Theoretical and Applied Mechanics
AIMETA 2022

Proceedings of the 25th AIMETA conference hosted by the Italian Association of Theoretical and Applied Mechanics in Palermo, Italy, September 4th - 8th, 2022.

https://pa22.aimeta.it/

Editors

Mario Di Paola, Livan Fratini, Fabrizio Micari, Antonina Pirrotta

Peer review statement

All papers published in this volume of "Materials Research Proceedings" have been peer reviewed. The process of peer review was initiated and overseen by the above proceedings editors. All reviews were conducted by expert referees in accordance with Materials Research Forum LLC high standards.

Published under License by **Materials Research Forum LLC**
Millersville, PA 17551, USA

Published as part of the proceedings series
Materials Research Proceedings
Volume 26 (2023)

ISSN 2474-3941 (Print)
ISSN 2474-395X (Online)

ISBN 978-1-64490-242-4 (Print)
ISBN 978-1-64490-243-1 (eBook)

Distributed worldwide by

Materials Research Forum LLC
105 Springdale Lane
Millersville, PA 17551
USA
https://www.mrforum.com

Manufactured in the United State of America
10 9 8 7 6 5 4 3 2 1

Table of Contents

Fluid Mechanics

Solid Mechanics

Structural Mechanics

Mechanics of Machine

Novel approaches in computational mechanics

Theoretical and applied biomechanics

Masonry modelling and analysis: from material to structures

Dynamical systems and applications in civil and mechanical structures

Control and experimental dynamics

Mechanical modelling of metamaterials and periodic structures

Novel stochastic dynamics methodologies & signal processing techniques for civil engineering applications

Open issues on procedures and methodologies for the vibration-based monitoring and dynamic identification of historic constructions

Mechanics of renewable energy systems

Advances in mathematical modeling and experimental techniques for quantification and prediction of fluid dynamic noise

Advanced process mechanics

Preface

The Organizing Committee of the XXV National Congress of the Italian Association of Theoretical and Applied Mechanics, in collaboration with the Engineering Department of the University of Palermo, organized the XXV National Congress of the Italian Association of Theoretical and Applied Mechanics, in Palermo, at the Department of Engineering, on 4-8 September 2022.

For fifty years, the AIMETA congress has been an opportunity for researchers working in the fields of General Mechanics, Fluids, Solids, Structures, Machines and Biomechanics to share advanced knowledge. The objectives of the congress are: to promote the exchange of ideas and experiences among the participants; foster stimuli towards new research topics; disseminate the latest scientific results. The topics dealt with concern theoretical, computational, experimental, technical-applicative aspects, and extend to multi-physics problems and between different sectors of scientific and technological knowledge.

AIMETA 2022, organized in Palermo, has been very successful and was developed in five days of meetings, symposia and work sessions, with participants among the most authoritative members of the Italian and International scientific community in the fields of theoretical and applied mechanics.

In particular, the Conference was composed of four invited talks, by outstanding scientists on the topics covered, presented in plenary sessions; of three hundred presentations of scientific papers, distributed over ten parallel sessions. In addition to general topics of relevant scientific interest in the field of basic and applied research, fourteen mini-symposia were organized on specific topics.

We wish to thank those many colleagues, young and less young, that took part in the organization of the congress; the scientific committee that faced the heavy task of examining the submitted abstracts and papers; and all those who contributed to the logistics and success of the event.

Mario Di Paola
Livan Fratini
Fabrizio Micari
Antonina Pirrotta

Acknowledgements

A special and sincere thank goes to all those people who reviewed the submitted abstracts and the following papers, thus helping to keep the quality of the congress high. Among them, we would like to quote S. Marfia, G. Garcea, S. de Miranda, P. Bisegna, V. Parenti Castelli, G. Pedrizzetti, M. D. De Tullio, M. Marino, N. Sancisi, G. Vairo, D. Addessi, G. Castellazzi, F. Clementi, G. Milani, F. D'Annibale, M. Ferretti, M. Romeo, F. Pellicano, G. Piccardo, D. Zulli, A. Bacigalupo, F. Dal Corso, M. L. De Bellis, A. Piccolroaz, V. Gusella, M. Betti, G. Boscato, N. Cavalagli, A. Cecchi, F. Clementi, F. P. Pinnola, M. S. Vaccaro, F. Davì, M. Paggi, A. Gizzi, F. Davì, M. Paggi, A. Gizzi, A. Cammarata, P.D. Maddìo, F. Garescì, M. Cammalleri, C. Baniotopoulos, C. Borri, C. L. Bottasso, L. Cappietti, E. Marino, V. Armenio; R. Camussi, M. Felli, M. Gennaretti, L.Filice, G. Buffa, A. Ghiotti.

We know that some reviewers might have been forgotten, hence we apologize; the success of the congress would not have been possible without the precious work of these people.

Mario Di Paola
Livan Fratini
Fabrizio Micari
Antonina Pirrotta

Committees

AIMETA Steering Committee

Walter D'Ambrogio (L'Aquila) - Presidente

Stefano Lenci (Politecnica delle Marche) - Vice-Presidente

Sandra Carillo (Roma Sapienza) - Segretario

Umberto Perego (Politecnico di Milano) - Tesoriere

Carlo Casciola (Roma Sapienza) - Responsabile del sito

Fernando Fraternali (Salerno) - Resp. rapporti internazionali

Elio Sacco (Federicoll Napoli) - Resp. gruppi Aimeta

Organizing Committee

Chairman: Mario Di Paola, Fabrizio Micari

C. Arico

S. Benfratello

E. Bologna

G. Borino

R. Di Lorenzo

A. Di Matteo

L. Fratini

G. Giambanco

E. La Malfa

V. Nigrelli

L. Palizzolo

F. Parrinello

A. Spada

M. Zingales

Scientific Committee

G. Rega (Roma Sapienza)

V. Armenio (Trieste)

A. Bottao (Genova)

A. Cecchi (Venezia)

C. Comi (Politecnico di Milano)

M. Fabrizio (Bologna)

L. Gambarotta (Genova)

F. Greco (Calabria)

M. Paggi (Lucca)

A. Pirrotta (Palermo)

F. Resta (Politecnico di Milano)

A. Rivola (Bologna)

M. Romeo (Genova)

E. Sacco (Federico Il Napoli)

R. Sinatra (Catania)

Fluid Mechanics

A novel one domain approach for free fluid-porous medium transport simulation - preliminary results

Costanza Aricò[1,2,a *], Martin Schneider[2,b], Tullio Tucciarelli[1,c] and Rainer Helmig[2,d]

[1]Department of Engineering, University of Palermo, Viale delle Scienze, 90128, Palermo, Italy

[2]Institute for Modelling Hydraulic and Environmental Systems, University of Stuttgart, Pfaffenwaldring 61, 70569 Stuttgart, Germany

[a]costanza.arico@unipa.it, [b]martin.schneider@iws.uni-stuttgart.de, [c]tullio.tucciarelli@unipa.it, [d]rainer.helmig@iws.uni-stuttgart.de

Keywords: Free Flow, Porous Medium, Transport Phenomena, Numerical Solver

Abstract. We present a new numerical solver for free-fluid flowing over and inside a porous medium. It is based over a macroscopic approach and one fictitious medium is assumed inside the domain, according to the One Domain Approach. Preliminary results are shown and compared with the ones provided by the well-known DuMux solver which applies a two Domain Approach.

Introduction

Transport phenomena of combined free fluid (ff) and porous medium (pm) flow occur in several industrial, environmental and biological applications (e.g., surface and groundwater flow, contaminant transport from lakes by groundwater, fuel cells, oil filters, blood flow in vessels and tissue, transfer of therapeutic agents).

The related mass, momentum and energy transport mechanisms at the ff–pm interface have been intensively investigated during the last decades applying two main approaches, the formulations at the micro- and macro-levels, respectively.

At the microscopical level, the porous medium is assumed as a connected domain of pore spaces filled with the fluid. The flow is governed by Navier–Stokes (or Stokes) equations, and no–slip boundary conditions are imposed on the interfaces between fluid and solid particles of the porous medium. The main drawback of this approach is its application for real problems, due to the huge amount of CPU time and computer memory storage, as well as the lack of an exact knowledge of the real porous geometry.

At the macroscopical level, the sets of the governing equations are obtained by averaging or upscaling the equations at the microscopic level over a Representative Elementary Volume (REV) [1]. The REV size is much larger than the characteristic size of the pore, but much smaller than the representative size of the domain. Two general different approaches have been derived at the macroscale, namely the Two-Domain Approach (TDA) and the One-Domain Approach (ODA) [1].

In TDA, the domain is split into two regions, and, in the most general case, the Navier–Stokes equations describe the fluid flow in the ff domain, while the Darcy's law is applied in the pm region. This is the most difficult approach from a mathematical point of view, since the two sets of governing equations are completely different systems of Partially Differential Equations (PDEs) and need interface conditions (IFCs). A sharp interface is assumed, where appropriate boundary conditions are imposed (typically, the conservation of mass, the balance of normal forces and the Beavers–Joseph condition (BJC) for the tangential velocity components [2]).

On the contrary, in ODA, the porous layer is regarded as a pseudo-fluid and the composite region free fluid-porous medium is treated as a continuum. One set of PDEs is assumed, typically the Brinkman or the (Navier)-Stokes-Brinkman equations (e.g., [1, 3]). The transition from the fluid to the porous medium is achieved through a continuous spatial variation of physical properties, such as permeability and porosity inside a transition layer (TL) which separates the homogeneous ff region from the bulk porous medium. The major drawback of the ODA is related to the knowledge of the size of the TL between the two homogeneous regions and how the physical parameters of the porous medium change inside it.

In the recent literature, some ODA have been proposed for the Brinkman equations [4] or a momentum equation having a Darcy form, including inertial and slip effects [5].

In the present paper, we present a novel numerical solver for ODA for incompressible and single-phase fluid over a saturated anisotropic porous medium. The governing equations, derived by averaging the pore scale microscopic Navier-Stokes equations, are the Incompressible Navier-Stokes-Brinkman equations (INSBEs), discretized over general unstructured meshes.

Preliminary model results are shown.

The Governing Equations and spatial discretization

The governing equations are a set of Partial Differential Equations (PDEs), the INSBEs,

$$\nabla \cdot \boldsymbol{u} = 0 \qquad (1),$$

$$\frac{\partial u}{\partial t} + \boldsymbol{u} \cdot \nabla \boldsymbol{u} + \nabla \Psi - \upsilon \Delta_2 \boldsymbol{u} + \frac{\upsilon \varepsilon}{\boldsymbol{K}} \boldsymbol{u} = 0 \qquad (2),$$

where t is the time, $\mathbf{u}$ is the fluid velocity vector (with u and v its x and y components), υ is the kinematic viscosity of the fluid, Ψ is the kinematic pressure of the fluid ($\Psi = p/\rho$ with p and ρ the fluid pressure and fluid constant density, respectively), $\mathbf{K}$ the permeability matrix of the porous medium, symmetric and positive-definite, and ε is the porosity of the porous medium. The last term on the l.h.s. of Eq. (2) takes into account the drag effects due to interaction of the fluid with the solid particles of the porous medium. The set of the governing equations has been derived from the Navier-Stokes equations at the pore-scale, applying average techniques [3, 6]

The computational domain Ω is discretized by means of unstructured triangulations of N_T non-overlapping triangles and N nodes. Inside each triangle e we assume the velocity vector be piecewise constant and $\boldsymbol{u}_e(\boldsymbol{x}) \in \mathbf{X}_e$, where $\mathbf{X}_e$ is the lowest-order Raviart-Thomas (RT-0) space. Ψ is assumed to be piecewise linear inside each triangle. The computational mesh satisfies the Delaunay Property.

In the present algorithm, INSBEs are solved by sequentially applying a fractional time step procedure, by solving consecutively a prediction and two correction steps. The prediction step is solved by applying the MAST (Marching in Space and Time) procedure, recently proposed for the solution of shallow waters, groundwater and incompressible Navier Stokes problems ([7] and cited works). MAST presents some important advantages: 1) explicit handling of the non-linear momentum terms due to a sequential solution of a small Ordinary Differential Equations (ODEs) system for each computational cell, 2) a computational effort proportional to the number of triangular elements and 3) numerical stability with respect to Courant-Friedrichs-Lewy (CFL) numbers also greater than one. The correction steps are solved by a Mixed Hybrid Finite Elements discretization that assumes a Generalized Delaunay mesh condition. They involve the solution of large linear systems, whose matrices are sparse, symmetric, positive definite and diagonally dominant, allowing a well performing condition number and a very fast solution of the associated systems. Strong reduction of the computational effort, in comparison with other numerical schemes (e.g., Lagrangian schemes), is due to the matrix coefficients, which are constant in time and are calculated only once, before the beginning of the numerical iteration loop. Flux continuity at each triangle side is guaranteed, as well as the local and global mass balance (more details in the cited works). There is no need to compute pressure at each time iteration, but only at target simulation times.

Model Results

1. Comparison with analytical solution and study of convergence order

We assume a 1D domain (see figure 1), with an isotropic porous medium and an analytical velocity profile given by Eqs. (3), continuous from the ff to the pm region.

$$u = 4U_M\left(\left(\frac{y-H}{y_{max}-H}-1\right)\left(\frac{y-H}{y_{max}-H}+\delta\right)\right) \text{ in ff, } u = 4U_M\left(\frac{\varepsilon(2\varepsilon^2-1)e^{\left(\frac{y-H}{\varepsilon(y_{max}-H)}-1\right)}}{1+\varepsilon}-\delta\right) \text{ in pm} \quad (3),$$

with $U_M = \frac{\nabla_x\Psi}{8\nu}(y_{max}-H)^2$, $\nabla_x\Psi$= 1d-08 m/s^2, $\nabla_y\Psi$= 0, $\varepsilon = \varepsilon\left(\sqrt{\frac{k_0}{\nu}}\Big/H\right)$, $\delta = \delta(\varepsilon)$, and k_0 the value of the permeability in the bulk porous medium. Vertical velocity component is zero. A source term vector is computed according to the expression of the velocity and the pressure gradient, in order to set to zero the momentum equation (2). We assume a hyperbolic variation of the permeability and porosity inside the TL, as in Eq. (4),

$$k(\mathbf{x}) = \frac{1}{2}\left((1-k_0)tanh(dist)+(1-k_0)\right) \quad \varepsilon(\mathbf{x}) = \frac{1}{2}\left((1-\varepsilon_0)tanh(dist)+(1-\varepsilon_0)\right) \quad (4),$$

where the distance *dist* from the midline of the TL is shown in Fig. 1 and k_0 = 1d-06 m^2. Two studies have been performed, assuming two values of the Reynolds number (*Re*), computed according to the maximum velocity in the ff region and the depth ($y_{max} - H$). The first study concerns the effect of the mesh refinement, without any special regard to the size and discretization of the TL. We set the size of the TL equal to y_{max}/1000. Results are summarized in tables 1,a-1,b of Fig. 2, in terms of L_2 norm of absolute and relative errors of the numerical solution compared to the analytical one in Eqs. (3). As expected, the convergence order is almost 1, due to the spatial approximation order of the velocity inside each triangle. The second study concerns the effect of the size of the TL, and the L_2 norms of the errors of the velocity have been reported in tables 2,a-2,b of Fig. 2. According to the definition of the velocity profile (in Eqs. (3)), function of the bulk value of the permeability k_0, we expect that the numerical solution converges to the exact one reducing the size of the TL. This is confirmed from the results in tables 2.

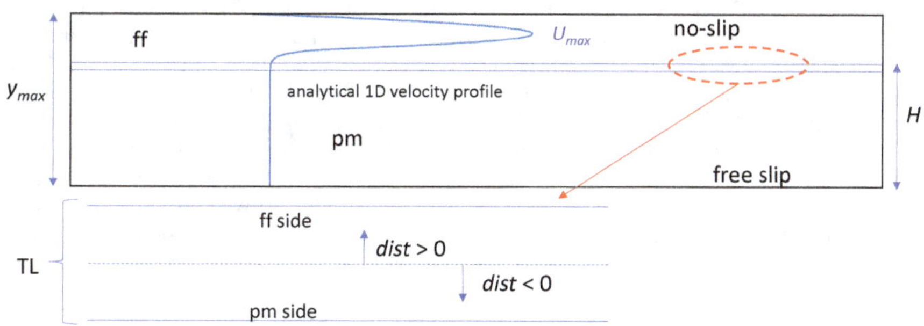

Figure 1. Test 1. Definition sketch

mesh linear size [m]	L2 ass err	L2 rel err	convergence	convergence
0.016	3.42E-05	1.57E+01		
0.008	1.71E-05	7.457924	1.00E+00	1.07E+00
0.004	8.26E-06	3.5568935	1.05E+00	1.07E+00
0.002	3.99E-06	1.70490455	1.05E+00	1.06E+00

mesh linear size [m]	L2 ass err	L2 rel err	convergence	convergence
0.016	3.61E-05	1.59E+01		
0.008	1.73E-05	7.90083	1.06E+00	1.01E+00
0.004	8.63E-06	3.789456	1.01E+00	1.06E+00
0.002	4.20E-06	1.833356788	1.04E+00	1.05E+00

Effect of the size of the transition layer

Table 2,a Re 1.5

TL size [m]	L2 ass err	L2 rel err	converger	converger
0.02	2.80E-05	8.863378		
0.005	1.30E-05	4.154056	1.11E+00	1.09E+00
0.00125	6.11E-06	2.022567	1.09E+00	1.04E+00
0.0003125	2.90E-06	0.9868	1.08E+00	1.04E+00

Table 2,b Re 150

TL size [m]	L2 ass err	L2 rel err	converger	converger
0.02	2.99E-05	9.054097		
0.005	1.40E-05	4.259854	1.09E+00	1.09E+00
0.00125	6.70E-06	2.129765	1.07E+00	1.00E+00
0.0003125	3.20E-06	1.030066	1.07E+00	1.05E+00

Figure 2. Test 1. Errors and study of the convergence order

2. Test 2. Free fluids over a porous obstacle. Isotropic and anisotropic cases

The geometry for this test is reported in Fig. 3. The working fluid is air. We set a pressure drop $\Psi_1 - \Psi_2$ = 1d-06 m^2/s^2. The permeability matrix **K** in the (bulk) porous medium is defined as

$$\mathbf{K} = \mathbf{RAR^{-1}} \qquad \mathbf{R} = \begin{pmatrix} \boldsymbol{cos\alpha} & \boldsymbol{-sin\alpha} \\ \boldsymbol{sin\alpha} & \boldsymbol{cos\alpha} \end{pmatrix} \qquad \mathbf{A} = \begin{pmatrix} k/\beta & \mathbf{0} \\ \mathbf{0} & k \end{pmatrix} \tag{5}$$

with angle α defined in figure 3. The porosity in the porous medium is 0.4. We compare the results of the ODA solver with the DuMux TDA solver [8]. DuMux applies a monolithic approach, where the Navier-Stokes equations for the ff region and the Darcy equations for the pm are solved, along with the interface conditions (IFCs), are solved in one large system. The IFCs impose the equality of fluxes and momentum fluxes in the ff and pm regions along the normal direction to the interface, as well as the BJC for the velocity along the tangential direction to the interface.

For the ODA simulation, the mesh sizes range from 1d-05 m inside the TL to 2.5 d-02 m in the ff and pm bulk regions (see in figure 3 the zoom close to the TL region). For the DuMux simulation, a structured mesh is used to discretize the domain, with constant size 1d-04 m. We assumed a hyperbolic variation of porosity and permeability parameters, as in the previous test. We performed simulations assuming both isotropic and anisotropic porous obstacle.

For the isotropic case, we set $\alpha = 0$ and k = 1d-08 m^2 in Eq. (5) and, for the ODA simulations, we assume two sizes of the TL, 1d-04 m and 5d-04 m. Due to the obstacle, a "channelized" flow is established above the porous medium, as shown in Fig. 4,a where we plot the flow field computed by the ODA solver. Similar overall results are given by the TDA solver (not shown for brevity). In Fig. 4,b-4,c, we compare, in the region close to the TL region, the velocity components along the vertical lines in the middle and at the upstream (left) side of the porous obstacle. Observe the differences, in the ODA solutions, due to the size of the TL, and the jump of the velocity component parallel to the interface in the DuMux results, due to the IFCs. Along the upstream side of the obstacle, the two solvers compute vertical velocity component v with opposite sign.

In the case of anisotropic obstacle, we set in Eq. (5) $\alpha = -\pi/4$, k = 1d-06 m^2 and coefficient β = 100. The size of the TL for the ODA model is 5d-04 m. In Fig. 5,a we plot the velocity vectors computed by the ODA solver. Similar overall results are given by the TDA solver. In Fig. 5,b we show the differences of the velocity components along the vertical right boundary of the obstacle in the TL region. Observe the different sign of the u component. Interesting is the comparison of the pressure fields (in Fig. 5.c), The ODA solver computes local maxima and minima along the upstream and downstream sides of the obstacle, respectively, while DuMux computes almost constant pressure values upstream and downstream the porous medium.

Theoretical and Applied Mechanics - AIMETA 2022 Materials Research Forum LLC
Materials Research Proceedings **26** (2023) 3-8 https://doi.org/10.21741/9781644902431-1

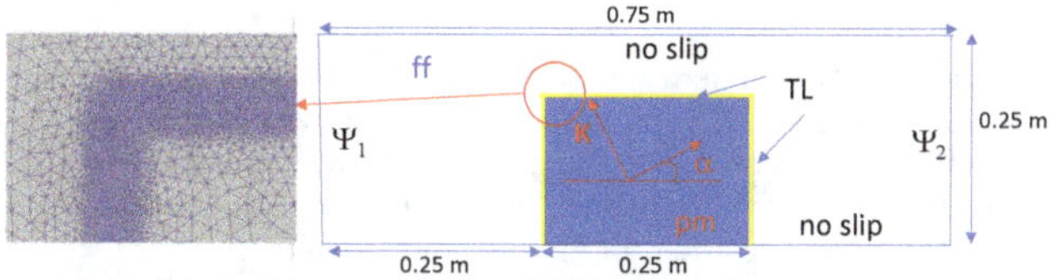

Figure 3. Test 2. Definition sketch. Boundary conditions, zoom of the mesh

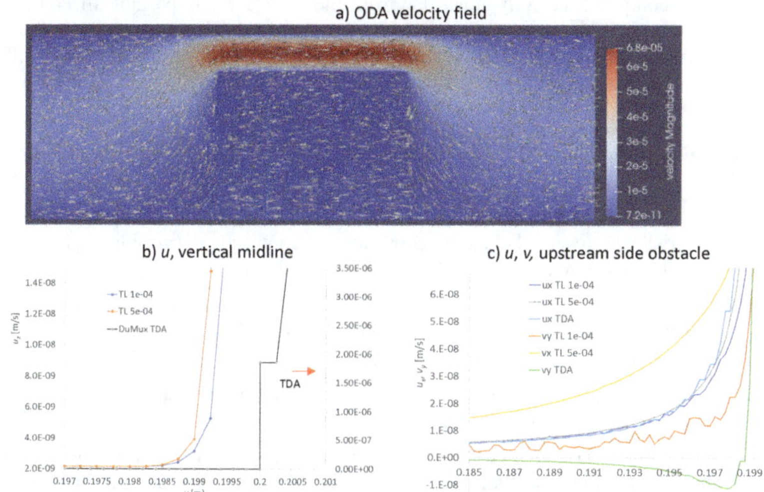

Figure 4. Test 2. Isotropic porous obstacle. a) velocity field, ODA solver. b) u component vertical midline of the obstacle. c) u and v components upstream side of the obstacle

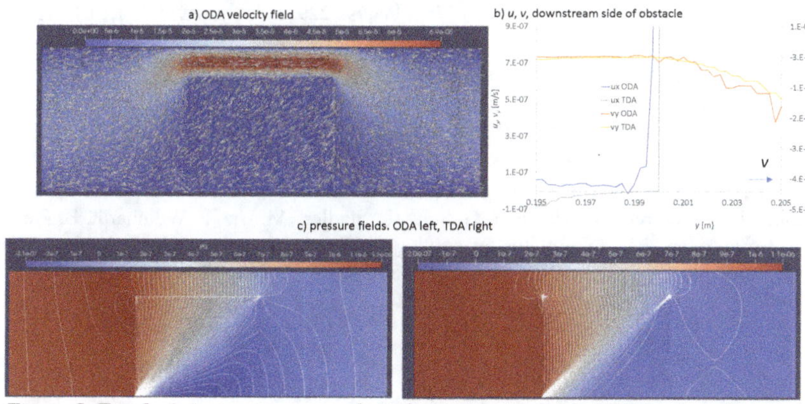

Figure 5. Test 2. Anisotropic porous obstacle. a) velocity field of ODA solver. b) u and v components at the downstream side of the obstacle, c) pressure fields, left ODA solver, right TDA solver

Conclusions

We have presented a new ODA numerical solver for the simulation of the interaction of free fluid over and inside a porous medium. This is a new research activity and we are aware that a lot of work have to be still done. According to the preliminary results shown in this paper, in-deep investigation deserves the role of the transition layer at the interface between the fluid and porous regions, along with its size and position. In the literature, there is a lack of studies regarding the comparison of the solutions provided by ODA and TDA models. Overall agreement has been obtained of the computed velocity fields far from the interfaces. Important differences of flow velocity and pressure fields have been obtained by the two approaches close to the transition layer. Prediction of transport phenomena of heat and species at interfaces could be strongly affected by the solutions of the velocity and pressure.

References

[1] B. Goyeau, D. Lhuillier, D. Gobina, M.G. Velarde, Momentum transport at a fluid–porous interface, Int. J. Heat and Mass Transfer 46 (2003) 4071–4081. https://doi.org/10.1016/S0017-9310(03)00241-2

[2] G.S. Beavers, D. D. Joseph, Boundary conditions at a naturally permeable wall, J. Fluid Mech. 30(01) (1967) 197–207. https://doi.org/10.1017/S0022112067001375

[3] L. Wang, L.-P. Wang, Z. Guo, J. Mi, Volume-averaged macroscopic equation for fluid flow in moving porous media, Int. J. Heat and Mass Transfer 82 (2015) 357–368. https://doi.org/10.1016/j.ijheatmasstransfer.2014.11.056

[4] H. Chen, X.-P. Wang, A one-domain approach for modeling and simulation of free fluid over a porous medium, J. Comp. Phys., 259 (2014) 650–671. https://doi.org/10.1016/j.jcp.2013.12.008

[5] F. J. Valdés-Parada, D. Lasseux, A novel one-domain approach for modeling flow in a fluid-porous system including inertia and slip effects, Phys. Fluids 33, 022106 (2021). https://doi.org/10.1063/5.0036812

[6] W. Gray, A derivation of the equations for multi-phase transport, Chemical Engineering Science Vol. 30, Issue 2, February 1975, Pages 229-233. https://doi.org/10.1016/0009-2509(75)80010-8

[7] C. Aricò, M. Sinagra, Z. Driss, T.Tucciarelli, A new solver for incompressible non-isothermal flows in natural and mixed convection over unstructured grids, Applied Mathematical Modelling 103 (2022), 445-474. https://doi.org/10.1016/j.apm.2021.10.042

[8] T. Koch, D. Gläser, K. Weishaupt, S. Ackermann, M. Beck, B. Becker, S. Burbulla, H. Class, E. Coltman, S. Emmert, T. Fetzer, C. Grüninger, K. Heck, J. Hommel, T. Kurz, M. Lipp, F. Mohammadi, S. Scherrer, M. Schneider, G. Seitz, L. Stadler, M. Utz, F. Weinhardt, B. Flemisch, DuMux 3 – an open-source simulator for solving flow and transport problems in porous media with a focus on model coupling, Comp. Math. with Applications 81 (2021) 423–443. https://doi.org/10.1016/j.camwa.2020.02.012

Solid Mechanics

Theoretical and Applied Mechanics - AIMETA 2022 Materials Research Forum LLC
Materials Research Proceedings **26** (2023) 11-16 https://doi.org/10.21741/9781644902431-2

Mechanical properties of cables made with helically wound carbon-nanotube fibers for advanced structural applications

Giovanni Migliaccio[1,a *], Reginald DesRoches[2,b], Gianni Royer-Carfagni[1,c]

[1]Department of Engineering and Architecture, University of Parma, Parma, Italy

[2]Department of Civil and Environmental Engineering, Rice University, Houston, USA

[a]giovanni.migliaccio.it@gmail.com, [b]provost@rice.edu, [c]gianni.royer@unipr.it

Keywords: Carbon-Nanotube Fibers, Effective Strength, Helically Wound Cable

Abstract. Big structures, such as super long suspension bridges, require materials that overcome the specific strength of steel, since there is a material-dependent limit size beyond which they shall collapse under their own weight. Carbon NanoTube Fibers (CNTFs) hold great promise for advanced applications, for their exceptional strength and stiffness per unit specific weight. We propose a theoretical model to describe the mechanical response of cables made of CNTFs. The mechanical response is studied via a variational approach and closed-form expressions are obtained for the stiffness parameters of the cable and the stress in the constituent CNTFs. This study shows that the stiffness and strength of the cable present opposite variations with respect to parameters associated with the geometric arrangement of the CNTFs, suggesting the existence of an optimal compromise yet to be experimentally verified.

Introduction

Challenging engineering applications require materials with specific strength much higher than metals [1,2]. Carbon NanoTubes (CNTs) may be used in new applications, yet to be fully appreciated, as they possess exceptional physical properties [3-4] not only to replace metals in current engineering applications, but also to design structures not even conceivable with the materials currently available [1,2]. The problem is how to employ very small CNTs (1nm diameter, 10μm length) in large structural elements. When assembled into a matrix, their strength is impaired by non-optimal orientations, defects and the intrinsic weakness of the matrix. CNT Fibers (CNTFs), obtained via solution spinning [3], are made of highly-aligned densely-packed CNTs and can currently be manufactured with diameter of about 100μm and, in principle, no limit in length. Since there is no matrix embedding the CNTs, the structural capacity of such CNTFs could theoretically approach that of the CNTs. Indeed, technological improvements have permitted an increase in specific strength of about 23% per year over the last 20 years [3], and a similar positive trend is expected for the years to come. Recent studies [5] have shown that stiffness and strength of CNTFs [4] depend on the lateral bond and on the longitudinal offset between the constituent CNTs. New production techniques are expected to allow denser packaging of longer CNTs with optimal offset and, consequently, higher performing CNTFs [6].

Here, we present a theoretical model, propaedeutic for an experimental campaign, to define the mechanical properties of cables made of CNTFs. The cable mechanical properties are derived via a variational approach, in which the CNTFs are rod-like elements with macroscopic properties derived by the micromechanics of the constituent CNTs. Some of the main results of this study are summarized in the Conclusion section, together with indications on future developments.

Theoretical and Applied Mechanics - AIMETA 2022 Materials Research Forum LLC
Materials Research Proceedings **26** (2023) 11-16 https://doi.org/10.21741/9781644902431-2

Mechanical model

Consider a cable made of M layers of CNTFs wound around a straight central fiber. Figure 1(a) shows the cases with $M = 1,2,3$. The tensile and torsional properties of the cable clearly depend upon the geometric arrangement of the CNTFs.

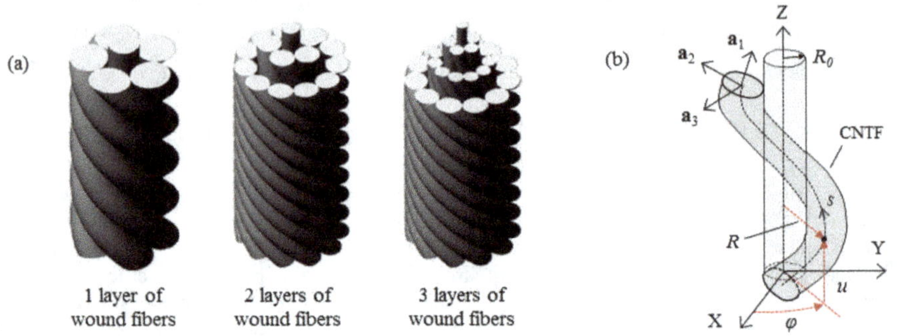

Figure 1: (a) cable with 1,2,3 layers of wound fibers; (b) problem of a fiber wound on a cylinder.

The basic problem, represented in Figure 1(b), is that of a fiber, modelled as a circular cross-sectioned rod of radius R_f, wound around a straight cylinder of radius R_0 ($R=R_0+R_f$), whose ends are displaced in a given position. The fiber deformation is described via classical indicators of strain [7-9], i.e., the axial extension ε_f, bending curvature k_{Bf}, and torsion curvature k_{Tf}, while the equilibrium state is found with a variational approach. The strain energy of the fiber is the line integral over the fiber length L_f of the strain energy density W_f, which depends upon the fiber axial stiffness $\mathcal{E}_f$, bending stiffness $\mathcal{B}_f$, and torsional stiffness $\mathcal{T}_f$ (derivable from a micro-mechanical model of the CNTFs [5], or from experimental tests [4]), and is written in the form

$$2W_f = \mathcal{E}_f \varepsilon_f^2 + \mathcal{B}_f k_{Bf}^2 + \mathcal{T}_f k_{Tf}^2. \tag{1}$$

For a stretched and twisted cable, we assume that the problem symmetry is respected: all the fibers deform together, cross-sections remain plane, and frictional forces are negligible. With respect to the fixed reference frame X,Y,Z in Figure 1(b), the actual position of a point of the fiber center-line at the arc-length $s \in (0,L_f)$ is expressed in terms of the cylindrical coordinates $\varphi(s)$ and $u(s)$, while $\theta(s)$ describes the rotation of the fiber cross-section about the fiber center-line, with tangent $\mathbf{a}_1$. When $\varphi(s)$, $u(s)$, $\theta(s)$ are prescribed at the fiber ends, s=0 and s=L_f, a stationary point for the fiber strain energy corresponds to a configuration in which the fiber center-line is a helix, the cross-sectional twist θ' is constant (prime denotes s-derivative), and

$$k_{Bf} = \varphi' cos\gamma, \;\; k_{Tf} = \theta' + \varphi' sin\gamma, \tag{2}$$

where $tan\,\gamma = u'/(R\varphi')$ defines the constant pitch angle γ in terms of the other variables.

It is important to observe that the torsion curvature k_{Tf} is composed of two contributions: the cross-sectional twist θ' and the tortuosity-induced twist $\varphi' sin\,\gamma$. The fiber is torsion-free when the cross-sectional twist balances the tortuosity-induced twist. This condition plays an important role in the manufacturing of a cable made of helically wound CNTFs, which consists in inducing a cross-sectional "pre-twist" in a straight fiber that is subsequently transformed in "tortuosity" for its helical configuration, as in the manufacturing of a hemp rope [6].

A 7-wire strand (1-layer cable) is represented in Figure 2. For its manufacturing, straight fibers are pulled and pre-twisted and, subsequently, are brought closer and wound together in such a way that section pre-twist turns into center-line tortuosity. Adding a second layer of fibers one obtains the 2-layer cable of Figure 1(a); repeating the construction, a M-layer cable can be manufactured.

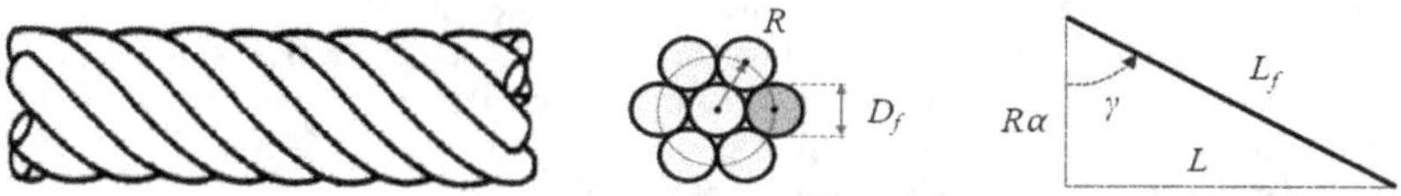

Figure 2: A 7-wire strand (1 layer of helically wound fibers) and geometric parameters.

The first layer of helical fibers in a 7-wire strand is the prototypical model for the *i-th* layer in a M-layer cable, for which each layer is wound around a cylinder (i.e., the straight central fiber for the strand). Assume that the strand has been manufactured: each fiber is arranged according to an helix of length L_f, helical radius R, slope γ, $L=L_f \sin\gamma$, and coordinate φ that spans the angle α from the first end to the second end of the fiber, such that $\alpha R = L_f \cos\gamma$. A system of forces, equipollent to an axial force and a twisting moment, are applied at the strand ends. All fibers deform together and their varied state is a new helix with new parameters $L_f+\Delta L_f$, $L+\Delta L$, $R+\Delta R$, $\alpha+\Delta\alpha$, $\gamma+\Delta\gamma$.

The macroscopic strain of an arrangement of this kind, when pulled and twisted, is described via the axial strain $\varepsilon=\Delta L/L$ and twist rate $\beta=\Delta\alpha/L$. For small variations, ΔL_f, ΔL, ΔR, $\Delta\alpha$, $\Delta\gamma$, it is possible to demonstrate that the extension, ε_f, and curvatures, k_{Bf} and k_{Tf}, of all fibers can be expressed as linear combinations of the strain indicators ε and β, via coefficients that depend on the slope γ and radius R of the unstrained state. Therefore, considering that the strain energy density W of the arrangement is the sum of those of the constituent CNTFs, W turns out to be a quadratic function of the strain indicators ε and β. This holds for a 7-wire strand as well as for a cable made of M concentric layers of helically wound CNTFs.

Denoting as $W_{(2)}$ the quadratic part of the cable strain energy density, one can write

$$2W_{(2)} = \mathcal{E}\varepsilon^2 + \mathcal{T}\beta^2 + 2X\varepsilon\beta\,, \tag{3}$$

where $\mathcal{E}$ represents the axial stiffness of the cable, $\mathcal{T}$ is the torsional stiffness, and X accounts for the extension-torsion coupling associated with the helical configuration. These parameters depend on the slope, diameter, and Poisson's ratio of the helical CNTFs, but their expressions are generally complicated. Compact analytical formulas can be obtained in paradigmatic cases, e.g., when all fibers have the same slope γ, diameter D_f, and Poisson's ratio ν_f, and the contribution of the fiber bending stiffness $\mathcal{B}_f$ and torsional stiffnesses $\mathcal{T}_f$ are negligible with respect to that of the fiber axial stiffness $\mathcal{E}_f$, as is expected for CNTFs [4,5].

Under the aforementioned hypotheses, the dependence of the cable axial stiffness $\mathcal{E}$ on the axial stiffness $\mathcal{E}_f$ and slope γ of the helical fibers can be expressed [6] as

$$\mathcal{E} = \rho\, N_f \mathcal{E}_f sin^3\gamma, \tag{4}$$

where N_f represents the total number of fibers in the cable and ρ is a coefficient (close to 1) that depends on ν_f. Similar expressions hold for the other stiffness parameters of the cable [6].

Observe that the cable axial stiffness $\mathcal{E}$ increases with the angle γ, and is maximized for $\gamma=90°$ (straight fibers). However, helical fibers are expected to provide a cable with a superior internal cohesion because of the radial contact forces that take place among the cable layers.

Results

Recent experiments [4] have measured the mechanical properties of CNTFs with D_f=22μm, made of CNTs of diameter d=1.5nm and lengths 2.2-6.3μm. The axial stiffness $\mathcal{E}_f$ of these fibers is in

 Materials Research Forum LLC
https://doi.org/10.21741/9781644902431-2

the range 27.5-54.4 Nm/m, and is dominant over the bending stiffness, being $\mathcal{B}_f/(\mathcal{E}_f D_f^2)$ within the interval 0.01-0.05. Relying on these measures, we have performed a parametric analysis with our theoretical model, to evaluate the mechanical properties of a cable made of M layers of helical CNTFs as a function of representative properties, i.e., fiber stiffnesses, Poisson's ratio, slope γ.

To illustrate, Figure 3 reports results obtained for the cable axial stiffness $\mathcal{E}$ normalized via the reference parameter $\mathcal{E}_{ref} = N_f \mathcal{E}_f$. It is found that the CNTF bending and torsional stiffnesses $\mathcal{B}_f$ and $\mathcal{T}_f$ have a negligible effect; Poisson's ratio has a mild influence for larger γ (say, above 65-70°), while the helical slope γ, the number of fibers N_f, and the fiber axial stiffness $\mathcal{E}_f$ play the major role. Similar results hold for the other stiffness parameters of the cable.

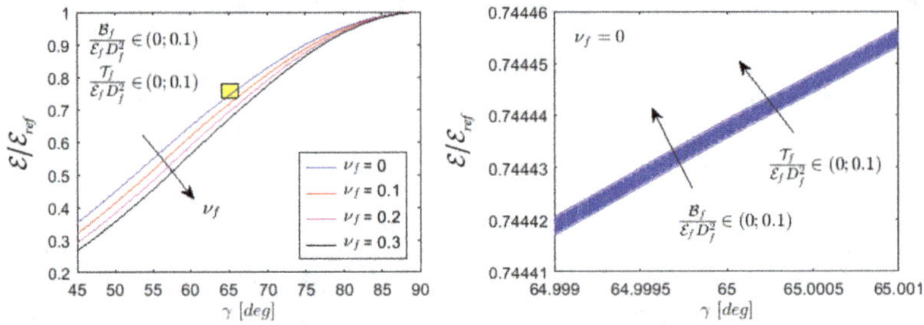

Figure 3: Cable axial stiffness as a function of the fiber stiffness, Poisson's ratio, and slope γ.

When the cable is pulled by a force F and the rotation of its end sections is prevented, the constituent helical fibers undergo a tensile force $\mathcal{E}_f \varepsilon_f$, which is independent of the position of the layer they belong to in the cable and is shown (normalized by F/N_f) in Figure 4. The conclusions about the dependence of the cable axial stiffness on the fiber parameters also hold true for the stress parameter $(\mathcal{E}_f \varepsilon_f)/(F/N_f)$, which represents the ratio between the tensile force in the helical fibers of the cable and the value of such force if all the fibers were straight.

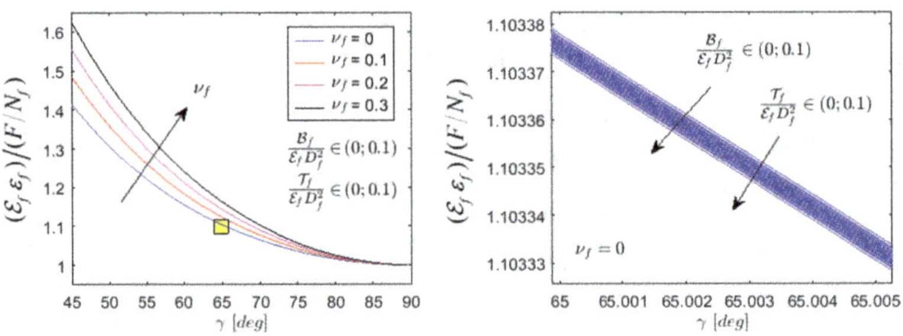

Figure 4: Tensile force in cable fibers as a function of fiber stiffness, Poisson's ratio, and slope γ.

For a cable made of M layers of helical CNTFs, another important stress indicator is represented by the radial contact forces among the cable layers. Indeed, a pressure orthogonal to the CNTFs can increase the lateral bond among the constituent CNTs and, consequently, the mechanical properties of the CNTFs. Furthermore, radial contact forces compress the CNTFs together, so that if a fiber breaks, the frictional forces associated with the compression allows to re-establish the load bearing capacity of the broken fiber at a certain distance from the failure point. Therefore, in a cable made of many fibers, a limited number of fiber failures occurring at different sections in the cable does not compromise the overall bearing capacity of the cable. In conclusion, the cable resilience is enhanced by the radial contact forces produced by a helical arrangement of fibers.

For the considered cable, the radial force that is exerted on each fiber of the *i-th* layer at the radial position R_i (distance between the *i-th* layer and the cable axis) can be evaluated as the sum of all actions that are exerted on this layer by the other layers $i+1,\ldots, M$. Ignoring the effects of the fiber bending/torsional stiffness and considering only the dominant contribution of the fiber axial stiffness, it is possible to show [6] that if the cable is pulled and the rotation of its end sections is prevented, the radial contact forces can be evaluated via the simplified formula

$$\frac{F_R D_f}{\mathcal{E}_f \varepsilon_f} = cos^2\gamma\left(\frac{R_M}{R_i} - 1\right), \tag{5}$$

where R_M represents the radius of the cable and F_R is the radial force at the *i-th* layer.

The dimensionless parameter $(F_R D_f)/(F/N_f)$ associated with the radial forces among the cable layers represents a measure of the resilience of the cable and of the increase of its tensile strength obtainable by helically arranged fibers. This parameter is reported in Figure 5 as a function of the slope γ. In the same Figure, for comparison, we juxtapose the graph of Figure 3, which is associated with the stiffness of the cable. The case shown corresponds to a cable composed of $M = 100$ layers of CNTFs and considers the state of stress at the layer $R_i = 30D_f$.

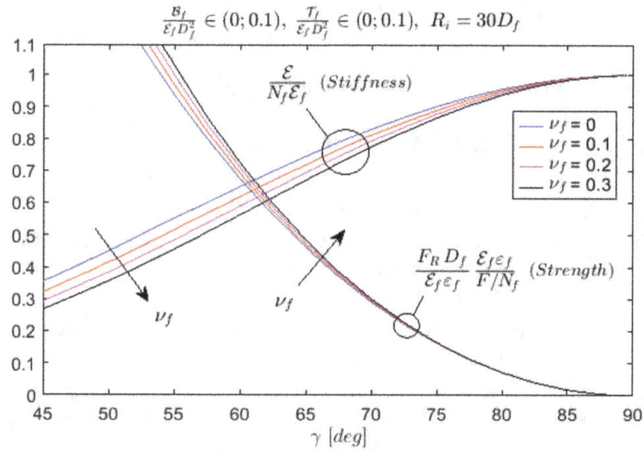

Figure 5: Cable stiffness and strength as a function of fiber stiffness, Poisson's ratio, and slope γ.

Remarkably, the cable axial *stiffness* increases with the slope γ, as opposed to the parameter related to the cable *strength*. Therefore, a value of $\gamma = 90°$, which optimizes the stiffness, may not optimize the strength. This suggests the existence of an optimal compromise between strength and stiffness of the cable, which needs to be further investigated and verified experimentally.

Conclusions

Thanks to their exceptional mechanical properties, CNTFs appear particularly suitable for being the basic constituents of cables for advanced structural applications. An ad-hoc theoretical model has been proposed to describe the mechanical properties of cables made of wound CNTFs.

The approach is variational and considers the minimization of the strain energy that additively accounts for the contribution of the individual CNTFs in the cable. Using symmetry, the problem is reduced to the paradigmatic case of an isolated fiber wound onto a cylinder, whose equilibrium states are found to be helixes with constant pitch. Analytical expressions have been found for the stiffness parameters of a cable composed by several concentric layers of helical CNTFs, as well as for the internal stresses in such CNTFs. A parametric analysis has been performed to evaluate the dependence of the cable mechanical response on the stiffness, Poisson's ratio, and helical pitch of the constituent CNTFs. The analysis has demonstrated the important influence of the fiber axial stiffness and helical pitch, the limited effect of the fiber Poisson's ratio for large helical pitch, and the negligible contribution of the fiber bending stiffness and torsional stiffness when dealing with the calculation of the stiffness and strength parameters of the cable.

The theoretical model has disclosed important aspects to be corroborated by experiments. If the characterization of the axial stiffness and bending stiffness of the CNTFs has been the subject of intense experimental tests, little has been done on Poisson's ratio and torsional stiffness. Another important issue concerns the radial contact forces between helically wound CNTFs, which result from the stretching or twisting of the cable. Such forces can increase the bond between the CNTs that form the CNTFs, enhancing their effective strength, and produce a friction constraint which improves the cable resilience in the event of rupture of one CNTF. This study has shown that the radial contact forces increase by reducing the helical pitch of the cable CNTFs, but the cable axial stiffness shows an opposite trend. This suggests the existence of an optimal compromise between the strength and stiffness of the cable. An extended experimental campaign has been planned to corroborate the theoretical findings.

References

[1] DesRoches R., Migliaccio G., Royer-Carfagni G., Structures that can be made with carbon nanotube fibers but not with other materials, J. Eng. Mech., 2022 (in press). https://doi.org/10.1061/(ASCE)EM.1943-7889.0002138

[2] Carpinteri A., Pugno. N.M., Super-bridges suspended over carbon nanotube cables, J. Phys. Condens. Matter, 20:1-8, 2008. https://doi.org/10.1088/0953-8984/20/47/474213

[3] Taylor L.W., Dewey O.S., et al., Improved properties, increased production, and the path to broad adoption of carbon nanotube fibers, Carbon, 171:689-694, 2021. https://doi.org/10.1016/j.carbon.2020.07.058

[4] Adnan M., Pinnick R.A., et al., Bending behavior of CNT fibers and their scaling laws, Soft Matter, 14, 2018. https://doi.org/10.1039/C8SM01129J

[5] Galuppi L., Pasquali M., Royer-Carfagni G., The effective tensile and bending stiffness of nanotube fibers, Int. J. Mech. Sci., 163, 2019. https://doi.org/10.1016/j.ijmecsci.2019.105089

[6] Migliaccio G., DesRoches R., Royer-Carfagni G., Theoretical mechanical properties of strands and cables made of wound carbon-nanotube fibers, Int. J. Mech. Sci., 2022 (submitted). https://doi.org/10.1016/j.ijmecsci.2022.107706

[7] Love A.E.H., A treatise on the mathematical theory of elasticity, 4th ed., Dover Pub., 1944.

[8] Migliaccio G., Ruta G., The influence of an initial twisting on tapered beams undergoing large displacements, Meccanica, 56, 7, 1831-1845, 2021. https://doi.org/10.1007/s11012-021-01334-2

[9] Costello G.A., Theory of wire rope, Mech. Eng. Series, 2nd ed., Springer-Verlag, 1997. https://doi.org/10.1007/978-1-4612-1970-5

Theoretical and Applied Mechanics - AIMETA 2022 Materials Research Forum LLC
Materials Research Proceedings **26** (2023) 17-22 https://doi.org/10.21741/9781644902431-3

A variational model for plastic reorientation in fibrous material: numerical experiments on phase segregation

Andrea Rodella[1, a *], Antonino Favata[1,b] and Stefano Vidoli[3,c]

[1]DISG, "Sapienza" University of Rome, Via Eudossiana 18, 00185 Rome, Italy

[a]andrea.rodella@unitoma1.it, [b]antonino.favata@uniroma1.it, [c]stefano.vidoli@uniroma1.it

Keywords: Phase Segregation, Fibrous Material, Plastic Remodeling, FEniCS

Abstract. We propose a continuum model of fibrous material that may undergo an internal reorganization, which turns out in a plastic change of the orientation of the fibers when the remodeling torque achieves a threshold. We have recently found that the reorientation may induce a complex scenario in the response of such materials. In a traction test, we show that the most general transversely isotropic material may evolve in three different ways; in particular, the fibers asymptotically tend (regularly or with jumps): (*A*) to a given angle; (*B*) to align perpendicularly to the load direction; (*C*) to align with the load direction if their initial orientation is less than a given value otherwise perpendicularly. We focus on the latter material response (*C*) which has all the ingredients to manifest a phase transition phenomenon. Finally, we provide a numerical investigation to demonstrate phase segregation.

Introduction

Due to their ubiquity, fibrous materials have gained a predominant role in the scientific community in the last few decades. Fibrous structures are everywhere; the human body is a glaring example: the connective tissue, the most abundant tissue in mammals, is composed mainly of extracellular matrix and collagen fibers (~6% of the total body weight). It is also well established that biological tissues may experience an internal reorganization, including segregation [1], due to external stimuli at chemo-mechanical levels. This phenomenon corresponds to a change in the fiber orientation that may be associated with a loss of the elastic energy content, resulting in irreversible deformations [2]. The effects of reorientation, particularly those occurring during the post-yield deformation, remain largely unexplored in biological and manufactured composites despite their fundamental role in the mechanical response [3]. We focus on the most general 2D transversely isotropic material to model the plastic reorientation in fibrous materials. Then, we consider a linearized framework concerning the strain measure; nevertheless, we admit finite rotation for the fibers. Within a thermodynamically consistent framework [4], we present a variational model taking into account the anisotropic response of the material, which depends on an internal variable, *e.g.*, the fiber orientation. The latter, similarly to plasticity [5], has the peculiarity to evolve till a threshold beyond which the reorientation is permanent. We recall the main analytical results of the incremental homogenous traction problem we recently found in [6]. In this work, we show three different asymptotic behavior for the orientation of the fibers, which tend to align, smoothly or not: *(A)* with a given angle; *(B)* perpendicularly with the load direction and *(C)* perpendicularly with the load direction or with the load direction itself in accordance with the initial orientation. The energy of the material in class *(C)* presents a double-well landscape which may induce a phase transition and give room for exploring the phase segregation.

We organize the work as follows: i) we present the assumptions to formulate a thermodynamically consistent phase-field model taking into account the anelastic response of the material; ii) we recall the main analytical results found in [6] for the traction test; iii) we provide some numerical results for a non-homogeneous initial fiber orientation distribution showing the evolution of the traction test leading to the segregation phenomenon.

Assumptions for the free energy and the reversibility domain

For the sake of simplicity, we confine our analysis to a two-dimensional body $\mathcal{B}$. The vector position of the point $x \in \mathcal{B}$ is defined as $\mathbf{x} = x - o = x_a \, \mathbf{e}_a$ with respect to the origin o of the Cartesian frame $\{o, \, \mathbf{e}_1, \, \mathbf{e}_2\}$. We chose to associate an orientated fiber to each material point in $\mathcal{B}$, whose direction is represented by the unit vector $\mathbf{n}(\vartheta) = \cos\vartheta \, \mathbf{e}_1 + \sin\vartheta \, \mathbf{e}_2$ as a function of the *internal state variable* ϑ . Recalling [4,7,8], the state variable influences the free energy and the dissipation rests upon the evolution of ϑ. The state for the body $\mathcal{B}$ at each time t is known if the list of functions of the spatial coordinate x, $\boldsymbol{\Lambda} = \{\mathbf{E}, \, \vartheta\}$ is known, where $\mathbf{E} = \mathrm{sym}\nabla\mathbf{u}$ is the linearized strain measure expressed in terms of the displacement field $\mathbf{u}$. The free energy density is taken as a quadratic form of the strain field

$$\psi = \hat{\psi}(\mathbf{E}, \vartheta) = \frac{1}{2}\mathbb{C}\big(\mathbf{n}(\vartheta)\big)\mathbf{E} \cdot \mathbf{E}, \tag{1}$$

where $\mathbb{C}\big(\mathbf{n}(\vartheta)\big)$ represents the elasticity tensor of a linearly elastic transversely isotropic material with respect to the direction $\mathbf{n}(\vartheta)$. For the most general two-dimensional case, we specify the free energy density as follows:

$$\hat{\psi}(\mathbf{E}, \vartheta) = \mu\|\mathbf{E}\|^2 + \frac{\lambda}{2}(\mathrm{tr}\mathbf{E})^2 + c_1(\mathrm{tr}\mathbf{E})\mathbf{E}\mathbf{n}(\vartheta) \cdot \mathbf{n}(\vartheta) + c_2\big(\mathbf{E}\mathbf{n}(\vartheta) \cdot \mathbf{n}(\vartheta)\big)^2, \tag{2}$$

where λ and μ are the Lamé coefficients and c_1 and c_2 are the material constants characterizing the transversely isotropy. By fixing the Lamé coefficients, which must respect the conditions $\mu > 0$ and $2\mu + \lambda > 0$, the pair of constants (c_1, c_2) must lie in a set $\mathcal{P}$, ensuring the positiveness of the energy Eq. (2)

$$\mathcal{P} = \left\{c_1, c_2 \in \mathbb{R}: c_2 > \frac{c_1^2 - 4\mu c_1 - 4\mu(\mu+\lambda)}{2(2\mu+\lambda)}\right\}, \tag{3}$$

which represents a parabola, see Fig. 1. In order to clarify the role of the constant c_1 and c_2 on the anisotropic material response, it is useful to recall the definitions of Young modulus and Poisson ratio, defined for a uniaxial traction test $\bar{\mathbf{T}} = \sigma\mathbf{t}\otimes\mathbf{t}$ in the direction $\mathbf{t} = \cos\alpha \, \mathbf{e}_1 + \sin\alpha \, \mathbf{e}_2$. The strain corresponding to $\bar{\mathbf{T}}$ is defined as $\bar{\mathbf{E}} = \mathbb{C}^{-1}\big(\mathbf{n}(\vartheta)\big)\bar{\mathbf{T}}$. Therefore, the Young modulus and the Poisson ratio are functions of the angle resulting from the difference between the testing direction and the fiber orientation $(\alpha - \vartheta)$ and follow the definitions:

$$E(\alpha, \vartheta) = \hat{E}(\alpha, \vartheta) := \frac{\bar{\mathbf{T}}\mathbf{t}\cdot\mathbf{t}}{\bar{\mathbf{E}}\mathbf{t}\cdot\mathbf{t}}; \qquad \nu(\alpha, \vartheta) = \hat{\nu}(\alpha, \vartheta) := \frac{\bar{\mathbf{E}}\mathbf{t}^{\perp}\cdot\mathbf{t}^{\perp}}{\bar{\mathbf{E}}\mathbf{t}\cdot\mathbf{t}}. \tag{4}$$

Polar plots are, therefore, a meaningful tool to understand how Young modulus and Poisson ratio change in function of the angle $(\alpha - \vartheta)$, see the insert in Fig. 1. We define the stiffening set $\mathcal{S}$ by imposing the ratio between the Young modulus in the parallel and perpendicular with respect to the fiber direction $E_{||} = \hat{E}(\alpha = \vartheta, \vartheta)$ and $E_{\perp} = \hat{E}(\alpha = \vartheta - \pi/2, \vartheta)$ to be greater than 1:

$$E_{||}/E_{\perp} = 1 + \frac{2(c_1+c_2)}{2\mu+\lambda} > 1 \quad \Rightarrow \quad \mathcal{S} = \{c_1, c_2 \in \mathcal{P}: c_1 + c_2 > 0\} \tag{5}$$

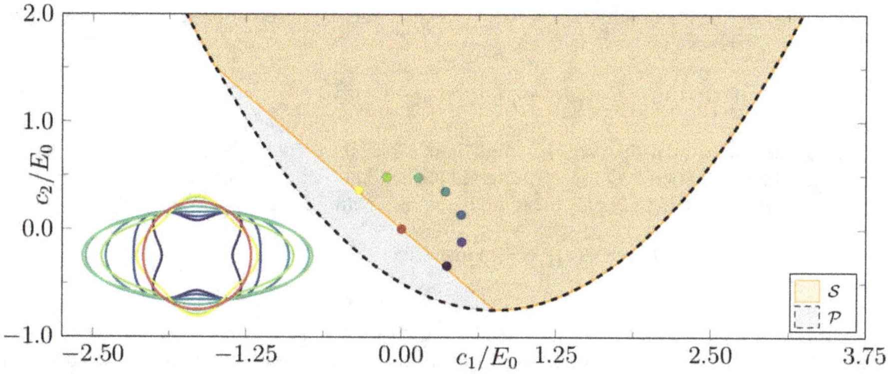

***Figure 1**: The gray area confined by the black dashed line represents the set where the pair $\{c_1, c_2\}/E_0$ ensures the positiveness of the energy, see Eq. (3), where E_0 is the Young modulus of the isotropic case. While the yellow area represents the stiffening set, a subset of $\mathcal{P}$, obeying Eq. (5). The insert figure is the polar plot of the Young modulus $\hat{E}(\alpha, \vartheta = 0)$ where $\alpha \in [0, 2\pi]$. Each polar plot corresponds to the colored dot picked from the stiffening set $\mathcal{S}$.*

In the framework of *Generalized Standard Materials* [4], the dissipative behavior is described through the dissipation potential, which is a convex and positive function of the internal state variables rates. Since $\mathbf{n} \cdot \mathbf{n} = 1$, we have $\dot{\mathbf{n}} \cdot \mathbf{n} = 0$ or, equivalently, $\dot{\mathbf{n}} = \beta \mathbf{n}^\perp$ with $\mathbf{n}^\perp := \partial_\vartheta \mathbf{n}(\vartheta) = -\sin\vartheta\, \mathbf{e}_1 + \cos\vartheta\, \mathbf{e}_2$, from Eq. (2), the energy release rate turns out to be

$$\begin{aligned} -\partial_\vartheta \hat{\psi}(\mathbf{E}, \vartheta)\dot{\vartheta} &= -\partial_\mathbf{n} \hat{\psi}(\mathbf{E}, \mathbf{n}(\vartheta)) \cdot \dot{\mathbf{n}}(\vartheta) \\ &= -2\big(c_1 \mathrm{tr}\mathbf{E} + 2c_2 \mathbf{E}\mathbf{n}(\vartheta) \cdot \mathbf{n}(\vartheta)\big)\mathbf{E}\mathbf{n}(\vartheta) \cdot \beta \mathbf{n}^\perp(\vartheta) \\ &= -\beta \gamma(\mathbf{E}, \vartheta) \end{aligned} \tag{6}$$

where $\gamma(\mathbf{E}, \vartheta)$ is the force thermodynamically associated to the change of fiber orientation ϑ, hereafter named *remodeling torque*. In order to respect the fundamental inequality [4], the energy release rate in Eq. (6) calculated at $\mathbf{E}$ must be greater than the energy release rate calculated at any other admissible $\tilde{\mathbf{E}}$:

$$-\beta\big[\gamma(\mathbf{E}, \vartheta) - \gamma(\tilde{\mathbf{E}}, \vartheta)\big] > 0 \tag{7}$$

for any $\mathbf{E}$ and $\tilde{\mathbf{E}}$ in the reversibility domain, $\mathcal{R}(\eta)$. This latter is the set of the symmetric strain tensors that make the remodeling torque stay below the critical material threshold $\eta > 0$:

$$\mathcal{R}(\eta) = \left\{ \mathbf{E} \in \mathbb{S}\mathrm{ym}: \sup_{\vartheta \in [-\pi/2,\ \pi/2]} |\gamma(\mathbf{E}, \vartheta)| \le \eta \right\}. \tag{8}$$

Equivalently, this set can also be interpreted as the strains for which the fiber orientation ϑ cannot evolve, and the material response is purely elastic. Finally, by following [4,5], the dissipation rate is given through the Legendre transform:

$$d(\vartheta)(\dot{\vartheta}) = \sup_{\mathbf{E} \in \mathcal{R}(\eta)} \big(-\partial_\vartheta \hat{\psi}(\mathbf{E}, \vartheta)\dot{\vartheta}\big) = \eta\beta|\dot{\vartheta}| = \bar{\eta}|\dot{\vartheta}|. \tag{9}$$

Now, by considering a quasi-static process over a time interval observation t, the total dissipation is defined as

$$\delta = \hat{\delta}(t) = \int_0^t d(\vartheta(\tau))(\dot{\vartheta}(\tau))d\tau = \bar{\eta}\int_0^t |\dot{\vartheta}(\tau)|d\tau =: \bar{\eta}\bar{\Theta}(t), \tag{10}$$

where the dependency with respect to the time t has been highlighted; $\bar{\Theta}(t)$ represents the accumulated fiber rotation over the time interval t. The total energy is, then, defined by integrating the free energy density Eq. (1) plus the remodeling dissipation Eq. (10) over the body

$$\mathcal{E}(\mathbf{E}(t), \vartheta(t)) = \int_\Omega \left(\hat{\psi}(\mathbf{E}(t), \vartheta(t)) + \hat{\delta}(t)\right)dA. \tag{11}$$

Traction problem

We examine the traction problem sketched in Fig. 2(a). A rectangular sample, of length L and height H, is left free on the upper and lower sides and free to slide on the left side; the horizontal displacement of the points on the right side is equal to $\bar{\mathbf{u}} = \varepsilon L\mathbf{e}_1$, whilst their vertical displacement is left free.

The present section consists of two parts: the first is dedicated to briefly recalling the analytical results obtained in our recent work [6], while the second focuses on the main objective of this paper: the numerical observation of the segregation phenomenon.

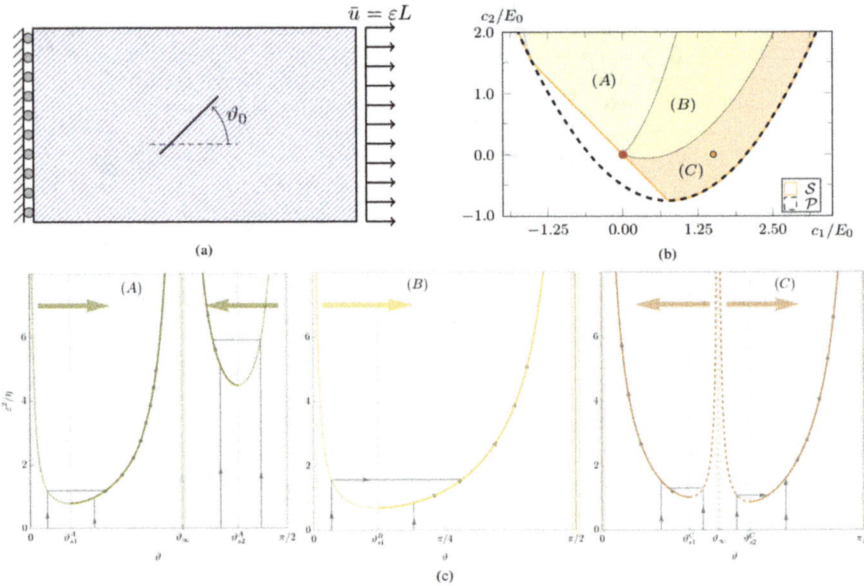

Figure 2: *(a) Schematics of the traction test, ϑ_0 represents the initial fiber orientation, $\bar{u}$ is the imposed displacement acting on the left side of the sample. (b) In the stiffening set $\mathcal{S}$ three different materials are highlighted. (c) ϑ evolution paths (gray continuos lines with arrows) for each case for the three different domains of (b): (A) left, (B) center, and (C) right. Dashed lines represent unstable branches that make the initial fiber orientation jump to the stable branches when the elastic limit is reached.*

Materials Research Proceedings **26** (2023) 17-22 https://doi.org/10.21741/9781644902431-3

Analytical solution for the homogeneous case In [6] it is presented a complete analytical characterization of the homogeneous case in terms of strain and fiber rotation, as the imposed horizontal displacement $\bar{u} = \varepsilon L$ monotonically increases starting from 0. Hence, it is possible to find an asymptotic value of the fiber orientation $\vartheta_\infty(\lambda, \mu, c_1, c_2)$ for $\varepsilon \to \infty$. The stiffening materials are identified in three classes, see Fig. 2(b):

- class (*A*) presents a minimum of the elastic energy ψ in correspondence of $0 \leq \vartheta_\infty \leq \pi/2$[1] and two maxima $\vartheta = \{0, \pi/2\}$. The evolution of the homogeneous traction problem with an initial uniform fiber orientation $\vartheta_0 \in [0, \pi/2]$ is depicted in Fig. 2(c) left. The fiber orientation evolves with jumps if ϑ_0 has been taken on an unstable branch ($0 \leq \vartheta_0 < \vartheta_{s1}^A$ or $\vartheta_{s2}^A \leq \vartheta_0 < \pi/2$), or continuously if ϑ_0 is initially on a stable branch ($\vartheta_{s1}^A \leq \vartheta_0 < \vartheta_\infty$ or $\vartheta_\infty < \vartheta_0 \leq \vartheta_{s2}^A$).
- class (*B*) presents the only minimum of ψ at $\vartheta = \pi/2$; $\vartheta = 0$ is, instead, the maximum. Moreover, ϑ_∞ does not exist in $\mathbb{R}$, see Fig. 2(c) center.
- class (*C*) presents two minima of ψ at $\vartheta = \{0, \pi/2\}$, while ϑ_∞ exists and corresponds to the maximum, see Fig. 2(c) right. In this scenario, it is then possible to observe the segregation in two phases of the fiber orientation, which is the objective of the numerical simulations presented in the next section.

Phase segregation The Lamé constants are chosen $\lambda = \mu = 3/8$ in order to have a unitary Young modulus $E_0 = 1$ and Poisson ratio $\nu_0 = 1/3$. The anisotropic parameters $c_1 = 1.5$ and $c_2 = 0$ are chosen in region (*C*); see the orange dot in Fig. 2(b). We present three study cases where we consider the evolutions of different non-uniform distributions of initial orientation.

- In Fig.s 3(a-d) are displayed the results obtained by tacking into account an initial linear distribution of orientation $\vartheta_0(\mathbf{x})$ that goes from 9° on the left side to 22° on the right one, Fig. 3(a). This specific non-homogeneous distribution is collocated below the ϑ_∞. The imposed displacement grows until the incipient touching of the yield surface Fig. 3(b). The orientation of the fiber starts to decrease with or without jumps; Fig. 3(c) shows one of those intermediate steps. Eventually, the fibers reach a uniform distribution decreasing toward 8°, Fig. 3(d). At this point, the analytical solution, see Fig. 2(c) left, describes the evolution of the orientation until 0°, that is the closest minimum of the elastic energy.
- Fig.s 3(e-h) describe a sample with a distribution of fiber orientated from 52° to 72°. The evolution is similar to the previous one, but in this case the initial distribution $\vartheta_0(\mathbf{x}) > \vartheta_\infty$. Therefore, the fibers tend to reach the minimum collocated at 90° once the elastic limit is overcome, Fig. 3(g). Again, once the field ϑ reaches homogeneity, see Fig. 3(h), the analytical solution provides the evolution.
- Finally, Fig.s 3 from (i) to (l) show the orientation evolution for a sample in which the initial distribution is $\vartheta_0(\mathbf{x}) = (\pi/2)(x_1/L)$. In this case, ϑ_∞ is a value comprised in the initial distribution, expecting the rotation of the fibers going toward the two minima at 0° and 90°. Snapshot (k) shows the beginning of this process, while snapshot (i) shows the left part of the sample completely segregate with respect to the right one. The left part has the fibers initially orientated above $\vartheta_\infty = 30°$.

Conclusions

We have presented a variational model describing the reorientation in a transversely isotropic material. Similarly to plasticity, the model considers the irreversibility of an internal variable, *e.g.*, the fiber orientation, when the remodeling torque reaches a threshold. Then, we focused on the traction problem by recalling the essential analytical findings for the homogeneous case.

[1] It suffices to consider $\vartheta \in [0, \pi/2]$ as $\vartheta \equiv \vartheta + \pi$ and, for the symmetry of the problem, $\vartheta \equiv -\vartheta$.

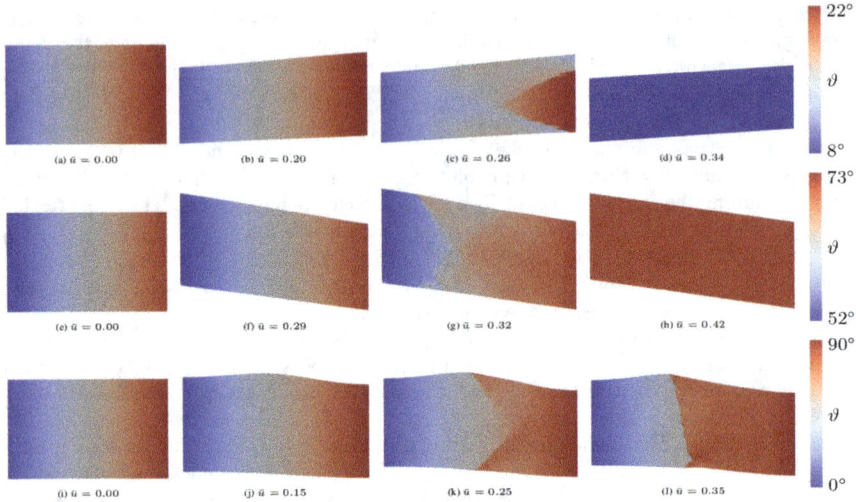

Figure 3: Sequences for a stiffening non-homogeneous material in class (C) characterized by $\boldsymbol{\vartheta}_\infty = \mathbf{30}^\circ$ *with different initial linear distribution of* $\boldsymbol{\vartheta}$. *Sequence from (a) to (d) has* $\boldsymbol{\vartheta}_0(\boldsymbol{x}) < \boldsymbol{\vartheta}_\infty \forall \boldsymbol{x} \in \boldsymbol{\mathcal{B}}$. *Sequence from (e) to (h) has* $\boldsymbol{\vartheta}_0(\boldsymbol{x}) > \boldsymbol{\vartheta}_\infty \forall \boldsymbol{x} \in \boldsymbol{\mathcal{B}}$. *Finally (i)-(l) has* $\boldsymbol{\vartheta}_0(\boldsymbol{x}) = (\boldsymbol{\pi}/\mathbf{2})(\boldsymbol{x}_1/\boldsymbol{L})$.

It has been possible to divide the material responses into three categories based on their energetic characteristics. Materials in class (*C*), characterized by double-well energy, are suitable for observing phase transitions and, eventually, segregation. In this context, we have presented a numerical case study where an initial inhomogeneous distribution of fiber orientation leads to the segregation of the internal variable.

References

[1] G. Grekas, M. Proestaki, P. Rosakis, J. Notbohm, C. Makridakis, and G. Ravichandran, Cells exploit a phase transition to mechanically remodel the fibrous extracellular matrix, Journal of the Royal Society Interface, vol. 18, no. 175, p. 20200823, (2021). https://doi.org/10.1098/rsif.2020.0823

[2] E. Ban, J. M. Franklin, S. Nam, L. R. Smith, H. Wang, R. G. Wells, O. Chaudhuri, J. T. Liphardt, and V. B. Shenoy, Mechanisms of Plastic Deformation in Collagen Networks Induced by Cellular Forces, Biophysical Journal, vol. 114, no. 2, pp. 450–461, (2018). https://doi.org/10.1016/j.bpj.2017.11.3739

[3] B. Buchmann, L. Engelbrecht, P. Fernandez, et al., Mechanical plasticity of collagen directs branch elongation in human mammary gland organoids, Nature Communication, vol. 12, no. 2759, (2021). https://doi.org/10.1038/s41467-021-22988-2

[4] B. Halphen and Q. Nguyen, Sur les matériaux standards généralisés, Journal de Mécanique, vol. 14, pp. 39–63, 01 (1975).

[5] J.-J. Marigo, Constitutive relations in plasticity, damage and fracture mechanics based on a work property, Nuclear Engineering and Design, vol. 114, pp. 249–272, (1989). https://doi.org/10.1016/0029-5493(89)90105-2

[6] A. Favata, A. Rodella, and S. Vidoli, An internal variable model for plastic remodeling in fibrous materials, European Journal of Mechanics - A/Solids, p. 104718, (2022). https://doi.org/10.1016/j.euromechsol.2022.104718

[7] B. Coleman and M. Gurtin, Thermodynamics with internal state variables, The Journal of Chemical Physics, vol. 47, no. 2, pp. 597–613, (1967). https://doi.org/10.1063/1.1711937

[8] P. Germain, The method of virtual power in continuum mechanics. Part 2: Microstructure, SIAM Journal on Applied Mathematics, vol. 25, no. 3, pp. 556–575, (1973). https://doi.org/10.1137/0125053

Theoretical and Applied Mechanics - AIMETA 2022 Materials Research Forum LLC
Materials Research Proceedings **26** (2023) 23-28 https://doi.org/10.21741/9781644902431-4

A coupled thermo-mechanical and neutron diffusion numerical model for irradiated concrete

Jiangkun Zhang[1, a *], Beatrice Pomaro [1, b], Gianluca Mazzucco [1, c], Beaudin Freinrich Dongmo[1, d], Carmelo Maiorana[1, e] and Valentina Salomoni[2, f]

[1]Department of Civil, Environmental and Architectural Engineering (DICEA), University of Padova, via F. Marzolo 9, 35131, Padova, Italy

[2]Department of Management and Engineering (DTG), University of Padova, Stradella S. Nicola 3, 36100, Vicenza, Italy

[a]jiangkun.zhang@phd.unipd.it, [b]beatrice.pomaro@unipd.it, [c]gianluca.mazzucco@unipd.it, [d]beaudinfreinrich.dongmo@phd.unipd.it, [e]carmelo.maiorana@unipd.it, [f]valentina.salomoni@unipd.it

Keywords: Finite Element Method, Neutron Diffusion, Irradiated Concrete, Coupled Problem

Abstract: Neutron irradiation plays an important role in nuclear-induced degradation for concrete shielding materials, specifically in determining the radiation induced volume expansion (RIVE) phenomenon driving its failure. When analyzing at the structural level the effects of nuclear radiation on concrete, a non-uniformed distribution of neutron radiation must be considered. This can be done via particle transport calculations preventive to the thermo-mechanic study, or by solving numerically the coupled set of governing equations of the problem. In this work the second approach is pursued in the theoretical framework of the Finite Element Method (FEM). The proposed formulation not only considers an accurate neutron transport model based on the two-group theory, but also it includes the effects induced by thermal neutrons to the temperature field. The formulation lends itself to include RIVE and the other relevant radiation induced effects on the mechanical field. The governing equations are presented and discussed, and some results obtained by using the general 3D numerical formulation proposed herein are compared to results from literature obtained via analytical methods addressing simplified 1D problems.

Introduction

The biological shielding walls undergo a coupling effect of irradiation and temperature rise during the operation of nuclear power plants (NPPs). Both, neutron and gamma radiation produce internal heating from absorption of radiation energy. Additionally, at high fluence levels, they affect the microstructure and induce changes in certain mechanical properties of concrete (e.g., compressive strength, tensile strength, modulus of elasticity) [1]. Since neutron radiation can deteriorate the mechanical properties of concrete materials, it is critical to obtain accurate neutron radiation levels in concrete structures during their service life. Often, this problem is addressed via analytical simplified 1D studies [2].

For more complex geometries, however, the Finite Element Method (FEM) is generally accepted as an effective approach for solving coupled transient problems, also when porous multiphase materials are involved [3-5]. Based on FEM, some diffusion-deformation and diffusion-reaction-deformation coupling models are proposed in literature [6, 7]. As far as the authors know, a FEM thermo-mechanical-neutron diffusion coupled model for irradiated concrete has not yet been presented.

The thermo-mechanical-neutron diffusion coupled model proposed herein combines an elastic mechanical constitutive model with the two-group neutron-transport theory, and thermal diffusion.

This model can be used to quantify a realistic radiation flux, and temperature field on the domain of interest, by considering both, the fast neutron and thermal neutron fractions. In fact, both temperature and the neutron flux account for relevant state variables to predict damage and durability of biological shields in the NPPs and nuclear facilities, in general.

Mathematical Model

The governing equations of the multi-physics problem affecting irradiated concrete are outlined in this section. A general three-dimensional body is considered, that occupies a domain $B \in \mathbb{R}^3$ and is bounded by the surface S. Vector $\boldsymbol{X}$ is used for representing the position of an arbitrary material point of the body in a Cartesian coordinate system, that is $\boldsymbol{X} = [x, y, z]^T$.

Mechanical Model. The constitutive model for concrete is defined based on the small strain theory. Thermal effects and radiation-induced effects, are assimilated to an additional mechanical contribution in the definition of the total strain, according to an additive decomposition of the same strain (in the following, ∇ stands for the Nabla operator and indicates the gradient operator, while ∇^{T} indicates the divergence). Let S_{t} and S_{u} be complementary sub-surfaces of the boundary S of the body B, then the mechanical boundary conditions complete the set of equations governing the mechanical field. According to them, the surface loads $\bar{\mathbf{t}}$ are specified on S_{t}, characterized by normal outward vector $\mathbf{n}$ to the surface, and the displacement $\bar{\mathbf{u}}$ is specified on S_{u} (t denotes time). Hence,

$$\begin{cases} \mathbf{L}^{\mathrm{T}}\boldsymbol{\sigma} + \boldsymbol{b} = \mathbf{0} \\ \boldsymbol{\sigma} = \mathbf{D}\boldsymbol{\varepsilon}_e = \mathbf{D}(\boldsymbol{\varepsilon} - \boldsymbol{\varepsilon}_T - \boldsymbol{\varepsilon}^{RIVE}) = \mathbf{D}\,\mathbf{L}(\boldsymbol{u} - \boldsymbol{u}_T - \boldsymbol{u}^{RIVE}) \\ \boldsymbol{\varepsilon}_T = \mathbf{I}\alpha(T - T_0) \\ \boldsymbol{\varepsilon}^{RIVE} = \mathbf{I}\kappa\varepsilon_{max}\frac{e^{\delta\Phi}-1}{\varepsilon_{max}+\kappa e^{\delta\Phi}} \\ \boldsymbol{\sigma}\,\mathbf{n} = \bar{\mathbf{t}}(\boldsymbol{X}, t) \quad \text{on } S_{\mathrm{t}} \\ \mathbf{u} = \bar{\mathbf{u}}(\boldsymbol{X}, t) \quad \text{on } S_{\mathrm{u}} \end{cases} \tag{1}$$

where $\boldsymbol{\sigma}$ is the Cauchy stress; $\boldsymbol{b}$ is the body force per unit volume; $\boldsymbol{u}$ is the total displacement; $\mathbf{L}$ is the differential operator such that $\boldsymbol{\varepsilon} = \mathbf{L}\,\boldsymbol{u}$; $\boldsymbol{u}_T$ is the displacement component due to temperature, and $\boldsymbol{u}^{RIVE}$ is the component due to radiation; $\boldsymbol{\varepsilon}$ is the total strain vector; $\boldsymbol{\varepsilon}_e$ is the purely mechanical strain vector; $\boldsymbol{\varepsilon}_T$ is the strain induced by temperature, and $\boldsymbol{\varepsilon}^{RIVE}$ the one induced by radiation, i.e. RIVE; $\mathbf{D}$ is the elastic constitutive matrix; α is the thermal expansion coefficient (assumed isotropic). This model accounts for an estimate of RIVE in function of the neutron fluence Φ [n/cm^2] in line with the sigmoidal growth model proposed in [8], where $\mathbf{I} = [1\ 1\ 1\ 0\ 0\ 0]^T$ is the identity vector, and ε_{max}, κ, and δ are material parameters calibrated over irradiated concrete specimen of different aggregate nature.

Multi-group Neutron Diffusion Problem. Neutrons are either scattered in, attenuated or stopped by the absorbing medium during their transport processes. In this work the two-speed neutron diffusion model proposed in [9-11] is considered. This model is more realistic with respect to the one-speed diffusion model in relation to the fact that neutrons have a wide range of energy spectrum, in reality, ranging from 10MeV down to 0.01MeV. The two-group model assumes that the secondary neutrons from scattering reaction occurring within a specific kinetic group (fast group) represent a neutron source in the diffusion process of the group of lower kinetic energy (thermal group). The governing equation for the two-group theory for the so-called fast and thermal neutrons are

$$\text{Fast group:} \quad \frac{1}{v_1}\frac{\partial\phi_1(\mathbf{X},t)}{\partial t} - \nabla^{\mathrm{T}}(\boldsymbol{D}_1(\mathbf{X},t)\nabla\phi_1(\mathbf{X},t)) + \Sigma_{R1}\phi_1(\mathbf{X},t) = 0 \tag{2}$$

$$\text{Thermal group:} \quad \frac{1}{v_2}\frac{\partial\phi_2(\boldsymbol{X},t)}{\partial t} - \nabla^{\mathrm{T}}(\boldsymbol{D}_2(\mathbf{X},t)\nabla\phi_2(\mathbf{X},t)) + \Sigma_{a2}\phi_2(\boldsymbol{X},t) = \Sigma_{s12}\phi_1(\boldsymbol{X},t) \tag{3}$$

where subscripts $i = 1, 2$ stand for the fast and the thermal group, respectively. In Eq. (2) and Eq. (3) ϕ_i is the scalar neutron flux [n/(cm^2·s)]; v_i is the neutron speed; $\boldsymbol{D}_i$ is the neutron diffusion coefficient matrix; Σ_a is the macroscopic absorption cross section; Σ_R is the macroscopic removal cross section; Σ_{S12} is the macroscopic fast to thermal group-transfer cross section.

Let $S_{\phi 1}$, $S_{\phi 2}$ be complementary sub-surfaces of the boundary S of the body B, then the problem is completed by proper Dirichlet conditions of the kind

$$\phi_1 = \overline{\phi_1}\,(\boldsymbol{X}, t) \qquad \text{on } S_{\phi 1} \tag{4}$$

$$\phi_2 = \overline{\phi_2}\,(\boldsymbol{X}, t) \qquad \text{on } S_{\phi 2.} \tag{5}$$

Heat Conduction Problem. The temperature at which the shielding concrete is working during normal operations, along with the gamma radiation heating, can cause non-uniform thermal strains in concrete, at the scale of its constituents, so driving deterioration [12]. On the other hand, the neutron radiation also generates heat during the transport and attenuation processes, which could alter the thermal field in the wall. The classical heat conduction equation is

$$\rho c_p \frac{\partial T(\boldsymbol{X},t)}{\partial t} = \nabla^{\mathrm{T}}\big(\mathbf{k}\nabla T(\boldsymbol{X},t)\big) + Q(\boldsymbol{X},t) \tag{6}$$

where c_p is the specific heat capacity; ρ is the mass density; T is temperature; $\mathbf{k}$ is the thermal conductivity matrix; Q is the volumetric heat source.

Let S_1, S_2, S_3 be complementary sub-surfaces of the boundary S of the body B. Then, proper boundary conditions to the problem include a specified temperature on S_1, a specified heat flux on S_2, and specified convection boundary conditions on S_3

$$T = \bar{T}(\boldsymbol{X}, t) \qquad \text{on } S_1 \tag{7}$$

$$-\big(\mathbf{k}\nabla T(\boldsymbol{X},t)\big)\,\mathbf{n} = \bar{q}\,(\boldsymbol{X},t) \qquad \text{on } S_2 \tag{8}$$

$$-\big(\mathbf{k}\nabla T(\boldsymbol{X},t)\big)\,\mathbf{n} = h(T(\boldsymbol{X},t) - T_e) \qquad \text{on } S_3 \tag{9}$$

where $\bar{T}$ is the prescribed temperature; $\bar{q}$ is the prescribed heat flux; h is the heat transfer coefficient; T_e is the environment temperature; and T is the unknown temperature at the boundary.

Based on the two-group neutron diffusion model, heat generated either by absorption or scattering of neutrons is in good approximation due to the thermal neutron fraction. This is true as long as one accepts that heat due to scattering is negligible, if compared to heat generated by the capture of thermal neutrons. The neutron radiation induced heating rate by absorption can be estimated as [2]

$$Q(\boldsymbol{X},t) = 1.6 \times 10^{-13}\, \Sigma_c \mathcal{E}_b \phi_2(\boldsymbol{X},t) \tag{10}$$

where Σ_c is the macroscopic neutron capture cross section, and $\mathcal{E}_b$ is the binding energy for the capture process.

Weak formulation

The global coupled system of governing equations yields

$$\begin{cases} \mathbf{L}^{\mathrm{T}}\boldsymbol{\sigma} + \boldsymbol{b} = \mathbf{0} \\ \rho c_p \frac{\partial T(\boldsymbol{X},t)}{\partial t} = \nabla^{\mathrm{T}}\left(\mathbf{k}\nabla T(\boldsymbol{X},t)\right) + 1.6 \times 10^{-13}\Sigma_c \mathcal{E}_b \phi_2(\boldsymbol{X},t) \\ \frac{1}{v_1}\frac{\partial \phi_1(\boldsymbol{X},t)}{\partial t} - \nabla^{\mathrm{T}}\left(\boldsymbol{D_1}(\mathbf{X},t)\nabla\phi_1(\mathbf{X},t)\right) + \Sigma_{R1}\phi_1(\boldsymbol{X},t) = 0 \\ \frac{1}{v_2}\frac{\partial \phi_2(\boldsymbol{X},t)}{\partial t} - \nabla^{\mathrm{T}}\left(\boldsymbol{D_2}(\mathbf{X},t)\nabla\phi_2(\mathbf{X},t)\right) + \Sigma_{a2}\phi_2(\boldsymbol{X},t) = \Sigma_{s12}\phi_1(\boldsymbol{X},t). \end{cases} \quad (11)$$

A numerical solution to the boundary value problem is sought in agreement with the weighted residual method, that minimizes the residual of the differential problem over the domain B and the boundary S. In line with the FEM approach the body B can be discretized into finite elements $B_e | B = \cup B_e$, and the approximate solution for the state variables $\boldsymbol{u}$, T, ϕ_1, and ϕ_2 within each element are sought as linear combination of the shape functions $\mathbf{N}$ defined at the nodes of the element, and the solution at the same nodes.

By applying the Galerkin variational approach. The coupled problem illustrated herein reduces, therefore, to the following system

$$\begin{bmatrix} 0 & 0 & 0 & 0 \\ 0 & \mathbf{M}_T & 0 & 0 \\ 0 & 0 & \mathbf{M}_{\phi_1} & 0 \\ 0 & 0 & 0 & \mathbf{M}_{\phi_2} \end{bmatrix} \begin{bmatrix} \dot{\boldsymbol{u}} \\ \dot{\boldsymbol{T}} \\ \dot{\boldsymbol{\phi}}_1 \\ \dot{\boldsymbol{\phi}}_2 \end{bmatrix} + \begin{bmatrix} \mathbf{K}_{uu} & 0 & 0 & 0 \\ 0 & \mathbf{K}_T & 0 & \mathbf{K}_{T\phi_2} \\ 0 & 0 & \mathbf{K}_{\phi_1} & 0 \\ 0 & 0 & \mathbf{K}_{\phi_1\phi_2} & \mathbf{K}_{\phi_2} \end{bmatrix} \begin{bmatrix} \boldsymbol{u} \\ \boldsymbol{T} \\ \boldsymbol{\phi}_1 \\ \boldsymbol{\phi}_2 \end{bmatrix} = \begin{bmatrix} \boldsymbol{f}_u \\ \boldsymbol{f}_T \\ \boldsymbol{f}_{\phi_1} \\ \boldsymbol{f}_{\phi_2} \end{bmatrix} \quad (12)$$

where $\boldsymbol{u}$, $\boldsymbol{T}$, $\boldsymbol{\phi}_1$, and $\boldsymbol{\phi}_2$ are the vectors of the nodal values of the state variables.
The coupled system can be solved by using a finite difference scheme in time, and the Newton–Raphson scheme.

Numerical Results

A slice of biological shield is considered in the numerical analysis of dimensions 100×100×80 cm. The temperature at the inner surface is set at 65°C, with an outer temperature of 20°C. Thermal convection boundary condition is imposed on the outer surface, by considering a heat transfer coefficient with air, while thermal symmetrical boundary conditions are imposed on the upper, down and lateral surfaces. On the inner surface a uniform constant value of 3.2×10^{10} n/(cm^2 s) for the fast neutron flux, and 4×10^{10} n/(cm^2 s) for the thermal neutron flux is assumed. The flux is directed outwards and involves the 80 cm thickness of the wall. The time of analysis is 1 year; the time step used in the analysis is 1×10^{-6} s. The concrete properties used in the analyses are listed in Table 1.

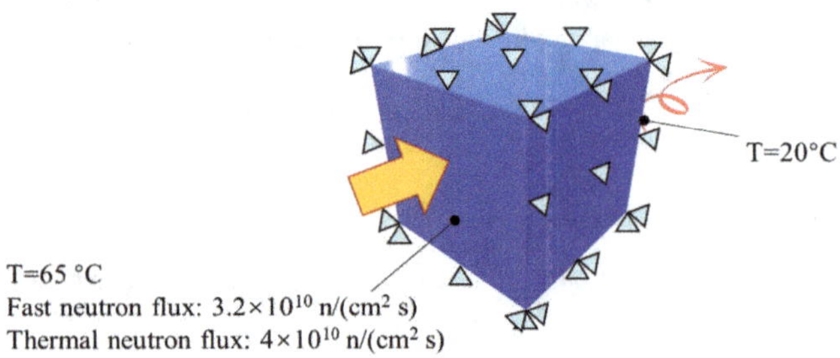

Figure 1: Studied portion of biological shielding wall with boundary conditions.

Materials Research Proceedings **26** (2023) 23-28 https://doi.org/10.21741/9781644902431-4

Table 1: Parameters used in the numerical example (thermal, and neutron diffusion parameters are taken from reference [2]).

E [MPa]	ν	α [°C^{-1}]	k [W/(cm·°K)]	ρ [kg/cm^3]	c_p [J/(kg·°K)]	h [W/(cm^2·°K)]
30000	0.2	$1\cdot10^{-5}$	8.7×10^{-3}	2.3×10^{-3}	650	0.0025

D_1 [cm]	D_2 [cm]	Σ_{R1} [cm^{-1}]	Σ_{s12} [cm^{-1}]	Σ_{a2} [cm^{-1}]	v_1 [cm/s]	v_2 [cm/s]	Σ_c [cm^{-1}]	$\mathcal{E}_b$ [MeV]
1.14	0.484	0.085	0.08	0.0094	4.37×10^8	2.2×10^5	0.0094	5.5

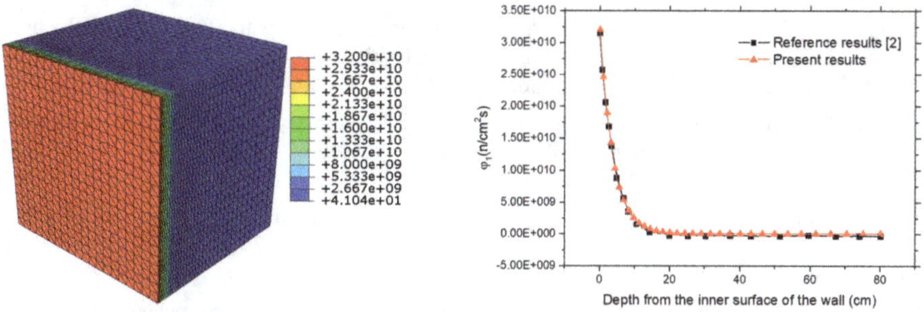

Figure 2: (a) Fast neutron flux field in the model; (b) fast neutron flux along the thickness of the wall.

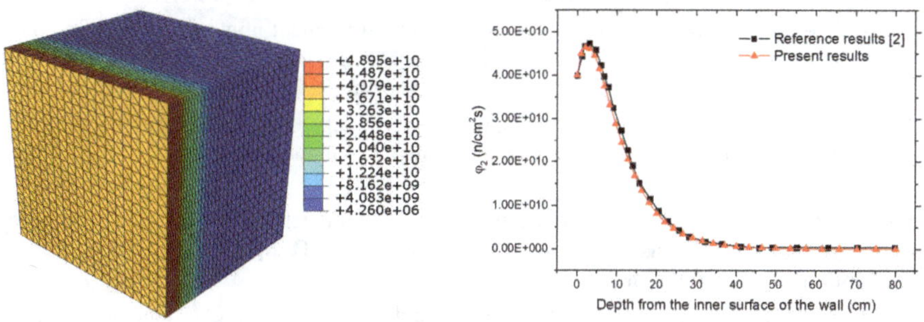

Figure 3: (a) Thermal neutron flux field in the model; (b) Thermal neutron flux along the thickness of the wall;

In Fig. 3 (b) a peculiar behavior of the thermal neutron flux at steady-state close to the hot surface is envisaged, showing a slight increase in this region, and a subsequent decrease in space, which indicates the attenuation of fast neutrons in the wall. On the other hand, the fast neutron flux (Fig. 2) shows a monotonic behavior, as expected.

Summary

Neutron radiation can deteriorate the mechanical properties of concrete materials at long term, therefore it is critical to obtain accurate neutron radiation levels in concrete structures during their service life.

A thermo-mechanical-neutron diffusion coupled formulation is proposed in this paper, which combines the mechanical constitutive model with the two-group neutron-transport theory, and the thermal diffusion theory. Some preliminary results are shown, that demonstrate the capability of the model to catch the main coupling mechanisms associated to the three problems, and, specifically the influence of the two kinetic groups of neutron fluxes to each other in the attenuation/absorption process within the shielding wall, as well as the local temperature increase due to neutron absorption.

References

[1] NRC, Standard review plan for review of subsequent license renewal applications for nuclear power plants, NUREG-2192, Office of Nuclear Reactor Regulation, Washington, DC, 2017.

[2] Y. Jing, Y. Xi, Modelling long-term distribution of fast and thermal neutron fluence in degraded concrete biological shielding walls, Construction and Building Materials, 292 (2021) 123379. https://doi.org/10.1016/j.conbuildmat.2021.123379

[3] C.E. Majorana, V.A. Salomoni, B.A. Schrefler, Hygrothermal and mechanical model of concrete at high temperature, Materials and Structures, 31(210) (1998) 378-386. https://doi.org/10.1007/BF02480710

[4] V.A. Salomoni, G. Mazzucco, C.E. Majorana, Mechanical and durability behaviour of growing concrete structures, Engineering Computations, 24 (2007) 536-561. https://doi.org/10.1108/02644400710755906

[5] B. Pomaro, G. Xotta, Salomoni V.A., Majorana C.E., A thermo-hydro-mechanical numerical model for plain irradiated concrete in nuclear shielding, Materials and Structures 55(1) (2022) 14. https://doi.org/10.1617/s11527-021-01844-1

[6] S. A. Chester, C. V. Di Leo, and L. A. Anand, Finite Element implementation of a coupled diffusion-deformation theory for elastomeric gels, International Journal of Solids and Structures, 52 (2015) 1-18. https://doi.org/10.1016/j.ijsolstr.2014.08.015

[7] B. Qin, Y. Zhou, and Z. Zhong, A computational model for diffusion-reaction-deformation coupled problems and its finite element implementation, Engineering Computations 39 (2021) 837-857. https://doi.org/10.1108/EC-02-2021-0084

[8] Y. Le Pape, Structural effects of radiation-induced volumetric expansion on unreinforced concrete biological shields, Nuclear Engineering and Design, 295 (2015) 534-548. https://doi.org/10.1016/j.nucengdes.2015.09.018

[9] B. T. Price, K. T. Horton Spinney, Radiation shielding, Pergamon Press, London, UK, 1957.

[10] R.G. Jaeger, Engineering compendium on radiation shielding, Volume 2, Springer, Berlin, Germany, 1970.

[11] M.F. Kaplan, Concrete radiation shielding: nuclear physics, concrete properties, design, and construction, John Wiley & Sons, NY, USA, 1989.

[12] Y. Khmurovska, P. Štemberk, T. Fekete, and T. Eurajoki, Numerical analysis of VVER-440/213 concrete biological shield under normal operation, Nuclear Engineering and Design, 350 (2019) 58-66. https://doi.org/10.1016/j.nucengdes.2019.05.004

Theoretical and Applied Mechanics - AIMETA 2022 Materials Research Forum LLC
Materials Research Proceedings **26** (2023) 29-34 https://doi.org/10.21741/9781644902431-5

Fracturing process in an anisotropic layered geomaterial: theoretical and computational predictions

Martina Rinaldi[1,a], Marco Trullo[1,b], Francesco Tornabene[1,c] and Rossana Dimitri[1,d*]

[1]Department of Innovation Engineering, Università del Salento, 73100, Lecce, Italy

[a]martina.rinaldi@unisalento.it, [b]marco.trullo@unisalento.it, [c]francesco.tornabene@unisalento.it, [d*]rossana.dimitri@unisalento.it

Keywords: Anisotropic Geomaterials, Fracture Mechanics, XFEM

Abstract. The present work investigates the fracturing behavior of an anisotropic layered geomaterial, known as Opalinus Clay (OPA). The formation of this rock is mainly related to a sedimentation process, where bedding planes correspond to planes of isotropy. OPA is here studied because of its good properties, primarily, the low permeability and high adsorption capability, which make it a perfect candidate for the storage of radioactive waste. The characterization of this rock takes place experimentally in the Mont Terri Rock Laboratory, in the northern Switzerland, with an increased attention to theoretical and computational predictions. In this context, this work aims at simulating the nonlinear crack behavior of OPA by using the eXtended Finite Element Method (XFEM) and damage mechanics. The study is applied on a Semi-Circular specimen under a Bending load (SCB), whose fracturing response is investigated in terms of peak load and direction of the cracking propagation for different notch dimensions and geometries.

Introduction

An underground capability in which radioactive waste can be permanently stored, is called deep geological repository. Based on studies conducted in the Mont Terri Rock laboratory, OPA has been identified as a capable material of hosting high-level radioactive Waste. In order to define the feasibility of such repository, the rock must be characterized in its fracturing behavior, which is highly affected by its mechanical anisotropy [1-3]. In the last years, Barbi et al. [4] applied on OPA the design parameters defined experimentally by Bock [5], i.e. deformability, strength, permeability and hydro-mechanically-coupled parameters, and studied the orientation-dependence of fracture mechanics parameters for SCB tests. Following the studies [4,6] , and based on the pioneering work by Bock [5], we here evaluate comparatively, the SCB behavior as provided by an advanced computational XFEM and damage concepts. This falls within a wider numerical investigation, as provided in Ref. [7], focusing on the force-CMOD response, together with the specimen toughness and kinematic behavior.

Due to the efficient kinematic approximation at the interfacial level, and the proper introduction of discontinuous enrichment functions, the XFEM is capable to avoid very fine mesh refinements ahead of the crack tip, to follow a crack propagation [8-12]. The main XFEM basics are, thus, applied to model the crack propagation in transversely-isotropic geomaterials, accounting for a standard SCB test, which is explored in its fracturing response for different notch shapes and sizes, in terms of global response, maximum strength, toughness and cracking propagation. After a brief description of the XFEM theoretical basics, we discuss about the numerical investigation and results, focusing on the overall sensitivity of the response.

Theoretical Formulation

The main basics of XFEM are briefly recalled in this section, before its application on the fracturing process of anisotropic solids as OPA. Based on a classical Galerkin method, in addition to the usual global shape functions applied to describe generic nodes, some enrichment functions are added for cracked elements to account for discontinuities in the displacement field, as singularities at the crack-tip. Starting from the general shape functions for the entire domain of nodes, the enriched approximation for the displacement function is defined as [8-12]

$$u^h(x)=\sum_{i\in I}u_iN_i+\sum_{j\in J}b_jN_jH(x)+\sum_{k\in K}N_k\left[\sum_{l=1}^{4}c_k^lF_l(x)\right] \tag{1}$$

in which x stands for the global coordinate, N_i are the shape functions referred to an arbitrary node i, u_i are the corresponding degrees of freedom, $H(x)$ is the Heaviside function which accounts for the displacement jump, N_j refers to the shape function at the discontinuity node j, and b_j represents the additional degrees of freedom associated with $H(x)$. Moreover, $F_l(x)$ are the enrichment functions at the crack-tip, N_k are the shape functions associated to the crack-tip functions at node k, while c_k^l are the other degrees of freedom for the elastic asymptotic crack-tip enrichment functions.

In an enriched element, the degradation process starts taking place, when some fixed crack initiation criteria are satisfied, which can be referred either to the stress and strain field. In this perspective, we select a maximum nominal stress criterion, i.e.

$$\max\left\{\frac{\langle p_n\rangle}{p_n^o},\frac{p_s}{p_s^o},\frac{p_t}{p_t^o}\right\}=1 \tag{2}$$

where $\langle\ \rangle$ is the Macauley bracket, and $\mathbf{p}$ is the nominal traction stress vector, defined by means of its normal component p_n, and shear components p_s, p_t, whereas p_n^o, p_s^o, p_t^o are the associated peak values. Based on relation (2), the damage process starts once the maximum nominal stress ratio reaches the unit value. Such stress components follow the material degradation defined by the damage variable D, i.e.

$$p_n=\begin{cases}(1-D)P_n & P_n\geq 0\\ P_n & \text{otherwise}\end{cases} \tag{3}$$

$$p_s=(1-D)P_s \tag{4}$$

$$p_t=(1-D)P_t \tag{5}$$

where P_n, P_s, P_t stand for the normal and shear stress components as predicted by an elastic traction-separation behavior for an undamaged strain field. On the contrary, for a compressive state $P_n<0$, the material behaves elastically, without any presence of damage. The scalar damage parameter assumes a null value before the damage initiation, and it increases monotonically up to the unit value, during the damage evolution until the complete damage. More specifically, the damage parameter is assumed to follow an exponential evolution of the type

$$D=1-\left\{\frac{\delta_m^0}{\delta_m^{\max}}\right\}\left\{1-\frac{1-\exp\left(-\alpha\left(\frac{\delta_m^{\max}-\delta_m^0}{\delta_m^f-\delta_m^0}\right)\right)}{1-\exp(-\alpha)}\right\} \tag{6}$$

where the damage evolution is represented by the dimensionless parameter α, while the effective displacement at the damage initiation, the maximum value of the effective displacement reached during the loading steps and the effective displacement at complete failure are, labeled, respectively, as $\delta_m^0, \delta_m^{\max}$ and δ_m^f.

Numerical Investigation

We start now the numerical analysis for the OPA-based SCB, implemented in the code as a transversely isotropic material with bedding planes aligned by 45° with respect to the borehole axis, as suggested in Refs. [5-7]. Elastic properties derived from in-situ and laboratory tests are herein assigned to OPA, according to a quasi-brittle and transversely isotropic constitutive law of the type

$$\begin{Bmatrix}\varepsilon_{xx}\\ \varepsilon_{yy}\\ \varepsilon_{zz}\\ \gamma_{xy}\\ \gamma_{xz}\\ \gamma_{yz}\end{Bmatrix}=\begin{bmatrix}\frac{1}{E_x} & \frac{-\nu_{xy}}{E_x} & \frac{-\nu_{xz}}{E_z} & 0 & 0 & 0\\ \frac{-\nu_{xy}}{E_x} & \frac{1}{E_x} & \frac{-\nu_{yz}}{E_z} & 0 & 0 & 0\\ \frac{-\nu_{xz}}{E_z} & \frac{-\nu_{yz}}{E_z} & \frac{1}{E_z} & 0 & 0 & 0\\ 0 & 0 & 0 & \frac{2(1+\nu_{xy})}{E_x}=\frac{1}{G_{xy}} & 0 & 0\\ 0 & 0 & 0 & 0 & \frac{1}{G_{xz}} & 0\\ 0 & 0 & 0 & 0 & 0 & \frac{1}{G_{xz}}\end{bmatrix}\begin{Bmatrix}\sigma_{xx}\\ \sigma_{yy}\\ \sigma_{zz}\\ \sigma_{xy}\\ \sigma_{xz}\\ \sigma_{yz}\end{Bmatrix} \tag{7}$$

where $E_z=4$ GPa, $\nu_{xy}=0.33$, $\nu_{xz}=0.24$, $G_{xz}=1.2$ GPa, with $G_{xy}\neq G_{yz}=G_{xz}$ and $\nu_{xz}=\nu_{yz}$ [7]. The specimen has diameter $D=80$ mm and thickness $S=6$ mm, and it is subjected to a downward displacement of 0.1 mm in the mid-side of the top face (Fig. 1). The specimen is discretized with tetrahedral elements of quadratic geometric order (C3D10), as also applied in the extended work by Dimitri et al. [7]. The damage evolution (defined in the same notation as in the Abaqus code) is controlled by a constitutive law with a softening branch of the following type

$$\frac{p_n}{p_n^0}=1-\frac{1-e^{\frac{-x\delta_m}{\delta_m^f}}}{1-e^{-\alpha}} \tag{8}$$

where p_n^0 is the parallel-to-bedding tensile strength of the material, here kept equal to 2 MPa; δ_m is the displacement discontinuity, while δ_m^f is its associated critical value, assumed as $\delta_m^f=0.1$mm; and α is kept equal to 5.

Materials Research Proceedings **26** (2023) 29-34 https://doi.org/10.21741/9781644902431-5

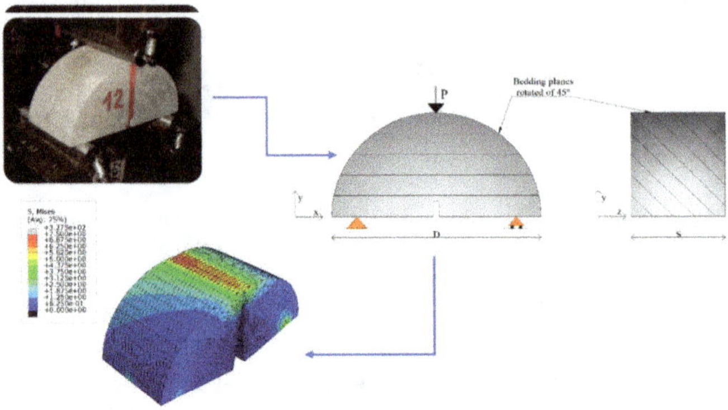

Figure 1. *An SCB test modeled starting from experimental results.*

Based on these assumptions, a sensitivity study is performed to investigate the influence of the notch shape and dimension on the response of the SCB specimen. To this end, four different notch shapes are taken into account as shown in Fig. 2.

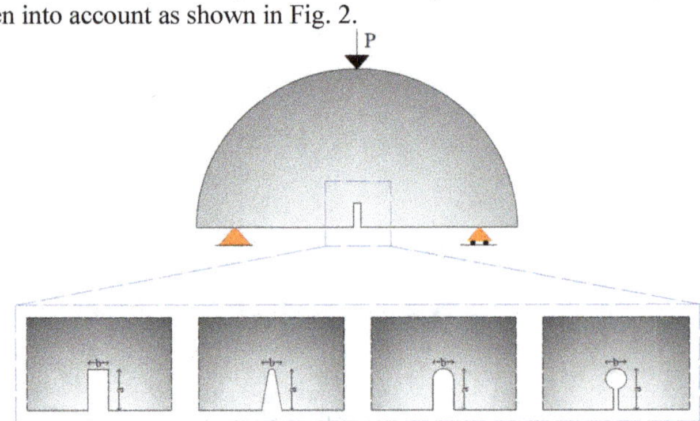

(a) (b) (c) (d)

Figure 2. *Notch shapes investigated: (a) rectangular notch, (b) V-notch, (c) U-notch, (d) key-hole like notch.*

The notch depth $a = 6$ mm is kept constant, while the notch width b is varied from 1 mm up to 4 mm by steps of 1 mm. The V-notch tip has a connecting radius of 0.5 mm for each value of b, so that, for $b = 1$ mm it reverts to the U-notch with the same dimensions. The global force-CMOD response of the specimens (Fig. 3) shows a slight dependence of the curves to the notch shape at least in the maximum strength and descending branches, with possible variations in the overall ductility, together with small variations in the crack path, around the vertical direction. For each selected notch shape with $b = 2$ mm, in Fig. 4 we display some principal stress contour plots at the same time step.

Materials Research Proceedings **26** (2023) 29-34 https://doi.org/10.21741/9781644902431-5

XFEM

Notch depth $a = 6\text{mm}$

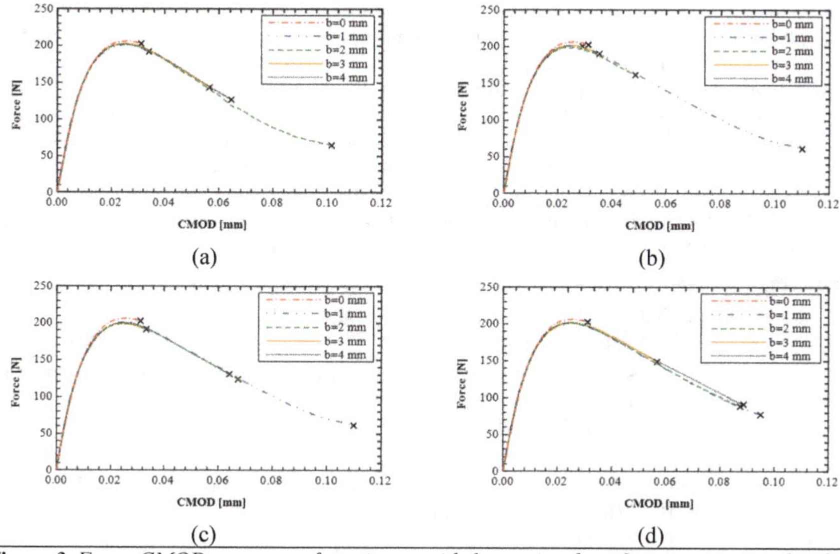

Figure 3. *Force-CMOD response of specimens with b varying from 0 mm to 4 mm and with: (a) rectangular notch, (b) U-notch, (c) V-notch, (d) key-hole-like notch.*

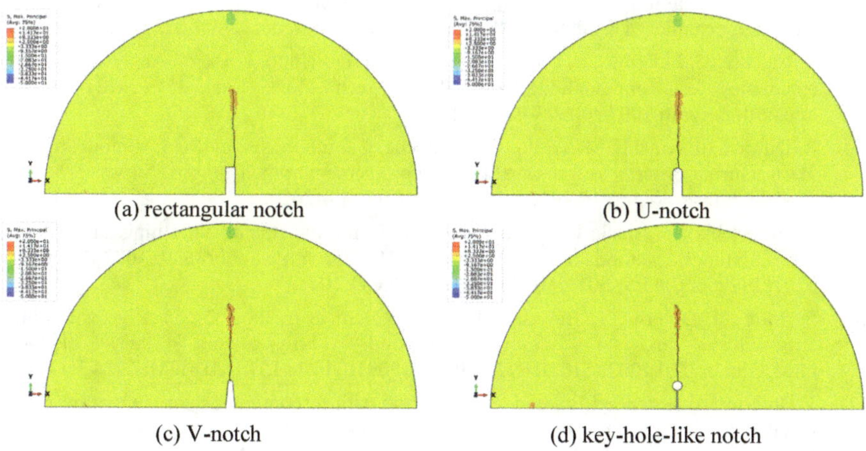

Figure 4. *Principal stress distributions for the specimen with* $a = 6\text{mm}$ *and* $b = 2\text{mm}$.

It is worth observing that the crack propagates in the same vertical direction in all specimens. Furthermore, the principal stress distributions in the entire solid are almost the same for an equal crack length. In addition, the position of the crack initiation corresponds to the notch vertex for the rectangular shape (Fig. 4a), while it is always located in the center of the circular notch tip for the other shapes (Fig. 4b-c-d).

Conclusion

In the present work, the XFEM-based approach is implemented together with damage mechanics to study the fracturing response of an OPA-based rock, with a transversely isotropic behavior. In line with an experimental investigation from literature [6], a notched Semi-Circular Bending test is explored to assess the influence of the notch shape and size on the overall fracturing response. Based on results, a negligible variation is observed in the ascending branch of the load-CMOD curves, with a more pronounced variation in the peak load and brittleness of the specimen. Such variations come with few deviations of the crack from the vertical direction. A further extension of the work will include the arbitrariness of the bedding properties or the influence of the notch inclination with respect to the bedding planes.

References

[1] R. McLamore, K.E. Gray, The mechanical behavior of Anisotropic Sedimentary Rocks, Am. Soc. Mechannical Eng. J. Eng. Ind. 89 (1967) 73–6. https://doi.org/10.1115/1.3610013

[2] H. Niandou, J. F. Shao, J. P. Henry, D. Fourmaintraux, Laboratory investigation of the mechanical behaviour of tournemire shale, Int. J. Rock Mech. Min. Sci. 34 (1997) 3–16. https://doi.org/10.1016/S1365-1609(97)80029-9

[3] M. Naumann, U. Hunsche, O. Schulze, Experimental investigations on anisotropy in dilatancy, failure and creep of Opalinus Clay, Phys. Chem. Earth. 32 (2007) 889–895. https://doi.org/10.1016/j.pce.2005.04.006

[4] F. Barpi, S. Valente, M. Cravero, G. Iabichino, C. Fidelibus, Fracture mechanics characterization of an anisotropic geomaterial, Eng. Fract. Mech. 84 (2012) 111–122. https://doi.org/10.1016/j.engfracmech.2012.01.010

[5] H. Bock, P. Blümling, RA Experiment Rock mechancis analysis and synthesis: data report on rock mechanics. TR 2000-02, Mont Terri Technical Report 2000-02. 52 (2001).

[6] S. Valente, C. Fidelibus, S. Loew, M. Cravero, G. Iabichino, F. Barpi, Analysis of fracture mechanics tests on opalinus clay, Rock Mech. Rock Eng. 45 (2012) 767–779. https://doi.org/10.1007/s00603-012-0225-2

[7] R. Dimitri, M. Rinaldi, M. Trullo, F. Tornabene, C. Fidelibus, FEM/XFEM modeling of the 3D fracturing process in transversely isotropic geomaterials, Compos. Struct. 276 (2021) 114502. https://doi.org/10.1016/j.compstruct.2021.114502

[8] T. Belytschko, T. Black, Elastic crack growth in finite elements with minimal remeshing, Int. J. Numer. Methods Eng. 45 (1999) 601–620. https://doi.org/10.1002/(SICI)1097-0207(19990620)45:5<601::AID-NME598>3.0.CO;2-S

[10] N. Moës, J. Dolbow, T. Belytschko, A finite element method for crack growth without remeshing, Int. J. Numer. Methods Eng. 46 (1999) 131–150. https://doi.org/10.1002/(SICI)1097-0207(19990910)46:1<131::AID-NME726>3.0.CO;2-J

[11] J. Dolbow, An extended finite element method with discontinuous enrichment for applied mechanics, (2000).

[12] J. Dolbow, N. Moës, T. Belytschko, Discontinuous enrichment in finite elements with a partition of unity method, Finite Elem. Anal. Des. 36 (2000) 235–260. https://doi.org/10.1016/S0168-874X(00)00035-4

Theoretical and Applied Mechanics - AIMETA 2022 | Materials Research Forum LLC
Materials Research Proceedings **26** (2023) 35-40 | https://doi.org/10.21741/9781644902431-6

A 3D visco-elasto-plasto damage constitutive model of concrete under long-term effects

Beaudin Freinrich Dongmo[1, a*], Gianluca Mazzucco[1, b], Beatrice Pomaro[1, c], Jiangkun Zhang[1, d], Carmelo Majorana[1, e], and Valentina Salomoni[2, f]

[1]Department of Civil, Environmental and Architectural Engineering (DICEA), University of Padova, via F. Marzolo 9, 35131, Padova, Italy

[2]Department of Management and Engineering (DTG), University of Padova, Stradella S. Nicola 3, 36100, Vicenza, Italy

[a]beaudinfreinrich.dongmo@phd.unipd.it, [b]gianluca.mazzucco@unipd.it, [c]beatrice.pomaro@unipd.it, [d]jiangkun.zhang@phd.unipd.it, [e]carmelo.maiorana@unipd.it, [f]valentina.salomoni@unipd.it

Keywords: Visco-Elasto-Plasto-Damage, Long-Term Effects, Concrete, Creep

Abstract. A comprehensive 3D visco-elasto-plasto-damage constitutive model of concrete is proposed to analyze its behaviour under long-term and cyclic loadings. This model combines the visco-elasticy and plasticity theories together with damage mechanics. The work aims at providing an efficient model capable of predicting the material behaviour, taking into account time-dependent effects at the mesoscale. The visco-elastic part is modeled within the framework of the linear visco-elasticity theory. The creep function is evaluated with the aid of the B3 model by Bažant and Baweja, and implemented via the exponential algorithm. The modified Menétrey-Willam pressure-dependent yield surface, and a non-associated flow rule are used for the plastic formulation of the model. The damage part of the model considers two exponential damage parameters: one in tension, and one in compression, that account for a realistic description of the transition from tensile to compressive failure. After discussing the numerical implementation, the proposed model is calibrated, and numerical results at the mesoscale level are compared to experimental results.

Introduction

The calculation of creep deformations under long-term loading is crucial to ensure the functionality and the durability of concrete structures. Moreover, for a comprehensive description of the mechanical behaviour of concrete, nonlinear characteristics, such as stiffness degradation and reduction of strength, must be considered altogether with creep deformations. In the past and recent years, many researchers have combined various plasticity and damage models with creep models in order to reproduce accurately the phenomenological behavior of concrete. In 1998, Majorana et al. [1] consider a Maxwell chain model to combine creep with a scalar isotropic damage model. Ren et al. [2] have proposed a creep-damage-plasticity model based on strain additivity, in which an energy-based damage-plasticity model and a modified ACI model are combined together. Yu et al. [3] have studied the nonlinear creep failure of concrete by combining a Mazars' damage model [4] with the B3 creep model proposed by Bažant and Baweja. However, the coupling mechanism between these various models has not yet been fully revealed, and there is a need to use the most recent models in literature which allow for a more realistic description of concrete nonlinear behaviour.

In this paper, a comprehensive 3D visco-plasto-damage model is proposed. The plasto-damage model based on Menétrey-Willam pressure-dependent yield surface [5,6], and a non-associated flow rule is coupled with the B3 creep model of Bažant and Baweja [7]. The damage part of the model considers two exponential damage parameters, one in tension, and one in compression, that help accounting for crack closure and stiffness increase during the transition from tensile to compressive loading [8]. After calibration of the unified model, numerical simulations are performed and compared with experimental data. It is proved that the new model is capable of reproducing with satisfactory precision the nonlinear characteristics of concrete under long-term loading.

The visco-plasto-damage model

The Plasto-Damage Model. The plasto-damage model implemented in this paper considers the Menétrey-Willam plastic surface [6] modified with a scalar damage variable ω that accounts for the change in shape of the plastic surface due to damage. The plastic surface is expressed through the unified coordinates (ξ,ρ,θ) in the Haigh–Westergaard (HW) stress space as

$$f = 1.5\frac{\rho^2}{(1-\omega^2)f_c^2} + \frac{q_h(\kappa)m}{(1-\omega)f_c}\left[\frac{\rho}{\sqrt{6}}r(\theta,e) + \frac{\xi}{\sqrt{3}}\right] - q_h(\kappa)q_s(\kappa) \leq 0, \tag{1}$$

where κ is the internal variable for plastic evolution, taken as the volumetric component of the plastic strain such that $\dot{\kappa}(\dot{\varepsilon}^p) = tr(\dot{\varepsilon}^p)$. The shape of the deviatoric section is controlled by the elliptic function r(θ,e), where θ is the Lode angle, and e is the eccentricity parameter; m is the cohesion parameter, and f_c the concrete strength in compression; q_h and q_s represent the hardening and softening laws, respectively, as defined in [9].

A non-associated flow rule has been adopted, and is defined in agreement with

$$\dot{\varepsilon}_p = \dot{\lambda}\frac{\partial g_p}{\partial \sigma}(\sigma,\kappa), \tag{2}$$

where $\dot{\varepsilon}_p$ is the rate of the plastic strain, and $\dot{\lambda}$ is the plastic multiplier; κ is the internal variable defined above. The function g_p represents the plastic potential, expressed using the coordinates in the HW stress space as [8]

$$g_p = -\frac{A}{f_c}\left(\frac{\rho}{\sqrt{q(\kappa)}}\right)^2 - B\left(\frac{\rho}{\sqrt{q(\kappa)}}\right) + \frac{\xi}{\sqrt{q(\kappa)}} = 0, \tag{3}$$

where A and B are experimentally derived coefficients; f_c is the concrete strength in compression, and q(κ) is the hardening/softening law, defined as $q(\kappa) = q_h(\kappa)q_s(\kappa)$.

Damage has been considered via a total damage parameter that includes a stiffness recovery function able to catch the increase of stiffness due to closure of cracks when the mechanical loading changes from tension to compression. It is defined as a combination of the compressive and tensile damage parameters as follows

$$\omega = 1 - [1 - \omega_c(\kappa_c)][1 - s(\sigma^{tr})\omega_t(\kappa_t)], \tag{4}$$

where the scalar damage variables ω_c and ω_t are defined using exponential functions as suggested by Pijaudier-Cabot and Mazars [4]; κ_c and κ_t are internal variables used to describe the damage evolution in compression and tension, respectively, and $s(\sigma^{tr})$ is the stiffness recovery function.

Materials Research Proceedings **26** (2023) 35-40 https://doi.org/10.21741/9781644902431-6

The Creep Model. Herein, the creep model used to describe the long-term behaviour of concrete is the linear viscoelastic B3 model proposed by Bažant and Baweja [7]. The compliance function of the B3 creep model is expressed as

$$J(t, t') = q_1 + C_0(t, t') + C_d(t, t', t_0), \tag{5}$$

where q_1 is the instantaneous strain due to unit stress; t and t′ represent the current age, and loading age, respectively.

In Eq. 5 $C_0(t,t')$ is the basic creep term of the B3 model, expressed as [7]

$$C_0(t,t') = q_2 Q(t,t') + q_3 \ln[1 + (t-t')^{0.1}] + q_4 \ln\left(\frac{t}{t'}\right), \tag{6}$$

while $C_d(t, t', t_0)$ is the drying creep term of the B3 model expressed as [7]

$$C_d(t,t',t_0) = q_5 \left[e^{-8H(t)} - e^{-8H(t_0')}\right]^{0.5}, \tag{7}$$

where t_0 is the time at which drying starts, and $t_0' = \max(t', t_0)$; H(t) and $H(t_0')$ are spatial averages of pore relative humidity [7].

The parameters q_1 to q_5 can be approximated via empirical formulas [7]. The creep model is implemented via the exponential algorithm [10].

Numerical results

In the scope of this paper, 3D mescoscale models of ordinary concrete made by calcareous aggregates have been reconstructed using 3D computed tomography. In the mesoscale models, the aggregates are modelled as elastic; the cement paste has been modelled using the above mentioned visco-plasto-damage model. A sketch of the adopted solid model is represented in Fig. 1 (left).

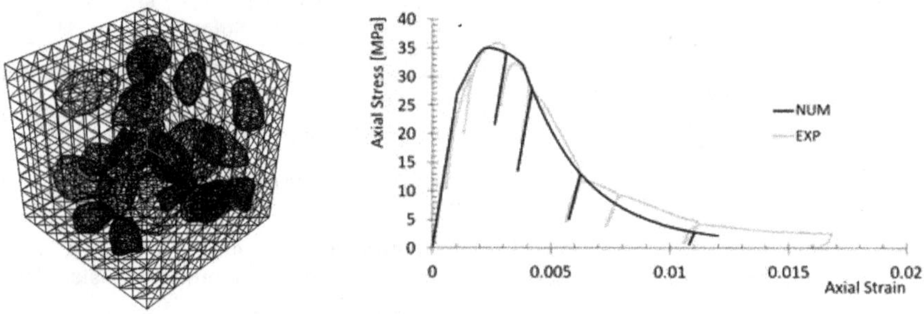

Figure 1: 3D sample reconstruction (left), and comparison between numerical and experimental results of the uniaxial compression test (right).

The calibration of the model has been conducted over experimental results obtained from a uniaxial compression test with loading and unloading cycles, on a cubic sample of same geometry and grading curve of the adopted model. Subsequently, similar experiments considering short- and long-term loadings have been considered for validation of the present model. In particular, the two parameters A and B of the plastic potential are determined by means of the axial strain in uniaxial compression at maximum stress and the axial strain in triaxial compression at maximum stress, as illustrated in [9]. The elasto-plasto-damaged parameters for cement matrix constituent are reported in Table 1.

Table 1: Elasto-pasto-damage parameters after calibration.

ELASTO-PLASTICITY			DAMAGE		
E	Young modulus at 28 days [MPa]	33,000	k_{c0}	1D initial inelastic strain for compression damage	0.015
ν	Poisson's ratio	0.2	A_c	Damage parameter in compression	1
f_c	Compressive strength at 28 days [MPa]	34.6	B_c	Damage parameter in compression	100
f_t	Traction strength at 28 days [MPa]	3.5	$\omega_{c\,max}$	Maximum damage in compression	0.95
e	Eccentricity	0.55	k_{t0}	1D initial inelastic strain for tension damage	0.0015
t	Slope for the softening function q_s	0.003	A_t	Damage parameter in tension	1
k_{1D}	Plastic strain in the compressive pick	0.1	B_t	Damage parameter in tension	1
q_{h0}	Initial hardening function q_h	0.4	$\omega_{t,max}$	Maximum damage in tension	0.95
A	Plastic potential parameter	-2.22	s_0	Maximum recovery value	0.2
B	Plastic potential parameter	-3.46	G_f	Fracture energy [$10^3 J/m^2$]	0.025
			w_c	Crack bands width [mm]	1

Fig. 1 (right) reports the comparison between numerical and experimental results for the short-term loading with loading and unloading cycles. The model proves to reproduce satisfactorily the post-peak behaviour and stiffness degradation during the softening regime.

The comparison between numerical and experimental results for the long-term loading is shown in Fig. 2. Also in this case the numerical results are in good agreement with the experimental ones.

Materials Research Proceedings **26** (2023) 35-40 https://doi.org/10.21741/9781644902431-6

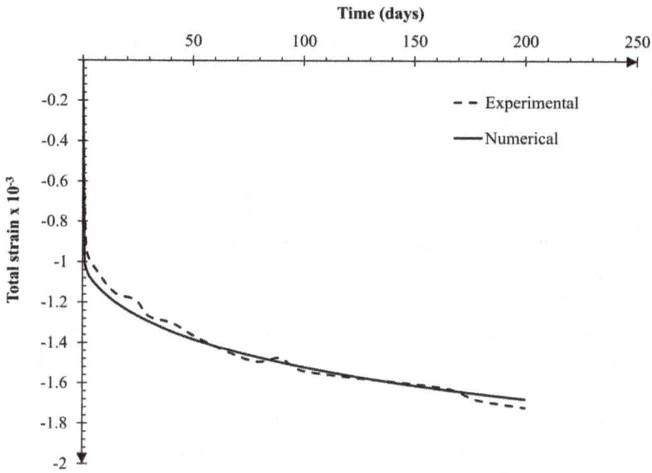

Figure 2: Comparison between numerical and experimental results of the creep test.

Fig. 3 shows the stress distribution in the direction of the load, across the sample. Overall, stress localizations in the regions of the cement paste surrounding the aggregates are envisaged. This is coherent with the expected triggering of damage at the interface zone between aggregates and cement paste, and subsequent failure following these preferential paths.

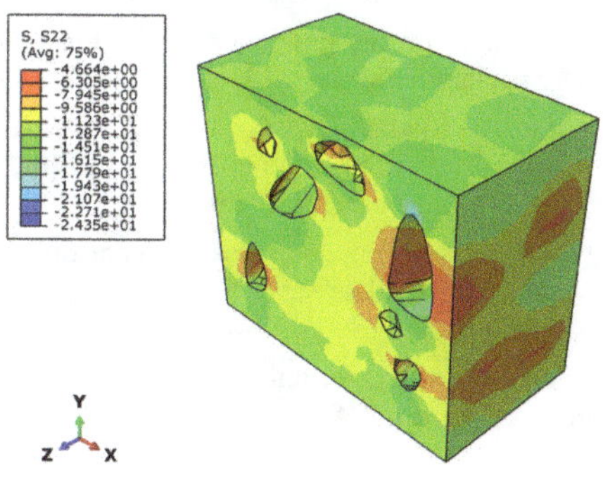

Figure 3: Stress contour map (in MPa) in the cement paste.

Summary

The comprehensive 3D visco-plasto-damage model presented herein combines a pressure-dependent plastic model extended to include damage with a linear visco-elastic creep model for the study of the coupling mechanism between creep and damage. Specifically, the plasto-damage model includes a pressure-dependent yield surface and a stiffness recovery function for a more realistic description of the transition from tensile to compressive failure. The creep model considers two contributions: a basic creep, and drying creep. After calibration and validation of the model, a numerical simulation has been conducted on a cubic concrete sample at the mesoscale, undergoing creep. The comparison between numerical and experimental results confirms that the proposed unified model is capable of characterizing accurately the nonlinear material behaviour of concrete under short and sustained loading.

References

[1] C. E. Majorana, V. Salomoni, B. A. Schrefler, Hygrothermal and mechanical model of concrete at high temperature, Materials and structures 31(1998) 378-386. https://doi.org/10.1007/BF02480710

[2] X. Ren, Q. Wang, R. Ballarini, X. Gao, Coupled creep-damage-plasticity model for concrete under long-term loading, Journal of Engineering Mechanics 146 (2020) 04020027. https://doi.org/10.1061/(ASCE)EM.1943-7889.0001748

[3] P. Yu, R. Li, D. Bie, X. Yao, X. Liu, Y. Duan, A coupled creep and damage model of concrete considering rate effect, Journal of Building Engineering 45 (2022) 103621. https://doi.org/10.1016/j.jobe.2021.103621

[4] G. Pijaudier-Cabot, J. Mazars, Handbook of materials behavior models, three-volume set, Elsevier, 2001.

[5] G. Mazzucco, B. Pomaro, G. Xotta, C. E. Maiorana, V. A. Salomoni, Tomography reconstruction of concrete materials for mesoscale modelling, Engineering Computations 37 (2020) 2275–2291. https://doi.org/10.1108/EC-10-2019-0448

[6] P. Menétrey, K.J. Willam, Triaxial failure criterion for concrete and its generalization, Structural Journal 92 (1995) 311–318. https://doi.org/10.14359/1132

[7] Z. P. Bažant, S. Baweja, Creep and shrinkage prediction model for analysis and design of concrete structures: Model b3, ACI Special Publications 194 (2000) 1–84.

[8] P. Grassl, D. Xenos, U. Nyström, R. Rempling, K. Gylltoft, Cdpm2: A damage-plasticity approach to modelling the failure of concrete, International Journal of Solids and Structures 50 (2013) 3805–3816. https://doi.org/10.1016/j.ijsolstr.2013.07.008

[9] P. Grassl, K. Lundgren, K. Gylltoft, Concrete in compression: a plasticity theory with a novel hardening law, International Journal of Solids and Structures 39 (2002) 5205–5223. https://doi.org/10.1016/S0020-7683(02)00408-0

[10] Q. Yu, Z. P. Bažant, R. Wendner, Improved algorithm for efficient and realistic creep analysis of large creep-sensitive concrete structures, ACI Structural Journal 109 (2012) 665. https://doi.org/10.14359/51684044

Theoretical and Applied Mechanics - AIMETA 2022 Materials Research Forum LLC
Materials Research Proceedings **26** (2023) 41-46 https://doi.org/10.21741/9781644902431-7

A multiscale model for anisotropic damage and hysteresis in biodegradable polymers

Vitucci Gennaro*[,a], De Tommasi Domenico, Di Stefano Salvatore, Puglisi Giuseppe and Trentadue Francesco

DICATECh, Politecnico di Bari, Via Re David 200, 70125 Bari, Italy

[a]gennaro.vitucci@poliba.it

Keywords: Mullins Effect, Multiscale Models, Microsphere Approach, Biodegradable Polymers

Abstract. Predicting the mechanical response of biological soft materials requires an understanding of the complex phenomena characterizing their microscale. In this work, we use an existing versatile framework, based on assumptions on the statistical distribution of biolpolymers at the network scale, for extending our previous entropic constitutive model of Worm-Like Chains networks to different deformation classes. Furthermore, we include the effect of molecules topological constraints by introducing an energy term depending on the second invariant of the Green-Cauchy tensor. In this way we are able to qualitatively reproduce, with a limited set of physically meaningful constitutive parameters, a range of observed phenomena such as induced anisotropy, stress softening, hardening, Mullins effect, evolution of permanent stretches.

Introduction

In this work, we extend our previous micromechanical model [1] by accounting for a fully three-dimensional distribution of polymeric chains, taking care of damage and contour length variation depending on loading history. The analysis is inscribed in the framework of statistical mechanics. The constitutive Helmholtz free energy is considered to stem from two terms: a microscopic Worm Like Chain type and a macroscale topological one describing network constraints. This novel model is described by a set of five parameters: one for each component of the WLC and topological free energy, one that relates contour length to end-to-end molecules variations, and two describing refolding and unfolding during unloading and reloading, respectively [2]. In this way we are able to capture both irreversible damage, onset of residual stretches (Mullins effect), hysteresis and induced anisotropy. With the single chain characterization in hand, the three dimensional model is directly obtained via a numerical integration of the energy on the microsphere [3]. The micro-to-macro shift in scale relies on the classical affinity hypothesis commonly used in rubberlike elasticity. We show the versatility and effectiveness of the present framework in describing the aforementioned effects by means of a parametric study. The constitutive parameters have a clear physical interpretation, which allows predicting the range of parameters variation. We believe that the present model is useful in describing the mechanical response of biopolymers exhibiting damage and permanent deformations of interest in material design and biomedical applications.

Methods

In this section, first we model the behavior of a single filament, then we use this model for passing to the macroscopic scale via the microsphere approach [3] and an additional macroscopic additional term.

Macromolecule model. In the previous works [1, 4] about absorbable suture threads, we modified the Helmholtz free energy of [5] in order to account for residual stretches. Here we use the same free energy of the macromolecule and the entropic force

$$\Psi(\lambda,\lambda_n)=\kappa\frac{(\lambda-\lambda_n)^2}{\lambda_c-\lambda},\quad f=\frac{\partial\Psi}{\partial\lambda}=\kappa\left[\left(\frac{\lambda_c-\lambda_n}{\lambda_c-\lambda}\right)^2-1\right] \tag{1}$$

In the formulas above, λ, λ_c and λ_n are the molecule stretch, contour stretch and natural stretch, respectively. The parameter κ accounts for temperature and number of free monomers. The natural and contour stretch are related as , based on well known statistical mechanics considerations such as e.g. in [6, 7], where λ_{co} is the initial contour stretch ($\lambda_{co}>1$). Observe that the natural stretch λn corresponds to a zero value of the force, thus the residual stretch λ_{res} is identified by the condition $\lambda=\lambda_n$. Simultaneously, the contour stretch λ_c is a limit stretch. Notice that the force tends to infinity as $\lambda\rightarrow\lambda_c$.

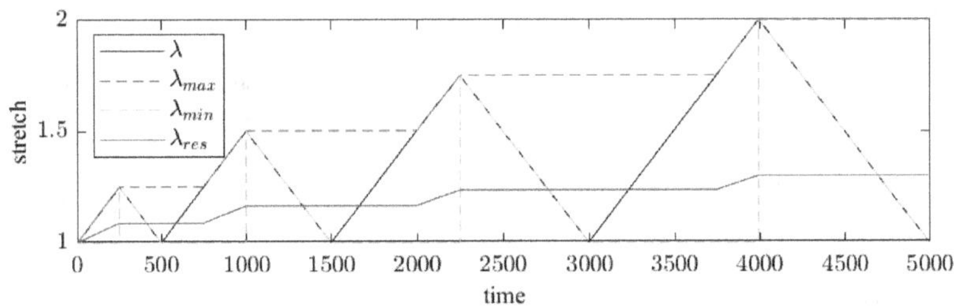

Figure 1: Evolution of the stretches under cyclic loading. The unloading parameter, describing partial refolding andthus residual stretches λ_{res} of copolymer domains is taken β_u = 0.2.

As in our previous work [4], in order to model folding and unfolding of the macromolecule under stretch, we assume the following phenomenological evolution law for the natural stretch

$$\lambda_n=\sqrt{\lambda_{max}}\left(\frac{\lambda}{\lambda_{max}}\right)^{-\beta_u+\beta_r\sqrt{\lambda-\lambda_{min}}}, \tag{2}$$

depending on the two constitutive parameters and . The former β_u describes the number of free monomers to crystallize during unloading, whereas the latter β_r models unfolding under reloading. Observe that, according to this evolution law, the maximum force exerted at the reach of λ_{max} on the primary path before unloading is exactly attained again at $\lambda=\lambda_{max}$ after reloading. In other words our model satisfies the so called Return Point Memory property [8]. Notice that the exponent term $\sqrt{\lambda-\lambda_{min}}$ is nonzero only in reloading as shown in Fig.1. Furthermore, (4) also identifies residual stretches. The filament force of Eq.(2) vanishes at $\lambda=\lambda_n$, thus the residual stretch must meet the condition $\lambda=\lambda_n=\lambda_{res}$ which returns the analytical dependency $\lambda_{res}=\lambda_{max}^{\frac{1}{2}\frac{1-2\beta_u}{1-\beta_u}}$ of filament residual stretches and maximum historical load in the simple form of power law.

Macroscopic deformation. For a material point, the rate of work carried out by the first Piola-Kirchhoff stress Σ must equal the rate of change of free energy, , where the macroscopic deformation is described through the deformation gradient $\mathbf{F}$. Entropic free energies terms like the one in Eq.(1) are usually able to describe the mechanics of polymers in solution. When the polymers are, as in solids, crosslinked in networks, additional energy terms are needed for considering self and mutual entanglement effects (see [9, 4]). Here we adopt a hybrid approach and, leaving untouched the microscopic energy, we add a topological energy term at the macroscopic level borrowing by solid mechanics (e.g. [10]). Denoting terms of microscopic and macroscopic origin with *m* and *M*, the stress tensor can be decomposed into

$$\mathbf{\Sigma} = \mathbf{\Sigma}_m + \mathbf{\Sigma}_M = \frac{\partial \Psi_m}{\partial \mathbf{F}} + \frac{\partial \Psi_M}{\partial \mathbf{F}}. \tag{3}$$

Notice that the quasi-static assumption allows treating time as a unitless order parameter. We assume that the stretches are applied along the principal directions and, as usual, the solid is incompressible, i.e. det**F**=1. Therefore, by looking at Fig.2, we are able to describe uniaxial extension, equibiaxial extension and pure shear via a unique macrostretch variable Λ as follows (see [10]):

$$\mathbf{F}(\Lambda) = \begin{cases} \mathrm{diag}(\Lambda, \Lambda^{-\frac{1}{2}}, \Lambda^{-\frac{1}{2}}) & \text{(uniaxial)} \\ \mathrm{diag}(\Lambda, \Lambda, \Lambda^{-2}) & \text{(equibiaxial)} \\ \mathrm{diag}(\Lambda, 1, \Lambda^{-1}) & \text{(pure shear)} \end{cases} \tag{4}$$

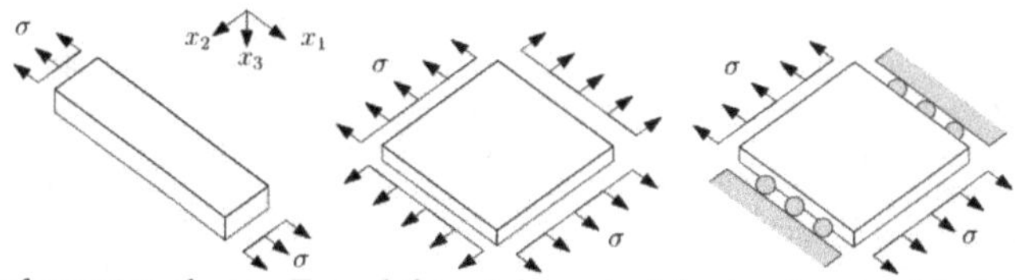

Figure 2: Deformation classes. From left to right: uniaxial extension, equibiaxial extension, pure shear.

In this way, one can find the components of $\mathbf{\Sigma}_m$ that contribute to the microscopic term work as

$$\sigma_m = \frac{1}{4\pi\delta} \int_{\text{sphere}} \frac{\partial \psi}{\partial \lambda} \frac{\mathrm{d}\lambda}{\mathrm{d}\Lambda} \,\mathrm{d}\Omega = \frac{1}{4\pi\delta} \int_{\text{sphere}} f(\lambda) \frac{\mathrm{d}\lambda}{\mathrm{d}\Lambda} \,\mathrm{d}\Omega. \tag{5}$$

with δ=1 for uniaxial extension and pure shear, δ=2 for equibiaxial extension. As in standard nonlinear elasticity [10], notice in Fig.2 that pure shear is obtained by imposed elongation in direction 1, zero displacement in direction 2 and free stretch, i.e. here resulting from incompressibility, across the thickness in direction 3. Moreover, assuming the classical affinity hypothesis, the stretch $\lambda(\Lambda, d)=\|\mathbf{F}(\Lambda)d\|$ of a filament oriented along the unit vector $\mathbf{d}=[d_1, d_2, d_3]$ and its derivative with respect to the independent stretch variable Λ have to be evaluated. A numerical quadrature rule for integrals of the kind Eq.(5) was discovered by [11] and successfully applied to polymer mechanics for the first time by [3]. This is the so-called microsphere approach. It is based on N quadrature orientations d_j such that, for a function v defined at the microscale, the macroscale counterpart can be computed as a simple weighted average. In order to consider the interaction between the network macromolecules we add a further term to the total energy, which enforces topological constraints. It has been repeatedly highlighted in literature that the second invariant I_2 of the right Cauchy-Green deformation tensor $\mathbf{C}=\mathbf{F}^T\mathbf{F}$, plays a crucial role in describing the energy, especially for large deformations of rubber-like and biological materials (e.g. see [12, 13] and references therein). Similarly as in the classical Mooney-Rivlin model of incompressible hyperelasticity, we assume that the energy depends linearly on I_2, which accounts for the square of the aerial deformation normal to the elongating fibers. Following the same thermodynamic reasoning as in Eq.6, we obtain

$$\sigma_M = \kappa C_2 \frac{\mathrm{d}I_2}{\mathrm{d}\Lambda}, \tag{5}$$

so that C2 is unitless.

Materials Research Forum LLC
https://doi.org/10.21741/9781644902431-7

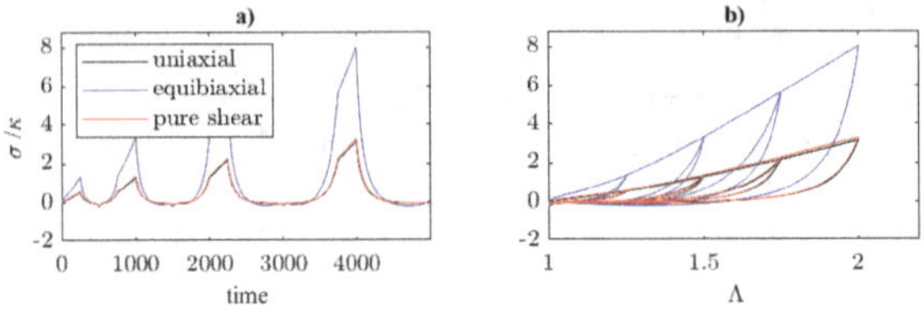

Figure 3: Mechanical response of the microscopic term. The material is cyclically tested inuniaxial extension,equibiaxial extension and pure shear. Material parameters: β_u=0.2, β_r=0.2, λ_{co}=1.1, C_2 = 0. a) Normalized stress vstime. b) Normalized stress vs macroscopic stretch.

Results and conclusions
In order to show the versatility of the present model, here we perform a parametric analysis which elucidates the role of the five parameters in capturing Mullins effect, internal hysteresis and permanent set in the three loading cases above considered. In Fig.3, we show the mechanical response in absence of the topological term. The material is cyclically tested in uniaxial extension, equibiaxial extension and pure shear.

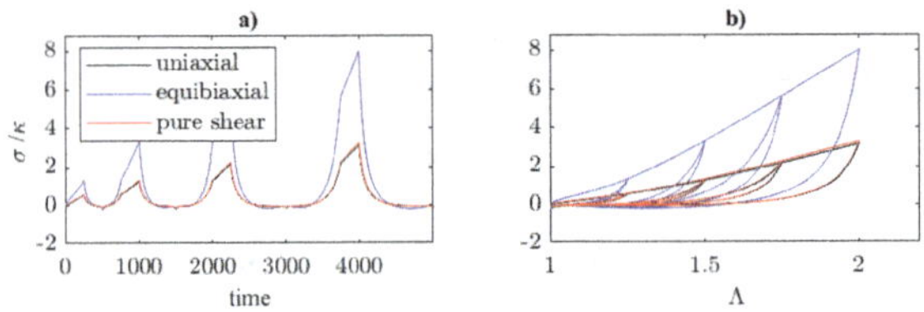

Figure 4: Mechanical response of the microscopic term. The material is cyclically tested inuniaxial extension,equibiaxial extension and pure shear. Material parameters: β_u=0.2, β_r=0.2, λ_{co}=1.1, C_2 = 0. a) Normalized stress vstime. b) Normalized stress vs macroscopic stretch.

In Fig.4 the effect of initial contour stretch is investigated. Looser chains, i.e. with a higher λ_{co}, show a lower response in terms of stresses for all loading scenarios.

The contribution of reversible folding is depicted in Fig.5. A higher unloading parameter β_u, i.e. a higher rate of refolding, causes lower stresses in unloading. A higher reloading parameter βr entails higher stresses in reloading. Finally, the contribution of the second invariant related macrostress is numerically analysed in Fig.6. Whereas the multiscale microsphere network predicts a very similar behavior for uniaxial extension and pure shear, the adding of a macrostress depending on the second invariant of the right Cauchy-Green deformation tensor allows the model to differentiate the two cases as measured in experiments such as in [14].

The main contribution of this work is a multiscale constitutive model which is able to predict, with only five physically motivated parameters, a wide range of phenomena and deformation classes.

Numerous biological materials are constituted by a network of stiff macromolecules, e.g.. collagen, immersed into a soft incompressible extracellular matrix (see [15,16]). This enables outstanding mechanical properties which, as in this work, can be explained via a multiscale approach. Future work will quantitatively investigate real biological materials by means of our novel two-scale energy function.

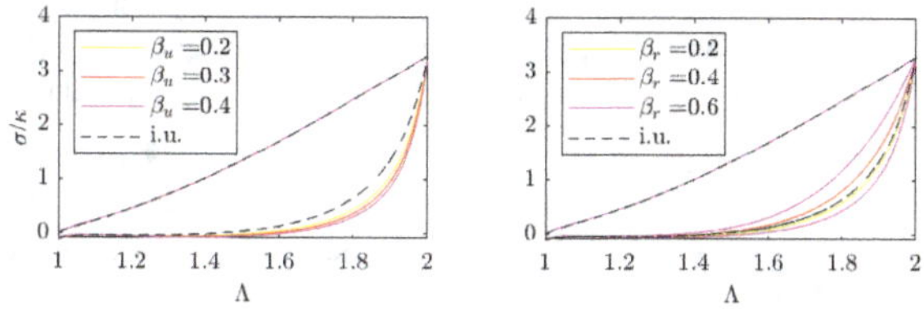

Figure 5: Effect of reversible folding. In dotted line the case of only irreversible unfolding,i.e. β_u and β_r are null and the only unfolding is along the primary path. Left-hand side)A higher unloading parameter β_u, i.e. a higher rate of refolding, causes lower stresses in unloading. Right-hand side) The lowest magenta curve is the common unloading stress forthe three cases. A higher reloading parameter βr entails higher stresses in reloading.

Acknowledgements
G. Vitucci is supported by the POR Puglia FESR-FSE project REFIN A1004.22, S. Di Stefano by Regione Puglia in the context of the research project "Riciclo di materiali e sostenibilità: modelli di delaminazione per dispositivi laminati" and Gruppo Nazionale per la Fisica Matematica (GNFM) under Istituto Nazionale di Alta Matematica (INdAM) in the context of "Progetto Giovani GNFM 2020-2022", F. Trentadue and D. De Tommasi by the Prin Project 2017J4EAYB, G. Puglisi by the Prin Project 2017KL4EF3 and GNFM.

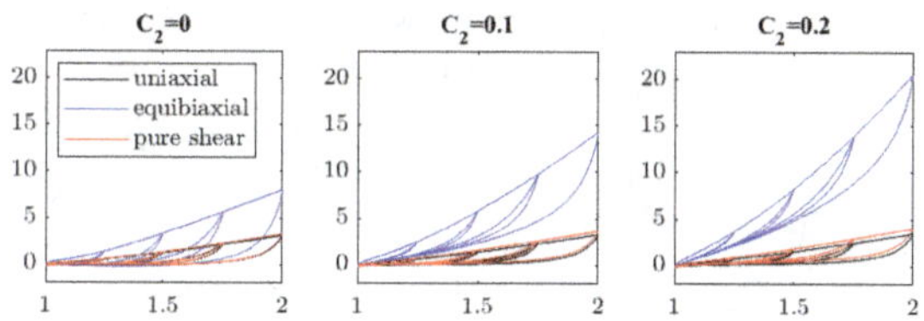

Figure 6: Effect of second invariant related macrostress. Whereas the multiscale microsphere network predicts a very similar behavior for uniaxial extension and pure shear, the adding of a macrostress depending on the second invariant of the right Cauchy-Green deformation tensor allows the model to differentiate the two as measured in published experiments.

References

[1] F. Trentadue, D. De Tommasi, and G. Puglisi, "A predictive micromechanically-based model for damage and permanent deformations in copolymer sutures," Journal of the Mechanical Behavior of Biomedical Materials, vol. 115, p. 104277, 2021. https://doi.org/10.1016/j.jmbbm.2020.104277

[2] D. De Tommasi, N. Millardi, G. Puglisi, and G. Saccomandi, "An energetic model for macromolecules unfolding in stretching experiments," Journal of The Royal Society Interface, vol. 10, no. 88, p. 20130651, 2013. https://doi.org/10.1098/rsif.2013.0651

[3] S. Goektepe and C. Miehe, "A micro–macro approach to rubber-like materials. part iii: The micro-sphere model of anisotropic mullins-type damage," Journal of the Mechanics and Physics of Solids, vol. 53, no. 10, pp. 2259–2283, 2005. https://doi.org/10.1016/j.jmps.2005.04.010

[4] G. Vitucci, D. De Tommasi, G. Puglisi, and F. Trentadue, "A predictive microstructure-based approach for the anisotropic damage, residual stretches and hysteresis in biodegradable sutures," arXiv preprint arXiv:2206.00345, 2022. https://doi.org/10.1021/ma00130a008

[5] J. F. Marko and E. D. Siggia, "Stretching dna," Macromolecules, vol. 28, no. 26, pp. 8759–8770, 1995.

[6] M. Rubinstein, R. H. Colby, et al., Polymer physics, vol. 23. Oxford university press New York, 2003.

[7] W. Kuhn and F. Gr¨un, "Beziehungen zwischen elastischen konstanten und dehnungsdoppelbrechung hochelastischer stoffe," Kolloid-Zeitschrift, vol. 101, no. 3, pp. 248–271, 1942. https://doi.org/10.1007/BF01793684

[8] G. Bertotti, Hysteresis in magnetism: for physicists, materials scientists, and engineers. Gulf Professional Publishing, 1998.

[9] M. Doi, S. F. Edwards, and S. F. Edwards, The theory of polymer dynamics, vol. 73. oxford university press, 1988.

[10] R. W. Ogden, Non-linear elastic deformations. Courier Corporation, 1997.

[11] P. Baˇzant and B. Oh, "Efficient numerical integration on the surface of a sphere," ZAMMJournal of Applied Mathematics and Mechanics/Zeitschrift f¨ur Angewandte Mathematik und Mechanik, vol. 66, no. 1, pp. 37–49, 1986. https://doi.org/10.1002/zamm.19860660108

[12] C. O. Horgan and G. Saccomandi, "Constitutive modelling of rubber-like and biological materials with limiting chain extensibility," Mathematics and mechanics of solids, vol. 7, no. 4, pp. 353–371, 2002. https://doi.org/10.1177/108128028477

[13] M. Destrade, J. G. Murphy, and G. Saccomandi, "Simple shear is not so simple," International Journal of Non-Linear Mechanics, vol. 47, no. 2, pp. 210–214, 2012. https://doi.org/10.1016/j.ijnonlinmec.2011.05.008

[14] B. Yohsuke, K. Urayama, T. Takigawa, and K. Ito, "Biaxial strain testing of extremely soft polymer gels," Soft Matter, vol. 7, no. 6, pp. 2632–2638, 2011. https://doi.org/10.1039/c0sm00955e

[15] V. Gennaro, I. Argatov, and G. Mishuris. "An asymptotic model for the deformation of a transversely isotropic, transversely homogeneous biphasic cartilage layer." Mathematical Methods in the Applied Sciences 40.9 (2017): 3333-3347. https://doi.org/10.1002/mma.3895

[16] G. Vitucci and G. Mishuris. "Three-dimensional contact of transversely isotropic transversely homogeneous cartilage layers: a closed-form solution." European Journal of Mechanics-A/Solids 65 (2017): 195-204. https://doi.org/10.1016/j.euromechsol.2017.04.004

Theoretical and Applied Mechanics - AIMETA 2022 Materials Research Forum LLC
Materials Research Proceedings **26** (2023) 47-52 https://doi.org/10.21741/9781644902431-8

Multiscale approach to decohesion in cell-matrix systems

Salvatore Di Stefano[1, a *], Ariel Ramirez-Torres[2, b], Luca Bellino[1, c], Vincenzo Fazio[3, d], Gennaro Vitucci[1, e], Giuseppe Florio[1,4, f]

[1]Dipartimento di Ingegneria Civile, Ambientale, del Territorio, Edile e di Chimica (DICATECh), Politecnico di Bari – via Edoardo Orabona, 4, 70125 Bari, Italy

[2]School of Mathematics and Statistics, Mathematics & Statistics Building, University Place, Glasgow, G12 8QQ, United Kingdom

[3]Laboratory for Bioinspired, Bionic, Nano, Meta Materials & Mechanics, University of Trento, Via Mesiano 77, 38123 Trento, Italy

[4]INFN, Sezione di Bari, I-70126, Bari, Italy

[a]salvatore.distefano@poliba.it, [b]ariel.ramireztorres@glasgow.ac.uk, [c]luca.bellino@poliba.it, [d]vincenzo.fazio@poliba.it, [e]gennaro.vitucci@poliba.it, [f]giuseppe.florio@poliba.it

* Corresponding Author

Keywords: Focal Adhesion, Decohesion, Two-Scale Asymptotic Homogenization

Abstract. We propose a model for cell-matrix decohesion that highlights the role of elasticity in this process. In doing this, we specialize our previous study of focal adhesion, an integrin mediated structure that oversees and guides the mutual interactions between cells and the extracellular matrix. Specifically, we consider a two-scale asymptotic homogenization technique to study the multi-scale nature of decohesion. Thus, we are able to use micro-structural information available at length scales smaller than those at which focal adhesions are observed. Based on classical two-scale asymptotic techniques the proposed approach allows to define effective elastic coefficients encoding the intrinsic heterogeneous properties of both focal adhesions and extracellular matrix.

Introduction

We focus on integrin-mediated structures of cell attachment, by specializing our study to focal adhesions (FAs) [1-8]. Generally speaking, FAs represent the basic sites by means of cells anchor to ECM and share mechanical forces and biomechanical signals with it [1-8]. It is possible to schematize the structure of a FA in terms of a three-component system, comprising the adhesion plaque, integrin receptors and stress fibers [3, 5, 6].

Experiments show the capability of cells of sensing forces and convert them in biochemical signals, and particular attention has been devoted to the role played by the structural properties of FAs [3, 5, 6, 9-12]. For example, it has been experimentally validated how the ECM's rigidity determines the length and the stability of FAs [3, 5, 6, 9-12]. Conversely, forces originated from cells and transmitted to the ECM via FAs can determine the onset of remodeling, i.e., the rearrangement of the ECM's internal structure and its consequent structural adaptation to cell-induced stimuli [6, 13, 14].

Following all these specific examples, we are interested in studying how the elastic properties of the FAs and of the ECM may influence the dynamics of FA-ECM systems, in the case in which the insurgence and propagation of decohesion is considered. In doing this, based on [5], we develop a multi-scale approach, based on Asymptotic Homogenization [15-18] in order to deduce how the inhomogeneous microstructure of the system comprising a FA and the ECM may affect its overall behavior.

Materials Research Proceedings **26** (2023) 47-52 https://doi.org/10.21741/9781644902431-8

The model

It is known that FA, ECM and integrins are typically non-homogeneous [19-22]. By considering the ECM at first, we remark that it is a complex medium comprising several species of biological elements, such as proteins and polysaccharides, which present remarkably different biochemical and mechanical properties, as well as topological and geometrical characteristics [19]. Analogous considerations are also valid for the adhesion plaque. In fact, all the protein substructures characterizing the adhesion plaque are quite heterogeneous, since they significantly vary in size, shape, distribution and, in general, in their molecular composition [20]. Several experimental procedures have been developed with the purpose of a comprehensive description of the structure of a FA [20], although the determination of the morphology and of the dynamics of the adhesion plaque represents an important open problem. This aspect results to be clearer if we consider the fact that FA experiences remodeling, which is induced by mechanical actions exchanged with the ECM [6, 13, 14]. The last aspect to mention is the heterogeneity associated with the integrin receptors [21]. In this respect, the integrins' heterogeneity is the result of the formation of interphase regions close to the FA and of the ECM, thereby determining the onset of micro-scale effects. which aims at affecting larger scales phenomena [21]. Furthermore, the attachment of the adhesion plaque to the ECM occurs via different types of integrins, so that, the overall generated forces depend on their composition and non-uniform distribution [21, 22].

Following the discussion reported so far, we can now clarify why we can speak of a multi-scale behavior in the case of FAs and ECM. In fact, the heterogeneity character of both the FA and the ECM are associated with length scales which are smaller than those characterizing the whole adhesion complex. In other words, the system under investigation exhibits at least the co-existence of two main scales: the first one, hereon referred to as *micro-scale*, is typical of the heteregeneities of the components constituting the adhesion plaque, the ECM and the integrins, whereas the *macro-scale* is representative of the system as a whole [15-18]. Such characteristic lengths are well separated. This implies that, by denoting with L_{M} and L_{m} the macro- and the micro-scales, their ratio is much smaller than one, i.e., $\varepsilon = \frac{L_m}{L_M} \ll 1$ [1-8, 15-18]. Such condition, known as *scale separation,* is the starting hypothesis which preludes the application of a multiscale analysis based on a two-scale homogenization technique [15-18].

In this work, we employ a two-scale homogenization procedure to deduce the overall mechanical behavior of a system comprising a FA, the ECM and the integrin receptors and accounting for their inhomogeneities. By referring to the mechanical picture outlined in [5] and by virtue of the scale separation condition discussed above, we re-interpret it in a two-scale fashion. This leads us to formulate constitutive relations that incorporate the information deducible from the micro-structure. In details, we compute *homogenized* or *effective* elastic coefficients, encoding the elastic properties of FAs, ECM and integrins at the scale of the heterogeneities. Moreover, we infer a system of local field equations, describing the point-wise equilibrium of the considered physical system at the micro- and micro-scale, respectively. With reference to [5], we consider a one-dimensional, continuum model, comprising a FA adhesion, the ECM and the integrin receptors.

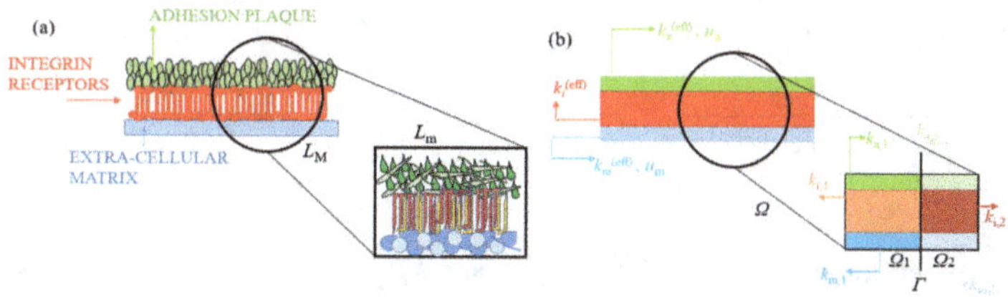

Figure 1: *Schematization of the system comprising FA, ECM and integrins and of its microstructure (a) and mechanical scheme employed in [14] and an example of periodic cell (b).*

The first two components are described as linear elastic fibers, with stiffness k_a and k_m, while the integrins are energetically represented by a linear elastic fragile potential, with stiffness k_i.

We remark that, with the term fragile, we mean that, when the force exerted by the integrins reaches a threshold value, it drops to zero. This way, as shown in [5], we model the onset and propagation of fracture and, in doing this, we illustrate how we adapt and generalize shear lag models to account for decohesion [5, 23, 24].

In this work, we retrace the same steps as in [5], but, instead of considering the elastic properties of the system to be homogeneous, we let them be inhomogeneous, thereby meaning that they vary because of the inhomogeneities arising from the microstructure. Following standard arguments of two-scale asymptotic homogenization, we assume the microstructure to be periodic [15-18]. Although such hypothesis seems to be too restrictive, it has shown to be sufficiently reliable in predicting the development of complex biological structures, e.g., breast cancer and bone remodeling [17, 18]. On the other hand, other biological media, such as spider silks, are known to possess a hierarchical periodic structure [25]. Such a hypothesis leads to the existence of an elementary cell, denoted by Ω, which can be assumed as representative of all elastic properties of the system at the micro-scale. In particular, it can be written as the union of N elements, i.e. $\Omega = \cup_{n=1,\dots,N} \Omega_n$, being Ω_n the portion of the elementary cell in which the elastic properties k_a, k_m and k_i assume the value $k_{a,n}$, $k_{m,n}$, and $k_{i,n}$, jumping across the boundary of Ω_n [15-18].

Results and discussion

We design the mechanical problem at hand in a mono-dimensional framework, and we introduce the scalar-valued functions u_a and u_m to represent the scalar displacements of the adhesion plaque and of the ECM. Moreover, u_a and u_m depend on the variable $x \in [0, L]$, being L the initial length of the system and $[0, L]$ its reference placement [5]. By applying two-scale asymptotic homogenization we obtain the following homogenized equations [26]

$$\frac{d}{dx}\left(k_a^{(eff)}\frac{du_a}{dx}\right) - \frac{k_i^{(eff)}}{d_i l_i}(u_a - u_m) = 0, \qquad x \ in \ ]0, L - L_{det}[, \qquad (1a)$$

$$\frac{d}{dx}\left(k_m^{(eff)}\frac{du_m}{dx}\right) + \frac{k_i^{(eff)}}{d_i l_i}(u_a - u_m) = 0, \qquad x \ in \ ]0, L - L_{det}[, \qquad (1b)$$

$$\frac{d}{dx}\left(k_a^{(eff)}\frac{du_a}{dx}\right) = 0, \qquad x \ in \ ]L - L_{det}, L[, \qquad (1c)$$

$$\frac{d}{dx}\left(k_m^{(eff)}\frac{du_m}{dx}\right) = 0, \qquad x \ in \ ]L - L_{det}, L[, \qquad (1d)$$

where L_{det} is the length of the detached portion of the system, while d_i and l_i are reference lengths associated with the distribution of integrins and their length, respectively [10], and with the effective coefficients $k_a^{(eff)}$, $k_m^{(eff)}$ and $k_i^{(eff)}$ computed as [15-18, 26]

Materials Research Proceedings **26** (2023) 47-52 https://doi.org/10.21741/9781644902431-8

$$k_a^{(eff)} = \frac{k_{a,1}k_{a,2}}{k_{a,2}\Gamma + k_{a,1}(1-\Gamma)}, \tag{2a}$$

$$k_m^{(eff)} = \frac{k_{m,1}k_{m,2}}{k_{m,2}\Gamma + k_{m,1}(1-\Gamma)}, \tag{2b}$$

$$k_i^{(eff)} = k_{i,1}\Gamma + k_{i,2}(1-\Gamma), \tag{2c}$$

The effective elastic parameters $k_a^{(eff)}$, $k_m^{(eff)}$ and $k_i^{(eff)}$ are, in fact, the solutions of suitable *cell equations*, in the jargon of Asymptotic Homogenization [15-18], obtained by employing standard arguments of two-scale periodic analysis. We emphasize that the functional form of such cell equations depends on the considered mechanical model, describing the macro-scale mechanical interactions exchanged at the scale of the FA-ECM system, and on the way in which inhomogeneities are distributed at the microstructure. In our framework, granted for the linearity of the problem, the obtained cell equations are linear and decoupled from each other, which help us in obtaining their analytical expressions as reported in Equations (2a) – (2c).

As an example, we show the analytical solutions u_a and u_m (Fig. 2, left) computed with the homogenized Equations (1a) – (1d), equipped with the same boundary conditions as in [5]. In particular, we assume that the ECM is clamped at $x = 0$ and traction free at $x = L$, while the FA is traction free at $x = 0$ and subjected to an imposed displacement $\bar{u}$= 2 nm applied at $x = L$. Moreover, we plot the trend of FA's traction, T_a, in the case of ductile rupture [10] (Fig. 2, right). In this case, we fix $k_{a,1}$ = 2 pN/nm, $k_{a,2}$ = 0.7 pN/nm, $k_{m,1}$ = 8.7 pN/nm, $k_{m,2}$ = 11.4 pN/nm, $k_{i,1}$ = 3.9 pN/nm, $k_{i,2}$ = 6.3 pN/nm and Γ = 0.6 (see Fig. 1(b)). The value L_{det} = 0.336 μm descends from a minimization analysis of the energy put forward in [5]. Although the micro-scale elastic parameters are chosen to give a proof of concept of our model, they are physically sound and taken within ranges of values experimentally predicted [1-8, 26-28].

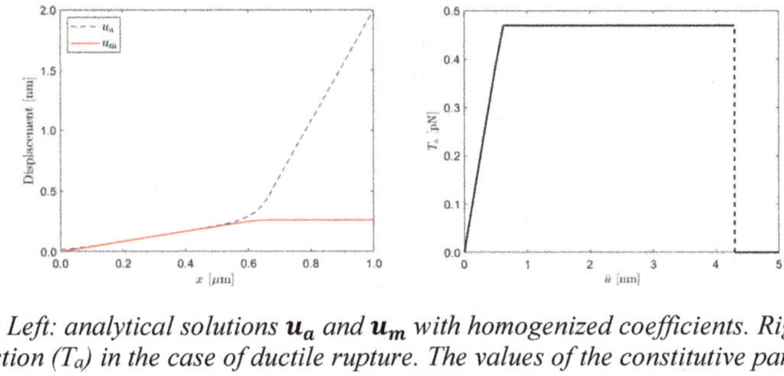

Figure 2: *Left: analytical solutions* $\mathbf{u_a}$ *and* $\mathbf{u_m}$ *with homogenized coefficients. Right: Trend of FA's traction* (T_a) *in the case of ductile rupture. The values of the constitutive parameters are from [10].*

We can motivate the difference in the trends of u_a and u_m by noticing that, while the adhesion plaque is subjected to the imposed displacement $\bar{u}$, the ECM is not. Moreover, the two components of the FA-ECM system respond differently to the forces exchanged through the integrins because of their diverse elastic properties. This behavior is coherent with that predicted by classical shear lag models [5, 23, 24]. Finally, these results constitute the first step towards a deeper investigation of the decohesive phenomenon and aiming, through a multi-scale approach, at furnishing a more detailed description of the overall nucleation and front advancement [26].

Acknowledgements

SDS is supported by Regione Puglia in the context of the research project *"Riciclo di materiali e sostenibilità: modelli di delaminazione per dispositivi laminati"* and Gruppo Nazionale per la

Fisica Matematica (GNFM) in the context of *"Progetto Giovani GNFM 2020-2022"*. GV is supported by the POR Puglia FESR-FSE project REFIN A1004.22. GF, SDS, LB, VF are supported by the Italian Ministry MIUR-PRIN project *Mathematics of active materials: From mechanobiology to smart devices* (2017KL4EF3) and by GNFM. GF is also supported by Istituto Nazionale di Alta Matematica (INdAM) and by Istituto Nazionale di Fisica Nucleare (INFN) through the project QUANTUM, by the FFABR research grant (MIUR) and the PON S.I.ADD.

References

[1] B.N.J Persson, On the mechanism of adhesion in biological systems. The Journal of chemical physics 118.16 (2003): 7614-7621. https://doi.org/10.1063/1.1562192

[2] A. Nicolas, B. Geiger, S.A. Safran. Cell mechanosensitivity controls the anisotropy of focal adhesions. PNAS 101.34 (2004): 12520-12525. https://doi.org/10.1073/pnas.0403539101

[3] X. Cao, et al. A chemomechanical model of matrix and nuclear rigidity regulation of focal adhesion size. Biophysical Journal 109.9 (2015): 1807-1817. https://doi.org/10.1016/j.bpj.2015.08.048

[4] X. Cao, et al. Multiscale model predicts increasing focal adhesion size with decreasing stiffness in fibrous matrices. Proceedings of the National Academy of Sciences 114.23 (2017): E4549-E4555. https://doi.org/10.1073/pnas.1620486114

[5] S. Di Stefano, et al. On the role of elasticity in focal adhesion stability within the passive regime. Int. J. Non Linear Mech. 10.1016/j.ijnonlinmec.2022.104157 (2022). https://doi.org/10.1016/j.ijnonlinmec.2022.104157

[6] S. Di Stefano, E. Benvenuti, V. Coscia. On the role of friction and remodelling in cell-matrix interactions: A continuum mechanical model. Int. J. Non Linear Mech.142 (2022): 103966. https://doi.org/10.1016/j.ijnonlinmec.2022.103966

[7] A. Besser, U. S. Schwarz. Coupling biochemistry and mechanics in cell adhesion: a model for inhomogeneous stress fiber contraction. New Journal of Physics 9.11 (2007): 425. https://doi.org/10.1088/1367-2630/9/11/425

[8] G. Rh Owen, et al. Focal adhesion quantification: A new assay of material biocompatibility? Review. European cells & materials 9 (2005): 85-96. https://doi.org/10.22203/eCM.v009a10

[9] M. Ben Amar, O.V. Manyuhina, and G. Napoli. Cell motility: a viscous fingering analysis of active gels. The European Physical Journal Plus 126.2 (2011): 1-15. https://doi.org/10.1140/epjp/i2011-11019-7

[10] L. Chun-Min, et al. Cell movement is guided by the rigidity of the substrate. Biophysical journal 79.1 (2000): 144-152. https://doi.org/10.1016/S0006-3495(00)76279-5

[11] D.E. Discher, P. Janmey, Y. Wang. Tissue cells feel and respond to the stiffness of their substrate. Science 310.5751 (2005): 1139-1143. https://doi.org/10.1126/science.1116995

[12] L. Bellino, et al. On the competition between interface energy and temperature in phase transition phenomena, Application of Engineering Science, 2, 100009, (2020). https://doi.org/10.1016/j.apples.2020.100009

[13] A. Grillo, et al. A study of growth and remodeling in isotropic tissues, based on the Anand-Aslan-Chester theory of strain-gradient plasticity. GAMM-Mitteilungen 42.4 (2019): e201900015. https://doi.org/10.1002/gamm.201900015

[14] C. Giverso, et al. A three dimensional model of multicellular aggregate compression. Soft Matter 15.48 (2019): 10005-10019. https://doi.org/10.1039/C9SM01628G

[15] N.S. Bakhvalov, G. Panasenko. Homogenisation: averaging processes in periodic media: mathematical problems in the mechanics of composite materials. Vol. 36. Springer Science & Business Media, 2012.

[16] S. Di Stefano, et al. Effective balance equations for electrostrictive composites. Zeitschrift für angewandte Mathematik und Physik 71.5 (2020): 1-36. https://doi.org/10.1007/s00033-020-01365-x

[17] A. Ramírez-Torres, et al. The role of malignant tissue on the thermal distribution of cancerous breast. Journal of Theoretical Biology 426 (2017): 152-161. https://doi.org/10.1016/j.jtbi.2017.05.031

[18] A. Ramírez-Torres, et al. An asymptotic homogenization approach to the microstructural evolution of heterogeneous media. Int. J. Non Linear Mech.106 (2018): 245-257. https://doi.org/10.1016/j.ijnonlinmec.2018.06.012

[19] C. Frantz, K.M. Stewart, V.M. Weaver. The extracellular matrix at a glance. Journal of cell science 123.24 (2010): 4195-4200. https://doi.org/10.1242/jcs.023820

[20] A. Gladkikh, et al. Heterogeneity of focal adhesions and focal contacts in motile fibroblasts. Cellular Heterogeneity. Humana Press, New York, NY, 2018. 205-218. https://doi.org/10.1007/978-1-4939-7680-5_12

[21] S. Diebels, A. Geringer. Modelling inhomogeneous mechanical properties in adhesive bonds. The Journal of Adhesion 88.11-12 (2012): 924-940. https://doi.org/10.1080/00218464.2012.725612

[22] A. Shuaib, et al. Heterogeneity in the mechanical properties of integrins determines mechanotransduction dynamics in bone osteoblasts. Scientific reports 9.1 (2019): 1-14. https://doi.org/10.1038/s41598-019-47958-z

[23] J.J. Marigo, L. Truskinovsky. Initiation and propagation of fracture in the models of Griffith and Barenblatt. Continuum Mechanics and Thermodynamics 16.4 (2004): 391-409. https://doi.org/10.1007/s00161-003-0164-y

[24] N.M. Pugno, Nicola M., et al. A generalization of the Coulomb's friction law: from graphene to macroscale. Meccanica 48.8 (2013): 1845-1851. https://doi.org/10.1007/s11012-013-9789-5

[25] V. Fazio, D. De Tommasi, N. M. Pugno, G. Puglisi, Spider Silks Mechanics: Predicting Humidity and Temperature Effects. JMPS, in print, (2022). https://doi.org/10.1016/j.jmps.2022.104857

[26] Di Stefano et al. A multiscale study of decohesion occurring in integrin-mediated biological systems. Preprint 2022

[27] M.J. Paszek, et al. Integrin clustering is driven by mechanical resistance from the glycocalyx and the substrate. PLoS computational biology 5.12 (2009): e1000604. https://doi.org/10.1371/journal.pcbi.1000604

[28] K.E. Caputo, D.A. Hammer. Effect of microvillus deformability on leukocyte adhesion explored using adhesive dynamics simulations. Biophysical journal 89.1 (2005): 187-200. https://doi.org/10.1529/biophysj.104.054171

Theoretical and Applied Mechanics - AIMETA 2022 | Materials Research Forum LLC
Materials Research Proceedings **26** (2023) 53-58 | https://doi.org/10.21741/9781644902431-9

One century of theoretical and applied mechanics of concrete and stone materials permeability. What have we learned?

Michela Monaco[1,a*], Roberto Serpieri[2,b*]

[1]DING - Department of Engineering, University of Sannio
Palazzo ex INPS, Piazza Roma, 21, 82100 Benevento, Italy

[2]Dipartimento di Architettura e Disegno Industriale, Università della Campania "L. Vanvitelli"
Via San Lorenzo - Abazia di San Lorenzo, 81031, Aversa (CE), Italy

[a]monaco@unisannio.it, [b]roberto.serpieri@unicampania.it

Keywords: Variational Poroelasticity, Multiphase Physics, Concrete Permeability

Abstract. Periodic and non-routine inspections on buildings, infrastructures, and sites of historical or architectural interest frequently involve a necessary deepening of the knowledge of natural or artificial stone materials of the built heritage. Insight into stone materials is of particular relevance when the problem is approached of fine-tuning best practices for assessment and maintenance within Life Cycle Assessment methodologies, in the light of an increased awareness that social, infrastructural and environmental issues and policies cannot be faced by single compartments but need comprehensive evaluation. Permeability has been early recognized as a basic property for the assessment of durability of concrete and stone materials. This contribution reports and discusses selected conclusions of studies which have scrutinized a large body of theoretical and technical scientific literature on permeability testing of concrete, as well as of regulatory literature, keeping a historical perspective and a focus in pinpointing fundamental open research issues of both theoretical and applied relevance in the mechanics of permeation of fluids through porous media. The fundamental experimental studies conducted in the first decades of last century are also the starting point for an assessment of the continuum mechanics and thermodynamics equations describing the flow of compressible fluids in porous media and for a critical review of the approach of current European standards to the determination of concrete watertightness.

Introduction

The first published documents reporting of studies on concrete permeability can be traced back to the first quarter of the last century and were mainly devoted to verifying the applicability of Darcy's law to water permeability for this porous construction material which was new, or considered new, at the time. The work by Withey and Wiepking [1] together with other remarkable experimental studies developed in the U.S.A., among others by McMillan and Lyse [2], Ruettgers et al., [3, 4], and Vidal and Samson [5], were all devoted to find a suitable test setup and methodology to correctly evaluate the intrinsic permeability and the material and testing parameters influencing experimental results. The porous nature of concrete was well understood, and its watertightness was considered essential for the efficiency of strategic constructions like dams and bridges. In the same years, in Europe, experimental works developed in the U.K. by Glanville [6] and Madgwick [7] and in France by Mary [8] were carried out on the subject of concrete and stone permeability. Although the testing apparatuses and the tested stone and concrete materials were different, all these research efforts were focused on the identification of regularly observable phenomena and material properties. The only common features shared by the different experimental procedures were the creation of a uniaxial steady-state flow through a porous medium specimen acting like a filter under open-flow conditions and having, in general, a tile-like shape with thickness generally lower than 5 cm. These common features had, among others, the objective of maximizing the fluid flow in order to facilitate measurement by conveniently reducing error factors, like the formation of air pockets, as well as testing duration.

A comprehensive treatise on the flow of homogeneous fluids through porous media, with a focus on large-scale geological problems also of multiaxial nature, was published by Muskat [9] while studies focused on the correlation of permeability to microstructural parameters of the stone filter were successfully developed by Kozeny [10] and Carman [11]. It is worth remarking that all these studies gave for granted that the experimental determination of permeability is to be obtained by set-ups enforcing uniaxial steady-state open-flow conditions. Additional phenomena introducing deviations from the basic Darcy's formula, determined by the use of gases as permeating fluids, were later investigated by Klinkenberg [12]. Studies dedicated to Portland cement pastes were carried out by Powers et al. [13, 14]. Hallmark of the subsequent period from the fifties of the past century onwards, concerning permeability testing in concrete, was the evolution of tests in the wake of standardization processes controlling the transfer to a broader public of knowledge and procedures from scientific, industrial and engineering corps' research. For instance, one important regulative publication in this standardization process was the Initial Surface Absorption Testing methodology (ISAT) [15]. The main aspects to be underlined of this subsequent period are a partial renouncement to pressure-cells and set-ups granting strict conditions of uniaxial open flow (see, for instance, Levitt [16]), the renouncement to thorough oven drying of all specimens before test, the attempt to accommodate nondestructive on-site testing (Figg [17]), and a general push towards practices for concrete watertightness testing having a much more conventional character. It is necessary to wait the end of the '80s to find a study by Dhir et al. [18] performing a review of the ISAT and Figg's techniques and a partial reconnection with past methods based on the quantitative determination of intrinsic permeability by steady state uniaxial flow transmission. The results of the above-mentioned trend of pushing towards more conventional practices for concrete watertightness testing can be observed at present times examining the codes actually in force in the EU, almost all derived from EN 12390-8:2019 [19]. Most codes and combined provisions are based on just measuring the penetration, after splitting test, of the depth of water in an almost 10 cm deep concrete specimen, after a 72 hrs constant pressure.

A comprehensive critical review of one hundred years of permeability tests has been recently carried out by Monaco et al. [20] with the objective of assessing a near-zero-invasiveness practice which is therein proposed for the Life Cycle Assessment (LCA) of concrete structures and building manufacts by integrating actions of surveying, sampling, and watertightness testing of a concrete cover object of appraisal for possible replacement.

This contribution reports and discusses selected conclusions by Monaco et al. [20], keeping a focus in pinpointing fundamental open research issues of both theoretical and applied relevance in the mechanics of permeation of fluids through porous media with some critical considerations on current European standards for the determination of concrete watertightness. Given the importance of uniaxial flow, the document also reports in brief the main conclusions of an assessment of the continuum mechanics and thermodynamics equations describing uniaxial open-flow permeation of compressible fluids in a porous filter (Serpieri and Monaco [21]).

Fundamental mechanical quantities and phenomena in stone permeability testing

As reviewed in Muskat's treatise [9], the permeation of an incompressible liquid through a cylindrical filter made of a macroscopically homogeneous porous medium saturated by a Newtonian fluid – under the usual hypotheses of stationary conditions, macroscopically uniform, uniaxial and nonturbulent flow – is well described by Darcy's linear relation, depending on the permeability K_P, a proportionality constant (in $[m/s]$) which is characteristic of the porous medium and which also depends on the fluid properties. For incompressible liquids, the pressure gradient and the velocity are uniform through the filter length and Darcy's relation can be conveniently expressed in the form containing the intrinsic permeability which has the dimension of square meters. As shown in the theoretical and experimental studies by Kozeny [10] and Carman [11], in complete saturation the intrinsic permeability is reasonably well predicted by a correlation with the specific surface area and the void volume fraction. It should be remarked that the valid

Materials Research Proceedings **26** (2023) 53-58
https://doi.org/10.21741/9781644902431-9

predictive character of Carman's correlation has been successfully verified not only for granular beds of highly controlled geometry, investigated in the original works, but also subsequently for hardened concrete pastes by Powers and Brownyard [13]. These Authors have shown that, by assuming porosity to be equal to the nonevaporable water and by applying the method by Brunauer, Emett and Teller [22] to acquire measures of the specific surface of solid boundaries from the experimental determination of adsorption isotherm curves, a reasonable prediction is obtained of lowest intrinsic permeability values, in the range $[10^{-18}, 10^{-19}]$ m^2, inferable from experimental data on cement pastes reported by Ruettgers et al. [4] concerning the concretes employed in the construction of the Colorado Boulder 'Hoover' dam. Water-adsorption-related phenomena entailing deviations from the fundamental equation were studied by Carman [11] and Zunker [23], while the following compact expression of the deviation of intrinsic permeability for Portland cement pastes due to adsorbed water represents the results tabulated by Powers and Brownyard [13]:

$$K_{Pi} = 5.96 \cdot 10^{-22} \cdot \left(\frac{w/c - 0.15}{0.038} - k_1\right)^3 \cdot \frac{1}{w/c + 1/3.16}, \tag{1}$$

where w/c is the water-cement ratio and k_1 is a parameter varying from 1 to approximately 4 tentatively accounting for the possible reduction of space available to transmission due to adsorbed water. A relation which predicts a deviation from Darcy's equation when the testing fluid is compressible, such as in air-permeability testing, and when conditions are isothermal was proposed by Muskat [9]:

$$\left(\bar{v}_{f1}\right)_{gas,Musk.} = \frac{dQ}{dt} \cdot \frac{1}{A} = \frac{K_{Pi}}{\mu L} \cdot \frac{(p_1^2 - p_2^2)}{2p_1}. \tag{2}$$

where p_1 and p_2 are the inlet and outlet velocity, respectively, A is the cross-sectional area, μ the viscosity and L the filter length. The close adherence of Eq. (2) to experimental data had been earlier verified by Madgwick [7] in studies on the permeability of several construction materials like building stones, bricks and plasters. It should be noticed, however, that while Madgwick derived (2) by never explicitly mentioning an isothermal hypothesis, Muskat [9] proceeded explicitly from such a hypothesis. A second deviation, observed only with gases, was shown by Klinkenberg [12] who theoretically derived and experimentally validated relations capable of accounting for a flow increment due to the so-called phenomenon of gas-slippage.

For water permeation, two further possible side-phenomena which, depending on the stone constitution and the test duration, may affect a regular steady permeation, and thus should be considered prior to an open-flow permeation test, are: 1) the decrease of permeability to water along test run time, termed by some authors "self-sealing", and 2) the opposite phenomenon of "self-unsealing" which is a gain of permeability during the test. The root causes for a same observed self-sealing or self-unsealing behavior can however be of very different mechanical or chemical nature. In concretes, the possible presence of unhydrated reactive cement fractions is a main common cause for self-sealing [2]. Hearn et al. [24] provided important insights on further factors, other than cement paste hydration, which might determine a permeability reduction, including: clogging, osmotic pressure, precipitation of soluble hydrates, chemical interaction between the water and the cement matrix, dissolution of air into permeating water, and incomplete saturation of the test specimen.

Self-unsealing can be originated by leaching of significant fractions of soluble components possibly present in the stone matrix and, although deserving the utmost attention for stone durability, appears instead to be a much less investigated phenomenon. El-Dieb and Hooton [25] recorded the variation of water-permeability as function of time in three concrete mixes. Their study exemplifies a typical graph of the recorded coefficient of permeability (in m/s) plotted versus the time at which permeability is measured, and shows that, during 80 hours testing, permeability undergoes an almost ten-fold increase probably ascribable to dissolution or transport of soluble components.

A general model of uniaxial permeation of liquids and gases through stone materials
A continuum mechanics theory of permeation of a fluid phase in a porous medium which is sufficiently general to address the case of a compressible fluid and which, proceeding from the consideration of a minimum possible number of kinematic descriptors, is also purely-mechanical, purely variational and purely macroscopic (in that it does not require a detailed knowledge or detailed assumptions on the small-scale features of the porous medium and of the solid-fluid interaction) has been derived by Serpieri and Travascio [26] on the basis of canonic variational arguments and of preceding studies by Serpieri and Rosati [27], Serpieri et al. [28]; and by Travascio et al. [29]. For compelling page limits, only the final set of equations governing the stationary thermomechanical variational statement of the uniaxial permeation of an ideal gas is synoptically reported below.

$$\hat{\rho}_f \bar{v}_f = \hat{\rho}_{f0} \bar{v}_{f0} \quad \text{Complete space saturation + stationary Eulerian mass balance} \tag{3}$$

$$\hat{J}_f = \frac{\hat{\rho}_{f0}}{\hat{\rho}_f} \quad \text{Lagrangian mass balance} \tag{4}$$

$$-p' - \frac{\mu}{K_{Pi}} \bar{v}_f = \hat{\rho}_f \bar{v}_f \bar{v}_f' \quad \text{Linear momentum balance under stationary conditions} \tag{5}$$

$$\frac{\hat{\rho}_{f0}}{M} C_v T' \bar{v}_f = k_T T'' - p \hat{J}_f' \bar{v}_f \quad \text{Energy balance under stationary conditions} \tag{6}$$

$$p = \frac{\hat{\rho}_{f0} R}{M} \frac{T}{\hat{J}_f} \quad \text{Ideal gas law} \tag{7}$$

In the equations above, $\hat{\rho}_f$ is the true fluid density, $\bar{v}_f$ is the seepage velocity ($\hat{\rho}_{f0}$ and $\bar{v}_{f0}$ being reference values), $\hat{J}_f$ is the finite volumetric strain of the fluid, M is the molar mass of the gas, T the absolute temperature, C_v is the constant-volume heat capacity of the fluid, k_T its thermal conductivity, R is the universal gas constant, and prime notation indicates space derivation as usual.

Serpieri and Monaco [21] have shown that the system above can be reduced to the following single differential equation in the only unknown p:

$$\frac{K_{Pi}}{\mu} \frac{1}{k_T} \frac{\hat{\rho}_{f0} R}{M} \left[\frac{C_v}{R} (pp')' + pp'' \right] p' + (pp')'' = 0 \tag{8}$$

which, upon integration and approximation of negligible terms, provides a temperature-dependent solution encompassing deviations from the basic isothermal formula (2). Such solution can be employed to derive correlations applicable to the technical problem of air-permeability measurement in thin concrete filters which can be obtained from specimens carved out from an ordinary concrete cover.

Discussion
In the light of the studies herein summarized, research issues having both theoretical and applied relevance in the mechanics of stone permeability – in particular for concrete –, which are judged to be still-open or deserving further elucidation are:

- the definition of a general permeability measurement protocol capable of discerning the presence of a self-unsealing, self-sealing or stationary behavior in a given concrete specimen of tile-like form from recordings of permeability during a test of suitable duration;

- the determination of bounds to experimental and theoretical uncertainties associated with the possible occurrence of phenomena of self-unsealing and self-sealing;
- the definition of an experimental protocol, based on permeability variations, capable of discerning the possible presence of fractions of unhydrated cement and of swelling constituents in a given concrete mix.

A final comment is devoted to most recent and currently active standards, such as EN 12390-8, [19], which are based on a conventional measurement of penetration depth. The literature review has shown that intrinsic permeability by open-flow tests should be preferred, as a diagnostic parameter, to penetration depth since the former is less affected than the latter by hygrothermal and environmental factors.

The invasiveness of the size of the specimens, even 10 cm deep, prescribed by some recent standards is also pointed out in the light of the evidence that many testing protocols of the past have successfully employed much thinner specimens setting a path for the definition of less invasive standards.

References

[1] Withey, M.O., Wiepking, C.A. (1923). The permeability of broken stone concrete, University of Wisconsin Bulletin, Engineering, Ser. 9, 1245.

[2] McMyllan, F.R., Lyse, I. (1929). Some permeability studies of concrete. Journal of the American Concrete Institute Proceedings, 101-124.

[3] Ruettgers, A., Vidal, W.N., Wing S.P (1935). An investigation of the permeability of mass concrete with particular reference to Boulder dam. American Concrete Institute Proceedings, 31, 382. https://doi.org/10.14359/8350

[4] Ruettgers, A., Vidal, W.N., Wing S.P (1936). Authors' closure in the discussion of paper: "An investigation of the permeability of mass concrete with particular reference to Boulder dam". American Concrete Institute Proceedings, 32, 378.

[5] Vidal, E. N., Samson, G. (1936). Apparatus and Technique for the Bureau of Reclamation Concrete Permeability Tests. In Proceedings of the American Society for Testing Materials, 36(1), 289.

[6] Glanville, W.H. (1931). The permeability of Portland cement concrete. Building Research Station, Department of Science and Industrial Research, London, Great Britain, Technical paper no. 3.

[7] Madgwick, E. (1932). An investigation into some of the properties of porous building materials. Published in six parts in The London, Edinburgh, and Dublin Philosophical Magazine and Journal of Science: Parts I-VII.

[8] Mary, M. (1933,1934). La perméabilité du béton. Annales des Ponts et Chaussées, VI, May-Juin 1933, 467-487, Novembre-Décembre 1934, 421-449.

[9] Muskat, M. (1937) The flow of homogeneous fluids through porous media, McGraw-Hill Book Company, Inc. https://doi.org/10.1063/1.1710292

[10] Kozeny, J. (1927). Über kapillare leitung des Wassers im Boden (Aufstieg, Versickerung und Anwendung auf die Bewässerung). Sitzungsberichte der Kaiserlichen Akademie der Wissenschaften in Wien – Mathematisch- NaturwissenSchaftliche Classe, 136 (Abt. IIa) 271-306.

[11] Carman, P.C. (1939). Permeability of saturated sands, soils, and clays. Journal Of Agricultural Science 29, 262-273. https://doi.org/10.1017/S0021859600051789

[12] Klinkenberg, L.J. (1941). The permeability of porous media to liquids and gases, Drilling and production practice, New York, American Petroleum Institute, 200-213.

[13] Powers, T.C., Brownyard, T.L. (1947). Studies on the physical properties of hardened cement pastes. Part 7. Permeability and absorptivity. Jour. of the American Concrete Institute. 43, 865-880. https://doi.org/10.14359/15306
[14] Powers, T. C., Copeland, L. E., Hayes, J. C., Mann, H. M. (1954): Permeability of Portland Cement Paste. Journal of the American Concrete Institute, 26(3), 285-298.
[15] BS 1881-5:1970. Methods of Testing Hardened Concrete for other than Strength. With Feb.1983 and March 1990 amendments. British Standard Institution.
[16] Levitt, M. (1970). Non-destructive testing of concrete by the initial surface absorption method. Research Committee for the Cast Stone and Cast Concrete Products Industry, In Non-destructive testing of concrete and timber, Thomas Telford Publishing, 23-26.
[17] Figg, J. W. (1973): Methods of measuring the air and water permeability of concrete. Magazine of Concrete Research, 25(85), p. 213–219. https://doi.org/10.1680/macr.1973.25.85.213
[18] Dhir, R.K., Hewlett, P.C., Chan, Y.N. (1989). Near surface characteristics of concrete: intrinsic permeability. Magazine of Concrete Research, 41(147), 87-97. https://doi.org/10.1680/macr.1989.41.147.87
[19] EN 12390-8:2019 - Testing hardened concrete - Part 8: Depth of penetration of water under pressure, CEN- European Committee for Standardization
[20] Monaco, M., Guadagnuolo, M., Frunzio, G., Serpieri, R. (2022) A critical review of one hundred years of permeability tests for the Life Cycle Assessment of concrete structures, under review, Construction Building Materials.
[21] Serpieri, R., Monaco, M. (2022). A general consistent variational model describing Darcian permeation of liquids and gases through stone materials. Submitted to Meccanica.
[22] Brunauer, S., Emmett, P. H., Teller, E. (1938). Adsorption of gases in multimolecular layers. Journal of the American Chemical Society, 60(2), 309-319. https://doi.org/10.1021/ja01269a023
[23] Zunker, F. (1932). Die Durchlässigkeit des Bodens. Zeitschrift für Pflanzenernährung, Düngung und Bodenkunde, 25 A, 1–24.
[24] Hearn, N., Detwiler, R.J., Sframeli, C. (1994). Water permeability and microstructure of three old concretes. Cement and concrete research, 24(4), 633-640. https://doi.org/10.1016/0008-8846(94)90187-2
[25] El-Dieb, A.S., Hotoon, R.D. (1995). Water-permeability measurement of high-performance concrete using a high-pressure triaxial cell. Cement and Concrete Research, 25(6), 1199-1208. https://doi.org/10.1016/0008-8846(95)00112-P
[26] Serpieri, R., Travascio, F. (2017). Variational Continuum Multiphase Poroelasticity. Springer. https://doi.org/10.1007/978-981-10-3452-7
[27] Serpieri, R., Rosati, L. (2011). Formulation of a finite deformation model for the dynamic response of open cell biphasic media. Journal of the Mechanics and Physics of Solids, 59(4), 841-862. https://doi.org/10.1016/j.jmps.2010.12.016
[28] Serpieri, R., Travascio, F., Asfour, S., Rosati, L. (2015). Variationally consistent derivation of the stress partitioning law in saturated porous media. Int. Journal of Solids and Structures, 56, 235-247. https://doi.org/10.1016/j.ijsolstr.2014.10.033
[29] Travascio, F., Asfour, S., Serpieri, R., Rosati, L. (2017). Analysis of the consolidation problem of compressible porous media by a macroscopic variational continuum approach. Mathematics and Mechanics of Solids, 22(5), 952-968. https://doi.org/10.1177/1081286515616049

Theoretical and Applied Mechanics - AIMETA 2022 Materials Research Forum LLC
Materials Research Proceedings **26** (2023) 59-64 https://doi.org/10.21741/9781644902431-10

Supercontraction of spider silks as a humidity-driven phase transition

FAZIO Vincenzo[1, a *], FLORIO Giuseppe[2,3,b], PUGNO Nicola Maria[1,4,c] and PUGLISI Giuseppe[2,d]

[1]Laboratory for Bioinspired, Bionic, Nano, Meta Materials & Mechanics, University of Trento, Via Mesiano 77, 38123 Trento, Italy

[2]Department of Civil Environmental Land Building Engineering and Chemistry, Polytechnic University of Bari, via Orabona 4, 70125 Bari, Italy

[3]INFN, Sezione di Bari, I-70126 Bari, Italy

[4]School of Engineering and Materials Science, Queen Mary University of London, Mile End Road, London E1 4NS, U.K.

[a]vincenzo.fazio@unitn.it, [b]giuseppe.florio@poliba.it, [c]nicola.pugno@unitn.it, [d]giuseppe.puglisi@poliba.it

Keywords: Spider Silk, Supercontraction, Phase Transition, Statistical Mechanics

Abstract. Spider silks have been intensively studied among natural materials for their extreme mechanical properties such as very high strength, ultimate strain, and toughness. Another striking phenomenon characterizing spider silk, known as supercontraction, is a substantial contraction, up to a half of the initial length, occurring when an unconstrained silk thread is exposed to a wet environment. We propose a multiscale model that deduces the hygro-dependent macroscopic behaviour of the spider silks starting from the nano and micro-structure properties of the material. In particular, we describe the influence of humidity at the macromolecular scale by considering the moisture effects disrupting hydrogen-bonds and enabling the decrease of the natural (zero force) end-to-end chains length due to entropic effects. The main novelty of our theoretical approach is a description in the field of solid-solid phase transitions, with the system undergoing a transition driven by humidity from an unfolded, hard dry to a folded, soft wet configuration. Based on a statistical mechanical approach, we are able to describe the temperature dependence of the supercontraction effects and its cooperative properties quantitatively predicting the observed experimental behaviour.

Introduction

Spider silk is the most studied natural material for his extreme mechanical properties, in particular the strength and the toughness overcoming also many high-performance man-made materials. Moreover, in the framework of biomimetics, spider silks are considered as the basis of a new class of high-performance fibers [1,2]. However, the behavior of the spider silk is very sensitive to external environmental conditions. More in detail, a Relative Humidity (RH) above a certain critical threshold, may give rise to an abrupt variation of material properties and a significant reduction in the fiber length, known as *supercontraction* and discovered by Work in 1977 [3]. This contraction is due the influence of the hydration water molecules at the macromolecular scale that disrupt Hydrogen-bonds, thus enabling the decrease of the natural end-to-end chain length due to entropic effects [4–6].

The observation of a stress reaching values of tens of MPa in experiments at fixed length is very interesting also in the perspective of humidity driven actuators and sensors [7–9]. The silk properties change is localized in a small range of RH (few points percent) and this is related to the transition from a stiff state to a rubbery state undergoing entropic coiling [10,11] with important analogies with the glass transition typically induced by temperature in polymers. As in the case of other solids undergoing phase transitions, such as shape memory nanowires, where one observes a switch from the austenitic to the martensitic phase or in protein materials undergoing unfolding from a stiff to a softer configurations [12], the evolution of the system is regulated by the transition strategy among different phase configurations as the external fields are modified [13]. In this perspective, starting from the pioneering work of Müller and Villaggio [13], extended to include the fundamental thermal effects based on a Statistical Mechanics approach, lattices of elements with non-convex energies and intrinsic discrete length-scale have been used to mimic many biological systems, such as the misfolding and refolding in proteins [14,15], the DNA denaturation and replication [16,17], the attachment and detachment of tau proteins in the neuronal axon, usually associated to neurodegenerative diseases [18,19], and in the study of focal adhesions [20,21].

Here we extend this approach to study the experimentally observed transition in spider silks, describing the mechanical properties of the soft and hard phases of the silks at the molecular length scale. Specifically, we extend the model proposed in [12,22] to account for the effect of the relative humidity inducing the supercontraction, here modelled as a phase transition. This effect is due to an internal change of the natural configuration of the hard phase, and it can be described, following the classical Landau approach to phase transition, by introducing a RH-dependent transition energy as show in FIG. 1. The behavior of the silks is also highly affected by temperature effects (see Refs. [11,23]). Thus, in this paper we introduce thermal fields in the framework of equilibrium Statistical Mechanics, that has been proved to be an effective tool for describing thermal effects in multistable systems [24–27]. We deduce an analytic approach quantitatively describing the humidity and temperature effects in the supercontraction phenomenon. Such analytic results are in our opinion fundamental also in the spirit of biomimetics.

Model

Following the theoretical model developed in [22], we analyse a discrete chain with N two-state elements (see Fig. 1), each described by a bistable potential energy having two wells characterized by different elastic constants k (unfolded state) and αk (folded state), where $0 < \alpha < 1$. Moreover, the two states have their equilibrium lengths equal to l and χl with $0 < \chi < 1$.
Following an Ising type approach the overall energy assumes the compact form

$$\Phi = \sum_{i=1}^{N}\left\{Q(S_i) + \frac{K(S_i)}{2}\left[(\lambda_i - \lambda_0(S_i))l\right]^2\right\} - J\sum_{i=1}^{N-1} S_i S_{i+1} \tag{1}$$

where the internal variables $S_i = \pm 1$ identify the phase (energy wells) of each spring. Here, λ_i is the stretch of the i-th spring and $\lambda_0(-1) = 1, \lambda_0(+1) = \chi$, are the natural configurations. The parameter $J > 0$ penalizes energy interfaces. It is useful to take into account the nondimensional energy φ

$$\varphi = \frac{\Phi}{J} = \sum_{i=1}^{N}\left\{\tilde{Q}(S_i) + \frac{\tilde{K}(S_i)}{2}\left[(\lambda_i - \lambda_0(S_i))\right]^2\right\} - \sum_{i=1}^{N-1} S_i S_{i+1} \tag{2}$$

where $\tilde{Q}(S_i) = Q(S_i)/J$ and $\tilde{K}(S_i) = K(S_i)l^2/J$, $\tilde{k} = kl^2/J$ and $\tilde{K}(-1) = \tilde{k}$ and $\tilde{K}(+1) = \alpha\tilde{k}$. The main novelty of the proposed approach is the microstructure-inspired assumption of Q depending on the relative humidity as

$$\tilde{Q}(RH) = \frac{1}{\mathrm{J\,h(RH)}} \tag{3}$$

where we included the dependence on the relative humidity in the quantity h(RH); this dependence will be specialized in the following to adapt the results of the model to the experimental results.

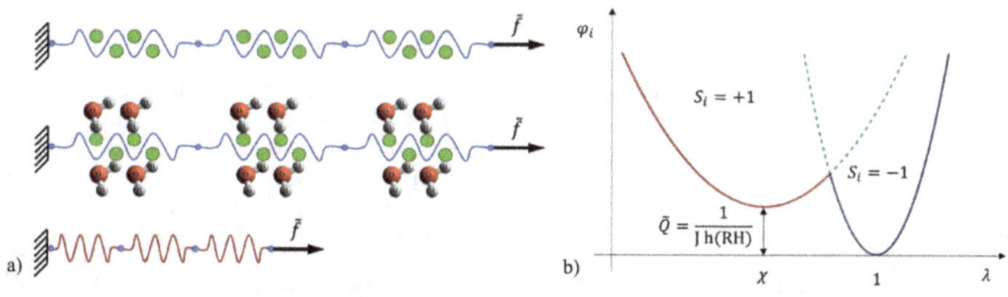

Figure 1: a) Scheme of a chain with three elements. Top: the chain is in the unfolded configuration ($\boldsymbol{S_i} = \mathbf{-1}$) with the green circles representing the H-bonds naturally present in the silk; middle: the hydration water molecules link and disrupt the H-bonds; bottom: final folded (supercontracted) configuration ($\boldsymbol{S_i} = \mathbf{+1}$). b) Scheme of the energy wells of a single element.

To introduce temperature effects in the framework of the equilibrium Statistical Mechanics, we consider the canonical partition function also including the fixed force acting on the last element of the chain (the so-called Gibbs ensemble):

$$Z_G(\tilde{f}) = l^N \sum_{\{S_i\}} \int_{\mathbb{R}^N} e^{-\tilde{\beta}(\varphi - \tilde{f}\sum_{i=1}^{N}\lambda_i)} d\lambda_1 \dots \lambda_N. \tag{4}$$

Here, the non-dimensional force is defined as $\tilde{f} = f\,l/J$ and

$$\tilde{\beta} = \beta\,J = \frac{J}{k_B T}. \tag{5}$$

The sums over $\{S_i\}$ must be considered extended to the values $+1$ and -1 for each spin variable $(i = 1,\dots,N)$.

By definition, the Gibbs free energy is

$$\mathcal{G} = -\frac{1}{\tilde{\beta}} \ln Z_G(\tilde{f}). \tag{6}$$

Thus, in order to evaluate the unrestrained supercontraction phenomenon, we compute the expectation value of the average deformation of the chain for $\tilde{f} = 0$, as

$$N\bar{\lambda} = \langle \sum_{i=1}^{N} \lambda_i \rangle = -\frac{d\mathcal{G}}{d\tilde{f}}\bigg|_{\tilde{f}=0}. \tag{7}$$

A direct evaluation of this quantity, (see [22] for detailed calculations) gives

$$\bar{\lambda} = \frac{1}{2}\left[\left(1 + \frac{c_- - c_+}{\sqrt{\Delta}}\right) + \chi\left(1 - \frac{c_- - c_+}{\sqrt{\Delta}}\right)\right], \tag{8}$$

where $c_- = c(-1)$ and $c_+ = c(+1)$ with

$$c(s_i) = \sqrt{\frac{2\pi}{\tilde{\beta}\tilde{k}(s_i)}} e^{-\tilde{\beta}\,\tilde{Q}(s_i)}, \Delta = (c_+ - c_-)^2 + 4\,c_+\,c_-\,e^{-4\tilde{\beta}}. \tag{9}$$

Experimental comparison

The effectiveness of the model in reproducing the experimental behavior of the phase transition occurring in spider silks, i.e. the supercontraction, is verified by quantitatively comparing the values of the silk thread stretch as a function of the external humidity at different temperatures. In Fig. 2 we represent by markers the experiments performed on *Argiope trifasciata* spider silk fibers [23], that exhibits a relevant contraction, reaching less than half of its length in dry conditions. The curves obtained by Eq. (7) are accurate in predicting the main physical phenomena observed during the phase transition at variable RH and temperature. We assumed $\mathrm{h}(RH) = \gamma\sqrt{\frac{RH}{100}}$ obtaining $\tilde{Q}(RH) = \frac{1}{\delta\sqrt{RH/100}}$, with $\delta = J\,\gamma$. We use the experimental data at the reference temperature ($T_r = 20°C$) to fit the parameters of the model δ and $\tilde{\beta}_r$, whereas from Eq. (4) we can compute the values $\tilde{\beta} = \tilde{\beta}_r\, T_r/T$ for other temperatures. We remark that for the main parameters of the model we employed the values measured during the experiments in [23], i.e. $\chi = 0.46$ corresponding to the stretch of the fiber at the saturation of the RH and $\alpha = 0.005$ corresponding to the ratio of the stiffness of the silk in the full supercontracted (soft, folded) state and pristine (hard, unfolded) state.

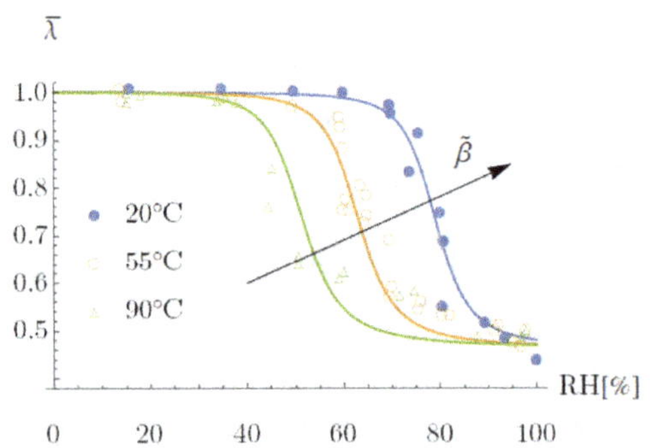

Figure 2: Comparison of the model with experiments [23]. Values of the parameters : $\boldsymbol{\alpha = 0.005;\ \chi = 0.46, \tilde{\beta}_r = 1.4, \delta = 0.59}$.

Thus, based on minimal assumptions on the microscopic structure, we can describe a solid-solid phase transition, with the system undergoing an unfolded-folded (hard→soft) transition driven by the humidity. Based on an equilibrium statistical mechanical approach, we can deduce analytically the temperature dependence of the supercontraction effect and its cooperative properties, quantitatively predicting the observed experimental behavior.

Acknowledgements

VF, GF and GP have been supported by "Gruppo Nazionale per la Fisica Matematica" (GNFM) under "Istituto Nazionale di Alta Matematica" (INdAM). GF and GP have been supported by the Italian Ministry MIUR-PRIN project "Mathematics of active materials: From mechanobiology to smart devices (2017KL4EF3) and GF is also supported by "Istituto Nazionale di Fisica Nucleare" (INFN) through the project QUANTUM, by the FFABR research grant (MIUR) and the PON

S.I.ADD. NMP is supported by the European Commission under the FET Open "Boheme" grant no. 863179 and by the Italian Ministry of Education MIUR, Italy under the PRIN-20177TTP3S.

References

[1] G. Greco et al., *Tyrosine Residues Mediate Supercontraction in Biomimetic Spider Silk*, Commun Mater **2**, 43 (2021). https://doi.org/10.1038/s43246-021-00147-w

[2] T. Arndt et al., *Engineered Spider Silk Proteins for Biomimetic Spinning of Fibers with Toughness Equal to Dragline Silks (Adv. Funct. Mater. 23/2022)*, Adv Funct Materials **32**, 2270134 (2022). https://doi.org/10.1002/adfm.202270134

[3] R. W. Work, *Dimensions, Birefringences, and Force-Elongation Behavior of Major and Minor Ampullate Silk Fibers from Orb-Web-Spinning Spiders—The Effects of Wetting on These Properties*, Textile Research Journal **47**, 650 (1977). https://doi.org/10.1177/004051757704701003

[4] N. Du, X. Y. Liu, J. Narayanan, L. Li, M. L. M. Lim, and D. Li, *Design of Superior Spider Silk: From Nanostructure to Mechanical Properties*, Biophysical Journal **91**, 4528 (2006). https://doi.org/10.1529/biophysj.106.089144

[5] M. Elices, G. V. Guinea, J. Pérez-Rigueiro, and G. R. Plaza, *Finding Inspiration in Argiope Trifasciata Spider Silk Fibers*, JOM **57**, 60 (2005). https://doi.org/10.1007/s11837-005-0218-7

[6] V. Fazio, D. De Tommasi, N. M. Pugno, and G. Puglisi, *Spider Silks Mechanics: Predicting Humidity and Temperature Effects*, Journal of the Mechanics and Physics of Solids **164**, 104857 (2022). https://doi.org/10.1016/j.jmps.2022.104857

[7] F. I. Bell, I. J. McEwen, and C. Viney, *Supercontraction Stress in Wet Spider Dragline*, Nature **416**, 37 (2002). https://doi.org/10.1038/416037a

[8] K. N. Savage, P. A. Guerette, and J. M. Gosline, *Supercontraction Stress in Spider Webs*, Biomacromolecules **5**, 675 (2004). https://doi.org/10.1021/bm034270w

[9] V. Fazio, N. M. Pugno, and G. Puglisi, *"Water to the Ropes": A Predictive Model for the Supercontraction Stress of Spider Silks*, Preprint (2022).

[10] M. Elices, G. R. Plaza, J. Pérez-Rigueiro, and G. V. Guinea, *The Hidden Link between Supercontraction and Mechanical Behavior of Spider Silks*, Journal of the Mechanical Behavior of Biomedical Materials **4**, 658 (2011). https://doi.org/10.1016/j.jmbbm.2010.09.008

[11] C. Fu, D. Porter, and Z. Shao, *Moisture Effects on* Antheraea Pernyi *Silk's Mechanical Property*, Macromolecules **42**, 7877 (2009). https://doi.org/10.1021/ma901321k

[12] L. Bellino, G. Florio, S. Giordano, and G. Puglisi, *On the Competition between Interface Energy and Temperature in Phase Transition Phenomena*, Applications in Engineering Science **2**, 100009 (2020). https://doi.org/10.1016/j.apples.2020.100009

[13] I. Müller and P. Villaggio, *A Model for an Elastic-Plastic Body*, Arch. Rational Mech. Anal. **65**, 25 (1977). https://doi.org/10.1007/BF00289355

[14] C. M. Dobson, *Protein Folding and Misfolding*, **426**, 7 (2003). https://doi.org/10.1038/nature02261

[15] G. Florio, N. M. Pugno, M. J. Buehler, and G. Puglisi, *A Coarse-Grained Mechanical Model for Folding and Unfolding of Tropoelastin with Possible Mutations*, Acta Biomaterialia **134**, 477 (2021). https://doi.org/10.1016/j.actbio.2021.07.032

[16] H. E. Gaub, M. Rief, and H. Clausen-Schaumann, *Sequence-Dependent Mechanics of Single DNA Molecules*, Nat. Struct Biol. **6**, 346 (1999). https://doi.org/10.1038/7582

[17] J. D. Watson and F. H. C. Crick, *Molecular Structure of Nucleic Acids: A Structure for Deoxyribose Nucleic Acid*, Nature **171**, 737 (1953). https://doi.org/10.1038/171737a0

[18] H. van den Bedem and E. Kuhl, *Tau-Ism: The Yin and Yang of Microtubule Sliding, Detachment, and Rupture*, Biophysical Journal **109**, 2215 (2015). https://doi.org/10.1016/j.bpj.2015.10.020

[19] L. Bellino, G. Florio, A. Goriely, and G. Puglisi, *Cooperative Melting in Double-Stranded Peptide Chains through Local Mechanical Interactions*, Preprint (2022).

[20] S. Di Stefano, G. Florio, G. Napoli, N. M. Pugno, and G. Puglisi, *On the Role of Elasticity in Focal Adhesion Stability within the Passive Regime*, International Journal of Non-Linear Mechanics 104157 (2022). https://doi.org/10.1016/j.ijnonlinmec.2022.104157

[21] X. Cao, Y. Lin, T. P. Driscoll, J. Franco-Barraza, E. Cukierman, R. L. Mauck, and V. B. Shenoy, *A Chemomechanical Model of Matrix and Nuclear Rigidity Regulation of Focal Adhesion Size*, Biophysical Journal **109**, 1807 (2015). https://doi.org/10.1016/j.bpj.2015.08.048

[22] A. Cannizzo, L. Bellino, G. Florio, G. Puglisi, and S. Giordano, *Thermal Control of Nucleation and Propagation Transition Stresses in Discrete Lattices with Non-Local Interactions and Non-Convex Energy*, Eur. Phys. J. Plus **137**, 569 (2022). https://doi.org/10.1140/epjp/s13360-022-02790-9

[23] G. R. Plaza, G. V. Guinea, J. Pérez-Rigueiro, and M. Elices, *Thermo-Hygro-Mechanical Behavior of Spider Dragline Silk: Glassy and Rubbery States*, J. Polym. Sci. B Polym. Phys. **44**, 994 (2006). https://doi.org/10.1002/polb.20751

[24] L. Bellino, G. Florio, and G. Puglisi, *The Influence of Device Handles in Single-Molecule Experiments*, Soft Matter **15**, 8680 (2019). https://doi.org/10.1039/C9SM01376H

[25] M. Benedito and S. Giordano, *Thermodynamics of Small Systems with Conformational Transitions: The Case of Two-State Freely Jointed Chains with Extensible Units*, The Journal of Chemical Physics **149**, 054901 (2018). https://doi.org/10.1063/1.5026386

[26] Y. R. Efendiev and L. Truskinovsky, *Thermalization of a Driven Bi-Stable FPU Chain*, Continuum Mech. Thermodyn. **22**, 679 (2010). https://doi.org/10.1007/s00161-010-0166-5

[27] D. E. Makarov, *A Theoretical Model for the Mechanical Unfolding of Repeat Proteins*, Biophysical Journal **96**, 2160 (2009). https://doi.org/10.1016/j.bpj.2008.12.3899

Theoretical and Applied Mechanics - AIMETA 2022 Materials Research Forum LLC
Materials Research Proceedings **26** (2023) 65-70 https://doi.org/10.21741/9781644902431-11

Rate-dependent response of axonal microtubules and tau proteins under shear forces

Luca Bellino[1, a*], Giuseppe Florio[1,3,b], Alain Goriely[2,c] and Giuseppe Puglisi[1,d]

[1]Dipartimento di Ingegneria Civile, Ambientale, del Territorio, Edile e di Chimica (Dicatech), Politecnico di Bari, Bari, Italy

[2]Mathematical Institute, University of Oxford, OX2 6GG, UK

[3]INFN, Sezione di Bari, I-70126 Bari, Italy

[a*] luca.bellino@poliba.it, [b] giuseppe.florio@poliba.it, [c] goriely@maths.ox.ac.uk, [d] giuseppe.puglisi@poliba.it

Keywords: Rate Effects, Brain Tissue, Microtubules, Tau Proteins

Abstract. The brain tissue is a very complex biological material exhibiting viscoelastic-type properties at the macroscopic level, arising from a hierarchical multiscale structure. Thus, to describe such interesting features at the molecular level, we introduce a model mimicking the coupling of microtubules and tau proteins inside the neuronal axon and we study the rate-dependent response under different conditions of applied load, rate, and temperature.

Introduction

One of the most complex biological systems is the brain tissue, which is made of billions of neurons, the fundamental component of nervous tissue, responsible for transmitting electrochemical signals. It has been addressed that mechanical fields have crucial importance for the correct functioning of these cells [1]. Specifically, it has been found that external forces, impacts or traumatic accidents acting on the brain can generate damage causing severe consequences at the level of the axons, also years after the event, a class of pathologies known as Traumatic Brain Injury (TBI) [2]. This phenomenon has first been observed in U.S.A. boxers because a great percentage of them suffered Alzheimer's or Parkinson's diseases even years after their retirement [3]. In these patients, there is evidence that the damage acting on the brain can cause scattered lesions at the level of the axons, a pathology known as Diffuse Axonal Injury (DAI) [4]. These lesions are responsible for the axonal failure, a phenomenon that can happen abruptly (primary axotomy) or with a progressive degradation of the internal components of the axons (secondary axotomy) [5], finally resulting in the malfunction of the whole tissue. As a matter of fact, these multiscale features are due to the damage mechanisms and to the structure of the complex energy landscape at the microscale. As a consequence, the comprehension of these phenomena requires to understand the behaviour of the material at this length scale.

In this paper, we study the effect of forces and loading rates at the level of the axons, considering two main components of the axonal cytoskeleton *i.e.,* the microtubules (MT) and the tau proteins [6]. MTs are arranged in bundles and are kept together by the crosslinking tau proteins. They do not cover the whole length of the axon and their length may vary in the range from 2 to 100 μm. There can be found almost 20 MTs per section radially spaced at about 23 – 38 nm, with a Young Modulus of about 1.9 GPa [7]. On the other hand, the main function of the tau proteins is to stabilize the microtubules and maintain the correct arrangement along the axonal length [8]. It is important to stress that another clinical evidence is that agglomerates of tau proteins are found inside the brain's white matter of patients affected by neurodegenerative diseases. This observation suggests the possibility that damages caused by mechanical fields may break these proteins depending mainly on the force and the loading rate. Indeed, three main regimes can be identified.

When the strain and the strain rate are low, the microtubules may slide through each other, and the tau proteins can detach and reattach in other vacant sites on the same MT or on adjacent ones. If the strain increases, the crosslink tau proteins start to break and are no longer able to reattach. When also the strain rate is large, there exists the possibility that the MTs break, causing malfunction and irreversible damage [9,10]. From a theoretical point of view, the challenge of describing such small systems considering both mechanical and thermal fields at varying rates is widely tackled using discrete or continuous models and numerical methods such as Molecular Dynamic (MD) simulations [11]. To mention some examples, it is possible to approach the direct modelling of the inertial dynamics [12,13] or to describe the evolution of the system using kinetic theories [14,15]. On the other hand, the possibility of investigating the microscopic characterization of the complex energy landscape of such systems obtaining analytical results is still to be thoroughly investigated, possibly considering phenomena such as hysteresis, damage, and residual stresses. In this spirit, we refer to recent papers where the study of different biological problems has been approached by analytical methods, such as the unfolding and misfolding of proteins materials [16-18], the effect of nucleation and propagation stresses in memory shape materials taking into account also thermal fields [19,20], focal adhesion [21,22], spider silks and new bio-inspired materials [23] or the mechanics of double-stranded polypeptide chains [24].

Specifically, following [24], we propose a micromechanical-based model describing the response of the microtubules and tau proteins to different loading rates (including temperature effects) when a shear force is applied to the system, as shown in Fig. 1. The analysis of rate effects can be performed by using different approaches such as the Fokker-Planck formulation [25], numerical FEM simulation [26] or Langevin dynamics [27]. To highlight the role of energy barriers, we will tackle the problem by adopting a Bell-type theory [28] based on Kramer's rate equation [29,30], as widely developed in recent works [24,31]. In particular, based on the possibility of evaluating the energy barriers separating the metastable states, we obtain the rate-dependent response of a system composed of two MTs linked by tau proteins and study the mechanical response of such a system under shear forces for different loading rates and temperatures.

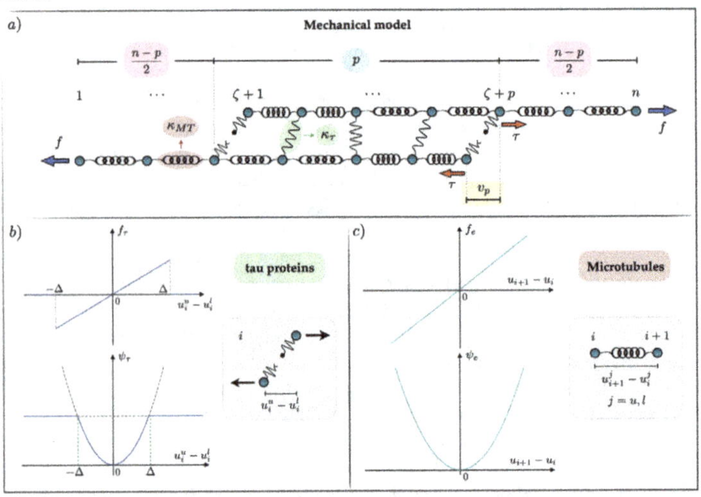

Fig. 1: Panel a): Mechanical model of the system describing microtubules and tau proteins. The energies used for the two microscopic elements are displayed in panel b) for the tau protein and in panel c) for the microtubule.

Mathematical framework

Following the theoretical model developed in [24], we consider a bundle of two elastic chains representing the microtubules connected by n breakable units spaced by a length ℓ, describing the tau proteins. The total length of the system is $L = n\ell$. The upper and the lower chains are described by elastic springs, as shown in Fig. 1c). On the other hand, the behaviour of each breakable unit is described by introducing a discrete variable χ_i (a "spin" variable) such that the condition $\chi_i = 1$ ($\chi_i = 0$) corresponds to an attached (detached) configuration. Thus, the quantity $p = \sum_{i=1}^{n} \chi_i$ represents the number of attached proteins (see Fig. 1b). By considering the rescaled end-points displacement $\delta = \frac{u_n^u}{\Delta} = -\frac{u_1^l}{\Delta}$ (Δ representing the breaking threshold displacement of the tau proteins, see Fig. 1b) and introducing the total stiffness of the system κ^t, the free energy and the force-displacement relation at the equilibrium read (see Ref. [24] for a detailed calculation)

$$g_{eq}(f) = -\frac{f^2}{\kappa^t} + \mu^2\left(1 - \frac{p}{n}\right), \qquad \delta(f) = \frac{f}{\kappa^t}, \tag{1}$$

where

$$\kappa^t = \frac{4n}{2n-p-1+4\gamma(p)}, \quad \gamma(p) = \frac{\sinh(\lambda)+\sinh(p\lambda)}{2\sinh[(p+1)\lambda]-\sinh(\lambda)-\sinh(p\lambda)}, \quad \lambda = \cosh^{-1}\left(1 + \frac{2\mu^2}{n^2}\right). \tag{2}$$

In the previous formula, we introduced the main non-dimensional parameter of the system

$$\mu^2 = \frac{\kappa_\tau}{2\kappa_{MT}}\frac{L^2}{\Delta^2}, \tag{3}$$

where κ_τ and κ_{MT} are the stiffnesses of the tau proteins and the of MTs, respectively,

Energy Barriers

In our approach, the loading rate affects the probability of jumping from a metastable equilibrium solution to another, and it depends on the height of the energy barriers. The ability of overcoming these barriers depends, among other effects, on the mechanical and thermal fields. This approach allows the possibility of exploring different locally stable solutions with partially detached elements. To evaluate $g_b(f)$, *i.e.* the height of the barrier for a certain value of the applied shear force f, we initially consider the elements of the system in equilibrium and we study the effect of changing the position of the last attached element p from its equilibrium value to the breaking threshold, by imposing a force τ (see Fig. 1a). The resulting mechanical quantities obtained by studying this variational problem represent the equilibrium conditions of the system when the energy is maximized (the analogous of a saddle point when smooth non-convex energies are considered [31,32]). Thus, we can evaluate the height of the barriers as

$$g_b(f) = g_\tau(f) - g_{eq}(f), \tag{4}$$

where $g_\tau(f)$ is the maximized energy and $g_{eq}(f)$ is given in Eq.(1) (see Ref. [31] for detailed analysis and calculation).

Rate effects

Following [29,32], we consider the time-dependent force $f(t)$. The loading rate reads

$$v(t) = v_0 e^{-\beta g_b(f(t))}, \tag{5}$$

where the energy barrier is given by Eq. (4), v_0 is a constitutive parameter and $\beta = 1/(K_B T)$, with T the absolute temperature and K_B the Boltzmann constant. Following [32], we define the probability $P_p(t)$ that the system exhibits p attached units a time t. The evolution equation for this probability is a first-order differential equation of the form

$$\frac{d\,P_p(t)}{d\,t} = -v_{p\to p-1}(t)P_p(t) + v_{p+1\to p}(t)P_{p+1}(t). \tag{6}$$

By Eq. (6) we can obtain the value of $P_p(t)$ at time t. In particular, it depends on the process where the configuration of the system can pass from the state with p attached elements to that with $p-1$ (with rate $v_{p\to p-1}(t)$) and the process where the system goes from a configuration with $p+1$ attached elements to p (with rate $v_{p+1\to p}(t)$). Two configurations must be considered separately *i.e.*, $v_{n+1\to n}(t)=0$ and $v_{0\to -1}(t)=0$. Starting the analysis at time $t=0$, the probability of being in the fully attached configuration ($p=n$) is $P_n(t)=1$. Following the action of the applied force, the tau proteins start to detach and the probability of finding the system in the initial state decreases while the probability of being in the configuration with $p=n-1$ increases. This process goes on until the system completely breaks (corresponding to the configuration with $p=0$).
Eventually, the total displacement is

$$\langle\delta(t)\rangle=\sum_{p=0}^{n}P_p(t)\delta(p), \tag{7}$$

with $\delta(p,t)$ given by Eq. (1).

Results and discussion

To approach the problem of the rate-dependent response of the microtubule bundles into the neuronal axon, we introduced a microscale-based model describing two MTs connected by crosslinking tau proteins. In Fig. 2a we show the effect of the loading rate v_f (with the applied shear force being $f=v_f t$). It is possible to observe that as the rate grows the force breaking thresholds increase and the system is not able to explore metastable configuration, being confined in the solution $p=n$. On the other hand, for low rates, solutions with partially detached elements are found before rupture, which occurs at smaller values of the force. Thus, at increasing rates, a "Maxwell" behaviour is preferred, with the overall collapse of the system attained without exploring other solutions whereas for decreasing rates the sequential detachment of the tau proteins is observed, as described in many papers [10,33,34]. In Fig. 2b we show the effect of changing temperatures, affecting the possibility of overcoming the energy barriers. As the temperature increases, the system exhibits a ductile type of fracture attained at low forces and explores different metastable solutions. Conversely, when the zero-temperature limit is approached, *i.e.* $\beta\to\infty$, the system is 'frozen' in the configuration with all the elements attached, following the "Maxwell" path.

Eventually, we obtained the mechanical response of a simple bundle composed of two MTs and crosslinking tau proteins and we observed the effect of the loading rate and the temperature. Through our model, we obtained results previously described in the literature, where other methods such as numerical FEM simulations, Molecular Dynamics or continuum models have been used to approach the analysis of similar systems [8-11,33,34].

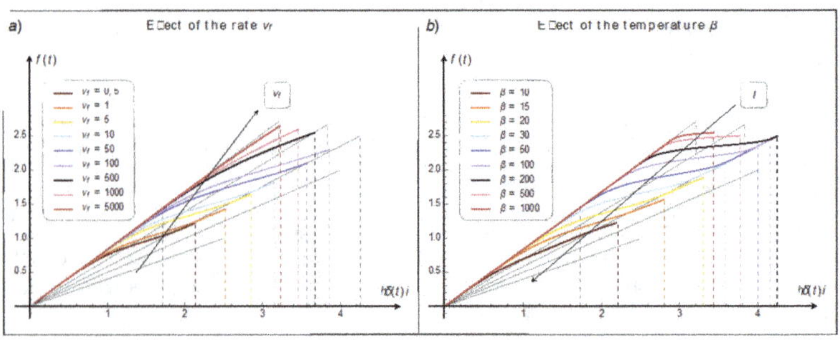

Fig. 2: Panel a) Effect of the loading rate. Panel b) Effect of the temperature.

Acknowledgements
LB, GF and GP have been supported by the Italian Ministry MIUR-PRIN project "Mathematics of active materials: From mechanobiology to smart devices (2017KL4EF3) and by "Gruppo Nazionale per la Fisica Matematica" (GNFM) under "Istituto Nazionale di Alta Matematica" (INdAM). GF is also supported by "Istituto Nazionale di Fisica Nucleare" (INFN) through the project QUANTUM, by the FFABR research grant (MIUR) and the PON S.I.ADD.

References

[1] S. B. Smith, D. F. Meaney, Axonal damage in traumatic brain injury, The neuroscientist, 6, 6, 483 – 495, (2000). https://doi.org/10.1177/107385840000600611
[2] M.D. Tang-Schomer, A. R. Patel, P. W. Baas, D. H. Smith, Mechanical breaking of microtubules in axons during stretch injury underlies delayed elasticity, microtubule disassembly and axon degeneration, The FASEB Journal, 24, 5, 1401 – 1410, (2010). https://doi.org/10.1096/fj.09-142844
[3] C. Ballatore, V. M. Y. Lee, J. Q. Trojanowski, Tau-mediated neurodegeneration in Alzheimer's disease and related disorders, Nature rev. neuroscience, 8, 9, 663 – 672, (2007). https://doi.org/10.1038/nrn2194
[4] Ya. Shi, W. Zhang, Y. Yang, A. G. Murzin, B. Falcon, A. Kotecha, M. van Beers, A. Tarutani, F. Kametani, H. J. Garringer et al, Structure-based classification of tauopathies, Nature, 598, 7880, 359 – 363, (2021). https://doi.org/10.1038/s41586-021-03911-7
[5] T. Hawkins, M. Mirigian, M. S. Yasar, J. L. Ross, Mechanics of microtubules, Journal of Biomechanics, 43, 1, 23 – 30, (2010). https://doi.org/10.1016/j.jbiomech.2009.09.005
[6] H. Ouyang, E. Neuman, R. Shi, Contribution of cytoskeletal elements to the axonal mechanical properties, Journal of biological engineering, 7, 1, 1 – 8, (2013). https://doi.org/10.1186/1754-1611-7-21
[7] D. S. Eisenberg, M. R. Sawaya, Taming tangled tau, Nature, 547, 7662, 170–171, (2017). https://doi.org/10.1038/nature23094
[8] K. J. J. Rosenberg, L. Ross, H. E. Feinstein, S. C: Feinstein, J. Israelachvili, Complementary dimerization of microtubule-associated tau protein: implications for microtubule bundling and tau-mediated pathogenesis, PNAS, 105, 21, 7445 – 7450, (2008). https://doi.org/10.1073/pnas.0802036105
[9] R. De Rooij, K. E. Miller, E. Kuhl, Modeling molecular mechanism in the axon, Computational mechanics, 59, 3, 523 – 537, (2017). https://doi.org/10.1007/s00466-016-1359-y
[10] H. Van den Bedem, E. Kuhl, Tau-ism: the Yin and Yang of microtubules sliding, detachment and rupture, Biophysical Journal, 109, 11, 2215, (2015). https://doi.org/10.1016/j.bpj.2015.10.020
[11] E. Evans, K. Ritchie, Dynamic strength of molecular adhesion bonds, Biophysical Journal, 72, 4, 1541 – 1555, (1997). https://doi.org/10.1016/S0006-3495(97)78802-7
[12] T. Cohen, S. Givli, Dynamics of a discrete chain of bi-stable elements: a biomimetic shock absorbing mechanism, JMPS, 64, 426 – 439, (2014). https://doi.org/10.1016/j.jmps.2013.12.010
[13] Y. R. Efendiev, L. Truskinovsky, Thermalization of a driven bi-stable FPU chain, Continuum mechanics and thermodynamics, 22, 6, 679 – 698, (2010). https://doi.org/10.1007/s00161-010-0166-5
[14] S. Givli, K. Bhattacharya, A coarse-grained model of the myofibril: overall dynamics and the evolution of sarcomere non-uniformities, JMPS, 57, 2, 221 – 243, (2009). https://doi.org/10.1016/j.jmps.2008.10.013
[15] R. Raj, P. K. Purohit, Phase boundaries as agents of structural change in macromolecules, JMPS, 59, 10, 2044 – 2069, (2011). https://doi.org/10.1016/j.jmps.2011.07.003

[16] L. Bellino, G. Florio, G. Puglisi, The influence of device handles in single molecule experiments, Soft Matter, 15, 43, 8680 – 8690, (2019). https://doi.org/10.1039/C9SM01376H
[17] M. Benedito, S. Giordano, Isotensional and isometric force-extension response of chains with bistable units, PRE, 98, 5, 052146, (2018). https://doi.org/10.1103/PhysRevE.98.052146
[18] I. Benichou, S. Givli, the hidden ingenuity of titin structure, Applied Physics Letters, 99, 091904, (2011). https://doi.org/10.1063/1.3558901
[19] L. Bellino, G. Florio, S. Giordano, G. Puglisi, On the competition between interface energy and temperature in phase transition phenomena, Application of Engineering Science, 2, 100009, (2020). https://doi.org/10.1016/j.apples.2020.100009
[20] A. Cannizzo, L. Bellino, G. Florio, G. Puglisi, S. Giordano, Thermal control of nucleation and propagation transition stresses in discrete lattices with non-local interactions and non-convex energy, EPJ Plus, 137, 569, (2022). https://doi.org/10.1140/epjp/s13360-022-02790-9
[21] A. Cannizzo, G. Florio, G. Puglisi, S. Giordano, Temperature controlled decohesion regimes of an elastic chain adhering to a fixed substrate by softening and breakable bonds, Journal of Physics A, 54, 445001, (2021). https://doi.org/10.1088/1751-8121/ac2a07
[22] S. Di Stefano, G. Florio, G. Napoli, N. M. Pugno, G. Puglisi, On the role of elasticity in focal adhesion within the passive regime, International Journal of Nonlinear Mechanics, accepted, (2022). https://doi.org/10.1016/j.ijnonlinmec.2022.104157
[23] V. Fazio, D. De Tommasi, N. M. Pugno, G. Puglisi, Spider Silks Mechanics: Predicting Humidity and Temperature Effects. JMPS, in press, (2022). https://doi.org/10.1016/j.jmps.2022.104857
[24] L. Bellino, G. Florio, A. Goriely, G. Puglisi, Role of interlinker stiffness in melting and unfolding cooperativity and toughness of double stranded peptide chains, Preprint, (2022).
[25] H. Herrmann, U. Aebi, Intermediate filaments: molecular structure, assembly mechanism, and integration into functionally distinct intracellular scaffolds, Annual review of biochemistry, 73, 1, 749 – 789, (2012). https://doi.org/10.1146/annurev.biochem.73.011303.073823
[26] R. De Rooji, K. E. Miller, E. Kuhl, Modeling molecular mechanism in the axon, computational mechanics, 59, 3, 523 – 537, (2017). https://doi.org/10.1007/s00466-016-1359-y
[27] H. B. Da Rocha, L. Truskinovsky, Rigidity-controlled crossover, from spinodal to critical failure, PRL, 124, 1, 015501, (2020). https://doi.org/10.1103/PhysRevLett.124.015501
[28] G. I. Bell, Models for the specific adhesion of cells to cells, Science, 200, 4342, 618 – 627, (1978). https://doi.org/10.1126/science.347575
[29] P. Hanggi, P. Talkner, M. Borkovec, Reaction-rate theory: fifty years after Kramers, Reviews of modern physics, 62, 2, 251, (1990). https://doi.org/10.1103/RevModPhys.62.251
[30] H. A. Kramers, Brownian motion in a field of force and the diffusion model of chemical reactions, Physica, 7, 4, 284 – 304, (1940). https://doi.org/10.1016/S0031-8914(40)90098-2
[31] L. Bellino, G. Florio, A. Goriely, G. Puglisi, Effects of rate and temperature on the mechanical response of microtubules and tau proteins, Preprint, (2022).
[32] I. Benichou, Y. Zhang, O. K. Dudko, S. Givli, The rate dependent response of a bistable chain at finite temperature, JMPS, 95, 44 – 63, (2016). https://doi.org/10.1016/j.jmps.2016.05.001
[33] H. Ahmadzadeh, D. H. Smith, V. B. Shenoy, Mechanical effects of dynamic binding between tau proteins on microtubules during axonal injury, Biophysical Journal, 106, 5, 1123 – 1133, (2015). https://doi.org/10.1016/j.bpj.2014.01.024
[34] R. De Rooij, E. Kuhl, Physical biology of axonal damage, Frontiers in cellular neuroscience, 12, 144, (2018). https://doi.org/10.3389/fncel.2018.00144

Theoretical and Applied Mechanics - AIMETA 2022 | Materials Research Forum LLC
Materials Research Proceedings **26** (2023) 71-76 | https://doi.org/10.21741/9781644902431-12

A moving cohesive interface model for brittle fracture propagation

Umberto De Maio[1, a], Fabrizio Greco[1, b *], Paolo Lonetti[1, c], Paolo Nevone Blasi [1, d], and Andrea Pranno [1, e]

[1]Department of Civil Engineering, University of Calabria, Rende, Italy

[a] umberto.demaio@unical.it, [b] fabrizio.greco@unical.it, [c] paolo.lonetti@unical.it, [d] paolo.nevoneblasi@unical.it, [e] andrea.pranno@unical.it

Keywords: Cohesive Zone Model, Moving Mesh Technique, Fracture Propagation, Quasi-Brittle Materials

Abstract. In this work, an advanced numerical model, based on the cohesive zone approach and the moving mesh technique, is proposed to simulate fracture propagation in quasi-brittle materials. In particular, the proposed numerical procedure consists of two stages. In the former one, according to the inter-element crack approach, once a suitable stress criterion for fracture onset is satisfied, a mesh boundary, representing the crack segment, is selected and aligned along the crack propagation direction by using the well-known moving mesh technique. In the latter one a zero-thickness interface cohesive element, equipped with a traction-separation law, is inserted on-the-fly along the previously selected mesh boundary, in order to describe the nonlinear fracture process. Comparisons with available experimental and numerical results have highlighted the effectiveness and the reliability of the proposed model in the prediction of the brittle fracture phenomenon.

Introduction

The accurate prediction of fracture propagation in quasi-brittle materials is one of the most challenging issues in the framework of structural mechanics because of the complex interplay of different irreversible processes such as defects nucleation, micro-cracks growth. In the literature, different numerical models, based on smeared and discrete fracture approaches, have been proposed to analyze fracture phenomena and their influence on the global mechanical response of the structures [1]. In particular, the discrete fracture models simulate the crack as an actual displacement discontinuity within or between the finite elements of a standard computational mesh, adopting a suitable cohesive traction-separation law to describe the crack propagation within the so-called fracture process zone (FPZ) [2,3]. Such a modeling approach results to be very efficient for predicting the fracture process in quasi-brittle materials with respect to the smeared crack approaches that often lead to ill-posed boundary value problems (BVPs) [4]. The discrete crack models are often employed to simulate the interfacial crack in adhesive-bonded ductile sheets by using different phenomenological-type traction-separation laws, to describe the mechanical behavior of the cohesive forces acting along a predefined crack path [5,6]. Recently, to predict unknown crack paths, a diffuse interface model has been introduced and employed to investigate the damage phenomena in quasi-brittle materials and masonry structures [7,8]. However, such a model suffers from the well-known artificial compliance effect and requires a suitable calibration of the cohesive properties of the embedded interfaces to ensure the desired numerical accuracy [9,10]. Moreover, the strongly nonlinear fracture behavior, including plate end debonding phenomena, has been properly simulated by cohesive crack models [11,12]. In some recent works, fracture models, based on discrete and smeared crack approaches, have been incorporated within advanced numerical homogenization to predict micro-cracking and contact evolution in composite materials [13–16]. However, fracture mechanics-based models result to be very efficient for predicting damage phenomena with a-priori known crack path but provide some divergences with respect to the experimental outcomes, in terms of crack patterns, in the case of unknown crack paths. To overcomes these drawbacks, the moving mesh technique has been recently introduced in the structural mechanics field to easily permit the moving of the computational mesh according to specific conditions [17–19].

In this work, an advanced numerical model, based on the cohesive zone approach and the moving mesh technique, is proposed to simulate fracture propagation in quasi-brittle materials. In particular, the cohesive fracture model, based on an inter-element crack approach, has been employed to describe the nonlinear behavior of the crack while the moving mesh technique allows aligning the crack element along the crack propagation direction.

Adopted numerical framework for cracking simulation

The numerical procedure proposed to simulate the crack propagation in quasi-brittle materials consists of two stages. In the former one, according to the inter-element crack approach, once a suitable stress criterion for fracture onset is satisfied, a mesh boundary, representing the crack segment, is selected and aligned along the crack propagation direction by using the well-known moving mesh technique. In the latter one a zero-thickness interface cohesive element, equipped with a traction-separation law, is inserted on-the-fly along the previously selected mesh boundary, in order to describe the nonlinear fracture process.

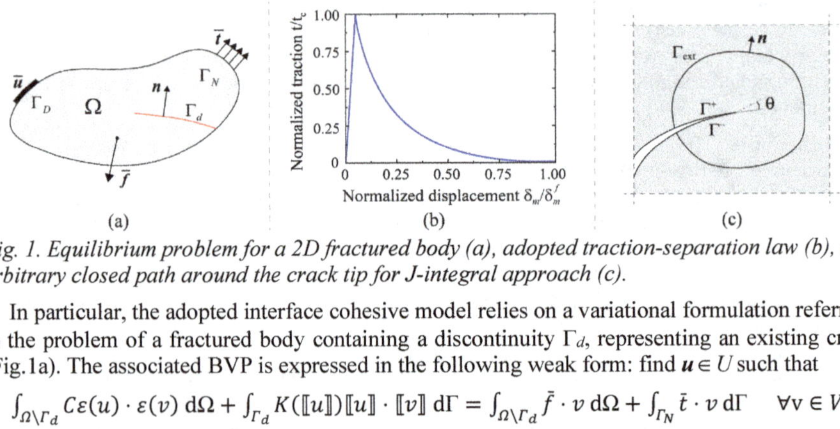

(a) (b) (c)

Fig. 1. Equilibrium problem for a 2D fractured body (a), adopted traction-separation law (b), and arbitrary closed path around the crack tip for J-integral approach (c).

In particular, the adopted interface cohesive model relies on a variational formulation referring to the problem of a fractured body containing a discontinuity Γ_d, representing an existing crack (Fig.1a). The associated BVP is expressed in the following weak form: find $\boldsymbol{u} \in U$ such that

$$\int_{\Omega\backslash\Gamma_d} C\varepsilon(u)\cdot\varepsilon(v)\,\mathrm{d}\Omega + \int_{\Gamma_d} K([\![u]\!])[\![u]\!]\cdot[\![v]\!]\,\mathrm{d}\Gamma = \int_{\Omega\backslash\Gamma_d} \bar{f}\cdot v\,\mathrm{d}\Omega + \int_{\Gamma_N} \bar{t}\cdot v\,\mathrm{d}\Gamma \quad \forall \mathrm{v}\in V, \tag{1}$$

where $\boldsymbol{C}$ and $\boldsymbol{\varepsilon}$ are the fourth-order elasticity tensor and the linear strain operator, respectively, whereas $\boldsymbol{u}$ and $\boldsymbol{v}$ are the unknown displacement field and arbitrary virtual displacement field, respectively, which belong to the set of kinematically admissible displacements U and to set of kinematically admissible variations of the approximated displacement field V, respectively. The second term in Eq. 1, represents the nonlinear constitutive behavior of the cohesive interfaces expressed by a traction-separation law, reported in Fig. 1b, which involves the following effective displacement jump:

$$\delta_m = \sqrt{\langle\delta_n\rangle^2 + \delta_s^2}\ . \tag{2}$$

A stress-based criterion has been employed to detect the onset of the fracture and to insert a cohesive crack element. In particular, when the normal stress along a mesh boundary exceeds the critical tensile strength of the material, such an element is selected and aligned with the theoretical crack propagation direction. The direction of the crack extension has been found by the maximum energy release rate (ERR). ERR is computed by using the J-integral-based approach with respect to a closed path containing the crack tip and a local coordinate system aligned with the crack direction:

$$J_k = \int_\Gamma \left[W n_k - \sigma_{ij} \frac{\partial u_i}{\partial x_k} n_j \right] d\Gamma, \tag{3}$$

with $\Gamma = \Gamma_{ext} + \Gamma^+ + \Gamma^-$ an arbitrary closed path around the crack tip singularity (see Fig. 1c). Is worth noting that, the Γ^+ and Γ^- represent the faces of the stress-free crack. The ERR can be expressed as a function of J vector components, J_1 and J_2, as follows: $G = J_1\cos(\theta)+J_2\sin(\theta)$. Maximization of G yields to evaluation of the kinking angle:

$$\theta = \arctan\left(\frac{J_2}{J_1}\right). \tag{4}$$

After that, the selected crack element is aligned according to the computed kinking angle (Eq. 4). The crack alignment procedure is accomplished by relocating nodes performed by the Arbitrary Lagrangian-Eulerian (ALE) description. To this end, a referential configuration $\Omega\chi$, whose points are identified by the reference coordinates $\boldsymbol{\chi}$, is introduced. The corresponding mesh displacement field $\boldsymbol{u}_{mesh} = \mathbf{X} - \boldsymbol{\chi}$ represents a structural change, which, in the context of this work, consists in the selected element rotation of the small kink emanated from the current crack tip.

Numerical results

In order to validate the proposed numerical fracture model, a numerical simulation of the three-point bending test, experimentally tested by Rots in [20], is performed, involving a plain concrete specimen. Both geometry and boundary conditions of the notched concrete beam are depicted in Fig. 2a. The Young's modulus and the Poisson's ratio of the bulk material are equal to 20 GPa and 0.20, respectively. The cohesive parameters, i.e. critical tensile stress, mode-I fracture energy, and initial cohesive stiffness, required by the adopted traction-separation law, are set equal to t_c = 2.4 MPa, G = 113 N/m, and K = 1E15 N/m³, respectively.

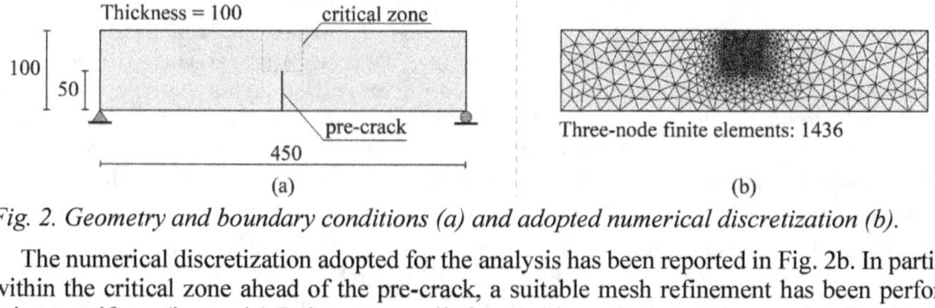

Fig. 2. Geometry and boundary conditions (a) and adopted numerical discretization (b).

The numerical discretization adopted for the analysis has been reported in Fig. 2b. In particular, within the critical zone ahead of the pre-crack, a suitable mesh refinement has been performed, using a uniform (isotropic) Delaunay tessellation and imposing a maximum edge length of 3.00 mm, which results in an average mesh size of about 1.46 mm.

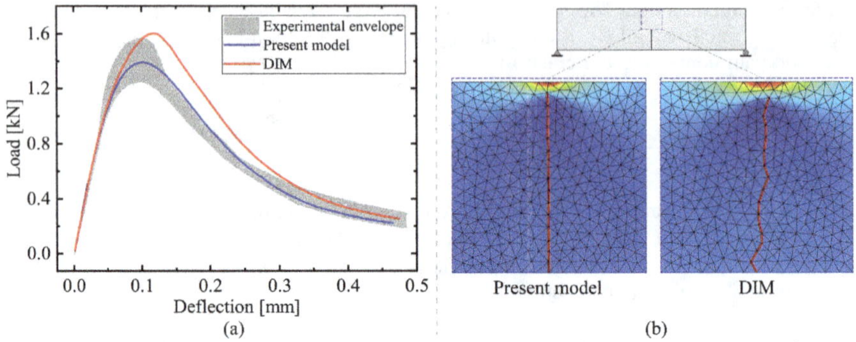

Fig. 3. Numerical loading curves together with the experimental envelope (a) and crack pattern obtained by the proposed model and the DIM model (b).

The numerical results, in terms of load carrying capacity and crack pattern, are reported in Fig. 3. In particular, Fig. 3a shows a comparison of the loading curves predicted by the proposed model, the Diffuse Interface Model (DIM), and the experimental test. The numerical model (DIM), taken for comparison purposes, has been recently introduced by some of the authors in [21,22] and relies on an inter-element fracture approach according to which cohesive interfaces are inserted at the beginning of the simulation along all mesh boundaries inside the critical zone. We can note that the proposed model predicts a loading curve within the experimental envelope providing a numerical prediction, in terms of peak load, better than the DIM model. This result is confirmed by the crack pattern (see Fig. 3b) predicted by the proposed model which results to be coincident with the exact crack path, assuming self-similar propagation.

Conclusions

In this work, a numerical model, based on the cohesive fracture approach and the moving mesh technique, has been proposed to simulate the crack propagation in quasi-brittle materials. In particular, once a suitable stress criterion for fracture onset is satisfied, a boundary of the adopted mesh is selected and aligned with the theoretical crack direction. This alignment procedure relies on the J-integral approach, to compute the kinking angle, and the moving mesh technique, to relocate nodes of the mesh due to the crack update. Mode-I fracture has been simulated by the proposed model involving quasi-brittle materials like concrete. Comparisons with available experimental and numerical results have highlighted the effectiveness and the reliability of the proposed model to predict the cracking behavior in concrete structures. As a future perspective of this work, the proposed model could be incorporated into a multiscale approach to improve the related computational performances, similarly to what has been suggested by [23–25].

Acknowledgments

Umberto De Maio and Paolo Lonetti gratefully acknowledge financial support from the Italian Ministry of University and Research (MUR) under the Research Grant PRIN 2020 "Opportunities and challenges of nanotechnology in advanced and green construction materials" (Project Code 2020EBLPLS, University of Calabria Research Unit).

References

[1] R. de Borst, J.J.C. Remmers, A. Needleman, M.-A. Abellan, Discretevs smeared crack models for concrete fracture: bridging the gap, Int. J. Numer. Anal. Meth. Geomech. 28 (2004) 583–607. https://doi.org/10.1002/nag.374

[2] A. Hillerborg, M. Modéer, P.-E. Petersson, Analysis of crack formation and crack growth in concrete by means of fracture mechanics and finite elements, Cement and Concrete Research. 6 (1976) 773–781. https://doi.org/10.1016/0008-8846(76)90007-7

[3] U. De Maio, D. Cendón, F. Greco, L. Leonetti, P. Nevone Blasi, J. Planas, Investigation of concrete cracking phenomena by using cohesive fracture-based techniques: A comparison between an embedded crack model and a refined diffuse interface model, Theoretical and Applied Fracture Mechanics. 115 (2021) 103062. https://doi.org/10.1016/j.tafmec.2021.103062

[4] J. Červenka, V. Červenka, S. Laserna, On crack band model in finite element analysis of concrete fracture in engineering practice, Engineering Fracture Mechanics. 197 (2018) 27–47. https://doi.org/10.1016/j.engfracmech.2018.04.010

[5] A. Pascuzzo, A. Yudhanto, M. Alfano, G. Lubineau, On the effect of interfacial patterns on energy dissipation in plastically deforming adhesive bonded ductile sheets, International Journal of Solids and Structures. 198 (2020) 31–40. https://doi.org/10.1016/j.ijsolstr.2020.04.001

[6] D. Bruno, F. Greco, P. Lonetti, Computation of Energy Release Rate and Mode Separation in Delaminated Composite Plates by Using Plate and Interface Variables, Mechanics of Advanced Materials and Structures. 12 (2005) 285–304. https://doi.org/10.1080/15376490590953563

[7] F. Greco, D. Gaetano, L. Leonetti, P. Lonetti, A. Pascuzzo, A. Skrame, Structural and seismic vulnerability assessment of the Santa Maria Assunta Cathedral in Catanzaro (Italy): classical and advanced approaches for the analysis of local and global failure mechanisms, Frattura Ed Integrità Strutturale. 60 (2022) 464–487. https://doi.org/10.3221/IGF-ESIS.60.32

[8] A. Pascuzzo, F. Greco, L. Leonetti, P. Lonetti, A. Pranno, C. Ronchei, Investigation of mesh dependency issues in the simulation of crack propagation in quasi-brittle materials by using a diffuse interface modeling approach, Fatigue Fract Eng Mat Struct. 45 (2022) 801–820. https://doi.org/10.1111/ffe.13635

[9] U. De Maio, F. Greco, L. Leonetti, R. Luciano, P. Nevone Blasi, S. Vantadori, A refined diffuse cohesive approach for the failure analysis in quasibrittle materials—part I: Theoretical formulation and numerical calibration, Fatigue Fract Eng Mater Struct. 43 (2020) 221–241. https://doi.org/10.1111/ffe.13107

[10] U. De Maio, F. Greco, L. Leonetti, R. Luciano, P. Nevone Blasi, S. Vantadori, A refined diffuse cohesive approach for the failure analysis in quasibrittle materials—part II: Application to plain and reinforced concrete structures, Fatigue Fract Eng Mater Struct. 42 (2019) 2764–2781. https://doi.org/10.1111/ffe.13115

[11] U. De Maio, F. Greco, L. Leonetti, P. Nevone Blasi, A. Pranno, An investigation about debonding mechanisms in FRP-strengthened RC structural elements by using a cohesive/volumetric modeling technique, Theoretical and Applied Fracture Mechanics. 117 (2022) 103199. https://doi.org/10.1016/j.tafmec.2021.103199

[12] E. Benvenuti, N. Orlando, Failure of FRP-strengthened SFRC beams through an effective mechanism-based regularized XFEM framework, Composite Structures. 172 (2017) 345–358. https://doi.org/10.1016/j.compstruct.2017.02.099

[13] F. Greco, L. Leonetti, U. De Maio, S. Rudykh, A. Pranno, Macro- and micro-instabilities in incompressible bioinspired composite materials with nacre-like microstructure, Composite Structures. 269 (2021) 114004. https://doi.org/10.1016/j.compstruct.2021.114004

[14] F. Greco, L. Leonetti, C.M. Medaglia, R. Penna, A. Pranno, Nonlinear compressive failure analysis of biaxially loaded fiber reinforced materials, Composites Part B: Engineering. 147 (2018) 240–251. https://doi.org/10.1016/j.compositesb.2018.04.006

[15] F. Greco, P. Lonetti, R. Luciano, P. Nevone Blasi, A. Pranno, Nonlinear effects in fracture induced failure of compressively loaded fiber reinforced composites, Composite Structures. 189 (2018) 688–699. https://doi.org/10.1016/j.compstruct.2018.01.014

[16] A. Pranno, F. Greco, L. Leonetti, P. Lonetti, R. Luciano, U. De Maio, Band gap tuning through microscopic instabilities of compressively loaded lightened nacre-like composite metamaterials, Composite Structures. 282 (2022) 115032. https://doi.org/10.1016/j.compstruct.2021.115032

[17] F. Zhang, W. Huang, X. Li, S. Zhang, Moving mesh finite element simulation for phase-field modeling of brittle fracture and convergence of Newton's iteration, Journal of Computational Physics. 356 (2018) 127–149. https://doi.org/10.1016/j.jcp.2017.11.033

[18] D. Ammendolea, F. Greco, P. Lonetti, R. Luciano, A. Pascuzzo, Crack propagation modeling in functionally graded materials using Moving Mesh technique and interaction integral approach, Composite Structures. 269 (2021) 114005. https://doi.org/10.1016/j.compstruct.2021.114005

[19] A. Pascuzzo, F. Greco, P. Lonetti, D. Ammendolea, Dynamic fracture analysis in quasi-brittle materials via a finite element approach based on the combination of the ALE formulation and M-integral method, Engineering Failure Analysis. (2022) 106627. https://doi.org/10.1016/j.engfailanal.2022.106627

[20] J.G. Rots, Computational modeling of concrete fracture, Doctoral thesis, TU Delft, 1988.

[21] A. Pranno, F. Greco, P. Lonetti, R. Luciano, U. De Maio, An improved fracture approach to investigate the degradation of vibration characteristics for reinforced concrete beams under progressive damage, International Journal of Fatigue. 163 (2022) 107032. https://doi.org/10.1016/j.ijfatigue.2022.107032

[22] U. De Maio, F. Greco, L. Leonetti, P. Nevone Blasi, A. Pranno, A cohesive fracture model for predicting crack spacing and crack width in reinforced concrete structures, Engineering Failure Analysis. 139 (2022) 106452. https://doi.org/10.1016/j.engfailanal.2022.106452

[23] F. Greco, L. Leonetti, A. Pranno, S. Rudykh, Mechanical behavior of bio-inspired nacre-like composites: A hybrid multiscale modeling approach, Composite Structures. 233 (2020) 111625. https://doi.org/10.1016/j.compstruct.2019.111625

[24] F. Greco, L. Leonetti, P. Lonetti, R. Luciano, A. Pranno, A multiscale analysis of instability-induced failure mechanisms in fiber-reinforced composite structures via alternative modeling approaches, Composite Structures. 251 (2020) 112529. https://doi.org/10.1016/j.compstruct.2020.112529

[25] F. Greco, L. Leonetti, R. Luciano, A multiscale model for the numerical simulation of the anchor bolt pull-out test in lightweight aggregate concrete, Construction and Building Materials. 95 (2015) 860–874. https://doi.org/10.1016/j.conbuildmat.2015.07.170

Materials Research Proceedings **26** (2023) 77-82 https://doi.org/10.21741/9781644902431-13

A fractional-order theory of phase transformation in presence of anomalous heat transfer

Gianmarco Nuzzo[1, a *], Fabiana Amiri[1], Salvatore Russotto[1], Emanuela Bologna[1], and Massimiliano Zingales[1]

[1]Department of Engineering, University of Palermo, Viale delle Scienze ed. 8, 90128, Palermo, Italy

[a]gianmarco.nuzzo@unipa.it

Keywords: Fractional Calculus, Anomalous Heat Transfer, Stefan Problem

Abstract. In this paper the solution for the one-dimensional Stefan problem of fractional order was examined, considering a generalization of Fourier's law, in which flux is related to temperature through Caputo's fractional derivative.

Introduction

The term phase transformation denotes the transition of a thermodynamic system from one aggregate state to another. Dealing with phase transformations, it is necessary to refer to the so-called processes at the interface. They are characterized by two distinct phases and a moving interface advancing and consequently by the transformation of one phase and the gradual decrease of the other. Typical examples may be liquid-solid phase changes, chemical reactions, liquid-vapor transformations etc.

These phenomena were observed in the late 19th century by Slovenian physicist Joseph Stefan, especially the formation of ice in the polar seas, to whom we owe the so-called formulation of *Stefan's problem* [1].

Stefan problems are systems of partial derivative differential equations comprising two governing equations, for the two phases, with respective boundary conditions and another differential equation called *Stefan condition*, valid at the interface and binding the two phases [2]. Given the high nonlinearity of these problems, depending on the choice of boundary conditions and because, as time goes on, one domain widens and one shrinks, it is not always possible to derive analytical solutions, so it is often necessary to proceed numerically.

The pivotal equation governing the heat diffusion process of both phases is the *Fourier law*, which links the flux to the temperature gradient. In previous years, however, it has been observed that in multiphase conductors or in the presence of high-frequency phenomena, the Fourier equation, in its the steady state, cannot fully describe such processes [3]. For this reason, modifications and generalizations of Fourier's law have been proposed in the last half of the past century.

Cattaneo suggested a modification of Fourier's law, introducing a new term that considers the rate of heat flow [4], but Fourier/Cattaneo law-based models are not appropriate for capturing the behavior of recent applications such as high-efficiency exchangers, microprocessor cooling or additive manufacturing.

So, in [5] a fractional formulation of the Fourier equation was proposed, replacing the standard integer-order derivative and integral with real-order operators.

The aim of this paper is to exploit the fractional formulation of Fourier's law to reach the solutions found in [6], starting with physical considerations.

Materials Research Forum LLC
https://doi.org/10.21741/9781644902431-13

Derivation of fractional Fourier law

Considering an inhomogeneous medium (Fig. 1), consisting of several masses, m_i, with different heat capacities $C^{(V)}(z)$, separated by conductive material with different conductivities $\chi^{(V)}(z)$ and subjected to a flow of heat $\bar{q}(t)$:

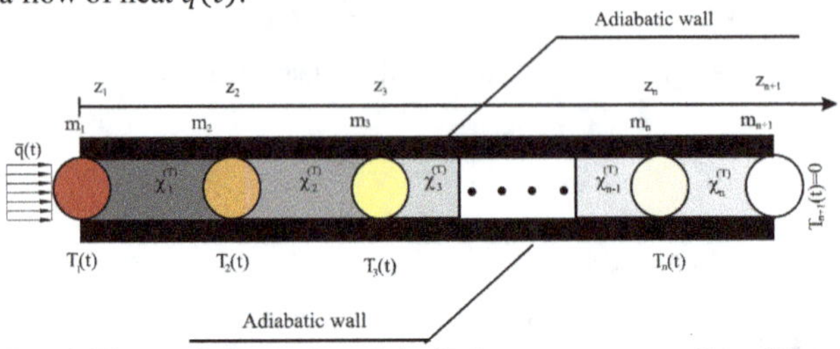

Fig. 1. Thermodynamical model of inhomogeneous medium [5].

Capacities and conductivities hold the following power laws: $C^{(V)}(z) = \frac{C_\beta^{(V)} z^{-\beta}}{\Gamma[1-\beta]}$, $\chi^{(V)}(z) = \frac{\chi_\beta^{(V)} z^{-\beta}}{\Gamma[1-\beta]}$ with dimensions $\left[C_\beta^{(V)}\right] = \frac{L^{2+\beta}T^2}{K}$, $\left[\chi_\beta^{(V)}\right] = \frac{ML^{1+\beta}}{KT^3}$. If we examine an infinitesimal part (Fig. 2) of the medium:

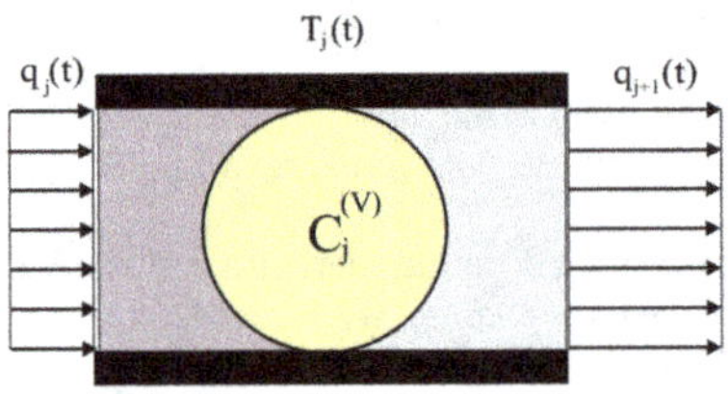

Fig. 2. Thermal energy balance of the jth-mass [5].

The energy balance for the generic mass m_j involves the rate of internal energy U_j of the mass and the incoming and outgoing energy fluxes:

$$\frac{dU_j(t)}{dt} = m_j \frac{du_j(t)}{dt} = m_j C_j^{(V)} \frac{dT_j(t)}{dt} = A_j q_j(t) - A_{j+1} q_{j+1}(t). \tag{1}$$

In which u_j is the internal density function of the mass m_j. The mass m_j occupies the volume of $A\Delta z$, so introducing the density ϱ, we can express the mass as $m_j = \varrho A \Delta z$. Simplifying the previous expression, we get (assuming $A_j = A_{j+1}$):

$$\varrho C_j^{(V)} \frac{dT_j(t)}{dt} = -\frac{q_{j+1}(t) - q_j(t)}{\Delta z}. \tag{2}$$

Introducing the temperature field $T_j(t) \to T(z_j, t)$, the thermal energy flux field $q_j(t) \to q(z_j, t)$, $C_j^{(V)} \to C^{(V)}(z_j)$ and performing the limit $\Delta z \to 0$ in (2) we obtain the following balance equation:

$$\varrho C^{(V)}(z)\frac{\partial T(z,t)}{\partial t} = -\frac{\partial q(z,t)}{\partial z}. \quad (3)$$

We can substitute to $q(z,t)$ the Fourier law: $q(z,t) = -\chi^{(V)}(z)\frac{\partial T(z,t)}{\partial z}$. Finally, we reach the diffusion temperature equation:

$$\varrho C^{(V)}(z)\frac{\partial T(z,t)}{\partial t} = \frac{\partial}{\partial z}\left[\chi^{(V)}(z)\frac{\partial T(z,t)}{\partial z}\right]. \quad (4)$$

It remains only to associate the (4) with the corresponding boundary conditions: $\bar{q}(t) = \lim_{z\to 0}\left(-\chi^{(V)}(z)\frac{\partial T(z,t)}{\partial z}\right)$, $\lim_{z\to\infty} T(z,t) = 0$. The solution of (4) was found by applying the Laplace transform, applying boundary conditions, and then anti-transforming (refer to [5] for more details, R_β is the anomalous thermal conductivity coefficient):

$$\bar{q}(t) = R_\beta\left(D^{\beta}_{0^+}T_0\right)(t). \quad (5)$$

Formulation of Fractional Stefan Problem

Before writing the fractional system, we derive Stefan's condition by the fractional Fourier law. Consider two infinitesimal times (highlighted in the figure below) consider two infinitesimal times (accentuated in the figure below), in which we will obtain two different values of the interface, respectively $s(t_0)$ and $s(t_1)$:

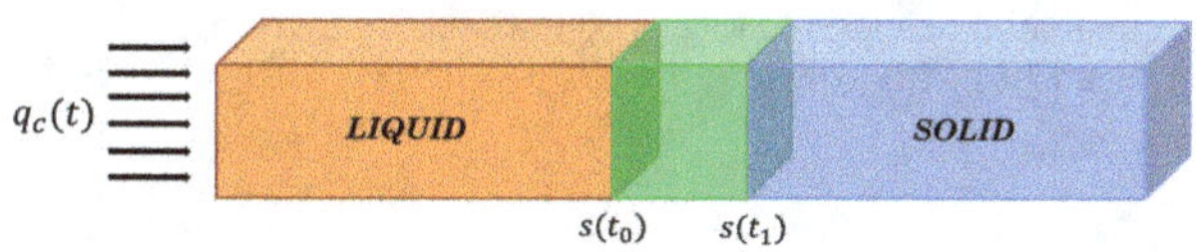

Fig. 3. Schematic representation of two generic heat diffusion times, at which we find two interface values.

We know from thermodynamics that to melt a piece of bar from $s(t_0)$ to $s(t_1)$ requires heat equal to:

$$Q_1 = V\varrho\lambda = A\big(s(t_1) - s(t_0)\big)\varrho\lambda. \quad (6)$$

Heat within both phases propagates through flux of fractional Fourier law, which takes this form:

$$q_{\beta,i} = -k_{\beta,i}\left({}_0^CD_t^\beta\frac{\partial T_i}{\partial x}\right)(x,t). \quad (7)$$

Materials Research Forum LLC
https://doi.org/10.21741/9781644902431-13

Where the dimensions of the anomalous diffusivity coefficient are $[k_{\beta,i}] = \frac{g\,cm}{s^{3-\beta}K}$. To employ the principle of energy conservation and equalize the fluxes of liquid and solid to heat Q_1, it is necessary to integrate them twice, first with respect to area and then with respect to time.

$$Q_2 = A\int_{t_o}^{t_1}\left(-k_{\beta,L}\left({}_0^C D_t^\beta \frac{\partial T_L}{\partial x}\right)(s(t),t) + k_{\beta,S}\left({}_0^C D_t^\beta \frac{\partial T_S}{\partial x}\right)(x,t)\right)dt. \quad (8)$$

Now we can equalize $Q_1 = Q_2$ and make some simplifications:

$$\left(s(t_1) - s(t_0)\right)\varrho\lambda = \int_{t_o}^{t_1}\left(-k_{\beta,L}\left({}_0^C D_t^\beta \frac{\partial T_L}{\partial x}\right)(s(t),t) + k_{\beta,S}\left({}_0^C D_t^\beta \frac{\partial T_S}{\partial x}\right)(s(t),t)\right)dt. \quad (9)$$

We divide both members by $(t_1 - t_0)$ and perform the limit for $t_1 \to t_0$. Then, since t_0 can be any instant, we substitute t:

$$\frac{ds}{dt}(t) = -\gamma_{\beta,L}\left({}_0^C D_t^\beta \frac{\partial T_L}{\partial x}\right)(s(t),t) + \gamma_{\beta,S}\left({}_0^C D_t^\beta \frac{\partial T_S}{\partial x}\right)(s(t),t). \quad (10)$$

Where $\gamma_{\beta,i} = \frac{k_{\beta,i}}{\varrho\lambda}$ with these dimensions $[\gamma_{\beta,i}] = \frac{cm^2}{s^{1-\beta}K}$. It becomes interesting to note that for $\beta \to 0$ in (10) we again obtain the classic Stefan's condition. Then writing the complete system in term of Caputo derivative [7,8] with $0 < \beta < 1$:

Liquid side:

$$\begin{cases} \frac{\partial T_L}{\partial t}(r,t) = \alpha_{\beta,L}\left({}_0^C D_t^\beta \frac{\partial^2 T_L}{\partial x^2}\right)(x,t) & 0 < x < s(t), t > 0. \quad (10.a) \\ \frac{\partial T_L}{\partial x}(0,t) = -\frac{q_c(t)}{k_L} = -\frac{q_0}{k_L\sqrt{t^{1-\beta}}}. & (10.b) \\ T_L(s(t),t) = T_{Melt}. & (10.c) \end{cases}$$

Solid side:

$$\begin{cases} \frac{\partial T_S}{\partial t}(r,t) = \alpha_{\beta,S}\left({}_0^C D_t^\beta \frac{\partial^2 T_S}{\partial x^2}\right)(x,t) & s(t) < x < \infty, t > 0. \quad (11.a) \\ T_S(s(t),t) = T_{Melt}. & (11.b) \\ T_S(\infty,t) = T_{Cold}. & (11.c) \\ T_S(x,0) = T_{Cold}. & (11.d) \end{cases}$$

Interface condition:

$$\begin{cases} \frac{ds}{dt}(t) = -\gamma_{\beta,L}\left({}_0^CD_t^\beta \frac{\partial T_L}{\partial x}\right)(s(t),t) + \gamma_{\beta,S}\left({}_0^CD_t^\beta \frac{\partial T_S}{\partial x}\right)(s(t),t). & (12.a) \\ s(0) = 0. & (12.b) \end{cases}$$

Where $\alpha_{\beta,i} = \frac{k_{\beta,i}}{\varrho\, cp_i}$ are the anomalous diffusivity coefficients: $[\alpha_{\beta,i}] = \frac{cm^2}{s^{1-\beta}}$. The solutions of $(11.a - 12.b)$ were found in [6], not shown here for brevity.

Numerical Application

Now some comparisons between the classical and fractional solutions will follow. The numerical data that will be used, for example purposes only, are from a Titanium alloy. For simplicity, parameters such as thermal conductivity, which depend on the order of derivation in the fractional case, will be taken in equal number value with respect to their integer-order counterparts. So, for example, $k_{\beta,L} = k_L$. Here are data for the Titanium alloy used: $\varrho = 2.67\ \frac{g}{cm^3}$, $\lambda = 171.6\ \frac{J}{g}$, $k_L = 0.3\ \frac{W}{cm\,K}$, $c_{p,L} = 1.18\ \frac{J}{g\,K}$, $k_S = 0.081\ \frac{W}{cm\,K}$, $c_{p,S} = 0.43\ \frac{J}{g\,K}$, $q_0 = 1000\ \frac{W}{cm^2}\sqrt{s}$, $T_{Melt} = 1923\ K$, $T_{Cold} = 298\ K$. In the next graph it is possible to observe, for a certain fixed time, different interface curves as β changes:

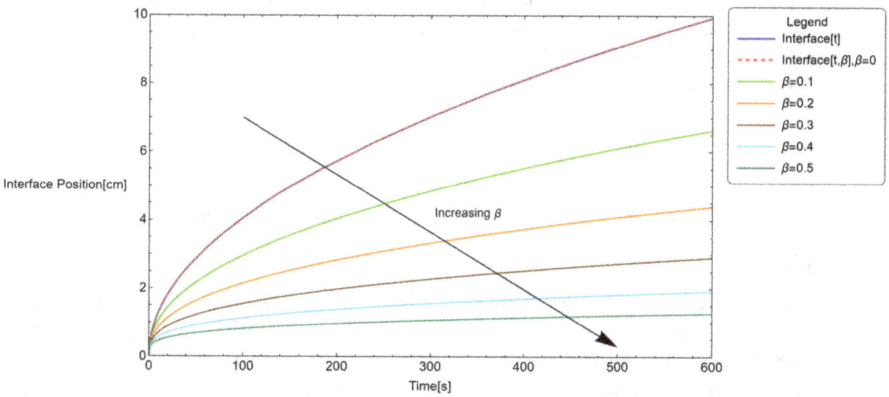

*Fig. 4. Interface functions in **β** dependence, while in blue the classic solution.*

For $\beta = 0$ the interface function derived from the problem in fractional form (dashed red curve) coincides with the classical solution (blue curve). Instead, as β increases there is a rapid decrease in the process, so fast that even just for a $\beta = 0.5$ the process reaches almost complete stationarity. Examples were not given for values above 0.5 because the curves were completely parallel to the x-axis, meaning that the interface remains in the same position for all time.

Lastly, in Fig. 4, the functions of temperature distributions at the same time fixed for the interface graph and at different values of β were plotted. Again, for $\beta = 0$, the fractional functions (black and grey) coincide with the classical functions (red and blue). The other liquid and solid curves have been plotted in the same colour for simplicity.

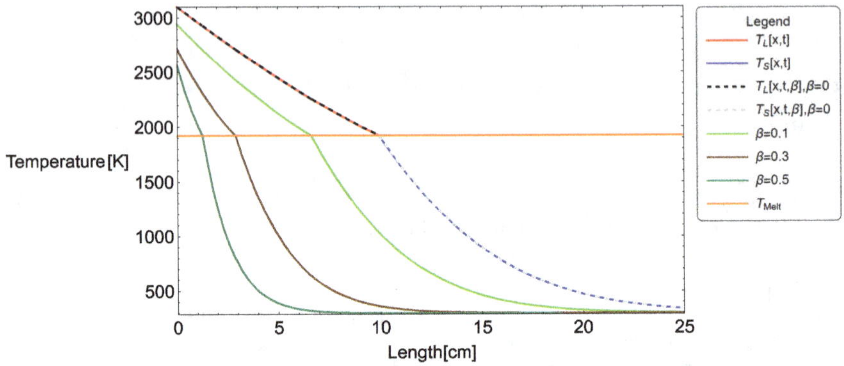

Fig. 5. Temperature distributions for liquid and solid in $\boldsymbol{\beta}$ *dependence. In red and blue the solutions of the classical problem, in orange the melting temperature. The point where the curves meet on the melting line is the value of the interface.*

Conclusions

In this paper, the solution for the one-dimensional stefan problem of fractional order was examined. It started from the fractional beta-order generalization of Fourier's law and then derived the formulation of the problem. By varying the order of the differintegral operator the process goes into *subdiffusion* and already for small values of beta slows down considerably, while for $\beta = 0$ we obtain to the solution of the classical problem.

References

[1] C. Vuik (1993). Some historical notes about the Stefan problem.

[2] S. C. Gupta (2017). *The classical Stefan problem: basic concepts, modelling and analysis with quasi-analytical solutions and methods* (Vol. 45). Elsevier. https://doi.org/10.1016/B978-0-444-63581-5.00012-9

[3] G. Baffou, & R. Quidant (2013). Thermo-plasmonics: using metallic nanostructures as nano-sources of heat. *Laser & Photonics Reviews*, *7*(2), 171-187. https://doi.org/10.1002/lpor.201200003

[4] C. Cattaneo (1948). Sulla conduzione del calore. *Atti Sem. Mat. Fis. Univ. Modena*, *3*, 83-101.

[5] M. Zingales (2016). An exact thermodynamical model of power-law temperature time scaling. *Annals of Physics*, *365*, 24-37. https://doi.org/10.1016/j.aop.2015.08.014

[6] S. D. Roscani, & D. A. Tarzia (2018). Explicit solution for a two-phase fractional Stefan problem with a heat flux condition at the fixed face. *Computational and Applied Mathematics*, *37*(4), 4757-4771. https://doi.org/10.1007/s40314-018-0600-z

[7] I. Podlubny (1999). Fractional differential equations. *Mathematics in science and engineering*, *198*, 41-119.

[8] R. Gorenflo, A. A. Kilbas, F. Mainardi, & S. V. Rogosin (2020). *Mittag-Leffler functions, related topics and applications* (p. 540). New York, NY, USA: Springer. https://doi.org/10.1007/978-3-662-61550-8

Structural Mechanics

Theoretical and Applied Mechanics - AIMETA 2022 Materials Research Forum LLC
Materials Research Proceedings **26** (2023) 85-90 https://doi.org/10.21741/9781644902431-14

An accurate and refined nonlinear beam model accounting for the Poisson effect

RUOCCO E[1, a *] and REDDY JN[2,b]

[1]Department of Engineering, University of Campania Luigi Vanvitelli, Aversa, Italy

[2]Department of Mechanical Engineering, Texas A&M University, College Station, TX 77843-3123, USA

[a]eugenio.ruocco@unicampania.it, [b]jnreddy@tamu.edu

Keywords: Nonlinear Beam Model, Poisson Effect, Large Deformation Analysis, Snap-Back/Through Behavior

Abstract. In this paper, an enhanced beam model based on a 5-parameter displacement field, recently proposed by the authors and able to reproduce the Poisson effect in transverse direction is presented, and utilized to simulate the fully geometrically nonlinear response of elastic beam structures. The adoption of the linear solution as approximation functions for the nonlinear case allows prediction of nonlinear response of problems involving complex geometries with a relatively small computational effort. Several numerical examples of benchmark problems are analyzed, highlighting the characteristic features of the proposed five-parameter model and comparing the results with those obtained using the classical Bernoulli beam model and 3D finite element model.

Introduction

The development of enriched beam models aiming to overcome the classical Euler-Bernoulli or Timoshenko-Ehrenfest hypotheses [1] is currently an active research subject. Among them, the cubic displacement model proposed by Reddy in the early 1980s [2] is certainly among the most popular [3-5], and has paved the way to many models proposed later on, generally motivated by the need to create a consistent and straightforward model able to simulate the mechanical behavior of the new materials and structures now available. For instance, in [6] the authors compare the high-order beam theories available in the literature to describe sandwich beams and based on polynomial expansions, whereas in [7] the authors propose a strain-based geometrically

nonlinear beam formulation for structural and aeroelastic modeling and analysis of slender wings of very flexible aircraft. A beam model suitable for meta-materials design in a dynamic setting and formulated directly in a discrete way is recently reported in [8]. A cubic displacement field for flexural analysis of two-layered composite beams is proposed in [9], and a higher-order beam model able to accurately represent the buckling behavior of three-dimensional thin-walled structures is proposed in [10]. The number of works dealing with this subject is very large, and the interested readers can find in [1,11] reviews of the literature containing hundreds of papers devoted to bending, buckling and free vibration analyses based on different beam models. A common feature of all this model, justified by experimental evidence, is adopting an axial displacement component enriched with various contributions while leaving the transverse component unchanged (i.e., assuming inextensibility of the transverse normal lines and, as a consequence, neglecting the Poisson effect). Nevertheless, such a hypothesis also affects the strain and the stress distribution over the beam, sometimes in a meaningful way. Moreover, the assumption of constant transverse displacement in the beam thickness does not allow to evaluate the Poisson effects which, in some cases, can be of considerable interest [12]. In this paper, the nonlinear version of the enhanced beam model based on a 5-parameter displacement field, recently proposed in [13] and able to reproduce the Poisson effect in transverse direction is presented. This work summarizes the results recently published by the authors in [14] and stems from the 7-parameter displacement field, proposed in [15] for large deformation analysis of composite shell structures, simplified by the one-dimensional nature inherent in the beam model. The loss of information in the passage from a two-dimensional to a one-dimensional model is rewarded by the identification of an analytical solution able to overcome the difficulties associated with the numerical implementation of the 2D model. As in [15], the new kinematic model is utilized to simulate the fully geometrically nonlinear response of elastic beam structures.

Governing equations

Structural theories can be developed by expanding the displacements in a power series of the transverse coordinates [1]. In the proposed model the displacement field is a linear expansion of the thickness coordinate around the mid-surface and the transverse displacement is parabolic through the thickness coordinate y of the beam as follow (Figure 1):

$$\begin{bmatrix} s_x \\ s_y \end{bmatrix} = \begin{bmatrix} u \\ v \end{bmatrix} + y \begin{bmatrix} \varphi_v \\ \varphi_u \end{bmatrix} + y^2 \begin{bmatrix} 0 \\ \psi \end{bmatrix} \tag{1}$$

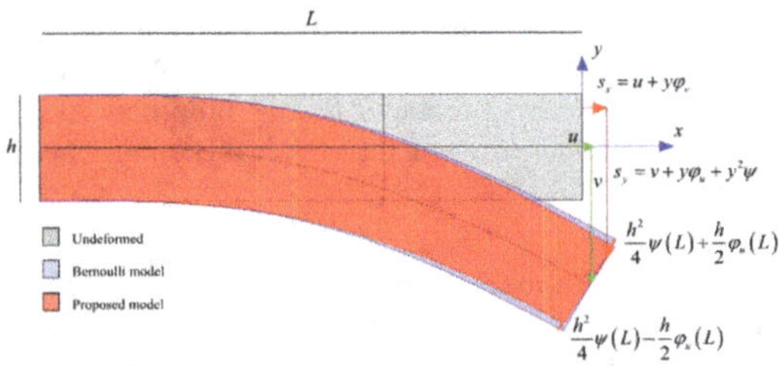

***Fig. 1**: Displacement field proposed in Eq. (1), compared with classical Bernoulli beam model. The Figure shows how both the new parameters ψ and φ_u contribute to the deformation in the direction perpendicular to the cross section area. The remaining kinematic parameters are common to the Timoshenko beam model.*

Given the displacement field (1), under the von Karman hypothesis the linear and nonlinear strain components can be accordingly calculated as:

$$\begin{bmatrix} \varepsilon_x \\ \varepsilon_y \\ \gamma_{xy} \end{bmatrix}^L + \begin{bmatrix} \varepsilon_x \\ \varepsilon_y \\ \gamma_{xy} \end{bmatrix}^{NL} = \begin{bmatrix} u_{,x} + y\varphi_{v,x} \\ \varphi_u + 2y\psi \\ \frac{1}{2}\left((v_{,x} + \varphi_v) + y\varphi_{u,x} + y^2\psi_{,x}\right) \end{bmatrix} + \frac{1}{2}\begin{bmatrix} u_{,x}^2 + v_{,x}^2 \\ 0 \\ 0 \end{bmatrix} = \begin{bmatrix} \eta + y\chi_v \\ \varphi_u + y\omega \\ \frac{1}{2}\left(t + y\chi_u + y^2\omega\right) \end{bmatrix} + \frac{1}{2}\begin{bmatrix} u_{,x}^2 + v_{,x}^2 \\ 0 \\ 0 \end{bmatrix} \tag{2}$$

the dual stress vector corresponding to the linear strain can be obtained by invoking the virtual work equivalence:

$$\begin{aligned} &\int_V \boldsymbol{\sigma}_L^T \delta\boldsymbol{\varepsilon}_L dV = \int_0^L \int_A \boldsymbol{\sigma}_L^T \delta\boldsymbol{\varepsilon}_L dA dx = \int_0^L \mathbf{Q}_L^T \delta\mathbf{q}_L dx \\ &\int_V \mathbf{X}^T \delta\mathbf{s} dV = \int_0^L \int_A \mathbf{X}^T \delta\mathbf{s} dA dx = \int_0^L \mathbf{P}^T \delta\mathbf{U} dx \end{aligned} \tag{3}$$

having indicate with $\boldsymbol{\sigma}^T = \begin{bmatrix}\sigma_x & \sigma_y & \tau_{xy}\end{bmatrix}$, $\mathbf{X}^T = \begin{bmatrix}X_x & X_y\end{bmatrix}$ the vectors containing the stress and the applied mass-forces, respectively, with:

$$\begin{aligned}&\mathbf{q} = \begin{bmatrix}\eta & \chi_v & \chi_u & \varphi_u & t & \phi & \omega\end{bmatrix}^T;\\ &\mathbf{U} = \begin{bmatrix}u & v & \varphi_u & \varphi_v & \psi\end{bmatrix}^T;\end{aligned} \tag{4}$$

the generalized strain and displacement vectors and with:

$$\begin{aligned}&\mathbf{Q} = \int_A \begin{bmatrix}\sigma_x & y\sigma_x & y\tau_{xy} & \sigma_y & \tau_{xy} & y\sigma_y & y^2\tau_{xy}\end{bmatrix}^T dA = \begin{bmatrix}N_x & M_x & M_{xy} & N_y & T & M_y & V\end{bmatrix};\\ &\mathbf{P} = \int_A \begin{bmatrix}X_x & X_y & yX_x & yX_y & y^2X_y\end{bmatrix}^T dA = \begin{bmatrix}n & p & m & r & q\end{bmatrix}^T;\end{aligned} \tag{5}$$

The derived vector of generalized stress and applied load, respectively.

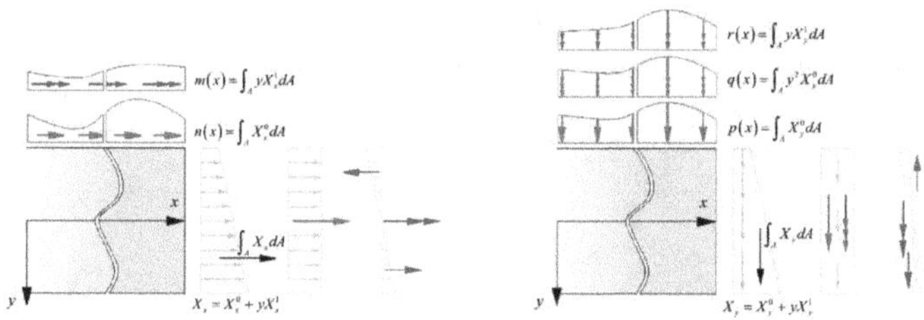

Fig. 2: Mechanical interpretation of the generalized loads deriving by the mass forces. In the proposed model, the axial load X_x returns the two distributed loads n(x) and m(x), the latter accounting fort he position of n(x). Both are considered in first and higher-order beam theories, whereas Bernoulli beam model only consider n(x). Regarding the out-of-plane contributes, the proposed model shares with ones available in the literature the generalized component p(x). To this it adds the two more contributes q(x) and r(x) invisible tot he classical models.

It is worth noticing that, compared with the Timoshenko beam, the proposed model presents the four more stress components (M_{xy}, N_y, V, M_y) dual to the strain components $(\chi_u, \varphi_u, \phi, \omega)$ and the two more applied loads (r, q) dual to the displacements (φ_u, ψ). In particular, the applied load $r = \int_A yX_y dA$ can be interpreted as a moment resulting from the mass force Xy applied in the direction perpendicular to the middle line of the beam, so that an applied load $X_y = X_y^0 + X_y^1 y$, varying linearly along y, returns in addition to the generalized load $p = I_0 X_y^0$ common to the classical beam models, an additional component $r = I_2 X_y^1$. If isotropic, in plane stress state the constitutive relationships between stress and strain components involved in the proposed model is:

Materials Research Proceedings **26** (2023) 85-90 https://doi.org/10.21741/9781644902431-14

$$
\begin{bmatrix} N_x \\ M_x \\ M_{xy} \\ N_y \\ T \\ M_y \\ V \end{bmatrix} = \begin{bmatrix} \bar{E}I_0 & 0 & 0 & \nu\bar{E}I_0 & 0 & 0 & 0 \\ 0 & \bar{E}I_2 & 0 & 0 & 0 & \nu\bar{E}I_2 & 0 \\ 0 & 0 & GI_2 & 0 & 0 & 0 & 0 \\ \nu\bar{E}I_0 & 0 & 0 & \bar{E}I_0 & 0 & 0 & 0 \\ 0 & 0 & 0 & 0 & GI_0 & 0 & GI_2 \\ 0 & \nu\bar{E}I_2 & 0 & 0 & 0 & \bar{E}I_2 & 0 \\ 0 & 0 & 0 & 0 & GI_2 & 0 & GI_4 \end{bmatrix} \begin{bmatrix} \eta \\ \chi_v \\ \chi_u \\ \varphi_u \\ t \\ \phi \\ \omega \end{bmatrix} \tag{6}
$$

The development of the nonlinear equilibrium equations lies on the application of the Principle of the Minimum Potential Energy, sum of the elastic stored energy and geometric potential energy, obtained neglecting terms of order higher than two [1]. The sum of external energy with the linear and non-linear contribution of the internal virtual energy assumes the form:

$$
W_E + W_I^L + W_I^{NL} = \int_0^L \left(nu + pv + m\varphi_v + r\varphi_u + q\psi\right)dx + \int_0^L \mathbf{q}^T \mathbf{Q} dx + \frac{1}{2}\int_0^L N_0 v_{,x}^2 \, dx \tag{7}
$$

The weak form of the equilibrium equations:

$$
\frac{\partial\left(W_E + W_I^L + W_I^{NL}\right)}{\partial \mathbf{U}} = 0 \tag{8}
$$

gives the following system of two in-plane:

$$
\begin{aligned}
&\bar{E}I_0\left(u_{,xx} + \varpi\varphi_{u,x}\right) + N_0 u_{,xx} + n = 0 \\
&GI_2\varphi_{u,xx} - \bar{E}I_0\left(\varphi_u + \nu u_{,x}\right) + N_2\varphi_{u,xx} + r = 0
\end{aligned} \tag{9}
$$

and three out-of-plane:

$$
\begin{aligned}
&G\left(I_0\left(v_{,x} + \varphi_v\right)_{,x} + I_2\psi_{,xx}\right) + N_0 v_{,xx} + N_2\psi_{,xx} + p = 0 \\
&\bar{E}I_2\left(\varphi_{v,xx} + 2\nu\psi_{,x}\right) + G\left(I_0\left(v_{,x} + \varphi_v\right) + I_2\psi_{,x}\right) + N_2\varphi_{v,xx} + m = 0 \\
&G\left(I_2\left(v_{,x} + \varphi_v\right)_{,x} + I_4\psi_{,xx}\right) - 2\bar{E}I_2\left(2\psi + \nu\varphi_{v,x}\right) + N_2 v_{,xx} + N_4\psi_{,xx} + q = 0
\end{aligned} \tag{10}
$$

equilibrium equations. It is interesting to note that the proposed kinematic model also affects the equations governing the in-plane behavior: assuming a Poisson's ratio $\nu = 0$, Eqs. (9) returns two uncoupled equations, the first of which related to the u-component and coincident with the classical in-plane equilibrium equation, the latter in terms of the φ_u-component. Both are affected by a nonlinear contribution (set to zero assuming the von Kármán hypothesis). The analytical solution of the linear expressions of Eqs. (9-10), reported in [13] and [14] are finally adopted in a displacement-based approach to solve structures with general geometries.

Examples

The proposed example is a simply supported, two-member L-shaped structure subjected to a single eccentric load shown in Fig. 3. Each member has a length $6L$ = 120 cm and a 2 cm by 3 cm rectangular cross section. Originally proposed by Lee et al. [16], and later studied by several researchers [17–19], this structure exhibits complicated equilibrium path, characterized by large displacement including both snap-through and snap-back phenomena. Fig. 3 contains a comparison of the present solution with the solution given in [17] in terms of u and v displacement components of the loaded point.

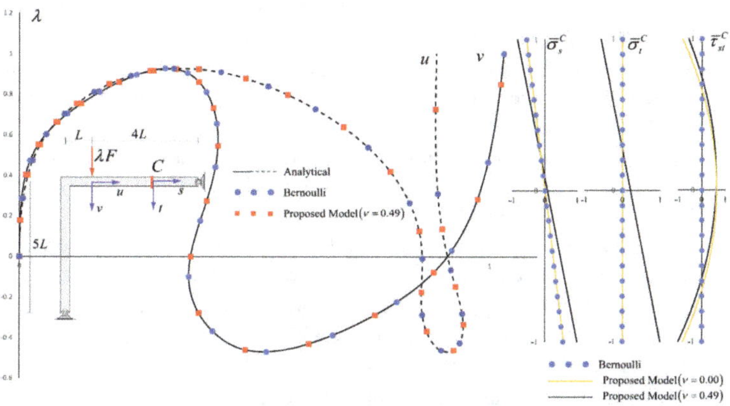

***Fig. 3**: Geometry, stress field and equilibrium path of a Lee's frame with $L = 20$ cm and $h = 0.3$ cm. The Figure shows as the proposed model still accurately traces the equilibrium path with negligible differences with respect to the analytical solution. However, the stress field obtained at section C still shows significant variations with the Poisson ratio ν, matching the Bernoulli's values only for $\nu = 0$.*

The proposed model accurately traces the equilibrium path, with negligible differences with respect to both Bernoulli and analytical solution for each value of the Poisson ratio; however, it shows an increasing value of any σ-component by increasing the Poisson ratio ν.

Summary

In this work an enhanced beam model suitable for nonlinear analysis of frame structures us developed. The model's ability to predict both displacements and stresses is tested by comparing the produced results with ones available in the literature for the classical Bernoulli beam model. The obtained results show that, for straight thin beams, the displacement field does not present appreciable differences but when the Poisson ratio is different from zero the related stress field show substantial increase in magnitude.

References

[1] J.N. Reddy, Theories and Analyses of Beams and Axisymmetric Circular Plates, CRC Press, Boca Raton, FL, 2022. https://doi.org/10.1201/9781003240846

[2] J.N. Reddy, A simple higher order theory for laminated composite plates, ASME J. Appl. Mech. 51 (4) (1984) 745–752. https://doi.org/10.1115/1.3167719

[3] C. Polizzotto, From the Euler–Bernoulli beam to the Timoshenko one through a sequence of Reddy-type shear deformable beam models of increasing order, Eur.J. Mech. A Solids 53 (2015) 62–74. https://doi.org/10.1016/j.euromechsol.2015.03.005

[4] E. Ruocco, J.N. Reddy, C. Wang, An enhanced hencky bar-chain model for bending, buckling and vibration analyses of Reddy beams, Eng. Struct. 221(2020) 111056, https://doi.org/10.1016/j.engstruct.2020.111056

[5] M. Di Paola, J.N. Reddy, E. Ruocco, On the application of fractional calculus for the formulation of viscoelastic reddy beam, Meccanica 55 (3) (2020) 1–14. https://doi.org/10.1007/s11012-020-01177-3

[6] W. Castañeda, J. Yarasca, J. Mantari, Best shear deformation theories based on polynomial expansions for sandwich beams, Eng. Struct. 190 (2019) 422–434. https://doi.org/10.1016/j.engstruct.2019.04.022

[7] W. Su, C.E. Cesnik, Strain-based geometrically nonlinear beam formulation for modeling very flexible aircraft, Int. J. Solids Struct. 48 (16) (2011) 2349–2360. https://doi.org/10.1016/j.ijsolstr.2011.04.012

[8] E. Turco, Modeling nonlinear beams for metamaterials design in a dynamic setting, Mech. Res. Commun. 117 (2021) 103786. https://doi.org/10.1016/j.mechrescom.2021.103786

[9] J. Wen, A. Hamid Sheikh, M. Uddin, B. Uy, Analytical model for flexural response of two-layered composite beams with interfacial shear slip using a higher order beam theory, Compos. Struct. 184 (2018) 789–799. https://doi.org/10.1016/j.compstruct.2017.10.023

[10] E.P.R. Vieira, F. Virtuoso, Buckling of thin-walled structures through a higher order beam model, Compos. Struct. 180 (2019) 104–116. https://doi.org/10.1016/j.compstruc.2016.01.005

[11] A.S. Sayyad, Y.M. Ghugal, Bending, buckling and free vibration of laminated composite and sandwich beams: A critical review of literature, Compos. Struct. 171 (2017) 486–504. https://doi.org/10.1016/j.compstruct.2017.03.053

[12] K. Evans, Auxetic polymers: a new range of materials, Endeavour 15 (4) (1991) 170–174. https://doi.org/10.1016/0160-9327(91)90123-S

[13] E. Ruocco, J.N. Reddy, E. Sacco, Analytical solution for a 5-parameter beam displacement model, Int. J. Mech. Sci. 201 (2021) 106496. https://doi.org/10.1016/j.ijmecsci.2021.106496

[14] E. Ruocco, J.N. Reddy, A new nonlinear 5-parameter beam model accounting for the Poisson effect. Int. J. of Non-Linear Mech. 142 (2022) 103996. https://doi.org/10.1016/j.ijnonlinmec.2022.103996

[15] M. Gutierrez Rivera, J.N. Reddy, Stress analysis of functionally graded shells using a 7-parameter shell element, Mech. Res. Commun. 78 (2016) 60–70. https://doi.org/10.1016/j.mechrescom.2016.02.009

[16] S. Lee, F. Manuel, E. Rossow, Large deflections and stability of elastic frames, J. Eng. Mech. 94 (1968) 521–547. https://doi.org/10.1061/JMCEA3.0000966

[17] C. Cichon, Large displacements in-plane analysis of elastic-plastic frames, Comput. Struct. 19 (5–6) (1984) 737–745, http://dx.doi.org/10.1016/0045-7949(84)90173-1. https://doi.org/10.1016/0045-7949(84)90173-1

[18] J. Simo, L. Vu-Quoc, A three-dimensional finite strain rod model: Part 2: Computational aspects, Comput. Methods Appl. Mech. Engrg. 58 (1986) 79–116. https://doi.org/10.1016/0045-7825(86)90079-4

[19] B. Coulter, R. Miller, Numerical analysis of a generalized plane 'elastica' with non-linear material behaviour, Int. J. Numer. Methods Eng. 26 (1988) 617–630. https://doi.org/10.1002/nme.1620260307

Theoretical and Applied Mechanics - AIMETA 2022 | Materials Research Forum LLC
Materials Research Proceedings **26** (2023) 91-96 | https://doi.org/10.21741/9781644902431-15

The extended membrane analogy for an engineered evaluation of the torsional properties of multi-material beams

GALUPPI Laura[1, a *] and ROYER CARFAGNI Gianni[1,2,b]

[1] Department of Engineering and Architecture, University of Parma, Parma, Italy

[2] Construction Technologies Institute - Italian National Research Council (ITC-CNR), San Giuliano Milanese, Milan, Italy

[a]laura.galuppi@unipr.it, [b]gianni.royer@unipr.it

* corresponding author

Keywords: De Saint Venant Torsion, Membrane Analogy, Multi-Material Cross Sections

Abstract. Prandtl's membrane analogy for cylindrical bars under torsion is extended to beams with multi-material cross section of any geometry. It was demonstrated that the problem is governed by equations formally analogue to those describing the deformation of an inflated membrane, differently tensioned in the regions corresponding to the cross-section domains hosting different materials. The analogy allows to evaluate not only the state of stress in the composite bar, but also its torsional stiffness. Here, the proposed method is used to evaluate the torsional response of a wide range of geometries, corresponding to glued multi-material bars, layered composites and laminated glass with reinforcements.

Introduction

There is a renewed and growing interest for non-homogeneous, multi-material, sandwich and laminated beams and bars. It is important to assess their response to torsion because they can be used as shafts, or at risk of flexural-torsional buckling. Elements of this type are usually verified using three-dimensional Finite Element (FE) models, requiring a consistent computational effort and a particular attention at the interface and boundary conditions at the end sections.

It is well known that the linear elastic problem à la De Saint Venant for a cylindrical bar under torsion is governed by differential equations and boundary conditions presenting a formal analogy with those describing the small deformation of an inflated initially-flat membrane stretched by an uniform equibiaxial stress, shaped as the cross section of the bar and constrained at the border. This *membrane* or *soap-film analogy*, ingeniously first proposed by Prandtl in 1903 [1], has been recently extended [2] to multi-material cross sections of any geometry. Specifically, it has been demonstrated that the problem is governed by the same equations describing the deformation of an inflated membrane, differently tensioned in the regions corresponding to the cross-section domains hosting different materials, under a stress state inversely proportional to shear moduli of the materials. The torsional stiffness is related to the volume enclosed between the deformed membrane and its reference flat configuration. The model allows to recover the case of multi-connected cross sections, by considering the holes as material domains with vanishing elastic stiffness. In the analogy, these correspond to infinitely taut regions of the membrane, that remain plane when pressurized.

Here, we present the practical use of the extended membrane analogy, to estimate the torsional stiffness in three different kind of composites: multi-material bars with circular symmetry, layered composites and laminated glass strengthened by stiff elements. For the simplest case, the multi-tensioned membrane problem can be solved in closed from, while for more complex geometries the membrane problem is implemented in a commercial FE code. In general, the numerical model of the membrane presents no difficulty, and it is certainly much easier than implementing a three-dimensional FE model for the whole bar.

The membrane analogy
Consider the De Saint–Venant's problem for a prismatic bar, whose cross section is composed by N different materials, with shear modulus G_i (i=1,…, N), connected by smooth interfaces, subjected to a load distribution at the ends equipollent to a torque T. With reference to Fig. 1, introduce the right-handed reference system (x,y,z), with z coincident with the bar axis, and denote by Ω_i the domain hosting the ith material, by Γ_{i0} the intersection of the boundary of Ω_i with the external boundary, and by Γ_{ij} the interface between materials i and j ($i \neq j$). To distinguish 'open' and 'closed' bi-material interfaces, the latter will be formally denoted by Γ_{ij}^0.

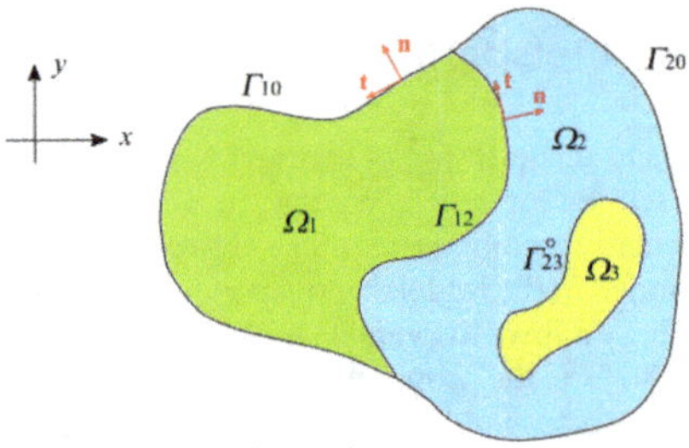

Figure 1. *Scheme of a multi-material cross section.*

In analogy with the classical approach by Prandtl, introduce, for each material domain, the stress function $F_i(x,y)$, such that the shear stresses take the form

$$\tau_{xz,i} = \frac{\partial F_i(x,y)}{\partial y} \text{ and } \tau_{yz,i} = \frac{\partial F_i(x,y)}{\partial x}. \tag{1}$$

Using the same arguments of the classical approach, for the ith material,

$$\Delta F_i(x,y) = -2G_i\Theta \ in \ \Omega_i \text{ and } F_i(x,y) = 0 \ on \ \Gamma_{i0}, \tag{2}$$

Where $\Delta(\cdot) = \partial^2(\cdot)/\partial x^2 + \partial^2(\cdot)/\partial y^2$, and Θ is the torsion angle per unit length. Bi-material matching conditions [3] require the continuity of the normal traction, and of the strain in tangential direction. In terms of stress function, this may be written as [2]

$$F_i(x,y) = F_j(x,y) \text{ and } \frac{1}{G_i}\frac{\partial F_i(x,y)}{\partial \mathbf{n}} = \frac{1}{G_j}\frac{\partial F_j(x,y)}{\partial \mathbf{n}} \ on \ \Gamma_{ij}. \tag{3}$$

where **n** denote outward normal vector on Γ_{ij}, as depicted in Figure 1. If the jth material domain Ω_j is enclosed by a closed curve Γ_{ij}^0, recalling (2), one obtains [2]

$$\frac{1}{G_j}\oint_{\Gamma_{ij}^0} \frac{\partial F_i(x,y)}{\partial \mathbf{n}} dl = -2A_j\Theta. \tag{4}$$

Once the stress function has been determined, the torque T and the torsional stiffness k are obtained as

$$T = 2\sum_{i=1}^{N}\int_{\Omega_i} F_i(x,y)\, dA_i, \quad k = \frac{2}{\Theta}\sum_{i=1}^{N}\int_{\Omega_i} F_i(x,y)\, dA_i. \tag{5}$$

Consider now an inflated membrane, defined by the same domain Ω, subjected to a uniform pressure p, whose (equibiaxial) tension is not uniform: the membrane portions corresponding to the domains Ω_i are subjected to the tension $s_i = 1/(2G_i)$. Denoting with $w_i(x, y)$ the out-of-plane displacement of the portion Ω_i, the governing equations read

$$\Delta w_i(x,y) = -p/s_i \quad in \ \Omega_i \quad \text{and} \quad w_i(x,y) = 0 \quad on \ \Gamma_{i0}, \tag{6}$$

with interface conditions

$$w_i(x,y) = w_j(x,y) \quad \text{and} \quad s_i \frac{\partial w_i(x,y)}{\partial \mathbf{n}} = s_j \frac{\partial w_j(x,y)}{\partial \mathbf{n}} \quad on \ \Gamma_{ij}, \tag{7}$$

The first of (7) states the continuity of the membrane displacement at the interface, while the second may be interpreted by considering that, under the hypothesis of small out-of-plane displacements, $s_i\, \partial w_i(x,y)/\partial \mathbf{n}$ represents the out-of-plane force per unit length exerted on the interface by the tension of the membrane portion Ω_i. Therefore, it corresponds to the *out-of-plane* equilibrium condition for the interface Γ_{ij}. Observe that the *in-plane* equilibrium of the interface cannot be satisfied (except for the trivial case $s_i = s_j$). Hence, the membrane necessitates constrains in correspondence of the interface, which allow out-of-plane movements but refrain in-plane displacements, providing a constraint reaction that re-establishes in-plane equilibrium.

If the contour Γ_{ij}^0 is closed, it may be verified that

$$s_j \oint_{\Gamma_{ij}^0} \frac{\partial w_i(x,y)}{\partial \mathbf{n}} \, dl = -pA_j, \tag{8}$$

which corresponds to the condition of global equilibrium in the out-of-plane direction of the whole membrane portion Ω_j, subjected to a total load pA_j.

There is a formal analogy between equations (2-4), governing the torsional response of the multi-material bar, and the "multi-tensioned" membrane equations (6-8). As in the classical Prandtl's analogy, the torque T and the torsional stiffeness k (5) correspond to the double of the volume V contained between the deformed membrane and its projection on the (x, y) plane, and to $2V/p$, respectively. As discussed in [2], the analogy may be extended to multiply connected cross sections, by considering the holes as domains hosting a material with vanishing elastic modulus, corresponding in the analogy to regions where the membrane is infinitely taut, i.e., $s_j \rightarrow \infty$.

Practical application

For simplest geometries, the problem can be solved in closed form, providing exact results. For most complicated cases, the analogy is very useful, but not for an experimental verification, as originally conceived by Prandtl, which is possible but not easy. More conveniently, the membrane problem can be directly implemented in a commercial FE code, allowing to consider membrane portions with different tautness. Therefore, one can determine the torsional stiffness for multi-material sections of any kind with a two-dimensional numerical model, at the same time evaluating the stress state, and use the result in a subsequent structural model based on 1D beam elements. This procedure simplifies the common engineering approach based on a 3D FE model, requiring a consistent effort for meshing and a particular attention at the interface and boundary conditions at the beam ends.

Working Examples

Example 1: Glued multi-material bar with circular symmetry.

Consider the multi-material bar whose cross section is shown in Fig. 2a, composed of an inner cylindrical bar of radius R_{23} and an external hollow cylinder with internal radius R_{12} and external radius R_{01}, both made of steel ($G_1 = G_3 = 80$ GPa), glued by a thin layer of glue, of thickness t and mean radius R_m, with variable shear modulus G_2. Thanks to the circular symmetry, both the torsional and the membrane problems, associated with eq,s (2-3) and (6-7), respectively, can be solved in closed form. The volume enclosed in the deformed membrane is given by

$$V = \frac{\pi p}{8}\left[\frac{\left(R_{01}^4 - R_{12}^4\right)}{s_1} + \frac{\left(R_{12}^4 - R_{23}^4\right)}{s_2} + \frac{R_{23}^4}{s_3}\right], \tag{9}$$

Hence, the torsional stiffness (5) is

$$k = \frac{2V}{p} = \frac{\pi}{4}\left[\frac{\left(R_{01}^4 - R_{12}^4\right)}{s_1} + \frac{\left(R_{12}^4 - R_{23}^4\right)}{s_2} + \frac{R_{23}^4}{s_3}\right] = \frac{\pi}{2}\left[G_1\left(R_{01}^4 - R_{12}^4\right) + G_2\left(R_{12}^4 - R_{23}^4\right) + G_3 R_{23}^4\right]. \tag{10}$$

Fig. 2b shows k as a function of G_2, for R_{01}= 20 mm and R_m=10 mm (obviously, R_{23}= R_m-t/2 and R_{12}= R_m+t/2), for different values of t, together with the qualitative deformed shape of the corresponding membranes, for different values of G_2. It is evident that the influence of the glue stiffness on the overall torsional stiffness is quite limited.

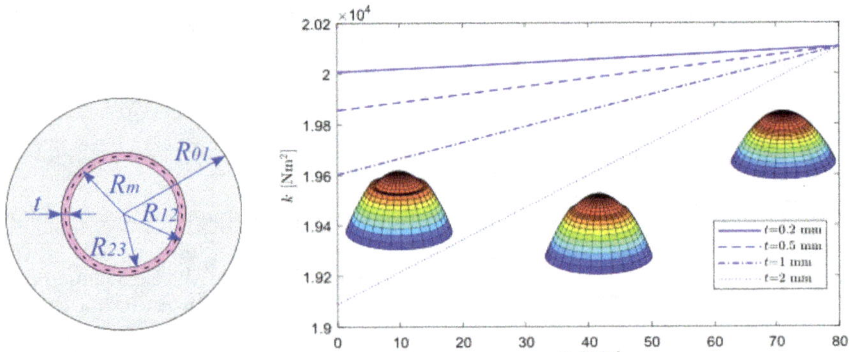

a) *b)*

Figure 2. *Example 1. a) Considered geometry, and b) torsional stiffness as a function of* G_2*, with qualitative membrane deformed shape for* $G_2=G_1/100$*,* $G_2=G_1/2$*, and* $G_2=G_1$*.*

<u>*Example 2: triple laminated glass element.*</u>

Consider now the triple laminated glass element of Fig. 3a, with b=200 mm, composed of three glass plies (h =12 mm, G_{glass} =28.69 GPa), bonded by two interlayers of thickness t = 1.52 mm. To evaluate the influence of the interlayer properties on the torsional response, its shear modulus G_{int} is made to vary between 0 (correspondent to infinitely taught membrane) and G_{glass}.

To numerically evaluate the torsional stiffness, a 3D model has been implemented with the software ABAQUS. A relative rigid rotation of 0.1 rad has been prescribed to the extreme cross sections of a 1m long beam, while allowing the warping of the cross section. The body has been discretized with quadratic solid elements with reduced integration. The membrane model has been implemented by using linear quadrilateral membrane elements with reduced integration, with the same mesh used for the beam cross section. Null displacements have been prescribed on the external boundary, while at the inner interfaces only the in-plane displacements are prevented.

Materials Research Proceedings **26** (2023) 91-96 https://doi.org/10.21741/9781644902431-15

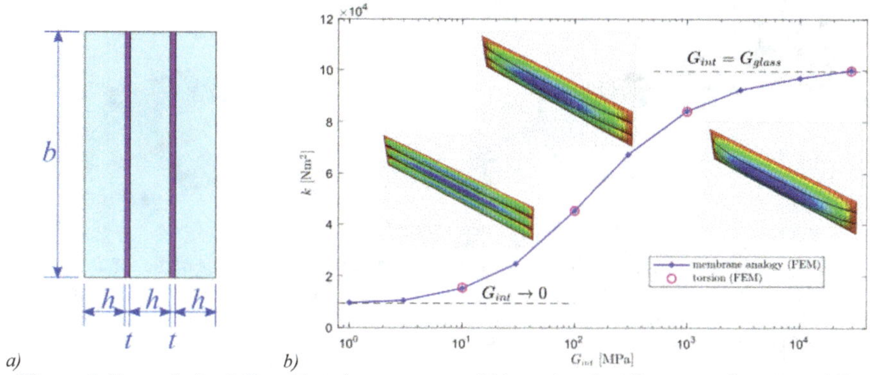

Figure 3. *Example 2. a) Considered geometry, and b) torsional stiffness as a function of G_{int}, with qualitative membrane deformed shape for G_{int}=10 MPa, G_{int}=100 MPa, and G_{int}=1000 MPa.*

Fig. 3b shows the comparison of the torsional stiffness evaluated by means of the torsion FE analyses (circles), and with the membrane analogy (stars). The results are in very good agreement (max error of the order of 1%). The analytically evaluated bounding values, corresponding to the cases of interlayer with vanishing stiffness ($G_{int} \to 0$) and of a full-glass section ($G_{int} = G_{glass}$) are plotted on the same graph. Remarkably, the computational effort required to numerically solve the membrane problem is of about 1% of that required to solve the 3D torsion problem.

Example 3: *triple laminated glass element, with steel reinforcement.*
Laminated glass beams and fins are often reinforced by steel elements, to avoid brittle response and to enhanced their failure behavior [4]. Here, the same triple laminated glass element studied in Example 2, reinforced by a steel element of width w=10 mm, as shown in Fig. 4, is considered.

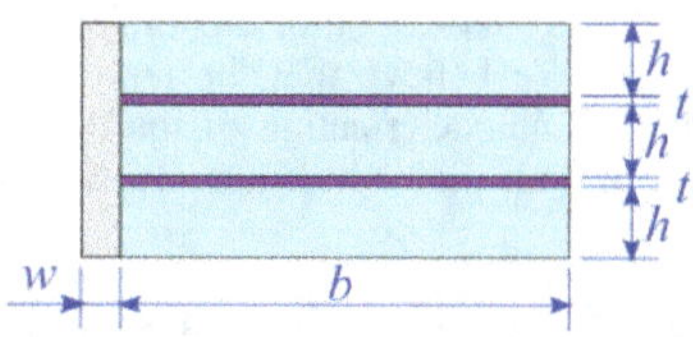

Figure 4. *Example 3. Considered geometry.*

The numerical evaluation of the torsional stiffness has been conducted as described for Example 2, by considering two different values for the interlayer stiffness, i.e., G_{int}=10 MPa, and G_{int}=100 MPa. The qualitative membrane deformed shape for these cases are shown in Fig. 5.

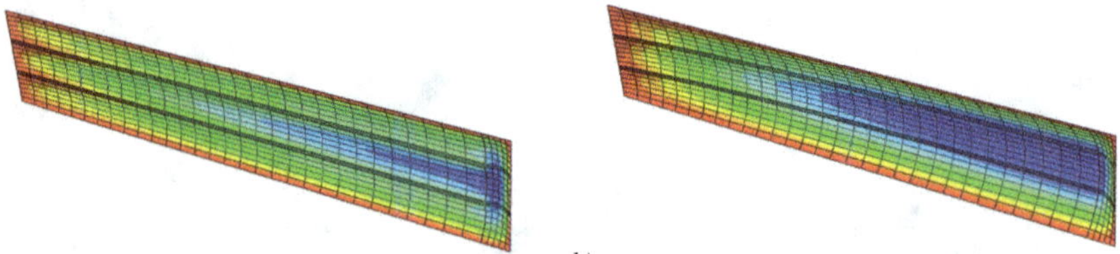

a) b)

Figure 5. *Example 3Membrane deformed shape for a) G_{int}=10 MPa, b) G_{int}=100 MPa.*

Table 1 records the torsional stiffness *k*, evaluated either with FE analysis of the torsion problem and with the membrane analogy. Again, the results are in very good agreement. The computational effort required to numerically solve the membrane problem is of about 0.6% of that required to solve the 3D torsion problem.

Table 1. *Example 3. Torsional stiffness for G_{int}=10 MPa and G_{int}=100 MPa.*

	k (FEM-torsion) [Nm^2]	*k* (FEM-membrane) [Nm^2]	Error [%]
G_{int}= 10 MPa	31615.95	31604.00	0.038
G_{int}= 100 MPa	76467.32	76558.25	-0.119

Conclusions

The semi-analytical method based on an extension of the classical Prandtl's membrane analogy, allows to readily evaluate the torsional properties of multi-material bars. This method replaces the 3D FE analysis of the torsion problem for the bar, which is time-consuming due to the complex shape and the high numbers of materials, with a FE model of the equivalent membrane structure. In the worked examples, this required about the 1% of the computational effort for the 3D case.

The practical use of this method has been here demonstrated for a wide range of geometries, corresponding to glued multi-material bars, layered composites and laminated glass beams with reinforcements. The worked examples confirm the potential of this approach in providing both a qualitative (intuitive) and quantitative indication of the torsional properties, and its advantages in the practice of structural design. We believe that the method here proposed may represent a powerful tool to engineers for the characterization of the torsional response of composite and layered structures, also in view of their verification against buckling under bending.

References

[1] Prandtl L., Zur torsion von prismatischen stäben. Physikalische Zeitschrift 4 (1903) 758–770.

[2] Galuppi, L., Royer-Carfagni, G., Membrane analogy for multi-material bars under torsion, Proc. R. Soc. A 475 (2019) 20190124. https://doi.org/10.1098/rspa.2019.0124

[3] Dundurs J, Sotiropoulos D. On stress conditions and computations at bi-material interfaces. Comput. Mech. 21 (1998) 300–305. https://doi.org/10.1007/s004660050306

[4] Martens, K., Caspeele, R., Belis, J. Development of reinforced and posttensioned glass beams: review of experimental research. J. Struct. Eng. 142.5 (2016) 04015173. https://doi.org/10.1061/(ASCE)ST.1943-541X.0001453

Theoretical and Applied Mechanics - AIMETA 2022 Materials Research Forum LLC
Materials Research Proceedings **26** (2023) 97-102 https://doi.org/10.21741/9781644902431-16

Flexural tensegrity: Field applications

Claudio Boni [1,a] * and Gianni Royer-Carfagni [1,2,b]

[1] Department of Engineering and Architecture, University of Parma, Parma, Italy

[2] Construction Technologies Institute - Italian National Research Council (ITC-CNR), San Giuliano Milanese, Milano, Italy

[a] claudio.boni@unipr.it, [b] gianni.royer@unipr.it

Keywords: Kinetic Structure, Segmental Beam, Nonlinear Nonlocal Response

Abstract. Flexural tensegrities represent a class of structural elements composed of segments in unilateral contact along properly-designed pitch profiles and held together by pre-tensioned cables. The cables are constrained in cavities inside the segments and can move within them, while the segments can roll one another along the pitch surfaces, thus straining the cables to an amount dictated by the shape of the contact profiles, and affecting the energy landscape of the assembly. A range of possible field applications, yet to be fully explored, is here presented.

Introduction

The term *tensegrity* was coined by Buckminster Füller [1] to indicate structures whose integrity is granted by tensioned cables, prestressed by matching a few floating compression struts. A novel type of beam-like structures was proposed in [2]. They have been named *flexural tensegrity* to indicate that they are formed by a chain of segments whose integrity under flexure is provided by prestressed tendons anchored at the beam ends and pressing the segments together. The key point is that the contact surfaces of adjacent segments are shaped according to properly designed pitch profiles, such that the relative rotation of the segments opens up the joints. There is a substantial difference with prestressed reinforced concrete segmented beams, for which the contact joints are approximately flat surfaces, designed to remain tight (no opening) at least in the serviceability limit state. Here, on the contrary, the opening of the joints produces the straining of the tendons, free to move in properly shaped cavities inside the segments (unbonded cables), to an amount dictated by the shape of the contact surfaces. This affects the elastic energy of the system, thus characterizing the constitutive bending properties, as a function of the shape of contact profiles.

The flexural-tensegrity concept can be explored in many forms. By changing the shape of the contact profiles, while the cables axially move inside tubular sheaths, either linear, or sub-linear or super-linear constitutive responses can be obtained [2]. As a function of the tendon stiffness, nonlinear Duffing-like vibrations are attained and can be controlled by varying the tendon force [3]. By enhancing the mobility of the tendon in large cavities, the bending energy can be made non-convex in type [4], possibly achieving complex snap-through sequential motions [5]. The limit, when the number of segments goes to infinity and their length to zero, corresponds to a particular type of Euler's elastica with nonlocal response [6], whose bent shape can be computed either analytically (with elliptic integrals), or numerically.

Here, we want to present a possible range of field applications, yet to be fully explored. The segmental beam can be used in archery, to manufacture a new type of bow [7], and in soft robotics, obtaining limbs controlled by internal/external cables. Multi-stable flexural tensegrities can be employed for propulsion in fluids in the form of flagellating tails [8], while larger scale applications in kinetic architecture, yet to be fully appreciated, can be found in the manufacturing of movable skeletons supporting envelopes for water collection and shielding.

Materials Research Proceedings **26** (2023) 97-102 https://doi.org/10.21741/9781644902431-16

The concept of flexural tensegrity

Flexural-tensegrity structures [2] consist of segments, supposed to be stiff, that are pierced and house tubular sheaths, or properly shaped larger cavities, through which an unbonded elastic cable can freely slide while keeping the segments together. Specifically, the contact surfaces are tailor-shaped, so that the segments can rotate one another along the design pitch profiles in pure rolling motion (no sliding). To illustrate, Figs. 1(a) and 1(b) report a theoretical scheme for a simply-supported flexural-tensegrity beam under transversal loads in the reference straight configuration and in the deformed state under applied loads, respectively. The cable (red solid line in Fig. 1) is tensioned and anchored only at the beam ends, where additional springs can be added in series to increase the cable compliance. A detail view of the segments in contact is shown in Fig. 1(c): the segments are pressed together and a certain stiffness against relative rotation is offered, similarly to a spring hinge, because their rotation causes the elongation of the cable. The elongation between segments i and $i+1$ in Fig. 1(c) is indicated as Λ_i , while the total elongation of the cable reads $\Lambda = \sum_{i=1}^{n-1} \Lambda_i$, being n the total number of segments in the chain. Note that, since the cable can freely slide within the segments, the increase of length due to the relative rotation of just one joint causes an increase of tensile force in the cable that is transmitted to *all* the other joints of the segmental beam. This renders the flexural response *nonlocal* in type.

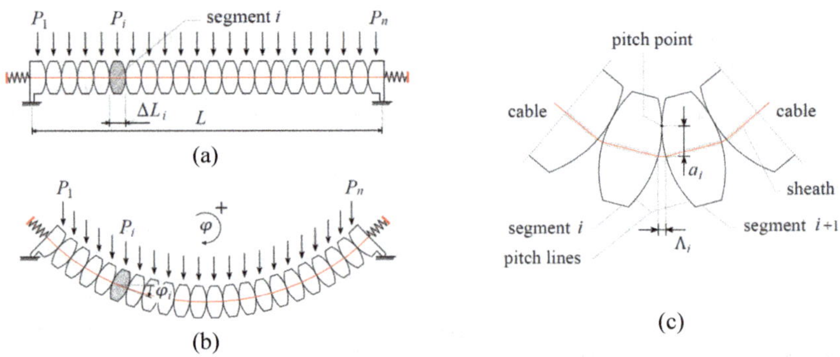

Fig. 1. Theoretical scheme for a simply-supported flexural-tensegrity beam under transversal loads: (a) reference straight configuration; (b) deformation under applied loads; (c) detail of the contact joint, with indication of the lever arm a_i and the local cable elongation Λ_i.

Another noteworthy parameter, reported in Fig. 1(c) is the lever arm a_i of the cable tensile force N with respect to the pitch point. Let N_0 denote the initial prestress in the cable and K its effective axial stiffness, accounting for the springs in series. Hence, the internal bending moment at joint i can be written as $M_i = Na_i = (N_0 + K\Lambda)a_i$. Note that both Λ_i and a_i depend upon the shape of the pitch profiles when the relative rotation of the segments is increased: the response of the joint can be tuned by simply changing the shape of contact profiles, kept fixed the cable properties.

The requirement that the segments are in pure rolling contact is fulfilled by means of the interlocking provided by conjugate profiles associated with the pitch profiles, for which various possibilities are reported in Fig. 2. Each segment is composed of three layers: the central one reproduces the pitch profile, while the two external ones are shaped according to the conjugate profiles. Each one of the external layers can be shaped according to smooth conjugate profiles, as per Fig. 2(a), or manufactured with multiple teeth similarly to spur gears, as in Fig. 2(b).

Under the hypothesis of stiff segments (rigid in the limit case), one can overlook their deformation energy, so that the only contribution remains that of the (compliant) tendon. In this case, with respect to the reference state under null external actions, the variation of the strain energy of the structure reads $\Delta U = N_0\Lambda + K\Lambda^2/2$, which depends upon the cable initial prestress N_0 and, specifically, the shape of contact profiles through the total cable elongation Λ. The Lagrangian variables for the bending problem can be set equal to the rotations φ_i of the segment i, $i = 1\ldots n$ (Fig. 1(b)). A variational approach, via Hamilton principle [9], can be used to find the set of nonlinear equations governing the static and dynamic response of the structural system.

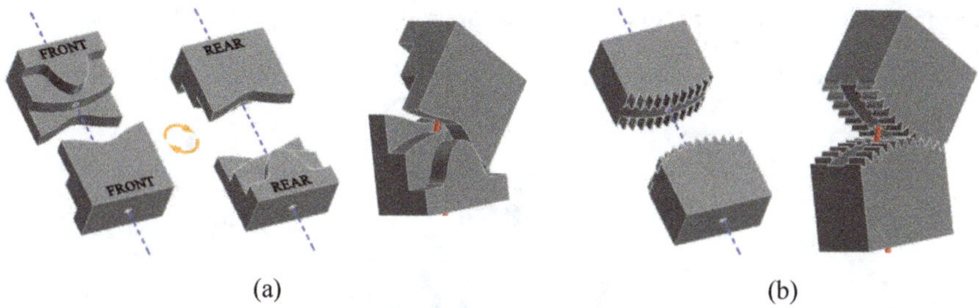

Fig. 2. Three-dimensional drawing of the joint. Exploded view and assembled joint in the rotated state for: example (a) with one pair of smooth conjugate profiles (external layers) and pitch profiles (central layer) and (b) with toothed profiles (external layers) and pitch profiles (central layer). Blue dashed lines indicate the cable axis; exposed portion of the cable is drawn in red in the rotated configurations.

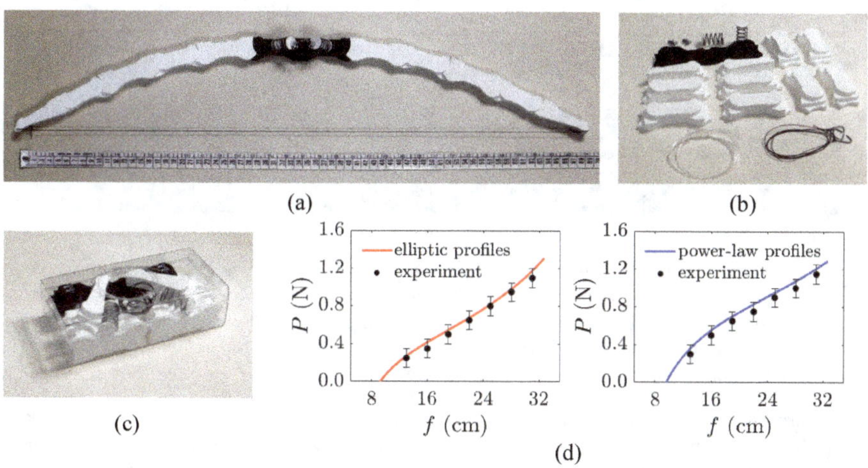

Fig. 3. Flexural-tensegrity bow: (a) photograph of the assembled bow; (b) components of the bow, including segments, bow string, prestressing tendons and additional springs; (c) the bow can be packed in a prismatic box for transportation. (d) Influence of the shape of pitch profiles on drawing force P as a function of drawing height f.

Examples

The beam-like segmental structure, with just one cable passing through the segments inside a tubular sheath, as per the schematic of Fig. 1, has been used to manufacture the elastic limbs of a new type of bow. The assembled bow is displayed in Fig. 3(a); the various components (segments, bow string, one prestressed tendon for each limb, and additional springs) are shown in Fig. 3(b). The segmental assembly permits to fold the bow by slackening the tendons; it can be totally disassembled and stored in a small box for easy transportation (Fig. 3(c)). The design is focused on the shape of contact profiles to attain different responses for the bow. Fig. 3(d) compares the static drawing of two bows in terms of drawing force P as a function of the drawing height f for two different shapes of the contact profiles; the theoretical description is corroborated by experimental dots in the chart. Note that power-law profiles of the type $y \propto x^4$ provide a response more concave than circular/elliptic profiles, for the same prestress $N_0 \approx 16.5$ N, so that a greater amount of energy is stored in the deformed limbs and can be later transmitted to the fired arrow.

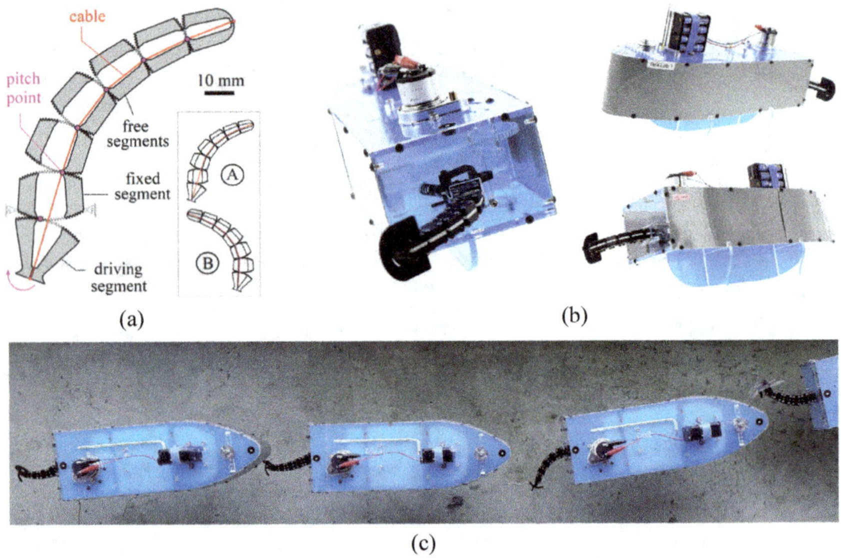

Fig. 4. Application to propulsion in fluids: flexural-tensegrity snapping flagellum. (a) Schematics of the structure with large cavities inside the segments where the cable can move; the snap occurs between configurations A and B. (b) Views of a toy boat propelled by the device, with evidence of the actuation via crank and crankshaft connected to a motor; the tail is equipped with a fin to enhance propulsion. (c) Toy boat sailing in a water tank.

The presence of large cavities inside the segments, instead of a tubular sheath, can provoke the snapping of a segmental cantilever, in response to a localized perturbation, represented by the relative rotation of the end segments close to the clamp restraint [5]. Hence, the cantilever can function as a tail that flagellates under a cyclic relative rotation of these segments, finding specific applications in propulsion devices for locomotion in fluids. Fig. 4(a) reports a schematic section of the cantilever segmental beam. Here, the internal cavities are represented by the non-hatched regions inside the segments. The cable (red solid line) can move inside these cavities: while rotating the driving segment, at a certain point the configuration becomes unstable and the beam reverses its shape, passing from state A to state B. By changing the sign of the driving rotation, the beam can snap back to the original state. Fig. 4(b) shows the application to a toy boat, where the actuation is obtained through a simple crank and crankshaft mechanism connected to an electric motor. In this example, the terminal segment is equipped with a fin, to enhance the propulsion [8]. Fig. 4(c) reports the final field experiment where the toy boat is sailing in a water tank.

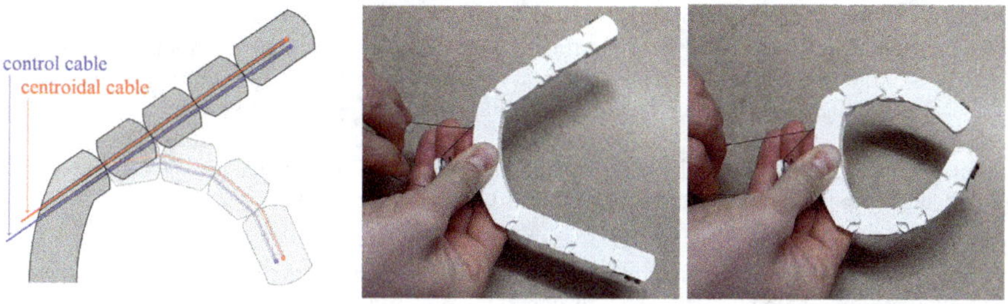

(a) (b)

Fig. 5. Applications to robotics. (a) Schematics of a cable-actuated limb: the centroidal tendon gives integrity to the assembly, while the eccentric one controls grasping when pulled. (b) Corresponding robotic picker manufactured according to this principle.

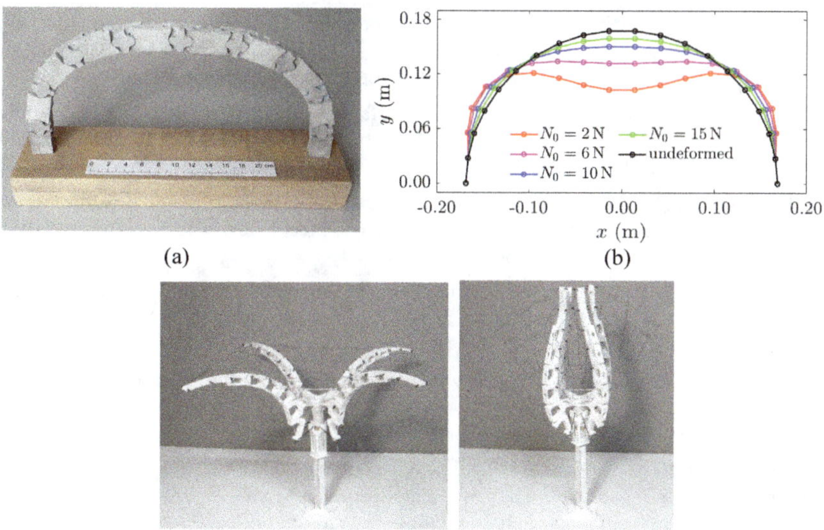

(a) (b)

(c)

Fig 6. Applications in architecture. Flexural-tensegrity arch: (a) photograph of a small-scale concrete arch and (b) corresponding mobility as the tendon prestress is varied. Flexural-tensegrity umbrella: (c) photograph of a small-scale polylactide structure in the open/closed configuration.

Using two cables a robotic segmental limb can be manufactured. As represented in Fig. 5(a), one tendon (red color) passes through the segments in a centroidal tubular sheath to provide integrity to the assembly, while another tubular sheath, eccentric with respect to the segment centroid, hosts the control cable (blue color): by pulling this second tendon, the limb folds.

Combining two of such limbs, a cable-actuated picker of the type shown in Fig. 5(b) can be obtained, which is capable of grasping and collecting also very soft objects.

Applications in kinetic architecture can be found in the manufacturing of movable skeletons supporting envelopes for water collection and shielding. In Fig. 6(a) a concrete flexural-tensegrity arch is proposed, with the mobility shown in Fig. 6(b) as the tension force in the cable is varied. With a functioning similar to that of the flagellating tail of Fig. 4(a), an umbrella can be designed that opens and closes like a flower, as shown in Fig. 6(c). Note that the structural bending stiffness of the segmental assembly can be tuned by pulling/releasing the prestressing tendons [3][6]; therefore, the force in the cable can be controlled to reduce structural vibrations. This can be very useful to manufacture seismic-resistant bracing systems, amenable to actively respond to earthquakes, especially where the seismic hazard is particularly high.

Conclusive remarks

A range of possible field applications has been presented within the class of flexural-tensegrity segmental structures, to show the versatility and the potentials of this recently proposed structural concept, yet to be fully explored and appreciated. Although the examples were limited to 2D planar mobility and small-scale structures, these represent a noteworthy rich scenario. The extension to full 3D mobility is the subject of current research.

References

[1] S.H. Juan, and J.M. Mirats Tur, Tensegrity frameworks: static analysis review, Mechanism and Machine Theory 43.7 (2008) 859-881. https://doi.org/10.1016/j.mechmachtheory.2007.06.010

[2] C. Boni, M. Silvestri, and G. Royer-Carfagni, Flexural tensegrity of segmental beams, Proceedings of the Royal Society A, 476.2237 (2020) 20200062. https://doi.org/10.1098/rspa.2020.0062

[3] C. Boni, and G. Royer-Carfagni, Nonlinear effects in the vibrations of flexural tensegrity beams, International Journal of Non-Linear Mechanics, 128 (2021) 103616. https://doi.org/10.1016/j.ijnonlinmec.2020.103616

[4] C. Boni, and G. Royer-Carfagni, Equilibrium of bi-stable flexural-tensegrity segmental beams, Journal of the Mechanics and Physics of Solids, 152 (2021) 104411. https://doi.org/10.1016/j.jmps.2021.104411

[5] C. Boni, and G. Royer-Carfagni, Energy harnessing in the snap-through motion of a flexural-tensegrity flagellum, Mechanism and Machine Theory, 173 (2022) 104845. https://doi.org/10.1016/j.mechmachtheory.2022.104845

[6] C. Boni, and G. Royer-Carfagni, A nonlocal elastica inspired by flexural tensegrity, International Journal of Engineering Science, 158 (2021) 103421. https://doi.org/10.1016/j.ijengsci.2020.103421

[7] C. Boni, and G. Royer-Carfagni, A new flexural-tensegrity bow, Mechanism and Machine Theory, 164 (2021) 104398. https://doi.org/10.1016/j.mechmachtheory.2021.104398

[8] C. Boni, P.M. Reis, and G. Royer-Carfagni, Flexural-tensegrity snapping tails for bio-inspired propulsion in fluids, submitted for publication (2022). https://doi.org/10.1016/j.eml.2022.101853

[9] R.W. Clough, and J. Penzien. Dynamics of structures, third ed., Computers and Structures, Berkeley, 2003.

Theoretical and Applied Mechanics - AIMETA 2022 Materials Research Forum LLC
Materials Research Proceedings **26** (2023) 103-108 https://doi.org/10.21741/9781644902431-17

Structural designs that required thinking

Federico Bosetti[1,a], Massimo Maffeis[1,b] and Gianni Royer-Carfagni [2,3,c*]

[1] Maffeis Engineering SpA, Solagna (Vi), Italy

[2] Department of Engineering and Architecture, University of Parma, Parma, Italy

[3] Construction Technologies Institute - Italian National Research Council (ITC-CNR), San Giuliano Milanese, Milano, Italy

[a] f.bosetti@maffeis.it, [b] m.maffeis@maffeis.it, [c] gianni.royer@unipr.it

Keywords: Structural Engineering, Genetic Optimization, Parametric Design

Abstract. We discussed the peculiarities of three structural projects. The *El Ferdan swing bridge* over the Suez Canal, the longest in the world, required the optimization of structural weight and deflection, pursed with genetic algorithms. The *Museum of the Future* in Dubai, is supported by a single-layer steel gridshell following a pseudo-toroidal double-curvature surface, which required the optimization of a great number of diagrid nodes, parametrically designed with a custom tool. The competition-winner project for the new *Stadio Milano* follows the vision of a Gothic Cathedral with slender buttresses, tall multi-storey arched frames, all covered by a first-class cable-supported glazed facade, which fulfils the clean "glass-box" architectural intent. Although very sophisticated calculation tools are available, these experiences suggest that the human contribution of the structural engineer is still, at least for now, of paramount importance.

Introduction

Advancements in numerical structural analysis have freed the engineer from the burden of calculations. Buildings and bridges of rare beauty, that thirty years ago could not have been conceived due to the difficulties in the calculations, are now reality. We are witnessing an actual Renaissance in structural engineering, to carry out the finest futuristic designs by imaginative architects. At the same time, we assist at the definition of increasingly complex technical regulations and standards: even the simplest structures (regular frames, truss bridges) shall comply with long verification formulas, where each uncertainty carries a coefficient, to be somehow calibrated. At the same time, the paradigm of artificial intelligence suggests that everything can be dealt with automatically, via modern calculation tools, making the human contribution of the structural engineer and the basic knowledge of structural mechanics superfluous. Does all this contribute to the Renaissance trend? On the one hand, the engineer is hindered in creative design, because the counterintuitive complexity of the rule, which can only be pursed with a numerical approach, prevents a synthetic vision: the structure is not governed, but governs! On the other hand, standards and codes are of little help, if not an obstacle, for complex or innovative designs, whereas the automatic decision making approach cannot fulfill all the requirements. Here we humbly present a few structures we know well enough from working on them, with the aim of highlighting the problems and the scientific advancements (modeling, structural optimization, customized parametric approach) they required, which could not be completely solved by standards and automatized numerical tools.

The swing bridges over the Suez Canal

The project for the two double swing bridges at El Ferdan, Ismailia over the Suez Canal [*en.wikipedia.org/wiki/El_Ferdan_Railway_Bridge*] consists in the retrofitting and upgrading of the existing bridge from one to double rail lane, and in the design of a new double-rail lane bridge, with geometry similar to the existing one. With a total length of 640 m, and a 340 m span between the central piers, this is the longest swing bridge in the world. Fig. 1(a) shows half portion of the new bridge in the rotated configuration. This is a metallic truss structure, assembled by welding on the bank of the canal as represented in Fig. 1(b). The design included the mechanical rotation and its locking system through electrical motors and hydraulic pistons. A detail of the cylindrical hinge with rollers that permit the rotation of the two halves of the bridge is indicated in Fig. 1(c).

For a bridge of this type, it is of paramount importance to minimize the weight, to facilitate the movement, as well as the deflection, both in the "open" and "closed" configurations, to fit the constraints. Derivative-free genetic optimization algorithms were implemented for minimum weight (objective 1) and minimum displacement (objective 2). These objectives are in contrast one other: to obtain the minimum weight the structure should be as thin as possible, but to reduce the displacement, the structure has to be strengthened.

(a)

(b) (c)

Fig. 1. The new swing bridges at El Ferdan over the Suez Canal. (a) View of half portion of the completed bridge in the rotated configuration. (b) The bridge during the construction phase on the bank of the canal. (c) Detail of the cylindrical hinge with rollers that permits the rotation.

Genetic algorithms are based on principles of natural genetics and selection. In derivate-free algorithms [1], to find the optimum it is not required to calculate the derivative of the *fitness function*, which is in general hard to define. The typical procedure is based on: a starting population of individuals (usually a "preset" configuration defined by experience is used as a starting point); each point is characterized by a set of variables, composing the pool of genes of each individual; objective functions are evaluated for each individual to obtain the fitness; a new generation is defined and includes the best individuals from the previous population and new individuals generated with reproduction, crossover and mutation techniques; objective function are evaluated for each new individual; new generations are defined up to convergence. For the case at hand, following [2] and [3], the workflow consisted in: 1) creation of FEM model based on the starting geometry; 2) random increment of area for each structural element to create a new individual; 3) structural analysis of this new individual from point 2; 4) determination of fitness functions for each individual; 5) use of the genetic algorithm to define the new individuals based on the elitism, mutuation and crossover between different genes; 6) structural analysis of this new individuals from point 5; 7) iteration from point 4 to 6 until the optimal solution is reached. The software chosen for the parametric structural analysis is Grasshopper associated with SAP2000.

Fig 2(a) describes, in the plane total weight vs. vertical displacement, the results of the first generation: the initial condition, corresponding to the configuration of Fig. 2(c) is highlighted as a red dot. The other points are the results of the structural analysis for random increments of the variables, represented by the area of each truss element. Observe in the graphs the *Pareto Front*, i.e., the locus of points of solution that dominate all the other solutions (*dominant solutions*).

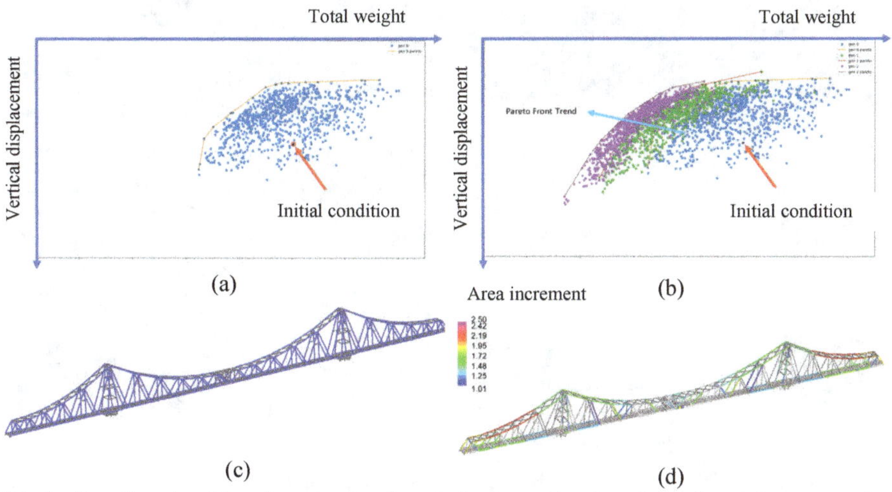

Fig.2. Genetic algorithm for structural optimization. Graphs of total weight vs. vertical displacement obtained with (a) the first generation of optimization and (b) after 12 simulations. Structural model for (c) the initial condition, highlighted with a red dot in the graphs (a), and (d) final optimal configuration at the end of the process, with indication of the area increments.

The typical analysis for each population is made on 60 individuals, subjected to a set of constraints (structural verification according to Egyptian codes, cost reliability-interaction [4]). Fig. 2(b) shows the distribution of fitness function for the successive generations, where each dot is associated with one individual. The first generation is represented by the blue dots on the right-hand-side of the graph. Following generations show that the genetic algorithm provides new individuals (genetic evolution of genes) with solutions better than the previous ones. The evolutive trend is clearly visible, because the Pareto front moves leftwards. The latest generation gives individuals placed on the best pareto front: part of them optimizes weight and the others minimize displacements and members ratio. The selected solution, after many simulations and verification, is represented in Fig. 2(d), with indication of the area increments with respect to the initial condition of Fig. 2(c). The results were systematically compared with hand-made optimization, in order to corroborate the results obtained with the complex genetic algorithm.

The Museum of the Future in Dubai

The Museum of the Future [*en.wikipedia.org/wiki/Museum_of_the_Future*] in Dubai, UAE, shown in Fig. 3(a), is a multistorey building, with steel-RC composite floors and steel floor trusses supported by following the feature pseudo-toroidal double-curvature surface; additional lateral stiffness is provided by an internal RC core. The tubular truss of the shell is visible in photograph of Fig. 3(b), taken before the installation of the metallic cladding,

(a) (b) (c) (d)

Fig. 3. Museum of the Future in Dubai. (a) The completed building and (b) the tubular truss structure before the installation of the metallic cladding. (c) The model for the definition of the geometry of the nodes and (d) their assembly during the construction phase.

The steelwork connections and the construction method were designed by analyzing the staged erection sequence, to meet the specified construction tolerances and to allow for the effect of the lock-in forces into the connections and members design. This included the optimization of the diagrid nodes in order to minimize the use of internal stiffeners and simplify the steel-contractor manufacturing operations. Since all nodes are different one another (different geometries, pipes and internal actions), more than 800 FEM models had to be developed, by using a dedicated internal parametric tool developed in Grasshopper [5]. Fig. 3(c) indicates the model that defines the geometry of the nodes; their assembly during the construction phase is shown in Fig. 3(d).

The new Stadio Milano

The competition-winner design for the new Stadio Milano, shown in Fig. 4(a), was developed following the architectural vision [*www.nuovostadiomilano.com/it*] of a "Gothic Cathedral". This is materialized by slender regularly-spaced buttresses and tall multi-storey arched frames, covered by a glass envelope, as per Fig. 4(b). In the structural design the arrangement of the members intuitively follows the flow of the forces, as in gothic architecture. A modular repetition of the structural frames reduces both the construction costs and the erection time, improving quality.

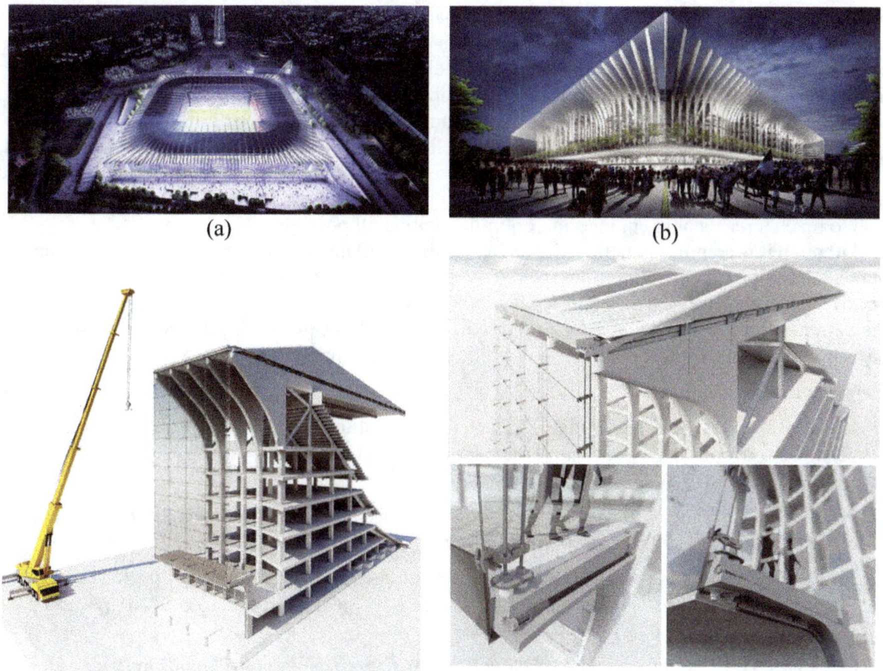

(c) (d)

Fig. 4. The new Stadio Milano. Rendering of (a) the completed building and (b) detail view of the structure with the glazed curtain wall. (c) Schematic section of the load-bearing structures and (d) details of the cable supported glass façade.

A section of the load bearing structure is shown in Fig. 4(c). A very innovative transparent façade is planned, where the glass panels are supported by a set of vertical stainless steel cables [6], tensioned between the upper buttresses at the roof level, and the lower RC cantilevers at the Galleria Mezzanine level. These synergically contribute to balance, at the roof level, the overturning moment due to the cantilever roof, spanning over the playfield, as well as the large cantilever at the corner of the Galleria Mezzanine (the upward forces from the cables reduce the long-term deflection of the concrete slab). Glass panels are heat-strengthened, stratified with high performance polymers, connected to the cables by custom-designed stainless steel clamped connectors, indicated in Fig. 4(d). Cable supported glazed surfaces have remarkable properties, including blast-resistance capacity [7]. The design will provide the Milan Stadium with a first-class, transparent and elegant glazed facade, to fulfill the clean "glass-box" architectural intent.

Conclusive remarks

The design process of the structures discussed here was quite complex, as it required the selection of geometry, static scheme and initial parameters, before proceeding with the structural analysis and verification of the suitability of members and connections. Modern calculation tools allow structural verifications to be carried out in a very short time on a large number of models, in which the geometry and sections can be varied until an optimal configuration is achieved. Although genetic algorithms based on parametric design tools apparently make it possible to automatically obtain the prefigured target according to the paradigm of the so-called Artificial Intelligence, nothing can replace, at least for now, the skill of a wise structural engineer. The designer has to make a rather difficult decision in selecting a preliminary, albeit rough, solution (preset), among all the possible alternatives, adequately satisfying all the requirements in terms of safety and serviceability imposed by design codes and a number of specified merit criteria such as cost and weight. Human, as opposed to artificial, intelligence is necessary to find an good preset, sufficiently closed to the ideal one, in order to direct the software operations and associated processing (crossover, migration, mutation), not only to reduce the required computation time, but also to reach a reasonable engineering optimum among all possible local optima.

Although it is commonly believed that the use of increasingly complex regulations may suggest universal design rules, these rules can only be used in the verification phase. Only a deep knowledge of structural theory and a wide experience on the really built constructions can allow the extreme synthesis of a complex structure in a simple but significant scheme, verifiable on the basis of elementary formulas, which will guide the preliminary, but necessary, basic choices.

References

[1] L.M. Rios, and N.V. Sahinidis, Derivative-free optimization: a review of algorithms and comparison of software implementation, J Glob Optm, 56.3 (2013) 1247-1293. https://doi.org/10.1007/s10898-012-9951-y

[2] C. Soh, and J. Yang, Optimal layout of bridge trusses by genetic algorithms, Comp Aided Civ Infract Eng, 13 (1998) 247-254. https://doi.org/10.1111/0885-9507.00103

[3] J. Park, Y. Chun, and J. Lee, Optimal design of an arch bridge with high performance steel for bridges using genetic algorithm, Int J Steel Struc, 16.2 (2016) 559-572. https://doi.org/10.1007/s13296-016-6024-y

[4] P.C. Das, D.M. Fragopol, and A.S. Novak (Eds.), Current and Future Trends in Bridge Design, Construction and Maintenance 2: Safety, Economy, Sustainability and Aesthetics , Thomas Telford, London, 2001.

[5] M. Maffeis, A. Biasi, F. Ceccato, E. Mazzarolo, and L. Michelini, One single model: A new parametric approach to megastructures, in Proceedings of IABSE Symposium, Nantes 2018: Tomorrow's Megastructures, 2018, pp. S13/35-S13/42. https://doi.org/10.2749/nantes.2018.s13-35

[6] J. Schlaich, H. Schober, and T. Moschner, Prestressed cable-net façade, Struct Eng Int, 15.1 (2005) 36-39. https://doi.org/10.2749/101686605777963332

[7] G. Royer-Carfagni, and L. Viviani, Basic design of cable-supported glazed surfaces under blast waves, Int J Non-Lin Mech, 123 (2020) 103489. https://doi.org/10.1016/j.ijnonlinmec.2020.103489

Theoretical and Applied Mechanics - AIMETA 2022 | Materials Research Forum LLC
Materials Research Proceedings **26** (2023) 109-114 | https://doi.org/10.21741/9781644902431-18

Fractional viscoelastic characterization of laminated glass

Luca Viviani[1, a *], Mario Di Paola[2, b] and Gianni Royer-Carfagni[1, c]

[1]Departement of Engineering and Architecture, University of Parma, Parma, Italy

[2]Departement of Engineering, University of Palermo, Palermo, Italy

[a]luca.viviani@unipr.it, [b]mario.dipaola@unipa.it, [c]gianni.royer@unipr.it

Keywords: Laminated Glass, Fractional Viscoelastic Modelling, Blast Loading

Abstract. Glass façades are often required to withstand against explosive events due to premeditated or accidental causes. Laminates made by glass plies bonded by thin polymeric foils (laminated glass) need to be used to avoid catastrophic breakage. The paradigmatic case study considered is that of a rectangular three-layer laminate, made of two glass plies, modelled as Kirchhoff-Love plates, sandwiching a thin viscoelastic polymeric interlayer. Its time-dependent response under the action of a blast wave is described via fractional calculus operators, whose main advantage is that only two material constants are needed for an exhaustive characterization. The dynamic equations are treated à la Galërkin and their integration in time relies on the Grünwald-Letnikov approach. The fractional characterization presents noteworthy advantages from a computational point of view. We find that the maximum stress peak is mildly affected by the viscosity of the interlayer, which instead dictates the subsequent rebounding oscillations.

Viscoelastic properties of polymeric interlayer

Laminated glass is a composite formed by two (or more) glass plies sandwiching one (or more) polymeric foil(s), permanently bonded with a process at high temperature and pressure in autoclave. The polymeric interlayer is too thin to present bending capacity, but it contributes to the stiffness and strength of the laminate by coupling the glass plies [1, 2]. Its viscoelastic behavior is characterized by the relaxation function $R(t)$. Since the lamination process in autoclave can modify the mechanical properties of the interlayer, the relaxation curve should be measured directly on laminated glass samples. Shear relaxation tests are usually performed but, for practical reasons, the observation period shall be in an interval varying from a minimum of a few seconds to a maximum of a few weeks. Further details about the experimental procedures can be in found in [3] and [4]. The extrapolation to smaller/larger time scales can be done on the basis of Time–Temperature Superposition (TTS) principle, according to which a variation of the testing temperature is associated with a variation of the time scale for the viscosity effects [5]. For most commercial polymers, the typical shape of $R(t)$, in the bi-logarithmic stress–time plane, is as indicated in Figure 1: it presents two pseudo-linear branches, connected by a transition zone. Figure 1 Qualitative plot of a typical relaxation function $R(t)$. The first branch is fitted by a power law (straight line), which allows to analytically extrapolate the values at time scales of the same order of the characteristic load duration TdUnder the impulsive action of a blast wave, the branch of interest is certainly that on the left hand side of the graph. As represented in the figure, this branch can be analytically fitted by a power law (linear trend in the bi-log scale) of the form

$$R(t) = \frac{C_\alpha}{\Gamma(1-\alpha)} t^{-\alpha}, \tag{1}$$

where $\Gamma(\cdot)$ is the Euler's Gamma function, C_α[MPa s$^\alpha$] is a dimensional coefficient, and α the number associated with the slope of the line in the bi-log graph.

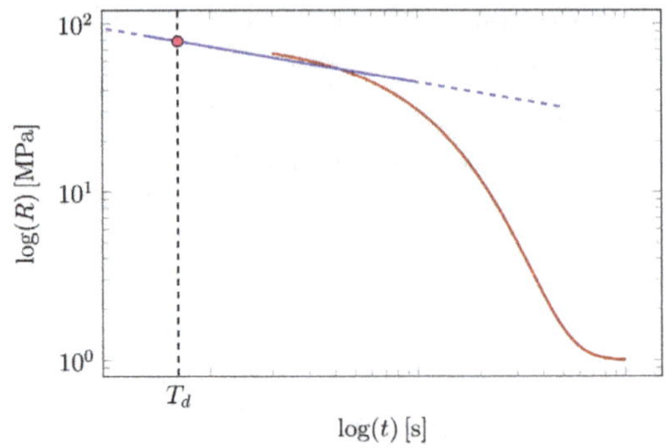

Figure 1 Qualitative plot of a typical relaxation function R(t). The first branch is fitted by a power law (straight line), which allows to analytically extrapolate the values at time scales of the same order of the characteristic load duration T_d of the blast wave.

The structural model for laminated plates

The schematic representation of the composite plate is shown in Figure 2. This is a three-layer plate composed by two thick elastic plies (layers "1" and "2") sandwiching one thin viscoelastic core (layer "0"). A reference frame is introduced: the (x, y) axes are located in-plane, while the z axis indicates the out-of-plane direction, with the mid-surface of each layer corresponding to $z = 0$. The external plies are subjected only to the bending contribution; while the core produces the shear coupling under the assumption that there is no sliding at the interfaces. The thickness of the external plies is denoted by h_i with $i = 1, 2$, so that the distance between the mid-planes reads $\bar{h} = h_0 + (h_1 + h_2)/2$, where h_0 represents the thickness of polymeric core. With reference to Figures 2(b) and 2(c), the interlayer "0" undergoes shear strains, which depend on the in-plane displacements $u_i = u_i(x, y, t)$ and $v_i = v_i(x, y, t)$ of the external layers, for $i = 1, 2$, as well as the out-of-plane displacement $w = w(x, y, t)$, which is the same for both layers. For a constitutive model based on fractional calculus, the shear forces per unit length $Q_{x,0}$ and $Q_{y,0}$ depend on the shear stresses $\tau_{xz,0} = \tau_{xz,0}(x, y, t)$ and $\tau_{yz,0} = \tau_{yz,0}(x, y, t)$, in the form

$$Q_{x,0} = Q_{x,0}(x, y, t) = \int_{-h_0/2}^{h_0/2} \tau_{xz,0}\, dz = C_\alpha {}_0^C\mathcal{D}_t^\alpha \left[u_2 - u_1 + \bar{h}\frac{\partial w}{\partial x}\right](t), \tag{2.1}$$

$$Q_{y,0} = Q_{y,0}(x, y, t) = \int_{-h_0/2}^{h_0/2} \tau_{yz,0}\, dz = C_\alpha {}_0^C\mathcal{D}_t^\alpha \left[v_2 - v_1 + \bar{h}\frac{\partial w}{\partial y}\right](t), \tag{2.2}$$

where ${}_0^C\mathcal{D}_t^\alpha[\dots]$ denotes Caputo's fractional derivative of order α, with $0 < \alpha < 1$, defined as

$${}_0^C\mathcal{D}_t^\alpha[f(\cdot)](t) = \frac{1}{\Gamma(1-\alpha)}\int_0^t (t-\bar{t})^{-\alpha}\, \dot{f}(\bar{t})\, d\bar{t}.$$

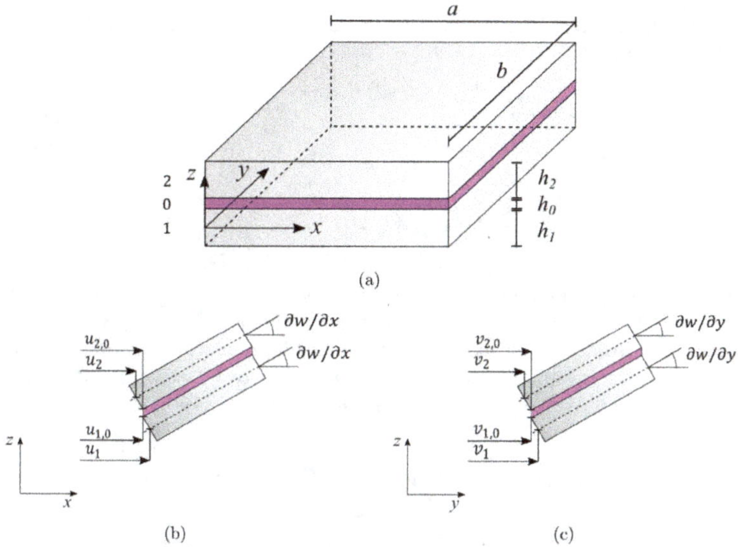

Figure 2 Schematic representation of a three-layer laminated glass plate. The glass plies are denoted as "1" and "2", while the core is the layer "0". For each layer, the reference system is located at one corner of the corresponding middle plane, with the z axis in the out-of-plane direction. The figure shows (a) the undistorted reference configuration and (b) the $\mathbf{x}-\mathbf{z}$ *view and the (c)* $\mathbf{y}-\mathbf{z}$ *view of the deformed configuration of the plate, with indication of the variables used to describe the displacement field.*

The glass plates are modelled according to Kirchhoff-Love plate theory, with Young's modulus $E_1 = E_2 \equiv E_g$ and Poisson's ratio $\nu_1 = \nu_2 \equiv \nu$, and with the same thickness ($h_2 = h_1 \equiv h$). Their dimensions are indicated as $a \times b \times h$. The inertia of the laminate is accounted by means of the global mass per unit area μ. Due to the geometric and load symmetry, $u_1 = -u_2$ and $v_1 = -v_2$. Hence, setting $\Delta u = u_2 - u_1$ and $\Delta v = v_2 - v_1$, the problem is described by the three dynamic equations [6]

$$\begin{aligned}\mu\frac{\partial^2 w(x,y,t)}{\partial t^2} + \frac{E_g h^3}{6(1-\nu^2)}\left(\frac{\partial^4 w(x,y,t)}{\partial x^4} + 2\frac{\partial^4 w(x,y,t)}{\partial x^2\,\partial y^2} + \frac{\partial^4 w(x,y,t)}{\partial y^4}\right) \\ = \frac{\bar{h}}{h_0} C_\alpha {}_0^C\mathcal{D}_t^\alpha \left[\frac{\partial \Delta u(x,y,\cdot)}{\partial x} + \bar{h}\frac{\partial^2 w(x,y,\cdot)}{\partial x^2}\right](t) \\ + \frac{\bar{h}}{h_0} C_\alpha {}_0^C\mathcal{D}_t^\alpha \left[\frac{\partial \Delta v(x,y,\cdot)}{\partial y} + \bar{h}\frac{\partial^2 w(x,y,\cdot)}{\partial y^2}\right](t) + p(t)\end{aligned} \tag{3.1}$$

$$\begin{aligned}\frac{E_g h}{1-\nu^2}\left[\frac{\partial^2 \Delta u(x,y,t)}{\partial x^2} + \frac{1}{2}(1-\nu)\frac{\partial^2 \Delta u(x,y,t)}{\partial y^2} + \frac{1}{2}(1+\nu)\frac{\partial^2 \Delta v(x,y,t)}{\partial x\,\partial y}\right] \\ = \frac{2}{h_0} C_\alpha {}_0^C\mathcal{D}_t^\alpha \left[\Delta u(x,y,\cdot) + \bar{h}\frac{\partial w(x,y,\cdot)}{\partial x}\right](t),\end{aligned} \tag{3.2}$$

Materials Research Forum LLC
https://doi.org/10.21741/9781644902431-18

$$\frac{E_g h}{1-\nu^2}\left[\frac{\partial^2 \Delta v(x,y,t)}{\partial y^2}+\frac{1}{2}(1-\nu)\frac{\partial^2 \Delta v(x,y,t)}{\partial x^2}+\frac{1}{2}(1+\nu)\frac{\partial^2 \Delta u(x,y,t)}{\partial x \partial y}\right] = \frac{2}{h_0} C_\alpha {}_0^C\mathcal{D}_t^\alpha \left[\Delta v(x,y,\cdot)+\bar{h}\frac{\partial w(x,y,\cdot)}{\partial y}\right](t), \tag{3.3}$$

where

$$p(t) = p_r\left(1-\frac{t}{T_d}\right)e^{-\beta\frac{t}{T_d}} \tag{4}$$

represents the loading action interpreted via Friedlander equation [7]. We set $p_r = 80$ kPa as peak overpressure, $T_d = 12.7 \cdot 10^{-3}$ s as time positive duration and $\beta = 0.95$ denotes the decay coefficient. The time history recalls that classified as EXV25, in accordance with the standard ISO 16933:2007 [8]. In order to solve the set of equilibrium equations for a simply supported plate, the unknown variables and the loading action are expressed in double Fourier sine series, in the form

$$w(x,y,t) = \sum_{m=1}^{M}\sum_{n=1}^{N} w_{mn}(t)\sin\left(\frac{m\pi x}{a}\right)\sin\left(\frac{n\pi y}{b}\right) \tag{5.1}$$

$$\Delta u(x,y,t) = \sum_{m=1}^{M}\sum_{n=1}^{N} \Delta u_{mn}(t)\cos\left(\frac{m\pi x}{a}\right)\sin\left(\frac{n\pi y}{b}\right) \tag{5.2}$$

$$\Delta v(x,y,t) = \sum_{m=1}^{M}\sum_{n=1}^{N} \Delta v_{mn}(t)\sin\left(\frac{m\pi x}{a}\right)\cos\left(\frac{n\pi y}{b}\right) \tag{5.3}$$

$$\Delta v(x,y,t) = \sum_{m=1}^{M}\sum_{n=1}^{N} \Delta v_{mn}(t)\sin\left(\frac{m\pi x}{a}\right)\cos\left(\frac{n\pi y}{b}\right) \tag{5.4}$$

This choice automatically fulfils the boundary conditions for a simply supported plate along the perimeter with in-plane free edges. By substituting them in the solving system (3.1)-(3.3), we can find the time dependent functions (modes $m-n$) $w_{mn}(t)$, $\Delta_{mn}(t)$ and $\Delta v_{mn}(t)$ with a step-by-step numerical integration, relying on the Grünwald-Letnikov approach [9] for fractional derivatives. The imposed initial conditions are $w_{mn}(t) = \Delta u_{mn}(t) = \Delta v_{mn}(t) = 0$ for $t < 0$.

Numerical experiments

The geometric and mechanical parameters for the considered laminated plate are $\boldsymbol{a \times b \times h_i} = \mathbf{1 \times 1 \times 0.01}$ m³, $\boldsymbol{i} = \mathbf{1, 2}$, $\boldsymbol{h_0} = \mathbf{2.28}$ mm, $\boldsymbol{E_g} = \mathbf{70}$ GPa, $\boldsymbol{\nu} = \mathbf{0.25}$, densities $\boldsymbol{\rho} = \mathbf{2500}$ kg/m³ and $\boldsymbol{\rho_0} = \mathbf{1000}$ kg/m³. The constitutive properties for three different types of interlayer, labelled as materials A B and C, according to the experimental campaign of [10] are: $\boldsymbol{\alpha} = \mathbf{0.155}$, $\boldsymbol{C_\alpha} = \mathbf{0.474}$ MPa s$^\alpha$ (material A); $\boldsymbol{\alpha} = \mathbf{0.117}$, $\boldsymbol{C_\alpha} = \mathbf{9.409}$ MPa s$^\alpha$ (material B); $\boldsymbol{\alpha} = \mathbf{0.117}$, $\boldsymbol{C_\alpha} = \mathbf{84.138}$ MPa s$^\alpha$ (material C). In order to understand how the hereditary memory of viscoelasticity contributes to the dynamic response, it is interesting to compare the bending stress $\boldsymbol{\sigma_{xx}}$ with the correspondent solution for perfectly elastic interlayers. The stress is evaluated at the plate center $(\boldsymbol{x}, \boldsymbol{y}) = (\boldsymbol{a/2}, \boldsymbol{b/2})$, on the external surface not directly invested by the pressure wave. The relaxation functions are assumed to be power laws and these are represented by inclined straight lines on bi-log plane; the equivalence with an elastic material (quasi-elastic approximation) is done by considering for it a shear modulus corresponding to the value of the relaxation curve at characteristic duration of the load $\boldsymbol{T_d}$. Graphically, the elastic behavior corresponds to a horizontal line ($\alpha = 0$), which intersects the relaxation function at $\boldsymbol{t} = \boldsymbol{T_d}$, where $\boldsymbol{T_d} = \mathbf{12.7 \cdot 10^{-3}}$ s is the time duration of the compression phase provided by Friedlander equation (4). Hence, the elastic limit corresponds to $\boldsymbol{R(t)} = \boldsymbol{C_0}$ which, for the cases at hand, corresponds to $\boldsymbol{C_0} = \mathbf{0.835}$ MPa for material A, $\boldsymbol{C_0} = \mathbf{14.484}$ MPa for material B, $\boldsymbol{C_0} = \mathbf{129.516}$ MPa for material C. The results to be compared are reported in Figure 3 for the three considered interlayers.

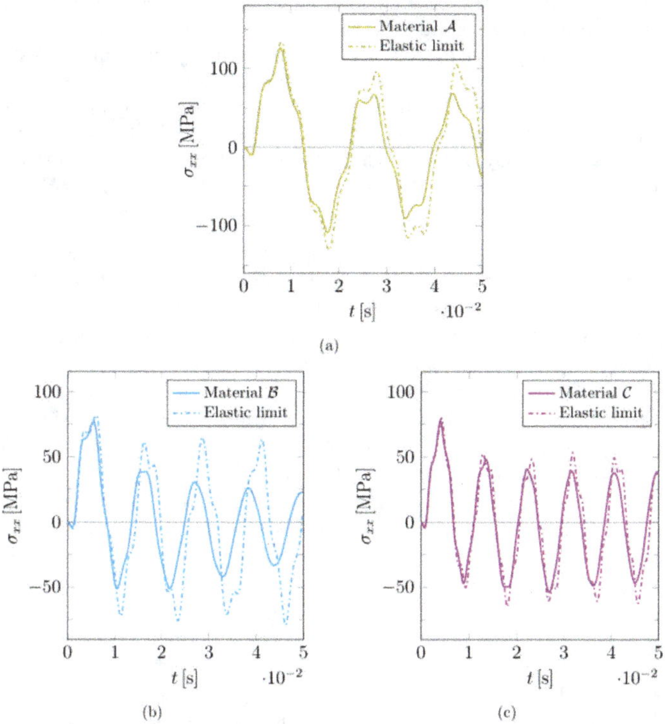

Figure 3 Stress component $\boldsymbol{\sigma_{xx}}$ evaluated at the centre of ply 2 at $\mathbf{z = -h/2}$. The plate is simply supported along its edges, and it is subjected to the blast pressure $\mathbf{p(t)}$. The corresponding solutions are compared with the quasi-elastic approximation. (a) Interlayer of type A; (b) interlayer B; (c) interlayer C.

In all the cases, there is a good superimposition between viscoelastic and elastic responses for what concerns the first peak. If we consider the response in a broader time interval, the difference is minimal when the interlayer is made of material C, the stiffest of all, inasmuch the monolithic limit (rigid shear coupling of the plies) is attained. Materials A and B provide a similar response: the dissipation due to the viscous component decreases the magnitude of the successive peaks and slightly increases the frequency of oscillations. With the interlayer is of type A, the sandwich is more compliant; hence, there are less oscillations within the considered time interval if compared with the other materials.

Conclusions

The case study is represented by a three-layer plate, where two elastic glass plies sandwich a thin polymeric foil without bending capacity, but sufficient to provide the shear coupling of the glass plies. The novelty of the presented analysis consists in the fractional constitutive model used for the interlayer. If compared with the more classic viscoelastic description via Prony series, the fractional approach is more straightforward and computationally advantageous, since only two parameters are needed to characterize the material response. The numerical experiments have accounted for three different polymeric interlayers, so to obtain a spectrum of dynamic responses in terms of stress. Under impulsive loading, the first (maximum) peak is not qualitatively affected by the viscosity of the interlayer, which however significantly lowers the values of the subsequent rebounding peaks. The greatest limitation of this model is represented by the assumption of linear elastic model à la Kirchhoff–Love for the glass plates, therefore neglecting the geometric non-linearities, which certainly play a role under high deformations. The extension to a non-linear model will represent the subject of future work.

References

[1] L. Galuppi G. Royer-Carfagni, "The post-breakage response of laminated heat-treated glass under in plane and out of plane loading," *Composites Part B: Engineering,* vol. 147, pp. 227-239, 2018. https://doi.org/10.1016/j.compositesb.2018.04.005

[2] I. V. Ivanov, "Analysis, modelling, and optimization of laminated glasses as plane beam," *International Journal of Solids and Structures,* vol. 43, pp. 6887-6907, 2'006. https://doi.org/10.1016/j.ijsolstr.2006.02.014

[3] M. López-Aenlle A. Noriega F. Pelayo, "Mechanical characterization of polyvinil butyral from static and modal tests on laminated glass beams," *Composites Part B: Engineering,* vol. 169, pp. 9-18, 2019. https://doi.org/10.1016/j.compositesb.2019.03.077

[4] X. Centelles F. Pelayo M. J. Lamela-Rey A. I. Fernandez R. Salgrado-Pizarro J. R. Castro L. F. Cabeza, "Viscoelastic characterization of seven laminated glass interlayer materials from static tests," *Construction and Building Materials,* vol. 279, p. 122503, 2021. https://doi.org/10.1016/j.conbuildmat.2021.122503

[5] M. L. Williams R. F. Landel J. D. Ferry, "The temperature dependence of relaxation mechanisms in amorphous polymers and other glass-forming liquids," *Journal of the American Chemical society,* vol. 77, pp. 3701-3707, 1955. https://doi.org/10.1021/ja01619a008

[6] L. Viviani M. Di Paola G. Royer-Carfagni, "A fractional viscoelastic model for laminated glass sandwich plates under blast actions," *International Journal of Mechanical Sciences,* vol. 222, p. 107204, 2022. https://doi.org/10.1016/j.ijmecsci.2022.107204

[7] P. S. Bulson, Explosive loading of engineering structures, CRC Press, 1997. https://doi.org/10.4324/9780203473863

[8] Iintenrational Organization of Standards (ISO 16933:2007), Glass in building—Explosion-resistant security glazing—Test and classification for arena air-blast loading, 2007.

[9] I. Podlubny, Fractional Differential Equations, Academic Press, 1998.

[10] L. Biolzi S. Cattaneo M. Orlando L. Ruggero Piscitelli P. Spinelli, "Constitutive relationships of different interlayer materials for laminated glass," *Composite Structures,* vol. 244, p. 112221, 2020. https://doi.org/10.1016/j.compstruct.2020.112221

Theoretical and Applied Mechanics - AIMETA 2022 | Materials Research Forum LLC
Materials Research Proceedings **26** (2023) 115-120 | https://doi.org/10.21741/9781644902431-19

A phase-field model for fracture in beams from asymptotic results in 2D elasticity

Giovanni Corsi[1, a *], Antonino Favata[1,b] and Stefano Vidoli[1,c]

[1]Dipartimento di Ingegneria Strutturale e Geotecnica, Sapienza Università di Roma, Rome, Italy

[a]giovanni.corsi@uniroma1.it, [b]antonino.favata@uniroma1.it, [c]stefano.vidoli@uniroma1.it

Keywords: Asymptotic Approach, Fracture Mechanics, Slender Structures

Abstract. We propose a derivation of a damage model in slender structures, focusing on the particular case of a rod. The peculiarity of the model is that it takes into account the changes in rigidity of the body, distinguishing between bending, traction and the possible mixed interactions between the two. The approach is based on a matched asymptotic expansion, taking the recent work of Baldelli et al [1] as starting point. Choosing the slenderness of the rod as small parameter for the asymptotic expansion, we determine the first order at which a correction occurs with respect to the Saint-Venant solution of the elastic problem, due to the presence of a crack. The results highlight that the presence of a defect affects in different ways the bending and traction rigidities of the rod, and that a coupling between the two deformation modes might occur, depending on the geometry of the crack. Moreover, the derivation allows to explicitly calculate the coefficients of this correction, for any given depth of the crack, by means of a simple numerical procedure. Application to the classic three-point bending problem is considered in order to highlight the predictive capabilities of the model. These results suggest ways in which state of the art phase-field models (e.g. [2]) for damage could be refined. This work goes in the direction of developing phase-field models suitable for application to slender structures, where the use of reduced dimensional models has proved promising [3].

Introduction

In modeling of fracture and damage of structures, a lot of attention has been devoted to the case of slender structures. A particular regime of loading is that when bending and membrane stresses interact in non-trivial ways. This happens in some important applications, such as thin shells that have bending and membrane deformation coupling due to geometric constraints (Gauss compatibility) [4, 5]. Such structures are characterized by their slenderness and give way to interesting applications taking advantage of their shape morphing features [6, 7]. The question then arises as to how to properly model damage for such cases. Currently in damage models, the state of the art of the classical phase-field approach is to introduce a unique damage parameter, which modulates the change of all stiffness constants that appear in the definition of the elastic energy. For example, considering a 1D beam:

$$E_e = \frac{1}{2}\int_{\Omega}(1-\alpha)^2\,(A_0\varepsilon^2 + D_0\chi^2), \tag{1}$$

where ε and χ are the stretching and curvature measures, A_0 and D_0 the relative stiffness constants, and α is a scalar field describing the damaging of the structure. In this work we consider the case when different effects on the coefficients are possible, as well as coupling effects between stretching and bending contributions:

$$E_e = \frac{1}{2}\int_{\Omega}(A(\alpha)\varepsilon^2 + D(\alpha)\chi^2 + 2C(\alpha)\varepsilon\chi). \tag{2}$$

Materials Research Proceedings **26** (2023) 115-120 https://doi.org/10.21741/9781644902431-19

For sake of simplicity, we will focus on the case when the crack starts at the middle of a beam, on one of the two lateral faces. The main features of this model are that the progressive damaging of the material could affect the stiffnesses differently, and also that the presence of a crack is expected to couple bending and stretching effects, since the reduction in thickness of the cross section causes the neutral axis of bending to be displaced. In order to do this, we consider a case of particular interest, that of a slender beam subject to traction force and bending load at the bases. In the domain of small displacements, the base solution of the 1D elastic problem is that of Saint-Venant. The presence (or the possible nucleation) of a crack causes a modification of the base solution, in that at the middle section the axial displacement and section rotation will show a jump discontinuity. The assumption of slenderness allows us to use an asymptotic approach, expanding the base solution in powers of a small parameter.

In the following, we will give an outline of the asymptotic procedure. The final result will be the calculation of the first order of correction (in the small parameter) of the solution, with respect to the base Saint-Venant solution. This correction coincides with the jumps in axial displacement and rotation due to the crack. These additional deformations can be modeled with an additional compliance, or, inversely, with a decrease in the stiffnesses. These factors will determine the crack-dependent coefficients of the energy, as in the example of Eq. 2, and allow us to formulate our damage model.

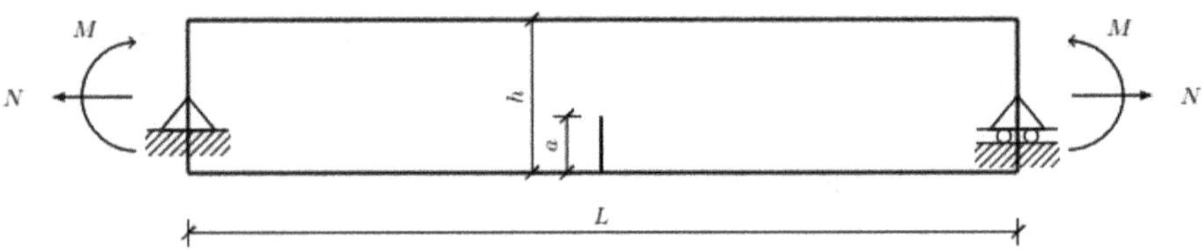

Figure 1. Schematic representation of the beam with a crack in the mid section, of the loading and other important geometric quantities.

Problem and non-dimensionalization

The solid domain is defined in the Cartesian space with coordinates x_1, x_2, x_3, and has large width, b, in the x_3 direction. In the other two directions, the domain of the beam occupies the rectangle $[-L/2, L/2] \times [-h/2, h/2]$, where $h \ll L \ll b$, and x_1 is in the direction of the axis, while x_2 is in the direction of the thickness. We introduce the dimensionless parameter $\eta = h/L$, and we will study the elasticity problem, in a small displacements setting, when the domain has a surface crack, with profile symmetric about the x_2 axis. The problem will be two dimensional, under the assumption of plane strain. The following non-dimensional coordinates are introduced: $\overline{x}_1 = x_1/L$ and $y = x_2/h$. We can then scale the displacement field

$$u(x) = Lu^{\eta}(\overline{x}_1, y), \tag{3}$$

the strains and stress tensors

$$E(x) = \eta E^{\eta}(\overline{x}_1, y),$$
$$S(x) = \eta Y S^{\eta}(\overline{x}_1, y), \tag{4}$$

and the traction and bending load along the beam

$$M_{\alpha\beta}(x) = \eta^3 bYL^2 M^{\eta}_{\alpha\beta}(\overline{x}_1, y),$$
$$N(x) = \eta^2 bYLN^{\eta}(\overline{x}_1, y), \tag{5}$$

noticing that the powers of η that appear in Eq. 5 are such that the traction and bending load scale differently with respect to the small parameter: their relative scaling is such that their contribution to the stress will be of the same order in η in the expansion, allowing for the possibility of interactions between bending and traction in the solution.

Outer and inner asymptotic expansion

We suppose that, far from the boundaries of the crack in the middle of the domain, the displacements, stresses and strains admit the following asymptotic expansion in the small parameter:

$$u^{\eta} = u^0(\overline{x}_1, y) + \eta u^1(\overline{x}_1, y) + \eta^2 u^2(\overline{x}_1, y) + \eta^3 u^3(\overline{x}_1, y) + \cdots,$$
$$S^{\eta} = \eta^{-2} S^{-2}(\overline{x}_1, y) + \eta^{-1} S^{-1}(\overline{x}_1, y) + S^0(\overline{x}_1, y) + \eta S^1(\overline{x}_1, y) + \cdots, \tag{6}$$
$$E^{\eta} = \eta^{-2} E^{-2}(\overline{x}_1, y) + \eta^{-1} E^{-1}(\overline{x}_1, y) + E^0(\overline{x}_1, y) + \eta E^1(\overline{x}_1, y) + \cdots.$$

This coincides with the power expansion of the Saint-Venant solution, and is not valid in the proximity of the crack region, since a boundary layer is present. Another asymptotic expansion has to be derived, after a rescaling of the coordinates. Then, once the two expansions have been determined, it will be enforced that the two coincide in a transition region, where both are supposed to be valid. The rescaling consists of a stretching of the domain in the axial direction, rewriting the elastic problem in the variable

$$x = \frac{\overline{x}_1}{\eta}, \tag{7}$$

and the inner expansion:

$$u^{\eta} = v^0(x, y) + \eta v^1(x, y) + \eta^2 v^2(x, y) + \eta^3 v^3(x, y) + \cdots,$$
$$S^{\eta} = \eta^{-2} \sigma^{-2}(x, y) + \eta^{-1} \sigma^{-1}(x, y) + \sigma^0(x, y) + \eta \sigma^1(x, y) + \cdots, \tag{8}$$
$$E^{\eta} = \eta^{-2} \gamma^{-2}(x, y) + \eta^{-1} \gamma^{-1}(x, y) + \gamma^0(x, y) + \eta \gamma^1(x, y) + \cdots.$$

Matching and energy of the new beam model

The matching of the two asymptotic expansions requires that these coincide when the outer solution is evaluated at $\overline{x}_1 = 0$, and the inner one for x tending to infinity. This yields three equations, choosing to stop at the first order at which a correction is obtained with respect to the Saint-Venant solution:

 Materials Research Forum LLC
https://doi.org/10.21741/9781644902431-19

$$\lim_{x\to\pm\infty} v^0(x,y) = u^0(0\pm,y),$$
$$\lim_{x\to\pm\infty} v^1(x,y) = u^1(0\pm,y) + \lim_{x\to\pm\infty} x u^0_{,1}(0\pm,y), \tag{9}$$
$$\lim_{x\to\pm\infty} v^2(x,y) = u^2(0\pm,y) + \lim_{x\to\pm\infty} \left(x u^1_{,1}(0\pm,y) + \tfrac{1}{2}x^2 u^0_{,11}(0\pm,y)\right),$$

and from the solution of these, the jumps of the solution (which is allowed to possibly be discontinuous at $\bar{x}_1 = 0$) in terms of the external loads are calculated

$$\left[\left[U^1_{,1}\right]\right] = C_t(1-\nu^2)N^0(0) + 12C_b(1-\nu^2)M^0_{11}(0),$$
$$[[U^2]] = K_{t2}(1-\nu^2)N^0(0) + 12K_{b2}(1-\nu^2)M^0_{11}(0), \tag{10}$$
$$[[W^2]] = K_{t1}(1-\nu^2)N^0(0) + 12K_{b1}(1-\nu^2)M^0_{11}(0).$$

Here, W and U are the axial and transversal components of the displacements at a certain order. The coefficients that relate these jumps and the external loads are the additional compliance of the beam due to the presence of a boundary layer, which in turn is due to the crack of the domain. To get back to a continuous model, we assume that the calculated jumps occur as a linear distribution in a small region of width $\bar{l}_c = l_c/L$. This allows us to calculate the derivatives of the approximated jumps, and thus the strains and the compliance coefficients. These are given by:

$$\begin{pmatrix} \bar{a} & \bar{c} \\ \bar{c} & \bar{d} \end{pmatrix} = (1-\nu^2)\begin{pmatrix} \left(1+\frac{\eta K_{t1}}{\bar{l}_c}\right) & 12\frac{\eta K_{b1}}{\bar{l}_c} \\ \frac{\eta C_t}{\bar{l}_c} & 12\left(1+\frac{\eta C_b}{\bar{l}_c}\right) \end{pmatrix} \tag{11}$$

The energy of the beam, in the interval $\left[-\bar{l}_c/2, \bar{l}_c/2\right]$, can then be written as:

$$P^{\eta,\bar{l}_c} = -\frac{1}{2}\int_{-\frac{\bar{l}_c}{2}}^{\frac{\bar{l}_c}{2}} \begin{pmatrix} \bar{a} & \bar{c} \\ \bar{c} & \bar{d} \end{pmatrix}\begin{pmatrix} N^\eta_e \\ M^\eta_e \end{pmatrix} \cdot \begin{pmatrix} N^\eta_e \\ M^\eta_e \end{pmatrix} \mathrm{d}\bar{x}_1. \tag{12}$$

Far from the crack, instead, the energy is unchanged with respect to the base solution.

Discussion and conclusions

Using an asymptotic expansion, we calculated the perturbation of the solution of an elastic beam due to a crack. This perturbation is determined by 4 scalar coefficients, representing the additional deformations with respect to the non-cracked beam, due to traction and bending, or to the interaction of the two. This latter effect will be present if the non-diagonal coefficient of the stiffness (or compliance) matrix is non-zero. These coefficients have to be calculated numerically, for any value of the non-dimensional crack depth in the interval (0,1). Once these are known, the problem is determined and one can solve it as a 1D damage model. The first thing to notice is that, from the numerical results, these coefficients are clearly different and have a different evolution when the crack propagates along the thickness of the structure. This confirms the starting idea that, in general, the effect of a crack on the bending and traction stiffnesses is different. Furthermore, the off-diagonal values of the stiffness (or compliance) matrix in general will be non-zero; that is, a coupling energy term between bending and traction will be generated by the presence of the crack. This opens up to interesting future directions of research: once the model is obtained, it can be applied to some cases of interest (e.g., the classic three-point bending problem), and compared with experimental data, if available. This initial work can then be extended to other cases where the interaction between traction and bending effects can be interesting, prominent among these, the study of thin shells [8-10], using asymptotic techniques which have been extensively studied [1, 11-14] in literature.

References

[1] Baldelli, A.L., Marigo, JJ. Pideri, C., Analysis of Boundary Layer Effects Due to Usual Boundary Conditions or Geometrical Defects in Elastic Plates Under Bending: An Improvement of the Love-Kirchhoff Model. J Elast 143, 31–84 (2021). https://doi.org/10.1007/s10659-020-09804-6

[2] Gerasimov, T., De Lorenzis, Laura, "On penalization in variational phase-field models of brittle fracture." Computer Methods in Applied Mechanics and Engineering 354 (2019): 990-102 6. https://doi.org/10.1016/j.cma.2019.05.038

[3] Brunetti M., Freddi F., Sacco E., "Layered Phase Field Approach to Shells". In: Carcate rra A., Paolone A., Graziani G. (eds) Proceedings of XXIV AIMETA Conference 2019. AIMETA 2019. Lecture Notes in Mechanical Engineering. Springer, Cham (2020). https://doi.org/10.1007/978-3-030-41057-5_36

[4] Brunetti, M., Vincenti, A., & Vidoli, S. (2016). A class of morphing shell structures satisfying clamped boundary conditions. *International Journal of Solids and Structures*, *82*, 47-55. https://doi.org/10.1016/j.ijsolstr.2015.12.017

[5] M. Brunetti, A. Favata, S. Vidoli (2020), Enhanced models for the nonlinear bending of planar rods: localization phenomena and multistability, *Proceedings of the Royal Society of London A*, **476**: 20200455. https://doi.org/10.1098/rspa.2020.0455

[6] Corsi, G., De Simone, A., Maurini, C., & Vidoli, S. (2019). A neutrally stable shell in a Stokes flow: a rotational Taylor's sheet. *Proceedings of the Royal Society A*, *475*(2227), 20190178. https://doi.org/10.1098/rspa.2019.0178

[7] Corsi, G. (2021). Asymptotic approach to a rotational Taylor swimming sheet. *Comptes Rendus. Mécanique*, *349*(1), 103-116. https://doi.org/10.5802/crmeca.75

[8] Kiendl, J., Ambati, M., De Lorenzis, L., Gomez, H., & Reali, A. (2016). Phase-field description of brittle fracture in plates and shells. *Computer Methods in Applied Mechanics and Engineering*, *312*, 374-394. https://doi.org/10.1016/j.cma.2016.09.011

[9] Proserpio, D., Ambati, M., De Lorenzis, L., & Kiendl, J. (2020). A framework for efficient isogeometric computations of phase-field brittle fracture in multipatch shell structures. *Computer Methods in Applied Mechanics and Engineering*, *372*, 113363. https://doi.org/10.1016/j.cma.2020.113363

[10] Proserpio, D., Ambati, M., De Lorenzis, L., & Kiendl, J. (2021). Phase-field simulation of ductile fracture in shell structures. *Computer Methods in Applied Mechanics and Engineering*, *385*, 114019. https://doi.org/10.1016/j.cma.2021.114019

[11] Bessoud, A. L., Krasucki, F., & Serpilli, M. (2011). Asymptotic analysis of shell-like inclusions with high rigidity. *Journal of Elasticity*, *103*(2), 153-172. https://doi.org/10.1007/s10659-010-9278-1

[12] Abdelmoula, R., & Marigo, J. J. (2000). The effective behavior of a fiber bridged crack. *Journal of the Mechanics and Physics of Solids*, *48*(11), 2419-2444. https://doi.org/10.1016/S0022-5096(00)00003-X

[13] Geymonat, G., Krasucki, F., Hendili, S., & Vidrascu, M. (2011). The matched asymptotic expansion for the computation of the effective behavior of an elastic structure with a thin layer of holes. *International Journal for Multiscale Computational Engineering*, *9*(5). https://doi.org/10.1615/IntJMultCompEng.2011002619

[14] Marigo, J. J., & Pideri, C. (2011). The effective behavior of elastic bodies containing microcracks or microholes localized on a surface. *International Journal of Damage Mechanics*, *20*(8), 1151-1177. https://doi.org/10.1177/1056789511406914

Theoretical and Applied Mechanics - AIMETA 2022 Materials Research Forum LLC
Materials Research Proceedings **26** (2023) 121-126 https://doi.org/10.21741/9781644902431-20

Static and free vibration analysis of anisotropic doubly-curved shells with general boundary conditions

TORNABENE Francesco[1,a], VISCOTI Matteo[1,b*] and DIMITRI Rossana[1,c]

[1]Department of Innovation Engineering, University of Salento, Lecce, Italy

[a]francesco.tornabene@unisalento.it, [b]matteo.viscoti@unisalento.it, [c]rossana.dimitri@unisalento.it

Keywords: Higher Order Theories, GDQ Method, General Boundary Conditions

Abstract. In the present work, a two-dimensional model based on a higher order Layer-Wise (LW) approach is presented for the static and dynamic analysis of doubly-curved anisotropic shell structures. The Equivalent Single Layer (ESL) methodology is also obtained as particular case of LW. Each lamina of the stacking sequence is modelled as an anisotropic continuum. The fundamental equations account for both surface and concentrated loads, as well as the effects of the Winkler-Pasternak foundation. Moreover, non-conventional boundary conditions are introduced, and the numerical solution is assessed from the Generalized Differential Quadrature (GDQ) method. The proposed formulation is validated with respect to refined three-dimensional simulations, pointing out its accuracy and computational efficiency.

Introduction

In many engineering applications, layered structures with complex shapes are very frequently adopted in many branches of engineering. In this context, novel design perspectives require more complicated models capable of providing accurate predictions in terms of structural response.

Among two-dimensional methodologies, the Layer-Wise (LW) formulation [1] seems to provide very accurate results with respect to three-dimensional solutions, accounting for the compatibility conditions at the interface between adjacent laminae. More specifically, the governing equations are solved directly within each lamina. On the other hand, when the Equivalent Single Layer (ESL) approach [2-3] is adopted, a reference surface is provided for the entire structure, and a higher order through-the-thickness expansion of the field variable is adopted taking into account a generalized formulation.

From literature, closed-form solutions can be derived only for a limited number of cases, such that numerical procedures like classical finite elements are more suitable to solve approximately more complicated cases. In this context, refined simulations can be very high computationally demanding, thus spectral collocation approaches like the Generalized Differential Quadrature (GDQ) are adopted [4] since they lead to very accurate results with a reduced number of Degrees of Freedom (DOFs).

In the present contribution, a generalized higher order two-dimensional formulation based on a LW approach is proposed to study the linear statics and dynamics of laminated shell structures featuring a double curvature, general lamination schemes, and enforced with unconventional external constrains [5]. Then, a unified higher order ESL theory is outlined as a particular case of the LW. A numerical solution of the fundamental equations is provided, taking into account the GDQ method. A doubly-curved shell structure is here selected as benchmark, characterized by a softcore lamination scheme, and unconventional boundary conditions. The results are compared to those ones obtained from a 3D Finite Element Method (FEM), pointing out the accuracy of the proposed formulation, and its computational efficiency. The present ESL and LW higher order formulations have been implemented in the DiQuMASPAB software [4], and all the material properties are obtained from its database.

Materials Research Forum LLC
https://doi.org/10.21741/9781644902431-20

Theoretical formulation

Let consider a laminated doubly-curved shell made of l laminae, described in a reference surface $\mathbf{r}(\alpha_1,\alpha_2)$ located in its middle thickness. In particular, a global coordinate system $O'\alpha_1,\alpha_2\zeta$ is introduced starting from the principal directions of $\mathbf{r}(\alpha_1,\alpha_2)$. Furthermore, a local coordinate system $O'\alpha_1,\alpha_2\zeta^{(k)}$ is assessed in each k-th layer of thickness h_k, with $k=1,...,l$. As a consequence, the position vector $\mathbf{R}^{(k)}(\alpha_1,\alpha_2,\zeta)$ of an arbitrary point of the shell can be described as [1]:

$$\mathbf{R}^{(k)}\left(\alpha_1,\alpha_2,\zeta^{(k)}\right)=\mathbf{r}\left(\alpha_1,\alpha_2\right)+\left(\frac{\zeta_{k+1}+\zeta_k}{2}+\frac{h_k}{2}z_k\right)\mathbf{n}\left(\alpha_1,\alpha_2\right) \tag{1}$$

where $(\alpha_1,\alpha_2)\in[\alpha_1^0,\alpha_1^1]\times[\alpha_2^0,\alpha_2^1]$ and $\zeta\in[\zeta_k,\zeta_{k+1}]$. In addition, ζ_k,ζ_{k+1} denote the locations of the intrados and the extrados of the lamina at issue in the global reference system, respectively, whereas the dimensionless local out-of-plane coordinate is defined as $z_k=2\zeta^{(k)}/h_k$. In other words, in Eq. (1) a midsurface is provided for each lamina, so that the global and the local out-of-plane coordinates ζ and $\zeta^{(k)}$ are related as:

$$d\zeta^{(k)}=d\zeta \tag{2}$$

On the other hand, the geometry of the structure can be described in the ESL framework in terms of the global thickness coordinate ζ, as follows [2]:

$$\mathbf{R}(\alpha_1,\alpha_2,\zeta)=\mathbf{r}(\alpha_1,\alpha_2)+\zeta\mathbf{n}(\alpha_1,\alpha_2) \tag{3}$$

Referring to the local geometric reference system $O'\alpha_1\alpha_2\zeta^{(k)}$, the three-dimensional displacement field vector $\mathbf{U}^{(k)}\left(\alpha_1,\alpha_2,\zeta^{(k)}\right)=\left[U_1^{(k)}\ U_2^{(k)}\ U_3^{(k)}\right]^T$ is described by means of generalized thickness functions $F_\tau^{\alpha_i(k)}$ for $i=1,...,3$ collected in the matrix $\mathbf{F}_\tau^{(k)}$ defined in each k-th layer for each $\tau=0,...,N+1$. Thus, $U_i^{(k)}$ is expressed in terms of the so-called generalized displacement field components $u_i^{(k\tau)}$, setting $i=1,2,3$:

$$\begin{bmatrix}U_1^{(k)}\\U_2^{(k)}\\U_3^{(k)}\end{bmatrix}=\sum_{\tau=0}^{N+1}\begin{bmatrix}F_\tau^{\alpha_1(k)}&0&0\\0&F_\tau^{\alpha_2(k)}&0\\0&0&F_\tau^{\alpha_3(k)}\end{bmatrix}\begin{bmatrix}u_1^{(k\tau)}\\u_2^{(k\tau)}\\u_3^{(k\tau)}\end{bmatrix}\leftrightarrow\mathbf{U}^{(k)}\left(\alpha_1,\alpha_2,\zeta^{(k)}\right)=\sum_{\tau=0}^{N+1}\mathbf{F}_\tau^{(k)}\mathbf{u}^{(k\tau)} \tag{4}$$

Based on the ESL approach, the relation $\mathbf{u}^{(k\tau)}=\mathbf{u}^{(\tau)}$ should be considered in Eq. (4). The constitutive equation considered in the problem is valid for generally anisotropic materials relating each component of the three-dimensional stress and strain vectors $\boldsymbol{\sigma}^{(k)}$ and $\boldsymbol{\varepsilon}^{(k)}$:

$$\boldsymbol{\sigma}^{(k)} = \overline{\mathbf{E}}^{(k)}\boldsymbol{\varepsilon}^{(k)} \leftrightarrow \begin{bmatrix} \sigma_1^{(k)} \\ \sigma_2^{(k)} \\ \tau_{12}^{(k)} \\ \tau_{13}^{(k)} \\ \tau_{23}^{(k)} \\ \sigma_3^{(k)} \end{bmatrix} = \begin{bmatrix} \overline{E}_{11}^{(k)} & \overline{E}_{12}^{(k)} & \overline{E}_{16}^{(k)} & \overline{E}_{14}^{(k)} & \overline{E}_{15}^{(k)} & \overline{E}_{13}^{(k)} \\ \overline{E}_{12}^{(k)} & \overline{E}_{22}^{(k)} & \overline{E}_{26}^{(k)} & \overline{E}_{24}^{(k)} & \overline{E}_{25}^{(k)} & \overline{E}_{23}^{(k)} \\ \overline{E}_{16}^{(k)} & \overline{E}_{26}^{(k)} & \overline{E}_{66}^{(k)} & \overline{E}_{46}^{(k)} & \overline{E}_{56}^{(k)} & \overline{E}_{36}^{(k)} \\ \overline{E}_{14}^{(k)} & \overline{E}_{24}^{(k)} & \overline{E}_{46}^{(k)} & \overline{E}_{44}^{(k)} & \overline{E}_{45}^{(k)} & \overline{E}_{34}^{(k)} \\ \overline{E}_{15}^{(k)} & \overline{E}_{25}^{(k)} & \overline{E}_{56}^{(k)} & \overline{E}_{45}^{(k)} & \overline{E}_{55}^{(k)} & \overline{E}_{35}^{(k)} \\ \overline{E}_{13}^{(k)} & \overline{E}_{23}^{(k)} & \overline{E}_{36}^{(k)} & \overline{E}_{34}^{(k)} & \overline{E}_{35}^{(k)} & \overline{E}_{33}^{(k)} \end{bmatrix} \begin{bmatrix} \varepsilon_1^{(k)} \\ \varepsilon_2^{(k)} \\ \gamma_{12}^{(k)} \\ \gamma_{13}^{(k)} \\ \gamma_{23}^{(k)} \\ \varepsilon_3^{(k)} \end{bmatrix} \quad \text{for} \quad k = 1,\ldots,l \tag{5}$$

The elastic stiffness matrix $\overline{\mathbf{E}}^{(k)} = \mathbf{T}^{(k)}\mathbf{E}^{(k)}\left(\mathbf{T}^{(k)}\right)^T$ of Eq. (5), referred to the geometric reference system $O'\alpha_1\alpha_2\zeta^{(k)}$, is derived from the rotation transformation $\mathbf{T}^{(k)}\left(\vartheta_k\right)$ of matrix $\mathbf{E}^{(k)}$, with coefficients $E_{ij}^{(k)}$ for $i,j = 1,\ldots,6$, associated to the reference system of the material [4]. The fundamental equations and the related boundary conditions are derived from the Hamiltonian Principle, accounting for the virtual variation of the elastic strain energy, kinetic energy and external work. Referring to an arbitrary τ-th kinematic expansion order, one gets:

$$\sum_{\eta=0}^{N+1} \mathbf{L}^{(k\tau\eta)}\mathbf{u}^{(k\eta)} - \sum_{\eta=0}^{N+1} \mathbf{M}^{(k\tau\eta)}\ddot{\mathbf{u}}^{(k\eta)} + \mathbf{q}^{(k\tau)} = \mathbf{0} \qquad \text{for} \quad \tau = 0,\ldots,N+1, \quad k = 1,\ldots,l \tag{6}$$

where $\mathbf{L}^{(k\tau\eta)}$ and $\mathbf{M}^{(k\tau\eta)}$ denote the fundamental and mass matrix, respectively. The symbol $\mathbf{q}^{(k\tau)} = \left[q_1^{(k\tau)}\ q_2^{(k\tau)}\ q_3^{(k\tau)}\right]^T$ accounts for the vector of generalized external loads, whose components are determined according to a static equivalence principle.

The higher order two-dimensional problem of Eq. (6) is solved with the GDQ method, starting from a discretization of the physical domain in $I_N \times I_M$ discrete points according to the Chebyshev-Gauss-Lobatto (CGL) distribution [4]. Referring to an arbitrary univariate function $f = f(x)$, the GDQ technique provides the following expression for the n-th order derivative evaluated at an arbitrary point x_i for $i = 1,\ldots,I_Q$:

$$f^{(n)}\left(x_i\right) = \left.\frac{\partial^n f(x)}{\partial x^n}\right|_{x=x_i} \cong \sum_{j=1}^{I_Q} \varsigma_{ij}^{(n)} f\left(x_j\right) \qquad i = 1, 2,\ldots, I_Q \tag{7}$$

where the weighting coefficients $\varsigma_{ij}^{(n)}$ are computed with a recursive procedure.

Applications and results

We now present some results from the statics and dynamics of a doubly-curved laminated panel with a softcore, made of generally anisotropic materials. A revolution hyperbolic hyperboloid is considered [4], whose reference surface can be described with principal coordinates α_1, α_2 according to the following relation:

$$\mathbf{r}(\alpha_1,\alpha_2) = a\cosh\alpha_1 \cos\alpha_2\, \mathbf{e}_1 - a\cosh\alpha_1 \sin\alpha_2\, \mathbf{e}_2 + c\sinh\alpha_1\, \mathbf{e}_3 \tag{8}$$

with $a = 2.00\,\text{m}$ and $c = 1.50\,\text{m}$. The lamination scheme consists of four different layers with general orientation $(30/70/70/45)$. The two external sheets of the structure are made of graphite-epoxy $\left(\rho^{(k)} = 1450\ \text{kg/m}^3\right)$, here modeled as orthotropic material with elastic moduli $E_1^{(k)} = 13.79\ \text{GPa}$, $E_2^{(k)} = E_3^{(k)} = 8.96\ \text{GPa}$, shear moduli $G_{12}^{(k)} = G_{13}^{(k)} = 7.10\ \text{GPa}$, $G_{23}^{(k)} = 6.21\ \text{GPa}$ and Poisson's ratios $\nu_{12}^{(k)} = \nu_{13}^{(k)} = 0.30$, $\nu_{23}^{(k)} = 0.49$. On the other hand, the central core is made of a triclinic material $\left(\rho^{(k)} = 7750\ \text{kg/m}^3\right)$, characterized by the following anisotropic stiffness matrix $\mathbf{E}^{(k)}$ [2]:

$$\mathbf{E}^{(k)} = \begin{bmatrix} E_{11}^{(k)} & E_{12}^{(k)} & E_{16}^{(k)} & E_{14}^{(k)} & E_{15}^{(k)} & E_{13}^{(k)} \\ E_{12}^{(k)} & E_{22}^{(k)} & E_{26}^{(k)} & E_{24}^{(k)} & E_{25}^{(k)} & E_{23}^{(k)} \\ E_{16}^{(k)} & E_{26}^{(k)} & E_{66}^{(k)} & E_{46}^{(k)} & E_{56}^{(k)} & E_{36}^{(k)} \\ E_{14}^{(k)} & E_{24}^{(k)} & E_{46}^{(k)} & E_{44}^{(k)} & E_{45}^{(k)} & E_{34}^{(k)} \\ E_{15}^{(k)} & E_{25}^{(k)} & E_{56}^{(k)} & E_{45}^{(k)} & E_{55}^{(k)} & E_{35}^{(k)} \\ E_{13}^{(k)} & E_{23}^{(k)} & E_{36}^{(k)} & E_{34}^{(k)} & E_{35}^{(k)} & E_{33}^{(k)} \end{bmatrix} = \begin{bmatrix} 98.84 & 53.92 & 0.03 & 1.05 & -0.1 & 50.78 \\ 53.92 & 99.19 & 0.03 & 0.55 & -0.18 & 50.87 \\ 0.03 & 0.03 & 22.55 & -0.04 & 0.25 & 0.02 \\ 1.05 & 0.55 & -0.04 & 21.1 & 0.07 & 1.03 \\ -0.1 & -0.18 & 0.25 & 0.07 & 21.14 & -0.18 \\ 50.78 & 50.87 & 0.02 & 1.03 & -0.18 & 87.23 \end{bmatrix} \text{GPa} \tag{9}$$

More specifically, the core is made of a lamina with triclinic-soft material, whose stiffness constants are equal to 1/1000 of those reported in Eq. (9), whereas the third layer follows exactly the triclinic material of Eq. (9). Unconventional boundary conditions have been enforced accounting for a Double-Weibull distribution of linear springs. For more details on the topic, the interested reader is referred to [5].

In Table 1 the first ten mode frequencies, calculated with different higher order ESL and LW theories, are compared to those ones resulting from a 3D FEM simulation with 20-node brick elements.

Table 1. *Free vibration analysis of a revolution hyperbolic hyperboloid laminated with generally anisotropic materials employing higher order theories with both the ESL and LW approaches.*

Mode f[Hz]						($\text{B}^{\text{K}}_{\text{sss}}$CFF)					
	3D FEM	**FSDT**	**TSDT**	**ED3**	**EDZ3**	**ED4**	**EDZ4**	**LD1**	**LD2**	**LD3**	**LD4**
DOFs	327402	5046	10092	10092	12615	12615	15138	20184	30276	40368	50460
1	11.53	19.21	18.38	18.37	17.26	14.79	14.76	12.17	11.42	11.37	11.39
2	14.96	28.02	26.70	26.69	24.60	20.24	20.16	15.30	15.06	15.09	15.24
3	21.42	32.11	31.05	31.07	29.77	26.62	26.57	22.40	21.17	21.13	21.13
4	28.53	43.04	41.27	41.31	39.21	35.50	35.45	29.76	28.74	28.75	28.90
5	31.71	52.75	50.13	50.16	46.65	38.95	38.81	32.65	31.50	31.42	31.49
6	33.12	54.80	52.27	52.35	48.62	40.82	40.73	33.80	32.96	32.95	33.18
7	40.15	64.36	61.41	61.38	57.92	51.33	51.18	41.83	40.08	40.08	40.15
8	43.33	66.00	63.11	63.20	59.75	52.09	51.95	44.51	43.01	42.96	42.98
9	44.38	69.84	66.35	66.44	62.07	53.68	53.56	45.51	43.93	43.98	43.99
10	47.04	80.76	75.26	75.23	67.94	57.89	57.77	48.89	46.85	46.86	46.93

Geometric Inputs: Revolution Hyperbolic Hyperboloid, $a = 2.00$ m, $c = 1.50$ m
$\alpha_1^0 = -1$, $\alpha_1^1 = 1$, $\alpha_2^0 = 0$ and $\alpha_2^1 = \pi/2$, $h_1 = h_4 = 0.01$ m, $h_2 = 0.10$ m, $h_3 = 0.03$ m

Lamination Scheme: 1st layer: graphite-epoxy, 2nd layer: triclinic, 3rd layer: triclinic-soft, 4th layer: graphite-epoxy

Computational Grid: CGL distribution with $I_N = I_M = 31$

When the Murakami's zigzag function [2] is adopted in Eq. (4) within an ESL framework, more accurate results are obtained. However, if LW simulations are performed, a perfect alignment between the 3D FEM-based predictions is outlined. Fig. 1 shows the first eight mode shapes of the structure, calculated by means of the LD4 theory, showing the three-dimensional capability of the proposed formulation.

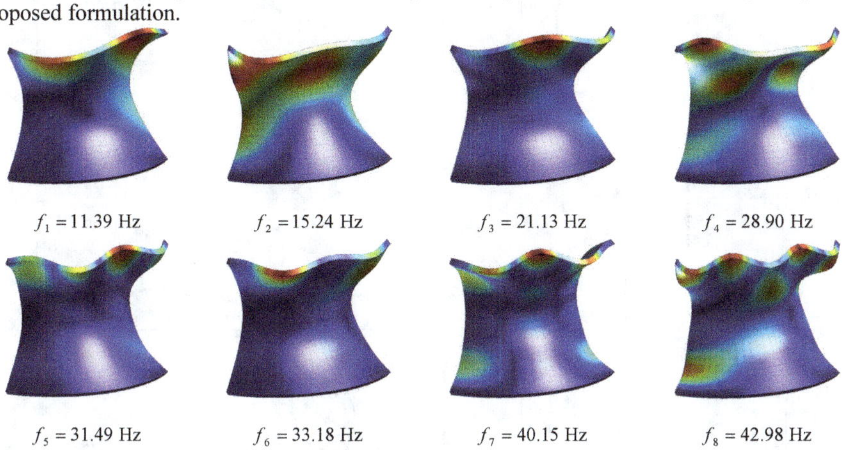

Figure 1. *First eight mode shapes of a laminated anisotropic revolution hyperbolic hyperboloid under general boundary conditions. The modal eigenvectors have been calculated employing the LD4 theory.*

The same structure has been investigated under a static load. In particular, a load $q_3^{(+)} = -2000$ N is applied on the structure, according to Ref. [5]. Taking into account a bivariate super elliptic distribution [5], the external load is distributed only in a limited area within the physical domain, setting $\alpha_{1m} = \alpha_{2m} = 0$ and $\delta_1 = \delta_2 = 0.53$ and $n = 1000$. The three-dimensional through-the-thickness stress distribution is depicted in Fig. 2, referring to the point of the physical domain located at $\left(0.25\left(\alpha_1^1 - \alpha_1^0\right), 0.75\left(\alpha_2^1 - \alpha_2^0\right)\right)$.

Classical approaches like FSDT and TSDT are not capable of predicting the three-dimensional finite element outcomes, as well as higher order ESL theories. The static response of the entire lamination scheme can be properly evaluated only with higher order LW theories for both in-plane and out-of-plane stress components. As can be seen from the three-dimensional solution, the abrupt change of stiffnesses between two adjacent layers leads to very complicated stress distributions, which requires a higher order LW approach among two-dimensional theories.

Conclusions

In the present work a generalized higher order two-dimensional theory has been presented for the static and modal analysis of shell structures made of generally anisotropic laminates. Following the LW approach, the fundamental equations are derived within each layer of the structure. As particular case, a unified ESL theory accounting for zigzag functions has been derived. The equations of motion have been discretized in a strong form via the GDQ method, together with the associated boundary conditions. The proposed methodology has been applied to a doubly-curved shell structure with a generally anisotropic lamination scheme and soft layers, showing the accuracy of the formulation, as well as its computational efficiency.

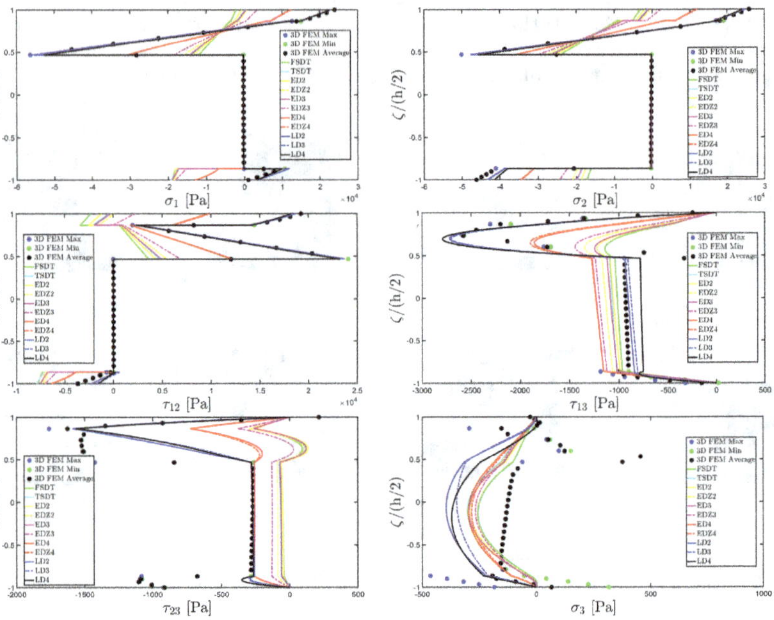

Figure 2. *Through-the-thickness distributions of the three-dimensional stress components calculated by means of various higher order ESL theories of a fully-clamped ellipsoid subjected to a uniform surface load applied at the top surface.*

References

[1] F. Tornabene, M. Viscoti, R. Dimitri, Generalized higher order layerwise theory for the dynamic study of anisotropic doubly-curved shells with a mapped geometry, Eng. Anal. Bound. Elem. 134 (2022) 147–183. https://doi.org/10.1016/j.enganabound.2021.09.017

[2] F. Tornabene, M. Viscoti, R. Dimitri, J.N. Reddy, Higher order theories for the vibration study of doubly-curved anisotropic shells with a variable thickness and isogeometric mapped geometry, Compos. Struct. 267 (2021) 113829. https://doi.org/10.1016/j.compstruct.2021.113829

[3] F. Tornabene, M. Viscoti, R. Dimitri, Equivalent single layer higher order theory based on a weak formulation for the dynamic analysis of anisotropic doubly-curved shells with arbitrary geometry and variable thickness, Thin-Walled Struct. 174 (2022) 109119. https://doi.org/10.1016/j.tws.2022.109119

[4] F. Tornabene F., and Bacciocchi M., Anisotropic Doubly-Curved Shells. Higher-Order Strong and Weak Formulations for Arbitrarily Shaped Shell Structures, Esculapio (Ed.), Bologna, 2018. https://doi.org/10.15651/978-88-938-5080-3

[5] F. Tornabene, M. Viscoti, R. Dimitri, Static analysis of anisotropic doubly-curved shells with arbitrary geometry and variable thickness resting on a Winkler-Pasternak support and subjected to general loads, Eng. Anal. Bound. Elem. 140 (2022) 618–673. https://doi.org/10.1016/j.enganabound.2022.02.021

Theoretical and Applied Mechanics - AIMETA 2022 Materials Research Forum LLC
Materials Research Proceedings **26** (2023) 127-132 https://doi.org/10.21741/9781644902431-21

R-Funicularity of shells and effective eccentricity: Influence of tensile strength

Gloria Rita Argento[1,a *], Stefano Gabriele[1,b] and Valerio Varano[1,c]

[1] Department of Architecture, Roma Tre University, Rome, Italy

[a]gloriarita.argento@uniroma3.it, [b]stefano.gabriele@uniroma3.it, [c]valerio.varano@uniroma3.it

Keywords: R-Funicularity, Generalized Eccentricity, Effective Eccentricity, Tensile Strength

Abstract. The concept of Relaxed Funicularity (R-Funicularity) has been introduced in [2] to deal with the difficulties of designing a shell with a funicular behavior. The R-Funicularity has been defined by using a parameter called generalized eccentricity: a shell is said to be R-Funicular if the generalized eccentricity belongs to an admissibility domain defined in terms of membrane forces N and bending moments M. Here we define an $N-M$ admissibility domain that includes the shell tensile limit force N_t, not previously taken into account. The proposed domain is well described by a new parameter, the effective eccentricity, and is used to control the tensile stresses occurring on shells designed to be R-Funicular and to work under compression.

Introduction

Shells with a funicular shape with respect to a given load are ideally able to bear it without introducing bending. They are an example of mechanical efficiency and subject of a fruitful research [1]. A funicular behavior is difficult to reach due to shell's bending stiffness, boundary conditions or multiple load cases. For this reason, an extension of the funicularity concept has been proposed defining the Relaxed Funicularity (R-Funicularity). A parameter called generalized eccentricity, i.e. the ratio between the generalized bending moments M and the generalized membrane forces N, has been used for this purpose: a shell is R-Funicular if the generalized eccentricity belongs to an admissibility domain [2]. The R-Funicularity has been used to study the modal and dynamic behavior of shells [3]. The velaroidal shells, example of funicular shells [4], have been studied by estimating their R-Funicularity [5,6]. Moreover, the R-Funicularity has been used to assess how the funicularity of timber gridshells is affected by the laths orientation [7]. A shells' shape optimization process aimed to minimize the generalized eccentricity extrema has been developed in [8]. A similar shape optimization process for 1D structures has been proposed in [9].

Here, we introduce an $N-M$ admissibility domain that includes the tensile limit force N_t related to the shell's material and thickness. The proposed domain is well described by the effective eccentricity, a new parameter that will be introduced by means of three main steps. First, the admissibility domain is shifted towards the right to include the tensile strength while showing that the generalized eccentricity is not a good parameter for spanning the new domain. Then, the effective eccentricity is introduced as a change of parameter to left shifting the ellipses. Finally, the mechanical meaning of the effective eccentricity is illustrated.

Generalized Eccentricity and R-Funicularity

Here, we briefly describe the concept of Relaxed Funicularity based on the generalized eccentricity; a complete discussion can be found in [2].

Let us consider a point P belonging to the surface S, endowed with its orthonormal basis $\boldsymbol{a}_1, \boldsymbol{a}_2, \boldsymbol{a}_3$, and a generic direction $\boldsymbol{u} = (\cos\theta, \sin\theta)$ (Fig. 1, Left). The membrane force $N(\theta)$ acting along $\boldsymbol{u}$ and the bending moment $M(\theta)$ acting in the plane $(\boldsymbol{u}, \boldsymbol{a}_3)$ are defined as follows:

$$N(\theta) = \boldsymbol{u}^T\boldsymbol{N}\boldsymbol{u},\ M(\theta) = \boldsymbol{u}^T\boldsymbol{M}\boldsymbol{u}, \tag{1a, b}$$

where $\boldsymbol{N}$ and $\boldsymbol{M}$ are the local tensors:

$$\boldsymbol{N} = \begin{pmatrix} N_{11} & N_{12} \\ N_{12} & N_{22} \end{pmatrix},\ \boldsymbol{M} = \begin{pmatrix} M_{11} & M_{12} \\ M_{12} & M_{22} \end{pmatrix}. \tag{2a, b}$$

The generalized eccentricity is defined as the following ratio:

$$e(\theta) = \frac{M(\theta)}{N(\theta)}. \tag{3}$$

Setting $\alpha = 2\theta$, the following explicit form for Eq. 1a, b has been obtained (Fig. 1, Right):

$$N(\alpha) = \overline{N} + \widehat{N}\cos\alpha + N_{12}\sin\alpha,\ \ M(\alpha) = \overline{M} + \widehat{M}\cos\alpha + M_{12}\sin\alpha, \tag{4a, b}$$

where $\overline{N} = \frac{N_{11}+N_{22}}{2}$, $\widehat{N} = \frac{N_{11}-N_{22}}{2}$, $\overline{M} = \frac{M_{11}+M_{22}}{2}$, $\widehat{M} = \frac{M_{11}-M_{22}}{2}$. Then, the generalized eccentricity is reparametrized:

$$e(\alpha) = \frac{M(\alpha)}{N(\alpha)}. \tag{5}$$

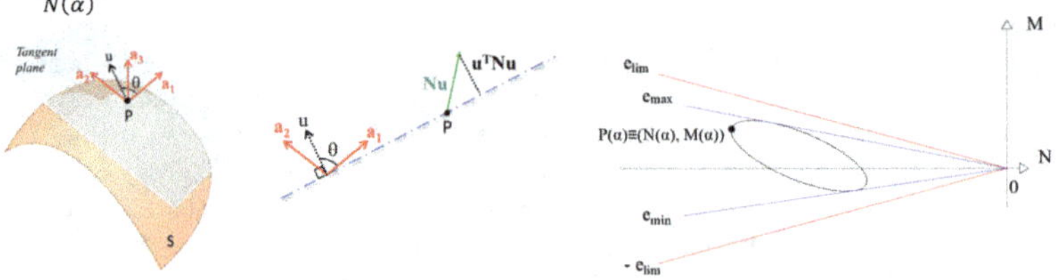

Figure 1 – Left: local basis and projection of internal forces [2]. Right: ellipse of eccentricity or RF ellipse. e_{max} and e_{min} are the slope of the upper and lower blue lines respectively; e_{lim} and $-e_{lim}$ are the slope of the red lines.

It is well known that a 2D structure is funicular if $M(\alpha) = 0\ \forall\ \alpha$ in each point of the surface. The actual behaviour exhibited by shells with non negligible thickness shows that the nullity of bending moments is an ideal that doesn't verify. The aim of R-Funicularity is to define a criterium to consider a structure funicular in a relaxed manner when $e(\alpha)$ belongs to an admissibility range $[-e_{lim}, e_{lim}]$, where $e_{lim} = \lambda h$, being h the shell thickness and $\lambda \in [0, 1/2]$ an admissibility coefficient. Thus, the Relaxed Funicularity (RF) is defined as follow: "a shell is R-Funicular if, $e(\alpha) \in [-e_{lim}, e_{lim}]\ \forall\ \alpha \in [0, 2\pi]$, for each point of the surface"(Fig. 1, Right).

In practice the problem of the R-Funicularity is that of verifying:

$$\left[min\left(e(\alpha)\right), max\left(e(\alpha)\right)\right] \subset [-e_{lim}, e_{lim}]. \tag{6}$$

Writing e as the Rayleigh quotient: $e(\theta) = \frac{\boldsymbol{u}^T\boldsymbol{M}\boldsymbol{u}}{\boldsymbol{u}^T\boldsymbol{N}\boldsymbol{u}}$, the maximum and minimum eccentricities (e_{max}, e_{min}) are found by solving the eigenvalue problem in Eq. 7.

Theoretical and Applied Mechanics - AIMETA 2022 Materials Research Forum LLC
Materials Research Proceedings **26** (2023) 127-132 https://doi.org/10.21741/9781644902431-21

$$(\boldsymbol{M} - e\boldsymbol{N})\,\boldsymbol{u} = 0. \quad (7)$$

The solution of Eq. 7 depends on the algebraic conditioning of ***M*** and ***N*** and has been discussed in [2]. The local extrema e_{max} and e_{min} of Eq. 7 have been calculated in the equation (17) of [2]. The global extrema of the eccentricities can be finally evaluated as follows:

$$min(e(\alpha)), max(e(\alpha)) = \begin{cases} -\infty, +\infty & if \det(\boldsymbol{N}) < 0 \\ e_{min}, e_{max} & if \det(\boldsymbol{N}) \geq 0 \end{cases}. \quad (8)$$

How to include tensile strength in the analysis
Let us consider a brittle material with a small tensile strength σ_t. This corresponds, in terms of the shell forces to the tensile limit force:

$$N_t = h\,\sigma_t. \quad (9)$$

A simple way to include tensile strength in the admissibility domain is to consider a Mohr-Coulomb like criterion in the plane N, M, by shifting the vertex of the cone from 0 to the positive value N_t. In this way the criterion becomes:

$$|M(\alpha)| \leq -\lambda h\,(N - N_t) = -\lambda h\,N + \lambda h N_t = -\lambda h\,N + M_t. \quad (10)$$

Where the flexural strength $M_t = \lambda h^2\,\sigma_t$ takes over the same role of the cohesion in the Mohr-Coulomb criterion (Fig. 2). The role of the quantity $(N - N_t)$, named effective normal force N_e, arises in Eq. 10.

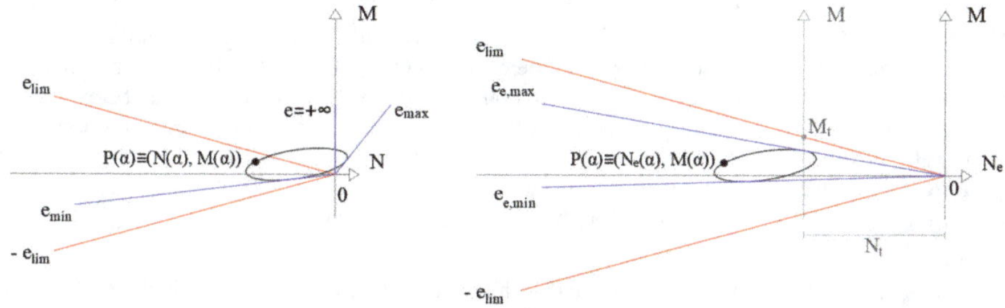

Figure 2 – Ellipse of eccentricities. The slopes of the blue lines and those of the red lines are, respectively, the eccentricity extrema and the eccentricity limits. Left: Eccentricity admissibility domain when no tensile strength is considered. Right: Eccentricity admissibility domain when the tensile limit force N_t is taken into account.

Effective Eccentricity as a change of parameter
In order to properly describe the admissibility domain, we perform a change of parameter, by plotting the ellipse of eccentricity in the plane $(N_e, M) \in \mathbb{R}^2$.

This is done by means of the following substitution:

$$\boldsymbol{N}_e = \boldsymbol{N} - N_t\boldsymbol{I}. \quad (11)$$

Consequently, the expression of the effective normal force in direction $\boldsymbol{u}$ can be calculated as:

$$N_e(\alpha) = \boldsymbol{u}^T\boldsymbol{N}_e\boldsymbol{u} = \boldsymbol{u}^T(\boldsymbol{N} - N_t\boldsymbol{I})\boldsymbol{u} = \boldsymbol{u}^T\boldsymbol{N}\boldsymbol{u} - N_t\boldsymbol{u}^T\boldsymbol{u} = N(\alpha) - N_t. \quad (12)$$

After this change of parameter the ellipse of eccentricity is shifted toward the left and the admissibility domain passes through the origin of the plane (N_e, M). In this way the lines starting from the origin span the admissibility domain. The slope of those that are tangent to the ellipse represent a new descriptor, named *effective eccentricity*:

$$e_e(\alpha) = \frac{M(\alpha)}{N_e(\alpha)} = \frac{M(\alpha)}{N(\alpha)-N_t} = \frac{N(\alpha)}{N_e(\alpha)} e(\alpha). \tag{13}$$

It is worth to notice that $|N_e| > |N|$, being $N < 0$ and $N_t > 0$; thus $e_e(\alpha)$ is proportional to $e(\alpha)$ with the ratio N/N_e and decreases as N_t increases.

Following the definition of the effective eccentricity $e_e(\alpha)$, then the problem of the R-Funicularity becomes to verify:

$$[min(e_e(\alpha)), max(e_e(\alpha))] \subset [-e_{lim}, e_{lim}]. \tag{14}$$

The mechanical meaning of the effective eccentricity is depicted in Fig. 3.

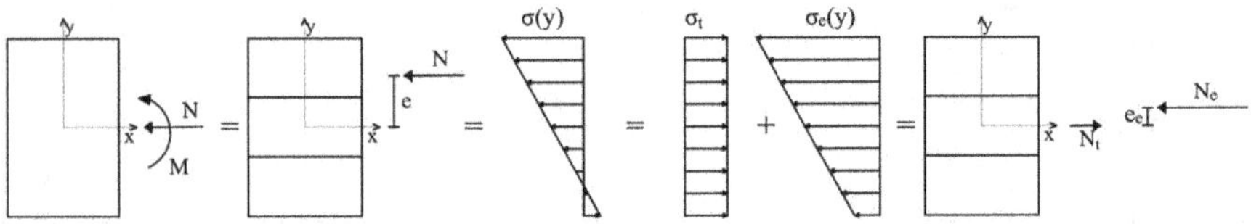

Figure 3 – Graphical representation of the mechanical meaning of the effective eccentricity.

With reference to Fig. 3, the left section is subject to N and M. These are substituted by the eccentric force N, located at distance e from the centre of the section. The stress distribution $\sigma(y)$ due to the eccentric force N goes from compression (top) to tension (bottom) and can be expressed as the sum of the uniform tensile stress distribution σ_t and a trapezoidal one of compression. This is equivalent to the stress state occurring on a section subject to N_t and N_e, the latter being an eccentric force located at distance e_e from the centre of the section. Thus, when the effective eccentricity e_e belongs to the middle-third, the tensile strength is not exceeded.

Example

The effective eccentricity has been evaluated with reference to a well-known funicular surface [4], the parabolic velaroidal surface, described by the following equation:

$$z = f(x, y) = c\left(\frac{x^2}{a^2} - 1\right)\left(\frac{y^2}{b^2} - 1\right), \tag{15}$$

where a and b are the measures of the spans and c is the rise of the surface (Fig. 4).

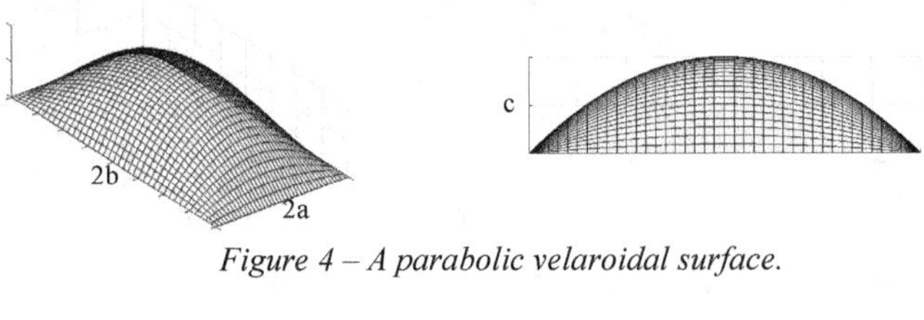

Figure 4 – A parabolic velaroidal surface.

Here we have set $a = b = 10\ m$, $c = 5\ m$; the shell material is concrete with a weight per unit volume $w = 25\ kN/m^3$ and a tensile strength $\sigma_t = 2\ MPa$, the shell thickness is $h = 0.1\ m$ and the admissibility coefficient $\lambda = 1/6$. A finite element model of the shell structure clamped along the entire perimeter has been built, a shell-thin element formulation has been used. Then, a linear static analysis of the shell subject to self-weight has been performed and the results have been used to evaluate the eccentricities and the effective eccentricities (Fig. 5).

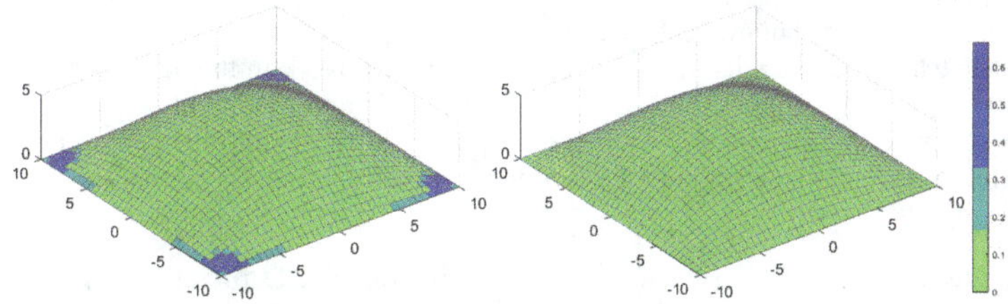

Figure 5 – Left: eccentricity distribution. Right: effective eccentricity distribution.

In Fig. 5, the colormap of the left graph represents the values of $e_{M,i} = max\ (|e_{max,i}|, |e_{min,i}|)$ that increases going from green to blue, the latter meaning that the eccentricity is $> h/2$, i.e. the resultant force N acts out of the cross section. The shell is R-Funicular in the light green areas. The effective eccentricities $e_{e_{M,i}} = max\ (|e_{e_{max,i}}|, |e_{e_{min,i}}|)$, are hence pictured with the colormap of the right graph, with the same color scale of the left one. In this case the effective eccentricity is everywhere included in the admissibility domain.

By exploiting the symmetry of the studied surface, the ellipse of eccentricities has been evaluated for the shell elements shown in Fig. 6.

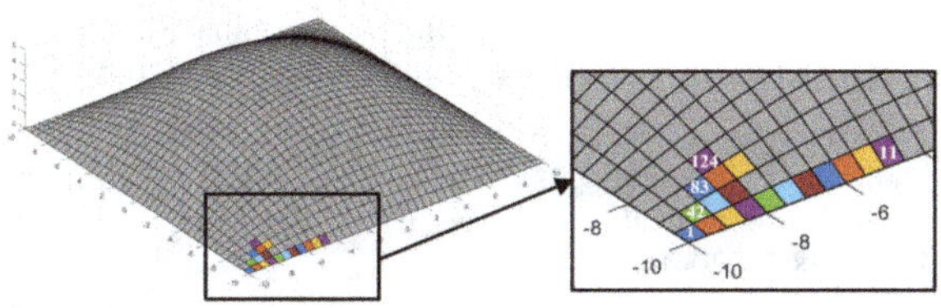

Figure 6 – Non R-Funicular shell elements that turn R-Funicular when the tensile strength is considered.

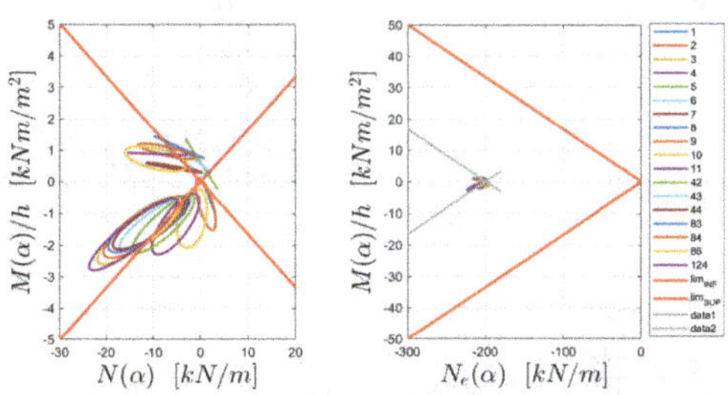

Figure 7 – Left: ellipse of eccentricity of the non R-Funicular shell elements. Right: the shell elements turn R-Funicular when the extended domain is considered.

As shown in Fig. 5 and in Fig. 7, the shell structure, originally non R-Funicular in the areas close to the boundaries, turns R-Funicular when the tensile strength is taken into account.

Conclusions

In this work has been developed the idea of how to formally extend the admissibility domain of shell's R-Funicularity by taking into account the tensile strength of the material. A new parameter called effective eccentricity has been introduced on purpose, in order to consider the left shifting of the R-Funicularity ellipses. The mechanical meaning of the effective eccentricity has been illustrated. The presented numerical example shown as a partially non R-Funicular shell turns fully R-Funicular when the extended domain is used. As a first work in this direction the authors are aware that taking into account the small tensile strength of shells, designed to work in compression, is a tricky task. For this reason, further development of the work will aim to account for the contribution of the prestress exploiting a formulation analogous to that here used.

References

[1] S. Adriaenssens, P. Block, D. Veenendaal, C. Williams, Shell structures for architecture: form finding and optimization, New York: Routledge, 2014. https://doi.org/10.4324/9781315849270

[2] S. Gabriele, V. Varano, G. Tomasello, D. Alfonsi, R-Funicularity of form found shell structures, Eng. Struct. 157 (2018) 157-169. https://doi.org/10.1016/j.engstruct.2017.12.014

[3] G. Tomasello, S. Adriaenssens, S. Gabriele, Dynamic behavior of form-found shell structures according to Modal and Dynamic Funicularity, Eng. Struct. 198 (2019) 109521. https://doi.org/10.1016/j.engstruct.2019.109521

[4] G. S. Ramaswamy, Design and construction of a new shell of double curvature, in: Proceedings of the Symposium on Shell Research, Delft, 1961, North Holland Publishing Co. Amsterdam, pp. 102-115.

[5] S. Gabriele, F. Marmo, V. Varano, About the funicularity of Velaroidal Shells, in: C. Lázaro, K.-U. Bletzinger, E. Oñate (Eds.), Proceedings of the IASS Annual Symposium 2019 - Structural Membranes 2019, Form and Force, Barcellona, Spain, 7-10 October 2019, 2019, pp. 45-52.

[6] F. Marmo , S. Gabriele, V. Varano, S. Adriaenssens, R-Funicularity of Analytical Shells, in: A. Carcaterra, A. Paolone, G. Graziani (Eds), Proceedings of XXIV AIMETA Conference 2019, Lecture Notes in Mechanical Engineering, Springer, 2020, pp. 947-957. https://doi.org/10.1007/978-3-030-41057-5_77

[7] M.L. Regalo, S. Gabriele, G. Salerno, V. Varano, Numerical methods for post-formed timber gridshells: Simulation of the forming process and assessment of R-Funicularity, Eng. Struct. 206 (2020) 110119. https://doi.org/10.1016/j.engstruct.2019.110119

[8] G.R. Argento, F. Marmo, V. Varano, S. Gabriele, Shells' shape optimization based on R-Funicularity, in: S. A. Behnejad, G. A. R. Parke, O. A. Samavati (Eds.), Inspiring the Next Generation - Proceedings of the International Conference on Spatial Structures 2020/21 (IASS2020/21-Surrey7), Guildford, UK, 23-27 August 2021, 2021, pp. 2548-2556.

[9] G. R. Argento, S. Gabriele, V. Varano, R-Funicularity for shells' shape optimization, in: C. Fivet, P. D'Acunto, M. Fernández Ruiz, P. O. Ohlbrock (Eds.), Proceedings of the International *fib* Symposium on the Conceptual Design of Structures, Attisholz Areal, Switzerland, 16-18 September 2021, 2021, pp. 503-510. https://doi.org/10.35789/fib.PROC.0055.2021.CDSymp.P062

Theoretical and Applied Mechanics - AIMETA 2022 Materials Research Forum LLC
Materials Research Proceedings **26** (2023) 133-138 https://doi.org/10.21741/9781644902431-22

Tensile behaviour of rayon cords in different conditions

Lucas Pires da Costa[1,a], Giorgio Novati[1,b], Paola Caracino[2,d] , Claudia Comi[1,c], Simone Agresti[2,e]

[1]Department of Civil and Environmental Engineering, Politecnico di Milano, Italy

[2]R&D Reinforcement Materials – Pirelli Tyre SpA, Italy

[a]lucas.piresdacosta@polimi.it, [b]giorgio.novati@polimi.it, [c]claudia.comi@polimi.it, [d]paola.caracino@pirelli.com, [e]simone.agresti@pirelli.com

Keywords: Rayon Twisted Yarns, Tensile Tests, Elasto-Viscoplastic Behaviour

Abstract. Different polymeric fibrous materials, such as polyester, nylon, aramid and rayon, can be employed as reinforcement in tyres or in other rubber composites. The present contribution focuses on the experimental characterization of rayon yarns and cords and makes some steps towards developing a constitutive model for cords, to be later integrated into a finite element procedure for the analysis of fibre-reinforced composites. Uniaxial tensile tests under monotonic and cyclic loading together with creep tests have been performed on rayon specimens in different conditions. A simple, preliminary viscoelastic-viscoplastic model is proposed which can describe the main experimental findings for untwisted yarns.

Introduction

Cord-reinforced rubber components, such as tyres, are complex systems and the development of cost-effective computational methods for their analysis and for efficient product development at industrial level is still a challenging goal, see e.g. [1,2,3]. The knowledge of the material properties of the individual constituents and the ability to numerically simulate their behaviour represent important intermediate objectives. For cords, such task is rather demanding in view of the nonlinear constitutive behaviour of the fibres combined with the geometrical nonlinearities linked to the twisted structure of the material. This work reports on research activities conducted on rayon cords within an industry-academia collaboration in which the authors are involved. An extensive experimental campaign has been conducted and the obtained results, which highlight an elastic-viscoplastic behaviour of the cords, are discussed. Attention is given to the irreversible aspect of such behaviour, related to the manifestation of inelastic permanent deformations, which is rarely taken into consideration in the literature concerning polymeric fibres (while it is more often considered in the case of polymer matrix materials, see e.g. [4]). A rheological model is also proposed for the simulation of the behaviour of untwisted yarns.

Experimental tests on rayon yarns and cords

Representative results of the experimental campaign carried out by the authors on the materials in point are reported in [5]. For this reason and also due to space limitations, only a selection of the obtained experimental results, complementary to those described in [5], are reported in the present work; in particular some creep tests that were carried out only recently are here discussed.

Materials. All tested materials were either rayon filament yarns or cords made from them. The single untwisted yarn, denoted by Y0, has a linear density of 1840 dtex (g/10000 m) and contains 1000 filaments. Using two different twist levels applied to the same filament bundle, yarns Y38 and Y48 are obtained, with 380 and 480 turns per meter (tpm), respectively. Two cords (also called multi-ply yarns) are considered: cord C2, produced by twisting two Y48 yarns together; and cord C3, obtained by twisting three Y38 yarns together. For both cords, the twist direction of the filaments in the single yarn and the ply twist direction are opposite, with the same tpm value. Note that only experimental results on greige materials are reported in the present work; some results for dipped cord can be found in [5].

Monotonic and non-monotonic tensile tests. A Zick-Roell testing machine with a 10 kN load cell was used to carry out the tests under displacement control, using different strain rates. The specimens were 500 mm long; bollard grips, suitable for filamentary materials, were employed. With reference to monotonic tests, we start showing some results that illustrate the influence of humidity conditions, see Fig. 1, and of strain rate, see Fig. 2. Note that "dry" conditions refer to samples that before being tested are kept in an oven at 100°C for at least two hours, while "humid" conditions apply to samples conditioned for 24 hours in the same climatic room where the test is run, at 20°C and at 65% of relative humidity. Each curve in Fig. 1 represents the mean response for ten experiments on nominally identical specimens. The adopted strain rate was 0.017 s^{-1}. It is apparent that rayon fibrous materials are very sensitive to moisture content: in dry conditions, the materials turn out to be generally stiffer and characterized by a higher maximum force and a lower maximum elongation than when humid. Note that yarn Y0 exhibits an almost bi-linear force-strain curve, interpretable as linear elastic in the first part and plastic with linear hardening in the second part. Figure 2 shows how the response of cords C2 and C3, in humid conditions, is affected by the adoption of different strain rates. Analogous experimental results for yarn Y0 are presented in Fig. 6 (b). As expected, both for the yarn and for the cords, the strength increases with the strain rate; the effect however is fairly limited, at least for the range of strain rates considered.

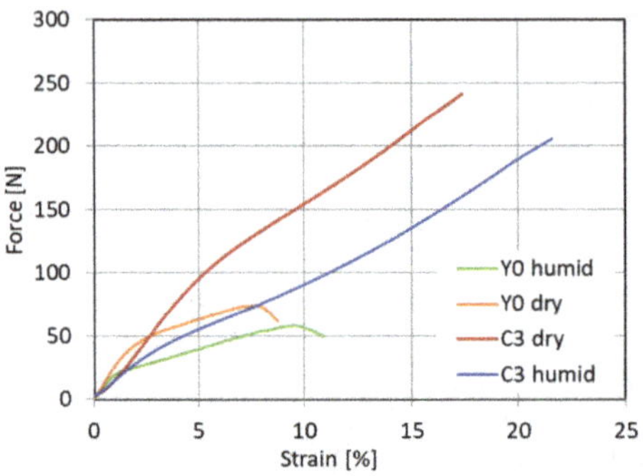

Fig. 1: Force-strain curves of uniaxial tensile tests in dry and humid conditions, for yarn Y0 and cord C3.

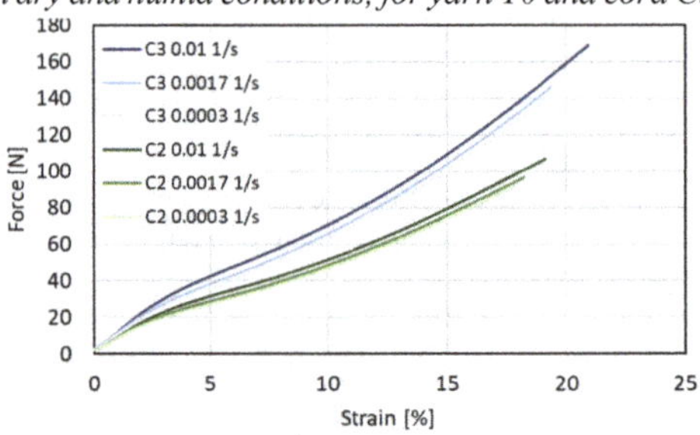

Fig. 2: Force-strain curves for monotonic tensile tests on cords C2 and C3.

Cyclic tests give important information on the inelastic behaviour of the materials under study. Such tests were performed under displacement control and using a constant strain rate ($\pm$ 0.003 s^{-1}), enforcing non-monotonic strain histories. Figure 3 shows the cyclic response of cords C2 and C3, under two different strain histories (in humid conditions). The two curves are plotted considering the average reactive force obtained from four tests on nominally identical samples. For cord C3, the same figure also depicts the curve obtained imposing an increasing strain (linear in time), using the same strain rate of the cyclic tests. One can note the presence of permanent deformation upon unloading and of hysteresis loops in the unloading-reloading phases. An analogous behaviour is observed for yarn Y0, see Fig. 7 (a). The above results indicate an elastic-viscoplastic behaviour of rayon fibres and cords.

Creep tests. Very recently the experimental campaign has been enriched with some creep tests that give important additional information on the time-dependent properties of the materials under

Materials Research Proceedings **26** (2023) 133-138 https://doi.org/10.21741/9781644902431-22

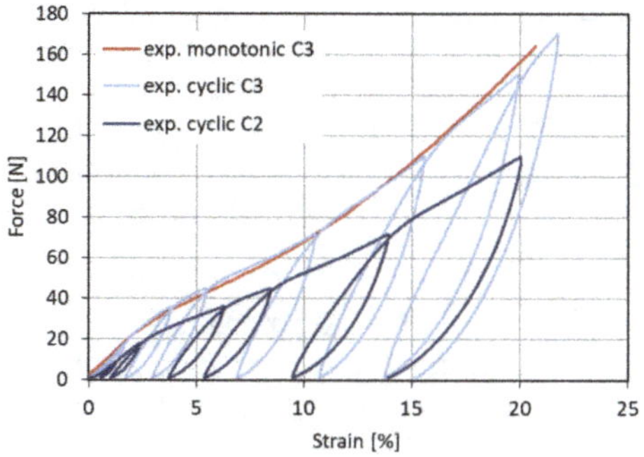

Fig. 3: Force-strain curves for cord C3 (cyclic and monotonic case) and for cord C2 (cyclic case only).

study. In all the creep tests, 200 mm long samples are used. Each test is carried out on two or three specimens and the reported result is the average obtained from the different specimens.

Figure 4 (a) shows the strain response to a creep test conducted on yarns Y0 and Y48 under a constant force of 10 N, holding this force for 24h (86400 s); the experiments are carried out at 20°C and at 65% of relative humidity. The tests were done by imposing increasing values of strain, for $0 < t < t_0$, up to the attainment of the given force; thereafter the force is kept constant. The strain at time t_0, denoted by ε_0, is called instantaneous strain. For the yarns Y0 and Y48 such strains (indicated by dotted horizontal lines in the graphs) turn out to be $\varepsilon_0^{Y0} = 0.7\%$ and $\varepsilon_0^{Y48} = 1.7\%$; note that the imposed load is such that the instantaneous response can be considered elastic. For both yarns, creep strain develops with an ever-decreasing strain rate. Denoting by ε_f^{Y0} and $\varepsilon_f^{Y48}$the final strains at the end of the experiment (i.e. after 24 h), the values obtained are $\varepsilon_f^{Y0} = 1.8\%$ and $\varepsilon_f^{Y48} = 4.5\%$; thus the ratios $\varepsilon_f^{Y0}/\varepsilon_0^{Y0} = 2.57$ and $\varepsilon_f^{Y48}/\varepsilon_0^{Y48} = 2.65$ turn out to be similar.

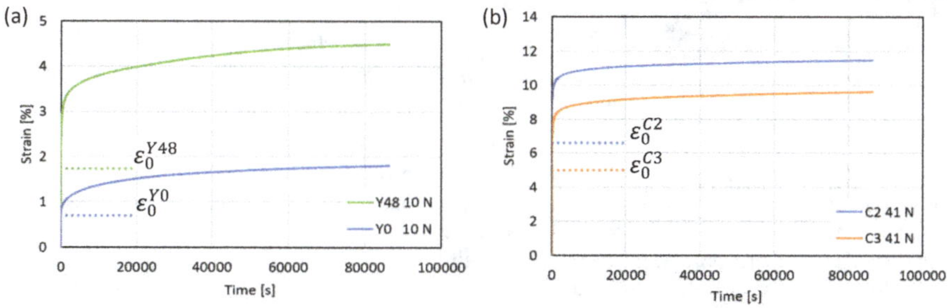

Fig. 4: Creep tests (a) for two yarns (Y0, Y48) and (b) for two cords (C2, C3) under different loads; the horizontal dotted lines mark the values of the so-called instantaneous strains.

Analogous results for creep test carried out for cords C2 and C3 are reported in Fig. 4 (b). The two cords were loaded by a (constant) force of 41 N (identical for C2 and C3); for both cords this load level is higher than the elastic limit (determined by monotonic tensile tests). Therefore, the instantaneous strain (denoted by ε_0) comprises elastic and inelastic contributions; it turns out that $\varepsilon_0^{C2} = 6.6\%$ and $\varepsilon_0^{C3} = 5\%$.

We also carried out creep tests at 100°C using the thermal chamber of Fig. 5 (a); note that these tests include load removal and a recovery phase. The results for cord C2 are shown in Fig. 5 (b) for two values of applied force (26 N, 83 N), held constant for 12600 s (3.5 h). The same figure shows also the creep response of cord C2 at 20°C and in humid condition (test conducted without the thermal chamber), under a load of 41 N. In Fig. 5 (b), the values of the "instantaneous" strains are marked by dotted horizontal segments, similarly to in Fig. 4.

Materials Research Proceedings **26** (2023) 133-138
https://doi.org/10.21741/9781644902431-22

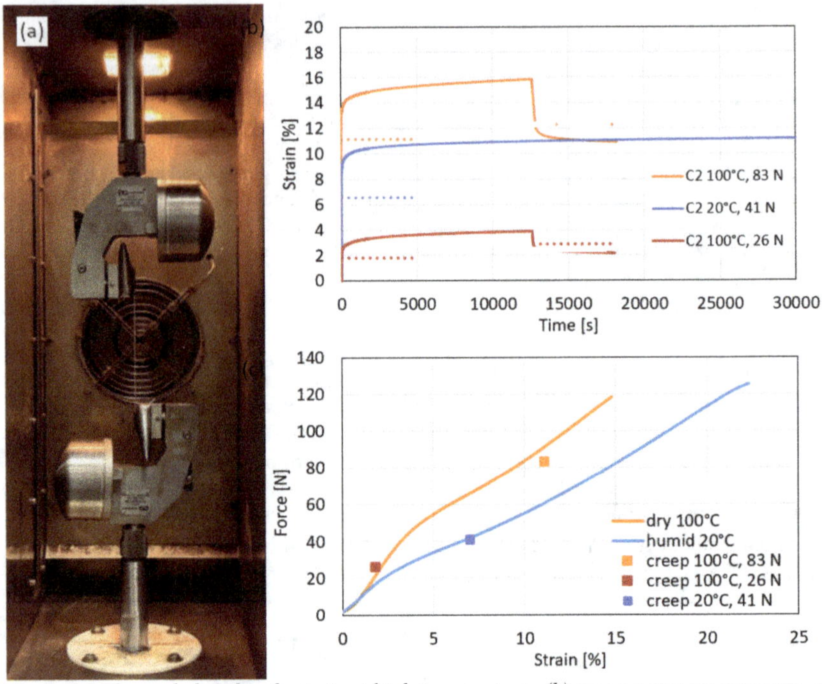

Fig. 5: (a) Thermal chamber for tests at high temperature; (b) creep-recovery response, at 100°C, for cord C2 and creep response for the same cord at 20°C (in humid conditions); (c) results from corresponding uniaxial tests on cord C2, performed at the same conditions.

Figure 5 (c) shows the result of the uniaxial tests on cord C2 performed at the same conditions employed in the creep tests. A square symbol is used to denote the force-strain values representing the "instantaneous response" at the beginning of the creep tests: in principle these square symbols should lie on the corresponding curves; the largest discrepancy can be noted for the creep instantaneous response at the higher load. One can note that among the three instantaneous strains shown in Fig. 5 (b-c), only the one corresponding to the lower load takes place in elastic regime.

Numerical simulation of the experimental behaviour of the untwisted yarn Y0

Viscoelastic-viscoplastic model. The one-dimensional rheological model depicted in Fig. 6 (a) consists of a linear spring (E) in series with a Kelvin unit ($\widetilde{\boldsymbol{E}}, \widetilde{\boldsymbol{\eta}}$) and with a viscoplastic unit comprising three elements in parallel: a frictional device ($\boldsymbol{\sigma_y}$), a hardening spring (H) and a viscous dashpot ($\boldsymbol{\eta}$). All elastic and viscous parameters are meant to be constant.

The total strain is hence the sum of three contributions: elastic, viscoelastic and viscoplastic:

$$\varepsilon = \varepsilon^{e} + \varepsilon^{ve} + \varepsilon^{vp} \tag{1}$$

Such 6-parameter model can represent the behaviour of a polymeric fibrous material, not accounting however for the geometric effects due to the twisted structure of yarns and cords; therefore, it is used only to simulate the behaviour of the untwisted yarn Y0.

By introducing a time discretization ($\Delta t = t_n - t_{n-1}$ denoting the time step) and assuming that (i) the state of the model is known at time t_{n-1} and that (ii) the strain ε_n at t_n is given (strain-driven process), it is possible to develop an integration algorithm over the time step for the updating of all the model variables to time t_n. The algorithm, based on Euler backward scheme, parallels the predictor/corrector scheme of classical rate-independent plasticity and makes use of "trial quantities" defined by "freezing" the viscoplastic strain ε^{vp} in the step. Thus, for any give strain history $\varepsilon(t)$, the response of the 6-parameter model can be obtained by an incremental method in time, with no iterations within each increment.

Note that for the above model, the creep response under a suddenly applied load $\bar{\sigma}$ can be obtained analytically and reads:

$$\varepsilon(t) = \bar{\sigma}/E + \bar{\sigma}/\tilde{E}\left(1 - exp\left(-\tilde{E}t/\tilde{\eta}\right)\right) + (\bar{\sigma} - \sigma_y)(1 - exp(-Ht/\eta))/H \qquad (2)$$

where the last term is present only for $\bar{\sigma} > \sigma_y$.

Comparison between experimental results and numerical simulations. The yarn Y0 experimental results in terms of force, for the monotonic and cyclic tests, have to be expressed in terms of stress in order to be compared with results obtained by the 6-parameter model using the same strain history. The force values are divided by a nominal cross-section area A of yarn Y0, setting $A = 0.17$ mm^2 according to a microscopic measurement reported in [5].

A parameter identification is conducted in order to obtain a best fit with the yarn Y0 experimental behaviour under different load conditions (monotonic, cyclic and creep), always at 20°C and 65% humidity. The chosen values for the model parameters are: E = 12 GPa, $\tilde{E}$ = 4.5 GPa, $\tilde{\eta} = 5 \cdot 10^4$ GPa·s, H = 3.2 GPa, η = 3.5 GPa·s, σ_y = 90 MPa. Figure 6 (b) shows the comparison between monotonic experimental results at different strain rates (continuous lines) and the numerical results obtained using the model (dotted lines). The latter is able to replicate the influence of different velocities, producing a reasonable agreement with the tests.

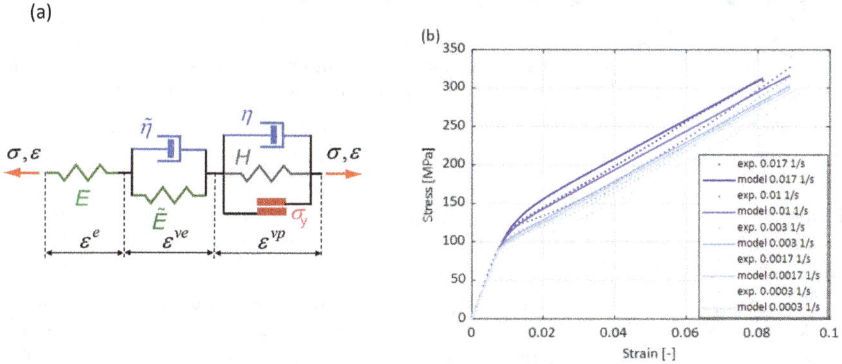

Fig. 6: (a) 6-parameter model; (b) comparison between experimental and numerical results for yarn Y0 under monotonic load at different strain rates (20°C, humid).

Figure 7 (a) compares the model and the experimental response to a given cyclic strain history (strain rate of 0.003 1/s) and shows fairly good agreement between the two curves. Figure 7 (b) shows the comparison between the creep test response in terms of strain and the model response based on Eq. 2, with the same set of parameters specified above.

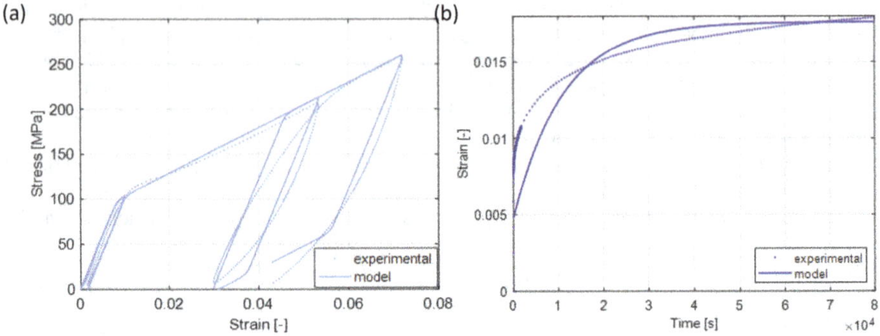

Fig. 7: Comparison between experimental data and results obtained with the 6-parameter model for yarn Y0 (at 20°C in humid conditions) for (a) cyclic load and (b) creep test.

Summary

In the present contribution, the elasto-viscoplastic behaviour of greige rayon yarns and cords was discussed on the basis of several experimental tests under monotonic and cyclic loading and under creep conditions. It was shown that the behaviour of untwisted rayon yarns can be fairly well represented by a 6-parameter viscoelastic-viscoplastic model. Current research is focusing on how to couple the fibre material behaviour and the geometrical effects due to the twisted structure of yarns and cords. With reference to such effects, it is worth noting that recent investigations using microtomography techniques [6] have led to a better understanding of fibre trajectories and of their evolution in twisted cords under tension.

References

[1] P. Helnwein, C. H. Liu, G. Meschke, H. A. Mang, A new 3-D finite element model for cord-reinforced rubber composites - application to analysis of automobile tires, Finite Elements in Analysis and Design 14 (1993) 1-16. https://doi.org/10.1016/0168-874X(93)90075-2

[2] M. G. Pastore Carbone, Investigating mechanical behavior of cord-rubber composites by multi-scale experimental and theoretical approach, PhD Thesis, Università degli Studi di Napoli Federico II, 2011.

[3] H. Donner, J. Ihlemann, On the efficient finite element modelling of cord-rubber composites, in: Constitutive Models for Rubber VIII – Proc. 8th European Conf. on Constitutive Models for Rubbers, ECCMR, 2013, pp. 149-155.

[4] R. K. Goldberg, G. D. Roberts, A. Gilat, Implementation of an associative flow rule including hydrostatic stress effects into the high strain rate deformation analysis of polymer matrix composites, J. of Aerosp. Eng., 18 (2005) 18-27. https://doi.org/10.1061/(ASCE)0893-1321(2005)18:1(18)

[5] P. Caracino, S. Agresti, L. Pires da Costa, G. Novati, C. Comi, The nonlinear behaviour of cords to be used in rayon-rubber composites, European Conference on Constitutive Models for Rubbers (ECCMR), Sept 7-9, 2022. https://doi.org/10.1201/9781003310266-77

[6] A. Sibellas, M. Rusinowicz, J. Adrien, D. Durville, E. Maire, The importance of a variable fibre packing density in modelling the tensile behaviour of single filament yarns, Journal of the Textile Institute, 112(5) (2021) 733-741. https://doi.org/10.1080/00405000.2020.1781347

Theoretical and Applied Mechanics - AIMETA 2022 Materials Research Forum LLC
Materials Research Proceedings **26** (2023) 139-144 https://doi.org/10.21741/9781644902431-23

Lower bound limit analysis through discontinuous finite elements and semi-analytical procedures

Zona Renato[1,a*], Esposito Luca[1,b], Ferla Paolo[1,c], Palladino Simone[1,d], Totaro Elena[1,e], and Vincenzo Minutolo[1,f]

[1]Department of Engineering, University of Campania "Luigi Vanvitelli", Aversa (CE), Italy

[a]renato.zona@unicampania.it, [b]esposito.luca@unicampania.it, [c]paolo.ferla@unicampania.it, [d]simone.palladino@unicampania.it, [e]elena.totaro@unicampania.it, [f]vincenzo.minutolo@unicampania.it

Keywords: Limit Analysis, Discontinuous Finite Element, Masonry Domes, Dislocations, Self-Equilibrated Stress

Abstract. This work deals with the limit analysis of structures through the lower-bound theorem, using dislocations based finite elements and eigenstress modelling. The lower bound approach is based on the knowledge of the self-equilibrated stresses that constitutes the basis of the domain where the optimal solution should be searched. A twofold strategy can be used to get self-equilibrated stresses, i.e., eigenstresses. The first one pursues the calculation of the self-equilibrated stress through the numerical approximation of the differential equilibrium equation in homogeneous form through an *a posteriori* discretization that used polynomial representation of finite degree. The second one consists of Finite Element implementation of the self-equilibrated stress calculation by discontinuous finite elements based on Volterra's dislocations theory. Both the formulations are written in terms of the strain and precisely in terms of the strain nodal displacement parameters. Consequently, it is possible to formulate an iterative procedure starting from the knowledge of the dislocation at the incoming collapse, in Melan's residual sense, and calculate the structural ductility requirement. Several numerical examples are presented to confirm the method's feasibility.

Introduction

The application of limit analysis and plasticity to structural safety concerns a wide range of engineering fields. Starting from the pioneer works of Prager, Drucker, and Greenberg's [1,2] and Massonnet and Save [3,4] that address the plastic response of structures introducing the collapse calculation for one-dimensional beams assembly as the main topic the matter has been formalized in the mathematical treatise of Hill [5], a two-fold approach is the way the limit analysis has been applied. The first, the kinematic method, consists of finding the collapse load as the load infimum among that in equilibrium with the stress linked to a compatible collapse mechanism. The second, the static approach, is based on the research of the supremum of the load in equilibrium with an admissible stress state that is a combination of self-equilibrated stress with the particular solution of the elastic equilibrium equation to the applied loads. An extensive chapter in the limit analysis is devoted to the mechanics of masonry structures following the work of Heyman [6], who extends the primary approach of limit analysis to the not tensile resistant materials (NTRM) modeling masonry. The limit analysis, and in a more general sense the plasticity modeling, is highly required when one must interpret the results of monitoring campaign since the structures during their life generally undergo permanent strain and cracks [7,8]. In the field of the biomechanics, the limit equilibrium is used as a primary tool to assess the fracture risk of prosthesis implants and relative optimization strategies as reported in [9,10,11].

The main topic of the limit analysis when addressed using the lower-bound approach is to define the field of the self-equilibrated stress that constitute the kernel of the equilibrium operator. The way the self-equilibrated stress is obtained characterizes the work here presented. Namely, a first strategy is based on mixed numerical-analytical solution of the equilibrium differential equations. The procedure evaluates the collapse load multiplier for masonry domes and vaults [12,13] and concrete caps. The results from the proposed formulation showed a good agreement with the experiments reported in [14]and with calculation presented in [15] that uses a kinematic approach based on the energy balance about crack lines which constitutes a typical pattern of collapse for plates and slender domes under radial load [16].
Alternatively, one can resolve the self-equilibrated stress, and consequently the domain within one must define the admissible stress state, with reference to finite element discretization of the model. The second procedure relates the permanent strain modeled as Volterra's dislocation to nodal parameters analogous to finite element nodal displacement. The formulated displacement base FEM gives the linear operator that relates the self-equilibrated stress domain to nodal parameters and boundary displacements. The dimensions of the self-equilibrated stress domain is the rank of the linear operator that in the case of truss and frames structures coincide with the redundancy degree of the structure.

Semi analytical method (SAM)

The first procedure consists of the search of the collapse multiplier through a mixed numerical and analytical procedure. The equilibrium equation for vaults and domes is solved through an optimization constrained problem. We have used a generalized stress formulation; hence the stress has been described through the internal forces, N, T, and M that are the resultant components, axial, and shear forces respectively, and the resultant bending moment of the stress acting on the section. Where the subscript $\{\ \}_{1,2}$ refers to the meridian or parallel cross section of the structure.

$$\begin{aligned} &\frac{d(N_1 r)}{d\theta} - N_2 R_1 \cos\theta - T_1 r = -X\, R_1 r \\ &N_1 r + N_2 R_1 \sin\theta + \frac{d(T_1 r)}{d\theta} = Z\, R_1 r \\ &\frac{d(M_1 r)}{d\theta} - M_2 R_1 \cos\theta - T_1\, R_1 r = 0 \end{aligned} \tag{1}$$

The set of equilibrium equation is solved numerically starting from a set of shape function and collocating the equation at the discrete colatitude angle θ_j .

$$\begin{cases} N_1^j = K_{N_1}^j c \\ N_2^j = K_{N_2}^j c \\ M_1^j = K_{M_1}^j c \\ M_2^j = K_{M_2}^j c \\ T_1^j = K_{T_1}^j c \end{cases} \tag{2}$$

The limit multiplier of prescribed load paths, either monotonically increasing or randomly variable, is obtained by maximizing the static multiplier of loads under the constraint that the sum of elastic response, plus any self-equilibrated time-independent stress solution, belongs to the admissible domain. Hence it results that the optimization program has the load multiplier as objective function and the parameters c as design variables. The optimization constraints are the linear inequalities representing the limit domain in terms of c. The elastic solution under the actual loads must be obtained from the equilibrium equation in any way. Namely, if closed form solution exists one can use it or can be obtained employing FEM analysis. The vectors N_i^e, M_i^e collected the effective generalized stress. Collocating the equations at discrete angles, i.e., at a finite number of θ_j with $j \in \{1, \dots \dots, m\}$, where m was the number of points along the meridian curve, one gets the desired solution. Finally, the optimization program, has the following discretized form:

$$\sup_{c} k \left| k \in \mathbb{R}^{+} : \begin{cases} h(N_i^r + kN_i^e) + \alpha(M_i^r + kM_i^e) < \beta h \\ h(N_i^r + kN_i^e) - \alpha(M_i^r + kM_i^e) < \beta h \\ (N_i^r + kN_i^e) < \beta \end{cases} \right|^{j} \tag{3}$$

Where β accounts for the presence of tensile resistance σ_0 such that axial limit stress is $N_0 = \sigma_0 h$ following the material constitutive properties.

Discontinuous finite element procedure (DFEP)

An alternative procedure uses finite elements in a discontinuous form to write the discrete operator that relates the set of self-equilibrated stress in the structure to discrete nodal displacement-like parameters. The approach is devoted to relating the permanent strain equivalent to Volterra's dislocation to nodal parameters. Moreover, the permanent strain and the corresponding self-equilibrated stress can be represented in terms of constraints displacements too. The representation can be assumed as a span of the eigenstress space. It can be seen that the span is not independent and that the base of the eigenstress is a proper subset of the parameters manifold.

In conclusion, any equilibrated stress under prescribed load can be calculated as the sum of the elastic response plus self-equilibrated stress, depending on the application of the linear operator V that maps the nodal dislocation parameter to the eigenstress.

The load collapse multiplier results as the 'sup' of the load multiplier in the constrained optimization program

$$s_\alpha = \sup k | f(k\sigma^* + V\delta) \leq 0, \alpha = \begin{cases} sd & \text{shakedown} \\ c & \text{collapse} \end{cases} \tag{4}$$

where σ^* is the stress in the structure calculated as being indefinitely elastic at any time during the load path, δ is the nodal dislocation parameter vector.

Results

Slender concrete caps have been analyzed following the *SAM* procedure and compared with experimental results obtained by [14,17] are reported in the following Fig.1 and Fig.2.

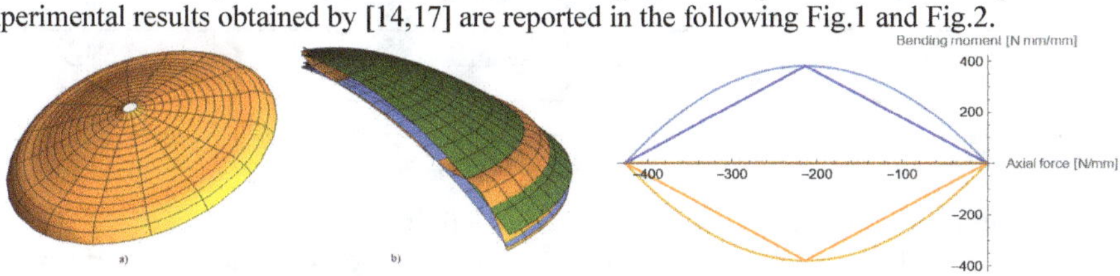

Figure 1 Thrust Surface and limit domain

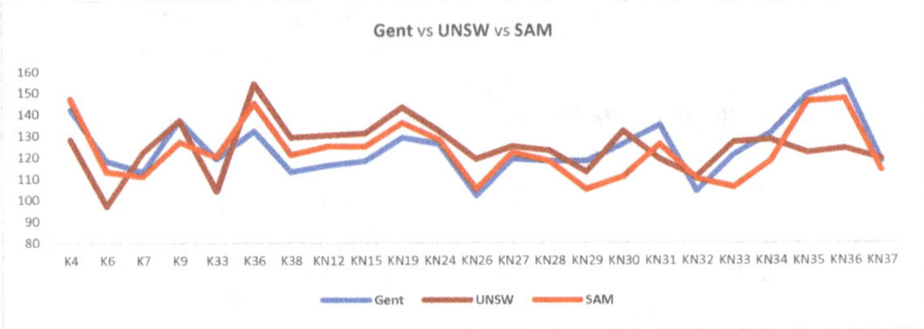

Figure 1. Result comparison among experimental results (Gent), analytical solution (UNSW) and proposed method (SAM) for different speciement $\boldsymbol{K_i}$

The *DFEP* has been applied to two-dimensional and three-dimensional examples. The first represent simple approximation of the Prestwood bridge Figure.3, whose collapse has been experimented in [18].

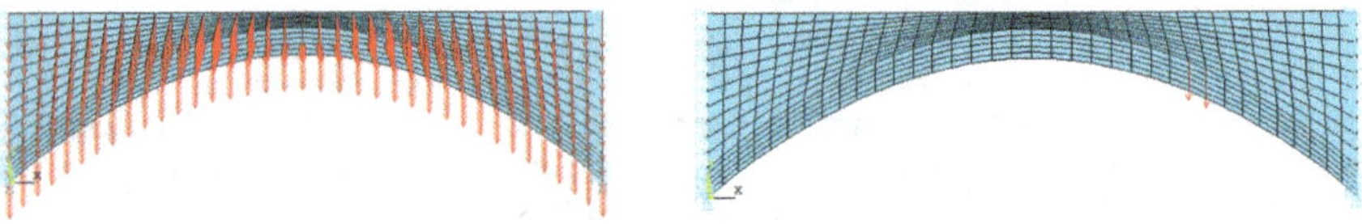

Figure 3 Bridge load condition

In Tab.1 the results for each mesh sizing are reported. The value of experimental load multiplier, in the destructive test conducted by Page is *228* KN.

Table 1 Load Multiplier values for each mesh size

Case	Element number along X	Element number along Y	Load Multiplier
1	18	5	701
2	36	5	423
3	72	5	312
4	72	10	258
5	88	12	**233**

The 3D example has modeled the plastic behavior of the cross section of a femur [9], modeled as a hollow cylinder, Fig.4, , after a hip prosthesis implant.The limit load multiplier has been compared with Ansys nonlinear analysis solution.

Table 2 Result comparison of femur limit load

Ansys Nonlinear solution	DFEP solution
13.64 MPa	12.86 Mpa

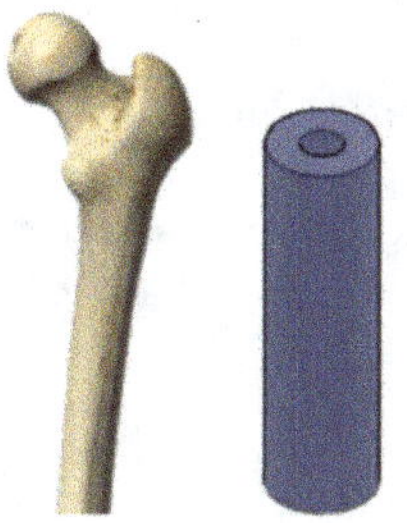

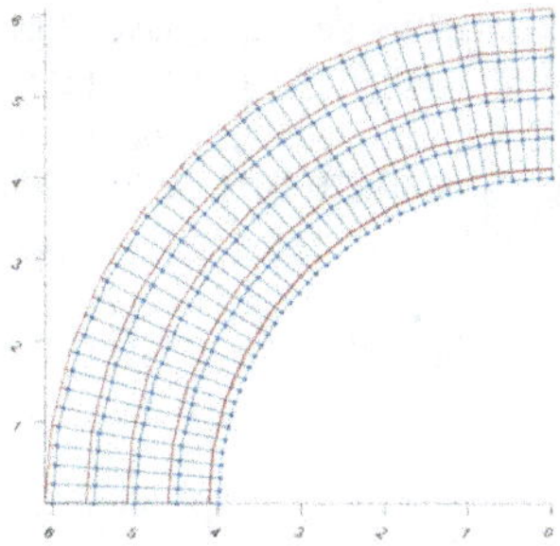

Figure 4 Approximation of the femur to a cave cylinder and elastic solution of the generic section (blue, undeformed shape/ red, deformed shape)

Both the examples confirm the accuracy of the procedure.

Conclusions

Two procedures have been set up. The first one, *SAM*, starts with the analytical solution of spherical domes to calculate any self-equilibrated stress. Moreover, a finite element elastic solution is obtained from the actual loads and was used as the purely elastic solution introduced into Melan's theorem. The analytical solution of the homogeneous equation of spherical domes is used to model the eigenstress of parabolical, conic, and slender domes through an approximate interpolation of the parabola with a sphere. The approximation allowed us to use the analytical solution to different non-spheric geometries. The proposed results were presented in terms of thrust lines. The results have been compared with numerical results obtained through commercial software of numerical analysis and experimental results. The thrust lines confirmed that the analyzed domes are safe under the applied loads or confirmed the collapse load multiplier under prescribed loads. The proposed method allowed us to calculate the safety factor under prescribed load patterns and assess the safety of the prescribed load level. Both presented strategies have shown the feasibility of the methods.

In second one, *DFEP*, the problem's schematization does not depend on its size.

The advantages of the procedure are manifold. First, the method allows the implementation of the load conditions, and the value of the collapse multiplier is computed without necessarily following the load path. Furthermore, being a FEM-based procedure, it can be borrowed from other commercial computing platforms for numerical analysis.

It should be noted that the *DFEP* procedure is structured in the deformation space. This fact has to be intended as a passage to an optimization problem in the displacement field, having chosen both the basic parameters of the self-stresses and the compatibility domain of the stresses in the displacement space. Therefore, in the proposed solution strategy, it is possible to control *a posteriori* the demand for structural ductility, which is essential for permanent deformations to unfold up to the desired load level without affecting the results obtained with a fragile local collapse.

References

[1] D. C. Drucker, W. Prager e H. J. Greenberg, Extended limit design theorems for continuous media, vol. 9, 1952. https://doi.org/10.1090/qam/45573

[2] W. Prager e B. G. Neal, Introduction to Plasticity, vol. 27, 1960. https://doi.org/10.1115/1.3643927

[3] C. E. Massonet e M. A. Save, Plastic Analysis and Design. Vol. 1, Beams and Frames, vol. 33, 1966. https://doi.org/10.1115/1.3625111

[4] M. Save, Limit analysis and design of containment vessels, vol. 79, 1984. https://doi.org/10.1016/0029-5493(84)90048-7

[5] R. Hill, The mathematical theory of plasticity, Oxford: Oxford University Press, 1950.

[6] J. Heyman, The stone skeleton: structural engineering of masonry architecture, 1995. https://doi.org/10.1017/CBO9781107050310

[7] V. Minutolo, E. Cerri, A. Coscetta, E. Damiano, M. D. Cristofaro, L. D. Gennaro, L. Esposito, P. Ferla, M. Mirabile, L. Olivares e R. Zona, NSHT: New smart hybrid transducer for structural and geotechnical applications, vol. 10, 2020. https://doi.org/10.3390/app10134498

[8] V. Minutolo, S. Di Ronza, C. Eramo, P. Ferla, S. Palladino e R. Zona, The use of destructive and non-destructive testing in concrete strength assessment for a school building, vol. 10, 2019. https://doi.org/10.34218/IJARET.10.6.2019.028

[9] L. Esposito, V. Minutolo, P. Gargiulo, H. Jonsson, M. K. Gislason e M. Fraldi, Towards an app to estimate patient-specific perioperative femur fracture risk, vol. 10, 2020. https://doi.org/10.3390/app10186409

[10] C. Ricciardi, J. Halldór, D. Jacob, G. Improta, M. Recenti, M. K. Gíslason, G. Cesarelli, L. Esposito, V. Minutolo, P. Bifulco e P. Gargiulo, Improving prosthetic selection and predicting BMD from biometric measurements in patients receiving total hip arthroplasty, vol. 10, 2020. https://doi.org/10.3390/diagnostics10100815

[11] V. Minutolo, L. Esposito, E. Sacco e M. Fraldi, Designing stress for optimizing and toughening truss-like structures, vol. 55, 2020. https://doi.org/10.1007/s11012-020-01189-z

[12] R. Zona, P. Ferla e V. Minutolo, «Limit analysis of conical and parabolic domes based on semi-analytical solution,» *Journal of Building Engineering,* vol. 44, p. 103271, 12 2021. https://doi.org/10.1016/j.jobe.2021.103271

[13] R. Zona, L. Esposito, P. Ferla, S. Palladino, E. Totaro e V. Minutolo, «Lower bound limit analysis of parabolic domes based on spherical analytical solution,» *International Journal of Advanced Research in Engineering and Technology,* vol. 11, p. 59–79, 2020.

[14] D. Vandepitte e G. Lagae, Buckling of Spherical Domes Made of Microconcrete and Creep Buckling of Such Domes Under Long-term Loading, 1986. https://doi.org/10.1007/978-3-642-82776-1_14

[15] Z. T. Chang, M. A. Bradford e R. I. Gilbert, Short-term behaviour of shallow thin-walled concrete dome under uniform external pressure, vol. 49, 2011. https://doi.org/10.1016/j.tws.2010.08.012

[16] S. Di Ronza, C. Eramo, V. Minutolo, S. Palladino, E. Totaro, P. Ferla, R. Zona, T. Ronga e C. C. Pomicino, Experimental tests on gully tops and manhole TOPS devices according to EN124 standard, vol. 11, 2020.

[17] Z. T. Chang, M. A. Bradford e R. Ian Gilbert, Limit analysis of local failure in shallow spherical concrete caps subjected to uniform radial pressure, vol. 48, 2010. https://doi.org/10.1016/j.tws.2010.01.014

[18] J. Page, «Load tests to collapse on two arch bridges at preston, shropshire and prestwood, staffordshire.,» 1987.

Finding damage in truss structures exploiting modal strains

MODESTI Martina[1,a *], PALERMO Antonio[1,b] and GENTILINI Cristina[2,c]

[1]DICAM Department, University of Bologna, Viale del Risorgimento 2, 40136 Bologna (Italy)

[2]DA Department, University of Bologna, Viale del Risorgimento 2, 40136 Bologna (Italy)

[a]martina.modesti2@unibo.it, [b]antonio.palermo6@unibo.it, [c]cristina.gentilini@unibo.it

*corresponding author

Keywords: Damage Detection, Damage Localization, Truss Structures, Structural Health Monitoring, Flexibility Matrix, Modal Strains

Abstract. The detection of potentially damaged elements in planar truss structures is a challenging task. Among the different methods proposed in literature, one promising procedure exploits the modal strains of the structure that are calculated from the flexibility matrix, which is, in turn, estimated from the lowest frequencies and corresponding modes of vibration. The benefit of this approach stems from the possibility of using a reduced number of mode shapes, usually available from the dynamic monitoring of the structure to perform the damage detection. In this work, a novel damage detection index based on modal strains is proposed, and its reliability in detecting stiffness reduction in elements of a planar truss is tested numerically.

Introduction

Damage detection is a key information for monitoring the status of structures and infrastructures. In the last decades, several procedures aimed at detecting structural damages and/or anomalies to prevent the degradation and avoid the collapse of the structures have been proposed. To reach this goal, a meaningful part of the available research literature makes use of the structure dynamic response. In fact, when a damage occurs, the reduction of the stiffness yields a change in the structural dynamic characteristics. Hence, vibration-based techniques able to detect the variation of such features are widely used to monitor the structural health.

Among others, the first proposed numerical strategies exploited the variation of natural frequencies and mode shapes to identify the presence of anomalies [1]. Although effective in identify the presence of damages, these methods suffer from the presence of noise which can conceal the variation of natural frequencies induced by damages. Hence, more sophisticated techniques were later proposed to locate structural damages, based, for example, on the use of modal strain energy, residual force vector, mode shape curvature and flexibility matrix [2,3,4]. Methods exploiting flexibility matrix allow to evaluate changes in the structural dynamic features using only the lowest eigenfrequencies and mode shapes [5]. This feature represents a great advantage for practical health monitoring where it is typically difficult to measure and estimate the highest modes of vibrations.

In this context, Montazer and Seyedpoor [6] recently proposed a damage index, called strain change based on flexibility index (SCBFI), which exploits the variation in strains computed from the flexibility matrix. The proposed index is proved to be particularly effective in locating the damage in truss structures.

In this paper, we stem from their approach to propose an alternative index based on the singular value decomposition of the difference between the strain matrix of the damaged and healthy structure. Following this approach, we can evaluate the bars which mostly contribute to the variation of strain in the damaged truss. The proposed index is tested numerically on a truss structure.

Damage index

Let us consider a generic planar truss, composed by N nodes, connected with e bar elements. The number of degrees of freedom of the structure is equal to $n = 2N - c$, with c being the number of constrained DOFs.

The modal matrix $\mathbf{\Phi}$ is composed by $\mathbf{\Phi}_i$ mass-normalized mode shapes. The flexibility matrix $\mathbf{f}$ of the structure can be approximated as, [5]:

$$\mathbf{f} \approx \sum_{i=1}^{m} \frac{1}{\omega_i^2} \mathbf{\Phi}_i \mathbf{\Phi}_i^{\mathrm{T}} \tag{1}$$

where ω_i is the i-th circular frequency and m is the number of considered modes. Given the inverse proportion between flexibility matrix and the square of the circular frequencies, the expression in Eq. (1) rapidly converges to the exact value as the number of considered frequencies increases. Therefore, an accurate approximation of the flexibility matrix can be obtained using the lowest m modes of vibration.

We remind that the i-th column of the flexibility matrix collects a vector of nodal displacements corresponding to a unitary force applied in the related i-th degree of freedom. Hence, we can compute the strain matrix **SM**, composed by j-th element strain values ϵ_{ji} caused by a unitary force applied to the i-th DOF:

$$\epsilon_{ji} = \mathbf{R}_j \mathbf{u}_{ji} \tag{2}$$

where $\mathbf{R}_j$ is the topological vector of the element j and $\mathbf{u}_{ji}$ is the 1×4 nodal displacement vector of j-th bar, associated with the i-th column of flexibility matrix.

As well known, the presence of one or more damaged bars in a truss produces a reduction in the structural stiffness and, in turn, an increase of flexibility and strain. For this reason, it is possible to evaluate and compare the strain matrix **SM** for both the healthy and for the damaged structure, hereinafter labelled as $\mathbf{SM}^{\mathrm{H}}$ and $\mathbf{SM}^{\mathrm{D}}$, respectively.

From the strain matrices, we compute the matrix collecting the strain variation between damaged and healthy structures as:

$$\mathbf{\Delta SM} = \mathbf{SM}^{\mathrm{D}} - \mathbf{SM}^{\mathrm{H}} \tag{3}$$

At this stage, we assume that only the damaged elements contribute significantly to the variation of the modal strain. Following this assumption, we propose to identify such damaged elements by performing a singular value decomposition (SVD) of the matrix $\mathbf{\Delta SM}$:

$$\mathbf{\Delta SM} = \mathbf{USV} \tag{4}$$

where $\mathbf{U}$ is a $e \times e$ matrix collecting the left-singular vectors of $\mathbf{\Delta SM}$, $\mathbf{S}$ is a $e \times n$ rectangular diagonal matrix whose diagonal entries are the singular values s_i of $\mathbf{\Delta SM}$ and $\mathbf{V}$ is a $n \times n$ matrix collecting the right-singular vectors of $\mathbf{\Delta SM}$. We remark that the norm of the matrix $\mathbf{\Delta SM}$ can be approximated by exploiting only the first $v \ll e$ singular values and vectors:

$$||\mathbf{\Delta SM}||^2 \approx \sum_{i=1}^{v} \mathbf{U}_i s_i^2 \mathbf{U}_i^* \tag{5}$$

where * indicate the complex conjugate.

In particular, the first singular value and the related left-singular vector provides the largest contribution to the strain difference $\mathbf{\Delta SM}$. Hence, we utilize the first left-singular vector $\mathbf{U}_1$ to define the damage index indicated as **SVD**$_1$:

$$\mathbf{SVD}_1 = |\mathbf{U}_1| \quad (6)$$

being | . | the absolute value. Similarly, taking into consideration both the first and the second singular values and the corresponding left-singular vectors, we can compute a second damage index defined as the weighted sum of the two lowest singular vectors, $\mathbf{U}_1$ and $\mathbf{U}_2$, of the $\mathbf{\Delta SM}$ matrix multiplied by the corresponding singular values, s_1 and s_2, as follows:

$$\mathbf{SVD}_2 = s_1|\mathbf{U}_1| + s_2|\mathbf{U}_2| \quad (7)$$

In the next section, the proposed indexes **SVD**$_1$ and **SVD**$_2$ are applied to detect and locate the presence of damages in a two-dimensional planar truss and their reliability in finding damage is compared.

Numerical results

Single damage scenario. For the numerical investigation, we consider the planar truss shown in Fig. 1, which was previously used in literature as a numerical benchmark for other damage identification methods [7]. The truss is organized in 6 square bays, for an overall of $e = 31$ bars and $n = 25$ degrees of freedom. Bars are made of aluminium, with an elastic modulus of 70 GPa and a density equal to 2770 kg/m^3.

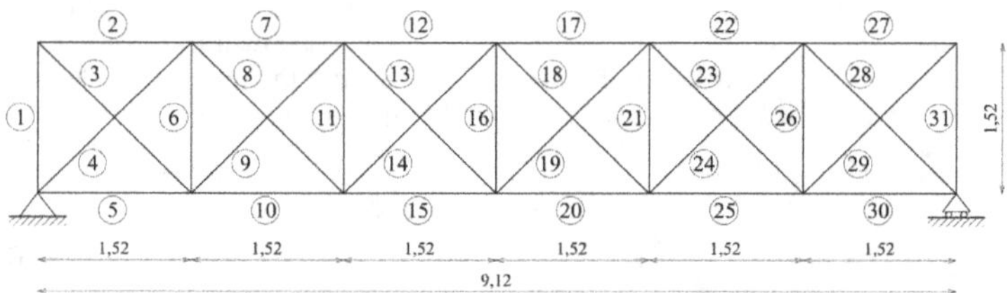

Fig. 1. Planar truss composed of 31 bars and 6 square bays.

The natural frequencies and mode shapes of the truss are computed using a standard Finite Element code where each element is modelled as linear elastic truss element. The damage in the bars is simulated reducing the element stiffness, namely by decreasing the Young's modulus.

To evaluate the capabilities and robustness of the proposed damage index, we have tested several configurations of single damaged bars. Our aim is to locate the presence of damage along the truss for increasing damage intensities. To this purpose, we computed natural frequencies and mode shapes of the truss structure considering 5 different levels of damage in each bar, with a damage intensity varying between 10% and 90%.

Additionally, we polluted each damaged mode shape components $\varphi_{mn}^{\mathrm{D}}$, related to the m-th mode of vibration and n-th DOF, considering a given level of noise p. This noise aims at replicating the inherent variability induced by environmental and operational conditions of the structure in the field. The components $\varphi_{mn}^{\mathrm{D}}$ utilized to build the flexibility matrix are thus obtained as the mean of 200 random polluted $\hat{\varphi}_{mn}$ as follows:

$$\hat{\varphi}_{mn} = \varphi_{mn}(1 + p \cdot \mathrm{rand}) \quad (8)$$

where p is the intensity of a uniformly distributed random noise rand in the interval [-1;1]. In the described analysis, we considered noise value p between 0% (without noise) and 3%. For the approximation of flexibility matrix, Eq. 1, we used the first $m = 4$ mode shapes.

Table 1. *Number of correct identifications over 100 attempts in single damage scenarios using* $\boldsymbol{SVD_1}$ *with noise* $p = 1\%$.

	1	2	3	4	5	6	7	8	9	10	11	12	13	14	15	16	17	18	19	20	21	22	23	24	25	26	27	28	29	30	31
10%	100	100	100	100	100	99	100	100	100	100	96	100	100	100	100	99	100	100	100	100	78	100	100	100	100	27	100	100	100	100	100
30%	100	100	100	100	100	100	100	100	100	100	100	100	100	100	100	100	100	100	100	100	100	100	100	100	100	100	100	100	100	100	100
50%	100	100	100	100	100	100	100	100	100	100	100	100	100	100	100	100	100	100	100	100	100	100	100	100	100	100	100	100	100	100	100
70%	100	100	100	100	100	100	100	100	100	100	100	100	100	100	100	100	100	100	100	100	100	100	100	100	100	100	100	100	100	100	100
90%	100	100	100	100	100	100	100	100	100	100	100	100	100	100	100	100	100	100	100	100	100	100	100	100	100	100	100	100	100	100	100

Table 2. *Number of correct identifications over 100 attempts in single damage scenarios using* $\boldsymbol{SVD_2}$ *with noise* $p = 1\%$.

	1	2	3	4	5	6	7	8	9	10	11	12	13	14	15	16	17	18	19	20	21	22	23	24	25	26	27	28	29	30	31
10%	100	100	100	100	100	97	100	99	98	100	88	100	92	100	100	86	100	86	92	100	69	100	98	100	100	58	100	100	99	100	100
30%	100	100	100	100	100	100	100	100	100	100	100	100	100	100	100	100	100	100	100	100	100	100	100	100	100	100	100	100	100	100	100
50%	100	100	100	100	100	100	100	100	100	100	100	100	100	100	100	100	100	100	100	100	100	100	100	100	100	100	100	100	100	100	100
70%	100	100	100	100	100	100	100	100	100	100	100	100	100	100	100	100	100	100	100	100	100	100	100	100	100	100	100	100	100	100	100
90%	100	100	100	100	100	100	100	100	100	100	100	100	100	100	100	100	100	100	100	100	100	100	100	100	100	100	100	100	100	100	100

Table 1 collects the number of correctly identified damage scenarios using the index $\mathbf{SVD_1}$, while Table 2 shows the same results obtained exploiting $\mathbf{SVD_2}$. Each column of the tables represents a different damaged bar of the truss and each row an increasing level of damage. Note that for each damage level and bar, we performed 100 simulations and damage identifications. The reader can appreciate that even with the presence of noise, the results with the proposed indexes are robust, since the greatest part of the damage cases are correctly identified by the algorithm.

Table 3. *Number of correct identifications over 100 attempts in single damage scenarios using* $\boldsymbol{SVD_1}$ *with noise* $p = 3\%$.

	1	2	3	4	5	6	7	8	9	10	11	12	13	14	15	16	17	18	19	20	21	22	23	24	25	26	27	28	29	30	31
10%	74	76	79	98	100	20	100	48	63	100	33	100	18	21	100	32	100	14	31	100	20	100	47	62	100	6	73	95	71	88	76
30%	100	100	100	100	100	100	100	100	100	100	100	100	100	100	100	100	100	100	100	100	97	100	100	100	100	47	100	100	100	100	100
50%	100	100	100	100	100	100	100	100	100	100	100	100	100	100	100	100	100	100	100	100	100	100	100	100	100	99	100	100	100	100	100
70%	100	100	100	100	100	100	100	100	100	100	100	100	100	100	100	100	100	100	100	100	100	100	100	100	100	100	100	100	100	100	100
90%	100	100	100	100	100	100	100	100	100	100	100	100	100	100	100	100	100	100	100	100	100	100	100	100	100	100	100	100	100	100	100

Table 4. *Number of correct identifications over 100 attempts in single damage scenarios using* $\boldsymbol{SVD_2}$ *with noise* $p = 3\%$.

	1	2	3	4	5	6	7	8	9	10	11	12	13	14	15	16	17	18	19	20	21	22	23	24	25	26	27	28	29	30	31
10%	43	32	71	56	100	42	100	29	48	100	46	100	44	41	100	45	100	38	34	100	34	100	42	64	100	12	54	74	66	68	58
30%	100	100	100	100	100	99	100	100	100	100	91	100	97	100	100	93	100	95	99	100	83	100	98	100	100	65	100	100	100	100	100
50%	100	100	100	100	100	100	100	100	100	100	100	100	100	100	100	100	100	100	100	100	100	100	100	100	100	99	100	100	100	100	100
70%	100	100	100	100	100	100	100	100	100	100	100	100	100	100	100	100	100	100	100	100	100	100	100	100	100	100	100	100	100	100	100
90%	100	100	100	100	100	100	100	100	100	100	100	100	100	100	100	100	100	100	100	100	100	100	100	100	100	100	100	100	100	100	100

In particular, to ease the interpretation of the tables, we report with a green background all the scenarios with a damage identification success rate greater than 85%.

Table 3 and Table 4 report the number of correct identifications with a level of noise p equal to 3%. As expected, using both $\mathbf{SVD_1}$ and $\mathbf{SVD_2}$ the success rate decreases as the level of noise increases and as the damage intensity decreases. However, the missed identified bars correspond to the lowest level of damage (10% reduction in the elastic modulus) and, likely, less harmful for the structure. We remark that the indexes $\mathbf{SVD_1}$ and $\mathbf{SVD_2}$ have similar results in terms of success rate in the analysed single damage scenarios, showing that the second singular value contributes marginally to the reconstruction of the strain matrix in Eq. (5).

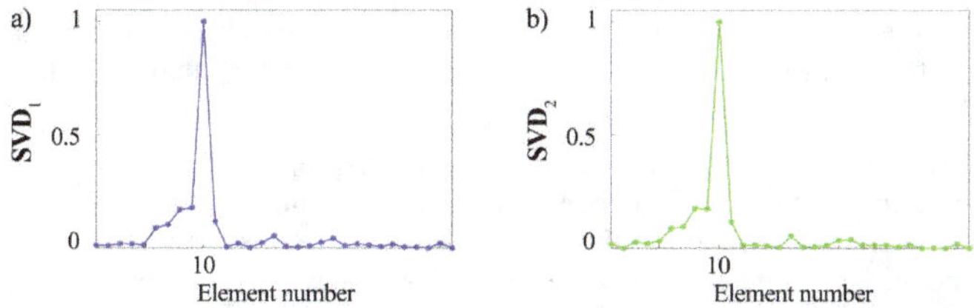

Fig. 2. *a)* ***SVD_1*** *and b)* ***SVD_2*** *graphs with a damage of 30% in bar 10 and noise level p of 3%.*

As an example, we report in Fig. 2 the value of the damage indices as computed from a single simulation of a damage scenario corresponding to a reduction of 30% in the elastic modulus of bar 10, with a noise level p equal to 3%. The indexes are normalized with respect to their maximum value. The reader can appreciate how the proposed damage indexes successfully identify the damaged bar, as highlighted by the peaks in the graphs.

Double damage scenario. Motivated by the satisfactory performance of the proposed indexes, we investigated the capabilities of the identification algorithm against damage scenarios with multiple damaged bars. In particular, we considered a stiffness reduction of 50% in bar 16 and of 30% in bar 5. In Fig. 3 and Fig. 4 a comparison between the indexes **SVD_1** and **SVD_2** as obtained for the considered damage scenario with noise level $p = 1\%$ and $p = 3\%$, respectively, is shown.

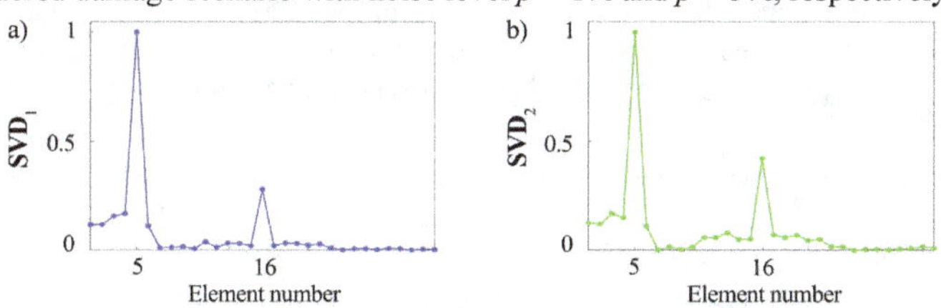

Fig. 3. *a)* ***SVD_1*** *and b)* ***SVD_2*** *graphs with an elastic modulus reduction of 50% in bar 16 and damage of 30% in bar 5, polluted by a noise p equal to 1%.*

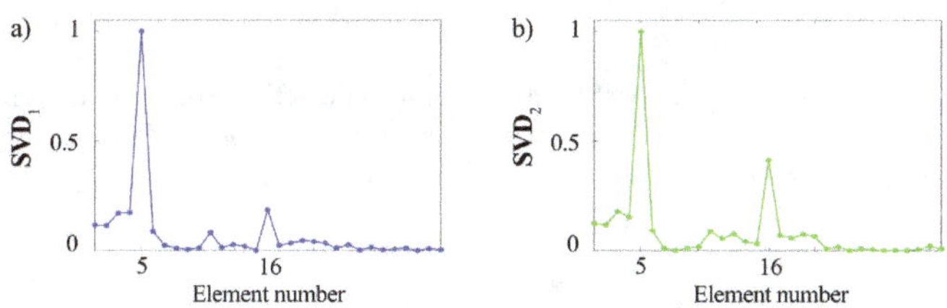

Fig. 4. *a)* ***SVD_1*** *and b)* ***SVD_2*** *graphs with an elastic modulus reduction of 50% in bar 16 and damage of 30% in bar 5, polluted by a noise p equal to 3%.*

Outputs in Fig. 3 and Fig. 4 show that the addition of the second singular value, as considered in **SVD_2** index, is beneficial in the multiple case scenario, as indicated by the peaks in Fig. 3b and Fig. 4b, considering both a level of noise p of 1% and of 3%. In fact, while the **SVD_1** still succeeds in locating properly the damaged bars with noise equal to 1%, Fig. 3a, it shows some limitations in the scenario with 3% of noise, where extra peaks are found by the damage index, Fig. 4a.

Finally, we remark that the proposed index cannot quantify the damage extent, namely there is no direct correlation between the value of the damage index and the intensity of the damage.

Conclusions

In this paper, we proposed a novel damage index which exploits the variation in modal strains to detect damages in truss structures. The index is based on the Singular Value Decomposition of the variation strain matrix $\mathbf{\Delta SM}$ between the healthy and damaged structure. The performance and reliability of the method in detecting the damaged elements were numerically tested on a 31 bars planar truss considering several single damage scenarios, even in presence of noise polluting the structures modal characteristics. As expected, the identification success rate increases as the damage extent increases and decreases with higher noise level. The proposed index is applied with promising results also to a double case scenario. Future research efforts will be devoted to extend the proposed algorithm to quantify the damage extent.

References

[1] J.-T. Kim, Y.-S. Ryu, H.-M. Cho, N. Stubbs, Damage identification in beam-type structures: frequency-based method vs mode-shape-based method, Engineering Structures 25 (2003) 57–67. https://doi.org/10.1016/S0141-0296(02)00118-9

[2] A. Alvandi, C. Cremona, Assessment of vibration-based damage identification techniques, Journal of Sound and Vibration 292 (1) (2006) 179–202. https://doi.org/10.1016/j.jsv.2005.07.036

[3] O. Avci, O. Abdejlaber, S. Kiranyaz, M. Hussein, M. Gabbouj, D.J. Inman, A review of vibration-based damage detection in civil structures: From traditional methods to machine learning and deep learning applications, Mechanical Systems and Signal Processing 147 (2021) 107077. https://doi.org/10.1016/j.ymssp.2020.107077

[4] S. Doebling, C. Farrar, M. Prime, A summary review of vibration-based damage identification methods, The Shock and Vibration Digest 30 (1998) 91–105. https://doi.org/10.1177/058310249803000201

[5] A. Pandey, M. Biswas, Damage detection in structures using changes in flexibility, Journal of Sound and Vibration 169 (1) (1994) 3–17. https://doi.org/10.1006/jsvi.1994.1002

[6] M. Montazer, S.M. Seyedpoor, A new flexibility based damage index for damage detection of truss structures, Shock and Vibration (2014) article ID 460692. https://doi.org/10.1155/2014/460692

[7] M. Nobahari and S. M. Seyedpoor, Structural damage detection using an efficient correlation-based index and a modified genetic algorithm, Mathematical and Computer Modelling (2011), vol. 53, no. 9-10, pp. 1798–1809. https://doi.org/10.1016/j.mcm.2010.12.058

Theoretical and Applied Mechanics - AIMETA 2022 Materials Research Forum LLC
Materials Research Proceedings **26** (2023) 151-156 https://doi.org/10.21741/9781644902431-25

Experimental analysis of new moment resisting steel connections

Salvatore Benfratello[1,a*], Luigi Palizzolo[1,b] and Santo Vazzano[1,c]

[1]Department of Engineering, University of Palermo, Viale delle Scienze, Ed. 8, I-90128, Palermo, Italy

[a]salvatore.benfratello@unipa.it, [b]luigi.palizzolo@unipa.it, [c]santo.vazzano@unipa.it

Keywords: Steel Connections, Experimental Analysis, Elastic and Limit Behaviour

Abstract. In the recent past, the authors proposed a new steel device devoted to representing an innovative moment resisting connection for steel frame elements called LRPD (Limited Resistance Plastic Device). It is a steel element characterized by symmetry with respect to three orthogonal barycentric planes and constituted by a sequence of three portions with abrupt cross section changes, each of one identifies a steel element of suitably designed geometry. LRPD possesses the following characterizing features: any elastic flexural stiffness variation with respect to the original selected member must be avoided; the bending moment resistance must be an appropriate reduced percentage of the original beam bending resistance; any local instability phenomenon must be avoided ensuring a full plastic deformation field. In previous papers the deep description of the geometrical and mechanical features of the device and the optimal design formulation are reported. In the present paper a first stage of experimental campaign on the mechanical behaviour of LRPD is presented. Specifically, the pure bending behaviour of LRPD is investigated by performing the four-point bending test. The test is performed monotonically until the selected ultimate plastic bending moment acts on the specimen. The mechanical response of LRPD, both in terms of deflections as well as of axial strains is evaluated by means of suitably positioned displacement and strain gauges. The experimental test is performed on LRPD designed for HEB240 cross section beams. The obtained results confirm the expected performance of LRPD constituting a fundamental step for the subsequent experimental steps mainly constituted by cyclic tests.

Introduction

The analysis and the design of the connections of steel structures is of particular interest in both the scientific and technical fields especially after the 1994 Northridge and 1995 Kobe earthquakes (see e.g. [1-4]). Two main types of connections are available: the rigid connections (also called moment connections) and the simple ones (also called shear connections). Among the moment connections, to avoid the brittle rupture of the connection, the most adopted approach proposes a reduction of the end sections of the frame beams, implementing the so-called "dog-bone" profiles, by cutting portions of the flanges of a I-beam profile (see e.g. [5-7]). The main drawbacks of this approach are that it is usable only with I-beam section elements and, furthermore, that the flanges cutting involves a reduction both in stiffness and strength of the treated element.

In the recent past, the authors proposed a new moment resisting connection for steel frame elements, called LRPD (Limited Resistance Plastic Device) [8-15]. Mainly, LRPD is thought as a steel device replacing a part of the structural elements with the aim to obtain a connection between structural members with preset mechanical and kinematical behaviour. The detailed description of the geometrical description of LRPD can be found in [15], while its leading features are those of the same elastic flexural stiffness as the original selected member, to possess a bending moment resistance reduced with respect to that of the original beam, to avoid any local instability phenomenon. In many papers [8-15] authors proposed different optimal design for LRPD as well performed a lot of numerical analyses to verify the LRPD behaviour and the related computational aspects. In the present paper a first step of experimental campaign on the mechanical behaviour of LRPD is presented by performing the four-point bending test to apply a pure bending moment without shear force in central zone of the specimen. The test is performed monotonically until the selected ultimate plastic bending moment acts on the specimen. The mechanical response of LRPD, both in terms of deflections as well as of axial strains is evaluated by suitably positioned displacement and strain gauges. The test is performed on LRPD designed for HEB240 beams. The obtained results confirm the expected performance of LRPD constituting a fundamental step for the subsequent experimental steps mainly constituted by cyclic tests and by a deep analysis of mechanical behavior of LRPD also adopting full-field contactless analysis method as Electronic Speckle Pattern Interferometry (ESPI) [17-19], to obtain highly accurate strain full-field.

Experimental investigations

Aim of the experimental investigations is the check of the bending behaviour of the LRPD by performing a four-point bending test on a suitable specimen. To this goal the first step has been to optimally design an LRPD following [3,4] to be adopted with a HEB240 profile and with an ultimate bending reduction equal to 0.67. The next step has been to design the specimen avoiding welding along the cross section which constitute a possible weakness. The final choice has been that of realizing a specimen simply by extending on both sides the flanges and the web of the LRPD to reach the desired length of the specimen. The experimental investigations have been performed on the specimen sketched in Fig. 1 where the corresponding main geometric characteristics and the location of displacement measuring point are also reported.

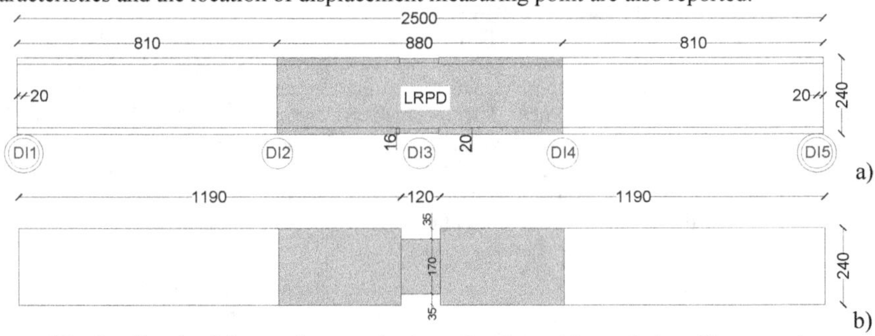

Fig. 1 – Sketch of the specimen under investigation: a) lateral view; b) upper view

The specimen is constituted by three different parts, two corresponding to the flanges (equal each other) and the other one for the web. Each part has been derived by a S275 steel slab, shaped following the results of the optimal design reported in [15] with an assigned ultimate bending moment equal to 0.67 of the corresponding one of HEB240 (dimensions in Fig. 1). The overall assembly has been obtained by factory welding the single parts and it has been performed by Tecnozinco s.r.l., whose contribution is gratefully acknowledged. The mechanical characterization of the material constituted the slab has been performed by tensile tests on suitable specimens following [15] and the results are reported in Table 1. Assuming the geometrical characteristics of the specimen, the mean value of f_y, the consequent ultimate bending moment of LRPD is equal to 241.8 kNm corresponding to a load of 306 kN. The experimental setup is sketched in Fig. 2

Materials Research Proceedings **26** (2023) 151-156 https://doi.org/10.21741/9781644902431-25

Table 1 – Results of the mechanical characterization of the material

Part	E [MPa]	f_y [MPa]	f_u [MPa]	A_g [%]	A_{gt} [%]
Web	208426	348.85	466.19	17.86	18.08
Flanges	231822	339.71	477.56	15.63	15.83

Fig. 2 – Sketch of the experimental setup

The experimental setup is constituted by a hydraulic jack (payload = 1 MN), driven by a ENERPAC pump; the acting load is measured by a CLF-1 class 1 AEP load cell (payload = 1 MN) connected to AEP MP-10 digital indicator, and the load is set on the specimen by means of a rigid beam and two steel semi-cylinders. Two steel cylinders are suitably positioned at the extreme of the specimen to obtain the simply supported scheme. The deflection is measured at three different locations (corresponding to the beginning, the middle section and the end section of the LRPD) by means of Mitutoyo DIGIMATIC Digital indicator (25 mm stroke) and two other indicators have been positioned in correspondence of the supports of the specimen. The strain acting on the LRPD has been measured by seven different HBM K-CLY-41/120 strain gages (6 mm length), arranged as sketched in Fig. 3, and connected to HBM MGCPlus. All the measurement values have been collected by a suitable virtual instrument developed in LabView 2020 environment.

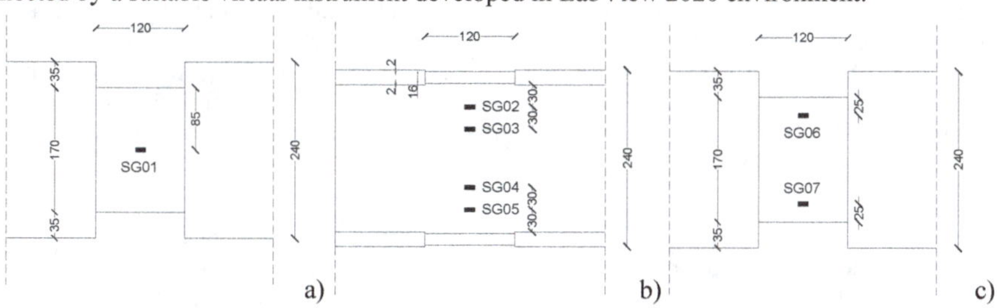

Fig. 3 – Arrangement of the strain gages: a) upper flange; b) web; c) lower flange.

Numerical model

The experimental test of the specimen described in the foregoing section has been analyzed by means of a suitable FEM model developed in ABAQUS environment and sketched in Fig. 4. The adopted mesh is equal to 4 mm, the mechanical model of the material is that of elastic perfectly plastic one with mechanical characteristics equal to that reported in Table 1 (i.e. two different materials have been adopted) while an elastic material with E 1000 times greater than that of the steel has been adopted for the support and loading cylinders. The support cylinders have been fully constrained (encastre condition), friction (0.125) has been assumed between the specimen and support and loading cylinders. The acting load is a concentrated one applied on the reference point for each loading cylinder, the reference points are rigidly linked to the surface of the cylinders.

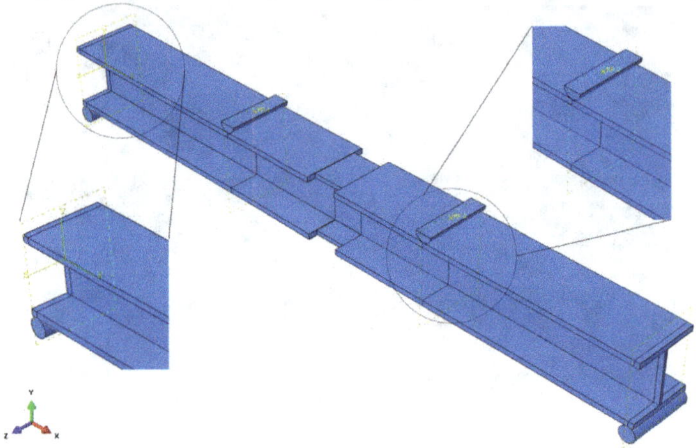

Fig. 4 – FE model of the experimental test in ABAQUS environment.

Results

The experimental results, in terms of axial stress σ_x evaluated from the axial strains ε_x at the locations indicated in Fig. 3, compared with those arising from the numerical analysis are reported in Fig. 5a. The kinematic behaviour of LRPD is evaluated defining as parameter $\Delta_{LRPD} = v_3 - \text{mean}(v_2; v_4)$ (being v_2, v_3 and v_4 the vertical displacement of DI1, DI2 and DI3 in Fig. 1a). In Fig. 5b the experimental Δ_{LRPD} is reported and compared with those arised from the numerical analysis and the theoretical one.

An examination of Fig. 5a confirms the expected mechanical behaviour of LRPD. The first points which undergo to yield are those corresponding to SG01, SG06 and SG07; the next points are SG02 and SG05, while the last points are SG03 and SG04. Finally, the numerical results are in well agreement with the experimental ones. The results of Fig. 5b confirm that the kinematic behaviour of LRPD is the expected one, being the overall behaviour practically coincident with the theoretical ones of the HEB240.

Conclusions

In the paper the experimental characterization of new moment resisting steel connections recently proposed by the authors and called LRPD has been proposed. The experimental results, both in terms of axial stress, axial strain and displacement confirm that the device behaves as expected following the constraints imposed in the optimal design. The experimental champaign will be enriched and completed soon studying other devices designed for different commercial profile.

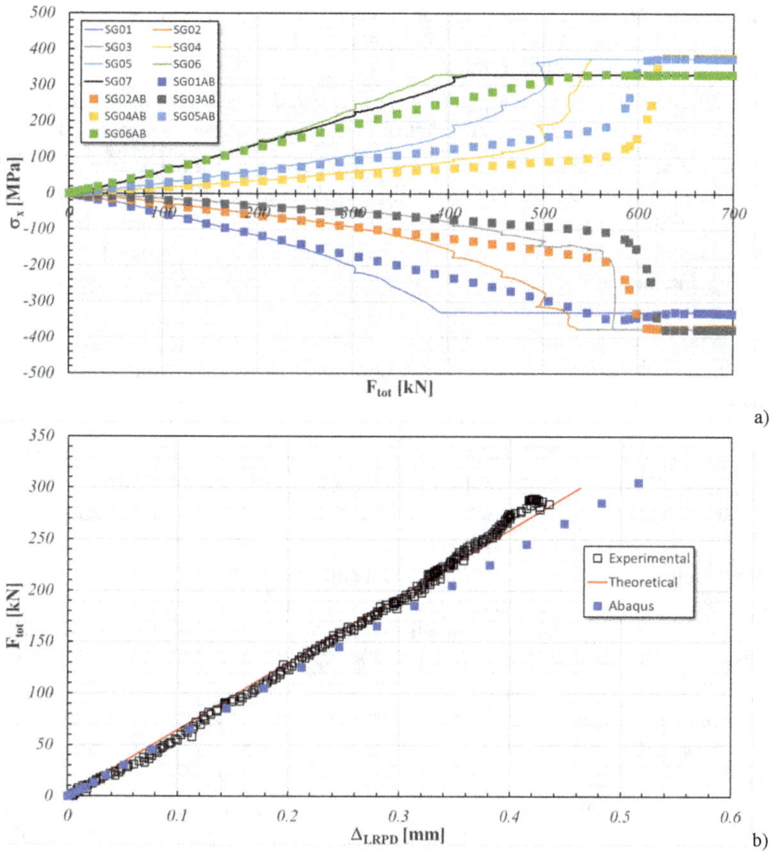

Fig. 5 – Experimental results compared with the numerical ones a) σ_x vs F_{tot}; b) Δ_{LRPD} vs F_{tot}.

References

[1] D.K. Miller, Lessons learned from the Northridge earthquake,Eng. Struc. 20 (1998), 249–260. https://doi.org/10.1016/S0141-0296(97)00031-X

[2] E.P. Popov, T.S. Yang, S.P. Chang, Design of steel MRF connections before and after 1994 Northridge earthquake. Eng Struct, 20 (1998) 1030-1038. https://doi.org/10.1016/S0141-0296(97)00200-9

[3] AISC 2016, Prequalified connections for special and intermediate steel moment frames for seismic applications. ANSI/AISC 358-16. Chicago.

[4] EN 1993-1-8:2006, Eurocode 3: Design of Steel Structures Part 1-8: Design of Joints, 2006.

[5] A. Plumier, The dogbone: back to the future. Eng. J. 34 (1997) 61-67.

[6] J. Shen, T. Kitjasateanphun, W. Srivanich, Seismic performance of steel moment frames with reduced beam sections, Eng. Struct. 22 (8) (2000) 968–983. https://doi.org/10.1016/S0141-0296(99)00048-6

[7] T. A. Horton, I. Hajirasouliha, B. Davison, Z. Ozdemir, More efficient design of reduced beam sections (RBS) for maximum seismic performance, J. of Constr. Steel Res., 183 (2021) 106728. https://doi.org/10.1016/j.jcsr.2021.106728

[8] S. Benfratello, L. Palizzolo, Limited resistance rigid perfectly plastic hinges for steel frames, Intern. Rev. Civ. Eng. 8 (6) (2017) 286–298. https://doi.org/10.15866/irece.v8i6.13190

[9] S. Benfratello, C. Cucchiara, P. Tabbuso, Fixed strength and stiffness hinges for steel frames, in: Feo L., Ascione L., Berardi V.P., Fraternali F., Tralli A.M. (Eds.), AIMETA 2017 – Proc. of the 23rd Conf. of the Italian Association of Theoretical and Applied Mechanics, pp. 1287–1296.

[10] S. Benfratello, L. Palizzolo, P. Tabbuso, S. Vazzano, On the post elastic behavior of LRPH connections, International Review on Modelling and Simulations, 12 (2019) 341-353. https://doi.org/10.15866/iremos.v12i6.18294

[11] L. Palizzolo, S. Benfratello, P. Tabbuso, S. Vazzano, Numerical validation of LRPH behaviour by fem analysis, Advances in Engineering Materials, Structures and Systems: Innovations, Mechanics and Applications - Proceedings of the 7th International Conference on Structural Engineering, Mechanics and Computation, 2019, 2019, 1224–1229. https://doi.org/10.1201/9780429426506-212

[12] S. Benfratello, L. Palizzolo, P. Tabbuso, S. Vazzano, LRPH device optimization for axial and shear stresses, Intern. Rev. Civ. Eng. 11 (4) (2020) 152–163. https://doi.org/10.15866/irece.v11i4.18100

[13] S. Benfratello, S. Caddemi, L. Palizzolo, B. Pantò, D. Rapicavoli, S. Vazzano, Smart beam element approach for lrph device, in: Carcaterra A., Graziani G., Paolone A. (Eds), 24th Conference of the Italian Association of Theoretical and Applied Mechanics, AIMETA 2019, Lecture Notes in Mechanical Engineering (2020), 197-213. https://doi.org/10.1007/978-3-030-41057-5_16

[14] S. Benfratello, S. Caddemi, L. Palizzolo, B. Pantò, D. Rapicavoli, S. Vazzano, Targeted steel frames by means of innovative moment resisting connections, Journal of Constructional Steel Research, 183 (2021) 106695. https://doi.org/10.1016/j.jcsr.2021.106695

[15] S. Benfratello, L. Palizzolo, S. Vazzano, A New Design Problem in the Formulation of a Special Moment Resisting Connection Device for Preventing Local Buckling, Journal of Applied Sciences, 12 (2022) 202. https://doi.org/10.3390/app12010202

[16] UNI EN 10002-1 (2004), Metallic materials - Tensile testing - Part 1: Method of test at ambient temperature.

[17] S. Benfratello, V. Fiore, L. Palizzolo, P. Tabbuso, Speckle Interferometry Analysis of Full-bending Behavior of GFRP Pultruded Material, Procedia Engineering, 161 (2016), 439-444. https://doi.org/10.1016/j.proeng.2016.08.587

[18] S. Benfratello, V. Fiore, L. Palizzolo, T. Scalici, Evaluation of continuous filament mat influence on the bending behaviour of GFRP pultruded material via Electronic Speckle Pattern Interferometry, Arch. Civ. Mech. Eng., 17 (2017), 169-177. https://doi.org/10.1016/j.acme.2016.09.009

[19] S. Benfratello, L. Palizzolo, P. Tabbuso, Evaluation of the bending behaviour of laminated glass beams via electronic speckle pattern interferometry, International Review on Modelling and Simulations, 10 (2017), 279-288. https://doi.org/10.15866/iremos.v10i4.12220

Theoretical and Applied Mechanics - AIMETA 2022 Materials Research Forum LLC
Materials Research Proceedings **26** (2023) 157-162 https://doi.org/10.21741/9781644902431-26

Innovative devices for the protection of welded sections in steel structures

Salvatore Benfratello[a*], Luigi Palizzolo[b] and Santo Vazzano[c]

Department of Engineering, University of Palermo, Viale delle Scienze, I-90128, Palermo, Italy

[a]salvatore.benfratello@unipa.it, [b]luigi.palizzolo@unipa.it, [c]santo.vazzano@unipa.it

Keywords: Steel Structures, Moment Resisting Connections, Innovative Devices

Abstract. The paper proposes the use of innovative devices devoted to the brittle collapse protection of welded steel sections, typically represented by the end beam cross-sections in framed structures. Reference is made to I-shaped cross-sections. At first, limiting to the case of plane stress, the relevant elastic domain is defined in the N, T, M space; then a plane frame equipped with the proposed devices and subjected to seismic load condition is studied, ensuring that the generalized stresses at the welded sections be within the relevant elastic domain.

Introduction

As it is well known, in the design of framed steel structures, special attention must be paid to the end beam sections connected to the columns. These cross-sections are usually welded to steel plates bolted to the flange columns. The welding can produce a modification of the material crystal lattice and, consequently, the transition from a ductile behaviour to a brittle one. Therefore, it is cautious to make suitable elastic check for the welded cross-sections adopting appropriate safety factors to avoid undesired brittle collapse.

A recommended strategy consists of limiting the stresses acting on the welded cross-sections by making use of special innovative devices [1-8] for beam-column connection, able to preserve the node integrity without modifying the elastic behaviour. As widely reported in [7-8], the proposed devices possess the property of suitably reducing the generalized stress conditions at the beam extreme maintaining unaltered the elastic bending stiffness. This latter feature makes the device different from the usual adopted ones available for structural designer [9-13] and approved by international codes (see e.g. [14-15]). In the present paper, referring to classical I-shaped profiles, the elastic domain of the relevant cross-sections is firstly defined, described in a suitable analytic form and represented in the N, T, M space. Then the improved optimal design problem is proposed and utilized for a simple plane frame, confirming the full reliability of the procedure and the goodness of the new optimal design formulation.

N, T, M elastic domain of a I-shaped cross-section

Referring to a typical I-shaped cross-section in the plane y, z of the principal axes of inertia (with y the greater inertia axis), the limit elastic axial force, shear force and pure bending moment have the form, respectively,

$$N^E = A\sigma_b; \quad T^E = \frac{\sigma_b I_y a}{2S'_y(G)}; \quad M^E = W_y^E \sigma_b \tag{1}$$

with A cross-section area, σ_b expected limit brittle stress evaluated as a fraction of the relevant material yield stress, I_y cross-section moment of inertia with respect to the y axis, a cross-section web thickness, $S'_y(G)$ statical moment of half cross-section with respect to the y axis, W_y^E cross-section elastic resistance modulus with respect to the y axis. In defining T^E, $\tau_b = \sigma_b/2$ has been assumed, being τ_b the chosen shear limit stress.

According with the Tresca yield criterion, the elastic domain boundary on the first quarter of the N, M plane is defined by the function

$$\frac{N}{A} + \frac{M}{W_y^E} = \sigma_b \tag{2}$$

holding for $0 \leq N \leq N^E$ e $0 \leq M \leq M^E$. On the other quarters the boundary can be defined imposing symmetry with respect to the N and M axes (Fig. 1a).

The elastic domain boundary on the first quarter of the N, T plane is defined by the function:

$$\left(\frac{N}{A}\right)^2 + 4\left[\frac{TS_y'(G)}{I_y a}\right]^2 = \sigma_b^2 \tag{3}$$

holding for $0 \leq N \leq N^E$ and $0 \leq T \leq T^E$. On the other quarters the boundary can be defined imposing symmetry with respect to the N and T axes (Fig. 1b).

The elastic domain boundary on the first quarter of T, M plane is defined by combining the function (flange maximum stress value):

$$\left(\frac{M}{W_y^E}\right)^2 + 4\left[\frac{TS_y'(E)}{I_y e}\right]^2 = \sigma_b^2 \tag{4}$$

holding for $0 \leq T \leq T^E$ and $0 \leq M \leq M^E$, and the discrete set of couples of values of shear and bending moment obtained by the solution to the minimum problem (web maximum stress value):

$$\min_{(z)} M \qquad \text{subjected to}$$

$$\left(\frac{M}{I_y} z\right)^2 + 4\left[\frac{\overline{T} S_y'(z)}{I_y a}\right]^2 \geq \sigma_b^2 \tag{5}$$

$$0 \leq z \leq \left(\frac{h}{2} - e\right)$$

for a prefixed discrete set of values of $\overline{T}$ $\left(0 \leq \overline{T} \leq T^E\right)$. On the other quarters the boundary can be defined imposing symmetry with respect to the T and M axes (Fig. 1c).

The elastic domain boundary surface on the first octant of N, T, M space is defined by combining the function (flange maximum stress value):

$$\left(\frac{N}{A} + \frac{M}{W_y^E}\right)^2 + 4\left[\frac{\overline{T} S_y'(E)}{I_y e}\right]^2 = \sigma_b^2 \tag{6}$$

which provides a discrete set of functions N, M in correspondence to an analogous discrete set of values of $\overline{T}$ $\left(0 \leq \overline{T} \leq T^E\right)$, and the discrete set of values of axial force, shear force and bending moment obtained by the solution to the minimum problem (web maximum stress value):

$$\min_{(z)} T \qquad \text{subjected to}$$

$$\left(\frac{\overline{N}}{A} + \frac{\overline{M}}{I_y} z\right)^2 + 4\left[\frac{TS_y'(z)}{I_y a}\right]^2 \geq \sigma_b^2 \tag{7}$$

$$0 \leq z \leq \left(\frac{h}{2} - e\right)$$

for a prefixed discrete set of couples of values of $\overline{N}$ and $\overline{M}$, with $0 \leq \overline{N} \leq N^E$ and $0 \leq \overline{M} \leq M^E$. The domain boundary is completed by symmetry with respect the coordinate planes (Fig. 1d).

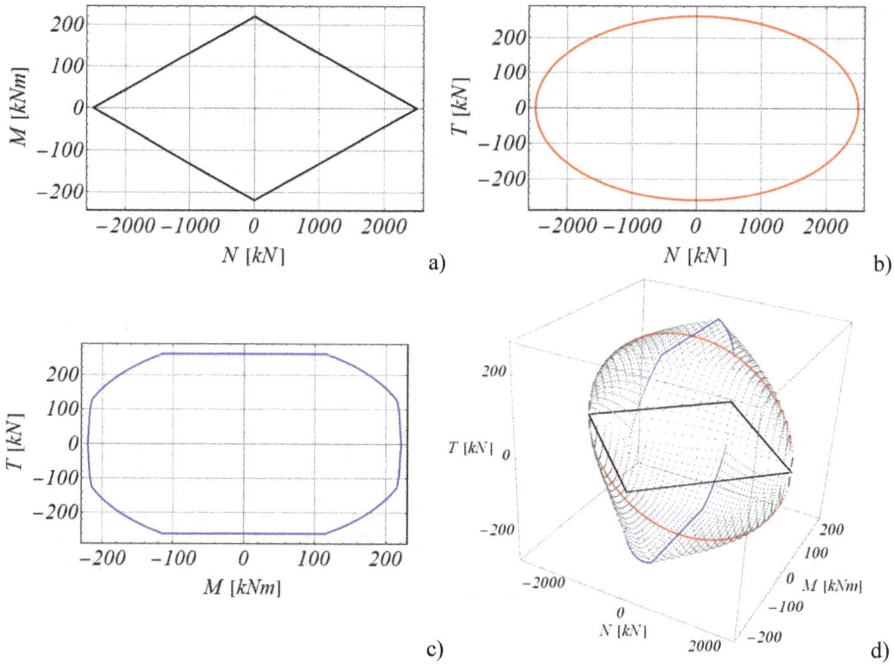

Fig. 1 – HEB240 profile (S235) elastic domain: a) N, M plane; b) N, T plane; c) M, T plane; d) N, M, T space.

Application

Let us consider the plane frame sketched in Fig. 2.

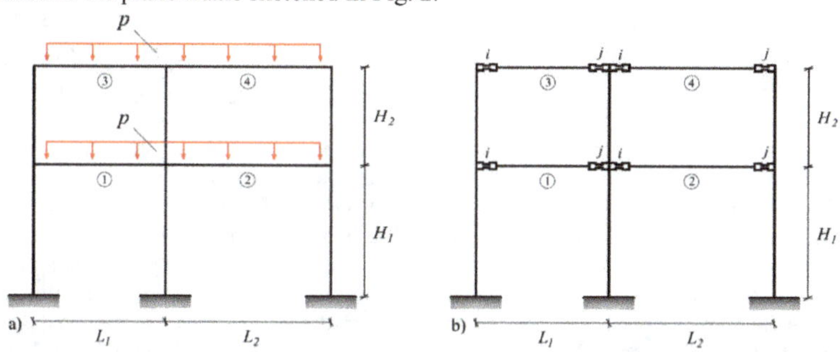

Fig. 2 – Steel plane frame: a) geometry and load condition; b) position of the LRPD.

The material constituting the frame is a steel S275 grade ($E = 210.000$ MPa). The geometrical data are: $L_1 = 4$ m; $L_2 = 5$ m; $H_1 = 4$ m; $H_2 = 3$ m. The distributed load $p = 45$ kN/m is the sum of $G_1 = 18$ kN/m (dead load), $G_2 = 12$ kN/m (permanent loads) and $Q_k = 15$ kN/m

Materials Research Proceedings **26** (2023) 157-162 https://doi.org/10.21741/9781644902431-26

(variable load). Beams 1 and 3 are constituted by HEA220 profiles, beams 2 and 4 are constituted by HEA260 profiles. The columns are constituted by HEB340 profiles.

At first, a modal dynamic analysis has been developed for the structure with behaviour factor $q = 4$, verifying the compliance of the ductile response of the element structure with the considered Italian code. Then the same analysis has been developed with behaviour factor $q = 1$. In this last case it has been verified, as expected, that the beam element extremes suffer load conditions above the related elastic domain. In Table I are reported the generalized stress values (N, T, M) in correspondence of all the beam element extremes and the related elastic limit bending moment value M^d for the assigned couple of N and T considering $\sigma_b = 0.9\,\sigma_0$.

***Table I**. Generalized stress response and limit bending moments for the beam end cross-section.*

	Beam internal forces							
	1 *i*	1 *j*	2 *i*	2 *j*	3 *i*	3 *j*	4 *i*	4 *j*
N	67.321	39.657	45.332	60.18	47.037	53.793	64.156	76.68
T	127.624	135.362	164.046	157.34	120.449	135.151	165.671	153.829
M	15,211.25	15,618.51	22,889.04	22,459.2	13,644.86	14,437.84	20,676.48	19,864.96
M^d	**10,672.07**	**10,939.92**	**17,911.30**	**18,306.07**	**11,661.63**	**10,359.52**	**17,691.75**	**18,270.61**

The values $N_{\alpha k}, T_{\alpha k}$ and $M^d_{\alpha k}$ for $\alpha = 1,2,3,4$ and $k = 1,2$ have been utilized for designing the LRPD devices (for the specific geometry details see [1-8]). Imposing, as usual, for the internal portion length $\ell_i = 0.5h$, with h indicating the height of the relevant cross-section, the optimal design problem results (see [7-8]) are listed in the Table II.

***Table II**. Optimal design results.*

	LRPD Optimal dimensions							
	1 *i*	1 *j*	2 *i*	2 *j*	3 *i*	3 *j*	4 *i*	4 *j*
h^*	18.0887	18.0725	21.8816	21.984	18.4181	17.5866	21.8009	22.0183
$t_{f,o}$	2.9113	2.9275	3.1184	3.016	2.5819	3.4134	3.1991	2.9817
$t_{f,i}$	1.1	1.1	1.25	1.25	1.1	1.1	1.25	1.25
ℓ_o	15.8804	15.4395	16.1619	14.8654	11.6727	19.5643	17.2592	14.4467
bi	10.3949	10.595	13.7983	14.5868	12.7731	8.9274	13.1891	14.8587

The considered frames were both studied by performing a non-linear dynamic analysis subjected to an assigned seismic time history compatible with the response spectrum (Fig. 3).

In Fig. 4 the bending moment response function related to the extremes of beam elements 2 and 3 is plotted where the dashed lines represent the limit bending moment imposed to avoid any undesired brittle behaviour. As it is possible to observe, the original frame exhibits a generalized stress response that in many instants is higher than the prescribed assigned limit value, resulting above the elastic domain previously defined. Whereas, as it was expected, the frame equipped with the suitably designed devices, exhibits a generalized stress response substantially brittle safe. The rare instants in which the prescribed limit is exceeded are due to the evident difference between the spectral analysis and compatible seismic time history and they can be accepted from a practical point of view or avoided by imposing more stringent brittle safety factor.

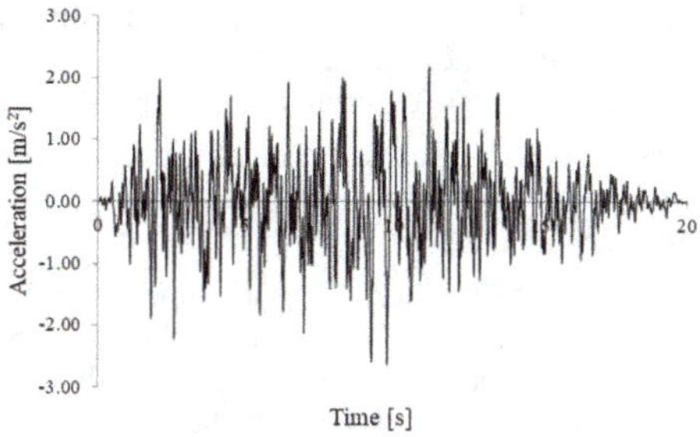

Fig. 3 – Accelerogram applied to the studied frames.

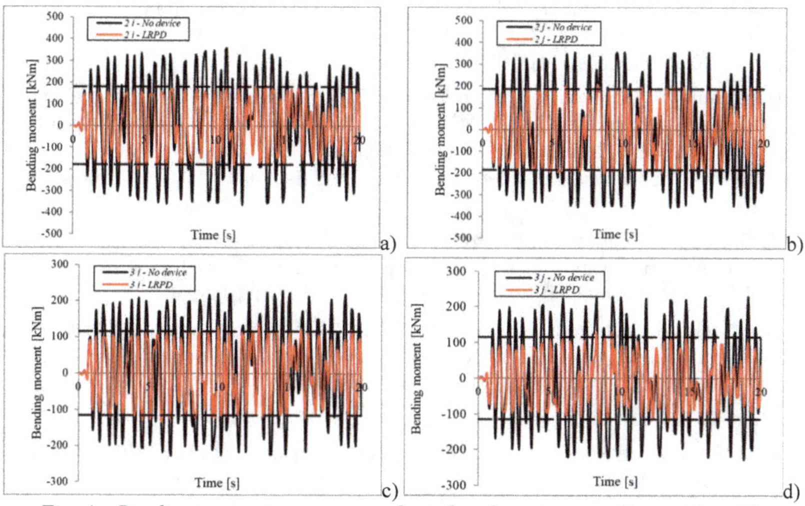

Fig. 4 – Bending moment response evaluated at the extremes of beams 2 and 3.

Conclusions

In the present paper, a special strategy devoted of limit the generalized stresses acting on the welded cross-sections of a frame structure by making use of some innovative devices for beam-column connections, able to preserve the node integrity without modifying the elastic behaviour, is proposed.

The computational procedure consists of evaluating the limit elastic bending moment on the relevant cross-sections complying the reference N, T, M domain and designing the LRPD devices for the assigned limit stress values able to prevent brittle behaviour as a percentage of elastic limit stress suitably selected depending on the welding methodology. The performed numerical applications, related to a simple plane frame, confirm the effectiveness of the proposed strategy.

References

[1] S. Benfratello, L. Palizzolo, Limited resistance rigid perfectly plastic hinges for steel frames, Intern. Rev. Civ. Eng. 8 (6) (2017) 286–298. https://doi.org/10.15866/irece.v8i6.13190

[2] S. Benfratello, C. Cucchiara, P. Tabbuso, Fixed strength and stiffness hinges for steel frames, in: Feo L., Ascione L., Berardi V.P., Fraternali F., Tralli A.M. (Eds.), AIMETA 2017 – Proc. of the 23rd Conf. of the Italian Association of Theoretical and Applied Mechanics, pp. 1287–1296.

[3] S. Benfratello, L. Palizzolo, P. Tabbuso, S. Vazzano, On the post elastic behavior of LRPH connections, International Review on Modelling and Simulations, 12 (2019) 341-353. https://doi.org/10.15866/iremos.v12i6.18294

[4] L. Palizzolo, S. Benfratello, P. Tabbuso, S. Vazzano, Numerical validation of LRPH behaviour by fem analysis, Advances in Engineering Materials, Structures and Systems: Innovations, Mechanics and Applications - Proceedings of the 7th International Conference on Structural Engineering, Mechanics and Computation, 2019, 2019, 1224–1229. https://doi.org/10.1201/9780429426506-212

[5] S. Benfratello, L. Palizzolo, P. Tabbuso, S. Vazzano, LRPH device optimization for axial and shear stresses, Intern. Rev. Civ. Eng. 11 (4) (2020) 152–163. https://doi.org/10.15866/irece.v11i4.18100

[6] S. Benfratello, S. Caddemi, L. Palizzolo, B. Pantò, D. Rapicavoli, S. Vazzano, Smart beam element approach for lrph device, in: Carcaterra A., Graziani G., Paolone A. (Eds), 24th Conference of the Italian Association of Theoretical and Applied Mechanics, AIMETA 2019, Lecture Notes in Mechanical Engineering (2020), 197-213. https://doi.org/10.1007/978-3-030-41057-5_16

[7] S. Benfratello, S. Caddemi, L. Palizzolo, B. Pantò, D. Rapicavoli, S. Vazzano, Targeted steel frames by means of innovative moment resisting connections, Journal of Constructional Steel Research, 183 (2021) 106695. https://doi.org/10.1016/j.jcsr.2021.106695

[8] S. Benfratello, L. Palizzolo, S. Vazzano, A New Design Problem in the Formulation of a Special Moment Resisting Connection Device for Preventing Local Buckling, Journal of Applied Sciences, 12 (2022) 202. https://doi.org/10.3390/app12010202

[9] A. Plumier, The dogbone: back to the future. Eng. J. 34 (1997) 61-67.

[10] D.K. Miller, Lessons learned from the Northridge earthquake,Eng. Struc. 20 (1998), 249–260. https://doi.org/10.1016/S0141-0296(97)00031-X

[11] J. Shen, T. Kitjasateanphun, W. Srivanich, Seismic performance of steel moment frames with reduced beam sections, Eng. Struct. 22 (8) (2000) 968–983. https://doi.org/10.1016/S0141-0296(99)00048-6

[12] A. Saleh, SR Mirghaderi, SM Zahrai, Cyclic testing of tubular web RBS connections in deep beams. J Constr Steel Res 117 (2016) 214-226. https://doi.org/10.1016/j.jcsr.2015.10.020

[13] T. A. Horton, I. Hajirasouliha, B. Davison, Z. Ozdemir, More efficient design of reduced beam sections (RBS) for maximum seismic performance, J. of Constr. Steel Res., 183 (2021) 106728. https://doi.org/10.1016/j.jcsr.2021.106728

[14] AISC 2016, Prequalified connections for special and intermediate steel moment frames for seismic applications. ANSI/AISC 358-16. Chicago.

[15] EN 1993-1-8:2006, Eurocode 3: Design of Steel Structures Part 1-8: Design of Joints, 2006.

Theoretical and Applied Mechanics - AIMETA 2022 Materials Research Forum LLC
Materials Research Proceedings **26** (2023) 163-168 https://doi.org/10.21741/9781644902431-27

Stress and strain fields in non-prismatic inhomogeneous beams

Giovanni Migliaccio

Department of Engineering and Architecture, University of Parma, Parma, Italy

giovanni.migliaccio.it@gmail.com

Keywords: Analytical Solutions, Numerically Efficient Method, Large Displacements

Abstract. Beamlike structures are commonly studied via 1D beam models, which are more efficient than 3D finite element methods, but do not permit accurate predictions of 3D stresses in non-prismatic cases. Despite the progress made in their modeling, either via direct 1D approaches or dimensional reductions from 3D formulations, the accurate analytical prediction of stresses and strains in beamlike yet 3D elements, with non-uniform properties both in terms of materials and cross-section shape, subject to large displacements, is an open problem. This work presents a model for such elements that is particularly suitable for efficient numerical implementations and that allows accurate analytical predictions of stresses and strains. A paradigmatic example shows the importance of non-trivial stress terms that are absent in prismatic homogeneous elements and the inadequacy of usual beam models and stepped-beam approaches when dealing with predictions of stresses and strains in non-prismatic inhomogeneous cases.

Introduction

The analytical prediction of stresses and strains in non-prismatic beamlike elements is very challenging. In the study of helicopter rotor blades, for instance, one needs to properly account for non-trivial couplings among bending, torsion, and traction associated with the cross-sectional pre-twist [1-3]. In addition to this, further complexities are present in wind turbine blades, which are also tapered [4-5]. Not to mention the complexities associated with large displacements and complex material properties. However, analytical predictions of stresses and strains are not simple even in prismatic homogenous isotropic cases. A paradigmatic example is represented by the classical Saint-Venant's problem [6-7], which is often addressed via approximated methods because of the difficulty to find closed-form solutions and the need for application-oriented formulas for cases of engineering interest. Jourawski's formula is one such application-oriented solutions [8], but it holds only for linear elastic prismatic beams whose material is homogeneous and isotropic, and provides erroneous predictions in non-prismatic cases [3,9,10].

Following Jourawski's method, several scholars have proposed formulas for shear stresses in tapered beams, e.g., Bleich [11], Pugsley and Weatherhead [12], Krahula [13], Bertolini et al. [14], Balduzzi et al. [10]. A review about critical issues and deficiencies of current engineering methods when dealing with tapered beams can be found in the works of Paglietti and Carta [9], Balduzzi et al. [10], and Migliaccio et al. [15]. Literature works usually address tapered beams undergoing small displacements, assume Navier's formula for the cross-sectional normal stresses, and derive formulas for the corresponding shear stresses by the static equilibrium of a beam slice in its reference (undeformed) state. Migliaccio's works [5,15,16] do not rely on such assumptions, account for the effects of large displacements, and is the basis of the present study.

Here, after introducing a mechanical model tailored to non-prismatic inhomogeneous beams susceptible to large deflections, an analytical method is proposed to accurately predict their stress and strain fields. Finally, the inadequacy of usual beam models when dealing with predictions of stresses and strains in the considered elements is discussed.

Mechanical model

The present beamlike body is a set of deformable plane figures (cross-sections) attached at a deformable line (center-line). For convenience, cross-sections in the reference state are orthogonal to the center-line. Changes from the reference state to the actual (deformed) state may take place with center-line large displacements and small strains, while cross-sections may undergo warping displacements, in and out of plane, which produce small deformation. See Figure 1.

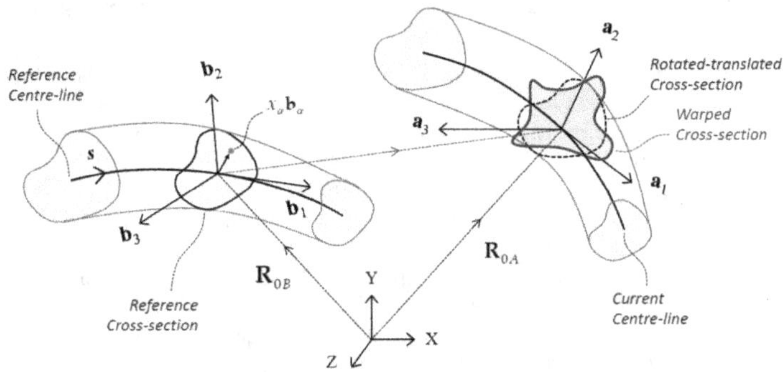

Figure 1: Schematic reference (left) and actual (right) states of non-prismatic beamlike element.

Using summation convention, with Latin (Greek) indices taking values on 1, 2, 3 (2, 3), the positions of the body points in the reference state, $\mathbf{R}_B$, and in the actual state, $\mathbf{R}_A$, are written as

$$\begin{aligned}\mathbf{R}_B(z_i) &= \mathbf{R}_{0B}(z_1) + x_\alpha(z_i)\mathbf{b}_\alpha(z_1),\\ \mathbf{R}_A(z_i,t) &= \mathbf{R}_{0A}(z_1,t) + x_\alpha(z_i)\mathbf{a}_\alpha(z_1,t) + w_k(z_i,t)\mathbf{a}_k(z_1,t),\end{aligned} \tag{1}$$

where $\mathbf{R}_{0B}$ and $\mathbf{R}_{0A}$ identify the center-line points in the reference and actual states with respect to a fixed Cartesian frame X,Y,Z; $\mathbf{b}_i=\mathbf{b}_i(s)$ is the local triad of orthogonal unit vectors in the reference state; $\mathbf{b}_1$ is tangent to the reference center-line of arc-length s; $\mathbf{a}_i=\mathbf{a}_i(s,t)$ is the image $\mathbf{b}_i$ in the current state and generally depends on the arc-length s and time t; x_α identify the cross-section points; w_i are warping displacements; finally, z_i are time-independent variables, with $z_1=s$ and z_α spanning a two-dimensional domain mapping the positions of the cross-section points. In this work, $x_i=\Lambda_{ij}z_j$, $\Lambda_{11}=1$, $\Lambda_{22}=\Lambda_2(z_1)$, $\Lambda_{33}=\Lambda_3(z_1)$, and coefficients Λ_{ij} with $i\neq j$ identically vanish.

The body strain state is described via the Green-Lagrange strain tensor $\mathbf{E}$ and the vector fields $\boldsymbol{\gamma} = \mathbf{T}^T\mathbf{R}'_{0A} - \mathbf{R}'_{0B}$ and $\mathbf{k} = \mathbf{T}^T\mathbf{k}_A - \mathbf{k}_B$, where $\mathbf{k}_B$ and $\mathbf{k}_A$ are curvature vectors in the reference and actual states, $\mathbf{T} = \mathbf{a}_i \otimes \mathbf{b}_i$, $\otimes$ is tensor product, and prime denotes s-derivative. Specifically, for small strain and warping fields, $\mathbf{E}$ is written as in [16], i.e., $2\mathbf{E} = \mathbf{T}^T\mathbf{H}+\mathbf{H}^T\mathbf{T}-2\mathbf{I}$, where $\mathbf{H}$ is the derivative of $\mathbf{R}_A$ with respect to $\mathbf{R}_B$ (deformation gradient) and $\mathbf{I}$ is the identity tensor.

The body stress state is described via the second (symmetric) Piola-Kirchhoff stress tensor $\mathbf{S} = S_{ij}\,\mathbf{b}_i \otimes \mathbf{b}_j$, which is given in terms of $\mathbf{E} = E_{ij}\,\mathbf{b}_i \otimes \mathbf{b}_j$ via a linear elastic constitutive model,

$$\mathbf{S} = \frac{Y}{1+\nu}\mathbf{E} + \frac{\nu Y}{(1+\nu)(1-2\nu)} tr\mathbf{E}\,\mathbf{I}, \tag{2}$$

where Y and ν are Young's modulus and Poisson's ratio, respectively. For completeness, it is useful introducing also the Cauchy stress tensor, $\mathbf{C} = C_{ij}\,\mathbf{a}_i \otimes \mathbf{a}_j$, which is related to $\mathbf{S}$ via the deformation gradient $\mathbf{H}$ and is associated with the stress in the body actual state [3,16].

Body stresses and strains can be obtained as solutions of balance equations (partial differential equations with boundary conditions) derivable from the principle of virtual power. To this aim, we introduce the external power functional Π_e (to describe the interactions between the body and the external environment), and the internal power functional Π_i (to describe the interactions among the body parts), as follows

$$\Pi_e = \int_V \mathbf{b}\cdot\mathbf{v} + \int_{\partial V} \mathbf{p}\cdot\mathbf{v}, \quad \Pi_i = \frac{d}{dt}\int_V \Phi, \tag{3}$$

where **b** represents body loads per unit body reference volume V, **p** denotes surface actions per unit area of the boundary ∂V, **v** is the time rate of the actual position of the body points, and the strain energy density $\Phi=\mathbf{S}\cdot\mathbf{E}/2$ is half the scalar product of **S** and **E**. Balance equations are finally obtainable by enforcing that for any velocity field attainable by the body its interactions with the external environment and among its parts are such that the total power vanishes (i.e., $\Pi_e=\Pi_i$) at any value of the evolution parameter t.

Following [16], and studying inhomogeneous beams with tapered, pre-twisted cross-sections and external actions (as in a Saint-Venant's beam) applied only at the end cross-sections, we write the cross-sectional strains, E_{11}, $E_{\alpha 1}$, in the form

$$\begin{aligned} E_{11} &= k_2x_3 - k_3x_2 + \gamma_1 + e_{1,1} + k_{B1}(x_3e_{1,2} - x_2e_{1,3}), \\ 2E_{21} &= e_{1,2} - k_1x_3 + e_2 + 2(1+\nu)(k_2x_3 - k_3x_2 + \gamma_1)(\Lambda_2^{-1}\Lambda_2'x_2 - k_{B1}x_3), \\ 2E_{31} &= e_{1,3} + k_1x_2 + e_3 + 2(1+\nu)(k_2x_3 - k_3x_2 + \gamma_1)(\Lambda_3^{-1}\Lambda_3'x_3 + k_{B1}x_2), \end{aligned} \tag{4}$$

where E_{11} and $E_{\alpha 1}$ are normal and shear strains in the cross-sections, γ_1 is center-line extension, k_α and k_1 are bending and torsion curvatures, k_{B1} is pre-twist function, Λ_α are taper functions, comma denotes x_i-derivate, and the scalar fields e_1, e_2, e_3 are solutions of the PDEs problem

$$\begin{aligned} & e_{1,22} + e_{1,33} = 0 \quad in\ \Sigma, \\ & e_{2,2} + e_{3,3} = d_{k_3'}k_3' + d_{k_3}k_3 + d_{k_2'}k_2' + d_{k_2}k_2 \quad in\ \Sigma, \\ & e_{3,2} - e_{2,3} = g_{k_3'}k_3' + g_{k_3}k_3 + g_{k_2'}k_2' + g_{k_2}k_2 + g_{\gamma_1}\gamma_1 \quad in\ \Sigma, \\ & (e_{1,2} - k_1x_3)n_2 + (e_{1,3} + k_1x_2)n_3 = 0 \quad on\ \partial\Sigma, \\ & e_2n_2 + e_3n_3 = 0 \quad on\ \partial\Sigma. \end{aligned} \tag{5}$$

In (5), Σ and $\partial\Sigma$ are cross-sectional domain and its boundary, n_α are components of the outward unit normal on $\partial\Sigma$, Λ_Y is the ratio between the material Young's modulus Y at a generic s and its reference value at s=0, and coefficients $d_{(\cdot)}$ and $g_{(\cdot)}$ are defined as follows

$$\begin{aligned} d_{k_3'} &= 2(1+\nu)x_2, \\ d_{k_3} &= 2(1+\nu)\left(\Lambda_3^{-1}\Lambda_3' + 2\Lambda_2^{-1}\Lambda_2' + \Lambda_Y^{-1}\Lambda_Y'\right)x_2, \\ d_{k_2'} &= -2(1+\nu)x_3, \\ d_{k_2} &= -2(1+\nu)\left(\Lambda_2^{-1}\Lambda_2' + 2\Lambda_3^{-1}\Lambda_3' + \Lambda_Y^{-1}\Lambda_Y'\right)x_3, \\ g_{\gamma_1} &= 2k_{B1}(2+2\nu), \\ g_{k_3'} &= -2\nu x_3, \\ g_{k_3} &= -2(1+\nu)\Lambda_3^{-1}\Lambda_3'x_3 - 2k_{B1}(3+2\nu)x_2, \\ g_{k_2'} &= -2\nu x_2, \\ g_{k_2} &= -2(1+\nu)\Lambda_2^{-1}\Lambda_2'x_2 + 2k_{B1}(3+2\nu)x_3. \end{aligned} \tag{6}$$

Equations (4)-(6) show that the cross-sectional strains E_{i1} explicitly depend on the geometric and material characteristics of the body (Λ_α, k_{B1}, Λ_Y) and can be expressed as linear combinations of the 1D strains γ_1 and k_i and their s-derivative. This result is interesting as it allows us to consider the individual effect of each 1D strain by solving PDEs whose solution only depends on the shape of the cross-sectional domain Σ. Such PDEs admit closed-form solutions only in a few cases (e.g., [3,16]), but can always be solved numerically. However, apart from directly determining the scalar fields e_i and the relevant strains and stresses, we can obtain analytical information on these latter without explicitly solving the PDEs (5)-(6), as is discussed in the following section.

Analytical results

Equations (4)-(6), together with (2), can be used to derive a closed-form expression of the flow of shear stress through the cross-sectional chords of non-prismatic inhomogeneous beams. To show this, let us consider the generic cross-section in Figure 2 (left): the boundary of the dashed domain Σ_q is oriented counter-clockwise and is composed of internal lines $\partial\Sigma_i$ and external lines $\partial\Sigma_e$, whose outward normal **n** and tangent **t** have components n_α and t_α, respectively.

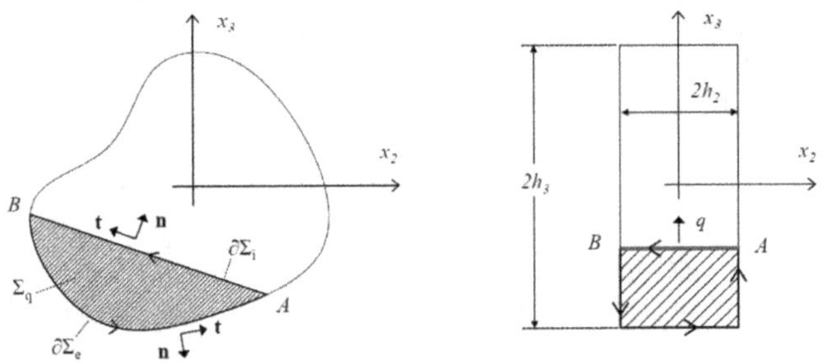

Figure 2: Generic reference cross-section of a beamlike solid (left) and rectangular shape (right).

The cross-sectional shear flow q through the internal lines $\partial\Sigma_i$ is defined via the line integral

$$q = \int_{\partial\Sigma_i} S_{1\alpha} n_\alpha \, , \tag{7}$$

and, by using (2), (4)-(6), and integration methods based on Green's formulas, takes the form

$$q = -YS_2 k_2' + YS_3 k_3' - YZ_2 k_2 + YZ_3 k_3 + YZ_1 \gamma_1 \, , \tag{8}$$

where coefficients S_α and Z_i are defined by the following surface and line integrals

$$S_2 = \int_{\Sigma_q} x_3, \quad S_3 = \int_{\Sigma_q} x_2 \, , \tag{9}$$

$$\begin{aligned} Z_2 &= \int_{\partial\Sigma_e} x_3 (\Lambda_2^{-1}\Lambda_2' x_2 n_2 + \Lambda_3^{-1}\Lambda_3' x_3 n_3) + k_{B1} \int_{\partial\Sigma_i} x_3 x_\alpha t_\alpha + \Lambda_Y^{-1}\Lambda_Y' \int_{\Sigma_q} x_3, \\ Z_3 &= \int_{\partial\Sigma_e} x_2 (\Lambda_2^{-1}\Lambda_2' x_2 n_2 + \Lambda_3^{-1}\Lambda_3' x_3 n_3) + k_{B1} \int_{\partial\Sigma_i} x_2 x_\alpha t_\alpha + \Lambda_Y^{-1}\Lambda_Y' \int_{\Sigma_q} x_2, \\ Z_1 &= \int_{\partial\Sigma_i} \Lambda_2^{-1}\Lambda_2' x_2 n_2 + \Lambda_3^{-1}\Lambda_3' x_3 n_3 + k_{B1} \int_{\partial\Sigma_e} x_\alpha t_\alpha . \end{aligned} \tag{10}$$

Equations (8)-(10) show the explicit dependence of the shear flow q on the body geometric and material characteristics (taper functions Λ_α, pre-twist function k_{B1}, material function Λ_Y). Note that coefficients S_α are classical area moments and are present also in prismatic cases. On the contrary, coefficients Z_i account for the effects of taper, pre-twist, and material inhomogeneity, which are absent in prismatic homogeneous cases and are unpredictable via usual linear theories of prismatic homogenous beams. Specifically, the shear formula (8) is a generalization of results derivable from the classical Saint-Venant's theory (such as Jourawski's formula), as well as of formulas presented in recent works [5,15]. Moreover, it is worth noting that if the s-derivative of the material and taper

Materials Research Proceedings **26** (2023) 163-168 https://doi.org/10.21741/9781644902431-27

functions vanish, together with the pre-twist function (i.e., in the prismatic homogenous case), the cross-sectional shear flow q will depend only on the s-derivative of the bending curvatures k_α via the area moments S_α. Furthermore, if the actual and reference states of the body can be considered almost coincident for equilibrium purposes (small displacements), the s-derivative of the bending curvatures k_α turn out to be proportional to the transverse shear forces calculated in the reference state of the body and equation (8) reduces exactly to the classical Jourawski's solution [8].

The shear formula (8) can be used for analytical predictions of shear stresses in thin-walled non-prismatic beams with spanwise variable material properties in the same way as Jourawski's formula is used in prismatic homogenous cases. We show this via a paradigmatic example in which the cross-sectional shape is rectangular, as in the original Jourawski's work [8], but its dimensions are varied along the beam length, together with the material properties.

Variable dimensions and material properties: a paradigmatic example.
Let us consider a straight beam with rectangular cross-sections, which are tapered from the root to the tip according to the taper functions $\Lambda_2=\Lambda_3=\Lambda(s)$. Let us assume that the material properties can vary as well: specifically, the Young's modulus is given in the form $Y(s)=Y_0\Lambda_Y(s)$, where Y_0 is the Young's modulus at the root section (clearly, for constant material properties $\Lambda_Y(s)=1$).

To illustrate the analytical procedure for predicting stresses and strains in elements of this kind, let us consider the case in which this beam is fixed at the root ($s=0$) and is bent within the plane x_1-x_3 by a tip load applied in the same plane. In such case, coefficients S_2, Z_2, Z_1 are

$$S_2 = h_2(x_3^2 - h_3^2),\quad Z_2 = h_2(x_3^2 - 3h_3^2)\Lambda^{-1}\Lambda',\quad Z_1 = 2h_2x_3\Lambda^{-1}\Lambda', \tag{11}$$

and the cross-sectional shear flow, from (8) and (11), turns out to be expressible in the form

$$\frac{q}{2h_2Y} = -\frac{x_3^2 - h_3^2}{2}k_2' - \frac{\Lambda'}{\Lambda}\left(\frac{x_3^2 - 3h_3^2}{2}k_2 - x_3\gamma_1\right) - \frac{\Lambda_Y'}{\Lambda_Y}\frac{x_3^2 - h_3^2}{2}k_2, \tag{12}$$

where $2h_2(s)$, $2h_3(s)$ are the dimensions of the rectangular cross-section in Figure 2 (right) at the axial coordinate s, and $q/(2h_2)$ represents the mean shear stress orthogonal to the chord AB.

As is apparent, equation (12) extends the well-known Jourawski's solution [8]: the mean shear stress over a chord parallel to the cross-section width $2h_2$ consists of a first Jourawski-like term (proportional to the s-derivative of the bending curvature k_2), plus additional terms proportional to the s-derivative of taper and material functions, which are needed to accurately predict the shear stresses in non-prismatic inhomogeneous cases, as is also confirmed by numerical analyses for the present beam shape (not reported here for brevity), as well as from the analyses performed on other beam shapes in recent works (e.g., [3,5,15,16]).

Conclusions

Non-prismatic inhomogeneous beams undergoing large displacements are characterized by stress distributions that can be quite different from those predictable by linear theories of prismatic homogeneous beams. The approach proposed in this work allows accurate analytical predictions of stresses and strains in such elements.

In a paradigmatic example, we have shown the importance of geometric and material functions that produce non-trivial stress distributions, which are unpredictable via usual beam models and which confirm the inadequacy of stepped-beam approaches when dealing with predictions of stresses and strains in non-prismatic inhomogeneous cases. These results are also corroborated by several numerical analyses, some of which are reported in recent works by the author.

The results of this paper focus on the cross-sectional out-of-plane deformations and the relevant stress fields. Investigations about the in-plane deformations and the associated stress fields, as well as analytical studies about the influence of other important parameters (e.g., an initial curvature of the beam), will be addressed in subsequent extended works.

References

[1] Krenk S., The torsion-extension coupling in pretwisted elastic beams, Int. J. Solids Struct., 19, 67-72, 1983. https://doi.org/10.1016/0020-7683(83)90038-0

[2] Rosen A., Structural and dynamic behavior of pretwisted rods and beams, ASME, 44, 483-515, 1991. https://doi.org/10.1115/1.3119490

[3] Migliaccio G., Ruta G., The influence of an initial twisting on tapered beams undergoing large displacements, Meccanica, 56, 7, 1831-1845, 2021. https://doi.org/10.1007/s11012-021-01334-2

[4] Griffith D.T., Ashwill T.D., et al., Large offshore rotor development: design and analysis of the Sandia 100-meter wind turbine blade, 53rd Structures, Structural Dynamics and Materials Conference, 23-26 April, 2012. https://doi.org/10.2514/6.2012-1499

[5] Migliaccio G., Analytical determination of the influence of geometric and material design parameters on the stress and strain fields in non-prismatic components of wind turbines, J. Phys. Conf. Ser., 2265, 032033, 2022. https://doi.org/10.1088/1742-6596/2265/3/032033

[6] Love A.E.H., A treatise on the mathematical theory of elasticity, 4th ed., Dover Publications, 1944.

[7] Muskhelishvili N.I., Some basic problems of the mathematical theory of elasticity, Springer Verlag, 1977. https://doi.org/10.1007/978-94-017-3034-1

[8] Jourawski D.I., Sur la résistance d'un corps prismatique et d'une piéce composée en bois ou on tôle de fer à une force perpendiculaire à leur longeur. Annales Des Ponts Et Chaussées, 12, 328-351, 1856.

[9] Paglietti A., Carta G., Remarks on the current theory of shear strength of variable depth beams, Open J. Civ. Eng., 3, 28-33, 2009. https://doi.org/10.2174/1874149500903010028

[10] Balduzzi G., Hochreiner G., Füssl J., Stress recovery from one dimensional models for tapered bi-symmetric thin-walled I beams: deficiencies in modern engineering tools and procedures, Thin-Walled Struct., 119, 934-945, 2017. https://doi.org/10.1016/j.tws.2017.06.031

[11] Bleich F., Stahlhochbauten Bd. 1, Springer, Berlin, 1932. https://doi.org/10.1007/978-3-642-99613-9

[12] Pugsley A.G., Weatherhead R.A., The shear stresses in tapered beams, Aircr. Eng., 46, 218-226, 1942. https://doi.org/10.1017/S0368393100099090

[13] Krahula J.L., Shear formula for beams of variable cross section, AIAA Journal, 13(10), 1390-1391, 1975. https://doi.org/10.2514/3.6996

[14] Bertolini P, Eder M.A., et al., Stresses in constant tapered beams with thin-walled rectangular and circular cross sections, Thin-Walled Struct., 137, 527-540, 2019. https://doi.org/10.1016/j.tws.2019.01.008

[15] Migliaccio G., Ruta G., et al., A new shear formula for tapered beamlike solids undergoing large displacements, Meccanica, 2022. https://doi.org/10.1007/s11012-022-01529-1

[16] Migliaccio G., Analytical evaluation of stresses and strains in inhomogeneous non-prismatic beams undergoing large deflections, Acta Mech., 2022. https://doi.org/10.1007/s00707-022-03247-x

Theoretical and Applied Mechanics - AIMETA 2022 Materials Research Forum LLC
Materials Research Proceedings **26** (2023) 169-174 https://doi.org/10.21741/9781644902431-28

A simple procedure for the non-linear optimization of cable tension for suspended bridges

MASCOLO Ida [1,a], MODANO Mariano[1,b] and GUARRACINO Federico[1,c]

[1]Department of Structures for Engineering and Architecture, University of Naples Federico II,

[a] ida.mascolo@unina.it, [b] modano@unina.it, [c] fguarrac@unina.it

Keywords: Suspended-Deck Bridge, Cable Tensioning, Bridge Optimisation

Abstract. Long-span suspended bridges rely upon networks of tensed cables that carry the weight of the deck. The networks of cables are generally connected to a number of vertical towers (pylons) that transfer the forces to the foundations as in the case of cable-stayed bridge. Their structural behaviour is highly influenced by the pretension forces on account of the redundancy of the structure. Several methods have been proposed for determining pretension the forces in the cables, such as Load-Balance Method, Iterative Unit Load Method, Force Equilibrium Method, Zero Displacement Method. The present study aims to investigate the influence of geometrical non-linearities on the optimization of the design. To this end, the Force Equilibrium Method is here extended and compared to the use of a Finite Element commercial package, since this is the standard method in everyday engineering practice. The comparison between the Force Equilibrium Method and FEM results shows that the first method, in spite of its simplicity, is able to provide a reasonable and reliable alternative to the more complex non-linear FE approaches.

Introduction

One of the fundamental problems with cable-sustained bridges is related to the determination of initial cable forces and loading sequence, which are needed to reach the correct geometrical configuration [1-2]. An optimal cable stress control allows for the stress and strain of the whole structure to be controlled both during and after the construction phase. Despite the fact that various approaches have been proposed and analysed, the whole problem still lacks concrete definition. There are no analytical frameworks to determine the stressing sequences necessary to obtain a predetermined final design configuration but a number of iterative algorithms that allow for the progressive adjustment of cable forces during the pre-construction phase have been proposed, some by the present authors, which can apply to both proportional and non-proportional loading [3-6]. Obviuosly, the cable tightening operations involve technological, structural, and economic problems.

The simple procedure proposed here is an original solution to the problem of cable pretension in cable-stayed structures. Unlike other methods, it makes reference to sensitivity analysis to identify the optimal configuration of stresses. More than a design tool, it can be seen as a structural behaviour optimization method that allows to determine the values of initial cable forces in order to give the structure a certain state of stress under prescribed loads.

Actually the illustrated method is an extension of the Influence Matrix method, already employed for other types of structures and suspended bridges [7-10]. In the present case it is applied to a suspension system consisting of a main cable and secondary hangers.

The proposed procedure

The procedure is based on a preliminary analysis of the structure subjected to external loads. By means of a sensitivity analysis, in which the stiffness of the load-bearing elements are changed, an optimal distribution of stress is identified, such as the bending moment in the deck.

Materials Research Proceedings **26** (2023) 169-174 https://doi.org/10.21741/9781644902431-28

The chosen optimal configuration constitutes the target solution to be attained by pre-tensioning the cables.

The proposed procedure relies upon the application of the Influence Matrix Method in order to attain a chosen optimum configuration previously identified by means of a suitable sensitivity analysis.

The novelty here with respect to [9] is that the geometry of the structure is updated step by step until the final configuration is reached. The optimization procedure is thus applied to each step. At each pre-tension step the influence matrix is updated according to the actual geometry of the structure.

Once the single step has been solved, the new geometric configuration of the structure is set. A new influence matrix is constructed on the updated configuration, the optimum solution is identified and the pre-tension stresses to be applied to each cable are calculated. The procedure is considered concluded when the variation of the pretension efforts from one step to another becomes negligible.

A worked example

In order to illustrate the procedure, a very simple system made of a main cable, three secondary cables, and a beam is taken into consideration, see Fig.1. The loading is applied to the horizontal beam and represents the self-weight load of the structure.

The sensitivity analysis showed that for this example reference can be made to a Low-stiffness deck (LSD) - High-stiffness cables model. Actually, the system can be considered as made of two macro systems, i.e. the suspension system and the loaded beam.

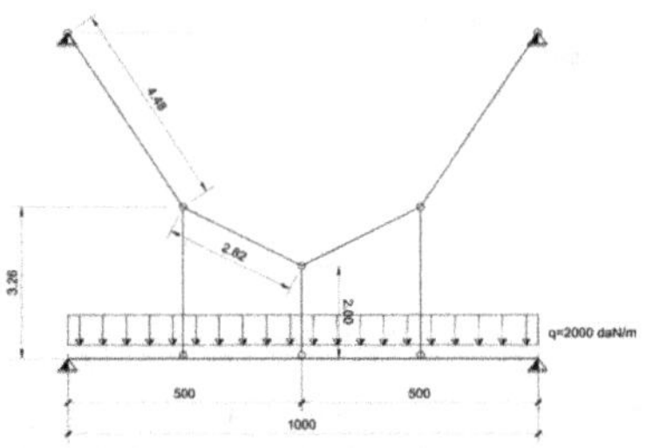

Fig.1. Geometry [m] and loads [daN].

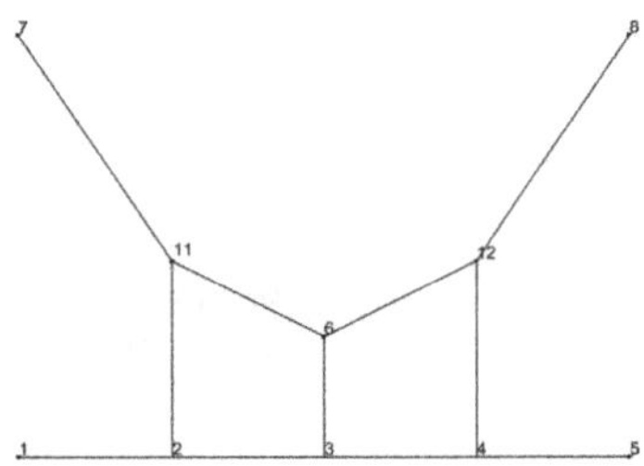

Fig.2. Point numbering.

In absence of any pre-tensioning of the cables, the bending moment diagram shown in Fig.2 is found. Fig.3 shows the diagram of the bending moment following the application of the first pre-stress loading in the cables and Figs. 4 and 5 the resulting optimised bending moment diagram and the deformed configuration at the end of the first step.

Materials Research Proceedings **26** (2023) 169-174 https://doi.org/10.21741/9781644902431-28

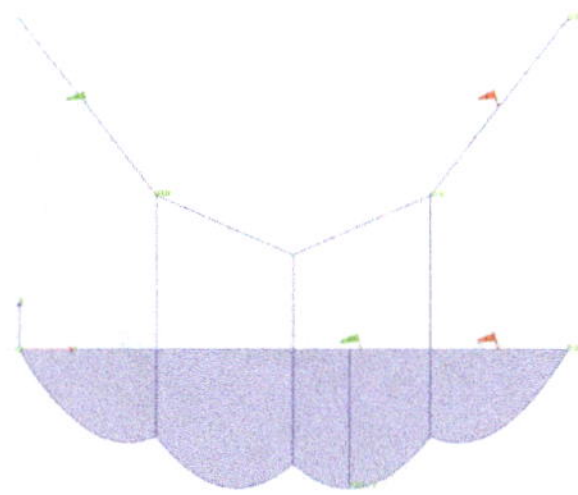

Fig.3. Bending moment diagram in absence of any cable pre-stress.

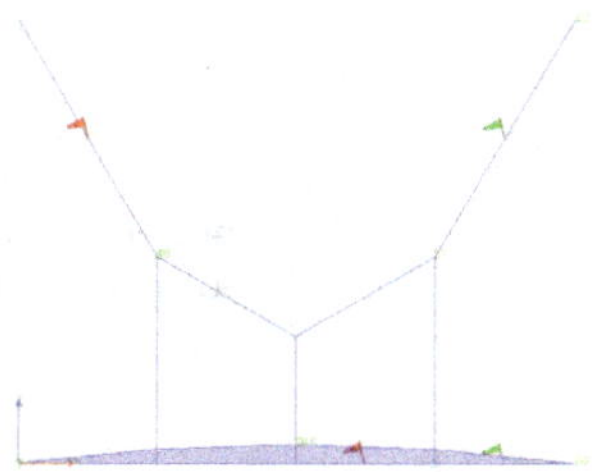

Fig.4. Bending moment diagram due to the first pre-stress step application.

Rather obviously, the configuration of the structure changes at the end of the first step, thus triggering a geometric non-linearity in the problem. Therefore, the optimization procedure needs to be repeated for the displaced structure.

By applying the optimization procedure iteratively, for each step a new influence matrix is calculated and updated pre-tension loads are found.

The optimization goal can be considered attained when the configuration differences between two subsequent steps become negligible.

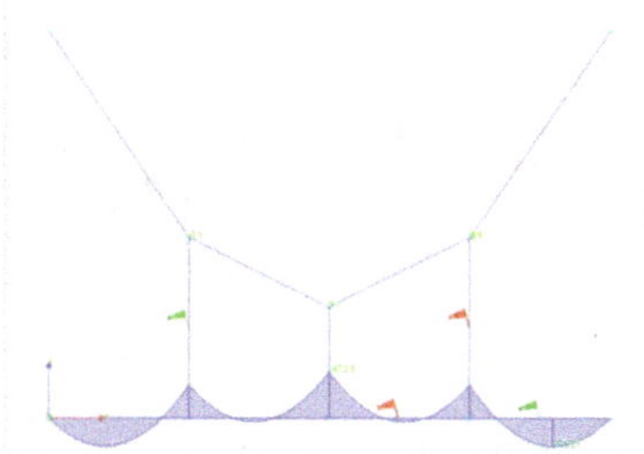

Fig.5. Optimized bending moment (first step).

Materials Research Proceedings **26** (2023) 169-174
https://doi.org/10.21741/9781644902431-28

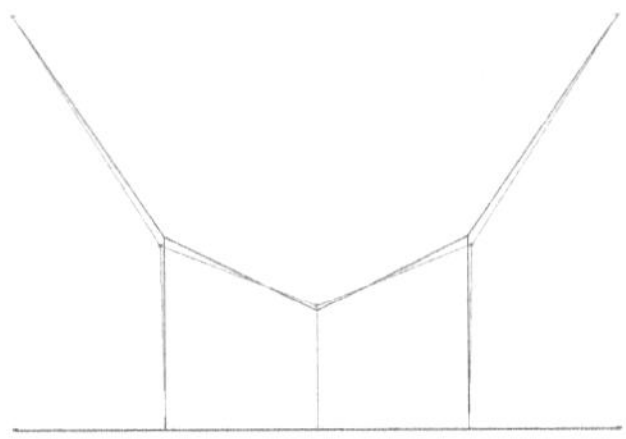

Fig.6. Deformed configuration after the first step optimization.

Rather intuitively, points 11 and 12 move outwards on account of the pretensions applied to the cables and elements (2-11-7) and (4-12-8) tend to align following the pre-tensioning. Overall, points 11 and 12 tend to line up with points 2-7 and 4-8. For a similar reason point 6 tends to move upwards. In the shift table, already in the first step, this phenomenon is found.

In the following steps, a new matrix of influence on the deformed configuration is defined and the forces to be applied to the cables in order to optimize the status of stress on the beam are calculated iteratively.

Discussion of results

The iterations were carried on until the configuration differences between two subsequent steps became close to zero. In the case at hand six steps were required in order to attain the final configuration of the structure. Figs. 6 and 7 show the evolution of the configuration of the structure along with the optimization steps and the initial and final ones, respectively.

With reference to Fig.8, which shows the point numbering, Table 1 collects the absolute and incremental displacements of the point at each step. It appears evident that after a few iterations, the configuration of the structure stabilises quite abruptly. This fact is typical of structures involving cables and is discussed in details in [11]. The plots relative to the displacements of points 6, 11 and 12 are shown in Figs. 8 and 9.

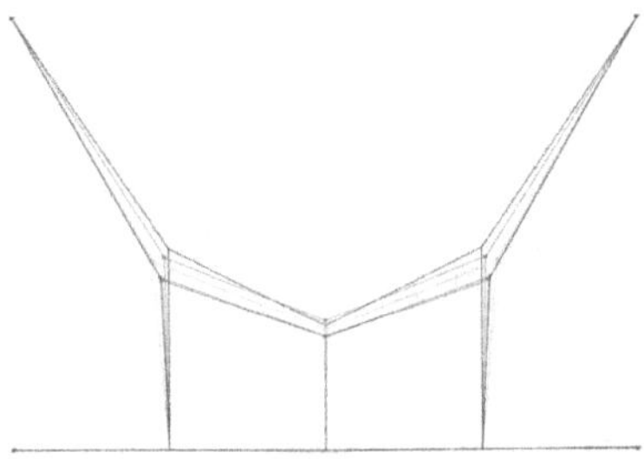

Fig.7. Deformed configuration evolution from step 0 to step 6.

Table 1. Point displacements [cm] vs. steps: absolute (left) and incremental (right).

Step	3	6	11	12
0	0	0	0	0
1	0,09	9,47	15,35	-15,35
2	0,36	15,02	18,95	-18,95
3	0,52	0,04	6,45	-6,45
4	0,64	12,12	31,49	-31,49
5	0,61	0,41	6,16	-6,16
6	0,85	0,68	6,22	-6,22

Step	3	6	11	12
0	0	0	0	0
1	0,09	9,47	15,35	-15,35
2	0,27	5,55	3,60	-3,60
3	0,16	-14,98	-12,50	12,50
4	0,12	12,08	25,04	-25,04
5	-0,03	-11,71	-25,34	25,34
6	0,24	0,27	0,06	-0,06

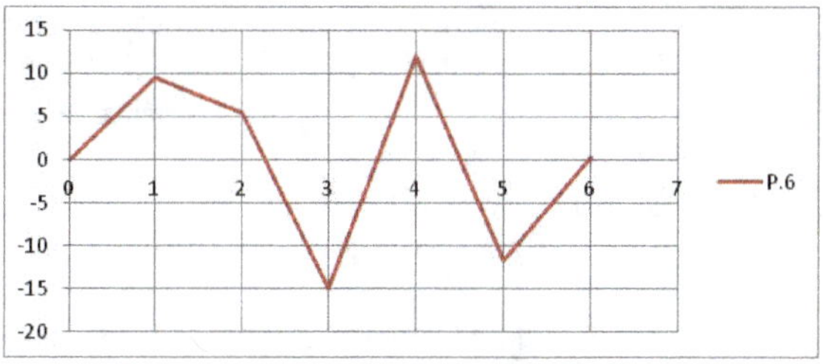

Fig.8. Displacement point n.6 [cm] from step 0 to step 6.

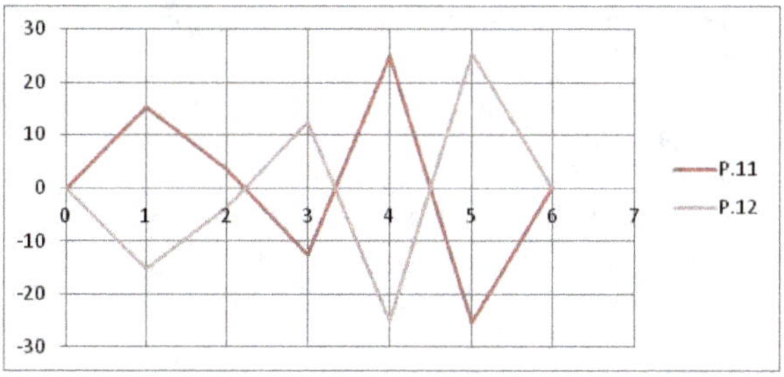

Fig.9. Displacement points n.11 and 12 [cm] from step 0 to step 6.

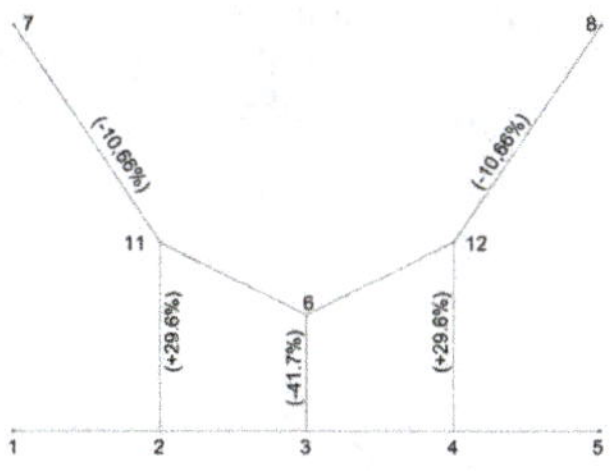

Fig.10. Different pretension between linear and non-linear optimization.

Conclusions

The proposed procedure is concerned with the application of the Influence Matrix Method to a simple suspended bridge scheme under large displacements. The iterative application of the method, taking into account the geometric non-linearity, allows to calculate quite easily the optimal pre-stress in the cables loading in order to minimise the bending moment in the lower deck. The procedure seems susceptible to be extended to real examples of suspension bridges with reasonable computational effort.

It is worth pointing out that, with respect to the linear optimization procedure, as the one performed at step 1, the main cable at the end of the process results tensioned on average by about 10% less than in the linear procedure. The side hangers, on the other hand, show a pretension increment of 29%, while the central hanger results less pretensioned than in the linear procedure.

References

[1] W. Kanok-Nukulchai (Ed.), Cable-Stayed Bridges. Experiences & Practice - Proceedings of the Int Conference on Cable-Stayed Bridges Bangkok, Thailand. November 18-20, 1987.

[2] H. Svensson, Cable-Stayed Bridges: 40 Years of Experience Worldwide, Ernst & Sohn, Berlin, 2012. https://doi.org/10.1002/9783433601044

[3] I. Mascolo, M. Pasquino, Lateral-torsional buckling of compressed and highly variable cross section beams. Curved and Layered Structures, 3(1): 146-153, 2016. https://doi.org/10.1515/cls-2016-0012

[4] I. Mascolo, M. Modano, A. Fiorillo, M. Fulgione, V. Pasquino, F. Fraternali, Experimental and numerical study on the lateral-torsional buckling of steel C-beams with variable cross-section. Metals, 8 (112018), 941, 2018. https://doi.org/10.3390/met8110941

[5] R. Shamass, G. Alfano, F. Guarracino, An investigation into the plastic buckling paradox for circular cylindrical shells under non-proportional loading, Thin-Walled Structures, 95: 347-362, 2015. https://doi.org/10.1016/j.tws.2015.07.020

[6] R. Shamass, G. Alfano, F. Guarracino, On Elastoplastic Buckling Analysis of Cylinders Under Nonproportional Loading by Differential Quadrature Method, International Journal of Structural Stability and Dynamics, 17(7), 1750072, 2017. https://doi.org/10.1142/S0219455417500729

[7] M. Modano, I. Mascolo, F. Fraternali, Staging and Pretensioning of Cable-Stayed Bridges. In: IOP Conference Series: Materials Science and Engineering, 473(1), IOP Publishing, 2019. https://doi.org/10.1088/1757-899X/473/1/012012

[8] M. Modano, I. Mascolo, F. Fraternali, Design and optimization of pre-tension forces in cable-stayed bridges. In: Advances in Engineering Materials, Structures and Systems: Innovations, Mechanics and Applications, CRC Press, 2019. https://doi.org/10.1201/9780429426506-183

[9] M. Modano, I. Mascolo, Optimisation of suspended-deck bridge design: a case study. Australian Journal of Structural Engineering, 21(3):244-53, 2020. https://doi.org/10.1080/13287982.2020.1778433

[10] Modano, M., Majumder, A., Santos, F., Luciano, R., Fraternali, F. (2020) Fast and Optimized Calculation of the Cable Pretension Forces in Arch Bridges With Suspended Deck. Frontiers in Built Environment, 2020, 6, 114 ISSN: 2297-3362. https://doi.org/10.3389/fbuil.2020.00114

[11] I. Orynyak, F. Guarracino, M. Modano, R. Mazuryk, An efficient iteration procedure for form finding of slack cables under concentrated forces, Archives of Civil Engineering, 68(2): 645-663, 2022. https://doi.org/10.24425/ace.2022.140664

Mechanics of Machine

Theoretical and Applied Mechanics - AIMETA 2022 Materials Research Forum LLC
Materials Research Proceedings **26** (2023) 177-182 https://doi.org/10.21741/9781644902431-29

Experimental identification of a pneumatic valve-cylinder system for attitude control

Michele Gabrio ANTONELLI[1,a], Jacopo BRUNETTI[1,b*], Walter D'AMBROGIO[1,c], Annalisa FREGOLENT[2,d], and Francesco LATINI[2,e]

[1] Dipartimento di Ingegneria Industriale e dell'Informazione e di Economia, Università degli Studi dell'Aquila, P.le Pontieri, Monteluco di Roio, 67100 L'Aquila, Italy

[2] Dipartimento di Ingegneria Meccanica e Aerospaziale, Università di Roma "La Sapienza", Via Eudossiana 18, 00184 Roma, Italy

[a]michelegabrioernesto.antonelli@univaq.it, [b]jacopo.brunetti@univaq.it, [c]walter.dambrogio@univaq.it, [d]annalisa.fregolent@uniroma1.it, [e]francesco.latini@uniroma1.it

Keywords: Active Suspensions, Pneumatics, Proportional Pressure Control Valve, Experimental Identification, Dynamic Response

Abstract. Vertical vibrations on agricultural tractors, due to soil irregularities, represent a major cause of diseases of agricultural operators. The control of noise and vibration on the operator is of interest to the (Italian) National Institute for Insurance against Accidents at Work (INAIL). A prototype of an active suspension system of the operator seat has been developed in the laboratories of INAIL. The prototype can be configured with hydraulic or pneumatic actuation. This paper focuses on the pneumatic solution powered by a proportional pressure control electro-valve. Since several proportional pressure control valves are commercially available, three of them have been experimentally tested to determine the most suitable one for control purposes. Numerical modeling of the three valves has been carried out and described. Experimental tests on the selected valves allow for identifying the main dynamic parameters of the numerical models, providing a reliable simulator to be adopted for the development and optimization of the control system. The performances of the valves are compared and discussed.

Introduction

A major cause of diseases of agricultural operators is represented by the vertical vibration exposure on agricultural tractors due to soil irregularities. The effects of the whole-body vibrations on operators of industrial and agricultural machines are extensively investigated in [1]. To reduce the vibrations that are transmitted to the operator, the cabin and the seat are generally provided with a suspension system that can exploit either passive, semi-active, or active solutions.

The first ones, typically mechanical or pneumatic [2], allow for an attenuation of the vibrations beyond a given cut-off frequency. Better performances are guaranteed by the semi-active systems that adjust the damping coefficient according to the soil irregularities [3]. Active solutions provide the best performances by the adoption of an actuation system generally based on either hydraulic, electromagnetic, or electromechanical technology [4].

The (Italian) National Institute for Insurance against Accidents at Work (INAIL) is interested in active vibration control (AVC) and active noise control (ANC) technology to reduce the vibration and noise exposure of operators and founded several research programs in the last years. A laboratory prototype of an active suspension system for the operator seat is developed. It can employ either hydraulic or pneumatic actuation.

The present study is focused on the pneumatic actuation made of a double-effect pneumatic cylinder powered by a proportional pressure control electro-valve. The attitude control system generates the command signal for the electro-valve according to the soil irregularities. This kind of valve converts an electrical input signal into an output pressure.

In this work, the pneumatic actuation is evaluated for adoption in an active control system. This technology is, in fact, very versatile and is widely used for its cheapness, cleanness, and high power-to-weight ratio. In fact, it allows for force control, variable stiffness actuation (VSA) [5], position control in several industrial sectors [6], and also for non-traditional applications such as deformable pneumatic soft actuators [7, 8].

To properly develop an active control system, the dynamic properties of the main components of the system must be known, including that of the electro-valve. Among all the commercially available valves, three have been experimentally tested. The dynamic parameters of each valve are experimentally identified, and performance comparisons of the valves are reported and discussed to find the most suitable one for control purposes. Starting from the identified dynamic parameters, numerical modeling of each valve is carried out. The models provide a reliable simulator of the dynamics of the valves to be included in a more complex model of the whole active suspension system for the development and optimization of its control system.

The prototype of the seat suspension

The prototype of the seat suspension is made of the actuation system placed on a vibrating plate. As shown in Fig. 1, the actuation system is mounted over the vibrating plate (1), and it is composed of the pneumatic actuator (2), the sliders (3), the suspended body (4), a set of springs (5), a couple of electro-valves (6), a linear position transducer and a couple of accelerometers.

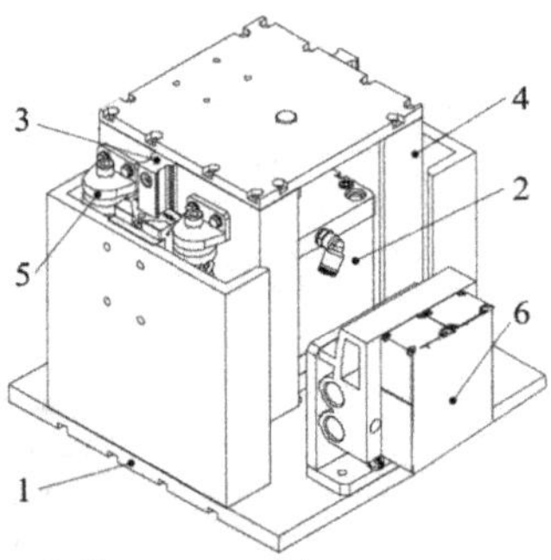

Fig. 1: Prototype of the seat suspension

The actuator (ISO 21287 double-acting, bore 80 mm, stroke 70 mm) is mounted along the vertical axis. The upper end of the rod is bolted to the suspended body and ensures its motion. The proportional pressure control valves are directly connected to the chambers of the cylinder. The outlet pressure of the valves must ensure the motion of the piston (rod) according to the soil irregularities. The dynamics of the suspended body is affected by the external disturbances in terms of platform vibrations and by the dynamics of the electro-valves.

Experimental activity

Three commercial pneumatic proportional pressure control valves are adopted and tested:
- Festo VPPN-6F-L-1-F-0L6H-A4N-S1 (hereinafter called Valve 1);
- MetalWork Regtronic 1/4" (hereinafter called Valve 2);
- SMC ITV2050-31F2BL3 (hereinafter called Valve 3)

The working principle of the three valves is the same: a pilot stage, made of two modulating pilot solenoid valves, ensures the pressure adjustment in a pilot chamber. Inside it, the pressure acts on the upper surface of a diaphragm connected to the air supply valve: changes in pressure cause the opening/closing adjustment of this valve, with a consequent adjustment of the outlet pressure.

Valve 1 provides multiple control loops instead of the conventional direct-acting controls. Although the outlet pressure is proportional to an input command current, which ranges from 4 to 20 mA, the valve is commanded by an equivalent tension command, which ranges from 1 to 5 Vdc. This valve can work according to three functional response modes (fast, universal and precise control behavior): in the present activity, the valve is set to the fast control behavior mode. No pressure full-scale adjustment is possible.

Valve 2 can work according to three functional modes: in the present activity, the fast regulation mode is set. In this case, the pressure full-scale can be adjusted.

Valve 3 can work in one only functional mode. No pressure full-scale adjustment is possible.

Regardless of the type, the dynamic behavior of these kinds of valves can be described, with a good approximation, by the following second-order equation [9]:

$$\ddot{p}_{out} + 2\cdot\zeta\cdot\omega_n\cdot\dot{p}_{out} + \omega_n^2\cdot p_{out} = \omega_n^2\cdot V_{in}\cdot k_v \quad (1)$$

where p_{out} is the outlet pressure from the valve (measured in bar), V_{in} is the input command signal (measured in V), k_v is the static gain (measured in bar/V), and ζ and ω_n are the damping ratio and the natural frequency, respectively.

All the valves are tested in the pressure range 0.0 - 6.0 bar. The k_v value of each valve is identified in specific experimental tests where V_{in} is increased in quasi-static conditions from 0.0 to 6.0 bar. The dynamic parameters ζ and ω_n of each valve are identified by imposing an offset sine wave V_{in} with three different amplitudes and the same offset at different frequencies. The corresponding pressure amplitudes are 1.80 bar, 1.20 bar, and 0.60 bar, representing 20-80%, 30-70%, and 40-60% of the outlet pressure range, respectively. The corresponding offset pressure is 3.0 bar. For a given amplitude, tests are carried out at different frequencies in the range starting from 0.1 Hz up to a frequency that does not provide any variation of the outlet pressure.

A signal generator (GW Instek AFG-2125) is used for V_{in}. A data acquisition system (made of a USB NI6001 DAQ-board and software developed in the NI LabView environment) is adopted to acquire V_{in}, converted into the corresponding pressure set-point, and the outlet pressure p_{out}. The sampling frequency is set to 1 kHz.

Fig. 2 shows the input pressure set-point and the associated output at the frequency 0.5 Hz. At the considered frequency, each valve exhibits a proper distortion pattern of the outlet pressure. In particular, Valve 1 exhibits a more visible distortion in the rising part of the curve. Such a behavior could be due to the multiple control loops performed by the valve controller. Valve 2 exhibits a high difference between the rising and falling parts of the curve. This might be due to the pneumatic resistances in the two airflow directions being very different. Valve 3 exhibits some distortions after the stationary points of the curve.

The magnitude and the phase shift of the outlet pressure with respect to the set-point, are achieved using the following criterion: at a given frequency, according to the nonlinear least-squares method, each of the two curves is described by an offset sine expressed by the equation:

$$y(t) = A + B\cdot\sin(2\pi f\cdot t+\varphi) \quad (2)$$

where A is the identified offset value of the input/output (i/o) signal (in bar), B is the identified amplitude of the signal (in bar), f is the frequency of the input signal (in Hz), t is the time (in s), and φ is the identified phase of the signal. Hence, for each frequency value, the magnitude $|Bo/Bi|$ and the phase shift $\varphi_f = \varphi_o - \varphi_i$ are computed. The whole procedure and the associated intermediate results are detailed in [10].

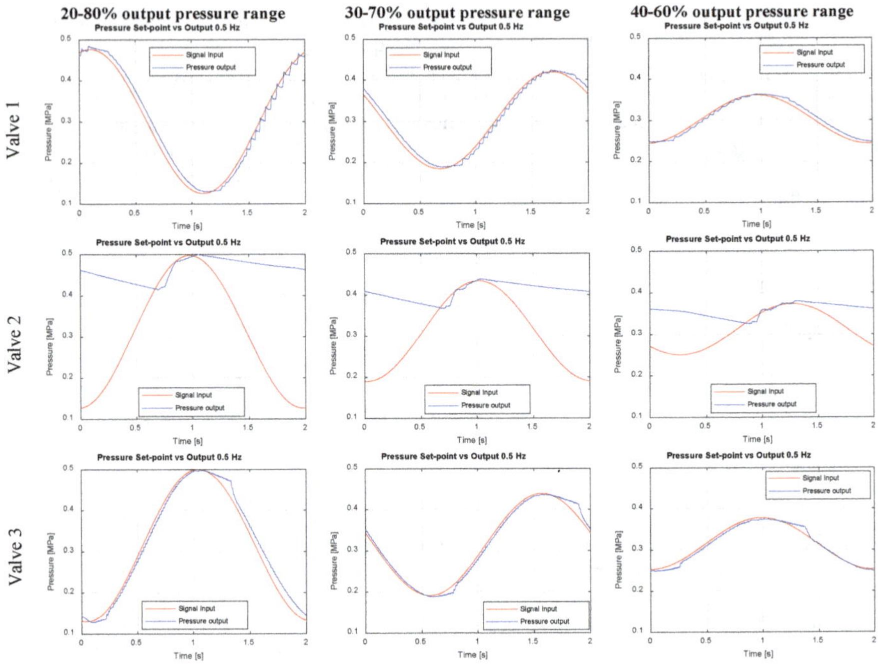

Fig. 2: Example of experimental input and output signals of the electro-valves

The identified magnitude and phase shift are gathered in Fig. 3, for the three valves. The identified k_v, ζ, and ω_n for each valve are reported in Table 1 together with the main characteristics. The identification aims to approximate the magnitude and the phase shift for the three amplitude values, representing a linear approximation of the non-linear behavior of the valves. Up to 5 Hz, Valve 1 behaves linearly and experimental data for different amplitudes provide the same values of magnitude and phase shift. On the contrary, for higher frequencies, different values are obtained. The high attenuation that is observed on the response of Valve 2 at 0.5 Hz (see Fig. 2) is confirmed by its magnitude plot. Valve 3 shows a non-linear behavior throughout the considered frequency range. In particular, also at low frequency, the experimental frequency response is amplitude-dependent.

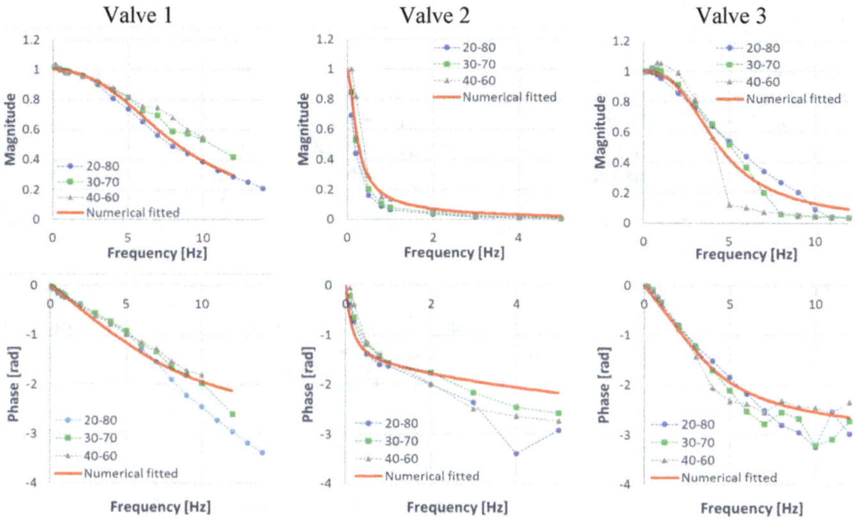

Fig. 3: Magnitude and phase of the three valves responses

Table 1 – Main characteristics of the tested proportional electro-valves

Valve	P_{out} *range [bar]*	V_{in} *range [Vdc]*	k_v *[bar/V]*	ζ	ω_n *[rad/s]*
1	0.00 – 6.0	1 – 5	1.43	0.85	45
2	0.05 – 6.0	0 – 10	0.60	3.5	6.28
3	0.05 – 9.0	0 – 10	0.90	0.8	23

Numerical modeling and experimental validation

To validate the model of the valves, experimental data are compared in Fig. 4 to numerical results obtained by imposing on the model the same experimental offset sine V_{in}. The model of Valve 1 fits the experimental data with a good approximation at 7 Hz and at the highest amplitude condition. As previously noted, the model of Valve 2 is not able to retrace the experimental data because the adopted model does not consider the nonlinear behavior of the valve, which produces a different offset of the response with respect to the expected one, in addition to different behavior with rising and falling pressures. Finally, also the model of Valve 3 does not properly retrace the experimental: some slight nonlinearities occur.

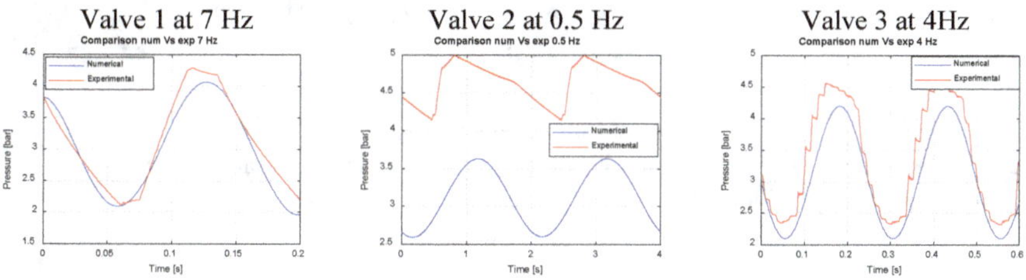

Fig. 4: Comparison between experimental and numerical output pressures of the three valves

Conclusions

In this paper, three commercial proportional pressure control pneumatic electrovalves are experimentally tested to identify their dynamic properties and to evaluate their use in a pneumatic vibration control system. The tests performed on the selected electrovalves allowed to identify the resonance frequencies and the damping factors of the second-order system adopted to model the valve behavior. However, for two of the three electrovalves, significant nonlinearities are present that introduce a relevant distortion in the pressure response of the valve, limiting their use for a control application. Although the model of Valve 1 fits the experimental data with a good approximation up to 7 Hz, the attenuation of the response can be considered suitable for control purposes only below 2 Hz.

References

[1] S. K. Essien, C. Trask, M. Khan, C. Boden, and B. Bath, Association between whole-body vibration and low-back disorders in farmers: A scoping review, Journal of Agromedicine, 23(1) (2018) 105–120. https://doi.org/10.1080/1059924X.2017.1383333

[2] I. Hostens, K. Deprez, and H. Ramon, An improved design of air suspension for seats of mobile agricultural machines, Journal of Sound and Vibration. 276(1) (2004) 141–156. https://doi.org/10.1016/j.jsv.2003.07.018

[3] A. Soliman, M. Kaldas, Semi-active suspension systems from research to mass-market – a review, Journal of Low Frequency Noise, Vibration and Active Control. 40(2) (2021) 1005–1023. https://doi.org/10.1177/1461348419876392

[4] W. East, J. Turcotte, J.-S. Plante, and G. Julio, Experimental assessment of a linear actuator driven by magnetorheological clutches for automotive active suspensions, Journal of Intelligent Material Systems and Structures. 32(9) (2021) 955–970. https://doi.org/10.1177/1045389X21991237

[5] M. G. Antonelli, P. Beomonte Zobel, F. Durante, and F. Gaj, Development and testing of a grasper for notes powered by variable stiffness pneumatic actuation, International Journal of Medical Robotics and Computer Assisted Surgery, 13(3) (2017). https://doi.org/10.1002/rcs.1796

[6] M. Sorli, G. Figliolini, S. Pastorelli, Dynamic Model and Experimental Investigation of a Pneumatic Proportional Pressure Valve, IEEE/ASME Transactions on Mechatronics. 9(1) (2004) 78-86. https://doi.org/10.1109/TMECH.2004.823880

[7] M.G. Antonelli, P. Beomonte Zobel, F. Durante, and M. Zeer, Modeling-Based EMG Signal (MBES) Classifier for Robotic Remote-Control Purposes, Actuators. 11(art. 65) (2022). https://doi.org/10.3390/act11030065

[8] M.G. Antonelli, W. D'Ambrogio, and F. Durante, Development of a pneumatic soft actuator as a hand finger for a collaborative robot, in Proceedings of ICMSCE 2018, Amsterdam, Netherland, February 2018, 67-71. https://doi.org/10.1145/3185066.3185079

[9] C. Ferraresi, and M. Sorli, Modeling and analysis of pressure control systems, Fluid. 357 (1994) 58–65.

[10] M.G. Antonelli, J. Brunetti, W. D'Ambrogio, A. Fregolent, and F. Latini, Multi-physics modeling of an active seat suspension with pneumatic or hydraulic actuation, in Proceedings of ISMA 2022, Leuven, Belgium, September 2022.

Theoretical and Applied Mechanics - AIMETA 2022 | Materials Research Forum LLC
Materials Research Proceedings **26** (2023) 183-188 | https://doi.org/10.21741/9781644902431-30

Performance evaluation of a Ball Screw mechanism through a multibody dynamic model

BERTOLINO Antonio Carlo[1, a *], DE MARTIN Andrea[1,b] and SORLI Massimo[1,c]

[1] Department of Mechanical and Aerospace Engineering, Politecnico di Torino, Torino, Italy

[a]antonio.bertolino@polito.it, [b]andrea.demartin@polito.it, [c]massimo.sorli@polito.it

Keywords: Ball Screw, Multibody Dynamics, Internal Recirculation

Abstract. Ball screws are mechanism to convert the rotational into linear motion and viceversa and are widespread in a variety of different sectors. A detailed high-fidelity dynamic mathematical model of such component is paramount in several fields and, in particular, in the definition of a PHM system for flight control EMAs in order to increase their reliability. In fact they can be used as a virtual test bench on which inject artificial defects and study their effect on specific indicators. This paper presents a MBD model of a single-nut ball screw with internal recirculation able of describing the full dynamic of each internal component allowing a more in-depth understanding of the system behavior and poses the basis for PHM-oriented analyses on different degradations.

Introduction

During last decades, a trend towards more electric aircraft has arisen. Several efforts have been made to replace currently wide-spread electro-hydraulic actuators with electro-mechanical ones (EMAs) for primary flight controls. However, the principal drawback is the increased jamming probability which makes this technology not suitable for safety critical applications in its base configuration. To overcome this issue a possible solution is to use a simple architecture EMA equipped with an efficient prognostic and health management system (PHM). To train this PHM algorithm and to understand the correlation between different parameters and the performance deterioration, a high-fidelity dynamic model is paramount: in fact it can be used as a virtual test rig on which inject artificial defects to study the system under nominal and degraded conditions and to extract meaningful features correlated with faults' extent, in a cost-effective way.

According to literature FMECAs, the ball screw is the most critical subcomponent of an EMA. For this reason, this component is the focus of the current research. The ball screw is the last and key component of the EMA's mechanical transmission, converting the rotational motion coming from the electric motor into a linear displacement. This component can reach extremely high efficiencies replacing the sliding with rolling friction. Its comprehension is essential to avoid performance degradations and catastrophic failures, such as jamming. However, experimental analyses to understand the internal behavior are fairly difficult because of its closed nature; therefore, several theoretical studies analyzed the system from different perspectives, such as internal kinematics [1,2], no-load drag torque [3,4] and recirculating units degradation [5,6].

Hitherto, most of the models used a quasi-static approach describing the motion of the element in steady state conditions. Unfortunately, these models are not suitable for PHM. To fill this gap, the authors developed a set of dynamic models with increasing complexity and realism in order to be used for analyses with different levels of required details. A unidimensional lumped parameter dynamic model was first created. Being the ball screw mainly a 3D mechanism, a MBD model was developed in Simscape Multibody environment [7] taking into account the full dynamic of each subcomponent as well as their mutual interactions through a dedicated contact model considering grease lubrication [8]. Several degradation model were inserted and investigated [9]. However, no recirculating channel model was present, hence the capability of the MSC ADAMS software to handle contacts between arbitrary-shaped bodies was exploited analyzing a ball screw with a simple end-cap type axial recirculation [10].

Differently from [10], an internal recirculation type with three liner circuits is considered in this work, representing a novel contribution in the modelling of recirculation, usually disregarded in literature studies. This paper presents a detailed three-dimensional multibody dynamic model of a ball screw with the aim to understand the underlying dynamic of internal elements and to make a further step towards a model-based PHM for this mechanism. The MBD model can accurately describe the kinematics and dynamics of each element within the mechanism and allows to understand the complex interaction between the gothic arch helical grooves and the spheres under time-variant speed and external loads. The spheres are restrained between the grooves only by the Hertzian contact with the grooves, while the motion is transmitted by friction. Each sphere can freely move within the grooves, creating either backlash or multiple contact points with variable contact angles. The capabilities of such model are presented by evaluating the overall performance and internal component motion. The considered ball screw refer to the one mounted on the experimental test bench which is being built in the laboratories of Politecnico di Torino [11,12] with the aim of validating the results of the current and previous models and getting more in-deep insights for prognostic purposes. The proposed model poses the basis for a complete understanding of the internal dynamics of the mechanism considering also the internal recirculating system.

Model definition

The developed model has been developed in the MSC ADAMS environments to exploit its general shape body contact feature. The ball screw under study has an only rotating screw shaft and a only translating not preloaded single nut, on which an external force is applied axially as disturbance. The motion is imposed to the screw shaft as speed law.

Geometry tessellation. The complex geometry of the screw shaft and nut bodies have initially been simplified by removing al the non-necessary features: the simpler the geometry, the less point are necessary for its accurate discretization, the faster the simulation. The resulting geometry, imported into ADAMS, is triangular tessellated using the RAPID algorithm in order to speed up the contact detection during dynamic simulation.

Contact detection. After a contact occurs, the contact point location is calculated. The solver computes the intersection volume, which may be single or, in general, multiple depending on the specific condition. Assuming an uniform density of the material, the centroid of the intersection volume is obtained, having the same center of mass of the intersection volume. This point represents the contact point, while the penetration depth, used as a penalization parameter to enforce the contact constraint, is calculated as the distance between the two points of each tessellated surface closest to the intersection volume centroid.

Normal contact model. The ADAMS solver contains a contact predictor that estimates the onset of contacts and modifies the integration time step accordingly. Theoretically, contacting bodies does not compenetrate. This can be usually obtained introducing a penalty regularization considering the penetration as a deformation of the bodies and enforcing the contact constraint with a reaction force, which can be calculated as the sum of an elastic and damping contributions where a smooth activation function modulates the damping contribution (of the order of 1 Ns/mm) with respect to the penetration itself, avoiding contact discontinuities due to non-null approaching speeds. The contact stiffness constant is calculated a priori from the knowledge of the geometry and exact curvature radii [13], and assuming a theoretical contact angle usually close to 45° [14], according to the explicit non-recursive method for Hertzian contacts proposed in [15].

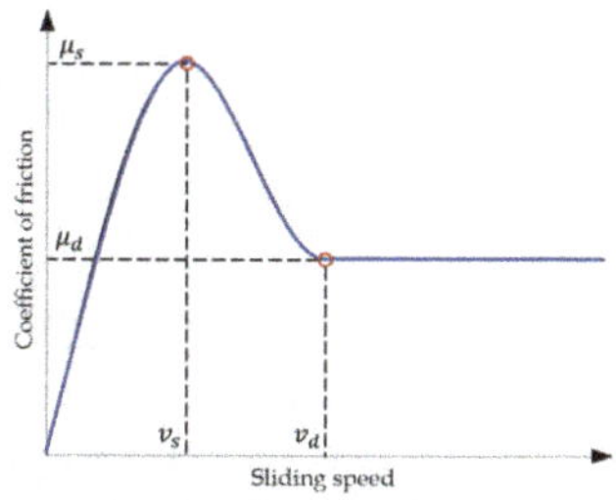

Figure 1 – Coefficient of friction model versus sliding speed.

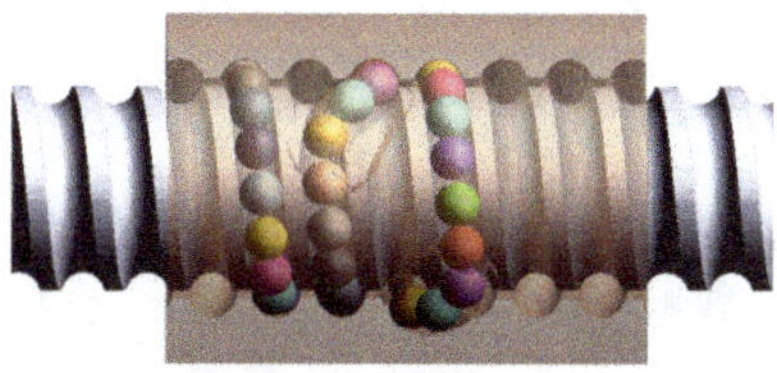

Figure 2 – ADAMS ball screw model with internal recirculating inserts.

Friction model. Once the contact point location is defined and the outward normal are identified, the normal and slip velocity can be easily obtained. The first is the previously cited approaching speed $\dot{g}$, while the second is the key parameter to calculate the tangential friction forces. ADAMS does not consider rolling friction resistance, but it only takes into account sliding and spin friction. There is no contact stiction in ADAMS, but a little amount of sliding speed is always required between the bodies in order to create a friction force: this is to ensure a zero force condition for the case of perfect adherence and to avoid numerical instabilities given by the abrupt transition between negative and positive speed values. Figure 1 depicts the coefficient of friction function versus the sliding speed: static and dynamic COF values and the relative threshold speeds must be chosen by the user to adapt the curve to the punctual problem. The shape of the COF function represents a lubricated contact with various lubricating regimes In this paper, the static and dynamic COFs have been set respectively to $\mu_s = 0.11$ ($v_s = 0.1$ mm/s) and $\mu_d = 0.06$ ($v_d = 0.15$ mm/s). The spin friction torque is applied to react to a relative angular speed about the contact normal axis and is calculated assuming an equivalent circular contact area.

Constraints. No ideal joints are imposed. Each rotation or translation to be constrained generate a reaction force/torque proportional to the constraint violation through an elastic-damping element, that has been constructed such that to allow also the presence of a little amount of backlash, considered null in the current paper. Two elastic joints have been imposed at the two extremities of the screw shaft to reflect the structural implementation of the experimental test bench which will be used to validate the current model [11,12].

Simulation results

This section presents the preliminary results of a first dynamic simulation of the presented model (Fig. 2), applied to a ball screw with a nominal diameter of 16 mm and lead 5 mm with internal recirculation. The three recirculating inserts are evenly spaced by $120°$ circumferentially.

Compliant joints have been considered for both the screw shaft and the nut. For the latter, only the rotational motion around the ball screw symmetry axis is constrained with an elasto-damping joint, while all the other rotations are not constrained. All the results will be presented referring to a fixed coordinate system, located on the ball screw symmetry axis and coincident with the lower face of the screw shaft, with the z axis correspondent to the ball screw symmetry axis.

The simulation has been performed imposing to the screw shaft a smooth speed step of -500 rpm from 0.15 s to 0.25 s while applying an axial external force as disturbance on the nut in the negative direction of the z axis, opposing to its imposed speed. Figure 3 depicts the nut linear speed and that of the screw shaft reported in the linear domain through the transmission ratio. Figure 4 shows the external force and the torque required on the screw shaft to realize the

Materials Research Proceedings **26** (2023) 183-188 https://doi.org/10.21741/9781644902431-30

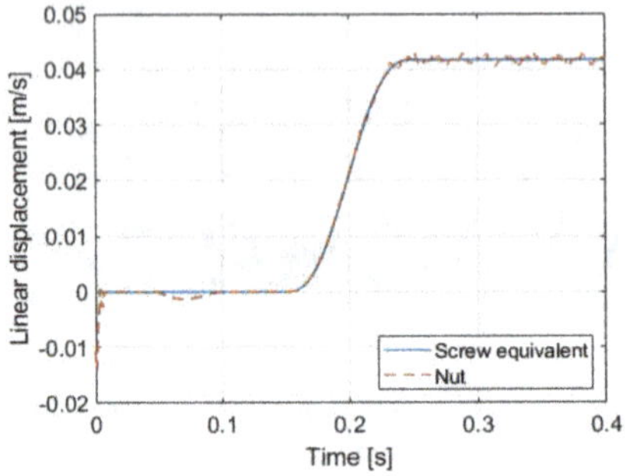

Figure 3 – Linear speed of the nut and screw

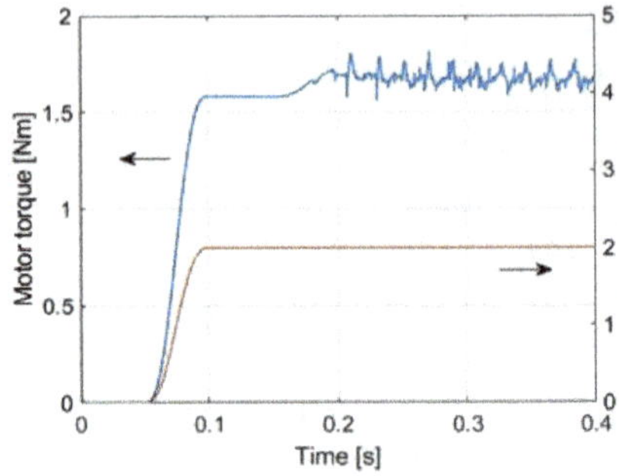

Figure 4 – External force on the nut and torque on the screw shaft.

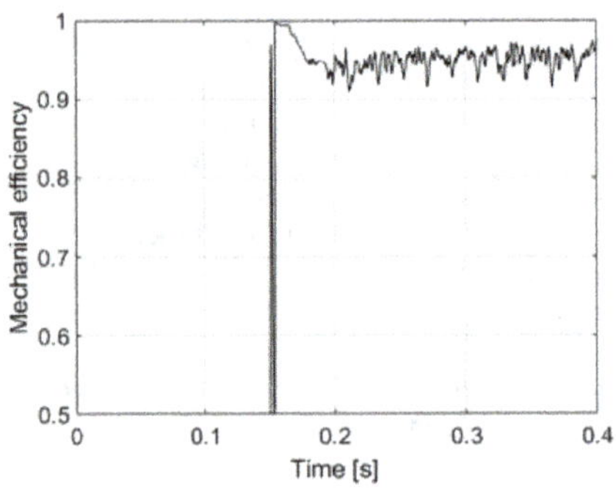

Figure 5 – Mechanical efficiency.

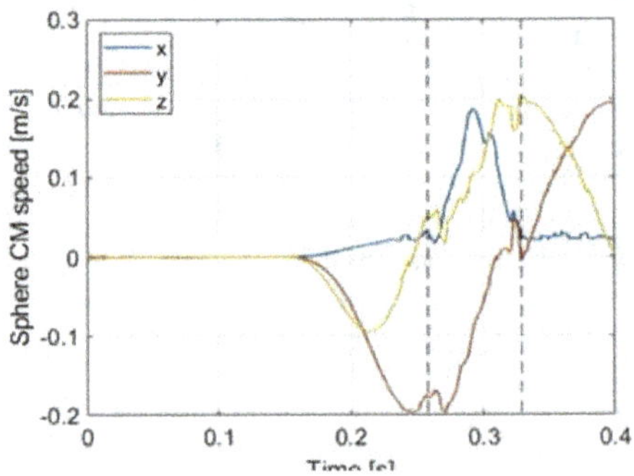

Figure 6 – Speed of the centre of mass of the selected sphere.

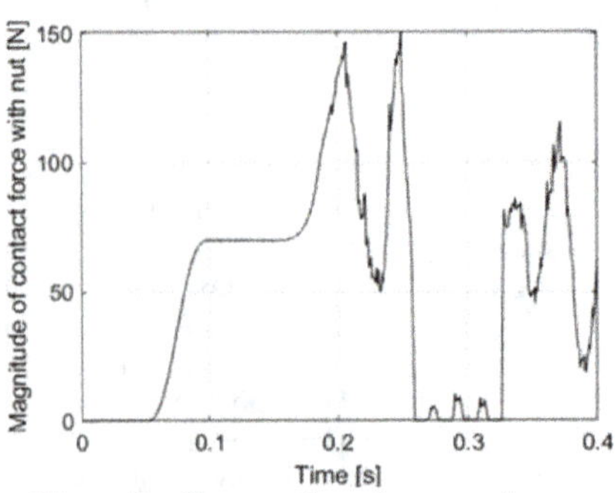

Figure 7 – Contact force between the nut and the analyzed sphere.

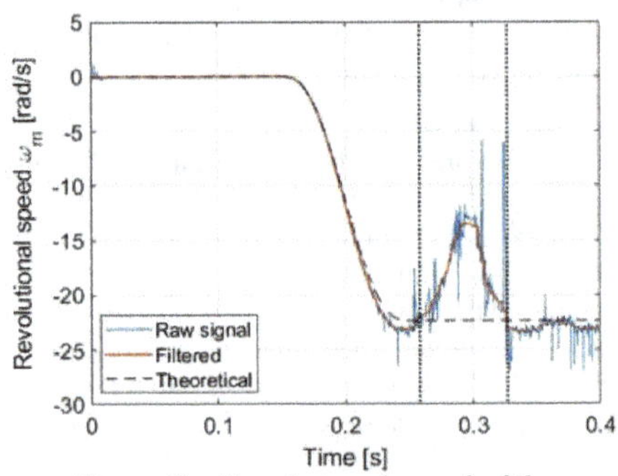

Figure 8 – Revolutional speed of the selected sphere.

imposed motion profile. It can be seen that when the external force is applied, though the screw shaft is steady, the nut slightly move because of the elastic deformations in the sphere/groove contacts. Furthermore, the presence of internal backlash is visible in the very beginning of the simulation on Fig. 3: in fact the initial configuration of the various bodies assumes the spheres and the nut in the nominal positions but, as the simulation starts, the gravity makes the system to reach a settle position. The motor torque rises to the level required to contrast the external force and after shows a little increment necessary to compensate the internal friction and guarantee the required motion. The system shows a realistic mechanical efficiency value of ≈ 93% (Fig. 5), typical of such mechanisms.

In order to clearly present the results, only the signals of one of the spheres are shown. Figure 6 illustrates the speed components of its center of mass. It is worth to be highlighted that the speeds of the spheres assume a sinusoidal shape when engaged in bearing load between the grooves while they varies when the spheres enters the recirculation. It can be seen around 0.3 s in the area between the two vertical dashed lines. A similar behavior is observable in Fig. 8, in which the revolution speed of the sphere with respect to the screw shaft around the ball screw axis is shown. The dashed lines represents the ideal revolution speed calculated according to [2]. When the sphere enters the recirculation, this speed sensibly decreases because its revolving motion is slowed down by the necessity of pass over the thread head to come back to the beginning of its motion path.

As observable in Fig. 3, because of the lack of other nut constraints but the anti-rotation around the symmetry axis combined with the slight asymmetry of the mass distribution and the presence of backlash, the nut oscillates and, hence, this reflects on the magnitude of the contact forces with the spheres. Figure 7 shows the contact force with the nut of the selected sphere. The contact forces originate with the application of the external force, remain steady until the motion starts and falls to near zero into the recirculation channel. The asymmetry of the nut causes a slight tilt of the nut, hence not every sphere bears the same load fraction. However, the mean value corresponds to the expected theoretical one [2]. The nut oscillation reflects also in the torque signal and finally in the mechanical efficiency. Because of the contact conditions, the spheres does not enter the recirculating inserts simultaneously on each recirculating circuit and this contribute to generate vibrations and nut oscillations.

Conclusions

A multibody dynamic model of a ball screw is presented in this paper. The innovation of such a model is the possibility to dynamically simulate the recirculation inserts analyzing the motion of the spheres within them, allowing the accuracy of the system to be sensibly increased. Firstly the model configuration is explained and finally the results of a preliminary simulation are presented. Further work includes the study of the friction increase in one or multiple recirculating circuits up to jamming, and the investigation on the effect of extraneous particles between the spheres. Last but not least, the experimental validation of the present model is scheduled as soon as the test bench will be fully functional. The current ball screw modeling is one of the activities included in the broader framework of PHM of electromechanical actuators ongoing within the authors' research group.

References

[1] Lin, M. C., Ravani, B., and Velinsky, S. A., 1994, "Kinematics of the Ball Screw Mechanism," J. Mech. Des., **116**(3), pp. 849–855. https://doi.org/10.1115/1.2919459

[2] Wei, C. C., and Lin, J. F., 2003, "Kinematic Analysis of the Ball Screw Mechanism Considering Variable Contact Angles and Elastic Deformations," J. Mech. Des., **125**(4), pp. 717–733. https://doi.org/10.1115/1.1623761

[3] Zhou, C. G., Feng, H. T., Chen, Z. T., and Ou, Y., 2016, "Correlation between Preload and No-Load Drag Torque of Ball Screws," Int. J. Mach. Tools Manuf., **102**, pp. 35–40. https://doi.org/10.1016/j.ijmachtools.2015.11.010

[4] Bertolino, A. C., Jacazio, G., Mauro, S., and Sorli, M., 2021, "Investigation on the Ball Screws No-Load Drag Torque in Presence of Lubrication through MBD Simulations," Mech. Mach. Theory, **161**(C), p. 104328. https://doi.org/10.1016/j.mechmachtheory.2021.104328

[5] Braccesi, C., and Landi, L., 2007, "A General Elastic-Plastic Approach to Impact Analisys for Stress State Limit Evaluation in Ball Screw Bearings Return System," Int. J. Impact Eng., **34**(7), pp. 1272–1285. https://doi.org/10.1016/j.ijimpeng.2006.06.005

[6] Hung, J. P., Shih-Shyn Wu, J., and Chiu, J. Y., 2004, "Impact Failure Analysis of Re-Circulating Mechanism in Ball Screw," Eng. Fail. Anal., **11**(4), pp. 561–573. https://doi.org/10.1016/j.engfailanal.2004.01.002

[7] Bertolino, A. C., Mauro, S., Jacazio, G., and Sorli, M., 2020, "Multibody Dynamic Model of a Double Nut Preloaded Ball Screw Mechanism With Lubrication," *Volume 7B: Dynamics, Vibration, and Control*, ASME, ed., American Society of Mechanical Engineers, Portland, OR, USA, pp. 1–11. https://doi.org/10.1115/IMECE2020-23619

[8] Bertolino, A. C., Jacazio, G., Mauro, S., and Sorli, M., 2019, "Developing of a Simscape Multibody Contact Library for Gothic Arc Ball Screws: A Three-Dimensional Model for Internal Sphere/Grooves Interactions," *IMECE Volume 4: Dynamics, Vibration, and Control*, ASME, ed., American Society of Mechanical Engineers, Salt Lake City, UT, USA, pp. 1–8. https://doi.org/10.1115/IMECE2019-10709

[9] Bertolino, A. C., De Martin, A., Fasiello, F., Mauro, S., and Sorli, M., 2022, "A Simulation Study on the Effect of Lubricant Ageing on Ball Screws Behaviour," *Proceedings of the International Conference on Electrical, Computer, Communications and Mechatronics Engineering, ICECCME*, IEEE, ed., IEEE, Maldives. https://doi.org/10.1109/ICECCME55909.2022.9987873

[10] Bertolino, A. C., De Martin, A., Mauro, S., and Sorli, M., 2021, "Multibody Dynamic ADAMS Model of a Ball Screw Mechanism with Recirculation Channel," *IMECE2021 Volume 7: Dynamics, Vibration, and Control*, pp. 1–10. https://doi.org/10.1115/IMECE2021-71121

[11] Bertolino, A. C., De Martin, A., Gaidano, M., Mauro, S., and Sorli, M., 2021, "A Fully Sensorized Test Bench for Prognostic Activities on Ball Screws," Int. Conf. Electr. Comput. Commun. Mechatronics Eng. ICECCME 2021, (October), pp. 7–8. https://doi.org/10.1109/ICECCME52200.2021.9591032

[12] Bertolino, A. C., De Martin, A., Jacazio, G., Mauro, S., and Sorli, M., 2019, "Robust Design of a Test Bench for PHM Study of Ball Screw Drives," *IMECE Volume 1: Advances in Aerospace Technology*, ASME, ed., American Society of Mechanical Engineers, Salt Lake City, UT, USA, pp. 1–9. https://doi.org/10.1115/IMECE2019-10713

[13] Bertolino, A. C., De Martin, A., Mauro, S., and Sorli, M., 2022, "Derivation of the Exact Curvature Formulation for Gothic Arch Ball Screw Grooves," *International Mechanical Engineering Congress and Exposition (IMECE2022)*, ASME, ed., ASME, Columbus, OH, pp. 1–9.

[14] Lin, M. C., Velinsky, S. A., and Ravani, B., 1994, "Design of the Ball Screw Mechanism for Optimal Efficiency," J. Mech. Des., **116**(3), pp. 856–861. https://doi.org/10.1115/1.2919460

[15] Antoine, J.-F., Visa, C., Sauvey, C., and Abba, G., 2006, "Approximate Analytical Model for Hertzian Elliptical Contact Problems," J. Tribol., **128**(3), p. 660. https://doi.org/10.1115/1.2197850

Theoretical and Applied Mechanics - AIMETA 2022 Materials Research Forum LLC
Materials Research Proceedings **26** (2023) 189-194 https://doi.org/10.21741/9781644902431-31

Dynamic performance of an aerostatic pad with internal pressure control

Federico Colombo[1, a], Luigi Lentini[1,b*], Terenziano Raparelli[1,c] and Andrea Trivella[1,d]

[1] Politecnico di Torino, DIMEAS department, Italy

[a]federico.colombo@polito.it, [b]luigi.lentini@polito.it, [c]terenziano.raparelli@polito.it, [d]andrea.trivella@polito.it

Keywords: Compensation, Diaphragm Valve, Aerostatic Bearings, Infinite Stiffness

Abstract. This paper presents a theoretical study on the dynamic performance of an aerostatic pad with an internal pressure control. The trend of the dynamic stiffness and damping over the frequency domain is analysed.

Introduction

Thanks to their zero friction, cleanness and infinite life, aerostatic pads are widely used in high-precision applications, e.g., machine tools, measuring machines and power board testing [1]. However, due to the compressibility of the lubricant, aerostatic bearings are characterized by relative low stiffness and poor damping. Much effort has been done in order to compensate for these drawbacks. A suitable selection and optimization of the feeding system of air pads is a common solution that makes it possible to obtain limited performance improvements. Several authors investigated to what extent the dimension, location and number of orifices affect aerostatic pad performance [2]. Compound restrictors, i.e., machining shallow grooves on the active surface of the pad, increase load capacity and stiffness but significantly reduce the bearing damping [3, 4]. The use of porous surfaces [5] is another solution that lead to performance improvement but it presents critical issues related to the choice of material that should have both suitable porosity and impact toughness.

Passive and active compensation methods makes it possible to significantly increase the static and dynamic performance of aerostatic bearings. In passive compensation methods bearings are integrated with devices that exploit only the energy associated with the supply pressure, e.g., pneumatic regulating valves and compliant elements. Conversely, actively compensated bearings are integrated with elements such as sensors, controllers and actuators that require external sources of energy to function. Active bearings have excellent static and dynamic performance [6–8] but the solutions are quite expensive. By contrast passive compensation solutions, notwithstanding their lower dynamics and effectiveness, represent an acceptable and cheap alternative to be integrated into current industrial applications [9]. In order to simplify the setting of the passive compensation, in work [10] the control signal and the supply pressure of the regulating valve are not dependent; the static behavior of the controlled pad is theoretically and experimentally analysed.

In this work the dynamic behavior of the system presented in [10] is theoretically analysed, evaluating the stiffness and damping of the air gap as the frequency of a sinusoidal load applied to the bearing varies. The numerical study is carried out by using a suitable lumped parameter model.

The compensated air pad

Figure 1 shows the geometry of the aerostatic pad considered in this numerical investigation. It has a rectangular base of 110x50 mm^2 and presents four supply holes with a diameter $d_p = 0.5$

Materials Research Proceedings **26** (2023) 189-194
https://doi.org/10.21741/9781644902431-31

mm which are located in the middle of a grooved rectangular supply line of dimensions 80x30 mm^2. These grooves have a triangular cross-section of depth h_g=60 μm and base w_g=0.3 mm.

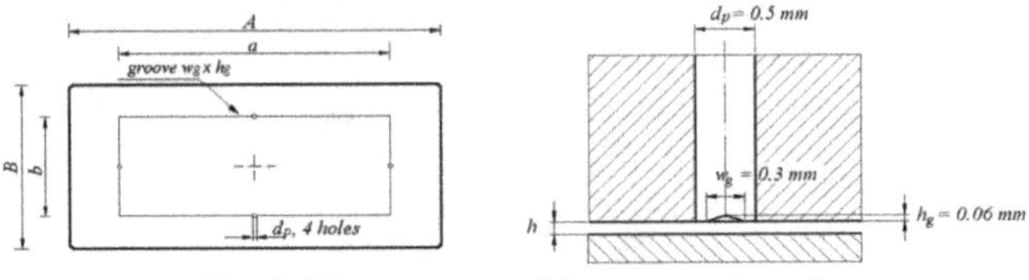

Fig. 1. The geometry of the aerostatic pad.

Figure 2a shows the scheme of the differential diaphragm valve that has been integrated with the aerostatic pad. The nozzle (1) has a diameter d_n= 0.5 mm and it is located in the chamber (2) that along with the chamber (3) is supplied with a constant supply pressure p_s. The other chambers (3) and (4) are supplied with a feedback $P_{Feedback}$ and a reference P_{Ref} pressures and they are separated through diaphragms (5 and 6) with different dimensions. The feedback pressure depends on the operating conditions of the pad (supply pressure and air gap height) and it is measured by means of back-pressure hole of diameter d_3=0.25 mm that is located on the active surface of the pad. The air flow supplied to the pad depends on the position of the shutter (7) that in turn depends on the difference between the feedback and supply pressure along with the diaphragm stiffnesses k_v.

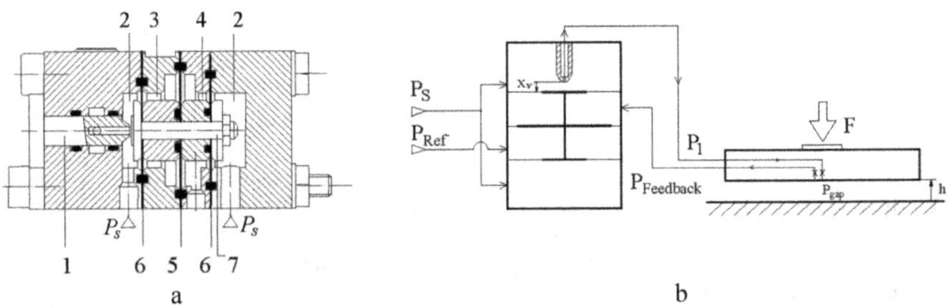

Fig. 2. a) scheme of the differential diaphragm valve. b) functional scheme of the control system.

Figure 2b shows a scheme useful to clarify the functioning of the system. Once the pressure P_S and P_{Ref} and the initial nozzle-shutter distance x_v are defined, the valve suitably regulates the air flow supplied to the pad depending on the applied load F. When the load increases, the pressures P_{gap} and $P_{Feedback}$ increase too producing a downward displacement of the shutter and in turn an increase of the air flow supplied to the pad. Consequently, the increase of the air flow produces an increase of the air gap height h that partially or completely compensated for the load variation. Conversely, the opposite holds in the presence of load reductions.

The numerical model

The considered compensated pad is modelled through a lumped parameter model consisting of pneumatic capacitances and resistances. Figure 3 shows the pneumatic model of the system.

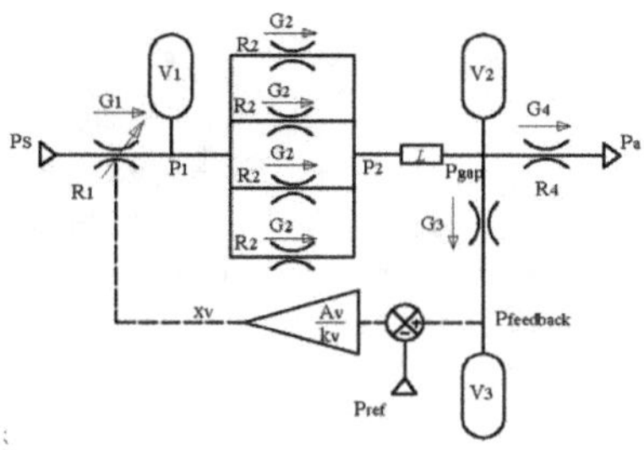

Fig. 3. Pneumatic model of the control system.

The nozzle and the air gap height are modelled through the variable pneumatic resistances R_1 and R_4, whereas the supply and back pressure hole of the pad are modelled through the constant resistances, R_2 and R_3 respectively. The air gap pressure (P_{gap}) distribution is considered uniform within the rectangle defined by the grooves (a x b) and linear in the region outside it. V_1, and V_3 are the volumes related to the ducts of the valve and the pad, whereas $V_2 = abh{+}V_g$ is the sum of the air gap and groove volume. The nozzle-shutter distance x_v is defined on the basis of the following equation:

$$\begin{cases} x_v = x_{by-pass} & x_v \le x_{by-pass} \\ x_v = x_0 + A_v \dfrac{(P_{feedback} - P_{Ref})}{k_v} & x_v > x_{by-pass} \end{cases} \tag{1}$$

where $x_{by-pass}$ is a threshold takes into account the air leakages experienced when the shutter is in contact with the nozzle (40 μm); x_0 is the initial distance between the nozzle and the shutter it is defined when $P_{feedback}$ and P_{Ref} are equal. A_v is the effective area of the valve diaphragms where the feedback and reference pressure are applied. The air mass flow rates passing through the resistances R_1, R_2, R_3 are modelled through the ISO formula 6358:

$$G_i = K_T\, cd_i\, \frac{0.685}{\sqrt{RT}} A_i P_{up} \sqrt{1 - \left(\frac{\frac{P_{down}}{P_{up}} - b_c}{1 - b_c}\right)^2} \tag{2}$$

where K_T is the square root of the ratio between the valve (293 K) and the reference absolute temperatures (273 K). A_i, cd_i, P_{up} and P_{down} are the annular area, discharge coefficient and the upstream and downstream pressures of the ith lumped resistance (i = 1, 2, 3); b_c=0.528 is the theoretical critical pressure ratio assuming an isentropic expansion. The discharge coefficients are assumed to be functions of the Reynolds' Number Re_i related to the flow of each resistance:

$$\begin{aligned} & cd_i = 1.05(1 - 0.3\, e^{-0.005\, Re_i}) \\ & Re_1 = \frac{G_1}{\pi \mu d_1};\ Re_2 = \frac{G_2\, h}{\mu A_2};\ Re_3 = \frac{G_3}{\pi \mu d_3};\ A_2 = \pi d_p h + w_g h_g\ ;\ d_1 = d_n \end{aligned} \tag{3}$$

Where $\mu = 18.89 \cdot 10^{-6}$ Pa s is the dynamic air viscosity. The mass flow rates G_4 exhausted from the air gap under isothermal conditions is:

$$G_4 = \frac{1}{6\mu RT}\left(\frac{b}{A-a} + \frac{a}{B-b}\right)\left(P_{gap}^2 - P_a^2\right) h^3 \tag{4}$$

Where R, T are the constant of gas ($R = 287.1\ \mathrm{J/(kg\ K)}$) and the absolute ambient temperature ($T = 293$ K). The pad's dimensions A, B, a, b are shown in figure 1. P_a is the ambient pressure, P_{gap} is the equivalent uniform pressure expressed by an empirical formula adopted for this air pad geometry [11]:

$$P_{gap} = L\,(h, P_2) = \left[1 - 0.02^{\left(\frac{5}{h}\right)}\right] \cdot (P_2 - P_a) + P_a \tag{5}$$

where h is expressed in μm. The load capacity of the pad is computed by integration of the pressure distribution on its area:

$$F_p = \left[ab + AB + \frac{(Ab+aB)}{2}\right]\frac{(P_{gap}-P_a)}{3} \tag{6}$$

The continuity equations related to the incoming and outgoing mass flow rates for each volume V_1, V_2, V_3 and to the corresponding pressures P_1, P_{gap}, $P_{feedback}$ are:

$$G_1 - 4G_2 = \frac{V_1}{RT}\frac{dP_1}{dt}$$

$$4G_2 - G_4 - G_3 = \frac{P_{gap}ab}{RT}\frac{dh}{dt} + \frac{V_2}{RT}\frac{dP_{gap}}{dt} \tag{7}$$

$$G_3 = \frac{V_3}{RT}\frac{dP_{feedback}}{dt}$$

The equilibrium equation of the pad (considered as a single degree of freedom system) is:

$$F - \mathrm{F}_P + M\frac{\mathrm{d}^2 h}{dt^2} = 0 \tag{8}$$

where F is the vertical external load applied to the pad and M the mass related to the supported payload (F=Mg).

Numerical results and Discussion

At the supply pressure $P_S = 0.5$ MPa the regulation parameters of the valve are identified with the numerical model: x_0= 55 μm, $P_{Ref} = P_{feedback} = 0.37$ MPa. In this way, the compensation works in a range of air gap near to h = 9.5 μm. The static behavior of the compensated air pad has been widely discussed in [10].

To determine the dynamic parameters K and c of the air bearing a sinusoidal load F of amplitude ΔF and frequency f is imposed around the initial point of static equilibrium. By applying a small amplitude load (ΔF is about 1% of the initial static load) it is possible to define with good accuracy the dynamic parameters of the air bearing at each air gap height. With the imposed sinusoidal load the air gap h is calculated and plotted as a function of the time; the amplitude Δh and the time delay Δt between the input signal F and the output signal h are therefore determined in stationary conditions. The pressure force of the air gap is the sum of the elastic component due to the stiffness

Materials Research Proceedings **26** (2023) 189-194
https://doi.org/10.21741/9781644902431-31

of the air gap and the viscous component due to the damping of the air gap. Considering all the acting forces as rotating vectors on a complex plane, the equilibrium of the forces decomposed in the direction of the elastic force and in the direction of the viscous force allows to obtain the values of K and c as a function of the frequency, according to the equations (9):

$$K(\omega) = \frac{\Delta F}{\Delta h}\cos(\phi) + M\omega^2 ; \quad c(\omega) = \frac{\Delta F}{\Delta h}\frac{\sin(\phi)}{\omega} \tag{9}$$

Where $\omega = 2\pi f$ is the angular frequency, $\phi = \omega\Delta t$ is the phase lag between the input and the output signals.

Figure 4 shows the trend of the dynamic stiffness K and the damping c over a frequency range from 0 to 30 Hz and considering initial static conditions near to $h = 9.5$ μm.

At low frequency, stiffness and damping of the air gap are much higher than those obtainable without the valve and they quickly reduce as ω increases. Outside the overcompensation zone, the stiffness is always positive, otherwise it can be positive or negative as a function of ω. Further studies will be involved to analyze the dynamic stability of the system, as already made for the solution studied in [12].

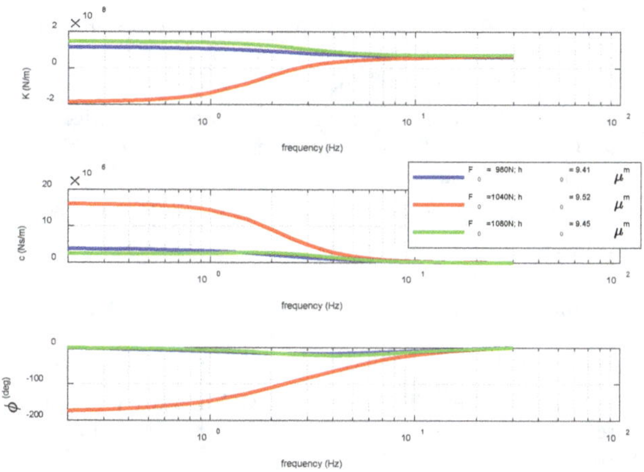

Fig. 4. Stiffness K, damping c and phase delay ϕ of the compensated air pad: numerical results. $P_S = 0.5\,MPa,\ k_v = 50\,N/mm,\ d_v = 0.5\,mm,\ d_P = 0.5\,mm.$

Conclusions

This paper presents a preliminary investigation on the dynamic performance of an aerostatic pad with internal pressure control. The dynamic performance of the compensated air pad dramatically decreases as the frequency increases due to the presence of the valve. For frequencies below about 40 Hz the phase lag is negative, which means the presence of the valve makes the displacement of the air pad in advance with respect to the applied external force. The results indicate that this compensation method represents an efficient and cost-effective method for low frequency air pad applications. Further studies will be aimed to compare numerical and experimental results and to analyze the stability of the proposed control system.

References

[1] Gao, Q., Chen, W., Lu, L., Huo, D., Cheng, K.: Aerostatic bearings design and analysis with the application to precision engineering: State-of-the-art and future perspectives. Trib. Intern. 139, (2019), 1-17. https://doi.org/10.1016/j.triboint.2019.02.020

[2] Charki, A., Diop, K., Champmartin, S., Ambari, A.: Numerical simulation and experimental study of thrust air bearings with multiple orifices. International Journal of Mechanical Sciences. 72, (2013), 28–38. https://doi.org/10.1016/j.ijmecsci.2013.03.006

[3] Ma, W., Cui, J., Liu, Y., Tan, J.: Improving the pneumatic hammer stability of aero-static thrust bearing with recess using damping orifices. Trib. Intern. 103, (2016), 281–288. https://doi.org/10.1016/j.triboint.2016.06.009

[4] Yoshimoto, S., Tamura, J., Nakamura, T.: Dynamic tilt characteristics of aerostatic rectangular double-pad thrust bearings with compound restrictors. Trib. Intern. 32, (1999), 731–738. https://doi.org/10.1016/S0301-679X(00)00004-9

[5] Kwan, Y.B.P., Corbett, J.: Porous aerostatic bearings: an updated review. Wear 222, 1998, 69–73. https://doi.org/10.1016/S0043-1648(98)00285-3

[6] Al-Bender, F.: On the modelling of the dynamic characteristics of aerostatic bearing films: From stability analysis to active compensation. Precision Engineering. 33, (2009), 117–126. https://doi.org/10.1016/j.precisioneng.2008.06.003

[7] Maamari, N., Krebs, A., Weikert, S., Wegener, K.: Centrally fed orifice based active aerostatic bearing with quasi-infinite static stiffness and high servo compliance. Trib. Intern. 129, (2019), 297–313. https://doi.org/10.1016/j.triboint.2018.08.024

[8] Colombo, F., Lentini, L., Raparelli, T., Viktorov, V.: Actively compensated aerostatic thrust bearing: design, modelling and experimental validation. Meccanica, 52, (2017), 1–16. https://doi.org/10.1007/s11012-017-0689-y

[9] Ghodsiyeh, D., Colombo, F., Lentini, L., Raparelli, T., Trivella, A., Viktorov, V.: An infinite stiffness aerostatic pad with a diaphragm valve. Tribology International. 141, (2020), 105964. https://doi.org/10.1016/j.triboint.2019.105964

[10] Lentini, L., Colombo, F., Raparelli, T., Trivella, A., Viktorov, V.: An aerostatic pad with an internal pressure control. E3S Web Conferences. 197, (2020), 07002. https://doi.org/10.1051/e3sconf/202019707002

[11] Colombo, F., Lentini, L., Raparelli, T., Trivella, A., Viktorov, V.: A Lumped Model for Grooved Aerostatic Pad. Advances in Service and Industrial Robotics. Springer (2019), 678–686. https://doi.org/10.1007/978-3-030-00232-9_71

[12] Colombo, F., Lentini, L., Raparelli, T., Trivella, A., Viktorov, V.: Dynamic behaviour and stability analysis of a compensated aerostatic pad. 76th Italian National Congress ATI, 15-17 Settembre 2021. https://doi.org/10.1051/e3sconf/202131205003

Theoretical and Applied Mechanics - AIMETA 2022 Materials Research Forum LLC
Materials Research Proceedings **26** (2023) 195-200 https://doi.org/10.21741/9781644902431-32

A novel prototype of diaphragm valve for passively compensated aerostatic pads

Federico Colombo [1, a], Luigi Lentini [1,b*], Terenziano Raparelli[1,c], Andrea Trivella[4c]

[1]Politecnico di Torino, Corso Duca degli Abruzzi 24, 10129, Turin, Italy

[a]federico.colombo@polito.it , [b]luigi.lentini@polito.it , [c]terenziano.raparelli@polito.it , [d]andrea.trivella@polito.it

*corresponding author

Keywords: Air Pad, Gas Bearings, Infinite Stiffness, Compensation, Diaphragm Valve

Abstract. Thanks to their low friction, aerostatic pads are currently used in many high precision applications. However, due to the air compressibility, aerostatic bearings suffer from relatively low stiffness and damping. Active and passive compensation methods represent a valuable solution to reduce these limitations. This paper presents a novel prototype of a diaphragm valve for passively compensated aerostatic pads. The proposed valve was obtained through the improvements of a previous prototype. The main goal of this new design of the valve was to improve the reliability, repeatability and accuracy of regulation. The novel prototype is modelled through the same lumped parameter model that was used for the previous prototype. The preliminary experimental results demonstrate the efficiency and effectiveness of the proposed valve.

Introduction

Because their zero friction, aerostatic pads are widely used in applications where high precision of positioning is required, e.g., machine tools and measuring machines [1,2]. However, this kind of bearings are characterized by low relative stiffness and they suffer from low damping due to the air compressibility. In this regard, different solutions have been proposed to improve aerostatic pad performance. In the attempt of reducing these limitations different kinds of feeding system were studied and experimented, e.g., multiples supply holes with different size, number and location [3], porous inserts or surfaces [4,5] and compound restrictors [6–8]. However, a suitable choice and design of the feeding system of aerostatic pads can lead only to limited performance improvement. Conversely, active and passive compensation methods make it possible to significantly improve both static and dynamic performance [9,10]. Passive compensation methods use components which require only the energy associated with the supply pressure of the bearing, e.g., pneumatic valves and compliant elements. By contrast, actively compensated bearings exploit elements that require external sources of energy to function, e.g., sensors, controllers and actuators. Although their high performance [11–13], active compensation solutions are still too expensive and thus result to be unsuitable for industrial applications. Notwithstanding their lower bandwidth, passive compensation solutions represent a cheaper alternative. This paper presents a novel prototype of passively compensated pad. This passive compensation solution consists in the integration of commercial pad and a custom-built diaphragm valve whose design is an improved version of that described in [14,15]. The novel prototype is modelled through the same lumped parameter model that was used for the previous prototype.

Materials Research Proceedings **26** (2023) 195-200 https://doi.org/10.21741/9781644902431-32

The Prototype

Figure 1a and 1b show a cross section of the designed diaphragm valve and the aerostatic pad.

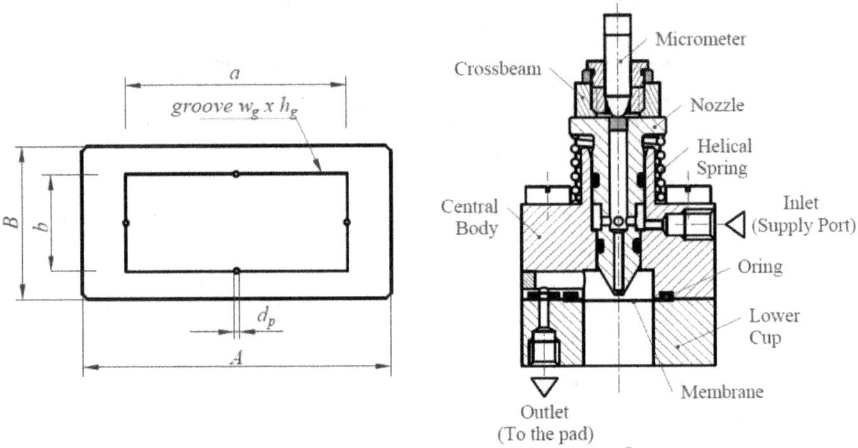

(a) Air pad. (b) Diaphragm valve.

Figure 1: The proposed prototype.

The diaphragm valve is composed of three main parts: an upper crossbeam, a central body and a lower cap. The chamber of the valve is located at the interface between the central body and the lower cap. The lower surface of the chamber consists of a membrane that is clamped between the central body and the lower cup thanks to the presence of an elastomeric o-ring. The chamber of the valve is supplied through a moving nozzle. The relative position from the nozzle and the membrane can be manually regulated through a micrometer. A coil spring is compressed between the nozzle and the central body to guarantee the presence of a restoring force when the nozzle must be moved upwards. The main goal of the novel design was to modify the mechanism to regulate the initial nozzle-membrane distance, i.e., when the system is not pressurized, since the accuracy, repeatability and reliability of the system is strictly related to this aspect. The air pad has rectangular base of 60x30 mm^2 (AxB) and four orifices with a diameter of 1 mm (d_p). Each restrictor is located in the middle of each side of a grooved rectangular supply line with a base of 45 mm (a) and a width of 20 mm (b). The grooves present a triangular cross-section with a base of 0.2 mm (w_g) and a height of 0.06 mm (h_g).

Numerical Model

The system is modelled through lumped pneumatic resistances and capacitances (see Figure 2).

Materials Research Forum LLC
https://doi.org/10.21741/9781644902431-32

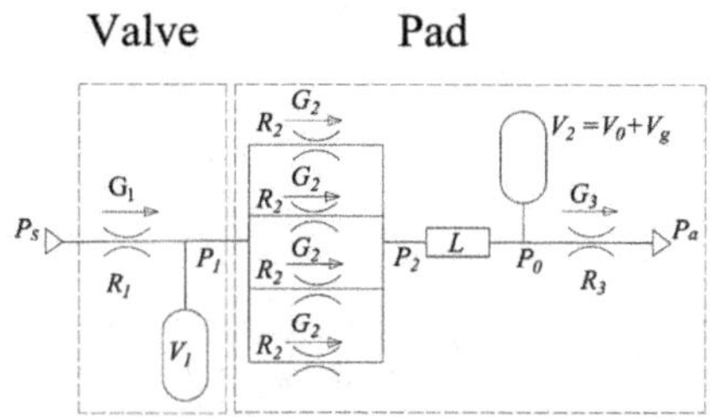

Figure 2: Pneumatic scheme of the prototype.

Starting from the upstream, the volume of the valve V_1 is supplied with a constant pressure P_s through the nozzle. The value of the air mass flow rate G_1 supplied by the nozzle (R1) depends on the initial position of the nozzle and its diameter d_n, the deflection of the membrane[1] x and its stiffness k_m, the supply pressure P_s and the pressure inside the valve chamber P_1. The air mass flow rate passing through the nozzle of the valve G_1 and the supply holes of the pad G_2 are computed by means of the ISO formula 6358 [16] as follows:

$$G_i = C_i P_s \sqrt{1-\left(\frac{\frac{P_{d,i}}{P_{up,i}}-0.5283}{1-0.5283}\right)^2} \qquad i=1,2.$$

$$C_1 = 1.05(1-0.3\,e^{-0.005\,Re_1})\frac{0.685}{\sqrt{R\,T}}\pi\,d_n\,x; \qquad Re_1=\frac{G_1}{\pi\mu d_n}$$

$$\begin{cases} x = 12\ \mu m; & x \le 12\ \mu m \\ x = x_n + \dfrac{\pi\,{d_m}^2}{4}\dfrac{(P_1-P_a)}{k_m}; & x > 12\mu m \end{cases} \tag{1}$$

$$C_2 = 1.05(1-0.3\,e^{-0.005\,Re_2})\frac{0.685}{\sqrt{R\,T}}\left(\pi\,d_p\,h+\,w_g\,h_g\right)$$

$$Re_2 = \frac{G_2\,h}{\mu\left(\pi d_p h + w_g\ h_g\right)}$$

where, P_{up} and P_d are the upstream and downstream pressure. C_1 and C_2 are the pneumatic conductance related to the valve nozzle and the pad supply holes. G_3 is the mass flow rate exhausted through the air gap (R3) and it computed through Equation 2, where, P_0 is the mean constant pressure inside the area of the grooved rectangular supply line that is obtained from the pressure downstream the supply holes of the pad P_2 through the semi-empirical formula of Equation 3 [8] (that in Figure 2 corresponds to the function L) [14]. R, T and μ are the gas constant temperature and dynamic viscosity of the air. Re_1 and Re_2 are the Reynolds numbers related to the air flow passing through the nozzle of the valve and the supply holes of the pad.

[1] That is a function of the membrane geometry: its diameter d_m .

$$G_3 = \frac{1}{6\mu RT}\left(\frac{b}{A-a} + \frac{a}{B-b}\right)(P_0^2 - P_a^2)h^3 \tag{2}$$

$$P_0 = \left[1 - 0.14^{\left(\frac{5}{h}\right)}\right](P_2 - P_a) + P_a \tag{3}$$

The load carrying capacity of the pad F_p is computed through Equation 4 by assuming that the surface of the pad is perfectly smooth and that the pressure distribution outside the rectangular area surrounded by the grooves is linear.

$$F_p = \left[ab + AB + \frac{(Ab + aB)}{2}\right]\frac{(P_0 - P_a)}{3} \tag{4}$$

The dynamic equations of the model (Equations 5) that are necessary to compute the pressures (P_1 and P_0) along with the air gap height of the pad (h) are the continuity equations applied to volumes V_1 and V_0 and the equilibrium equation of the pad respectively:

$$G_1 - 4G_2 = \frac{V_1}{RT}\frac{dP_1}{dt} \qquad 4G_2 - G_3 = \frac{1}{RT}\left(P_0\frac{dV_2}{dt} + V_2\frac{dP_0}{dt}\right)$$

$$F^{ext} = F_p - M\ddot{h} \tag{5}$$

These equations are discretized by the Euler explicit method.

Numerical and Experimental Results

As it can be seen from Figures 3 the numerical model was experimentally validated through static characterizations. Moreover, the equations of the lumped model were exploited to implement a numerical design procedure in order to define the valve parameters that optimize the system performance around a specific air gap height h_{des}. To verify the accuracy and reliability of the proposed design procedure, it was used to compute the optimal operating parameters of the valve for an air gap height of 21 μm.

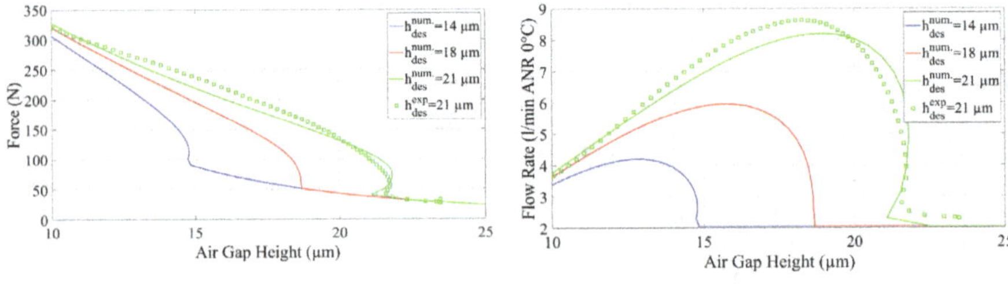

(a) Load Capacity (b) Air consumption

Figure 3: Experimental and numerical static feature of the prototype.

Figures 3 also reports the numerical curves that corresponds to operating air gap height $\boldsymbol{h_{des}}$ of 14 and 18 μm. Eventually, the stability of the prototype was experimentally verified around the operating air gap value (h_{des}=21 μm) through the application of a negative step force of about 20 N (see Figure 4).

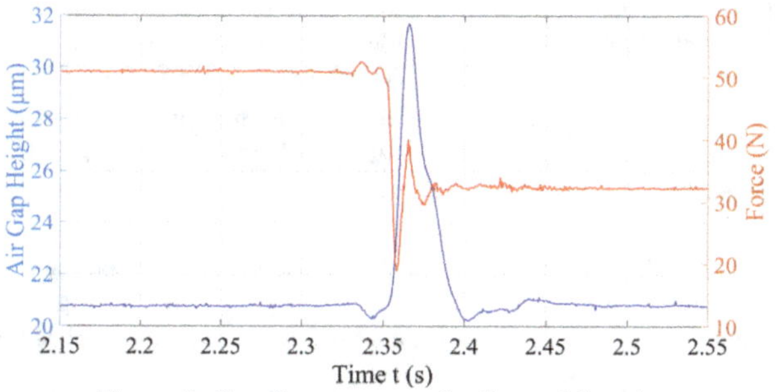

Figure 5: Step force test results (h_{des}=21 μm)

Step force test demonstrates the stability of the proposed prototype and even quasi-static infinite stiffness around the considered operating air gap height (h_{des}= 21 μm).

Conclusions

This paper presented a novel prototype of a passively compensated aerostatic pad consisting in the integration of a commercial pad and custom-built diaphragm valve. Compared to a previous solution, the new valve design was thought to obtain a simpler mechanical structure and higher repeatability. The repeatability was assessed by the fact that, conversely to the previous prototype, the initial position of the nozzle does not change after many experimental test repetitions. Moreover, the novel valve prototypes was built by exploiting a design algorithm (similar to those presented in [15]) based on the equation of the lumped parameter model of the system [14]. The static characterization of the prototype makes it possible to verify the accuracy of the lumped model and the proposed design algorithm. Step force tests were used to verify the stability of the system. Other than verifying the stability, step force tests demonstrated that the system is characterized by quasi-static infinite stiffness around the considered operating air gap height.

References

[1] Lentini L, Moradi M, Colombo F. A Historical Review of Gas Lubrication: From Reynolds to Active Compensations. Tribology in Industry 2018;40:165–82. https://doi.org/10.24874/ti.2018.40.02.01

[2] Gao Q, Chen W, Lu L, Huo D, Cheng K. Aerostatic bearings design and analysis with the application to precision engineering: State-of-the-art and future perspectives. Tribology International 2019;135:1–17. https://doi.org/10.1016/j.triboint.2019.02.020

[3] Colombo F, Lentini L, Raparelli T, Trivella A, Viktorov V. Dynamic Characterisation of Rectangular Aerostatic Pads with Multiple Inherent Orifices. Tribology Letters 2018;66. https://doi.org/10.1007/s11249-018-1087-x

[4] Luong TS, Potze W, Post JB, Van Ostayen RAJ, Van Beek A. Numerical and experimental analysis of aerostatic thrust bearings with porous restrictors. Tribology International 2004;37:825–32. https://doi.org/10.1016/j.triboint.2004.05.004

[5] Witelski T, Schwendeman D, Evans P. Analysis of Pressurized Porous Air Bearings. New Way Precision Air Bearings Inc; 2008.

[6] Colombo F, Lentini L, Raparelli T, Trivella A, Viktorov V. A Lumped Model for Grooved Aerostatic Pad. Advances in Service and Industrial Robotics, vol. 67, Springer International Publishing; 2019, p. 678–86. https://doi.org/10.1007/978-3-030-00232-9_71

[7] Colombo F, Lentini L, Raparelli T, Trivella A, Viktorov V. Dynamic model of a grooved thrust bearing: Numerical model and experimental validation. Proceeding of the 23rd Conference of the Italian Association of Theoretical and Applied Mechanics, AIMETA, vol. 4, 2017, p. 506–17.

[8] Colombo F, Lentini L, Raparelli T, Trivella A, Viktorov V. A nonlinear lumped parameter model of an externally pressurized rectangular grooved air pad bearing. Advances in Italian Mechanism Science, vol. 68, Springer; 2019, p. 490–7. https://doi.org/10.1007/978-3-030-03320-0_54

[9] Al-Bender F. Air Bearings: Theory, Design and Applications. John Wiley & Sons; 2021.

[10] Al-Bender F. On the modelling of the dynamic characteristics of aerostatic bearing films: From stability analysis to active compensation. Precision Engineering 2009;33:117–26. http://dx.doi.org/10.1016/j.precisioneng.2008.06.003

[11] Aguirre G, Al-Bender F, Van Brussel H. A multiphysics model for optimizing the design of active aerostatic thrust bearings. Precision Engineering 2010;34:507–15. https://doi.org/10.1016/j.precisioneng.2010.01.004

[12] Morosi S, Santos IF. Active lubrication applied to radial gas journal bearings. Part 1: Modeling. Tribology International 2011;44:1949–58. http://dx.doi.org/10.1016/j.triboint.2011.08.007

[13] Morosi S, Santos IF. On the modelling of hybrid aerostatic-gas journal bearings. Proceedings of the Institution of Mechanical Engineers, Part J: Journal of Engineering Tribology 2011;225:641–53. https://doi.org/10.1177/1350650111399845

[14] Ghodsiyeh D, Colombo F, Lentini L, Raparelli T, Trivella A, Viktorov V. An infinite stiffness aerostatic pad with a diaphragm valve. Tribology International 2020;141:105964. https://doi.org/10.1016/j.triboint.2019.105964

[15] Colombo F, Lentini L, Raparelli T, Trivella A, Viktorov V. Design and Analysis of an Aerostatic Pad Controlled by a Diaphragm Valve. Lubricants 2021;9:47. https://doi.org/10.3390/lubricants9050047

[16] Belforte G, Raparelli T, Viktorov V, Trivella A. Discharge coefficients of orifice-type restrictor for aerostatic bearings. Tribology International 2007;40:512–21. https://doi.org/10.1016/j.triboint.2006.05.003

Theoretical and Applied Mechanics - AIMETA 2022
Materials Research Proceedings **26** (2023) 201-206

Materials Research Forum LLC
https://doi.org/10.21741/9781644902431-33

Characterization of finger joints with underactuated modular structure

Gabriele Maria Achilli[1,a], Silvia Logozzo[1,b], Monica Malvezzi[2,c], Domenico Prattichizzo[2], Gionata Salvietti[2], Maria Cristina Valigi[1,d] *

[1]Department of Engineering, University of Perugia, Perugia, Italy

[2]Department of Information Engineering and Mathematics, University of Siena, Siena, Italy

[a]gabrielemaria.achilli@studenti.unipg.it, [b]silvia.logozzo@unipg.it, [c]monica.malvezzi@unisi.it, [d]mariacristina.valigi@unipg.it

Keywords: Underactuated Finger, Joint Stiffness, Modular Gripper, Soft Robotics

Abstract. The characterization of flexible joints of robotic underactuated fingers allows the investigation of the finger flexion trajectories using a tendon-driven actuation. In this paper, the characterization of TPU 3D printed joints used in underactuated robotic fingers, is experimentally and numerically performed. Since the mechanical properties of 3D printed materials are uncertain, this research helps to define the characteristics of robotic fingers in terms of stiffness. The role of the stiffness for the control of fingers' trajectories is fundamental and the obtained results are very useful for improving the method to track a certain predefined trajectory. Experimental and theoretical results evaluate the stiffness as a function of the infill density percentage of the material.

Introduction

In the last years, a novel generation of underactuated compliant hands is growing in the robotics community [1]. In these soft hands the simplification of actuation and control is central [2-7], as well as the reduction of the degrees of actuation with respect to the degrees of freedom (DoFs) and the intrinsic passive compliance at joint level. Underactuated compliant hands can be implemented using deformable materials that can be processed with Additive Manufacturing (AM) [8, 9]. Adding deformable elements allows the storage of elastic energy avoiding tendon slackness, passive compliance, the distribution of forces over a large contact area avoiding grasped object damage. Shape adaptation increases the grasp performance by compensating the uncertainties and stabilizing the grasp [10]. Soft grippers are commonly prototyped using different materials, e.g., elastomeric materials whose mechanical properties often present anisotropic, inhomogeneous, and nonlinear behaviours [11, 12]. Some recent development further exploits Interpenetrating Phase Composites [13] and topology optimization techniques for obtaining a multimaterial-like behaviour varying the 3D printing infill density [14, 15]. A proper mechanical characterisation is very important to predict the dynamic behaviour of the soft robotic component [16]. This work is focused on the mechanical characterisation of compliant joints of robotic modular fingers, providing methods and results that can be exploited in the design of new soft hands, improving dexterity [17]. A possible way to design a desired fingertip trajectory or a specific performance is to opportunely tune the stiffness of the compliant joints. For this purpose, this paper presents the characterization of a flexible joint for modular robotic fingers in terms of variation of stiffness depending on the density of the material, represented by the infill rate of the 3D printing process.

Joint characterization

Soft-rigid robotic fingers are modular structures composed of rigid or semi-rigid phalanges and soft flexible joints which reproduce the articulation function (Fig. 1(a)) [3, 8, 9, 18]. This work is focused on the analysis of the mechanical properties of flexible joints made in TPU (thermoplastic polyurethane, Ninjaflex Semiflex 85A, Lulzbot, USA) and rigid parts made in PLA (Polylactic Acid, ecoPLA, niceshops GmbH, Austria) by 3D printing techniques using a FDM (fused deposition modeling) 3D printer (Lulzbot TAZ 5, Lulzbot, USA). A TPU joint sample is represented in Fig. 1(b, c).

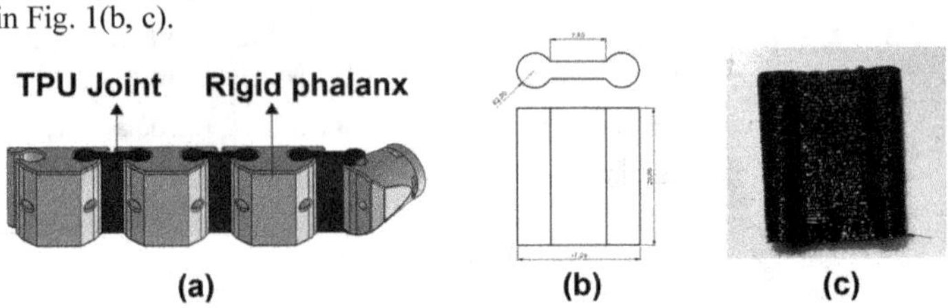

Fig. 1 – Modular finger: (a) assembly; (b) joint dimensions; (c) 3D printed joint

The objective of the joint characterization was to study the stiffness of the flexible component for different values of density. In fact, the stiffness of the deformable passive joint affects the trajectory of the fingers subjected to a tendon-driven actuation and thus the effectiveness of the grasping. For this purpose, both experimental and numerical analyses were performed. The experimental part of the work was based on the realization of 15 TPU joints with five different infill percentages. The stiffness was measured by using a digital force-displacement gauge. The numerical approach implied the use of FEM (finite element modelling) software. As the finger is composed of rigid modules and the length of the joint is comparable with the length of the rigid modules, the deflection is moderate and the problem can be modelled considering the theory of beams [19-20]. Analytical and experimental results can be compared to create a simple and validated numerical tool to design this kind of flexible joints with specific stiffnesses.

Experimental characterization

The experimental characterization of stiffness was performed for 15 joint samples with the same geometrical dimensions and five different infill percentages (20, 30, 40, 50 and 60%), in groups of 3 samples. A 3D printed rigid constraint made in PLA was used as support for the TPU samples, which were mounted between this rigid support and a finger phalanx (Fig. 2(a)). The phalanx was subjected to a vertical load by a flat tip to simulate the behavior of the joint during the flexion of the finger. For each flexion test the force-displacement curve was obtained and the stiffness value was extracted accordingly in two significative displacement ranges: [0 – 0.8 mm] and [0 – 4 mm]. In fact, a 2-grade polynomial approximation of the force-displacement curves demonstrates that the curve has an approximately parabolic trend with a quasi-linear increase in the first interval 0 – 0.8 mm. The maximum displacement was fixed at 4 mm as after this value the finger phalanx collides with the support.

Numerical analysis

The FEM analysis was performed analyzing the joint behavior when it is subjected to a static stress. A 3D CAD (Computer Aided Design) model of two simplified phalanges connected by a TPU joint was drawn and discretized and loads and constraints were applied as represented in Fig. 2(b).

Different simulations were performed by varying the infill conditions with the aim of observing the force required to obtain a fixed displacement. The stiffness and the force evaluated are compared to the experimental force-displacement curve. The TPU characteristics were extracted by the datasheets according with the infill percentages and a linear elastic model was used to study the deformations. Simulations were performed in displacement control, in large displacement hypothesis. In particular, the simulations results were compared to the following loads: 7.4, 13.7, 15.0, 16.9, 18.3 N. These values represent the mean values of the maximum loads measured during the experiments for the TPU joints with infill percentage equal to 20, 30, 40, 50 and 60 %.

Materials Research Proceedings **26** (2023) 201-206 https://doi.org/10.21741/9781644902431-33

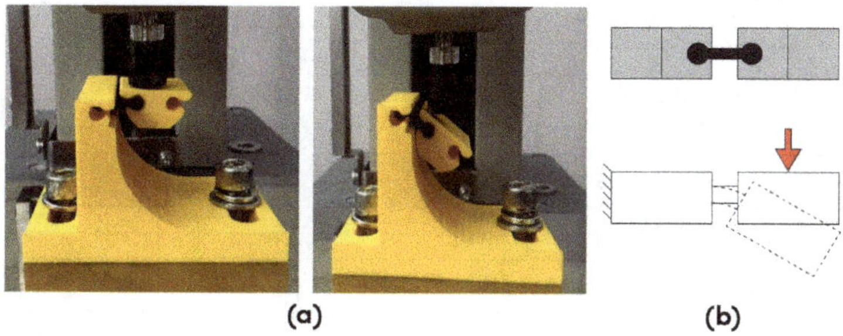

Fig. 2 – Experimental layout (a); 3D CAD model for FEM and load configuration (b).

Results

The mean force-displacement curves obtained by experiments are represented in Fig. 3. The fitting curves obtained by 2-grade polynomial approximation are also represented. The deformation map resulting by the simulations is reported in Fig. 4: setting the prescribed displacement as a boundary condition, the resulting plot is the same for all the infill percentages. Tab. 1 reports all the stiffness results obtained by experimental tests. The parameter K_{mean} represents the mean value of the stiffness. The force applied in the numerical analysis to allow the maximum displacement and the stiffness are reported in Tab. 2. Fig. 5 shows the comparison of numerical and experimental results.

Discussion

Results show that all the curves present similar shapes, with different load/deformation values. All the curves are characterized by a steep, almost linear part at the beginning (displacements 0-0.8 mm). In the 20% infill case, due to the higher flexibility, also in the first part of the curve a less stiff, nonlinear behavior is evident. For all the samples, a linear load/deformation behavior can be considered only for limited displacement values, and this is an important aspect that must be considered for the applications of these components as flexible joints of soft robotic fingers, which frequently undergo large deformations. This aspect must be considered both in the mechanical design phase and in the control system implementation.

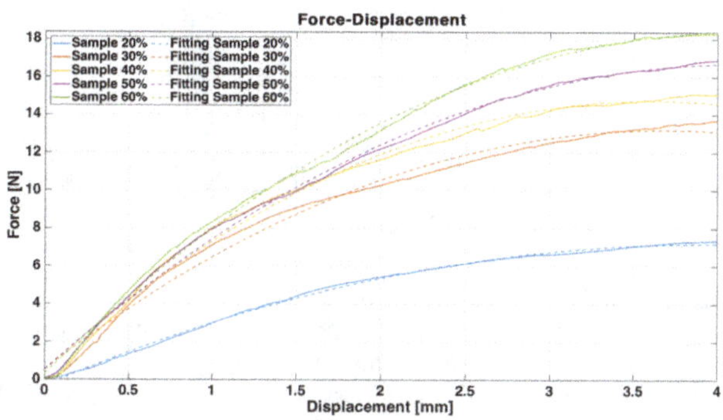

Fig. 3 – Mean force-displacement curves of the TPU joint with different infill percentages

Tab. 1 – Stiffness experimental results

		Stiffness [N/mm]			
Infill percentage	Displacement range	Sample 1	Sample 2	Sample 3	K_{mean}
20%	0 – 0.8 mm	3.19	3.39	2.19	2,92
	0 – 4 mm	2.39	1.55	1.59	1,84
30%	0 – 0.8 mm	7.06	7.87	7.68	7,53
	0 – 4 mm	3.25	3.50	3.54	3,43
40%	0 – 0.8 mm	8.00	9.23	8.10	8,44
	0 – 4 mm	3.56	4.02	3.73	3,77
50%	0 – 0.8 mm	9.05	7.10	9.13	8,43
	0 – 4 mm	4.24	4.00	4.43	4,22
60%	0 – 0.8 mm	6.57	10.78	9.25	8,87
	0 – 4 mm	4.06	6.97	4.69	4,57

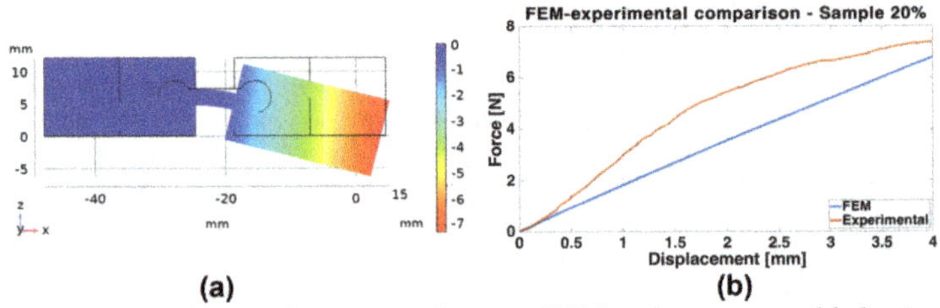

Fig. 4 – FEM results (a); Comparison between FEM and experimental behavior (b).

Tab. 2 – Stiffness numerical results

Infill percentage	Force measured at 4mm [N]	K [N/mm]
20%	7.07	1.77
30%	13.33	3.33
40%	14.26	3.56
50%	15.95	3.99
60%	17.35	4.34

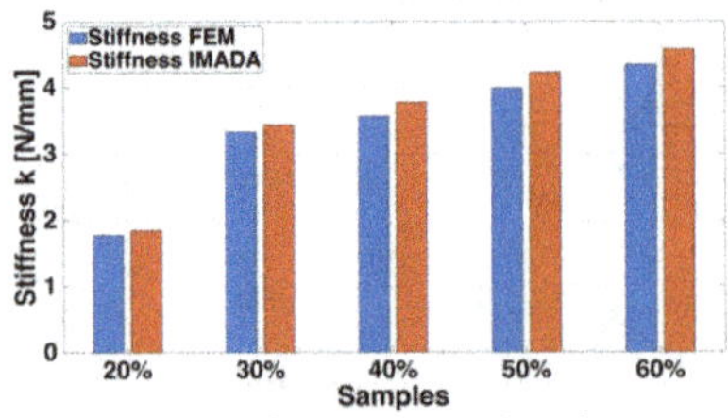

Fig. 5 – Comparison of experimental and numerical results

As expected, joints with higher infill densities present higher stiffness, but the infill density vs stiffness relationship is not linear: for lower infill percentages (20-30%) the stiffness sensitivity with respect to infill is much higher than for higher percentages. One can observe that all the samples show a higher stiffness in the linear elasticity interval with respect to the entire force-displacement curve. As the infill percentage affects the mechanical properties and consequently the joint behavior, it is necessary to find a compromise between desired mechanical properties, 3D printing time and material amounts. This suggests that it is not recommended to produce joints with too low (20%) or too high (>>60%) infill percentages to allow a correct joint work and a sustainable production process. Fig. 5 shows a correspondence between experimental and numerical outcomes. Differences between theoretical and experimental stiffnesses are more evident when the infill percentage increases. This discrepancy can be justified by the use of a linear elasticity model for the FEM analysis. For higher infill percentages the linear elasticity model is no more able to precisely approximate the joint deformation.

Conclusions

The results of the study show the relevant impact of manufacturing parameters, and in particular the infill density percentage, on the overall flexural stiffness of flexible joints for soft modular robotic fingers. The results show a relationship between load and displacement in flexible joints. Such behavior was both experimentally and numerically validated, and results will contribute at defining a more reliable characterization of deformable components of soft robotic hands. Future works can be focused on the improvement of the approximation of deformations through FEM simulations, introducing a comparison between different hyperelastic material models, also considering viscoelasticity and anisotropy. Furthermore, different load conditions and specimen shapes will be evaluated for experimental and numerical characterization. These models can be obtained from the force-displacement curves, to evaluate the behavior of the TPU joint.

References

[1] Shintake J, Cacucciolo V, Floreano D et al. (2018) Soft Robotic Grippers. Advanced Materials 30(29):art. 1707035. https://doi.org/10.1002/adma.201707035

[2] Eppner C, Deimel R, Álvarez-Ruiz J et al. (2015) Exploitation of environmental constraints in human and robotic grasping. International Journal of Robotics Research 34(7):1021-1038. https://doi.org/10.1177/0278364914559753

[3] Achilli GM, Valigi MC, Salvietti G et al. (2020) Design of soft grippers with modular actuated embedded constraints. Robotics 9(4):1-15, art. 105. https://doi.org/10.3390/robotics9040105

[4] Logozzo S, Valigi MC, Malvezzi M (2022) A methodology to evaluate contact areas and indentations of human fingertips based on 3D techniques for haptic purposes. MethodsX 9, art. 101781. https://doi.org/10.1016/j.mex.2022.101781

[5] Logozzo S, Valigi MC, Malvezzi M (2022) Modelling the human touch: A basic study for haptic technology. Tribology international 166:art. 107352. https://doi.org/10.1016/j.triboint.2021.107352

[6] Lotti F, Tiezzi P, Vassura G et al. (2002) Mechanical Structures for Robotic Hands based on the "Compliant Mechanism" Concept. In:7th ESA Workshop on Advanced Space Technologies for Robotics and Automation 'ASTRA 2002'.

[7] Odhner LU, Jentoft LP, Claffee MR et al. (2014) A compliant, underactuated hand for robust manipulation. International Journal of Robotics Research 33(5):736-752. https://doi.org/10.1177/0278364913514466

[8] Salvietti G, Hussain I, Malvezzi M et al. (2017) Design of the passive joints of underactuated modular soft hands for fingertip trajectory tracking. IEEE ROBOTICS AND AUTOMATION LETTERS 2(4):2008-2015, art. 7954638. https://doi.org/10.1109/LRA.2017.2718099

[9] Hussain I, Renda F, Iqbal Z et al. (2018) Modeling and Prototyping of an Underactuated Gripper Exploiting Joint Compliance and Modularity. IEEE ROBOTICS AND AUTOMATION LETTERS 3(4):2854-2861. https://doi.org/10.1109/LRA.2018.2845906

[10] Achilli GM, Logozzo S, Valigi MC et al. (2022) Underactuated Soft Gripper for Helping Humans in Harmful Works. In:Mechanisms and Machine Science 108 MMS:264-272. https://doi.org/10.1007/978-3-030-87383-7_29

[11] Martins PaLS, Jorge RMN, Ferreira AJM (2006) A comparative study of several material models for prediction of hyperelastic properties: Application to silicone-rubber and soft tissues. Strain 42(3):135-147. https://doi.org/10.1111/j.1475-1305.2006.00257.x

[12] Marechal L, Balland P, Lindenroth L et al. (2021) Toward a Common Framework and Database of Materials for Soft Robotics. Soft Robotics 8(3):284-297. https://doi.org/10.1089/soro.2019.0115

[13] Hussain I, Al-Ketan O, Renda F et al. (2020) Design and prototyping soft–rigid tendon-driven modular grippers using interpenetrating phase composites materials. International Journal of Robotics Research 39(14):1635-1646. https://doi.org/10.1177/0278364920907697

[14] Dragusanu M, Achilli GM, Valigi MC et al. (2022) The WaveJoints: A Novel Methodology to Design Soft-Rigid Grippers Made by Monolithic 3D Printed Fingers with Adjustable Joint Stiffness.In: 2022 International Conference on Robotics and Automation (ICRA),6173-6179. https://doi.org/10.1109/ICRA46639.2022.9811548

[15] Hussain I, Iqbal Z, Malvezzi M et al. (2020) How to 3D-Print Compliant Joints with a Selected Stiffness for Cooperative Underactuated Soft Grippers. In: Springer Proceedings in Advanced Robotics. pp 139-153. https://doi.org/10.1007/978-3-030-42026-0_11

[16] Achilli GM, Logozzo S, Valigi MC et al. (2021) Preliminary study on multibody modeling and simulation of an underactuated gripper with differential transmission. In:Proceedings of the ASME Design Engineering Technical Conference 9:art. V009t009a005. https://doi.org/10.1115/DETC2021-72162

[17] Malvezzi M, Valigi MC, Salvietti G et al. (2019) Design criteria for wearable robotic extra–fingers with underactuated modular structure. In:Mechanisms and Machine Science 68:509-517. https://doi.org/10.1007/978-3-030-03320-0_56

[18] Malvezzi M, Iqbal Z, Valigi MC et al. (2019) Design of multiplewearable robotic extra fingers for human hand augmentation. Robotics 8(4):art. 102. https://doi.org/10.3390/robotics8040102

[19] Achilli GM, Logozzo S, Valigi MC, et al. (2022) Theoretical and Experimental Characterization of a New Robotic gripper's Joint. In:Mechanisms and Machine Science 122 :738 - 745. https://doi.org/10.1007/978-3-031-10776-4_85

[20] Hussain I, Salvietti G, Malvezzi M et al. (2017) On the role of stiffness design for fingertip trajectories of underactuated modular soft hands. In:Proceedings - IEEE International Conference on Robotics and Automation:3096-3101, art. 7989354. https://doi.org/10.1109/ICRA.2017.7989354

Theoretical and Applied Mechanics - AIMETA 2022 | Materials Research Forum LLC
Materials Research Proceedings **26** (2023) 207-212 | https://doi.org/10.21741/9781644902431-34

On the effects of strain wave gear kinematic errors on the behaviour of an electro-mechanical flight control actuator for eVTOL aircrafts

Roberto Guida[1,a,*], Antonio C. Bertolino[1,b], Andrea De Martin[1,c], Andrea Raviola[1,d], Giovanni Jacazio[1,e] and Massimo Sorli[1,f]

[1]Department of Mechanical and Aerospace Engineering, Politecnico di Torino, Torino, Italy

[a]roberto.guida@polito.it, [b]antonio.bertolino@polito.it, [c]andrea.demartin@polito.it, [d]andrea.raviola@polito.it, [e]giovanni.jacazio@formerfaculty.polito.it, [f]massimo.sorli@polito.it

Keywords: Strain Wave Gear, Kinematic Error, eVTOL Aircrafts EMA

Abstract. In recent years, the increasingly growing overcrowding of urban environments and the resulting road traffic congestion have pushed toward the search for alternative mobility solutions, among which there are novel Urban Air Mobility (UAM) technologies. The UAM, together with the development of electric actuation systems, would allow decongesting the streets by exploiting the sky using electric Vertical Take-Off and Landing (eVTOL) aircrafts. Urban air mobility vehicles are primarily based on fully electrical flight control systems with rotary output. Since such technology is relatively new and unproven, Prognostic and Health Management (PHM) algorithms, able to continuously monitor the health status of such systems, are of particular interest. The diffusion of these systems strongly depends on the general confidence of possible customers. The present paper proposes a preliminary study on the effects of the kinematic error of a Strain Wave Gear (SWG), the most used reducer for this kind of applications, on the behaviour of an Electro-Mechanical Actuator (EMA) used as a flight control actuator for an eVTOL aircraft. The simulation results show how the unavoidable kinematic error affects the EMA performances and how its presence can be detected and quantified in strain wave gears.

Introduction

Overcrowded cities and road traffic congestion, together with their problems related to environmental and noise pollution, have pushed the search for alternative mobility solutions. A possible solution to this problem is the Urban Air Mobility. This term refers to the use of urban airspace for intracity passenger transportation, package delivery, healthcare applications, and emergency services. Since environmental pollution in cities is one of the main issues of our time, UAM converged to all-electric aircrafts with compact Electro-Mechanical Actuators (cEMAs) with rotary output for both lift devices and flight controls.

Fatal accidents and low reliability were the main obstacles to previous UAM development. For this reason, and for the fact that this technology is still relatively new, the application of Prognostic and Health Management logics to this system may be a valuable and efficient solution [1]. Nevertheless, the design of novel PHM routines requires a significant amount of data representative of both nominal and off-nominal health conditions. This information can come from historical series, laboratory tests, or simulation campaigns. Although the first two options are usually preferable, they are often not feasible for newly developed systems and require a significant amount of time and monetary investment. Within this framework, researchers often rely on High-Fidelity (HF) models to generate datasets required to train PHM algorithms. In traditional EMAs, a lot of work has been done with this aim for what concerns electrical [2] and mechanical components [3,4] as well as flight control connections [5].

One of the possible actuation and control systems for UAM applications consists of a fault-tolerant brushless DC motor, a Strain Wave Gear, control and power electronics, and sensors. Despite traditional flight commands, the actuator presents a rotative output and not a linear one

Materials Research Proceedings **26** (2023) 207-212 https://doi.org/10.21741/9781644902431-34

[3]. In this kind of application, the use of SWGs is mainly justified by the need for a high transmission ratio (τ) within a small volume. This solution also presents several advantages over traditional gearboxes like very low backlash, good precision, and excellent repeatability. For these reasons, it is commonly used in robotics [6] and aerospace [7] in which compactness and high precision are fundamental. As highlighted in [8], strain wave gears are subjected to peculiar fault and failure modes that must be taken into account for a proper design and implementation of such technology.

Through a first-approximation multibody model [9], the present paper describes a preliminary representation of the dynamics of a strain wave gear in an EMA for eVTOL flight control systems. With the aim of providing the basis for the definition of future PHM algorithms, the kinematic error, modelled according to a physics-based approach [10], has also been considered in the simulation environment to study its influence on the performance of a flight control actuator.

Dynamic Model of an EMA for eVTOL Aircraft

Even though eVTOL aircrafts are based on a specific design [11], they are very similar in almost all their configurations from an actuation point of view. In this paper, the actuator analyzed is the one that allows obtaining a vectoring of the thrust device to change from take-off to cruise configuration. The specific aim of the present model is to study the effect of the kinematic error of the strain wave gear. For this reason, all the components of the EMA, except the gearbox, are represented using simplified models. The three main elements of the model are the Electric Motor (EM) and its control logic, the strain wave gear, and the sensors. As in the real system, the only input to the model is the position set value required by the flight control computer. All the other quantities (i.e., motor voltage and current, angular velocity, and torque) are internally calculated according to the classic EMA scheme [12].

Electric Motor. To build a high-fidelity model of a mechanical system, all its components must be properly modelled. However, since this would lead to high simulation times, simplifications have been adopted. Within this framework, in the present research, a model of a DC brushless motor has been implemented.

Strain Wave Gear. To describe the kinematic and the dynamic behaviour of the SWG, an equivalent model is exploited. This model is a good compromise between the need for high-fidelity representation and limited simulation time. The proposed model is based on the work done in [9], in which, the interactions among the three main components of the SWG were analysed.

This approach allows describing the cyclic deformation of the flexspline caused by the motion of the Wave Generator (WG) and the interaction during the meshing between the flexspline (FS) and circular spline (CS) teeth. The operating mode of a SWG can be dual, depending on which element, between the FS and the CS, is locked. In the present paper, the FS is fixed, while the CS is connected to the output shaft.

Differently from [9], the mechanical transmission between the electric motor shaft and the WG and between the CS and the position transducer have been considered. In so doing, it is possible to obtain a three degree-of-freedom model distinguishing the three inertias, one for each main component of the SWG.

In Figure 1.a a schematic representation of the equivalent model is reported. The physical and geometric parameters associated with it are the primitive equivalent radius (r_g), the wave generator angular position (θ_{WG}), the circular spline angular position (θ_{CS}), the teeth pressure angle (α_t), the elliptical cam equivalent angle ($\alpha_n = \arctan\left(\frac{1}{\tau * \tan(\alpha_t)}\right)$), the stiffness ($K_b$) and the viscous damping coefficient (c_b) of the elliptical bearing, the flexspline torsional stiffness (K_t), the meshing stiffness between FS and CS (K_m) and the pure kinematic error between input and output of the strain wave gear (θ_e). The displacements of the components that occur during the transmission are the radial displacement of the bearing outer race (x_1), the radial displacement of the FS during elliptical deformation (x_2), the tangential displacement of the FS free edge caused by torsional deformation (y_1) and the tangential displacement of the CS (y_2).

By analyzing the free-body diagrams of the single components, it is possible to obtain the equations (Eq. 1) describing the dynamics of the system.

$$\begin{cases} T_{WG} = K_{WG}(\theta_{in} - \theta_{WG}) + c_{WG}(\dot{\theta}_{in} - \dot{\theta}_{WG}) = T_{FS\to WG} + J_{WG}\ddot{\theta}_{WG} \\ F_b = K_b(e_{WG,mis} + x_1 - x_2) + c_b(\dot{e}_{WG,mis} + \dot{x}_1 - \dot{x}_2) \\ F_b - \frac{J_{FS}\ddot{x}_2}{r_g^2} = F_m \sin(\alpha_t) \\ \frac{K_t y_1}{r_g^2} + \frac{c_t \dot{y}_1}{r_g^2} = -\frac{J_{FS}\ddot{y}_1}{r_g^2} + F_m \cos(\alpha_t) \\ F_m = K_m(x_2 \sin(\alpha_t) - (y_1 + y_2)\cos(\alpha_t)) + c_m(\dot{x}_2 - (\dot{y}_1 + \dot{y}_2)\cos(\alpha_t)) \\ T_{FS\to CS} - J_{CS}\ddot{\theta}_{CS} = K_{CS}(\theta_{CS} - \theta_{out}) + c_{CS}(\dot{\theta}_{CS} - \dot{\theta}_{out}) = T_{CS} \\ T_{FS\to WG} = F_b r_g \sin(\alpha_n)\cos(\alpha_n) \\ T_{FS\to CS} = F_m r_g \cos(\alpha_t) \end{cases} \quad (1)$$

The equations were implemented in a Simulink model to obtain a description of the SWG dynamics.

Kinematic error. The kinematic error is always present in SWGs, due to the FS flexibility, tolerances, and tooth machining errors. It is defined as:

$$\tilde{\theta} = \frac{\theta_{in}}{\tau} - \theta_{out} = (\theta_{CS} + \theta_e + \Delta\theta) - \theta_{out} = \theta_e + \Delta\theta \quad (2)$$

The term $\Delta\theta$ of the Eq. 2 is related to the torsional compliance, while θ_e is called pure kinematic error [13] and it depends on the kinematic structure of the SWG. The parameter $e_{WG,mis} = \theta_{e,mis} r_g \cong \theta_e r_g$ is the kinematic error caused by misalignment expressed as a radial displacement.

The pure kinematic error contributions can be classified into a constant component, related to clearance, and a variable one, related to misalignment and tooth machining error [10]. In [9,14,15], the presence of the kinematic error is modelled by means of finite Fourier series based on experimental data. On the contrary, the present study aims to describe such a phenomenon from its root causes such as a misalignment of the wave generator due to assembly and machining imperfections.

The modelling of the presence of misalignment among the three main components of the SWG is obtained by means of a roto-translation matrix depending on the wave generator angular position. In this way, it is possible to define a planar equivalent WG profile, depicted in Figure 1.b, changing, at each instant, with the input shaft position.

The input parameters of the kinematic error model are the eccentricity (e_1 and e_2) and the angular deviations ($\bar{\beta} = [\beta_1 \beta_2 \beta_3]$ and $\bar{\alpha} = [\alpha_1 \alpha_2 \alpha_3]$) shown in Figure 2.

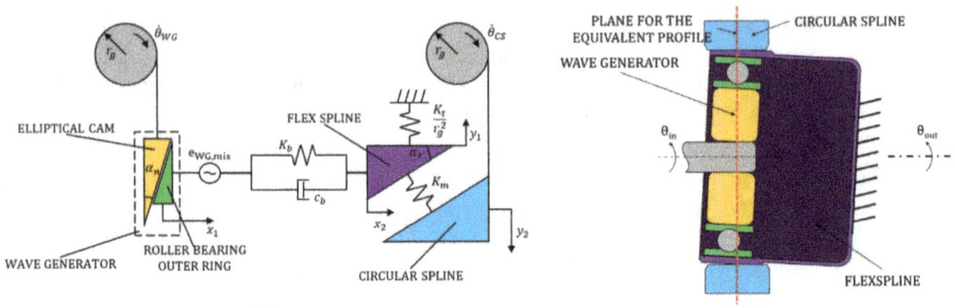

(a) (b)

Figure 1 (a) Diagram of the SWG dynamic equivalent model. (b) Schematic representation of a misaligned strain wave gear.

The two eccentricity values are measured, respectively, between the CS center (O_{CS}) and the nominal center of the reference frame (O), and between the nominal center (O) and the WG center (O_{WG}). The angular deviations are the angles between eccentricities (e_1 and e_2) and x-axes (β_1 and α_1), between CS Z-axis and nominal Z-axis, nominal Z-axis and WG Z-axis on XOZ-plane (β_2 and α_2) and between CS Z-axis and nominal Z-axis, nominal Z-axis and WG Z-axis on YOZ-plane (β_3 and α_3). By knowing these parameters, it is possible to define two rotation matrices linking the nominal position of the WG to its angular position in presence of misalignment.

The obtained 3D profile of the wave generator is intersected with the plane perpendicular to the input shaft obtaining its equivalent planar profile. This is compared to the nominal one to get the radial displacement and, as a direct consequence, the kinematic error associated with it. In Figure 3.a, a comparison of the pure kinematic error between experimental [13] and proposed model simulated data is reported.

For more accurate results, proper modelling of the other kinematic error sources, such as clearance and tooth machining error, is necessary.

Friction. Due to the presence of friction, a passive torque, modeled by an approximation of the Stribeck curve as in [12], must be subtracted to T_{CS}, the output torque of the circular spline, to obtain the actual torque (T) at the output shaft of the reducer [16].

Sensors. Before being used to close the relative control loops, the signals of the flap surface angular position, the motor angular velocity and current pass-through blocks simulating the presence of three sensors. Zero-order transfer functions have been adopted to simulate the current sensors, while first and second-order ones have been implemented for the speed and the position sensors. In addition, to replicate a real device, noise has been added.

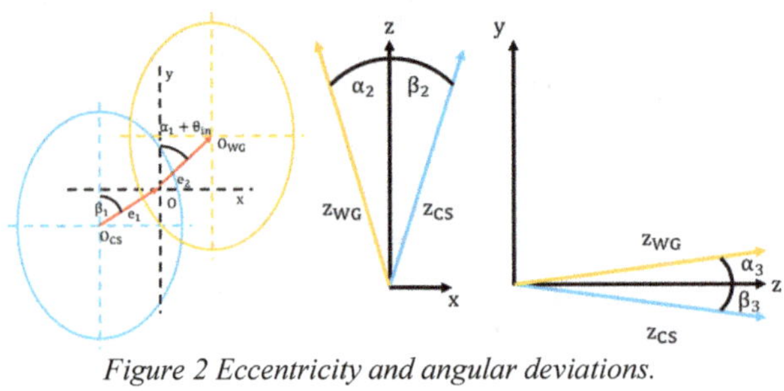

Figure 2 Eccentricity and angular deviations.

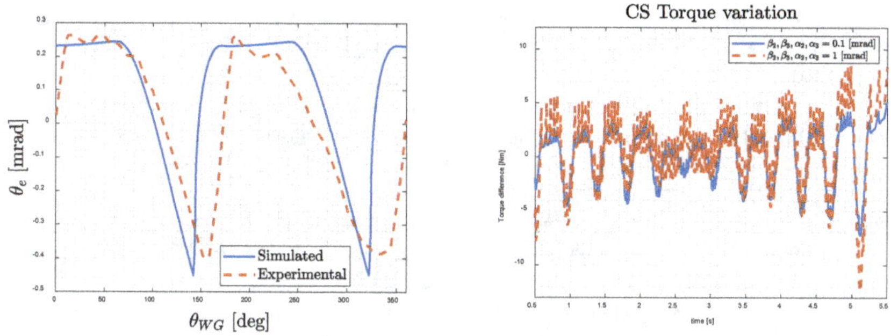

(a)

Figure 3 Analysis of the kinematic error model: (a) comparison of the pure kinematic error between experimental and simulated data and (b) angular deviation misalignment kinematic error.

Simulations Results

To study the effect of the presence of the kinematic error caused by misalignment, different simulations, performed by changing the input parameters (eccentricity, e_1 and e_2, and angular deviations, $\overline{\beta}$ and $\overline{\alpha}$), were run.

The angular position set is a ramp with a slope such that the maximum stroke of the actuator is reached in 5 s. In so doing, the speed of the actuator is constant, and the torque oscillations are primarily linked to the control logic. The simulations are performed by analyzing the effects of angular deviations ($\overline{\beta}$ and $\overline{\alpha}$). Because the position control loop operates to minimize the relative error the effect of the kinematic error on the flap surface angular position is negligible. Nevertheless, an effect could be observed in torques, and consequentially, in current signal, as reported in Figure 3.b in which the presence of the kinematic error caused a noisy signal with several oscillations. In Figure 3.b, are reported the differences between the nominal torque and the one in the presence only of angular deviations. As predictable, the effect of the kinematic error increases together with the misalignment level. A similar effect can be found in presence of misalignments.

Conclusions

In the present paper, a preliminary study on the effects of a strain wave gear pure kinematic error in an electro-mechanical actuator for UAM applications was proposed. For this purpose, a mathematical model of the SWG and relative misalignment was developed to study and quantify the effects on the main signals of the EMA.

In the analyzed case study, the results of the simulation made it possible to investigate one of the causes of vibrations between the components of a SWG. The presence of this error causes the appearance of oscillating torques which are instantaneously significantly greater than the nominal ones.

In the future, the proposed model will be validated with experimental results and integrated with more accurate ones of the other components of the EMA to study the effects of different phenomena and to obtain datasets suitable for PHM purposes.

References

[1] Ranasinghe, K., Sabatini, R., Gardi, A., Bijjahalli, S., Kapoor, R., Fahey, T., and Thangavel, K., Advances in Integrated System Health Management for Mission-Essential and Safety-Critical Aerospace Applications, Prog. Aerosp. Sci., 128 (2021), p. 100758. https://doi.org/10.1016/j.paerosci.2021.100758

[2] De Martin, A., Jacazio, G., and Vachtsevanos, G., Anomaly Detection and Prognosis for Primary Flight Control EMAs, European Conference of the Prognostic and Health Management Society, Bilbao, 2016, pp. 1–9.

[3] Bertolino, A. C., Sorli, M., Jacazio, G., and Mauro, S., Lumped Parameters Modelling of the EMAs' Ball Screw Drive with Special Consideration to Ball/Grooves Interactions to Support Model-Based Health Monitoring, Mech. Mach. Theory, 137 (2019) pp. 188–210. https://doi.org/10.1016/j.mechmachtheory.2019.03.022

[4] Bertolino, A. C., De Martin, A., Mauro, S., and Sorli, M., Multibody Dynamic Adams Model of a Ball Screw Mechanism with Recirculation Channel, ASME Int. Mech. Eng. Congr. Expo. Proc., (2021). https://doi.org/10.1115/IMECE2021-71121

[5] Bacci, A., Bertolino, A. C., De Martin, A., and Sorli, M., Multiphysics Modeling of a Faulty Rod-End and Its Interaction with a Flight Control Actuator to Support Phm Activities, ASME Int. Mech. Eng. Congr. Expo. Proc., (2021) pp. 1–10. https://doi.org/10.1115/IMECE2021-71095

[6] Li, Z., Melek, W. W., and Clark, C., Decentralized Robust Control of Robot Manipulators with Harmonic Drive Transmission and Application to Modular and Reconfigurable Serial Arms, Robotica, 27(2) (2021) pp. 291–302. https://doi.org/10.1017/S0263574708004712

[7] Ueura, K., Kiyosawaa, Y., Kurogi, J., Kanai, S., Miyaba, H., Maniwa, K., Suzuki, M., and Obara, S., Tribological Aspects of a Strain Wave Gearing System with Specific Reference to Its Space Application, Proc. Inst. Mech. Eng. Part J J. Eng. Tribol., 2008. https://doi.org/10.1243/13506501JET415

[8] Raviola, A., Martin, A. De, Guida, R., Jacazio, G., Mauro, S., and Sorli, M., Harmonic Drive Gear Failures in Industrial Robots Applications : An Overview, European Conference of the Prognostic and Health Management Society, 2021, pp. 350–360.

[9] Zhang, X., Tao, T., Jiang, G., Mei, X., and Zou, C., A Refined Dynamic Model of Harmonic Drive and Its Dynamic Response Analysis, Shock Vib., (2020). https://doi.org/10.1155/2020/1841724

[10] Jia, H., Li, J., Xiang, G., Wang, J., Xiao, K., and Han, Y., Modeling and Analysis of Pure Kinematic Error in Harmonic Drive, Mech. Mach. Theory, 155 (2021). https://doi.org/10.1016/j.mechmachtheory.2020.104122

[11] Bacchini, A., and Cestino, E., Electric VTOL Configurations Comparison, Aerospace, 6(3) (2019). https://doi.org/10.3390/aerospace6030026

[12] Guida, R., Raviola, A., Migliore, D. F., De Martin, A., Mauro, S., and Sorli, M., Simulation of the Effects of Backlash on the Performance of a Collaborative Robot: A Preliminary Case Study, 31st International Conference on Robotics in Alpe-Adria-Danube Region, Klagenfurt am Wörthersee, 2022, pp. 28–35. https://doi.org/10.1007/978-3-031-04870-8_4

[13] Ghorbel, F. H., Gandhi, P. S., and Alpeter, F., On the Kinematic Error in Harmonic Drive Gears, J. Mech. Des. Trans. ASME, 123(1) (2001) pp. 90–97. https://doi.org/10.1115/1.1334379

[14] Preissner, C., Royston, T. J., and Shu, D., A High-Fidelity Harmonic Drive Model, J. Dyn. Syst. Meas. Control. Trans. ASME, 134(1) (2012). https://doi.org/10.1115/1.4005041

[15] Zou, C., Tao, T., Jiang, G., Mei, X., and Wu, J., A Harmonic Drive Model Considering Geometry and Internal Interaction, Proc. Inst. Mech. Eng. Part C J. Mech. Eng. Sci., 231(4), 2017, pp. 728–743. https://doi.org/10.1177/0954406215621097

[16] Ma, D., Yan, S., Yin, Z., and Yang, Y., Investigation of the Friction Behavior of Harmonic Drive Gears at Low Speed Operation, Proceedings of 2018 IEEE International Conference on Mechatronics and Automation, ICMA 2018, IEEE, (2018) pp. 1382–1388. https://doi.org/10.1109/ICMA.2018.8484324

Braking torque estimation through machine learning algorithms

BONINI Federico[1,a], RIVOLA Alessandro[1,b], MARTINI Alberto[1,c*]

[1]DIN – Department of Industrial Engineering, University of Bologna, Bologna, Italy

[a]federico.bonini3@unibo.it, [b]alessandro.rivola@unibo.it, [c]alberto.martini6@unibo.it

Keywords: Machine Learning, Carbon Brakes, Friction Coefficient

Abstract. MotoGP class motorcycles rely on carbon braking system to cope with their incredible acceleration capability and high speed. Hence, assessing the torque generated by the front discs is a key to improve the vehicle performance. As direct measurement of the braking torque is not allowed during races, its value may be estimated through a physical model, using as inputs the brake fluid pressure (monitored on board), the braking system geometry and the friction coefficient (μ). However, the results obtained with this method are highly limited by the knowledge of the instantaneous friction coefficient between the disc rotor and the pads. Since the value of μ is a highly nonlinear function of many variables (namely temperature, pressure and angular velocity of the disc), an analytical model appears impractical to establish. This work aims to implement an innovative algorithm, based on machine learning, for determining μ from the signals regularly available in races, to enable accurate breaking torque computation. The proposed method consists of two main tools. An *artificial neural network* (ANN) is developed to approximate the unknown function that relates the input variables to μ, while a *Kalman filter* (KF) is implemented to estimate the real temperature distribution on the disc surface that constitutes one of the most important ANN inputs. The proposed algorithm has been successfully validated with real data collected from extensive tests in racetracks, with a special sensor setup.

Introduction

This research activity is carried out in collaboration with Ducati Corse (a division of Ducati Motor Holding S.p.A., Bologna, Italy) and investigates new methods for estimating the braking torque generated by carbon brakes mounted on MotoGP class motorcycles.

Currently, the company is able to estimate the braking torque delivered by the front brakes through an analytical model of the braking system. Given the pressure in the hydraulic circuit (available from telemetry) and the braking system geometry, such model calculates the braking torque assuming a constant value for the friction coefficient, μ. The selected value for μ has been defined on the basis of the company database and experience to fit a large portion of the breaking maneuvers performed in the racetracks of the MotoGP championship. However, measuring the braking torque during some specific racetrack tests revealed significant discrepancies between the real and the predicted values. Every time the operating conditions of the brake differed from the most common values, the constant friction coefficient approach turned out to be inaccurate for the calculation of the torque provided by the front brakes. Determining experimentally the instantaneous value of μ through measurements performed on a test rig is too expensive and time consuming. Indeed this approach implies that for every new brake configuration or variant developed all the possible operating conditions must be covered. Therefore, the research is aimed at achieving a more accurate calculation of the braking torque on the front wheel by estimating the instantaneous μ through a machine learning (ML) algorithm.

An essential requirement for accurately estimating the friction coefficient is reliably assessing the thermal dynamics of the system. Originally, the monitoring of the brake temperature was performed directly through the sensor present on board. However, it soon became clear that, besides being quite noisy, a single spot acquisition was not representative of the whole temperature distribution along the disc. Therefore, a two-dimensional (2D) finite element (FE) model of the disc is implemented along with a Kalman filter (KF) to predict with higher accuracy and reliability the temperature distribution on the disc-pads contact patch [1, 2]

The second part of the study focuses on the definition of a ML-based approach suitable for the disc-pad friction coefficient prediction. Some works investigated ML algorithms to estimate the friction coefficient of steel disc in test rig experiments [3]. To the Authors' best knowledge, no studies can be found concerning carbon brakes in real operating condition. A preliminary investigation [4] permitted to identify Artificial Neural Networks (ANN) and Decision Trees as the most promising algorithms. In particular, they could provide similar accuracy, with decision trees appearing faster, but also more prone to overfitting. In this paper, both algorithms are further investigated for possibly enhancing their performance. A new ANN is trained using the k-fold cross validation method [5] and with new inputs that improve its prediction capabilities. In addition, a new decision tree is developed by using the Minimal Cost-Complexity Pruning technique, which is an effective way to optimize the algorithm hyperparameters, to reduce overfitting issues.

Materials and Methods

The first stage of the research is aimed at implementing the FE model in Matlab, which will constitute the physical model used by the Bayesian filter for the posterior optimal temperature estimation. To assess the performance of the proposed approach, racetrack tests were performed by installing on the motorcycles a multi-spot sensor able to measure the disc temperature in four different radial positions (Fig. 1). Such sensor (not allowed during races) was employed for calibration and validation of the algorithm. Indeed, the final goal is achieving an accurate estimate of the temperature distribution based on the onboard (single-spot) sensor measurements.

As for the machine learning algorithms, the true friction coefficient values that represent the target for each input vector in the training dataset were obtained by performing specific track tests in which the motorcycle front wheel was equipped with a wheel torque transducer. Given the measured torque and the corresponding value of hydraulic pressure it is possible to calculate the friction coefficient values (i.e., to generate the ground truth for the network training dataset) by inverting the analytical model of the brake mentioned previously.

Once the temperature distribution is correctly estimated, it can be processed by the machine learning algorithm that outputs the actual value of the friction coefficient. Then, the latter information is fed into the original analytical model of the brake that is expected to compute the braking torque with enhanced accuracy.

Analytical model of the brake. Knowing the current value of the friction coefficient, the pressure in the hydraulic circuit measured by the sensor and the brake disc/caliper geometry, the model outputs the braking torque on the front wheel T_{bf}, by using the following expression:

$$T_{bf} = 2 \cdot \mu \cdot F_n \cdot R_{eff} \cdot K_\mu \quad (1)$$

where μ is the friction coefficient; R_{eff} is the effective radial distance at which the tangential force ($F_t = \mu \cdot F_n$) is applied; the parameter K_μ is a scaling factor, which is a rational function of μ related to the specific caliper geometry (its exact expression cannot be provided due to non-disclosure agreements – NDA); the factor 2 is related to the presence of two brake discs (with two distinct calipers) on the front wheel; finally F_n represents the normal force applied by the pads on the disc, calculated as

$$F_n = A_p \cdot p, \quad (2)$$

where A_p is the total cross section of the four pistons and p is the pressure of the brake fluid.

It is worth noting that training ML tools to directly estimate the value of the braking torque (instead of μ) is not deemed convenient. Indeed, calculating the braking torque through the analytical model (Eq. 1) permits to assess the effects of different design parameters of the braking system without the need of re-training the algorithm. In fact, MotoGP teams are allowed to choose different brake discs and pads according to the characteristics of each track. With the proposed approach, a different setup can be addressed simply by adjusting the parameters of the physical model (in particular the effective radius and the amplification coefficient K_μ).

2D FE model and KF. Measuring the disc temperature using the infrared sensor available on board is not robust enough to allow the machine learning algorithms to catch the correct temperature dependency of the friction coefficient. In fact, this kind of sensor detects the disc temperature at a fixed radial distance in the disc-pads contact patch, disregarding the temperature gradient and the overall thermal behavior of the disc. From the measurements performed with the multi-spot sensor (Fig. 1) it turned out that the heat generated by the friction with the pads is not always distributed uniformly along the contact patch. Conversely, the pads wear and the pressure distribution inside the caliper can induce local temperature peaks on the disc and very high temperature gradient that cannot be recorded by the single-spot sensor.

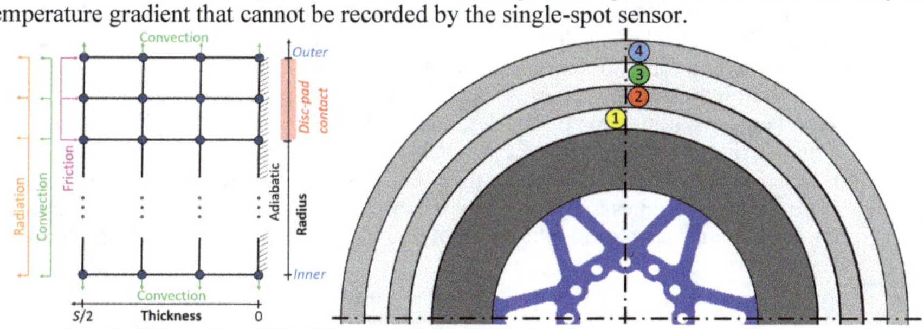

Fig. 1: Schematic of 2D FE model (left – l) and multi-spot sensor setup (right – r).

Building a thermal model of the carbon disc allows to take into account the whole temperature distribution on the disc hence giving a more robust information to the ML algorithms. The temperature on the disc evolves according to the energy transferred by convection, conduction, radiation and friction with the pads. In the 2D FE model implemented, temperature is a function of the radius, the thickness and time. As shown in Fig.1, thanks to the disc symmetry, only half of its thickness can be simulated, by imposing an adiabatic boundary condition in the median plane. Conversely, the opposite boundary represents the lateral surface of the disc in contact with the brake pad, hence being affected by convection, radiation and friction. The remaining two boundaries represent the frontal and internal surface of the disc in which only convection is simulated, for sake of simplicity.

Since, during the race, the actual temperature of the disc may drift away from the simulated results due to many external factors (e.g., the presence of slip stream or pads wearing), running the FE model alone can not grant reliable results. Therefore, a KF is developed to make up for the approximation error introduced by the physical model. At each time step, the filter updates the theoretical prediction performed by the thermal model with the empirical information coming from the single spot sensor and outputs the a posteriori temperature estimation.

Artificial Intelligence Algorithms. To preserve the ability to generalize to non-training data, hence avoiding overfitting, it is common practice to divide the dataset into training, validation and testing sets. However, partitioning the available data into three sets drastically reduces the number of samples which can be used for learning the model. Moreover, the results obtained depend on the particular choice for the three sets. To solve this problem the cross-validation (CV) procedure is implemented. The CV approach consists in splitting the training set into k smaller sets and iteratively training the ML model using the data belonging to k-1 folds while the remaining part of the data is kept for validation. The performance of the algorithm is computed by averaging the values obtained for each fold. This approach can be computationally expensive, but it allows to assess the performance of any ML algorithm across the whole dataset without wasting data. As for the implemented algorithms, both the ANN and the decision tree were trained using CV technique dividing the dataset into 10 equally populated folds.

The decision tree was pruned using Minimal Cost-Complexity Pruning technique (MCCP). MCCP involves the selective removal of certain leaves of a tree to optimize its hyperparameters (namely the maximum number of leaves, the minimum number of samples per leaf and the depth of the tree). This process is parametrized by the complexity parameter, α, that indicates a particular tree dimension. How α is calculated is beyond the scope of this work, but more information on the MCCP can be found at [6]. The more leaf nodes a tree has, the higher its complexity becomes and the lower the value of α. A null α corresponds to the full-size tree that is grown until the sum of the squared residual for each leaf is minimized (high overfitting). By increasing the value of α the weakest link of the original tree is pruned and a new optimal subtree is generated. This process is repeated until only one leaf is left, leading to a set of optimal subtrees (Fig. 2). For each of the ten folds the MCCP algorithm trains all these trees and tests them on the data kept for validation. As α increases the error on the training set grows because the less leaf nodes the tree contains (lower complexity) the less possibilities for fitting the training data. However, it can be observed that limiting the complexity of the tree may increase the accuracy on the validation set because the overfitting is reduced, and the algorithm generalizes better to unseen data. On the other hand, for high values of α the tree becomes too simple and the error on the validation set starts increasing again (Fig. 2). MCCP aims at selecting the simplest tree that gives the best average performance across the whole dataset.

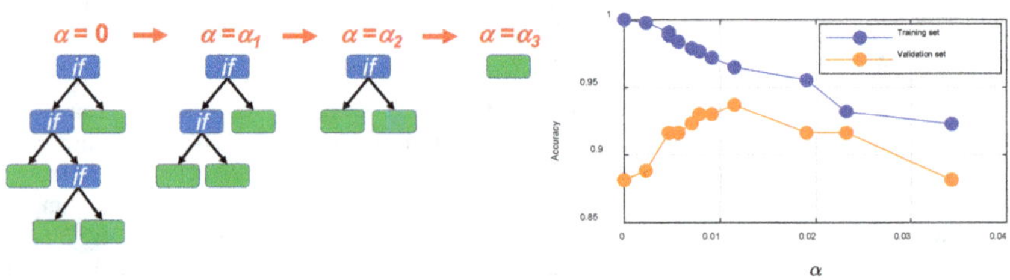

Fig. 2: MCCP subtrees creation (l) and qualitative trend of tree accuracy with respect to α (r).

As for the ANN, in [4] the most promising results were provided by a four-layered network with two hidden layers. The first layer consisted of the input signals, while both hidden layers had ten neurons (each one with the hyperbolic tangent as activation function). The same network architecture here considered, but new inputs are introduced, which help the algorithm to identify the actual friction coefficient. In addition to the pressure in the hydraulic circuit, the angular speed of the front wheel and the average disc temperature (calculated using the 2D-FE model and the KF) the ML algorithms were provided also with the temperature of the brake at the beginning of every braking maneuver, the duration of the braking maneuver and the difference between the average disc temperature and the temperature measured by the infrared sensor mounted on board (ΔT).

The latter is introduced to take into account how a non-uniform power distribution in the disc affects the friction coefficient value. Figure 3 shows two braking maneuvers with similar average temperature on the disc and almost identical fluid brake pressure. The one at the bottom shows a very high temperature gradient on the disc-pad contact region while the one at the top presents

Materials Research Proceedings **26** (2023) 213-218 https://doi.org/10.21741/9781644902431-35

almost uniform temperature distribution (the single-spot temperature measurement is equal to the average temperature on the disc estimated by the FE model). In the first scenario the region in which the sensor is located is contributing the most in the generation of the braking torque. Hence, due to the local temperature peak, the friction coefficient is drastically lower than the other case in which the local temperatures are cooler and uniform. This results in a significant discrepancy between the breaking torque measured experimentally (black line) and the values predicted by the constant friction coefficient approach (dashed). Since these breaking maneuvers present similar pressure values and the braking system geometry is the same (Eq. 1) the braking torque estimated by assuming a constant μ is very similar for both cases. However, in case of a local temperature peak, this prediction overestimates the measured torque by more than 20%.

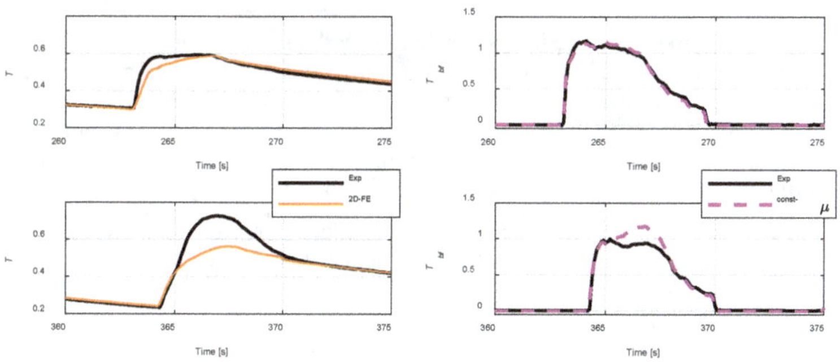

Fig. 3: Effect of power distribution on torque estimation.

Main Results

Table 1 compares the two tested ML algorithms. Their performance is expressed in terms of percent variation in the root mean square error ($\Delta\%_{RMSE}$) with respect to the constant-μ approach, and prediction speed (namely, computational time for predicting 10^4 samples). The former is calculated averaging the values obtained for each fold. Even though the decision tree is one order of magnitude faster in making new prediction, the ANN is deemed the best solution for the friction coefficient recognition because it almost halves the RMSE obtained with constant μ.

The implementation of the ANN as a preliminary stage for the torque calculation allows to enhance the results of the analytical model of the brake whenever the friction coefficient diverge significantly from the reference value, particularly in case of a very aggressive braking maneuver or a quite gentle one. For instance, Fig. 4 clearly shows that the proposed algorithm permits to significantly improve the accuracy of the braking torque prediction in case of hard braking (the same maneuver previously reported in Fig. 3, bottom, is considered).

Combining the thermal model of the brake and the new ANN allowed the creation of the colormap showed in Fig. 4, which provides a glimpse at the brake performance in many working conditions. The *X* axis is related to a synthetic coefficient that takes into account the dependency of μ with respect to both pressure and angular velocity while the *Y* axis is referred to the average disc temperature. Finally, the vertical axis and the color of the map indicate the values of μ.

Table 1: performance of the tested approaches

Algorithm	**$\Delta\%_{RMSE}$**	**Computational time [s]**
ANN [10;10]	– 42.7 %	0.40
MCCP Tree	– 17.1 %	0.02

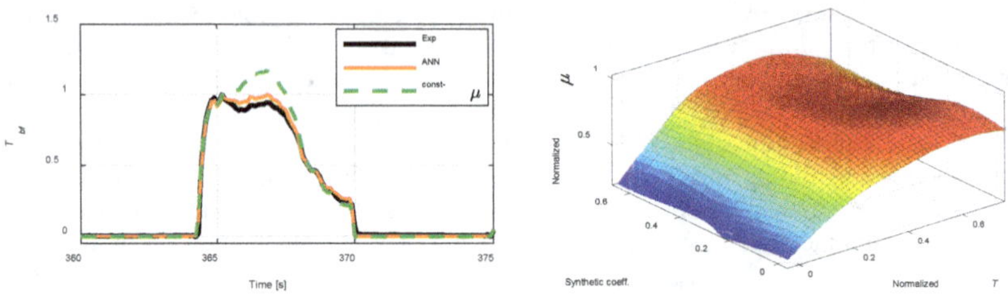

Figure 4: braking torque estimation (l) and normalized μ colormap (r).

Conclusions

A novel method to estimate the friction coefficient of carbon brakes for racing motorcycles has been presented. It combines a thermal model of the brake, consisting of a Kalman filter and a two-dimensional finite element model, with an Artificial Neural Network. The proposed method has been validated with a large dataset collected from an experimental campaign conducted in many different racetracks. The developed algorithm permitted to map the friction coefficient over most of the operating conditions that may be experienced over a race.

Acknowledgements. Ducati Corse is kindly acknowledged for providing real data to test the developed methods. The Authors wish to thank particularly N. Mancinelli and G. Manduchi for active collaboration and support.

References

[1] H. Sakamoto, Heat convection and design of brake discs, Proceedings of the Institution of Mechanical Engineers, Part F: Journal of Rail and Rapid Transit 218(3) (2004) 203–212. https://doi.org/10.1243/0954409042389436

[2] S. Sarkka, Bayesian Filtering and Smoothing, CUP, Cambridge, 2013. https://doi.org/10.1017/CBO9781139344203

[3] V. Ricciardi, K. Augsburg, S. Gramstat, V. Schreiber, V. Ivanov, Survey on Modelling and Techniques for Friction Estimation in Automotive Brakes, Applied Sciences 7(9):873 (2017). https://doi.org/10.3390/app7090873

[4] F. Bonini, G. Manduchi, N. Mancinelli, A. Martini, Estimation of the braking torque for MotoGP class motorcycles with carbon braking systems through machine learning algorithms, In Proc. of the 2021 IEEE International Workshop on Metrology for Automotive (MetroAutomotive), 2021, pp. 1-6. https://doi.org/10.1109/MetroAutomotive50197.2021.9502878

[5] I. Goodfellow, Y. Bengio and A. Courville, Deep Learning, MIT press, Cambridge, 2016.

[6] B.R. Kiran, J. Serra, Cost-Complexity Pruning of Random Forests, in: J. Angulo, S. Velasco-Forero, F. Meyer (Eds.), Mathematical Morphology and Its Applications to Signal and Image Processing (ISMM2017) Lecture Notes in Computer Science vol. 10225, Springer, Cham, 2017. https://doi.org/10.1007/978-3-319-57240-6_18

Theoretical and Applied Mechanics - AIMETA 2022 Materials Research Forum LLC
Materials Research Proceedings **26** (2023) 219-224 https://doi.org/10.21741/9781644902431-36

Sizing and control system definition of an intelligent facility for qualification tests and prognostic research activities for electrical landing gear systems

Antonio Carlo Bertolino[a], Andrea De Martin[b,*]
Giovanni Jacazio[c] and Massimo Sorli[d]

[1]Politecnico di Torino, Corso Duca degli Abruzzi 24, Torino, Italy

[a]antonio.bertolino@polito.it, [b]andrea.demartin@polito.it,
[c]giovanni.jacazio@formerfaculty.polito.it, [d]massimo.sorli@polito.it

Keywords: Electric-Brakes, Landing Gear System, LGS, Test-Rig, Prognostics, PHM

Abstract. The evolution towards "more electric" aircrafts has seen a decisive push in the last decade affecting both the propulsion components and the aircraft systems. An interesting and fast-developing application field of electro-mechanical actuators is that of the aeronautical brakes. The E-LISA research project under way within the Clean Sky 2 framework has the objective of developing an innovative iron bird dedicated to executing tests on the landing gear of a small aircraft transport equipped with electro-mechanical landing gear and electrical brake. The paper details the process required to the definition of such iron bird, with a particular focus on the control system.

Introduction

The evolution towards "more electric" aircrafts has seen a decisive push in the last decade, due to growing environmental concerns and the development of new market segments, such as the flying taxis. Such trend affects both the propulsion components and the aircraft systems, with the latter seeing a progressive trend in replacing the traditional solutions based on hydraulic power with electrical or electro-mechanical devices. Although more attention is usually devised towards the flight control actuation, an interesting and fast-developing application field for electro-mechanical systems is that of the aeronautical brakes. Electro-mechanical brakes, or E-Brakes hereby onwards, would present several advantages over their hydraulic counterparts, mainly related to the avoidance of leakage issues and the simplification of the system architecture. The E-LISA research project under way within the Clean Sky 2 framework has the objective of developing an innovative iron bird dedicated to executing tests on the landing gear of a small aircraft transport equipped with electro-mechanical landing gear and electrical brake. Literature on landing gear test rig is rather limited and focused on the different aspects of the certification process. A test rig dedicated to the study of the fatigue behavior of a landing gear is reported in [1], while a hydraulic solution to execute the drop-test is re-ported in [2]. In [3], the authors present the implementation of a high-dynamics force control loop to test the actuation system of a main landing gear within the M-346 iron bird. In [4] the authors performed the experimental validation of a simulation model of a landing gear leg through a comparison with the results obtained with a dedicated experimental setup. Literature on the development of PHM systems for landing gear legs is thus far limited to preliminary simulation studies for the actuation system [5], or to certain components of the electrical brakes [6], while a preliminary comparative analysis of data-driven methods was re-ported in [7]. The proposed iron bird is set to provide meaningful experimental data by collecting signals difficult to extract from the currently in-service aircraft. To this end, the E-LISA iron bird is able to reproduce fully comprehensive test scenarios and conditions, following the workflow described in [8]. The E-LISA iron bird, firstly presented in [9], aims at providing a significant advancement on the current state-of-the-art by defining a new test-rig able to cover different tests procedures without resorting to different test-benches and acts as a technological demonstrator for new prognostic functionalities for the electrical brake system. The paper is organized as follows. At first, the test-rig architecture is introduced, while requirements are detailed and discussed. Hence the sizing process is presented, along with the definition of the control system. The control system is at first designed through the classical linear-dynamics approach and then verified against a high-fidelity, non-linear model. Results are finally presented, highlighting the system capabilities.

Design Requirements and Test-Rig Architecture

The main purposes of the proposed iron-bird are to support the testing and certification of a novel E-Brake system and to foster the definition of dedicated prognostic logic. The iron bird is organized as depicted in in Figure 1 [9]. The mechanical structure can be divided between a fixed part and a moving platform integral with the landing gear leg, complete with a wheel and electrical brake. The moving platform slides along low friction vertical guides according to the force provided by an electro-hydraulic servoactuator controlled through a dedicated servovalve, reproducing the portion of the 6 tons aircraft weight acting on one landing gear leg. A calibrated by-pass orifice connects the two hydraulic lines serving the actuator to improve the dynamic response of the force-controlled system. The test-rig behavior is continuously monitored through one linear variable differential transformer (LVDT) sensor measuring the hydraulic actuator travel, a load cell measuring the force exchanged between the actuator and the moving platform and a differential pressure transducer sensing the pressure drop across the two actuator's chambers. The hydraulic power available for the test-rig operation is that of the facilities in which the rig will be installed and is limited at 207 bar. The contact between the landing gear wheel and the runway is represented through a runway simulator, a rotating disk, connected to a selected number of inertia disks, representative of the aircraft inertia, through a gearbox. A different solution, based on a novel hydraulic system, was considered in [10] but discarded due to cost concerns. A gearbox is interposed to significantly reduce the mass and the encumbrance of the flywheels, the number of which can be increased or decreased to scale-up or scale-down the weight of the simulated aircraft. The runway simulator includes the possibility to change the external coating to achieve the variation of the friction forces between the wheel and the runway and allow the verification of the anti-skid logic behavior in different operating conditions, while a sprinkler can be activated to reproduce the wet-runway conditions. An electric motor is used to accelerate the runway simulator up to the angular frequency corresponding to the aircraft horizontal speed given the diameter of the rotating disk. An emergency brake is installed in-line with the rotating cylinder, allowing to bring the full system to a complete stop in less than 60 s. The iron bird operation is managed by an engineering test station (ETS), which accepts the inputs from a central control unit (CCU) that in turn receives the commands from an operator via a user interface. The input signals are then sent together with rig measurements to a dedicated computer running a real-time (RT) representation of the aircraft dynamics during landing. Such real-time model is then used, along with a model of the runway and a model of the landing gear dynamics, to compute in real-time the load that must be applied to the test-article. The ETS also include the rig control logic, which is designed to manage both the position of the moving platform and the force exerted by the hydraulic

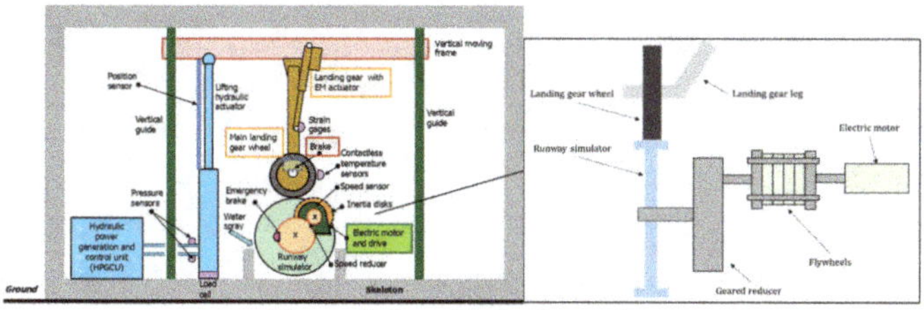

Figure 1. Iron bird schematics.

actuator and include a safety routine check to limit the damages to the rig and to the test article in case of a failure of the anti-skid system during the execution of a test. The structure of the control system is depicted in Figure 2, where three main modules can be identified. The Simulation Module involves the real-time representation of the landing dynamics, including a real-time representation of the aircraft dynamics [9], a runway model, which allows to describe the presence of periodical or localized runway irregularities, and the model of the landing gear legs, each modelled as a two-degrees of freedom vibrating systems, where the mechanical characteristics of the shock-absorber, of the tires and of the mechanical structure are provided by the industrial partners of the project. The control system is based on two control loops. The position control loop operates as a modulating element. It is based on a simple Proportional regulation which output does not act directly on the power lines of the block diagram, instead operating on the dynamic saturation defining the minimum and maximum current absorbed by the servovalve. The main, or basic, control loop is that used to manage the force exerted by the hydraulic actuator and is similar to the control systems of other demanding applications such as iron birds for flight control systems [11] and test rigs for actuators under PHM experimentation [12]. The final component of the control system is the safety module, designed to react to the eventual failure of the anti-skid. A failure of such system during the test can be potentially dangerous to both the rig and the test article, since blocked wheels tends to rapidly fail, hence causing a direct contact under load between the wheel structure and the runway simulator. To avoid catastrophic damages the safety module commands a positive (extract) force set, quickly lowering the load on the test article and achieving the separation between the wheel and the runway simulator.

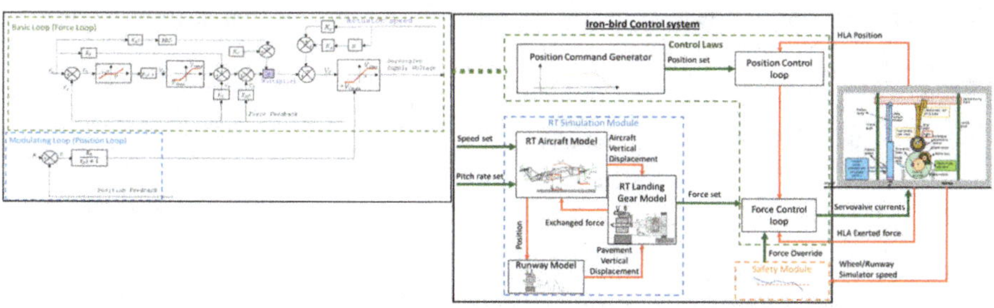

Figure 2. Control system structure.

Materials Research Proceedings **26** (2023) 219-224 https://doi.org/10.21741/9781644902431-36

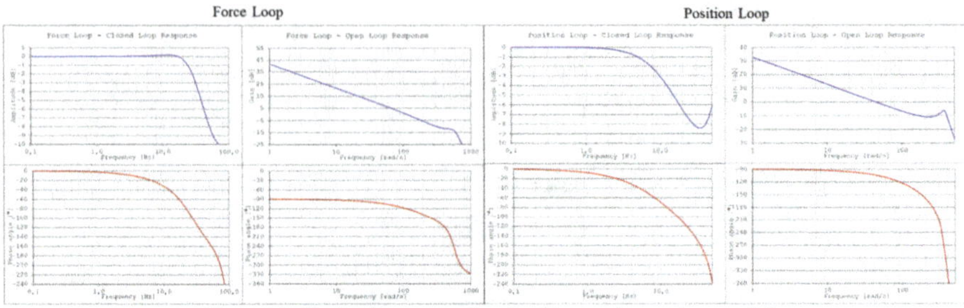

Figure 3. Bode diagrams for the linear dynamics model

Sizing process and Control System Definition

Given the geometrical and physical properties of each component of the loading system, it is possible to perform a preliminary assessment of the control system gains by considering separately the behavior of the force control loop and that of the position control loop. Each control loop can be then at first described as two linear electro-hydraulic servosystems, neglecting the presence of friction, the occurrence of flow-rate saturation of the servovalve and the non-infinite stiffness of the mechanical attachments. As shown in Figure 3, the chosen gain values ensure the system stability and high bandwidth. The latter however is expected to be significantly affected by the flowrate saturation of the servovalve and is expected to be significantly lower for high-amplitude force oscillations.

To check the results of the linear model and provide a more accurate assessment of the test-rig performances we resolved to the definition of a high-fidelity dynamic model. The model, implemented in Matlab/Simulink, is representative of the whole test-rig system, including both the hardware and the software components, and has been already published in [9]. Such model is used to test the control system performances both in the frequency and in the time domain and to verify the test-rig response to an emergency lift of the moving platform following the occurrence of blocked wheel conditions. Since the role of the position control loop is limited to the descent/lift phase of the test procedure, the frequency analysis is limited to the force control loop. Results for both the closed loop analysis are provided in Figure 4 for two different cases: on the left side, the behavior of the system in presence of the 160 l/min defined through the design analysis is depicted,

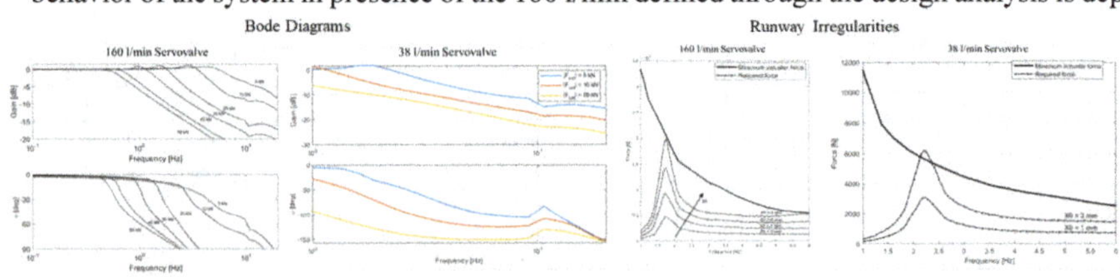

Figure 4. Closed Loop response of the force control loop with variable force amplitude

while on the right-hand side results with a 38 l/min servovalve with similar flow gain made available for the project are presented. It is evident that the use of the 38 l/min servovalve involves a significant downgrade from the 160 l/min case and from the linear analysis expectations. The impact of the servovalve limitations on the test bench performances can be best explained looking at the rig capabilities of reproducing the effects of periodical runway irregularities by comparing the maximum dynamic force applied by the hydraulic servo actuator against the dynamic force required to reproduce the effect of such pavement defects within a certain frequency range. The test-rig equipped with the 160 l/min servovalve is able to fully replicate the occurrence of periodical runway irregularities up to 8 mm deep, while the 38 l/min one is limited to just 2 mm oscillations of the runway pavement. The behavior of the safety system was also investigated. As shown in Figure 5, the force applied to the landing gear leg is cut to 0 N in 0.5 s, becoming positive and hence allowing the separation between the wheel and the runway simulator in less than 0.7 s. Please notice that this simulation was performed considering the 38 l/min servovalve, thus the slower option.

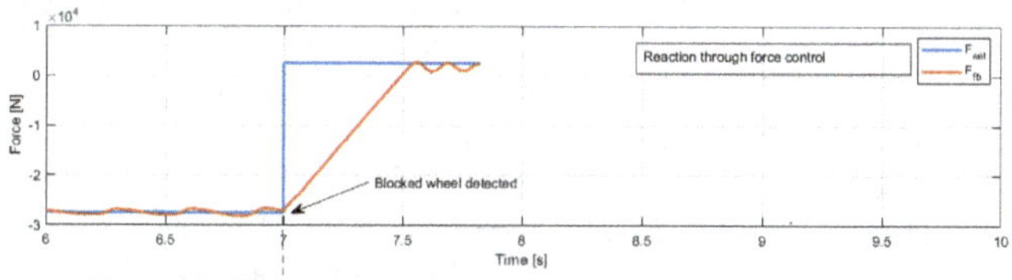

Figure 5. Test-rig reaction to blocked wheel occurrence

Conclusions

The E-LISA iron bird has been designed with to fulfill two main purposes. To support the certification procedures for a new, fully electrical landing gear system and act as a technological demonstrator for prognostic techniques developed for a few key components. As such, the need to closely reproduce a wide array of possible operating conditions including the presence of runway irregularities was assessed as a necessary feature. These considerations apply directly to the definition of the test-rig control system, which involves a real-time simulation model of the aircraft to continuously compute the force command used by the test rig to reproduce a realistic loading pattern on the landing gear under analysis. The definition of a novel control scheme, based on the combination of a simple position control loop with an advanced force control loop was also required. To support the design of the iron bird and contribute to the definition of its control laws, a high-fidelity model of the system was prepared according to well established equations. Simulation results showed that the iron bird control system was stable, although its expected performances are currently limited with respect to the original expectations. Such limitations are however mainly caused by the servovalve made available for the project and can be overcome through the adoption of servovalves with higher flowrate performances. Further work is required to properly characterize the test rig, through the proper identification of the dynamic parameters of the simulation model, the validation of the proposed control scheme, and the experimental verification of the iron bird performance.

References

[1] R. K. Schmidt and P. Eng, "an Integrated Modular Test Rig for Landing Gear Fatigue and Strength Testing," in *ICAS2002 CONGRESS*, 2002, pp. 1–6.

[2] J. Han, Y.-S. Lee, and O.-S. Ahn, "Design & Fabrication for Small Aircraft Landing Gear Drop Test Rig System," *J. Korean Soc. Aeronaut. Sp. Sci.*, vol. 36, no. 11, pp. 1121–1125, 2008. https://doi.org/10.5139/JKSAS.2008.36.11.1121

[3] G. Jacazio and G. Balossini, "A Mechatronic Active Force Control System," in *Proceedings of the ASME 2009 International Design Engineering Technical Conferences & Proceedings of the ASME 2009 International Design Engineering Technical Conferences & Computers and Information in Engineering Conference August 30 -*

September 2, 2009, 2009, pp. 1–8.

[4] L. Dravico, M. Tanelli, and S. M. Savaresia, "Experimental validation of landing-gear dynamics for anti-skid control design," *2018 Eur. Control Conf. ECC 2018*, pp. 2751–2756, 2018. https://doi.org/10.23919/ECC.2018.8550258

[5] C. S. Byington, M. Watson, and D. Edwards, "Data-driven neural network methodology to remaining life predictions for aircraft actuator components," 2004. https://doi.org/10.1109/AERO.2004.1368175

[6] G. Ramesh, P. Garza, and S. Perinpanayagam, "Digital simulation and identification of faults with neural network reasoners in brushed actuators employed in an e-brake system," *Appl. Sci.*, vol. 11, no. 19, 2021. https://doi.org/10.3390/app11199171

[7] A. Oikonomou, N. Eleftheroglou, F. Freeman, T. Loutas, and D. Zarouchas, "Remaining Useful Life Prognosis of Aircraft Brakes," *Int. J. Progn. Heal. Manag.*, vol. 13, no. 1, pp. 1–11, 2022. https://doi.org/10.36001/ijphm.2022.v13i1.3072

[8] S. Autin, A. De Martin, G. Jacazio, J. Socheleau, and G. J. Vachtsevanos, "Results of a feasibility study of a Prognostic System for Electro-Hydraulic Flight Control Actuators," *Int. J. Progn. Heal. Manag.*, vol. 12, no. 3, pp. 1–18, 2021. https://doi.org/10.36001/ijphm.2021.v12i3.2935

[9] A. De Martin, G. Jacazio, and M. Sorli, "Simulation of Runway Irregularities in a Novel Test Rig for Fully Electrical Landing Gear Systems," *Aerospace*, vol. 9, no. 2, p. 114, 2022. https://doi.org/10.3390/aerospace9020114

[10] A. De Martin, G. Jacazio, A. Ruffinatto, and M. Sorli, "A Novel Hydraulic Solution to Simulate Inertial Forces on a Landing Gear Qualification Test Rig," in *Proceedings of the ASME/BATH Symposium on Fluid Power and Motion Control, FPMC2022*, 2022, pp. 1–9. https://doi.org/10.1115/FPMC2022-89900

[11] G. Jacazio and G. Balossini, "Real-time loading actuator control for an advanced aerospace test rig," *Proc. Inst. Mech. Eng. Part I J. Syst. Control Eng.*, 2007. https://doi.org/10.1243/09596518JSCE269

[12] P. Chiavaroli, A. De Martin, G. Evangelista, G. Jacazio, and M. Sorli, "Real Time Loading Test Rig for Flight Control Actuators Under PHM Experimentation," in *ASME International Mechanical Engineering Congress and Exposition, Proceedings (IMECE)*, 2018, vol. 1, p. V001T03A032. https://doi.org/10.1115/IMECE2018-86967

Novel approaches in computational mechanics

Theoretical and Applied Mechanics - AIMETA 2022 Materials Research Forum LLC
Materials Research Proceedings **26** (2023) 227-232 https://doi.org/10.21741/9781644902431-37

A mixed finite-element formulation for the elasto-plastic analysis of shell structures

LIGUORI Francesco S.[1, a*], MADEO Antonio [1,b] and GARCEA Giovanni [1,c]

[1]Dipartimento di Ingegneria Informatica, Modellistica, Elettronica e Sistemistica Università della Calabria 87036 Rende (Cosenza), Italy

[a]francesco.liguori@unical.it, [b]antonio.madeo81@unical.it, [c]giovanni.garcea@unical.it

Keywords: Mixed Finite Element, Elasto-Plastic Analysis, Equilibrated Assumed Stresses, Dual Decomposition Strategy

Abstract. Mixed assumed stress finite elements for elastic-perfectly plastic materials require the solution of a Closest Point Projection (CPP) involving all the element stress parameters for the integration of the constitutive equation. Here, a dual decomposition strategy is adopted to split the CPP at the element level into a series of CPPs at the integration points level and in a nonlinear system of equations over the element. The strategy is tested with a four nodes mixed shell finite element, named MISS-4, characterised by an equilibrated stress interpolation and a displacement field assumed only along its boundaries. The recovered elasto-plastic solution preserves all the advantages of MISS-4, namely it is accurate for coarse meshes in recovering the equilibrium path and evaluating the limit load showing a quadratic rate of convergence.

Introduction

Shell structures are largely used in different areas of engineering and constitute the primary elements of many mechanical, aerospace, marine, and civil constructions. Their nonlinear material behaviour plays a main role, for instance, in masonry and reinforced concrete structures [1] in presence of seismic actions. In such cases, and so in many others, an effective evaluation of the safety factor against plastic collapse [2] is of paramount importance to ensure reliable design choices and wise retrofitting strategies. The most frequently employed numerical approach for modelling shell structures is based on the Finite Element (FE) analysis. Beside more traditional displacement-based FEs, mixed formulations are nowadays widely employed thanks to their effectiveness in eliminating locking effects and improving accuracy. In mixed formulations, in fact, stresses or strains are assumed as independent variables together with displacements, and this gives the chance to design optimised FE characterised by high performance in both linear and nonlinear problems [3]. This, generally speaking, means low error in recovering the solution for coarse mesh grids, elevated rate of convergence and equivalent convergence properties for all the unknown fields. Recently, a wide number of mixed FE has been proposed [4]. Among these, a bright choice is represented by the so-called hybrid FEs where the assumed stress interpolation a-priori satisfies the equilibrium equations.

The family of FEs denoted with the acronym MISS (Mixed Isostatic Self Equilibrated Stresses) represents an interesting option among the state-of-art mixed hybrid FE [5, 6]. In particular, the element called MISS-4 is a quadrilateral four noded flat shell FE with 24 degrees of freedom (DOFs) [5]. It is developed within the Hellinger-Reissner variational formulation and is based on equilibrated and isostatic assumed stress fields. MISS-4 has demonstrated good performance in elasticity, being characterised by accurate solutions for coarse meshes, low sensitivity to mesh distortion, quadratic rate of converge for both displacement and stress fields.

In addition, an enhanced variant, called MISS-4c, improves the performance of MISS-4 [7] by using assumed stress fields that a-priori satisfy also compatibility equations for symmetric composite materials.

Despite being widespread in linear and geometrically nonlinear problems, mixed FEs based on assumed stresses are not a prior choice in incremental elastoplastic analysis, even though their aforementioned advantages hold. The reason for this resides in the fact that the weak imposition of the constitutive laws should be performed at the element level and the consequent return mapping process results into a Closest Point Projection (CPP) of the elastic predictor over the multi-surfaces admissibility domain of the element. This convex optimisation problem requires appropriate mathematical programming methods to be solved.

In this work we propose an alternative strategy for solving the CPP problem based on a dual decomposition method which splits the CPP over the element into two sub-problems. The first one is a series of standard CPP problems over the integration points (IP) which provide the IP stresses, while the second is a nonlinear system of equations over the element that recovers the assumed stress interpolation. The advantage of this procedure resides in the fact that the CPP problem at the IP level can be solved using the same strain-driven integration return map- ping schemes adopted by displacement-based formulations. The additional system of nonlinear equations can be readily solved by the Newton method. Within this strategy, it is possible to efficiently extend the field of application of MISS-4 FE to elasto-plastic problems, obtaining good accuracy even at coarse meshes thanks to the equilibrated assumed stress fields. In is worth noting that the proposed decomposed formulation furnishes the same discrete equations as in other works [8, 9, 10]. Further details on the proposed approach can be found in [11].

Elasto-plastic FE formulation

The FE used in this work is named MISS-4. We refer elsewhere [5, 7] for all the details on its formulation in the linear-elastic range. For the linear elastic problem, using a Hellinger-Reissner approach and the assumed displacement and generalised stresses interpolation one obtains

$$\int_{Be} \boldsymbol{t}^T \boldsymbol{\rho} d\Omega_e - \frac{1}{2}\int_{Be} \boldsymbol{t}^T \boldsymbol{F} \boldsymbol{t} d\Omega_e - L_{ext} = \boldsymbol{\beta}_e^T \boldsymbol{Q}_e \boldsymbol{d}_e - \frac{1}{2}\boldsymbol{\beta}_e^T \boldsymbol{H}_e \boldsymbol{\beta}_e - \boldsymbol{d}_e^T \boldsymbol{p}_e \tag{1}$$

where L_{ext} is the work of the element external loads pe, $\boldsymbol{H}_e$ and $\boldsymbol{Q}_e$ are the element compliance and the compatibility/equilibrium matrices, $\boldsymbol{t}$ and $\boldsymbol{\rho}$ are the generalised stresses and strains, respectively, $\boldsymbol{\beta}_e$ and $\boldsymbol{d}_e$ are the stress and displacement parameters, respectively, and $\boldsymbol{F}$ is the inverse of the first order shear deformation theory elasticity matrix. The key problem of an elasto-plastic analysis is represented by the integration of the constitutive equations. A common approach is based on the use of a Backward-Euler integration scheme of the incremental constitutive equations, where the stresses become an implicit function of the assigned strain (displacement) increment. For mixed FE, the weak form of the finite step constitutive relations over the element becomes

$$\begin{cases} \boldsymbol{H}_e\left(\boldsymbol{\beta}_e - \boldsymbol{\beta}_e^{(n)}\right) + \sum_g \mu_g \dfrac{\partial f_g}{\partial \boldsymbol{\beta}_e} = \boldsymbol{Q}_e \Delta \boldsymbol{d}_e \\ \mu_g \geq 0, \; f_g \leq 0, \; \mu_g f_g = 0 \end{cases} \tag{2}$$

where we assume to test the plastic admissibility conditions in a discrete number of IP identified by the subscript g, $\boldsymbol{\Delta} = (\cdot)^{(\boldsymbol{n}+\boldsymbol{1})} - (\cdot)^{(\boldsymbol{n})}$ represents the difference between quantities at step $\boldsymbol{n}+\boldsymbol{1}$ and $\boldsymbol{n}$, $\boldsymbol{f}$ is the yield function, and $\boldsymbol{\mu}_g$ are the positive plastic multipliers. The superscript $(+\boldsymbol{1})$ is omitted to simplify the notation.

It is easy to show that Eq. (2) are the first order conditions of the following convex optimisation problem

$$\begin{aligned} &\min_{\boldsymbol{\beta}_e} && \frac{1}{2}(\boldsymbol{\beta}_e - \boldsymbol{\beta}_e^*)^T \boldsymbol{H}_e (\boldsymbol{\beta}_e - \boldsymbol{\beta}_e^*) \\ &subject\ to && f_g = f\left[\boldsymbol{N}_{tg}\boldsymbol{\beta}_e\right] \leq 0, \;\; \forall g \end{aligned} \tag{3}$$

which represents a CPP of the predictor $\boldsymbol{\beta}_e^* = \boldsymbol{\beta}_e^{(n)} + \boldsymbol{H}_e^{-1}\boldsymbol{Q}_e\Delta\boldsymbol{d}_e$ over the admissible domain of the element which, in this case, is a multi-surface yield domain. The solution of this problem is unique when the quantities on the initial step are known and $\Delta\boldsymbol{d}_e$ assigned.

Equations (2) furnish the implicit constitutive step equation we are searching for as $\beta_e = \boldsymbol{\beta}_e[\boldsymbol{\beta}_e^{(n)}, \Delta\boldsymbol{d}_e]$. The internal force vector of the element, for a given $\Delta\boldsymbol{d}_e$ and starting from a known state $\boldsymbol{\beta}_e^{(n)}$, is expressed as

$$\boldsymbol{s}_e[\boldsymbol{d}_e] = \boldsymbol{Q}^T\boldsymbol{\beta}_e\left[\boldsymbol{\beta}_e^{(n)}, \Delta\boldsymbol{d}_e\right] \tag{4}$$

while the tangent stiffness matrix becomes

$$\boldsymbol{K}_{et} = \frac{\partial \boldsymbol{s}_e[\boldsymbol{d}_e]}{\partial \boldsymbol{d}_e} = \boldsymbol{Q}_e^T\boldsymbol{H}_{et}^{-1}\boldsymbol{Q}_e, \qquad \boldsymbol{H}_{et}^{-1} = \frac{\partial \boldsymbol{\beta}_e}{\partial \boldsymbol{\eta}_e} \tag{5}$$

And $\boldsymbol{\eta}_e = \boldsymbol{Q}_e\boldsymbol{d}_e$ is the generalized strain.

The incremental elastoplastic analysis

The equilibrium path of the structures is usually evaluated by means of a continuation method [11]. Starting from a known equilibrium point $z_0 = z^{(n)}$, where $z = [d, \lambda]$, a new state z^{k+1} is obtained by solving, through a Newton iteration, the following system representing the equilibrium equations plus the arc-length constraint for an assigned value of $\Delta\xi$.

$$\begin{cases} \boldsymbol{r}[\boldsymbol{d}, \lambda] = \boldsymbol{s}[\boldsymbol{d}] - \lambda\boldsymbol{p} = \boldsymbol{0} \\ r_\lambda[\boldsymbol{d}, \lambda] = \boldsymbol{n}_z^T\boldsymbol{M}_z(\boldsymbol{z} - \boldsymbol{z}_0) - \Delta\xi = 0 \end{cases} \quad with \;\; \Delta\xi = \boldsymbol{n}_z^T\boldsymbol{M}_z(\boldsymbol{z}_1 - \boldsymbol{z}_0), \tag{6}$$

where $\boldsymbol{z}_1$ is the first predictor. The most used and effective arc-length equation is a moving hyperplane with $\boldsymbol{M}_z = diag(\boldsymbol{M}, \mu)$ a suitable metric matrix and $\boldsymbol{n}_z = [\boldsymbol{n}, n\_\lambda]$ its normal vector. Letting

$$\boldsymbol{K} = \frac{\partial \boldsymbol{s}[\boldsymbol{d}]}{\partial \boldsymbol{d}} \tag{7}$$

evaluated as assemblage of element matrices $\boldsymbol{K}_{et}$ in Eq. (5), Newton iterations are applied to the extended nonlinear system of equations (6), giving a sequence of estimates

$$\Delta\boldsymbol{z}_{j+1} = \Delta\boldsymbol{z}_j + \dot{\boldsymbol{z}}_j \;\; and \;\; \boldsymbol{z}_{j+1} = \boldsymbol{z}_0 + \Delta\boldsymbol{z}_j$$

Where the correction $\dot{\boldsymbol{z}}_j$ is evaluated as

$$\begin{cases} \boldsymbol{K}_j\dot{\boldsymbol{d}} - \dot{\lambda}\,\boldsymbol{p} - \boldsymbol{r}_j \\ \boldsymbol{n}^T\boldsymbol{M}\dot{\boldsymbol{d}} + \mu n_\lambda\dot{\lambda} = -r_{\lambda j} \end{cases}$$

At each iteration, the internal forces vector s_j, and then the residual r_j, is evaluated by assembling element quantities $s_e^j = Q^T\beta_e[\beta^{(n)}, \Delta d_e^j]$ with the stresses obtained using the CPP problem in Eq. (3).

The solution of the Closest Point Projection scheme over the element

The mathematical programming problem in Eq.(3) represents the main difference between mixed and displacement-based FE formulations.

Herein, the CPP problem is alternatively solved by means of a dual decomposition technique. The dual decomposition produces a series of CPP over the IPs, as in usual displacement formulations, therefore simplifying the solution which can be performed using standard numerical tools based on the application of the Newton method.

The dual decomposition of the CPP problem over the element

Two kind of dual decomposition approaches are investigated, depending on the way the plastic admissibility is checked. In a first case, we assume to express the plastic-admissibility condition in terms of the generalised stresses $\boldsymbol{t}$

 Materials Research Forum LLC
https://doi.org/10.21741/9781644902431-37

$$f_g[\boldsymbol{t}_g] = f[\boldsymbol{t}_g, \boldsymbol{x}_g] \leq 0, \ \ \forall g \tag{8}$$

and to control it in a finite number of IPs having middle-surface coordinate x_g and generalised stresses $\boldsymbol{t}_g = \boldsymbol{N}_t[\boldsymbol{x}_g]\boldsymbol{\beta}_e$. The decomposition process is performed at the level of these midplane shell IPs.

The second approach further decomposes the problem at level of material point. In this case, the plastic admissibility is expresses by standard yield functions in terms of the Cauchy 3D stresses $\boldsymbol{\sigma}$ evaluated in a discrete number of control points distributed also in the thickness direction $(\boldsymbol{x}_g, z_{gm})$

$$f_{gm}[\boldsymbol{\sigma}_{gm}] = f[\boldsymbol{\sigma}_{gm}, \boldsymbol{x}_g, z_m] \leq 0 \ \ \forall g, m \tag{9}$$

Where $\boldsymbol{\sigma}_{gm} = \boldsymbol{E}[\boldsymbol{z}_m]^T \boldsymbol{N}_t[\boldsymbol{x}_g]\boldsymbol{\beta}_e$ is the stress at the $(g,m)IP$.

A dual decomposition scheme for the CPP problem over the element

Assuming a numerical integration scheme to evaluate the complementary energy terms so that

$$t_g = N_{tg}\beta_e \rightarrow \sum_g t_g^T F t_g w_g = \beta_e^T H_e \beta_e \tag{10}$$

The minimization problem in Eq.(3) becomes

$$\min_{t_e} \ \ \frac{1}{2}\sum_g \left(\boldsymbol{t}_g - \boldsymbol{t}_g^{(n)}\right)^T \boldsymbol{F}\left(\boldsymbol{t}_g - \boldsymbol{t}_g^{(n)}\right) w_g - \boldsymbol{\beta}_e^T \boldsymbol{Q}_e \Delta \boldsymbol{d}_e$$
$$subject\ to\ \ f_g[t_g] \leq 0, \ \ \forall g = 1, \dots, N_g, \qquad t_g = N_{tg}\beta_e \tag{11}$$

where the objective function of problem in Eq.(3), since a constant with βe term is inessential, is rewritten as

$$\frac{1}{2}\left(\boldsymbol{\beta}_e - \boldsymbol{\beta}_e^{(n)}\right)^T \boldsymbol{H}_e\left(\boldsymbol{\beta}_e - \boldsymbol{\beta}_e^{(n)}\right) - \boldsymbol{\beta}_e^T \boldsymbol{Q}_e \Delta \boldsymbol{d}_e \tag{12}$$

The Lagrangian associated to problem in Eq.(11), under the assumption $\mu_g \geq 0$, is

$$L = \frac{1}{2}\sum_g \left(\boldsymbol{t}_g - \boldsymbol{t}_g^{(n)}\right)^T \boldsymbol{F}\left(\boldsymbol{t}_g - \boldsymbol{t}_g^{(n)}\right) w_g - \boldsymbol{\beta}_e^T \boldsymbol{Q}_e \Delta \boldsymbol{d}_e + \sum_g \mu_g f_g[\boldsymbol{t}_g] w_g - \sum_g \Delta\boldsymbol{\rho}_g^T (\boldsymbol{t}_g - \boldsymbol{N}_{tg}\boldsymbol{\beta}_e) w_g$$

Having relaxed the element interpolation equations, the first order equations can be subdivided into those involving only IPs variables and those regarding elemental quantities. The firsts are

$$\frac{\partial L}{\partial t_g} \rightarrow \begin{cases} \boldsymbol{F}\left(\boldsymbol{t}_g - \boldsymbol{t}_g^{(n)}\right) + \mu_g \frac{\partial f_g[\boldsymbol{t}_g]}{\partial \boldsymbol{t}_g} - \Delta\boldsymbol{\rho}_g = \boldsymbol{0} \\ \mu_g f_g[\boldsymbol{t}_g] = 0 \ \ \mu_g \geq 0 \ \ f_g[\boldsymbol{t}_g] \leq 0 \end{cases} \forall g, \tag{13}$$

while the equations regarding elemental variables are

$$\boldsymbol{r}_g = \frac{\partial L}{\partial \boldsymbol{\rho}_g} = \boldsymbol{t}_g - \boldsymbol{N}_{tg}\boldsymbol{\beta}_e = \boldsymbol{0} \ \forall g$$
$$\boldsymbol{r}_e = \frac{\partial L}{\partial \boldsymbol{\beta}_e} = \boldsymbol{Q}_e \Delta \boldsymbol{d}_e - \sum_g \boldsymbol{N}_{tg}^T \Delta\boldsymbol{\rho}_g w_g = \boldsymbol{0} \tag{14}$$

A sequence of IP state determination and element state determination problems allows the evaluation of a plastically-admissible $\boldsymbol{\beta}_e$ for an assigned $\Delta\boldsymbol{d}_e$, as explained in [11].

Numerical results

The proposed FE solution for the elasto-plastic analysis of shell structures is herein tested in a numerical example. The results obtained with MISS-4 are compared with those provided by the commercial software Abaqus using S4 and S8r FEs. The plastic admissibility is tested using a grid of 2×2 Gauss Points GP grid over the FE mid-plane and m = 8 GP along the thickness. In all tests the material is elastic perfectly plastic and a von Mises yield criterion is assumed. The test regards a the square plate of length L. The thickness is h = L/100. Due to the symmetry of the problem, only a quarter of the plate is analysed. Two boundary conditions are studied, namely simply supported and clamped where in both cases the normal rotation to the side is constrained. A uniformly distributed out-of-plane load is applied, having amplitude $p = \lambda M_0/(2L)^2$, with $M_0 = \sigma_0 t^2/4$ and σ_0 is Mises stress limit.

Table 1: Square plate: limit load for different models and mesh densities, clamped condition.

FE	*m*	mesh density				
		2×2	4×4	8×8	16×1	32×3
MISS-4	8	44.164	46.146	46.624	46.215	45.643
MISS-4	g	43.664	45.629	46.104	45.705	45.143
Abaqus	8	61.975	54.797	49.840	47.267	45.965

The limit loads for different FE and meshes are shown in Tables 1 for the clamped case. It shows that MISS-4 is accurate also for the coarsest mesh. The good convergence properties of MISS-4 are highlighted in Fig. 1 which shows that for both the boundary conditions it gives lower error than Abaqus S4.

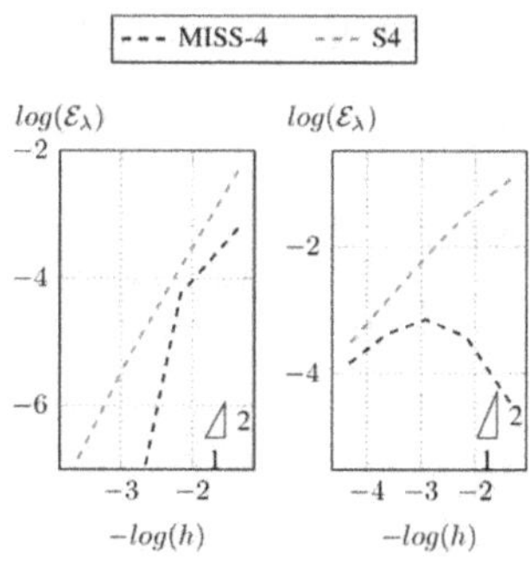

Figure 1: Square plate: convergence of the limit load.

References

[1] Nicola A. Nodargi. An overview of mixed finite elements for the analysis of inelastic bidimensional structures. Arch Computat Methods. Eng., 26:1117–1151, 2019. https://doi.org/10.1007/s11831-018-9293-0

[2] J. Bleyer and P. de Buhan. On the performance of non-conforming finite elements for the upper bound limit analysis of plates. International Journal for Numerical Methods in Engineering, 94(3):308–330, 2013. https://doi.org/10.1002/nme.4460

[3] G. Castellazzi and P. Krysl. A nine-node displacement-based finite element for Reissner-Mindlin plates based on an improved formulation of the NIPE approach. Finite Elements in Analysis and Design, 58:31–43, 2012. https://doi.org/10.1016/j.finel.2012.04.004

[4] L. Leonetti and Canh V. Le. Plastic collapse analysis of Mindlin–Reissner plates using a composite mixed finite element. International Journal for Numerical Methods in Engineering, 105(12):915– 935, 2016. https://doi.org/10.1002/nme.4997

[5] A. Madeo, G. Zagari, R. Casciaro, and S. De Miranda. A mixed 4-node 3d plate element based on self-equilibrated isostatic stresses. International Journal of Structural Stability and Dynamics, 15(4), 2015. https://doi.org/10.1142/S0219455414500667

[6] Antonio Madeo, Francesco S. Liguori, Giovanni Zucco, and Stefania Fiore. An efficient isostatic mixed shell element for coarse mesh solution. International Journal for Numerical Methods in Engineering, 122:82–12, 2021. https://doi.org/10.1002/nme.6526

[7] Francesco S. Liguori and Antonio Madeo. A corotational mixed flat shell finite element for the efficient geometrically nonlinear analysis of laminated composite structures. International Journal for Numerical Methods in Engineering, 122(17):4575–4608, 2021. https://doi.org/10.1002/nme.6714

[8] R. L. Taylor, F. C. Filippou, A. Saritas, and F. Auricchio. A mixed finite element method for beam and frame problems. Computational Mechanics, 31(1):192–203, May 2003. https://doi.org/10.1007/s00466-003-0410-y

[9] Nicola A. Nodargi and Paolo Bisegna. A novel high-performance mixed membrane finite element for the analysis of inelastic structures. Computers and Structures, 182:337–353, 2017. https://doi.org/10.1016/j.compstruc.2016.10.002

[10] D. Magisano and G. Garcea. Fiber-based shakedown analysis of three-dimensional frames under multiple load combinations: Mixed finite elements and incremental-iterative solution. International Journal for Numerical Methods in Engineering, 121(17):3743–3767, 2020. https://doi.org/10.1002/nme.6380

[11] Francesco S. Liguori, Antonio Madeo, and Giovanni Garcea. A dual decomposition of the closest point projection in incremental elasto-plasticity using a mixed shell finite element. International Journal for Numerical Methods in Engineering; 123:6243–6266, 2022. https://doi.org/10.1002/nme.7112

Theoretical and Applied Mechanics - AIMETA 2022 Materials Research Forum LLC
Materials Research Proceedings **26** (2023) 233-238 https://doi.org/10.21741/9781644902431-38

Geometrically nonlinear thermoelastic analysis of shells: modelling, incremental-iterative solution and reduction technique

LIGUORI F. S.[1, a *], MAGISANO D.[1,b], LEONETTI L. [1,c], MADEO A. [1,d] and GARCEA G. [1,e]

[1]University of Calabria, DIMES, Ponte P. Bucci, cubo 39 C, 87030, Rende (CS), Italy

[a]francesco.liguori@unical.it, [b]domenico.magisano@unical.it, [c]leonardo.leonetti@unical.it, [d]antonio.madeo81@unical.it, [e]giovanni.garcea@unical.it

Keywords: Shell Structures, Thermoelastic Analysis, Geometric Nonlinearity, Buckling, Newton Method, Isogeometric Analysis

Abstract. This work presents an accurate and efficient numerical tool for geometrically nonlinear thermoelastic analyses of thin-walled structures. The structure is discretized by an isogeometric solid-shell model avoiding the parameterization of finite rotations. An efficient modeling of thermal strains, temperature-dependent materials and general temperature profiles is proposed. Then, a generalized path-following method is developed for solving the discrete equations with the temperature amplifier as additional unknown. Finally, a reduction technique based on Koiter theory is derived for a quick estimate of the nonlinear thermal buckling.

Introduction

Civil, aeronautical and mechanical structures can be affected by high temperature rise [1]. The consequent high thermally-induced stresses and degradation of the thermoelastic properties can lead to buckling. Both aspects can significantly influence the buckling point and the postbuckling response. The nonlinear systems of equations defining the structural response can be solved step-by-step by using arc-length continuation method. In geometrically nonlinear analysis for mechanical loads, it has been shown that the use of a mixed formulation with independent stress variables makes the solution method able to withstand much larger step sizes with fewer iterations compared to purely displacement-based formulations [2]. It is also worth mentioning reduced order models based on the FEM implementation of Koiter's theory of elastic stability [3]. This work presents an accurate and efficient numerical tool for geometrically nonlinear thermoelastic analyses of thin-walled structures. The starting point is an accurate isogeometric solid-shell discretization without rotational DOFs. Focus is given to an efficient modelling of thermal strains, temperature-dependent materials and general temperature profiles based on a numerical pre-integration through the shell thickness in a pre-processing stage. Then, a generalized path-following method is developed for solving the discrete equations with the temperature amplifier as additional unknown, in order to trace the temperature-displacement nonlinear curve of the structure also in case of limit points and unstable paths. A consistent definition of the tangent operators and a mixed integration point strategy give a robust analysis with a reduced iterative burden, even in case of complex nonlinear behaviors where well-known simulation codes are outperformed or even fail. Finally, a reduction technique is derived for a quick estimate of the nonlinear thermal buckling point of structures with a high number of discrete degrees of freedom. It is a generalization of the so-called Koiter method to thermoelasticity. Numerical application are given to validate the proposal.

Solid-shell model with thermal strains and temperature-dependent material proprieties
This section introduces the main equations of the solid-shell model [4,5] extended to thermoelastic nonlinear analyses. Assuming a linear through-the-thickness interpolation, the position vector $\boldsymbol{X}$ and the displacement field $\boldsymbol{d}$ are expressed as

$$\boldsymbol{X} = \boldsymbol{X}_0\,[\xi,\eta] + \frac{2h}{\zeta}\boldsymbol{X}_n[\xi,\eta], \qquad \boldsymbol{d} = \boldsymbol{d}_0\,[\xi,\eta] + \frac{2h}{\zeta}\boldsymbol{d}_n\,[\xi,\eta]\,, \tag{1}$$

with h the shell thickness. The Green-Lagrange strain components are collected in vector $\boldsymbol{E} = [E_{11}, E_{22}, 2E_{12}, E_{33}, 2E_{23}, 2E_{13}]^T$ and linearized with respect to ζ as

$$\boldsymbol{E} \approx \begin{bmatrix} \boldsymbol{e}[\xi,\eta] + \zeta\boldsymbol{\chi}\,[\xi,\eta] \\ \boldsymbol{E}_{33}[\xi,\eta,0] \\ \boldsymbol{\gamma}\,[\xi,\eta] \end{bmatrix} \tag{2}$$

Generalised strains (3) presents the same format than in the Mindlin-Reissner model plus the thickness strain E_{33} but does not requires a direct parameterisation of 3D finite rotations.
Constitutive law, thermal strains and strain energy. We assume only in-plane thermal strains:

$$\boldsymbol{E}_T = \begin{bmatrix} \boldsymbol{\alpha}_p \\ 0 \\ \boldsymbol{0} \end{bmatrix} \boldsymbol{T} \;\text{ with }\; \boldsymbol{\alpha}_p = \begin{bmatrix} \alpha_1 \\ \alpha_2 \\ 0 \end{bmatrix} \tag{3}$$

where α_1 and α_2 are the thermal expansion coefficients. The strain energy of the solid-shell model can be written in compact notation as

$$\boldsymbol{\Phi}[\boldsymbol{d}] \equiv \frac{1}{2}\int_\Omega \int_{-\frac{h}{2}}^{\frac{h}{2}} (\mathbf{E} - \mathbf{E}_T[T])^T\, \mathbf{C}[\xi,T](\mathbf{E} - \mathbf{E}_T[\xi,T])\; d\zeta\, d\Omega \;= \int_\Omega \left(\frac{1}{2}\boldsymbol{\varepsilon}^T\, \boldsymbol{C}[T]\boldsymbol{\varepsilon} - \boldsymbol{\varepsilon}^T\boldsymbol{\sigma}_T\right) d\Omega + \text{cost} \tag{4}$$

Let us consider a distribution of temperature increment over the structure $T[\xi,\eta,\zeta]$ with respect to the ambient value. We can assume

$$T[\xi,\eta,\zeta] = \lambda\,\hat{T}\,[\xi,\eta,\zeta]$$

where $\hat{T}\,[\xi,\eta,\zeta]$ is the given temperature distribution and λ an amplifier factor. In case of temperature-dependent material properties, the elastic moduli and the thermal expansion coefficients depend on T. Their laws of variability can be assumed on the basis of experimental results or using formulas available in standards.
From a computational point of view, it is more handy to express $\mathbf{S}_T = \mathbf{C}[T]\mathbf{E}_T[T]$ by means of a polynomial interpolation in T at some temperatures T_k, which can be rewritten as

$$\mathbf{C}[\lambda] = \sum_{k=0}^{n} \lambda^k\, \hat{\mathbf{C}}_k\, \hat{T}^k, \qquad \mathbf{S}_T[\lambda] = \sum_{k=0}^{n} \lambda^k\, \hat{\mathbf{S}}_{Tk}\, \hat{T}^k \tag{5}$$

This choice makes it possible to express the generalised quantities of the solid-shell model $\boldsymbol{C}[T]$ and $\boldsymbol{\sigma}_T$ as a function of λ by a numerical through-the-thickness integration carried out once and for all in a pre-processing stage, regardless of the complexity of $\hat{T}\,[\xi,\eta,\zeta]$.

The isogeometric solid-shell discretisation
The continuum solid-shell model is approximated over the shell mid-surface by using NURBS functions, following the approach proposed in [4,5] to which we refer for further details.
Geometry and displacement fields are described as

$$\boldsymbol{X}[\xi,\eta,\zeta] = \boldsymbol{N}_d[\xi,\eta,\zeta]\boldsymbol{X}_e\,, \quad \boldsymbol{d}[\xi,\eta,\zeta] = \boldsymbol{N}_d[\xi,\eta,\zeta]\boldsymbol{d}_e. \tag{6}$$

where $\boldsymbol{X}_e = [\boldsymbol{X}_{0e}, \boldsymbol{X}_{ne}]$ and $\boldsymbol{d}_e$ collects the control points. The matrix $\boldsymbol{N}_d[\xi,\eta,\zeta]$ collects the interpolation functions

$$\boldsymbol{N_d}[\xi,\eta,\zeta] = \left[\boldsymbol{N}[\xi,\eta] \quad 2\frac{\zeta}{h}\boldsymbol{N}[\xi,\eta]\right] \tag{7}$$

where $\boldsymbol{N}[\xi,\eta]$ are 2D NURBS functions of the middle surface coordinates only. By exploiting Eq.(3) we obtain the generalised strain-displacement relationship. An efficient patch-wise integration schemes [4,5] to avoid membrane and shear locking is adopted for numerical integration. In this work, we adopt cubic C^2 NURBS with the integration scheme named S_1^4 .

The MIP arc-length continuation method for thermoelastic nonlinear analyses

We consider a slender hyperelastic structure subject to a conservative load p and to a temperature rise defined as $\mathrm{T}[\xi,\eta,\zeta] = \lambda\,\hat{\mathrm{T}}\,[\xi,\eta,\zeta]$. The equilibrium condition is given by the virtual work equation in discrete form

$$\boldsymbol{r}[\boldsymbol{u},\lambda] \equiv \boldsymbol{s}[\boldsymbol{u},\lambda] - \boldsymbol{p} = \boldsymbol{0} \tag{8}$$

where $\boldsymbol{r}: \mathbb{R}^{n+1} \to \mathbb{R}^n$ is a nonlinear vectorial function of the vector $[\boldsymbol{u},\lambda] \in \mathbb{R}^{n+1}$, collecting the discrete DOFs $\boldsymbol{u} \in \mathbb{R}^n$ and the multiplier $\lambda \in \mathbb{R}$. Eq.(8) represents a system of n-equations and n+1 unknowns and defines the equilibrium path as a curve in the $[\boldsymbol{u},\lambda]$ space.

Letting $\boldsymbol{x} = \{\boldsymbol{u},\lambda\}$, a new equilibrium point $\boldsymbol{x}^{k+1}$ is obtained starting from the previous known point $\boldsymbol{x}_0 = \boldsymbol{x}^{(k)}$ by solving iteratively the equilibrium equations plus the arc-length constraint for an assigned value of $\Delta\xi$

$$\begin{cases} \boldsymbol{r}[\boldsymbol{u},\lambda] \equiv \boldsymbol{s}[\boldsymbol{u},\lambda] - \boldsymbol{p} = \boldsymbol{0} \\ r_\lambda[\boldsymbol{u},\lambda] \equiv \boldsymbol{n}_x^T\boldsymbol{M}_x(\boldsymbol{x}-\boldsymbol{x}_0) - \Delta\xi = \boldsymbol{0} \end{cases} \qquad \Delta\xi = \boldsymbol{n}_x^T\boldsymbol{M}(\boldsymbol{x}_1 - \boldsymbol{x}_0) \tag{9}$$

with $\boldsymbol{x}_1$ the first predictor. The most simple arc-length equation is an adaptive hyperplane with $\boldsymbol{M}_x$ a suitable metric matrix and $\boldsymbol{n}_x$ its normal vector. Introducing the following tangent operators

$$\boldsymbol{K} \equiv \frac{\partial \boldsymbol{s}[\boldsymbol{u},\lambda]}{\partial \boldsymbol{u}} \qquad \hat{\boldsymbol{s}} \equiv \frac{\partial \boldsymbol{s}[\boldsymbol{u},\lambda]}{\partial \lambda} \tag{10}$$

Newton iterations can be then used to solve the extended nonlinear system of equations (9).

As shown in [2], the Newton method convergence depends on how rapidly the iteration matrix changes with respect to the configuration variables $\boldsymbol{u}$. The MIP Newton method using displacements and independent stress DOFs, makes the Newton method able to withstand very large steps with a very small iterative effort.

The strain energy is rewritten in a Hellinger-Reissner mixed format using a MIP strategy. The stress field is not interpolated, but the stresses at each integration point $\boldsymbol{\sigma}_g$ are taken directly as independent variables. In particular, by relaxing the constitutive law at each integration point one obtains the pseudo Hellinger-Reissner energy

$$\boldsymbol{\Phi}_e[\boldsymbol{u}_e,\lambda] = \left\{\textstyle\sum_g \boldsymbol{\sigma}_g^T\left(\boldsymbol{\varepsilon}_g - \boldsymbol{C}_g^{-1}[\lambda]\boldsymbol{\sigma}_{Tg}\right) - \frac{1}{2}\boldsymbol{\sigma}_g^T\boldsymbol{C}_g^{-1}[\lambda]\boldsymbol{\sigma}_g\right\}\mathrm{w}_g \tag{11}$$

Due to the local nature of the stress variables, decoupled at each IP, they can be condensed out when solving the linear systems and are then not involved in the global operations.

A Koiter-inspired reduction technique

A reduced order model (ROM) is proposed for an efficient nonlinear thermoelastic analysis of structure prone to buckling using a generalisation of the so-called Koiter method [3,5] to include thermal effects. The difficulty of this extension is due to the nonlinear dependence of the strain energy from the parameter λ that regulates the temperature rise. In particular, an accurate and coherent ROM requires the solution of an accurate 2-point linearized buckling problem in mixed (stress-displacement) form.

The space of admissible configuration, following a Galerkin approach, is limited to the following reduced model

$$\boldsymbol{u}_d = \boldsymbol{u}_f[\lambda] + \boldsymbol{v}[\boldsymbol{t}] + \frac{1}{2}\boldsymbol{w}[\lambda, \boldsymbol{t}] \qquad (12)$$

where $\boldsymbol{u}_f[\lambda] = \lambda\,\hat{\boldsymbol{u}}$ is the solution along the initial path tangent, $\boldsymbol{v}[\boldsymbol{t}] = \sum_i t_i \dot{\boldsymbol{v}}_i$ is a combination of buckling modes corresponding to the first m linearized critical temperatures and $\boldsymbol{w} = \sum_i \sum_j t_i t_j \boldsymbol{w}_{ij} + \lambda^2\,\widehat{\boldsymbol{w}}$ is the corresponding orthogonal second order correction.

Introducing the approximation of $\boldsymbol{u}_d$ in the discrete equilibrium equation using $\dot{\boldsymbol{v}}_i$ as test functions according to a Ritz-Galerkin approach and expanding in Taylor series up to the 3rd order in λ and t_i, we obtain the nonlinear reduced equations

$$\mathrm{r_i}[\boldsymbol{t}, \lambda] = 0 \quad \mathrm{i} = 1 \cdots \mathrm{n} \qquad (13)$$

Equations (13) are an algebraic nonlinear system of m equations in the m+1 variables $\lambda, t_1 \cdots t_m$ with known coefficients and can be solved using a path--following algorithm. Because of the small size of the system, the solution is very fast.

Numerical results

The performance of the proposed formulation is herein tested. The efficiency of the MIP strategy is compared with a displacement-based Newton method. Additionally, comparison is made with the nonlinear analysis implemented in the commercial code Abaqus and shell finite element S8R. A linear temperature distribution $\hat{T}(\zeta) = T_M + \frac{2\zeta}{h} T_L$ is assumed through the shell thickness. Additional tests are reported in [6,7].

A simply-supported Euler beam-like structure, reported in Fig.1, is considered. The length is $L = 10^3 mm$ and different values of L/h are analysed. Temperature-dependent material properties are considered and reported in details in [6,7,8]. The structure is subject to an initial compression load $q = 0.95\, k_c$, with $k_c = \pi^2\, h^3\, E_0/(12\, L^2)$. Subsequently, a temperature change is applied, with $T_M = 1$ and different values of the through-the-thickness gradient T_L.

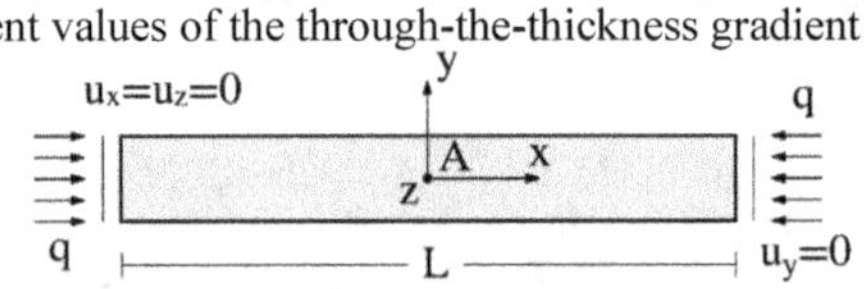

Figure 1 *Euler beam: geometry, loads and boundary conditions.*

It is worth noting that the structure does not exhibit any additional stress due to the temperature rise. Conversely, since the material properties are temperature-dependent, the temperature rise reduces the beam stiffness leading to buckling. Figure 2 depicts the equilibrium paths for three different values of T_L obtained though the proposed model and Abaqus which fails to reach convergence beyond the bifurcation load. Interestingly, its last converged point is characterized by a decreasing temperature for increasing values of T_L. This is likely due to an approximated evaluation of the tangent operators for TD material properties.

Table 1 compares the number of iterations required by Newton and MIP Newton for different values of L/h. The Newton method performances drastically deteriorate for increasing values of L/h and convergence is no longer possible starting from L/h=10^3. Conversely, MIP Newton is not influenced by the slenderness ratio.

Table 1. Euler beam: steps and iterations for Newton and MIP Newton with $T_L = 10^{-4}$.

Materials Research Forum LLC
https://doi.org/10.21741/9781644902431-38

L/h	Newton		MIP Newton		
	steps	iterations	steps	iterations	elapsed time*
30	110	368	52	131	0.33
100	158	577	48	121	0.21
1000	fails		36	95	-

* normalised with respect to Newton's elapsed time

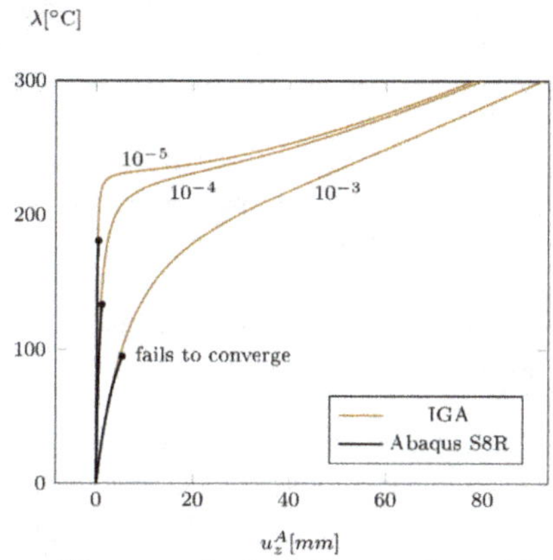

Figure 2. Euler beam: equilibrium paths for different values of the through-the-thickness temperature gradient and L/h=100.

Furthermore, it is possible to observe that the proposed Koiter reduced formulation with just one mode (and relative corrective) is pretty accurate in all cases up to significant post-buckling deformations.

Conclusions

The paper presented a robust and efficient simulation tool for the nonlinear thermoelastic analysis of thin-walled structures. The first part of the work consisted in developing a solid-shell isogeometric discretisation which requires a low number of displacement DOFs for accurately describing the shell kinematics. Focus was given to the modelling of thermal strains and temperature-dependent materials within the solid-shell concept. A reduction technique based on the Koiter method was also proposed and applied to buckling problems. Tests on composite shells and thin-walled structures are available in [6,7].

References

[1] M. Bhatia, E. Livne (2008), Design-oriented thermostructural analysis with internal and external radiation, part 1: Steady state, AIAA Journal 46 (3), 578–590. https://doi.org/10.2514/1.26236

[2] D. Magisano, L. Leonetti, G. Garcea (2017), How to improve efficiency and robustness of the Newton method in geometrically non-linear structural problem discretized via displacement-based finite elements, Computer Methods in Applied Mechanics and Engineering 313, 986-1005. https://doi.org/10.1016/j.cma.2016.10.023

[3] G. Garcea, F. S. Liguori, L. Leonetti, D. Magisano, A. Madeo, (2017), Accurate and efficient a posteriori account of geometrical imperfections in Koiter finite element analysis, International Journal for Numerical Methods in Engineering 112, 1154–1174. https://doi.org/10.1002/nme.5550

[4] L. Leonetti, F. Liguori, D. Magisano, G. Garcea (2018), An efficient isogeometric solid-shell formulation for geometrically nonlinear analysis of elastic shells, Computer Methods in Applied Mechanics and Engineering 331, 159-183. https://doi.org/10.1016/j.cma.2017.11.025

[5] L. Leonetti, D. Magisano, F. Liguori, G. Garcea (2018), An isogeometric formulation of the Koiter's theory for buckling and initial post-buckling analysis of composite shells, Computer Methods in Applied Mechanics and Engineering 337, 387-410. https://doi.org/10.1016/j.cma.2018.03.037

[6] F. S. Liguori, D. Magisano, L. Leonetti, G. Garcea (2021), Nonlinear thermoelastic analysis of shell structures: solid-shell modelling and high-performing continuation method, Composite Structures 266, 113734. https://doi.org/10.1016/j.compstruct.2021.113734

[7] F. S. Liguori, D. Magisano, A. Madeo, L. Leonetti, G. Garcea (2022), A Koiter reduction technique for the nonlinear thermoelastic analysis of shell structures prone to buckling, International Journal for Numerical Methods in Engineering 123 (2), 547-576. https://doi.org/10.1002/nme.6868

[8] J. Lee, M. Bhatia (2019), Impact of corrugations on bifurcation and thermoelastic responses of hat-stiffened panels, Thin-Walled Structures 140, 209-221. https://doi.org/10.1016/j.tws.2019.03.027

Theoretical and Applied Mechanics - AIMETA 2022 Materials Research Forum LLC
Materials Research Proceedings **26** (2023) 239-244 https://doi.org/10.21741/9781644902431-39

Extended virtual element method for elliptic problems with singularities and discontinuities in mechanics

CHIOZZI Andrea[1, a *], BENVENUTI Elena[2,b], MANZINI Gianmarco[3,c] and SUKUMAR, N.[4,d]

[1]Department of Environmental and Prevention Sciences, University of Ferrara, C.so Ercole I D'Este 32, 44121 Ferrara, Italy

[2]Department of Engineering, University of Ferrara, Via Saragat 1, 44122 Ferrara, Italy

[3] IMATI, Consiglio Nazionale delle Ricerche, Via Ferrata 4, 27100 Pavia, Italy

[4] Department of Civil and Environmental Engineering, University of California, One Shields Avenue, 95616 Davis, USA

[a]andrea.chiozzi@unife.it, [b]bnvlne@unife.it, [c]marco.manzini@imati.cnr.it, [d]n.sukumar@ucdavis.edu

* Corresponding author

Keywords: Virtual Element Method, Extended Finite Element Method, Singularities, Discontinuities

Abstract. Drawing inspiration from the extended finite element method (X-FEM), we propose for two-dimensional elastic fracture problems, an extended virtual element method (X-VEM). In the X-VEM, we extend the standard virtual element space with the product of vector-valued virtual nodal shape functions and suitable enrichment fields, which reproduce the singularities of the exact solution. We define an extended projection operator that maps functions in the extended virtual element space onto a set spanned by the space of linear polynomials augmented with the enrichment fields. Several numerical examples are adopted to illustrate the convergence and accuracy of the proposed method, for both quadrilateral and general polygonal meshes.

Introduction

Numerical techniques for the solution of problems that admit singular or discontinuous solutions such as fracture propagation in solids have attracted significant attention in the last two decades. Among these, enriched finite element approximations based on the partition-of-unity concept [1] and the eXtended Finite Element Method (X-FEM) [2] have become widely popular. More recently, extended finite element formulations for polygonal meshes have been proposed [3] even though, on polygonal elements, the construction of shape functions is generally cumbersome and additional numerical integration issues must be carefully addressed [4]. The Virtual Element Method (VEM) is a stabilized Galerkin scheme proposed in [5] to solve partial differential equations on general polygonal meshes that overcomes many of the difficulties related to standard polygonal finite element formulations. The VEM can be looked at as a generalization of the Finite Element Method (FEM) in which the explicit knowledge of the basis functions is not needed, since the bilinear form and the continuous linear functional deriving from the variational formulation, are approximated by means of elliptic projections of the basis functions onto suitable polynomial spaces, which turn out to be computable from the degrees of freedom of the method. More recently, taking inspiration from the X-FEM, an eXtended Virtual Element Method (X-VEM) has been proposed in [6,7], for the scalar Laplace problem with singularities and discontinuities, and in [8] for fracture problems in two-dimensional linear elasticity. Here, we summarize the main finding related to the extended virtual element formulation for linear elastic fracture problems proposed in

[8], in which the displacement field exhibits both crack-tip singularities and discontinuities. An enriched virtual element space is constructed by resorting to an additional set of virtual basis functions, starting from suitably chosen vectorial enrichment fields which allow to incorporate additional information about the exact solution, improving numerical accuracy in the presence of singularities. The X-VEM for elastic fracture proves to be more flexible with respect to the X-FEM since it is applicable to arbitrary polygonal meshes, while using a very simple one-dimensional quadrature rule for the computation of all the integrals involved.

Two-dimensional elasticity model

Let us consider a linear elastic body occupying the two-dimensional domain $\Omega \in \mathbb{R}^2$, bounded by Γ cut by a traction-free internal crack Γ_c. We denote the displacement field on Ω by u(x) and assume small strains and displacements. The boundary is such that $\Gamma = \Gamma_u \cup \Gamma_t \cup \Gamma_c$. Prescribed displacements $\mathbf{g} \in C^0(\Gamma_u)$ are imposed on Γ_u, whereas tractions $\overline{\mathbf{t}} \in C^0(\Gamma_t)$ are imposed on Γ_t. Let $\boldsymbol{\sigma}$ be the Cauchy stress tensor. In the absence of body forces, equilibrium equations read

$$\nabla \cdot \boldsymbol{\sigma} = 0 \quad \text{in } \Omega, \tag{1}$$

with the natural boundary conditions

$$\begin{aligned} \boldsymbol{\sigma} \cdot \mathbf{n} &= \overline{\mathbf{t}} \quad \text{on } \Gamma_t, \\ \boldsymbol{\sigma} \cdot \mathbf{n} &= \mathbf{0} \quad \text{on } \Gamma_c, \end{aligned} \tag{2}$$

where n is the unit outward normal, and the essential boundary condition

$$\mathbf{u} = \mathbf{g} \quad \text{on } \Gamma_u. \tag{3}$$

The small strain tensor $\boldsymbol{\varepsilon}$ is related to the displacement field u by the compatibility equation

$$\varepsilon(\mathbf{u}) = \frac{1}{2}\left(\nabla(\mathbf{u}) + \nabla^T(\mathbf{u})\right). \tag{4}$$

Lastly, the isotropic linear elastic constitutive for a homogeneous material reads

$$\boldsymbol{\sigma}(\mathbf{u}) = \mathbf{C} : \varepsilon(\mathbf{u}), \tag{5}$$

Where $\mathbf{C}$ is the fourth-order elasticity tensor.
To state the weak form of the problem we define the test function space as:

$$U_0 = \left\{ \mathbf{v} \in [H^1(\Omega)]^2 : \mathbf{v} = 0 \text{ on } \Gamma_u, \ \mathbf{v} \text{ discontinuous on } \Gamma_c \right\}. \tag{6}$$

The weak form of the equilibrium equation reads as: Find the admissible displacement field $\mathbf{u}$ such that

$$a(\mathbf{u}, \mathbf{v}) := \int_\Omega \boldsymbol{\sigma}(\mathbf{v}) : \varepsilon(\mathbf{u}) dx = \int_{\Gamma_t} \overline{\mathbf{t}} \cdot \mathbf{v} \, d\Gamma =: b(\mathbf{v}) \quad \forall \mathbf{v} \in U_0. \tag{7}$$

Extended virtual element formulation

We now summarize the formulation of the extended virtual element method for fracture problems in two-dimensional elasticity presented in . Let $\mathrm{T} = \{\Omega_h\}_h$ be a family of decompositions of Ω into nonoverlapping polygonal elements E with nonintersecting boundary ∂E, barycenter $\mathbf{x}_E \equiv (x_E, y_E)^T$, area $|E|$, and diameter $h_E = \sup_{\mathbf{x},\mathbf{y}\in E} |\mathbf{x}-\mathbf{y}|$.

Enrichment with singular fields. The main concept of the X-VEM is to enrich the standard virtual element space by means of independent fields carrying information about the singularities affecting the exact solution. For the problem at hand, we choose the enrichment fields $\hat{\mathbf{u}}^I = \mathbf{u}^I / h^{1/2}$ and $\hat{\mathbf{u}}^{II} = \mathbf{u}^{II} / h^{1/2}$, where $\mathbf{u}^I$ and $\mathbf{u}^{II}$ are the exact asymptotic crack-tip displacement fields for mode I and mode II crack opening respectively, and h the maximum elemental diameter of the mesh [8]. We observe that these fields satisfy equilibrium. In order to define the extended virtual element space, we first introduce the local virtual element space $\mathbf{V}^{h,*}(E)$:

$$\mathbf{V}^{h,*}(E) = \left\{ \mathbf{v}^h = (v_x^h, v_y^h)^T \in \mathbf{V}^h(E) : v_x^h = v_y^h \right\}, \tag{8}$$

where $\mathbf{V}^h(E) = \left[V^h(E) \right]^2$ with $V^h(E)$ the standard virtual element space, spanned by the scalar virtual basis functions $\{\varphi_i\}_{i=1}^{N_E}$. Hence, the space $\mathbf{V}^{h,*}(E)$ is generated by the linear combination of the basis functions $\left\{ \varphi_i^* = (\varphi_i, \varphi_i)^T \right\}_{i=1}^{N_E}$. Then, we define the matrices ψ^I and ψ^{II} as

$$\psi^I \equiv \begin{bmatrix} \hat{u}_x^I & 0 \\ 0 & \hat{u}_y^I \end{bmatrix}, \quad \psi^{II} \equiv \begin{bmatrix} \hat{u}_x^{II} & 0 \\ 0 & \hat{u}_y^{II} \end{bmatrix}, \tag{9}$$

so that the local extended virtual element space $\mathbf{V}_X^h(E)$ reads as

$$\mathbf{V}_X^h(E) \equiv \mathbf{V}^h(E) \oplus \psi^I \mathbf{V}^{h,*}(E) \oplus \psi^{II} \mathbf{V}^{h,*}(E). \tag{10}$$

A basis of this space can be obtained as the union of the basis functions of $\mathbf{V}_X^h(E)$, $\psi^I \mathbf{V}^{h,*}(E)$ and $\psi^{II} \mathbf{V}^{h,*}(E)$. Therefore, at every enriched node the vector-valued field $\mathbf{v}_X^h(\mathbf{x})$ that belongs to the extended virtual element space $\mathbf{V}_X^h(E)$ is characterized by four values and for an element whose nodes are all enriched, we have $4N_E$ degrees of freedom. We denote the basis functions of $\mathbf{V}_X^h(E)$ by the symbol φ_i, $i = 1, \dots, 4N_E$, where

$$\varphi_i = \begin{cases} (\varphi_i, 0)^T & \text{for } 1 \le i \le 2N_E,\ i \text{ odd}, \\ (0, \varphi_i)^T & \text{for } 1 \le i \le 2N_E,\ i \text{ even}, \\ (\hat{u}_x^I \varphi_i, \hat{u}_y^I \varphi_i)^T & \text{for } 1 + 2N_E \le i \le 3N_E, \\ (\hat{u}_x^{II} \varphi_i, \hat{u}_y^{II} \varphi_i)^T & \text{for } 1 + 3N_E \le i \le 4N_E. \end{cases} \tag{11}$$

Finally, the extended global virtual element space $\mathbf{V}_X^h$ reads:

$$\mathbf{V}_X^h = \left\{ v_X^h \in \left[H^1(\Omega) \right]^2 : v_{X|E}^h \in \mathbf{V}_X^h(E)\ \forall E \in \Omega_h \right\}. \tag{12}$$

Since $\{\varphi_i\}_{i=1}^{4N_E}$ are not known in the interior of the element, we construct a convenient projection operator that allows to compute the approximations $a_X^h(\cdot,\cdot): \boldsymbol{V}_X^h(E) \times \boldsymbol{V}_X^h(E) \to \mathbb{R}$ and

$b_X^h(\cdot): \boldsymbol{V}_X^h(E) \to \mathbb{R}$ of the exact bilinear form $a(\cdot,\cdot)$ and the linear functional $b(\cdot)$ appearing in (7). The extended virtual element formulation then reads: Find $\mathbf{u}_X^h \in \mathbf{V}_{X,g}^h$ such that

$$a_X^h(\mathbf{u}_X^h, \mathbf{v}_X^h) = b_X^h(\mathbf{v}_X^h) \quad \forall \mathbf{v}_X^h \in \mathbf{V}_{X,0}^h \tag{13}$$

where the bilinear form $a_X^h(\cdot,\cdot)$ is built element-wise as

$$a_X^h(\mathbf{u}_X^h, \mathbf{v}_X^h) = \sum_{E\in\Omega} a_X^{h,E}(\mathbf{u}_X^h, \mathbf{v}_X^h) \quad \forall \mathbf{u}_X^h, \mathbf{v}_X^h \in \mathbf{V}_X^h, \tag{14}$$

and we set $b_X^h(v_X^h) = b(v_X^h)$. To construct a bilinear form $a_X^{h,E}(\cdot,\cdot)$ which is computable from the degrees of freedom, we extend the vector-valued linear polynomial space $\wp^1(E)$ to a subspace $\wp_X$ of $\mathbf{V}_X^h(E)$ which includes the linear polynomials and the additional enrichment functions $\hat{\mathbf{u}}^I$ and $\hat{\mathbf{u}}^{II}$. Such space is spanned by the eight linearly independent vector fields representing the three fundamental rigid body motions, the three independent deformation modes and the two enrichment fields:

$$\wp_X(E) = \text{span}\left\{ \begin{pmatrix}1\\0\end{pmatrix}, \begin{pmatrix}0\\1\end{pmatrix}, \begin{pmatrix}\eta\\-\xi\end{pmatrix}, \begin{pmatrix}\xi\\0\end{pmatrix}, \begin{pmatrix}0\\\eta\end{pmatrix}, \begin{pmatrix}\eta\\\xi\end{pmatrix}, \begin{pmatrix}\hat{u}_x^I\\\hat{u}_y^I\end{pmatrix}, \begin{pmatrix}\hat{u}_x^{II}\\\hat{u}_y^{II}\end{pmatrix} \right\}. \tag{15}$$

We then define the extended elliptic projection operator $\Pi_X^a : \mathbf{V}_X^h(E) \to \wp_X(E)$ for each element E, which is the solution of the following variational problem:

$$\int_E \boldsymbol{\sigma}(\mathbf{q}_X) : \boldsymbol{\varepsilon}(\Pi_X^a \mathbf{v}_X^h)\, d\mathbf{x} = \int_E \boldsymbol{\sigma}(\mathbf{q}_X) : \boldsymbol{\varepsilon}(\mathbf{v}_X^h)\, dx \quad \forall \mathbf{q}_X \in \wp_X(E), \tag{16}$$

with the additional conditions

$$\begin{aligned} \overline{\Pi_X^a v_X^h} &= \overline{v_X^h}, \\ \overline{(\Pi_X^a v_X^h)_R} &= \overline{(v_X^h)_R}, \end{aligned} \tag{17}$$

where $\overline{(\cdot)}$ and $\overline{(\cdot)}_R$ represent the average translation and rotation. Then, the local extended bilinear form can be computed as:

$$\begin{aligned} a_X^{h,E}(\mathbf{v}_X^h, \mathbf{w}_X^h) &\equiv a^E\left(\Pi_X^a(\mathbf{v}_X^h), \Pi_X^a(\mathbf{w}_X^h)\right) + S_X^E\left(\mathbf{v}_X^h - \Pi_X^a(\mathbf{v}_X^h), \mathbf{w}_X^h - \Pi_X^a(\mathbf{w}_X^h)\right) \\ &= \int_E \boldsymbol{\sigma}(\Pi_X^a(\mathbf{v}_X^h)) : \boldsymbol{\varepsilon}(\Pi_X^a(\mathbf{w}_X^h))\, d\mathbf{x} + S_X^E\left(\mathbf{v}_X^h - \Pi_X^a(\mathbf{v}_X^h), \mathbf{w}_X^h - \Pi_X^a(\mathbf{w}_X^h)\right), \end{aligned}$$

(18)

where $S_X^E(\cdot,\cdot)$ is a suitable stabilization term needed to guarantee linear consistency and stability of the method. According to the virtual element methodology, $S_X^E(\cdot,\cdot)$ can be any symmetric, positive definite, continuous bilinear form defined on the kernel of the extended projection operator Π_X^a. In [8], we provide two possible choices of the stabilization term by considering the standard dofi-dofi and D-recipe formulations in our extended setting. Such choices are widely accepted in the VEM literature and in some cases they were theoretically proved to be effective to guarantee stability.

Numerical examples

Patch test. We first conduct an extended patch test, addressing the enrichment with singular fields. The extended patch test ensures that the singular enrichment fields can be exactly reproduced using the X-VEM. To this aim, we consider a square elastic plate that occupies the region $(-1,1)^2$ under plane strain conditions, with a horizontal crack of unit length that extends from $(-1,0)$ to $(0,0)$. A coarse mesh of 64 polygonal elements are considered, where all the nodes in the domain are enriched the near-tip displacement fields are imposed on the boundary of the domain by requiring that all the enriched boundary degrees of freedom are equal to 1 and all the standard boundary degrees of freedom are equal to 0. The relative error in strain energy for the extended patch tests is of the order of 10^{-12}.

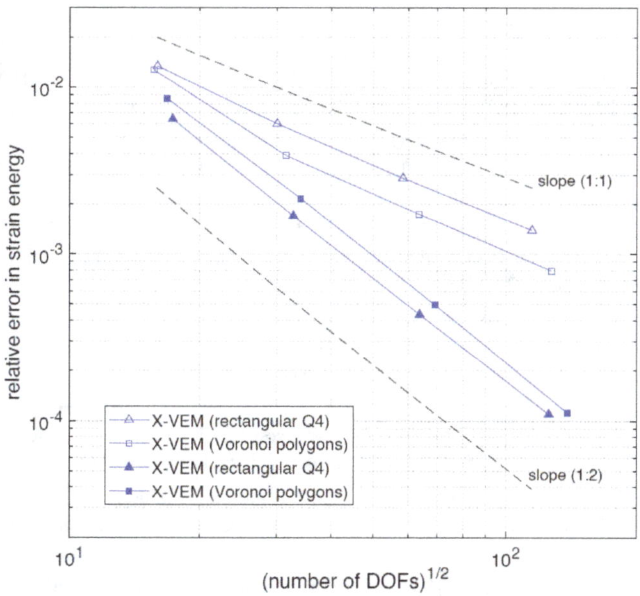

Figure 1: X-VEM convergence in strain energy for the mixed-mode benchmark problem with both topological and enrichment strategies enrichment.

Convergence study. We investigate the convergence of the X-VEM for the problem of a twodimensional square plate under plain strain conditions in the presence of a horizontal crack, extending from the boundary to the center of the specimen. The geometry of the domain is the same adopted as that for the extended patch test. On the boundary of the domain, we apply the exact near-tip mixed mode I and mode II displacement fields, which are also employed as enrichment fields for the X-VEM and represent the exact solution for the problem at hand. Both quadrilateral and general polygonal meshes are considered. To compute the element stiffness matrix, we follow two different strategies: topological enrichment and geometric enrichment. In the topological enrichment, we only enrich the node located at the singularity of the solution whereas in geometric enrichment we enrich all the nodes within a given radius from the origin. As in extended finite element methods, due to the presence of the singularity in the crack tip, the theoretical convergence rate for this problem is R = 1 that is non-optimal. As shown in Fig. 1, both VEM and X-VEM with topological enrichment converge in strain energy with a rate close to 1, in agreement with theory. It turns out that the X-VEM is insensitive to the type of mesh (quadrilaterals or polygons), and the results from the X-VEM are consistently more accurate than those from standard VEM. In order to establish if the proposed X-VEM can deliver the optimal convergence rate R = 2 that is predicted by theory, we enrich all nodes that are located within a ball of radius $r_e = 0.5$ from the origin. As shown in Fig. 1, in the case of geometric enrichment, the convergence rate is close to 2, which is consistent with theory.

References

[1] J. M. Melenk and I. Babŭska. The partition of unity finite element method: Basic theory and applications. Computer Methods in Applied Mechanics and Engineering, 139:289–314, 1996. https://doi.org/10.1016/S0045-7825(96)01087-0

[2] N. Möes, J. Dolbow, and T. Belytschko. A finite element method for crack growth without remeshing. International Journal for Numerical Methods in Engineering, 46(1):131–150, 1999. https://doi.org/10.1002/(SICI)1097-0207(19990910)46:1<131::AID-NME726>3.0.CO;2-J

[3] A. Tabarraei and N. Sukumar. Extended finite element method on polygonal and quadtree meshes. Computer Methods in Applied Mechanics and Engineering, 197(5):425–438, 2008. https://doi.org/10.1016/j.cma.2007.08.013

[4] E. B. Chin, J. B. Lasserre, and N. Sukumar. Modeling crack discontinuities without element partitioning in the extended finite element method. International Journal for Numerical Methods in Engineering, 86(11):1021–1048, 2017.

[5] L. Beirão da Veiga, F. Brezzi, A. Cangiani, G. Manzini, L. D. Marini, and A. Russo. Basic principles of virtual element methods. Mathematical Models & Methods in Applied Sciences, 23:119–214, 2013. https://doi.org/10.1142/S0218202512500492

[6] E. Benvenuti, A. Chiozzi, G. Manzini, and N. Sukumar. Extended virtual element method for the Laplace problem with singularities and discontinuities. Computer Methods in Applied Mechanics and Engineering, 356:571–597, 2019. https://doi.org/10.1016/j.cma.2019.07.028

[7] A. Chiozzi and E. Benvenuti. Extended virtual element method for the torsion problem of cracked prismatic beams. Meccanica, 55:637–648, 2020. https://doi.org/10.1007/s11012-019-01073-5

[8] E. Benvenuti, A. Chiozzi, G. Manzini, and N. Sukumar. Extended virtual element method for two-dimensional linear elastic fracture. Computer Methods in Applied Mechanics and Engineering, 390:114352, 2022. https://doi.org/10.1016/j.cma.2021.114352

Theoretical and Applied Mechanics - AIMETA 2022 | Materials Research Forum LLC
Materials Research Proceedings **26** (2023) 245-250 | https://doi.org/10.21741/9781644902431-40

An efficient plasticity-based model for reinforced concrete flat shells by a 4-nodes mixed finite element

LIGUORI Francesco S.[1, a*], CORRADO Antonella[1,b], BILOTTA Antonio[1, c] and MADEO Antonio[1,d]

[1]Dipartimento di Ingegneria Informatica, Modellistica, Elettronica e Sistemistica Università della Calabria 87036 Rende (Cosenza), Italy

[a]francesco.liguori@unical.it, [b]antonella.corrado@unical.it, [c]antonio.bilotta@unical.it, [d]antonio.madeo81@unical.it

Keywords: Mixed Shell Elements, Plasticity, Concrete-Like Materials

Abstract. A 4-nodes flat shell Finite Element (FE) is formulated on the basis of an assumed interpolation of the stress and displacement fields in order to perform the elastoplastic analysis of reinforced concrete structures. The nonlinear material behaviour of the concrete is described through a confinement-sensitive plasticity constitutive model not requiring the often difficult calibration of several parameters but it is expressed in terms of only the uniaxial compressive strength of the concrete. The presence of steel reinforcement bars is accounted by defining additional material layers with a uniaxial von Mises elastoplastic behaviour. These different material responses are assembled through the shell thickness by numerical integration. Computational efficiency and accuracy are assessed by comparing the proposed shell strategy and other shell FE models by performing the step-by-step nonlinear analyses on a numerical test.

Introduction

In the solution of many engineering problems the nonlinear behaviour of structure due to the inelastic response of materials are often described using elastoplastic models which are capable to accurately simulate the behaviour of many kinds of engineering materials, not only metals but also stone-like materials and soils, [1]. Moreover, the complexity of the problem to be solved, involving highly nonlinear and non smooth elastoplastic constitutive equations, [2], and the need for application of complex design rules, for example the application of several loading conditions with the check of different limit states, require the adoption of robust and very efficient solution algorithms.

In this work a finite element for the modelling of concrete flat shell embodying steel reinforcement bars is proposed. The element formulation is based on assumed stress and displacement field interpolations, a kind of formulation certainly enjoying the feature of providing high performance models as shown in [3] in the elastic field, in [4] for the analysis of slender structures and in [5] for 2D and 3D applications involving standard elastoplastic materials, i.e. von Mises-like models. The analysis of shell structures constituted by concrete, a material requiring more challenging elastoplastic modelling choices, in presence of steel reinforcement bars, see for example [7,8], constitutes a new context still not explored.

However, the proposed formulation is very general and capable to describe layered shells composed by different materials. The kinematic description is based on a first order shear deformable shell model whose primary kinematic fields are interpolated on the basis of a four nodes plane element, [9], whose mixed formulation contains also the discrete representation of the components of generalised stresses of the shell. The control of stress history and of the related evolution of plastic strain is however performed at 3D-level thanks to a numerical integration along the thickness of the shell. In this way is possible to accommodate a layered description of the materials composing the shell where concrete behaviour is modelled through a confinement-

Materials Research Proceedings **26** (2023) 245-250 https://doi.org/10.21741/9781644902431-40

sensitive plasticity yield surface depending on all the three stress invariants, see [10], and expressed in terms of only the uniaxial compressive strength of the concrete. At negligible computational costs the presence of steel rebars is computed by defining additional material layers contributing through a von Mises one-dimensional material response. It worth to observe how the proposed approach can be readily used for the modelling of additional material layers as in the case of retrofitting of existing structures. Computational efficiency and accuracy are assessed by comparing the proposed shell model with the solution obtained by standard displacement-based FEM formulations and the solutions provided by the commercial code Abaqus.

Layer-wise shell formulation of the MISS-4 finite element

The MISS-4 shell finite element proposed in [9] is now reformulated in order to describe the different layers of materials composing the shell. The MISS-4 element interpolates both stress and displacement fields by choosing them in order to achieve high accuracy on very coarse meshes. In particular, the assumed interpolation for the components of the generalised stresses are selected in order to *a-priori* satisfy the internal equilibrium equations. As a consequence, the displacement field needs to be interpolated only along the element boundary. All other details, the assumed interpolations etc., can be found in [9].

Shell element response

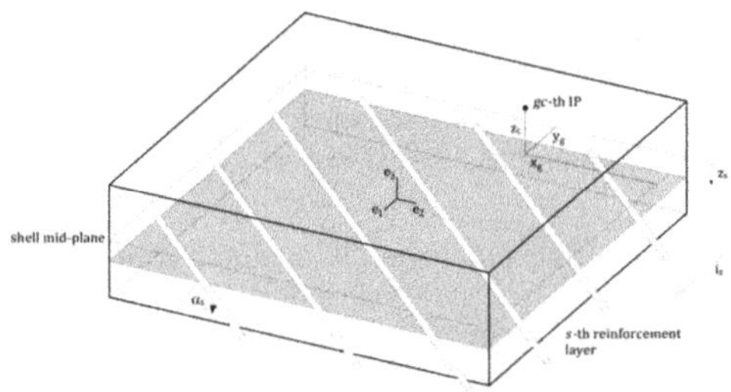

Figure 1: reinforced concrete shell.

Figure 1 shows the generic material composition of the shell finite element. In particular, it is made of concrete and a certain number of reinforcement layers. Each reinforcement layer, identified by the subscript s, has orientation α_s with respect to the element local Cartesian axis $\boldsymbol{e}_1$, rebar area A_s, spacing i_s and depth z_s with respect to the shell mid-surface.

As shown in Fig. 1, a discrete number of IP is considered located in position $\boldsymbol{x}_{gc} = [\boldsymbol{x}_g, z_c]$, where $\boldsymbol{x}_g$is the position of a generic IP over the mid-surface of the element and z_c indicates the depth of the generic IP along the thickness direction.

The vector of the generalised shell stresses at the g-th mid-surface IP is evaluated as

$$\boldsymbol{t}_g = \boldsymbol{t}[\boldsymbol{x}_g] = \int_{-h/2}^{h/2} \boldsymbol{E}[z]^T \boldsymbol{\sigma}[\boldsymbol{x}_g, z]\, dz = \sum_{c=1}^{n_c} \boldsymbol{E}[z_c]^T \boldsymbol{\sigma}_{gc} w_c + \sum_{s=n_c+1}^{n_c+n_s} \boldsymbol{E}[z_s]^T \boldsymbol{\sigma}_{gs} \frac{A_s}{i_s} \tag{1}$$

where n_c is the number of concrete IPs along the thickness, n_s is the number of steel layers, $\boldsymbol{\sigma}_{gc}$ are the concrete stresses at the c-th integration point along the thickness direction, w_c is the corresponding weight and $\boldsymbol{\sigma}_{gs}$ are the stresses of the s-th reinforcement layer.

The stresses in both concrete and steel must satisfy the plastic admissibility in each IP, expressed by

Materials Research Forum LLC
https://doi.org/10.21741/9781644902431-40

$$f_m[\boldsymbol{\sigma}_{gm}] \leq 0, \quad m = 1, \ldots, n_c + n_s \tag{2}$$

where, with m we identify the generic IP and f_m is the related yield function. The stresses $\boldsymbol{\sigma}_{gc}$ and $\boldsymbol{\sigma}_{gs}$ are independently evaluated by solving the IP state determination scheme at level of material point

$$\begin{cases} F_m\left(\boldsymbol{\sigma}_{gm} - \boldsymbol{\sigma}_{gm}^{(n)}\right) + \mu_{gm} \dfrac{\partial f_m[\boldsymbol{\sigma}_{gm}]}{\partial \boldsymbol{\sigma}_{gm}} - \Delta\boldsymbol{\varepsilon}_{gm} = \mathbf{0} \\ \mu_{gm} f_m[\boldsymbol{\sigma}_{gm}] = 0 \quad \mu_{gm} \geq 0 \quad f_m[\boldsymbol{\sigma}_{gm}] \leq 0 \end{cases} \qquad \forall m = 1, \ldots, n_c + n_s. \tag{3}$$

where $\Delta = (\cdot)^{(n+1)} - (\cdot)^{(n)}$ represents the difference between quantities at step $n+1$ and n, and μ_{gm} are the positive plastic multipliers, $\Delta\boldsymbol{\varepsilon}_{gm} = \boldsymbol{E}[z_m]\Delta\boldsymbol{\rho}_g$ represents the strain on the mth IP, while the assigned $\Delta\boldsymbol{\rho}_g$ is constant for each m. The superscript $(n+1)$ is omitted to simplify the notation.

The stress parameters $\beta_e = \beta_e\left[\beta_e^{(n)}, \Delta\boldsymbol{d}_e\right]$ for an assigned value of $\Delta\boldsymbol{d}_e$ are obtained using the element state determination algorithm defined by the following additional equations

$$\begin{cases} r_g \equiv \boldsymbol{t}_g - \boldsymbol{N}_{tg}\,\beta_e = \mathbf{0}, \quad \forall g \\ r_e \equiv \boldsymbol{Q}_e \Delta\boldsymbol{d}_g - \sum_g \boldsymbol{N}_{tg}^T \Delta\boldsymbol{\rho}_g w_g = \mathbf{0} \end{cases} \tag{4}$$

where first one imposes that the generalised stresses $\boldsymbol{t}_g$ coming from the solution of the IP state determination in Eq. (3) and evaluated using Eq. (1), are the same than those furnished by the assumed stress interpolation. The second equation imposes, in a weak form, the strain/displacement relation over the element.

Concrete mechanical response

The stress response of concrete, $\boldsymbol{\sigma}_{gc}$, relative to the generic IP along the thickness is evaluated by solving problem (3) on the basis of an elastic isotropic behaviour defined by the Young modulus, E_c, Poisson ratio, ν_c, of concrete and the yield function proposed in [10] and here reformulated with respect to the stress invariants I_1 and J_2 and the Lode angle θ as follows

$$f_c \equiv \frac{3}{\bar{\sigma}_c^2} J_2 + \frac{m}{\sqrt{3}\,\bar{\sigma}_c} \sqrt{J_2}\, r(\theta, e) + \frac{m}{3\,\bar{\sigma}_c} I_1 - 1. \tag{5}$$

The exact meaning of the material parameters, m and e, is defined in the already cited paper [10] where their dependence from $\bar{\sigma}_c$, the compressive strength of concrete, is formulated and discussed. In this way an apparently three-parameter hydrostatic-pressure-sensitive loading surface, capable to describe parabolic meridians and a variable shape on the deviatoric plane, is simply defined on the basis of $\bar{\sigma}_c$ only.

The Newton-Raphson solution of Equation (3) provides the required stress $\boldsymbol{\sigma}_{gc}$ but also the consistent tangent operator F_{tc}^{-1} relative to the concrete IP formulated as

$$F_{tc}^{-1} = \Xi - \frac{\Xi \dfrac{\partial f_c}{\partial \boldsymbol{\sigma}} \otimes \Xi \dfrac{\partial f_c}{\partial \boldsymbol{\sigma}}}{\dfrac{\partial f_c}{\partial \boldsymbol{\sigma}} \cdot \Xi \dfrac{\partial f_c}{\partial \boldsymbol{\sigma}}}, \quad \Xi^{-1} = F_c + \mu_{gc} \frac{\partial^2 f_c}{\partial \boldsymbol{\sigma}^2}, \tag{6}$$

and ready to be used in the assembly of the global tangent stiffness matrix.

Steel reinforcement layer mechanical response

Problem (3) is also solved for the generic steel reinforcement layer whose mono-axial mechanical response is now simply defined by the Young modulus E_s, the yield stress $\bar{\sigma}_s$ and the vector components $c = \cos\alpha_s$ and $s = \sin\alpha_s$ defined by rebar orientation, α_s, with respect to the axes describing the shell mid-surface. In particular, in the elastic phase, the mechanical response is defined by

$$\boldsymbol{\sigma}_{gs} = \boldsymbol{\sigma}_{gs}^{(n)} + F_s^{-1}\,\Delta\boldsymbol{\varepsilon}_{gs}, \quad F_s^{-1} = E_s\begin{bmatrix} c^4 & c^2s^2 & c^3s & 0 & 0 \\ c^2s^2 & s^4 & s^3c & 0 & 0 \\ c^3s & s^3c & c^2s^2 & 0 & 0 \\ 0 & 0 & 0 & 0 & 0 \\ 0 & 0 & 0 & 0 & 0 \end{bmatrix} \tag{7}$$

In the plastic phase the mechanical response is given by

$$\boldsymbol{\sigma}_{gs} = \pm\bar{\sigma}_s[c^2, s^2, -cs, 0, 0]^T, \quad F_{ts}^{-1} = \mathbf{0}_{5\times5}\,. \tag{8}$$

Numerical results

Efficiency and accuracy of the proposed FE strategy have been tested performing some numerical tests consisting in the step-by-step nonlinear analysis of structures subjected to a system of loads amplified through a multiplier λ used to describe the loading history. The results provided by the layer-wise MISS-4 element are compared with those provided by the commercial software Abaqus. In particular, all the analyses are also performed by using the Abaqus S4 shell element and describing the nonlinear behaviour of concrete through the Concrete Damage Plasticity (CDP) model parameterised in order to obtain a perfectly plastic behaviour.

The response of MISS-4 element is evaluated by using a 2×2 IP grid over the finite element mid-surface and 12 IPs through the shell thickness. The results compare the obtained equilibrium curves, i.e. the plot of the load multiplier λ versus a reference displacement component.

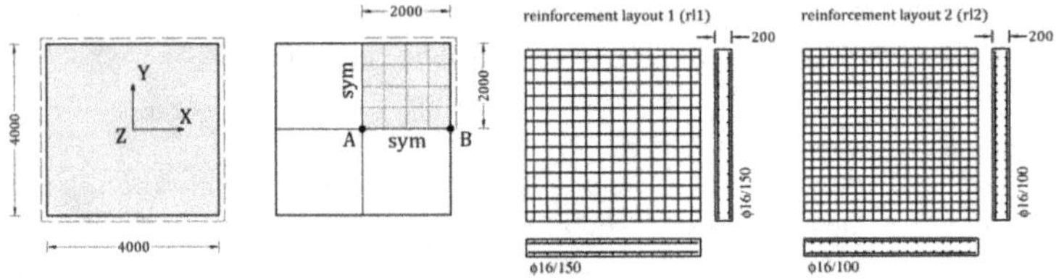

Figure 2: Square plate: geometry, boundary conditions and a 4x4 mesh (all the measures are given in [mm]).

The test regards the reinforced concrete square plate of length $L = 4$ [m] and thickness $s = 0.2$ [m] represented in Fig. 2. Two different schemes of reinforcement are considered.

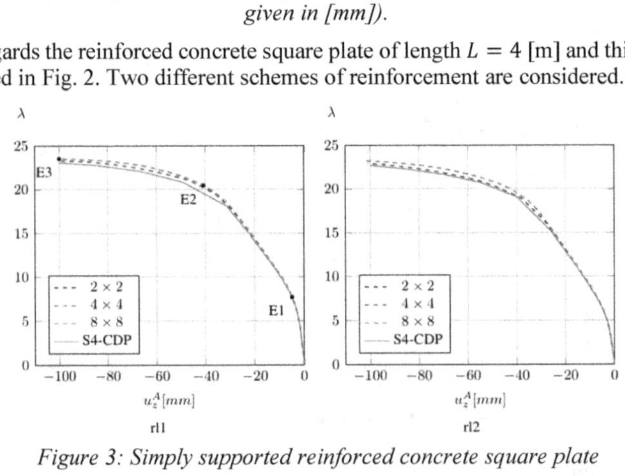

Figure 3: Simply supported reinforced concrete square plate

The concrete parameters are: $\overline{\sigma}_c = 30$ [MPa], the compressive strength; $\overline{\sigma}_t = 0.1\,\overline{\sigma}_c$ [MPa], the tensile strength; $\overline{\sigma}_b = 1.14\,\overline{\sigma}_c$, the biaxial compressive strength; $E_c = 2200\,(\overline{\sigma}_c/10)^{0.3}$ and $\nu_c = 0.2$, the Young modulus and the Poisson ratio, respectively. The parameters of steel reinforcement bars are $\overline{\sigma}_s = 450$ [MPa] and $E_s = 2.1 \times 10^5$ [MPa].

According to the symmetry of the problem, only a quarter of the plate is analysed, see Fig 2. Two boundary conditions are taken into consideration, namely simply supported condition and clamped condition. The plate is subjected to a uniformly distributed out-of-plane uniform pressure of amplitude $q = \lambda\,M_p/L^2$, where λ is the load multiplier and M_p is the plastic bending moment considering a mono-axial yield condition only for the concrete subjected to compressive state and steel rebars subjected to tensile state.

Table 1: Square plate: normalised load multiplier

B.C.	Reinf. layout	2×2	4×4	8×8	16×16	S4-CDP
Simply supported	rl1	23.2644	23.4824	23.5056	23.5282	23.051
	rl2	22.8898	23.1895	23.2061	23.2100	22.676
clamped	rl1	45.6447	44.3623	42.8930	41.7100	43.401
	rl2	45.4340	43.8796	43.3289	41.0107	43.107

Table 1 reports the load multipliers obtained for a maximum value of the displacement at the centre of plate equal to 100 [mm]. The solutions obtained by several meshes of MISS-4 elements are compared to the result provided by Abaqus with a very fine mesh of S4 elements. The same analyses are compared in Figs. 3 and 4 where the equilibrium curves are reported. The results show a good agreement between the proposed formulation and Abaqus code, even in the case of the coarsest mesh.

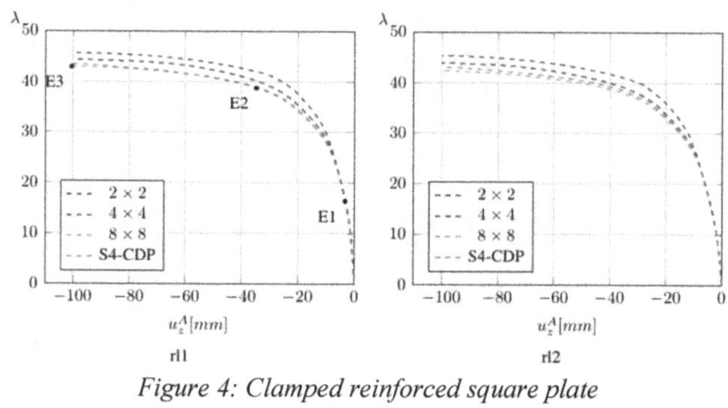

Figure 4: Clamped reinforced square plate

Concluding remarks

A mixed 4-nodes finite element has been formulated for the analysis of reinforced concrete shells. The element is based on the assumed interpolation of both displacement and stress fields, the latter is chosen in order to *a-priori* satisfy the equilibrium. The material mechanical response of the concrete is numerically integrated along the shell's thickness assuming a plasticity yield surface depending on all the three stress invariants which allows the description of the confinement-sensitive behaviour. The steel reinforcement bars are modelled as additional layers, embedded in the shell thickness, with von Mises mechanical response. The accuracy and efficiency of the proposed formulation are confirmed by the numerical test performed.

References

[1] Jirasek, M., Bazant, Z. P., 2002. Inelastic analysis of structures. John Wiley & Sons.

[2] Simo, J. C., Hughes, T. J. R., 1998. Computational inelasticity. Springer-Verlag.

[3] Bilotta A., Casciaro R., 2002. Assumed stress formulation of high order quadrilateral elements with an improved in-plane bending behaviour. Comput. Methods Appl. Mech. Engrg. 191/15-16, 4365–4378. https://doi.org/10.1016/S0045-7825(01)00334-6

[4] Liguori, F. S., Madeo, A., 2021. A corotational mixed flat shell finite element for the efficient geometrically nonlinear analysis of laminated composite structures. Int J Numer Methods Eng 122 (17), 4575–4608. https://doi.org/10.1002/nme.6714

[5] Bilotta A., Casciaro R., 2007. A high performance element for the analysis of 2D elastoplastic continua. Comput. Methods Appl. Mech. Engrg. 196, 818–828. https://doi.org/10.1016/j.cma.2006.06.009

[6] Bilotta A., Garcea G., Leonetti L., 2016. A composite mixed finite element model for the elastoplastic analysis of 3D structural problems, Finite Elem. Anal. Des. 113, 43–53. https://doi.org/10.1016/j.finel.2016.01.002

[7] Hofstetter G., Mang H. A., 1996. Computational plasticity of reinforced and prestressed concrete structures, Computational Mechanics 17, 243–254. https://doi.org/10.1007/BF00364827

[8] Haussler-Combe, U., 2014. Computational Methods for Reinforced Concrete Structures. John Wiley & Sons. https://doi.org/10.1002/9783433603611

[9] Madeo, A., Liguori, F. S., Zucco, G., Fiore, S., 2021. An efficient isostatic mixed shell element for coarse mesh solution. Int J Numer Methods Eng 122 (1), 82–121. https://doi.org/10.1002/nme.6526

[10] Papanikolaou, V. K., Kappos, A. J., 2007. Confinement-sensitive plasticity constitutive model for concrete in triaxial compression. Int. J. Solids Struct. 44, 7021–7048. https://doi.org/10.1016/j.ijsolstr.2007.03.022

Theoretical and Applied Mechanics - AIMETA 2022 Materials Research Forum LLC
Materials Research Proceedings **26** (2023) 251-256 https://doi.org/10.21741/9781644902431-41

Automatic construction of structural meshes from photographic and laser surveys

PADUANO Ivan[1,a], MILETO Andrea[1,b] and LOFRANO Egidio[1,c]*

[1]Department of Structural and Geotechnical Engineering, "Sapienza" University of Rome, Italy

[a]ivan.paduano@uniroma1.it, [b]andrea.mileto@uniroma1.it, [c]egidio.lofrano@uniroma1.it

* corresponding author

Keywords: Work Pipelines, Photographic/Laser Surveys, Structural Meshes

Abstract. The focus of this paper is the analyses of various work pipelines, allowing to manage the transition from *in-situ* surveys to cloud of points and geometric meshes optimized for structural purposes. This topic is very challenging, and today is almost always performed through homemade, uncontrolled, approaches, requiring the passage of information between numerous codes. These unsupervised workflows often compromise the integrity and the reliability of the results. Here two experimental case studies are reported to check the performances of two pipelines, based on photographic and laser surveys, respectively. The proposed comparison is used to outline significant indications on how properly manage the transformation, in order to create a "true" digital twin of the given structure.

Introduction

One of the most important issue in the field of structural surveying is the capability of modeling real-life constructions. The recent increasing interest given by conservation purposes put the topic as a crucial task in modern engineering. In particular, the introduction of BIM (Building Information Modeling) procedures among the fundamental goals of civil engineering [1], has shifted attention to a whole series of expeditious techniques aimed at making the process sustainable in terms of timing and costs. Among these we can mention, for our theme, photo modeling (not to be confused with photogrammetry) and laser scanning. Even if these two technologies are not exhaustive of the problem, we can confidently state that their proper combination could cover most of the artifacts in our area of interest [2].

During the last two decades some field experiments on this topic have been developed; the best results come from the field of chemical plants constructions, where the simplicity and very high standardization of the elements has allowed to create pipelines commonly accepted by a large sector of the operators working in the field. In the area of structural engineering, this process did not take place for a whole series of reasons, some of which merely ascribable to the IT (Information Technology) part of the procedure, as the high computational burden, the complexity of the elements composing the geometry and the complications related to the interoperability among different software. On the other hand, the rapid evolution of drone-mounted cameras has actually solved some important problems due to inaccessibility. However, the possibility of quickly acquiring a geometrical model does not allow to immediately pass to the FEM (Finite Element Model) work phase [3].

The rising of the BIM environments for executive design has made interoperability among models a central step. Indeed, within a BIM environment, the models intended as geometrical mesh must have a high degree of interoperability and must preserve the required LOD (Level Of Detail) [4]. It is therefore necessary to correctly work on the meshes to ensure high values of their main quality indicators (such as: aspect ratio, skewness and so on). The worst enemy of a correct procedure is the enormous amount of data stored during the detection procedures. Then, relying on the LOD required by the project, only the strictly necessary data should be acquired [5]: overabundance of data not only does not improve the quality of the output mesh, but also introduces some critical issues (e.g., in the case of photo modeling, a critical issue could be the processing of photos taken in conditions of very different light or with lenses distorting the images). It follows that the choice of the method for the survey must be made once the LOD and the level of interoperability have been set [6].

The paper focuses on the transition from photographic and laser surveys to cloud of points and geometric meshes optimized for structural purposes. At first, the relevant work pipelines are proposed, then, two experimental case studies are discussed to check their performances. The main objective of the manuscript is a comparison among the results provided by the two kinds surveys, when controlled and supervised approaches are adopted.

Work pipelines

To have an efficient pipeline ranging from data acquisition to model finalization, it is necessary to divide the process into at least three main blocks. The proposed pipeline is in Figure 1.

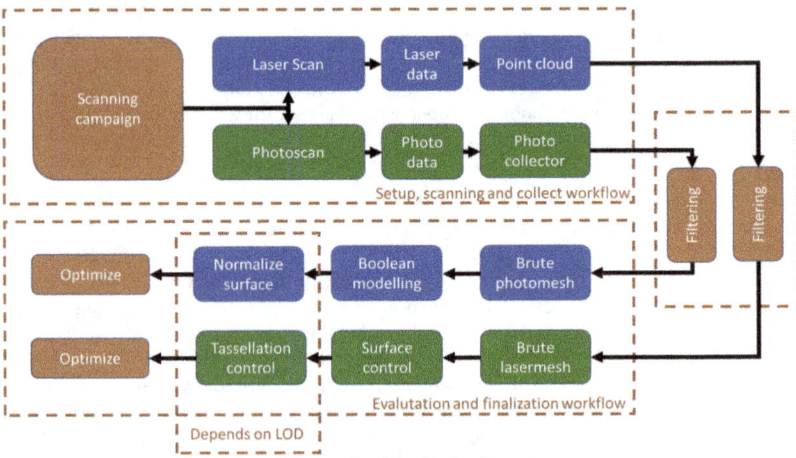

Figure 1 – Work pipelines.

In the first block there is the data acquisition phase with the choice of the survey method and a reasoned storage of these. Starting from the experience of the Authors, the following empirical laws are here suggested:

- for photo scanning, considering a LOD of at least 400 and congruently with the geometric nature of the artifact, at least 80 full-frame 1200 DPI (Dots Per Inch) photos per m^3 taken from different camera points (in similar light conditions) are required;

- for the laser scan, it will be necessary to place the stations so that there are as few data redundancies as possible, but ensuring that a functional overlap between the various stations is maintained. However, an estimation of the amount of stations is not possible in this case, since it strongly depend on the specific case study.

The data must then be carefully filtered in order to obtain 'brute' meshes that have the right LOD and are free of noise surfaces and all those biased clouds created by the limits of the instruments, such as the color block for photo modeling (occurring when two surfaces with the same color are associated with the same plane, even though they do not belong to it) and reflection for the laser scanner. Specifically, three types of filters are here adopted: remove duplicate points filter, SOR (Statistical Outlier Removal) filter and noise filter.

Moving towards the third block, leading to the evaluation and finalization of the model, different techniques can be used to make the model as efficient as possible with respect to the use of interest. According to the nature of the element, the topology (tri, quad, Delaunay, etc.) will then be chosen and the number of iterations of the tessellation (depending on the LOD) must be set. At the end of the pipeline the optimization step is fund, the process eliminating all the constructive problems of the model and reducing its computational burden on the basis of given criteria. The quality of the mesh must be checked in every operation that is done in the third block as some of these are irreversible. To speed up this phase, several retopologization tools have recently been produced that allow to automatically restore meshes whose topology is compromised. Nowadays, retopology becomes crucial to manage the big data coming from experimental surveys. Here, basing on the given case studies and relevant hot spots, a specific scheme is calibrated (see the following Section for the details).

The software stack used in this paper to develop the pipelines of Figure 1 is composed of several tools, all belonging to the Autodesk environment: ReCap Photo, Meshmixer, Inventor, Inventor Nastran, Retopology Tool, Autocad. Moreover, the software Cloud Compare has been used to improve the filter phase and, then, the mesh quality.

Experimental and numerical findings

Two cases studies were considered, a concrete slab and a small-scale steel frame, Figure 2. The slab, measuring 100 x 80 x 15 cm, is made of a C28/35 concrete. The second structure is a shear-type steel frame, with one span and four floors. The overall height is 800 mm (inter-floor of 200 mm), and the plant is square with a side of 300 mm. The columns have a rectangular section of 50 x 4 mm, while the beams have an L-shaped section 50 x 50 x 4 mm. All beam-column joints are bolted (Figure 2.b). The steel class is S235.

Figure 2 – Case studies: concrete slab 100 x 80 x 15 cm (left) and shear-type steel frame, height of 80 cm (right).

The two structures were detected with a photographic and laser scanner survey from different station points, as shown in Figure 3.

A CANON EOS 5D camera (resolution 4368x2912 pixels) with natural light was used to perform the photo survey. For the concrete slab, 179 photos have been gathered, with 2 different lenses (CANON EF 16-35mm f / 4L IS USM and CANON EF 75-300mm f / 4-5.6 III); the duration of the survey was 8 minutes. The same camera and lenses plus one (CANON EF 20mm f / 2.8 USM) were used for the steel frame, gathering a total of 231 photos, in 14 minutes.

The laser surveys were performed with a Leica BLK360 (properties can be found here: https://shop.leica-geosystems.com/au/leica-blk/blk360-g1/product-details), with 5 and 6 station points, for the concrete and steel specimen, respectively. The timing required for the acquisition was 5 minutes for each station.

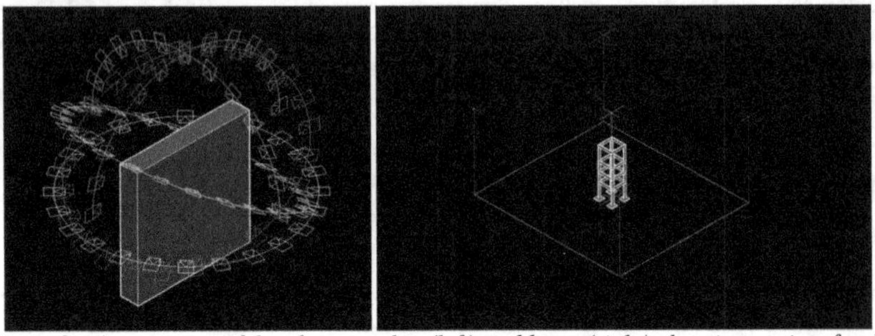

Figure 3 - Axonometry of the photographic (left) and laser (right) shooting project for the concrete slab and the steel frame.

The pipelines described in the previous Section have been developed to build up four geometrical meshes, one for each sample (concrete slab and steel frame) and each survey (photo and laser). For the sake of brevity, only the results obtained for the concrete slab are here discussed. The two meshes provided by the photo and laser surveys have been processed through a retopology procedure; the main settings are: 40'000 faces, quad tolerance of 10%, subdivision factor 1, regularize 0.7, anisotropy 0.65, adaptivity 0.45.

The two meshes were imported into the finite element calculation software Midas FEA NX, via an Iges CAD file. Once the geometry was imported, the auto-mesh tool of Midas was used to generate a mesh of three-dimensional elements, hexahedrons and tetrahedrons. A mesh size equal to 10 mm was imposed, small enough not to lose the obtained LOD (Table 1); for instance, it was even possible to consider the roughness present on the surfaces. To facilitate the comparisons, the models were also scaled imposing the same average depth, 150 mm.

The images of Table 1 show a counterintuitive result: the geometry created from the laser survey (column Laser 1) has less detailed than the one based on the photographic survey (column Photo). This is due to some issues of the import procedure. Even if the laser device was capable of generating a much more refined mesh than the photographic survey, the retopology operation performed with the same settings on both meshes results average surfaces with less asperities in the case of laser (basically because there are more asperities to be averaged).

In order to obtain a geometry that is more similar to the 'real' concrete slab (in the meaning of the required LOD) even with the laser survey, the retopology scheme was repeated after optimizing the model with a smoothing procedure. This allowed to remove the counterintuitive results (see the new column Laser 2).

Materials Research Proceedings **26** (2023) 251-256 https://doi.org/10.21741/9781644902431-41

Table 1 - Geometry and mesh obtained for the concrete slab.

	Photo	**Laser 1**	**Laser 2**
Imported geometry			
Height (average)	1090 mm	1070 mm	1030 mm
Width (average)	881 mm	849 mm	820 mm
Depth (average)	150 mm	150 mm	150 mm
Structural mesh			
Mesh elements	143'329	161'340	154'350
Mesh nodes	143'242	153'206	92'939
Mesh gen. timing	108 s	283 s	500 s

From a geometrical standpoint, the results of Table 1 show that the various models may carry out discrepancies in the dimensions of almost 6 cm among them, both in height and in width (remember that, to facilitate the comparisons, the models have been scaled by imposing the same average depth). The maximum percentage discrepancies are 7%, for the width, and 6%, for the height.

To evaluate the structural performance of the relevant three meshes, two static and one dynamic analyses were performed, Table 2 and Table 3 (elastic parameters are 32.31 GPa and 0.20, for the Young's modulus and the coefficient of Poisson, respectively, and density is 24 kN/m^3):

- the static analyses were conducted considering only the self-weight, the first, and considering the self-weight plus a uniformly distributed load of 1 MPa applied to the upper face, the second. The bottom face was constrained preventing the vertical displacement direction; to enable rigid motions, but allowing the Poison's effect, two vertices of the same face were also respectively constrained with a spherical hinge and a roller preventing the displacement along the short edge;
- the dynamic simulation was a modal analysis. In this case the slab was constrained preventing on the bottom face all the displacements.

Table 2 - Linear static analyses: resultant of the vertical reactions and maximum vertical displacement.

	Photo		**Laser 1**		**Laser 2**	
	Reaction	Displacement	Reaction	Displacement	Reaction	Displacement
Self-weight	3'539 N	$0.45 \cdot 10^{-3}$ mm	3'752 N	$0.45 \cdot 10^{-3}$ mm	3'318 N	$0.40 \cdot 10^{-3}$ mm
Self-weight + Load	137'773 N	$34.19 \cdot 10^{-3}$ mm	138'996 N	$32.63 \cdot 10^{-3}$ mm	144'984 N	$34.40 \cdot 10^{-3}$ mm

All the static analyses (Table 2) were completed in less than 2 minutes and showed results in fairly good agreement among them; the maximum percentage discrepancies (equal for reactions and displacements) are 13%, for the case considering only the self-weight, and 5%, for the case considering the self-weight and the applied load. These important discrepancies are mainly due to the differences obtained in terms of geometry.

Table 3 - Modal analyses: frequencies of the first ten mode-shapes.

Mode	**Photo**	**Laser 1**	**Laser 2**	Mode	**Photo**	**Laser 1**	**Laser 2**
1	6.1 Hz	6.7 Hz	6.7 Hz	6	95.0 Hz	101.8 Hz	100.1 Hz
2	19.0 Hz	21.5 Hz	21.1 Hz	7	96.9 Hz	106.0 Hz	103.1 Hz
3	28.2 Hz	28.9 Hz	28.5 Hz	8	126.4 Hz	131.6 Hz	130.0 Hz
4	56.0 Hz	63.5 Hz	61.9 Hz	9	131.0 Hz	144.5 Hz	140.5 Hz
5	86.3 Hz	89.4 Hz	88.4 Hz	10	143.3 Hz	151.6 Hz	147.8 Hz

All the modal analyses (Table 3) were finalized in less than 6 minutes and did not reveal any critical issue. Even the comparison among the frequencies of the first ten mode-shapes clearly highlights important discrepancies, up to 13%.

Conclusions

In this study two work pipelines have been proposed and tested for the automatic construction of finite element models starting from site surveys. The two different types of surveys taken into consideration are the photographic and the laser scanner. The software stack used to develop the pipelines belongs to the Autodesk environment; the software Cloud Compare has been also used to improve the filter phase.

Two cases studies were considered, a concrete slab and a small-scale steel frame. For the concrete slab, the LOD was set to even detect the roughness of the surfaces: despite the simplicity of the considered specimen, some important discrepancies among the results are present, up to 7% for the geometry, and up to 13% for the results of the mechanical (both static and dynamic) simulations. These results point out that, even for simple case study and controlled and supervised pipelines, the state-of-the-art techniques may provide significant discrepancies, thus justifying the need of further investigations and studies. More extended results (including those related to the small-scale steel frame) will be presented during the conference and published in due course.

References

[1] Y. Pan, A. Braun, A. Borrmann, I. Brilakis, Void-growing: a novel Scan-to-BIM method for Manhattan world buildings from point cloud, in European Conference on Computing in Construction, Online eConference, July 26-28 (2021). https://doi.org/10.35490/EC3.2021.162

[2] J.W. Ma, B. Han, F. Leite, An automated framework for generating synthetic point clouds from as-built BIM with semantic annotation for scan-to-BIM, in Winter Simulation Conference (WSC), IEEE (2021).

[3] Y. Alshawabkeh, A. Baik, Y. Miky, Integration of laser scanner and photogrammetry for heritage BIM enhancement, ISPRS International Journal of Geo-Information 10 (2021) 316.

[4] S. Mostafa, H. Villamor, R.A. Stewart, K. Sturm, E. Suprun, S. Vohland, Costs and benefits data mapping of BIM laser scan integration: a case study in Australia, in Joint International Conference on Design and Construction of Smart City Components, Springer, Cham (2019). https://doi.org/10.3390/ijgi10050316

[5] M.-K. Kim, Q. Wang, J.-W. Park, J.C.P. Cheng, H. Sohn, C.-C. Chang, Automated dimensional quality assurance of full-scale precast concrete elements using laser scanning and BIM, Automation in construction 72 (2016) 102-114. https://doi.org/10.1016/j.autcon.2016.08.035

[6] N. Ham, S.-H. Lee, Empirical study on structural safety diagnosis of large-scale civil infrastructure using laser scanning and BIM, Sustainability 10 (2018) 4024. https://doi.org/10.3390/su10114024

Theoretical and Applied Mechanics - AIMETA 2022 Materials Research Forum LLC
Materials Research Proceedings **26** (2023) 257-262 https://doi.org/10.21741/9781644902431-42

An unconditionally stable time integration for the dynamics of elastic beams and shells in finite motions

MAGISANO Domenico[1,a*], LEONETTI Leonardo[1,b] and GARCEA Giovanni[1,c]

[1] Dipartimento di Ingegneria Informatica, Modellistica, Elettronica e Sistemistica Universita` della Calabria 87036 Rende (Cosenza), Italy

[a]domenico.magisano@unical.it, [b]leonardo.leonetti@unical.it, [c]giovanni.garcea@unical.it

Keywords: Nonlinear Dynamics, Large Deformation, Stability, Kirchhoff-Love

Abstract. This work presents a numerical framework for long dynamic simulations of structures made of multiple thin shells undergoing large deformations. The C1-continuity requirement of the Kirchhoff-Love theory is met in the interior of patches by cubic NURBS approximation functions with membrane locking avoided by patch-wise reduced integration. A simple penalty approach for coupling adjacent patches, applicable also to non-smooth interfaces and non-matching discretization is adopted to impose translational and rotational continuity. A time-stepping scheme is proposed to achieve energy conservation and unconditional stability for general nonlinear strain measures and penalty coupling terms, like the nonlinear rotational one for thin shells. The method is a modified mid-point rule with the internal forces evaluated using the average value of the stress at the step end-points and an integral mean of the strain-displacement tangent operator over the step computed by time integration points.

Introduction

One-step implicit time integration methods such as Newmark's schemes are only conditionally stable when used in large deformation analyses [1]. Simo and Tarnow proposed a simple method that guarantees unconditional stability by conserving the algorithmic energy in elastodynamics [2]. However, energy conservation is lost for other structural models as the Kirchhoff-Love theory, more efficient in the terms of spatial DOFs for thin shell problems, where the strain-displacement relationship is no longer quadratic. This work presents a numerical framework for long term dynamic simulations of structures made of multiple thin shells undergoing large deformations. The C1-continuity requirement of the Kirchhoff-Love theory is met in the interior of patches by cubic NURBS approximation functions, according to the isogeometric concept, with membrane locking avoided by patch-wise reduced integration [4]. A simple penalty approach for coupling adjacent patches, applicable to either smooth or non- smooth interfaces and either matching or non-matching discretizations is adopted to impose translational and rotational continuity [5]. The time-stepping scheme of Simo and Tarnow is generalized to achieve energy conservation for generally nonlinear strain measures and penalty coupling terms, like the nonlinear rotational one for thin shells. The method is based on a particular integral mean of the internal forces over the step, that includes Simo and Tarnow's method as a reduced quadrature rule, and has unconditional stability.

Large deformation dynamic problem with general nonlinear strain measure

Let us consider a generic elastic body characterized by a linear elastic constitutive law

$$\boldsymbol{\sigma} = \boldsymbol{C\varepsilon}$$

where $\boldsymbol{\sigma}$ is the vector collecting the stress/generalized stress components, $\boldsymbol{\varepsilon}$ is the vector collecting the strain/generalized strain components and C is the constitutive matrix. The strain is linked to the displacement field u by means of a differential operator generally nonlinear in u:

$$\boldsymbol{\varepsilon} = \boldsymbol{\varepsilon}(\boldsymbol{u}) \tag{1}$$

Applying a spatial discretization technique, for instance the finite element method or the isogeometric analysis, the displacement is approximated at element level as

$$\boldsymbol{u} = \mathbf{N}_{\mathrm{u}}\mathbf{u}_{\mathrm{e}}$$

where matrix $\mathbf{N}_{\mathrm{u}}$ collect the spatial shape functions. The strain energy of the body can be expressed as a sum of element contributions as $\Phi \equiv \sum_e \Phi_{\mathrm{e}}(\mathbf{u}_{\mathrm{e}})$, with

$$\Phi_{\mathrm{e}}(\mathbf{u}_{\mathrm{e}}) \equiv \int_{\Omega_e} \frac{1}{2}\boldsymbol{\varepsilon}^T\mathbf{C}\boldsymbol{\varepsilon}\, \mathrm{d}\Omega_{\mathrm{e}} \tag{2}$$

where Ω_{e} is the element domain. The first variation of the strains in Eq.(1) can be written as

$$\delta\boldsymbol{\varepsilon} = \mathbf{B}_{\mathrm{e}}\delta\mathbf{u}_{\mathrm{e}} \tag{3}$$

where matrix B is the strain-displacement tangent operator. The first variation of the element strain energy

$$\Phi'_{\mathrm{e}}(\mathbf{u}_{\mathrm{e}})\delta\mathbf{u}_{\mathrm{e}} \equiv \int_{\Omega_e} \delta\boldsymbol{\varepsilon}^T\mathbf{C}\boldsymbol{\varepsilon}\, \mathrm{d}\Omega_{\mathrm{e}} = \delta\mathbf{u}_{\mathrm{e}}^{\mathrm{T}} \int_{\Omega_e} \mathbf{B}_{\mathrm{e}}{}^T(\mathbf{u}_{\mathrm{e}})\boldsymbol{\sigma}(\mathbf{u}_{\mathrm{e}})\mathrm{d}\Omega_{\mathrm{e}} = \delta\mathbf{u}_{\mathrm{e}}^{\mathrm{T}}\boldsymbol{s}_e(\mathbf{u}_{\mathrm{e}}) \tag{4}$$

allows us to define the element internal force vector

$$\boldsymbol{s}_e(\mathbf{u}_{\mathrm{e}}) \equiv \int_{\Omega_e} \mathbf{B}_{\mathrm{e}}{}^T(\mathbf{u}_{\mathrm{e}})\boldsymbol{\sigma}(\mathbf{u}_{\mathrm{e}})\mathrm{d}\Omega_{\mathrm{e}} \quad \text{with} \quad \boldsymbol{\sigma}(\mathbf{u}_{\mathrm{e}}) = \mathbf{C}\boldsymbol{\varepsilon}(\mathbf{u}_{\mathrm{e}}) \tag{5}$$

Similarly, the velocity field v is approximated consistently as

$$\boldsymbol{v} = \mathbf{N}_{\mathrm{u}}\mathbf{v}_{\mathrm{e}}$$

where $\mathbf{v}_{\mathrm{e}}$ are the discrete velocity DOFs. The kinetic energy is then sum of element contributions $\mathrm{T} \equiv \sum_e \mathrm{T}_{\mathrm{e}}(\mathbf{v}_{\mathrm{e}})$

$$\mathrm{T}_e(\mathbf{v}_{\mathrm{e}}) \equiv \int_{\Omega_e} \frac{1}{2}\boldsymbol{\varepsilon}^T\Xi\boldsymbol{\varepsilon}\, \mathrm{d}\Omega_{\mathrm{e}} = \frac{1}{2}\boldsymbol{v}_e^T\boldsymbol{M}_e\boldsymbol{v}_e \tag{6}$$

where Ξ is a diagonal matrix with the mass density associated to each component of $\boldsymbol{v}$, which can be different for examples when v collects both translational and rotational velocities, and $\mathbf{M}_{\mathrm{e}}$ is the element mass matrix.

After a standard assemblage process, the semi-discrete equations of motion for the discretized body can be written, neglecting the damping, in terms of the global DOFs u and v as

$$\begin{cases} \mathbf{v} = \dot{\mathbf{u}} \\ \mathbf{M}\dot{\mathbf{v}} + \mathbf{s}(\mathbf{u}) = \mathbf{f} \end{cases} \tag{7}$$

where a dot denotes time derivative and **f** is the discrete load vector, while the total energy of the system is

$$\Pi(\mathbf{u},\mathbf{v}) \equiv T(\mathbf{v}) + \Phi(\mathbf{u}) - \mathbf{u}^{\mathrm{T}}\mathbf{f} \tag{8}$$

One-step time integration

Letting $\alpha = \frac{t-t_0}{t_1-t_0}$ a one-step time integration scheme can be obtained introducing the approximation in time for u and v over the time step $[t_0, t_1]$

$$\begin{cases} \mathbf{v}(\alpha) = \mathbf{v}_0 + \alpha(\mathbf{v}_1 - \mathbf{v}_0) \\ \mathbf{u}(\alpha) = \mathbf{u}_0 + \alpha(\mathbf{u}_1 - \mathbf{u}_0) \end{cases} \tag{9}$$

The semi-discrete equations of motion (7) can be then rewritten in discrete form as

$$\begin{cases} \overline{\mathbf{v}} = \frac{\mathbf{u}_1 - \mathbf{u}_0}{\Delta t} \\ \mathbf{M}\frac{\mathbf{v}_1 - \mathbf{v}_0}{\Delta t} + \overline{\mathbf{s}}(\mathbf{u}) = \overline{\mathbf{p}} \end{cases} \tag{10}$$

where $\overline{\mathbf{v}}, \overline{\mathbf{s}}$ and $\overline{\mathbf{p}}$ are representative value of $\mathbf{v}(\alpha)$, $\mathbf{s}(\alpha)$ and $\mathbf{p}(\alpha)$ over the step. For $\overline{\mathbf{v}}$ and $\overline{\mathbf{p}}$, a simple and natural choice is to define them as the integral mean of the corresponding function, so that we have

$$\begin{cases} \frac{\mathbf{v}_1+\mathbf{v}_0}{2} = \frac{\mathbf{u}_1-\mathbf{u}_0}{\Delta t} \\ \mathbf{M}\frac{\mathbf{v}_1-\mathbf{v}_0}{\Delta t} + \overline{\mathbf{s}}(\mathbf{u}) = \overline{\mathbf{p}} \quad \text{with } \mathbf{p} \equiv \int_0^1 p(\alpha)d\alpha \end{cases} \tag{10}$$

We can get $\mathbf{v}_1$ from the first of Eq. (11) and substitute it into the second, so obtaining the final form of the discrete equations of motion in terms of the only unknowns $\mathbf{u}_1$:

$$2\mathbf{M}\frac{\mathbf{u}_1+\mathbf{u}_0-\mathbf{v}_0\Delta t}{\Delta t^2} + \overline{\mathbf{s}}(\mathbf{u}) = -\mathbf{u}^{\mathrm{T}}\mathbf{f} \tag{8}$$

Different choices are possible for $\overline{\mathbf{s}}(\mathbf{u})$.

The new integration scheme

Now, we present a new one-step integration method aimed at preserving energy as well as the strain-displacement compatibility at the step end-point for generally nonlinear strain measures. Starting for the time interpolation in Eq.(9), the idea is to generalize the Simo and Tarnow method as follows. Firstly, at each spatial integration point, a representative value of the strain-displacement tangent operator is computed as integral mean of B(u) over the step:

$$\overline{\mathbf{B}} = \int_0^1 \mathbf{B}(\alpha)d\alpha \approx \sum_{n=1}^{N_t} \mathbf{B}(\mathbf{u}(\alpha_n))\,\overline{\omega}_n \tag{13}$$

Then the representative internal force vector of the step is evaluated as

$$\overline{\mathbf{s}}(\mathbf{u}) = \int_0^1 \overline{\mathbf{B}}^{\mathrm{T}}\boldsymbol{\sigma} dV \quad \text{with} \quad \boldsymbol{\sigma} = \mathbf{C}\frac{\varepsilon(u_1)+\varepsilon(u_0)}{2} \tag{14}$$

with α_n and $\overline{\omega}_n$ temporal position and weight of the nth time integration point respectively. So, the scheme is very similar to Simo and Tarnow's one [2]. The difference is the integral mean $\overline{\mathbf{B}}$ instead of the mean value $\mathbf{B}_{\frac{1}{2}}$. This choice is the crucial idea to achieve energy conservation for arbitrarily nonlinear strain-displacement laws. Indeed, it is easy to verify that

$$\boldsymbol{\varepsilon}_1 - \boldsymbol{\varepsilon}_0 = \Delta t\int_0^1 \dot{\boldsymbol{\varepsilon}}d\alpha = \Delta t\int_0^1 \mathbf{B}(\mathbf{u}(\alpha))\dot{\mathbf{u}}d\alpha = \overline{\mathbf{B}}\,(\mathbf{u}_1 - \mathbf{u}_0) \tag{15}$$

and then energy conservation holds independently of the strain measure nonlinearity:

$$\begin{aligned}(\mathbf{u}_1 - \mathbf{u}_0)^{\mathrm{T}}\overline{\mathbf{s}}(\mathbf{u}) &= \int_V(\mathbf{u}_1 - \mathbf{u}_0)^{\mathrm{T}}\,\overline{\mathbf{B}}^{\mathrm{T}}\,\mathbf{C}\frac{(\boldsymbol{\varepsilon}_1+\boldsymbol{\varepsilon}_0)}{2}dV \\ &= \frac{1}{2}\int_V\{\boldsymbol{\varepsilon}_1^{\mathrm{T}}\,\mathbf{C}\boldsymbol{\varepsilon}_1 - \boldsymbol{\varepsilon}_0^{\mathrm{T}}\,\mathbf{C}\boldsymbol{\varepsilon}_0\}dV = \Phi(\mathbf{u}_1) - \Phi(\mathbf{u}_1)\end{aligned} \tag{15}$$

Clearly, the energy conservation tends to be exact by increasing the number of time integration points in (13). For example the Simo and Tarnow method, that is only approximately conserving for nonlinear measures other than quadratic, can be seen as a reduced integration of the proposed method where the integral mean of **B** is approximated by a single Gauss point in time. Although the number of time integration point for a converged $\overline{\mathbf{B}}$ is a-priori unknown for general problems, a few time points, as shown in the numerical examples, gives energy conservation for practical applications and time steps. Compared to the full integration of $\mathbf{s}(\mathbf{u})$ discussed in [1], the number of time integration points required is usually lower, because only $\mathbf{B}(\mathbf{u})$ in now integrated instead of the more nonlinear term $\mathbf{B}(\mathbf{u})^{\mathrm{T}}\boldsymbol{\sigma}(\mathbf{u})$. More importantly, in our proposal the internal forces are evaluated using only the stress at the end-points of the step, avoiding the inaccuracies caused by the inner stress derived from the displacement. With respect to the Sansour et al. method [3], the proposal fulfill exactly the strain-displacement relation at the step end-points and results to be unconditionally stable.

Materials Research Proceedings **26** (2023) 257-262 https://doi.org/10.21741/9781644902431-42

Extension to nonlinear multi-body coupling laws

Multi-body coupling laws can be generically written as

$$\boldsymbol{\varepsilon}_c(\boldsymbol{u}) = \mathbf{0} \quad \forall \boldsymbol{x} \in \boldsymbol{\ell} \tag{17}$$

where ℓ denotes the boundary of the bodies where the coupling occurs. Without introducing additional DOFs for the Lagrange multipliers and upsetting the structure of existing finite element packages, general couplings expressed by Eq. (17) can be easily imposed in penalty form by adding to the strain energy of the system the penalty term

$$\Phi_c(\mathbf{u}_c) \equiv \int_\ell \frac{1}{2} \boldsymbol{\varepsilon}_c(\boldsymbol{u})^T \mathbf{C}_c \boldsymbol{\varepsilon}_c(\boldsymbol{u})\, \mathrm{ds} \tag{18}$$

where $\boldsymbol{\varepsilon}_c(\boldsymbol{u})$ is interpreted as a pseudo-strain that is constrained to negligible values by a stiff pseudo-constitutive matrix $\mathbf{C}_c = a\, \hat{\mathbf{C}}_c$ with matrix $\hat{\mathbf{C}}_c$ chosen on the basis of the actual materi- als and geometry of the coupled bodies [6], so that the penalty factor a defines how small we want the penalty energy to be with respect to strain energy of the system. For general structural problems and coupling laws, the constraint tends to be satisfied point-wise with the mesh refinement as in the Lagrange multipliers approach for a sufficiently high penalty factor. The penalty coupling is useful to model real complex structures with generic interfaces and non- matching discretizations [5]. The gradient of $\Phi_c(\mathbf{u}_c)$ provides the equivalent internal forces due to the coupling

$$\boldsymbol{s}_c(\mathbf{u}) \equiv \int_\ell \mathbf{B}_c{}^T(\mathbf{u}) \boldsymbol{\sigma}\, \mathrm{ds} \quad \text{with} \quad \boldsymbol{\sigma}_c = \mathbf{C}\boldsymbol{\varepsilon}_c(\mathbf{u}) \tag{19}$$

that has the same form of the element internal force vector with $\mathbf{B}_c(\mathbf{u})$ the tangent pseudo-strain/displacement operator. This is then assembled together with the element ones to obtain the global internal force vector to be used in global equations of motion Eq. (7). In this frame-
work, the time integration method developed in the previous subsection can be directly applied to problems with generally nonlinear multi-body couplings, guarantying energy conservation and unconditional stability.

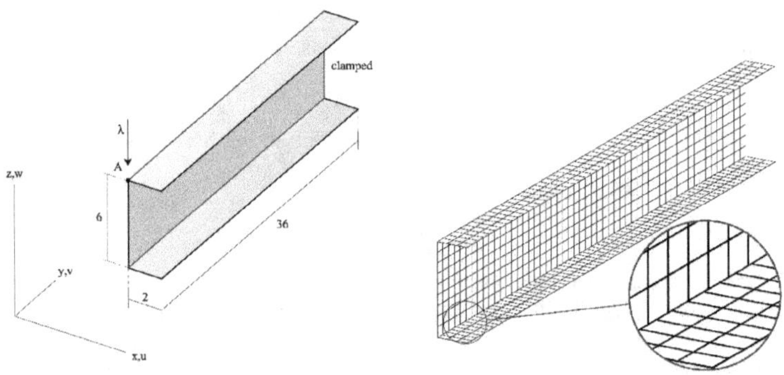

Figure 1: Thin-walled cantilever beam: geometry, boundary conditions and mesh

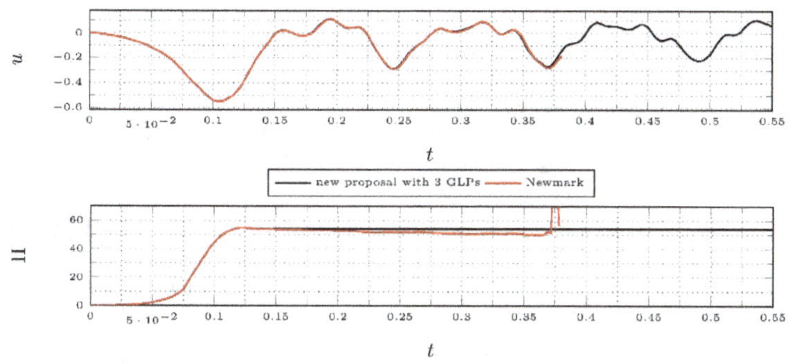

Figure 2: Thin-walled cantilever beam in composite material: Newmark vs new proposal

Numerical examples: thin-walled cantilever beam with local buckling

A thin-walled cantilever beam with a U cross section is considered in this test. Geometry, load, boundary conditions and mesh are reported in Fig. 1. Both isotropic and composite materials are considered. The isotropic material is characterized by $E = 1.0 \times 10^7$ and $\nu = 0.3$, while for the composite material we have $E_1 = 3.06 \times 10^7, E_2 = E_3 = 8.70 \times 10^6, \nu_{12} = 0.29$, $\nu_{23} = \nu_{13} = 0.3$ and $G_{12} = G_{13} = 3.24 \times 10^6, G_{23} = 2.90 \times 10^6$, with the material direction 1 corresponding to the longitudinal beam axis. The thickness of the walls is 0.05, while the density per unit of volume is 10^{-2}. The approximation is based on cubic NURBS functions. In such a structure, a patch coupling strategy is necessary for the Kirchhoff-Love shell model along the red interfaces. The load amplitude varies linearly from 0 to 150, a little higher value than the static buckling load, in 0.075 seconds. Afterwards, the load is removed linearly in 0.075 seconds and the simulation proceeds without load. The time step of the analysis is $\Delta t = 0.003125$ seconds. The results in terms of deflection of the loaded point and of total energy of the system are reported in Fig. 2 for the composite case. The Newmark method fails to preserve the energy after the load removal and its iterative solution fails when the energy oscillation becomes relevant. This is even more marked in the composite case with a dramatic blow-up. Instead, the proposal is perfectly stable with constant energy for a zero load and without any difficulty in the iterative solution.

Conclusions

A novel and very simple one-step time integration scheme for large deformation dynamics of elastic structures was presented. It is a generalization of the Simo and Tarnow method [2], designed to achieve unconditional stability for a quadratic strain, to arbitrarily nonlinear strain measures. The method is also suitable for models with finite 3D rotations, parametrized by the pseudo-rotation vector [9, 10].

References

[1] Betsch, P. and Steinmann, P. Conservation properties of a time FE method. Part I: time-stepping schemes for N-body problems. Int. J. Numer. Meth. Engng. (2000) 49: 599–638. https://doi.org/10.1002/1097-0207(20001020)49:5<599::AID-NME960>3.0.CO;2-9

[2] Simo, J.C. and Tarnow, N. The discrete energy-momentum method. Conserving algorithms for nonlinear elastodynamics. Z. angew. Math. Phys. (1992) 43: 757–792. https://doi.org/10.1007/BF00913408

[3] Sansour, C. and Wriggers, P. and Sansour, J. Nonlinear Dynamics of Shells: Theory, Finite Ele- ment Formulation, and Integration Schemes. Nonlinear Dynamics (1997) 17: 279–305. https://doi.org/10.1023/A:1008251113479

[4] Leonetti, L. and Magisano, D. and Madeo, A. and Garcea, G. and Kiendl, J and Reali, A. A sim- plified Kirchhoff–Love large deformation model for elastic shells and its effective isogeometric formulation. Computer Methods in Applied Mechanics and Engineering (2019) 354: 369–396. https://doi.org/10.1016/j.cma.2019.05.025

[5] Leonetti, L. and Liguori, F. S. and Magisano, D. and Kiendl, J and Reali, A. and Garcea, G. A robust penalty coupling of non-matching isogeometric Kirchhoff–Love shell patches in large deformations. Computer Methods in Applied Mechanics and Engineering (2020) 371: 113289. https://doi.org/10.1016/j.cma.2020.113289

[6] Austin J. Herrema, Emily L. Johnson, Davide Proserpio, Michael C.H. Wu, Josef Kiendl, Ming- Chen Hsu, Penalty coupling of non-matching isogeometric Kirchhoff–Love shell patches with application to composite wind turbine blades. Computer Methods in Applied Mechanics and En- gineering (2019) 346: 810–840. https://doi.org/10.1016/j.cma.2018.08.038

[7] Leonetti, L. and Liguori, F. S. and Magisano, D. and Garcea, G. An efficient isogeometric solid- shell formulation for geometrically nonlinear analysis of elastic shells. Computer Methods in Ap- plied Mechanics and Engineering (2018) 331: 159–183. https://doi.org/10.1016/j.cma.2017.11.025

[8] Magisano, D. and Leonetti, L. and Garcea, G. Unconditional stability in large deformation dynamic analysis of elastic structures with arbitrary nonlinear strain measure and multi-body cou- pling. Computer Methods in Applied Mechanics and Engineering (2022) 393: 114776. https://doi.org/10.1016/j.cma.2022.114776

[9] Magisano, D. and Leonetti, L. and Madeo, A. and Garcea, G. A large rotation finite element analysis of 3D beams by incremental rotation vector and exact strain measure with all the desirable features. Computer Methods in Applied Mechanics and Engineering (2020) 361: 112811. https://doi.org/10.1016/j.cma.2019.112811

[10] Magisano, D. and Leonetti, L. and Garcea, G. Isogeometric analysis of 3D beams for arbitrarily large rotations: Locking-free and path-independent solution without displacement DOFs inside the patch. Computer Methods in Applied Mechanics and Engineering (2021) 373: 113437. https://doi.org/10.1016/j.cma.2020.113437

Theoretical and Applied Mechanics - AIMETA 2022 Materials Research Forum LLC
Materials Research Proceedings **26** (2023) 263-266 https://doi.org/10.21741/9781644902431-43

Development of a multi-field computational tool for high-fidelity static aeroelastic simulations

Marco Grifò[1,a], Andrea Da Ronch[2,b], Alberto Milazzo[1,c], Ivano Benedetti[1,d*]

[1] Department of Engineering, University of Palermo, Viale delle Scienze, Edificio 8, 90128, Palermo, Italy

[2] Faculty of Engineering and Physical Sciences, University of Southampton, England, UK

[a]marco.grifo01@unipa.it, [b]A.Da-Ronch@soton.ac.uk, [c]alberto.milazzo@unipa.it, [d]ivano.benedetti@unipa.it

*Corresponding author

Keywords: Aeroelasticity, Composite Materials, FEM, CFD

Abstract. A new method for high-fidelity aeroelastic static analysis of composite laminated wings is proposed. The structural analysis and the fluid-dynamic analysis are coupled in a heterogeneous staggered process. The Finite Element Method (FEM), the Carrera Unified Formulation (CUF) and Equivalent Plate Modelling (EPM) are combined to model complex three-dimensional geometries in a bi-dimensional framework; Computational Fluid Dynamics (CFD) is employed to solve the Navier-Stokes equations and different turbulence models (i.e. Spalart-Allmaras) through SU2, an open-source software that implements C++ routines for 3D fluid-dynamics analysis. The Moving Least Square patch technique is adopted to manage the fluid-structure interaction. The use of an equivalent plate model, as opposite to 1D models often employed in the literature, shows competitive performances in terms of number of degrees of freedom. High-fidelity aerodynamics allows studying non-linear phenomena associated to irregularities of the fluid-structure interaction, showing a level of accuracy that low-fidelity methods such as Vortex Lattice Method (VLM) and Doublet Lattice Method (DLM) are unable to provide. Such advantages are balanced by the need to elaborate a staggered iterative method for the resolution of static aeroelastic problems, which leads to higher computational costs.

Introduction

Aeroelasticity has a fundamental role in many fields of engineering. Multiple strategies are adopted in the literature to investigate the behaviour of structures subjected to the action of a fluid, so to avoid catastrophic phenomena such as divergence or flutter. Besides providing insights into potentially catastrophic phenomena, aeroelastic methods allow to investigate the aerodynamic loads redistribution due to the structural deformation, thus providing a valuable tool for the assessment of the structural performance. When composite materials are involved, Aeroelastic Tailoring aims at selecting suitable material arrangements so to provide reasonable margins with respect to dangerous operating conditions. Such tasks generally require reliable computational models for representing the aero-structural interaction.

CUF [1] provides a general framework for the generation of variable order structural theories and it has been successfully employed for the generalised analysis of composite structures. In this work, EPM [2] is combined with CUF and FEM to generate variable order structural models for aeroelastic analysis.

The fluid properties and phenomena are studied through the CFD open-source software SU2 [3], from which pressure distribution is obtained. This high-fidelity approach prevents the loss of information inherent to low fidelity theories such as VLM or DLM, especially when the drag and

Materials Research Proceedings **26** (2023) 263-266
https://doi.org/10.21741/9781644902431-43

possible non-linearities coming from irregular geometries or high-speed regimes play important roles.

After a brief introduction to the main items of the developed framework, some representative results are reported to illustrate its scope and potential.

Theoretical Background

In this work an EPM+CUF+FEM approach is chosen to model generally complex composite structures within a static aeroelastic context. The strategy shows to be computationally competitive with respect to classical CUF beam models employed for aeroelastic analysis. In the CUF, the kinematic model may be represented as

$$\boldsymbol{u}(\hat{\boldsymbol{x}}) = \boldsymbol{F}(\hat{x}_3)\boldsymbol{U}(\hat{x}_1, \hat{x}_2) \tag{1}$$

where $\boldsymbol{u}(\hat{\boldsymbol{x}})$ collects the displacement components at the space point $\hat{\boldsymbol{x}} = (\hat{x}_1, \hat{x}_2, \hat{x}_3)$ which spans a three-dimensional structural domain, $\boldsymbol{F}(\hat{x}_3)$ is the matrix containing through-the-thickness functions and $\boldsymbol{U}(\hat{x}_1, \hat{x}_2)$ are the generalized displacement components.

In the proposed static aeroelastic framework, the kinematic model is employed in conjunction with the Principle of Virtual Displacements (PVDs) for addressing generally complex structural components of aeronautic interest, e.g. a wing with skins or spars or ribs. The volume integration needed for computing the stiffness contributions takes into account the material distribution of the considered structure, separating through-the-thickness integration from the integration over the reference plane, thus providing an equivalent plate representation of the analysed structure.

The CFD analysis is based on Reynolds-averaged Navier-Stokes equations and turbulence models such as the Spalart-Allmaras or $k - \omega$ models. SU2 solves these equations through a discretization of the 3D domain and applying the Finite Volumes Method once the freestream conditions (ρ_∞, V_∞) are given. If $\breve{\boldsymbol{x}} = (\breve{x}_1, \breve{x}_2, \breve{x}_3)$ is the three-dimensional CFD domain, tractions are defined by

$$\boldsymbol{t}(\breve{\boldsymbol{x}}) = \frac{1}{2}\rho_\infty V_\infty \boldsymbol{c}(\breve{\boldsymbol{x}})\boldsymbol{n}(\breve{\boldsymbol{x}}) \tag{2}$$

where $\boldsymbol{c}(\breve{\boldsymbol{x}})$ is the vector of pressure and skin friction coefficients c_p, $c_{f,\breve{x}_1}$ and $c_{f,\breve{x}_2}$ and $\boldsymbol{n}(\breve{\boldsymbol{x}})$ contains the normal vectors to the surface defined as a boundary wall.

These tractions are transferred to the structural model through the Moving Least Square patch technique, which is based on the conservation of energy and minimisation of the mean square error on two displacement fields defined over the aerodynamic and structural domains, weighted by radial basis functions.

$$\sum_{k=1}^{\hat{N}} \boldsymbol{F}_k (\delta\hat{\boldsymbol{U}})_k = \int_{\breve{\Omega}} -\boldsymbol{t}(\breve{\boldsymbol{x}}) \sum_{m=1}^{\breve{N}} S_i \sum_{k=1}^{\hat{N}} h_{ij} (\delta\hat{\boldsymbol{U}})_m \, d\boldsymbol{A} \tag{3}$$

$$\text{Minimise} \int_{\Gamma} \chi \left(\text{Tr}(\delta\breve{\boldsymbol{U}})|_{\Gamma} - \text{Tr}(\delta\hat{\boldsymbol{U}})|_{\Gamma}\right)^2 d\boldsymbol{A} \tag{4}$$

where $\breve{N}$ and $\hat{N}$ are the number of CFD mesh nodes and structural mesh nodes respectively (with indexes k and m), S_i are base functions for the aerodynamic surface domain approximation, h_{ij} is the single element of the interpolation matrix $\boldsymbol{H}$, χ is the radial basis function and Γ is the virtual surface domains on which the traces of $\delta\breve{\boldsymbol{U}}$ and $\delta\hat{\boldsymbol{U}}$ are projected.

The staggered iterative process for static aeroelastic analysis consists of the recalculation of pressure loads for every change in the geometry of the structure caused by deformation. $\boldsymbol{H}$ is employed to transfer information from the structural to aerodynamic domain and vice versa, and it is computed once and re-used in every staggered iteration. An overall schematic representation of the method is reported in Fig.(1).

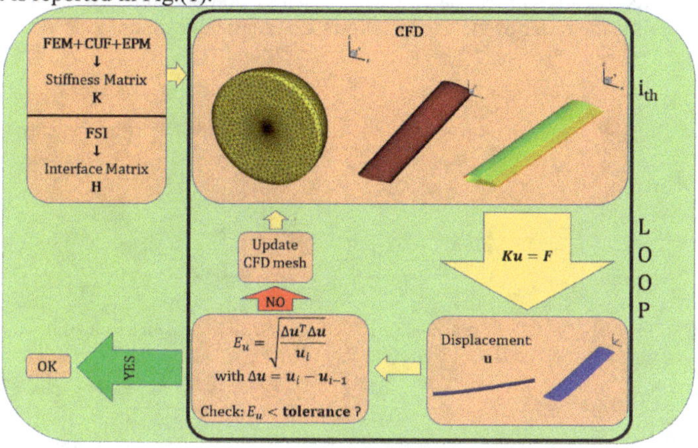

Figure 1: Scheme of the staggered iterative static aeroelastic analysis

Preliminary computational results

Some preliminary results are reported here. First, the EPM is assessed in terms of accuracy considering a torsional load scheme for a realistic composite wing configuration. Second, the validation of the static aeroelastic response of an isotropic wing with a NACA2415 airfoil transverse section and two spars is shown and compared to the solutions reported in Ref.[4].

Structural validation: torsion of a composite [0/90] laminate wing

A NACA2415 straight wing with span $b = 5$ m, chord $c = 1$ m, skin thickness of $t_{sk} = 0.04h$, being h the height of the airfoil, and two spars positioned at $x_{sp1} = 0.25c$ and $x_{sp2} = 0.75c$ respectively with thickness $t_{sp1} = 0.1h$ and $t_{sp2} = 0.07h$ is considered. The wing is subjected to a torsional load applied through a suitable fictitious distribution of pressures on portions of the lower and upper sides of the wing, as shown in Fig.(2). The value of the pressure distribution is $p = 1$ Pa. The results presented in Fig.(3a-3b) show good accuracy with respect to the results obtained from ABAQUS through $\sim 10^6$ quad-structured 2D shell elements. The error $e_{\%}$ is measured on the maximum displacement, located at the leading edge of the tip section, and on the tip-section twist $\Delta u_z = u_z(0, b, 0) - u_z(c, b, 0)$.

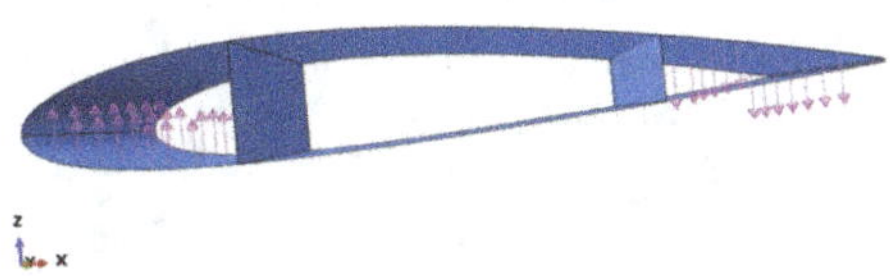

Figure 2: Load scheme for the structural validation.

Aeroelastic deformation of a straight wing in subsonic conditions

The same geometry from Section 1.1 is adopted for a static aeroelastic analysis with isotropic aluminium, considering a freestream at $V_\infty = 50\ \mathrm{m/s}$ in standard conditions. The result is compared with that reported in Ref.[4] and it shows a 7% difference due to the difference between the employed aerodynamic theories. In Fig.(4a-4b) the step values of $u_{z,max}$ and percentual error from the staggered iterative process are presented, with a comparison between $N_u = 1$ and $N_u = 3$, which represents the order of the expansion in the kinematic assumption.

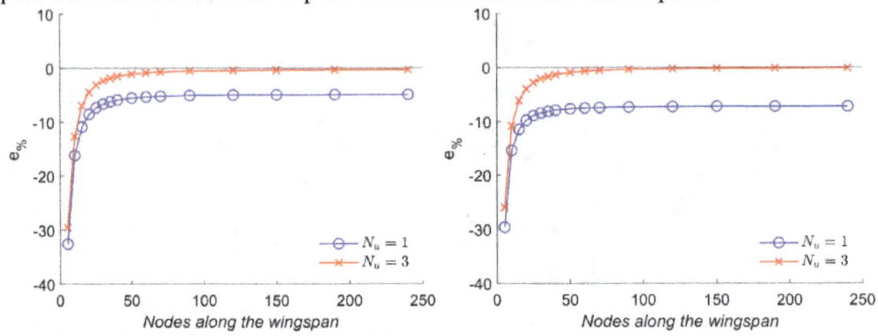

Figure 3: Results of structural validation for a torsional load on a composite wing

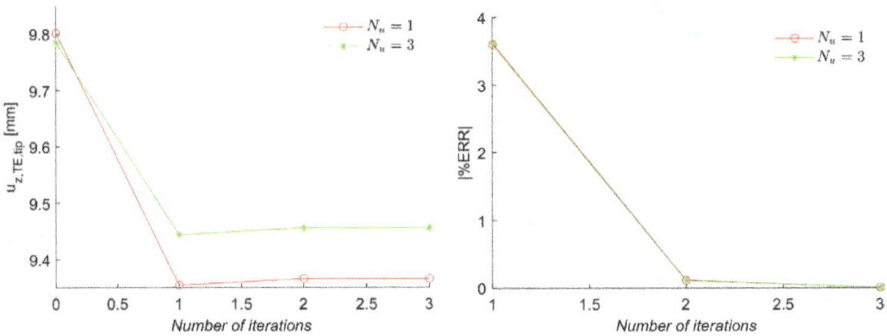

Figure 4: Convergence of the reference analysis variable in the staggered aeroelastic solution scheme

Conclusions

Some preliminary results from a new CUF+FEM+EPM+CFD approach for static aeroelastic analysis are presented, showing the flexibility of the proposed tool for the analysis of complex geometries and different materials. The main advantage of the method is the accuracy in the fluid-structure coupling, which allows the study of complex aerodynamic conditions without changing anything in the structural approach.

References

[1] E. Carrera, M. Cinefra, M. Petrolo, E. Zappino. "Finite Element Analysis of Structures through Unified Formulation". Wiley, 1st edition (2014). https://doi.org/10.1002/9781118536643

[2] R. Kapania, Y. Liu, Static and vibration analysis of general wing structures using equivalent plate models, in: 41st AIAA/ASME/ASCE/AHS/ASC Structures, Structural Dynamics and Materials Conference, 2000. https://doi.org/10.2514/6.2000-1434

[3] F. Palacios, T.D. Economon et al. Stanford University Unstructured (SU2): Open-source Analysis and Design Technology for Turbulent Flows in "Proceeding of *AIAA SciTech, Aerospace Sciences Meeting*", (2014). https://doi.org/10.2514/6.2014-0243

[4] E. Carrera, A. Varello, L. Demasi. A refined structural model for static aeroelastic response and divergence of metallic and composite wings. *CEAS Aeronaut J*, **4**, pp. 175-189 (2013). https://doi.org/10.1007/s13272-013-0063-2

Theoretical and Applied Mechanics - AIMETA 2022 Materials Research Forum LLC
Materials Research Proceedings **26** (2023) 267-272 https://doi.org/10.21741/9781644902431-44

Numerical modeling of dynamic crack propagation mechanisms using a moving mesh technique based on the ALE formulation

Arturo Pascuzzo[1,a*], Fabrizio Greco[1,b], Paolo Lonetti[1,c], Domenico Ammendolea[1,d], Giulia Sansone[1,e]

[1]Departement of Civil Engineering, University of Calabria, Via P. Bucci, Cubo 39B, 87036, Rende, Cosenza, Italy

[a]arturo.pascuzzo@unical.it, [b]fabrizio.greco@unical.it, [c]paolo.lonetti@unical.it, [d]domenico.ammendolea@unical.it, [e]giulia.sansone@unical.it

Keywords: Crack Propagation, Moving Mesh Technique, Finite Element Method

Abstract. This work proposes a new FE model to predict dynamic crack propagation mechanisms in quasi-brittle materials. The numerical strategy uses a Moving Mesh (MM) technique consistent with the Arbitrary Lagrangian-Eulerian (ALE) formulation to reproduce the variation of the geometry of the computational domain caused by dynamically growing cracks. Specifically, the motion of the mesh nodes takes place consistently with conditions dictated by classic Fracture Mechanics, which provide conditions concerning the direction of propagation and the velocity of advancing cracks. As a remarkable key novelty, the proposed method introduces the ALE formulation of the M-integral for extracting Dynamic Stress Intensity Factors (DSIFs) at a moving crack front. This formulation allows extracting fracture variables on deforming elements without losing accuracy. Comparisons with analytical and numerical data are proposed to assess the validity and efficiency of the proposed strategy.

Introduction

Recent developments in computational fracture mechanics have led to a proliferation of numerical methodologies for analyzing the failure behavior of most materials typically used in civil engineering and mechanical applications. Many numerical methodologies have been developed in the context of the Finite Element method because of its widespread diffusion and flexibility in modeling complex structures (see, for instance, [1-10]). Unfortunately, each method presents advantages and disadvantages, so it is relatively complicated to identify the best one.

Currently, there is an increasing request for numerical methods that combine high accuracy and computational efficiency [11, 12].

Consistently with such a necessity, this work presents an effective numerical model for reproducing dynamic fracture phenomena in quasi-brittle materials. The proposed model consists of an FE code enhanced by a Moving Mesh (MM) technique based on the Arbitrary Lagrangian-Eulerian (ALE) formulation [13, 14]. Specifically, the MM is used to reproduce the geometric variation of the computational domain because of the dynamically advancing cracks. Besides, it offers a reliable evaluation of key fracture variables (*i.e.*, Dynamic Stress Intensity Factors (DSIFs)) owing to the use of the Interaction Integral method (*i.e.*, *M*-integral). In particular, this work proposes as a key novelty the ALE formulation of the dynamic *M*-integral, which permits using the Interaction integral in the context of the ALE formulation.

The validity of the present approach is assessed by comparing the proposed method's previsions with experimental data and numerical results available in the literature.

Theoretical background

Arbitrary Lagrangian-Eulerian formulation

Traditional problems of continuum mechanics are formulated either in a spatial or a material system of coordinates, denoted as R_x and R_X, respectively. The ALE formulation uses a third

Materials Research Proceedings **26** (2023) 267-272 https://doi.org/10.21741/9781644902431-44

coordinate system, that is the mesh (or referential) coordinates system R_χ, in which the coordinate $\boldsymbol{\chi}$ identifies the nodes of the computational mesh. A proper set of mapping functions serve to link R_x, R_X, and R_χ (see Fig. 1). The use of the ALE formulation requires that the governing equations of the problem must be expressed in R_χ instead of R_X. According to the ALE, the gradient of a vectorial field $\mathbf{v}^M(\mathbf{X},t)$ in the material domain can be stated in the referential one using the following expression:

$$\nabla^M \mathbf{v}^M = \frac{\partial \mathbf{v}^M}{\partial \mathbf{X}} = \frac{\partial \mathbf{v}^R}{\partial \boldsymbol{\chi}} \frac{\partial \boldsymbol{\chi}}{\partial \boldsymbol{\Psi}} = \nabla^R \mathbf{v}^R \mathbf{J}_{\boldsymbol{\Psi}}^{-1} \tag{1}$$

where $\nabla^M(\cdot) = \partial(\cdot)/\partial\mathbf{X}$, $\nabla^R(\cdot) = \partial(\cdot)/\partial\boldsymbol{\chi}$, and $\mathbf{J}_{\boldsymbol{\Psi}}$ is the Jacobian of the mapping function $\boldsymbol{\Psi}(\boldsymbol{\chi},t)$.

Governing equations

The governing equations at the base of the proposed method are twofold: the fundamental equations of solid mechanics and those concerned with the MM problem. In particular, the latter is consistent with the Laplace regularization approach, which smoothly moves mesh nodes, avoiding excessive distortions for the Finite Element. Besides, additional equations of Fracture Mechanics are included for (i) identifying crack onset conditions, (ii) evaluating the direction of propagation, and (iii) quantifying the velocity of the crack tip.

The Interaction Integral method

The Interaction Integral method, also known as *M*-integral, is a strategy used in numerical methods to extract DSIFs at the crack front. The M-integral expression derives from the J-integral applied to a superimposed state formed by two admissible fields. In particular, the first regards the problem under investigation (denoted as the actual state), while the second is an auxiliary state (aux) with DSIFs known. The expression of the *M*-integral assumes the following form:

$$\begin{aligned} M = &\int_A \left[\left(\sigma_{ij}^{act} u_{j,1}^{aux} + \sigma_{ij}^{aux} u_{j,1}^{act} \right) - \left(\sigma_{ij}^{act} \varepsilon_{ij}^{aux} + \rho \dot{u}_j^{act} \dot{u}_j^{aux} \right) \delta_{1i} \right] q_{,i} dA + \\ &+ \int_A \left[\rho \left(\ddot{u}_j^{act} u_{j,1}^{aux} + \ddot{u}_j^{aux} u_{j,1}^{act} \right) - \rho \left(\dot{u}_{j,1}^{act} \dot{u}_j^{aux} + \dot{u}_{j,1}^{aux} \dot{u}_j^{act} \right) \right] q dA \end{aligned} \tag{2}$$

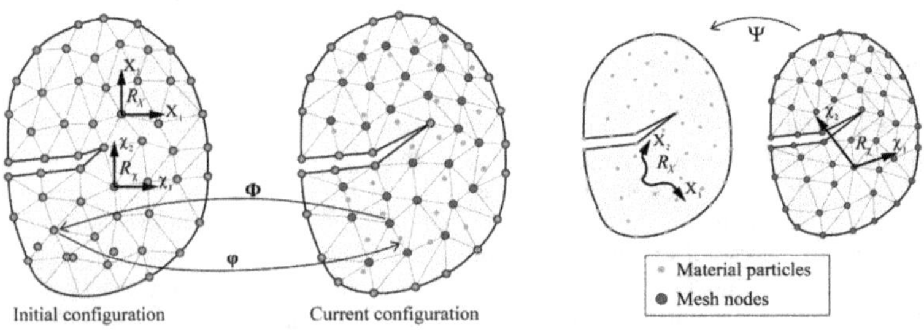

Fig. 1 – A schematic representation of Spatial, Material, and Referential coordinate systems.

Numerical implementation

The proposed strategy has been implemented in COMSOL Multiphysics [15], a commercially available software. COMSOL offers a helpful tool that connect it with MATLAB software, thereby managing COMSOL functions through script codes [16, 17]. This feature has served to develop a user-made script that manages the propagation process automatically. The fundamental steps of the script are reported in Table 1. For further clarification, see [18-20].

Table 1 – Main steps of the MATLAB script

START	1.	Set data of the problem: Geometry, material properties and boundary conditions
	2.	Define stretching segment and initial mesh configuration
	3.	Loading process: Evaluate DSIFs and assess crack initiation conditions
	4.	Propagation process: Evaluate crack tip velocity and check mesh quality
STOP	5.	Check tolerance to the angle variation, crack arrest or collapse conditions

Results

Fig. 2-a shows a rectangular pre-cracked plate of length L=10 m, height 2H=4 m, and thickness b=1 m. The plate is subjected to a uniform and distributed traction σ_0 = 500 MPa at the upper boundary. Externally, line constraints on vertical boundaries limit the horizontal displacements of the plate against horizontal displacements. The Young's Modulus, Poisson's ratio, and mass density are equal to E=210 GPa, ν=0.3, ρ=8000 kg/m^3, respectively. Besides, the dilatational and Rayleigh wave speeds are c_d=5944.5 m/s and c_r=2942.8 m/s, respectively.

Fig. 2-b depicts the mesh configuration used in the analysis. It comprises 432 triangular elements densely arranged around the crack tip and somewhat coarser in the remaining zones.

The analytical solutions associated with this fracture problem concerning a stationary crack and a moving crack have been developed by Freund [21]. In particular, the mode I DSIF analytical expressions gained by Freund for stationary and moving cracks assume these form:

$$K_I(0,t)=\frac{2\sigma_0}{1-\nu}\sqrt{\frac{c_d(t-t_c)(1-2\nu)}{\pi}}\quad \text{(Stationary crack)}$$
$$K_I(\dot{a},t)=\frac{2\sigma_0}{1-\nu}\sqrt{\frac{c_d(t-t_c)(1-2\nu)}{\pi}}\,\frac{1-\dot{a}/c_r}{1-\dot{a}/(2c_r)}\quad \text{(Moving crack)} \tag{3}$$

Note that such solutions are valid until the stress wave generated by the action of the external loads on the upper boundary of the plate, and reflected by the bottom boundary, reaches the crack tip. This behavior is experienced in the interval of time estimated to be $0<t<3t_c = 3H/c_d$.

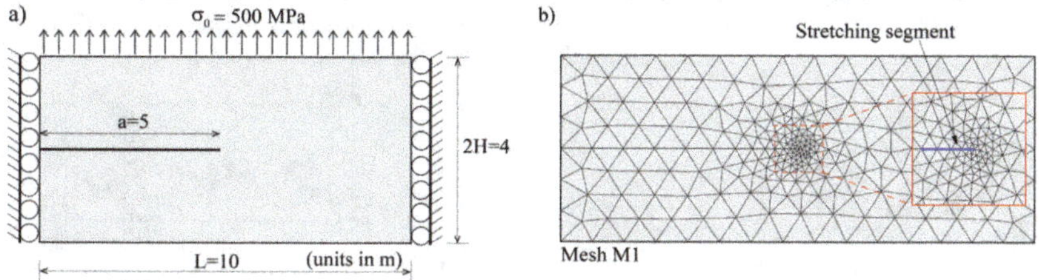

Fig. 2 – (a) Geometry and boundary conditions; (b) Mesh configuration adopted in numerical simulations.

Besides analytical solutions developed by Freund, there are several numerical results in the literature. Among these, Menouillard et al. [22] and Chen et al. [23] have investigated the fracture behavior of the plate using an XFEM strategy and an advanced numerical procedure based on the Singular Edge-based Smoothed Finite Element Method (SE-FEM), respectively.

In order to assess the reliability of the proposed model, it is supposed a failure behavior of the plate characterized by an initial stationary crack that starts moving for t/t_c>1.5.

Fig. 3-a compares the normalized Mode-I DSIFs K_I/K_0 ($K_0 = \sigma_0\sqrt{H}$) versus normalized time t/t_c predicted by the proposed method with the analytical solution of Freund and numerical results of Menouillard et al. [22] and Chen et al. [23]. The results denote that the proposed method agrees well with both analytical and numerical solutions since no appreciable differences in the curves of K_I are observed.

A parametric study in terms of mesh configurations is now performed to assess the computational efficiency of the numerical model. In particular, the study compares the mesh in Fig. 2-b (Mesh M1), with the Mesh M2 (entirely coarse with 175 triangular elements) and the Mesh M3 (refined configuration of 1113 triangular elements) reported in Fig. 4-b.

Fig. 4-a shows that the mesh discretization does not influence Mode-I DSIFs results as there are no significant differences between the curves. In addition, Fig. 4-a reports the percentage error (ERR) between numerical results and analytical solutions defined as follows:

$$\mathrm{ERR}(\%) = \frac{\left|\mathrm{K}_{\mathrm{I}}^{\mathrm{Analytical}} - \mathrm{K}_{\mathrm{I}}^{\mathrm{Numerical}}\right|}{\mathrm{K}_{\mathrm{I}}^{\mathrm{Analytical}}} \tag{4}$$

From the curves of ERR, it is possible to note that the proposed approach extracts K_I quite well for each mesh configuration because the percentage variation is less than 10%, even if peaks that exceed this threshold just before a re-meshing event occurs.

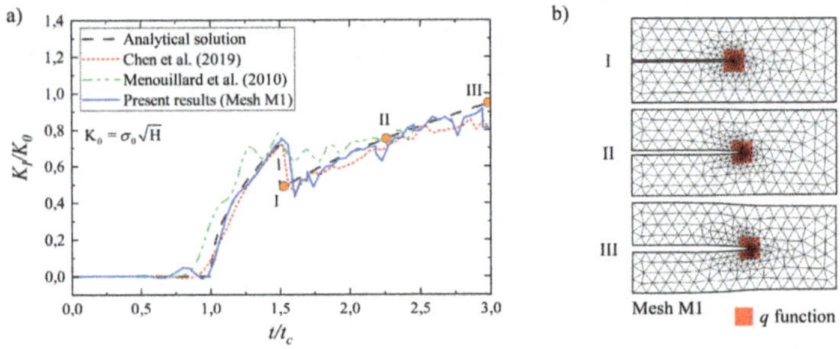

Fig. 3 – (a) Comparison in terms of normalized Mode-I DSIF between the analytical solutions proposed by Freund [21], the numerical results obtained by Chen et al. [23] and Menouillard et al. [22] and the present approach; (b) A schematic of the crack propagation process simulated by the proposed approach.

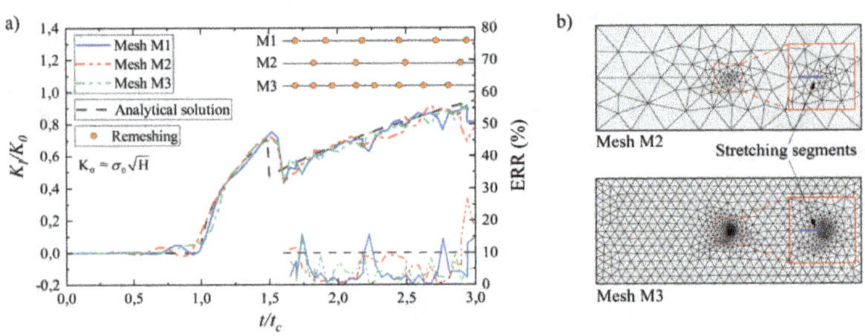

Fig. 4 – (a) Comparison in terms of normalized Mode-I DSIF and percentage error between numerical results and analytical solutions; (b) Mesh configurations used in the parametric study.

Conclusions

This work has proposed a new FE modeling strategy for reproducing dynamic crack propagation phenomena in quasi-brittle materials. The results denote that the proposed model ensures accurate fracture variable prediction, thus representing a powerful numerical tool for analyzing the failure behavior of quasi-brittle materials under the action of dynamic loadings.

Acknowledgements

Paolo Lonetti acknowledges financial support from the Italian Ministry of Education, University and Research (MIUR) under the P.R.I.N. 2020 National Grant "Opportunities and challenges of nanotechnology in advanced and green construction materials" (Project Code 2020EBLPLS; University of Calabria Research Unit). Arturo Pascuzzo acknowledges financial support from the (MIUR) under the National Grant "PON R&I 2014-2020, Attraction and International Mobility (AIM)", Project n° AIM1810287, University of Calabria". Domenico Ammendolea gratefully acknowledges financial support from "Programma Operativo Regione (POR) Calabria FESR-FSE 2014/2020". Giulia Sansone acknowledges financial support from the (MIUR) under the National Grant "PON R&I 2014-2020, Action IV.4" – M.D. n°. 1062 of 10/08/2021.

References

[1] A. Pranno, F. Greco, P. Lonetti, R. Luciano, U. De Maio, An improved fracture approach to investigate the degradation of vibration characteristics for reinforced concrete beams under progressive damage, International Journal of Fatigue 163 (2022) 107032. https://doi.org/10.1016/j.ijfatigue.2022.107032

[2] U. De Maio, F. Greco, L. Leonetti, P. Nevone Blasi, A. Pranno, A cohesive fracture model for predicting crack spacing and crack width in reinforced concrete structures, Engineering Failure Analysis 139 (2022) 106452. https://doi.org/10.1016/j.engfailanal.2022.106452

[3] U. De Maio, F. Greco, L. Leonetti, P. Nevone Blasi, A. Pranno, An investigation about debonding mechanisms in FRP-strengthened RC structural elements by using a cohesive/volumetric modeling technique, Theoretical and Applied Fracture Mechanics 117 (2022) 103199. https://doi.org/10.1016/j.tafmec.2021.103199

[4] U. De Maio, N. Fantuzzi, F. Greco, L. Leonetti, A. Pranno, Failure Analysis of Ultra High-Performance Fiber-Reinforced Concrete Structures Enhanced with Nanomaterials by Using a Diffuse Cohesive Interface Approach, 10(9) (2020) 1792. https://doi.org/10.3390/nano10091792

[5] A. Pranno, F. Greco, L. Leonetti, P. Lonetti, R. Luciano, U. De Maio, Band gap tuning through microscopic instabilities of compressively loaded lightened nacre-like composite metamaterials, Composite Structures 282 (2022) 115032. https://doi.org/10.1016/j.compstruct.2021.115032

[6] A. Pascuzzo, F. Greco, L. Leonetti, P. Lonetti, A. Pranno, C. Ronchei, Investigation of mesh dependency issues in the simulation of crack propagation in quasi-brittle materials by using a diffuse interface modeling approach, 45(3) (2022) 801-820. https://doi.org/10.1111/ffe.13635

[7] F. Greco, L. Leonetti, R. Luciano, A. Pascuzzo, C. Ronchei, A detailed micro-model for brick masonry structures based on a diffuse cohesive-frictional interface fracture approach, Procedia Structural Integrity 25 (2020) 334-347. https://doi.org/10.1016/j.prostr.2020.04.038

[8] F. Greco, D. Gaetano, L. Leonetti, P. Lonetti, A. Pascuzzo, Structural and seismic vulnerability assessment of the Santa Maria Assunta Cathedral in Catanzaro (Italy): classical and advanced approaches for the analysis of local and global failure mechanisms, Frat. Integrita Strutr. 16(60) (2022). https://doi.org/10.3221/IGF-ESIS.60.32

[9] U. De Maio, D. Cendón, F. Greco, L. Leonetti, P. Nevone Blasi, J. Planas, Investigation of concrete cracking phenomena by using cohesive fracture-based techniques: A comparison

between an embedded crack model and a refined diffuse interface model, Theoretical and Applied Fracture Mechanics 115 (2021) 103062. https://doi.org/10.1016/j.tafmec.2021.103062

[10] D. Bruno, F. Greco, P. Lonetti, Computation of Energy Release Rate and Mode Separation in Delaminated Composite Plates by Using Plate and Interface Variables, Mechanics of Advanced Materials and Structures 12(4) (2005) 285-304. https://doi.org/10.1080/15376490590953563

[11] M. Pingaro, E. Reccia, P. Trovalusci, R. Masiani, Fast statistical homogenization procedure (FSHP) for particle random composites using virtual element method, Computational Mechanics 64(1) (2019) 197-210. https://doi.org/10.1007/s00466-018-1665-7

[12] M. Pingaro, E. Reccia, P. Trovalusci, Homogenization of Random Porous Materials With Low-Order Virtual Elements, ASCE-ASME J Risk and Uncert in Engrg Sys Part B Mech Engrg 5(3) (2019). https://doi.org/10.1115/1.4043475

[13] F. Greco, P. Lonetti, A. Pascuzzo, A moving mesh FE methodology for vehicle–bridge interaction modeling, Mechanics of Advanced Materials and Structures 27(14) (2020) 1256-1268. https://doi.org/10.1080/15376494.2018.1506955

[14] F. Greco, L. Leonetti, P. Lonetti, A novel approach based on ALE and delamination fracture mechanics for multilayered composite beams, Composites Part B: Engineering 78 (2015) 447-458. https://doi.org/10.1016/j.compositesb.2015.04.004

[15] COMSOL, COMSOL Multiphysics® v. 5.4, Stockholm, Sweden, 2018.

[16] P. Lonetti, A. Pascuzzo, Design analysis of the optimum configuration of self-anchored cable-stayed suspension bridges, Struct Eng Mech 51(5) (2014) 847-866. https://doi.org/10.12989/sem.2014.51.5.847

[17] D. Bruno, P. Lonetti, A. Pascuzzo, An optimization model for the design of network arch bridges, Computers & Structures 170 (2016) 13-25. https://doi.org/10.1016/j.compstruc.2016.03.011

[18] F. Greco, D. Ammendolea, P. Lonetti, A. Pascuzzo, Crack propagation under thermo-mechanical loadings based on moving mesh strategy, Theoretical and Applied Fracture Mechanics 114 (2021) 103033. https://doi.org/10.1016/j.tafmec.2021.103033

[19] D. Ammendolea, F. Greco, P. Lonetti, R. Luciano, A. Pascuzzo, Crack propagation modeling in functionally graded materials using Moving Mesh technique and interaction integral approach, Composite Structures 269 (2021) 114005. https://doi.org/10.1016/j.compstruct.2021.114005

[20] A. Pascuzzo, F. Greco, P. Lonetti, D. Ammendolea, Dynamic fracture analysis in quasi-brittle materials via a finite element approach based on the combination of the ALE formulation and M-integral method, Engineering Failure Analysis (2022) 106627. https://doi.org/10.1016/j.engfailanal.2022.106627

[21] L.B. Freund, Dynamic fracture mechanics, Cambridge university press1998.

[22] T. Menouillard, J.-H. Song, Q. Duan, T. Belytschko, Time dependent crack tip enrichment for dynamic crack propagation, International Journal of Fracture 162 (2010) 33-49. https://doi.org/10.1007/s10704-009-9405-9

[23] H. Chen, Q. Wang, W. Zeng, G.R. Liu, J. Sun, L. He, T.Q. Bui, Dynamic brittle crack propagation modeling using singular edge-based smoothed finite element method with local mesh rezoning, European Journal of Mechanics - A/Solids 76 (2019) 208-223. https://doi.org/10.1016/j.euromechsol.2019.04.010

Theoretical and applied biomechanics

Theoretical and Applied Mechanics - AIMETA 2022 Materials Research Forum LLC
Materials Research Proceedings **26** (2023) 275-280 https://doi.org/10.21741/9781644902431-45

A 3-year follow-up study on bone structure elastic quality

Francesca Cosmi[1, a *], Simona Gentile[1,b] and Sergio Carrato[1,c]

[1]Dipartimento di Ingegneria e Architettura, Università degli Studi di Trieste , Trieste, Italy

[a*]cosmi@units.it, [b]simonagentile214@gmail.com, [c]carrato@units.it

Keywords: Bone, Osteoporosis, Medical Devices, BES TEST®

Abstract. Osteoporosis is called a silent disease because bone fragility manifests itself to the patient only in an advanced state, through fracture and pain. Medical and industry leaders recognize that the current golden standard diagnostic method, densitometry, or Dual Energy X-ray Absorptiometry (DEXA), may not always be sufficient to assess the patient's real risk of fragility fracture [1]. Indeed, pathological alterations affect not only the mineral content (quantity) of the bone, but also its "quality", which can be measured from the elastic properties of the bone internal trabecular structure by means of the Bone Elastic Structure Test, BES TEST®. In this study, the incidence of fragility fractures was assessed after a 3-year follow-up period in the women enrolled for a population study in 2015. The BES TEST® resulted an effective estimator of bone health, and can improve the assessment of the patient's fracture risk map.

Introduction

Osteoporosis is called a silent disease, due to the fact that bone fragility often manifests itself to the patient in an advanced state, through a major fracture (such as that of femur or wrist) or pain due to trabecular micro-fractures (often in the vertebrae).

The current golden standard diagnostic method is the densitometry, or Dual Energy X-ray Absorptiometry (DEXA). The DEXA results are usually expressed in terms of T-score, a statistical value indicating the number of standard deviations below the average value of young Caucasian women. According to WHO [2], there are four general diagnostic categories for women:

- *Normal.* A value of BMD within 1 standard deviation of the young adult reference mean (T-score >= -1).
- *Low bone mass (osteopenia).* A value of BMD 1 or more standard deviations below the young adult mean, but less than 2.5 standard deviations below this value (-2.5 < T-score < -1).
- *Osteoporosis.* A value of BMD 2.5 or more standard deviations below the young adult mean (T-score <= -2.5).
- *Severe osteoporosis (established osteoporosis).* A value of BMD 2.5 or more standard deviations below the young adult mean value in the presence of one or more fragility fractures.

However, medical and industry leaders recognize that the DEXA examination may not always be sufficient to assess the patient's real risk of fragility fracture: 62% of fragility fractures that could be prevented fail to be diagnosed in time [1]. Indeed, bone alterations are not usually detected at an early stage by DEXA scans, as they firstly affect the elastic properties of the internal structure of the bone [3]. This translates into a cost of care on health budgets that is far from negligible: in 2010 the EU6 countries (France, Germany, Italy, Spain, Sweden, and UK) spent € 37 billion, and this cost is projected to reach €47 billion by 2030 [4].

The Bone Elastic Structure Test, BES TEST®, is a software medical device for the assessment of the bone micro-architecture elasticity, i.e. its ability to withstand loads. It is registered by the Italian Ministry of Health and has CE mark and ISO 13485 qualification.

The purpose of this study is to investigate the effectiveness of BES TEST® in the identification of patients at risk for incident osteoporotic fractures.

Theoretical and Applied Mechanics - AIMETA 2022 Materials Research Forum LLC
Materials Research Proceedings **26** (2023) 275-280 https://doi.org/10.21741/9781644902431-45

Methods

BES TEST®

The test is based on an application of the Cell Method, a direct discrete numerical method, effective from the point of view of computation time, memory requirements and accuracy of the results [5, 6]. A radiographic virtual biopsy of the patient, acquired in the proximal phalanges of the non-dominant hand, is converted into a structural model and the response to compressive loads along the orthogonal axes is computed.

The simulation results are synthetized in a numerical score, which combines the elastic response of the reconstructed structure in several directions, purified of the normalized sum of gray tones, indicative of mineralization in the region under examination: the *Bone Structure Index* (*BSI*) summarizes the elastic behavior of the trabecular structure in different directions and, therefore, the ability to withstand the loads of the trabecular part of the bone.

The BES TEST® results are not correlated with the DEXA ones [7], and its precision parameters are in line with the golden standard requirements [8]. The test has shown potential in a variety of fields like rheumatology [9], oncology [10] and nephrology [11], in which clinical trials are currently being carried on. The procedure is schematically represented in Fig.1.

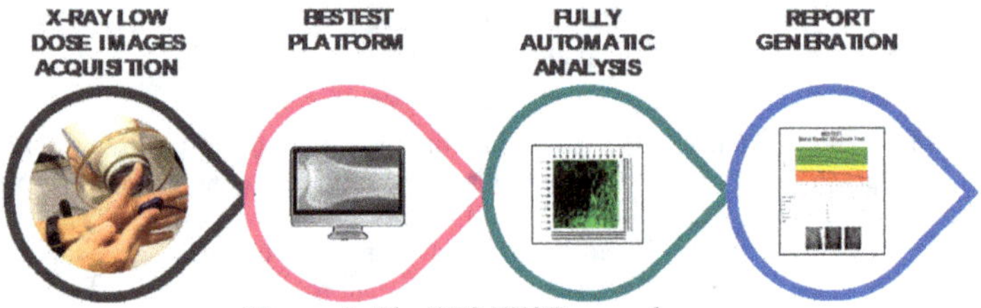

Figure 1. *The BES TEST procedure.*

Analogously to DEXA, the BESTEST results are presented to the doctor also in terms of *BSI_T-score*, which compares the patient's *BSI* with the mean value for young Caucasian women (age 20-45) and measures this difference in number of standard deviations [7].

Given the statistical nature of the T-score parameter, the same guidelines that are normally used in assessing bone density can be used for *BSI_T-score*, with reference to the clinical assessment of bone quality:

- A T *BSI_T-score* = 0 indicates that a patient's *BSI* is equal to the average *BSI* found in young Caucasian women.
- A *BSI_T-score* = -1 or above indicates that a patient's bone quality is normal.
- A T *BSI_T-score* between -1 and -2.5 indicates a first level of bone elasticity deficiency.
- A *BSI_T-score* < -2.5 indicates a second level of bone elastic deficiency.

Population

A population of 351 Caucasian women was enrolled for the BES TEST® population study in 2015 [7], and incidence of fragility fractures has been assessed after a 3-year follow-up period. For the purpose of the present research, two groups were considered (Table1):

- *Non-Fractured*, who had not sustained a fragility fracture in 2015 nor in the following 3 years.
- *Fractured*, who had sustained a fragility fracture at the time of BES TEST® assessment.

The *BSI-Tscore*, *Age,* and the *Bone Mass Index (BMI),* an indirect index of adiposity calculated as the patient's weight in kilograms divided by the square of height in meters, were collected for each subject at the time of the test.

Materials Research Proceedings **26** (2023) 275-280 https://doi.org/10.21741/9781644902431-45

Table 1. *Population descriptors: Age, BMI and BSI-Tscore Mean (Min – Max) values*

	n	*Age* [years]	*BMI* [kg/m^2]	*BSI-Tscore* [St. Dev.]
Fractured	91	63 (27 – 93)	22.8 (14.7 – 36.3)	- 1.5 (-3.4 – 0.8)
Non-Fractured	75	58 (25 – 84)	23.7 (17.4 – 34.0)	-0.4 (-3.2 – -2.4)

Diagnostic accuracy
True positive (*TP*) is an outcome where the test correctly predicts the positive class. True negative (*TN*), false positive (*FP*) and false negative (*FN*) have analogous definitions.

The *Accuracy* of a diagnostic test is the percentage of correct classifications, giving equal importance to positives and negatives [12]:

$$Accuracy = (TP + TN)/(TP + TN + FP + FN). \tag{1}$$

The *Receiving Operator Characteristic (ROC)* curve is created by plotting at various threshold settings the *True Positives Rate* (*TPR*), or *Sensitivity*, against the *False Positive Rate* (*FPR*), or probability of false alarm, which can be calculated as *(1 – Specificity)*.

Lowering the threshold classifies more items as positive, increasing both *FP* and *TP*. The *Area Under the ROC Curve*, *AUC*, provides an aggregate measure of performance of a diagnostic test across all possible thresholds [13].

Results
Population
The Age distribution of the population is depicted in Fig.1. The *Fractured* group appears to be slightly older than the *Non-Fractured* one (Student's test p =0.00485).

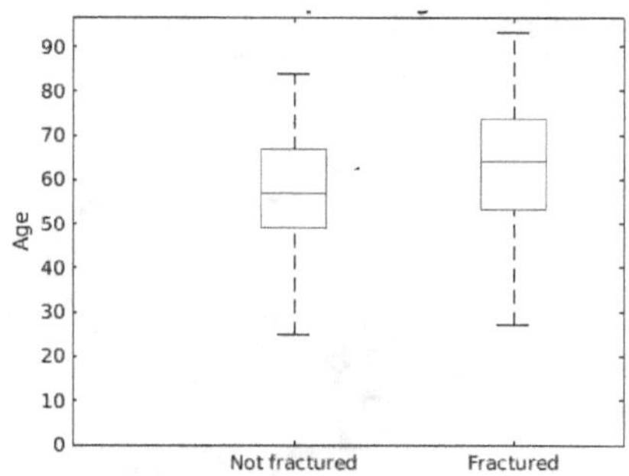

Figure 1. *Box plot of Age distribution.*

The *BMI* distribution for the two groups, classified according to the WHO guidelines, is reported in Table 1. The *BMI* values for the two groups appear to be similar (p = 0.1243).

Table1. *BMI distribution for the two groups, according to the WHO classification.*

BMI	*CLASSIFICATION*	*FRACTURED*	*NON-FRACTURED*
< 18.5	*UNDERWEIGHT*	8%	3%
18.5 – < 25.0	*NORMAL RANGE*	74%	64%
>= 25.0	*OVERWEIGHT*	13%	27%
>= 30.0	*OBESE*	5%	5%

The distribution of *BSI_T-score* in the two groups is shown in Fig.2. The *Non-Fractured* patients are characterized by a median of -0.5 and a first and third quartile of -1.2 and 0.1, respectively, and are very satisfactorily discriminated from the *Fractured* group ($p = 3.18e\text{-}10$).

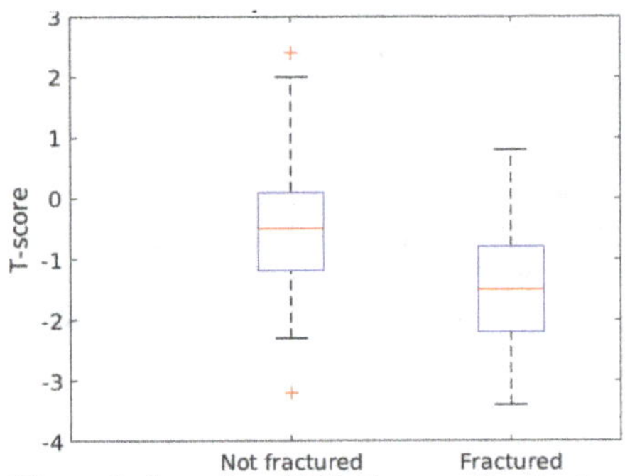

Figure 2. *Box plot of BSI_T-score distribution.*

As shown in Fig.3, no correlation appears to be present between the *BMI* and the *BSI_T-score* values for each of the two groups ($R^2 = 0.0032$ in the *Fractured* and $R^2 = 0.0188$ in the *Non-Fractured* group). The fitted ROC curve [14] is plotted in Fig.4, and the *AUC* value is 0.78.

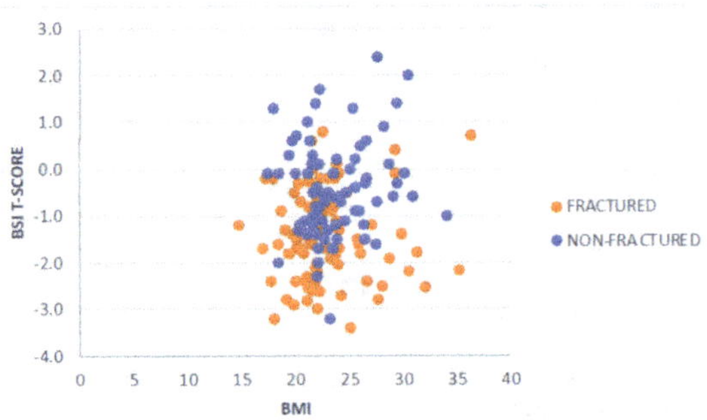

Figure 3. *BSI_T-score vs. BMI distribution in the two groups.*

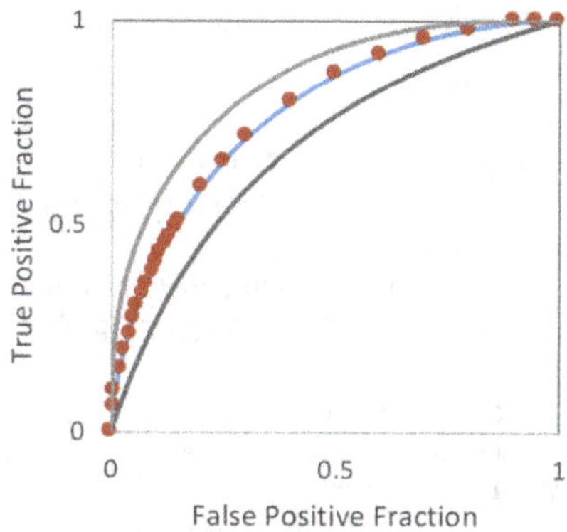

Figure 4. *Red symbols and blue line: Fitted ROC curve for BSI_T-score. Gray lines: 95% confidence interval.*

An optimal threshold value could be identified: the accuracy trend shown in Fig. 5 is characterized by a peak at the *BSI_T-score* = -1.4 cut-off value that maximizes *Accuracy*, corresponding to 78% of correct results, i.e. a diagnostic capacity of the BES TEST® of 0.78. *Accuracy* for the common threshold values of diagnostic significance for fracture risk assessment resulted respectively 53% for a cut-off value of *BSI T-score* =-2.5, and 31% for *BSI T-score* =1.

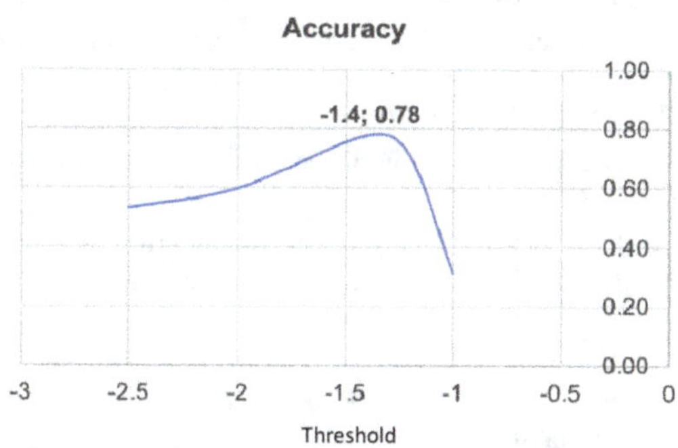

Figure 5. *Accuracy plotted against threshold values.*

Conclusions

The *BSI T-score* resulted an effective predictor for the risk of incident fragility fractures, and an optimal cut-off value for fragility fracture risk was identified.

The results achieved are certainly helpful in the diagnostic and preventive practice: the clinical portrait of a patient who obtained a *BSI T-score* lower than -1.4 should be carefully investigated because there is a high probability of fracturing in the following 3 years.

Given the DEXA reduced ability in assessing the load bearing capability of the internal part of the bone, it appears that BES TEST®, also in association with DEXA, could be helpful in completing the fracture risk map and reducing the clinical and economic burden of fragility fractures.

References

[1] Pasco, J. A. et al., The population burden of fractures originates in women with osteopenia, not osteoporosis, Osteoporosis International, 2006;17(9):1404-9. https://doi.org/10.1007/s00198-006-0135-9

[2] WHO Study Group Assessment of fracture risk and its application to screening for postmenopausal osteoporosis. Report of a WHO Study Group. World Health Organ Tech Rep Ser, 1994, 843:1–129.

[3] Brear K., Currey J.D., Raines S., Smith K.J., Density and temperature effects on some mechanical properties of cancellous bone, Engineering in Medicine, 1988, 17, 163-167. https://doi.org/10.1243/EMED_JOUR_1988_017_043_02

[4] Borgström B. et al., Fragility fractures in Europe: burden, management and opportunities Arch Osteoporos, 2020, 15, 17(9), 1404-9. https://doi.org/10.1007/s11657-020-0706-y

[5] Tonti E., A Direct Discrete Formulation of Field Laws: The Cell Method, CMES 2001,1(1), 11-26.

[6] Cosmi F., Numerical Evaluation of Trabecular Bone Alterations, MCB, 2015, 201(12), 87-105.

[7] Cosmi F. et al., Osteoporosis risk factors and bone microstructure evaluation: a population breakdown, Mater Today-Proc, 2018, 5, 26772–26777. https://doi.org/10.1016/j.matpr.2018.08.150

[8] Cosmi F., Dal Maso A., Zatta G. 3D-printing for the precision assessment of a new medical device, Proc. Inst. Mech. Eng. Part C, 2021, 236, 1–9. https://doi.org/10.1177/09544062211003623

[9] Saviola G. et al., BES TEST analysis in patients with normal or osteopenic DEXA, Minerva Medica, 2019, 10, 597-9.

[10] Saracchini S. et al. BES TEST®: a new diagnostic opportunity for bone structure evaluation in oncology, 2019, The Breast, 44S1, S79–S142. https://doi.org/10.1016/S0960-9776(19)30428-X

[11] Ferraro P et al. Evaluation of the Bone Elastic Structure in Persons with and without Chronic Kidney Disease on Dialysis, Dial. Transplant., 2021, 36S1 ,445. https://doi.org/10.1093/ndt/gfab096.003

[12] Umemneku Chikere CM et al. Diagnostic test evaluation methodology: A systematic review of methods employed to evaluate diagnostic tests in the absence of gold standard – An update. PLoS. ONE 2019, 14(10): e0223832. https://doi.org/10.1371/journal.pone.0223832

[13] Fawcet T., An introduction to ROC analysis. Pattern Recognit. Lett. 2006, 27, 861–874. https://doi.org/10.1016/j.patrec.2005.10.010

[14] ROC analysis: web-based calculator for ROC curves. Retrieved May 2022 from http://www.jrocfit.org

Theoretical and Applied Mechanics - AIMETA 2022
Materials Research Forum LLC
Materials Research Proceedings **26** (2023) 281-286
https://doi.org/10.21741/9781644902431-46

Numerical simulation of coronary arteries blood flow: effects of the aortic valve and boundary conditions

Seyyed Mahmoud Mousavi[1, a*], Gianluca Zitti[1, b], Marco Pozzi[2,c] and Maurizio Brocchini[1, d]

[1]Department of Construction, Civil Engineering, and Architecture, Polytechnic University of Marche, Ancona, Italy

[2]Department of Pediatric and Congenital Cardiac Surgery and Cardiology, United Hospital, Ancona, Italy

[a]s.m.mousavi@pm.univpm.it, [b]g.zitti@staff.univpm.it, [c]marco.pozzi@ospedaliriuniti.marche.it, [d]m.brocchini@staff.univpm.it

* Corresponding author, (s.m.mousavi@pm.univpm.it)

Keywords: Coronary Perfusion, Aortic Valve, Boundary Conditions, Numerical Simulation

Abstract. Sudden cardiac death in athletes is often related to anomalies in coronary origin, which may affect how coronary arteries supply blood to the heart; this highlights the importance of understanding coronary perfusion. Studies for the simulation and predictions of coronary blood flow under normal or disease conditions are many. Numerical simulations can provide rich information about the coronary blood flow with reduced costs in relation to physical models. Correct numerical simulation of the coronary blood flow is still challenging due to the complex interactions with the upstream, e.g., the aortic root, and downstream, e.g., the ventricular contraction, parts of the coronary arteries. Among all, the intrinsic ability of coronary artery autoregulation, intramyocardial resistance, cardiac frequency, and aortic valve functioning could significantly affect the coronary artery perfusion. In the present study, we aim at investigating the effects of the aortic valve and boundary conditions on coronary perfusion. We numerically modeled, by means of the ANSYS-Fluent software, the blood flow inside the proximal parts of the left and right coronary arteries, aortic sinuses of Valsalva, and ascending aorta with and without the aortic valve; the geometry of the computational model represents the average healthy person. Physiological boundary values have been applied at the computational domain boundaries to achieve a stable solution of the Navier-Stokes equations, which govern the incompressible, laminar, Newtonian blood flow. Our numerical results give insight into a proper numerical setup to predict coronary blood flow.

Introduction

The main medical cause of death in athletes is sudden cardiac death (SCD) [1]. High risks of SCD have been observed among patients with a history of coronary artery disease (CAD) [2]. There are multiple research studies in the literature indicating the possible relations between coronary abnormal functioning and myocardial ischemia which might lead to SCD [3,4]. Thus, knowing about coronary artery perfusion could help coronary artery disease diagnosis and surgical planning prior to SCD.

While medical imaging is applied widely for cardiovascular and coronary artery diagnosis, they show no predictive capabilities particularly in pathology of the anomalous course of the coronary artery in SCD [5,6]. Alternatively, computational methods provide this opportunity to virtually simulate the cardiovascular system including the coronary arteries. Lumped parameter or one-dimensional computational models have been successfully hired to study the variety of patient-

Materials Research Proceedings **26** (2023) 281-286 https://doi.org/10.21741/9781644902431-46

specific blood flow simulations [7–9]. However, these methods cannot suitably model pressure wave propagations or fluid-structure interaction phenomena, which are important factors in arterial blood flow simulations [10], such as in the aortic root and proximal part of the coronary arteries. Thus, 3D computational fluid dynamics (CFD) simulations, particularly when they are coupled with the structural models, can provide a detailed analysis of the blood flow in coronary arteries.

In this paper, we simulate the blood flow in a three-dimensional model of the proximal parts of the left and right coronary arteries, aortic sinuses of Valsalva, and ascending aorta of average healthy people. We applied physiological pressure and flow rate boundary values at the boundary locations of the computational domain. Subsequently, a numerical flexible aortic valve model has been integrated into the computational domain of the blood volume to study the effects of the aortic valve functioning on coronary artery perfusion.

Methods

The computational domain is twofold. One is the blood volume, and the other is the aortic valve structural model.

We follow the descriptions of works by [11–13] to design the blood flow computational volume that represents average healthy people. The blood volume consists of the inlet volume (included to manage the inflow), the aortic sinuses of Valsalva, proximal parts of the left and right coronary arteries, and the ascending aorta which are shown in Fig. 1; the dimensions of the model elements are reported in Table 1.

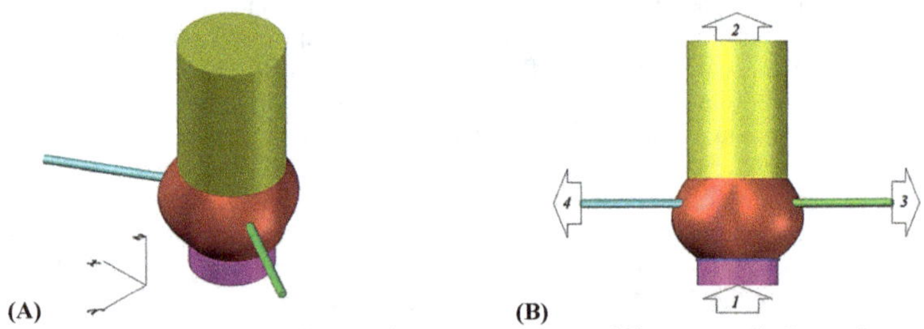

Fig. 1 (A) 3D schematic of the blood volume computational domain including inlet aorta (magenta), aortic sinuses of Valsalva (red), ascending aorta (yellow), left (green), and right (blue) coronary arteries. (B) Numbers showing the flow domain boundary locations, (1) inlet, (2) aorta outlet, (3) left coronary outlet, and (4) right coronary outlet.

Table 1 The dimensions of the flow volume computational domain elements as shown in Fig. 1.

No	Part	Length $[mm]$	Diameter $[mm]$
1	Ascending aorta	50	36
2	Left and right coronary arteries	40	3
3	Sinuses of Valsalva	29	29 (aortic annulus) 36 (sinotubular junction)
4	Inlet aorta	10	29

Moreover, the computational flow domain is integrated into the aortic valve structural model that is shown in Fig. 2-A. The aortic valve is located right after the aortic inlet downstream at the location of the aortic annulus, see Fig. 2-B. The design of the numerical model of the aortic valve mimics fairly well the geometry of the Medtronic Avalus 400 bioprosthetic aortic valve that is shown in Fig. 2-C; the corresponding dimensions of the aortic valve model are reported in Table 2. The dynamics of the fluid and solid parts of the domain have been simulated with the numerical software Ansys® Fluent and Ansys® Transient Structural, respectively, coupled in a solver package that iteratively solves the discretized forms of the governing equations of the fluid flow and solid motion at each computational mesh point. The mesh characteristics of the computational domain are summarized in Table 3.

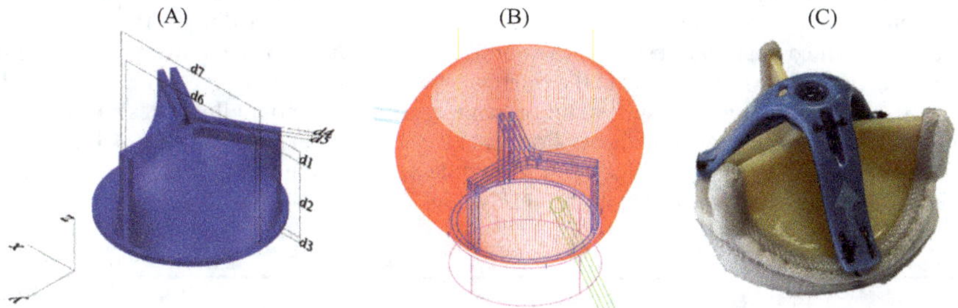

Fig. 2. The aortic valve computational domain (A) is located downstream of the aortic inlet (magenta) at the aortic annulus (red) center (B), which has been designed according to the Medtronic Avalus 400 bioprosthetic aortic valve (C).

The fluid flow governing equations are the principle of mass conservation and the Navier-Stokes equations for incompressible fluid, characterized by density of $\rho = 1060[kg/m^3]$ and dynamic viscosity of $\mu = \nu \times \rho = 3.5 \times 10^{-3}$ $[Pa.s]$ [14]. The motion of the aortic valve with flexible leaflets is governed by the linear momentum balance equation, considering a linear elastic and isotropic material, with the Young's modulus of $1[MPa]$, Poisson's ratio of 0.45, and density $\rho_s = 1000[kg/m^3]$, in line with previous studies [15,16]. The equilibrium of surface forces is applied at the interface between the solid and fluid parts of the domain.

Table 2 Dimensions of the computational aortic valve structural model elements according to the parameters shown in Fig. 2-A.

No	Part	Dimension $[mm]$
1	$d1$	2
2	$d2$	14
3	$d3$	1
4	$d4$	1
5	$d5$	1
6	$d6$	28
7	$d7$	30

Table 3 Summary of the computational domain mesh characteristics generated in ANSYS.

Computational domain	Element type	Element order	Number of nodes	Number of elements	Min element size $[m]$	Max element size $[m]$
Blood volume	Tetrahedral	Linear	71248	389719	4.9063e-05	0.00887179
Aortic valve	Tetrahedral	Linear	665	1617	4.2214e-04	4.5706e-03

Results and Discussion

To evaluate the effect of inlet boundary value, we applied two sets of boundary conditions: the first set of boundary conditions (called *inlet flow*), taken from reference [17], are a physiological aortic flow rate at the inlet volume, and physiological pressure values at three other outlets; the second set (called *inlet pressure*), evaluated from reference [14], are the physiological left ventricular pressure at the inlet volume and three physiological pressure values at three outlet boundary locations. Moreover, the role of the aortic valve is studied considering two numerical model configurations: one configuration simulates merely the blood volume computational domain, without the valve (called *without valve*) and the other is the fluid-structure interaction of the blood volume with the aortic valve structural model (called *with valve*).

Fig. 3-A and Fig. 3-B, show respectively the calculated numerical flow rate values of the left and right coronary arteries when we hired the *inlet flow* boundary conditions. In each panel of the figure, we can compare the results of the configuration *without valve* (shown in blue) and of the one *with valve* (shown in yellow) to the reference values (shown in orange) borrowed from [17].

Moreover, the numerical results of the left and right coronary artery flow rates when we applied the *inlet pressure* boundary conditions from [14] are shown in Fig. 4-A and Fig. 4-B respectively. Similar to the former case, the blue color shows the results of the configuration *without valve* and yellow indicates the results of the model *with valve*.

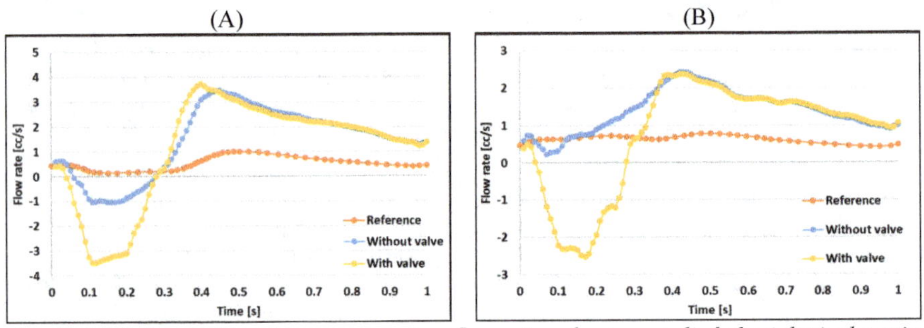

Fig. 3. (A) Left and (B) right coronary artery flow rates when we applied physiological aortic flow rate at the inlet volume boundary location and physiological pressure values at three outlet boundary locations; all boundary values are hired from reference [17].

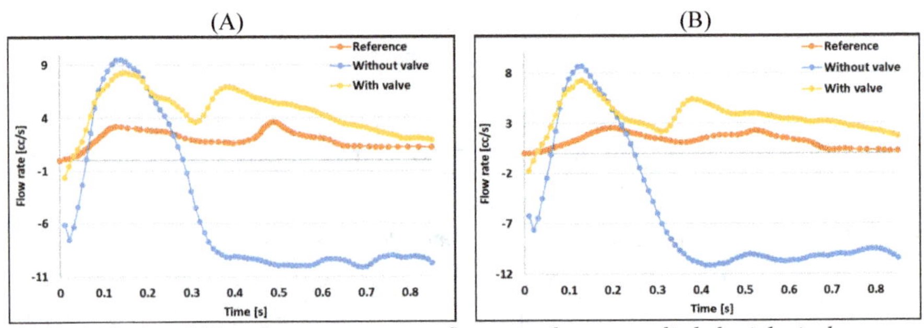

Fig. 4. (A) Left and (B) right coronary artery flow rate when we applied physiological pressure values at the aorta inlet boundary location as well as physiological pressure values at three outlet boundary locations; all boundary values are hired from reference [14].

By comparing the numerical results of the models *without valve* in Fig. 3 and Fig. 4, we can observe that the model with the *inlet flow* boundary conditions could better simulate the coronary artery perfusion compared to the model with the *inlet pressure* boundary conditions. In other words, the calculated coronary perfusion waveforms in the former model are mostly positive values following, though not perfectly, the waveforms of the reference values. In contrast, the numerical coronary artery flow rates in the latter model show large negative values, particularly in the diastolic phase. However, the results of the model *without valve* by applying either set of the boundary conditions show deviation from their corresponding reference values. Thus, we run the simulations *with valve* to study the role of the aortic valve functioning in coronary artery perfusion. The comparison of the results obtained in models with and without valve in Fig. 3 indicates that the inclusion of the aortic valve in the numerical model does not improve the calculated coronary flow rates when the flow rate inlet boundary condition is used. This might be due to the fact that the inlet volume flow rate boundary condition reported in reference [17] is located downstream of the aortic valve. On contrary, in our numerical model, the inlet volume boundary location is upstream of the aortic valve, see Fig. 1-B and Fig. 1-B. Nonetheless, the results of the simulation *with valve* in Fig. 4 show that the left and right coronary artery flow rates better resemble the reference values compared to the corresponding simulations *without valve*, if pressure inlet boundary condition is hired.

Conclusion

We numerically studied the effects of boundary conditions and aortic valve functioning on coronary artery flow rates. We could observe that the choice of upstream boundary condition plays a significant role in coronary perfusion. Particularly, considering the model without valve, the *inlet flow* boundary condition provides better results than the *inlet pressure* boundary conditions. Instead, if the aortic valve is included, the *inlet pressure* boundary condition helps better simulate the coronary flow rates. In fact, in real human hearts, the left ventricular pressure acts on the upstream side of the aortic valve.

References

[1] K. G. Harmon, J. A. Drezner, M. G. Wilson, and S. Sharma, "Incidence of sudden cardiac death in athletes: a state-of-the-art review," *Heart*, vol. 100, no. 16, pp. 1227–1234, 2014. https://doi.org/10.1136/heartjnl-2014-093872.rep

[2] A. Rohila and A. Sharma, "Detection of sudden cardiac death by a comparative study of heart rate variability in normal and abnormal heart conditions," *Biocybern. Biomed. Eng.*, vol. 40, no. 3, pp. 1140–1154, 2020. https://doi.org/10.1016/j.bbe.2020.06.003

[3] A. P. Burke, A. Farb, R. Virmani, J. Goodin, and J. E. Smialek, "Sports-related and non-sports-related sudden cardiac death in young adults," *Am. Heart J.*, vol. 121, no. 2, pp. 568–575, 1991. https://doi.org/10.1016/0002-8703(91)90727-Y

[4] A. J. Taylor, K. M. Rogan, and R. Virmani, "Sudden cardiac death associated with isolated congenital coronary artery anomalies," *J. Am. Coll. Cardiol.*, vol. 20, no. 3, pp. 640–647, 1992. https://doi.org/10.1016/0735-1097(92)90019-J

[5] M. C. de Jong, T. S. S. Genders, R.-J. van Geuns, A. Moelker, and M. G. Hunink, "Diagnostic performance of stress myocardial perfusion imaging for coronary artery disease: a systematic review and meta-analysis," *Eur. Radiol.*, vol. 22, no. 9, pp. 1881–1895, 2012. https://doi.org/10.1007/s00330-012-2434-1

[6] M. Kadem, L. Garber, M. Abdelkhalek, B. K. Al-Khazraji, and Z. Keshavarz-Motamed, "Hemodynamic modeling, medical imaging, and machine learning and their applications to cardiovascular interventions," *IEEE Rev. Biomed. Eng.*, p. 1, 2022. https://doi.org/10.1109/RBME.2022.3142058

[7] I. E. Vignon-Clementel, C. Alberto Figueroa, K. E. Jansen, and C. A. Taylor, "Outflow boundary conditions for three-dimensional finite element modeling of blood flow and pressure in

arteries," *Comput. Methods Appl. Mech. Eng.*, vol. 195, no. 29, pp. 3776–3796, 2006. https://doi.org/10.1016/j.cma.2005.04.014

[8] H. J. Kim, I. E. Vignon-Clementel, J. S. Coogan, C. A. Figueroa, K. E. Jansen, and C. A. Taylor, "Patient-specific modeling of blood flow and pressure in human coronary arteries," *Ann. Biomed. Eng.*, vol. 38, no. 10, pp. 3195–3209, 2010. https://doi.org/10.1007/s10439-010-0083-6

[9] M. Mirramezani and S. C. Shadden, "A Distributed Lumped Parameter Model of Blood Flow," *Ann. Biomed. Eng.*, vol. 48, no. 12, pp. 2870–2886, 2020, doi: 10.1007/s10439-020-02545-6. https://doi.org/10.1007/s10439-020-02545-6

[10] M. Mirramezani and S. C. Shadden, "Distributed lumped parameter modeling of blood flow in compliant vessels," *J. Biomech.*, vol. 140, p. 111161, 2022. https://doi.org/10.1016/j.jbiomech.2022.111161

[11] H. Reul, A. Vahlbruch, M. Giersiepen, T. H. Schmitz-Rode, V. Hirtz, and S. Effert, "The geometry of the aortic root in health, at valve disease and after valve replacement," *J. Biomech.*, vol. 23, no. 2, pp. 181–191, 1990. https://doi.org/10.1016/0021-9290(90)90351-3

[12] J. T. Dodge Jr, B. G. Brown, E. L. Bolson, and H. T. Dodge, "Intrathoracic spatial location of specified coronary segments on the normal human heart. Applications in quantitative arteriography, assessment of regional risk and contraction, and anatomic display.," *Circulation*, vol. 78, no. 5, pp. 1167–1180, 1988. https://doi.org/10.1161/01.CIR.78.5.1167

[13] J. T. Dodge, B. G. Brown, E. L. Bolson, and H. T. Dodge, "Lumen diameter of normal human coronary arteries. Influence of age, sex, anatomic variation, and left ventricular hypertrophy or dilation.," *Circulation*, vol. 86, no. 1, pp. 232–246, Jul. 1992. https://doi.org/10.1161/01.CIR.86.1.232

[14] S. Nobari, R. Mongrain, R. Leask, and R. Cartier, "The effect of aortic wall and aortic leaflet stiffening on coronary hemodynamic: a fluid–structure interaction study," *Med. Biol. Eng. Comput.*, vol. 51, no. 8, pp. 923–936, 2013. https://doi.org/10.1007/s11517-013-1066-1

[15] G. Marom, R. Haj-Ali, E. Raanani, H.-J. Schäfers, and M. Rosenfeld, "A fluid–structure interaction model of the aortic valve with coaptation and compliant aortic root," *Med. Biol. Eng. Comput.*, vol. 50, no. 2, pp. 173–182, 2012. https://doi.org/10.1007/s11517-011-0849-5

[16] H. E. Salman, L. Saltik, and H. C. Yalcin, "Computational Analysis of Wall Shear Stress Patterns on Calcified and Bicuspid Aortic Valves: Focus on Radial and Coaptation Patterns," *Fluids* , vol. 6, no. 8. 2021. https://doi.org/10.3390/fluids6080287

[17] M. Yin, A. Yazdani, and G. E. Karniadakis, "One-dimensional modeling of fractional flow reserve in coronary artery disease: Uncertainty quantification and Bayesian optimization," *Comput. Methods Appl. Mech. Eng.*, vol. 353, pp. 66–85, 2019. https://doi.org/10.1016/j.cma.2019.05.005

Theoretical and Applied Mechanics - AIMETA 2022 Materials Research Forum LLC
Materials Research Proceedings **26** (2023) 287-292 https://doi.org/10.21741/9781644902431-47

A limit analysis approach for the prediction of the human proximal femur ultimate load

Aurora Angela Pisano[1,a*], Paolo Fuschi[1,b]

[1]University Mediterranea of Reggio Calabria, Reggio Calabria, Italy

[a]aurora.pisano@unirc.it, [b]paolo.fuschi@unirc.it

Keywords: Static Limit Analysis, Human Proximal Femur, Tsai-Wu Constitutive Criterion

Abstract. A limit analysis numerical procedure for the determination of a lower bound on the ultimate load of the human proximal femur is presented. The procedure, already applied by the authors in different contexts, is based on a simplified 3D geometrical model of the human femur and on the assumption of a few geometric and material data available in the relevant literature. The perfectly plastic behaviour of the human bone, due to phenomena starting at molecular scale, and the orthotropic behaviour of the main human femur tissues, *trabecular* and *cortical*, allows to assume a yield surface of Tsai-Wu-type for its constitutive description. The effectiveness of the promoted numerical approach is validated by comparison of the obtained results with experimental findings on in-vitro tests of human femurs.

Introduction

The limit analysis theory, in a numerical formulation oriented to the determination of a lower bound to the collapse load of a structural element and already employed by the authors for the solution of several engineering structural problems [1] is applied in the context of human long bones. The study is focused on of the collapse load of the human proximal femur and this, in the spirit of the static approach of the limit analysis theory, without following the load history that has produced it, excluding as well all the post elastic phenomena, damage or fracture propagation in the bone.

In the promoted approach, the main pre-collapse phenomenon arising within the human bone material, that is the stress redistribution process, is numerically simulated by means of an appropriate adjustment of the elastic moduli of the main constituent materials of the bone, namely the cortical and the trabecular bone tissues. For both the above main constituents it is assumed a criterion of Tsai-Wu-type formulated in principal stress space which, in the 3D context, proves to be a strictly convex domain for the admissible stresses. The few material parameters needed to define the constitutive criterion and to carry on the whole limit analysis procedure represents a very attractive feature of the proposed method which can be easily connected with molecular imaging and/or other diagnostic tools of prevention.

The simulation of the stress redistribution process, arising within the bone element until collapse, is carried on iteratively with reference to a simplified model of the proximal human femur, which approximate the geometry of the bone and that, discretized in 3D solids finite elements, is used to analyse two well known mechanical configurations producing the human proximal femur collapse, i.e. the so-called stance and side configurations, [2,3]. The former is aimed to predict bone fracture under single-leg-stance loading, the latter under sideways-fall.

Few numerical examples are carried out for comparison with the results of experimental tests available in the relevant literature for in-vitro tests on human femurs, [3], also with the purpose to define the sensitivity of the results by varying some geometric and material parameters. The obtained numerical findings definitively show a good capacity of the employed numerical method to predict the real limit load of the proximal human femur under prescribed load conditions.

Materials Research Proceedings **26** (2023) 287-292 https://doi.org/10.21741/9781644902431-47

Theoretical background

The static approach of limit analysis theory, as said in the introductory section, is aimed to the determination of a lower bound to the collapse load of a structure or of a structural element. This goal is classically pursued by applying the Lower-Bound Theorem of limit analysis that leads to a maximization problem in which, a fixed reference load distribution, defining the loading conditions, is amplified by a single scalar load multiplier, say P_{LB}, that produces loads that are in equilibrium with a stress field, that nowhere violates the yield criterion of the constituent material, and that do not exceed the effective collapse load multiplier, say P_U . P_{LB} is then a Lower Bound (LB) to the load multiplier whose maximum, obtainable by increasing the acting loads, defines the collapse, or peak, load value of the structure. It is worth to remind that no matters the elastic behaviour before collapse or any type of elasticity law followed by the constituent material.

On the other hand, the existence of a yield criterion for bone material is assured by the circumstance that plasticity in bones starts at molecular scale and is due to breaking of hydrogen bonds within single collagen molecules followed by breaking of bonds and intermolecular sliding within collagen fibrils, see e.g. [4]. The irrelevance of the elastic behaviour before collapse is a key point for the numerical procedure hereafter promoted, namely the Elastic Compensation Method (ECM) [1,2]. Indeed, the ECM varies, within a discretized FE model of the bone, the elastic moduli of the "bone material" to simulate the process of stress re-distribution arising within the bone suffering an increasing load until the attainment of its strength threshold.

Limit analysis on human proximal femur

The simplified geometrical model of Fig.1 has been assumed for the human proximal femur. The external dimensions of the bone, borrowed from the relevant literature, are given in the same figure in tabular form.

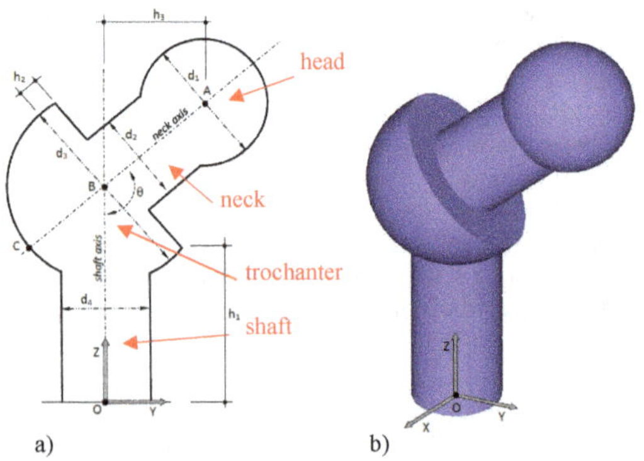

Dimension		Reference
d_1 [mm]	51.60	[5]
d_2	37.40	[5]
d_3	65.64	[6]
d_4	35.40	[6]
h_1	80.00	[3]
h_2	5.30	[6]
h_3	44.00	[6]
AC	90.26	[6]
θ [grad]	130	[7]

***Figure 1**. Human proximal femur model: a) geometry referred to the medio-lateral section; b) 3D model and main reference orthogonal coordinate system; dimensions taken from the relevant literature in tabular form.*

It should be specified that the bone has an external part made of cortical tissue and an internal part made of trabecular and marrow tissues, the latter is only present in the shaft zone. The thickness of the cortical tissue is here considered variable in the different zones into which the simplified proximal femur model has been divided. In particular, from now on S1, S2 and S3 indicate the cortical thickness in the head zone, in the neck and trochanter zones and in the shaft zone, respectively.

As already said, the cortical and trabecular tissues are assumed to obey a Tsai-Wu-type constitutive criterion, which in principal stress space can be given the shape:

$$G_{11}\sigma_{11}+G_{22}\sigma_{22}+G_{33}\sigma_{33}+F_{1111}\sigma_{11}^2+F_{2222}\sigma_{22}^2+F_{3333}\sigma_{33}^2+ \\ 2F_{1122}\sigma_{11}\sigma_{22}+2F_{1133}\sigma_{11}\sigma_{33}+2F_{2233}\sigma_{22}\sigma_{33}=1 \quad (1)$$

where:

$$G_{ii}=\frac{1}{\sigma_i^+}-\frac{1}{\sigma_i^-};\quad F_{iiii}=\frac{1}{\sigma_i^+\sigma_i^-};\quad F_{iijj}=\frac{1}{2}\left(\frac{1}{\sigma_i^+\sigma_i^-}+\frac{1}{\sigma_j^+\sigma_j^-}-\frac{1}{\sigma_{ij}^2}\right) \quad (2)$$

In Eq.(1) σ_{11}, σ_{22}, σ_{33} are principal stresses and repeated indices in Eqs.(2) do not imply summation. Moreover, σ_i^+, σ_i^-, σ_{ij} are the moduli of ultimate strengths in tension, compression and shear, respectively; finally i, j=1, 2, 3 refer to direction and plane of orthotropy.

Typical Tsai-Wu-type yield surfaces for the two main bone tissues are shown in Fig.2 and the ECM limit analysis procedure is applied with reference to both these yield surfaces.

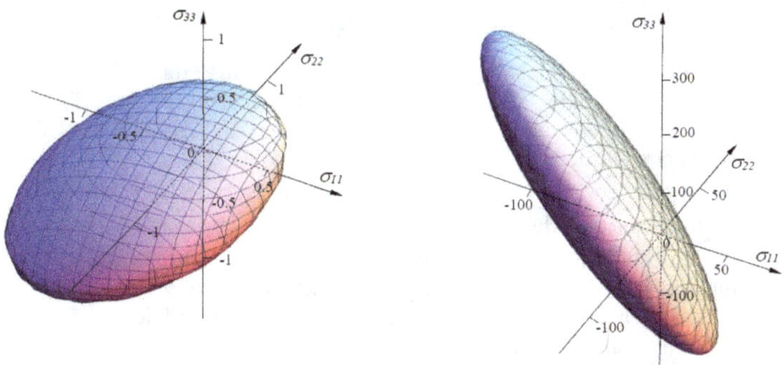

a) b)

***Figure 2**. Typical Tsai-Wu-type yield surface for trabecular a) and cortical b) proximal femur tissues.*

For sake of brevity, the Reader can refer to [1,2] for the details on the ECM iterative procedure. The latter performs a sequence of FE elastic analyses and, within each sequence carried on for a fixed loads distribution, applies an iterative procedure. It is worth noting that, at current iteration, the elastic moduli within the discretized structural model, are updated (reduced) only within those FEs where the elastic element solution, in terms of average stress within the element, is "greater" than the corresponding stress at yield. This produces, at the subsequent iteration (within the current sequence of elastic FE analyses), a redistribution of the stresses, which migrate from the weakened elements to the neighbour ones, so miming the redistribution of the stresses arising before the incipient state of collapse. The latter is reached when, for the applied loads (increased sequence by sequence by increasing the acting loads multiplier), such redistribution is not possible any more.

In-silico *vs* in-vitro tests

Two experimental tests on real human proximal femurs, after D'Allara et al. 2013, [3], have been numerically reproduced, precisely, the so-called STANCE and SIDE configurations which are shown in Fig.3(a-c) together with the assumed FE models.

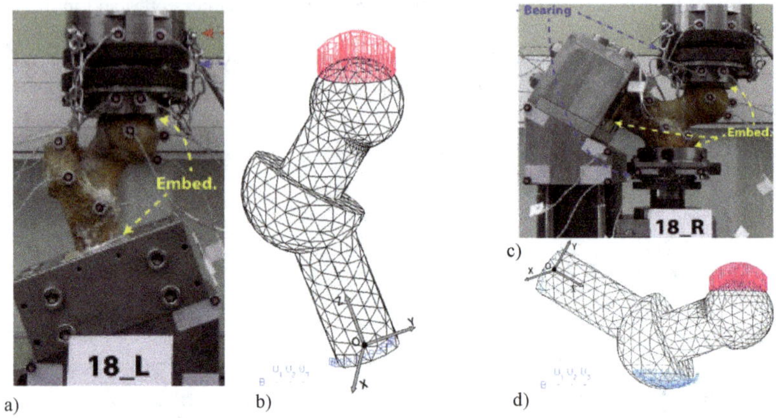

Figure 3. *Human proximal femur in-silico configurations: STANCE a) Experimental set up (after [3]), b) FE mesh of 6408 3D-Solids (10nodes, 17GPs); SIDE c) Experimental set up (after [3]), d) FE mesh of 7111 3D-Solids (10nodes, 17GPs)*

In order to evaluate the influence on the collapse load of material and geometric parameters, three different material strength values (for cortical and trabecular) and three different geometrical models have been considered and this for each configuration, STANCE and SIDE.

Three pairs of apparent density values are assumed leading to the definition of three different materials, hereinafter referred to as Mat_1, Mat_2 and Mat_3. In particular, for Mat_1 $\rho_{cortical} = 1.5$ g/cm^3 and $\rho_{trabecular} = 0.1$ g/cm^3; for Mat_2 $\rho_{cortical} = 1.75$ g/cm^3 and $\rho_{trabecular} = 0.4$ g/cm^3 ; finally for Mat_3 $\rho_{cortical} = 2.0$ g/cm^3 and $\rho_{trabecular} = 0.7$ g/cm^3. These values, corresponding to the minimum, average and maximum values of apparent density either in cortical and in trabecular, give rise to the strength material values reported in Table1. Finally, the contribution of the marrow to the overall strength of the femur is neglected.

Concerning the geometry, three geometrical models, hereinafter named GM_1, GM_2 and GM_3, are obtained by assuming for the thickness of the cortical in the shaft (S3) the values 5.7 mm, 7 mm and 8.3 mm, respectively. Moreover, following [16] the thickness in other zones of the bone are derived by assuming the ratios S1 / S2 = 0.5 and S1 / S3 = 0.25. In summary, the adopted ECM FE-based numerical predictive procedure has been applied nine times for each configuration.

The obtained results, in terms of P_{LB} are reported in Table 2. It can be observed that for fixed geometry (values reported in columns) there is a great variability of the P_{LB} as the strength characteristics of the material vary, while the variability for fixed material strengths (values reported in rows) is less relevant on varying the geometry. It is then essential to calibrate the strength parameters as accurately as possible to increase the predictive capabilities of the whole procedure. Another important remark is that the ECM is able to reproduce the experimental data and, in fact, for the average value of $\rho_{cortical}$, corresponding to the row Mat_2, the values of the collapse loads are very close to each other and to the ones given in the quoted experiments. More precisely, in the STANCE configuration the experimental collapse load value is 8.7 ± 2.9 kN and the one obtained from the numerical simulations is 9 ± 1.5 kN, in the SIDE configuration the experimental collapse load is 3.1 ± 1.1 kN and the one obtained applying the ECM procedure is 4 ± 0.5 kN.

***Table 1.** Assumed strength values [MPa] for the two main bone tissues.*

Strength	Cortical	Trabecular	Cortical	Trabecular
	Mat1		Mat2	
$\sigma_1^- = \sigma_2^-$	68.249(1)	0.913(1)	86.137(1)	6.099(1)
σ_3^-	155.164(1)	0.526(1)	207.324(1)	7.220(1)
$\sigma_1^+ = \sigma_2^+$	23.887(2)	0.602(3)	30.148(2)	4.025(3)
σ_3^+	93.098(4)	0.347(5)	124.394(4)	4.765(5)
$\sigma_{13} = \sigma_{23}$	53.50(6)	0.315(7)	53.50(6)	4.332(7)
σ_{12}	67.80(6)	0.548(8)	67.80(6)	3.659(8)
	Mat3		Reference	
$\sigma_1^- = \sigma_2^-$	105.380(1)	13.128(1)	[8,9,10] (1)	
σ_3^-	266.486(1)	20.792(1)		
$\sigma_1^+ = \sigma_2^+$	36.883(2)	8.664(3)	[11,12] (2)	$(\sigma_1^+ = \sigma_2^+ = 35\%\ \sigma_1^-)$
			[13] (3)	$(\sigma_1^+ = \sigma_2^+ = 66\%\ \sigma_1^-)$
σ_3^+	159.892(4)	13.722(5)	[11,12] (4)	$(\sigma_3^+ = 60\%\ \sigma_3^-)$
			[13] (5)	$(\sigma_3^+ = 66\%\ \sigma_3^-)$
$\sigma_{13} = \sigma_{23}$	53.50(6)	12.475(7)	[14] (6), [15] (7)	$(\sigma_{13} = \sigma_{23} = 0.6\ \sigma_3^-)$
σ_{12}	67.80(6)	7.877(8)	[15] (8)	$(\sigma_{12} = 0.6\ \sigma_1^-)$

***Table 2.** P_{LB} values for the three geometrical models (GM_1, GM_2, GM_3) and for the three considered materials (Mat_1, Mat_2, Mat_3): a) STANCE configuration, b) SIDE configuration.*

Geometrical Model/ Material Model	GM1	GM2	GM3
Mat1	3	3.5	5
Mat2	7.5	9	10.5
Mat3	14	17	18

a)

Geometrical Model/ Material Model	GM1	GM2	GM3
Mat1	1.7	2.2	2.7
Mat2	3.5	4	4.5
Mat3	6.2	7	7.3

b)

Concluding remarks

The numerical iterative procedure, known in the literature as Elastic Compensation Method, has been applied to determine a lower bound on the ultimate load of the human proximal femur. The specification of a few geometrical and material parameters suffices to reach a very good prediction of the ultimate load. In particular, rather than the geometrical parameters, the needed crucial parameters are the strengths of the cortical and trabecular tissues, which are related to the bone mineral density. The obtained results seem very promising as witnessed by comparison with available experimental data on the collapse load of in-vitro specimens for two typical conditions driving to human femur rupture. On taking into account that the bone mineral density and, in general, the actual bones' strengths of a patient affected by orthopedic diseases are nowadays easily detectable via molecular imaging, a predictive tool as the one here promoted could be easily integrated within diagnostic tools of prevention.

References

[1] A.A. Pisano, P. Fuschi, D. De Domenico. Numerical limit analysis of steel-reinforced concrete walls and slabs, Computers and Structures, 160, 42-55, (2015). https://doi.org/10.1016/j.compstruc.2015.08.004

[2] A.A. Pisano, P. Fuschi. Limit Analysis of Human Proximal Femur. Journal of the Mechanical Behavior of Biomedical Materials. 124, 104844, (2021). https://doi.org/10.1016/j.jmbbm.2021.104844

[3] E. Dall'Ara, B. Luisier , R. Schmidt, M. Pretterklieber, F. Kainberger, P. Zysset , D. Pahr. DXA predictions of human femoral mechanical properties depend on the load configuration, Medical Engineering & Physics, 35, 1564-1572, (2013). https://doi.org/10.1016/j.medengphy.2013.04.008

[4] R.O. Ritchie, M.J. Buehler, P. Hansma. Plasticity and toughness in bone, Physics Today, 62(6), 41-47, (2009). https://doi.org/10.1063/1.3156332

[5] G. Holzer, G. von Skrbensky, L.A. Holzer, W. Pichl. Hip Fractures and the Contribution of Cortical Versus Trabecular Bone to Femoral Neck Strength, Journal of Bone and Mineral Research, 24(3), 468-474, (2009). https://doi.org/10.1359/jbmr.081108

[6] J. Michelotti, J. Clark. Femoral Neck Length and Hip Fracture Risk, Journal of Bone and Mineral Research, 14(10), 1714-1720, (1999). https://doi.org/10.1359/jbmr.1999.14.10.1714

[7] Z. Yang, W. Jian, L. Zhi-han, X. Jun, Z. Liang, Y. Ge, S. Zhan-jun. The Geometry of the Bone Structure Associated with Total Hip Arthroplasty. PLOS ONE, 9(3):e91058, doi: 10.1371/journal.pone. 0091058, (2014). https://doi.org/10.1371/journal.pone.0091058

[8] D.C. Wirtz, N. Schiffers, T. Pandorf, K. Radermacher, D. Weichert, R. Forst. Critical evaluation of known bone material properties to realize anisotropic FE-simulation of proximal femur, Journal of Biomechanics, 33, 1325-1330, (2000). https://doi.org/10.1016/S0021-9290(00)00069-5

[9] J.C. Lotz, T.N. Gerhart, W.C. Hayes. Mechanical properties of trabecular bone from the proximal femur: a quantitative CT study, Journal of Computer Assisted Tomography, 14, 107-114, (1990). https://doi.org/10.1097/00004728-199001000-00020

[10] J.C. Lotz, T.N. Gerhart, W.C. Hayes. Mechanical properties of metaphyseal bone in the proximal femur, Journal of Biomechanics, 24, 317-329, (1991). https://doi.org/10.1016/0021-9290(91)90350-V

[11] J. Currey. Cortical Bone. In: Handbook of Biomaterial Properties, (Chapter A1) Springer Science + Business Media New York 2016, W. Murphy et al. (eds.), (2016).

[12] M. Doblaré, J.M. García, M.J. Gómez. Modelling bone tissue fracture and healing: a review, Engineering Fracture Mechanics, 71, 1809-1840, (2004). https://doi.org/10.1016/j.engfracmech.2003.08.003

[13] L. Rincón-Kohli, P.K. Zysset. Multi-axial mechanical properties of human trabecular bone. Biomechanics and Modeling in Mechanobiology, 8(3), 195-208, (2009). https://doi.org/10.1007/s10237-008-0128-z

[14] C.H. Turner, T. Wang, D.B. Burr. Shear Strength and Fatigue Properties of Human Cortical Bone Determined from Pure Shear Tests, Calcified Tissue International, 69, 373-378, (2001). https://doi.org/10.1007/s00223-001-1006-1

[15] A. Sanyal, A. Gupta, H.H. Bayraktar, R.Y. Kwon, T.M. Keaveny. Shear strength behavior of human trabecular bone. Journal of Biomechanics, 45, 2513-2519, (2012). https://doi.org/10.1016/j.jbiomech.2012.07.023

[16] S.P. Väänänen, L. Grassid, G. Lorenzo, G. J.S. Flivike, J.S, Jurvelina., H. Isaksson, Generation of 3D shape, density, cortical thickness and finite element mesh of proximal femur froma DXA image, Medical Image Analysis,24, 125-134, (2015). https://doi.org/10.1016/j.media.2015.06.001

Theoretical and Applied Mechanics - AIMETA 2022 Materials Research Forum LLC
Materials Research Proceedings **26** (2023) 293-298 https://doi.org/10.21741/9781644902431-48

The role of the interstitial fluid content in bone remodeling

Esposito Luca[1,a*], Zona Renato[1,b], Palladino Simone[1,c], Minutolo Vincenzo[1,d], Fraldi Massimiliano[2,e].

[1] Department of Engineering, University of Campania Luigi Vanvitelli, Aversa (CE), Italy

[2] Department of Structures for Engineering and Architecture, University of Naples Federico II, Napoli, Italy

[a]luca.esposito@unicampania.it, [b]renato.zona@unicampania.it, [c]simone.palladino@unicampania.it [d]vincenzo.minutolo@unicampania.it , [e]fraldi@unina.it

Abstract. Bone is an extraordinary biological material able to modify dynamically its outer shape and inner microstructure in response to chemo-mechanical stimuli coming from the environment adapting its hierarchical microstructure to respond to static and dynamic loads for offering optimal mechanical features, in terms of stiffness and toughness. To date, many theories and mathematical models have been proposed by several authors to describe the remodeling phenomenon, all approaches starting from the adaptive elasticity and the bone maintenance theories. Within this framework, one of the most classical strategies employed in the studies is the so-called Stanford's law, which allows uploading the effect of the time-dependent load-induced stress stimulus into a biomechanical model to guess the bone structure evolution. In the present work, we generalize this approach by introducing the bone poroelasticity, thus incorporating in the model the role of the fluid content that, by driving nutrients and contributing to the removal of wastes of bone tissue cells, synergistically interacts with the classical stress fields, in this way affecting growth and remodeling of the bone tissue. Two paradigmatic example applications, i.e. a cylindrical slice with internal prescribed displacements idealizing a tract of femoral diaphysis pushed out by the pressure exerted by a femur prosthesis and a bone element in a form of a bent beam, and a real study case of a patient subject to total hip replacement, and CT scanned at 24 hours after surgery and at 1 year post-surgery have been considered. It has to note that the proposed model is capable to catch more realistically both the transition between spongy and cortical regions and the expected non-symmetrical evolution of bone tissue density in the medium-long term, unpredictable with the standard approach. Although limitations still characterize some hypotheses at the basis of the present approach, the proposed model overcomes the intrinsic - and unrealistic - independence of the bone remodeling from the stress sign and from the indirect effect of stress gradients driving nutrients through the flow of the fluid content in the tissue, allowing to predict important spatial asymmetries in bone mass density, so paving the way to more reliable mechanobiological strategies and engineering tools for the faithful prediction of bone remodeling, with implications in diagnosis of risk fracture, optimal design of scaffolds and bone prostheses.

Introduction

The bone tissue is a dynamic system able to modify dynamically its outer shape and inner microstructure in response to chemo-mechanical stimuli coming from the environment, through several processes such as growth (mass change), remodeling (material properties change) and morphogenesis (shape and structural changes). The first studies were based on clinical observations. Galileo Galilei compared the dimensions of bones from animals of different sizes and suggested that their forms were determined by their functions, gravity and environment [1]. To date, many theories and mathematical models have been proposed by several authors to describe the remodeling phenomenon [2]. Two general approaches are proposed in the literature: the *adaptive elasticity theory* and the *bone maintenance theory*. The adaptive elasticity theory, is a continuum mechanical-based formulation, in which the remodeling equations relate the change

Materials Research Proceedings **26** (2023) 293-298 https://doi.org/10.21741/9781644902431-48

of the bone tissue density to mechanical stimuli [3] In particular, a mathematically rigorous theory for the remodeling of the internal architecture of the cancellous bone, called the *evolutionary Wolff's law*, was proposed by [4]. On the other hand, the bone maintenance theory, developed by [5], is based on optimizing a local remodeling objective function depending on the apparent density, the orientation of material axes, and the stress tensor. Expanding these ideas, Carter [6] introduced the daily remodeling stimulus. Beauprè [7] proposed the so-called *Stanford law* by extending the bone maintenance theory including time-dependent remodeling and adding surface growth, with internal remodeling considered as surface growth in internal cavities.

The influence of interstitial fluid on bone remodeling plays a crucial role, often highlighted in literature with respect to processes occurring at cellular level [8], and is believed to play important roles in providing nutrients and removing wastes, being directly involved in cellular mechano-transduction [9]. By starting from this framework, the present study is aimed to extend the Stanford law proposed by [10] by including the dimensionless fluid content, in a way to affect the relation between stress stimulus at the tissue level and the rate of density responsible for the remodeling, regulating the nutrients supply and thus the activity of bone tissue cells, being the stress stimulus at tissue level supposed as a function of the density, the stress, and the fluid content as well.

Materials and Methods

A time-dependent bone tissue remodeling theory – the Stanford law

Beauprè [7] proposed a unified time-dependent approach – known as the Stanford law – for periosteal and internal bone. The essence of this approach is that the bone tissue needs a certain level of mechanical stimulus to maintain itself. At the macroscopic scale, the bone adaptation process is described on a daily basis by relating the remodeling rate to a set of stresses corresponding to successive loading conditions. At the microscopic level, the bone remodeling response is expressed by the daily stress stimulus at the tissue level, and the remodeling response is measured in terms of the bone resorption/formation rate, which gives the net tissue volume formed or resorbed per unit time, with a homeostatic region corresponding to a range of "normal" activities, and absorption and formation zones associated with decreased and increased levels of daily stress stimulus, respectively.

Moreover the density change rate, is directly related to the bone surface area per unit volume. Summing up, the internal and external remodeling are treated in a unified surface-mediated phenomenon.

The role of the interstitial fluid content in bone remodeling – the proposal

In this work, a new remodeling formulation, aimed to extend the Stanford law by including the dimensionless fluid content [11], is proposed. We thus assumed that the stress stimulus at tissue level is a function of the density, the stress, and the fluid content. In this way, the fluid content can be interpreted as a triggering factor influencing both the support of nutrients and the removal of wastes of bone tissue cells, in turn affecting the relation between stress stimulus at the tissue level and remodeling rate of density. This more realistic interpretation of the bone remodeling process is translated mathematically by means of an *ad hoc* function f_ζ, related to the dimensionless fluid content ζ, as follows

$$\begin{aligned} &\text{if} \quad \zeta \geq 0 \quad f_\zeta = (1+\zeta)^\gamma \\ &\text{if} \quad \zeta < 0 \quad f_\zeta = (1+|\zeta|)^{-\gamma} \end{aligned} \tag{1}$$

where the quantity $(1+\zeta)$ works as an activator factor and γ is the power to be set for properly describing the effectiveness of the action. The power value γ depends in fact on the level of activities of the bone tissue cells, the concentration and quality of the fluid and the possible presence of growth factors, and needs to be determined by experimental tests. Hence, the stress tissue at tissue level is supposed to be proportional to the fluid content through the function f_ζ

with effects on the bone formation and resorption rate, and the , modifying the DZ window width in the case of homeostasis.

The proposed examples

Two paradigmatic examples of direct interest in biomechanical applications have been chosen to verify the feasibility and the effectiveness of the proposed remodeling approach, that is a beam with a rectangular section undergoing bending moment and a bi-phase hollow cylinder subject to prescribed radial displacements at the innermost boundary [12, 13]. To this purpose, both the classical Stanford Law and our proposal have been implemented by means of customized algorithms in the numerical Finite Element code ANSYS® Multiphysics (ANSYS Inc., Canonsburg, PA, USA) and the obtained results have been compared. The adopted poroelastic material parameters are provided by [14]

Discussion and Results

The numerical simulations have been performed with the aim of simulating the remodeling process during a period of 360 days. At the end of each iteration, once calculated the current value of the density for each element, the corresponding material properties have been consequently updated. As an effect to the applied bending moment the material grows on the outer surfaces and absorbs in the central part of the beam. The density distribution produced by using the Stanford law (continuous red line) remains symmetrical during the time with highest and equal values at the outermost fibers. On the contrary, the proposed formulation leads to have asymmetry of the results in terms of spatial distribution of density during remodeling (continuous blue line), see Figure 1 (left). Both formulations gave a thickening of the material up to the cortical value at the internal radius, however in a faster way for the poroelastic model, see Figure 2 (right). Additionally, when implementing our formulation (continuous blue line), the density became quite symmetrical at 180 days with cortical density values at the internal and external radius. Conversely, when implementing the Stanford law, the density reached the cortical value for most of the section.

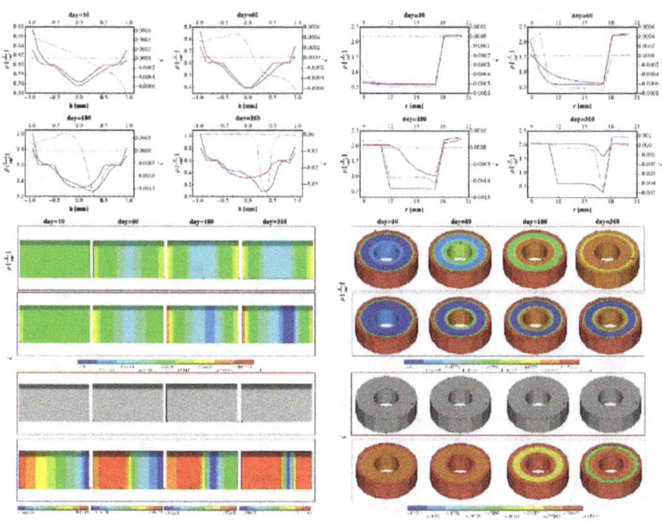

Figure 1: Plot of the density (continuous) and fluid content (dotted) when implemented Stanford law (red) and our proposal (blue), along with the height of the beam (left) and the radius of the

cylinder (right) at 10 days 60 days, 180 and 360 days. Contour plots of the density and fluid content when implemented Stanford law (upper row) and our proposal (lower row), along with the height of the beam (left) and the radius of the cylinder (right) at 10 days 60 days, 180 and 360 days.

A real study case for testing the proposed model

A real patient from the clinical trial at the Healthcare Center of the Icelandic National Hospital has been considered, and the related CT data at 24 hours have been used to build-up the corresponding in silico model by means of the voxel-based approach [15]. Due to the high computational times related to poroelastic-based remodeling analyses, the geometry-based meshing strategy was chosen, and bone material properties, both elastic and porous parameters, were estimated from CT data. The model was loaded by forces on the prosthesis cup by considering the actions related to the living daily activities [16] and constrained in the distal part of the model. Both the classical Stanford's law and our proposal were implemented in the FE-based algorithm, performing the numerical simulations for a period of 360 days, finally comparing the results of the two models. Figure 2 shows the whole three-dimensional femur obtained by filtering CT data (left), the maps of the density of the considered transverse section at 24H (top-middle) and 1Y (bottom-middle), and the corresponding plot of the densities obtained by CT data along a line from the innermost to outermost points across a femur region actually interested by significant changes in bone density, with clear spatially inhomogeneous transition from cortical to trabecular tissues.

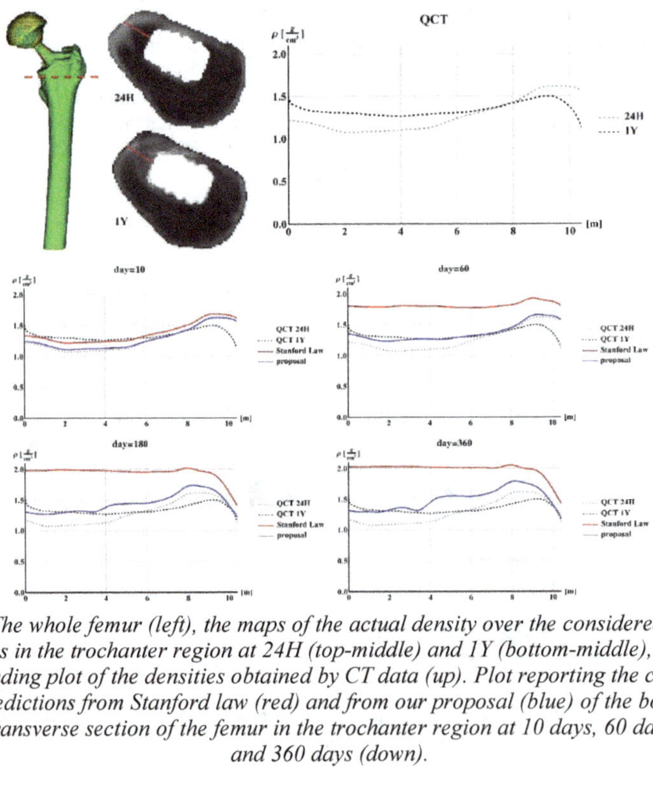

Figure 2: The whole femur (left), the maps of the actual density over the considered transverse sections in the trochanter region at 24H (top-middle) and 1Y (bottom-middle), and the corresponding plot of the densities obtained by CT data (up). Plot reporting the comparison between predictions from Stanford law (red) and from our proposal (blue) of the bone densities along the transverse section of the femur in the trochanter region at 10 days, 60 days, 180 days and 360 days (down).

The comparison between the predictions from Stanford law (red) and from our proposal (blue) as densities along the transverse section of the femur in the trochanter region at 10 days, 60 days, 180 days and 360 days is shown in Figure 2 (down). Both the model approaches produced a denser bone at the interface with the prosthesis. However, when implementing the Stanford law, the simulated remodeling process predicts values of density close to those of the cortical bone in the whole section, starting from the outcomes at 60 days and up to one year, in contrast to the actual density distribution one year after the implant. On the contrary, the proposed strategy seems instead affect growth and remodeling, the densities curves replicating more faithfully the actual bone density profile measured at one year along the selected line of the femur slice of the benchmark.

A sensitivity analysis to set the activation factor

The function f_ζ is a measure of the surplus of fluid content strictly related to the applied loads to which the bone tissue is subjected to. If the fluid content is positive, the bone tissue has a surplus of fluid and, as a consequence, a greater supply of nutrients and a better capability to remove wastes of bone tissue cells. Conversely, when negative, the dearth of the fluid reduces the amount of nutrients available for bone tissue cells and helps the stack of wastes. When null, the bone tissue does not involve a strain due to proroelastic effect and thus the bone tissue is soaked by basal fluid, and the effect of the activator factor vanishes. Due to the key role of the fluid content on the remodeling phenomenon, a in silico sensitivity analysis to set the activator factor was performed on the real patient model. It must be highlighted that, despite the very encouraging obtained results, the *a priori* setting of the crucial parameter p of the proposed model would deserve a deeper discussion, for example by designing an experiment to be performed in vitro, a topic that is however beyond the scope of the present work.

Conclusions

The work presented an improved version of the classical so-called Stanford's law by coupling the direct effect of the stress stimulus on bone growth with the spatially inhomogeneous nutrient supply kindled by pressure gradients inside the bone, modeled as a poroelastic medium. Two study cases, i.e. a ring under axis-symmetrical conditions and a plate under pure bending, were built up and numerically solved, which gave symmetrical results when adopting standard approaches based on Stanford's law, giving instead non-symmetrical and biophysically coherent results in terms of bone density spatial distribution, if the classical Stanford's law was enriched by taking into account the role of the fluid transporting nutrients throughout the poroelastic bone medium. Moreover, a real patient case was considered, and once again, the results obtained with the proposed affect growth and remodeling, the densities curves replicating more faithfully the actual bone density profile measured at one year along the selected line of the femur slice of the benchmark. Although limitations still characterize some hypotheses at the basis of the present approach, the proposed model overcomes the intrinsic – and unexpected – independence of the bone remodeling from the stress sign and from the indirect effect of stress gradients driving nutrients through the flow of the fluid content in the tissue, allowing to predict important spatial asymmetries in bone mass density, so paving the way to more reliable mechanobiological strategies and engineering tools for the faithful prediction of bone remodeling, with implications in diagnosis of risk fracture, optimal design of bone prostheses and precise medicine.

References

[1] Minutolo V, Esposito L, Sacco E, Fraldi M. 2020. Designing stress for optimizing and toughening truss-like structures. Meccanica. 55:1603-1622. https://doi.org/10.1007/s11012-020-01189-z

[2] Della Corte A, Giorgio I, Scerrato D. 2020. A review of recent developments in mathematical modeling of bone remodeling. Proc Inst Mech Eng H: J Eng Med. 234(3):273-281. https://doi.org/10.1177/0954411919857599

[3] Cowin SC, Hegedus DH. 1976. Bone remodeling I: Theory of adaptive elasticity. J Elasticity. 6:313-326. https://doi.org/10.1007/BF00041724

[4] Cowin SC, Sadegh AM, Luo GM. 1992. An evolutionary Wolff's law for trabecular architecture. J Biomech Eng. 114:129-136. https://doi.org/10.1115/1.2895436

[5] Fyhrie DP, Carter DR. 1986. A unifying principle relating stress to trabecular bone morphology. J Orthop Res. 4:304-317. https://doi.org/10.1002/jor.1100040307

[6] Carter DR. 1987. Mechanical loading history and skeletal biology. J Biomech. 20:1095-1109. https://doi.org/10.1016/0021-9290(87)90027-3

[7] Beaupré GS, Orr TE, Carter DR. 1990a. An approach for time-dependent bone modelling and remodeling-theoretical development. J Orthop Res. 8:651-661. https://doi.org/10.1002/jor.1100080506

[8] Fornells P, Garcia-Aznar JM, Doblaré, M. 2007. A finite element dual porosity approach to model deformation-induced fluid flow in cortical bone. Ann Biomed Eng. 35:1687-1698. https://doi.org/10.1007/s10439-007-9351-5

[9] Fritton SP, Weinbaum S. 2009. Fluid and solute transport in bone: flow-induced mechanotransduction. Annu Rev Fluid Mech. 41:347-374. https://doi.org/10.1146/annurev.fluid.010908.165136

[10] Beaupré GS, Orr TE, Carter DR. 1990b. An approach for time-dependent bone modeling and remodeling--Application: a preliminary remodeling simulation, J Orthop Res. 8:662-670. https://doi.org/10.1002/jor.1100080507

[11] Esposito L, Minutolo V, Fraldi M, Sacco E. 2022. Stress peaks, stiffening and back-flow in bilayer poro-elastic metamaterials. Int J Solids Struct. Volumes 236-237:111334. https://doi.org/10.1016/j.ijsolstr.2021.111334

[12] Esposito, L.; Bifulco, P.; Gargiulo, P.; Gislason, M.K.; Cesarelli, C.; Iuppariello, L.; Jonsson, H.J.; Cutolo, A.; Fraldi, M. 2018. Towards a patient-specific estimation of intraoperative femoral fracture risk. Comput. Methods Biomech. Biomed. Eng. 21:663-672. https://doi.org/10.1080/10255842.2018.1508570

[13] Esposito L, Minutolo V, Gargiulo P, Jonsson, H, Gislason MK, Fraldi M. 2020. Towards an App to Estimate Patient-Specific Perioperative Femur Fracture Risk. Appl. Sci. 10:6409. https://doi.org/10.3390/app10186409

[14] Cowin SC. 1999. Bone poroelasticity. J Biomech. 32:217-238. https://doi.org/10.1016/S0021-9290(98)00161-4

[15] Esposito L, Bifulco P, Gargiulo P, Fraldi M. 2016. Singularity-free finite element model of bone through automated voxel-based reconstruction Comput Methods Biomech Biomed Engin. 19(3):257-262. https://doi.org/10.1080/10255842.2015.1014347

[16] Bergmann G, Deuretzbacherb G, Hellerc M, Graichena F, Rohlmanna A, Strauss J, Duda GN. 2001. Hip contact forces and gait patterns from routine activities. J Biomech. 34:859-871. https://doi.org/10.1016/S0021-9290(01)00040-9

Fluid-structure interaction (FSI) analysis of 3D printing personalized stent-graft for aortic endovascular aneurysm repair (EVAR)

Sara Ragusa[1, a], Katia Siciliano[2,b], Francesco P. Di Simone[2,b], Salvatore Russotto[1,c] Emanuela Bologna[1d *] and Massimiliano Zingales[1,e]

[1]Department of Engineering, University of Palermo, 90128 Palermo, Italy

[2]Amed s.r.l., 90128 Palermo, Italy

[a] sara.ragusa01@community.unipa.it, [b] info@amedsrl.com, [c]salvatore.russotto@unipa.it
[d] emanuela.bologna@unipa.it, [e] massimiliano.zingales@unipa.it

Keywords: Stent Graft Patient Specific, EVAR, Fluid-Structure Interaction

Abstract. Abdominal aortic aneurysm (AAA) is an irreversible dilation of abdominal aorta, which may rupture if not surgically treated. To date, most aorta stent-graft used in clinical practice are batch manufactured devices with a uniform diameter. Custom abdominal aortic stent-grafts are able to overcome standard stents limitations. In this study, a customized aortic stent-graft (NiTi - Dacron) for the treatment of AAA has been proposed. Fluid dynamics analyses were performed to deepen the hemodynamic of aneurysm vessel and the proposed patient-specific graft. By means this study, the authors have shown the real benefits of the device for the patient and the possibility to apply this new stent-graft in the near future.

Introduction

Abdominal aortic aneurysms (AAAs) are irreversible dilations of the abdominal aortic wall due to gradual weakening and remodeling, which alter physiologic blood flow patterns. AAA is defined as a permanent localized dilation of an artery with a 50% increase in diameter compared with expected normal diameter [1, 2]. The risk of AAAs increases of above 60 years population with a major incidence in men (four to six times more common in men than in women [3]), and in patients who present cardiovascular diseases, history of smoking or family history of aneurysmal disease. Progression of the disease can ultimately result in rupture of the abdominal aorta, which has mortality rates of 50-75% [4]. The surgical options for the aneurysm treatment include the conventional Open Surgical Repair (OSR) and the Endovascular Aneurysm Repair (EVAR). OSR is an invasive procedure consisting of a large incision on the site of aneurysm in order to remove and replace it by a synthetic graft. Treatments vary depending on the patient's condition and on a case-to-case basis. Choices include open-surgery procedures which are largely invasive or endovascular aneurysm repair (EVAR). Size and structure mismatch between standardized stent and aorta may cause internal leakage or rupture of blood vessels. With the development of precision medicine, personalized vascular stents that conform to the patient's true vascular structure have become the purpose of the next generation stents. EVAR has clear benefits in terms of less trauma, reduced mortality and lower morbidity.

Nevertheless, it has also limitations related to the anatomy and the morphology of the patient vessel and major post-operative complications may still occur due to stent graft migration and endoleaks. The conventional geometry are, in some cases, not compliant with the patient's blood vessels and the mismatch may cause internal leakage, rupture and the failure of EVAR treatment. For these reasons, the patient-specific stent grafts are considered the next generation of device, able to accommodate the patient's anatomical vascular structure. Hemodynamics is believed to play a key role in the formation and the progression of the dissecting aneurysm [7]. These numerical methods can provide detailed information about wall stresses and fluid flow. Several

study of AAA by means of computational models have been reported in literature [8–11]. Over the years, Computational Fluid Dynamics (CFD) analysis and, in the last decade, Fluid-Structure Interaction (FSI) are used to simulate the complex cardiovascular system.
The aim of the work is to evaluate the advantages of using a patient-specific stent graft for EVAR treatment by means FSI analysis. In this preliminary study, a case of abdominal aortic aneurysm was examined by means of CFD in order to compare hemodynamics in the proposed patient-specific graft with respect to fluid dynamics in abdominal aortic aneurismatic vessel and in a conventional graft, ideally implanted after aneurysm resection by means OSR.

Materials and Methods

Image Segmentation and Model Construction.

The three-dimensional model of the patient's aneurismatic aorta was reconstructed from the patient's CT images using the software Mimics 21.0 (Materialise). The first step in the reconstruction process is the image segmentation to highlight the region of interest, the aorta in this specific case. Based on this, to perform the CFD analyses, the AAA model was obtained and conventional and the patient-specific grafts models were reconstructed in Solidworks®.

Geometry Discretization.

In the CFD approach, the continuous domain volume was discretized in approximate discrete domain. The three fluid models (i.e. aneurysmatic aorta, conventional and patient-specific grafts) were imported in .STL format in ICEM CFD 19.2 (Ansys Inc.) to discretize the volume in small elements and realize the mesh. For each model, *tetrahedral* elements in the core region and *prismatic* elements in the boundary layer (3 layers) were used. The mesh created consist of about 1 million elements with a mesh quality above 0.3, ratio between the volume of the element and the cube of the radius of the sphere that circumscribes it. In addition, meshes were verified in order to eliminate the unconnected vertices.

Computational Fluid Dynamics Analysis.

Fluid dynamics was evaluated by a transient CFD simulation performed using FLUENT v19.2 (Ansys Inc.). CFD solver is used to obtain an approximate solution of the Navier-Stokes differential equations.

$$\frac{D\rho}{Dt} + \rho\nabla \cdot \boldsymbol{u} = 0 \tag{1}$$

$$\rho\left[\frac{\partial \boldsymbol{u}}{\partial t} + \boldsymbol{u}(\nabla \cdot \boldsymbol{u})\right] = -\nabla p + \mu\nabla^2\boldsymbol{u} + \rho g \tag{2}$$

where ρ is the density, ∇ is defined as $\nabla = \mathbf{i}\frac{\partial}{\partial x}+\mathbf{j}\frac{\partial}{\partial y}+\mathbf{k}\frac{\partial}{\partial z}$, u is the velocity, p is the pressure, μ is the viscosity, g the gravitational acceleration.

The analyses set-up are the same for all three fluid models.

The blood was assumed to be an incompressible and Newtonian fluid with a density ρ of 1060 kg/m^3 and viscosity μ of 0.003 Pa*s. Assuming a diameter of 0.5mm it is correct to assume a Newtonian flow through the aorta, in fact the blood viscosity is relatively constant at the high rates of shear (100 s^{-1}) typically found in the aorta [12]. The boundary conditions were imposed on inlet, outlets (abdominal, renal - branch1 and branch2 - and mesenteric) and the wall surfaces (Fig. 1). Mixed-type boundary condition represents physiological conditions more accurately because when the heart ventricles contract, it induces a change in volume that causes a pressure gradient and forces the blood out [13]. In addition, rigid and no slip conditions were assumed at wall. Two cardiac cycles have been simulated in order to achieve a solution independent from the initial conditions.

The investigated hemodynamic variables are Wall Shear Stress (WSS) and velocity. WSS in a blood vessel is the force per unit area created when tangential force (blood flow) acts on a surface (endothelium). The magnitude of WSS is proportional to the velocity gradient near the wall of the vessel, that is called wall shear rate:

$$\tau_w = \mu \left(\frac{\partial \gamma}{\partial t}\right)_{y=0} \tag{3}$$

Where μ is the dynamic viscosity, $\dot{\gamma}$ is shear rate and y the distance from the wall.

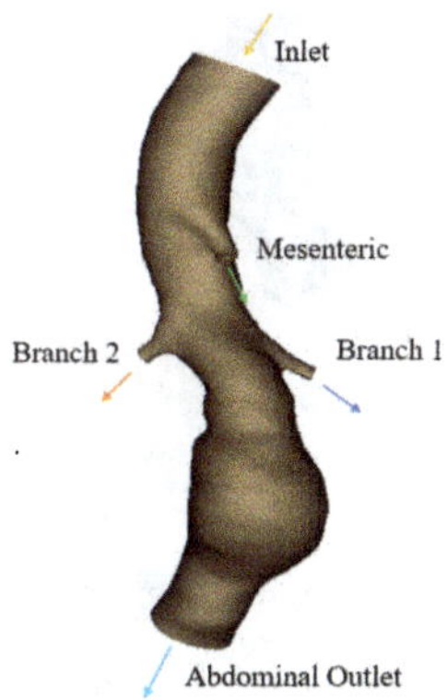

***Figure 1** CFD analysis set-up. The same surfaces are defined for both conventional graft fluid model and patient-specific graft fluid model*

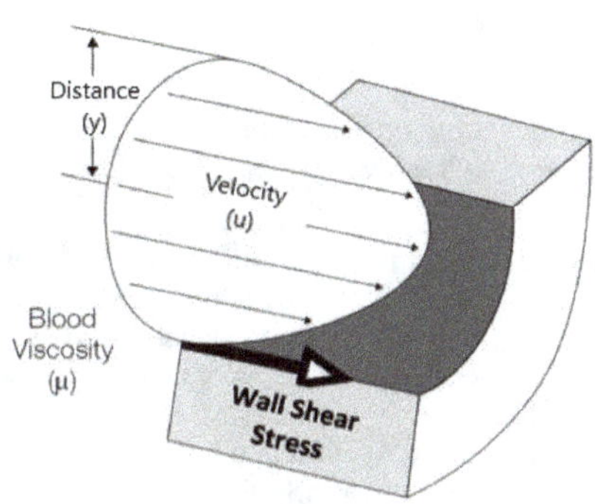

***Figure 2** Wall shear stress. The figure shows schematic illustration of the velocity profile experienced by inner surfaces of a vessel because of flowing blood*

Results and Discussion

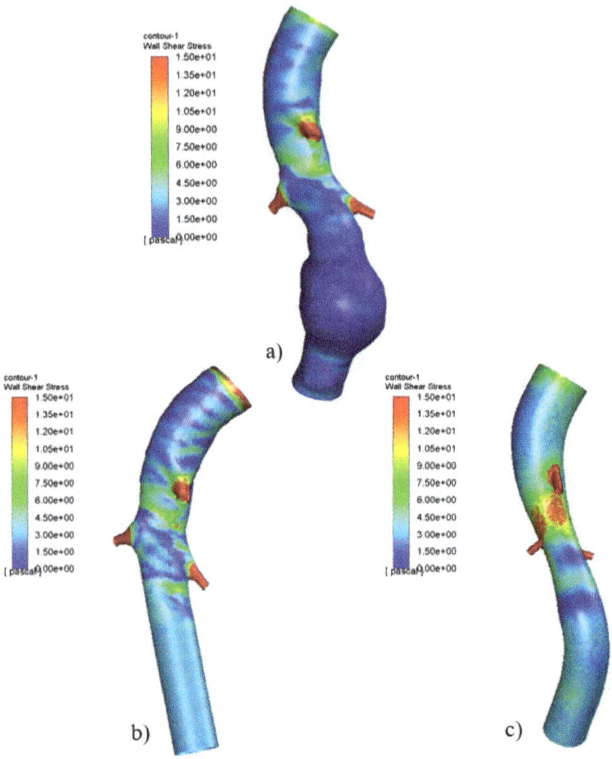

Figure 6 Wall Shear Stress at systole peak in a) aneurismatic aorta, b) conventional graft, c) patient-specific graft.

Wall shear stress (WSS) and velocity at peak systole are the hemodynamic in-depth variables in all three fluid models by CFD analyses. Fig. 6 shows the WSS in three fluid models. At the site of aneurysm, the WSS is zero (Fig. 6a). This indicates that blood flow does not exert tangential stresses on the wall due to the presence of low-velocity areas in the site as shown in Fig. 6a. The maximum WSS is located in branches: its value is of 184.6 Pa in aneurismatic vessel and the conventional graft (Fig. 6 a-b) compared to 142 Pa in patient-specific graft (Fig. 6c). The use of patient-specific graft reduces the maximum WSS of about 23% to the vessel and exhibits a distribution of wall shear stress more uniform than other models.

In Figure 7, the velocity magnitude of selected cross-sections is visualized at peak systole (0.313 s). Cross-section A is at aneurysm site, while cross-sections B and C are located in graft models at the same place of cross-section A. The use of both conventional and patient-specific graft removes the low-velocity area present in the aneurismatic vessel (Fig. 7a). The flow pattern in conventional graft (Fig. 7b) shows that the blood flow is not full developed and it is very different to the flow pattern of patient-specific graft (Fig. 7c). This aspect is associated to the conventional graft morphology and alters the vascular physiology. On the other hand, the flow pattern of patient-specific graft (Fig. 7c) is consistent and accommodates the vessel curvatures. The maximum of velocity is located in the branches in all three models.

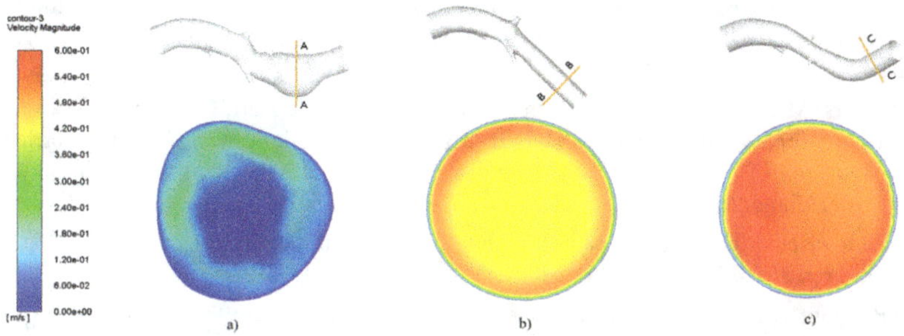

Figure 7 Velocity at systole peak in a) aneurismatic aorta, b) conventional graft, c) patient-specific graft.

Summary

In this work, we introduce a patient specific case study of the aortic aneurysm with the consequent design of a stent graft that can accommodate the complex anatomical structure of the patient. The device was modeled from abdominal aorta segmentation derived by CT scan images of abdomen. CFD, structural analysis and, more recently, the FSI are the numerical methods used for the study of cardiovascular disorders such as aneurysm even in patient specific cases. Computational methods were used in order to evaluate its hemodynamic advantages with respect to conventional graft. The fluid dynamics analysis carried out on the aneurysm vessel allowed to highlight salient aspects such as, first of all, the presence of flow recirculation and a high pressure profile. At the same time, it's worth to note how the insertion of a patient specific graft is beneficial.

The blood flow inside does not present the zones at low speed, not altering the physiology and fluid-dynamic of the vessel (unlike conventional graft) with a consistent fluid pattern. In addition, patient-specific graft reduces the maximum WSS of about 23% to the vessel and conventional graft and exhibits a distribution of Wall Shear Stress more uniform than aneurismatic vessel. The future goal is to implement the fluid structure interaction of the patient specific stent graft in order to evaluate the behavior of the two materials and their interactions when subjected to the fluid forces and, consequently, of the device itself.

References

[1] R. Erbel, V. Aboyans, C. Boileau, E. Bossone, R. D. Bartolomeo, H. Eggebrecht, A. Evangelista, V. Falk, H. Frank, O. Gaemperli, M. Grabenwöger, A. Haverich, B. Iung, A. J. Manolis, F. Meijboom, C. A. Nienaber, M. Roffi, H. Rousseau, U. Sechtem, P. A. Sirnes, … ESC Committee for Practice Guidelines. 2014 ESC Guidelines on the diagnosis and treatment of aortic diseases: Document covering acute and chronic aortic diseases of the thoracic and abdominal aorta of the adult. The Task Force for the Diagnosis and Treatment of Aortic Diseases of the European Society of Cardiology (ESC). European heart journal. *35*(41), (2014), 2873–2926. https://doi.org/10.5603/KP.2014.0225

[2] A. Daugherty, L.A. Cassis, Mechanisms of abdominal aortic aneurysm formation. *Current atherosclerosis reports* 4, (2002), 222–227. https://doi.org/10.1007/s11883-002-0023-5

[3] S. Aggarwal, A. Qamar, V. Sharma, A. Sharma, Abdominal aortic aneurysm: A comprehensive review. Experimental and clinical cardiology. 16(1), (2011) 11–15

[4] M. Xenos, S. H. Rambhia, Y. Alemu, S. Einav, N. Labropoulos, A. Tassiopoulos, J. J. Ricotta, & D. Bluestein, Patient-based abdominal aortic aneurysm rupture risk prediction with fluid structure interaction modeling. Annals of biomedical engineering. 38(11), (2010) 3323–3337. https://doi.org/10.1007/s10439-010-0094-3

[5] B.J.B.M. Wolters, M.C.M. Rutten, G.W.H. Schurink, U. Kose, J. de Hart, F.N. van de Vosse, A patient-specific computational model of fluid–structure interaction in abdominal aortic aneurysms. Medical Engineering and Physics. 27 (2005), pp. 871–883. https://doi.org/10.1016/j.medengphy.2005.06.008

[6] S. Jaunoo, Endovascular aneurysm repair (EVAR). International journal of surgery (London, England). 6 (2008), pp. 266–269. https://doi.org/10.1016/j.ijsu.2008.02.001

[7] K.M. Tse, P. Chiu, H.P. Lee, P. Ho, Investigation of hemodynamics in the development of dissecting aneurysm within patient-specific dissecting aneurismal aortas using computational fluid dynamics (CFD) simulations. Journal of Biomechanics. 44 (2011), pp. 827–836. https://doi.org/10.1016/j.jbiomech.2010.12.014

[8] E. S. Di Martino, G. Guadagni, A. Fumero, G. Ballerini, R. Spirito, P. Biglioli, & A. Redaelli, Fluid-structure interaction within realistic three-dimensional models of the aneurysmatic aorta as a guidance to assess the risk of rupture of the aneurysm. Medical Engineering and Physics. 9 (2001), pp. 647–655. https://doi.org/10.1016/S1350-4533(01)00093-5

[9] D.A. Vorp, M.L. Raghavan, M.W. Webster, Mechanical wall stress in abdominal aortic aneurysm: influence of diameter and asymmetry, J Vasc Surg. 27 (1998), pp. 632–63. https://doi.org/10.1016/S0741-5214(98)70227-7

[10] D.H. Wang, M.S. Makaroun, M.W. Webster, D.A. Vorp, Effect of intraluminal thrombus on wall stress in patient-specific models of abdominal aortic aneurysm. In: J Vasc Surg. 36 (2002), pp. 598–604. https://doi.org/10.1067/mva.2002.126087

[12] E. Soudah, E.Y. Ng, T.H. Loong, M. Bordone, U. Pua, S. Narayanan, CFD Modelling of Abdominal Aortic Aneurysm on Hemodynamic Loads Using a Realistic Geometry with CT, Computational and Mathematical Methods in Medicine (2013). https://doi.org/10.1155/2013/472564

[13] Christine M. Scotti and Ender A. Finol. 2007. Compliant biomechanics of abdominal aortic aneurysms: A fluid-structure interaction study. Comput. Struct. 85 (2007) 1097–1113. https://doi.org/10.1016/j.compstruc.2006.08.041

Theoretical and Applied Mechanics - AIMETA 2022 Materials Research Forum LLC
Materials Research Proceedings **26** (2023) 305-310 https://doi.org/10.21741/9781644902431-50

Fractional diffusion of membrane receptors in endocytosis pathway

Gianmarco Nuzzo[1, a *], Emanuela Bologna[1], Kaushik Dayal[2] and Massimiliano Zingales[1]

[1]Department of Engineering, University of Palermo, Viale delle Scienze ed. 8, 90128, Palermo, Italy

[2]Center for Nonlinear Analysis, Carnegie Mellon University, Pittsburgh, PA 15213, USA

[a]gianmarco.nuzzo@unipa.it

Keywords: Endocytosis, Biomechanics, Fractional Calculus, Stefan Problem, Phase Transformation

Abstract. In this paper the diffusion model representing the motion of membrane receptors with respect to virus endocytosis is considered in the context of applied mechanics. The unexpected behaviour of the receptor density that moves from higher concentrations in the unbound phase to lower concentration at the right of the virus surface is accounted for introducing a mechanical drift term in the governing equation so that the difference of concentrations, higher in the bounded phase and lower in the unbounded phase is accounted for in the receptor motion. Additionally, a non-gaussian model of diffusion has been introduced in terms of fractional generalization of the Fick law.

Introduction

Endocytosis is one of the main processes by which the cells of the human body receive substances and nutrients from the extracellular matrix, but it is also the route whereby viruses enter host cells to reproduce [1,2]. Generally, viruses have protuberances, small spikes, called ligands, on their outer surface. Ligands play a major role in the first phase of endocytosis, because they are responsible for the first contact by the virus with the cell membrane. The ligands form bonds with membrane receptors, which diffuse under the virus, to the area where endocytosis will occur, to allow the formation of various ligand-receptor pairs. As the virus descends toward the membrane, and more ligand-receptor pairs are formed, it exerts pressure on the membrane, deforming it. So, when the entire surface of the virus is in contact with the membrane, a concavity will have been formed in which the virus will reside. This last mechanism is called invagination and is completed as soon as the entire virus is surrounded by the cell membrane and passes the plasmalemma finding itself in the cytosol. Such a foreign body in the cell is called an endosome. This process will be followed by dissolution of the endosomatic capsule and biosynthesis of the viral components.

It's very important to know the time it takes viruses or nanoparticles to permeate the membrane, because on the one hand a virus that spends too much time to cross membrane will not be able to infect the host body, and on the other hand drug-containing nanoparticles can be functionalized more optimally to form ligand-receptor bonds effectively and act faster.

Recently the fractional calculus tool has provided excellent results to describe tissue biomechanics [3,4], proving its usefulness in describing tissue behaviour.

The aim of this paper is to propose a model that describes receptor diffusion, through a fractional form of Smoluchowski's fractional equation, which considers not only the chemical potential, given by Fick's law, but also an external potential in term of Morse potential.

The gaussian diffusion of membrane receptors

The mathematical formulation of the Stefan' problem proposed in previous section is not satisfactorily, from a mechanical perspective for the following reasons: i) The motion of the receptors following diffusion is always in the direction of the lower concentrations; ii) In order to achieve condition i) a first-order singularity in concentration appears at the moving boundary separating the bounded-unbounded phases of the receptors; iii) The force-flux relation presented by the Fick relation involves a random motion of diffusion of particles, non-mutually interacting, in a viscous media. This condition is hardly to find in receptors floating in the cell-membrane since the presence of other larger molecules of cholesterol, rafts and large protein-channels may be observed in real cellular membranes.

The aforementioned considerations pushed the authors, toward a different model of receptor diffusion that make use of the generalization of the Fick law with the introduction of fractional calculus [5,6,7,8,9]. In this regard, in the present section the first two issues i) and ii) will be considered introducing a mechanical drift in the model, derived from a specific potential, that represent the driving force that allows to overcome the hole of concentrations among the bounded-unbounded phases of the receptors. This potential represents an electrostatic attraction among the OH- sites of the receptors and the H+ protons free to connect in the virus ligands and, for them, a specific expression of potential has been chosen in the following. Let us consider a generic instant when the virus has already made first contact with the membrane:

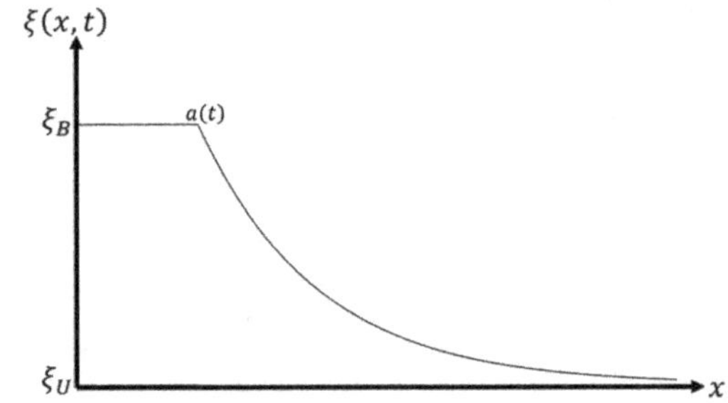

Fig. 1. Schematic representation of the receptor density profile of the proposed model.

Receptor diffusion follows Fick's classic law, in which there is an additional term:

$$j(x,t) = -\lambda \frac{\partial \xi(x,t)}{\partial x} - \frac{\xi(x,t)}{\zeta} \frac{dU(x)}{dx}. \tag{1}$$

Where $\lambda = \frac{k_B T}{\zeta}$ is the diffusivity coefficient $[\lambda] = \frac{L^2}{T}$, $\zeta = 4\pi\eta r$ friction factor $[\zeta] = \frac{M}{T}$, η membrane viscosity $[\eta] = \frac{M}{L\,T}$ e r radius of the receptor $[r] = L$. While $U(x)$ is the potential energy, which we decided to express through the following Morse potential:

$$U(x) = k_B T \left(1 - e^{-\sqrt{\frac{k_e}{2k_B T}}\,(x - a(t))}\right)^2. \tag{2}$$

Where k_e is the force at the minimum of the well. Contrary to the model [10], we assumed that we know the density of receptors at the interface. This density is the same as that found under the virus with regard to ligand-bound receptors, and we denote it by ξ_B. We also know the density of unbound receptors in areas far from where the process of endocytosis is occurring. We denote this density by ξ_U. Noting that $\xi_B > \xi_U$. Summarizing in mathematical terms, with reference to Fig. 1:

Materials Research Forum LLC
https://doi.org/10.21741/9781644902431-50

$$\xi(x,t) = \begin{cases} j(x,t) = 0, \xi(x,t) = \xi_B & if\ x \le a(t) \\ j(x,t) \to 0, \xi(x,t) = \xi_U & if\ x \to \infty \end{cases} . \quad (3)$$

To find the condition at the interface, we assume that over time the change in receptor density over the domain remains constant:

$$\frac{d}{dt}\left[\int_0^{a(t)} \xi(x,t)\, dx + \int_{a(t)}^{\infty} \xi(x,t)\, dx\right] = 0. \quad (4)$$

We apply Leibniz rule to (4) and taking into account the boundary conditions (3), we get:

$$\int_0^{a(t)} \frac{\partial \xi(x,t)}{\partial t} dx + \xi(a(t),t)\frac{da(t)}{dt} + \int_{a(t)}^{\infty} \frac{\partial \xi(x,t)}{\partial t} dx - \xi(a(t),t)\frac{da(t)}{dt} = 0. \quad (5)$$

We can elide the equal terms of (5) and since the density of receptors under the virus over time remains constant, we can erase the first integral, then:

$$\int_{a(t)}^{\infty} \frac{\partial \xi(x,t)}{\partial t} dx = 0. \quad (6)$$

Recalling the continuity equation: $\frac{\partial \xi(x,t)}{\partial t} = -\frac{\partial j(x,t)}{\partial x}$. We can substitute it into (6) and integrate:

$$[j(x,t)]_{a(t)}^{\infty} = 0. \quad (7)$$

The first term validated to infinity is null, while in the second term we have that the potential present a minin in $a(t)$ and therefore its derivative will be zero, so there remains only a single term:

$$\lambda \frac{\partial \xi(a(t),t)}{\partial x} = 0. \quad (8)$$

By deriving with respect to time the (8), the speed of the interface can be obtained:

$$\dot{a}(t) = -\frac{\frac{\partial \xi(a(t),t)}{\partial t \partial x}}{\frac{\partial^2 \xi(a(t),t)}{\partial x^2}}. \quad (9)$$

It is interesting to note that by adopting the form (2) of the Morse potential the diffusion of receptors is from higher to lower potentials, in this case that at the interface. Finally, we can write the complete system of the set of equations, in the form of Stefan's problem [11], describing the flow of receptors:

$$\begin{cases} \dfrac{\partial \xi(x,t)}{\partial t} = \lambda_\beta \dfrac{\partial^2 \xi(x,t)}{\partial x^2} + \dfrac{1}{\zeta}\left(\dfrac{\partial \xi(x,t)}{\partial x}\dfrac{dU(x)}{dx} + \xi(x,t)\dfrac{d^2U(x)}{dx^2}\right) & (10.a) \\ \xi(a(t),t) = c_B. & (10.b) \\ \xi(x \to \infty, t) = c_U. & (10.c) \\ \dot{a}(t) = -\dfrac{\frac{\partial \xi(a(t),t)}{\partial t \partial x}}{\frac{\partial^2 \xi(a(t),t)}{\partial x^2}}. & (10.d) \end{cases}$$

The domain of the $(10.a)$ is $a(t) < x < \infty, t > 0$.

The fractional-order model of membrane receptors diffusion

We introduce the fractional Fick law in terms of the Caputo derivative [12] with $0 < \beta < 1$:

$$j(x,t) = -\lambda_\beta \left({}_0^C D_t^\beta \frac{\partial \xi(x,t)}{\partial x}\right)(x,t) - \frac{\xi(x,t)}{\zeta}\frac{dU(x)}{dx}. \tag{11}$$

In which λ_β is the anomalous diffusivity coefficient with dimensions: $[\lambda_\beta] = \frac{L^2}{T^{1-\beta}}$. $U(x)$ is as previously the Morse potential. For the condition at the interface, we can follow the steps given above up to equation (7). So, we get: $\lambda\left({}_0^C D_t^\beta \frac{\partial \xi}{\partial x}\right)(a(t),t) = 0$. Using property ${}_0I_t^\beta\left({}_0^C D_t^\beta f\right)(t) = f(t) - f(0)$ and then deriving with respect to time and isolating the interface velocity, we obtain the same equation for the non-fractional system proposed before. To obtain the time-fractional diffusion equation, we substitute (11) into the continuity equation. The complete system will be:

$$\begin{cases} \dfrac{\partial \xi(x,t)}{\partial t} = \lambda_\beta \left({}_0^C D_t^\beta \dfrac{\partial^2 \xi}{\partial x^2}\right)(x,t) + \dfrac{1}{\zeta}\left(\dfrac{\partial \xi(x,t)}{\partial x}\dfrac{dU(x)}{dx} + \xi(x,t)\dfrac{d^2U(x)}{dx^2}\right) & (12.a) \\ \xi(a(t),t) = c_B. & (12.b) \\ \xi(x \to \infty, t) = c_U. & (12.c) \\ \dot{a}(t) = -\dfrac{\frac{\partial \xi(a(t),t)}{\partial t \partial x}}{\frac{\partial^2 \xi(a(t),t)}{\partial x^2}}. & (12.d) \end{cases}$$

Numerical investigation

For both system $(10.a - 10.d)$ and $(12.a - 12.d)$, it is not possible to obtain an analytical solution, so we applied the finite difference method. For simplicity we will report the case where $\beta = 0$ and then we get the non-fractional system again. Considering a generic particle with a diameter $d \cong 38\,nm$, the data used follow: $\eta = 10^{-3}\frac{pN\,\mu s}{nm^2}$, $r = 5\,nm$, $\zeta = 0.0628\frac{pN\,\mu s}{nm}$, $\lambda = 67.8\,\frac{nm^2}{s}$, $k_e = 0.01\,pN$, $c_u = 5 \times 10^{-4}\,nm^{-2}$,$c_B = 5 \times 10^{-3} nm^{-2}$. In the following graph, one can observe the curves obtained from the numerical solution for the receptor densities as a function of the radius of a generic virus for different times:

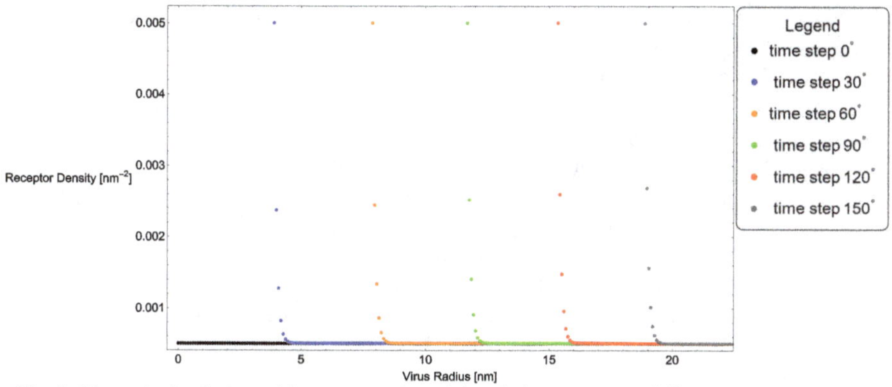

Fig. 2. Numerical solution of the proposed model with five curves at different time instants. The first point at the top of each curve represents the particle-membrane interface.

While in this second graph the corresponding interface position as a function of time. Each first point of the receptor density curves corresponds to one of the red curve points of the interface position.

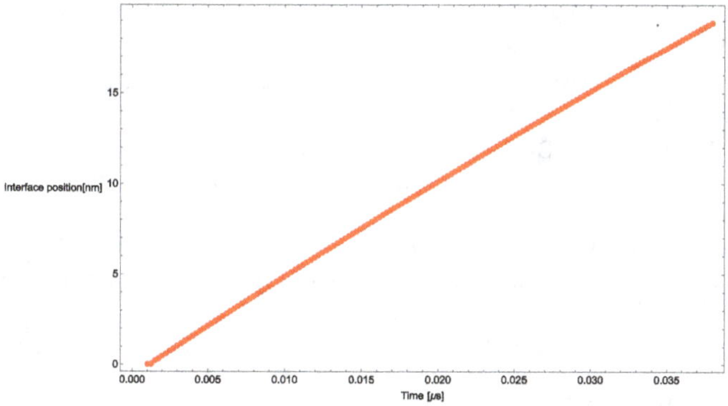

Fig. 3. Position of the particle-cell membrane interface as a function of time.

Conclusions

In this paper we analyzed the process of endocytosis and specifically the diffusion of receptors under the virus, an important step that allows the virus to bind to the cell membrane. We proposed a new model of receptor diffusion by considering an external potential as a function of the Morse potential. And finally proposed a generalized time-fractional form of the model, representing a typical sub-diffusive process, in which for $\beta = 0$ we regain the classical model.

References

[1] J. G. Joseph, & A. P. Liu (2020). Mechanical regulation of endocytosis: new insights and recent advances. *Advanced biosystems*, *4*(5), 1900278. https://doi.org/10.1002/adbi.201900278

[2] J. Shang, Y. Wan, C. Luo, G. Ye, Q. Geng, A. Auerbach, & F. Li (2020). Cell entry mechanisms of SARS-CoV-2. *Proceedings of the National Academy of Sciences*, *117*(21), 11727-11734. https://doi.org/10.1073/pnas.2003138117

[3] L. Deseri, P. Pollaci, M. Zingales, & K. Dayal (2016). Fractional hereditariness of lipid membranes: Instabilities and linearized evolution. *journal of the mechanical behavior of biomedical materials*, *58*, 11-27. https://doi.org/10.1016/j.jmbbm.2015.09.021

[4] E. Bologna, M. Di Paola, K. Dayal, L. Deseri, & M. Zingales (2020). Fractional-order nonlinear hereditariness of tendons and ligaments of the human knee. *Philosophical Transactions of the Royal Society A*, *378*(2172), 20190294. https://doi.org/10.1098/rsta.2019.0294

[5] G. Alaimo, V. Piccolo, A. Chiappini, M. Ferrari, D. Zonta, L. Deseri, & M. Zingales (2018). Fractional-order theory of thermoelasticicty. I: Generalization of the Fourier equation. *Journal of Engineering Mechanics*, *144*(2), 04017164. https://doi.org/10.1061/(ASCE)EM.1943-7889.0001394

[6] G. Alaimo, & M. Zingales (2015). Laminar flow through fractal porous materials: the fractional-order transport equation. *Communications in Nonlinear Science and Numerical Simulation*, *22*(1-3), 889-902. https://doi.org/10.1016/j.cnsns.2014.10.005

[7] M. Zingales (2016). An exact thermodynamical model of power-law temperature time scaling. *Annals of Physics*, *365*, 24-37. https://doi.org/10.1016/j.aop.2015.08.014

[8] M. Di Paola, & M. Zingales (2012). Exact mechanical models of fractional hereditary materials. *Journal of Rheology*, *56*(5), 983-1004. https://doi.org/10.1122/1.4717492

[9] G. Alaimo, V. Piccolo, A. Cutolo, L. Deseri, M. Fraldi, & M. Zingales (2019). A fractional order theory of poroelasticity. *Mechanics Research Communications*, *100*, 103395. https://doi.org/10.1016/j.mechrescom.2019.103395

[10] H. Gao, W. Shi, & L. B. Freund (2005). Mechanics of receptor-mediated endocytosis. *Proceedings of the National Academy of Sciences*, *102*(27), 9469-9474. https://doi.org/10.1073/pnas.0503879102

[11] S. C. Gupta (2017). *The classical Stefan problem: basic concepts, modelling and analysis with quasi-analytical solutions and methods* (Vol. 45). Elsevier. https://doi.org/10.1016/B978-0-444-63581-5.00012-9

[12] R. Gorenflo, A. A. Kilbas, F. Mainardi, & S. V. Rogosin (2020). *Mittag-Leffler functions, related topics and applications* (p. 540). New York, NY, USA: Springer. https://doi.org/10.1007/978-3-662-61550-8

Masonry modelling and analysis: from material to structures

Theoretical and Applied Mechanics - AIMETA 2022 Materials Research Forum LLC
Materials Research Proceedings **26** (2023) 313-318 https://doi.org/10.21741/9781644902431-51

Some recent advances and applications in Distinct Element modelling of masonry structures

Mattia Schiavoni[1,a], Ersilia Giordano[1,b] and Francesco Clementi[1,c *]

[1]Department of Civil and Building Engineering, and Architecture, Polytechnic University of Marche, Ancona, Italy

[a]m.schiavoni@pm.univpm.it, [b]e.giordano@pm.univpm.it, [c]francesco.clementi@univpm.it

Keywords: Masonry, Distinct Element Method, Nonlinear Dynamics

Abstract. In this paper, advanced numerical models are used to study the progressive damage of a historic building, namely the Palazzo of Podestà and the Civic Tower of Accumoli (central Italy). The dynamic behaviour of the structure is analyzed following important seismic events such as those that occurred in 2016-2017. Discontinuous and continuous approaches are used. In the formers, the masonry response is represented both with Discrete Element Method (DEM) and the Non-Smooth Contact Dynamic (NSCD) method; in the latter the masonry nonlinearity is replicated using the Concrete Damage Plasticity (CDP) model. The numerical results showed a good correspondence of all the approaches with the real damage suffered by the structure after the seismic sequence.

Introduction

In 2016, a series of catastrophic seismic shocks caused victims and considerable damage to the heritage structures in the Central Italy regions [1,2]. These earthquakes heavily stroked the Norcia, Visso, Arquata del Tronto, Amatrice and Accumoli villages [3,4].

Figure 1. The geographical location of Accumoli village.

The case study of this paper is located in the Accumoli village, in the Lazio region (central Italy). The historic center of Accumoli, before this disaster (Fig. 1), was the tangible testimony of a troubled history.

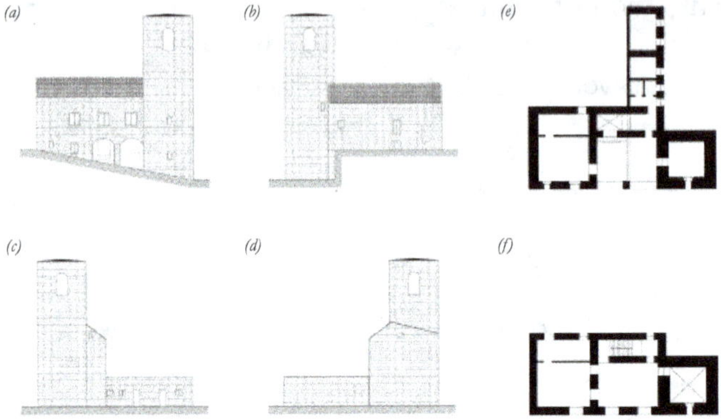

Figure 2. North (a), South (b), East f (c) and West façade (d), Ground (e) and first floor (f).

Accumoli dates to 1211, the territory was initially much larger than the current ones including several municipalities. At the heart of this small village there is the case study of this paper, i.e. the Civic Tower of Accumoli and the adjacent Podestà palace. The Tower dates to the twelfth century, it is unique in its kind in the entire Tronto's valley. It is in via Tommasi, featuring a square plan measuring 6.15x6.15 m^2 with walls of 1.20 m thickness at the base and 1.00 m at the top. In elevation, the maximum height exceeds 20 m (Fig. 2). The bearing structure consists of multi-leaf walls [5,6], the perimeter curtains are in cut stones and the inner core is in irregular stones. The structure ends with a pitched roof in reinforced concrete.

Next to the tower, there is the Palace of Podestà. It dates to the thirteenth century, and it is the oldest structure in the Accumoli village. The palace has a rectangular plan of 8.6x16 m^2 dimensions and a maximum height of 10 m. It consists of ground and noble floors. In the latter, there are architrave windows, instead on the ground floor there are two arched openings, typical of medieval public buildings. The bearing structure is made of square and smooth sandstone ashlars. (Fig. 2). On the north side of the palace, there is a little annex of one floor. It has a rectangular plant of 10x3.8 m.

After the 2016 earthquakes, the complex exhibited visible cracks, especially on the tower as visible in Figure 3.

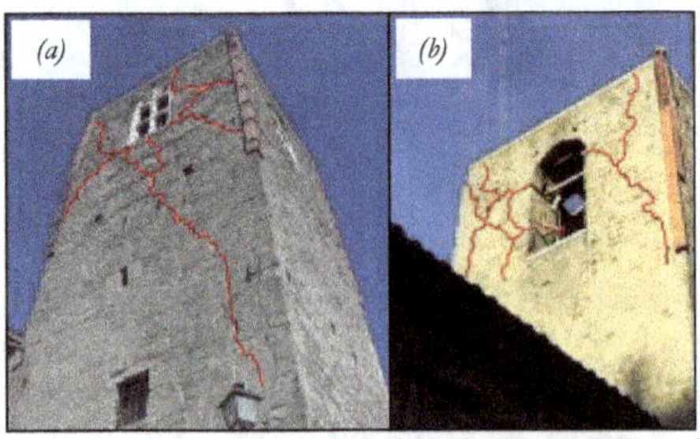

Figure 3. Real damage of the North façade (a) and West façade (b).

Discontinuous and continuous approaches

Numerical models were created to have a complete picture of the progressive damage that the structure undergoes when subjected to a seismic sequence. A comparison was made between the

Materials Research Proceedings **26** (2023) 313-318 https://doi.org/10.21741/9781644902431-51

discontinuous and the continuous approach [7–9].

In the discontinuous approach, the dynamic response of the structure is studied with the Non-Smooth Contact Dynamic (NSCD) method [10–12] implemented in LMGC90© open-source code and with the Distinct Element Method (DEM) [13–15] implemented in 3DEC© code. A 3D numerical model is created, and the masonry is discretized in different individual rigid blocks, reproducing the real behaviour of the structure, and analysing the progressive damage under a seismic action [16–18]. The size of the blocks has been approximated to obtain a fair compromise between a good degree of detail and a not too expensive computational burden (Fig. 4) [12]. Furthermore, it was decided to model the tower also considering its internal filling (Figure 4.b-c-d) as can be seen in Figure 4c, made with larger blocks than the two external leaves in order to limit the computational burden.

The blocks in the NSCD method are subjected to Signorini's law (i.e., impenetrability condition) and to the dry-friction Coulomb's law. The contacts have than a non-smooth nature.

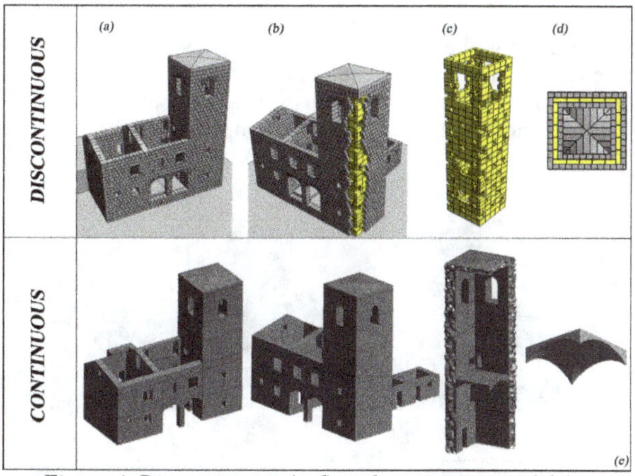

Figure 4. Discontinuous (a-d) and continuous model (e).

DEM models use smooth functions to represent interactions between blocks. A Mohr-Coulomb constitutive model is used.

Differently, in the continuous approach, the masonry is approximated to a fictitious and homogeneous isotropic medium [19]. To reproduce the nonlinear behaviour of the masonry, the Concrete Damage Plasticity (CDP) model [20,21] is used. Such a model is one of the most used to analyse the masonry behaviour under horizontal loads [8]. Indeed, it can simultaneously provide the compression and tension damages and consider the recovery of stiffness due to cracks closure.

Numerical results

The structural behaviour of the structure was first studied under gravitational loads, then the records of the three most important seismic events that occurred in Central Italy in 2016 were applied in the three main directions to obtain a first comparison between the numerical and the real damages.

The three events taken into consideration are summarized in Table 1. The events were applied in sequence considering 10 seconds of the peak amplitude of each and two seconds of rest between one event and another, the total time histories apply were of 34 seconds.

*Table 1. Characteristic of the three main shocks of the Central Italy seismic sequence of 2016 recorded in Amatrice (AMT) and Accumoli (ACC) station, where * indicates that the site classification is not based on a direct $V_{s,30}$ measurements.*

Seismic event		M_L	Depth [km]	Station	Class EC8	R [km]	R [km]	R [km]	Channel NS PGA [cm/s²]	Channel EW PGA [cm/s²]	Channel UD PGA [cm/s²]
1st	24/08/2016	6	8.1	AMT	B*	1.38	4.62	8.5	368.39	-850.8	391.37
2nd	26/10/2016	5.9	7.5	AMT	B*	25.93	26.09	33.3	-58.55	90.74	-49.11
3rd	30/10/2016	6.1	9.2	ACC	B*	35.33	35.32	47.10	-122.44	75.95	-44.07

Figure 5 shows an excellent correspondence of results between DEM and NSCD. The part of the structure that is most affected by the seismic sequence is the belfry; in fact, following the first two events the formation of crack patterns on it is visible. Following the earthquake of 30th October 2016, it is possible to notice the propagation of crack patterns between the palace and the tower. Furthermore, the palace after the last shock is damaged near the openings on the ground floor. The tower instead highlights the formation of vertical crack patterns on all the façades. On the other hand, the FE model also shows the vulnerability of the bell-cell. After the first event, horizontal cracks at its base are visible; instead, at the end of the sequence, a crack appeared also in the upper corner of the single arched windows. In contrast to the discrete approaches, the continuous showed important cracks in the connection between the tower and the palace just after the first events (Fig. 5).

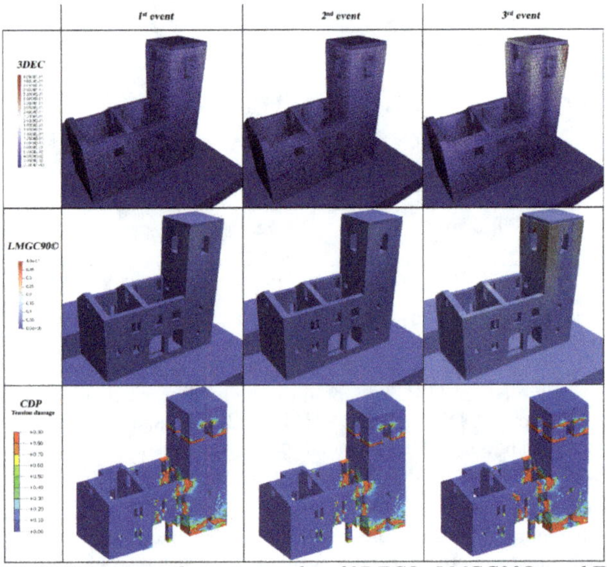

Figure 5. Comparison between results of 3DEC©, LMGC90©, and FEM.

Finally, it is important to stress the fact that all the approaches were able to identify the bell-cell as the most vulnerable element, as incidentally demonstrated by the 2016 seismic sequence. However, the cracks' pattern is more accurate with the discontinuous approaches instead than with continuous one. It is certainly linked to a loss of the resistant properties of the mortar, but also to a masonry's texture with irregularities, at least in the filling. With these boundary conditions, the masonry exhibits crumbling instead of a monolithic behaviour, the first one impossible to catch with continuous approaches.

Conclusions

The Palazzo of Podestà and the Civic Tower were studied through continuous and discontinuous approaches. The latter allows to represent the discontinuities of the masonry, capturing the dynamic behaviour of the structure following the seismic sequence of Central Italy in 2016. It is possible to note a good correspondence between the models with the NSCD and DEM methods and the real damage. In particular, the global and local behaviour of the structure is obtained. Furthermore, good results are obtained with the continuous approach, which, however, unlike the discontinuous approach, cannot faithfully reproduce the cracks because the texture of the masonry has a high influence on the structural response.

Acknowledgments

The authors express their gratitude to Itasca and Harpaceas for providing access and support to 3DEC under the IEP program. The authors gratefully acknowledge the support of the Italian Ministry of University (Italy), under the program "Dipartimento di Eccellenza" of the Department of Civil and Building Engineering, and Architecture - Polytechnic University of Marche (Ancona, Italy).

References

[1] G. Milani, Lesson learned after the Emilia-Romagna, Italy, 20–29 May 2012 earthquakes: A limit analysis insight on three masonry churches, Engineering Failure Analysis. 34 (2013) 761–778. https://doi.org/10.1016/j.engfailanal.2013.01.001

[2] M. Valente, G. Milani, Damage assessment and collapse investigation of three historical masonry palaces under seismic actions, Engineering Failure Analysis. 98 (2019) 10–37. https://doi.org/10.1016/j.engfailanal.2019.01.066

[3] F. Clementi, G. Milani, A. Ferrante, M. Valente, S. Lenci, Crumbling of amatrice clock tower during 2016 central Italy seismic sequence: Advanced numerical insights, Frattura Ed Integrita Strutturale. 14 (2020). https://doi.org/10.3221/IGF-ESIS.51.24

[4] F. Clementi, Failure Analysis of Apennine Masonry Churches Severely Damaged during the 2016 Central Italy Seismic Sequence, Buildings. 11 (2021) 58. https://doi.org/10.3390/buildings11020058

[5] A.M. D'Altri, G. Castellazzi, S. de Miranda, Collapse investigation of the Arquata del Tronto medieval fortress after the 2016 Central Italy seismic sequence, Journal of Building Engineering. 18 (2018) 245–251. https://doi.org/10.1016/j.jobe.2018.03.021.

[6] F. Clementi, A. Ferrante, S. Lenci, The Non-smooth Dynamics of Multiple Leaf Masonry Walls of the Arquata Del Tronto Fortress, in: 2020: pp. 1798–1807. https://doi.org/10.1007/978-3-030-41057-5_145

[7] A. Ferrante, E. Giordano, F. Clementi, G. Milani, A. Formisano, FE vs. DE Modeling for the Nonlinear Dynamics of a Historic Church in Central Italy, Geosciences (Basel). 11 (2021) 189. https://doi.org/10.3390/geosciences11050189

[8] E. Giordano, E. Bertolesi, F. Clementi, M. Buitrago, J.M. Adam, S. Ivorra, Unreinforced and TRM-Reinforced Masonry Building Subjected to Pseudodynamic Excitations: Numerical and Experimental Insights, Journal of Engineering Mechanics. 147 (2021). https://doi.org/10.1061/(ASCE)EM.1943-7889.0002017

[9] G. Standoli, G.P. Salachoris, M.G. Masciotta, F. Clementi, Modal-based FE model updating via genetic algorithms: Exploiting artificial intelligence to build realistic numerical

models of historical structures, Construction and Building Materials. 303 (2021) 124393. https://doi.org/10.1016/j.conbuildmat.2021.124393

[10] M. Jean, The non-smooth contact dynamics method, Computer Methods in Applied Mechanics and Engineering. 177 (1999) 235–257. https://doi.org/10.1016/S0045-7825(98)00383-1

[11] J.J. Moreau, Unilateral Contact and Dry Friction in Finite Freedom Dynamics, in: Nonsmooth Mechanics and Applications, Springer Vienna, Vienna, 1988: pp. 1–82. https://doi.org/10.1007/978-3-7091-2624-0_1

[12] A. Ferrante, M. Schiavoni, F. Bianconi, G. Milani, F. Clementi, Influence of Stereotomy on Discrete Approaches Applied to an Ancient Church in Muccia, Italy, Journal of Engineering Mechanics. 147 (2021) 04021103. https://doi.org/10.1061/(ASCE)EM.1943-7889.0002000

[13] P.A. Cundall, A computer model for simulating progressive large-scale movements in blocky rock systems, in: Proocedings of the Symposio of the International Society of Rock Mechanics, Nancy 2, 1971

[14] J. V Lemos, Discrete Element Modeling of Masonry Structures, International Journal of Architectural Heritage. 1 (2007) 190–213. https://doi.org/10.1080/15583050601176868

[15] S. Chen, A. Ferrante, F. Clementi, K. Bagi, DEM analysis of the effect of bond pattern on the load bearing capacity of barrel vaults under vertical loads, International Journal of Masonry Research and Innovation. 6 (2021) 346. https://doi.org/10.1504/IJMRI.2021.116234

[16] N. Mendes, S. Zanotti, J. V. Lemos, Seismic Performance of Historical Buildings Based on Discrete Element Method: An Adobe Church, Journal of Earthquake Engineering. 24 (2020) 1270–1289. https://doi.org/10.1080/13632469.2018.1463879

[17] A. Ferrante, F. Clementi, G. Milani, Advanced numerical analyses by the Non-Smooth Contact Dynamics method of an ancient masonry bell tower, Mathematical Methods in the Applied Sciences. (2020) mma.6113. https://doi.org/10.1002/mma.6113

[18] A. Ferrante, F. Clementi, G. Milani, Dynamic Behavior of an Inclined Existing Masonry Tower in Italy, Frontiers in Built Environment. 5 (2019). https://doi.org/10.3389/fbuil.2019.00033

[19] F. Bianconi, G.P. Salachoris, F. Clementi, S. Lenci, A Genetic Algorithm Procedure for the Automatic Updating of FEM Based on Ambient Vibration Tests, Sensors. 20 (2020) 3315. https://doi.org/10.3390/s20113315

[20] J. Lee, G.L. Fenves, Plastic-Damage Model for Cyclic Loading of Concrete Structures, Journal of Engineering Mechanics. 124 (1998) 892–900. https://doi.org/10.1061/(ASCE)0733-9399(1998)124:8(892)

[21] J. Lubliner, J. Oliver, S. Oller, E. Oñate, A plastic-damage model for concrete, International Journal of Solids and Structures. 25 (1989) 299–326. https://doi.org/10.1016/0020-7683(89)90050-4

Theoretical and Applied Mechanics - AIMETA 2022 Materials Research Forum LLC
Materials Research Proceedings **26** (2023) 319-324 https://doi.org/10.21741/9781644902431-52

Sequential linear analysis of no-tension masonry structures

Grigor Angjeliu [a] *, Matteo Bruggi [b] and Alberto Taliercio [c]

Department of Civil and Environmental Engineering, Politecnico di Milano
Piazza Leonardo da Vinci, 32, 20133, Milan, Italy

[a]grigor.angjeliu@polimi.it, [b]matteo.bruggi@polimi.it, [c]alberto.taliercio@polimi.it

Keywords: Masonry, Elastic No-Tension Material, Sequential Linear Analysis

Abstract. A new application of the elastic no-tension material model is developed through Sequential Linear Analysis (SLA) to analyze masonry structures. The approach has been demonstrated to be more robust compared to incremental analysis procedures. In the SLA framework, the equilibrium state of a masonry-like material is sought through a series of linear elastic analyses. In the loading process, cracking strains are simulated by sequential reduction of the directional stiffness upon violation of the no-tension constraint in terms of principal stresses. Some applications are presented to show the effectiveness of the proposed method in analyzing masonry structures under the effect of gravity, lateral loading, and ground settlements.

Introduction

In the last decades a number of refined models have been developed for masonry structures based on the theory of plasticity and damage mechanics [1]. Although they have been shown to be quite accurate in simulating the mechanical behavior of masonry specimens and small-scale buildings tested in the laboratory, these models require many input parameters, which are hardy available in practical situations. A comparison between limit analysis solutions and finite element methods in the stability assessment of masonry structures was addressed in [2]. An alternative is the so-called no-tension material model, which completely neglects the limited tensile strength of masonry [3][4]. The reduced number of parameters required by the no-tension model makes it very appealing for unreinforced masonry constructions, compared to more refined constitutive models.

Despite the apparent simplicity of the linear elastic no tension (ENT) model, numerical issues arise as discontinuities in the stress and displacement fields have to be dealt with, leading to convergence issues. In fact, the application of ENT material models remains still limited. Readers may refer in particular to Angelillo [5], who proposed a finite element solution based on a complementary energy theorem for ENT bodies, or to Bruggi and Taliercio [6], who reformulated the analysis of no-tension structures as an energy-based problem introducing an equivalent orthotropic material.

In this contribution, the analysis of no-tension structures in plane stress conditions is dealt with through the application of a procedure of Sequential Linear Analysis (SLA). This technique provides a robust alternative to traditional incremental-iterative methods for finite element simulations, since it transforms the problem into a series of linear elastic analyses, see in particular [7]. The implementation presented here is developed through the combined application of a user subroutine in Abaqus, Python and Matlab scripts.

Two benchmark cases have been analyzed to assess the capabilities of the proposed model under vertical loads, lateral loads, and ground settlements.

Governing equations

According to [3][8], a no-tension masonry-like material has to fulfill a prescribed set of conditions. The stress tensor must be negative semidefinite:

$$\sigma_{ij} \in Sym^{-}, \tag{1}$$

where Sym^- is the closed cone of negative semi-definite symmetric second-order tensors.

The strain tensor is assumed to consist of two parts, an elastic part ε_{ij}^e and a latent part ε_{ij}^c accounting for cracking:

$$\varepsilon_{ij} = \varepsilon_{ij}^e + \varepsilon_{ij}^c. \tag{2}$$

Finally, the elastic strain is related to the stress σ_{ij} through the elasticity tensor C_{ijhk}, while the latent strain, ε_{ij}^c, follows the normality condition:

$$\sigma_{ij}\varepsilon_{ij}^c = 0 \quad \text{and} \quad \varepsilon_{ij}^c \in Sym^+, \tag{3}$$

where Sym^+ is the closed cone of the positive semi-definite symmetric second order tensors.

Sequential Linear Analysis Methodology

The domain occupied by the structure is discretized by 4-node plane-stress finite elements with one integration point (Fig. 1a). The SLA consists in a series of elastic analyses with modified stiffness and orientation of each element, until a stable value of the total strain energy is reached. This criterion is used as convergence condition for the iterative procedure.

Upon violation of the condition that states that all the principal stresses must be negative, the original isotropic material is replaced by an equivalent orthotropic one, with vanishing stiffness in tension. The symmetry axes, $\tilde{z}_1$, and $\tilde{z}_2$, of the equivalent orthotropic material and the principal stress directions, z_I and z_{II}, of the no-tension medium share the same orientation with respect to the general reference system Oz_1z_2 (Fig. 1b). This is achieved through a process of alignment of the symmetry axes of the equivalent orthotropic material with the principal stress directions, z_α, $\alpha = I, II$, detected in the no-tension medium.

The principal stresses are computed as the eigenvalues of the stress tensor at the Gauss points, whereas the principal directions are found as the relevant eigenvectors.

In order to track cracking strains, at each Gauss point, two nondimensional material densities $\rho_i \in (0,1]$ are introduced along the material axes; these parameters govern the stiffness penalization of the orthotropic material along $\tilde{z}_1$ and $\tilde{z}_2$. The material density ρ_i is related to the damage variable D_i through the expression $D_i = (1 - \rho_i)$. If any of the principal stresses becomes positive, the material density variable is initialized with a value, $\rho_0 = 0.25$, whereas further reductions are performed using a quadratic reduction factor, i.e.:

$$\rho_{k+1,i} = 1 - \left(1 - \rho_{k,i}\right)^2, i = 1,2. \tag{4}$$

In each material direction, the model can capture elastic (negative) compressive strains, or positive strains, which correspond to cracking strains. In the latter case, a scaled stiffness is computed in the direction where the principal stress becomes positive, to account for cracking.

The stress-strain law written in matrix form in the general reference system reads:

$$\boldsymbol{\sigma} = \boldsymbol{D}\,\boldsymbol{\varepsilon} \tag{5}$$

where:

$$\boldsymbol{D} = \boldsymbol{T}(\theta)\ \widetilde{\boldsymbol{D}}\ \boldsymbol{T}(\theta)^T \tag{6}$$

being T a transformation matrix and θ the angle between z_1 and $\boldsymbol{z}_I$. In plane stress, the stiffness matrix in the material (principal stress) reference system reads:

$$\widetilde{\boldsymbol{D}} = \frac{1}{1-\tilde{\nu}_{12}\tilde{\nu}_{21}}\begin{bmatrix} \tilde{E}_1 & \tilde{\nu}_{12}\tilde{E}_2 & 0 \\ \tilde{\nu}_{21}\tilde{E}_1 & \tilde{E}_2 & 0 \\ 0 & 0 & \tilde{G}_{12}(1-\tilde{\nu}_{12}\tilde{\nu}_{21}) \end{bmatrix} \tag{7}$$

being $\tilde{E}_1$ and $\tilde{E}_2$ the elastic moduli of the equivalent orthotropic material along the axis $\tilde{z}_1$ and $\tilde{z}_2$, respectively, $\tilde{G}_{12}$ the in-plane shear modulus, and $\tilde{\nu}_{12}$ $\tilde{\nu}_{21}$ the material Poisson's ratios such that $\tilde{\nu}_{21}\tilde{E}_1 = \tilde{\nu}_{12}\tilde{E}_2$. When the negative stress criterion is violated at any Gauss point, the equivalent moduli are reduced with respect to those of the original material as follows:

$$\tilde{E}_1 = \rho_1 E_0, \quad \tilde{E}_2 = \rho_2 E_0, \quad \tilde{G}_{12} = \rho_1\rho_2 \frac{E_0}{2(1+\nu)} \tag{8}$$

being E_0 and ν the elastic constants of the original isotropic material. At the same time, the Poisson's ratios of the equivalent orthotropic material are given by:

$$\tilde{\nu}_{12} = \rho_2 \nu, \quad \tilde{\nu}_{21} = \rho_1 \nu. \tag{9}$$

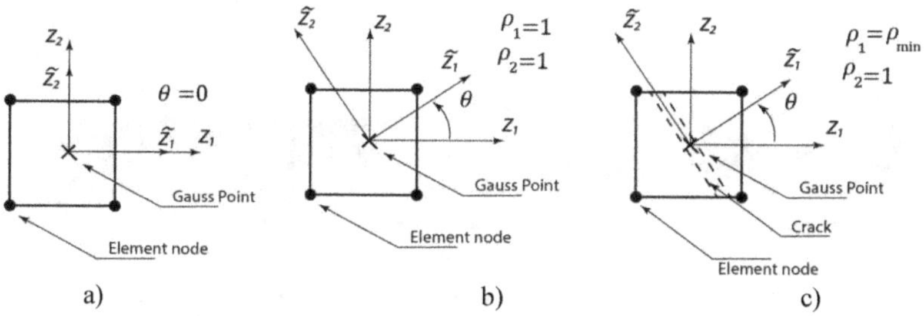

Fig. 1. State of stress in a plane element: a) initial state, b) alignement of the material axes with the principal stress directions, c) damage onset normal to a nonnegative principal stress.

Applications no. 1: Masonry panel under gravity loads and soil settlements

The aim is to test the ability of the ENT model to simulate the behavior of a masonry wall over a strip foundation subjected to ground settlements along a portion of its constrained boundary. The wall is supposed to be 3.2 m long, 0.6 m high and 0.1 m thick; only the central part, for an extension of 1.2 m, is affected by settlements (Fig. 2). The initial modulus of elasticity, E_0, is assumed to be equal to 1020 MPa. The panel is discretized with a mesh 75 x 75 mm^2. The wall is analyzed under self-weight condition, while settlements are simulated by removing all the constraints at the nodes of the central region.

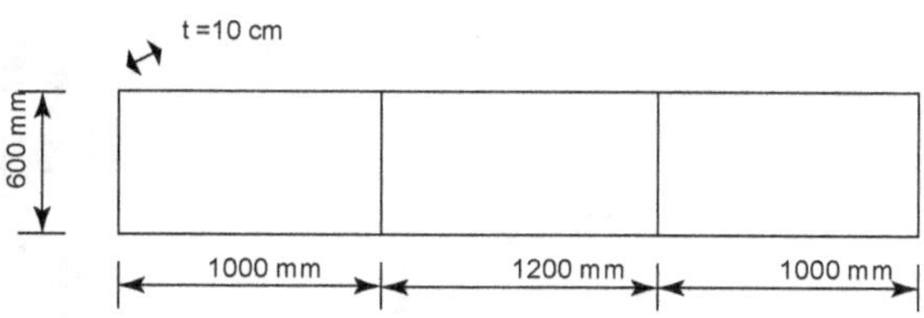

Fig. 2. Scheme of a masonry wall experiencing ground settlements in the central region.

The numerical simulations show the formation of a relieving arch (Fig. 3), which is typical of no-tension bodies. Starting from the ends of the fixed boundary, the direction of the principal compressive stress decreases and becomes horizontal at the middle of the unconstrained region. The rest of the model is not affected by settlements, and only nearly vertical compressive stresses are found.

The formation of the arch-like compressive stress path affects the displacement field significantly. In fact, compared to the conventional elastic case (Fig. 4a), where settlements affect also the neighboring regions, in the case of the ENT material displacements of non-negligible amplitude arise only beneath the inner arch (Fig. 4b). The zone above experiences much smaller displacements, whereas the rest of the model is virtually undeformed. This is in line with the schemes proposed by Mastrodicasa [9] regarding the effects of settlements in masonry walls.

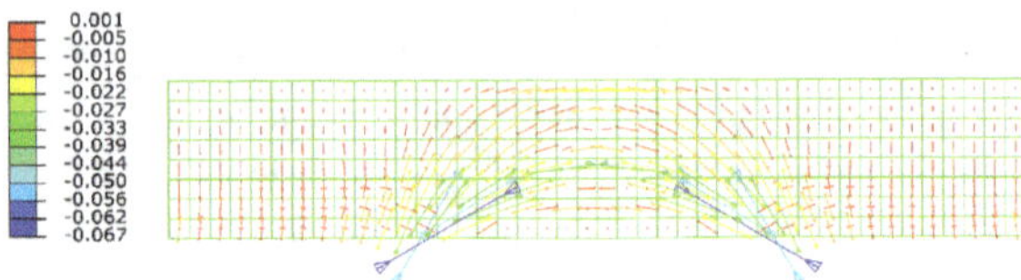

Fig. 3. Wall with settlements. Principal stresses from the ENT simulation.

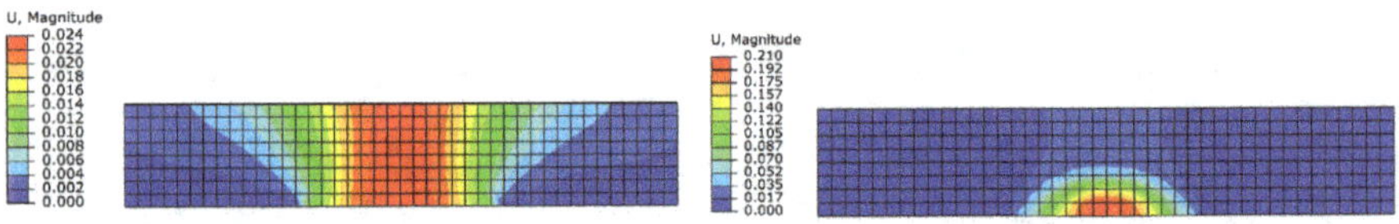

Fig. 4. Wall with settlements. Displacement contour plots: a) elastic model with symmetric behaviour in tension/compression, b) Elastic No-Tension model.

Application no. 2: Masonry panel under vertical pressure and horizontal loading
A slender masonry panel with dimensions 2.7m x 1.1m x 0.102m was tested at TU Delft [10] (experiment TUDCOMP-20): this experiment will be now adopted to validate the proposed numerical procedure. The top of the panel is loaded with a uniform vertical pressure of 0.63MPa. The pressure is applied through a horizontal steel beam and is kept constant during the experiment.

The domain is discretized by 100 mm x 100 mm square elements. The top steel beam is modelled though a row of elastic elements in the mesh, and is allowed to rotate similarly to the experimental setup. The material properties are: E_0=4972 MPa, ν=0.16. The compressive strength of the material, f_c, reported by the experimenters, is of 6.35 MPa.

In Fig. 5 the distribution of the principal stresses and that of the "void" phase (elements with biaxial damage, in blue) are both shown at the first converged stage in which any of the principal compressive stresses reaches the strength limit f_c.

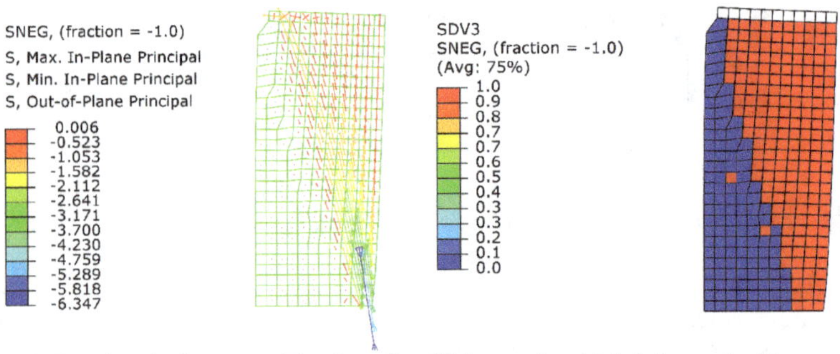

Fig. 5. Panel under horizontal loading, first SLA stage in which f_c is reached in some element: a) distribution of the principal stresses in the masonry wall, b) distribution of "void" elements, i.e. those with biaxial damage (in blue).

Materials Research Proceedings **26** (2023) 319-324 https://doi.org/10.21741/9781644902431-52

The collapse load is estimated in 12.77 kN. This value matches quite well the horizontal branch of the experimental curve whose maximum value is 12.8 kN.

The analytical calculation considering a simple panel overturning under vertical and horizontal loads gives a value of 14.4 kN, if the hinge is formed at the corner, or a value of 13.1 kN, if the hinge is considered 50 mm far away. The second value seems to be more representative of the considered case, since in the numerical model the hinge is allowed to occur at the midpoint of the side of a finite element, that is, at 50mm from the corner. Therefore, the difference between the numerical failure load and the analytical model is about 2.4%.

Regarding the displacement capacity, it must be remarked that limit analysis can predict only the limit load. However, if we consider a finite strength in compression, the predicted displacements when f_c is reached is 13.2 mm. Additional converged steps were obtained, but the compressive strength of masonry is exceeded.

Conclusions

In this contribution, the ENT model has been implemented in the framework of sequential linear analysis, which consists in a number of elastic analyses sequentially launched. The properties of the equivalent orthotropic material that replaces the ENT material to avoid the occurrence of tensile stresses are obtained through a penalization procedure.

The proposed approach was used to analyze masonry structures subjected to prescribed loads or ground settlements. In the latter case, the cracking strains computed by the SLA procedure match the real crack pattern expected at incipient collapse fairly well. Also, the limit load of a panel tested under increasing horizontal load and dead vertical load was predicted with good accuracy.

An advantage of the proposed approach compared to traditional incremental approaches is its inherent robustness.

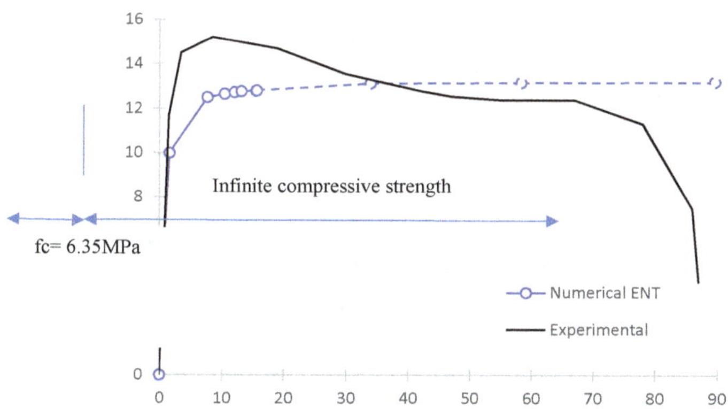

Fig. 6. Panel under horizontal loading. Comparison of experimental and numerical shear vs displacement capacity curve. Displacement are in mm, loads in kN.

References

[1] P.B. Lourenço, J.G. Rots, J. Blaauwendraad, Continuum Model for Masonry: Parameter Estimation and Validation, Journal of Structural Engineering, 124 (1998) 642-652. https://doi.org/10.1061/(ASCE)0733-9445(1998)124:6(642)

[2] C. Cusano, G. Angjeliu, A. Montanino, G. Zuccaro, C. Cennamo, Considerations about the static response of masonry domes: a comparison between limit analysis and finite element method, International Journal of Masonry Research and Innovation, 6 (2021) 502-528. https://doi.org/10.1504/IJMRI.2021.118835

[3] E. Sacco, Modellazione e calcolo di strutture in materiale non resistente a trazione, Rend. Mat. Acc. Lincei, 1 (1990) 235-258. https://doi.org/10.1007/BF03001757

[4] M. Angelillo, Constitutive relations for no-tension materials, Meccanica, 28 (1993) 195-202. https://doi.org/10.1007/BF00989121

[5] M. Angelillo, L. Cardamone, A. Fortunato, A numerical model for masonry-like structures, Journal of Mechanics of Materials and Structures, Volume 5, No. 4 (2010). https://doi.org/10.2140/jomms.2010.5.583

[6] M. Bruggi, A. Taliercio, Analysis of no-tension structures under monotonic loading through an energy-based method, Computers & Structures, 159 (2015) 14-25. https://doi.org/10.1016/j.compstruc.2015.07.002

[7] M. Pari, M. Hendriks, J.G. Rots, Non-proportional loading in sequentially linear solution procedures for quasi-brittle fracture: A comparison and perspective on the mechanism of stress redistribution, Engineering Fracture Mechanics, 230 (2020) 106960. https://doi.org/10.1016/j.engfracmech.2020.106960

[8] G. Del Piero, Constitutive equation and compatibility of the external loads for linear elastic masonry-like materials, Meccanica, 24 (1989) 150-162. https://doi.org/10.1007/BF01559418

[9] S. Mastrodicasa, Dissesti statici delle strutture edilizie. Diagnosi e consolidamento, (1958).

[10] R. Esposito, G.J.P. Ravenshorst, Quasi-static cyclic in-plane tests on masonry components, Report C31B67WP3-4, Delft University of Technology, 2017.

Gaussian process emulation for rapid in-plane mechanical homogenization of periodic masonry

LUIS C. M. da Silva [1, a *], ANDRÉ Jesus[2,b] and GABRIELE Milani[1,c]

[1] Department of Architecture, Built environment and Construction engineering (A.B.C.), Politecnico di Milano, Piazza Leonardo da Vinci 32, 20133, Milan, Italy

[2] School of Architecture, Building and Civil Engineering, Loughborough University, Leicestershire, England, United Kingdom

[a]luiscarlos.martinsdasilva@polimi.it, [b]a.jesus@lboro.ac.uk, [c]gabriele.milani@polimi.it

*corresponding author

Keywords: Masonry Homogenization, Gaussian Process, Surrogate Modelling

Abstract. Numerical homogenization strategies can provide average mechanical responses, either in stress or coupled-stress quantities, which include many phenomenological features. Nonetheless, a direct application of numerical homogenization in sensitivity analysis in which uncertainty is propagated becomes impractical, as hundreds or thousands of simulations are conventionally required. In this study, a reliable and rapid predictive surrogate model is developed aiming to characterize the homogenized response of masonry. The case of English-bond arrangement is addressed, and the following steps are considered: (1) creation of a synthetic database through numerical homogenization based on a Finite-Element method, generated by randomization of model parameters; (2) training of a nonlinear Gaussian process; and (3) approximation of homogenized stress-strain curves for a masonry wall and for both linear and nonlinear ranges. The performance of the proposed technique is evaluated using training-validation-test in terms of computational accuracy. Results indicate that computational time is lessened 1200% while relative errors remain below 5-10%.

Introduction

Including uncertainty in the mechanical analysis of masonry structures is generally achieved through forward propagation of uncertainty. Input variables are assumed to be random and uncertainty is propagated aiming to evaluate its effect in the mechanical response [1]. Inverse propagation is the opposite process and it is receiving attention due to the advance of surrogate modelling, as for instance to predict cement mortar strength [2], but also in dynamic identification problems [3]–[5]. Inverse problems allow predicting the parameters of a given system through a data-driven framework and within a probabilistic prediction interval. Although these are receiving a growing attention [6], its extension to the mechanical analysis of masonry structures is still limited.

In such a context, the present study tries to tackle such opportunity. Classical Finite-Element (FE) and Discrete-Element (DE) strategies are expensive to run when analyzing masonry structures in the inelastic range and when parameter uncertainty is considered. This study addresses such difficulty by reducing the high computational time-cost required. The provision of a mesoscopic numerical model through a homogenization technique [7]–[11] and within a probabilistic framework has a paramount role, as it can provide data for training and testing of a surrogate model. In specific, a Gaussian process [12] is adopted since it offers promising accuracy when applied to non-linear structural mechanics problems [6], [13]. The presented strategy is thus based on numerical tools only, which is convenient as it precludes the need of other data sources such experimental tests. Hence, the study develops a simple-to-use and fast model based on a

Gaussian process to predict the homogenized in-plane mechanical response of an English-bond masonry wall.

Mesoscopic boundary value problem

A meso-mechanical problem is solved to compute average field variables for the in-plane mechanical response of masonry. The problem is formulated for the case of regular and periodic masonries. A single Representative Volume Element Ω_m (RVE) is defined. The kinematical description of the in-plane case assumes that the macroscopic strain tensor E is obtained as the volume average of the mesoscopic strain field $\grave{o}_m=\grave{o}_m(y)$ over the associated volume V_m of the RVE (Eq. (1)). The mesoscopic strain field can be decomposed into a macro-scale and meso-scale contributions. The homogenized generalized stress is computed considering the mesoscopic stress field $\boldsymbol{\sigma}_m$ upon RVE equilibrium following the Hill-Mandell principle. The homogeneous macroscopic stress tensor $\boldsymbol{\Sigma}$ can be written as the volume average of the mesoscopic stress field $\boldsymbol{\sigma}_m=\boldsymbol{\sigma}_m(y)$ over the RVE (Eq. (1)).

$$E=\frac{1}{V_m}\int_{\Omega_m}\grave{o}_m\,dV \quad , \quad \boldsymbol{\Sigma}=\frac{1}{V_m}\int_{\Omega_m}\boldsymbol{\sigma}_m\,d\Omega \;. \tag{1}$$

A 2D unit-cell FE model based on a Kirchhoff-plate theory method is adopted to solve the meso-scale BVP. Brick units ($235\times115\times70$ mm^3) are modelled as elastic and through quadrilateral FEs and material nonlinearity is lumped on mortar joints that are represented through interface FEs. Such assumption is particularly adequate for strong unit masonry structures [14]. The so-called composite interface model [15] is adopted, which is able to reproduce fracture, frictional slip and crushing along the interface elements of the joints. It is defined by a convex composite yield criterion with three individual functions able to represent softening behavior, i.e. a (i) a tension cut-off criterion designated as $f_{criterion,1}$ and defined in Eq. (3); (ii) a Mohr-Coulomb shear criterion designated as $f_{criterion,2}$ and defined in Eq. (4); and (iii) a cap in compression designated as $f_{criterion,3}$ and defined in Eq. (5).

$$f_{criterion,1}(\boldsymbol{\sigma},k_1)=\sigma-\bar{\sigma}_1(\mathrm{k}_1) \quad and \quad \bar{\sigma}_1=f_t\,\mathrm{e}^{\left(-\frac{f_t}{G_{ft}^{I}}k_1\right)} \;. \tag{3}$$

$$f_{criterion,2}(\boldsymbol{\sigma},k_2)=|\,\tau\,|+\sigma\tan(\phi)(\mathrm{k}_2)-\bar{\sigma}_s(k_2) \quad and \quad \bar{\sigma}_2=c\,\mathrm{e}^{\left(-\frac{c}{G_{ft}^{II}}k_2\right)} \;. \tag{4}$$

$$f_{criterion,3}(\boldsymbol{\sigma},k_3)=\frac{1}{2}\left(\boldsymbol{\sigma}^T\mathbf{P}\boldsymbol{\sigma}\right)+\mathbf{p}^T\boldsymbol{\sigma}-\bar{\sigma}_3^2(k_3) \;. \tag{5}$$

Here, $\boldsymbol{\sigma}$ is the generalized stresses, f_t is the interface bond strength, c is the interface cohesion, ϕ is the friction angle; $\mathbf{P}$ is a projection diagonal matrix and $\mathbf{p}$ a projection vector based on material parameters [15]; G_{ft}^{f}, G_f^{II} are the mode-I and mode-II fracture energy terms, respectively; $\bar{\sigma}_1$, $\bar{\sigma}_2$ and $\bar{\sigma}_3$ are the effective stresses of each the adopted yield functions governed by the internal scalar variables κ_1,κ_2 and κ_3, respectively. The deterministic strategy given in [14] was enriched by attributing probability distribution functions to each input parameter; specifically to both geometric and material properties. If X defines the random input variables X_i, then one can write the following:

$$X=\{E_{brick}\,,E_{mortar}\,,f_t\,,G_{ft}^{I}\,,f_c\,,G_{fc}^{I}\,,c\,,G_f^{II}\,,f_c\,,t_{joint}\,,t\} \;. \tag{6}$$

in which $\forall X_i \in X : X_i = f(x)$ whereas $f(x)$ is assumed to be an uniform function. The sampling of random variables (RVs) is achieved by the Latin Hypercube method. The solution of the mesoscopic BVP is processed N_{simul} times, hence finding N_{simul} homogenized quantities.

Gaussian process emulation

A Gaussian process is a non-parametric probabilistic model that can be used for nonlinear regression of an existent model, hence approximating the structural response containing noise for classification problems [16]. For the present study, the nonlinear mapping between the input parameters and the corresponding homogenized stress-strain curves of the masonry will be approximated using a G_p. A G_p is fully defined by a mean and covariance function that is dependent on a set of parameters – known as hyperparameters – that must be estimated from a training data set. The mean m and covariance V are given by:

$$m(x) = \mathbf{H}(x)\boldsymbol{\beta} \qquad V(x,x) = s^2\mathbf{R}(x,x). \tag{7}$$

in which $\mathbf{H} \in \mathrm{R}^{N\times p}$ is a regression matrix, $\boldsymbol{\beta} \in \mathbb{R}^p$ is a vector of regression coefficients, x represents the strain and parameters described in the previous section that define the stress-strain curve, s^2 is a scalar response variance and $\boldsymbol{R} \in \mathbb{R}^{N\times N}$ is a correlation matrix. The matrix $\mathbf{H}$ contains N first order polynomial functions $h_1(x) = [1\, x_1]$ of degree $p = 2$. The matrix $\mathbf{R}$ contains the entries for a correlation function, which has been assumed to be linear:

$$R(\omega, x, x') = \max\{0, 1-\omega\,|\,x-x'\,|\}. \tag{8}$$

in which ω is a roughness parameter [17] that represents how roughly the response changes between training data points. The hyperparameters that define the above functions, $\vartheta = \{\boldsymbol{\beta}, s^2, \boldsymbol{\omega}\}$, must be estimated from training data. Summarily, a likelihood function that describes how well the model fits the data is established and maximized. To simplify it, analytical expressions can be found for s^2 and $\boldsymbol{\beta}$. For further theoretical details see [16].

Training and testing data

The homogenized $\Sigma - E$ curves have been generated through the meso-mechanical model. The input parameters are given in Table 1. Both friction angle ϕ and masonry wall thickness t are assumed to be deterministic variables. An elastic response has been considered in compression, which is acceptable for a strong-unit-weak-joint type of masonry.

Table 1 – Random input for the meso-mechanical model.

	Random parameters						
X_i	E_{brick} (N/mm²)	E_{mortar} (N/mm²)	$f_{t,mortar}$ (N/mm²)	G_{ft}^{I} (N/mm)	G_f^{II} (N/mm)	c (N/mm²)	t_{joint} (mm)
$E[X_i]$	[1000;15000]	[500;4000]	[0.05;0.40]	[0.005;0.015]	[0.02;0.055]	[0.05;0.80]	[10;15]

The generated data comprises 2697 curves for each associated Cauchy's deformation mode: (i) in-plane tension xx, (ii) in-plane tension yy, and (iii) in-plane shear xy. Homogenized curves are presented in Fig. 1. Each curve is composed of 100 points with a total of $N_{samples} = 269700$ points. Nonetheless, only 2.9% have been randomly selected for the training ($N_{training} = 7600$ points) and 20% for the testing ($N_{testing}$).

Materials Research Proceedings **26** (2023) 325-330 https://doi.org/10.21741/9781644902431-53

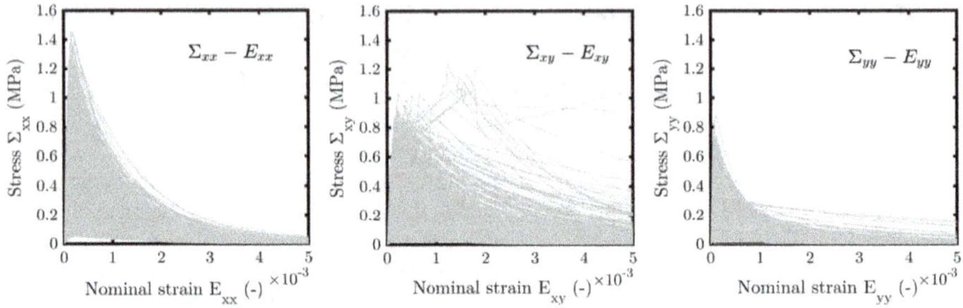

Fig. 1 – Probabilistic homogenized stress-strain curves used in the Gaussian process emulation.

Discussion of results

The elastic and inelastic homogenized responses have been emulated separately since the curves have a discontinuity between these ranges. Note that such division is easily achieved by scanning the peak of each curve. The mean function and prediction interval of two curves are hence predicted – elastic and inelastic ranges –, being subsequently merged to form the full predicted in-plane response of the periodic masonry arrangement. The mean function is used for the validation or testing of the predicted results. In the elastic part, both the estimated Young's modulus and elastic strain energy are validated against the reference data, as given in Fig. 2. Results indicate that relative errors below 10% are expected for the homogenized Young's modulus when the training data are equal to 2697 points. From an energetic standpoint, a total of 1500 points are enough to respect such threshold of 10%, as given in Fig.2(b).

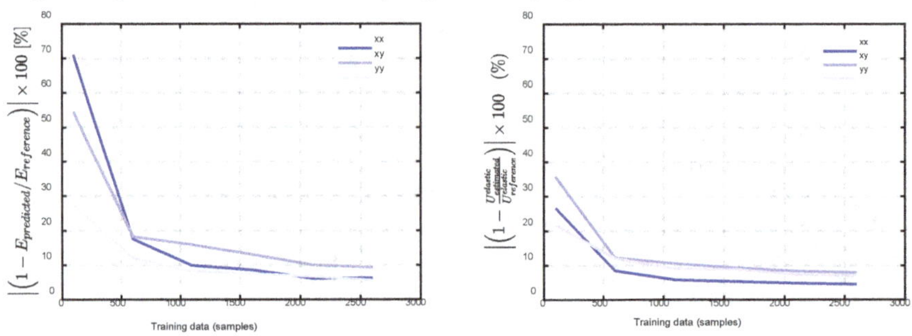

(a) Young's modulus (b) elastic strain energy

Fig. 2 – Relative errors of the predicted response in the elastic range.

For the inelastic range, the estimated peak and plastic strain energy, based on the Gp predicted mean function, are evaluated and compared with the reference testing data. The relative error is plotted in Fig. 3(a) and Fig. 3(b), respectively. In this case, all the training samples ($N_{training}$ = 7900 points) have been used, for which relative errors below 5% for the peak stress are found; and around 10% for the plastic strain energy in shear (xy curve) and vertical tension (yy curve) and around 4% for horizontal tension (curve xx). We can also highlight the existence of a plateau for $N_{training}$ > 7900 points, meaning that the improvement of the solution accuracy in the inelastic range may require a significant number of more data.

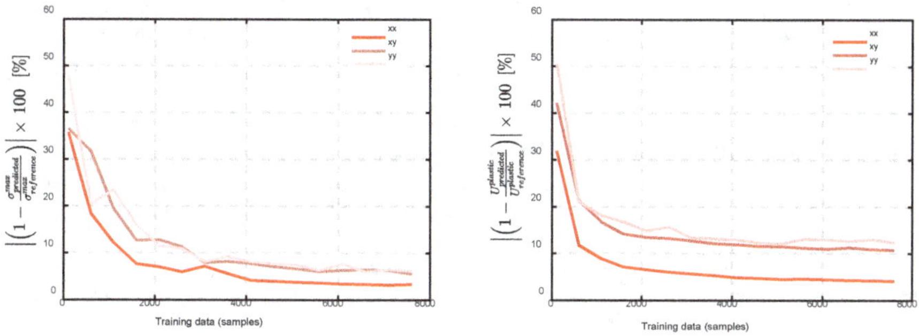

(a) peak stress (b) plastic strain energy

Fig. 3 – Relative errors of the predicted response in the inelastic range.

Conclusions

A Gaussian process has been used to predict the in-plane mechanical homogenized response of an English-bond masonry. The framework is purely numerical data-driven and proved its efficiency to compute the elastic and inelastic ranges of the in-plane tension (horizontal and vertical directions) and shear responses has been demonstrated. The homogenized Young's modulus, the elastic strain energy, the strength value, and the inelastic strain energy were considered to demonstrate the accuracy level. Generally, the Gaussian process is effective and accurate even for a small number of training data. After the training step, it requires in average 4 seconds to compute each homogenized curve, i.e. ~1200% lower than the time required by the meso-mechanical simulation. Its generalizability is guaranteed, as the topological input is independent on the type of masonry bond. Lastly, the study gives an important contribution to the application of inverse methods for the structural analysis of masonry structures. The results are promising and may lead to a future publication, in which case studies at a structural level will be provided.

References

[1] B. Pulatsu, S. Gonen, E. Erdogmus, P. B. Lourenço, J. V Lemos, and R. Prakash, "In-plane structural performance of dry-joint stone masonry Walls: A spatial and non-spatial stochastic discontinuum analysis," *Eng. Struct.*, vol. 242, p. 112620, 2021, https://www.doi.org/10.1016/j.engstruct.2021.112620

[2] P. G. Asteris, L. Cavaleri, H.-B. Ly, and B. T. Pham, "Surrogate models for the compressive strength mapping of cement mortar materials," *Soft Comput.*, vol. 25, no. 8, pp. 6347–6372, 2021, https://www.doi.org/10.1007/s00500-021-05626-3

[3] A. Jesus, P. Brommer, R. Westgate, K. Koo, J. Brownjohn, and I. Laory, "Modular Bayesian damage detection for complex civil infrastructure," *J. Civ. Struct. Heal. Monit.*, vol. 9, no. 2, pp. 201–215, 2019, https://www.doi.org/10.1007/s13349-018-00321-8

[4] A. Jesus, P. Brommer, R. Westgate, K. Koo, J. Brownjohn, and I. Laory, "Bayesian structural identification of a long suspension bridge considering temperature and traffic load effects," *Struct. Heal. Monit.*, vol. 18, no. 4, pp. 1310–1323, Sep. 2018, https://www.doi.org/10.1177/1475921718794299

[5] A. Jesus, P. Brommer, Y. Zhu, and I. Laory, "Comprehensive Bayesian structural identification using temperature variation," *Eng. Struct.*, vol. 141, pp. 75–82, 2017, https://www.doi.org/10.1016/j.engstruct.2017.01.060

[6] P. D. Arendt, D. W. Apley, and W. Chen, "Quantification of Model Uncertainty: Calibration, Model Discrepancy, and Identifiability," *J. Mech. Des.*, vol. 134, no. 10, Sep. 2012, https://www.doi.org/10.1115/1.4007390

[7] P. B. Lourenço, M. F. Funari, and L. C. Silva, "Building resilience and masonry structures : How

can computational modelling help ?," in *Computational Modelling of Concrete and Concrete Structures*, 1st Editio., G. Meschke, B. Pichler, and J. G. Rots, Eds. London: CRC Press, 2022, pp. 30–37.

[8] M. F. Funari, L. C. Silva, N. Savalle, and P. B. Lourenço, "A concurrent micro/macro FE-model optimized with a limit analysis tool for the assessment of dry-joint masonry structures," *Int. J. Multiscale Comput. Eng.*, vol. 20, no. 5, pp. 65–85, 2022, https://www.doi.org/10.1615/IntJMultCompEng.2021040212

[9] M. F. Funari, L. C. Silva, E. Mousavian, and P. B. Lourenço, "Real-time Structural Stability of Domes through Limit Analysis: Application to St. Peter's Dome," *Int. J. Archit. Herit.*, pp. 1–23, Oct. 2021, https://www.doi.org/10.1080/15583058.2021.1992539

[10] D. Addessi and E. Sacco, "Enriched plane state formulation for nonlinear homogenization of in-plane masonry wall," *Meccanica*, vol. 51, no. 11, pp. 2891–2907, 2016, https://www.doi.org/10.1007/s11012-016-0484-1

[11] J. A. Dauda, L. C. Silva, P. B. Lourenço, and O. Iuorio, "Out-of-plane loaded masonry walls retrofitted with oriented strand boards: Numerical analysis and influencing parameters," *Eng. Struct.*, vol. 243, p. 112683, 2021, https://www.doi.org/https://doi.org/10.1016/j.engstruct.2021.112683

[12] S. Conti, J. P. Gosling, J. E. Oakley, and A. O'Hagan, "Gaussian process emulation of dynamic computer codes," *Biometrika*, vol. 96, no. 3, pp. 663–676, Jul. 2009, [Online]. Available: http://www.jstor.org/stable/27798855

[13] L. Portelette, J.-C. Roux, V. Robin, and E. Feulvarch, "A Gaussian surrogate model for residual stresses induced by orbital multi-pass TIG welding," *Comput. Struct.*, vol. 183, pp. 27–37, 2017, https://www.doi.org/10.1016/j.compstruc.2017.01.009

[14] S. Sharma, L. C. Silva, F. Graziotti, G. Magenes, and G. Milani, "Modelling the experimental seismic out-of-plane two-way bending response of unreinforced periodic masonry panels using a non-linear discrete homogenized strategy," *Eng. Struct.*, vol. 242, no. December 2020, 2021, https://www.doi.org/10.1016/j.engstruct.2021.112524

[15] P. B. Lourenço and J. G. Rots, "Multisurface Interface Model for Analysis of Masonry Structures," *J. Eng. Mech.*, vol. 123, no. 7, pp. 660–668, Jul. 1997, https://www.doi.org/10.1061/(ASCE)0733-9399(1997)123:7(660)

[16] S. Conti and A. O'Hagan, "Bayesian emulation of complex multi-output and dynamic computer models," *J. Stat. Plan. Inference*, vol. 140, no. 3, pp. 640–651, 2010, https://www.doi.org/10.1016/j.jspi.2009.08.006

[17] N. Cressie, *Statistics for Spatial Data. Wiley series in probability and mathematical statistics.* New York, USA: Wiley, 1993.

Impacts analysis in the rocking of masonry circular arches

Paolo Bisegna[a], Simona Coccia[b], Mario Como[c], Nicola Nodargi[d]

University of Rome "Tor Vergata", Italy

[a]bisegna@uniroma2.it, [b]coccia@ing.uniroma2.it, [c]como@uniroma2.it, [d]nodargi@ing.uniroma2.it

Keywords: Masonry, Circular Arch, No Tension Material, Rocking, Restitution Coefficient

Abstract. The in-plane rocking of a masonry arch is studied. The starting motion of the arch is assumed to take place along the four-link failure mechanism ***u*** corresponding to a constant distribution of horizontal loads proportional to the arch self-weight. During such a motion, if overturning does not occur, the arch returns to the undeformed configuration, with all the hinges of the mechanism ***u*** simultaneously closing. At such an instant, impacts occur, implying a motion exchange and a subsequent movement of the arch along a new four-link mechanism ***u'***. A novel impact model, different from the one available in literature, is here proposed. The restitution coefficient is computed by applying the linear and angular momentum impulse theorems. The rocking mechanism ***u'*** of the arch after the impact is determined under the condition of minimum energy expense at the impact.

Introduction

Fig. 1 shows a free-standing stone column over a supporting plane that, hit by a horizontal impulse from left to right, puts itself in a rocking motion. The friction between the column and its supporting plane is sufficiently high, and the column is sufficiently slender, to avoid occurrence of sliding or bouncing during the motion. The column thus rotates around the point O of its basis up to the attainment of a maximum rotation angle; then, if overturning does not occur, it inverts its motion until an impact occurs at the opposite point O' of its base. Afterwards, the column continues to rotate with non-harmonic oscillations of reduced amplitudes and periods until it stops. This type of motion is called "rocking".

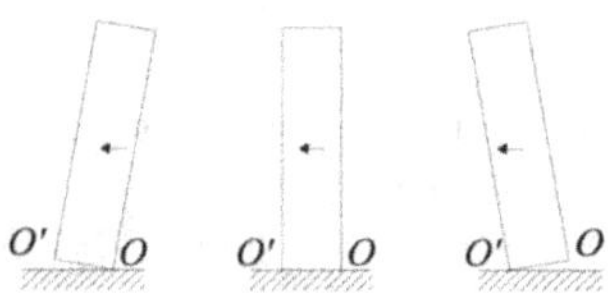

Figure 1. Rocking motion of a free-standing masonry wall

A fundamental paper by Housner [1], concerning the study of the behaviour of a free-standing stone column placed upon a rigid horizontal base and undergoing rocking motion, set the basis for the application of these concepts in the assessment of the dynamical behaviour of masonry constructions. The first description of the in-plane motion of a masonry circular arch, without any impact analysis, can be found in [2]. More recently, energy dissipation models of the impacts occurring in the rocking of the arch, considered as a rigid body four-link mechanism, have been given in [3, 4, 5]. In spite of the numerous researches on the subject, the definition of the mechanical model of the impacts in the rocking motion of the masonry arch is still debated and a new model is here proposed.

The rocking mechanism of the arch

A horizontal ground acceleration pulse, acting from right to left, is assumed to hit a masonry arch. If the magnitude of the acceleration pulse exceeds the incipient rocking value A_L, the arch starts to move along a four-link mechanism ***u***. Both the incipient rocking acceleration A_L and the mechanism ***u*** can be straightforwardly computed by the Limit Analysis theorems [6]. During such a motion, if overturning does not occur, the arch, moving from right to left, returns to the undeformed configuration, with all the hinges of the mechanism ***u*** simultaneously closing. At that instant, a motion exchange takes place and a new motion, still from right to left, develops along a new four-link mechanism ***u'***.

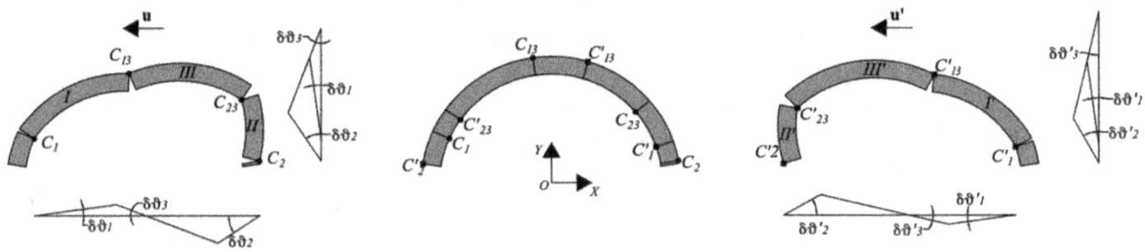

*Figure 2. Hinge locations in mechanisms **u** and **u'***

In the mechanism ***u*** the rotations $\delta\theta_2$ and $\delta\theta_3$ can be expressed as a function of $\delta\theta_1$ as:

$$\delta\theta_3 = k_{31}\delta\theta_1; \qquad \delta\theta_2 = k_{21}\delta\theta_1 \tag{1}$$

where the coefficients k_{i1} are evaluated in the undeformed configuration as function of the coordinates of the centres of the rotation:

$$k_{31} = \frac{x_{C_{13}}-x_{C_1}}{x_{C_3}-x_{C_{13}}}; \quad k_{21} = k_{31}\frac{x_{C_{23}}-x_{C_3}}{x_{C_2}-x_{C_{23}}} \tag{2}$$

Analogous expressions hold for the mechanism ***u'***.

A new impact model

An essential problem to be addressed for computing the mechanism ***u'*** after the impact is the localization of the internal impulsive forces arising at the instant when impacts occur. A first step in solving that problem requires to recognize that the locations of the hinges of ***u'*** are impact points, thus transmitting impulsive internal forces.

Some researchers, basing on the analogy with the behaviour of Housner's column, account for impacts occurring at the opposite side of the sections containing the hinges of the mechanism ***u***, which are *closing* [2, 3]. However, besides those unquestionable impact points, also the locations of the hinges of ***u'***, i.e. the points around which the arch segments pivot in the new motion of the arch after the impact, should be recognized as impact points.

In fact, consider the instant when the hinges of the mechanism ***u*** begin to close while the new mechanism ***u'*** is taking place. If the hinges of ***u'*** develop at the same sections of the arch where the hinges of former mechanism ***u*** are located (Fig. 3), the passage from ***u*** to ***u'*** is similar to what happens in the Housner column (Fig. 1) and it is recognized that impacts occur at points where the hinges of the new mechanism are located.

Consider then the case in which the hinges of ***u'*** develop at sections different from those of the former mechanism ***u***. While the hinges of ***u*** are closing, inertial effects force the arch to develop the new mechanism ***u'***. Impulsive compressive forces flow across the closed sections of ***u*** while the sections including the new hinges of ***u'*** begin to open and start to reciprocally rotate, thus

Materials Research Proceedings **26** (2023) 331-336 https://doi.org/10.21741/9781644902431-54

impacting one against the other. So, an impact occurs at the points where the hinges of the mechanism ***u'*** are located and therein compressive impulse forces are transmitted (Fig. 4).

The impulsive forces can be determined by using the linear and angular momentum impulse theorems, as shown below.

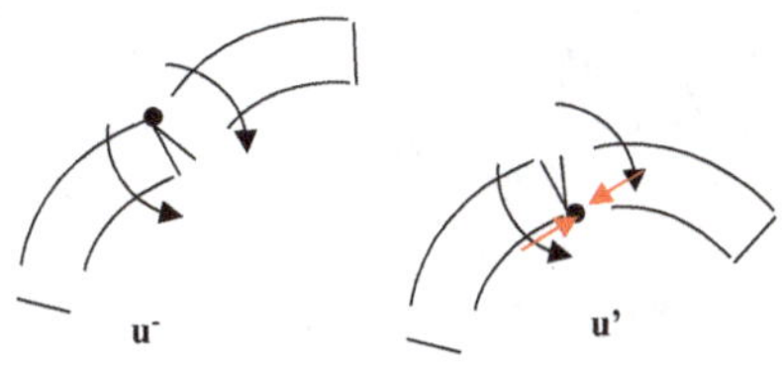

4

*Figure 3. The case, similar to the Housner column, in which the hinges of **u'** occur at the same sections where the hinges of **u** are located*

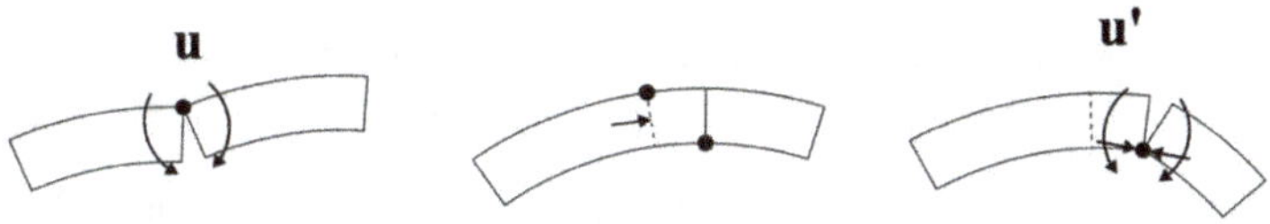

*Figure 4. The case in which hinges of **u'** develop at sections different from those of the former mechanism **u***

However, the issue to define the locations of the hinges of the mechanism ***u'*** developing in the arch after the impact remains to be solved. Here it is postulated that those hinges are formed at locations producing the minimum energy expense at the impacts, i.e. maximizing the restitution coefficient *r*.

In passing, some authors in the literature assume that the locations of the hinges of the mechanism ***u'*** are at the mirrored locations of the hinges of the mechanism ***u***. However, that assumption does not generally imply the minimum energy expense at the impact.

To that purpose, it is recalled that *I, II, III* denote the arch segments included between the sections through the hinges C_1, C_{13}, C_{23}, C_2 of the mechanism ***u***, whereas *I', II', III'* denote the arch segments included between the sections through the hinges C_1', C_{13}', C_{23}', C_2' of the mechanism ***u'*** (Fig. 2). If considered altogether, the sections through the hinges of the mechanisms ***u*** and ***u'*** pinpoint seven bodies, labelled as 1, …, 7, such that the arch segments of those mechanisms can be recovered as (Fig. 5):

$$\begin{array}{lll} I = 2 \cup 3 & II = 6 \cup 7 & III = 4 \cup 5 \\ I' = 5 \cup 6 & II' = 1 \cup 2 & III' = 3 \cup 4 \end{array} \tag{3}$$

Let m_i and $I_{O,i}$ respectively denote the mass and the polar moment of inertia around the centre O of the ith of those bodies. It is noticed that the kinematics of that body before [resp., after] the impact is completely described by its angular velocity $\dot{\theta}_i\mathbf{k}$ [resp., $\dot{\theta}_i'\mathbf{k}$], with $\mathbf{k}$ as a unit vector orthogonal to the plane of the arch, and by its absolute centre of rotation C_i [resp., C_i'].

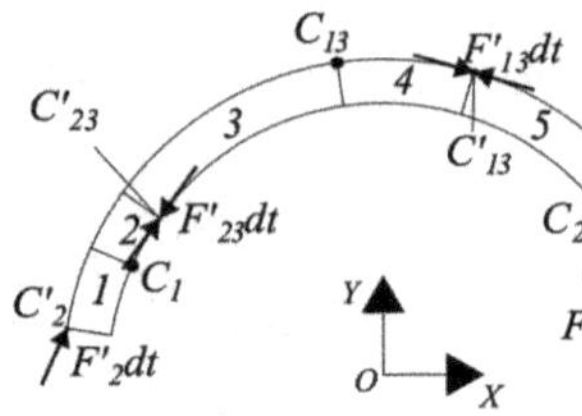

*Figure 5. The seven bodies composing the mechanisms **u** and **u'***

Indeed, due to the compatibility conditions of the mechanism ***u*** [resp., ***u'***], the angular velocities can be expressed in terms of a single free parameter $\dot{q}$ [resp., $\dot{q}'$] by:

$$\dot{\theta}_i = \Theta_i\dot{q}; \quad \dot{\theta}'_i = \Theta'_i\dot{q}' \tag{4}$$

where Θ_i [resp., Θ'_i] are known geometrical coefficients (e.g., if the choice $\dot{q} = \dot{\theta}_1$ [resp., $\dot{q}' = \dot{\theta}'_1$] is made, it follows that $\Theta_1 = 1$, $\Theta_2 = k_{21}$ and $\Theta_3 = k_{31}$ [resp., $\Theta'_1 = 1$, $\Theta'_2 = k'_{21}$ and $\Theta'_3 = k'_{31}$], consistently with Equation (1)).

Accordingly, the linear momentum $\boldsymbol{p}_i$ [resp., $\boldsymbol{p}'_i$] and the angular momentum $\boldsymbol{L}_{O,i}$ [resp., $\boldsymbol{L}'_{O,i}$] around the pole O of the ith body before [resp., after] the impact are given by (equations to be read without [resp., with] apex):

$$\boldsymbol{p}_i = m_i\boldsymbol{V}_{Gi} = [m_i\boldsymbol{k} \times (G_i - C_i)]\Theta_i\dot{q} \tag{5}$$

$$\boldsymbol{L}_{O,i} = I_{G_i,i}\dot{\theta}_i\mathbf{k} + (G_i - O) \times \boldsymbol{p}_i = \left[I_{G_i,i} + m_i(G_i - O) \cdot (G_i - C_i)\right]\Theta_i\dot{q}\boldsymbol{k} \tag{6}$$

where $\boldsymbol{V}_{Gi}$ denotes the velocity of the centre of gravity G_i of the ith body.

As the impact is of very short duration, very large impulsive forces act on the arch, producing finite changes of velocity of its various parts without significant changes in configuration. Accordingly, during the impact, all ordinary forces (such as gravity) are negligible, and no change of configuration of the arch needs to be considered.

On such a basis, the computation of the unknown parameter $\dot{q}'$ can be achieved by including among the unknowns the impulsive forces $\boldsymbol{F}'_1dt$ and $\boldsymbol{F}'_2dt$ carried out by the hinges C'_1 and C'_2 at the impact, and by formulating a system of five equations in five unknowns.

In detail, those equations are:

- the linear momentum impulse theorem for the portion of the arch between C'_1 and C'_2 along x and y directions (Fig. 6):

$$\textstyle\sum_{i=1}^{6}\boldsymbol{p}_i + (\boldsymbol{F}'_1 + \boldsymbol{F}'_2)dt = \sum_{i=1}^{6}\boldsymbol{p}'_i \tag{7}$$

- the angular momentum impulse theorem for the portion of the arch between C'_1 and C'_2 about the center O (Fig. 6):

$$\textstyle\sum_{i=1}^{6}\boldsymbol{L}_{O,i} + (C'_1 - O) \times \boldsymbol{F}'_1dt + (C'_2 - O) \times \boldsymbol{F}'_2dt = \sum_{i=1}^{6}\boldsymbol{L}'_{O,i} \tag{8}$$

- the angular momentum impulse theorem for the portion of the arch between C'_{13} and C'_1 about the point C'_{13} (Fig. 7):

$$\textstyle\sum_{i=5}^{6}\boldsymbol{L}_{C'_{13},i} + (C'_1 - C'_{13}) \times \boldsymbol{F}'_1dt = \sum_{i=5}^{6}\boldsymbol{L}'_{C'_{13},i} \tag{9}$$

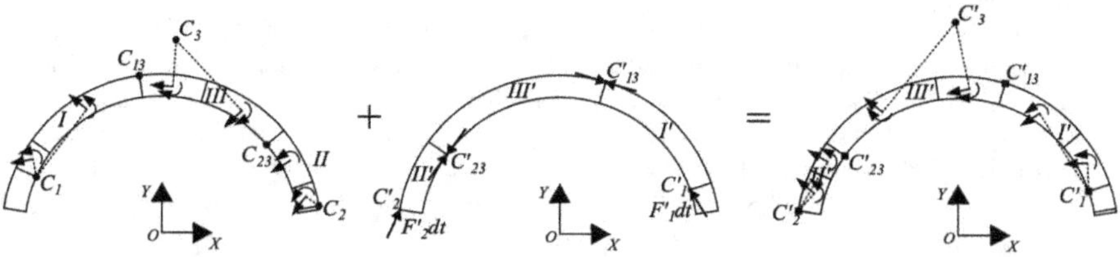

Figure 6. Linear and angular momentum impulse theorems during the impact

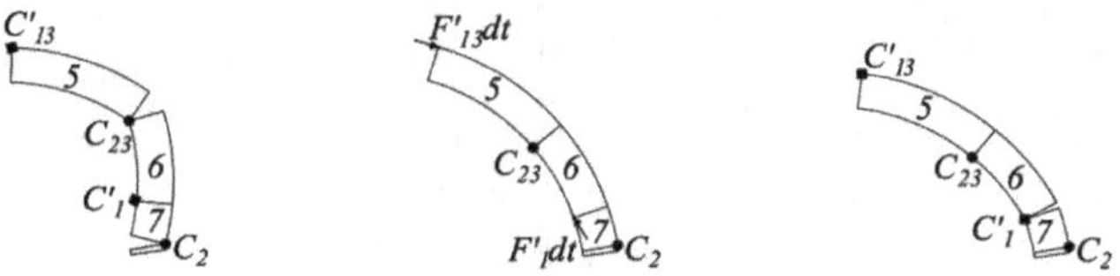

Figure 7. Angular momentum impulse theorem around the point C'_{13}

- the angular momentum impulse theorem for the portion of the arch between C'_2 and C'_{23} about the point C'_{23} (Fig. 8):

$$\sum_{i=1}^{2} \boldsymbol{L}_{C'_{23},i} + (C'_2 - C'_{23}) \times \boldsymbol{F}'_2 dt = \sum_{i=1}^{2} \boldsymbol{L}'_{C'_{23},i} \tag{10}$$

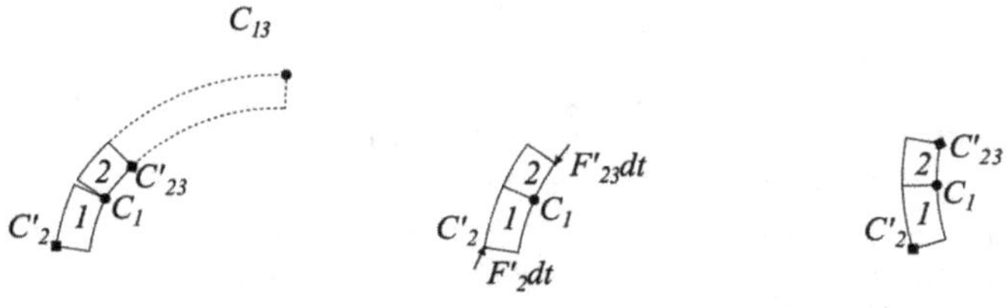

Figure 8. Angular momentum impulse theorem around the point C'_{23}

It is remarked that the angular momentum impulse theorem equations (9) and (10) are purposely written in such a way that the impulsive forces explicated by the internal hinges C'_{23} and C'_{13} are not explicitly involved (Fig. 5).

After solving Eqs. (7)–(10), and hence computing the free parameter $\dot{q}'$ of the arch mechanism ***u'***, the angular velocities $\dot{\theta}'_i$ follow from Eq. (4), whence it is a simple matter to compute the restitution coefficient of the arch as the ratio of the kinetic energy after the impact to the kinetic energy before the impact [1].

Furthermore, it is highlighted that r depends on the arch geometry and location of the hinges before the impact, which are known in advance, and by the location of the hinges after the impact, which are not. Hence, the latter are assumed as the unknowns of a nonlinear optimization problem, aimed at determining the maximum value of r, i.e. the minimum value of energy dissipation, among the various possible mechanisms ***u'***.

Numerical applications

A circular arch with an angle of embrace equal to 180°, a ratio t/R_m equal to 0.15 is considered for a numerical application of the proposed model. The mean radius R_m does not influence the results. The failure acceleration A_L results to be 0.1443. The mechanisms ***u*** before the impact, and the mechanism ***u'*** after the impact that maximize the restitution coefficient *r*, are shown in Fig. 10. It is remarked that the hinges of the mechanism ***u'*** are not in the mirrored location of the hinges of the mechanism ***u***. The corresponding maximum value of the restitution coefficient *r* results to be 0.6812.

Finally, the new concept of impulse line is introduced. Analogously to the thrust line, which graphically depicts the equilibrium between applied loads (including inertia forces) and constraint reactions at regular instants of motion, the impulse line depicts the equivalence between impulsive forces and variations of linear and angular momenta at impact instants. The impulse line at the impact instant relevant to the present case study is depicted in Fig. 10. It passes through the indrados and the extrados of the arch at sections where the hinges of ***u'*** take place. It can be shown that the impulse line is statically admissible, i.e. it is entirely contained within the thickness of the arch, if and only if the post-impact mechanism ***u'*** maximizes the restitution coefficient *r* over the space of kinematically admissible post-impact mechanisms.

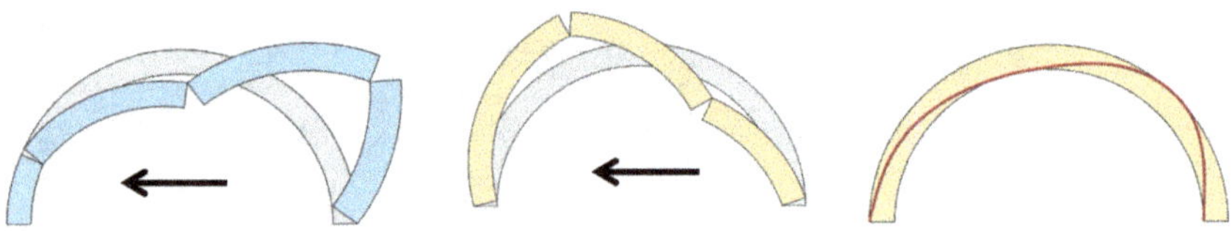

*Figure 10. Mechanisms **u** and **u'** and impulse line at the impact instant*

References

[1] G.W. Housner, The behaviour of inverted pendulum structures during earthquakes. Bulletin of the Seismological Society of America, 53(2), 403-417, 1963. https://doi.org/10.1785/BSSA0530020403

[2] I.J. Oppenheim, The masonry arch as a de mechanism under base motion. Earthquake Engineering and Structural Dynamics, 21, 1005-1017, 1992. https://doi.org/10.1002/eqe.4290211105

[3] L. De Lorenzis, M. DeJong, J. Ochsendorf, Failure of masonry arches under impulse base motion. Earthquake Engineering and Structural Dynamics, 36, 2119–2136, 2007. https://doi.org/10.1002/eqe.719

[4] M.J. DeJong, L. De Lorenzis, S. Adams, J.A. Ochsendorf, Rocking Stability of Masonry Arches in Seismic Regions. Earthquake Spectra, 24(4), 847–865, 2008. https://doi.org/10.1193/1.2985763

[5] L.P. Kollár, T. Ther, Numerical model and dynamic analysis of multi degree of freedom masonry arches, Earthquake Engineering & Structural Dynamics, 48(7), 709-730, 2019. https://doi.org/10.1002/eqe.3158

[6] M. Como, Statics of Historic masonry Constructions, 3rd ed., Springer, Heidelberg, 2017. https://doi.org/10.1007/978-3-319-54738-1

Theoretical and Applied Mechanics - AIMETA 2022 | Materials Research Forum LLC
Materials Research Proceedings **26** (2023) 337-342 | https://doi.org/10.21741/9781644902431-55

Numerical strategies for modelling masonry arch bridges strengthened with PBO-FRCM composites

Enrico Compagnone[1, a], Salvatore Gazzo[2, b], Leopoldo Greco[2, c], Massimo Cuomo[2, d], Loredana Contrafatto[2, e *]

[1] Giniu Ingegneria e Architettura, Spinoff Università degli Studi di Catania, Italy

[2] Dipartimento di Ingegneria Civile e Architettura, Università degli Studi di Catania, Italy

[a]enrico.compagnone@giniu.it, [b]salvatore.gazzo@dica.unict.it, [c]leopoldo.greco@unict.it, [d]mcuomo@dica.unict.it, [e]loredana.contrafatto@unict.it

Keywords: Masonry Arch Bridge, FRCM, PBO

Abstract. The work reports some numerical tests concerning the use of the Embedded Truss elements implemented in the Midas FEA NX software in modeling a specific FRCM reinforcement system for masonry structures. The efficacy of the finite element is demonstrated through the comparison of the numerical results with experimental results obtained by other authors relating both to the mechanical characterization of the FRCM system materials and to the response of masonry arch models. The Embedded Truss element appears to be promising for application to real large structures, such as masonry arch bridges of large span or multi-span masonry bridges.

Introduction

New reinforcement systems have recently been proposed for retrofitting of existing masonry structures, consisting of cementitious matrices reinforced by fiber networks (Fiber Reinforced Concrete Matrix). These materials have high strength and durability, higher compatibility with the masonry support than other polymeric matrix reinforcement systems, originally conceived for the reinforcement of reinforced concrete works. FRCM are therefore finding extensive use in civil, infrastructural and historical masonry structures. The design of the reinforcement intervention and ex-post verification of the strengthened structure require the modeling of the reinforcement system through the software commonly used in structural analysis. The need for a detailed model of the reinforcement, capable of adequately simulating the crisis mechanisms, is countered by the need for streamlined numerical models, which do not burden the non-linear calculation in terms of size of the problem and computation time. The paper concerns modeling methods of a specific type of FRCM with PBO mesh. The method is based on the FEM technology named Embedded Truss and implemented in Midas FEA NX software [1]. The purpose is the definition of an effective modeling methodology of the FRCM reinforcement, which is susceptible of application in the structural analysis of masonry multi-arched bridges, even with very large spans. The method, applied to a specific commercial product, allows the accurate simulation of experimental tests for the characterization of the Ruregold PBO-Mesh 22/22 system [2] and reproduce the response of fiber reinforced masonry arches. Non-linear analyses based on the Smeared Crack Model for the masonry support and for the matrix of the reinforcement are presented. Specific constitutive models are used to define the fibers behaviour and the support-reinforcement interface. Sensitivity analyses show the importance that certain constitutive parameters have on the numerical solution. The numerical results are compared with the experimental ones to validate the method.

FEM Modeling of FRCM composites in the analysis of reinforced structures by Embedded Truss elements

In the literature of the last 10 years there are some attempts to model FRCM reinforcements in FEM codes [3]. This study proposes the use of the Embedded Truss element, implemented in

Materials Research Proceedings **26** (2023) 337-342 https://doi.org/10.21741/9781644902431-55

Midas FEA NX software [1], for modeling the mesh of the FRCM composite. The element consists in a truss embedded in a matrix. The lattice, which represents the fibers, and the matrix are attributed distinct nonlinear constitutive behaviour that reproduce the phenomenological aspects typical of the two materials.

The input method, coordinate system, material property, etc. of the Embedded Truss elements are identical to those of the standard Truss elements. Embedded Truss elements are generally used to model structural elements such as anchors, nails, and rock bolts, which ignore flexural behavior.

The elements (Fig. 1a) do not require the sharing of discretization nodes (Fig. 1b) and are therefore convenient for modeling and analyzing systems with fiber networks embedded in a continuum (Fig. 1c). The parent element in which the truss system is incorporated can be a flat deformation element or a solid element. The parent element includes each node of the Embedded Truss element. A multipoint constraint is used to automatically constrain the nodal displacement of the Embedded Truss element to the internal displacement of the parent element.

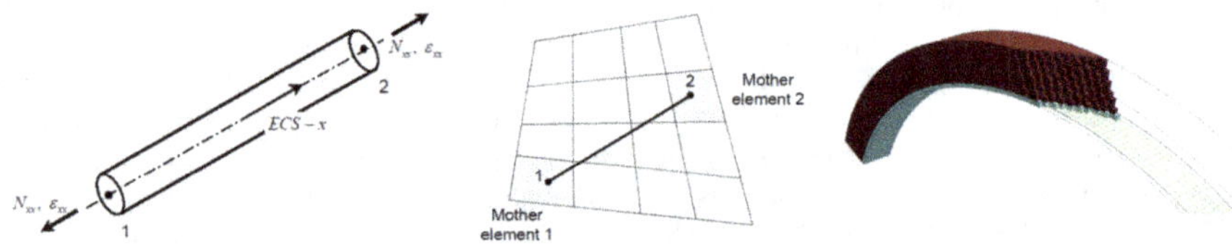

(a) (b) (c)

Figure 1 (a) Mother elements; (b) Embedded Truss element and (c) FEM model of a reinforced masonry arch

Both an elastic behaviour and a non-linear uniaxial constitutive behaviour with limited strength in compression and tension can be attributed to the Embedded Truss element.

The compressive non-linear behaviour of the matrix of the FRCM composite and of the masonry structure is ruled according to the Smeared Crack Model by a specific plasticity function. A Linearly Elastic-Perfectly Plastic stress-strain relationship defined by the elastic modulus E_{matrix} and compressive strength $f_{c,matrix}$ has been considered to model the FRCM matrix compressive behavior. A parabolic softening, defined by the size of the mesh h_{mesh}, the fracture energy $G_{c,matrix}$ and the compressive strength $f_{c,matrix}$ has been selected to define the compressive response of the masonry. The tensile behaviour is ruled by a crack detection surface. The FRCM matrix tensile behaviour is defined by Linear or Hordijk softening [1] defined by the size of the mesh h_{mesh}, the fracture energy $G_{f,matrix}$ and the tensile strength $f_{t,matrix}$. The tensile behaviour of the masonry is described by means of a bi-linear function defined by the size of the mesh h_{mesh}, the fracture energy $G_{f,masonry}$ and the tensile strength $f_{t,masonry}$. The shear behaviour of all fragile elements is defined by means of a linear τ-γ function, being β the coefficient regulating the slope.

Mechanical characterization of FRCM composites

The results of laboratory tests carried out on Ruregold PBO-Mesh 22/22 net, Ruregold MX-PBO mortar and on FRCM specimens reported in the work [4] have been considered for comparison.

Based on these experimental results, different numerical models have been calibrated for the various materials constituting the FRCM composite and for the composite itself.

Numerical results of the simulation of the uniaxial tensile test on a network specimen of PBO-Mesh 22/22 (Fig. 2a) are reported in the stress-strain graph shown in Fig. 2b. The net model is made of standard truss elements whose tensile behaviour is ruled by an elastoplastic function. Parameters of elastic modulus E =306924 [Mpa], tensile strength $\sigma_t = 3031$ [Mpa] and compressive strength $\sigma_c = 0$ [Mpa] allows a very good fitting of the experimental test.

Materials Research Forum LLC
https://doi.org/10.21741/9781644902431-55

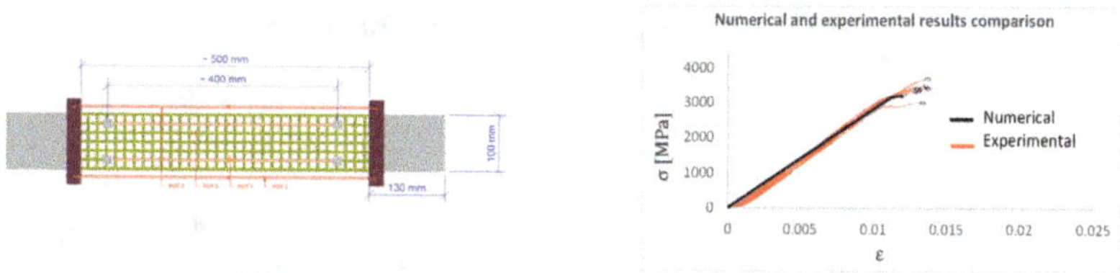

(a) (b)

Figure 2 (a) PBO mesh 22/22 specimen scheme and (b) Numerical and experimental result comparison

A second numerical test refers to the experimental tensile test on the FRCM composite panel reported in Fig. 3a [4]. The finite element model is illustrated in Figs. 3b, 3c, 3d.

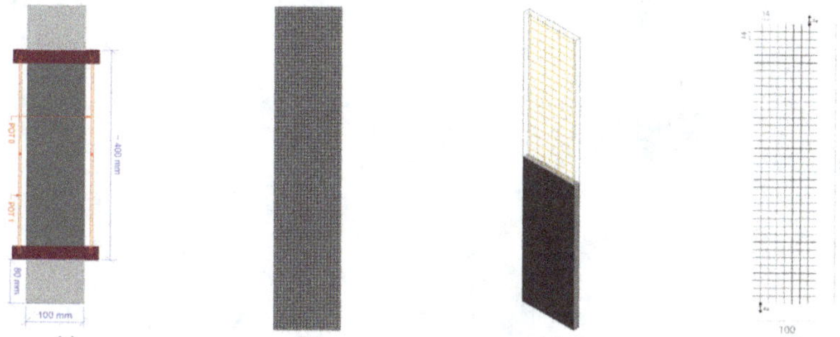

(a) (b) (c) (d)

Figure 3 (a) Geometric Scheme of FRCM Specimen; FRCM panel: FEM model (b) Frontal view, (c) cut plane Z and (d) the net.

The result of the numerical simulation is compared with the envelope of the stress-strain curves derived from the experimental tensile tests performed on 14 specimens of fiber-reinforced panels [4], according to the set-up shown in Fig. 3a. In Fig. 4 it is possible to appreciate the influence of the value of the fracture energy that defines the tensile behavior of the matrix of the FRCM system, governed by the linear softening. The matrix parameters are: elastic modulus E_{matrix} =2875 MPa, tensile strength $f_{t,matrix}$=2.84 MPa , fracture energy $G_{f,matrix}$ =0.05 N/mm.

A final test concerns the phenomenon of detachment of the FRCM from the support. Fig. 5a shows the discrete model. The dimensions of the surface of the reinforcement adhered to the tuff block are 100 x 300 mm. The longitudinal fibers of the mesh are extended to a length of 150 mm and then connected to a rigid bar. An interface has been introduced between the matrix and the support (in tuff), which approximates the stress redistribution in the matrix. Fig. 5 shows respectively the solid stress of all solid elements (b), the tangential tension along the interface (c) and the crack status at the interface (d) at a rate of the load of 85 %. It is clear that the detachment of the interface in the upper part involves a redistribution of the stresses in the matrix. The plot in Fig. 5a show that the loss of the interface bond strength happens about at the maximum force applied in experimental tests. The model based on the Embedded Truss is therefore able to adequately reproduce both the tensile response of the composite system and delamination tests, provided appropriate interface laws are adopted to simulate the behavior between the support and the matrix.

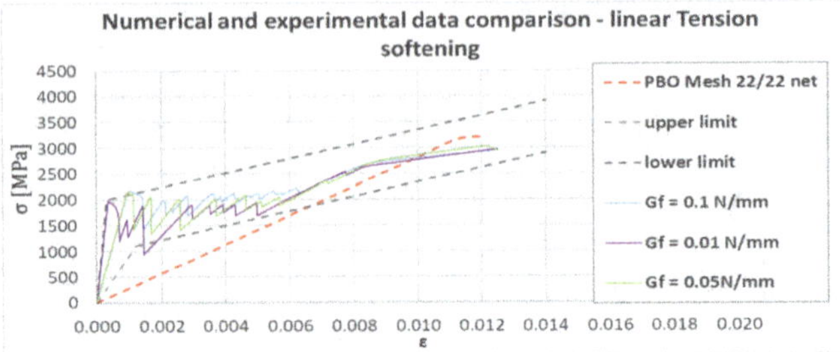

Figure 4 Comparison of numerical ad experimental results.

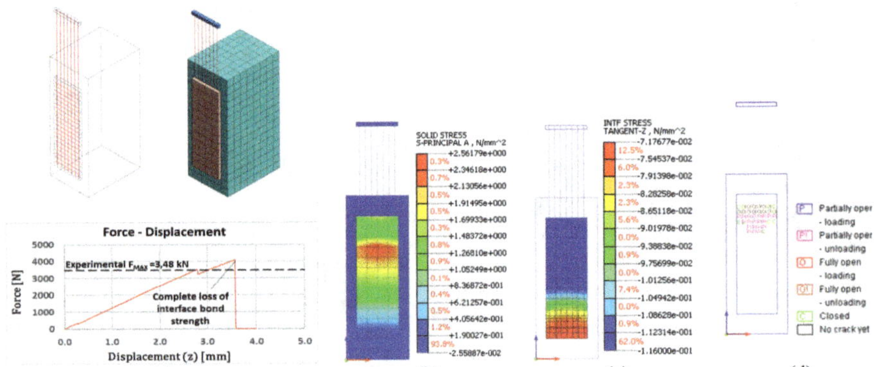

(a) (b) (c) (d)

Figure 5 Debonding FEM model. (a) Wireframe view, Solid View and Interface resistance; Results at 85% max load: (b) Matrix tensile stress; (c) Interface tangential stress and (d) Matrix crack status.

Behaviour of a single span FRCM strengthened masonry arch. Results and discussion.
In order to validate the proposed numerical technique, the experiments developed in [5] [6] [7] have been considered as benchmark test. The works experimentally investigate the structural behavior of masonry arches with different types of reinforcement, including PBO-Mesh 22/22 reinforcement. The tests were performed on 1:2 scale models with the dimensions depicted in Fig. 6a. The load is applied to a quarter of the span (378 mm from the left of the stump). Two samples were tested in the non-reinforced configuration (1-US and 2-US samples) and two samples with intrados reinforcement made of PBO Mesh 22/22 [5]; the control point of the displacements matches the loaded nodes. A nonlinear analysis was performed, in which the imposed displacement is assigned incrementally up to the last value of 6 mm with a step of 3.0 10^{-3}; the solution convergence criterion of the nonlinear problem was on the norm in energy (Work (W)) with a tolerance of 1. 0 10^{-3}. The elastic mechanical properties, from technical sheets and experimental identification, are reported in Table 1. The Smeared Crack model was used for both the matrix and the masonry. The input parameters of the non-linear model used in the numerical simulation were calibrated on the previous tests. Ruregold MX matrix: Linearly Elasic-Perfectly Plastic in compression, $f_{c,matrix}$= 26.4 MPa; Hordjik tensile softening, $f_{t,matrix}$= 2.64 MPa, $G_{f,matrix}$= 0.05 N/mm, internal length h_{mesh}=10 mm, shear retention factor β=0.01. Masonry: Linearly Elasic-

Materials Research Proceedings **26** (2023) 337-342 https://doi.org/10.21741/9781644902431-55

Perfectly Plastic in compression, $f_{c,masonry}$= 8.53 MPa; Linear tensile softening, $f_{t,masonry}$= 0.27 MPa, $G_{f,masonry}$= 0.001 N/mm, internal length h_{mesh} =20 mm, shear retention factor β=0.01.

Table 1 Material Parameters

	Ruregold MX-PBO	Masonry Parameter	PBO mesh 22-22
E (MPa)	7500	4500	270000
ν	0.3	0.27	0.3
G (MPa)	2885	1771.654	103846
f_t (MPa)			3031
ε_{el}			0.013

The graph in Fig. 6b shows the numerical response of both the unreinforced (N-UNR) and reinforced (N-RF) arch in terms of the Applied Force vs. Control Point Displacement curve. The model satisfactorily reproduces the experimental trend (Exp), both in the pre-peak and in the post-peak phase. In the case of the reinforced arch model, an increase of about 480% in the maximum strength was obtained compared to the non-reinforced model, in line with the experimental results.

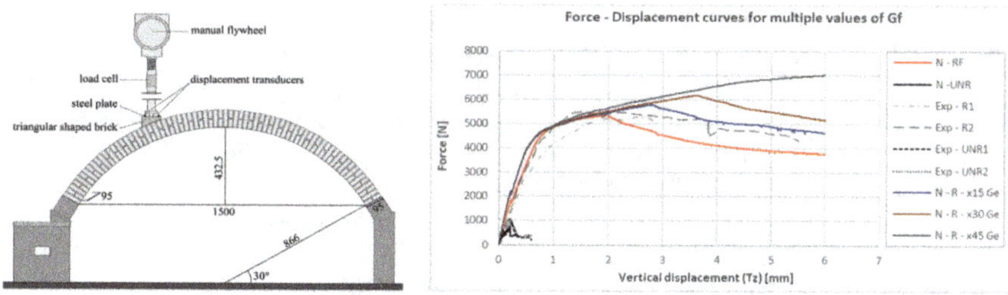

(a) (b)

Figure 6 (a) Geometric representation of the masonry arc specimen and (b) Comparison between experimental and numerical results of the strengthened and unstrengthened arch

In the graph (b), the three upper experimental curves (N-R) correspond to simulations in which the Hordjik tension function of the matrix was used with fracture energy $G_{f,matrix}$ chosen as multiple of the elastic energy E_e, that is, of the area underlying the strain curve in the elastic range. It is clear that, among the various parameters characterizing the Smeared Crack Model, the fracture energy plays a fundamental role and significantly affects the definition of the tensile behaviour of the matrix and the arch material. This implies that G_f must be carefully assessed. In the absence of certain data, the use of simpler models that adequately reproduce the salient aspects in the initial phase of the deformation history of the reinforced system, seems more prudential, even if they are not able to capture the entire evolutionary path of the damage.

From numerous numerical tests it has been observed that the shear retention function also plays a significant role, tuning the loss of shear stiffness (βG) and the evolutionary state of the cracks. On the contrary, not very significant, for the applications investigated, is the detailed definition of the compression damage function of both the masonry and the matrix of the FRCM. Indeed, the collapse mechanisms occur by the formation of hinges (cracks) due to the achievement of ultimate tensile stresses in specific sections, while the arch is divided into blocks whose compression stress state is well below the values at the elastic limit. For this reason, in the numerical applications, a simple Linearly Elastic-Perfectly Plastic model was assigned in compression to the brittle materials, that never exhibit excursions in the inelastic range.

The analysis of the crack pattern of the reinforced arch shows a widespread cracking in the masonry due to the transfer of the stresses performed by the FRCM, otherwise concentrated in the inelastic hinges in the case of the unreinforced arch.

(a) (b)

Figure 7 Crack pattern at the failure load of the unstrengthened arch (Tz=0.6 mm) (a) Unstrengthened arch and (b) strengthened arch

The above applications show the reliability of the numerical modeling methodology through Embedded Truss. The method appears suitable of application even to large structures, such as large-span and multi-span masonry arch bridges. Indeed, the reinforcement can be easily added to existent FEM models, with a limited increase of the size of the discrete structural model.

References

[1] MIDAS FEA NX, https://www.midasoft.com/

[2] RUREGOLD PBO-MESH 22/22, https://www.ruregold.com/it/

[3] D. V. Oliveira, B. Ghiassi, R. Allahvirdizadeh, X. Wang, G. Mininno e R. A. Silva, Macromodeling approach for pushover analysis of textile-reinforced mortar-strengthened masonry, Numerical Modeling of Masonry and Historical Structures (2019) 745-778. https://doi.org/10.1016/B978-0-08-102439-3.00021-X

[4] V. A. Alecci, F. Focacci, L. Rovero, G. Stipo e M. De Stefano, Extrados strengthening of brick masonry arches with PBO–FRCM composites: Experimental and analytical investigations, Composite Structures 149 (2016) 184-196. https://doi.org/10.1016/j.compstruct.2016.04.030

[5] V. A. Alecci, F. Focacci, L. Rovero, G. Stipo e M. De Stefano, Intrados strengthening of brick masonry arches with different FRCM composites: Experimental and analytical investigations, Composite Structures, 176 (2017) 898-909. https://doi.org/10.1016/j.compstruct.2017.06.023

[6] V. A. Alecci, G. Misseri, L. Rovero, G. Stipo, M. De Stefano, L. Feo e L. Raimondo, Experimental investigation on masonry arches strengthened with PBO-FRCM composite, Composites Part B, 100 (2016), 228-239. https://doi.org/10.1016/j.compositesb.2016.05.063

[7] E. Grande e G. Milani, Numerical simulation of the tensile behavior of FRCM strengthening systems, Composite, Composites Part B Engineering, 189 (2020) 107-886. https://doi.org/10.1016/j.compositesb.2020.107886

Theoretical and Applied Mechanics - AIMETA 2022 Materials Research Forum LLC
Materials Research Proceedings **26** (2023) 343-348 https://doi.org/10.21741/9781644902431-56

Numerical procedure for detecting the optimal stress state within the profile of a cracked arch

Stefano Galassi[1, a *] and Giacomo Tempesta[2, b]

[1]Department of Architecture, University of Florence, Florence, Italy

[2]Department of Architecture, UKZKM, Nostra Signora del Buon Consiglio, Tirana, Albania

[a]stefano.galassi@unifi.it, [b]g.tempesta@prof.unizkm.al

Keywords: Masonry, Cracked Arch, Optimal Stress State

Abstract. The method for identifying the line of thrust closest to the geometrical axis of an undamaged arch, developed by the authors in a publication in 2019 for the first time, is extended in this paper to the case of cracked arches. Cracks subdivide the arch into rigid portions linked in-between in correspondence to points at the extrados or the intrados, which can be interpreted as internal hinges. Under this assumption, the procedure is reformulated and extended to the cases of the one and two-hinged arch. In the paper, the procedure is presented in detail for the case of the one-hinged arch.

Introduction

The static indeterminacy of the equilibrium problem of the masonry arch is a well-known concept. In the 19th century, a great debate arose on this topic and the eighteenth-century theories for which the safety of an arch could be evaluated based on the condition of the existence of any line of thrust within the profile of the arch was replaced, little by little, by the emerging topic of detecting the true line of thrust within the arch, i.e. the actual state of stress. By removing the rigidity assumption, the true line of thrust can be identified by considering the elastic and inertial properties of the arch to formulate the congruency conditions.

However, some authors have proposed methodologies to work around the indeterminacy of the equilibrium problem by avoiding considering the masonry properties and its deformability. In this context it is worth mentioning the graphic procedure proposed by E. Mèry [1], where the elasticity theory affects his methodology only in the choice of the centers of pressure.

To work around the indeterminacy, particularly interesting is the research carried out by Winkler in [2], who, referring to the *"principle of least pressure"* enunciated by Moseley [3], stated that *"of all the thrust lines that can be built for the external loads, the true thrust line is the one that deviates least, on average, from the geometrical axis of the arch"*. Although it cannot be proved that this is the true one, what is certain is that the thrust line closest to the geometrical axis is the optimal one because it provides the lowest bending and shearing stresses and the arch is subjected to a fairly uniform compression stress. The topic of detecting the thrust line closest to the geometrical axis was addressed, firstly, by Heyman [4] who proposed an iterative procedure by trials and errors and, later on, by the authors of this contribution who provided the closed form solution to the problem stating it as a minimization problem of the quadratic deviation of the centroids of the discrete elements composing the arch from the nodes of the thrust line [5,6]. The minimum of this function is obtained expressing the zeroing of its partial derivatives with respect to three redundant unknowns, chosen as the horizontal thrust and the ordinate of the first and the last node of the thrust line. A system of three equations is obtained, which provides the value of the three redundant unknowns and, backwards, the ordinate of all the nodes of the thrust line. It is worth noting that the development of closed-form equations, capable of accurately estimating the load-bearing capacity of structures, is strongly recommended as it is a useful tool for practitioners

Materials Research Forum LLC
https://doi.org/10.21741/9781644902431-56

to perform a preliminary assessment of structure capacity [7].

In all the procedures mentioned above, only undamaged arches are considered, i.e. structures with a degree of statically indeterminacy equal to 3. However, arches can also be cracked or constructed with geometric defects, and, at the cracked or imperfect joints, hinges can have occurred. Nevertheless, no one, until now, has addressed this topic considering cracked arches. Leaving aside the three-hinged arch, which is a statically determined structure for which only one line of thrust exists, in this paper the numerical procedure developed by the authors for identifying the thrust line closest to the geometrical axis of an undamaged arch is extended to the case of the one and two-hinged arch and is presented, in detail, for the former case.

Procedure extended to cracked arches

Let us consider a continuous arch, subjected to a set of *n* vertical loads $F_1, F_2, F_3, \ldots, F_n$ (Fig. 1). Let us subdivide the arch into *n* elements in such a way as to apply these loads at the element centroids G_i. Let us also consider the occurrence of one or two hinges, C_1 and C_2, at the intrados/extrados of some lines of separation between the elements. The line of thrust is a polyline passing through the points P_i, vertically aligned to the element centroids, and through the hinges C_1 and C_2. It is worth noting that the line of thrust forms a cusp in correspondence to the nodes P_i and it is straight in correspondence to the hinges, where no external load is applied. For this reason, hereafter, nodes P_i, C_1 and C_2 are referred to as "cusp" nodes and "hinge" nodes, respectively.

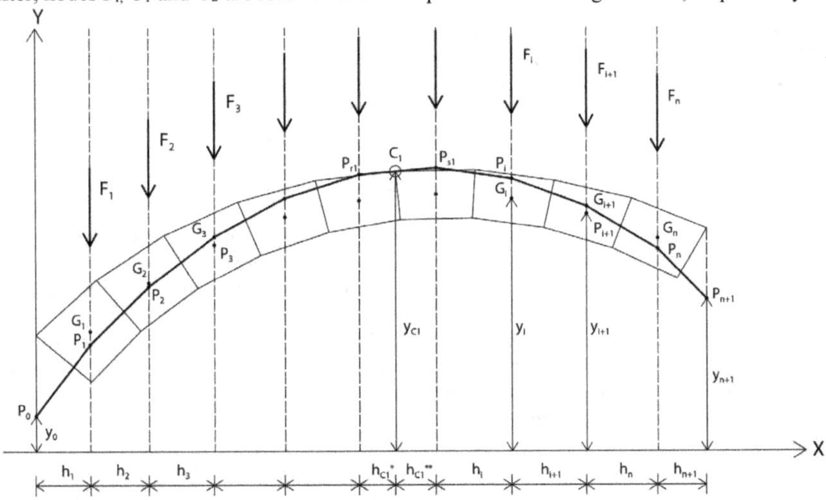

Fig. 1. Reference one-hinged arch.

To address the problem and identify the position of the nodes of the thrust line, the equilibrium conditions of both the "cusp" nodes and the "hinge" nodes are written. The equilibrium condition between the vertical components of the internal and external forces acting on the generic "cusp" node P_i of the thrust line is given by Eq. 1:

$$\left(\frac{-1}{h_i}\right) \cdot y_{i-1} + \left(\frac{h_i + h_{i+1}}{h_i \cdot h_{i+1}}\right) \cdot y_i + \left(\frac{-1}{h_{i+1}}\right) \cdot y_{i+1} = \frac{F_i}{H} \tag{1}$$

The equilibrium condition between the vertical components of the internal forces acting on the generic "hinge" node C_j ($1 \leq j \leq 2$) of the thrust line, located in-between the "cusp" nodes P_{r_j} and P_{s_j}, is given by Eq. 2:

$$\left(\frac{-1}{h_{c_j^*}}\right)\cdot y_{r_j}+\left(\frac{h_{c_j^*}+h_{c_j^{**}}}{h_{c_j^*}\cdot h_{c_j^{**}}}\right)\cdot y_{C_j}+\left(\frac{-1}{h_{c_j^{**}}}\right)\cdot y_{s_j}=0 \tag{2}$$

By collecting the equilibrium conditions of all nodes of the thrust line, a system of ($n+n_c$) linear equations in ($n+3$) unknowns is obtained (Eq. 3), where n_c is the number of internal hinges:

$$\begin{cases}\left(\frac{-1}{h_1}\right)\cdot y_0+\left(\frac{h_1+h_2}{h_1\cdot h_2}\right)\cdot y_1+\left(\frac{-1}{h_2}\right)\cdot y_2=\frac{F_1}{H}\\ \left(\frac{-1}{h_2}\right)\cdot y_1+\left(\frac{h_2+h_3}{h_2\cdot h_3}\right)\cdot y_2+\left(\frac{-1}{h_3}\right)\cdot y_3=\frac{F_2}{H}\\ \dots\dots\dots\dots\dots\dots\\ \left(\frac{-1}{h_i}\right)\cdot y_{i-1}+\left(\frac{h_i+h_{i+1}}{h_i\cdot h_{i+1}}\right)\cdot y_i+\left(\frac{-1}{h_{i+1}}\right)\cdot y_{i+1}=\frac{F_i}{H}\\ \dots\dots\dots\dots\dots\dots\\ \left(\frac{-1}{h_{r_j}}\right)\cdot y_{r_j-1}+\left(\frac{h_{r_j}+h_{c_j^*}}{h_{r_j}\cdot h_{c_j^*}}\right)\cdot y_{r_j}=\frac{F_{r_j}}{H}+\left(\frac{1}{h_{c_j^*}}\right)\cdot y_{C_j}\\ \left(\frac{-1}{h_{c_j^*}}\right)\cdot y_{r_j}+\left(\frac{h_{c_j^*}+h_{c_j^{**}}}{h_{c_j^*}\cdot h_{c_j^{**}}}\right)\cdot y_{C_j}+\left(\frac{-1}{h_{c_j^{**}}}\right)\cdot y_{s_j}=0\\ \left(\frac{h_{c_j^{**}}+h_{s_j+1}}{h_{c_j^{**}}\cdot h_{s_j+1}}\right)\cdot y_{s_j}+\left(\frac{-1}{h_{s_j+1}}\right)\cdot y_{s_j+1}=\frac{F_{s_j}}{H}+\left(\frac{1}{h_{c_j^{**}}}\right)\cdot y_{C_j}\\ \dots\dots\dots\dots\dots\dots\\ \left(\frac{-1}{h_n}\right)\cdot y_{n-1}+\left(\frac{h_n+h_{n+1}}{h_n\cdot h_{n+1}}\right)\cdot y_n+\left(\frac{-1}{h_{n+1}}\right)\cdot y_{n+1}=\frac{F_n}{H}\end{cases} \tag{3}$$

In the system of Eq. 3 the unknowns are the ordinates y_i of the n+2 cusp nodes P_i of the thrust line, comprising the former one, P_0, and the latter one, P_{n+1}, and the constant thrust, H. Therefore, as expected, the problem is statically undetermined to ($3-n_c$) degree, because ∞^{3-n_c} lines of thrust in equilibrium with the external loads can be identified. In fact, the requirement that the line of thrust passes through one or two hinges reduces the indeterminacy of the problem by one (n_c=1) or two (n_c=2) degrees.

The numerical technique developed to solve the problem requires the choice of three parameters, as follows: primarily, the n_c ordinates y_{C_j} of the hinges (known terms) are selected; successively, the remaining ($3-n_c$) parameters (unknowns) that are chosen are the horizontal thrust H and the ordinate y_0 of the first node of the thrust line. These parameters are then isolated within the system of Eq. 3, which is rewritten, in matrix form, rearranging the system putting the equation(s) of the hinge node(s) in the last row(s).

Referring to the case of the one-hinged arch (n_c=1) and to the case of the two-hinged arch (n_c=2), the matrix system takes the form of Eq. 4 and Eq. 5, respectively:

$$[D]\{Y\}=\{T_1\}\cdot k+\{T_2\}\cdot y_0+\{T_3\}\cdot y_{C_1} \tag{4}$$

$$[D]\{Y\}=\{T_1\}\cdot k+\{T_2\}\cdot y_{C_1}+\{T_3\}\cdot y_{C_2} \tag{5}$$

where k=1/H.

For lack of space and for the sake of clarity, in the following lines the numerical procedure is presented and discussed only for the one-hinged arch. Hence, referring to Eq. 4, entries of vectors and matrices are shown below:

$$D=\begin{bmatrix}
\left(\frac{h_1+h_2}{h_1\cdot h_2}\right) & \left(\frac{-1}{h_2}\right) & 0 & 0 & 0 & 0 & 0 & 0 & 0 & 0\\
\left(\frac{-1}{h_2}\right) & \left(\frac{h_2+h_3}{h_2\cdot h_3}\right) & \left(\frac{-1}{h_3}\right) & 0 & 0 & 0 & 0 & 0 & 0 & 0\\
\cdots & \cdots & \cdots & \cdots & \cdots & \cdots & \cdots & \cdots & \cdots & \cdots\\
0 & 0 & \left(\frac{-1}{h_i}\right) & \left(\frac{h_i+h_{i+1}}{h_i\cdot h_{i+1}}\right) & \left(\frac{-1}{h_{i+1}}\right) & 0 & 0 & 0 & 0 & 0\\
\cdots & \cdots & \cdots & \cdots & \cdots & \cdots & \cdots & \cdots & \cdots & \cdots\\
0 & 0 & 0 & 0 & \left(\frac{-1}{h_{r_j}}\right) & \left(\frac{h_{r_j}+h_{c_j^*}}{h_{r_j}\cdot h_{c_j^*}}\right) & 0 & 0 & 0 & 0\\
0 & 0 & 0 & 0 & 0 & 0 & \left(\frac{h_{c_j^{**}}+h_{s_j+1}}{h_{c_j^{**}}\cdot h_{s_j+1}}\right) & \left(\frac{-1}{h_{s_j+1}}\right) & 0 & 0\\
\cdots & \cdots & \cdots & \cdots & \cdots & \cdots & \cdots & \cdots & \cdots & \cdots\\
0 & 0 & 0 & 0 & 0 & 0 & 0 & \left(\frac{-1}{h_n}\right) & \left(\frac{h_n+h_{n+1}}{h_n\cdot h_{n+1}}\right) & \left(\frac{-1}{h_{n+1}}\right)\\
0 & 0 & 0 & 0 & 0 & \left(\frac{-1}{h_{c_j^*}}\right) & \left(\frac{-1}{h_{c_j^{**}}}\right) & 0 & 0 & 0
\end{bmatrix},$$

$$Y=\begin{Bmatrix} y_1\\ y_2\\ \cdots\\ y_i\\ \cdots\\ y_{r_j}\\ y_{s_j}\\ \cdots\\ y_n\\ y_{n+j}\end{Bmatrix},\qquad T_1=\begin{Bmatrix} F_1\\ F_2\\ \cdots\\ F_i\\ \cdots\\ F_{r_j}\\ F_{s_j}\\ \cdots\\ F_n\\ 0\end{Bmatrix}.\qquad T_2=\begin{Bmatrix} 1/h_1\\ 0\\ \cdots\\ 0\\ \cdots\\ 0\\ 0\\ \cdots\\ 0\\ 0\end{Bmatrix},\qquad T_3=\begin{Bmatrix} 0\\ 0\\ \cdots\\ 0\\ \cdots\\ \frac{1}{h_{c_j^*}}\\ \frac{1}{h_{c_j^{**}}}\\ \cdots\\ 0\\ -\left(\frac{h_{c_j^*}+h_{c_j^{**}}}{h_{c_j^*}\cdot h_{c_j^{**}}}\right)\end{Bmatrix}. \tag{6}$$

The solution of Eq. 4 is provided by Eq. 7:

$$\{Y\}=\{R_1\}\cdot k+\{R_2\}\cdot y_0+\{R_3\}\cdot y_{C_1} \tag{7}$$

where:

$$\{R_1\}=[D]^{-1}\cdot\{T_1\},\qquad \{R_2\}=[D]^{-1}\cdot\{T_2\},\qquad \{R_3\}=[D]^{-1}\cdot\{T_3\}. \tag{8}$$

In order to identify, among the ∞^{3-n_c} lines of thrust (n_c=1 in the case at hand), the one closest to the geometrical axis and passing through the hinge node, the squared distance function between the ordinates of the nodes of the thrust line and the ordinates of the centroids of the elements is computed and minimized with respect to the (3-n_c) redundant unknowns, n_c parameters being known terms:

$$S=\sum_{i=1}^{n+n_c}(\Delta y_i)^2=\sum_{i=1}^{n+n_c}(y_i-y_{Gi})^2=\left(\{R_1\}\cdot k+\{R_2\}\cdot y_0+\{R_3\}\cdot y_{C_1}-\{Y_G\}\right)^2 \tag{9}$$

Function S is minimized by expressing the partial derivatives of it, with respect to the two unknowns, equal to zero:

$$\begin{aligned}\frac{\partial S}{\partial k}(y_0,k)&=0\\ \frac{\partial S}{\partial y_0}(y_0,k)&=0\end{aligned} \tag{10}$$

By doing so, the system of two linear equations that follows is obtained:

$$\begin{cases} \{R_1\}^2 \cdot k + \{R_1\}\{R_2\} \cdot y_0 = \{R_1\}\left(\{Y_G\} - \{R_3\} \cdot y_{C_1}\right) \\ \{R_1\}\{R_2\} \cdot k + \{R_2\}^2 \cdot y_0 = \{R_2\}\left(\{Y_G\} - \{R_3\} \cdot y_{C_1}\right) \end{cases} \tag{11}$$

and it is then put in matrix form:

$$\begin{bmatrix} \{R_1\}^2 & \{R_1\}\{R_2\} \\ \{R_1\}\{R_2\} & \{R_2\}^2 \end{bmatrix} \cdot \begin{Bmatrix} k \\ y_0 \end{Bmatrix} = \begin{Bmatrix} \{R_1\}\left(\{Y_G\} - \{R_3\} \cdot y_{C_1}\right) \\ \{R_2\}\left(\{Y_G\} - \{R_3\} \cdot y_{C_1}\right) \end{Bmatrix} \tag{12}$$

and, more compactly, re-written as:

$$[N]\{P\} = \{W\} \tag{13}$$

The solution of Eq. 13 provides the value of the two unknowns, k and y_0, stored in vector P. The ordinates of the nodes of the line of thrust are finally obtained by back-substituting these values into Eq. 7.

Nevertheless, it is worth noting that the solution of Eq. 13 is not unique because it depends on the squared distance function in Eq. 9, which is affected by the value of the ordinate $y_{G(n+1)}$ of the centroid of the element *n+1*, which does not exist within the arch. It is possible to explore different lines of thrust passing through the hinge, with different shape and distance from the geometrical axis of the arch, by setting different values of $y_{G(n+1)}$. To identify the line of thrust passing through the hinge and closest to the geometrical axis, a procedure for trials and errors is conceived which converges when the value of $y_{G(n+1)}$ corresponding to the minimum of the squared distance function, computed for the *n* elements composing the arch, is obtained. At convergence, the two unknowns in Eq. 13 and the ordinates of the nodes of the thrust line are obtained.

Numerical explanatory example

The simple case of a portion of a segmental barrel vault, 100 cm deep, with a span length of 600 cm, a rise of 210.1 cm, 45 cm thick and an angle of embrace of 140° (t/R=0.13) is analyzed in this section. The vault is subdivided into six elements of the same size (*n=6*), with a unit weight of $18 kN/m^3$. Except for the self-weight, no additional load is applied on the vault.

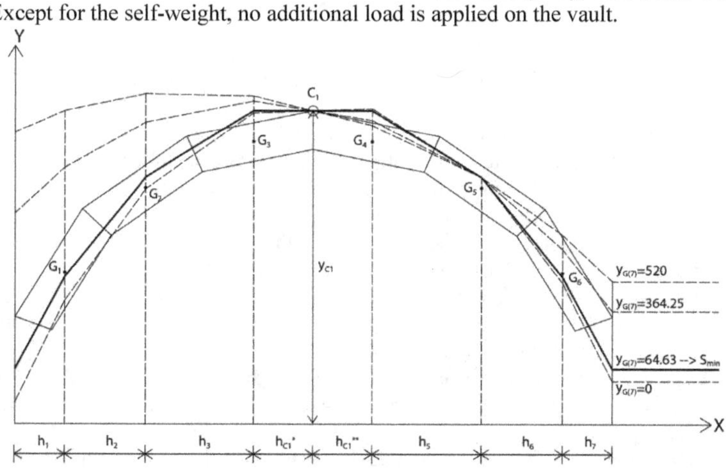

Fig. 2. Explanatory example; set of lines of thrust passing through the extrados hinge at the key section and identification of the one which is closest to the geometrical axis.

The numerical procedure proposed is used to identify the line of thrust closest to the geometrical axis of the structure passing through the extrados hinge at the key section (Fig. 2).

Fig. 2 allows for the exploring of the set of lines of thrust passing through the hinge at the key, corresponding to different values of the squared distance function, and the one which is also closest to the geometrical axis, corresponding to the minimum of the squared distance function which is found to be equal to *min(S)*=64.63 cm^2 by applying Eq. 9 iteratively. For that value, the ordinate $y_{G(7)}$ = 64.63 cm of the centroid of the fictitious 7th element is obtained.

In detail, in order to allow the interested reader the possibility of reproducing the example numerically on his own, we report the main data and entries of the input vectors built by the procedure:

$$h_1 = h_7 = 57.01\text{ cm}, h_2 = h_6 = 93.02\text{ cm}, h_3 = h_5 = 124.47\text{ cm}, h_{c_1^*} = h_{c_1^{**}} = 67.78\text{ cm}, y_{C_1} = 364.25\text{ cm}$$

$$\{T_1\}^t = \{16.96|16.96|16.96|16.96|16.96|16.96|0\}\text{ kN}$$

$$\{T_2\}^t = \{0.0175|0|0|0|0|0|0\}\text{ cm}^{-1}$$

$$\{T_3\}^t = \{0|0|0.0148|0.0148|0|0|-0.02951\}\text{ cm}^{-1}$$

The ordinates of the nodes of the line of thrust closest to the geometrical axis and passing through the hinge at the key, computed for the centroid ordinate $y_{G(7)}$ = 64.63 cm of the fictitious 7th element are finally shown in the vector that follows, obtained by applying Eq. 7:

$$\{Y\}^t = \{64.63|171.05|286.80|364.25|286.80|171.05|64.63\}\text{ cm}$$

The corresponding thrust line is drawn in Fig. 2 with the thick continuous polyline.

Conclusions

In this paper the numerical procedure developed by the authors for identifying the line of thrust closest to the geometrical axis of an undamaged arch has been reformulated and extended to the case of cracked arches. The case of the one-hinged arch has been addressed and the related numerical procedure presented in detail. The case of the two-hinged arch will be addressed in an extended paper in the near future and, most certainly, presented, in Palermo, at the Aimeta 2022 conference in person by the corresponding author of this contribution.

References

[1] E. Mèry. Mémoire sur l'équilibre des voutes en berceau, Annales des Ponts et Chaussées, X,1 (1840) 50-70.

[2] E. Winkler. Die Lebre von der Elasticitat und Festigkeit, Dominicus, Prague (1867).

[3] H. Moseley. On a new principle in statics called the principle of least pressure, The London, Edinburgh, and Dublin Philosophical Magazine and Journal of Science, 3,16 (1833) 285-88. https://doi.org/10.1080/14786443308648178

[4] J. Heyman. The safety of masonry arches, Int. J. Mech. Sci., 11,4 (1969) 363-85. https://doi.org/10.1016/0020-7403(69)90070-8

[5] G. Tempesta, S. Galassi. Safety evaluation of masonry arches. A numerical procedure based on the thrust line closest to the geometrical axis, Int. J. Mech. Sci., 155 (2019) 206-21. https://doi.org/10.1016/0020-7403(69)90070-8

[6] S. Galassi, G. Tempesta. The Matlab code of the method based on the Full Range Factor for assessing the safety of masonry arches, MethodsX, 6 (2019) 1521-42. https://doi.org/10.1016/j.mex.2019.05.033

[7] F. Gusella. Effect of the plastic rotation randomness on the moment redistribution in reinforced concrete structures, Eng. Struct., 252 (2022) 113652. https://doi.org/10.1016/j.engstruct.2021.113652

Theoretical and Applied Mechanics - AIMETA 2022 Materials Research Forum LLC
Materials Research Proceedings **26** (2023) 349-353 https://doi.org/10.21741/9781644902431-57

Multiscale analysis of masonry vaults coupling shell elements to 3D-Cauchy continuum

Daniela Addessi[1,a], Paolo Di Re[1,b], Cristina Gatta[1,c*] and Elio Sacco[2,d]

[1]Department of Structural and Geotechnical Engineering, Sapienza University of Rome, Rome, Italy

[2]Department of Structure for Engineering and Architecture, University of Naples Federico II, Naples, Italy

[a]daniela.addessi@uniroma1.it, [b]paolo.dire@uniroma1.it, [c]cristina.gatta@uniroma1.it, [d]elio.sacco@unina.it

Keywords: Multiscale, Masonry, Vault

Abstract. This study adopts an enhanced multiscale approach to investigate the effects of the damaging process on the structural behavior of masonry vaults with regular texture, in view of their safety assessment. The model, recently developed by the authors, links two different structural models at macro and microscale, exploiting the advantages of each formulation. At the macroscopic level a homogeneous Mindlin-Reissner shell is modeled and its constitutive response is derived by the detailed analysis of a three-dimensional (3D) masonry Unit Cell (UC) studied at microlevel. The UC is considered as the assembly of elastic bricks and damage-plastic zero-thickness interfaces, representative of both mortar and mortar-unit interaction, thus accounting for the actual geometry, arrangement and constitutive response of each constituent material. A Transformation Field Analysis procedure is used to link the two scales, speeding up the numerical simulations. Structural response of a masonry vault under differential settlements is investigated, determining its load-bearing capacity and the damaging path evolving in the structure up to collapse. The reliability of the results is proved by comparison with outcomes derived by detailed micromechanical analysis, interpreting and arguing similarities and differences.

Introduction

Vaulted structures are the key element of many historical and monumental masonry constructions and require a special attention to maintain their structural integrity and preserve cultural identity. Many efforts were devoted to developing efficient procedures, both analytical and numerical, to accurately study the response of masonry curved elements, like arches and vaults. An extensive review specially dedicated to masonry vaults can be found in [1].

The limit analysis is one of the oldest methods, widely adopted in the past [2]. However, recently some interesting contributions proposed modern numerical procedures relying on limit analysis concepts [3].

Among the computational approaches, the discrete element [4] and the finite element method (FEM) are today the most selected to model in detail general geometry, mechanical properties, boundary conditions and accurately reproduce the masonry structural response. Limiting the attention to the FEM, a possible classification of the proposed procedures relies on the scale of the analysis [5] and distinguishes between macromechanical, micromechanical and multiscale approaches [6].

This paper investigates the effects of the damaging process on the structural response of an experimental masonry vault, by adopting the multiscale procedure proposed by the authors in [7]. This relies on a Reduced Order Model (ROM) based on the Transformation Field Analysis (TFA)

to address the nonlinear homogenization problem of the regular masonry considered at the microscale level.

Two different models are coupled at the macroscopic structural scale and microscopic material level. Indeed, a thick shell formulation is adopted at the higher level for the vault, whereas a 3D Cauchy continuum is used for the masonry repetitive Unit Cell (UC) at the lower material scale. Here, possible nonlinear damage-plastic mechanisms are only concentrated at the mortar joints modeled as zero-thickness interfaces. The two scales communicate by means of a kinematic map, defined as function of the macroscopic strain measures, that represents the displacement field in the UC, in the framework of a computational homogenization procedure. The overall micromechanical response of the UC is evaluated under the effects of the imposed macroscopic strain components and inelastic strains and proper localization operators are computed. Based on these, the evolution problem of the damage and plasticity at each macroscopic point is solved.

The presented procedure is implemented in the finite element program FEAP and used to study the structural response of the masonry vault under differential settlements, investigating both the global response features and the damaging paths.

Shell-3D multiscale formulation

The multiscale strategy proposed in [7] is used to analyze response of masonry vaults. The model adopts different formulations at the two analysis scales (Fig. 1): at macroscale, a homogeneous thick shell is considered, based on Mindlin-Reissner plate theory. To derive the constitutive response of the shell, at each macroscopic point a UC, modeling the actual masonry texture, is linked. This is studied at microscale adopting a three-dimensional Cauchy continuum. The UC is formed by assembling linear elastic bricks and nonlinear mortar joints, modeled as interfaces. The relationship between tractions, **t**, and displacement jumps, **s**, at the interface results as:

$$\mathbf{t} = \mathbf{C}\left[\mathbf{s} - \boldsymbol{\pi}\right] = \mathbf{C}\left[\mathbf{s} - D(\mathbf{c} + \mathbf{p})\right] \tag{1}$$

with **C** the diagonal stiffness matrix and $\boldsymbol{\pi}$ the inelastic vector accounting for damage, D, unilateral contact, **c**, and sliding friction, **p** [8].

The linking procedure involves the definition of a suitable kinematic map. Indeed, the UC microscopic strain field in the bricks, where the displacements are continuous, is expressed as:

$$\boldsymbol{\varepsilon} = \mathbf{B}\mathbf{E} + \boldsymbol{\varepsilon}^{*} \tag{2}$$

being **E** the vector collecting the generalized strains of the shell, $\boldsymbol{\varepsilon}^{*}$ the strain field derived by the periodic displacement perturbation, and **B** the matrix governing the kinematic map.

The upscaling process is performed by invoking a generalized Hill-Mandel principle. Accordingly, the shell stresses, $\boldsymbol{\Sigma}$, result as:

$$\boldsymbol{\Sigma} = \frac{1}{A}\int \int_{-t/2}^{t/2} \mathbf{B}^{T} \boldsymbol{\sigma} \, dx_3 \, dA \tag{3}$$

being A the area of the UC mid-plane, t the UC thickness and x_3 the axis running along t. In the right side of Eq. 3, only the brick stresses, $\boldsymbol{\sigma}$, are considered, as the interface terms give null contributions.

The Transformation Field Analysis technique is exploited to determine the homogenized response of the composite material. This consists in the subdivision of the interfaces into regions, called subsets, where the inelastic quantities are assumed as uniform. The effects of the macroscopic strains and that of the inelastic ones in each subset of the UC are determined by

Materials Research Forum LLC
https://doi.org/10.21741/9781644902431-57

preliminary elastic analyses and, then, the results obtained in terms of localization matrices are used during the multiscale analysis to evaluate the constitutive response of the homogenized shell.

The model is implemented in the finite element code FEAP using a nonlocal integral formulation to overcome the mesh-dependency numerical issue.

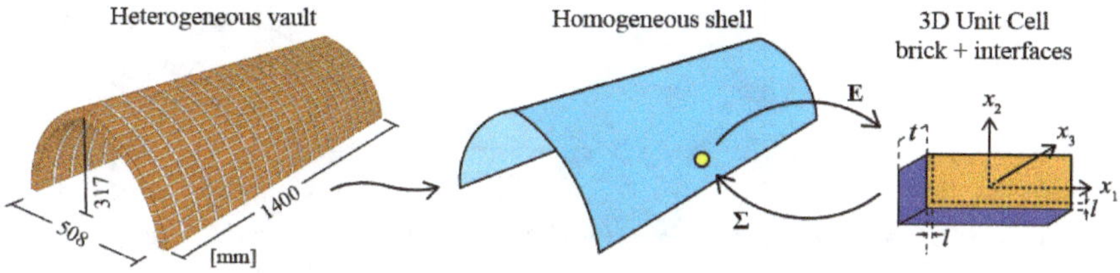

Figure 1: Multiscale modeling approach for masonry vaults.

Masonry vault under differential settlements

The response of the masonry barrel vault experimentally tested in [9] (Fig. 1) is numerically reproduced. In [10], same specimen is tested under vertical differential settlements to investigate the performance of the proposed approach in reproducing the behavior of curved elements. To extend this study to different loading conditions and analyze the response under different stress-strain states, horizontal displacements of the abutments are here considered, as indicated in Fig. 2, simulating the eventual out-of-plane mechanisms of the walls supporting the vault. Fig. 1 shows the adopted UC, made of a single brick, one head and one bed mortar joint. As opposed to the real specimen geometry that assumes running bond texture, for this test stacking bond arrangement is considered, resulting in a simpler and computationally less time-consuming description. In fact, under such loading conditions, structural response is expected to be governed by bed mortar joint opening, which is equally described by both masonry layouts.

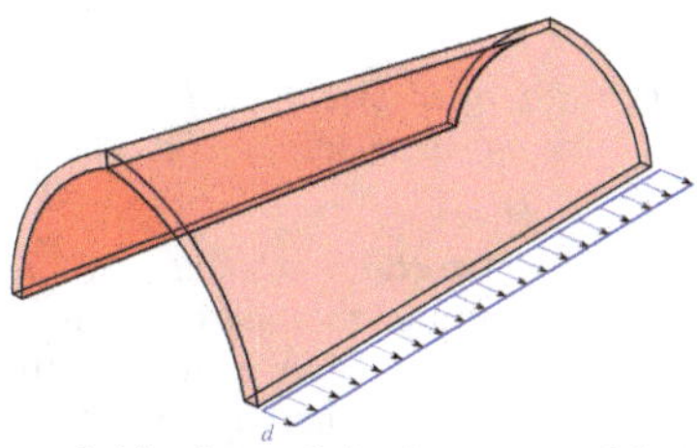

Figure 2: Differential horizontal displacement of the vault abutments.

Fig. 3 shows the global response curve of the vault in terms of total horizontal reaction of the fixed abutment (outward direction assumed as positive) versus imposed horizontal displacement. Blue curve refers to the solution obtained with the proposed model by dividing the UC into 16 subsets, i.e. 8 for the head and 8 for the bed joint, uniformly spaced across the masonry thickness [10]. This solution well agrees with that obtained by a reference micromechanical model (dashed red curve) made of linear elastic 8-node finite elements for the bricks and nonlinear zero-thickness interfaces for mortar joints, proving that the proposed model can correctly predict the vault nonlinear structural behavior.

Materials Research Forum LLC
https://doi.org/10.21741/9781644902431-57

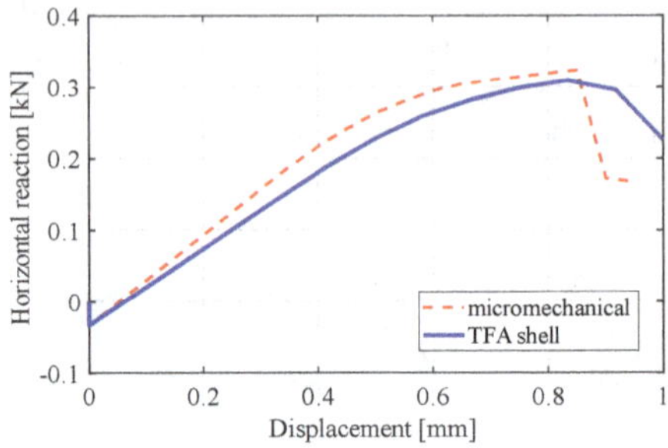

Figure 3: Global response of the vault under differential horizontal displacement of the abutments.

As expected, under the imposed displacement, vault walls uniformly bend inducing opening of the bed mortar joints at the abutments, on the extrados side, and at the key, on the intrados side. The vault deformed shape at $d = 0.85$ mm is shown in Fig. 4, for the two models. For the proposed shell model, contour plot indicates the values of the damage occurring in the bed mortar joints, averaged over the thickness of the UC. Damage appears more diffused than in the micromechanical model due to the homogenization and nonlocal nature of the multiscale model.

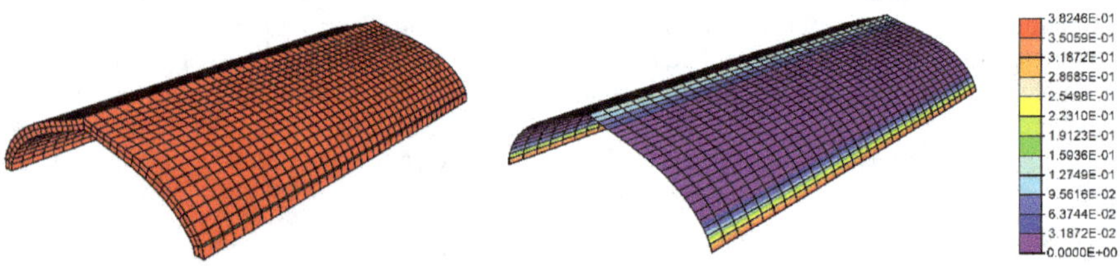

Figure 4: Deformed shape of the vault: micromechanical FEM (left) and multiscale model with damage distribution (right).

Conclusions

The study presented investigated the structural response of masonry vaults, representative elements of many components of the world cultural heritage, mainly focusing on the effects of damage evolution on the response global features. A multiscale procedure, relying on a reduced order model, was adopted, that consider a thick shell model at the structural scale and a Cauchy continuum at the real material scale. Here, the assumption of nonlinear damage-plasticity mechanisms only concentrated at mortar joints turned out to be accurate enough and suitable to reproduce the main characteristics of the masonry vault response, under simple distributions of the ground settlements.

Moreover, the use of the TFA technique, based on the assumption of piece-wise uniform distribution of nonlinearities over the mortar joints at the UC level, has allowed to avoid the high computational burden related to a full FE^2 multiscale procedure. Indeed, the obtained numerical results, as compared to those evaluated by a cumbersome micromechanical analysis, were accurate and reliable. Of course, the adopted method cannot describe localized cracking processes, although giving a correct indication of the region where damage concentrates but in a more diffused fashion.

Therefore, the multiscale model adopted, due to its computational efficiency and stability, has resulted to be a very promising tool and encourages to extend its application to other types of masonry elements characterized by different, more complex, geometries and fabric, as well as different boundary conditions. Further developments will be devoted to enhancing the description of the nonlinearity distributions in the TFA procedure to overcome the limits related to the piecewise uniform assumption. Moreover, also the possible introduction of strengthening interventions will be accounted for, by properly modifying the presented multiscale approach.

Acknowledgements

E.S. acknowledges the financial support of the ReLUIS project from the Italian Department of the Civil Protection.

D.A., P.D.R. and C.G. acknowledge the grant PRIN-2017 20173C478N.

References

[1] A. Tralli, C. Alessandri, G. Milani, Computational methods for masonry vaults: a review of recent results, Open Civ. Eng. J. 8 (2014) 272-287. https://doi.org/10.2174/1874149501408010272

[2] M. Como, Statics of historic masonry constructions, Volume 1, Springer, 2013. https://doi.org/10.1007/978-3-642-30132-2_1

[3] A. Chiozzi, G. Milani, A. Tralli, A genetic algorithm NURBS-based new approach for fast kinematic limit analysis of masonry vaults, Comput. Struct. 182 (2017) 187-204. https://doi.org/10.1016/j.compstruc.2016.11.003

[4] V. Sarhosis, J. Lemos, K. Bagi, Discrete element modeling, in: B. Ghiassi and G. Milani, (Eds.), Numerical Modeling of Masonry and Historical Structures, Woodhead Publishing. Series in Civil and Structural Engineering, Chapter 13, 469-501, 2019. https://doi.org/10.1016/B978-0-08-102439-3.00013-0

[5] E. Sacco, D. Addessi, K. Sab, New trends in mechanics of masonry, Meccanica 53(7) (2018) 1565-1569. https://doi.org/10.1007/s11012-018-0839-x

[6] P. Di Re, D. Addessi, E. Sacco, A multiscale force-based curved beam element for masonry arches, Comput. Struct. 208 (2018) 17-31. https://doi.org/10.1016/j.compstruc.2018.06.009

[7] D. Addessi, P. Di Re, C. Gatta, E. Sacco, Multiscale analysis of out-of-plane masonry elements using different structural models at macro and microscale, Comput. Struct. 247 (2021) 106477. https://doi.org/10.1016/j.compstruc.2020.106477

[8] D. Addessi, C. Gatta, S. Marfia, E. Sacco, Multiscale analysis of in-plane masonry walls accounting for degradation and frictional effects, Int. J. Multiscale Comput. Eng. 18 (2) (2020) 159-180. https://doi.org/10.1615/IntJMultCompEng.2020031235

[9] J. Hudson, D. Theodossopoulos, Gothic barrel vaults under differential settlement: The effects of boundary conditions and FRP on structural behaviour, Key Eng. Mater. 747 (2017) 496-503. https://doi.org/10.4028/www.scientific.net/KEM.747.496

[10] D. Addessi, P. Di Re, C. Gatta, E. Sacco, Shell-3D multiscale modeling of masonry vaults based on the TFA procedure, Int. J. Multiscale Comput. Eng. in press, 2022. https://doi.org/10.1615/IntJMultCompEng.2022041774

Theoretical and Applied Mechanics - AIMETA 2022 | Materials Research Forum LLC
Materials Research Proceedings **26** (2023) 355-360 | https://doi.org/10.21741/9781644902431-58

Vulnerability assessment of historical masonry buildings to excavation-induced settlements: *palazzo assicurazioni generali*

ADDESSI Daniela, DI RE Paolo* and PAOLONE Achille

Department of Structural and Geotechnical Engineering, Sapienza University of Rome, Italy

daniela.addessi@uniroma1.it, paolo.dire@uniroma1.it, achille.paolone@uniroma1.it

Keywords: Masonry, Settlement, Numerical Modelling, Nonlinear Analysis

Abstract. Historical masonry structures represent a conspicuous part of the European architectural heritage. However, these often are more vulnerable to damage risk than other constructions due to their peculiar mechanical properties. This justifies the lively interest in the development of efficient and reliable approaches for their structural capacity assessment. This work presents the vulnerability assessment of the historical masonry building, "Palazzo Assicurazioni Generali", for the effects induced by the underlying tunnel excavation required for line C of the Rome underground. The structure is a 12-floor building, built in 1902-1906 in Piazza Venezia, one of the most prestigious areas of the city. A detailed macromechanical three-dimensional numerical model of the entire structure is developed and two kinds of static analyses are conducted, assuming linear elastic and nonlinear constitutive response for masonry.

Introduction

Masonry structures compose most of the architectural heritage of many European countries. However, due to the poor tensile strength, these usually exhibit high vulnerability against natural loadings but also anthropic phenomena, such as excavation-induced ground settlement.

The assessment of structural response to differential settlements has always been a felt concern, especially in metropolitan areas. Indeed, several works have been dedicated to the estimation of damage in masonry buildings due to settlement effects. Meaningful examples and extensive reviews are reported in [1,2]. Most adopted techniques usually rely either on simplified approaches, where elastic strains in the material are compared with limit values, e.g. [3,4], or on more complex methods, based on nonlinear Finite Element (FE) analyses [5-7]. For the latter, soil-structure interaction can be neglected [8], for simpler and more conservative analyses, or included [9], for more accurate solutions.

This work is part of the studies conducted to estimate the effects induced on historical buildings by the tunnelling operations required for the construction of the new Line C of the Rome underground in Italy. These involve the realization of two tunnels that connect the station located near the Colosseum to that to be built under Venice Plaza (*Piazza Venezia*). In particular, this work presents the vulnerability assessment of the Assicurazioni Generali building (*Palazzo Assicurazioni Generali*), an important historical masonry building located in Venice Plaza, on the northeast side of the underground station.

A three-dimensional (3D) numerical model of the building is defined and structural analyses are carried out to estimate the damage distribution in the masonry due to the settlement-induced deformations. These are conducted by modeling the structure as decoupled from the soil. However, assumed vertical displacement profiles of the foundation, corresponding to the settlements, are defined according to two different approaches, namely level 1, based on the assumption that interaction between soil and building is negligible (free field), and level 2, based on numerical simulations capable of accounting for this interaction.

Two types of structural analyses are performed. In the first, the response of the building is assumed as linear elastic. Hence, the level of strains induced by the application of the subsidence fields is determined and used to evaluate the potential crack patterns occurring in the masonry. In the second, a smeared crack non-linear constitutive model (SCM) [10,11] is used to simulate the material damage and, thus, the evolution of cracks produced by the differential settlements.

Structure geometry

The Assicurazioni Generali building is a 12-floor masonry structure, occupying a large area of almost 3500 m^2 in plan. It was built in 1902-1906 by the Architect G. Sacconi, who decided to use high-quality solid brick masonry for almost the entire construction.

In the main portion of the building, the structure consists of 9 floors of different heights, 2 of which are underground, for a total height of about 32 m from the ground level (Figure 1: left). The building has a pseudo-quadrilateral layout, with one of the vertices beveled. The main façade overlooks Venice Plaza. On the southern corner of the building, a square tower rises for almost 17 m, reaching a height of approximately 48 m.

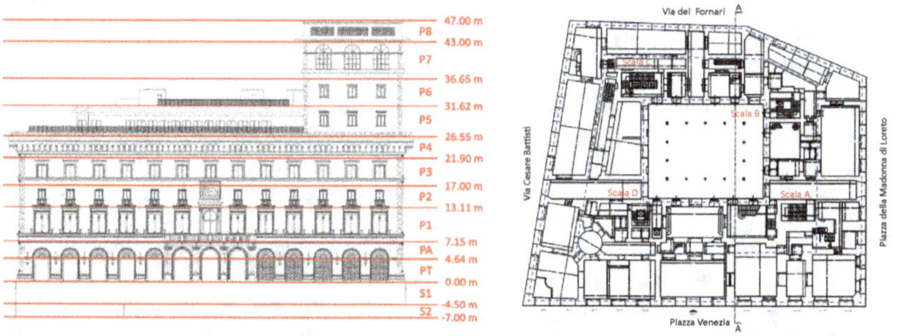

Figure 1: Façade on Piazza Venezia with indication of the building floors (left) and ground floor plan with indication of the main walls as blue dotted lines (right)

The structural plan is organized according to a "ring" system (Figure 1: right), consisting of the perimetral façades and central spine walls, connected by transverse walls and horizontal slabs. All the main walls have a large thickness, reaching at some points the value of 2.5 m. The horizontal slabs and the 4 main stairways, located at the internal courtyard vertices, have various natures, ranging from barrel and cross vaults to wood, steel or RC elements.

On the ground floor, the center of the building hosts a cloister and a travertine colonnade made with 18 cross vaults. These stand above a vast basement, covered by a reinforced concrete (RC) vault that is supported by 14 RC pillars.

The foundation is an RC slab supported by wooden piles that go beyond the first layer (about 11 m) of fill material soil and reach the underlying layer of compact silty clay (about 43.5 m).

Excavation-Induced Settlements

The evaluation of the settlement profile is carried out by the geotechnical engineering research group led by Prof. S. Rampello at the Sapienza University of Rome. The excavation operations are grouped into two main construction phases:

- phase 1: mechanized excavation of the two tunnels and the station shaft;
- phase 2: traditional enlargement necessary for the quay tunnels and the rail junction.

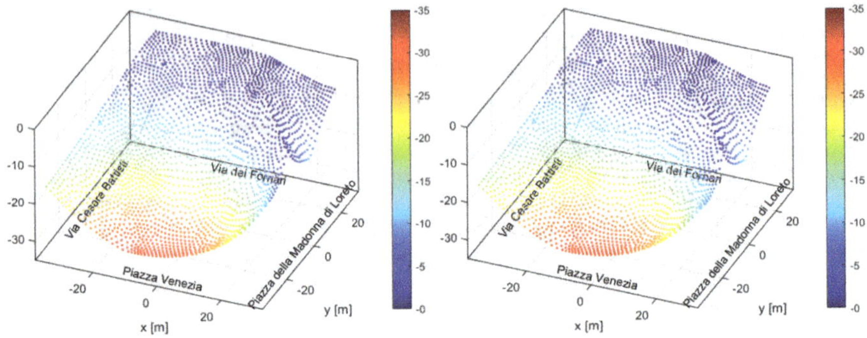

Figure 2: Settlement profile for phases 1 (left) and 2 (right), obtained from level 2 analyses (the color bar and the vertical axis indicate the value of the settlements in mm)

Two distinct procedures are followed, referred to as level 1 and level 2 analyses. Level 1 analyses are based on a semi-empirical approach that neglects the interaction between the soil and the building (free field). By contrast, level 2 analyses use FE numerical simulations where the ground is modeled as a continuous medium with non-linear behavior and the ground-building interaction is included.

For example, Figure 2 shows the settlement profile at the foundation intrados obtained from the level 2 analyses, for phases 1 (left) and 2 (right). The maximum settlement value is almost equal to 31 mm. These mainly develop during phase 1 in the central area of the façade overlooking Venice Plaza and at the corner between this façade and that overlooking Piazza della Madonna di Loreto. Very small increments (less than 2 mm) are observed during phase 2.

Finite Element numerical model

To carry out the structural analysis of the building, a 3D FE model is created (Figure 3) by using the software MIDAS FEA NX [12].

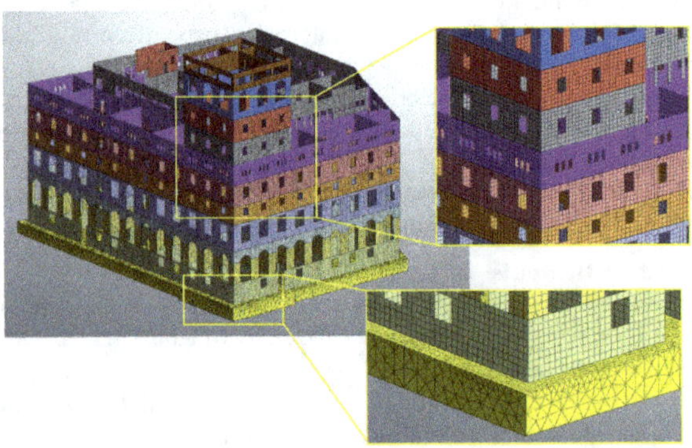

Figure 3: FE model of the building defined in MIDAS FEA NX

Walls are modeled with 3- and 4-node shell elements based on the Mindlin-Reissner plate theory and a mixed formulation (*Assumed Strain Elements* with in-compatible modes [12]) that avoids shear locking issues. The foundation is modeled through tetrahedral 4-node solid FEs, based on a classical displacement formulation. Columns, pillars and beams are modeled through 2-node

Materials Research Proceedings **26** (2023) 355-360 https://doi.org/10.21741/9781644902431-58

frame elements, based on a force formulation. Given the uncertainty about the actual mechanical characteristics of the horizontal slabs, these and the stairs are included in the model only as loads and by applying equivalent rigid diaphragm constraints at the node of the FE mesh.

Damage distribution assessment through linear elastic analyses

In the linear elastic analyses, the structure behavior is studied by imposing at the foundation intrados the settlement vertical displacements. Hence, stress and strain distribution attained in the structural elements are analyzed. To quantitatively define the vulnerability of the structure against the settlements, the results are interpreted based on the criterion proposed by Burland et al. [3], in the version developed by Boscardin & Cording [4]. This evaluates the possible damage to the masonry from the principal tensile strains. Given the historical and architectural interest of the building, a more conservative definition of the damage levels than that provided by Boscardin & Cording is assumed as indicated in Table 1.

Table 1: Adopted damage level classification

Damage level	Severity of the damage	Tensile strain	
		min	max
0	Negligible	0.0	$4.0\ 10^{-4}$
1	Very low	$4.0\ 10^{-4}$	$6.0\ 10^{-4}$
2	Low	$6.0\ 10^{-4}$	$1.0\ 10^{-3}$
3	Medium	$1.0\ 10^{-3}$	$2.0\ 10^{-3}$
4	Severe	$2.0\ 10^{-3}$	∞

Figure 4 shows, for example, the deformed shape (left) and the distribution of the principal tensile strains (right) obtained by imposing the vertical displacements computed for phase 2 through the level 2 analyses. These represent the worst condition for potential cracks in the masonry, although no significant difference is observed with respect to phase 1. Indeed, during phase 1, significant tensile strains arise in the main façades on Venice and Madonna di Loreto Plazas. These reach values up to 3.5×10^{-4} on the upper floors, corresponding to damage level 0, and 6.0×10^{-4} in very limited areas of the lower floors, corresponding to damage level 1.

Figure 4: Deformed shape (left) and maximum tensile principal strains (right) obtained for phase 2 settlements (level 2 analyses)

Nonlinear analyses

To obtain a more accurate assessment of the damage patterns potentially occurring in the masonry, nonlinear static analyses are conducted. To limit the computational cost, a reduced model of the building is employed. This includes only the two façades on Venice and Madonna di Loreto Plazas mainly affected by the settlement effects and portions of the transverse walls, as reported in Figure 5. Moreover, the foundation is not explicitly included in the model and the settlement displacements are directly imposed at the base of the walls.

Figure 5: Reduced model of the structure adopted for the nonlinear analyses

To verify the accuracy of the reduced model, preliminary linear elastic analyses under the vertical settlements are conducted and the results are checked with those of the full model.

In the nonlinear model, the SCM formulation [10,11] is assumed for the masonry and the following mechanical parameters are adopted: Young's modulus $E = 1714$ MPa; Poisson ratio $\upsilon = 0.15$; compressive strength $f_c = -2.7$ MPa and fracture energy density $g_c = G_c/h = 0.864$ MPa; tensile strength $f_t = |f_c|/10 \simeq 0.30$ MPa and fracture energy density $g_f = G_f/h = 0.0035$ MPa; crack band width $h = 500$ mm; and shear factor $\beta = 0.01$ [5].

Loads are applied to the structure in 3 successive phases, i.e. gravity loads are applied in a single step, assuming that these instantly act at the initial time of the analysis (phase 0); then the settlement vertical displacements are imposed, according to the two phases 1 and 2, by carrying out an incremental step-by-step analysis. Level 2 analyses values are considered.

Figure 6: Maximum tensile strains in the Venice Plaza (left) and Madonna di Loreto Plaza (right) façades obtained at the final step of phase 2

Tensile strain distributions for the last step of the analysis are reported in Figure 6. As expected, for gravity loads (phase 0), strains in the structure remain in the elastic limit, although the masts of the ground floor reach values close to this limit. During phase 1, the façade on Venice Plaza shows a significant increase of tensile strains in correspondence with its two vertical edges. Indeed, this mainly bends in its plane (see also Figure 4, left) with maximum vertical displacements in the middle part and minimum displacements at the edges (clamped-clamped-beam-like deformation). By contrast, the façade on Madonna di Loreto Plaza undergoes in-plane bending with maximum displacements near the tower and minimum displacements at the opposite edge (cantilever-beam-like deformation); the maximum tensile strains are observed in the central part of the lower floors. During phase 2, there are no significant increases in the deformations, with maximum tensile strain equal to 1.4×10^{-3}, corresponding to a crack width of about 0.60 mm. These results confirm those obtained in the linear elastic analyses, showing a maximum damage level equal to 1 (very low).

Summary

This work presented the vulnerability assessment of the Assicurazioni Generali masonry building for the settlements induced by the tunneling operation required for the new Line C of the Rome underground. Linear elastic and nonlinear static analyses are conducted by employing a detailed macromechanical model. These showed that the damage risk against the excavation-induced settlement is negligible (damage level 0) in almost the entire structure and reaches very low values (damage level 1) in limited zones of the main façades on Venice and Madonna di Loreto Plazas. Hence, the tunnelling operations do not generate significant stress-strain states in the structures that require specific interventions or attention.

References

[1] L. Masini, D. Gaudio, S. Rampello, E. Romani, Observed performance of a deep excavation in the historical center of Rome. J. Geotech. Geoenv. Eng., 147 (2021) 05020015. https://doi.org/10.1061/(ASCE)GT.1943-5606.0002465

[2] A. Amorosi, M. Sangirardi, Coupled three-dimensional analysis of the progressive tunnelling-induced damage to masonry buildings: is it always worth it?. Tunn. Undergr. Sp. Tech., 118 (2021) 104173. https://doi.org/10.1016/j.tust.2021.104173

[3] J. B. Burland, C. P. Worth, Settlement of buildings and associated damage, in British Geotechnical Society's Conference on the Settlement of Structures (1975).

[4] M. D. Boscardin, E. J. Cording, Building response to exacavtion-induced settlement, J. Geotech. Eng., 115 (1989) 1-21. https://doi.org/10.1061/(ASCE)0733-9410(1989)115:1(1)

[5] P. B. Lourenço, Recent advances in masonry modelling: micromodelling and homogenisation, Multiscale Model. Solid Mech.: Comp. App., (2010) 251-294. https://doi.org/10.1142/9781848163089_0006

[6] D. Addessi, P. Di Re, C. Gatta, E. Sacco, Multiscale analysis of out-of-plane masonry elements using different structural models at macro and microscale. Comput. Struct., 247 (2021) 106477. https://doi.org/10.1016/j.compstruc.2020.106477

[7] D. Addessi, P. Di Re, E. Sacco, Micromechanical and multiscale computational modeling for stability analysis of masonry elements. Eng. Struct., 211 (2020) 110428. https://doi.org/10.1016/j.engstruct.2020.110428

[8] D. Addessi, P. Di Re, C. Gatta, E. Sacco, Shell-3D multiscale modeling of masonry vaults based on the TFA procedure, Int. J. Multiscale Comp. Eng., 20 (2022) 41774. https://doi.org/10.1615/IntJMultCompEng.2022041774

[9] A. Amorosi, M. Sangirardi, G. De Felice, S. Rampello, Assessment of Tunneling Induced Damage on Historical Constructions Through a Fully Coupled Structural and Geotechnical Approach. In 12th International Conference on Structural Analysis of Historical Constructions – SAHC, (2021). https://doi.org/10.23967/sahc.2021.037

[10] H. D. Hibbitt, A simplified model for concrete at low confining pressures, Nuclear Engineering and Design, 104 (1987) 313-320. https://doi.org/10.1016/0029-5493(87)90209-3

[11] M. Jirásek, Damage and smeared crack models, Numerical Modeling of Concrete Cracking, (2011) 1-49. https://doi.org/10.1007/978-3-7091-0897-0_1

[12] MIDAS, MIDAS FEA NX - Analysis reference manual, (2020), https://www.midasoft.com/bridge-library/civil/products/midasfeanx.

Theoretical and Applied Mechanics - AIMETA 2022 | Materials Research Forum LLC
Materials Research Proceedings **26** (2023) 361-366 | https://doi.org/10.21741/9781644902431-59

Pure compressive solutions for masonry domes under gravity loads

Arsenio Cutolo[1, a], Enrico Babilio,[1,b], Ida Mascolo,[1,c *] and Elio Sacco [1,d]

[1]Department of Structures for Engineering and Architecture (DiSt), University of Naples "Federico II", Naples, Italy

[a]arsenio.cutolo@unina.it, [b]enrico.babilio@unina.it, [c]ida.mascolo@unina.it, [d]elio.sacco@unina.it

Keywords: Dome, No-Tension Material, Membrane Equilibrium Problem, Thrust Surface Analysis, Finite Element Method

Abstract. In recent years, efforts have been made to compare equilibrium solutions of masonry domes obtained through different methods. The research summarized in the present paper focuses on the evaluation of the equilibrium of masonry domes in the context of the classical membrane theory, in comparison with more sophisticated finite element models. In particular, we consider solutions, with vanishing tensile stresses, for domes under gravity loads. The equilibrium problem is traced back to the equilibrium of a thrust surface under the assumption of no-tension material in the sense of Heyman (i.e., no tensile strength, infinite resistance in compression, and no slide along fracture lines). A finite difference discretization of the equilibrium equation is proposed and the obtained solution is compared with results from a nonlinear finite element analysis. The good agreement of the results shows that both finite difference and finite element approaches represent reasonable and reliable alternative tools.

Introduction

Masonry domes, widely used in monumental buildings, are great works of architecture and engineering with a long history dating back to the dawn of civilization. Their static behavior has been widely studied through different methods which can be broadly classified into equilibrium methods [1–3] and deformation methods [4–6]. The present work aims at comparing equilibrium solutions for masonry domes under gravity loads obtained by applying a membrane approach and a Finite Element (FE) analysis. For the sake of comparison, we will consider very simple examples for which a reliable membrane solution can be easily obtained and then used to validate the more sophisticated FE results.

The paper is organized as follows. Section 2 presents the theoretical framework of the membrane approach and the Finite Difference (FD) algorithm used for the computations. Section 3 focuses on the fully three-dimensional, nonlinear FE analysis. Section 4 presents the findings of the two numerical-based analyses, remarking on the substantial agreement between the two approaches. This highlights the consistency and robustness of the nonlinear FE analysis for masonry domes characterized by very reduced tensile strength, as emphasized in the concluding remarks in Section 5.

Membrane approach and FD solution

Let us consider a masonry dome subjected to its self-weight and assume its equilibrium may be written in agreement with Pucher's approach to the classical membrane theory [7-10]. We define an axial-symmetric (with respect to the direction of the load) membrane as an unknown function $g = g(r)$, where r is the radius of the membrane projection Ω on the support plane, orthogonal to the direction of the load. The generalized stresses along the meridian and parallel curves on the membrane are S_{rr} and $S_{\theta\theta}$, respectively, and S_{11} and S_{22} are the corresponding projections on Ω. By introducing an unknown Airy's stress function $\varphi = \varphi(r)$ such that

$$\varphi' = \frac{r\,S_{rr}}{\sqrt{1+g'^2}} = rS_{11}\,, \quad \varphi'' = S_{\theta\theta}\sqrt{1+g'^2} = S_{22}\,, \tag{1}$$

the equilibrium along the direction of the load is given by the differential Pucher's equation

$$\varphi' g'' + \varphi'' g' = \gamma\, s\, r\, \sqrt{1+g'^2}\,, \tag{2}$$

where the prime symbol (′) denotes the derivative with respect to the membrane radius r, and γ and s stand for the specific weight of the masonry and the thickness of the dome, respectively.

In the problem we are dealing with, the function g is admissible if it corresponds to a membrane fully contained within the dome thickness and the function φ is admissible if leads to non-positive stresses. Since in Eq. (2) both g and φ are unknowns, one can assign any tentative g_0 or φ_0 function and compute the other. Choosing $n+1$ not uniformly spaced points along the radius r on the platform Ω, we transform Eq. (2) in a set of $n+1$ algebraic equations with $2(n+1)$ unknowns $g(i)$ and $\varphi(i)$, $i = 0,1,\dots,n$. In a finite difference fashion, to compute the discrete derivatives of functions we are dealing with, we use three-point stencils and adopt central differences for any point $1 \le i \le n$, forward difference scheme for the point 0, backward difference scheme for the point n, thus making as simple as possible the setting of the boundary conditions.

To determine the solution to the discretized problem, a two-stage algorithm is implemented in *Mathematica*®, taking advantage of the possibilities of that high-level, general-purpose programming environment.

Fully three-dimensional, nonlinear FE analysis

Modelling, simulations, and FE analysis are performed by means of the FE package Ansys®, adopting the concrete constitutive model (CCM) for masonry, which uses the William–Warnke constitutive equations [11] with a failure surface which is smooth and convex with noncircular and no affine sections in the deviatoric plane.

The performed analysis takes into account the geometric nonlinearities by means of an incremental approach based on a modified Newton-Raphson method [12-14]. An *ad hoc* algorithm is written in the Ansys Parametric Design Language (APDL) [15] to compute the coordinates of some points of the thrust surface on (any) vertical cross-section of the dome.

Numerical results and method comparison

In the following numerical applications, we consider here a hemispherical dome with a mean radius of 12.5 m, a constant thickness of 2 m, and self-weight $\gamma = 16\ \mathrm{kN/m^3}$.

Finite difference solution

The FD solution of Eq. (2) is summarized in Figures 1a and 1b reporting the computed function $\varphi(r)$ and the stresses per unit length, respectively. We observe that at the end of the first stage of the FD algorithm, in the compressed part of the dome, the numerical solution is in agreement with the predictions of the classical theory of shells (dashed lines in Figure 1b), while at the end of the second stage, the generalized stresses tend to relax (see, in particular, the hoop stress and the horizontal thrust drawn, respectively, in red and black in Figure 1b).

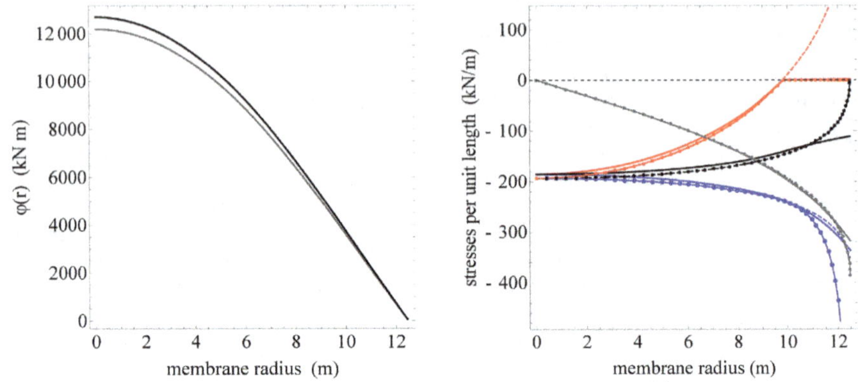

(a) (b)

Figure 1: (a) Stress function $\varphi(r)$: FD solutions from the first (black) and second (gray) stages of the algorithm. (b) Generalized stresses S_{rr} (blue), $S_{\vartheta\vartheta}$ (red), horizontal thrust S_{11} (black), vertical thrust $S_v = S_{11}\, g'$ (gray), theoretical solution for hemispherical membrane (dashed lines): FD solutions from the first (dotted lines) and second (thick lines) stages of the algorithm

Finite element solution

For the FE model, we consider a Cartesian reference frame $(0, x_1, x_2, x_3)$, with the origin O corresponding to the center of the projection of the dome onto the support plane $x_2 = 0$, being x_2 the gravity direction. On account of the hemispherical symmetry of the analyzed problem, the computational effort in the analysis is reduced by considering only half of the dome, where $x_1 = 0$ is the cutting plane. The dome is loaded only by gravity loads and mechanical parameters of masonry, namely Young's modulus, Poisson's ratio, and density, are set as $E = 2$GPa, $\nu = 0.1$, $\rho = 1630.99$kg/m^3. For the CCM, only the ultimate uniaxial tensile strength and the ultimate uniaxial compressive strength are assumed as failure parameters, set as $\sigma_t = 1$MPa and $\sigma_c = -20$ MPa, respectively. Two models which differ in terms of boundary conditions are considered:

Sample 1 discretized by means of 75,896 SOLID65 linear tetrahedral elements, resulting in 15,706 nodes with three translational degrees of freedom each, adopting the following boundary conditions are used in the simulations:

• displacements along the direction x_1 are prevented, i.e. $u_1 = 0$, for all nodes which belong to the plane of equation $x_1 = 0$;

• displacements along the radial direction, i.e. along the lines from the origin of the axes are allowed, for all nodes which belong to the plane of equation $x_2 = 0$;

• displacements along the direction x_3 are prevented, i.e. $u_3 = 0$, for all nodes which belong to the plane of equation $x_3 = 0$.

Sample 2 discretized by means of 264,800 SOLID65 linear tetrahedral elements, resulting in 48,853 nodes with three translational degrees of freedom each. With the purpose of mimicking as accurately as possible the real boundary conditions of the dome, the drum is also partially modeled. The drum, 2.8 m high and 3.3 m wide, is fully-constrained at the base. Moreover, displacements along the horizontal direction x_1 are prevented for all nodes which belong to the plane of equation $x_1 = 0$.

 https://doi.org/10.21741/9781644902431-59

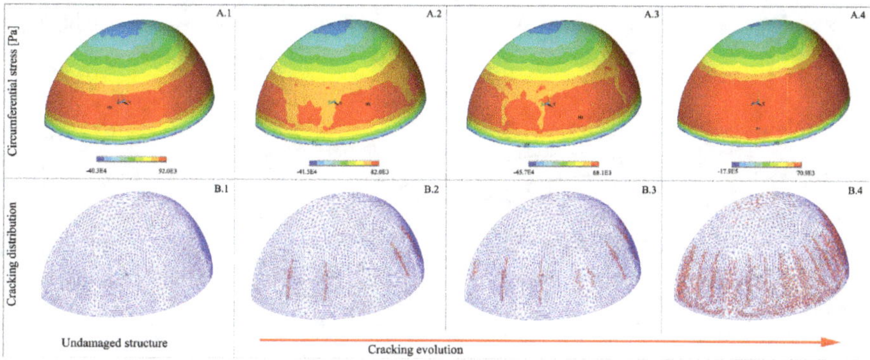

(a) Sample 1

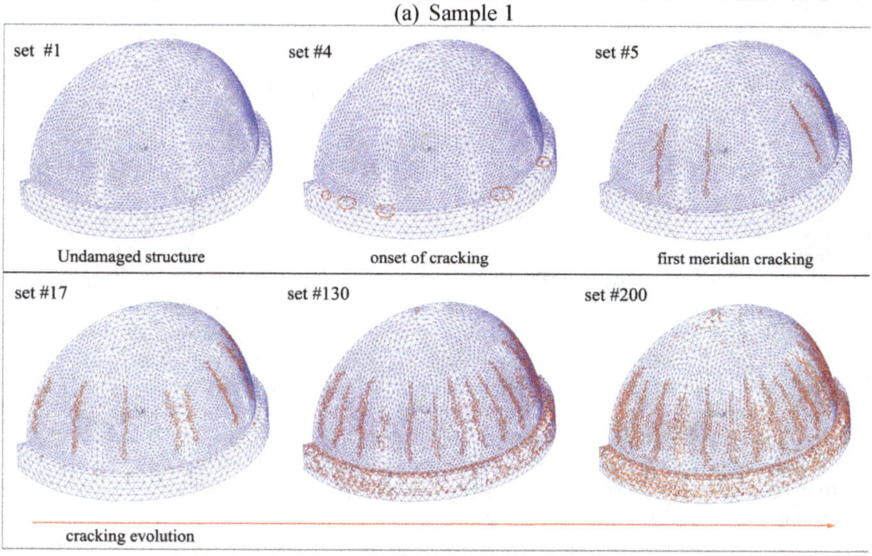

(b) Sample 2

Figure 2: Nonlinear FE simulation of cracking

The nonlinear FE simulation of cracking for the *Sample 1* and *Sample 2* are reported in Figures 2a and 2b, respectively. We can observe that the two samples exhibit a fairly similar cracking evolution with the development of typical meridian cracks due to the masonry weakness to sustain hoop stresses. The cracking arises at the springing subsequently propagating vertically upwards toward the crown (*Sample 1* and *Sample 2*) and downwards into the drum (*Sample 2*) and arresting at the haunches, as expected [16]. Figure 3 reports a superimposition of the thrust surface cross-section obtained by means of the membrane approach and the two FE samples. We emphasize that the lines computed from FE solutions are in agreement with the thrust surface cross-section obtained with the membrane approach, by imposing the initial point of the thrust line at the crown of the dome and the final point at the springing of the dome coincident with the corresponding points evaluated with FE simulations. We argue that the comparison shows the reliability of the performed FE analysis.

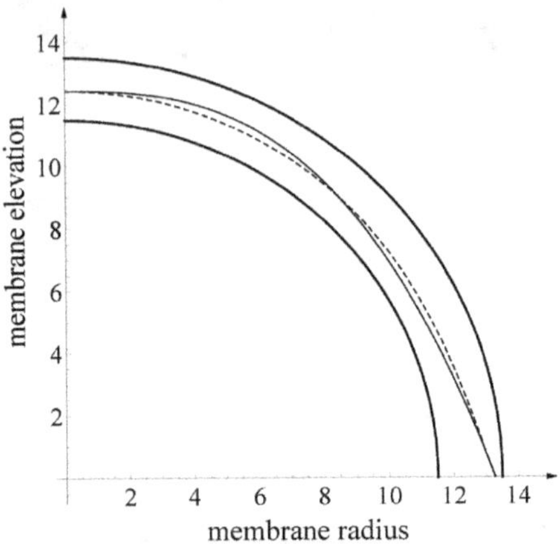

Figure 3: Superposition of the thrust surface to the half cross-section of the dome: membrane approach (dashed line) vs. FE approach (solid black line)

Conclusions

In the present contribution, two thrust surfaces in the thickness of a simple hemispherical masonry dome, under its self-weight, are computed: one integrating the Pucher's equilibrium equation through a finite difference approach, the other obtained from a more sophisticated constitutive model accounting for damage and smeared crack implemented into a finite element model. The solution from the membrane Pucher's approach, applied to a very simple example, is here assumed as the benchmark to assess the FE results. However, while the membrane approach is flexible enough to solve a large class of problems, its FD discretization becomes cumbersome in the case of complex geometries. In such cases, finite element models are instead suitable. In this respect, the utility of a Pucher-based solution represents a simple and physically significant benchmark for the FE model characterized by a large number of parameters that need to be carefully set up. In fact, the agreement between the thrust surface cross-section obtained with the nonlinear elastic analysis and the membrane approach demonstrates the reliability and usefulness of the proposed FE approach.

References

[1] M. Como, Statics of historic masonry constructions: An essay, in Masonry Structures: Between Mechanics and Architecture. Springer, 2015, pp. 49–72. https://doi.org/10.1007/978-3-319-13003-3_3

[2] N. A. Nodargi and P. Bisegna, A new computational framework for the minimum thrust analysis of axisymmetric masonry domes, Engineering Structures, vol. 234, p. 111962, 2021. https://doi.org/10.1016/j.engstruct.2021.111962

[3] A. M. D'Altri, V. Sarhosis, G. Milani, J. Rots, S. Cattari, S. Lagomarsino, E. Sacco, A. Tralli, G. Castellazzi, and S. de Miranda, Modeling strategies for the computational analysis of

unreinforced masonry structures: review and classification, Archives of Computational Methods in Engineering, vol. 27, no. 4, pp. 1153–1185, 2020. https://doi.org/10.1007/s11831-019-09351-x

[4] C. Olivieri, C. Cennamo, C. Cusano, A. Cutolo, A. Fortunato, and I. Mascolo, Masonry spiral stairs: A comparison between analytical and numerical approaches, Applied Sciences, vol. 12, no. 9, p. 4274, 2022. https://doi.org/10.3390/app12094274

[5] N. A. Nodargi and P. Bisegna, A finite difference method for the static limit analysis of masonry domes under seismic loads, Meccanica, vol. 57, no. 1, pp. 121–141, 2022. https://doi.org/10.1007/s11012-021-01414-3

[6] M. Angelillo, E. Babilio, and A. Fortunato, Singular stress fields for masonry-like vaults, Continuum Mechanics and Thermodynamics, vol. 25, no. 2-4, pp. 423–441, 2013. https://doi.org/10.1007/s00161-012-0270-9

[7] M. Angelillo, E. Babilio, A. Fortunato, M. Lippiello, and A. Montanino, Equilibrium of masonry-like vaults treated as unilateral membranes: Where mathematics meets history, Civil-Comp Proceedings, vol. 108, 2015.

[8] E. Babilio, C. Ceraldi, M. Lippiello, F. Portioli, and E. Sacco, Static analysis of a doublecap masonry dome, in Proceedings of XXIV AIMETA Conference 2019, A. Carcaterra, A. Paolone, and G. Graziani, Eds. Springer International Publishing, 2020, pp. 2082–2093. https://doi.org/10.1007/978-3-030-41057-5_165

[9] C. Olivieri, A. Castellano, I. Elia, A. Fortunato, and I. Mascolo, Horizontal force capacity of a hemi-spherical dome, in 8th International Conference on Computational Methods in Structural Dynamics and Earthquake Engineering, COMPDYN 2021, 2021, pp. 614–625. https://doi.org/10.7712/120121.8513.19613

[10] M. De Piano, M. Modano, G. Benzoni, V.P. Berardi, F.Fraternali, A numerical approach to the mechanical modeling of masonry vaults under seismic loading, Ingegneria Sismica, 34(4): 103–119, 2017.

[11] K. J.Willam and E. P.Warnke, Constitutive model for the triaxial behaviour of concrete, Proceedings of IABSE, International Association for Bridge and Structural Engineering, vol. 19, pp. 1–30, 1975.

[12] R. Shamass, G. Alfano, F. Guarracino, An Analytical Insight into the Buckling Paradox for Circular Cylindrical Shells under Axial and Lateral Loading, Mathematical Problems in Engineering, 2015, Article ID 514267. https://doi.org/10.1155/2015/514267

[13] F. Guarracino, M.G. Simonelli, The torsional instability of a cruciform column in the plastic range: Analysis of an old conundrum, Thin-Walled Structes, 113: 273-286, 2017. https://doi.org/10.1016/j.tws.2016.11.007

[14] I. Mascolo, M. Fulgione, M. Pasquino, Lateral torsional buckling of compressed open thin walled beams: Experimental confirmations, International Journal of Masonry Research and Innovation, 4(1-2), 150-158, 2019. https://doi.org/10.1504/IJMRI.2019.096829

[15] M. K. Thompson and J. M. Thompson, ANSYS mechanical APDL for finite element analysis. Butterworth-Heinemann, 2017.

[16] Z. Nanhai and Y. Jihong, Structural vulnerability of a single-layer dome based on its form, Journal of Engineering Mechanics, vol. 140, no. 1, pp. 112–127, 2014. https://doi.org/10.1061/(ASCE)EM.1943-7889.0000636

Dynamical systems and applications in civil and mechanical structures

Theoretical and Applied Mechanics - AIMETA 2022 Materials Research Forum LLC
Materials Research Proceedings **26** (2023) 369-374 https://doi.org/10.21741/9781644902431-60

Some remarks on the evaluation of work and dissipated energy associated with rate-independent hysteretic forces

Raffaele Capuano[1, a *], Nicolò Vaiana[1, b] and Luciano Rosati[1, c]

[1]Department of Structures for Engineering and Architecture, University of Naples Federico II, Via Claudio, 21, 80124, Napoli, Italy

[a] raffaele.capuano@unina.it, [b] nicolo.vaiana@unina.it, [c] rosati@unina.it

Keywords: Path-Dependent Work, Dissipated Energy, Vaiana-Rosati Model

Abstract. We provide closed-form expressions to compute the path-dependent work performed by generalized rate-independent hysteretic forces, simulated by using a brand-new model denominated Vaiana-Rosati Model (VRM). In particular, such expressions are valid over a generic generalized displacement interval. Furthermore, we provide a closed-form expression for evaluating the dissipated energy associated with the work done by a rate-independent hysteretic force when a full cycle of periodic generalized displacement is applied.

Introduction

In structural dynamics, the problem of evaluating the amount of mechanical energy dissipated by rate-independent hysteretic mechanical systems and materials plays a crucial role since such an issue plays a paramount role in vibrations control. In particular, with reference to the seismic protection of structures and objects contained in them [1], the evaluation of the dissipated energy is necessary for the modeling and design of isolation bearings [2-4] and energy dissipation devices [5].

The complexity of the nonlinear response exhibited by rate-independent hysteretic mechanical systems strongly influences the shape of generalized force-displacement hysteresis loops, which can be *symmetric* or *asymmetric*. Over the years, many researchers have proposed phenomenological models for evaluating the generalized force by means of different types of equations.

As regards the simulation of *symmetric* hysteresis loops, the Bouc-Wen model [6-7] and its subsequent extensions [8] are the most employed differential models. On the other hand, non-differential models, such as algebraic or transcendental ones, have been recently formulated by Vaiana et al. [9]. As far as the simulation of *asymmetric* hysteresis loops is concerned, it is possible to employ modified versions of the Bouc-Wen model [10] or, alternatively, recently developed models belonging to the generalized class proposed by Vaiana et al. [11].

In general, hysteresis phenomena are so complex that it is quite complicated to describe them by means of a single model. Conversely, Vaiana and Rosati [12] have recently proposed a preliminary formulation of a novel rate-independent hysteretic model capable of simulating a wide range of hysteresis loops in a unified manner. This model provides a closed-form expression to evaluate the output variable, thus allowing for significant advantages in terms of computational efficiency [13-14].

In particular, in this paper, after a short description of the Vaiana-Rosati model, we illustrate the closed-form expressions that have been derived to compute the incremental generalized work performed by an hysteretic force; on the basis of this result, we derive the closed-form expression for the evaluation of the dissipated energy associated with a full hysteresis loop.

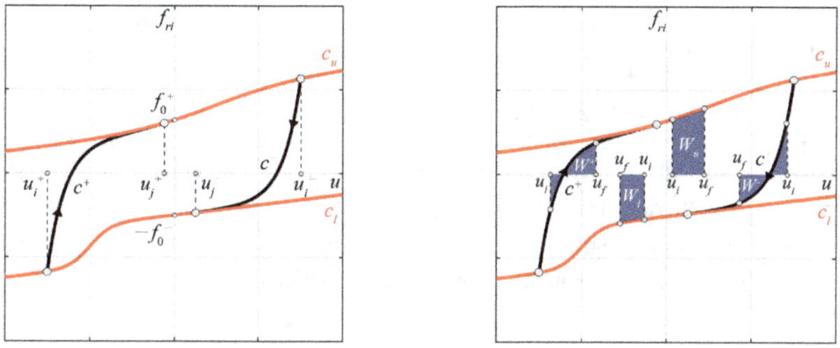

(a) (b)

***Figure 1**: VRM formulation: curves c^+, c^-, c_u, c_l, (a) and incremental generalized work associated with each curve (b).*

Brief Review of the VRM

In this section, we briefly describe the Vaiana-Rosati Model (VRM). The model can accurately simulate the response of a great variety of rate-independent hysteretic mechanical systems and materials. In particular, it can reproduce the complex hysteretic behavior characterized by *asymmetric, pinched, S-shaped, and flag-shaped* hysteresis loops, or by a combination of them, providing *closed-form expressions* for the computation of the generalized rate-independent force. Moreover, the model allows for uncoupled modeling of the generic loading and unloading phases thanks to a set of parameters that can be simply calibrated due to their clear mechanical meaning and direct relationship with experimental loops.

Fig. 1a shows a typical hysteresis loop that the VRM can reproduce. The generalized rate-independent force f_{ri}, during a generic loading phase $(\dot{u} > 0)$, can be evaluated as:

$$f_{ri}\left(u, u_j^+\right) = \begin{cases} c^+\left(u, u_j^+\right) & \text{if} \quad u_i^+ \leq u < u_j^+ \\ c_u\left(u\right) & \text{if} \quad u > u_j^+, \end{cases} \tag{1}$$

whereas, during the generic unloading one $(\dot{u} < 0)$, it can be computed as:

$$f_{ri}\left(u, u_j^-\right) = \begin{cases} c^-\left(u, u_j^-\right) & \text{if} \quad u_j^- < u \leq u_i^- \\ c_l\left(u\right) & \text{if} \quad u < u_j^-, \end{cases} \tag{2}$$

In Eq. (1), c^+ and c_u describe, respectively, the generic loading curve and the upper limiting curve, which have the following expressions:

$$c^+\left(u, u_j^+\right) = \beta_1^+ e^{\beta_2^+ u} - \beta_1^+ + \frac{4\gamma_1^+}{1+e^{-\gamma_2^+\left(u-\gamma_3^+\right)}} - 2\gamma_1^+ + k_b^+ u + f_0^+ - \frac{1}{\alpha^+}\left[e^{-\alpha^+\left(+u-u_j^+ + \bar{u}^+\right)} - e^{-\alpha^+ \bar{u}^+}\right], \tag{3}$$

$$c_u\left(u\right) = \beta_1^+ e^{\beta_2^+ u} - \beta_1^+ + \frac{4\gamma_1^+}{1+e^{-\gamma_2^+\left(u-\gamma_3^+\right)}} - 2\gamma_1^+ + k_b^+ u + f_0^+, \tag{4}$$

whereas, in Eq. (2), c^- and c_l define, respectively, the generic unloading curve and the lower limiting curve; their expressions read:

$$c^-\left(u, u_j^-\right) = \beta_1^- e^{\beta_2^- u} - \beta_1^- + \frac{4\gamma_1^-}{1+e^{-\gamma_2^-\left(u-\gamma_3^-\right)}} - 2\gamma_1^- + k_b^- u - f_0^- + \frac{1}{\alpha^-}\left[e^{-\alpha^-\left(-u+u_j^- + \bar{u}^-\right)} - e^{-\alpha^- \bar{u}^-}\right], \tag{5}$$

Materials Research Forum LLC
https://doi.org/10.21741/9781644902431-60

$$c_l\left(u\right)=\beta_1^- e^{\beta_2^- u}-\beta_1^- +\frac{4\gamma_1^-}{1+e^{-\gamma_2^-\left(u-\gamma_3^-\right)}}-2\gamma_1^- +k_b^- u-f_0^- . \tag{6}$$

The parameters k_b^+, f_0^+, α^+, β_1^+, β_2^+, γ_1^+, γ_2^+, γ_3^+ $\left(k_b^-, f_0^-, \alpha^-, \beta_1^-, \beta_2^-, \gamma_1^-, \gamma_2^-, \gamma_3^-\right)$, that appear in the expressions of the generalized rate-independent hysteretic force, represent the set of eight parameters that control the generic loading (unloading) phase and need to be calibrated from experimental or numerical tests. Concerning the internal variable $u_j^+\left(u_j^-\right)$ characterizing the generic loading (unloading) phase, its expression is reported in Vaiana and Rosati [12], whereas $\bar{u}^+\left(\bar{u}^-\right)$ is an internal model parameter expressed as a function of $\alpha^+\left(\alpha^-\right)$.

Generalized Work and Dissipated Energy

In the sequel, we show that the closed-form expressions for the generalized rate-independent hysteretic force, described in the previous Section, allow us to derive closed-form expressions for the evaluation of the area under the four curves characterizing the VRM (Fig. 1b) as:

$$W_{ri}=\int_{u_i}^{u_f} f_{ri}\left(u\right)du. \tag{7}$$

In fact, the quantity in Eq. (7) represents the path-dependent work performed by the generalized force f_{ri} over a generic generalized displacement interval $\left[u_i, u_f\right]$, and will be referred to as generalized rate-independent hysteretic work W_{ri}. The path-dependent work will be first derived for the four curves defining the model; subsequently, the relevant expressions will be used to evaluate the closed-form expression of the energy dissipated in a full hysteresis loop.

Incremental Generalized Work

The closed-form expressions of the incremental generalized rate-independent hysteretic work associated with the four VRM curves are obtained by recalling Eqs. (1) and (2) and using them in Eq. (7). In this way, during the generic loading phase $\left(\dot{u}>0\right)$, the generalized work is:

$$W_{ri}=\begin{cases} W^+ & \text{if} \quad u_i^+ \le u < u_j^+ \\ W_u & \text{if} \quad u > u_j^+, \end{cases} \tag{8}$$

whereas, in the generic unloading phase $\left(\dot{u}<0\right)$, it can be computed as:

$$W_{ri}=\begin{cases} W^- & \text{if} \quad u_j^- < u \le u_i^+ \\ W_l & \text{if} \quad u < u_j^-, \end{cases} \tag{9}$$

where W_u $\left(W^+\right)$ represents, on the basis of Eq. (7), the area under the curve c_u $\left(c^+\right)$ over a generic generalized displacement interval $\left[u_i, u_f\right]$, whereas W_l $\left(W^-\right)$ is the area under the curve c_l $\left(c^-\right)$ over a generic generalized displacement interval $\left[u_f, u_i\right]$, as shown in Fig. 1b.

To evaluate the expressions of W_u, W^+, W_l, and W^-, it is more convenient to rewrite the expressions of the four functions defining the curves, as the sum of terms that depend, respectively, upon suitable subsets of the model parameters, namely:

$$c_u\left(u\right)=c_a^+\left(\beta_1^+,\beta_2^+\right)+c_b^+\left(\gamma_1^+,\gamma_2^+,\gamma_3^+\right)+c_c^+\left(k_b^+,f_0^+\right), \tag{10}$$

$$c^+\left(u,u_j^+\right)=c_a^+\left(\beta_1^+,\beta_2^+\right)+c_b^+\left(\gamma_1^+,\gamma_2^+,\gamma_3^+\right)+c_c^+\left(k_b^+,f_0^+\right)+c_d^+\left(\alpha^+\right), \tag{11}$$

$$c_l(u)=c_a^-(\beta_1^-,\beta_2^-)+c_b^-(\gamma_1^-,\gamma_2^-,\gamma_3^-)+c_c^-(k_b^-,f_0^-), \tag{12}$$

$$c^-(u,u_j^-)=c_a^-(\beta_1^-,\beta_2^-)+c_b^-(\gamma_1^-,\gamma_2^-,\gamma_3^-)+c_c^-(k_b^-,f_0^-)+c_d^-(\alpha^-). \tag{13}$$

Thus, the expressions of W_u and W^+ can be derived by integrating Eqs. (10)-(11) over $\left[u_i,u_f\right]$:

$$W_u=\int_{u_i}^{u_f} c_u(u)\,du=\int_{u_i}^{u_f}\left(c_a^+ + c_b^+ + c_c^+\right)du=W_a^+ + W_b^+ + W_c^+, \tag{15}$$

$$W^+=\int_{u_i}^{u_f} c^+(u,u_j^+)\,du=\int_{u_i}^{u_f}\left(c_a^+ + c_b^+ + c_c^+ + c_d^+\right)du=W_a^+ + W_b^+ + W_c^+ + W_d^+, \tag{14}$$

where the terms W_a^+, W_b^+, W_c^+, and W_d^+ are given by:

$$W_a^+=\frac{\beta_1^+}{\beta_2^+}\left(e^{\beta_2^+u_f}-e^{\beta_2^+u_i}\right)-\beta_1^+\left(u_f-u_i\right), \tag{18}$$

$$W_b^+=2\gamma_1^+\left\{\frac{2\ln\left[e^{-\gamma_2^+\left(u_f-\gamma_3^+\right)}+1\right]}{\gamma_2^+}-\frac{2\ln\left[e^{-\gamma_2^+\left(u_i-\gamma_3^+\right)}+1\right]}{\gamma_2^+}+\left(u_f-u_i\right)\right\}, \tag{19}$$

$$W_c^+=\frac{k_b^+}{2}\left(u_f^2-u_i^2\right)+f_0^+\left(u_f-u_i\right), \tag{20}$$

$$W_d^+=\frac{1}{\left(\alpha^+\right)^2}e^{\alpha^+\left(\bar{u}^+-u_j^+\right)}\left[\alpha^+\left(u_f-u_i\right)e^{-\alpha^+u_j^+}+e^{-\alpha^+u_f}-e^{-\alpha^+u_i}\right]. \tag{21}$$

It is important to note that Eq. (18) assumes finite values only if $\beta_2^+\neq 0$. Similarly, Eq. (19) is defined only if $\gamma_2^+\neq 0$. However, if $\beta_2^+=0$ it turns out to be $c_a^+=0$, and, consequently, $W_a^+=0$. Similarly, if $\gamma_2^+=0$ then $c_b^+=0$ and $W_b^+=0$. Eq. (21) does not present the same problem for $\alpha^+=0$ since, according to the conditions indicated by Vaiana and Rosati [12], this parameter needs to be always greater than zero. Similarly, the expressions of W_l and W^- are obtained by integrating Eqs. (12)-(13) over $\left[u_f,u_i\right]$; the relevant expressions W_a^-, W_b^-, W_c^-, and W_d^-, are omitted for brevity.

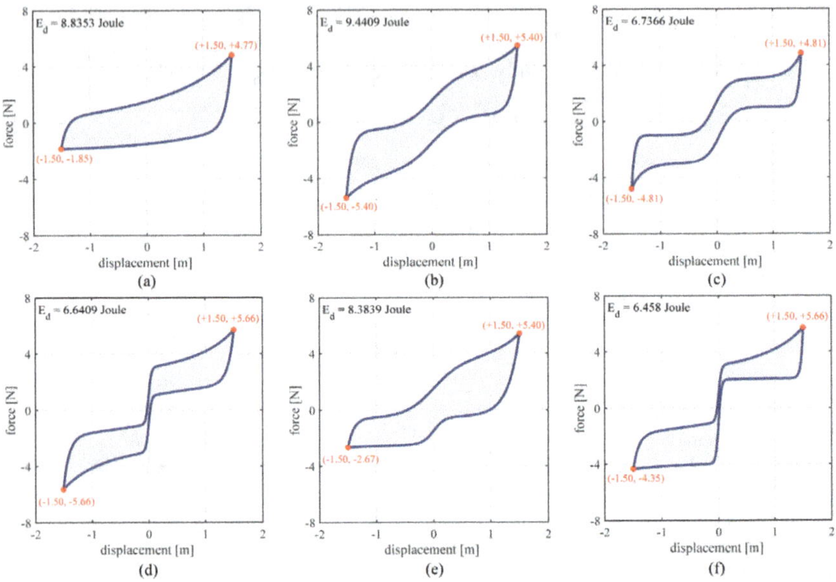

***Figure 2**: Dissipated energy associated with asymmetric (a), pinched (b), S-shaped (c), flag-shaped (d) hysteresis loops and two combinations of them (e)-(f), simulated by the VRM.*

<u>*Energy Dissipated in a Full Hysteresis Loop*</u>

Finally, the closed-form expression of the energy dissipated in a full hysteresis loop is obtained by summing the areas under the four curves (Fig. 1b). Recalling Eqs. (10)-(13), the energy dissipated in a full loop E_d can be obtained, by exploiting the additivity rule of definite integrals, as:

$$E_d = \int_{u_i^+}^{u_i^-} \left(c_a^+ + c_b^+ + c_c^+\right) du + \int_{u_i^+}^{u_j^+} c_d^+ \, du + \int_{u_i^-}^{u_i^+} \left(c_a^- + c_b^- + c_c^-\right) du + \int_{u_i^-}^{u_j^-} c_d^- \, du, \tag{22}$$

where all the quantities appearing in Eq. (22) can be calculated using the general expressions of the incremental work previously calculated, paying attention to set the proper displacement interval. Fig. 2 shows the values of the *dissipated energy* E_d, calculated using Eq. (22), for different hysteresis loops simulated by the VRM and obtained upon application of a full cycle of harmonic (sinusoidal) generalized displacement.

Conclusions

We have exploited one of the peculiar features of the VRM, that is, the analytical expressions assumed for the functions ruling the rate-independent hysteretic behavior, to derive closed-form formulas providing the path-dependent work done by an arbitrary rate-independent hysteretic force. Furthermore, analytical expressions of the energy dissipated in a full hysteresis loop have been explicitly computed. They represent the prerequisite to model the evolution of the hysteretic loop shape as a function of the energy dissipated in previous loading histories of cyclic nature, an issue that will be pursued in forthcoming papers.

References

[1] D. Pellecchia, S. Lo Feudo, N. Vaiana, JL. Dion, and L. Rosati, A procedure to model and design elastomeric-based isolation systems for the seismic protection of rocking art objects, Comput. Aided. Civ. Inf, (2021) 1– 18. https://doi.org/10.1111/mice.12775

[2] N. Vaiana, F. Marmo, S. Sessa, and L. Rosati, Modeling of the hysteretic behavior of wire rope isolators using a novel rate-independent model, Nonlinear Dyn. (2020) 309–317. https://doi.org/10.1007/978-3-030-34713-0_31

[3] N. Vaiana, R. Capuano, S. Sessa, F. Marmo, and L. Rosati, Nonlinear dynamic analysis of seismically base-isolated structures by a novel OpenSees hysteretic material model, Applied Sciences 11.3 (2021) 900. https://doi.org/10.3390/app11030900

[4] D. Losanno, D. De Domenico, and IE. Madera-Sierra, Experimental testing of full-scale fiber reinforced elastomeric isolators (FREIs) in unbounded configuration, Eng. Struct. 260 (2022) 114234. https://doi.org/10.1016/j.engstruct.2022.114234

[5] I. Nuzzo, D. Losanno, F. Cilento, and N. Caterino Analytical and numerical modelling of shear-link device for seismic energy dissipation in frame structures, Eng. Struct. 214 (2020) 110630. https://doi.org/10.1016/j.engstruct.2020.110630

[6] R. Bouc, Model mathematique d'hysteresis, Acustica 24(1) (1971) 16-25.

[7] Y. Wen, Method for random vibration of hysteretic systems, J. Eng. Mech. Div. ASCE 102(2) (1976) 249-263. https://doi.org/10.1061/JMCEA3.0002106

[8] T. Baber, M. Noori, Modeling general hysteresis behavior and random vibration application, J. Vib. Acoust. Stress Reliab. Des. ASME 108(4) (1986) 411-420. https://doi.org/10.1115/1.3269364

[9] N. Vaiana, S. Sessa, F. Marmo, and L. Rosati, A class of uniaxial phenomenological models for simulating hysteretic phenomena in rate-independent mechanical systems and materials, Nonlinear Dyn. 93.3 (2018) 1647–1669. https://doi.org/10.1007/s11071-018-4282-2

[10] J. Song, and A. Der Kiureghian, Generalized Bouc-Wen model for highly asymmetric hysteresis, J. Eng. Mech. 132.6 (2006) 610–618. https://doi.org/10.1061/(ASCE)0733-9399(2006)132:6(610)

[11] N. Vaiana, S. Sessa, and L. Rosati, A generalized class of uniaxial rate-independent models for simulating asymmetric mechanical hysteresis phenomena, Mech. Syst. Sig. Process. 146 (2021) 106984. https://doi.org/10.1016/j.ymssp.2020.106984

[12] N. Vaiana, and L. Rosati, Classification and modeling of uniaxial rate-independent hysteresis phenomena: some preliminary results, AIMETA (2022).

[13] N. Vaiana, S. Sessa, M. Paradiso, F. Marmo, and L. Rosati, An efficient computational strategy for nonlinear time history analysis of seismically base-isolated structures, Lecture Notes in Mech. Eng. (2020) 1340–1353. https://doi.org/10.1007/978-3-030-41057-5_108

[14] N. Vaiana, S. Sessa, F. Marmo, and L. Rosati, Nonlinear dynamic analysis of hysteretic mechanical systems by combining a novel rate-independent model and an explicit time integration method, Nonlinear Dyn. 98.4 (2019) 2879–2901. https://doi.org/10.1007/s11071-019-05022-5

Classification and modeling of uniaxial rate-independent hysteresis phenomena: some preliminary results

Nicolò Vaiana[1,a*] and Luciano Rosati[1,b]

[1]Department of Structures for Engineering and Architecture, University of Naples Federico II, via Claudio, 21, 80124, Napoli, Italy

[a] nicolo.vaiana@unina.it, [b] rosati@unina.it

Keywords: Complex Hysteresis Loop, Mechanical Hysteresis, Rate-Independent Model

Abstract. In this paper, we propose a preliminary classification of complex generalized force-displacement hysteresis loops, that is, asymmetric, pinched, S-shaped, flag-shaped hysteresis loops or those obtained as arbitrary combination of them, that generally characterize the nonlinear response of mechanical systems and materials. Subsequently, we present a preliminary formulation of a novel phenomenological model capable of reproducing such complex hysteresis phenomena and we illustrate the different types of hysteresis loop shapes that can be simulated. Finally, we perform an experimental verification of the proposed model by adopting some experimental hysteresis loops retrieved from the literature.

Introduction

The nonlinear response exhibited by mechanical systems and materials may be characterized by generalized force-displacement hysteresis loops having quite complex shapes such as the *asymmetric* [1], *pinched* [2], *S-shaped* [3], *flag-shaped* [4] ones, or those obtained as arbitrary combination of them [5].

Many uniaxial phenomenological models have been formulated and proposed in the literature to reproduce the above-described hysteresis loops [6-9]. Typically, such models require the numerical solution of a differential equation to evaluate the output variable, can be only adopted to reproduce a specific type of hysteresis loop shape, employ the same set of parameters to simulate both loading and unloading phases, and adopt parameters that not always have a clear theoretical or experimental significance.

With the aim of developing a unified approach for the modeling of complex rate-independent hysteretic behavior, we first present a preliminary classification of hysteresis phenomena and then we illustrate a preliminary formulation of a novel model that has been derived on the basis of a generalized class of uniaxial phenomenological models proposed by Vaiana et al. [10-14].

Such a model offers a series of advantages with respect those currently available in the literature since it i) employs closed-form expressions to evaluate the output variable, ii) allows one to simulate the generic loading phase by using a set of parameters different from the one adopted for the generic unloading one, iii) needs a simple identification procedure because of the clear meaning characterizing the adopted parameters, and iv) can be easily implemented in a computer program.

After the model formulation, we briefly describe the different types of hysteresis loops that can be reproduced and we show its capability to simulate complex hysteresis phenomena that characterize the experimental response of some mechanical systems and materials retrieved from the literature.

Materials Research Forum LLC
https://doi.org/10.21741/9781644902431-61

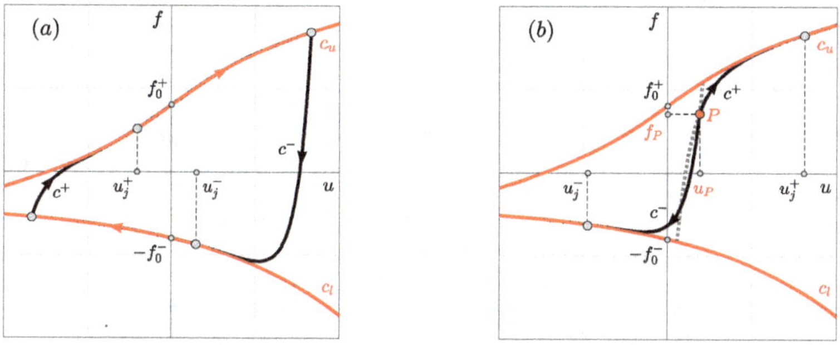

Figure 1. VRM: curves (a) and internal variables (b) characterizing a generic hysteresis loop.

Classification of Complex Uniaxial Hysteresis Phenomena

Mechanical systems and materials, typically employed in the aerospace, civil, and mechanical engineering fields, may exhibit uniaxial rate-independent hysteretic behavior characterized by quite complex generalized force-displacement hysteresis loops. In particular, the latter may be *asymmetric, pinched, S-shaped, flag-shaped*, or an arbitrary combination of them.

As proposed by Vaiana et al. [10,11], a generalized force-displacement hysteresis loop can be described by means of four different curves: a generic loading curve, a generic unloading curve, an upper limiting curve, and a lower limiting curve.

By looking at the analytical properties of the two limiting curves, it is possible to classify the hysteresis loops into four main categories:

- hysteresis loops limited by two straight lines;
- hysteresis loops limited by two curves with no inflection point;
- hysteresis loops limited by two curves with one inflection point;
- hysteresis loops limited by two curves with two inflection points.

In the sequel, we propose a novel model, denominated Vaiana-Rosati Model (VRM), that is capable of simulating the above-described complex hysteretic behavior by using parameters having a clear theoretical and/or experimental interpretation.

Proposed Model

Fig. 1a presents a typical generalized force-displacement hysteresis loop that can be reproduced by the proposed VRM. In particular, during a generic loading phase, the generalized force f is evaluated as:

$$f^+\left(u,u_j^+\right)=\begin{cases} c^+\left(u,u_j^+\right) & \text{if} \quad u<u_j^+ \\ c_u\left(u\right) & \text{if} \quad u>u_j^+, \end{cases} \tag{1}$$

whereas, during a generic unloading phase, it is obtained as:

$$f^-\left(u,u_j^-\right)=\begin{cases} c^-\left(u,u_j^-\right) & \text{if} \quad u>u_j^- \\ c_l\left(u\right) & \text{if} \quad u<u_j^-. \end{cases} \tag{2}$$

in which u is the generalized displacement, that is, the input variable of the model.

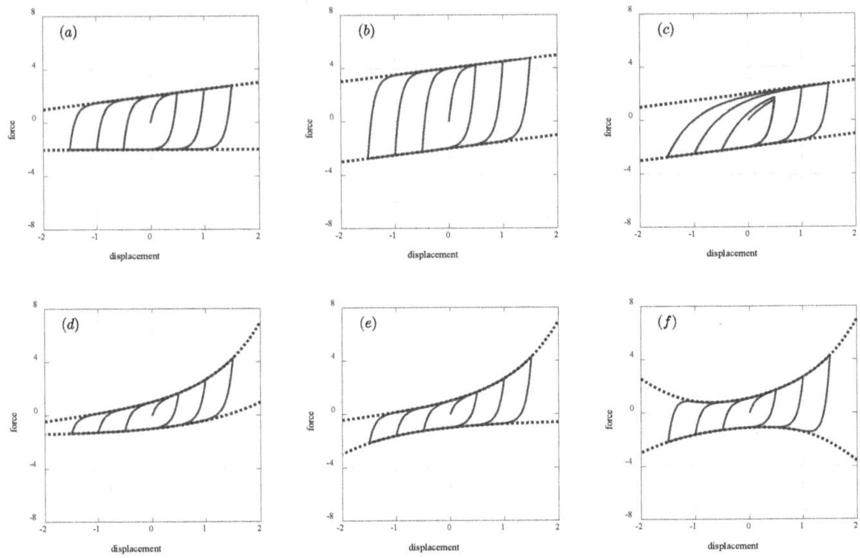

***Figure 2**. Types of hysteresis loop shapes simulated by the VRM (part 1).*

In Eqs. (1) and (2), c^+ and c^- represent, respectively, the generic loading and unloading curves, evaluated as:

$$c^+\left(u,u_j^+\right)=\beta_1^+ e^{\beta_2^+ u}-\beta_1^+ +\frac{4\gamma_1^+}{1+e^{-\gamma_2^+\left(u-\gamma_3^+\right)}}-2\gamma_1^+ +k_b^+ u+f_0^+ -\frac{1}{\alpha^+}\left[e^{-\alpha^+\left(+u-u_j^+ +\bar{u}^+\right)}-e^{-\alpha^+\bar{u}^+}\right], \tag{3}$$

$$c^-\left(u,u_j^-\right)=\beta_1^- e^{\beta_2^- u}-\beta_1^- +\frac{4\gamma_1^-}{1+e^{-\gamma_2^-\left(u-\gamma_3^-\right)}}-2\gamma_1^- +k_b^- u-f_0^- +\frac{1}{\alpha^-}\left[e^{-\alpha^-\left(-u+u_j^- +\bar{u}^-\right)}-e^{-\alpha^-\bar{u}^-}\right], \tag{4}$$

whereas c_u and c_l are, respectively, the upper and lower limiting curves, computed as:

$$c_u\left(u\right)=\beta_1^+ e^{\beta_2^+ u}-\beta_1^+ +\frac{4\gamma_1^+}{1+e^{-\gamma_2^+\left(u-\gamma_3^+\right)}}-2\gamma_1^+ +k_b^+ u+f_0^+, \tag{5}$$

$$c_l\left(u\right)=\beta_1^- e^{\beta_2^- u}-\beta_1^- +\frac{4\gamma_1^-}{1+e^{-\gamma_2^-\left(u-\gamma_3^-\right)}}-2\gamma_1^- +k_b^- u-f_0^-. \tag{6}$$

Finally, u_j^+ and u_j^- represent the internal variables having expressions:

$$u_j^+=u_P+\bar{u}^+ +\frac{1}{\alpha^+}\ln\left\{+\alpha^+\left[\beta_1^+ e^{\beta_2^+ u_P}-\beta_1^+ +\frac{4\gamma_1^+}{1+e^{-\gamma_2^+\left(u_P-\gamma_3^+\right)}}-2\gamma_1^+ +k_b^+ u_P+f_0^+ +\frac{1}{\alpha^+}e^{-\alpha^+\bar{u}^+}-f_P\right]\right\}, \tag{7}$$

$$u_j^-=u_P-\bar{u}^- -\frac{1}{\alpha^-}\ln\left\{-\alpha^-\left[\beta_1^- e^{\beta_2^- u_P}-\beta_1^- +\frac{4\gamma_1^-}{1+e^{-\gamma_2^-\left(u_P-\gamma_3^-\right)}}-2\gamma_1^- +k_b^- u_P-f_0^- -\frac{1}{\alpha^-}e^{-\alpha^-\bar{u}^-}-f_P\right]\right\}, \tag{8}$$

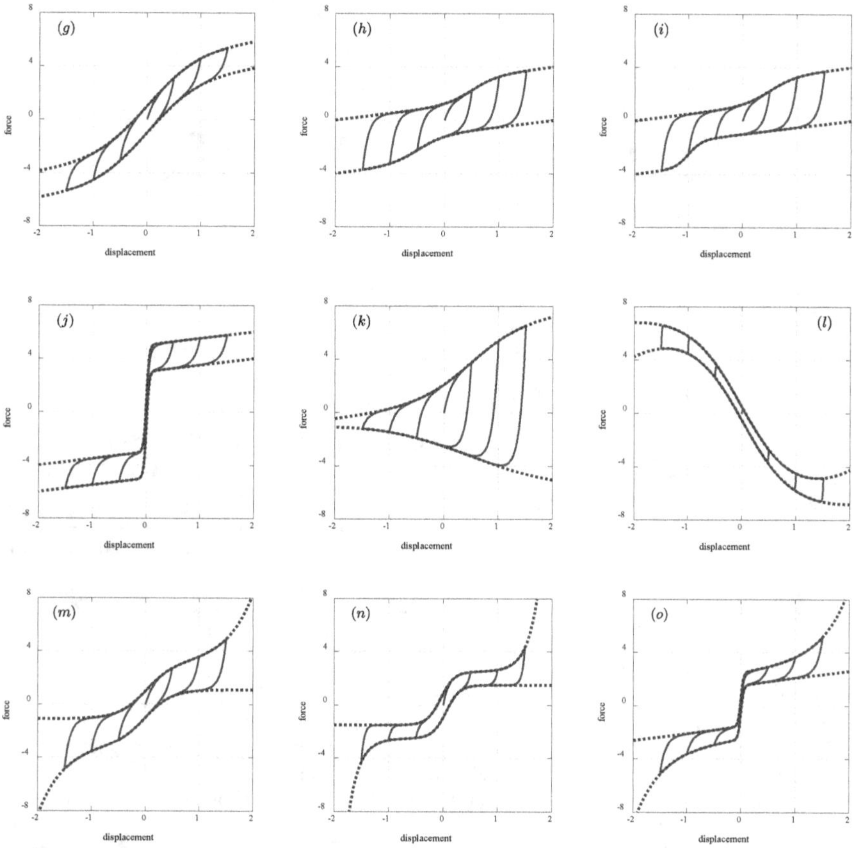

***Figure 2**. Types of hysteresis loop shapes simulated by the VRM (part 2).*

where u_P and f_P are the coordinates of a generic point P belonging to c^+ or c^-, as illustrated in Fig. 1b.

Finally, the following quantities $k_b^+, f_0^+, \alpha^+, \beta_1^+, \beta_2^+, \gamma_1^+, \gamma_2^+, \gamma_3^+$ $\left(k_b^-, f_0^-, \alpha^-, \beta_1^-, \beta_2^-, \gamma_1^-, \gamma_2^-, \gamma_3^-\right)$ represent the eight model parameters governing the generic loading (unloading) phase, whereas $\overline{u}^+$ $\left(\overline{u}^-\right)$ is an internal model parameter, whose expression will be presented in a future paper.

Fig. 2 presents several types of hysteresis loops that can be simulated by using the VRM (the parameters are omitted for brevity); it is possible to note that the hysteresis loops illustrated in:

- Figs. 2a-c are limited by two straight lines;
- Figs. 2d-f are limited by two curves with no inflection point;
- Figs. 2g-l are limited by two curves with one inflection point;
- Figs. 2m-o are limited by two curves with two inflection points.

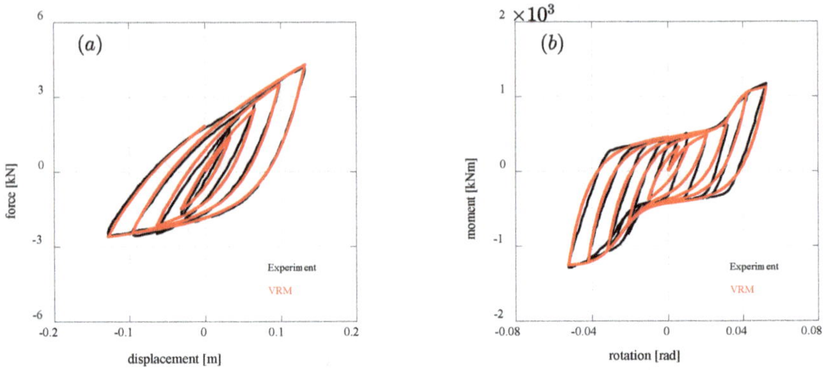

Figure 3. VRM experimental verification: asymmetric (a) and pinched (b) hysteresis loops.

***Table 1**. VRM parameters adopted to simulate the hysteresis loops in Fig. 3.*

Figure	sgn$(\dot{u})$	k_b	f_0	α	β_1	β_2	γ_1	γ_2	γ_3
3a	+	16	2.3	14	0	0	0	0	0
	−	4	2.1	24	0	0	0	0	0
3b	+	3.5	0.70	100	0	0	0.12	300	0.035
	−	0.0	0.85	90	0	0	0.21	160	-0.025

Experimental Verification

Fig. 3a (3b) compares the asymmetric (pinched) experimental hysteresis loops, characterizing the cyclic response of a spring connector (steel bolted flange plate connection) tested by Filiatrault and Kremmidas [1] (Schneider and Teeraparbwong [2]), with those simulated by using the VRM parameters listed in Table 1.

Note that in Table 1 the parameters governing the generic loading (unloading) phase are the ones put in the row where the sign of the generalized velocity $\dot{u}$, given by the second column, is positive (negative).

Looking at Fig. 3 it is possible to observe a satisfactory agreement between the experimental and simulated responses and, consequently, the capability of the model to accurately reproduce quite complex hysteresis loops is verified.

Conclusions

A preliminary classification of complex hysteresis phenomena has been presented together with a preliminary formulation of a novel uniaxial phenomenological model. Such a model can simulate different types of hysteresis loops, as shown in Fig. 2, and is capable of accurately reproducing complex experimental hysteresis responses, as illustrated in Fig. 3.

In future papers, the model will be employed to perform nonlinear dynamic analyses of mechanical systems and materials typically employed in the field of aerospace, civil, and mechanical engineering [15-16].

References

[1] A. Filiatrault, S. Kremmidas, Seismic interaction of interconnected electrical substation equipment, J. Struct. Eng., ASCE 126 (2000) 1140-1149. https://doi.org/10.1061/(ASCE)0733-9445(2000)126:10(1140)

[2] S. P. Schneider, I. Teeraparbwong, Inelastic behavior of bolted flange plate connections, J. Struct. Eng., ASCE 128 (2002) 492-500. https://doi.org/10.1061/(ASCE)0733-9445(2002)128:4(492)

[3] A. Manzoori, H. Toopchi-Nezhad, Application of an extended Bouc-Wen model in seismic response prediction of unbonded fiber-reinforced isolators, J. Earthquake Eng. 21 (2017) 87-104. https://doi.org/10.1080/13632469.2016.1138166

[4] A. Hashemi, P. Zarnani, P. Quenneville, Seismic assessment of rocking timber walls with energy dissipation devices, Eng. Struct. 221 (2020) 111053. https://doi.org/10.1016/j.engstruct.2020.111053

[5] H. Lim, F. Lam, R. Foschi, M. Li, Modeling load-displacement hysteresis relationship of a single-shear nail connection, J. Eng. Mech., ASCE 143 (2017) 04017015. https://doi.org/10.1061/(ASCE)EM.1943-7889.0001204

[6] G. Demetriades, M. Constantinou, A. Reinhorn, Study of wire rope systems for seismic protection of equipment in buildings, Eng. Struct. 15 (1993) 321-334. https://doi.org/10.1016/0141-0296(93)90036-4

[7] T. Baber, M. Noori, Random vibration of degrading, pinching systems, J. Eng. Mech., ASCE 111 (1985) 1010-1026. https://doi.org/10.1061/(ASCE)0733-9399(1985)111:8(1010)

[8] Y. Chen, G. Ahmadi, Wind effects on base-isolated structures. J. Eng. Mech., ASCE 118 (1992) 1708-1727. https://doi.org/10.1061/(ASCE)0733-9445(1992)118:9(2641.2)

[9] E. Graesser, F. Cozzarelli, Shape-memory alloys as new materials for aseismic isolation, J. Eng. Mech., ASCE 117 (1991) 2590-2608. https://doi.org/10.1061/(ASCE)0733-9399(1991)117:11(2590)

[10] N. Vaiana, S. Sessa, F. Marmo, L. Rosati, A class of uniaxial phenomenological models for simulating hysteretic phenomena in rate-independent mechanical systems and materials. Nonlinear Dyn. 93 (2018) 1647-1669. https://doi.org/10.1007/s11071-018-4282-2

[11] N. Vaiana, S. Sessa, L. Rosati, A generalized class of uniaxial rate-independent models for simulating asymmetric mechanical hysteresis phenomena. Mech. Syst. Sig. Process. 146 (2021) 106984. https://doi.org/10.1016/j.ymssp.2020.106984

[12] N. Vaiana, S. Sessa, F. Marmo, L. Rosati, An accurate and computationally efficient uniaxial phenomenological model for steel and fiber reinforced elastomeric bearings. Compos. Struct. 211 (2019) 196-212. https://doi.org/10.1016/j.compstruct.2018.12.017

[13] N. Vaiana, S. Sessa, F. Marmo, L. Rosati, Nonlinear dynamic analysis of hysteretic mechanical systems by combining a novel rate-independent model and an explicit time integration method. Nonlinear Dyn. 98 (2019) 2879-2901. https://doi.org/10.1007/s11071-019-05022-5

[14] N. Vaiana, D. Losanno, N. Ravichandran, A novel family of multiple springs models suitable for biaxial rate-independent hysteretic behavior, Comput. Struct. 244 (2021) 106403. https://doi.org/10.1016/j.compstruc.2020.106403

[15] D. Losanno, F. Palumbo, A. Calabrese, T. Barrasso, N. Vaiana, Preliminary investigation of aging effects on Recycled Rubber Fiber Reinforced Bearings (RR-FRBs), J. Earthquake Eng. (2021). https://doi.org/10.1080/13632469.2021.1871683

[16] S. Sessa, F. Marmo, N. Vaiana, D. De Gregorio, L. Rosati, Strength hierarchy provisions for transverse confinement systems of shell structural elements, Composites, Part B. 163 (2019) 413-423. https://doi.org/10.1016/j.compositesb.2019.01.018

Materials Research Proceedings **26** (2023) 381-385 https://doi.org/10.21741/9781644902431-62

Materials with memory: some new results in viscoelastic models

CARILLO Sandra[1,2,a]

[1]Dipartimento di Scienze di Base e Applicate per l'Ingegneria, Università di Roma LA SAPIENZA, Via Antonio Scarpa 16, 00161 Rome, Italy

[2]I.N.F.N. - Sezione Roma1, Gr. IV - M.M.N.L.P., Rome, Italy

[a]sandra.Carillo@uniroma1.it

Keywords: Materials with Memory, Viscoelasticity, Magneto-Viscoelasticity, Aging

Abstract. The classical model of viscoelastic body is reconsidered. As well known the deformation response of a material is termed viscoelastic when it does not depend only on the status of the material in the considered time, but also on its deformation history. Furthermore, the system is dissipative: such dissipative nature induce the use of viscoelastic materials in devising anti-seismic dissipators. Aiming to model new and innovative materials different forms of the relaxation modulus which characterises the response of the material are considered. Cases of a non-classical relaxation modulus are studied. Thus, a relaxation modulus which may be unbounded or less regular or modified to describe a material whose mechanical response is changed when the material with time, this phenomenon is usually termed "aging" are investigated. Finally, the viscoelastic response can be controlled on devising magneto-viscoelastic materials via injection of micro or nano particles magnetically sensible.

Introduction

The model of viscoelastic body, according to Fabrizio and Morro [18, 2] are the background for the present investigation. The viscoelastic body is assumed homogeneous and isotropic so that the dependence on the spatial variable can be omitted. Conversely, the dependence with respect to time is not only via the present time, but also through the whole *deformation history* of the material. Accordingly, the quantities of interest are:

$\mathbf{E}=\mathbf{E}(t)$	*strain tensor*
$\mathbf{T}=\mathbf{T}(t)$	*stress tensor*
$\mathbf{G}=\mathbf{G}(t)$	*relaxation modulus*
$\mathbf{G}_0=\mathbf{G}(0)$	*initial relaxation modulus*

wherein no space dependence is indicated; these quantities are connected via the *constitutive assumptions*

$$\mathbf{T}(t) = \int_0^\infty \mathbf{G}(\tau)\, \dot{\mathbf{E}}(t-\tau)\, d\tau \quad , \quad \mathbf{G}(t) = \mathbf{G}_0 + \int_0^t \dot{\mathbf{G}}(s)\, ds \qquad (1)$$

or equivalently, when $\mathrm{E}_t(\tau)$ denotes the **strain past history**

$$\mathbf{T}(t) = \mathbf{G}_0\mathbf{E}(t) + \int_0^\infty \dot{\mathbf{G}}(\tau)\, \mathbf{E}^t(\tau)\, d\tau \quad , \quad \mathbf{E}^t(\tau) := \mathbf{E}(t-\tau) \qquad (2)$$

Classical problem: regular kernel

The previous assumptions imply that, in the case the relaxation functions satisfies all the written regularity requirements, the problem, termed *classical*, in the one-dimensional case can be written as

$$u_{tt} = G(0)u_{xx} + \int_0^t \dot{G}(t-\tau)u_{xx}(\tau)d\tau + f \tag{3}$$

where, respectively, u and f denote the displacement and the external force which takes into account also the history of the material and, hence, is not zero. In addition, in the one-dimansional case, the tensor G is represented by a real valued function. The problem is assigned prescribing the following initial and boundary conditions:

$$u(\cdot,0) = u_0, \ \ u_t(\cdot,0) = u_1 \text{ in } \Omega; \ \ \ u = 0 \text{ on } \Sigma = \partial\Omega \times (0,T) \ . \tag{4}$$

Then, the existence and uniqueness result by Dafermos [17] can be applied. The *relaxation modulus* is the quantity, in this model, whose properties describe the behaviour of the material under investigation. The classical requirements on the relaxation modulus, the kernel in the integro-differential equation (3) are

$$\dot{G} \in L^1(\mathbb{R}^+) \ \ , \ \ G(t) = G_0 + \int_0^t \dot{G}(s) \ ds \ , \ G(\infty) := \lim_{t\to\infty} G(t) \ , \ G_0 := \lim_{t\to 0} G(t) \tag{5}$$

Hence, G enjoys the *fading memory property* that is

$$\forall \epsilon > 0 \, \exists \tilde{a} = a\left(\epsilon, E^t\right) \in \mathbb{R}^+ s.t. \ \forall a > \tilde{a}, \left| \int_0^\infty \dot{G}(s+a)E^t(s) \ ds \right| < \epsilon \tag{6}$$

whose physical meaning is that the effects very far in the past are negligible. In the following a list of different generalisations are given: they are devised to describe different materials for which the classical assumptions on the relaxation modulus cannot be adopted. The assumptions (5), which correspond to G continuous and differentiable positive valued with a negative derivative for all positive times, approaching to zero as t goes to infinity. In the case G twice differentiable thermodynamical compatibility implies also that

$$\dot{G} < 0, \ddot{G} > 0, \forall t \in (0.T), \forall T > 0$$

Non classical problems: Singular Kernel and "aging"

When we consider the case of a relaxation modulus $\lim_{t\to 0} G(t) = +\infty$. In this case [11, 4] as well the 3-dimensional generalisation [6], the problem cannot be formulated via (3) since both G(0) is not defined and $\dot{G} \notin L^1$, hence a different approach is adopted. Specifically, approximated problems are introduced and, via a suitable limit procedure, existence and uniqueness of the solution of an initial boundary value problem is proved [11].

On the other hand, another generalisation which is suggested by applications consists in taking into account the fact that, in general, the response of the material changes over time due to the natural deterioration of the material itself. These effects are studied in [15, 16] an overview is given in [5].

Magneto-viscoelastic materials

The model we adopted to describe the interaction between the viscoelastic body and an external magnetic field goes back to in [19] later revisited in [1], coupling between viscoelasticity and magnetisation.

Materials Research Forum LLC
https://doi.org/10.21741/9781644902431-62

Regular kernel

The regular kernel problem, which corresponds to consider the viscoelastic solid characterised by a relaxation function which satisfies the classical assumptions is studied in [12], in the one dimensional case and in [14] in the three dimensional case. The one-dimensional problem reads

$$\begin{cases} u_{tt} - G(0)u_{xx} - \int_0^t G'(t-\tau)u_{xx}(\tau)d\tau - \frac{\lambda}{2}(\Lambda(\mathbf{m})\cdot\mathbf{m})_x = f\,, \\ \qquad\qquad\qquad\qquad\qquad\qquad\qquad\qquad\qquad\qquad \text{in } \Omega\times(0,T) \\ \mathbf{m}_t + \mathbf{m}\frac{|\mathbf{m}|^2-1}{\varepsilon} + \lambda\Lambda(\mathbf{m})u_x - \mathbf{m}_{xx} = 0\,, \end{cases} \tag{7}$$

where $\Omega = (0, 1)$ and $\mathbf{m} = (0, m_1, m_2)$, denotes the magnetization vector, orthogonal to the conductor, since $\mathbf{u} = (u, 0, 0)$, when both quantities are written in $\mathbb{R}^3$. In addition, Λ is a linear operator defined by $\Lambda(\mathbf{m}) = (0, m_2, m_1)$, the scalar function u is the displacement in the direction of the conductor itself, here identified with the x−axis and λ is a positive parameter. In addition, the term f represents an external force which also includes the deformation history up to t = 0. Letting ν be the outer unit normal at $\partial\Omega$, initial and boundary conditions are given as follows

$$u(\cdot,0) = u_0 = 0,\ \ \mathbf{m}(\cdot,0) = \mathbf{m}_0,\quad |\mathbf{m}_0| = 1 \quad \text{in } \Omega\,,\ \ , \tag{8}$$

$$u = 0, \qquad \frac{\partial\mathbf{m}}{\partial\nu} = 0 \qquad \text{on} \quad \Sigma = \partial\Omega\times(0,T)\,, \tag{9}$$

Under the assumptions

$G(t) \times C^2(0, T)$, $u_0 \times H^1_0(\Omega)$, $u_1 \times L^2(\Omega)$, $m_0 \times H^1(\Omega)$ and $f \times L^2(\Omega\times(0, T))$,

the existence and uniqueness of the solution to the problem given by (7)-(8), is proved in [12]. The corresponding 3-dimensional problem is studied in [14].

Non classical problems: Singular Kernel and "aging"

The same problem, when the relaxation function is assumed to be unbounded at the initial time t = 0, that is coupling, now, the singular viscoelastic behaviour with the magnetisation effects is studied in [9] wherein a singular viscoelastic behaviour is coupled with the magnetic filed. Again, the introduction of suitable approximated problems and a limit procedure, allow to prove [9] existence of the solution of an initial boundary value problem.

A different generalisation is considered in [10] wherein the one-dimensional viscoelasticity model is modified to take into account the so called aging effects. The term aging is adopted to indicate that the response of the material is not unchanged with time. That is, if a long term use of the material is considered, then the response of the viscoelastic material is, in general, different from the initial one, after a long time. To model this behaviour the relaxation function is assumed to depend on the two time variables t and τ not only through their difference as in the classical model, see formulae (1) and (2) but we consider G a function of the two different time variables t and τ here the two time variables are independent. The regular magneto-viscoelasticity problem with aging in [10] is proved to admit an unique solution.

Conclusions

The wide variety of applications of viscoelastic materials for instance in the area of the study of attenuators which are devised to possibly prevent damages via dissipation of extra energy in the case of seismic events as mentioned in [3].

On the other hand, the more and more widely spread use of magnetically sensible particles for instance in gels which can be modelled as viscoelastic materials induces to further investigate this subject, see, for instance, [21], [22] and [23] to have an idea of the different kinds of applications which go from rheology to biomedical applications.

References

[1] M. Bertsch, P. Podio Guidugli and V. Valente, *On the dynamics of deformable ferromagnets. I. Global weak solutions for soft ferromagnets at rest*, Ann. Mat. Pura Appl. (4) 179 (2001), 331–360. https://doi.org/10.1007/BF02505962

[2] C. Giorgi, A. Morro, *Viscoelastic solids with unbounded relaxation function*, Continuum Mech. Thermodyn., 4, pp. 151-165, 1992. https://doi.org/10.1007/BF01125696

[3] B. Carboni, S. Carillo, W. Lacarbonara Shock dinamico in strutture in ECONOMIA E INGEGNERIA, Dinamiche comparate, Osservatorio sulle Imprese Facoltà ICI Sapienza Università di Roma, R. Gallo Editor, 2022.

[4] S. Carillo, Singular Kernel Problems in Materials with Memory, Meccanica, 50, pp. 603-615, 2015. https://doi.org/10.1007/s11012-014-0083-y

[5] S. Carillo, C. Giorgi, Non-classical memory kernels in linear viscoelasticity, Chap 13, Viscoelas- tic and Viscoplastic Materials, M.F.El-Amin Ed, ISBN 978-953-51-2602-7, pp. 295-331 (2016) InTech.

[6] S. Carillo, A 3-dimensional singular kernel problem in viscoelasticity: an existence result, Atti della Accademia Peloritana dei Pericolanti, Classe di Scienze Fisiche, Matematiche e Naturali 97 (S1), A3, 13 pp. (2019).

[7] S. Carillo, M. Chipot, V. Valente, G. Vergara Caffarelli, On weak regularity requirements of the relaxation modulus in viscoelasticity, CAIM, 10 (1), pp. 78-87, (2019). https://doi.org/10.2478/caim-2019-0014

[8] S. Carillo, M. Chipot, V. Valente, G. Vergara Caffarelli, A viscoelastic integro-differential equation with a discountinuos memory kernel, preprint 2022.

[9] S. Carillo, M. Chipot, V. Valente, G. Vergara Caffarelli, A magneto-viscoelasticity problem with a singular memory kernel, NONRWA, 35C, pp. 200-210, (2017). https://doi.org/10.1016/j.nonrwa.2016.10.014

[10] S. Carillo, C. Giorgi, A magneto-viscoelasticity problem with aging, Materi- als Physics, MDPI, Basel, (2022). doi:10.3390/ma15217810, 2022. https://doi.org/10.3390/ma15217810

[11] S. Carillo, V. Valente, G. Vergara Caffarelli *A linear viscoelasticity problem with a singular memory kernel: an existence and uniqueness result*, Differential and Integral Equations, **26**, (9/10), 1115– 1125, (2013). https://doi.org/10.57262/die/1372858565

[12] S. Carillo, V. Valente, G. Vergara Caffarelli, *A result of existence and uniqueness for an integro- differential system in magneto-viscoelasticity*, Applicable Analisys, 1791-1802, **90** (12), 2011. https://doi.org/10.1080/00036811003735832

[13] S. Carillo, *Regular and singular kernel problems in magneto-viscoelasticity*, Meccanica S.I. *New trends in Dynamics and Stability*, **52**, (13), 2017, pp 3053–3060. https://doi.org/10.1007/s11012-017-0722-1

[14] S. Carillo, V. Valente, G. Vergara Caffarelli, *An existence theorem for the magneto-viscoelastic problem* Discrete and Continuous Dynamical Systems Series S., 435 – 447, **5** (3) (2012). https://doi.org/10.3934/dcdss.2012.5.435

[15] M. Conti, V. Danese, C. Giorgi, V. Pata, *A model of viscoelasticity with time-dependent*

memory kernels, Amer. J. Math., **140** (2018) 349–389. https://doi.org/10.1353/ajm.2018.0008

[16] M. Conti, V. Danese, V. Pata, *Aging of viscoelastic materials: a mathematical model. Mathematical modeling in cultural heritage*, MACH2019, 135146, Springer INdAM Ser., 41, Springer, Cham, [2021]. https://doi.org/10.1007/978-3-030-58077-3_9

[17] C.M. Dafermos, An abstract Volterra equation with applications to linear viscoelasticity, *J. Diff. Equations*, **7**, 554–569, 1970. https://doi.org/10.1016/0022-0396(70)90101-4

[18] M. Fabrizio, A. Morro, *Mathematical problems in linear viscoelasticity*, SIAM Studies in Applied Mathematics, 12. Society for Industrial and Applied Mathematics (SIAM), Philadelphia, PA, 1992. x+203 pp. ISBN: 0-89871-266-1

[19] L. Gilbert, *A Lagrangian formulation of the gyromagnetic equation of the magnetization field*, Phys. Rev. 100 (1955) 1243.

[20] C. Giorgi, A. Morro, Viscoelastic solids with unbounded relaxation function, *Continuum Mech. Thermodyn.*, **4**, 151–165, 1992. https://doi.org/10.1007/BF01125696

[21] Xu Y., Gong X., Wan Q., Liu T., Xuan S. *Magneto-sensitive smart soft material and magnetorheo- logical mechanism* (2015) Advances in Mechanics, 45 (1), pp. 461 - 495.

[22] B. Wang, L. Kari, *A visco-elastic-plastic constitutive model of isotropic magneto-sensitive rubber with amplitude, frequency and magnetic dependency*, International Journal of Plasticity, 2020, 132, 102756. https://doi.org/10.1016/j.ijplas.2020.102756

[23] X.D. Zuo, H. Tang,X.Q. Zhu, D.M. Zhang, W. Gao, *Injectable magnetic hydrogels for self- regulating magnetic hyperthermia and drug release*, Modern Physics Letters B, Vol.35, 10, 2150169 2021. https://doi.org/10.1142/S0217984921501694

Theoretical and Applied Mechanics - AIMETA 2022 Materials Research Forum LLC
Materials Research Proceedings **26** (2023) 387-392 https://doi.org/10.21741/9781644902431-63

Direct dynamics of 2D cable-driven parallel robots including cables mass effect and its influence in the control performance

Guillermo Rubio Gómez[1,a *], Andrea Arena[2,b], Erika Ottaviano[3,c] and Vincenzo Gattulli[2,d]

[1]Industrial and Aerospace Engineering School, University of Castilla-La Mancha, Av. Carlos III s/n, 45004 Toledo, Spain

[2]Department of Structural and Geotechnical Engineering, Sapienza University of Rome, Via Eudossiana 18, Rome, 00184, Italy

[3]Department of Civil and Mechanical Engineering, University of Cassino and Southern Lazio, Via G. Di Biasio 43, Cassino, 03043, Italy

[a]guillermo.rubiogomez@uclm.es , [b]andrea.arena@uniroma1.it , [c]ottaviano@unicas.it, [d]vincenzo.gattulli@uniroma1.it

Keywords: Cable-Driven Parallel Robot, Nonlinear Modeling, Direct Dynamics, Distributed Mass Cable

Abstract. Cable-driven parallel robots are a type of parallel manipulators where rigid links are replaced by actuated cables. Although in many cases dynamic models that neglect the cables mass and elasticity are employed to simulate the robot behavior and test the control approach to be used, there are several situations in which their effect cannot be disregarded, especially when large span cables are used, and the cable mass density generates important cable sagging. This work proposes a dynamic model for planar cable-driven parallel robots with 3 degrees-of-freedom considering cables mass and elasticity. Furthermore, the effect of using control approaches based on massless inelastic cables dynamic models on robots with non-negligible cable mass and elasticity is finally assessed.

Introduction

Cable driven parallel robots (CDPR) are a special kind of parallel manipulators where rigid links are replaced by cables. By controlling the cables length, the position and orientation of the end-effector (EE) can also be controlled. These robots offer several advantages as the potential to cover large working areas and a very good power to weight ratio [1,2]. On the other hand, they also present disadvantages, one of them related to the complexity of the robot dynamic model. One common dynamic model used to analyze the dynamic behavior of CDPRs considers massless, inelastic cables, as in [3]. In this model, cables are considered as straight strings that transmit directly the cable tension generated by the winches to the end-effector, significantly reducing the complexity of the model. However, in some situations, these models do not represent the dynamic behavior of the robot with the required level of accuracy. These scenarios include large robots where, due to the cross-section and length of the cables, the cables mass cannot be neglected [4]. Static analyses were already performed by means of a geometrically exact model proposed in [5]. In this work, a dynamic model that considers cable mass and elasticity is proposed for planar CDPRs with 3 degrees-of-freedom (DOF) and n cables. The model proposed in this work is derived from a more comprehensive 3D model developed by the same authors of the present work in [6] where all the modeling details can be found. The model consists of a set of partial differential equations with boundary conditions modeling the cables behavior coupled with ordinary differential equations (ODE) that model the EE. A methodology for obtaining a solution to the system is proposed, it is validated through numerical simulations and the influence of the cable mass is assessed in two case-study.

Nonlinear parametric modeling

Mechanical formulation

Let us consider $s_i \in [0, L_i(t)]$ $(i = 1, \dots, n)$ the arclength of the unstretched configuration, $L_i(t)$ is the total unstretched length and $l_i(t)$ the linear distance between the boundaries of the i-th cable. Vectors $\boldsymbol{x}_i = [y_i\ z_i]^{\top}$represent the positions of the cable origins in the robot structure w.r.t. the fixed reference frame $(\boldsymbol{e}_y, \boldsymbol{e}_z)$ located at $\boldsymbol{x}_1 \equiv \boldsymbol{0}$. The vector $\boldsymbol{r}_i(t) = \boldsymbol{R}_x(\theta_x(t)) \cdot \boldsymbol{r}_i^0$ represents the position of the i-th cable attachment point to the end-effector (EE) w.r.t the fixed reference frame $(\boldsymbol{e}_y, \boldsymbol{e}_z)$ while $\boldsymbol{r}_i^0$ is the vector describing the EE geometry in the mass-fixed local axes $(\boldsymbol{b}_y(t), \boldsymbol{b}_z(t))$ centered in the EE center of mass O_M, while $\theta_x(t)$ is the rotation angle about the axis $\boldsymbol{e}_x \equiv \boldsymbol{b}_x$ orthogonal to the working plane, and $\boldsymbol{R}_x$ is the rotation matrix around the x-axis. The position of O_M can be calculated as

$$\boldsymbol{p}_M(t) = \boldsymbol{p}_{M,i}(t) - \boldsymbol{r}_i(t) + \boldsymbol{x}_i (i = 1, \dots, n). \qquad (1)$$

The position of the point on the i-th cable is given by the vector $\boldsymbol{p}_i(s_i, t)$ w.r.t. $\boldsymbol{x}_i$. The strain state of the i-th cable can be described by introducing the stretch vector $\boldsymbol{v}_i(s_i, t)$ obtained as

$$\boldsymbol{v}_i(s_i, t) = \frac{d}{ds_i} \boldsymbol{p}_i(s_i, t). \qquad (2)$$

The vector of the cable axial force can be expressed as: $\boldsymbol{n}_i(s_i, t) = N_i(s_i, t)\boldsymbol{v}_i(s_i, t)/v_i(s_i, t)$, where $N_i(s_i, t) = EA(v_i(s_i, t) - 1)$ represents the cable tension, being EA the cable axial stiffness.

Nondimensional form

The distance $l_{0,1} = l_1(0)$ is adopted as characteristic length while the characteristic frequency $\omega_c = \sqrt{EA/(\rho A\, l_{0,1}^2)}$ is adopted to nondimensionalize the time. The following nondimensional parameters can be introduced

$$\lambda_{0,i} = \frac{l_{0,i}}{l_{0,1}}, \bar{x}_i = \frac{x_i}{l_{0,1}}, \bar{r}_i^0 = \frac{r_i^0}{l_{0,1}}, \bar{r}_i = \frac{r_i}{l_{0,1}}$$
$$\bar{\boldsymbol{p}}_M = \frac{\boldsymbol{p}_M}{l_{0,1}}, \bar{\boldsymbol{p}}_i = \frac{\boldsymbol{p}_i}{l_{0,1}}, \bar{\boldsymbol{n}}_i = \frac{\boldsymbol{n}_i}{\rho A\ \omega_c^2\ l_{0,1}^2}, \tau = \omega_c \cdot t \qquad (3)$$
$$\lambda_i(\tau) = \frac{l_i(\tau)}{l_{0,i}}, \Lambda_i(\tau) = \frac{L_i(\tau)}{l_i(\tau)}, \sigma = \frac{s_i}{L_i(t)}, (i = 1 \dots n).$$

By defining $(\cdot)' := \frac{d}{d\sigma}(\cdot)$ and $\dot{(\cdot)} := \frac{d}{d\tau}(\cdot)$, as the derivatives in terms of the nondimensional arclength σ and time τ , respectively, the stretch vector can be then calculated as $\boldsymbol{v}_i(\sigma, \tau) = \bar{\boldsymbol{p}}_i'(\sigma, \tau)/(\Lambda_i(\tau)\lambda_i(\tau)\lambda_{0,i})$. Due to the definition of ω_c, it turns out that the i-th axial force is given in nondimensional form as $\bar{N}_i(\sigma, \tau) = (v_i(\sigma, \tau) - 1)$; therefore, the corresponding vector of the nondimensional axial force can be written as

$$\bar{\boldsymbol{n}}_i(\sigma, \tau) = \frac{1}{\Lambda_i(\tau)\lambda_i(\tau)\lambda_{0,i}} \frac{(v_i(\sigma, \tau) - 1)}{v_i(\sigma, \tau)} \bar{\boldsymbol{p}}_i'(\sigma, \tau). \qquad (4)$$

where $v_i = ||\boldsymbol{v}_i||$. Finally, the equation of motion of the i-th cable in nondimensional form reads

$$\frac{\bar{\boldsymbol{n}}_i'(\sigma, \tau)}{\Lambda_i(\tau)\lambda_i(\tau)\lambda_{0,i}} + \bar{\boldsymbol{f}}_i(\sigma, \tau) = \ddot{\bar{\boldsymbol{p}}}_i(\sigma, \tau) + \bar{c}_i \dot{\bar{\boldsymbol{p}}}_i(\sigma, \tau)\ (i = 1, \dots, n), \qquad (5)$$

where $\bar{c}_i = c_i/(\rho A_1 \omega_c)$. The nondimensional distributed load, calculated as $\bar{\boldsymbol{f}}_i = \boldsymbol{f}_i/(\rho A_1 \omega_c^2 l_{0,1})$, is here considered to be the only cable self-weight, therefore $\bar{\boldsymbol{f}}_i = -\gamma\, \boldsymbol{e}_z$, where $\gamma = g/(\omega_c^2\, l_{0,1})$ and $g = 9.81\ m/s^2$ is the gravity acceleration. Furthermore, the kinematic boundary conditions associated with Eq. (5) can be written in nondimensional form as

$$\bar{\boldsymbol{p}}_i(0,\tau) = 0 \quad (i = 1, \dots, n), \qquad (6)$$

$$\bar{\boldsymbol{x}}_i + \bar{\boldsymbol{p}}_i(1,\tau) - \bar{\boldsymbol{r}}_i(\tau) = \bar{\boldsymbol{p}}_1(1,\tau) - \bar{\boldsymbol{r}}_1(\tau) \quad (i = 2, \dots, n), \qquad (7)$$

while the nondimensional form of the EE balance equations can be written as

$$-\sum_{i=1}^{n} \bar{\boldsymbol{n}}_i(1,\tau) - \mu\gamma\, \boldsymbol{e}_z = \mu\left(\ddot{\bar{\boldsymbol{p}}}_1(1,\tau) - \ddot{\bar{\boldsymbol{r}}}_1(\tau)\right), \qquad (8)$$

$$-\sum_{i=1}^{n} \left(\bar{\boldsymbol{r}}_i(\tau) \times \bar{\boldsymbol{n}}_i(1,\tau)\right) \cdot \boldsymbol{e}_x = \bar{J}_\mu\, \ddot{\theta}_x(\tau), \qquad (9)$$

respectively, where $\mu = M/(\rho A_1\, l_{0,1})$ and $\bar{J}_\mu = J_M\left[1/(\rho A_1\, l_{0,1}^3)\right]$, being M and J_M the end-effector mass and mass moment of inertia, respectively.

Discretization procedure

The discretization technique based on the Galerkin method is adopted to reduce the space-dependence of the cables equations of motion so as to reduce them into a set of ordinary differential equations (ODEs), in τ. In this work $m + 1$ trial functions are chosen so as to satisfy the kinematic boundary conditions (6) and (7); therefore, the approximate solution of Eq. (5) is given by the 2-by-1 vector $\tilde{\boldsymbol{p}}_i(\sigma,\tau)$ $(i = 1, \dots, n)$ expressed as the linear combination of the $m + 1$ trial functions as

$$\tilde{\boldsymbol{p}}_i(\sigma,\tau) = \boldsymbol{q}_{i,0}(\tau)\, \sigma + \sum_{j=1}^{m} \phi_{i,j}(\sigma)\ \boldsymbol{q}_{i,j}(\tau), \qquad (10)$$

where $\phi_{i,j}(\sigma) = \mathrm{diag}(sin(j\pi\sigma), sin(j\pi\sigma))$ is the ij-th 2-by-2 diagonal matrix collecting the j-th trial functions and $q_{i,0}$ and $q_{i,j}$ are the vectors collecting the unknown generalized coordinates. Since $\tilde{p}_i(0,\tau) = 0$, Eq. (10) satisfies (6), whereas, to satisfy Eq. (7), the following relationship must hold:

$$\boldsymbol{q}_{i,0}(\tau) = \boldsymbol{q}_{1,0}(\tau) - \bar{\boldsymbol{r}}_1 + \bar{\boldsymbol{r}}_i - \bar{\boldsymbol{x}}_i, \ (i = 2, \dots, n), \qquad (11)$$

since $\tilde{\boldsymbol{p}}_i(1,\tau) = \boldsymbol{q}_{i,0}$. Therefore, only one out of n vectors $\boldsymbol{q}_{i,0}(\tau)$ is an effective set of unknown coordinates. By now substituting (10) into (5) one obtains the following unbalanced residual of the equations of motion

$$\tilde{\boldsymbol{\eta}}_i(\sigma,\tau) = \frac{1}{\Lambda_i^2 \lambda_i^2 \lambda_{0,i}^2}\left[\frac{(\tilde{\nu}_i - 1)}{\tilde{\nu}_i}\tilde{\boldsymbol{p}}_i'\right]' + \bar{\boldsymbol{f}}_i - \ddot{\tilde{\boldsymbol{p}}}_i - \bar{c}_i\dot{\tilde{\boldsymbol{p}}}_i \quad (i = 1, \dots, n), \qquad (12)$$

where $\tilde{\nu}_i$ corresponds to $\bar{\nu}_i$ in terms of the approximate solution. The unbalanced residuals can be minimized by ensuring that they are orthogonal in the nondimensional domain $[0,1]$ to the trial functions adopted in the discretization. Finally, the approximate form of the vector-valued balance equations of the end effector can be obtained by substituting Eqns. (10) and (11) into Eqns. (8) and (9). The resulting set of ODEs is solved taking a direct approach, i.e., the force applied to the cables is considered the input while the EE trajectory/orientation and the cable shapes are the outputs. In this sense, the values of $\Lambda_i(t)$ are assigned by means of $\Lambda_i(\tau) = 1 - \left|\left|\hat{\boldsymbol{n}}_i\right|\right|$. To perform the time-integration of the ODEs, the system of algebraic equations resulting from the evaluation of the governing equations of motion at $\tau = 0$ and considering zero initial acceleration and velocity is solved to obtain the initial values $\boldsymbol{q}_{i,j}(0)$. The initial values of $\Lambda_i(0)$ are computed so that the fulfill the equilibrium of the EE.

Simulation results and influence of the cable mass

To assess the impact of the cable mass in the behavior of a CDPR under real operational conditions, the following procedure is employed. First, a position control approach in the Cartesian space is designed using a very simple dynamic model for the robot, that assumes massless, inelastic cables, for simulating its behavior. Then, employing the same controller, the behavior of the robot is simulated using the dynamic model exposed and solved by the procedure described in the previous Sections. The control architecture is shown in Fig. 1, where $(\boldsymbol{p}_M, \theta_x)^*$ and $(\boldsymbol{p}_M, \theta_x)$ stand for the desired and the simulated EE trajectory, respectively. The redundancy of the actuation, i.e., the conversion between the desired EE acceleration $(\ddot{\boldsymbol{p}}_M, \ddot{\theta}_x)$ generated by the controller to the corresponding cable forces applied to the robot (T), is solved by means of the well-known Improved Closed Form proposed by Pott [7].

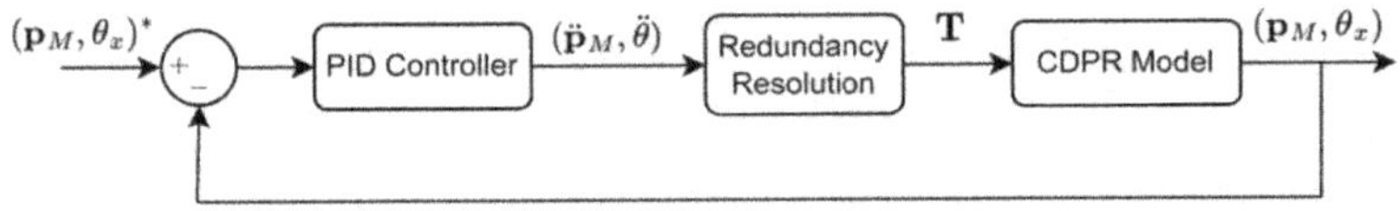

Figure 1. Block diagram of the control architecture

The PID gains are set by trial-and-error so that the position error committed with the massless cable model is less than 10mm. The set of ODEs is solved in Simulink using the classical Runge-Kutta (ODE4) method, with a sampling time of 1ms. The cable characteristics are $EA = 150360\,(\mathrm{N})$ and $\rho A = 4.19 \cdot 10^{-2}\,(\mathrm{kg/m})$ that correspond to those employed in [8]. An over-constrained CDPR with 4 cables and 3 DOF is considered. For the first case of study the robot frame is a rectangle of sizes $40 \times 10\,(\mathrm{m})$ and the EE geometry corresponds to a square of $0.5 \times 0.5\,(\mathrm{m})$. In case of study 2, the robot frame and EE are scaled by $1/4$. The EE mass is $M = 20\,\mathrm{kg}$ and its inertia $J = 0.833\,(\mathrm{kg{\cdot}m^2})$. The cable tension limits used for solving the redundancy by the Improved Closed Form are $[500{,}4000]\,(\mathrm{N})$. The employed trajectory is a circumference and the time evolution of $\boldsymbol{p}_M^*$ is generated by the interpolation algorithm detailed in [9]. For case of study 1 the radius is $r = 2.5\,(\mathrm{m})$ which is scaled proportionally for case of study 2. The maximum values for the EE linear speed, acceleration and jerk are, respectively $v = 12\,(\mathrm{m/s})$, $a = 27\,(\mathrm{m/s^2})$ and $j = 122\,(\mathrm{m/s^3})$, while for the second case these values are $v = 3\,(\mathrm{m/s})$, $a = 6.9\,(\mathrm{m/s^2})$ and $j = 30.6\,(\mathrm{m/s^3})$, which correspond to the same angular velocity of the EE describing the circumference in both cases. Fig. 2 shows the desired trajectory and the real one obtained for both cases of study. The legends MSC and MC stand for the results obtained with the massless and mass cable dynamic models, respectively. Analogously, Fig. 3 shows the tracking error and Fig. 4 shows the cable tension distributions generated.

Materials Research Proceedings **26** (2023) 387-392 https://doi.org/10.21741/9781644902431-63

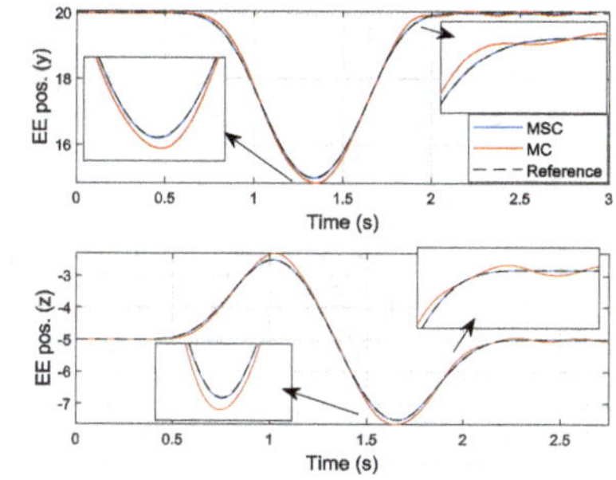

(a) Case 1

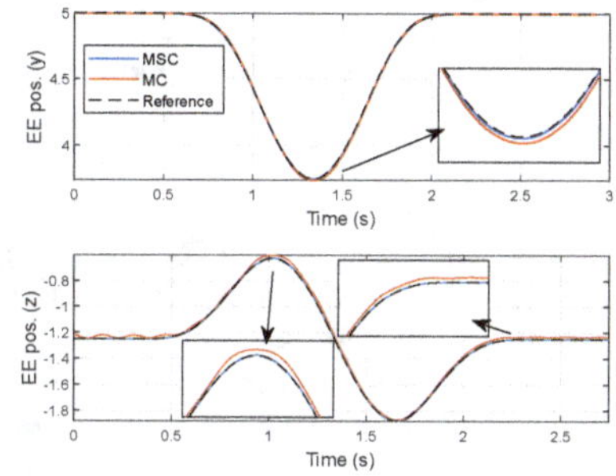

(b) Case 2

Figure 2. Position tracking

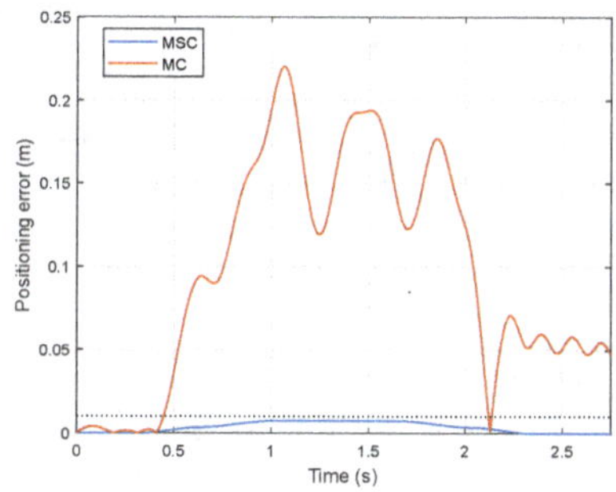

(a) Case 1

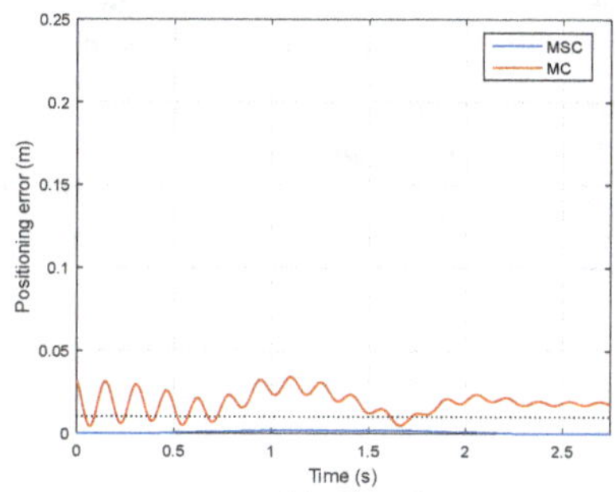

(b) Case 2

Figure 3. Tracking error

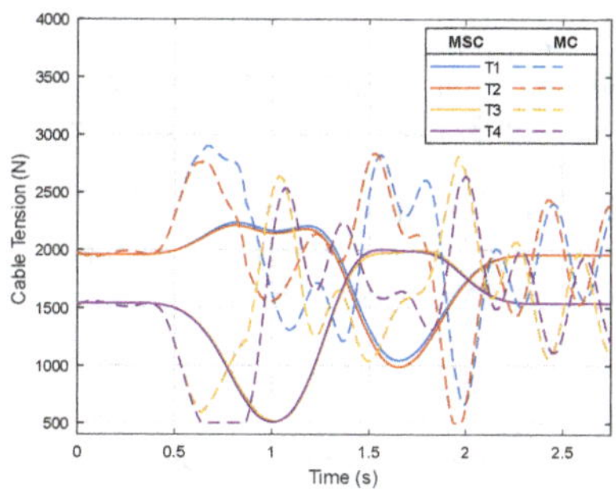

(a) Case 1

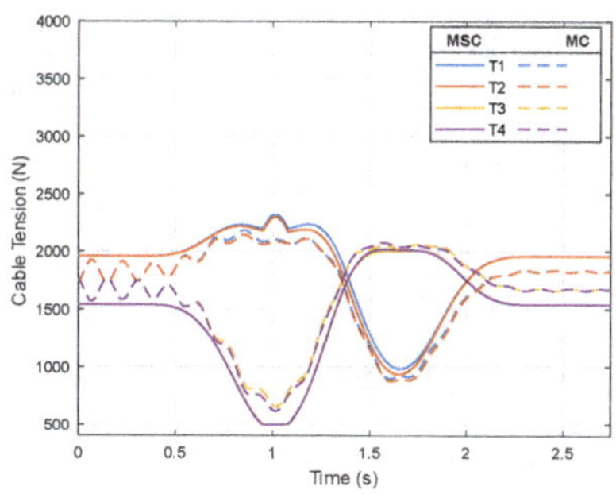

(b) Case 2

Figure 4. Cable tension distribution

In the case of study 1, it can be observed that using the controller designed considering a dynamic model with massless cables, yields to non-negligible differences between the expected and the real behavior of the robot, which is simulated by means of the proposed dynamic model when considering cable mass and elasticity. The tracking performance suffers a great deterioration as the maximum tracking error during the trajectory grows from less than 1 cm to more than 20 cm. Furthermore, the real cable force distribution presents an important change compared to the expected one, with considerable oscillations and values in the lower limit. On the other hand, we can see how that difference between the expected and the real behavior is considerably less in the case of study 2, as the robot size is 4 times smaller, therefore, cables total weight is reduced and therefore its effect on the robot dynamics, as expected.

Conclusions

In this work, a dynamic model for a planar CDPR with 3 DOF and n cables is proposed taking into account the effect of cables mass and elasticity. The system of partial differential equations corresponding to the cables motion is formulated, together with the boundary conditions imposed by the robot geometry and the ordinary differential equations that model the EE movement. A methodology based on the Garlekin discretization method is proposed to solve the system. Finally, the model is validated under a closed-loop position control strategy in the Cartesian space. The effect of the cables mass is assessed by comparing the results obtained under the same control strategy using the proposed model and a typical massless model. It can be concluded that the geometry of the robot is a key aspect to consider when deciding the level of complexity required for the CDPR model to be employed in model-based control strategies.

Acknowledgements

This research was in part sponsored by the NATO Science for Peace and Security Programme under grant id. G5924. Dr. Guillermo Rubio Gómez wishes to thank the University of Castilla-La Mancha for the financial support provided by the post-doctoral grant 2021-POST-20486.

References

[1] Pott, A.: Introduction. In: Cable-Driven Parallel Robots. Springer Tracts in Advanced Robotics, pp. 1-13. Springer, Cham (2018). https://doi.org/ 10.1007/978-3-319-76138-1 1

[2] Castillo-Garcia, F. J., Juárez-Pérez, S., González-Rodriguez, A., Rubio-Gómez, G., Rodríguez-Rosa, D., Ottaviano, E. (2021) Closed loop cable robot for large horizontal workspaces. Smart Structures and Systems, An International Journal, 27(2), 397-406.

[3] Santos, J.C., Chemori, A., Gouttefarde, M.: Model predictive control of large-dimension cable-driven parallel robots. In: International Conference on Cable-Driven Parallel Robots, pp. 221-232 (2019). Springer. https://doi.org/10.1007/978-3-030-20751-9_19

[4] Merlet, J.-P., Tissot, R.: A panorama of methods for dealing with sagging cables in cable-driven parallel robots. In: Advances in Robots Kinematics (2022). https://doi.org/10.1007/978-3-031-08140-8_14

[5] Ottaviano, E., Arena, A., & Gattulli, V. (2021). Geometrically exact three-dimensional modeling of cable-driven parallel manipulators for end-effector positioning. Mechanism and Machine Theory, 155, 104102. https://doi.org/10.1016/j.mechmachtheory.2020.104102

[6] Arena, A., Ottaviano, E., Gattulli, V. (2023). Dynamics of Cable-Driven Parallel Manipulators with Variable Length Vibrating Cables. Journal of Sound and Vibration. *Under review*

[7] Pott, A.: An improved force distribution algorithm for over-constrained cable-driven parallel robots. In: Computational Kinematics, pp. 139-146. Springer (2014). https://doi.org/10.1007/978-94-007-7214-4_16

[8] Du, J., Agrawal, S.K.: Dynamic modeling of cable-driven parallel manipulators with distributed mass flexible cables. Journal of Vibration and Acoustics 137(021020), 1-8 (2015). https://doi.org/10.1115/1.4029486

[9] Fang, Y., Hu, J., Liu, W., Shao, Q., Qi, J., Peng, Y.: Smooth and time-optimal s-curve trajectory planning for automated robots and machines. Mechanism and Machine Theory 137, 127-153 (2019). https://doi.org/10.1016/j.mechmachtheory.2019.03.019

Theoretical and Applied Mechanics - AIMETA 2022 Materials Research Forum LLC
Materials Research Proceedings **26** (2023) 393-398 https://doi.org/10.21741/9781644902431-64

On the dynamic stability of elastic structures subjected to follower forces

Francesca Levi[1,a*], Angelo Carini[1,b]

[1] Department of civil, environmental, architectural engineering and mathematics, University of Brescia, Brescia, Italy

[a] f.levi001@unibs,it, [b] angelo.carini@unibs.it

* corresponding author

Keywords: Dynamic Stability, Follower Forces, Divergence at Infinity

Abstract. Within the framework of the second-order theory, some classical stability problems, whose critical load corresponded to dynamic instability, were considered in the paper [1]. The main focus was on systems with just one lumped mass. This idealization, together with the assumption of negligible axial strain and the adoption of the second-order theory, reduced the considered systems to a single Lagrangian coordinate. In this way, static methods could be applied to derive the analytical expression of the stiffness coefficient and to study the dynamic stability, starting with a well-known example, namely, a cantilever beam with a lumped mass at the free end subjected to a follower load [2]. In this paper, a new lumped mass system is studied: a straight-axis beam with constant cross-sectional area and stiffness, mass-free, hinged at one end, simply supported at an intermediate point (with a sliding plane parallel to the beam axis) and with the other end free, where a lumped mass is present and a follower force is applied. As in the examples shown in [1], in this example the first asymptote of the stiffness coefficient corresponds to the critical load, due to divergence at infinity. It is shown that this critical load is equal to the buckling load due to divergence of an auxiliary structure, which differs from the original one in that the concentrated mass is replaced by a constraint that blocks the corresponding Lagrangian coordinate.

Introduction

A classical stability problem involving a straight-axis Euler Bernoulli beam without damping, with a concentrated mass M, inflecting in a plane, subjected to follower loads is discussed. No allowance is made for axial strain, and displacements are assumed to be small. This idealization reduces the system to a single Lagrangian coordinate u. Therefore, the equation of motion due to an initial perturbation can be written in the form:

$$M\ddot{u} + Ku = 0 \tag{1}$$

where the stiffness coefficient K takes second-order effects into account. It follows that stability depends on the sign of the stiffness coefficient. If it is positive, the motion is oscillatory and bounded. If it is negative, the motion is unbounded and non-oscillatory (divergence). Since the system possesses just one Lagrangian coordinate, unbounded oscillatory motion (flutter) cannot occur. In fact, a dynamical system that can be schematized with a lumped mass constrained by a linear spring and for which motion is governed by Eq. 1 is a conservative system with total energy $\mathcal{E}$ (kinetic plus potential) equal to

$$\mathcal{E} = \frac{1}{2}M\dot{u}^2 + \frac{1}{2}Ku^2 \tag{2}$$

which is conserved during the motion. The stability of such a system can therefore be studied statically by using Dirichlet's theorem and then going on to determine the sign of K. If $K > 0$ the potential energy in the initial undeformed configuration is minimum and the system is stable whereas if $K < 0$ the potential energy in the initial undeformed configuration is maximum and the system is unstable.

In [1] systems depicted in Fig. 1a, 1b and 1c have already been studied. Referring, for simplicity, to the system in Fig. 1a, the dependence of the stiffness coefficient on the applied load is nonlinear, due to the adopted mass modeling, so the change of sign of the stiffness coefficient may occur due to the presence of an asymptote. As the compressive load slowly increases from zero, the stiffness coefficient increases and consequently the vibration frequency increases. When the load tends to the critical value, the stiffness coefficient tends to infinity and the corresponding mass displacement tends to zero [2]. If the load is slightly higher than the critical load, the other branch of the function is involved and the stiffness coefficient is negative, corresponding to non-oscillatory unbounded motion.

This phenomenon is named divergence at infinity [3]. It is a dynamic instability, since it depends on the mass properties of the structure, and it can occur in systems that possess just one Lagrangian coordinate and are subjected to follower loads. In these cases, divergence at infinity can be studied by analyzing the sign of the stiffness coefficient, i.e., although divergence at infinity is a dynamic instability, a static method can be applied. The phenomenon of divergence at infinity also occurs for the structures in Fig. 1b and 1c.

In this paper, a new lumped mass system is studied: a straight-axis beam with constant cross-sectional area and stiffness, massless, hinged at one end, simply supported at an intermediate point at a distance a from the hinge and with the other end free in which there is a lumped mass and a follower force is applied (Fig. 2a). In the case of uniformly distributed mass, the system was studied by Zorii and Chernukha [4] and later by Elishakoff and Hollkamp [5].

Intuitively, for $a \to 0$ one should find the results already obtained in [1], while for $a \to l$ one should find again the Eulerian critical load. This implies (as indeed already pointed out in [4] and later in [5]) that there should exist a transition value $a = a^*$ such that when $a < a^*$ there is instability by divergence at infinity, while for $a > a^*$ there is Eulerian instability by divergence. Other lumped mass systems have been studied (Fig. 2b and 2c) but not discussed here for brevity of exposition.

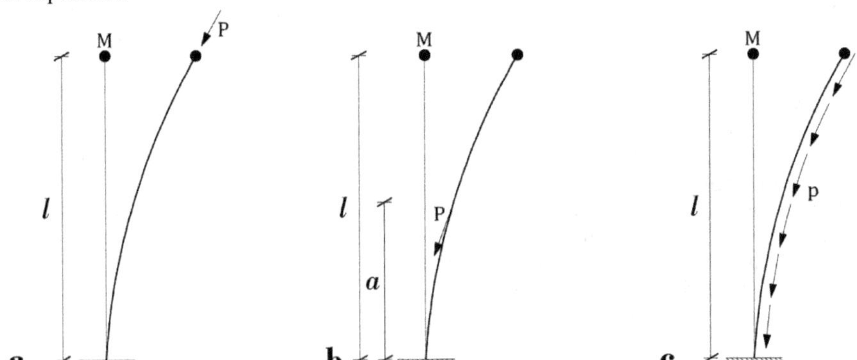

Fig. 1: Cantilever beam with lumped mass subject to (a) a follower force P at the free end, (b) to a follower force P applied at an intermediate point, and (c) to a uniformly distributed follower force p.

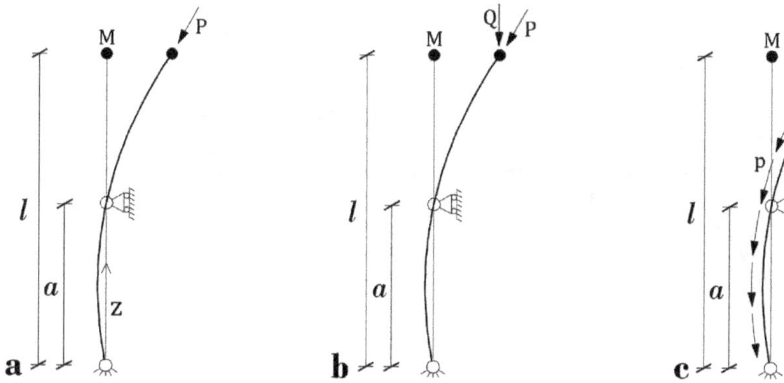

Fig. 2: Beam hinged at one end, free at the other end, simply supported at an intermediate point, with a mass concentrated at the free end and subject to (a) a follower force P at the free end, (b) a conservative force Q and a follower force P at the free end, and (c) a uniformly distributed follower force p.

Applying a static method to detect divergence at infinity

Consider the example in Fig. 2a, i.e., a straight-axis beam of length l, with a constant cross-section and moment of inertia J, with no distributed mass, hinged at one end, simply supported at an intermediate point at distance a from the hinge and with the other end free in which there is a concentrated mass M and a follower force P is applied. The stiffness coefficient K is determined by using the direct method, that is, neglecting the mass and applying a transverse static force F acting transversely to the beam at the free end such as to produce a unit displacement in the direction of F.

Having chosen, as shown in Fig. 2a, the hinged end as the origin of the z-coordinate along the axis of the beam and denoting by E the Young's modulus, the deflected curve $v(z)$, can be obtained by prescribing

$$EJv_i^{IV} + Pv_i^{II} = 0 \text{ with } i = 1 \text{ for } 0 \leq z \leq a, \quad i = 2 \text{ for } a \leq z \leq l \tag{3}$$

with boundary conditions:

$$v_1(0) = 0, \quad v_1^{II}(0) = 0, \quad v_2(l) = 0, \quad v_2^{II}(l) = 0 \tag{4}$$

and with the continuity conditions:

$$v_1(a) = v_2(a) = 0, \quad v_1^{I}(a) = v_2^{I}(a), \quad v_1^{II}(a) = v_2^{II}(a). \tag{5}$$

By solving the equation of the deflected curve, it is possible to determine K:

$$K = -EJv_2^{III}(l) = P\alpha \cdot \frac{a\sin(\alpha a)}{\alpha a^2 \sin(\alpha l) - \alpha a l \sin(\alpha l) + l\sin(\alpha a)\cdot[\cos(\alpha a)\sin(\alpha l) - \cos(\alpha l)\sin(\alpha a)]} \tag{6}$$

with $\alpha^2 = P/EJ$. Assuming $c = a/l$, Fig. 3 shows the dimensionless stiffness coefficient Kl^3/EJ as a function of αl for different c. For $c = 0$ a well-known case is found [2]. In this case, asymptotes occur as

$$\tan(\alpha l) = \alpha l \tag{7}$$

and the first asymptote (divergence at infinity) corresponds to $Kl^3/EJ = 20.1934$. If $c = 1$ another well-known case is obtained: the pinned-pinned Euler rod to which corresponds the Eulerian critical load $Kl^3/EJ = 9.8696$. Analyzing the same graph as c increases from zero to one, we move from instability by divergence at infinity to Eulerian instability by divergence. Such a transition occurs when $c = 0.5$.

Table 1: Critical loads for different values of c for (a) the real beam, (b) the auxiliary beam, and (c) the real beam without lumped mass.

$c = \frac{a}{l}$	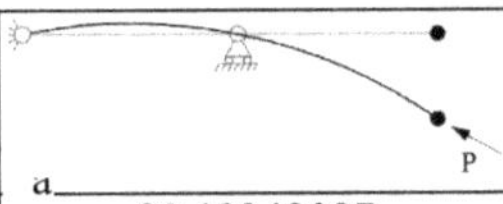	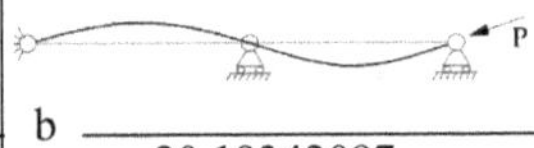	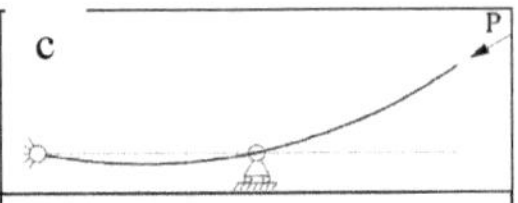
0	20.19342097	20.19342097	∞
0.1	23.22477944	23.22477944	986.96044011
0.2	27.05331941	27.05331941	246.74011003
0.3	31.75504645	31.75504645	109.66227112
0.4	36.79994680	36.79994680	61.68502751
0.5	39.47841374	39.47841760	39.47841760
0.6	27.41556778	36.79994680	27.41556778
0.7	20.14204980	31.75504645	20.14204980
0.8	15.42125688	27.05331941	15.42125688
0.9	12.18469679	23.22477944	12.18469679
1	9.86960440	20.19342097	9.86960440

Comparing the critical load by divergence at infinity with the Eulerian critical load by divergence of an auxiliary structure. Consider $c = 0$. As it is well known, Eq. 7 coincides with the equation that determines the Euler buckling load of a clamped-pinned beam. This coincidence is due to the fact that when the critical load is approached by slowly increasing the applied follower load, the stiffness coefficient tends to infinity (Fig. 3a) and the displacement of the free end tends to zero. This fact happens for all values of c between 0 and 0.5, that is, when there is instability by divergence at infinity. A general hypothesis already highlighted in [1] is confirmed: when instability is due to divergence at infinity, if one considers an auxiliary structure, different from the original one in that the lumped mass is replaced by a constraint that blocks the corresponding Lagrangian coordinate, and if the critical load of this new structure is due to divergence, that is, it does not depend on the mass distribution, then it coincides with the critical load by divergence at infinity of the original structure. It is then noticed that for $c \geq 0.5$ the critical load coincides with the Eulerian critical load of the massless beam. This circumstance is also noticed for the beams in Fig. 2b and 2c. Therefore, it can be assumed that, in general, if one wants to determine the dynamic critical load of a linear elastic structure subjected to follower forces in the small displacements regime, it suffices to determine the minimum critical load between the one determined with the auxiliary structure first introduced in Fig. 4a and the one determined statically (looking for equilibrium configurations other than the trivial) starting from the assigned structure in Fig. 4b. Table 1 shows the values of the critical loads thus deduced. It can be seen that for $0 \leq c \leq 0.5$ the correct critical loads are those of the auxiliary structure while for $0 \leq c \leq 1$ the correct critical loads are those derived statically from the actual structure.

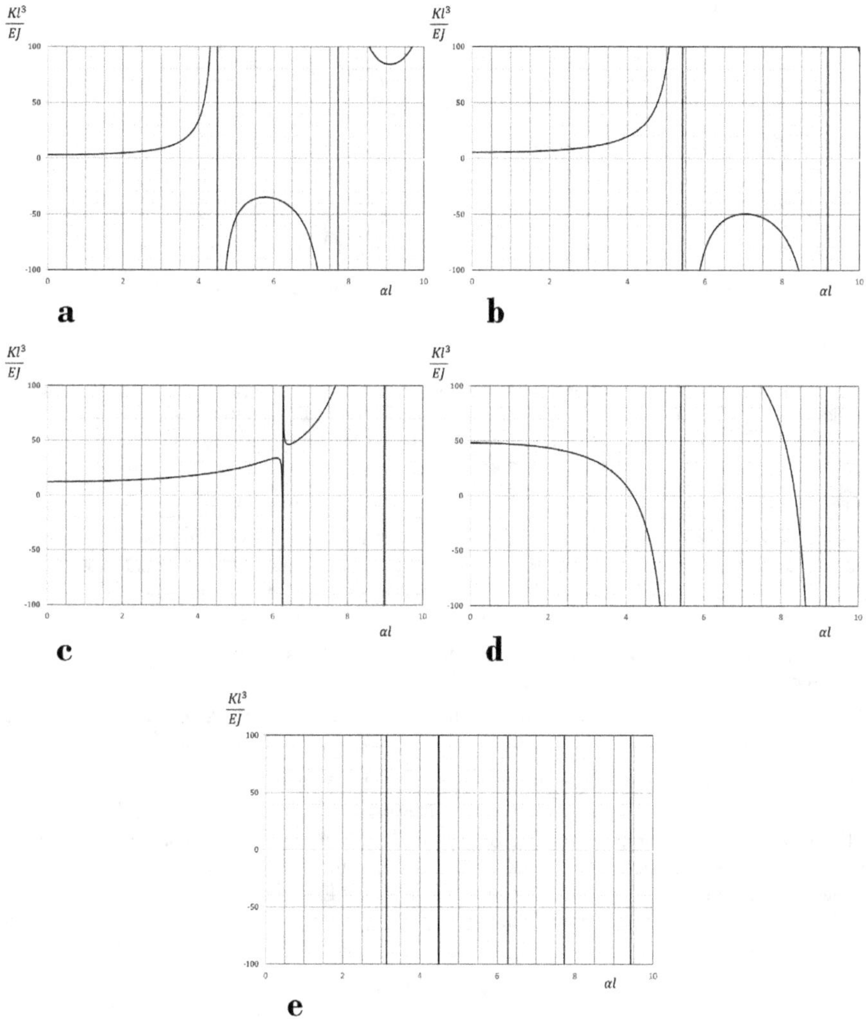

Fig. 3: Dimensionless stiffness coefficient Kl^3/EJ *versus* αl *when (a)* $c = 0$, *(b)* $c = 0.25$, *(c)* $c = 0.5$, *(d)* $c = 0.75$, *(e)* $c = 1$.

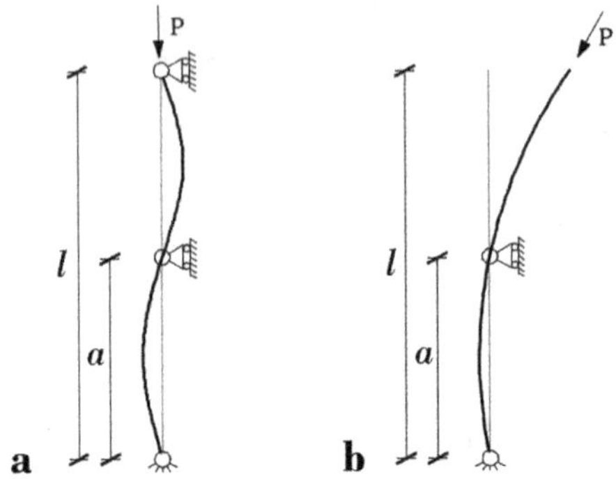

Fig. 4: (a) Auxiliary structure, (b) real beam without lumped mass.

Conclusions

Within the second order theory, a static method was detected to determine the critical load due to divergence at infinity in systems subjected to follower loads where the mass is modelled just as a lumped mass. The critical load is determined statically by taking the minimum value between the Eulerian critical load of an auxiliary structure and the Eulerian critical load of the given structure.

References

[1] A. Feriani, A. Carini, Some considerations on dynamic stability, in A. Zingoni (Ed.), Insights and Innovations in Structural Engineering, Mechanics and Computation, Proceedings of Sixth International Conference on Structural Engineering, Mechanics and Computation, Cape Town (South Africa), 2016-09-05 / 2016-09-07 (2016).

[2] Y. Panovko, I. Gubanova, Stability and oscillations of elastic systems: Models, paradoxes and errors, second ed., "Nunka" Press, Moscow, 1967.

[3] C. Felippa, Nonlinear Finite Element Methods. Department of Aerospace Engineering Sciences, University of Colorado at Boulder, Spring 2014, www.colorado.edu.

[4] L.M. Zorii, Yu. A. Chernukha, Influence of supports on the dynamic stability of elastic column, Prikl. Makh. 7(2) (1971) 134-136 (in Russian). https://doi.org/10.1007/BF00885991

[5] I. Elishakoff, J. Hollkamp, Computerized symbolic solution for a nonconservative system in which instability occurs by flutter in one range of a parameter and by divergence in another, Computer Methods in Applied Mechanics and Engineering 62 (1987) 27-46. https://doi.org/10.1016/0045-7825(87)90088-0

Theoretical and Applied Mechanics - AIMETA 2022 Materials Research Forum LLC
Materials Research Proceedings **26** (2023) 399-404 https://doi.org/10.21741/9781644902431-65

An inhomogeneous inelastic beam-like model for dynamic analyses of multistorey buildings

FIORE Ilaria[1, a*], GRECO Annalisa[1,b], CADDEMI Salvatore[1,c] and CALIO' Ivo[1,d]

[1] Department of Civil Engineering and Architecture, University of Catania, via Santa Sofia 64, 95123 Catania, Italy

[a] ilaria.fiore@phd.unict.it, [b] annalisa.greco@unict.it, [c] salvatore.caddemi@unict.it, [d] ivo.calio@unict.it

Keywords: Inelastic Dynamic Behaviour; Beam-Like Model; Shear Deformability

Abstract. This study presents an original inelastic beam-like model for the simulation of the nonlinear dynamic behaviour of three-dimensional buildings subjected to earthquake loadings. The proposed model is defined as a step-wise shear only cantilever beam, whose segments have the same mechanical properties of the corresponding inter-storeys of the building. The inter-storey constitutive laws are calibrated by means of the results of a static nonlinear analysis performed on the 3D FEM model of the building. An inverse nonlinear static identification procedure allows obtaining an inelastic beam-like model equivalent to the more demanding 3D FEM model, related to a specific direction and under a precise distribution of horizontal loadings. Dynamic nonlinear analyses performed on this simplified model allow a drastic reduction of the required computational effort and time with respect to the more demanding 3D FEM model. The reliability of the proposed beam-like model is validated through numerical applications on an irregular multi-storey RC frame representative of residential buildings in Catania designed to resist only gravity loadings.

Introduction

The simulation of the seismic response of multi-storey buildings is nowadays mostly performed by means of 3D FEM models and requires a relevant computational effort. In some contexts, such as preliminary design or seismic vulnerability assessment at urban scale, simplified models are anyway more desirable since they allow reducing the required computational burden.

Several MDOF simplified models have been presented in the scientific literature for the evaluation of both linear and nonlinear dynamic response of multi-storey buildings and particular interest has been devoted to beam-like equivalent models. Most of these studies deal with linear-elastic models [1, 2, 3, 4, 5, 6, 7].

Since the response of buildings subjected to seismic loads is predominantly inelastic, some researchers have proposed simplified inelastic equivalent MDOF models. In the context of beam-like models Kuang and Huang [8] modelled a wall-frame structure with uniform stiffness as an equivalent continuum system consisting of a combination of a flexural cantilever and a shear cantilever. In the latter case, a bilinear hysteretic model is used for the material properties of flexural and shear cantilevers. Ragni et al. [9] proposed a displacement-based method, particularly devoted to the seismic design of steel frames equipped with dissipative braces, by using an equivalent continuous beam-like model where flexural deformability and shear deformability are related to columns and diagonals of the bracing system respectively.

In the present paper an original inelastic beam-like model endowed with shear deformability only is presented. A cantilever beam with along axis multi-step discontinuous variations of the cross section, each uniform beam segment representing a building inter-storey, is considered. The calibration of the nonlinear shear-displacement relationship of each uniform beam is obtained by means of a pushover analysis performed, along a considered direction, on a FEM model of the analysed building. The displacement time history at each storey, as well as the inter-storey drifts, can be then obtained by means of the equivalent beam model. The proposed beam-like model has a number of degrees of freedom equal to the number of floors and therefore allows a drastic reduction of the computational burden with respect to the full FEM model. Therefore, it could be very useful in the seismic vulnerability assessments procedures that require several nonlinear dynamic analyses such as those expressed in terms of probability of failure by means of fragility curves. The numerical applications here reported refer to a reinforced concrete case study and show the comparison between the response time histories to an assigned seismic accelerogram of both a nonlinear FEM model and the proposed equivalent beam allowing highlighting the good accuracy of the described procedure.

The proposed inelastic beam-like model

In this section a new shear-only cantilever beam-like model, suitable to simulate the seismic inelastic response of a 3D framed building, is presented. The beam properties allow representing the inhomogeneous and the inelastic properties of the building along its height, assumed variable from one floor to the other. The mass of the beam-like model is assumed to be consistent with the actual mass distribution of the 3D structural model, and can be concentrated at the floor levels as well as, when required, partially distributed along the height.

Nonlinear constitutive laws are assumed to represent the inter-storey inelastic behaviour of the building. The beam-like model is calibrated considering a pushover analysis on the full 3D FEM model to reproduce the nonlinear behaviour of the three-dimensional building along a specific loading direction including torsional effects. It is worth highlighting that the inelastic beam model will exhibit a constitutive law that rigorously depends not only on the load direction but also on the adopted distribution of forces. Different load distributions have been tested in the numerical applications but only one of them has been reported in the following for the sake of brevity.

The beam element is divided into a number N_f sub-beam shear deformable elements whose length h_i is equal to the inter-storey height. The characterisation of the inelastic response of each i-th uniform beam sub-element can be defined by an appropriate uniaxial inelastic constitutive law, in terms of shear force T_i and inter-storey drift Δs_i. Fig. 1.a and 1.b summarise the equivalent shear beam-like model approach described so far.

The equivalence between the considered building and the corresponding beam-like model is enforced by calibrating the latter to predict the same pushover curve obtained by performing nonlinear static analysis on the 3D FEM model in a specific direction and under a precise distribution of horizontal loadings. Therefore, the tangent shear stiffness of each inter-storey beam segment (which provides the same inter-storey displacements obtained by the 3D FEM model) is calculated according to a step-by-step procedure. In particular, the inter-storey shear force is equal to the sum of the shear forces of the structural vertical elements (columns and walls) of the considered inter-storey, while the displacement is equal to the difference between the mean values of the displacements of the nodes situated at the top and at the bottom of the considered inter-storey in the direction of loading. These values, collected for each step of the pushover analysis, allow drawing the nonlinear "inter-storey capacity curve" and relate the shear force to the corresponding relative displacement.

Materials Research Forum LLC
https://doi.org/10.21741/9781644902431-65

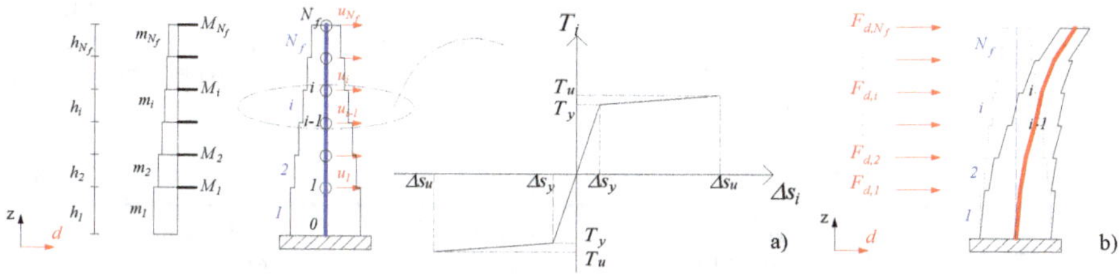

Figure 1 - a) Discretization of the equivalent beam-like model with nonlinear constitutive law of the i-th sub-element and b) the deflection of the equivalent beam under the considered load distribution

In this study the inter-storey constitutive laws have been defined according to bilinear energy-equivalent elasto-plastic models. In the numerical applications, reported in the next section, elasto-plastic behaviour with kinematic hardening has been adopted.

It is worth noting that the equivalent multi-degree-of-freedom beam-like model, as reported in Fig. 1, leads to the evaluation of all the floor displacements allowing modelling global as well as partial failure mechanisms. In Fig. 1, in each *i*-th inter-storey m_i denotes the distributed mass, M_i the lumped mass, T_y and T_u the yielding and ultimate shear force, respectively, and Δs_y and Δs_u the yielding and ultimate inter-storey drifts, respectively.

The results reported in the following section show how the proposed inelastic beam-like model can be able to reproduce the nonlinear static and dynamic behaviour of an entire building with sufficient accuracy drastically reducing the required computational time.

Numerical application

In this section the proposed inelastic beam-like model is applied to a multi-storey RC frame representative of residential buildings designed to resist only gravity loads. A detailed description of the building can be found in [10].

After being opportunely calibrated, as described in the previous paragraph, the proposed multi-stepped beam was adopted for simulating the static and dynamic inelastic behaviour of the building and the results, in terms of static and seismic response, were compared to those obtained by means of the conventional 3D FEM model performed using the software SAP2000 v.23.

The capacity curves depend on the applied distribution of forces, therefore in [10] different loading patterns have been taken into account in order to investigate the corresponding differences in predicting the inelastic responses. In the following, for the sake of brevity, only the force distribution associated with the fundamental natural mode is used.

The forces in the FEM and in the beam-like models are applied to the center of gravity (centroid) of each floor of the building, separately in the *x* and *y* directions. The values of the applied forces from the bottom to the top floor are assumed as follows: [0.1712 0.3654 0.5809 0.8087 1.0000 0.9502] in *x* direction; [0.1934 0.3828 0.5847 0.8114 1.0000 0.9466] in *y* direction.

The results of the nonlinear static analysis performed on the FEM model have been obtained in terms of inter-storey capacity curves in both *x* and *y* directions. The obtained inter-storey capacity curves have been transformed into equivalent bilinear elasto-plastic force-displacement laws with positive kinematic hardening. The equivalence has been obtained by imposing the equality between the areas below the nonlinear and bilinear capacity curves and assuming as initial inter-storey stiffness the tangent to the origin of the inter-storey capacity curve.

The stiffness values of each beam segment, assumed as initial stiffness in the successive nonlinear analyses, the inelastic limits and the hardening parameters for each beam segment are reported in Table 1. In Fig. 2 only the inter-storey inelastic equivalent force-displacement laws in x direction are reported, analogous curves have been obtained in y direction.

Once each inter-storey nonlinear behaviour has been defined, it is possible to perform nonlinear pushover analyses by making use of the beam-like model.

Table 1 - Initial stiffness, inelastic limits and hardening parameters for each beam segment

Beam segment	Height h [m]	Initial stiffness R [N/m] · 10^8		Yielding force Fy [N] · 10^5		Post yielding stiffness R_T [N/m] · 10^6	
		x	y	x	y	x	y
1	4.30	2.532	2.664	12.907	11.748	2.757	8.067
2	3.30	2.033	2.286	11.569	9.820	4.390	10.653
3	3.30	1.646	1.853	10.253	8.705	3.796	8.708
4	3.30	1.263	1.358	8.727	7.565	2.658	7.003
5	3.30	1.086	1.160	6.550	5.906	1.372	5.181
6	3.30	1.008	1.084	2.953	2.679	21.823	21.300

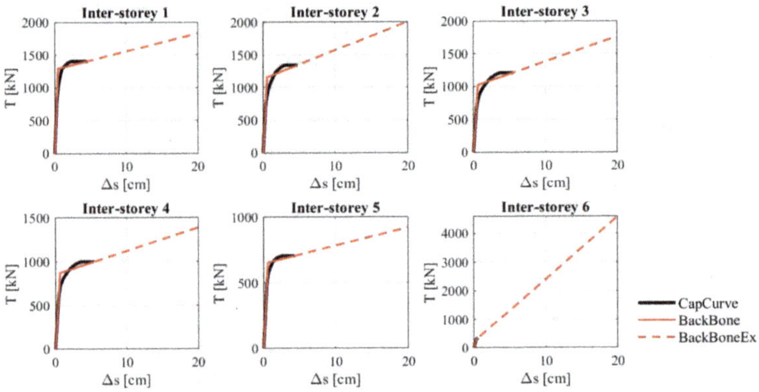

Figure 2 - Inter-storey capacity curves (CapCurve in black), the corresponding Back-Bone curves (red) and the extended Back-Bone without limits of the ductile behaviour (BackBoneEx dashed in red) in x direction

The nonlinear pushover analyses of the equivalent beam-like model consider the same horizontal load distribution adopted for the FEM model but refer to simplified inter-storey constitutive laws. The forces are applied on the beam axis at the floor level and are proportionally increased until conventional values of 0.25 m or 0.2 m top displacements are obtained, respectively in x and y directions.

For each considered direction, the capacity curves representative of the global behaviour of the building have been retrieved by considering the above-described beam-like models and expressed in terms of base shear force F and top floor displacement U. In Fig. 3.a-b the comparisons between the global capacity curves, obtained by means of the FEM and the beam-like models in x and y directions, are reported showing a satisfactory agreement.

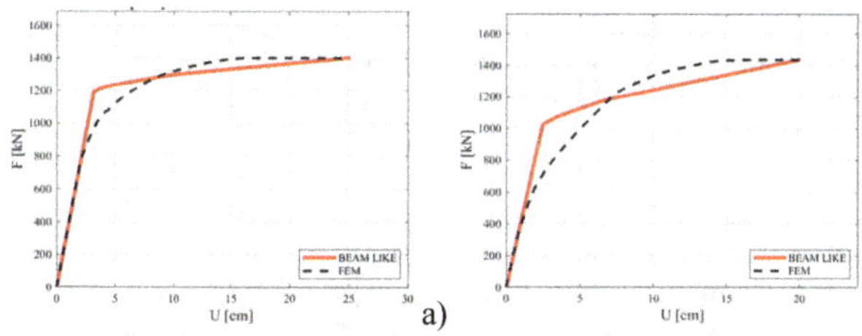

a) b)

Figure 3 - Capacity curves in a) x and b) y direction

In the following the beam-like model is adopted to predict the nonlinear dynamic response of the considered building by means of nonlinear dynamic analyses. Aiming to simulate representative seismic inputs that may occur on the Italian peninsula, nonlinear dynamic analyses were performed by considering the real seismic records in the x and y directions that occurred in L'Aquila (2009).

The reliability of the nonlinear dynamic response of the beam-like model was evaluated by comparing the time histories of the displacements of the centre of gravity at each floor in the considered direction. In the FEM model, the displacements of the centre of gravity have been calculated as the mean values of the four corner nodes of the same floor. Fig. 4 reports the time histories, expressed in terms of floor level displacements U_i with i=1, …,6, in x and y directions compared to the results of the FEM model, for L'Aquila earthquake.

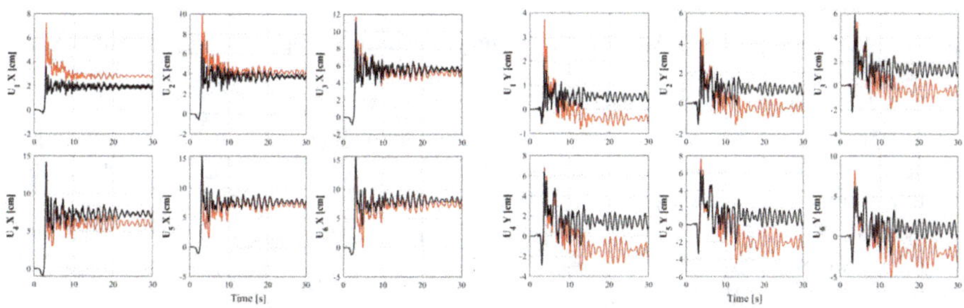

a) b)

Figure 4 - Dynamic response to L'Aquila earthquake in a) x and b) y directions obtained by means of the beam-like (red curve) and FEM (black curve) models

Due to the planar irregularity of the building the results are slightly different in x and y direction, anyway it can be seen that the proposed inelastic beam model is able to reproduce the nonlinear dynamic behaviour of the entire building with a good accuracy. The proposed beam-like model allows reducing drastically the computational burden, in fact, the computational time required by the beam-like model for each dynamic nonlinear analysis is about 60 times lower than the one required by a FEM model.

Conclusions

An original inhomogeneous and inelastic beam-like model suitable for the evaluation of the nonlinear dynamic response of multi-storey buildings subjected to seismic excitations is here presented. The proposed model is conceived to be representative of the nonlinear behaviour of the building when subjected to a prescribed distribution of horizontal forces in a specific direction. The beam-like model is calibrated on the results of pushover analyses performed on a FEM model of the entire building when subjected to an assigned horizontal load distribution. The equivalent beam-like model allows reproducing the nonlinear dynamic response of the building with a drastic reduction of the computational burden compared to the full FEM model. The model has been validated with reference to a 6-storey reinforced concrete structure, representative of residential buildings designed to resist only gravity loads. The results obtained for the beam-like model, in terms of response time histories to a recorded earthquake, have been compared to the corresponding full FEM model, showing a very good agreement. Considering its drastically reduced computation effort, the use of the presented beam-like model could be particularly advantageous for simplified seismic vulnerability approaches at urban scale and in the seismic assessment analyses expressed in terms of statistical probability of failure such as the fragility curves.

Acknowledgements
This research was funded by the University of Catania, with the project "PIA.CE.RI 2020/2022" and by the Italian Ministry of University and Research (MUR) with the project PRIN2020 #20209F3A37.

References

[1] E. Miranda, "Approximate seismic lateral deformation demands in multistory buildings.," *Journal of Structural Engineering ASCE,* pp. 417-425, 1999. https://doi.org/10.1061/(ASCE)0733-9445(1999)125:4(417)

[2] K. Zalka, "Maximum deflection of asymmetric wall-frame buildings under horizontal load," *Periodica Polytechnica: Civil Engineering,* pp. 387-396, 2014. https://doi.org/10.3311/PPci.7084

[3] G. Potza and L. Kollar, "Analysis of building structures by replacement sandwich beams," *International Journal of Solids and Structures,* pp. 535-553, 2003. https://doi.org/10.1016/S0020-7683(02)00622-4

[4] R. Rahgozar, A. Ahmadi and Y. Sharifi, "A simple mathematical model for approximate analysis of tall buildings," *Applied Mathematical Modelling,* pp. 2437-2451, 2010. https://doi.org/10.1016/j.apm.2009.11.009

[5] M. Malekinejad and R. Rahgozar, "An analytical model for dynamic response analysis of tubular tall buildings," *The Structural Design of Tall and Special Buildings,* pp. 67-80, 2014. https://doi.org/10.1002/tal.1039

[6] G. Piccardo, F. Tubino and A. Luongo, "Equivalent Timoshenko linear beam model for the static and dynamic analysis of tower buildings," *Applied Mathematical Modelling,* vol. 71, pp. 77-95, 2019. https://doi.org/10.1016/j.apm.2019.02.005

[7] A. Greco, I. Fiore, G. Occhipinti, S. Caddemi, D. Spina and I. Caliò, "An Equivalent Non-Uniform Beam-Like Model for Dynamic Analysis of Multi-Storey Irregular Buildings," *Applied Sciences,* vol. 10, p. 3212, 2020. https://doi.org/10.3390/app10093212

[8] J. S. Kuang and K. Huang, "Simplified multi-degree-of-freedom model for estimation of seismic response of regular wall-frame structures," *The structural design of tall and special buildings,* vol. 20, pp. 418-432, 2011. https://doi.org/10.1002/tal.538

[9] L. Ragni, A. Zona and A. Dall'Asta, "Analytical expressions for preliminary design of dissipative bracing systems in steel frames," *Journl of Constructional Steel Frames* , vol. 67, no. 102-113, 2011. https://doi.org/10.1016/j.jcsr.2010.07.006

[10] I. Fiore, *An equivalent non-uniform inelastic beam-like model for the seismic vulnerability assessment of buildings,* Catania: PhD thesis, University of Catania, 2022.

Theoretical and Applied Mechanics - AIMETA 2022 Materials Research Forum LLC
Materials Research Proceedings **26** (2023) 405-410 https://doi.org/10.21741/9781644902431-66

Using multiple singular values in topology optimization of dynamic systems

Paolo Venini

Department of Civil Engineering and Architecture, Via Ferrata 3, 27100 Pavia, ITALY

paolo.venini@unipv.it

Keywords: Topology Optimization, Dynamic Systems, Matrix Norms, SVD

Abstract. Within the general framework of frequency-domain topology optimization of Multi Input-Multi Output (MI-MO) dynamic systems, suitable norms of the input/output transfer matrix are introduced as possible merit functions to be minimized. Among them, the by now classical H_∞-norm (i.e. the supremum of the maximum singular value over the whole frequency range), and the so-called *nuclear norm* (i.e. the sum of all the positive singular values are considered. Heuristic motivations are given that suggest which norm should one choose according to the practical objective to be pursued alongside a few numerical examples on topology optimization of 2D linear-elastic multiload SI-SO and MI-MO dynamic systems.

Introduction and Motivation

Topology optimization of dynamical systems presents a few peculiarities among which the most important ones may be listed as follows:

- Time versus frequency domain approaches. This preliminary choice gives rise to two different families of solving approaches that do not share much but, possibly but not certainly, the final result. As a matter of fact, time domain approaches lead to optimal designs that depend strongly on the specific choice of the time-dependent loading functions [1]. On the other side, frequency domain methods work on the frequency response function that is expected to enjoy a few desirable features, at least in the frequency range of interest.
- Focusing the attention on frequency-domain approaches, [2] minimizes the dynamic compliance of the system via an incremental frequency approach that operates at low or high value of the excitation frequency whereas [3] sets the problem as a minimum H_∞-norm of the frequency response function, in a sense broadening to open-loop systems the well-established H_∞-norm-based active control strategy in a closed-loop feedback framework.

From a practical point of view, (for SI-SO systems) the H_∞-norm of the frequency response function is the peak of the function itself and is therefore crystal clear the motivation behind the adoption of such a design approach. However, looking at the methodology from a more algebraic perspective is likely to shed new light onto the method itself and opens the way to a few potentially useful extensions that are in fact the object of this contribution. The idea is then to define a novel goal function that depends on a few singular values (and not only on the first one as is the case when the minimization of the H_∞-norm is pursued). *Mutatis mutandis*, there are similarities with those eigenvalue optimization strategies that, at least to overcome the singularity of the min-max eigenvalue problem, introduce a goal function that depends on a few eigenvalues, see e.g. [4]. By so doing, eigenvalue crossing in the design space is no longer an issue, loss of regularity does not show up and standard gradient-based approaches are shown to work properly.

Having in mind a MI-MO rectangular frequency-response matrix, the usefulness of its Singular Value Decomposition (SVD) for a full understanding of the dynamic features of the system is highlighted next. The physical meaning of the singular values and associated left and right singular vectors is described along with the algebraic relation between singular values and H_∞-norm.

Reference is made to [5, 6] for a comprehensive exposition of such concepts from an algebraic as well computational point of view.

The Singular Value Decomposition
To start off, a formal definition of SVD of a rectangular, possibly non-symmetric, matrix is given next.

Singular Value Decomposition. Let G be an m by n (possibly complex valued) matrix. Two sets of singular vectors exist such that:

- n right singular vectors $v_1, \dots, v_n$ are orthogonal to each other in $\mathbb{R}^n$;
- m left singular vectors $u_1, \dots, u_m$ are orthogonal to each other in $\mathbb{R}^m$;
- left and right singular vectors are connected by an "eigen-like" relation $Av = \sigma u$ that may be written component wise as

$$Gv_1 = \sigma u_1, \dots, Gv_r = \sigma_r; \quad Gv_{r+1} = 0, \dots, Gv_n = 0, \tag{1}$$

where r is the *rank* of G and one may show that there exist r non-negative singular values that are usually cast in descending order: $\sigma_1 \leq \sigma_2 \leq \cdots \leq \sigma_r$.

The singular value decomposition of G finally reads:

$$A = U\Sigma V^T. \tag{2}$$

More explicitly, the matrix G may be written as a finite sum of rank-1 matrices as:

$$G = U\Sigma V^T = \sigma_1 u_1 v_1^T + \cdots + \sigma_r u_r v_r^T. \tag{3}$$

The following two propositions show a few reasons why the first singular values (and not only the largest one) are worth being investigated.

Theorem of Eckart-Young. Let $G_k = \sigma_1 u_1 v_1^T + \cdots \sigma_k u_k v_k^T$ be the *rank-k* SVD approximant to G. Then G_k is the overall best *rank-k* approximant to G, i.e.

$$||G - G_k|| \leq ||G - B|| \quad \forall B \text{ with rank } k. \tag{4}$$

On the properties and computation of the second-largest and further singular values. The solution of the Problem:

$$\max_x \frac{||Gx||}{||x||} \text{ under the condition } v_1^T x = 0, \tag{5}$$

is σ_2 at $x = v_2$.

Matrix Norms, Singular Values and H_∞-Norm of a Transfer Function Matrix
Let p be the design variable vector (that contains the element densities in a topology optimization framework) and $G(i\omega, p)$ the frequency response matrix function. For the sake of introducing a few key concepts in algebra of non-square and non-symmetric matrices, it is convenient to fix both the frequency ω and the design variable vector p, say $\omega = \omega^*$ and $p = p^*$, so that $G^* = G(i\omega^*, p^*)$ is any complex-valued rectangular matrix.

The 2-norm of a matrix G^*. The *"largest growth factor"* concept [5] appears to be the most natural to introduce the norm of a transfer function matrix that governs the input/output relation of a dynamical systems. In a more analytic format we may also refer to matrix norms induced by underlying (and previously defined) vector norms. One may write

$$||G^*|| = \sup_{v\neq 0}\frac{||G^*v||}{||v||}, \tag{6}$$

where each vector norm at the right hand side induces a corresponding matrix norm at the left hand side. If the 2-norm is used, one may show that

$$||G^*||_2 = \sigma_1, \tag{7}$$

i.e. the 2-norm of a matrix is its largest singular value. If now the dependence of G^* on the frequency ω is recovered, one may quickly realize that the H_∞-norm of a frequency-response matrix (that from an engineering point of view is the maximum amplification factor over the whole frequency axis, i.e. the *"largest growth factor"*) is defined as the supremum of the first singular value of $G(i\omega, p^*)$ with respect to the whole frequency axis, i.e.

$$||G(i\omega, p^*)||_\infty = \sup_{\omega\in(0,\infty)} \sigma_1(\omega). \tag{8}$$

It is interesting to note that the H_∞-norm of a matrix transfer function is in fact a 2-norm maximized over the frequency axis. From a numerical point of view, computing the H_∞-norm is quite a hard task as is the evaluation of the frequency at which the norm itself is attained. Alongside the H_∞-norm that depends exclusively on the largest (first) singular value, two more norms are theoretically suitable as goal functions when optimizing the topology of dynamical systems, i.e. the Frobenius norm and the Nuclear (the one adopted herein) norm that are respectively defined as:

$$||G||_F = \sqrt{\sigma_1^2 + \sigma_2^2 + \cdots + \sigma_r^2}, \quad ||G||_N = \sigma_1 + \sigma_2 + \cdots + \sigma_r. \tag{9}$$

Numerical Studies

Input data. As for the geometry of the problems investigated herein reference is made to Fig. 1 that also shows loads and constraints. By now standard aspects of topology optimization such as the SIMP idealization to handle intermediate materials, the nonlocal filters that are adopted to avoid checkerboarding, the finite elements that are used to derive the discrete version of the problem, not to mention the numerical scheme that is used to solve the optimization problem may be found in [7] among others and are not explicitly described herein.

Materials Research Forum LLC
https://doi.org/10.21741/9781644902431-66

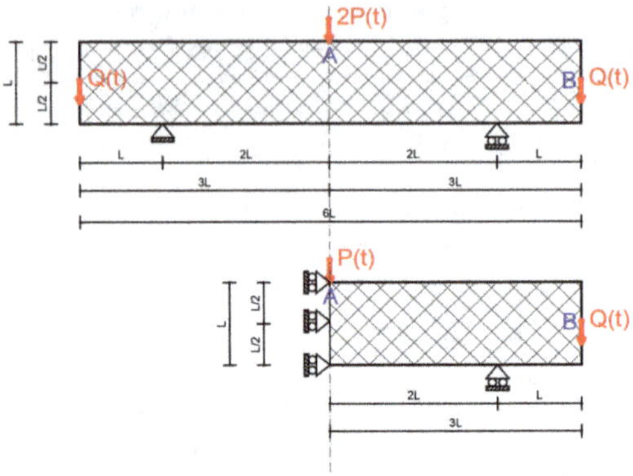

Figure 1 – The two-loaded beam: overall representation and symmetric problem

The problems under investigation. Three different optimization problems are considered and solved that are best defined referring to the so-called *descriptor state-space* formulation that reads:

$$\begin{cases} E\dot{x} = Ax + B\,u \\ y = Cx + D\,u \end{cases}, \tag{10}$$

where Eq. 10.1 is in fact the equation of motion in which x is the state vector that piles nodal displacements and velocities, u is the load vector, E is the *descriptor* (mass) matrix. A is the structural matrix that depends on the stiffness and damping matrices, and B is a topological matrix that distributes the loads to the degrees-of-freedom. Equation 10.2 is classically referred to as output equation and in fact y is the output vector that encompasses all the quantities that the optimization process should explicitly take care of (i.e. minimize). What is actually minimized is a suitable norm of the transfer-function matrix $G(s)$ that defines the Laplace-domain input/output relation $Y(s) = G(s)U(s)$ that may be shown to be equivalent to Eq. 10 and is graphically interpreted by the block in Fig. 2.

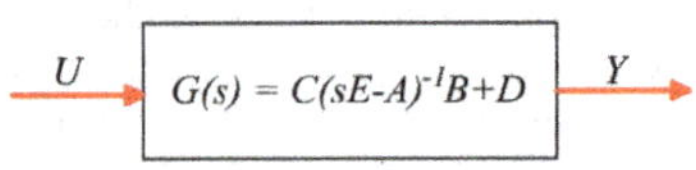

Figure 2 – Input-output Laplace-domain version of the system dynamics.

Given the system dynamics of Eq. 10.1, the specific optimization problem is fully defined by:

- The choice of the output vector y through a proper selection of the matrices C and D. At this regard both single-output and multi-output systems may inherently be handled by the proposed formulation. It should be noted that both single-output and multi-output cases give rise to a scalar minimization problem since a suitable norm of the resulting transfer function is the actual goal function that is minimized. Alternative approaches of vector minimization in a Pareto framework are left to future investigations.
- The selection of a proper matrix norm that is expected to address from a system-theoretic point of view the engineering goals that are expected to be reached by the designer.

Given this general scenario, the following optimization problems are considered.

1. Dynamic compliance H_∞-*norm* minimization (SI-SO). The two loads $P(t)$ and $Q(t)$ are supposed to belong to the same load combination (Single-Input), whereas the output is the so-called dynamic compliance [3];
2. The two loads $P(t)$, $Q(t)$ and the dual displacements $u_v(A)$, $u_v(B)$ define a 2-inputs, 2-outputs (MI-MO) transfer function of which the H_∞-*norm* is minimized;
3. Same as previous Problem 2 (MI-MO) but for the choice of the system norm to be minimized. A weighted version of the nuclear norm is chosen, i.e. $a_1\sigma_1 + a_2\sigma_2$.

Main results are briefly showcased next.

Problem 1: dynamic compliance H_∞-norm minimization. Figure 3 shows the optimal topology that solves the classical dynamic compliance problem along with the maximum-singular-value vs frequency curve (that for SI-SO systems is the same as the amplitude of the frequency response). The value of the goal function at convergence is 7.36 dB that is attained for $\omega = 0.28$ rad/s.

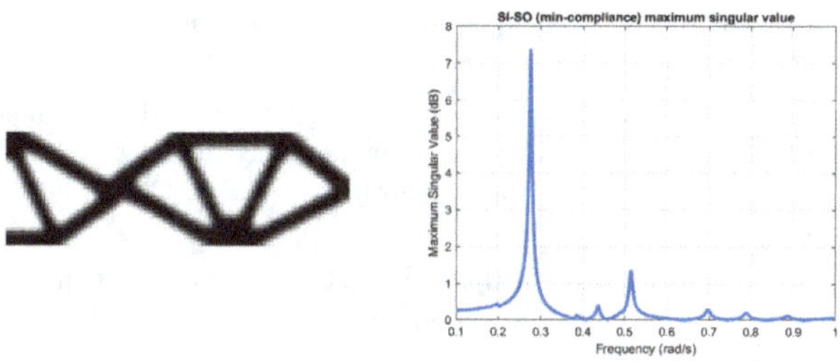

Figure 3 – Design case 1 – Optimal topology and maximum singular value vs frequency

Problem 2: H_∞-norm minimization of the 2×2 transfer function. Figure 4 shows the optimal topology that solves the MI-MO min H_∞-norm problem along with the maximum-singular-value vs frequency curve. The value of the goal function at convergence is 15.52 dB that is attained for $\omega = 0.048$ rad/s.

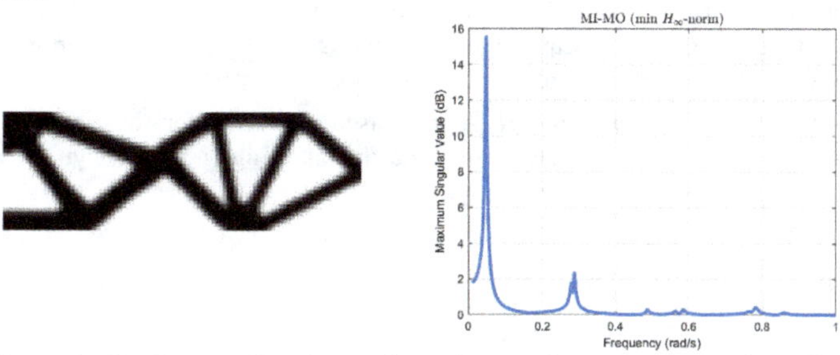

Figure 4 – Design case 2 – Optimal topology and maximum singular value

Problem 3: Nuclear-norm minimization of the 2×2 transfer function. Figure 5 shows the optimal topology that solves the MI-MO min $nuclear$ –norm problem along with the maximum-singular-value vs frequency curve. The value of the goal function at convergence is 53.87 dB that is attained for $\omega = 0.028$ rad/s.

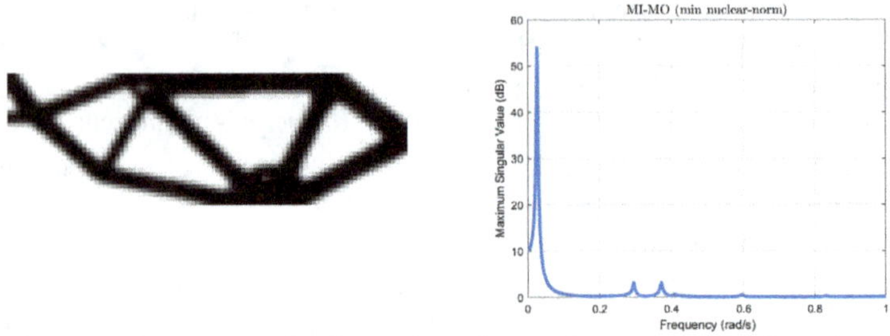

Figure 5 – Design case 3 – Optimal topology and maximum singular value

References

[1] J.S. Jensen, P.B. Nakshatrala, D.A. Tortorelli, On the consistency of adjoint sensitivity analysis for structural optimization of linear dynamic problems, Struct. Multidisc. Optim. 49 (2014) 831-837. https://doi.org/10.1007/s00158-013-1024-4

[2] N. Olhoff, J. Du, Generalized incremental frequency method for topological design of continuum structures for minimum dynamic compliance subject to forced vibration at a prescribed low or high value of the excitation frequency, Struct. Multidisc. Optim. 54 (2016) 1113-1141. https://doi.org/10.1007/s00158-016-1574-3

[3] P. Venini, Topology optimization of dynamic systems uncer uncertain loads: an H_∞-norm-based approach, J. Comput. Nonlinear Dynam. 14 (2019) 021007. https://doi.org/10.1115/1.4042140

[4] J. Du, N. Olhoff, Topological design of freely vibrating continuum structures for maximum values of simple and multiple eigenfrequencies and frequency gaps, Struct. Multidisc. Optim. 34 (2007) 91-110. https://doi.org/10.1007/s00158-007-0101-y

[5] G. Strang, Linear algebra and learning from data, Wellesley – Cambridge Press, Wellesley, 2019.

[6] S.L. Brunton, J.N. Kutz, Data-driven science and engineering – Machine learning, dynamical systems and control, Cambrige University Press, Cambridge, 2019. https://doi.org/10.1017/9781108380690

[7] P. Venini, P. Ceresa, A rational H_∞-norm-based approach for the optimal design of of sesmically excited reinforced concrete frames, Earthquake Engng Struct. Dyn. 47 (2018) 1522-1543. https://doi.org/10.1002/eqe.3028

Theoretical and Applied Mechanics - AIMETA 2022 | Materials Research Forum LLC
Materials Research Proceedings **26** (2023) 411-416 | https://doi.org/10.21741/9781644902431-67

The dynamics of circular arches with multiple damage

CANNIZZARO Francesco[1,a *], FIORE Ilaria[1,b], GRECO Annalisa[1,c], CADDEMI Salvatore[1,d] and CALIO' Ivo[1,e]

[1] Department of Civil Engineering and Architecture, University of Catania, via Santa Sofia 64, 95123 Catania, Italy

[a] francesco.cannizzaro@unict.it, [b] ilaria.fiore@phd.unict.it, [c] annalisa.greco@unict.it, [d] salvatore.caddemi@unict.it, [e] ivo.calio@unict.it

Keywords: Closed form Solution, Circular Cracked Arch, Vibration Analysis

Abstract. The governing equation of the dynamics of the planar inextensible Euler-Bernoulli arch with multiple damage is tackled in this study by employing the distributional approach. Precisely, the presence of impairments is modelled via cracks that can be effectively embedded in the governing equation by means of the Dirac's delta generalised functions. The governing equation is defined over a unique integration domain. The proposed integration strategy leads to closed form expressions of the displacement mode shapes maintaining the size of the problem as that of the undamaged arch regardless the number of cracks located along the span. The latter advantage avoids the enforcement of continuity conditions at the discontinuous sections. The proposed solution extends the integration procedure proposed in the static context to the vibration analysis and allows determining the modal characteristics of damaged circular arches. The versatility of the obtained closed form solution allows a straightforward execution of parametric analyses and is here adopted to evaluate the sensitivity of the eigenproperties of the multi-cracked circular arch to the change of meaningful geometric and mechanical parameters.

Introduction

The static and dynamic behaviour of the arches was studied by the researchers both in the linear and nonlinear context by means of different approaches over the years, such as limit analysis, Finite Element analysis and the Discrete Macro-Element Model. Arches may present circular shape [1] or a variable curvature [2, 3], and are often studied considering only the flexural deformability, neglecting both shear and axial effects. With regard to damaged arches, the presence of impairments is usually tackled with concentrated models [4]. The concentrated damage can be modelled according to various strategies, and in this paper the well-known equivalent spring model [5] is adopted. This leads to the insertion of a rotational spring with constant stiffness in each cracked section, allowing dealing with linear governing equations.

Many studies analysed the in-plane or out-of-plane vibrations of damaged circular arches [6, 7, 8] . The presence of concentrated cracks along the axis of straight or curved beams requires the subdivision of the element into uniform sub-elements connected by a rotational spring. Therefore, it is necessary to enforce continuity conditions or, alternatively, the assemblage of finite elements, thus increasing the computational burden. Alternatively, several authors proposed new approaches which require solely the enforcement of the external boundary conditions irrespectively of the number of internal cracks. These solutions are available for both the static and dynamic behaviour of multi-cracked straight beams [9, 10], whilst solutions for curved beams are limited to the static problem only [11].

In this paper, a free vibration six order differential governing equation of a multi-cracked circular inextensible Euler-Bernoulli arch is presented, and the associated closed form solution of the vibration modes is derived by means of the application of the Laplace transform. It is worth highlighting that the equation of motion is defined over a unique integration domain, even if in presence of multiple cracks, and the evaluation of the natural frequencies requires the enforcement of six boundary conditions only, as in the continuous arch case. The proposed solution is numerically validated by comparing the natural frequencies and the mode shapes of a multi-cracked circular arch with the results obtained by means of a finite element model in which the cracks are modeled by means of equivalent internal rotational springs. Some parametric studies referred to multi-cracked Euler arches with variable intensity or position of the concentrated crack are also presented and discussed.

Closed-form integration of the free vibration equation of the multi-cracked circular arch

The differential equation governing the in plane free vibrations of a Euler-Bernoulli circular arch with radius r in presence of n concentrated cracks, is presented in this section, and the relevant closed form solution in terms of mode shapes is obtained.

The integration domain is represented by $\vartheta \in [\vartheta_{o1}, \vartheta_{o2}]$, being ϑ a zenithal angular coordinate, and $\vartheta_{o1}, \vartheta_{o2}$ the left and right end cross sections, respectively. The kinematics of the arch involves the following components depending on the angular coordinate ϑ and the time t: the radial $u_r(\vartheta,t)$ and tangential $u_t(\vartheta,t)$ displacements, as well as the rotation of the centroidal axis $\varphi(\vartheta,t)$. For the case of inextensible Euler-Bernoulli arch model, the following kinematic constraints among the bending curvature $\chi(\vartheta,t)$, $\varphi(\vartheta,t)$, $u_r(\vartheta,t)$ and $u_t(\vartheta,t)$ hold:

$$\chi(\vartheta,t) = \frac{1}{r}\varphi^{I}(\vartheta,t) \quad , \quad u_t^{I}(\vartheta,t) = u_r(\vartheta,t) \quad , \quad \varphi(\vartheta,t) = -\frac{1}{r}\left[u_t(\vartheta,t) + u_t^{II}(\vartheta,t)\right] \tag{1}$$

where the Roman numbers superscripts indicate derivatives with respect to ϑ. Considering an Euler-Bernoulli circular arch characterized by n double-edge concentrated cracks, assumed to be always open, at cross sections $\vartheta_i, i = 1,...,n$, its free vibration motion is ruled by the following sixth order governing equation, expressed in terms of tangential displacement $u_t(\vartheta,t)$:

$$\left[\frac{\partial^6}{\partial\vartheta^6} + 2\frac{\partial^4}{\partial\vartheta^4} + \frac{\partial^2}{\partial\vartheta^2} + \frac{mr^4}{EJ_r}\frac{\partial^2}{\partial t^2}\left(\frac{\partial^2}{\partial\vartheta^2} - 1\right)\right]u_t(\vartheta,t) + r\sum_{i=1}^{n}\Delta\varphi_i(t)\left[\frac{\partial}{\partial\vartheta} + \frac{\partial^3}{\partial\vartheta^3}\right]\delta(\vartheta - \vartheta_i) = 0 \tag{2}$$

where the rotary inertia has been neglected in the dynamic equilibrium and denoting with m the distributed mass. The presence of n concentrated cracks implies rotation discontinuities $\Delta\varphi_i(t)$ at the cracked cross-sections $\vartheta_i, i = 1,...,n$. These singularities are taken into account in the summation in Eq. 2, characterized by sequences of n occurrences of first and third derivatives of Dirac's deltas. This strategy allows preserving the definition of the governing differential equation over the entire domain $\vartheta \in [\vartheta_{o1}, \vartheta_{o2}]$ as already proposed in the static case [11]. The governing free vibration equation provided by Eq. 2 can be reformulated as follows:

$$\left[\frac{\partial^6}{\partial\vartheta^6} + 2\frac{\partial^4}{\partial\vartheta^4} + \frac{\partial^2}{\partial\vartheta^2} + \mu^4\left(1 - \frac{\partial^2}{\partial\vartheta^2}\right)\right]\phi_t(\vartheta) = \sum_{i=1}^{n}\lambda_i\left[\phi_t^{I}(\vartheta_i^-) + \phi_t^{III}(\vartheta_i^-)\right]\left(\frac{\partial}{\partial\vartheta} + \frac{\partial^3}{\partial\vartheta^3}\right)\delta(\vartheta - \vartheta_i) \tag{3}$$

where the frequency parameter $\mu^4 = \omega^2 mr^4 / EJ_r$ is introduced and the tangential displacement $u_t(\vartheta,t) = \phi_t(\vartheta)\sin\omega t$ is expressed as the product of two functions, namely $\phi_t(\vartheta)$, depending solely on the non-dimensional angular coordinate ϑ, and a harmonic function $\sin\omega t$, being ω the radial frequency. Analogously, the unknown rotation discontinuities $\Delta\varphi_i(t) = \Delta\phi_{\varphi,i}\sin\omega t$, appearing in Eq. 2, can be expressed in terms of the tangential displacement mode shape $\phi_t(\vartheta)$ as follows:

$$\Delta\phi_{\varphi,i} = -\frac{\lambda_i}{r}\left[\phi_t^I\left(\vartheta_i^-\right)+\phi_t^{III}\left(\vartheta_i^-\right)\right] \tag{4}$$

where $\lambda_i = EJ_r / r K_i^{eq}$ represents the dimensionless crack compliance related to the stiffness of the equivalent rotational spring.

The closed form solution of Eq. 3 can be retrieved by means of the application of Laplace transform, leading to the expression of the tangential displacement mode shape as follows:

$$\phi_t(\vartheta) = \sum_{k=1}^{6} C_k\, h_k(\vartheta) + \sum_{i=1}^{n} \lambda_i\left[\phi_t^I\left(\vartheta_i^-\right)+\phi_t^{III}\left(\vartheta_i^-\right)\right]\overline{h}_i(\vartheta) \tag{5}$$

where:

$$\begin{aligned} &h_k(\vartheta) = e^{\alpha_k \vartheta}, \quad k=1,\ldots,6;\\ &\overline{h}_i(\vartheta) = \left[\sum_{k=1}^{6}\frac{\alpha_k^3+\alpha_k}{6\alpha_k^5+8\alpha_k^3+2(1-\mu^4)\alpha_k}\, e^{\alpha_k(\vartheta-\vartheta_i)}\right]U(\vartheta-\vartheta_i), \quad i=1,\ldots,n \end{aligned} \tag{6}$$

$$\begin{aligned} C_k = \frac{1}{6\alpha_k^5+8\alpha_k^3+2(1-\mu^4)\alpha_k}\Big[&\left(\alpha_k^5+2\alpha_k^3+(1-\mu^4)\alpha_k\right)\phi_t(0)+\left(\alpha_k^4+2\alpha_k^2+1-\mu^4\right)\phi_t^I(0)+\\ &+\left(\alpha_k^3+2\alpha_k\right)\phi_t^{II}(0)+\left(\alpha_k^2+2\right)\phi_t^{III}(0)+\alpha_k\phi_t^{IV}(0)+\phi_t^V(0)\Big] \end{aligned} \tag{7}$$

being $\alpha_k, k=1,\ldots,6$ the roots of the sixth order polynomial $p(s;\mu^4) = s^6+2s^4+\left(1-\mu^4\right)s^2+\mu^4$ that can be inferred in closed form.

The solution, as it stands in Eq. 5, does not provide an explicit expression for the tangential displacement mode shape $\phi_t(\vartheta)$ since it depends on the values of its first and third distributional derivatives $\phi_t^I(\vartheta)$ and $\phi_t^{III}(\vartheta)$, evaluated at ϑ_i^-. The latter can be expressed in explicit form and substituted into Eq. 5, providing the following explicit expression of the tangential displacement mode shape function:

$$\phi_t(\vartheta) = \sum_{k=1}^{6} C_k f_k(\vartheta) \tag{8}$$

where the functions $f_k(\vartheta)$, $k=1,\ldots,6$, are defined by the following expressions:

$$f_k(\vartheta) = h_k(\vartheta) + \sum_{i=1}^{n}\lambda_i\left[g_{I,k}(\vartheta_i^-)+g_{III,k}(\vartheta_i^-)\right]\overline{h}_i(\vartheta) \tag{9}$$

and the terms $g_{D,k}(\vartheta_i^-)$, $D=0,I,\ldots,V$, $k=1,\ldots,6$, are defined by the following expressions:

$$g_{D,k}(\vartheta_i^-) = h_k^{(D)}(\vartheta_i^-) + \sum_{m=1}^{i-1}\lambda_m\left[g_{I,k}(\vartheta_m^-)+g_{III,k}(\vartheta_m^-)\right]\overline{h}_m^{(D)}(\vartheta_i^-). \tag{10}$$

The radial displacement and rotation mode shape functions can be easily obtained, in view of the solution provided in Eq. 8, by means of the kinematic constraints expressed by Eq. 1. Each of the two end cross-sections provides three boundary conditions depending on the kinematic and mechanical conditions of the restraint. It has to be pointed out that the proper integration constants are the values of the derivatives of ϕ_t at $\vartheta=0$ of the arch denoted as $\phi_t^{(D)}(0)$, D indicating the order of the derivative $D=0,I,\ldots,V$, appearing in Eq. 7. The latter are to be determined according to specified boundary conditions dependent on the external constraints acting on the circular arch getting rid of the dependency of the frequency parameter μ^4 appearing in Eq. 7.

Numerical applications

Double cracked arch (FEM comparison)

In this sub-section an application regarding a clamped-clamped double cracked full circular arch (the angular span of the arch is 180°) is reported, aiming at validating the proposed closed-form solution. The radius of the axis of the arch is $r = 2$ m, the cross section is rectangular with a base $b = 40$ mm and a height $h = 50$ mm. The mass density is equal to $\rho = 7850$ kg/m^3, whilst the Young's modulus is $E = 210000$ MPa. The two cracks are localised in two sections placed at $\vartheta_1 = -\pi/4$ and $\vartheta_2 = \pi/6$ with respect to the axis of symmetry of the arch, and with intensity parameters $\lambda_1 = 0.0289$ and $\lambda_2 = 0.0833$, respectively . The intensity parameters are related to crack depth/cross section height ratios $\beta_1 = 0.3$, $\beta_2 = 0.5$, according to the model proposed in [12]. The analytical natural frequencies and modes of vibration have been obtained by means of the proposed model and compared to the results obtained by means of a FEM model implemented in the software environment SAP2000 in accordance with the assumptions of infinite axial and shear stiffness and where the cracks have been modeled by considering an equivalent rotational spring for each cracked cross section. The first three natural frequencies for the damaged configuration of the arch are shown in Table 1, whilst the first three displacement mode shapes for the proposed (dashed black line) and the FEM (solid grey line) models have been reported in Fig. 1.

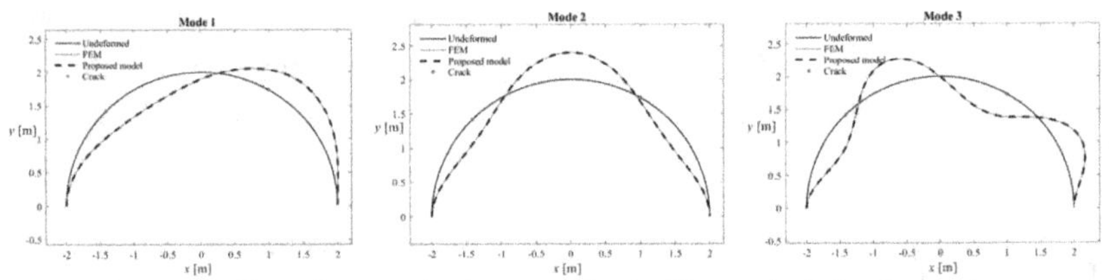

Figure 1 - First three mode shapes of the double cracked circular arch

Table 1 - Modal frequencies comparison

	Mode 1	Mode 2	Mode 3
Proposed model	12.7331 Hz	28.4654 Hz	52.4530 Hz
FEM model	12.7372 Hz	28.4786 Hz	52.4725 Hz

The agreement between the proposed and the FEM models, both in terms of frequency and mode shapes, is excellent. The subtle differences might be due to the discretized approach of the FEM model compared to the proposed continuous approach.

Parametric study

In this sub-section a parametric study is presented considering a clamped-clamped full circular arch with three cracks (at the two ends and at the midspan sections) with increasing intensities. The crack intensity parameters λ assume for all the cracks the following values: 0, 0.2, 0.4, 0.6, 0.8, 1, 2, 3, 4, 6, 8, 10. Furthermore, in view of the parametric nature of the study the cross section of the arch has not been specified, hence the crack intensity parameters can be related to specific crack depths in accordance with the case study at hand.

In Fig. 2 the variation of the frequency parameter μ^4 against the crack intensity parameter λ, for the first two vibration modes, is reported. In the same Figure, the frequency parameters μ^4 for a clamped-clamped and a three-hinged full circular arch are also reported for comparison.

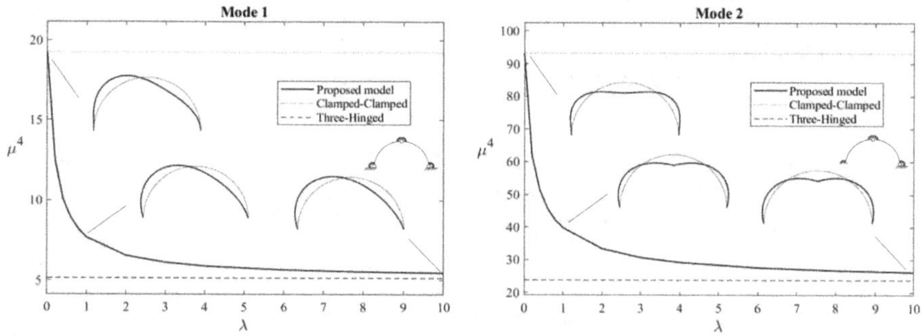

Figure 2 - Frequency parameter against the crack intensity parameter for the first and second mode of vibration

The results obtained in terms of frequency parameter show a monotonic trend as the value of intensity parameter increases, as expected. Furthermore, it can be observed that the frequency parameter is equal to the value of the clamped-clamped arch for $\lambda = 0$ and tends to the value of the three-hinged arch for higher values of the intensity parameter.

Conclusions

This work presented a free vibration six order differential governing equation of a multi-cracked circular inextensible Euler-Bernoulli arch and the associated closed form solution in terms of mode shapes. The proposed governing equation is defined over a unique integration domain regardless of the number of cracks, and the evaluation of the natural frequencies requires the enforcement of six boundary conditions only at the two ends of the arch, as in the continuous arch case, avoiding the introduction of continuity conditions at the cracked sections. The proposed procedure was validated by comparison with numerical solutions obtained with a classic FEM approach and the closed form solution was used for a representative parametric study.

Acknowledgements

This research was funded by the University of Catania with the project "PIA.CE.RI 2020/2022" and by the Italian Ministry of University and Research (MUR) with the project PRIN2020 #20209F3A37.

References

[1] P. Chidamparam and A. W. Leissa, "Vibration of planar curved beams, rings and arches," *Applied Mechanics Review,* vol. 46, no. 9, pp. 467-483, 1993. https://doi.org/10.1115/1.3120374

[2] A. Greco and A. Pau, "Detection of a concentrated damage in a parabolic arch by measured static displacements," *Structural Engineering and Mechanics,* vol. 39, no. 6, pp. 751-765, 2011. https://doi.org/10.12989/sem.2011.39.6.751

[3] U. Eroglu, G. Ruta and E. Tufekci, "Natural frequencies of parabolic arches with a single crack on opposite cross-section sides," *Journal of Vibration and Control,* vol. 25, no. 7, pp. 1313-1325, 2019. https://doi.org/10.1177/1077546319825681

[4] A. D. Dimarogonas, "Vibration of cracked structures: a state of the art review," *Engineering Fracture Mechanics,* vol. 55, no. 5, pp. 831-857, 1996. https://doi.org/10.1016/0013-7944(94)00175-8

[5] G. R. Irwin, "Relation of stresses near a crack to the crack extension force," in *9th Congress of Applied Mechanics*, Brussels, 1957.

[6] M. Krawczuk and W. Ostachowicz, "Natural vibrations of a clamped-clamped arch with an open transverse crack," *Journal of Vibration and Acoustics, Transactions of the ASME,* vol. 119, no. 2, pp. 145-150, 1997. https://doi.org/10.1115/1.2889695

[7] I. Caliò, D. D'Urso and A. Greco, "The influence of damage on the eigen-properties of Timoshenko spatial arches," *Computers & Structures,* vol. 190, pp. 13-24, 2017. https://doi.org/10.1016/j.compstruc.2017.04.012

[8] C. S. Huang, Y. P. Tseng, S. H. Chang and C. L. Hung, "Out-of-plane dynamic analysis of beams with arbitrarily varying curvature and cross-section by dynamic stiffness matrix method," *International Journal of Solids and Structures,* vol. 37, pp. 495-513, 2000. https://doi.org/10.1016/S0020-7683(99)00017-7

[9] S. Caddemi, I. Caliò and F. Cannizzaro, "Closed-form solutions for stepped Timoshenko beams with internal singularities and along-axis external supports," *Archive of Applied Mechanics,* vol. 83, no. 4, pp. 559-577, 2013. https://doi.org/10.1007/s00419-012-0704-7

[10] A. Palmeri and A. Cicirello, "Physically-based Dirac's delta functions in the static analysis of multi-cracked Euler-Bernoulli and Timoshenko beams," *International Journal of Solids and Structures,* vol. 48, no. 14-15, pp. 2184-2195, 2011. https://doi.org/10.1016/j.ijsolstr.2011.03.024

[11] F. Cannizzaro, A. Greco, S. Caddemi and I. Caliò, "Closed form solutions of a multi-cracked circular arch under static loads," *International Journal of Solids and Structures,* vol. 121, pp. 191-200, 2017. https://doi.org/10.1016/j.ijsolstr.2017.05.026

[12] C. Bilello, *Theoretical and experimental investigation on damaged beams under moving systems,* University of Palermo: PhD Thesis, 2001.

Theoretical and Applied Mechanics - AIMETA 2022 Materials Research Forum LLC
Materials Research Proceedings **26** (2023) 417-422 https://doi.org/10.21741/9781644902431-68

A reduced hysteretic model of stockbridge dampers

Francesco Bogani[1, a], Alex Sosio[1,b], Francesco Foti[1,c *]and Luca Martinelli[1,d]

[1]Department of Civil and Environmental Engineering (DICA), Politecnico di Milano, Milano, Italy

[a]francesco.bogani@mail.polimi.it, [b]alex.sosio@mail.polimi.it, [c]francesco.foti@polimi.it, [d]luca.martinelli@polimi.it

* corresponding author

Keywords: Dampers, Cable Dynamics, Vortex-Induced-Vibrations

Abstract. This paper presents a reduced model to describe the nonlinear dynamic behavior of Stockbridge dampers. The proposed model is based on the classic Bouc-Wen hysteretic law and requires the identification of a small number of model parameters. The proposed formulation is used within an extended version of the classic Energy Balance Method to assess the role of the damper nonlinearities in mitigating aeolian vibrations of a reference overhead transmission line.

Introduction

Conductors and guard wires in overhead transmission line (OHL) are prone to vortex-induced-vibrations, also known as aeolian vibrations in the electrical engineering community. Whenever not properly controlled, aeolian vibrations can induce wear damage and fatigue failures of the cables. Stockbridge dampers are often used to mitigate the severity of such vibrations. Stockbridge dampers are characterized by a markedly nonlinear dynamic response, which is related to the intrinsic properties of their components, namely the hysteretic bending behavior of messenger cables.

This paper presents a reduced model to describe the non-linear dynamic behavior of one of such dampers. The proposed model, which is based on the classic Bouc-Wen hysteretic law, is then used within an extended version of the classic Energy Balance Method (EBM) [1] to assess the role of the damper nonlinearities in mitigating aeolian vibrations of a transmission line.

The Stockbridge damper model

As it is schematically depicted in Fig. 1, Stockbridge dampers are made of a metallic clamp, two inertial bodies and a short metallic strand, also known as "messenger cable". The messenger cable is typically made of a straight core wire, which is surrounded by one or two concentric layers of helical wires. The clamp allows to rigidly connect the damper to OHL conductors and guard wires. Vibrations of the clamp set in motion the two branches of the messenger cable, which behave as two uncoupled cantilevers with lumped translational and rotational masses attached to their end sections.

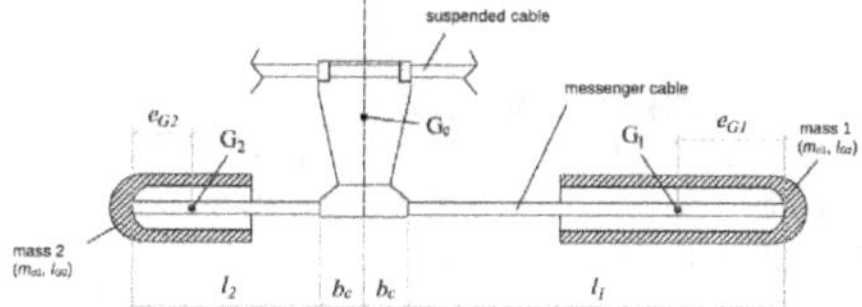

Figure 1. Geometry and inertial properties of a Stockbridge damper. Masses and centroidal mass moments of inertia are denoted respectively as M_{di} and I_{Gi} $(i = 1, 2)$.

The dynamic response of Stockbridge dampers is markedly non-linear and hysteretic, due to frictional dissipation of energy localized on the contact surfaces between adjacent wires of the messenger cable. Non-linearity may be in principle exploited to increase the range of frequencies over which the damper can effectively mitigate aeolian vibrations. Only few models, however, can be found in the literature that specifically address the issue of the characterization of the nonlinear dynamic response of Stockbridge dampers [2-7]. Most of these formulations are based on computationally expensive strategies to model the hysteretic behavior of the messenger cable and /or require the identification of a significant number of parameters from experimental tests. A remarkable exception, within this context, is the model initially proposed by Pivovarov and Vinogradov [2]. The messenger cable is herein modelled as an equivalent single-degree-of-freedom (SDOF) system, with restoring forces defined through the phenomenological hysteretic model proposed by Bouc [8]. The model of Pivovarov and Vinogradov has been recently re-considered by Foti et al. [7], which adopted a formulation of the Bouc-Wen model based on a minimal set of parameters [9].

As a major drawback, SDOF phenomenological models only allow to reproduce the dynamic response associated to the translation of the masses attached at the end sections of the messenger cable. In the present work, the formulation proposed by Foti et al. [7] is generalized to account for the rotational motion of the damper masses.

To keep the presentation of the model as straight as possible, let us herein focus on the special case of symmetric Stockbridge dampers (i.e. $m_{d1}=m_{d2}=m_d$, $I_{G1}=I_{G2}=I_G$, $l_1=l_2=l_d$, $e_{G1}=e_{G2}=e_G$ - see Fig. 1) subject to a pure vertical translation of the clamp. Details on the general case of non-symmetric dampers subject to a combined rotational and translational motion of the clamp can be found in [10].

The force F_d exerted by the damper on the cable (cable-damper interaction force) can be expressed as:

$$F_d = (2m_d + m_c)\,\dot{v}_c + 2m_d\,\ddot{u}_1 - 2\,m_d\,e_G\,\ddot{u}_2 \tag{1}$$

where m_c is the mass of the clamp, v_c is the velocity of the clamp, u_1 and u_2 are, respectively, the relative displacement of the centroid with respect to the clamp and the rotation of the rigid body attached at each end of the messenger cable (see also the formulation presented in [3]).

The dynamics of the two degrees-of-freedom u_1 and u_2 is assumed to be approximately described by the following decoupled equations of motion:

$$M_i\,\ddot{u}_i + F_i = Q_i, i = 1,2 \tag{2}$$

where the generalized masses M_i read: $M_1 = m_d$ and $M_2 = I_G + m_d\,e_G^2$; the generalized external forces Q_i read: $Q_1 = -\,m_d\,\dot{v}_c$ and $Q_2 = m_d\,e_G\,\dot{v}_c$; and the generalized restoring forces F_i are described through the following equations:

$$\begin{cases} F_i = k_{min,i}\,u_i + \left(k_{max,i} - k_{min,i}\right)u_{0,i}\,z_i \\ \quad u_{0,i}\,\dot{z}_i = \dot{u}_i - |\dot{u}_i|z_i \end{cases} \tag{3}$$

where z_i is a non-dimensional hysteretic variable with values in the range [-1, 1], while $k_{min,i}$ $k_{max,i}$ and u_{0i} are the parameters of the model. On the overall, hence, the total number of parameters of the proposed model that need to be identified from experimental tests is equal to six (three for each degree-of-freedom of the masses).

The Energy Balance Method

The technical approach currently adopted to assess the severity of aeolian vibrations relies on an application of the Energy Balance Method (EBM), e.g. [1]. Without loss of generality, the EBM will be presented in this section with reference to a taut cable of length l, suspended to rigid horizontal supports and equipped with a single Stockbridge damper. The damper is assumed to be attached at a distance x_d from one of the supports.

The EBM is based on the assumption of steady-state mono-modal vibrations of the cable. For each vibration frequency f, the single-peak antinode vibration amplitude y_{max} is obtained by imposing the balance between the average power per vibration cycle imparted by the wind to the cable (P_w) and the one dissipated in the coupled cable-damper system, that can be expressed as: $P_c + P_d$, where the term P_c accounts for the internal dissipation within the cable (also known as cable "self-damping"), while P_d is the average power per vibration cycle dissipated by the damper. The equation, hence, reads:

$$P_w(y_{max}, f) - P_c(y_{max}, f) - P_d(y_{max}, f) = 0, \text{for any } f \tag{4}$$

The wind input power P_w is typically obtained from wind-tunnel tests. Empirical results for laminar wind conditions can be expressed in the following general form:

$$\frac{P_w}{l} = D^4 f^3 \text{fnc}\left(\frac{y_{max}}{D}\right) \tag{5}$$

where D is the diameter of the cable and "fnc" is a nonlinear function of the non-dimensional vibration amplitude y_{max}/D. Different expressions have been proposed in the literature for the function "fnc" (see e.g. [1, 11]) along with correction coefficients that allows to modify Eq. (5) to account for the effect of turbulence (see e.g. [12]). Without loss of generality, the present work will focus only on laminar wind conditions and will adopt the definition of "fnc" recommended in [11].

The cable self-damping P_c can be modeled by means of both empirical and theoretical expressions (see e.g. [13] for a detailed discussion on this topic). Without loss of generality, in the present work the theoretical expression derived by Foti and Martinelli [13] under the assumption of micro-slip conditions on the contact surfaces between adjacent wires of the vibrating cable will be adopted:

$$\frac{P_c}{l} = \frac{128\pi^5 \gamma^3 RTS\, EI_{max,ef}}{3\, c_0 \mu} \frac{y_{max}^3 f^7}{T^4} \tag{6}$$

where T is the tensile force (approximately assumed as constant along the cable span); γ, RTS and $EI_{max,ef}$ are, respectively, the mass per unit of length, the Rated Tensile Strength (RTS) and the maximum bending stiffness of the cable; the coefficient c_0 is a construction parameter that only depends on the geometry of the cable cross section; and μ is the friction coefficient adopted to describe the contact conditions between the wires of the cable (see [13] for further details).

The power dissipated by the damper, P_d, reads:

$$P_d = \tfrac{1}{2}\cos(\zeta)|Z_d| v_c^2 \tag{7}$$

where $|Z_d|$ and ζ are, respectively, the module and phase of the mechanical impedance function (Z_d) of the damper. By adopting the classic undamped taut string model to describe the cable vibrations, the clamp velocity v_c can be related to the vibration frequency f and antinode vibration amplitude y_{max} through the equation:

$$v_c = f\, y_{max} \left|\sin\left(\frac{2\pi f}{\Omega_1}(1-\alpha)\right)\right| \tag{8}$$

where $\alpha = x_d/l$ and Ω_1 is the fundamental circular vibration frequency of the bare cable, i.e. $\Omega_1 = \frac{1}{l}\sqrt{\frac{T}{\gamma}}$. Substitution of Eqs. (5)-(8) in the balance equation (4) yields a non-linear algebraic equation that can be solved to get the maximum expected nondimensional aeolian vibration amplitude y_{max}/D as a function of the vibration frequency f.

It is worth noting that the procedure described up to this point applies as presented strictly to linear dampers i.e., in other terms, to dampers whose dynamic behavior can be characterized by a single impedance function. Whenever dealing with nonlinear damper models, the impedance function of the damper will depend on the motion amplitude in addition to the frequency, hence an iterative approach is required.

Example of application

The proposed modeling strategy is applied to investigate the aeolian vibrations of a benchmark OHL span already studied elsewhere (see e.g. [7, 14]). The length of the span is 450 m and the cable is a ACSR Bersfort 48/7 conductor (diameter D=35.6 mm, mass per unit of length γ=2.375 kg/m, Rated Tensile Strength RTS=180 kN, c_0 =0.139, μ =0.3). The cable is strung at T=0.4RTS.

A single symmetric Stockbridge damper that was experimentally tested by Sauter [15] is assumed to be present at the 1% of the span (i.e. at the non-dimensional arc-length coordinate $\alpha=x_d/l$=0.01). Geometrical and inertial parameters of the damper are fully reported in [10, 15]. Sauter [15] performed sweep tests on a shaker at two different constant values of the clamp velocity: v_c=50mm/s and v_c=200 mm/s. From these tests the parameters of the proposed non-linear model of the damper were identified to match the experimental data obtained for the clamp velocity v_c=200 mm/s. The identified model parameters are: $k_{min,1}$ = 2000 N/m, $k_{max,1}$ = 8000 N/m, $u_{0,1}$ = 0.002 m, $k_{min,2}$ = 68.5 N/m, $k_{max,2}$ = 275 N/m, $u_{0,2}$ = 0.0075 rad.

The required simulations were performed by applying a sinusoidal motion of the clamp with a frequency sweep from 0 to 100 Hz during a total duration of 200 s. The non-linear equations of motion of the damper were then numerically integrated.

Fig. 2 shows the matching between experimental results and the damper model output in terms of the real part of the damper impedance function. The predictions of the proposed damper model match very well the experimental results, not only for the clamp velocity v_c=200 mm/s, used in the identification process, but also for v_c = 50mm/s.

Fig. 3 shows the comparison of the OHL aeolian vibration amplitude computed with a linear model of the damper (Fig. 3a) and with a nonlinear model of the same (Fig. 3b). The vibration amplitude of the OHL without the damper is reported as well for comparison purposes. As it can be appreciated, the damper (no matter the model) is effective in the range of frequency approximately 10-20 Hz. The curve for the nonlinear damper model tends to follow the curve associated to higher vibration velocites at the lower end of the cited frequency range, while it follows the lower velocity one at the higher ends.

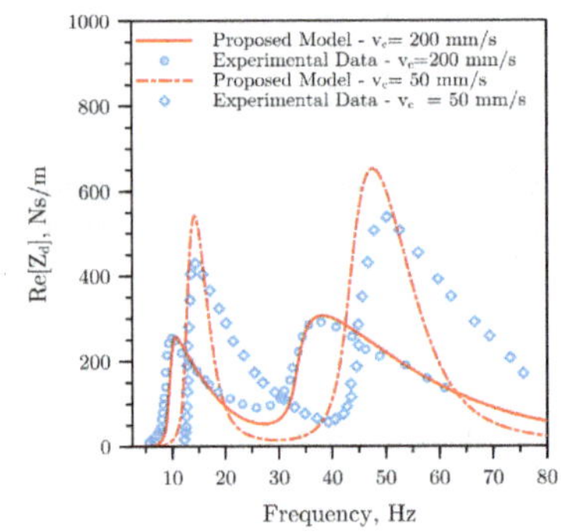

Figure 2. Absolute value of the real part of the impedance function of the identified model of the damper, compared to the experimental values in [15].

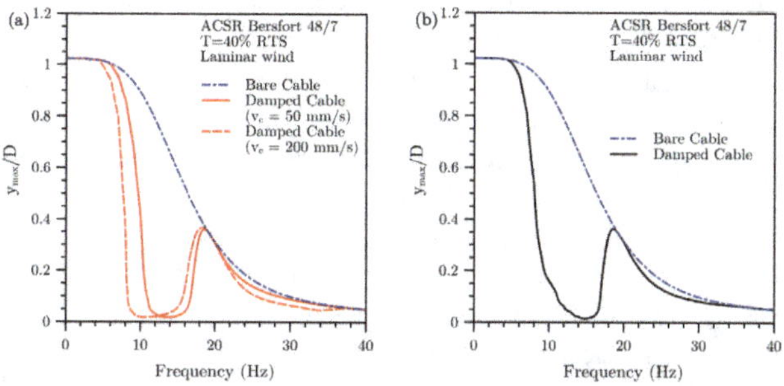

Figure 3. (a) Nondimensional vibration amplitudes of the line with and without linear damper strung at T= 0.4RTS. (b) Same for nonlinear damper.

Conclusions

A novel model for the nonlinear dynamic response of Stockbridge dampers has been presented. The model is based on a classic Bouc-Wen hysteretic law, and depends on a small number of physical parameters that can be identified from dynamic tests. A procedure to account for the Stockbridge damper nonlinear response on the assessment of aeolian vibrations of OHL has been briefly presented. Even though only one damper and one line has been considered in this work, numerical results suggest that an appropriate linear model is adequate to get meaningful estimates of vibrations amplitudes. Availability of a practical nonlinear model of the Stockbrigde damper paves the way to enhanced integrated design strategies of the dampers and the overhead lines that fully exploit the damper nonlinearities.

References

[1] EPRI Electric Research Power Institute, Transmission Line Reference Book: Wind-Induced Conductor Motion, 2006.

[2] I. Pivovarov, O.G. Vinogradov, One application of Bouc's model for non-linear hysteresis, Journal of Sound and Vibration, vol. 118, no. 2, pp. 209–216, 1987. https://doi.org/10.1016/0022-460X(87)90521-9

[3] F. Foti, L. Martinelli, Hysteretic Behaviour of Stockbridge Dampers: Modelling and Parameter Identification, Mathematical Problems in Engineering, 2018, article id: 8925121. https://doi.org/10.1155/2018/8925121

[4] S. Langlois, F. Legeron, Prediction of aeolian vibration on transmission-line conductors using a nonlinear time history model - Part I: Damper model, IEEE Transactions on Power Delivery, vol. 29, no. 3, pp. 1168–1175, 2014. https://doi.org/10.1109/TPWRD.2013.2291361

[5] N. Barbieri, R. Barbieri, R. A. da Silva, M. J. Mannala, and L. D. S. V. Barbieri, Nonlinear dynamic analysis of wire-rope isolator and Stockbridge damper, Nonlinear Dynamics, vol. 86, no. 1, pp. 501–512, 2016. https://doi.org/10.1007/s11071-016-2903-1

[6] X. Luo, L. Wang, and Y. Zhang, Nonlinear numerical model with contact for Stockbridge vibration damper and experimental validation, Journal of Vibration and Control, vol. 22, no. 5, pp. 1217–1227, 2014. https://doi.org/10.1177/1077546314535647

[7] F. Foti, V. Denoël, L. Martinelli, F. Perotti, A stochastic and continuous model of aeolian vibrations of conductors equipped with stockbridge dampers, Proceedings of the International

Conference on Structural Dynamic , EURODYN, 2020, 1, pp. 2088–2102. https://doi.org/10.47964/1120.9169.20304

[8] R. Bouc, "Modèle mathèmatique d'hystérésis" Acustica, vol. 21, pp. 16–25, 1971.

[9] F. Ikhouane, J. Rodellar, and J. E. Hurtado, "Analytical characterization of hysteresis loops described by the Bouc-Wen model", Mechanics of Advanced Materials and Structures, vol. 13, no. 6, pp. 463–472, 2006. https://doi.org/10.1080/15376490600862830

[10] F. Bogani and A. Sosio, Modellazione di dissipatori Stockbridge per la mitigazione delle vibrazioni eolica dei cavi sospesi, Politecnico di Milano (MSc. Thesis), Milano, Italy, 2021.

[11] G. Diana et al., "Modelling of aeolian vibrations of a single conductor plus damper: assessment of technology", Electra, vol. 223, pp. 28–36, 2005.

[12] M.L. Lu, The effect of turbulence on the wind power imparted to a vibrating conductor, (2003) 11th Conf. Wind Engineering, Lubbock, Texas (U.S.), June 2-5.

[13] F. Foti, L. Martinelli, An enhanced unified model for the self-damping of stranded cables under aeolian vibrations, Journal of Wind Engineering and Industrial Aerodynamics, 2018, 182, pp. 72–86. https://doi.org/10.1016/j.jweia.2018.09.005

[14] C. Gazzola, F. Foti, L. Martinelli, F. Perotti, An appraisal of modelling strategies for assessing aeolian vibrations of transmission lines, Lecture Notes in Mechanical Engineering, 2020, pp. 1522–1534. https://doi.org/10.1007/978-3-030-41057-5_123

[15] D. Sauter, Modeling the dynamic characteristics of slack wire cables in Stockbridge dampers, Technische Universitat Darmstadt (Dissertation), Darmstadt, Germany, 2003.

Theoretical and Applied Mechanics - AIMETA 2022 | Materials Research Forum LLC
Materials Research Proceedings **26** (2023) 423-428 | https://doi.org/10.21741/9781644902431-69

Explicit expressions of the eigenfrequencies of damaged frames

CANNIZZARO Francesco[1,a *], CADDEMI Salvatore[1,b], CALIO' Ivo[1,c] and IMPOLLONIA Nicola[1,d]

[1] Department of Civil Engineering and Architecture, University of Catania, via Santa Sofia 64, 95123 Catania, Italy

[a] francesco.cannizzaro@unict.it, [b] salvatore.caddemi@unict.it, [c] ivo.calio@unict.it, [d] nicola.impollonia@unict.it

Keywords: Cracked Frame, Dynamic Stiffness Matrix, Wittrick and Williams Algorithm, Explicit Solutions

Abstract. The presence of damage can strongly affect the residual carrying capacity and the dynamic properties of frame structures. The uncertainty in the position and intensity of damage implies the difficulty in adopting for practical purposes deterministic analyses, which rigorously require complicated calculations. In this paper, multi-cracked frames are studied considering the crack positions as deterministic whereas the intensities are uncertain. Explicit, although approximated, formulas of the main modal parameters as a function of the damage intensities are proposed. The latter expressions, which extend a previous study on beam-like structures, are built on the basis of detailed analyses, here computed combining the Dynamic Stiffness Matrix (DSM) approach with an efficient solution employing the distribution theory to treat the presence of cracks, and applying the Wittrick and Williams algorithm. An extremely low number of configurations of the frame is adopted to build the approximated solution is adopted. The proposed explicit formulas, which are duly verified for several meaningful cases, are then applied for the dynamic analysis of multi-cracked frames.

Introduction

The adoption of explicit expressions for the solution of structural problem can be very useful in many fields of engineering problems, e.g. when different design solutions have to be assessed according to the change of one or more parameters (parametric design), or when sensitivity analyses are needed. In addition, explicit expressions can be easily exploited for solving inverse problems. When closed form solutions have been obtained, the exact parametric expressions can be directly applied without running many numerical simulations. However, there is a wide variety of structural problems for which only numerical solutions are available, then the employment of re-analysis can become cumbersome and practically impossible to pursue. In such cases the availability of explicit solutions can be a powerful tool.

Within this framework, and with regard to free vibrations of damaged beams, in this paper a methodology to provide explicit although approximated expressions of the eigenfrequencies of damaged frames is proposed.

Damaged frame structures are here treated as continuous Euler-Bernoulli models and therefore handled with the Dynamic Stiffness Matrix (DSM) approach [1]. Damage is often assumed as an uncertain parameter [2], since its presence in structures is often hidden and, even when its presence is known, its intensity is hard to quantify. The presence of damage is considered with concentrated patterns, which is approximately realistic, especially for an early damage detection. The modelling of concentrated damage is usually approached with the so called equivalent rotational hinge model [3] where cracks correspond to rotational springs connecting two chunks of 1D elements. The stiffness of the spring is calibrated according to the crack depth following various strategies [4]. To optimize the required effort, in this study cracks are treated by means of a distributional approach rather than enforcing continuity conditions. This approach was successfully adopted for a wide variety of problems [5,6] and is able to provide closed form solutions as a function of the boundary conditions only, irrespectively of the number of along-axis discontinuities. For the specific case at hand, both the dynamics of multi-cracked beam [7] and frame [8], were already solved.

Indeed, when a structure is treated as a continuous system a frequency equation is inferred and then numerically solved to find its roots. One of the strategies that proved to be particularly effective in the frequency equation solution is the Wittrick and Williams (WW) algorithm [9], usually combined with the DSM.

This work aims at an explicit evaluation of the eigenfrequencies of damaged frames by considering deterministic crack locations, whereas a model of uncertainty for their severities is accounted for. The eigenfrequencies are approximated with a simple but effective explicit expression, calibrated on a limited obtained of reference solutions obtained combining the DSM with the WW algorithm for specific damage configurations. The proposed procedure represents a nontrivial extension of a previous study conducted on damaged beams [10]. This explicit solution can be applied to systems with multiple cracks and can be both employed to assess the response variability in direct problems and for damage detection purposes. In spite of the reduction of the computational effort for the parametric analysis with respect to a classic approach, it is shown how the proposed procedure is able to retrieve with good accuracy the results obtained with the DSM approach. A numerical application on a cracked portal frame exploring the potentiality of the method is presented, and the accuracy of the proposed methodology is evaluated.

The DSM approach for the multi-cracked Euler-Bernoulli beam

Reference solutions of the eigenfrequencies of multi-cracked frames are here obtained by means of the approach proposed in [7]. Each beam of a frame, characterised by length L, mass per unit length m, Young's modulus E_o, moment of inertia I_o, n cracks (where the i-th crack is located at the dimensionless abscissa ξ_i and is characterized by intensity λ_i, to be related to the crack depth according to one of the models available in the literature [5]). The Euler-Bernoulli multi-cracked beam model can be elaborated to infer the corresponding planar 4x4 DSM as a function of the frequency parameter $\alpha^4 = \omega^2 (mL^4 / E_o I_o)$, irrespectively of the number of cracks along each beam as follows [8]:

$$\mathbf{K} = \frac{E_o I_o}{L^2} \begin{bmatrix} \delta & \vartheta & -\bar{\delta} & \bar{\vartheta} \\ \vartheta & \rho & -\vartheta^* & \varepsilon \\ -\bar{\delta} & -\vartheta^* & \delta^* & \hat{\vartheta} \\ \bar{\vartheta} & \varepsilon & \hat{\vartheta} & \hat{\rho} \end{bmatrix} \tag{1}$$

where

$$\delta = 2\alpha^3 \frac{\alpha(cS + sC) + \alpha(Sg_{0,2} + sg_{0,4}) + g_{0,2}g_{I,4} - g_{0,4}g_{I,2} + cg_{I,4} - Cg_{I,2}}{D};$$

$$\vartheta=\alpha^2\frac{\left\{\begin{array}{l}\alpha\left[2sS+(c+C)(g_{0,2}-g_{0,4})+(s+S)(g_{0,1}+g_{0,3})\right]+\\+(g_{I,1}+g_{I,3})(c-C+g_{0,2}-g_{0,4})+(g_{I,4}-g_{I,2})(s+S+g_{0,1}+g_{0,3})\end{array}\right\}}{D};$$

$$\begin{aligned}
&\bar{\delta}=2\alpha^3\frac{\alpha(s+S)+g_{I,4}-g_{I,2}}{D};\qquad \bar{\vartheta}=2\alpha^3\frac{C-c+g_{0,4}-g_{0,2}}{D};\\
&\rho=2\alpha\frac{(\alpha C+g_{I,3})(g_{0,1}+s)-(\alpha c+g_{I,1})(g_{0,3}+S)}{D};\\
&\vartheta^*=2\alpha^2\frac{\alpha(C-c)+g_{I,3}-g_{I,1}}{D};\qquad \varepsilon=2\alpha^2\frac{s-S+g_{0,1}-g_{0,3}}{D};\\
&\delta^*=-\frac{\vartheta^*}{2\alpha^2}\left[\alpha^3(s-S)+g_{III,2}-g_{III,4}\right]-\frac{\bar{\delta}}{2\alpha^3}\left[\alpha^3(C+c)-g_{III,1}+g_{III,3}\right];\\
&\hat{\vartheta}=\frac{\bar{\vartheta}}{2\alpha^3}\left[\alpha^3(C+c)-g_{III,1}+g_{III,3}\right]+\frac{\varepsilon}{2\alpha^2}\left[\alpha^3(S-s)-g_{III,2}+g_{III,4}\right];\\
&\hat{\rho}=\frac{\bar{\vartheta}}{2\alpha^3}\left[-\alpha^2(s+S)+g_{II,1}-g_{II,3}\right]-\frac{\varepsilon}{2\alpha^2}\left[\alpha^2(C+c)-g_{II,2}+g_{II,4}\right]
\end{aligned}\tag{2}$$

with

$$\begin{aligned}
D=&2\alpha(1-cC)+\alpha\left[(C-c)(g_{0,4}-g_{0,2})-(s+S)(g_{0,3}-g_{0,1})\right]+\\
&+(g_{0,1}-g_{0,3}+s-S)(g_{I,4}-g_{I,2})+(g_{0,2}-g_{0,4}+c-C)(g_{I,1}-g_{I,3});\\
g_{P,k}=&\frac{1}{2\alpha}\sum_{i=1}^{n}\lambda_i\varepsilon_{ki}(\alpha)S_i^P(\alpha);\quad c=\cos\alpha;\quad s=\sin\alpha;\quad C=\cosh\alpha;\quad S=\sin\alpha
\end{aligned}\tag{3}$$

where $P=0,I,II,III$, $k=1,\ldots,4$. The values $S_i^P(\alpha)$, obtained by evaluating the P-th derivative of the function $S_i(\alpha,\xi)$ at the dimensionless abscissa $\xi=1$, are finally given, together with the terms $\varepsilon_{ki}(\alpha)$ and b_{ki}, by:

$$\begin{aligned}
&S_i(\xi)=\sin\alpha(\xi-\xi_i)+\sinh\alpha(\xi-\xi_i);\\
&\varepsilon_{ki}(\alpha)=\frac{\alpha}{2}\sum_{j=1}^{i-1}\lambda_j\varepsilon_{kj}(\alpha)\left[-\sin\alpha(\xi_i-\xi_j)+\sinh\alpha(\xi_i-\xi_j)\right]+\alpha^2 b_{ki};\\
&b_{1i}=-\sin\alpha\xi_i;\;\; b_{2i}=-\cos\alpha\xi_i;\; b_{3i}=\sinh\alpha\xi_i;\;\; b_{4i}=\cosh\alpha\xi_i
\end{aligned}\tag{4}$$

Once the global dynamic stiffness matrix of the entire structure is derived, the evaluation of the eigenfrequencies can be conveniently obtained via WW algorithm by evaluating the number J of natural frequencies that are lower than a specified dimensional frequency value ω^*. J is the sum of two terms $J=J_k+J_o$. J_k is the number of negative eigenvalues of the global dynamic stiffness matrix evaluated at ω^*. J_o is the sum of the number J_r of vibration frequencies lower than ω^* of the generic r-th beam with both ends clamped. Considering the number N_b of beams that compose the system, J_o can be obtained that $J_o=\sum_{r=1}^{N_b}J_r$. It is worth noting that J_r can be either evaluated by means of closed form expressions for undamaged beams or, alternatively, with substructuring approach.

Explicit approximate approach for the evaluation of the modal parameters

The procedure summarized in the previous section, although optimized with regard to the numerical algorithm and the size of the problem, significantly increases its computational burden when a considerable number of damage configurations are analyzed. In this section an approximated explicit formulation is proposed to assess the variability of the eigenfrequencies of a damaged frame with deterministic crack locations and uncertain intensities. The procedure is an extension of a study previously applied to cracked beams [10]. The generic crack severity λ_i is treated as a variable parameter ranging in the interval $\lambda_i \in \left[\underline{\lambda}_i, \overline{\lambda}_i\right]$ and expressed as $\lambda_i = \lambda_{o,i} + \beta_i$, where $\beta_i \in \left[-\Delta\lambda_i, \Delta\lambda_i\right]$ being $\Delta\lambda_i$ the deviation amplitude and $\lambda_{o,i}$ a reference damage intensity. The damage configuration of the frame is associated to the crack severities vector $\boldsymbol{\lambda} = \left[\lambda_1, \lambda_2, ..., \lambda_i, ...\lambda_n\right]$, considering the cracks located in one or more beams of the frame. In order to provide an approximated evaluation of the eigenfrequencies for a generic damage configuration of the frame, the correspondence between exact and approximated computations of the considered eigenfrequency parameter $\alpha_p^4(\boldsymbol{\lambda})$ is enforced for $2n$ significant configurations associated to the following damage intensity distributions

$$\boldsymbol{\lambda}_o = \left[\lambda_{o,1}, \lambda_{o,2}, ..., \lambda_{o,i}, ...\lambda_{o,n}\right];\ \overline{\boldsymbol{\lambda}}_s = \left[\lambda_{o,1}, \lambda_{o,2}, ..., \overline{\lambda}_s, ...\lambda_{o,n}\right];\ \underline{\boldsymbol{\lambda}}_s = \left[\lambda_{o,1}, \lambda_{o,2}, ..., \underline{\lambda}_s, ...\lambda_{o,n}\right] \tag{5}$$

with s=1,…,n. The interpolating formula proposed in [11] and already employed in [12] for the inversion of matrices, is adapted for the case at hand providing an estimation of frequency parameters as functions of the damage configuration in the form:

$$\alpha_p^4(\boldsymbol{\lambda}) \cong \alpha_{o,p}^4 + \sum_{i=1}^{n} \beta_i a_{p,i} \big/ \left(1 + \beta_i b_{p,i}\right), \quad p = 1, ..., n, \qquad \beta_i \in \left[-\Delta\lambda_i, \Delta\lambda_i\right] \tag{6}$$

being $\alpha_{o,p}^4$ the p-th eigenfrequency of the reference damage distribution $\boldsymbol{\lambda}_o = \left[\lambda_{o,1}, \lambda_{o,2}, ..., \lambda_{o,n}\right]$, whereas $a_{p,i}$, $b_{p,i}$ represent a set of coefficients obtained enforcing the correspondence between exact and approximated frequency parameters for the $2n$ damage configurations in Eq. (5):

$$\begin{aligned} a_{p,s} &= \frac{2}{\Delta\lambda_s} \frac{\left[\alpha_p^4(\overline{\boldsymbol{\lambda}}_s) - \alpha_{o,p}^4\right]\left[\alpha_{o,p}^4 - \alpha_p^4(\underline{\boldsymbol{\lambda}}_s)\right]}{\alpha_p^4(\overline{\boldsymbol{\lambda}}_s) - \alpha_p^4(\underline{\boldsymbol{\lambda}}_s)} \\ b_{p,s} &= \frac{1}{\Delta\lambda_s} \frac{2\alpha_{o,p}^4 - \alpha_p^4(\underline{\boldsymbol{\lambda}}_s) - \alpha_p^4(\overline{\boldsymbol{\lambda}}_s)}{\alpha_p^4(\overline{\boldsymbol{\lambda}}_s) - \alpha_p^4(\underline{\boldsymbol{\lambda}}_s)} \end{aligned} \qquad p = 1, ..., n_m; \quad s = 1, ..., n \tag{7}$$

Numerical applications

The accuracy of the approximated solution is tested on the cracked steel portal frame reported in Fig. 1 [13], composed of two columns of length L_c=800 mm and a beam of length L_b=1000 mm. Constant 40x8 mm rectangular cross section for all the elements, Young's modulus E=2*10^5 N/mm^2, mass density per unit volume m=7.849*10^{-9} N s^2/mm^4, have been assigned. Here, the crack is located at a dimensionless abscissa $\xi_1 = 0.05$ of the left column (40 mm from the clamp), considering a variable crack severity with reference intensity $\lambda_{o,1}$=0.1 and deviation amplitude $\Delta\lambda_1$=0.1.

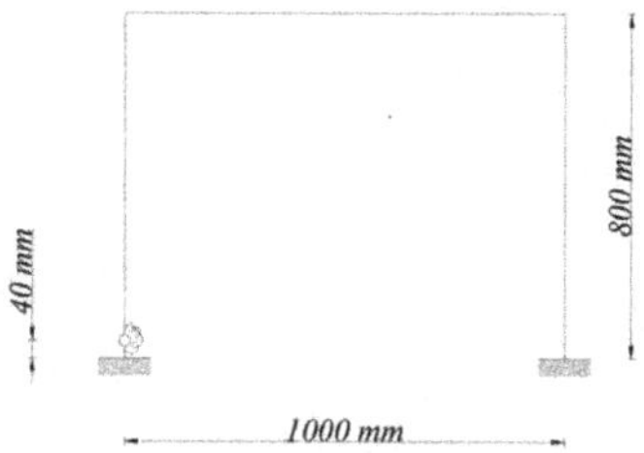

Figure 1 – Investigated damaged portal frame [14].

In Fig. 2 the comparisons between reference and approximated frequency parameters are shown for the first four frequencies; all the terms are normalized by the corresponding reference value. The continuous lines correspond to the exact properties, whereas the dashed lines are relative to the approximated solution. In Fig. 3 the corresponding relative errors are reported.

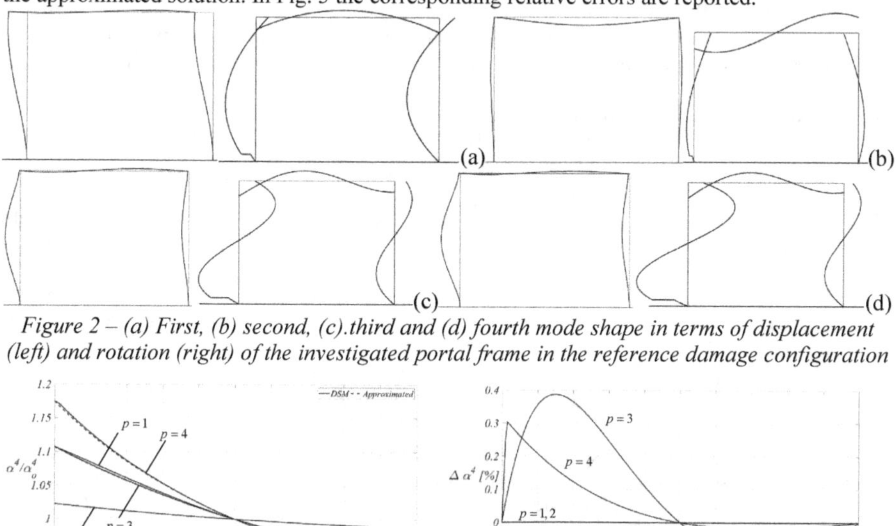

Figure 2 – (a) First, (b) second, (c).third and (d) fourth mode shape in terms of displacement (left) and rotation (right) of the investigated portal frame in the reference damage configuration

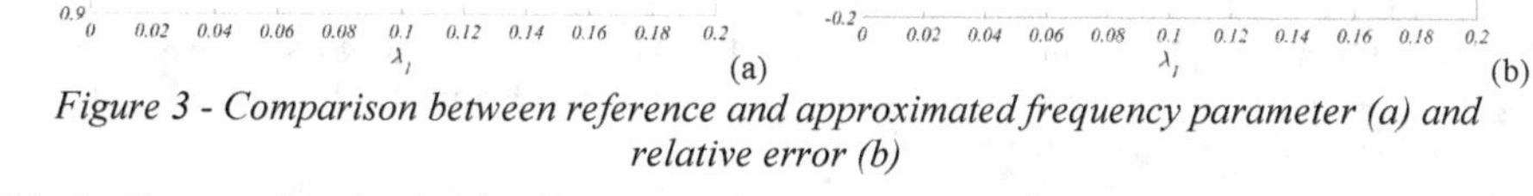

Figure 3 - Comparison between reference and approximated frequency parameter (a) and relative error (b)

For the first two frequencies the discrepancy between exact and approximated values is almost unappreciable (maximum error lower than 0.002%), while for third and fourth frequencies it is higher but always lower than 0.4%.

Conclusions

An explicit although approximated expression for the eigenfrequency of damaged frames was proposed in this study. The damaged was modelled via cracks with known position and uncertain severity, thus the frequencies are expressed as functions of the damage intensities. The proposed expressions, which are inspired to the Sherman and Morrison formula, are calibrated on reference results obtained combining the construction of the Dynamic Stiffness Matrix of multi-cracked Euler-Bernoulli beams with a formulation which takes advantage of the distributional theory with the Wittrick and Williams algorithm; a limited number of reference damage configurations is employed to calibrate the approximated solution. Validation of the proposed formulas, which can be exploited for damage identification problems as well as for sensitivity analyses, was provided for the first four frequencies with regard to a single cracked frame. The accuracy of the proposed approximated solution is satisfactory and assures a considerable computational advantage.

Acknowledgements
This research was funded by the Italian Ministry of University and Research (MUR) with the project PRIN2020 #20209F3A37.

References

[1] J.R. Banerjee, F.W. Williams, Exact Bernoulli–Euler dynamic stiffness matrix for a range of tapered beams, Int. J. for Numer. Meth. Eng. 21(12) (1985) 2289-2302. https://doi.org/10.1002/nme.1620211212

[2] G. Muscolino, R. Santoro, Dynamics of multiple cracked prismatic beams with uncertain-but-bounded depths under deterministic and stochastic loads J. Sound Vib. 443 (2019) 717-731. https://doi.org/10.1016/j.jsv.2018.11.029

[3] A.D. Dimarogonas, Vibration of cracked structures: a state of the art review Eng. Frac. Mech. 55(5) (1996) 831-857. https://doi.org/10.1016/0013-7944(94)00175-8

[4] C. Bilello, Theoretical and experimental investigation on damaged beams under moving systems, University of Palermo: PhD Thesis, 2001.

[5] S. Caddemi, I. Caliò, F. Cannizzaro, Tensile and compressive buckling of columns with shear deformation singularities Meccanica 50 (2015) 707-720. https://doi.org/10.1007/s11012-014-9964-3

[6] S. Caddemi, I. Caliò, F. Cannizzaro, Influence of an elastic end support on the dynamic stability of Beck's column with multiple weak sections Int. J. Nonin. Mech. 69 (2015) 14-28. https://doi.org/10.1016/j.ijnonlinmec.2014.10.016

[7] S. Caddemi, I. Caliò, Exact closed-form solution for the vibration modes of the Euler-Bernoulli beam with multiple open cracks J. Sound Vib. 327 (2009) 473-489. https://doi.org/10.1016/j.jsv.2009.07.008

[8] S. Caddemi, I. Caliò, The exact explicit dynamic stiffness matrix of multi-cracked Euler-Bernoulli beam and applications to damaged frame structures J. Sound Vib. 332(12) (2013) 3049-3063. https://doi.org/10.1016/j.jsv.2013.01.003

[9] F.W. Williams, W.H. Wittrick, An automatic computational procedure for calculating natural f requencies of skeletal structures, Int. J. Mech. Sci. 12(9) (1970) 781-791. https://doi.org/10.1016/0020-7403(70)90053-6

[10] F. Cannizzaro, N. Impollonia, S. Caddemi, I. Caliò, Explicit dynamic response of damaged beams with application to uncertain and identification problems J. Sound Vib. 487 (2020) 115608. https://doi.org/10.1016/j.jsv.2020.115608

[11] J. Sherman, W. Morrison, Adjustment of an inverse matrix corresponding to changes in the elements of a given column or a given row of the original matrix (abstract) Ann. Math. Stat. 20 (1949) 621.

[12] N. Impollonia, A method to derive approximate explicit solutions for structural mechanics problems Int. J. Solids Struct. 43 (2006) 7082-7098. https://doi.org/10.1016/j.ijsolstr.2006.03.003

[13] A. Greco, A. Pau, Damage identification in Euler frames Comput. Struct. 92-93 (2012) 328-336. https://doi.org/10.1016/j.compstruc.2011.10.007

Optimal design of single-degree-of-freedom vibro-impact system under harmonic base excitation

Giuseppe Perna[1,a*], Maurizio De Angelis[1,b] and Ugo Andreaus[1,c]

[1] Department of Structural and Geotechnical Engineering, Sapienza University of Rome, Via Eudossiana 18, Rome, 00148, Italy

[a]giuseppe.perna@uniroma1.it, [b]maurizio.deangelis@uniroma1.it, [c]ugo.andreaus@uniroma1.it

Keywords: Vibro-Impact Systems, Dynamic Response Control, Bumper, Gap, Numerical Modeling, Harmonic Base Excitation

Abstract. Strong seismic excitations can induce considerable displacements in base-isolated structures. These may in turn cause damage to the isolation system or impacts against adjacent structures in the event of an insufficient seismic gap. A strategy of reducing displacements can be achieved through the interposition, between the isolated structure and adjacent structures, of deformable and dissipative devices, called bumpers. In this paper, the response of single-degree-of-freedom (SDOF) base-isolated systems subjected to sine excitations and whose displacements are limited by optimally designed bumpers are analyzed, the bumpers being designed according to an optimal design criterion.

Introduction

One of the most widely used strategies for passive control of the dynamic response of sensitive structures and equipment is base isolation. This strategy consists of interposing a highly horizontally deformable element between the base floor and the structure (or equipment), so that the period of the system is increased significantly resulting in a reduction of the transmitted acceleration. However, strong earthquakes can cause large displacements at the isolated floor level, exceeding its limit deformation, or induce impacts with adjacent elements. Rigid impact causes significant increases in acceleration and floor drifts that can cause serious damage to both structural and nonstructural elements [1].

An effective strategy to reduce and control the adverse effects due to large isolation floor displacements is the interposition of deformable and dissipative devices, called bumpers. Various experimental investigations have been conducted by the authors on the influence of bumpers on the dynamic response of single-degree-of-freedom (SDOF) systems subjected to sine base excitation [2-4]. From these studies, it was observed that the parameters governing the impact between system and bumper can be identified by three elements: gap (distance between mass and bumper), stiffness and damping of the bumper. The authors, moreover, based on the experimental investigations defined a numerical model that allowed the identification of optimality relations between bumper stiffness, bumper damping coefficient and gap, reducing the design of such a control strategy to a single parameter [3-7].

This paper analyzes the response obtained from a numerical model of SDOF system constrained by two bumpers, arranged symmetrically on both sides of the mass of the system with an initial gap, subjected to sine excitation. The design of bumpers is done in accordance with the optimality relationship [6] and the optimal design curve [7]. The absolute maximum values of the response in absolute mass acceleration and relative mass displacement are represented as a function of frequency ratio and for six initial gaps. All results shown are represented in dimensionless terms so that they can be generalized.

The text is organized as follows: the first paragraph introduces the model and its equations of motion; the second paragraph explains the optimal design criterion that is used to identify the

Materials Research Forum LLC
https://doi.org/10.21741/9781644902431-70

parameters governing the impact; the third paragraph reports the results of the numerical analysis; and finally, the fourth paragraph reports the conclusions and possible future developments of the work.

Model and equations of motion

The model adopted in the numerical analyses is shown in Fig. 1. The figure represents a vibro-impact single-degree-of-freedom (SDOF) system that consists of a mass M, a damper D and two deformable and dissipative obstacles, arranged symmetrically on both sides of the mass with an initial gap G_{0j} (*j=R* right side, *j=L* left side) and denoted as right bumper B_R and left bumper B_L, respectively. The damper and the bumpers are modeled by a linear elastic element, with stiffness K and K_j *(j=R,L)*, respectively, and a linear viscous dashpot, with damping coefficient C and C_j *(j=R,L)*, respectively, arranged in parallel. The system is subjected by a sine base acceleration $A_t(t) = A_G sin(\Omega t)$, where A_G is the amplitude and Ω in the circular frequency of this excitation. Finally, in figure 1 $u(t)$ refers to the relative displacement of the mass with respect to the ground, and $u_j(t)$ *(j=R,L)* refers to the deformation of the bumper.

The equations of motion were written in dimensionless form to make them as general as possible. The components of the equation were normalized with respect to $F^* = M\omega^2 u^*$, which represents the maximum force in the SDOF system in free flight (without obstacles, FF): the quantity $u^* = u_{st} R_{d,max}$ represents the maximum relative displacement in FF, where $u_{st} = A_G/\omega^2$ is the static displacement and $R_{d,max} = 1/(2\xi\sqrt{1-\xi^2})$ is the maximum value of the dynamic amplification factor $R_d(\xi,\beta)$; while $\omega = \sqrt{K/M}$ denotes the frequency of the system. The frequency ratio $\beta = \Omega/\omega$ and the damping ratio $\xi = C/(2M\omega)$, are introduced and the dimensionless time $\tau = \omega t$ is defined. In the dimensionless equations, the quantities $q = u/u^*$ and $q_j = u_j/u^*$ *(j=R,L)* are the dimensionless displacement of the mass and dimensionless deformation of the bumper, respectively. Similarly, the dimensionless gap is $\delta_{0j} = G_{0j}/u^*$ *(j=R,L)* and can take values comprised within the range $0 \le \delta_{0j} \le 1$. Finally, $f(\tau) = 2\xi q'(\tau) + q(\tau)$ is the damper dimensionless force, $f_j(\tau) = 2\xi\gamma_j q'_j(\tau) + \lambda_j q_j(\tau)$ with *(j=R,L)* –where $\gamma_j = C_j/C$ and $\lambda_j = K_j/K$– as the dimensionless contact forces, and $a_G = 2\xi\sqrt{1-\xi^2}$ as the dimensionless amplification of the dimensionless sine excitation $a_t(\tau)$.

Thus, the motion equations of the model can be written in the following dimensionless form:

$$\begin{cases} q''(\tau) + f(\tau) + f_j(\tau) \cdot \psi_1[\delta_j(\tau)] \cdot \psi_2[f_j(\tau)] = -a_G sin\beta\tau \\ f_i(\tau) = 0 \end{cases}; \quad (1)$$

where it is assumed that if the mass is in contact with the left bumper, then *j=L* and *i=R*, or if the mass is in contact with the right bumper, then *j=R* and *i=L*. In the equations the Heaviside functions ψ_1 and ψ_2 are defined as follows:

$$\text{contact} \quad \psi_1[\delta_j(\tau)] = \begin{cases} 0, & \delta_j(\tau) > 0 \\ 1, & \delta_j(\tau) = 0 \end{cases} \quad (j = R, L); \quad (2)$$

$$\text{separation} \quad \psi_2[f_j(\tau)] = \begin{cases} 0, & f_R(\tau) \le 0 \ \ or \ \ f_L(\tau) \ge 0 \\ 1, & f_R(\tau) > 0 \ \ or \ \ f_L(\tau) < 0 \end{cases}; \quad (3)$$

where $\delta_j(\tau)$ *(j=R,L)* represents the gap function in terms of the dimensionless time τ and, if *j=R*, is equal to $\delta_R(\tau) = \delta_{0R} + q_R(\tau) - q(\tau)$, and, if *j=L*, is equal to $\delta_L(\tau) = \delta_{0L} - q_L(\tau) + q(\tau)$. In all the equations introduced the apex (') denotes differentiation with respect to the dimensionless time τ.

Because the bumpers are equal and arranged symmetrically on both sides of the mass $\gamma_R = \gamma_L = \gamma$, $\lambda_R = \lambda_L = \lambda$ and $\delta_{0R} = \delta_{0L} = \delta_0$.

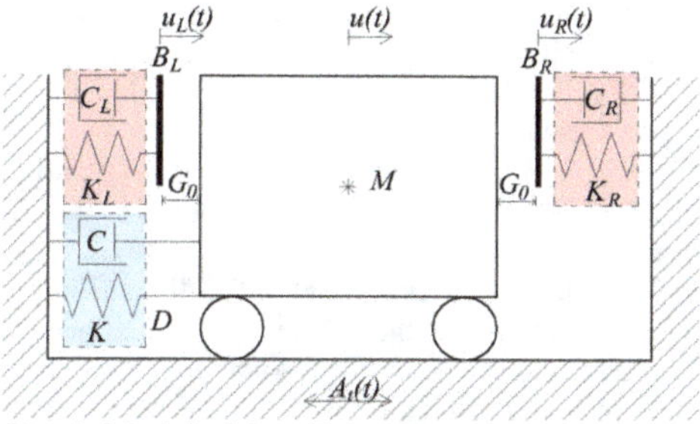

Fig. 1. ***Model of the vibro-impact system.***

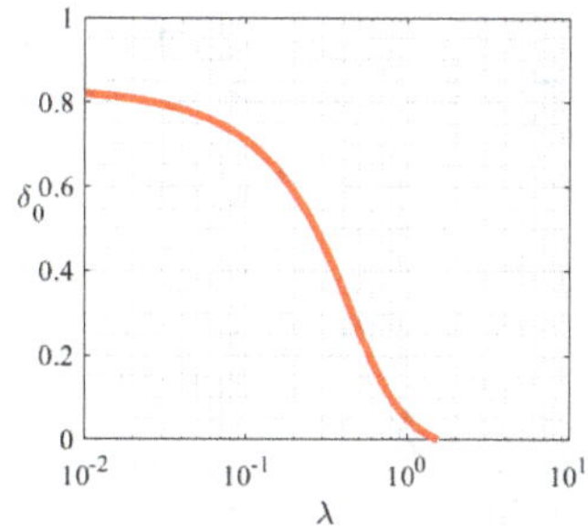

Fig. 2. ***Optimal design curve of the bumper.***

Optimal design of the bumpers

In order to optimize the control of the dynamic response of the system, the physical parameters governing the impact (gap δ_0, stiffness λ and damping γ of the bumper) to be adopted are obtained in relation to an optimal design criterion based on a relationship that links damping ratio ξ of the system with parameters λ and γ of the bumpers [6], and on a curve in which bumper stiffness λ is identified in relation to the gap δ_0 [7].

The optimality relationship is defined as follows:

$$\frac{\gamma}{\lambda} = \frac{1}{2\xi}. \tag{4}$$

This relationship was obtained by a parametric analysis in which, for each investigated δ_0, and for fixed values of ξ and of γ, the value of λ was searched such that the peak acceleration of the mass in the primary resonance is minimized. This is possible because the bumper is fully exploited: the bumper has sufficient time to recover its deformation before the next impact, dissipating all the deformation energy accumulated up to then, and it does not remain inactive because the next impact occurs practically immediately after recovery.

The optimality relationship (4) is independent of the gap and, through this relationship, it is possible to reduce the number of impact parameters from three, which in dimensionless terms can be identified as δ_0, λ and γ, to two, since by choosing λ we obtain the value of the corresponding γ.

With the introduction of the optimal curve an additional constraint is introduced, reducing the bumper design to one parameter, the dimensionless gap δ_0. The optimal design curve, shown in Fig. 2, associates to each δ_0 a pair of values λ and γ (Eq. 4), which minimizes the peak absolute acceleration of the mass in primary resonance and therefore the force acting on the mass.

Looking at the red curve in figure 2, it is clear how the control over the response of vibro-impact system with a dimensionless gap $\delta_0 > 0.82$, is ineffective. This is because, for these values of δ_0, the optimal stiffness ratio λ is so small that the presence of the bumpers is negligible.

Analysis of numerical response

In this paragraph, the numerical responses of the model characterized as follows are analyzed: the damping ratio of the damper is $\xi = 0.10$, the bumper parameters λ and γ are designed in accordance with the optimality relationship (4) and the optimal design curve (Fig.2). The represented response quantities are dimensionless excursion of the absolute acceleration and of the relative displacement, of the mass, both normalized with respect to two different factors:

$$\eta_{a1} = \frac{\Delta\alpha}{\Delta\alpha_0},\ \eta_{a2} = \frac{\Delta\alpha}{\Delta\alpha_t},\ \eta_{d1} = \frac{\Delta q}{\Delta q_0},\ \eta_{d2} = \frac{\Delta q}{\Delta q_t}, \tag{5}$$

where $\alpha = a_t + q''$, the dimensionless quantities whit subscript ($_o$) are related to the case of free flight in resonance (FF) and the dimensionless quantities whit subscript ($_t$) are related to the ground. The excursion ($\Delta i, i = \alpha, q$) was calculated as the difference between the maximum and minimum values recorded at steady state of each sub-frequency range.

Figures 3 show the normalized quantities analyzed as a function of the frequency ratio β and in the terms of six dimensionless gaps $\delta_0 = 1, 0.5, 0.3, 0.2, 0.1, 0$, where $\delta_0 = 1$ represents the FF and $\delta_0 = 0$ the case of mass adjacent with bumper.

The curves in Fig. 3a-3b represent the Pseudo-Resonance Curves (PRCs) of the dimensionless absolute acceleration of the mass, normalized in (a) with respect to the FF, and in (b) with respect to the ground. The black curves reproduce dimensionless excursion of the absolute acceleration of the mass in FF, subjected to the dimensionless sine excitation a_t. The curve $\eta_{a1_{FF}}(\beta)$ starts from the value $\eta_{a1_{FF}}(0) \simeq 0.20$ and then attains a peak at resonance, for $\beta = \sqrt{1-2\xi^2} \simeq 0.99$, equal to 1. Once the peak has been passed, for increasing β the acceleration $\eta_{a1_{FF}}$ tends to 0. The curve $\eta_{a2_{FF}}(\beta)$ starts from value 1 ($\beta = 0$), reaches the resonance peak at $\beta = \sqrt{1-2\xi^2} = 5.12$, and, for increasing values of β, $\eta_{a2_{FF}}$ decreases tending to 0 for β tending to infinity. The red dashed curves represented the cases with impact (thinner dashes imply smaller values δ_0). The study of Fig. 3a-3b shows that the curves $\eta_{a1}(\beta)$ are scaled by TR_{max}, that is the maximum value of the resonance transmissibility, relative to curves $\eta_{a2}(\beta)$, so the comments will be unique for both two types of curves, with percentage changes. In Fig. 3a-3b, as δ_0 decreases, a reduction and a shift toward larger β of the acceleration peaks are observed. The lowest value of the peak occurs in the case $\delta_0 = 0$, where a 62% reduction of the maximum acceleration value is obtained with respect to FF. The other cases show a reduction of 48% for $\delta_0 = 0.1$, 30% for $\delta_0 = 0.2$, 22% for $\delta_0 = 0.3$ and 9% for $\delta_0 = 0.5$. The β of resonance increases from 0.99 in the FF to 1.02 for $\delta_0 = 0.5$, 1.09 for $\delta_0 = 0.3$, 1.15 for $\delta_0 = 0.2$, 1.24 for $\delta_0 = 0.1$ up to 1.66 for $\delta_0 = 0$. The dashed curves, for some values of β, overlap with the FF curve: this is because for those values of β and for the value of δ_0 relative to that dashed curve, impact does not occur. The only case in which impact always occurs is the case of mass adjacent to bumpers ($\delta_0 = 0$), which, for $\beta > 1.25$, reports larger accelerations than FF and, for $\beta > 1.52$, larger than the other cases. For the other δ_0 investigated, the acceleration turns out to be greater than FF in a small range of β, between 1.04 and 1.72.

Fig. 3c-3d shows the PRCs of the dimensionless relative displacement of the mass, normalized with respect to FF, in (c), and with respect to the ground, in (d). The black curves represent the dimensionless excursion of the relative displacement of the mass in FF, subjected to the dimensionless sine excitation a_t. The dimensionless displacement curve in FF $\eta_{d1_{FF}}(\beta)$, reports the maximum at $\beta = \sqrt{1-2\xi^2} \simeq 0.99$, in which it takes value 1. For β between 0, in which $\eta_{d1_{FF}}(0) \simeq 0.20$, and $\sqrt{1-2\xi^2}$, $\eta_{d1_{FF}}$ grows as β increases. Passing the peak value, for increasing β the displacement $\eta_{d1_{FF}}$ decreases, tending, in the limit, to the value 0. The dimensionless displacement in FF, $\eta_{d2_{FF}}(\beta)$, reports the maximum in $\beta = 1/\sqrt{1-2\xi^2} \simeq 1.01$ of intensity equal to 5.03. In the first range between 0 and β relative to the maximum, $\eta_{d2_{FF}}$ has an increasing monotonic trend, after which, as β increases, it decreases tending, for infinite β, to the value 1 (maximum ground displacement). In contrast to $\eta_{d1_{FF}}$, $\eta_{d2_{FF}}$ at $\beta = 0$ assuming zero value.

The red dashed curves show the displacement responses of the impacted cases (thinner dashes imply smaller values of δ_0). Referring to Fig. 3c, the peaks of displacements, as δ_0 decreases, move toward larger β and decrease. With respect to FF ($\delta_0 = 1$), η_{d1} decreases by 90% for $\delta_0 = 0$, by 69% for $\delta_0 = 0.1$, by 55% for $\delta_0 = 0.2$, by 42% for $\delta_0 = 0.3$, and by 22% for $\delta_0 = 0.5$, while the β relative to these maximum values turns out to be 1.59, 1.19, 1.13, 1.09, and 1.03, respectively. Each curve related to $\delta_0 \geq 2\,\xi\sqrt{1-\xi^2}$ ($\simeq 0.2$) is initially coincident with the FF case, until η_{d1}, growing, reaches the value δ_0, causing the impact to occur because the displacement in FF is greater than the gap. Conversely, once the value of β has been exceeded beyond which $\eta_{d1_{FF}} \leq \delta_0$, the curve for that value of δ_0 turns out to coincide with the FF case. The only curve that shows a single point of coincidence with free flight is the one related to $\delta_0 = 0$, for $\beta = 1.75$; this curve also, for $\beta < 1.75$, shows smaller values than for FF, while for $\beta > 1.75$ larger values, with a maximum increase of 28% in displacement value in FF case. The PRCs for the other values of $\delta_0 > 0$ exhibit larger values than FF for $1.27 < \beta < 1.72$ with $\delta_0 = 0.1$, for $1.18 < \beta < 1.40$ with $\delta_0 = 0.2$, for $1.13 < \beta < 1.27$ with $\delta_0 = 0.3$, and for $1.07 < \beta < 1.15$ with $\delta_0 = 0.5$ with maximum increases of 33% (for $\beta = 1.5$), 24% (for $\beta = 1.3$), 18% (for $\beta = 1.2$), and 7% (for $\beta = 1.12$). In Fig. 3c, as δ_0 decreases, there is, as in Fig. 3d, a shift to the right and a reduction of peak displacement whit respect to FF, but at different dimensionless frequencies β and with different percentage reduction: for $\delta_0 = 0$ the peak is for $\beta = 2.04$ with 68% reduction, for $\delta_0 = 0.1$ the peak is for $\beta = 1.35$ with 50% reduction, for $\delta_0 = 0.2$ the peak is for $\beta = 1.22$ with 38% reduction, for $\delta_0 = 0.3$ the peak is for $\beta = 1.15$ with 28% reduction, and for $\delta_0 = 0.5$ the peak is for $\beta = 1.07$ with 13% reduction. As observed in Fig. 3c-3d, the response for $\delta_0 = 0$ never coincides with the FF curve except at $\beta = 1.75$.

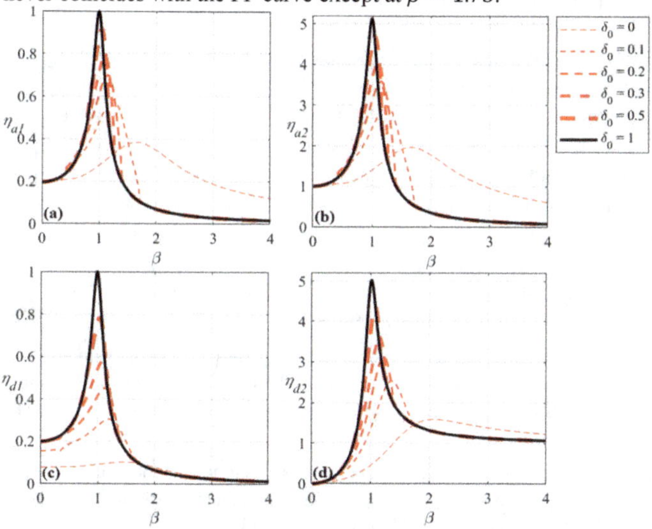

Fig. 3. PRCs of dimensionless acceleration: (a) PRC of $\eta_{a1}(\beta)$; (b) PRC of $\eta_{a2}(\beta)$; and PRCs of dimensionless displacement: (c) PRC of $\eta_{d1}(\beta)$; (d) PRC of $\eta_{d2}(\beta)$. The curves are related to different values of δ_0: in black δ_0=1 (FF), in dashed red δ_0=0.5, 0.3, 0.2, 0.1, 0 (lower thicknesses represent lower δ_0).

Conclusions and future developments

In this paper, the numerical dynamic response of the SDOF system, subjected to sine excitation, constrained by two symmetrically arranged deformable and dissipative bumpers designed with optimality relation (4) and optimal design curve (Fig. 2) is analyzed. Six different values of the design parameter δ_0 (dimensionless gap) are considered: two extreme cases, $\delta_0 = 1$, a touching condition between mass and bumpers representative of FF, and $\delta_0 = 0$, a condition with mass adjacent to the bumpers; and intermediate cases, $\delta_0 = 0.5$, $\delta_0 = 0.3$, $\delta_0 = 0.2$ and $\delta_0 = 0.1$.

The response quantities which have been analyzed are the absolute acceleration of the mass and the relative displacement of the mass with respect to the ground, normalized with respect to both (i) the peak values of acceleration and displacement in FF, and (ii) the peak values of base acceleration and ground displacement. It was found that as δ_0 decreases, the maximum values of peak acceleration and displacement undergo greater reductions with respect to FF, reporting in the case $\delta_0 = 0$ the greatest reductions (peak of acceleration reduced by 62% for both $\eta_{a1}(\beta)$ and $\eta_{a2}(\beta)$, peak of displacement reduced by 90% for $\eta_{d1}(\beta)$, and by 68% for $\eta_{d2}(\beta)$). However, the case $\delta_0 = 0$ reports larger values of acceleration for $\beta > 1.52$, compared with the other cases examined. A phenomenon of increased displacement, due to the presence of bumper, was also observed for some β following the primary resonance in FF ($1.08 < \beta < 1.15$ for $\delta_0 = 0.5$, $1.13 < \beta < 1.27$ for $\delta_0 = 0.3$, $1.18 < \beta < 1.4$ for $\delta_0 = 0.2$, $1.27 < \beta < 1.71$ for $\delta_0 = 0.1$, $\beta > 1.72$ for $\delta_0 = 0$). This phenomenon is named bouncing-effect in the literature.

These results show that the adoption of this optimal design criterion and an appropriate choice of the gap parameter δ_0 allow beneficial effects in mitigating the dynamic response to be gained. Thus, the need emerges to extend this study to cases of real dynamic actions, such as earthquakes, and evaluate the effects of appropriately designed dissipative and deformable bumpers on systems subjected to different seismic excitations.

References

[1] P.C. Polycarpou, P. Komodromos, On poundings of a seismically isolated building with adjacent structures during strong earthquakes, Earthq. Eng. Struct. Dyn. 39 (2010). https://doi.org/10.1002/eqe.975

[2] U. Andreaus, M. De Angelis, Influence of the characteristics of isolation and mitigation devices on the response of single-degree-of-freedom vibro-impact systems with two-sided bumpers and gaps via shaking table test, Struct. Control Health Monit. 27(2020). https://doi.org/10.1002/stc.2517

[3] G. Stefani, M. De Angelis, U. Andreaus, Scenarios in the experimental response of a vibro impact single-degree-of freedom systems and numerical simulations, Nonlinear Dyn. 103 (2021). https://doi.org/10.1007/s11071-020-05791-4

[4] G. Stefani, M. De Angelis, U. Andreaus, Influence of the gap size on the response of a single-degree-of-freedom vibro-impact system with two-sided constraints: Experimental tests and numerical modeling, Int. J. Mech. Sci. 206 (2021). https://doi.org/10.1016/j.ijmecsci.2021.106617

[5] G. Stefani, M. De Angelis, U. Andreaus, Numerical study on the response scenarios in a vibro-impact single-degree-of-freedom oscillator with two unilateral dissipative and deformable constraints, Commun. Nonlinear Sci. Numer. Simul. 99 (2021). https://doi.org/10.1016/j.cnsns.2021.105818

[6] G. Stefani, M. De Angelis, U. Andreaus, The effect of the presence of obstacles on the dynamic response of single-degree-of-freedom systems: Study of the scenarios aimed at vibration control, J. Sound Vib. 531 (2022). https://doi.org/10.1016/j.jsv.2022.116949

[7] G. Stefani, M. De Angelis, U. Andreaus, Exploit the study of the scenarios for the control of the response of single-degree-of-freedom systems with bumpers, Submitted to Eng. Struct. (2022).

Control and experimental dynamics

Theoretical and Applied Mechanics - AIMETA 2022 | Materials Research Forum LLC
Materials Research Proceedings **26** (2023) 437-442 | https://doi.org/10.21741/9781644902431-71

Friction-induced parametric oscillations in automotive drivelines: experimental analysis and modelling

Manuel Tentarelli[1,a*], Stefano Cantelli[2,b], Silvio Sorrentino[1,c] and Alessandro De Felice[1,d]

[1] Dipartimento di Ingegneria Enzo Ferrari, Università di Modena e Reggio Emilia, Via Pietro Vivarelli, 10, 41125 Modena, Italy

[2] CNH Industrial Italia, Viale delle Nazioni, 55, 41122 Modena, Italy

[a] manuel.tentarelli@unimore.it, [b] stefano.cantelli@cnhind.com, [c] silvio.sorrentino@unimore.it, [d] alessandro.defelice@unimore.it

Keywords: Clutch Judder, Friction-Induced Vibration, Parametric Oscillation

Abstract. Clutch judder is a friction-induced self-excited vibration occurring in automotive drivelines, an NVH issue studied for more than forty years and attributed by the scientific community to three possible causes: stick-slip, negative gradient of the coefficient of friction and geometric disturbances. However, these explanations fail to describe the kind of judder studied in this contribution, arising in presence of an oscillating component (dither) in the clutch actuation pressure. The analysis of experimental data collected on a dual-clutch transmission mounted on a specific test bench suggested the presence of a parametric resonance, generated by the dither. A specific 4 degrees of freedom model was then developed, able to predict with good accuracy the unstable parametric region in which judder occurs and useful in the design stage.

Introduction

Among various NVH issues affecting automotive drivelines, the friction-induced self-excited vibration called clutch judder holds particular importance.

To the best of the authors' knowledge, the current scientific literature on the subject identifies three main factors responsible for clutch judder. In several cases, it is attributed to a negative gradient of the coefficient of friction between the clutch plates (in this case referred to as cold judder) [1-7]. In fact, if the friction coefficient decreases with increasing slip speed, the system becomes unstable in most operating conditions. The second identified factor is stick-slip [7-11], associated with the alternate sticking and slipping motion which occurs due to the discontinuity of the friction curve at very low relative speeds. Finally, the third factor is given by geometric disturbances in terms of both asperity on the disc surfaces and misalignments between the components of the transmission. In the first case, the asperities of the discs can give rise to hot spots and thermoelastic instability [12,13]. In the second case, the misalignments between some components of the transmission give rise to the periodic fluctuation of the clamp load and to the possible excitation of a frequency of vibration of the system [7,14].

Although it has been shown that clutch judder can have different characteristics and causes, the phenomenon here under analysis has not been investigated yet in the scientific literature. It occurs in dual-clutch transmissions of heavy vehicles (tractors), in which the clutch actuation pressure is made up of a constant main component plus a small oscillating component due to the vibration of the circuit valve (dither). The experimental data collected on a specific test bench, designed in partnership with CNH Industrial, show the onset of a cold judder (total absence of hot spots) at a frequency approximately equal to the dither frequency, as displayed in fig. 1. Furthermore, judder's magnitude varies depending on whether odd or even gears are engaged.

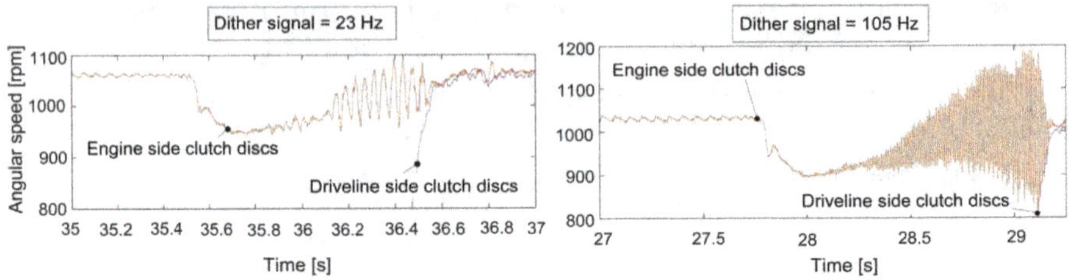

Fig. 1 - Experimental data: clutch judder arises at a high slip speed.

For a first qualitative investigation of this phenomenon, the authors developed a specific 3 degrees-of-freedom (dofs) minimal model, which was studied by means of Floquet theory [15], identifying parametric excitation due to dither as a new and unexplored source of clutch judder. For fitting experimental data and predicting the occurrence of the phenomenon, models with increasing complexity are here proposed, considering a larger number of dofs.

Methods

The system under analysis consists of four electro-hydraulically actuated wet clutches, eight gear ratios and three reduction ranges, as represented in fig. 2. Two clutches engage odd and even gears, respectively, while the other two engage forward and reverse. To improve the response of the hydraulic valves, they are actuated by dithering the main signal, consisting of a constant main pressure component p_m and an oscillating component p_f.

The hypothesis of parametric resonance as the cause of this kind of clutch judder was first assessed by the authors in [15], with the aid of a 3 dofs minimal model. Alongside the theoretical study of the phenomenon, an experimental campaign was conducted on a test bench. The input parameters selected for each test were the angular speed of the electric motor (replacing the engine of the tractor), the main actuation pressure, the dither (fluctuating) pressure (p_f) and frequency (ω_p) for each of the four clutches. Considering an adequate range of variation for ω_p and p_f, clutch engagements were performed for each pair (ω_p, p_f). The output parameter selected as an indicator for the occurrence of instability was the angular velocity upstream of the clutch (see fig. 1). After computing the relative spectrum of each experimental time history, the maximum value (peak) of each spectrum was reported on (ω_p, p_f) contour plots, as displayed in fig. 4b, which can be considered as potential and experimentally estimated stability charts. These charts showed that the essential model used in [15] cannot reproduce the identified unstable region. Therefore, models with higher complexity have to be considered.

Results and discussion

A 10 dofs model of the transmission is developed, as shown in Fig. 2. This model is described by a system of second order differential equations with time-periodic coefficients within the damping matrix.

Stability can be studied by either the Monodromy Matrix Method (MMM), or the Harmonic Balance Method (HBM). Both methods are based on Floquet theory [16], with the difference that the computational cost of MMM lies in the integration of the equations of the system, while the computational cost of HBM lies in the solution of a quadratic eigenvalue problem [17,18,19].

As a result, stability maps are drawn, characterized by the presence of unstable tongue-shaped regions (Arnold tongues). The stability map of the 10 dofs model is quite complex, with Arnold tongues of different kinds, as displayed in fig. 3. Tongues 1, 2 and 3 are regions of 'single-period instability, in which the system oscillates with a period equal to that of the parametric excitation (dither). On the other hand, in tongues 4 and 5 the unstable oscillation of the system has a double

Materials Research Proceedings **26** (2023) 437-442 https://doi.org/10.21741/9781644902431-71

period with respect to that of the parametric excitation, yielding a double-period instability. Finally, tongues 6, 7 and 8 are due to complex conjugate Floquet multipliers, coming out of the unit circle in the Argand plane [16], thus in general yielding non-periodic unstable solutions.

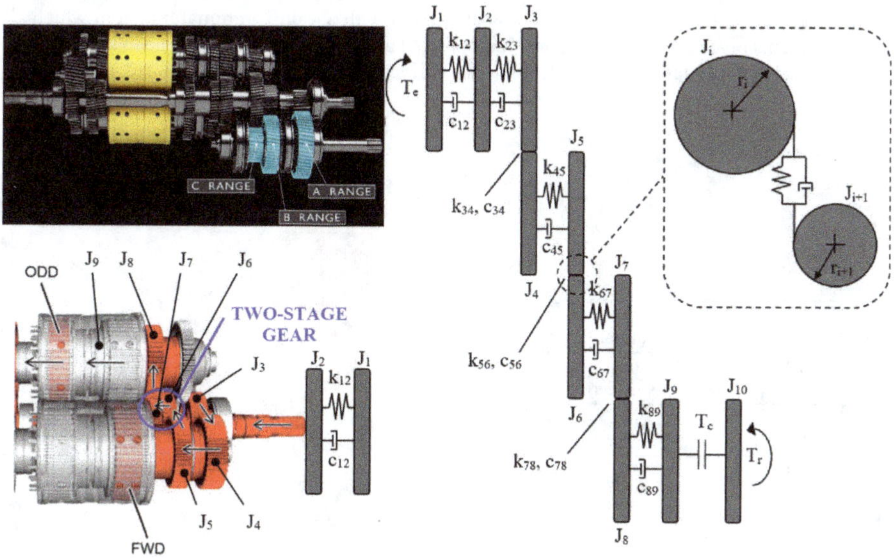

Fig. 2 – Dual-clutch transmission under analysis and 10 dofs model.

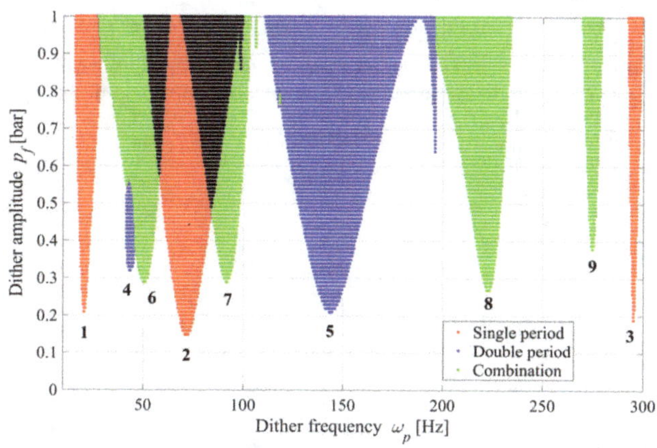

Fig. 3 – Stability map obtained with the 10 dofs model.

Given that the experimental campaign identified only one region of single-period instability, as shown in Fig. 4b, this suggests that the adopted 10 dofs model can actually be simplified into a more convenient minimal model. The stability map in fig. 3 shows only two single-period tongues at low frequencies (1 and 2), while the others are located at much higher frequencies (other single-period tongues appear at frequencies outside the upper limit of the x-axis in fig. 3).

Analyzing the model, it was found that the part of the transmission between the engine and the first gear wheel is the one that has a preponderant influence on unstable tongues 1 and 2.

Therefore, an excellent compromise has been found with a 4 dofs minimal model in which all the components placed between the clutch damper (J_2) and the clutch pack have been incorporated into a single rotating mass and a single torsional stiffness (fig. 4a).

(a)

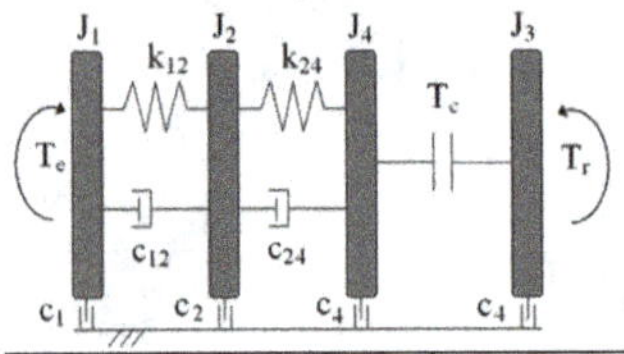

J_1	1.12 [kgm^2]	c_{12}	10 [Nms/rad]
J_2	0.231 [kgm^2]	c_{24}	0.1 [Nms/rad]
J_3	4.07 [kgm^2]	c_i	0.1 [Nms/rad]
J_4	0.122 [kgm^2]	μ_g	0.017
k_{12}	3856 [Nm/rad]	μ_0	0.09
k_{24}	32885 [Nm/rad]	p_m	0.1 [bar]

(b)

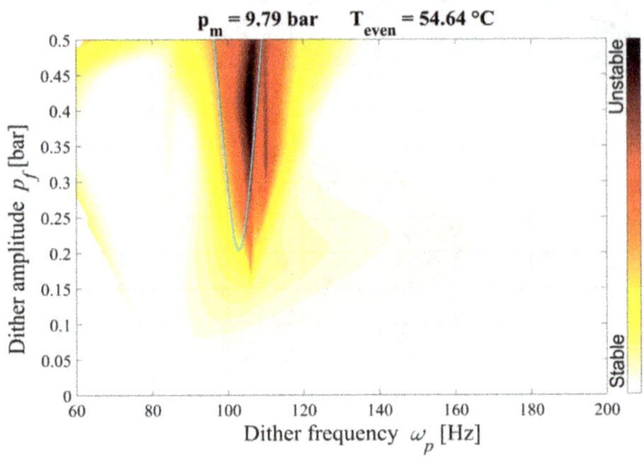

Fig. 4 – Four dofs model (a) and comparison with the experimental stability chart (b).

A very good correspondence was found between the theoretical stability map obtained with the 4 dofs model, as shown in fig. 4b (single-period tongue superimposed on the experimental chart), adopting the data reported below fig. 4a.

The equations of motion of the 4 dofs model shown in fig. 4a read:

$$\mathbf{M}\ddot{\boldsymbol{\theta}}(t)+\mathbf{C}(t)\dot{\boldsymbol{\theta}}(t)+\mathbf{K}\boldsymbol{\theta}(t)=\mathbf{g}(t) \tag{1}$$

with:

$$\mathbf{M}=\begin{bmatrix} J_1 & 0 & 0 & 0 \\ 0 & J_2 & 0 & 0 \\ 0 & 0 & J_4 & 0 \\ 0 & 0 & 0 & J_3 \end{bmatrix},\quad \mathbf{C}(t)=\begin{bmatrix} c_1+c_{12} & -c_{12} & 0 & 0 \\ -c_{12} & c_2+c_{12}+c_{24} & -c_{24} & 0 \\ 0 & -c_{24} & c_4+c_{24}+\mu_g f(t) & -\mu_g f(t) \\ 0 & 0 & -\mu_g f(t) & c_3+\mu_g f(t) \end{bmatrix}$$
$$\mathbf{K}=\begin{bmatrix} k_{12} & -k_{12} & 0 & 0 \\ -k_{12} & k_{12}+k_{24} & -k_{24} & 0 \\ 0 & -k_{24} & k_{24} & 0 \\ 0 & 0 & 0 & 0 \end{bmatrix},\quad \boldsymbol{\theta}(t)=\begin{bmatrix} \theta_1(t) \\ \theta_2(t) \\ \theta_4(t) \\ \theta_3(t) \end{bmatrix},\quad \mathbf{g}(t)=\begin{bmatrix} T_c \\ 0 \\ -\mu_0 f(t) \\ \mu_0 f(t)-T_r \end{bmatrix} \tag{2}$$

and:

$$f(t)=N_c R_m S\,p(t),\quad T_c=\mu(t)N_c R_m S\,p(t),\quad p(t)=p_m+p_f\cos(\omega_p t) \tag{3}$$

where J_i are moments of inertia, k_{ij} are torsional stiffness coefficients, c_{ij} are internal damping coefficients, c_i are external damping coefficients, N_c is the number of pairs of clutch plates, R_m is the mean radius of the clutch friction surfaces, $p(t)$ is the actuation pressure, S is the area of action of $p(t)$, T_c is the friction torque and μ is the friction coefficient.

As a first approximation, the friction coefficient is considered linearly dependent on the relative speed of the clutch plates and is expressed as:

$$\mu(t) = \mu_0 + \mu_g(\dot{\theta}_4 - \dot{\theta}_3) \tag{4}$$

The values adopted for the parameters in the 4 dofs model are reported below fig 4a.

Sensitivity analysis showed that the experimentally identified single-period tongue is mainly influenced by the equivalent stiffness k_{24} and moment of inertia J_4 of the components between the damper and the clutch. Furthermore, the damping coefficient of the clutch damper c_{12} has a great impact on the stability map because, if considering lower values, it gives rise to two combination regions which are symmetrical with respect to the identified single-period tongue, giving rise to a sort of tricuspid tongue. This tricuspid tongue is also visible in the map of the 10 dofs model, composed of unstable regions 6, 2 and 7: by reducing c_{12} the tips of tongues 6 and 7 are lowered and thus the unstable region broadens.

Finally, it should pointed out that the adopted model does not take into account several aspects of the complexity of the real transmission, such as nonlinear effects and asymmetries, able to produce relevant modifications in the stability thresholds, as for instance 'merging' effects between adjacent instability tongues [18,19].

Summary

The analysis of clutch judder on a dual-clutch gearbox for tractors has led to the identification of parametric resonance due to dither as a new possible cause for the onset of vibration.

Models with different levels of complexity have been developed, and a 4 dofs minimal model was identified, able to predict with good accuracy the unstable parametric region in which judder occurs and representing a useful tool at the design stage.

References

[1] D. Centea, H. Rahnejat, M.T. Menday, The influence of the interface coefficient of friction upon the propensity to judder in automotive clutches, Proceedings of the Institution of Mechanical Engineers, Part D: Journal of Automobile Engineering 213(5) (1999) 245-258. https://doi.org/10.1243/0954407991526847

[2] C. Bostwick, A. Szadkowski, Self-Excited Vibrations during Engagements of Dry Friction Clutches, SAE Technical Paper No. 982846 (1998). https://doi.org/10.4271/982846

[3] T. Kugimiya, N. Yoshimura, J. Mitsui, Tribology of Automatic Transmission Fluid, Tribology Letters 5(1) (1998) 49-56. https://doi.org/10.1023/A:1019156716891

[4] D. Centea, H. Rahnejat, M.T. Menday, Non-Linear Multi-Body Dynamic Analysis for the Study of Clutch Torsional Vibrations (Judder), Applied Mathematical Modelling 25(3) (2001) 177-192. https://doi.org/10.1016/S0307-904X(00)00051-2

[5] A. Crowther, N. Zhang, D. Liu, J. Jeyakumaran, Analysis and Simulation of Clutch Engagement Judder and Stick-Slip in Automotive Power Train System, Proceedings of the Institution of Mechanical Engineers Part D Journal of Automobile Engineering 218(12) (2004) 1427-1446. https://doi.org/10.1243/0954407042707731

[6] S. Iqbal, F. Al-Bender, A.P. Ompusunggu, B. Pluymers, W. Desmet, Modeling and Analysis of Wet Friction Clutch Engagement Dynamics, Mechanical Systems and Signal Processing (60) (2015) 420-436. https://doi.org/10.1016/j.ymssp.2014.12.024

[7] S. Mohire, R. Kapse, V. Tendulkar, Study of Clutch Judder Phenomenon in Manual Transmission Vehicle and Its Analysis Approach, SAE Technical Paper No. 2019-26-0215 (2019). https://doi.org/10.4271/2019-26-0215

[8] T. Gkinis, R. Rahmani, H. Rahnejat, Effect of clutch lining frictional characteristics on take-up judder, Proceedings of the Institution of Mechanical Engineers, Part K: Journal of Multibody Dynamics 231(3) (2017) 493-503. https://doi.org/10.1177/1464419317708946

[9] R.P. Jarvis, B. Mills, Vibrations Induced by Dry Friction, Proceedings of the Institution of Mechanical Engineers 178(1) (1963) 847-857. https://doi.org/10.1177/0020348363178001124

[10] M. Pandey, R.H. Rand, A.T. Zehnder, Frequency Locking in a Forced Mathieu-van Der Pol-Duffing System, Nonlinear Dynamics 54(1-2) (2008) 3-12. https://doi.org/10.1007/s11071-007-9238-x

[11] K. Berglund, P. Marklund, H. Lundh, R. Larsson, Prediction of Driveline Vibrations Caused by Ageing the Limited Slip Coupling, Proceedings of the Institution of Mechanical Engineers, Part D: Journal of Automobile Engineering 230(12) (2016) 1687-1698. https://doi.org/10.1177/0954407015619505

[12] J.R. Barber, Thermoelastic instabilities in the sliding of conforming solids, Proceedings of the Royal Society A 312(1510) (1969) 381-394. https://doi.org/10.1098/rspa.1969.0165

[13] Y.B. Yi, J.R. Barber, P. Zagrodzki, Eigenvalue solution of thermoelastic instability problems using Fourier reduction, Proceedings of the Royal Society A 456(2003) (2000) 2799-2821. https://doi.org/10.1098/rspa.2000.0641

[14] T. Paygude, R. Joshi, Modeling and Analysis of Clutch Engagement Judder in Commercial Vehicle Powertrain Systems, SAE Technical Paper No. 2019-01-0784 (2019). https://doi.org/10.4271/2019-01-0784

[15] M. Tentarelli, S. Cantelli, A. De Felice, S. Sorrentino, Cold Judder in Tractor Drivelines: an Essential Model for Stability Analysis, in: Proceedings of the 27th IAVSD Symposium on Dynamics of Vehicles on Roads and Tracks, 2021 (in press). https://doi.org/10.1007/978-3-031-07305-2_69

[16] V.A. Yakubovich, V.M. Starzhinski, Linear Differential Equations with Periodic Coefficients, John Wiley and Sons, New York, 1975.

[17] A. De Felice, S. Sorrentino, Stability Analysis of Parametrically Excited Gyroscopic Systems, in: Proceedings of XXIV Conference of the Italian Association of Theoretical and Applied Mechanics (AIMETA XXIV), 2019, pp. 1316-1331. https://doi.org/10.1007/978-3-030-41057-5_106

[18] A. De Felice, S. Sorrentino, Damping and Gyroscopic Effects on the Stability of Parametrically Excited Continuous Rotor Systems, Nonlinear Dynamics 103(4) (2021) 3529-3555. https://doi.org/10.1007/s11071-020-06106-3

[19] A. De Felice, S. Sorrentino, Effects of anisotropic supports on the stability of parametrically excited slender rotors, Nonlinear Dynamics, online 28 May 2022 (DOI 10.1007/s11071-022-07487-3). https://doi.org/10.1007/s11071-022-07487-3

Theoretical and Applied Mechanics - AIMETA 2022 Materials Research Forum LLC
Materials Research Proceedings **26** (2023) 443-448 https://doi.org/10.21741/9781644902431-72

Identification of normal modes of a set of strongly nonlinear springs

Francesco Latini[1, a *], Jacopo Brunetti[2, b], Walter D'Ambrogio[2, c], and Annalisa Fregolent[1, d]

[1]Dipartimento di Ingegneria Meccanica e Aerospaziale, Università di Roma "La Sapienza", Rome, Italy

[2]Dipartimento di Ingegneria Industriale e dell'Informazione e di Economia, Università dell'Aquila, L'Aquila, Italy

[a]francesco.latini@uniroma1.it, [b]jacopo.brunetti@univaq.it, [c]walter.dambrogio@univaq.it, [d]annalisa.fregolent@uniroma1.it

Keywords: Nonlinear Identification, Nonlinear Normal Modes, Nonlinear Time-Series Analysis

Abstract. The dynamics of connecting elements should be identified to evaluate their effects on the assemblies where they are employed. In general, a linear analysis is sufficient to determine their dynamics. However, in some cases, their responses depend on the amplitude and frequency of the excitation, thus nonlinear analyses must be carried out. Following a modal approach, Nonlinear Normal Modes (NNMs) can be used. The aim of this work is to identify the NNMs of a set of nonlinear connecting elements properly designed to be considered as strongly nonlinear springs. These elements have been manufactured and tested to measure some of their NNMs. The time series of the oscillation on some points is recorded using a laser vibrometer for different amplitude and frequencies of excitation. Then, the Virtual Point Transformation (VPT) is used to reduce the obtained data on the two physical points through which the spring is connected to other subsystems. The procedure is repeated for each NNM and the modal basis of each element is expressed as function of the level of excitation.

Introduction

In the last decades the dynamic characterization of joints has become of fundamental importance [1]. Many works have been done to address the problem of bolted joints and lap joints to evaluate their effect in the dynamics of coupled systems. On the other hand, nonlinearities are employed in energy-harvesting applications [2] and in dynamic vibration absorbers [3]. Different studies are focused on the effects that nonlinear connections produce on coupled assemblies. In general, nonlinearities arise when the stiffness and damping properties of the connection depend on some parameters, like the exchanged force or the relative motion between the connected parts. Those dependencies are usually represented using nonlinear laws to approximate the dynamic behavior of the connection. However, in some cases those laws might not represent properly the dynamic behavior of the joint, thus providing misleading results. In those cases, an experimental characterization is the only way to correctly estimate the nonlinear behavior of the connection.

The aim of this work is to characterize the dynamic behavior of a nonlinear element that can be coupled to any subsystem such that it can be regarded as a localized source of nonlinear phenomena. The connecting element considered in this work is suitably designed to exhibit a strong nonlinear behavior. It is derived from the one used in [4] but some adjustments in the design are performed to make it more robust and allow to couple/decouple it to/from the connected parts without affecting its dynamic behavior. Two connecting elements are assembled and tested, following the same procedure described in [5]. They are going to be used as nonlinear connecting elements to couple two beams and obtain an assembly like the one described in [4].

Description of the Nonlinear Connecting Element

In the current Section, the new design for the Nonlinear Connecting Element (NLCE) is presented. The connecting element is similar to the one described in [4]. However, some design aspects are improved to make the element more robust and ensure that the junction with other subsystems is localized. Fig. 1 shows the drawings of the NLCE and its 3D model.

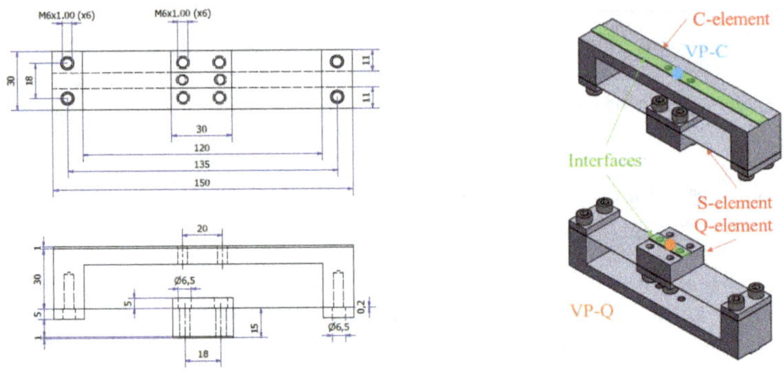

a) Design of the NLCE b) 3D model of the NLCE

Fig. 1– Design of the NLCE; the dimensions are expressed in millimeters

It is composed of three main parts: a C shaped element and a block element (referred as C-element and Q-element hereinafter, respectively) connected through a 0.2 mm thick spring-steel sheet (referred as S-element). During the assembly a thermal cycle has been introduced to guarantee a pre-tensile stress on the S-element to avoid buckling phenomena. Fig. 2 shows the first three modes of vibration of the NLCE obtained by performing a linear modal analysis on ANSYS. These are modes of interest because there is a large deformation of the S-element that leads to a significant nonlinear behavior of the NLCE.

a) First mode b) Second mode c) Third mode

Fig. 2 – First three modes of vibration of the NLCE

This design guarantees that once the NLCE is assembled, it can be treated as a unique subsystem, and can be jointed/disjointed to/from other subsystems making it robust and reliable for different tests. As it can be seen from Fig. 2, the three modes are associated with a relative motion between the C- and Q-element due to the deformation of the S-element. For this reason, it is possible to represent the dynamic behavior of the NLCE as the behavior of two masses connected through one longitudinal spring along the z-axis and two rotational springs acting along the x- and y-axis, respectively.

Since the internal deformation of the C- and Q-elements is negligible, a Virtual Point Transformation (VPT) [6] can be applied on their measurement points to obtain the time response on VP-C and VP-Q shown in Fig. 1b. For a given measurement point, the VPT is expressed by:

$$\begin{bmatrix} u_x \\ u_y \\ u_z \end{bmatrix} = \begin{bmatrix} e_{x,X} & e_{x,Y} & e_{x,Z} \\ e_{y,X} & e_{y,Y} & e_{y,Z} \\ e_{z,X} & e_{z,Y} & e_{z,Z} \end{bmatrix} \begin{bmatrix} 1 & 0 & 0 & 0 & r_Z & -r_Y \\ 0 & 1 & 0 & -r_Z & 0 & r_X \\ 0 & 0 & 1 & r_Y & -r_X & 0 \end{bmatrix} \begin{bmatrix} q_X \\ q_Y \\ q_Z \\ q_{\theta X} \\ q_{\theta Y} \\ q_{\theta Z} \end{bmatrix} + \begin{bmatrix} \mu_x \\ \mu_y \\ \mu_z \end{bmatrix} \tag{1}$$

where $\boldsymbol{u}=[u_x, u_y, u_z]^T$ is the array containing the displacements in the local frame (x,y,z) of the measurement point accounting for the sensor orientation; $\boldsymbol{q}=[q_X, q_Y, q_Z, q_{\theta X}, q_{\theta Y}, q_{\theta Z}]^T$ is the array containing the 6 DoFs of the virtual point in the global frame (X,Y,Z). Moreover, coefficients (r_X, r_Y, r_Z) represent the distances along the three axes between the measurement point and the virtual point; the matrix of coefficients $e_{i,j}$ with $i=x,y,z$ and $j=X,Y,Z$ represent the transformation matrix between the two frames and the vector of coefficients μ_i is the residual of displacements associated with the deformation. Note that, only the three DoFs $\boldsymbol{q}=[q_Z, q_{\theta X}, q_{\theta Y}]^T$ are meaningful to characterize the dynamics of the NLCE and only the displacements along the z-axis are measured. Then, the VPT is performed to obtain $\boldsymbol{q}_C$ and $\boldsymbol{q}_Q$. The deformation of the NLCE $\boldsymbol{q}_N$ is:

$$\boldsymbol{q}_N = \boldsymbol{q}_C - \boldsymbol{q}_Q \tag{2}$$

Since the NLCEs are suitably designed to be nonlinear, the resonance frequencies ω_N vary for increasing levels of the excitation. Moreover, the response along the DoFs $\boldsymbol{q}_N$ to harmonic excitation at a given resonance frequency can be nonlinear and contain super-harmonic components. Thus, each term $q_i(t)$ of $\boldsymbol{q}_N$ can be approximated using a truncated Fourier series:

$$q_i(t) \approx a_{0,i} + \sum_{k=1}^{n} a_{k,i}\cos(k\omega_0 t) + b_{k,i}\sin(k\omega_0 t) \tag{3}$$

This allows to evaluate how the coefficients $a_{0,i}$, $a_{k,i}$ and $b_{k,i}$ vary as the excitation level increases, highlighting the presence of super-harmonic components in the nonlinear modes of both NLCEs.

Experimental tests

Two NLCEs are manufactured and assembled following two different thermal cycles such that their dynamics are distinct. Both the NLCEs are tested following the same steps to obtain their linear and nonlinear behavior. In this Section the corresponding test procedures are described. The 1D scanner laser vibrometer Polytec PSV-500 is used to measure the velocity of the grid of points shown in Fig. 3.

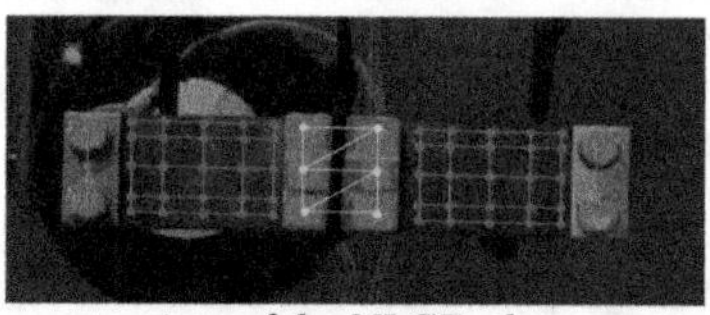

Fig. 3 – Grid of measurement points of the NLCE: the green points are referred to the Q-element, the red points to the C-element and the blue points to the S-element

Both the components are tested first using a low amplitude burst random excitation to estimate their linear behavior and then with increasing amplitude harmonic excitation signals to characterize their nonlinear dynamics [5]. In the linear tests, the Polytec software generates the excitation signal for the shaker (2025E Modal shaker, Modal Shop). In the nonlinear tests, instead, the Dewesoft acquisition system is used. The shaker is provided with an impedance head PCB 208 that measures both the force and the acceleration at the excitation point.

Materials Research Proceedings **26** (2023) 443-448 https://doi.org/10.21741/9781644902431-72

Results of linear tests

Fig. 4 shows the average spectrum of both NLCEs obtained by performing a linear test. This spectrum allows to localize the natural frequencies of the component in a given frequency range. The different thermal cycle followed for the second NLCE makes its natural frequencies higher than the first NLCE's. The dashed red lines in Fig. 4 highlight the frequencies of the modes of interest identified in the numerical model (Fig. 2) that are considered hereinafter. The resonance frequencies of the three modes are listed in Table 1. Fig. 5 shows the operational deflection shapes performing a Fast Scan analysis by exciting the NLCE with a harmonic force at the frequencies shown in Table 1. The obtained operational deflection shapes comply with the numerical results shown in Fig. 2.

a) First NLCE b) Second NLCE

Fig. 4 – Average spectra of the two NLCEs

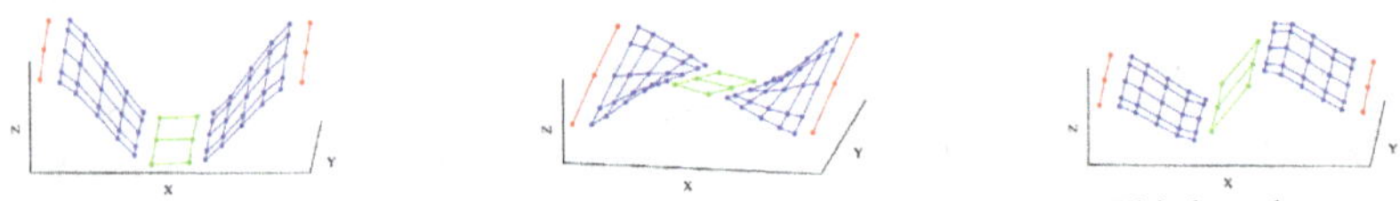

a) First mode b) Second mode c) Third mode

Fig. 5 – Operational deflection shape of the three modes obtained through Fast Scan analysis

Table 1 – Resonance frequencies of three modes of the two NLCEs

	Resonance frequency 1st NLCE	Resonance frequency 2nd NLCE	Δ%
First mode	28.3 Hz	38.4 Hz	35.7
Second mode	53.9 Hz	57.8 Hz	7.2
Third mode	76.1 Hz	101.8 Hz	33.8

Results of nonlinear tests

Once the linear resonance frequencies of the modes of interest are found, it is possible to perform the nonlinear analysis to reconstruct the corresponding Nonlinear Normal Modes (NNMs). The measurement of the NNMs is carried out by setting the amplitude and by adjusting the frequency of the sinusoidal excitation to reach the resonance condition, i.e., when the velocity is in-phase with respect to the excitation force. Once the resonance condition is attained, the velocity of all the points in the grid is recorded and it is integrated to obtain the displacement time series. By setting several excitation levels, it is possible to find the corresponding resonance frequency, thus reconstructing the considered NNM. This procedure is repeated for the three modes of each NLCE, always starting from the low value of amplitude at the identified linear natural frequencies, listed in Table 1.

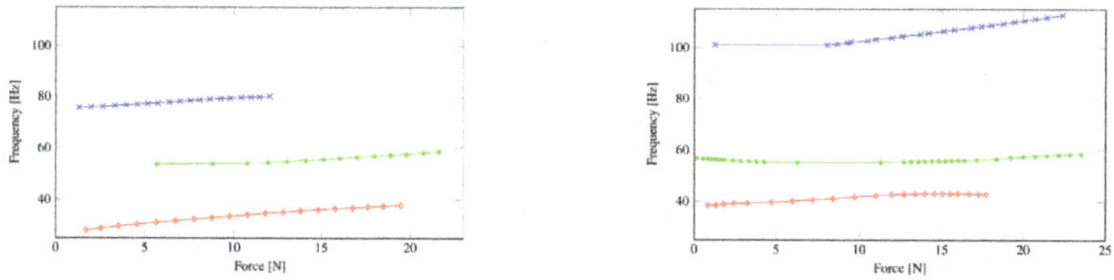

a) First NLCE b) Second NLCE

Fig. 6 – Frequency-Force Plot of both NLCEs: -◇- first mode; -●- second mode; -x- third mode

Fig. 6a and 6b show the resonance frequency of the three NNMs versus the excitation force for the first and second NLCE, respectively. Note that the resonance frequencies of the second NLCE are higher than those of the first one, especially for the first and third NNMs. As mentioned before, the internal deformation of the C- and Q-elements is negligible, thus Virtual Points can be introduced to describe their time response. According to the procedure described before, the analysis of the time response of VP-C and VP-Q allows to identify the mode shape of the NNMs associated with the deformation of the S-element. Fig. 7 shows the time responses of VP-C and VP-Q for the first NLCE along the three relevant DoFs (Z, θ_X, θ_Y). These displacement signals are obtained applying an excitation force of about 12 N. It is evident that, for each mode, the time response along one DoF is predominant with respect to the others. Nevertheless, the other DoFs have a non-null contribution in the time response, thus they cannot be neglected when estimating the mode shape of the corresponding modes. For example, the time responses along the DoFs θ_X and θ_Y of the first mode (Fig. 7 and 7g) show a non-negligible oscillation. Furthermore, it is possible to observe the presence of super-harmonic components. For this reason, the responses can be fitted using a Fourier series according to Eq. (3) truncated at $n=4$. Fig. 8 shows the variation of the Fourier coefficients for increasing levels of excitation on the three DoFs of each mode of the first NLCE. The results highlight that, for each mode, the contribution of the first harmonic of the response along one DoF is predominant with respect to all the other terms. Note that, for the first two modes, there is a non-negligible response for the other DoFs, while for the third mode only the DoF θ_Y is involved. The Fourier fitting highlights that super-harmonic components are non-negligible, especially in the non-predominant DoFs, e.g. DoF Z for the second mode and DoF θ_X for the first mode. The results of the second NLCE are similar to those of the first NLCE but they are not shown for the sake of brevity.

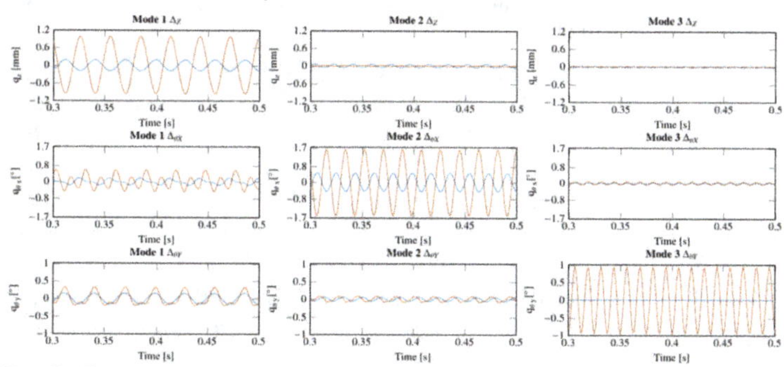

Fig. 7– Time response of VP-C (-) and VP-Q (-) of the first NLCE for F=12 N.

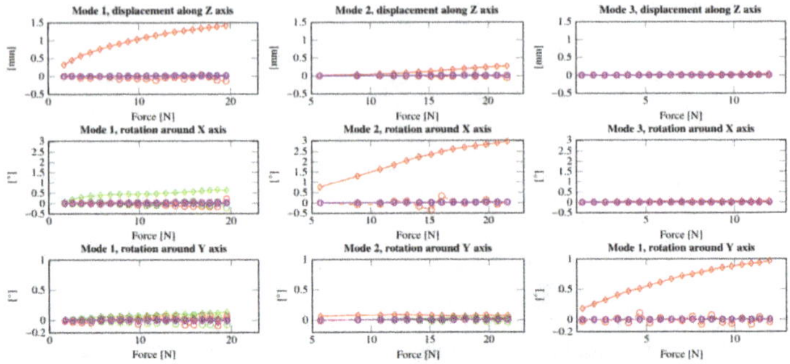

Fig. 8 – Fourier coefficients of the differential displacements for the three modes of the first NLCE: -a_0; -◊ a_1; -○- b_1; -◊ a_2; -○- b_2; -◊ a_3; -○- b_3; -◊ a_4; -○- b_4.

Summary

In this work, the dynamic behavior of a pair of Nonlinear Connecting Elements is experimentally evaluated. A linear experimental modal analysis is performed and three linear normal modes of interest are identified. Subsequently, nonlinear tests are carried out to identify the corresponding nonlinear normal modes in terms of variation of resonance frequency and mode shape. Since the nonlinearity introduces super-harmonics in the resonant response of the system, their contribution has been evaluated in the nonlinear normal mode identification. The identification procedure carried out on two similar nonlinear connecting elements shows that a different pre-tensioning of the spring-steel sheet significantly affects the nonlinear dynamics of these systems.

References

[1] Z. Saeed, S.W.B. Klaassen, C.M. Firrone, T. M. Berruti, D.J. Rixen, Experimental Joint Identification Using System Equivalent Model Mixing in a Bladed-Disk, J. Vib. Acoust. 142 (2020) 1-29. https://doi.org/10.1115/1.4047361

[2] G. Gatti, Optimizing elastic potential energy via geometric nonlinear stiffness, Commun. Nonlinear Sci. Numer. Simul. 103 (2021) 106035. https://doi.org/10.1016/j.cnsns.2021.106035

[3] F.S. Samani, F. Pellicano, Vibration reduction on beams subjected to moving loads using linear and nonlinear dynamic absorbers, J. Sound and Vib. 325 (2009) 742-754. https://doi.org/10.1016/j.jsv.2009.04.011

[4] F. Latini, J. Brunetti, W. D'Ambrogio, M.S. Allen, A. Fregolent, Nonlinear substructuring in the modal domain: numerical validation and experimental verification in presence of localized nonlinearities, Nonlinear Dyn. 104 (2021) 1043-1067. https://doi.org/10.1007/s11071-021-06363-w

[5] F. Latini, J. Brunetti, M. Kwarta, M.S. Allen, W. D'Ambrogio, A. Fregolent, Experimental results of nonlinear structure coupled through nonlinear connecting elements, Proceedings of ISMA 2020, Leuven, Belgium, September 2020, pp. 3011-3022.

[6] M.V. Van der Seijs, D.D. Van den Bosch, D.J. Rixen, D. De Klerk, An improved methodology for the virtual point transformation of measured frequency response functions in dynamic substructuring, Proc. COMPDYN 2013, Kos Island, Greece, June 2013, pp. 4334-4347. https://doi.org/10.7712/120113.4816.C1539

Theoretical and Applied Mechanics - AIMETA 2022 Materials Research Forum LLC
Materials Research Proceedings **26** (2023) 449-454 https://doi.org/10.21741/9781644902431-73

Complex dynamics in non-Newtonian fluid-structure interaction

F. Pellicano[1, a *], A. Zippo[1,b] and G. Iarriccio[1,c]

[1]University of Modena and Reggio Emilia, Dept. of Engineering Enzo Ferrari, InterMech MoRe Centre, 41121 Modena, Italy

[a]francesco.pellicano@unimore.it, [b]antonio.zippo@unimore.it, [c]giovanni.iarriccio@unimore.it

Keywords: Nonlinear Dynamics, Complexity, Fluid Structure Interaction, Experiments

Abstract. This paper presents the results of an extensive experimental campaign focused on the analysis of the dynamic interactions between an elastic structure and a non-Newtonian fluid. The structure is a circular cylindrical shell clamped in one end to a shaking table and in the other end to a heavy rigid disk. The shell has been investigated both in presence and absence of fluid. The fluid is a mixture of water and corn starch flour, commonly called Oobleck. The experiments were carried out at low and high vibrating energy, in order to clarify the influence of the fluid in different conditions: changing of modal properties, onset of complex dynamics when the fluid-solid transitions take place in the fluid.

Introduction

Fluid structure interaction (FSI) phenomena are of interest for several engineering fields as well as in medical science, and, of course in bioengineering or biomechanics. One can find countless examples of FSI problems in engineering, e.g. flutter of airplane wings, galloping in powerlines and bridge cables, supersonic panel flutter, pipes flutter, fully or partially filled tanks, heat exchangers. In the field of Medical Sciences an important example is the human aorta, where the fluid is highly viscous and non-Newtonian and the artery wall is hyper-elastic, this is a combination of exceptionally difficult problems.

At the end of the previous century (1983) Babcock [1] published an interesting review paper, where he counted, at that time, about 50,000 papers published on the topic of shell stability; he pointed out the attention on open topics such as post-buckling and imperfection sensitivity; dynamic buckling; plastic buckling; experiments. Further reviews, see Ref.[2-4], confirmed the need of further research on the dynamics and stability of shells, which is a topic not well understood.

In the sixties and seventies a series of studies regarding dynamic shells instabilities were published. Further studies addressing the parametric excitation of thin circular cylindrical shells, using the Donnell's shallow shell theory, can be found in Refs. [5-9]. More recent studies on dynamic instabilities of cylindrical shells considered more refined theories such as the Sanders-Koiter theory [13-18] validated from lab experiments; in such series of studies it was clarified the role of axial loads generated by inertial effects on the onset of chaotic vibrations, very high amplitudes of vibration with associated extreme noise production.

In order to have a comprehensive description of fluid structure interaction phenomena, models and applications, the reader is suggested to read the monumental work of Paidoussis [19], [20]; in such treatises the main fluid-interaction models are described as well as methods of analysis. Another interesting paper to be mentioned is a review published in 2003 by Amabili and Païdoussis [21], where more than 300 papers on the topic of nonlinear vibrations of shells with or without FSI were listed; among the deep and interesting comments on the literature the authors pointed out the attention on two aspects, we report their full sentences: "only 23 of the more than 350 papers discussed in the present review give experimental data on large-amplitude vibrations of complete shells", "most of the papers reviewed are dedicated to various theoretical aspects of the problem, with very few experimental results, although more experimental data are available for supersonic flutter of shells".

In the following, the bibliographic analysis is focused on FSI papers strictly related to the present study; it is worthwhile to stress that almost all studies deal with inviscid or Newtonian fluids.

In the 1967 Chu & Kana [22] published a paper focused on nonlinear vibrations of partially filled tanks; the nonlinearity was attributed to free surface waves. In the 1979 Ramachandran [23] analyzed large amplitude nonlinear vibrations of circular cylindrical shells in contact with a dense fluid, no free surface was present. The study was theoretical using the Donnell's shell theory and the potential theory for the inviscid incompressible fluid.

Using a semi analytical model based on the Donnell shallow shell theory for the structure and the potential flow theory for the inviscid and incompressible fluid, Amabili et al. [24]-[29] published a series of papers where the effect of a quiescent or flowing heavy fluid was investigated. Further investigations focused on the interactions of circular shells with supersonic flows, compressible, annular and unbounded flows can be found in Refs. [30]-[34]; in such series of studies the presence of compressive forces was accounted for as well.

In 2012 Girchenko et al studied numerically the interaction of a nonlinear viscous fluid, having a pseudoplastic nature, with a helical shell. They combined two commercial software FlowVision (finite volumes) and Simulia ABAQUS (finite elements). They showed the differences between Newtonian and Non-Newtonian flows in terms of stresses caused on the helical structure. Another study regarding FSI and non-Newtonian fluid was focused on arterial bypass [36], the effects of wall elasticity and non Newtonian rheology were investigated numerically through the commercial software ANSYS.

An experimental study on the rheology and processing of solvent-free core shell "polymer opals", see Ref. [37], analyzed an elastic shell grafted to hard colloidal polymer core particles in order to study the optical properties under deformation.

In 2019 Wu et al. [38] presented a numerical study on interaction between elastic multilayered spheres and a non-Newtonian fluid. They analyzed gold nanospheres immersed in water and calculated theoretically the natural frequencies and the quality factors.

The bibliographic analysis clearly shows that, even though a huge number of publications can be found about FSI problems, and many papers are available about non-Newtonian fluids, the interactions between vibrating structures and non-Newtonian fluids appear to be an almost unexplored field.

Experimental setup and Specimen definition

The studied structure is a polyethylene terephthalate (PET) shell, vertically placed, wedged in an aluminum base rigidly connected to the shaker at the bottom, and constrained at the top by a C-40 steel disc, which is called top mass.

The internal face of the shell was fixed to the external face of both the top mass and the basement by means of an instant cyano-acrylic glue. In addition, the lower end of the shell was secured to the basement via an aluminum ring, lying on the basement and tightened to it with screws. This ring was inserted to guarantee the interlocking constraint. The realized setup leads to a lower end of the shell solidary oscillating with the shaker, since the basement is anchored to the shaker horizontal plate with screws, whilst the top mass induces a rigid body motion at the upper end of the shell, preventing this end from moving freely.

The material of the shell was chosen to be PET, since, as a polymer, it can reach high oscillation amplitudes without addressing plastic deformation phenomena [16]. The shell is a center-holed cylinder characterized by a length (L = 135 mm) greater than the diameter (D = 80 mm).

The steel disc acting as a top mass is highly compact to avoid not wanted side effects, such as the spectral proximity of two vibration modes, especially the one to be excited (axial-symmetrical) and the tilt modes.

The tests were performed using a Dongling ES-340 shaker; a dynamic characterization of the shaker was realized in order to study the coupling with the shell.

The accelerometer on the shaker plate was dedicated to the measurement of the excitation produced by the shaker on the system basement-shell-top mass. Three accelerometers were placed on the top face of the top-mass, close to its circular edge and angularly spaced by 120°. Each sensor was positioned with the X-axis tangent to the circular edge and the Z-axis perpendicular to the top mass face and pointing upward (opposite to the shell). The goal of these three sensors is to measure the acceleration of the disc in the Z-direction, i.e., the axial (along L) oscillation of the shell, since under non-linearity conditions this acceleration/oscillation is not directly linked to the shaker excitation which is measured via the accelerometer on the shaker plate. However, the realized setup also allows to detect possible rotations of the structure around axes lying on a plane perpendicular to L and passing through either the structure mass center (tilt modes) or the basement fixing (beam modes). The detection of these modes is performed by correlating the signal acquired in the three directions by the three accelerometers.

The vibrometer and the telemeter are used for the study of the shell radial modes, namely possible vibrations of the shell along the radial direction (planes perpendicular to the shell length). As shown in Figure 3, the laser ray of both sensors focuses on points at half-length of the shell. Specifically, in order to obtain a proper refracted signal by the shell, needed for the vibrometer and telemeter measurements, a refractive sticker and a black mark were applied on the shell in the points where the rays imping the shell, respectively. The angle between the vibrometer and the telemeter was chosen to be 57°, because this value allows to avoid a redundancy in the measurements of the drift velocity (vibrometer) and the displacement (telemeter) of the shell wall in the radial direction for a wide range of possible radial modes which can be excited in the non-linear dynamic regime.

Figure 3 – a) Setup: (1) shaker, (2) shell, (3) measurement accelerometers, (4) telemeter and (5) control accelerometer on the shaker plate. b) Overall experimental setup.

Nonlinear dynamic scenario

Experiments are now carried out considering an empty and a fluid filled shell. The shell is now excited from the base, the excitation signal provided to the amplifier is harmonic with different amplitudes (0.01-0.08V) and frequencies (150-310Hz for the empty shell and 150-270Hz for the fluid filled shell).

Figure 4 shows a bifurcation diagram obtained from Poincaré sections obtained by using a single signal (lateral displacement). The dynamic scenario is extremely rich, we observed different sub-harmonic responses: ½, 1/3, ¼, 1/8 as well as quasiperiodic and chaotic dynamics in a very wide frequency range. For the sake of brevity bifurcation diagrams obtained by the other signals (accelerometers, Laser Doppler) and by changing the excitation levels are not reported for the sake of brevity.

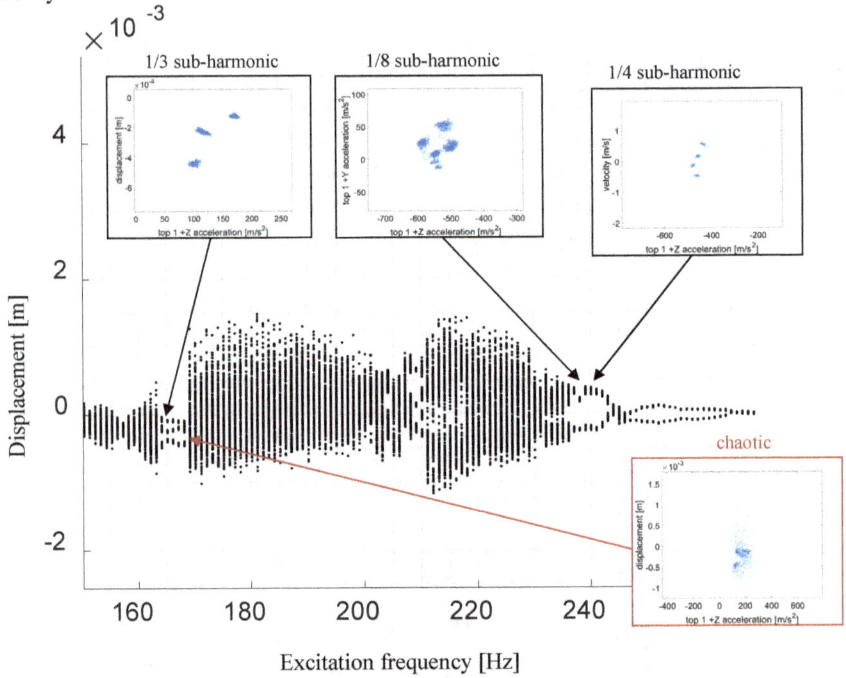

Figure 4: Bifurcation diagram of Poincaré maps: excitation level 0.06V, downward frequency sweep.

References

[1] C.D. Babcock, Shell stability, Journal of Applied Mechanics 50 (1983) 935–940. https://doi.org/10.1115/1.3167206

[2] C.R. Calladine, Understanding imperfection-sensitivity in the buckling of thin-walled shells, Thin-Walled Structures 23 (1995). https://doi.org/10.1016/0263-8231(95)00013-4

[3] J.G. Teng, Buckling of thin shells: recent advances and trends, Applied Mechanics Reviews 49 (4) (1996) 263–274. https://doi.org/10.1115/1.3101927

[4] Alijani, F., Amabili, M.: Non-linear vibrations of shells: a literature review from 2003 to 2013. Int. J. Non-linear Mech. **58**, 233–257 (2014). https://doi.org/10.1016/j.ijnonlinmec.2013.09.012

[5] F. Pellicano, M. Amabili, Stability and vibration of empty and fluid-filled circular cylindrical shells under static and periodic axial loads, International Journal of Solids and Structures 40 (2003) 3229–3251. https://doi.org/10.1016/S0020-7683(03)00120-3

[6] P.B. Goncalves, ZJGN Del Prado, Nonlinear oscillations and stability of parametrically excited cylindrical shells, Meccanica 37 (2002) 569–597. https://doi.org/10.1023/A:1020972109600

[7] A.A. Popov, Parametric resonance in cylindrical shells: a case study in the nonlinear vibration of structural shells, Engineering Structures 25 (2003) 789–799. https://doi.org/10.1016/S0141-0296(03)00006-3

[8] E. Jansen, Dynamic stability problems of anisotropic cylindrical shells via a simplified analysis, Nonlinear Dynamics 39 (2005) 349–367. https://doi.org/10.1007/s11071-005-4343-1

[9] Zippo A., Iarriccio G., Pellicano F., Shmatko T., Vibrations of plates with complex shape: Experimental modal analysis, finite element method, and R-functions method, (2020) Shock and Vibration, 2020. https://doi.org/10.1155/2020/8882867

[10] Iarriccio G., Zippo A., Pellicano F., Asymmetric vibrations and chaos in spherical caps under uniform time-varying pressure fields, (2022) Nonlinear Dynamics, 107 (1), pp. 313 – 329. https://doi.org/10.1007/s11071-021-07033-7

[11] Hemmatnezhad M., Iarriccio G., Zippo A., Pellicano F., Modal localization in vibrations of circular cylindrical shells with geometric imperfections (2022) Thin-Walled Structures, 181, art. no. 110079. https://doi.org/10.1016/j.tws.2022.110079

[12] G. Catellani, F. Pellicano, D. Dall'Asta, M. Amabili, Parametric instability of a circular cylindrical shell with geometric imperfections, Computers & Structures 82 (2004) 2635–2645. https://doi.org/10.1016/j.compstruc.2004.07.006

[13] F. Pellicano and K. V. Avramov, "Linear and nonlinear dynamics of a circular cylindrical shell connected to a rigid disk", *Communications in Nonlinear Science and Numerical Simulation*, 2007, **12(4)**, 496-518. https://doi.org/10.1016/j.cnsns.2005.04.004

[14] F. Pellicano, "VIBRATIONS OF CIRCULAR CYLINDRICAL SHELLS: THEORY AND EXPERIMENTS", J. of Sound and Vibration, 2007, **303**, 154–170. https://doi.org/10.1016/j.jsv.2007.01.022

[15] F. Pellicano Dynamic stability and sensitivity to geometric imperfections of strongly compressed circular cylindrical shells under dynamic axial loads, Communications in Nonlinear Science and Numerical Simulations, 2009, **14(8)** (2009), 3449-3462. https://doi.org/10.1016/j.cnsns.2009.01.018

[16] F. Pellicano, Dynamic instability of a circular cylindrical shell carrying a top mass under seismic excitation: experiments and theory, Int. J. of Solids and Structures, 48 (2011) 408–427. https://doi.org/10.1016/j.ijsolstr.2010.09.024

[17] Iarriccio G., Zippo A., Pellicano F., Barbieri M., Resonances and nonlinear vibrations of circular cylindrical shells, effects of thermal gradients, (2021) Proceedings of the Institution of Mechanical Engineers, Part C: Journal of Mechanical Engineering Science, 235 (20), pp. 4818 – 4832. https://doi.org/10.1177/0954406220907616

[18] Pellicano F., Zippo A., Iarriccio G., Barbieri M., Experimental Study on Nonlinear Random Excitation, (2020) Lecture Notes in Mechanical Engineering, pp. 637 – 648. https://doi.org/10.1007/978-3-030-31154-4_54

[19] M.P. Païdoussis, *Fluid-Structure Interactions, Volume 1: Slender Structures and Axial Flow, Second Edition*, ISBN: 9780123973122, Academic Press, 2014.

[20] M.P. Païdoussis, *Fluid-Structure Interactions, Volume 2: Slender Structures and Axial Flow, Second Edition*, ISBN: 9780123973122, Academic Press, 2016.

[21] M. Amabili and M.P. Païdoussis, Review of studies on geometrically nonlinear vibrations and dynamics of circular cylindrical shells and panels, with and without fluid-structure interaction, ASME Applied Mechanics Review **56(4)**, 2003. https://doi.org/10.1115/1.1565084

[22] W. Chu and D. Kana. 1967 A theory for nonlinear transverse vibrations of a partially filled elastic tank. AIAA Journal 5, 1828—1835. https://doi.org/10.2514/3.4312

[23] J. Ramachandran, 1979, Non-linear vibrations of cylindrical shells of varying thickness in an incompressible fluid. Journal of Sound and Vibration 64, 97—106. https://doi.org/10.1016/0022-460X(79)90575-3

[24] M. Amabili, F. Pellicano and M.P. Païdoussis, "Nonlinear vibrations of Simply Supported, Circular Cylindrical Shells, Coupled to Quiescent Fluid". J. of Fluids and Structures, 12, 883-918, 1998. https://doi.org/10.1006/jfls.1998.0173

[25] M. Amabili, F. Pellicano and M.P. Païdoussis, Letter to the Editor, "Further Comments on Nonlinear Vibrations of Shells", J. of Fluids and Structures, 13(1), 159-160, 1999. https://doi.org/10.1006/jfls.1998.0193

[26] M. Amabili, F. Pellicano and M.P. Païdoussis, "Non-Linear Dynamics and Stability of Circular Cylindrical Shells Containing Flowing Fluid. Part I Stability". J. of Sound and Vibration, 225(4), 655-699, 1999. https://doi.org/10.1006/jsvi.1999.2255

[27] M. Amabili, F. Pellicano and M.P. Païdoussis, "Non-Linear Dynamics and Stability of Circular Cylindrical Shells Containing Flowing Fluid, Part II: Large-Amplitude Vibrations without Flow". J. of Sound and Vibration, 228(5), 1103-1124, 1999. https://doi.org/10.1006/jsvi.1999.2476

[28] M. Amabili, F. Pellicano and M.P. Païdoussis, "Non-Linear Dynamics and Stability of Circular Cylindrical Shells Containing Flowing Fluid. Part III: Truncation Effect without Flow and Experiments". J. of Sound and Vibration, 237 (4), 617-640, 2000. https://doi.org/10.1006/jsvi.2000.3071

[29] M. Amabili, F. Pellicano and M.P. Païdoussis, "Non-Linear Dynamics and Stability of Circular Cylindrical Shells Containing Flowing Fluid. Part IV: Large Amplitude of Vibrations with Flow". J. of Sound and Vibration, 237(4), 641-666, 2000. https://doi.org/10.1006/jsvi.2000.3070

[30] M. Amabili and F. Pellicano, "Nonlinear Supersonic Flutter of Circular Cylindrical Shells". AIAA Journal, 39 (4), 564-573, 2001. https://doi.org/10.2514/2.1365

[31] M. Amabili, F. Pellicano and M.P. Païdoussis, "Nonlinear Stability of Circular Cylindrical Shells in Annular and Unbounded Axial Flow". ASME J. Applied Mechanics, 68, 827-834, 2001. https://doi.org/10.1115/1.1406957

[32] M. Amabili and F. Pellicano, "Multimode Approach to Nonlinear Supersonic Flutter of Imperfect Circular Cylindrical Shells", ASME J. of Applied Mechanics , 69, 117-129, 2002. https://doi.org/10.1115/1.1435366

[33] M. Amabili ,F. Pellicano and M. P. Païdoussis, "Non-linear dynamics and stability of circular cylindrical shells conveying Ô¨Çowing Ô¨Çuid", Computers and Structures, 2002, 80, 899-906. https://doi.org/10.1016/S0045-7949(02)00055-X

[34] F. Pellicano, M. Amabili, "Stability and vibration of empty and fluid-filled circular cylindrical shells subjected to dynamic axial loads" Int. J. of Solids and Structures, 2003, 40, 3229-3251. https://doi.org/10.1016/S0020-7683(03)00120-3

[35] A. A. Girchenko, V. A. Eremeyev, H. Altenbach, Interaction of a helical shell with a nonlinear viscous fluid, *International Journal of Engineering Science* **61** (2012) 53–58. https://doi.org/10.1016/j.ijengsci.2012.06.009

[36] F. Kabinejadiana, D. N. Ghista, Compliant model of a coupled sequential coronary arterial bypass graft: Effects of vessel wall elasticity and non-Newtonian rheology on blood flow regime and hemodynamic parameters distribution. *Medical Engineering & Physics* **34** (2012) 860–872. https://doi.org/10.1016/j.medengphy.2011.10.001

[37] H. S. Wong, M. Mackley, S. Butler, J. Baumberg, D. Snoswell, C. Finlayson, and Q. Zhao, The rheology and processing of "edge sheared" colloidal polymer opals, J. Rheol. **58(2)**, 397-409 (2014). https://doi.org/10.1122/1.4862920

[38] B. Wu, Y. Gan, E. Carrera and W. Q. Chen, Three-dimensional vibrations of multilayered hollow spheres submerged in a complex fluid, J. Fluid Mech. (2019), **879**, 682–715. doi:10.1017/jfm.2019.681. https://doi.org/10.1017/jfm.2019.681

Theoretical and Applied Mechanics - AIMETA 2022 Materials Research Forum LLC
Materials Research Proceedings **26** (2023) 455-460 https://doi.org/10.21741/9781644902431-74

The minimum variance distortionless response beamformer for damage identification using modal curvatures

Annamaria Pau[1, a *] and Ugurcan Eroglu[2,b]

[1]Department of Structural and Geotechnical Engineering, Sapienza University of Rome, Italy

[2]Department of Mechanical Engineering, MEF University, Istanbul, Turkey

[a]annamaria.pau@uniroma1.it, [b]erogluug@mef.edu.tr

* corresponding author

Keywords: Damage Identification, Modal Curvatures, Inverse Problems

Abstract. This study presents a damage identification procedure in beams based on the use of beamforming algorithms, which are mostly utilized in inverse problems of source identification and image reconstruction. We choose the modal curvatures as observed quantities and compare the performance of the Bartlett beamformer, minimum variance distortionless response (MVDR) processor, and of a conventional objective function based on the modal curvatures. By means of a set of experiments, we show that the MVDR processor can overcome some of the difficulties encountered with other estimators, especially in cases of slight damage, or damage located between two sensors.

Introduction

Numerous studies have proven that the use of modal quantities is a very effective strategy to locate damage [1]. Among different modal quantities, modal curvatures stand out for their low sensitivity to environmental and operational conditions, and better performance in case of slight damage [2,3]. When solving the related inverse problem, it is important to apply a robust estimator (or processor), which may improve the quality of the solution, especially in the presence of disturbance due to operating conditions, noisy data, and modelling errors.

The Minimum Variance Distortionless Response (MVDR) data processor [4], also named beamformer, is an estimator successfully applied in image-reconstruction techniques based on the wave response, for instance, to locate sources in oceans, but also to reconstruct the image of defects in plates and solids [5]. It is based on the comparison of modelled responses (replica vector) to data received by an array of sensors. Compared to other beamformers, the MVDR has proved its effectiveness in minimizing noise, thus providing images of very good quality. The use of the MVDR processor can be viewed as an approach to the solution of an inverse problem, with potential applications to a broader range of problems in which the forward solutions are computable and measurable, especially regarding the field of structural vibrations, where the use of such estimator has not received attention in the literature to date. We aim at applying the MVDR beamformer to damage identification in beams, using modal curvatures as measured response quantities. To this aim, we assess the performance of the MVDR processor in identifying damage in pseudo-experimental and experimental cases with different location and intensity. Comparisons of the MVDR estimates to those of the Bartlett beamformer, and of a conventional objective function based on the modal curvatures are carried out. For the sake of brevity, this paper only reports some experimental results, while the full analysis is presented in [7].

Numerical Model

The direct problem is solved by means of a 2D finite element (FE) plane stress analysis. This model is used to obtain natural frequencies and mode shapes of a free-free beam of length L, with a notch of small extension b^D, located at position S^D, measured from left end of the beam (see Fig.1). The cross-section of the beam is rectangular with height H^U and width b. The notch is modelled as a height reduction to the value H^D in a stretch b^D. We define damage by two non-dimensional variables, namely, its location $s^D=S^D/L$, and the residual height of the cross-section, $h^D=H^D/H^U$, collected in the vector, $\bar{\boldsymbol{x}} = \{s^D, h^D\}^T$.

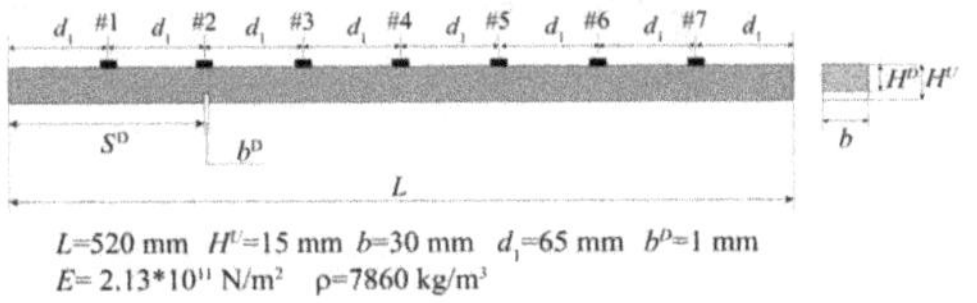

Figure 1. *Schematic representation of the beam used in benchmark case*

The numerical model is used for the construction of the replica vectors via sweeping the damage location from $s^D_{min} = 0.04$ to $s^D_{max} = 0.96$ in 48 steps, and the residual height from $h^D_{min} = 0.06$ to $h^D_{max} = 0.97$ in 30 steps. For all the analyses we evaluate the longitudinal modal strains ε at m=7 equally spaced points, as numbered in Fig.1, and derive modal curvatures using the kinematic relation $\chi=\varepsilon/H^U/2$ of the Euler-Bernoulli beam theory.

Experimental Study

The experimental setup is illustrated in Figure 2. The beam was excited by an instrumented hammer, at points #8 or #9, and the response was measured using seven strain gauges and two accelerometers. The tests were repeated ten times per each point of application of the forcing function. The applied impulsive force is capable of exciting frequencies up to 3000 Hz, which enables determining the first three natural frequencies. Accelerometers at points #8 and #9 are used for modal curvature normalization [6].

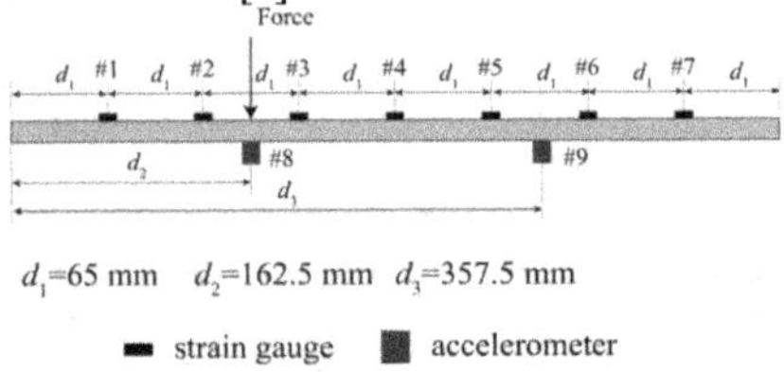

Figure 2. *Schematic of the experimental setup*

We consider two cases which we label Case A, and Case B. In the former, damage is located at $s^D = 0.25$, that is, under sensor #2, while in the latter damage is between sensors #3 and #4, with $s^D = 0.4375$. In both cases we consider two damage severities which correspond to nominal residual heights $h^D = 0.75$ (D1) and $h^D = 0.5$ (D2). In all cases the damage extension is fixed, with $b^D = 1$ mm.

The numerical and experimental curvatures of the first three modes for Case A and Case B are presented in Fig.3. Overall, a very good agreement is observed in both cases regarding the form of the curves. In Case A, the localized effect of damage can be captured by the sensor #2 and the curvature exhibits a distinct change, providing a strong hint at the location of damage. In Case B, instead, the localized effect cannot be captured, which makes locating the damage more difficult.

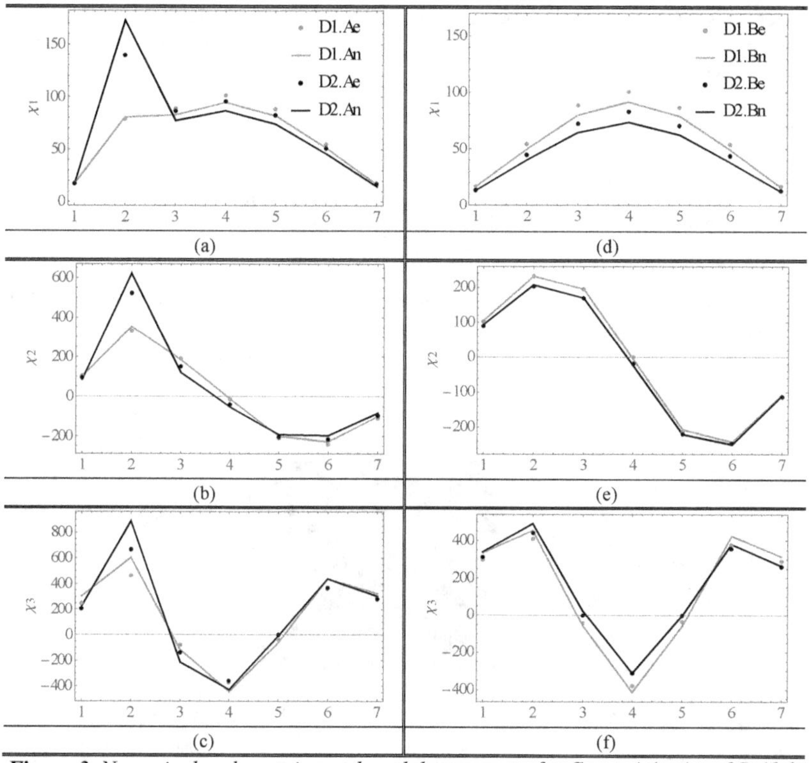

***Figure 3.** Numerical and experimental modal curvatures for Cases A (a-c) and B (d-f).*

Inverse Problem via MFP Algorithms

The direct solution of the free-vibration problem provides the replica vectors $\boldsymbol{\chi}_i(\overline{\boldsymbol{x}}) \in \mathbb{R}^m$, which list the mass-normalized modal curvatures of the ith mode at m sensor points. The beamformers evaluate the correlation between the normalized replica vector,

$$\overline{\boldsymbol{\chi}}_i(\overline{\boldsymbol{x}}) = \boldsymbol{\chi}_i(\overline{\boldsymbol{x}})/\|\boldsymbol{\chi}_i(\overline{\boldsymbol{x}})\|, \tag{1}$$

and the data vector which lists mass-normalized modal curvatures obtained by experimental or pseudo-experimental measurements, $\boldsymbol{d}_i(\overline{\boldsymbol{x}}) \in \mathbb{R}^m$, for the ith mode. Finding the damage parameter set which provides the best correlation with the measured data requires collecting the values of response data in the parameter space $[s^D_{min}, s^D_{max}] \times [h^D_{min}, h^D_{max}]$, which is performed by the 2D FE procedure described beforehand.

Bartlett Beamformer The Bartlett beamformer is a basic processor which has been used almost in all the studies about matched field processors (MFP) for comparative purposes. Simply, it is the average of the projection of the data vector onto the replica vector, and it can be written in terms of the cross-spectral density matrix (CSDM) of data, $\boldsymbol{K}_i = \boldsymbol{d}_i\boldsymbol{d}_i^T$. The expression for the Bartlett Beamformer for individual modes and their superposition are given below.

$$B_{Bart,i}(\overline{\boldsymbol{x}}) = \overline{\boldsymbol{\chi}}_i^T(\overline{\boldsymbol{x}})\boldsymbol{K}_i\overline{\boldsymbol{\chi}}_i\ (\overline{\boldsymbol{x}}), \quad B_{Bartlett}(\overline{\boldsymbol{x}}) = \sum_{i=1}^{3} B_{Bart,i}(\overline{\boldsymbol{x}}). \tag{2}$$

Materials Research Proceedings **26** (2023) 455-460 https://doi.org/10.21741/9781644902431-74

This processor has many side lobes even in case of perfect pseudo-experimental data. This drawback is overcome by the adaptive filtering of data, which leads to different processors, one of which is the MVDR beamformer.

Minimum Variance Distortionless Response (MVDR) Beamformer Trying to minimize the projection except for the best match, we have [4]:

$$B_{MVDR,i}(\overline{\boldsymbol{x}}) = \left[\overline{\chi}_i^T(\overline{\boldsymbol{x}})\boldsymbol{K}_i^{-1}\overline{\chi}_i\ (\overline{\boldsymbol{x}})\right]^{-1}, \quad B_{MVDR}(\overline{\boldsymbol{x}}) = \sum_{i=1}^{3} B_{MVDR,i}(\overline{\boldsymbol{x}}). \tag{3}$$

Here we note that the CSDM is usually ill-conditioned which makes its inversion not a straight-forward task. We improve its conditioning using a diagonal loading technique, with a magnitude of $10^{-6}\mathrm{tr}\boldsymbol{K}_i$.

Objective Function A widely used objective function which measures the difference between replica vector and the measurement is:

$$H(\overline{\boldsymbol{x}}) = \sum_{i=1}^{3}\left\|\overline{\chi}_i(\overline{\boldsymbol{x}}) - \overline{\boldsymbol{d}}_i(\overline{\boldsymbol{x}})\right\|, \quad \overline{\boldsymbol{d}}_i(\overline{\boldsymbol{x}}) = \frac{\boldsymbol{d}_i(\overline{\boldsymbol{x}})}{\|\boldsymbol{d}_i(\overline{\boldsymbol{x}})\|}. \tag{4}$$

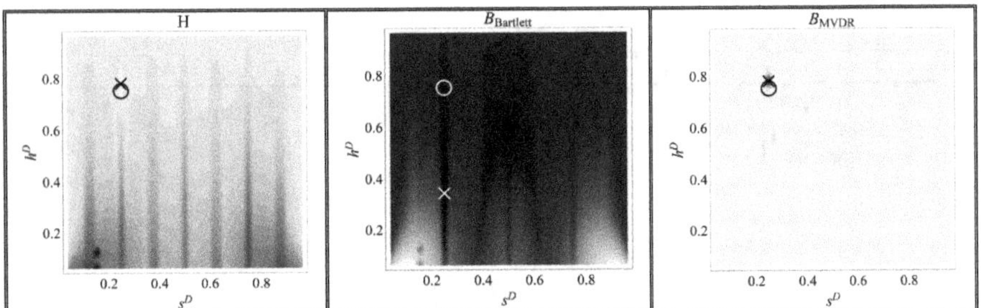

(a)	(b)	(c)
○ Actual $\mathbf{x}_e = \{0.25, 0.753\}^T$ × Estimate $\mathbf{x}_e = \{0.25, 0.781\}^T$	○ Actual $\mathbf{x}_e = \{0.25, 0.753\}^T$ × Estimate $\mathbf{x}_e = \{0.25, 0.344\}^T$	○ Actual $\mathbf{x}_e = \{0.25, 0.753\}^T$ × Estimate $\mathbf{x}_e = \{0.25, 0.781\}^T$

Figure 4. *Contour plots and estimates of the objective function a), Bartlett beamformer b) and MVDR c) for D1.A.*

Fig.4 shows the contour plots of the different estimators and reports the related estimates of the damage parameters. Note also that the reported values of h^D slightly differ from the nominal ones for they were updated with their experimental measurement with a caliber. The correct value is denoted by a circle and the estimate by a cross. Overall, we see that the objective function and MVDR are very accurate in determining both location and severity of damage, while the Bartlett beamformer overestimates the severity but captures the correct location. Objective function and Bartlett beamformer admit many local extrema and side lobes, while MVDR remarkably surpasses them especially in terms of damage intensity resolution.

The estimates in the case of a more severe damage are reported in Fig. 5. We see a pattern very similar to the previous case in which the estimates of the objective function and MVDR are accurate, but the Bartlett beamformer overestimates damage severity.

In case of damage in between sensors #3 and #4, the identification is more difficult since the local effects of damage cannot be captured by one of the sensors (Fig. 3). This is reflected in Fig. 6; the objective function has wider flat valleys, many local maxima are apparent in Bartlett beamformer, and MVDR has many side lobes even though their order is smaller than the global

Materials Research Proceedings **26** (2023) 455-460 https://doi.org/10.21741/9781644902431-74

maxima. When damage is more severe (Fig. 7), the estimate of Bartlett beamformer is improved, however, it is still less accurate than those of the objective function and MVDR.

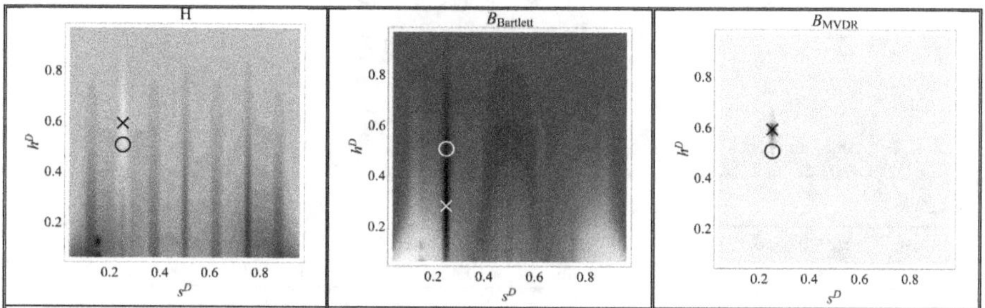

(a)		(b)		(c)	
○ Actual	$\mathbf{x}_e = \{0.250, 0.507\}^T$	○ Actual	$\mathbf{x}_e = \{0.250, 0.507\}^T$	○ Actual	$\mathbf{x}_e = \{0.250, 0.507\}^T$
× Estimate	$\mathbf{x}_e = \{0.250, 0.594\}^T$	× Estimate	$\mathbf{x}_e = \{0.250, 0.281\}^T$	× Estimate	$\mathbf{x}_e = \{0.250, 0.594\}^T$

Figure 5. *Contour plots and estimates of the objective function a), Bartlett beamformer b) and MVDR c) for D2.A.*

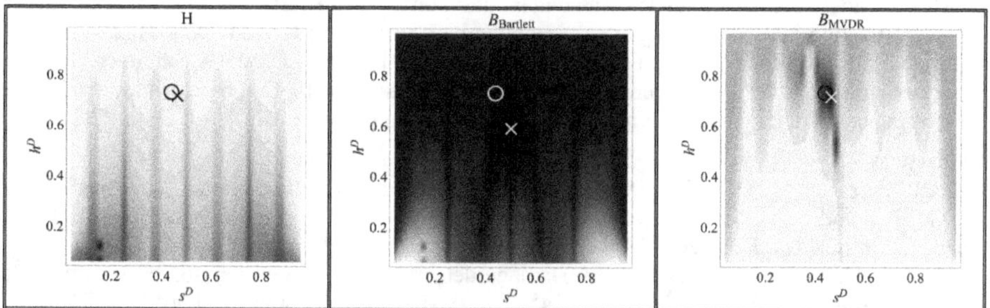

(a)		(b)		(c)	
○ Actual	$\mathbf{x}_e = \{0.438, 0.733\}^T$	○ Actual	$\mathbf{x}_e = \{0.438, 0.733\}^T$	○ Actual	$\mathbf{x}_e = \{0.438, 0.733\}^T$
× Estimate	$\mathbf{x}_e = \{0.462, 0.719\}^T$	× Estimate	$\mathbf{x}_e = \{0.500, 0.539\}^T$	× Estimate	$\mathbf{x}_e = \{0.462, 0.719\}^T$

Figure 6. *Contour plot and estimates of the objective function a), Bartlett beamformer b) and MVDR c) for D1.B.*

The overall estimates of different techniques for all the cases we considered here are summarized in Table 1. We see that the minimization of a suitable objective function and the MVDR provides identical results in these cases; however, the extrema of the latter are always more distinct than the former, which is a strong hint on higher reliability of MVDR.

Table 1. *Absolute errors of damage parameters for all the cases considered.*

	D1.A		D2.A		D1.B		D2.B	
	$\lvert s_e^D - s^D\rvert$	$\lvert h_e^D - h^D\rvert$	$\lvert s_e^D - s^D\rvert$	$\lvert h_e^D - h^D\rvert$	$\lvert s_e^D - s^D\rvert$	$\lvert h_e^D - h^D\rvert$	$\lvert s_e^D - s^D\rvert$	$\lvert h_e^D - h^D\rvert$
H	0	0.028	0	0.087	0.024	0.014	0.004	0.031
Bartlett	0	0.409	0	0.226	0.062	0.194	0.023	0.063
MVDR	0	0.028	0	0.087	0.024	0.014	0.004	0.031

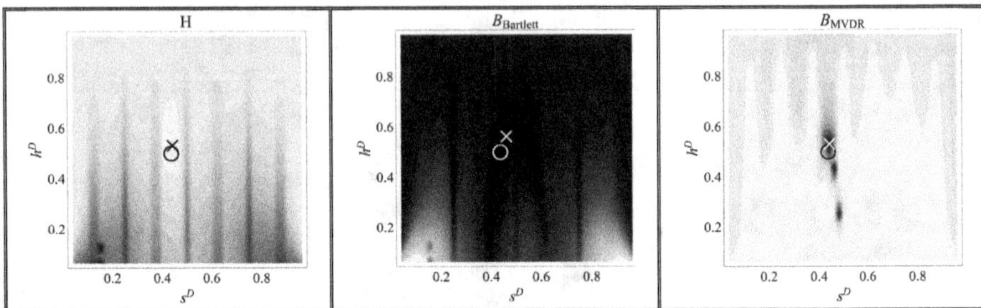

○ Actual $\mathbf{x}_e = \{0.438, 0.500\}^T$	○ Actual $\mathbf{x}_e = \{0.438, 0.500\}^T$	○ Actual $\mathbf{x}_e = \{0.438, 0.500\}^T$
× Estimate $\mathbf{x}_e = \{0.442, 0.531\}^T$	× Estimate $\mathbf{x}_e = \{0.461, 0.563\}^T$	× Estimate $\mathbf{x}_e = \{0.442, 0.531\}^T$
(a)	*(b)*	*(c)*

Figure 7. *Contour plots and estimates of the objective function a), Bartlett beamformer b) and MVDR c) for D2.B.*

Conclusions

We examined the possibility of applying to damage identification some beamforming algorithms which are widely used in fields such as underwater source identification and image processing. The identification of damage using experimental modal curvatures revealed that the use of MVDR is reliable and it suppresses local extrema which may lead to erroneous estimates. All the processors under investigation have exhibited high sensitivity to damage location, although MVDR appeared to be the most sensitive with regard to damage intensity.

References

[1] C. R. Farrar, S. W. Doebling, Damage detection II: field applications to large structures. In: J. M. M. Silva and N. M. M. Maia (eds.), Modal analysis and testing, Nato Science Series, Kluwer Academic Publishers, Dordrecht, Netherlands, 1999.

[2] A. Pandey, M. Biswas, M. Samman, Damage detection from changes in 465 curvature mode shapes, Journal of Sound and Vibration 145 (2) (1991) 321–332. https://doi.org/10.1016/0022-460X(91)90595-B

[3] M. Dilena, A. Morassi, M. Perin, Dynamic identification of a reinforced concrete damaged bridge, Mechanical Systems and Signal Processing 25 (8) (2011) 2990–3009. https://doi.org/10.1016/j.ymssp.2011.05.016

[4] G. Turek, W. Kuperman, Applications of matched-field processing to structural vibration problems, The Journal of the Acoustical Society of America 495 101 (3) (1997) 1430–1440. https://doi.org/10.1121/1.418168

[5] S. Sternini, A. Pau and F. Lanza di Scalea, "Minimum Variance Imaging in Plates Using Guided Wave Mode Beamforming," IEEE Transactions on Ultrasonics, Ferroelectrics and Frequency Control, 66 (12), 1906-1919 (2019). https://doi.org/10.1109/TUFFC.2019.2935139

[6] F. Vestroni, A. Pau, J. Ciambella, "The role of curvatures in damage identification" in IABMAS Proceedings, Barcelona, Spain, 11-15 July 2022, page 1-8, (2022).

[7] A. Pau, U. Eroglu, Identification of damage in beams using the Minimum Variance Distorsionless Response beamformer, submitted

Theoretical and Applied Mechanics - AIMETA 2022 Materials Research Forum LLC
Materials Research Proceedings **26** (2023) 461-466 https://doi.org/10.21741/9781644902431-75

Experimental characterisation and numerical modelling of axially loaded wire rope isolators

Davide Pellecchia[1, a *], Nicolò Vaiana[1, b], Salvatore Sessa[1, c], Francesco Marmo[1, d] and Luciano Rosati[1, e]

[1]Department of Structures for Engineering and Architecture, University of Naples Federico II, Naples, Italy

[a]davide.pellecchia@unina.it, [b]nicolo.vaiana@unina.it, [c]salvatore.sessa2@unina.it, [d]f.marmo@unina.it, [e]rosati@unina.it

Keywords: Wire Rope Isolators, Experimental Tests, Hysteretic Model

Abstract. We present the main outcomes of a recent experimental activity denoted to studying the dynamic behaviour of Wire Rope Isolators loaded along the axial direction. The experimental activity has been carried out at the laboratory of the Department of Structures for Engineering and Architecture of the University of Naples Federico II (Italy). Furthermore, a new hysteretic model derived from a recently developed class of asymmetric hysteretic models is illustrated. The main advantages of the proposed model with respect to the classical ones, typically having a differential nature, are: 1) the output variable is computed in closed-form, i.e. without numerically solving a differential equation; 2) the proposed model is based on two different sets of parameters that allows one to independently model the loading and unloading curves. Specifically, a close matching between the experimental asymmetric loop and the numerically one has been obtained.

Introduction

Wire Rope Isolators (WRIs) are widely used in controlling both vibration and shock in order to protect sensitive equipment in the areas of aerospace and mechanical engineering [1] as well as for seismic protection of buildings [2] and their contents/equipment [3]. These devices have also been used to reduce the seismic vibration in protecting high voltage ceramic circuit breakers [4] and to protect the two Riace bronzes towards vertical components of seismic ground motions [5,6].

The excellent damping properties of such a device are due to the sliding friction developed between strands and wires that allows one to dissipate a large amount of energy. In addition, the WRIs behaviour depends on the device's geometrical characteristics, e.g. the device length (l), width (w), height (h) and the wire rope diameter (ϕ).

The WRIs force-displacement curve is affected by geometrical and mechanical non-linearities resulting in a complex shape of hysteretic nature that depends on the direction along which the devices are loaded, i.e., typically, the shear, roll and axial directions [7], see, e.g., Fig. 1. As a matter of fact, the hysteretic loop is symmetric in both shear and roll directions, while along the axial direction is asymmetric, showing a remarkable difference between tension and compression behaviour [8].

The complex WRI behaviour is typically predicted by using hysteretic models of differential nature [9,10] including the well-known Bouc-Wen model [11,12]. These kinds of models show some issues, such as: 1) the evaluation of the output variable is performed by numerically solving a differential equation; 2) generally, the models' parameters have a not clear physically meaning; 3) the loading and unloading branches in the force-displacement curve are coupled; 4) they must be suitably modified to reproduce specific force-displacement curves such as asymmetric loops.

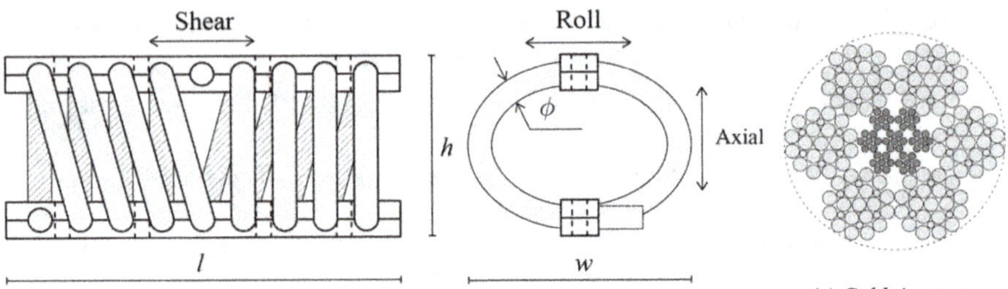

(a) Front view *(b) Transverse view* *(c) Cable's cross-section*

Figure 1 Geometrical characteristics of WRIs and load directions.

Being interested to characterise the dynamic behaviour of WRIs along the axial direction, we illustrate some preliminary results obtained for the PWHS16040 device manufactured by Powerflex S.r.l (Limatola, Italy) during an experimental campaign conducted at the laboratory of the Department of Structures for Engineering and Architecture of the University of Naples Federico II (Italy). Finally, the mathematical characterisation of such a behaviour is shown by adopting a new hysteretic model.

Experimental activity and results

The preliminary results of the experimental campaign carried out on the WRI PWHS16040 loaded along the axial direction in both small and large displacements regimes are described in this Section. Specifically, the force-displacement (*f-u*) curves are compared, aiming to investigate the dynamic behaviour of the above-mentioned device by studying the influence of the displacement amplitude and frequency of a sinusoidal displacement input as well as of the axial preload and of the wire rope diameter.

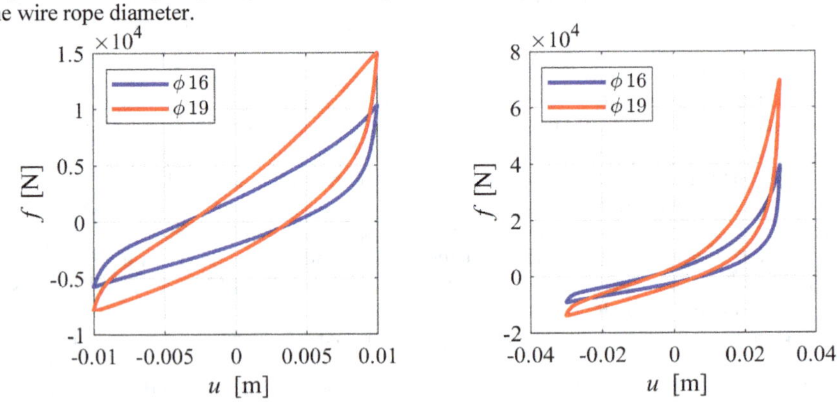

(a) Displacement amplitude ± 0.01 meters *(b) Displacement amplitude ± 0.03 meters*

Figure 2 Hysteresis loops exhibited by two PWHS16040 devices characterised by two different diameters of the wire rope.

Fig. 2 shows the hysteresis loops obtained for two PWHS16040 devices equipped with two different wire ropes having diameters equal to 16 mm (blue curves) and 19 mm (red curves), and by applying, in absence of axial preload, sinusoidal displacements having a frequency of 0.1 Hz and amplitudes equal to 10 mm (left) and 30 mm (right). One can observe that the use of a large wire rope diameter provides a clockwise rotation of the hysteresis loops both in small and large displacements.

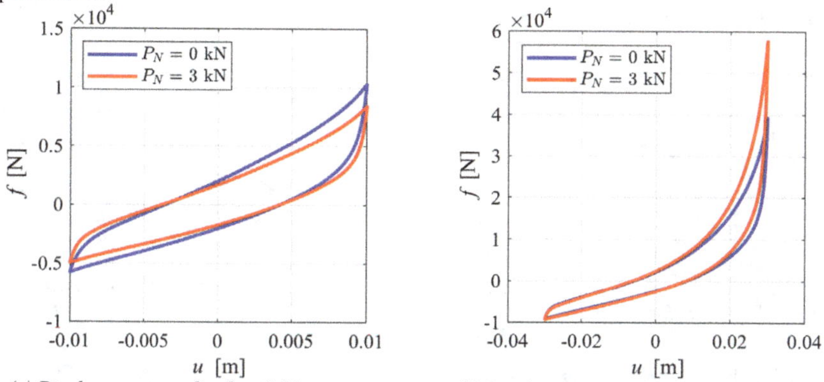

(a) Displacement amplitude ± 0.01 meters *(b) Displacement amplitude ± 0.03 meters*

Figure 3 Hysteresis loops obtained in absence of axial preload from cyclic tests characterised by a frequency of 0.1 Hz.

Hence, by increasing the wire rope diameters, the tangent stiffnesses in the state of tension and compression increase. In addition, a larger amount of dissipated energy, which is represented by the area enclosed in the hysteresis loop, is shown using a large wire rope diameter, mainly in the large displacements range.

The hysteresis loops of the WRI PWHS16040 are plotted in Fig. 3 as a function of the axial preload by applying sinusoidal displacements having a frequency of 0.1 Hz and amplitudes equal to 10 mm (left) and 30 mm (right). Notably, two values of axial preload are considered, namely 0 (blue curve) and 3 kN (red curves). The hysteresis loop exhibits a clockwise rotation, without changing its area, when the displacement amplitude is equal to 10 mm and the axial preload increases (see Fig.3(a)). On the other hand, in the range of large displacements (Fig. 3(b)), a stiffening behaviour in the state of tension is displayed when the axial preload is increased, whereas the tangent stiffness variation in the state of compression is negligible.

Fig. 4 shows the hysteresis loops obtained for the WRI PWHS16040 device, in absence of axial preload, when subjected to a sinusoidal displacement having displacements amplitude equal to 10 mm and 30 mm, see, e.g., Fig. 4(a) and (b), respectively, and increasing frequency. By increasing the input frequency, a pronounced hardening behaviour in the state of tension is shown for large displacements, while the change of the tangent stiffness in the state of compression remains negligible, see Fig. 4(b). In contrast, the frequency effect on the tested WRI is much lower at small displacements, see Fig. 4(a).

Experimental validation

In this Section, we derive a specific hysteretic model from a more general class of models [13]; subsequently the new model is validated by comparing the experimental hysteresis loops of the PWHS16040 device with the simulated one.

The general idea of the class of models consists of splitting the loop into four distinct curves that describe two phases, namely the generic loading and unloading phases in which the generalised velocity is, respectively, greater and smaller than zero.

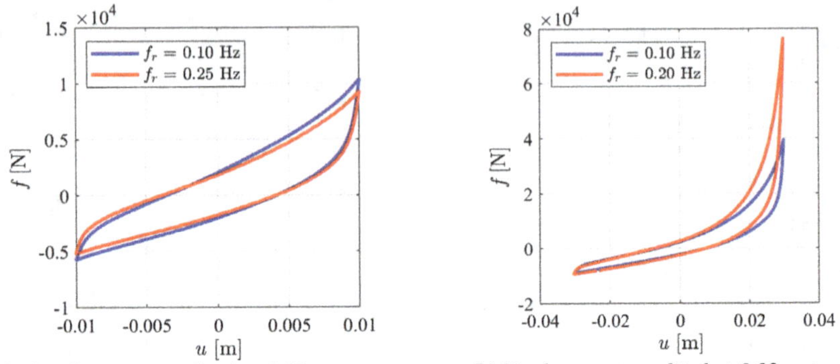

(a) Displacement amplitude ± 0.01 meters *(b) Displacement amplitude ± 0.03 meters*

Figure 4 Hysteresis loops obtained in absence of axial preload from cyclic tests characterised by increasing values of frequency f_r.

Notably, the generic loading phase is described by the loading and upper limiting curves denoted as c^+ and c_u, respectively; whereas the generic unloading phase is defined by the unloading and lower limiting curves referred to, respectively, as c^- and c_l, see, e.g., Fig. 5.

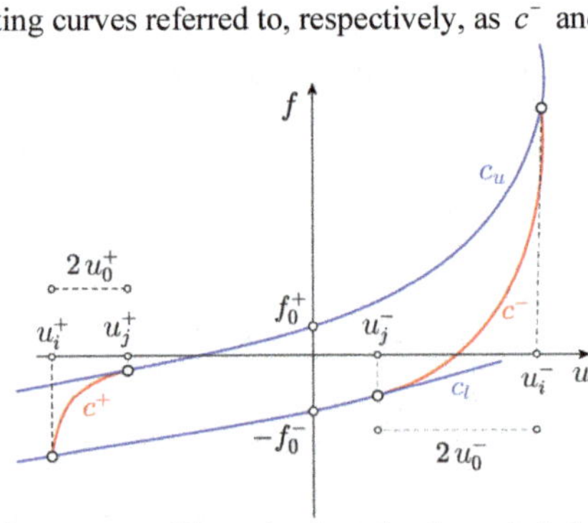

Figure 5 The four curves able to simulate the dynamic behaviour of WRIs.

The experimental validation is carried out by taking into account a hysteretic model belonging to the above-mentioned class. Specifically, the proposed model corresponds to a simplified version of the recently proposed Vaiana-Rosati model (VRM) [14] in which only hysteresis loops characterised by upper and lower limiting curves with no inflexion points are modelled.

Due to space limitations we just report the mathematical expressions of the functions describing the curves relating force f and displacement u along the axial direction

$$c_u(u) = \beta_1^+ e^{\beta_2^+ u} - \beta_1^+ + k_b^+ u + f_0^+,$$
$$c_l(u) = \beta_1^- e^{\beta_2^- u} - \beta_1^- + k_b^- u - f_0^-, \tag{1}$$

whereas the loading and unloading limiting curves

$$c^+(u,u_j^+) = \beta_1^+ e^{\beta_2^+ u} - \beta_1^+ + k_b^+ u + f_0^+ - \frac{1}{\alpha^+}\left[e^{-\alpha^+\left(+u-u_j^+ +\bar{u}^+\right)} - e^{-\alpha^+\bar{u}^+}\right],$$
$$c^-(u,u_j^-) = \beta_1^- e^{\beta_2^- u} - \beta_1^- + k_b^- u - f_0^- + \frac{1}{\alpha^-}\left[e^{-\alpha^-\left(-u+u_j^- +\bar{u}^-\right)} - e^{-\alpha^-\bar{u}^-}\right]. \tag{2}$$

The quantities $k_b^+, \alpha^+, \beta_1^+, \beta_2^+, f_0^+ (k_b^-, \alpha^-, \beta_1^-, \beta_2^-, f_0^-)$ represent the constitutive model parameters to be identified, whereas $\bar{d}^+, d_0^+ (\bar{d}^-, d_0^-)$ can be expressed as a function of the above-mentioned parameters. The quantities u_j^+ and u_j^- are illustrated in Fig. 5 and can be evaluated as a function of the coordinates of a generic initial point that lies on the upper or lower limiting curve. The relevant expressions and their mechanical significance are omitted for brevity.

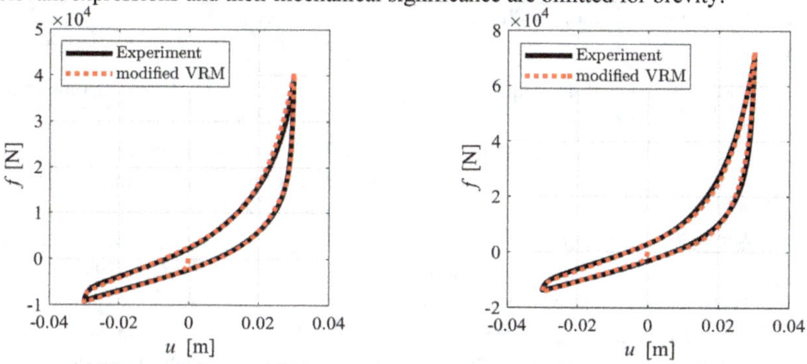

(a) Wire rope diameter ϕ 16 *(b) Wire rope diameter ϕ 19*

Figure 6 Experimental hysteresis loops vs. simulated ones obtained by using the model parameters listed in Tab. 1.

Fig. 6 compares the experimental and numerical hysteresis loops obtained for the PWHS16040 device by applying, in absence of preload, a sinusoidal displacement input having a frequency of 0.1 Hz and an amplitude of 0.03 m. Specifically, the validation of the modified VRM model has been carried out by assuming two wire ropes having different diameters, i.e. 16 mm (Fig. 6(a)) and 19 mm (Fig. 6(b)) and the constitutive model parameters listed in Tab. 1.

The numerical hysteresis loops obtained by means of the proposed hysteretic model are depicted by the dotted red lines, whereas the solid black lines represent hysteresis loops obtained by experimental tests. Such a comparison shows an excellent match between the two hysteresis loops, demonstrating that the proposed hysteretic model can satisfactorily predict the WRI's dynamic response.

Table 1 Model parameters used for reproducing hysteresis loops in Fig. 6.

ϕ [mm]	$\text{sgn}(\dot{u})$	k_b [N m^{-1}]	f_0 [N]	α [m^{-1}]	β_1 [N]	β_2 [m^{-1}]
16	+	300 000	2 400	1 800	750	125
	-	225 000	2 400	1 800	250	140
19	+	450 000	3 015	1 500	850	140
	-	380 000	3 015	1 500	60	220

Conclusions

We have presented some preliminary results of an experimental campaign, on the PWHS16040 device manufactured by Powerflex S.r.l. loaded along the axial direction, that has been carried out at the laboratory of the Department of Structures for Engineering and Architecture of the University of Naples Federico II (Italy). Specifically, aim of experimental campaign was to investigate the dynamic behaviour of the device by studying the influence of the amplitude and frequency of a sinusoidal displacement input as well as of the axial preload and of the wire rope diameter.

The experimental outcomes have shown that the effects of the frequency input and the axial preload do not significantly change the WRI dynamic behaviour in the field of small displacements. On the contrary, a stiffening behaviour in the state of tension has been shown in the field of large displacements.

Finally, in order to predict the WRI dynamic behaviour along the axial direction, a new hysteretic model has been proposed. Such a model has been calibrated and validated by comparing the experimental hysteresis loops with the numerically simulated ones.

References

[1] M.L. Tinker, M.A. Cutchins, Damping phenomena in a wire rope vibration isolation system, J. Sound and Vib. 157 (1992) 7-18. https://doi.org/10.1016/0022-460X(92)90564-E

[2] M. Spizzuoco, V. Quaglini, A. Calabrese, G. Serino, C. Zambrano, Study of wire rope devices for improving the re-centering capability of base isolated buildings, Struct. Control. Health. 24(6) (2017) e1928. https://doi.org/10.1002/stc.1928

[3] G.F. Demetriades, M.C. Constantinou, A.M. Reinhorn, Study of wire rope systems for seismic protection of equipment in buildings, Eng. Struct. 15 (1993) 321-334. https://doi.org/10.1016/0141-0296(93)90036-4

[4] S. Alessandri, R. Giannini, F. Paolacci, M. Malena, Seismic retrofitting of an HV circuit breaker using base isolation with wire ropes. part 1: Preliminary tests and analyses, Eng. Struct. 98 (2015) 251-262. https://doi.org/10.1016/j.engstruct.2015.03.032

[5] G. De Canio, Marble devices for the base isolation of the two bronzes of Riace: a proposal for the David of Michelangelo, P. WCEE (2012).

[6] D. Pellecchia, S. Lo Feudo, N. Vaiana, J.-L. Dion, L. Rosati, A procedure to model and design elastomeric-based isolation systems for the seismic protection of rocking art objects, Comput. Aided Civ. Inf. (2022) 1-18.

[7] N. Vaiana, M. Spizzuoco, G. Serino, Wire rope isolators for seismically base-isolated lightweight structures: Experimental characterization and mathematical modeling, Eng. Struct. 140 (2017) 498-514. https://doi.org/10.1016/j.engstruct.2017.02.057

[8] F. Foti, J. Galeazzi, L. Martinelli, On the modelling of the Hysteretic Behaviour of Wire Rope Isolators, P. XXIV AIMETA (2020) 1535–1542. https://doi.org/10.1007/978-3-030-41057-5_124

[9] S. Rashidi, S. Ziaei-Rad, Experimental and numerical vibration analysis of wire rope isolators under quasi-static and dynamic loadings, Eng. Struct. 148 (2017) 328-339. https://doi.org/10.1016/j.engstruct.2017.06.061

[10] A. Salvatore, B. Carboni, L.Q. Chen, W. Lacarbonara, Nonlinear dynamic response of a wire rope isolator: Experiment, identification and validation, Eng. Struct. 238 (2021) 112121. https://doi.org/10.1016/j.engstruct.2021.112121

[11] R. Bouc, Modèle mathématique d'hystérésis, Acustica 24 (1) (1971) 16-25.

[12] Y. K. Wen, Method for random vibration of hysteretic systems, J. Eng. Mech. Div. 102 (2) (1976) 249-263. https://doi.org/10.1061/JMCEA3.0002106

[13] N. Vaiana, S. Sessa, L. Rosati, A generalized class of uniaxial rate-independent models for simulating asymmetric mechanical hysteresis phenomena, Mech. Syst. Signal Pr. 146 (2021) 106984. https://doi.org/10.1016/j.ymssp.2020.106984

[14] N. Vaiana, L. Rosati, Classification and Modeling of Uniaxial Rate-Independent Hysteresis Phenomena: Some Preliminary Results, P. XXV AIMETA (2022).

Materials Research Proceedings 26 (2023) 467-472 https://doi.org/10.21741/9781644902431-76

Continuous particle swarm optimization for model updating of structures from experimental modal analysis

LO IACONO Francesco[1,a *], NAVARRA Giacomo[1,b], ORLANDO Calogero[1,c], RICCIARDELLO Angela[1,d]

[1]University of Enna Kore, Cittadella Universitaria, 94100 Enna

[a]francesco.loiacono@unikore.it, [b]giacomo.navarra@unikore.it, [c]calogero.orlando@unikore.it, [d]angela.ricciardello@unikore.it

Keywords: Model Updating, Particle Swarm Optimization, Experimental Modal Analysis

Abstract. The aim of this work is to present a structural model updating procedure based on a recent variant of Particle Swarm Optimization, called Continuous Particle Swarm Optimization. In order to validate the proposed model, the modal parameters of a real scale large electrical device have been identified by measuring the dynamical response of the system subjected to a measured input. Numerical results confirm the robustness and reliability of the proposed method.

Introduction

The discrepancy between the design of an artifact and its actual construction generates uncertainty on the mechanical behavior of the structures which depends on a large number of variables such as geometrical and mechanical properties of materials and structural parts, boundary conditions and so on.

In the past decades, the scientific literature proposed several model updating methods [1] mostly based on heuristic optimization techniques and on the experimental characterization of the structures in order to fill up the aforementioned gap.

Among the classical optimization techniques, the Continuous Particle Swarm Optimization (CPSO) [2, 3] has been recently developed showing good performance in terms of accuracy and computational time reduction in comparison with the others.

Moreover, the dynamical behavior of the artifact needs to be identified through experimental campaigns performed by means of two different approaches. The first one is the so-called Operational Modal Analysis (OMA) in which the modal parameters are obtained only from the measured data using environmental vibrations as unknown input (i.e., wind load, micro-tremors, traffic), without any artificial excitations applied on the structure. The second approach is the Experimental Modal Analysis (EMA) in which the identification of modal parameter is evaluated by applying a measured input on the system and measuring its response.

In the present paper, the CPSO has been adopted with the aim to develop a robust and fast structural model updating procedure. The method has been validated on a real scale large electrical device whose dynamical properties have been characterized by using EMA approach.

The experimental seismic response of the structure under investigation has been acquired by taking advantage of the shaking table system of the Laboratory of Earthquake engineering and Dynamic Analysis (L.E.D.A.) [4].

The paper is organized in three sections: in the first one the experimental model of a high voltage Current Transformer is presented as well as its numerical twin model defined by means of Finite Element Method (F.E.M.); the second section deals in detail with an original model updating procedure achieved through CPSO. Last, the comparison in terms of displacement between acquired data and dynamical response reproduced by the optimized numerical model shows a good agreement validating the proposed method.

Experimental Model

It is common practice that the design and construction process of industrial artifacts need to be verified by means of experimental test to guarantee the goodness of the actual objects to be placed in the market and their reliability during the operative lifetime. For such a reason, a great number of national and international standard have been delivered for example to qualify the artifacts under seismic events [5-6].

In order to foresee the behavior of these objects and their components under several input motions without the necessity to run many expensive experimental campaigns, it is often convenient to develop a reliable twin numerical model starting from the mechanical and dynamical data acquired during the mandatory qualification tests.

In this paper a model updating procedure is discussed to develop a numerical model able to reproduce the dynamic response of a high voltage Current Transformer which have been tested at the Laboratory of Experimental Dynamic of the L.E.D.A. Research Institute.

The Transformer has been mounted on a support structure by means of a six-spring damper and instrumented by several accelerometers and strain gauges to acquire the response in terms of accelerations and strains, respectively. Moreover, an infrared optoelectronic system has been deployed to measure the absolute 3D displacements of the structure under test. In Fig. 1a a picture of the experimental setup is reported and, for the sake of shortness, only the sensors A1 (tri-axial accelerometer) and D1 (markers for displacement measure) at the top of the transformer are highlighted.

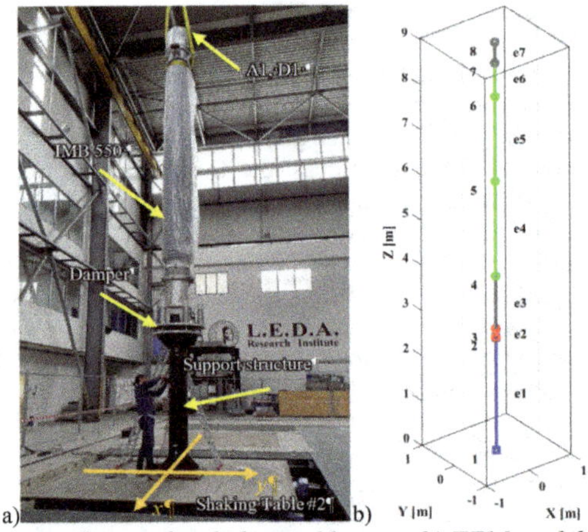

Figure 1. a) Experimental setup for shaking table tests; b) FEM model: nodes and element geometry.

With the aim to define a numerical twin model of the unit under test, a simplified Finite Element Model has been built by means of a Matlab® self-developed FEM software (Fig. 1b). The FEM model to be updated consists of n. 8 nodes (node n. 1 is fully restrained while nodes 2 to 8 are not constrained) and n. 7 elements among which elements 1, 3, 4, 5, 6, 7 have been discretized as 3D lumped-mass frame elements whose bending properties depend on the stiffness coefficients $EI_{x,i}$ and $EI_{y,i}$ [7], while element n. 2 has been modeled as a 3D spring with axial stiffness k_z and rotational springs $k_{\varphi,x}$ and $k_{\varphi,y}$ (lateral k_x, k_y and torsional $k_{\varphi,z}$ are herein considered 10 times bigger than the others in order to simulate the actual behavior of the spring). Elements n. 3 and 7

Materials Research Forum LLC
https://doi.org/10.21741/9781644902431-76

(corresponding to the massive oil tanks of the transformer) are much stiffer than the other elements and are not considered in the optimization process.

On each frame element with parameters to be identified (element n. 1, 4, 5 and 6) the damping has been modeled according to Raileigh formulation in such a way that their internal damping matrix is

$$\mathbf{C}_i = \alpha_{0,i}\mathbf{M}_i + \alpha_{1,i}\mathbf{K}_i,\ i = 1,4,5,6 \tag{1}$$

while for the spring element n. 2 the damping coefficients to be optimized are those corresponding to the axial and rotational degrees of freedom, namely c_z, $c_{\varphi,x}$ and $c_{\varphi,y}$, respectively.

Nominal values of the materials and geometrical properties of each element have been obtained by the design information available from the constructor and from the static tests performed before the seismic qualification tests.

Modal Updating by means of Continuous Particle Swarm Optimization

In this paper, a model updating strategy is developed by assuming that the simplified FEM model, previously described, is linear and the difficulties in modeling joints, flanges, the presence of non-structural components and other complicated boundary conditions could be compensated for by adjusting the stiffness and damping coefficients of some relevant elements of the structure.

As customary [1], a Finite Element Model Updating procedure can be defined as an iterative method where some physical parameters are updated until the FEM model reproduces the measured data to a sufficient degree of accuracy. This can be achieved by minimizing a so-called objective function with respect to the variation of the selected physical parameters:

$$\mathbf{x}_{opt} = \min_{\mathbf{x}}\left[J(\mathbf{x})\right],\ \underline{\mathbf{x}} < \mathbf{x} < \overline{\mathbf{x}} \tag{2}$$

where $J(\mathbf{x})$ is the objective function, $\mathbf{x}$ is the array of the physical parameters to be updated that can assume values in the interval $(\underline{\mathbf{x}}, \overline{\mathbf{x}})$.

In the following, the experimental data acquired during seismic tests will be assumed correct and considered as the refence signals to use for the computation of the objective function. Since the aim is to obtain a numerical model able to reproduce the dynamic response of the structure, the objective function adopted in this work is computed as:

$$J(\mathbf{x}) = w_1\left[\frac{\text{rms}\left(u_{8,x}^{\text{num}}(\mathbf{x},t_i,t_f) - u_{8,x}^{\text{exp}}(t_i,t_f)\right)}{\text{rms}\left(u_{8,x}^{\text{exp}}(t_i,t_f)\right)}\right] + w_2\left[\frac{\text{rms}\left(u_{8,y}^{\text{num}}(\mathbf{x},t_i,t_f) - u_{8,y}^{\text{exp}}(t_i,t_f)\right)}{\text{rms}\left(u_{8,y}^{\text{exp}}(t_i,t_f)\right)}\right] \tag{3}$$

where $w_1 = w_2 = 0.5$ are weighting parameters, rms is the root mean square function, $u_{8,x}$, $u_{8,y}$ are the displacements with respect to ground at the top of the transformer (node n. 8) in x and y global direction, respectively, for the numerical (num) and the experimental (exp) system. The time window $\left[t_i, t_f\right]$ have been chosen to be the time interval between the 2% and 98% contribution of the experimental shaking table input acceleration time histories $\ddot{\mathbf{u}}_g^{\text{exp}}(t)$ to the Arias intensity defined as:

$$A = \int_0^{T_f} \ddot{\mathbf{u}}_g^{\text{exp}}(t)dt \tag{4}$$

where $\left[0, T_f\right]$ is the whole signal duration.

Materials Research Forum LLC
https://doi.org/10.21741/9781644902431-76

At each step of the iterative procedure adopted, the FEM software computes the global Mass **M**, Damping **D** and Stiffness **K** matrices of the system and solves the numerical time integration problem

$$\begin{cases} \mathbf{M}\ddot{\mathbf{u}}(t) + \mathbf{C}\dot{\mathbf{u}}(t) + \mathbf{K}\mathbf{u}(t) = -\mathbf{M}\boldsymbol{\tau}\ddot{\mathbf{u}}_g^{\exp}(t) \\ \mathbf{u}_0(0) = \mathbf{0};\ \dot{\mathbf{u}}_0(0) = \mathbf{0} \end{cases} \tag{5}$$

where $\mathbf{u}(t)$ is the displacement vector of the nodes of the structure for all the 6 degrees of freedom in space with respect to ground, the upper dots indicate time derivative, $\boldsymbol{\tau}$ is the load incidence matrix and $\ddot{\mathbf{u}}_g^{\exp}(t)$ is the array of the *x*, *y* and *z* direction experimental acceleration time histories recorded at the base of the structure during the shaking table seismic tests.

To compute the optimal minimum value of the objective function $J(\mathbf{x})$ and the corresponding optimal parameter set $\mathbf{x}_{opt}$, the latest variant of PSO, the so called CPSO, introduced in [2,3], has been applied.

The discrete formulation of PSO has been extended in the continuous time space by means of an integration over the time interval, $[0,T]$, or rather, the bird flying motion evolution is defined by a continuous time function that is the integral of a given Cauchy problem associated with a second order ordinary differential equation describing the dynamics of a damped harmonic oscillator. In details, let μ denote the inertia weight, c_c and c_s the cognitive and social constant, p_{ib} the best position of the *i*-th bird history, p_{gb} the best position in the swarm history, r_1 and $r_2 \in [0,1]$ random values.

Eventually, let $\omega(t),\ \zeta(t):[0,T]\to[0,+\infty)$ be the angular frequency and the damping ratio of the oscillator, respectively, such that, $\forall t \in [0,T]$,

$$\begin{cases} 2\zeta(t)\omega(t) = 1-\mu(t) \\ \omega^2(t) = c_c r_1(t) + c_s r_2(t) \end{cases} \tag{6}$$

thus, if the time interval is split in *N* sub-interval of range $dt = T/N$, the particle position updates

$$p(t) = \cup_{k=0}^{N-1} p_{[(k-1)dt,kdt]}(t) \tag{7}$$

where $p_{[(k-1)dt,kdt]}(t)$ is the particle movement computed as follows.
Let be $\omega(t)=\omega_k$, $\zeta(t)=\zeta_k$, $f_k = c_c p_{ib}(kdt) + c_s p_{gb}(kdt)\ \ \forall k \in [kdt,(k+1)dt], k=1,\ldots,N$ and

$$\lambda_1^k = -\omega_k\left[\zeta_k + \sqrt{(\zeta_k^2-1)}\right] \quad \lambda_2^k = -\omega_k\left[\zeta_k - \sqrt{(\zeta_k^2-1)}\right]. \tag{8}$$

If $\zeta \neq 1$ then

$$p_{[(k-1)dt,kdt]}(t) = c_1^k \exp(\lambda_1^k t) + c_2^k \exp(\lambda_2^k t) + \frac{f_k}{\omega_k^2} \tag{9}$$

with

$$c_1^k = \frac{\left(p_k - \dfrac{f_k}{\omega_k^2}\right)\lambda_2^k - \dot{p}_k}{\left(\lambda_2^k - \lambda_1^k\right)\exp\left(\lambda_1^k kdt\right)} \quad \text{and} \quad c_2^k = \frac{\dot{p}_k - \left(p_k - \dfrac{f_k}{\omega_k^2}\right)\lambda_1^k}{\left(\lambda_2^k - \lambda_1^k\right)\exp\left(\lambda_2^k kdt\right)}. \tag{10}$$

Instead, if $\zeta = 1$ that is $\lambda_1^k = \lambda_2^k = \overline{\lambda^k}$, it results

$$p_{[(k-1)dt,kdt]}(t) = \left(c_1^k + c_2^k t\right)\exp(\overline{\lambda^k}t) + \frac{f_k}{\omega_k^2} \tag{11}$$

where

$$c_1^k = \left[\left(p_k - \frac{f_k}{\omega_k^2}\right)\left(1 + \overline{\lambda^k}kdt\right) - \dot{p}_k kdt\right]\exp-\left(\overline{\lambda^k}kdt\right) \tag{12}$$

$$c_2^k = \dot{p}_k - \overline{\lambda^k}\left(p_k - \frac{f_k}{\omega_k^2}\right). \tag{13}$$

Numerical results

The procedure described in the previous section has been applied for the computation of the optimal parameters able to reproduce the dynamical response of the numerical twin model.

The parameters **x** to be optimized are summarized in Table 1.

Table 1. Optimization parameters.

x_1	x_2	x_3	x_4	x_5	x_6	x_7	x_8	x_9	x_{10}	x_{11}	x_{12}	x_{13}	x_{14}
k_z	$k_{\varphi,x}$	$k_{\varphi,y}$	c_z	$c_{\varphi,x}$	$c_{\varphi,y}$	$EI_{x,1}$	$EI_{y,1}$	$\alpha_{0,1}$	$\alpha_{1,1}$	$EI_{x,(4,5,6)}$	$EI_{y,(4,5,6)}$	$\alpha_{0,(4,5,6)}$	$\alpha_{1,(4,5,6)}$

In order to achieve the optimum value of the parameters in Table 1, the CPSO has been executed by choosing: $c_c = 1.49$, $c_s = 1.49$, number of population $n_{pop} = 100$, $T = 200$, $N = 200$ and maximum number of iterations 200.

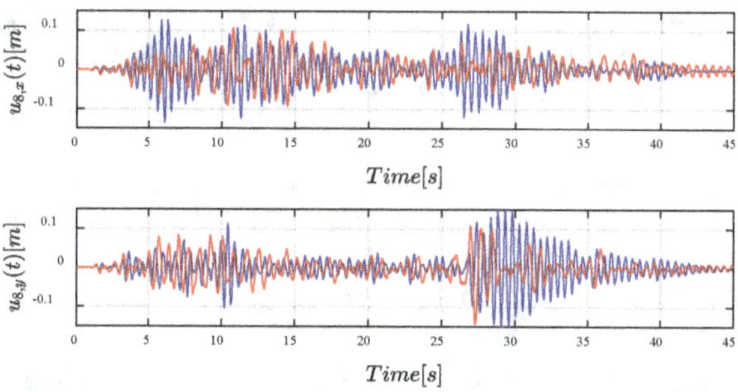

Figure 2. Displacements at the top of the structure (node 8) in x (upper) and y (lower) direction vs time: red line represents experimental data; blue line represents numerical simulation by using nominal parameters.

The Fig. 2 and Fig. 3 depict the displacement versus time of node 8 in the x and y direction considering the experimental data (red line) and numerical simulation (blue line) concerning the nominal parameters (Fig. 2) and the optimized parameters (Fig. 3). The comparison of the time signals in the latter figure shows a good match and it is possible to state that the proposed methodology allows to define a set of structural parameters able to correctly simulate the dynamic of the structure under investigation.

Several runs of the optimization procedure have been executed and a good agreement among them has been found in terms of both the optimum points and the corresponding objective function values.

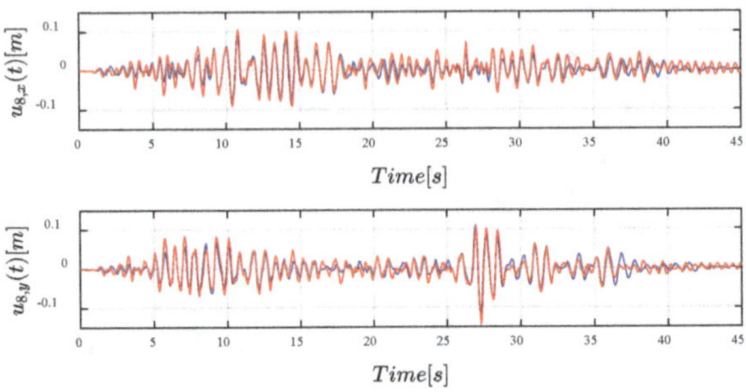

Figure 3. Displacements at the top of the structure (node 8) in x (upper) and y (lower) direction vs time: red line represents experimental data; blue line represents numerical simulation by using optimal parameters.

Summary

A model updating analysis based on CPSO has been presented and verified to numerically reproduce the dynamical response of a high voltage Current Transformer.

Future developments of this work will consider the comparison of the proposed model updating procedure with other classical optimization algorithms such as Genetic Algorithms and PSO, other objective functions depending on different components of the structural response and a modified version of the FEM model to investigate the role played by geometrical and mechanical nonlinearities.

Acknowledgement

This work is part of a national research project coordinated by Prof. F. Praticò and supported by M.I.U.R. through grant PRIN2017 prot. 2017XYM8KC. Local Unit Università di Enna Kore, coordinator Prof. F. Lo Iacono, prot. 2017XYM8KC_008.

References

[1] M. Friswell, J. E. Mottershead, Finite element model updating in structural dynamics, Springer Science & Business Media, 1995. https://doi.org/10.1007/978-94-015-8508-8

[2] C. Orlando, A. Ricciardello, Analytic solution of the continuous particle swarm optimization problem, Optimization Letters, 15 (6), (2021) 2005-2015. https://doi.org/10.1007/s11590-020-01671-3

[3] C. Orlando, A. Ricciardello, Continuous particle swarm optimization, AIP Conference Proceedings, 2293 (2020). https://doi.org/10.1063/5.0026570

[4] Lo Iacono, F., Navarra, G., Oliva, M., Cascone, D., "Experimental Investigation of the Dynamic Performances of the LEDA Shaking Tables System" in Proceedings of the 23rd Conference of the Italian Association of Theoretical and Applied Mechanics, Aimeta 2017, Salerno, 4-7 September 2017, 5, pages. 897–915 (2017).

[5] IEEE Recommended Practice for Seismic Design of Substations, in IEEE Std 693-2018 (Revision of IEEE Std 693-2005), vol., no., pp.1-220, 10 April 2019, doi: 10.1109/IEEESTD.2019.8686442. https://doi.org/10.1109/IEEESTD.2019.8686442

[6] EN 62271-207:2012, High-voltage switchgear and controlgear - Part 207: Seismic qualification for gas-insulated switchgear assemblies for rated voltages above 52 kV.

[7] K.J. Bathe, Finite Element Procedures, Second Edition, Watertown, MA, 2014.

Theoretical and Applied Mechanics - AIMETA 2022 Materials Research Forum LLC
Materials Research Proceedings **26** (2023) 473-478 https://doi.org/10.21741/9781644902431-77

Shake-table test assessment of a base-isolation device for the seismic protection of the Goddess of Morgantina statue

ALBERTI Elena[1, a *], LO IACONO Francesco[1,b] and NAVARRA Giacomo[1,c*]

[1] Facoltà di Ingegneria e Architettura, Università degli Studi di Enna "Kore", Enna, Italy

[a]mariagraziaelena.alberti@unikorestudent.it, [b]francesco.loiacono@unikore.it, [c]giacomo.navarra@unikore.it

Keywords: Shake-Table Test, Seismic Protection, Experimental Dynamics

Abstract. Monuments and museums contents in the Mediterranean area have revealed a poor dynamic behavior and suffered large damages during past severe seismic events. Their protection deserved great research efforts due to their inestimable values and high cultural significance. From an analytic point of view, art and museum objects, statues and displays are generally modelled as rigid blocks and have been extensively studied in the non-linear dynamic context. On the other hand, only limited experimental researches have been carried out. Base isolation devices have proven to be highly effective in control the dynamic behavior of for statue-like items. This work presents an analytical and experimental study for the dynamical characterization of a bidirectional base-isolation device. The ability of the device in reducing the dynamic response of statue-like objects has been investigated by means of full-scale shaking table tests carried out at L.E.D.A. Research Institute at the "Kore" University of Enna. Experimental results are then used to calibrate an analytical model of the device. The Goddess of Morgantina statue has been selected as case-study due to its significance and since it is equipped with a base-isolating device in its present placement.

Introduction

In the last decades, monuments and historic buildings located in the Mediterranean area have been exposed to several major earthquakes and extensive damages occurred. While the seismic protection of historic buildings is a well investigated research field with a large amount of numerical and experimental activities, also museums contents, artefacts and statues deserves the attention of scholars due to the high cultural significance and the inestimable value of most of them. Museum artefacts and statues are usually modelled as rigid bodies and no specific provisions are indicated in seismic codes, since they are considered as non-structural elements. Generally, they are analyzed in the non-linear dynamics context [1-3], but the experimental works remain limited. Base isolation techniques revealed to be very effective in seismic protection, also due to their small size, simplicity and affordability.

This paper presents a research conducted in the framework of the Research Project *eWAS - An Early-Warning System for Cultural Heritage*, funded by the MUR (Italian Ministry of University and Research), which aimed to evaluate the natural hazards to which Sicilian monumental and historical heritage are exposed. An extensive analytical and experimental study to characterize the dynamic performance of a bidirectional base-isolation device is presented. The isolator is based on an element patented as KSJ (Kinematic Steel Joint), consisting of a multiple articulated quadrilateral mechanism with crossing rods, entirely made of cut and folded steel sheets, thus reducing production and maintenance costs [4] and extending the applicability of the base-isolation techniques also to non-structural elements or cultural heritage items. Depending on the item to be protected, KSJ elements can be arranged in several configurations. In order to realize a base-isolation device for statue-like items, as illustrated in Fig. 1-a, KSJ elements can be assembled as

Materials Research Proceedings **26** (2023) 473-478 https://doi.org/10.21741/9781644902431-77

to form two double-rail stacked systems attached to a rigid base. In this way, the statue can undergo to horizontal bidirectional relative displacements with respect to the ground.

Figure 1 . The statue of the Goddess of Morgantina: a) the full scale replica with the isolation device on the shaking tables at L.E.D.A.; b) in its present placement;

The Goddess of Morgantina statue (depicted in Fig.1-b) has been selected as case-study. This 2.24m tall statue, carved in the 5th century BC in Sicily, was illegally excavated from the archaeological site of Morgantina (Sicily) in late 1970s, and exposed at the Paul Getty Museum in for several years. When the statue was finally returned to Italy on 2011, the base-isolating device, designed and realized in California, has been maintained.

Basing on an accurate 3D laser-scanner survey and 3D-printed form-work, a full-scale replica of the Goddess of Morgantina statue has been constructed. A fibre-reinforced cementitious mortar has been selected for its highly fluidity and for its unit mass, in order to reproduce as close as possible the mass and inertial properties of the actual statue. Full scale shaking table tests have been carried out at L.E.D.A. Research Institute of the "Kore" University of Enna [5]. A great amount of experimental data has been collected in order to prove that the design characteristics of the device are adequate to reduce the acceleration transmitted to the statue and to calibrate an analytical model of the isolated base.

Modelling of the dynamical behavior of the isolation device

Due to the kinematic characteristics of KSJ elements, the base-isolation device under study can be classified as a friction pendulum bearing, since it produces vertical upward displacements as it moves away from the central rest position. The protected item is then constrained to have trajectories contained in a double-curved concave surface. However, unlike conventional bearings, this device can resist uplift and provide reactions to the overturning moment but it does not protect against the earthquake vertical component.

In the base-isolation device at hand, the bi-directional behavior results from the composition of two orthogonal motions at each rail level, which depends on only one co-ordinate. The restoring and frictional forces exerted by the isolator and its self-centering capabilities depend on the slope of the trajectories. By considering, the rails in x direction, and by assuming that the surface can be expressed as z(x), it can be shown that the horizontal component of the restoring and frictional forces can me modelled, respectively, as [6]:

$$F_R(x) = mg\frac{z'(x)}{1+z'(x)^2}; \qquad F_{F,sl}(x,\dot{x}) = -mg\frac{\mu(\dot{x})\text{sgn}(\dot{x})}{1+z'(x)^2} \tag{1}$$

being m the mass of the protected item, g the gravity acceleration, z'(x) the first spatial derivative of the trajectory function, sgn(·) the signum function. Moreover, a velocity-variable friction coefficient can be set as [7]:

$$\mu(\dot{x}) = f_{max} - (f_{max} - f_{min})\exp(-\alpha|\dot{x}|) \tag{2}$$

where f_{min} and f_{max} are the minimum and maximum mobilized coefficient of friction, respectively, α is a parameter that control the friction coefficient variation rate and the over dot means time differentiation. Classical models of friction-based systems comprise two cases, sliding and sticking. In the sliding case the relative velocity $\dot{x}$ is non-zero and the friction force is expressed as in Eq. 1. In the sticking phase, the relative velocity is equal to zero and the friction force assumes, for equilibrium conditions, the following expression:

$$F_{F,st}(x,\dot{x}) = -\min\left(\left|F_{F,sl}(x,\dot{x})\right|, \left|F_{eq}\right|\right)\text{sgn}(F_{eq}) \tag{3}$$

where F_{eq} is the sum of all other forces acting on the protected object in the direction tangential to the trajectory. Similar equations apply also for the rails acting in y direction and, finally, the equation of motion of the bi-directional base isolation device subjected to a seismic loading can be written in state-space form as

$$\dot{\mathbf{Z}} = \mathbf{DZ} - \mathbf{G}(\mathbf{Z},\dot{\mathbf{Z}}) - \boldsymbol{\tau}\ddot{\mathbf{u}}_g(t), \quad \mathbf{Z}^T = [x \quad y \quad \dot{x} \quad \dot{y}] \tag{4}$$

where $\ddot{\mathbf{u}}_g(t)$ collects the ground acceleration components and $\boldsymbol{\tau}^T = [\mathbf{0}_{(2\times2)} \quad \mathbf{I}_{(2\times2)}]$ is the load location matrix. In Eq. 4, the system matrix **D** and the non-linear force vector **G** are expressed as:

$$\mathbf{D} = \begin{bmatrix} \mathbf{0}_{(2\times2)} & \mathbf{I}_{(2\times2)} \\ \mathbf{0}_{(2\times2)} & -\mathbf{M}^{-1}\mathbf{C} \end{bmatrix}; \quad \mathbf{G}(\mathbf{Z},\dot{\mathbf{Z}}) = \begin{bmatrix} \mathbf{0}_{(2\times1)} \\ \mathbf{M}^{-1}(\mathbf{F}_R + \mathbf{F}_F) \end{bmatrix} \tag{5}$$

being **M** and **C** diagonal matrices containing the mass and damping parameters in x and y direction respectively, whereas $\mathbf{F_R}$ and $\mathbf{F_F}$ are (2x1) vectors collecting the components in x and y direction of the restoring forces and friction force, respectively, as defined in the previous equations.

Eq. 4 can be easily solved in a step-by-step integration scheme. In the following applications, a fourth-order Runge-Kutta algorithm has been used in which, at the beginning of each step, the sliding and sticking conditions are separately assessed for each direction.

Results of shaking table tests

Several dynamic shaking table tests have been conducted to characterize the dynamical behavior of the device and to assess the seismic protection level [8]. In a first set of tests, broadband random noise acceleration time-histories have been imposed to measure the Frequency Response Function (FRF) between an accelerometer mounted on the shaking table (R1, input) and one on the rigid base (A1, output) of the device, under four increasing load conditions (from C0 to C3) realized by fixing heavy steel plates to the upper rigid base of the device. In the highest load level (C3), the same mass of the full-scale replica of the Goddess of Morgantina statue (about 900 kg) is applied. Fig. 2 reports the FRFs amplitude plots for each load level, for each direction and for an input intensity of 1.0 m/s^2 RMS. Results show that the device is able to strongly reduce the horizontal accelerations for each load level. In particular, it is to be noted that the load level C0 (no added masses) does not represent the normal operating configuration of the device. Moreover, a narrow magnification frequency range between 1.0 Hz and 1.15 Hz has been observed.

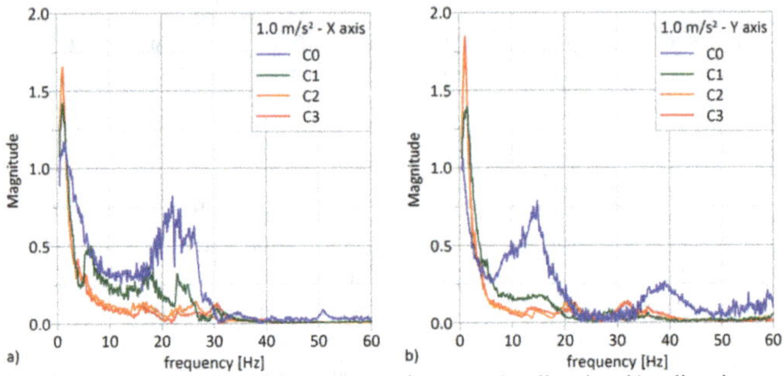

Figure 2 . FRF computed for random noise tests: a) x direction; b) y direction

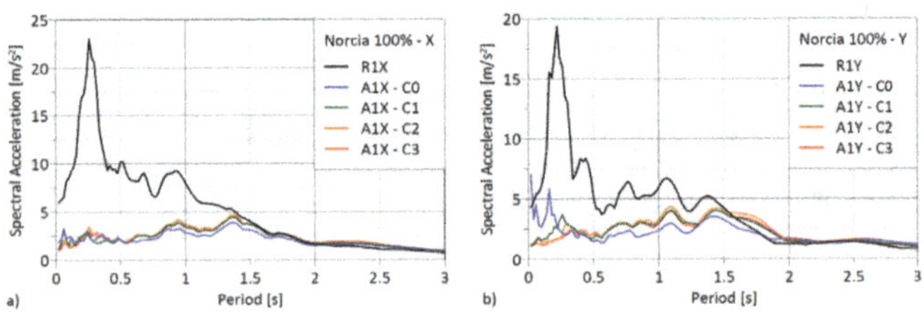

Figure 3 . Response Spectra comparison for Norcia signal.

A second set of shaking table tests has been executed by applying the horizontal components of four natural historical earthquakes acceleration time-histories (Kobe, Irpinia, Norcia and L'Aquila), selected for their different frequency content, impulsive behavior, duration and intensity. Firstly, the tests have been performed at the same load levels of previous tests (with steel plates) and, secondly, other tests have been carried out with the full-scale replica of the Goddess of Morgantina statue fixed on the device. In Fig. 3, the comparison between the Response Spectra (RS) computed 5% damping of the input and of the accelerations at the rigid base for Norcia signal and for all load levels is reported. The RS curves prove that the device strongly reduce the accelerations transmitted to the protected item independently of the load level.

The seismic tests performed with the statue mounted on the isolating device shown the effectiveness in the seismic protection of the statue. Fig. 4 illustrates the trends of peak accelerations at shaking table and on the rigid base with seven increasing input intensity (from 20% to 120% of the actual one) for both directions and with reference to Kobe and L'Aquila signals.

Materials Research Forum LLC
https://doi.org/10.21741/9781644902431-77

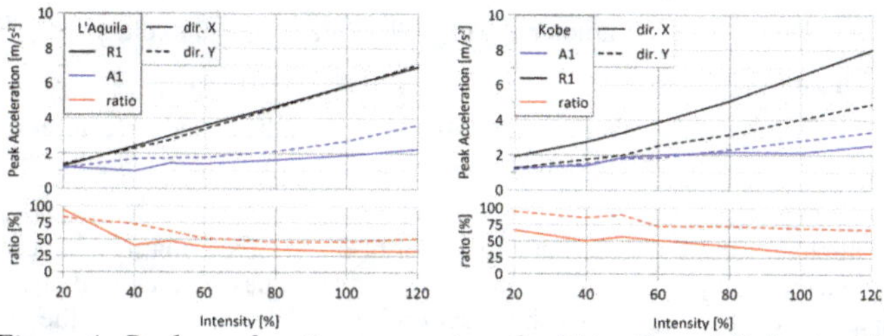

Figure 4 . Peak acceleration comparison for L'Aquila and Kobe signals.

Model calibration

The first step in characterizing the device was to identify the geometric properties of the double-curvature surface. Bi-axial horizontal displacements have been imposed to the rigid base, the trajectories have been acquired by a 3-D motion capture system [8] and it has been found that the surface containing the trajectories can be modelled as an elliptical paraboloid, expressed as:

$$z = ax^2 + by^2 \qquad (a > 0, b > 0) \tag{6}$$

where the parameters $a = 0.3391\ m^{-1}$ and $b = 0.3190\ m^{-1}$ have been evaluated by a fitting procedure. The equivalent oscillation periods, slightly variable along the stroke, have been estimated to be close to 2.5 s, typical values of base-isolation devices.

The remaining parameters of the analytical model, namely the masses, the damping coefficients, the friction at high and low velocity and their variation rate, have been evaluated by means of an optimization procedure, by minimizing the following objective function:

$$J(\mathbf{X}) = \sum_{j=x,y} \frac{1}{2} \sum_{i=1}^{N} w(i,j) \frac{\left[x_{exp}(i,j) - x_{th}(\mathbf{X},i,j)\right]^2}{x_{exp}(i,j)^2} \tag{7}$$

being **X** the vector that collects the optimization parameters, x_{exp} and x_{th} the (N x 2) matrices of experimental and analytical horizontal components of displacements, respectively, and w a weighting function which is calculated as the time derivative of the Arias's Intensity function, in order to better reproduce the strong motion phase of the response.

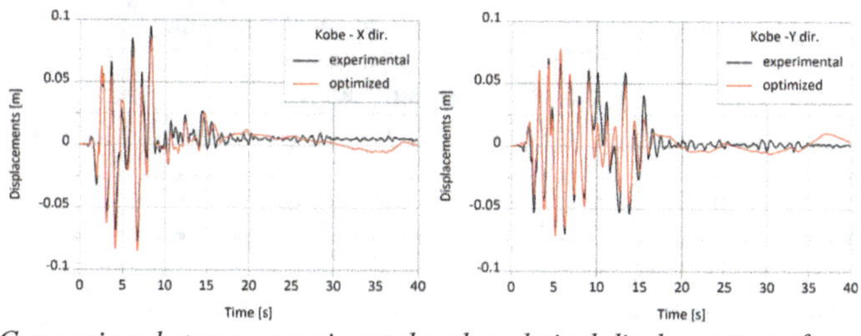

Figure 5 . Comparison between experimental and analytical displacements after optimization procedure.

Fig. 5 depicts the results of the proposed procedure for the seismic test conducted with Kobe signal at 100% of its actual intensity. The comparison among the acceleration time history recorded at the rigid base and the analytical one is reported. This latter, is obtained in correspondence of the following optimal set of parameters: $m_x = 900kg$, $m_y = 980.6kg$, $c_x = 133,4Ns/m$, $c_y = 137,4Ns/m$, $a_x = 0.3070m^{-1}$, $a_y = 0.3693m^{-1}$, $f_{max} = 0.88\%$, $f_{min} = 4.54\%$, $\alpha = 5.43$. It appears that the proposed analytical model is able to capture the essential features of the response in terms of accelerations, but further in-depth investigation is needed to enhance the model.

Conclusions

In order to characterize the dynamic behavior of a bi-directional isolating device based on the patented KSJ system, an experimental and analytical research has been carried out.

An analytical model of the device has been set up and calibrated, basing on the results of a full-scale shaking table tests campaign performed at L.E.D.A. Research Institute of Kore University of Enna. At this aim, a full-scale replica of the Goddess of Morgantina statue, chosen as case-study, has been constructed, placed on the isolation device and subjected to broadband random noise and seismic tests with increasing intensities. Test results demonstrated the ability of the device in highly reducing the seismic acceleration transmitted to statue-like art objects.

Finally, the calibration parameters of the analytical model have been evaluated by an optimization procedure, so that the model is able to capture the essential feature of the response.

Acknowledgements

This work has been carried out in the framework of the research project "eWAS An Early WArning System for Cultural Heritage" (ARS01 00926), funded by Italian Ministry for University and Research, whose financial support is gratefully acknowledged.

References

[1] G.W. Housner. The behaviour of inverted pendulum structures during earthquakes. Bulletin of the Seismological Society of America. 53 (1963) 403-417. https://doi.org/10.1785/BSSA0530020403

[2] P.D. Spanos, A.S. Koh. Rocking of rigid blocks due to harmonic shaking. Journal of Engineering Mechanics 110 (1984) 1627-1642. https://doi.org/10.1061/(ASCE)0733-9399(1984)110:11(1627)

[3] A. Palmeri, N. Makris. Response of rigid structures rocking on viscoelastic foundation. Earthquake Engineering and Structural Dynamics. 37 (2008) 1039-1063. https://doi.org/10.1002/eqe.800

[4] G. Guerrini, G. Ausenda, F. Graziotti, A. Penna. An innovative seismic isolation device based on multiple articulated quadrilateral mechanisms: analytical study and shake-table test. In: Proc. of XVIII ANIDIS Conference, Ascoli Piceno, Italy, 2019.

[5] F. Lo Iacono, G. Navarra, M. Oliva, D. Cascone. Experimental investigation of the dynamic performances of the L.E.D.A. shaking tables system. In: Proc. of the 23rd AIMeTA Conference, Salerno, Italy, 5 (2017) 897-915.

[6] L.Y. Lu, M.H. Shih, C.Y. Wu. Near fault seismic isolation using sliding bearings with variable curvatures. In: Proc. of 13th World Conference in Earthquake Engineering, paper no. 3264, Vancouver, Canada, 2004.

[7] A. Mokha, M.C. Constantinou, A.M. Reinhorn, V.A. Zayas. Experimental Study on Friction-Pendulum Isolation System. Journal of Structural Engineering. 117 (1991) 1201-1217. https://doi.org/10.1061/(ASCE)0733-9445(1991)117:4(1201)

[8] F. Castelli, V. Lentini, F. Lo Iacono, G. Navarra. Shake-table tests for the dynamic characterisation of an innovative isolator for seismic protection of statues. In: Geotechnical Engineering for the Preservation of Monuments and Historic Sites III, 389-400, CRC Press, London, 2022. https://doi.org/10.1201/9781003308867-25

Theoretical and Applied Mechanics - AIMETA 2022 Materials Research Forum LLC
Materials Research Proceedings **26** (2023) 479-484 https://doi.org/10.21741/9781644902431-78

Microcontroller design for active vibration control

Antonio Zippo[1,2, a *], Francesco Pellicano[1,2,b] ,Giovanni Iarriccio[1,c]

[1] University of Modena and Reggio Emilia - DIEF, 41121 Modena, Italy

[2] InterMech MoRe Centre, 41121 Modena, Italy

[a]antonio.zippo@unimore.it, [b]francesco.pellicano@unimore.it, [c]giovanni.iarriccio@unimore.it

Keywords: Active Vibration Control, Positive Position Feedback, Experiments

Abstract. Active Vibration Control (AVC) problem is successfully studied considering all the characteristics of a control problem. The controlled system consists of a honeycomb panel of carbon fiber. Expensive devices are in general used for setting up and designing AVC this due to limitations for practical implementation. In our work we point to a low cost and practical solution using a microcontroller that has been verified. The honeycomb plate has been forces with out of plane load using an electrodynamic shaker at resonances that has been identified by experimental modal analysis. Piezoceramic patches are used as sensor and actuator for the control. Multiple analog signal processing circuits were developed to scale and shift the signal at the input and output of the MCU. The Positive Position Feedback (PPF) control algorithm is proposed, and a campaign of tests are carried out with harmonic excitations at resonance frequencies. Experimental results show an amplitude velocity reduction from 50% to 77% less and Power Spectral Density (PSD) attenuation up to 12.8 dB. The size and structural properties of the MFC patches, the control unit and structure under control are suitable for automobile and aerospace applications.

Introduction

In the last years, more and more Active Vibration Control has been increasing a lot of consideration, problem related to uncontrolled vibrations finds its application in wide range of industries, as for example: body plates of automobiles, preventing the bending of aerodynamic components in motorsports or reducing unwanted noise and resonance response; also in the aerospace sector, active vibration control finds its application as in [1] where Ye-Lin Lee describe a reduction of vibration up to 98% in simulations of a lift-offset compound helicopter using two active control techniques. Moreover, studies regarding experimental activities and problems of complex systems, as in [2], [3] and [4] demonstrate the importance of experimental investigations.

Different techniques and algorithms can be used to control the vibration of a structure as the author in [5] devised an energy-fuzzy adaptive PD control to stabilize the structures in subsonic and transonic regime, or as in [6], Balakrishna used piezoceramic actuation embedded to the structure to reduce the response, producing a practical solution for the sting vibration. Another method is explained in [7] where Oveisi and Nestorović used a frequency domain subspace identification for robust control design, or in [8] Electrically Controlled Rotor (ECR) has been controlled developing a Kalman filter-based control algorithm.

Example of microcontroller for active vibration control applications are present as in [9] using an Arduino Uno, with clock speed of 16 MHz.

In [10], Moon K.Kwak developed a MIMO PPF controller realized by a dSpace DSP board, with the sampling frequency of ADC and DAC being 50 kHz, A.Zippo et alt., in [11], showed experiments implementing a variety of configurations: SISO, multiSISO and MIMO on a free-free composite panel with PPF method and MFC patches both for sensors and actuators in linear and nonlinear range. Moreover, analytically define chaotic structures is stimulating for improving

Materials Research Forum LLC
https://doi.org/10.21741/9781644902431-78

our knowledge in the occurrence of vibrations related to complex systems [12] with quasiperiodic and chaotic motion [13].

In [14], Isabelle Bruant et alt., used genetic algorithm for optimal placing patches for ACV.

Experimental modal analysis

In figure 1 is shown the experimental setup, while in figure 2 the TestLab sum of FRF is shown, this graph help to experimental identify modes of vibration.

Eleven modes have been identified in total and in figure 3 and 4 are represented the reconstructed modal shapes at 25.13Hz and at 132.31Hz, the respective identified damping is 0.33% and 0.18%

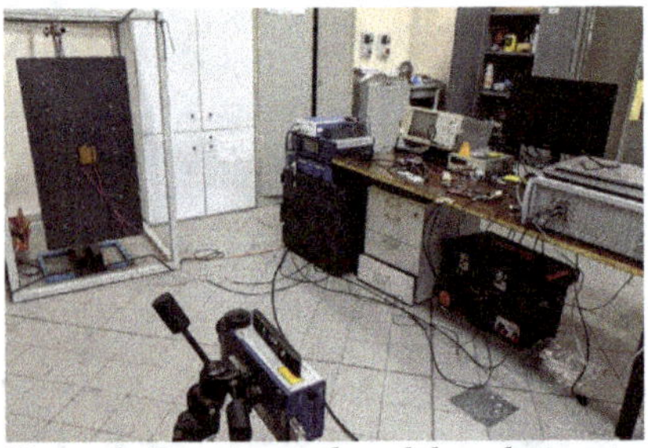

Figure 1. Experimental modal analysis setup

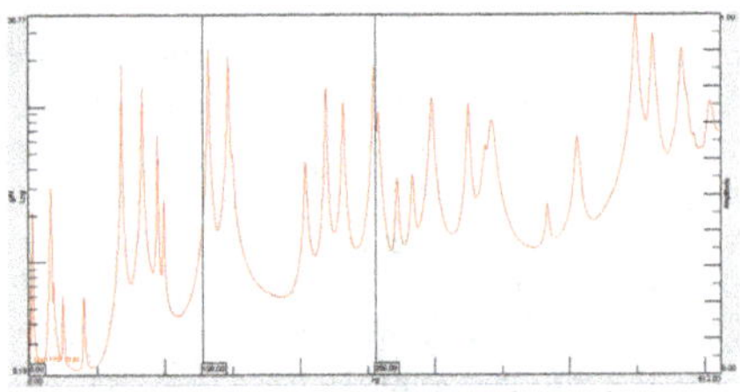

Figure 2: Sum of FRFs of the plate

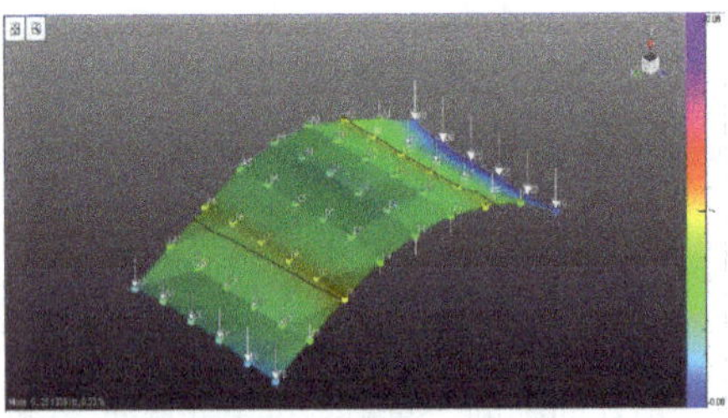

Figure 3: 2nd mode of vibration at 25.13 Hz

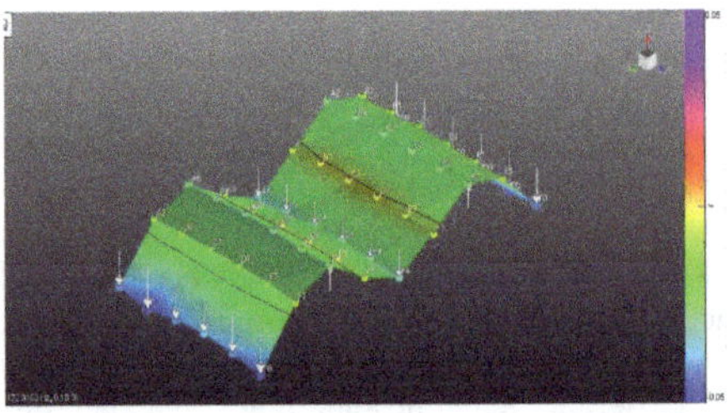

Figure 4: Vibration mode at 132.31 Hz

Sensor and actuator selection

In Table 1 are presented the selected. MFC patches used as sensor and actuator, the Macro Fiber Composites (MFC) can be used both as sensor or actuator, depending on the voltage is measured or powered from or to the patch, they consist of piezo-ceramic rods sandwiched in a design of multi-layers of electrodes and polyimide film. In figure 5 their location on the plate is showed. The physical effect used by this materials is the d33, achieved by the stack arrangement of the piezoelectric material, that are configured so that when a voltage is applied across the electrodes, the stack expands in length (commonly considered as 3-3 axis) and for a certain voltage, the net static displacement δ of the piezoelectric ceramic actuator is:

$$\delta = \frac{d_{33}V+\frac{F}{K_a}}{1+\frac{K}{K_a}} \tag{1}$$

where d33 is the strain constant of the piezoelectric material in the 3-3 axis, K is the external spring stiffness, Ka is the actuator stiffness ($K_a = E_aA_a/L_a$, where E_a, A_a, and L_a are the actuator's young's modulus, cross-sectional area, and length, respectively). F is the external load force.

Table 1. Dimensions and electrical properties of MFC patches

	M8507-P1 (sensor)	M8557-P1 (actuator)
Active length [mm]	85	85
Active width [mm]	7	57
Overall length [mm]	101	103
Overall width [mm]	13	64
Capacitance [nF]	3.1	16
Free strain [ppm]	1035	1350
Blocking force [N]	65	693

Figure 5: patches location

Micro controller unit selection

The application of AVC in certain industrial application as automotive adds more constraints for the implementation as compact and light weight. Indeed, a 32-bit Micro Controller Unit (MCU) from Texas Instrument is used for the study and experiment performing on a composite sandwich plate the control.

The C2000 32-bit microcontrollers are optimized for processing, sensing and actuation to improve closed-loop performance in real time.

The main feature are:

- 32-bit C28x architecture floating point MCU
- 200 MHz processing speed
- Selectable 16-bit & 12-bit SAR ADC, up to 14 MSPS system throughput
- 1MB flash & 204KB RAM
- USB communication with Simulink interface

Analog circuits design

The analog signal processing circuits are needed to scale and shift the signal at output of the sensors and MCU to match the input of the signal requirement at the succeeding element to improve resolution, hence the performance and avoid signal saturation. In figure 6 is showed the electric scheme of the system.

Materials Research Forum LLC
https://doi.org/10.21741/9781644902431-78

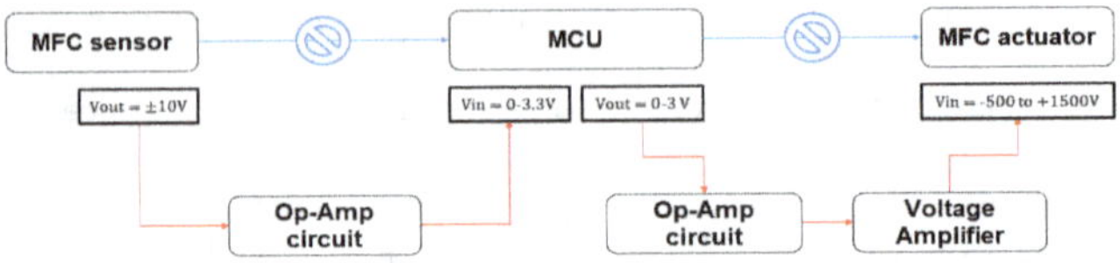

Figure 6: Electric schemes

The circuits are first prototyped in a breadboard, and after the components are implemented to a PCB to make it more compact and lightweight. BNC connectors has been used at the signal input and output for all the boards.

In the manufacture of the signal processing circuits, few of the resistors were replaced by trimmers to add the flexibility of tuning the gain and bias which can account for any errors during assembly and soldering.

Control algorithm

The PPF controller is a filter of second order that tends to maximize damping in the frequency specified without destabilizing other modes, and thus reducing spillover. PPF controller is modelled as a second-order dynamic system, which is represented by the following equation,

$$\ddot{\eta}(t) + 2\zeta_c\omega_c\dot{\eta}(t) + \omega_c^2\eta(t) = kq\omega_c^2 \tag{2}$$

Where $\zeta c$is the damping ratio of the compensator, ωc is the frequency of the compensator and k is the compensator gain. The transfer function takes the form:

$$H(s) = \frac{\omega_c^2}{s^2+2\zeta_c\omega_c s+\omega_c^2} \tag{3}$$

At frequencies above ω_c the slope of the transfer function amplitude is negative and very steep (-40 dB/octave), so it reduces the problem of spillover for vibrations at higher frequencies. On the contrary, at frequencies below ω_c the bode diagram of H(s) has an amplitude of 0 dB, the simulink model is showed in figure 7.

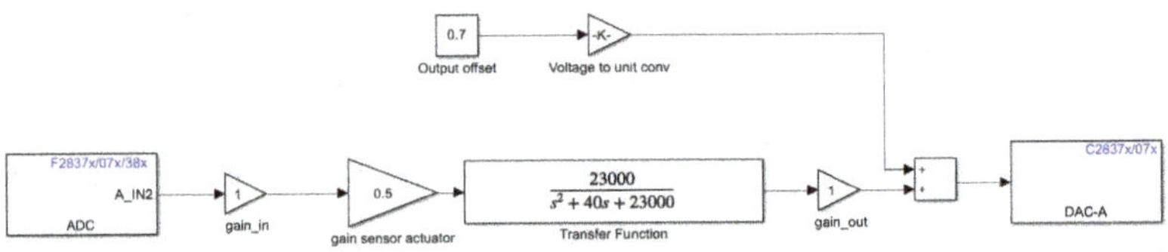

Figure 7. PPF controller modelled in simulink

Setup and results

The plate is excited using a traditional electrodynamic shaker, placed behind the plate. The force applied by the shaker is measured by a dynamic force transducer and the PPF algorithm is compiled and uploaded in the flash of MCU. The actuator and sensor are placed to interact optimally with the selected modes.

Results of the second vibration mode control at 25.17Hz are a Velocity reduction of 74.18%, a PSD amplitude attenuation of 11.76 dB inn a Settling time of about 1 second, in figure 8 is shown respectively the time history and PSD of the shaker input signal, and the PSD of the control, while in figure 9 is presented the time history while the control is activated.

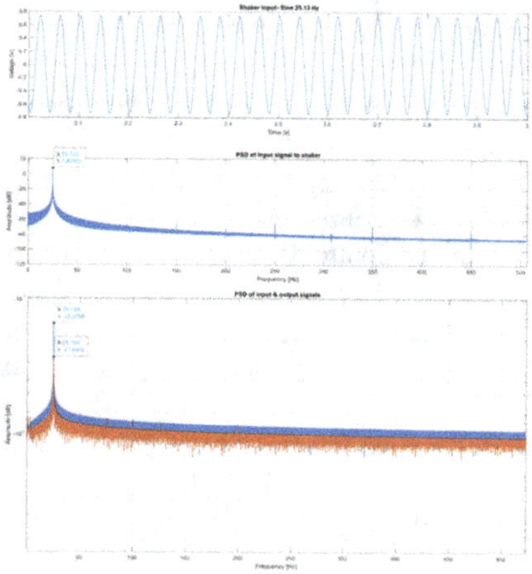

Figure 8: PPF controller modelled in simulink

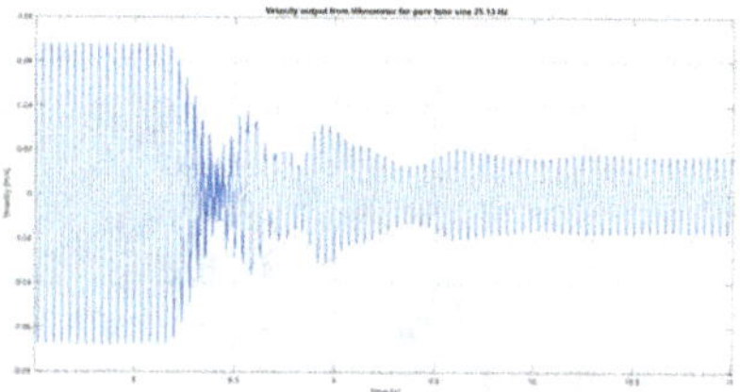

Figure 9: TH during control activation

Conclusions

This investigation designed an Active Vibration Control System on the base of a C2000 32-bit microcontroller unit by means of MFC patches used for sensing the disturbances and actuate a controlling force by means of a second order filter algorithm (Positive Position Feedback method). This work shows more practical solution can be implemented also with low-cost central unit and allow a direct implementation in automotive industrial problems. The control techniques discussed, and the developed hardware was proved effective in suppressing the amplitude of vibration when subjected to harmonic excitations. Nevertheless, harmonic disturbances are not common, forcing alone a system in a practical application, further studies will be carry out applying to the panel random excitations and complex signals to better reproducing and testing real case studies.

References

[1] Ye-Lin Lee, Do-Hyung Kim, Jae-Sang Park, Sung-Boo Hong. Vibration reduction simulations of a lift-offset compound helicopter using two active control techniques. Aerospace Science and Technology 2020; 106181. https://doi.org/10.1016/j.ast.2020.106181

[2] Zippo A., Iarriccio G., Pellicano F., Shmatko T., Vibrations of plates with complex shape: Experimental modal analysis, finite element method, and R-functions method, (2020) Shock and Vibration, 2020. https://doi.org/10.1155/2020/8882867

[3] Iarriccio G., Zippo A., Pellicano F., Barbieri M., Resonances and nonlinear vibrations of circular cylindrical shells, effects of thermal gradients, (2021) Proceedings of the Institution of Mechanical Engineers, Part C: Journal of Mechanical Engineering Science, 235 (20), pp. 4818 – 4832. https://doi.org/10.1177/0954406220907616

[4] Pellicano F., Zippo A., Iarriccio G., Barbieri M., Experimental Study on Nonlinear Random Excitation, (2020) Lecture Notes in Mechanical Engineering, pp. 637 – 648. https://doi.org/10.1007/978-3-030-31154-4_54

[5] Wei LIU, Weixiao LIU, Mengde ZHOU, Linlin TANG, Qinqin WANG, Zhengquan WEN, Zhuang YAO, Xiaojing YUAN. An active vibration control method based on energy-fuzzy for

cantilever structures excited by aerodynamic loads. Chinese Journal of Aeronautics,Volume 34, Issue 9, 2021; 1000-9361. https://doi.org/10.1016/j.cja.2020.07.001

[6] Balakrishna, Sundareswara & Houlden, Heather & Butler, David & White, Richard. Development of a Wind Tunnel Active Vibration Reduction System. 2007; 10.2514/6.2007-961. https://doi.org/10.2514/6.2007-961

[7] Atta Oveisi, Tamara Nestorović, Allahyar Montazeri. Frequency Domain Subspace Identification of Multivariable Dynamical Systems for Robust Control Design. IFAC, Volume 51, Issue 15, 2018; 2405-8963. https://doi.org/10.1016/j.ifacol.2018.09.065

[8] Jinchao MA, Yang LU, Taoyong SU, Shujun GUAN. Experimental research of active vibration and noise control of electrically controlled rotor. Chinese Journal of Aeronautics, Volume 34, Issue 11. 2021; 1000-9361. https://doi.org/10.1016/j.cja.2020.10.027

[9] Wang, Z, & Keogh, P. Active vibration control for robotic machining. Proceedings of the ASME 2017 International Mechanical Engineering Congress and Exposition. Volume 2: Advanced Manufacturing. Tampa, Florida, USA. November 3–9, 2017. V002T02A079. ASME. https://doi.org/10.1115/IMECE2017-71670

[10] Moon K. Kwak, Seok Heo. Active vibration control of smart grid structure by multi- input and multioutput positive position feedback controller. Journal of Sound and Vibration, Volume 304, Issues 1–2, 2007. https://doi.org/10.1016/j.jsv.2007.02.021

[11] Zippo, Antonio & Ferrari, Giovanni & Amabili, Marco & Barbieri, Marco & Pellicano, Francesco. Active vibration control of a composite sandwich plate. Composite Structures. 2015. 128. https://doi.org/10.1016/j.compstruct.2015.03.037

[12] Hemmatnezhad M., Iarriccio G., Zippo A., Pellicano F., Modal localization in vibrations of circular cylindrical shells with geometric imperfections (2022) Thin-Walled Structures, 181, art. no. 110079. https://doi.org/10.1016/j.tws.2022.110079

[13] Iarriccio G., Zippo A., Pellicano F., Asymmetric vibrations and chaos in spherical caps under uniform time-varying pressure fields, (2022) Nonlinear Dynamics, 107 (1), pp. 313 – 329. https://doi.org/10.1007/s11071-021-07033-7

[14] Isabelle Bruant, Laurent Gallimard, Shahram Nikoukar. Optimal piezoelectric actuator and sensor location for active vibration control, using genetic algorithm. Journal of Sound and Vibration, Volume 329, Issue 10, 2010. https://doi.org/10.1016/j.jsv.2009.12.001

Static and dynamic response analysis of stay cables using terrestrial laser scanning and vibration measurements

Cecilia Rinaldi[1, a *], Marco Lepidi[2, b] and Vincenzo Gattulli[1,c]

[1]Dept. of Structural and Geotechnical Engineering, Sapienza University of Rome, Rome, Italy

[2]Dept. of Civil, Chemical and Environmental Engineering, University of Genova, Genova, Italy

[a]cecilia.rinaldi@uniroma1.it, [b]marco.lepidi@unige.it, [c]vincenzo.gattulli@uniroma1.it

Keywords: Inclined Suspended Cables, Modal Analysis, Point Cloud Model, Vision-Based Measurement, Cable Tension Estimation

Abstract. Nowadays, high accuracy measurements provided by terrestrial laser scanner and vision sensors allow to collect useful and exhaustive information about the conditions of the existing structures, useful to detect defects and geometry anomalies and to better understand their mechanical behavior. These avant-garde technologies were found to be particularly effective for the structural health assessment of the cable-stayed pedestrian bridge described in this paper. Considering a continuous mono-dimensional model of an inclined perfectly flexible cable, the axial tension is locally tangent to the cable profile. Thus, determining the cable static response under self-weight consists of a geometric shape-finding problem. Through terrestrial laser scanning, a 3D point cloud model of the bridge was acquired, including a data-abundant description of the actual static configuration of the stays. Therefore, cable configuration was no longer an unknown of the static problem, which can be inverted to assess the static tension. Furthermore, modal analysis was conducted also through image-based vibrations measurements to identify the fundamental frequencies of the cables. The independent identification of the axial forces from static (geometric) and dynamic (spectral) data provided results in good agreement.

Introduction

Structural cables are primary load bearing elements in different civil engineering systems, such as cable-stayed and suspension bridge. The assessment of their integrity condition and stress level are key tasks in the health monitoring of such structures. Among the techniques used to identity cable forces, direct methods make use of load sensors, while indirect methods estimate cable tension through different quantities such as strain, or natural frequencies [1],[2].

The dynamic identification method based on determining the unknown cable tension from the geometric stiffness that can be evaluated for measured frequencies (known the cable mass) is the most used, because it is a nondestructive method which guarantees high efficiency [3],[4],[5]. Similarly to other model based methods, this approach requires refined mechanical formulations describing the dependence of the spectral properties on the structural parameters [6],[7]. In addition to traditional accelerometer sensors, different technologies are used to record cable vibrations. Microwave interferometry has been used to measure the vibration response of cables in cable-stayed bridges [8],[9]. Frequencies and cable tensions have been identified through radar technique with the same accuracy of the ones obtained with conventional sensor measurements. Modal analysis and axial force identification of stay cables are also performed through vision-based measurement, with motion magnification methods used to amplify microvibration of stay cable captured by video camera [10]. If not based sufficiently refined mechanical methods, frequency-based approaches often neglect or underrate the effects of the chord inclination angle, cable extensibility, flexural stiffness, complex boundary conditions and cable-beam interactions. All these effects could lead to significant errors in tension identification.

In this work, the 3D point cloud model of a cable-stayed pedestrian bridge was exploited to evaluate the actual static configuration of the stays and to identify their axial forces by comparing the quadratic term coefficient of the cubic function interpolating the acquired geometry data with the quadratic term of the cubic equation obtained from an approximate solution of the static problem. The quality of the estimation of cable force was evaluated through the definition of an error function based on the cubic term the interpolating function and thorough the comparison with the tension identified through frequency-based method.

Bridge description and data acquisition

The bridge is a steel truss structure located in Beinasco, Turin province, in Italy. It is a cycle and pedestrian cable-stayed bridge and consists of an access ramp and a deck both supported by steel columns and cables anchored to a steel tubular pylon (see Fig. 1) with star reinforcements in 15 mm thick shaped sheets, hinged at the base. The deck (80 m long, 3 m wide and 1.65 m height) is made of three main tubular members and smaller cross-members and it is supported by 7 cables (42 mm diameter), all prestressed during assembly to avoid loosening. The first part of the access ramp is simply supported on columns while the second part is suspended by other stays (28 mm diameter) converging on the same pylon that supports the deck. Several cables (40 and 42 mm diameter) connect the pylon to the ground, to ensure stability and counterbalance the loads of the deck and ramp. On the side of the ramp, the steel truss-type beam of the deck is also supported by an access concrete staircase which is located at the center of the ramp.

Experimental data were acquired by different technologies to collect both geometry and vibration information of the structure. The terrestrial laser scanner Cam2 Laser Scanner Focus Faro x 130 (130 m unambiguity interval, 0.6 m - 130 m Range Focus3D X 130 HDR, 122000/244000/ 488000/976000 points/sec measurement speed, ±2 mm distance error) was used to acquire actual geometry configuration of the whole structure; during the scans, it acquired measurements by the appropriate laser and photographs in both color and black and white. The given color allowed to obtain three-dimensional RGB models (see Fig. 1b).

Wireless and wired accelerometers and a high-speed camera were used to record structural dynamic response under ambient excitation. The vibrations of the cables were recorded by using 6 uniaxial piezoelectric wired accelerometers, PCB 393B31 with ICP technology (10.0 V/g sensitivity, 0.5 g pk measurement range, 200 Hz sampling frequency). Furthermore, the high-speed camera IO Industries Flare 12M125xCL (monochrome, F-mount, 4096 x 3072 resolution, 5.5 x 5.5 μm pixel size, 100-200 fps sampling frequency) was installed on the concrete cantilever of the stairs located in the central area of the ramp to record the vibration of cables SP15 and SP16 (Fig. 4a). The displacement time histories were obtained by processing the captured images with Digital Image Correlation (DIC) technique.

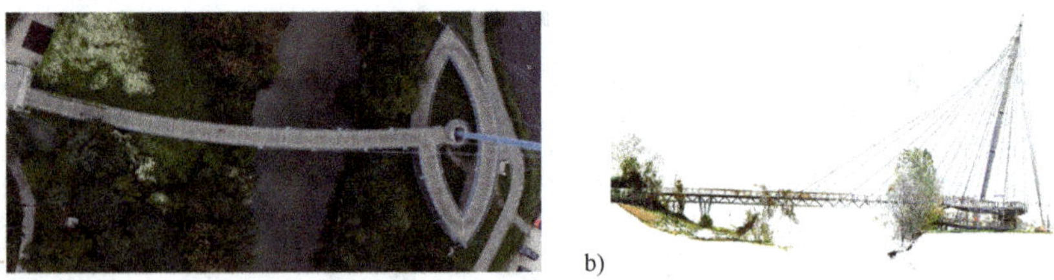

a) b)

Fig. 1. Beinasco bridge: a) plan view of the cable-stayed; b) side view of the 3D point cloud model.

Cable tension identification from static configuration

Consider a suspended cable hanging on a vertical plane under self-weight. Let S be the set of N laser-scanned points belonging to the cable, pointed by the configuration vector $\mathbf{X}_i = (X_i, Y_i, Z_i)$,

 Materials Research Forum LLC
https://doi.org/10.21741/9781644902431-79

within an orthogonal Cartesian reference system in a three-dimensional space (with $i = 1,...,N$). The vertical midplane of the N coordinate points is sought through the equations

$$y = ax + b, \quad a = \frac{1}{M}\sum a_{ij}, \quad b = \frac{1}{M}\sum b_{ij}, \tag{1}$$

where the coefficients a_{ij} and b_{ij} are determined by taking $M = N_r^2$ pairs of points (i, j) from the cable point cloud model, according to the formulas

$$a_{ij} = \frac{Y_i - Y_j}{X_i - X_j} = \frac{Y_j - Y_i}{X_j - X_i} = a_{ji}, \quad b_{ij} = \frac{X_i Y_j - X_j Y_i}{X_i - X_j} = \frac{X_j Y_i - X_i Y_j}{X_j - X_i} = b_{ji}, \tag{2}$$

where a subset of $2N_r < N$ points has been selected, namely N_r points $\mathbf{X}_i$ and N_r points $\mathbf{X}_j$ close to the lower and upper cable support – respectively – with the aim of maximizing the accuracy of the coefficient estimates by maximizing the denominators in equations (2). The coefficients a_{ij} and b_{ij}, obtained for each pair of points considered, are represented in the plane (a, b) in Fig. 2. It can be noted that the results associated to the same point $\mathbf{X}_i$ are approximately aligned. This behaviour can be easily justified by considering that, with respect to a fixed lower point $\mathbf{X}_i$, the positions of the upper points $\mathbf{X}_j$ (that are close to each other) determine small variations of the coefficients a_{ij} and b_{ij}, that can be estimated through their first variations

$$\delta a_{ij} = \frac{\partial a_{ij}(X_j, Y_j)}{\partial X_j}\delta X_j + \frac{\partial a_{ij}(X_j, Y_j)}{\partial Y_j}\delta Y_j, \quad \delta b_{ij} = \frac{\partial b_{ij}(X_j, Y_j)}{\partial X_j}\delta X_j + \frac{\partial a_{ij}(X_j, Y_j)}{\partial Y_j}\delta Y_j, \tag{3}$$

so that it can be demonstrated that the ratio $\delta b_{ij} / \delta a_{ij} = -X_i$ is a constant. The midplane identified with this procedure is represented in Fig. 2b. A roto-translation of the reference system are performed to make the Cartesian plane $\Pi(X, Z)$ coincident with the midplane of the cable.

The geometric curve to be identified based on the cable points is postulated to be describable by a cubic function $Z(X)$. Suited dimensionless variables and parameters are introduced

$$z = \frac{Z}{L}, \quad x = \frac{X}{L}, \quad \theta = \text{arctg}(B/L), \tag{4}$$

where the lengths L and B are the horizontal and vertical distance between the supports of the cable (assumed known) and θ is the angle of the cable chord with respect to the horizontal.

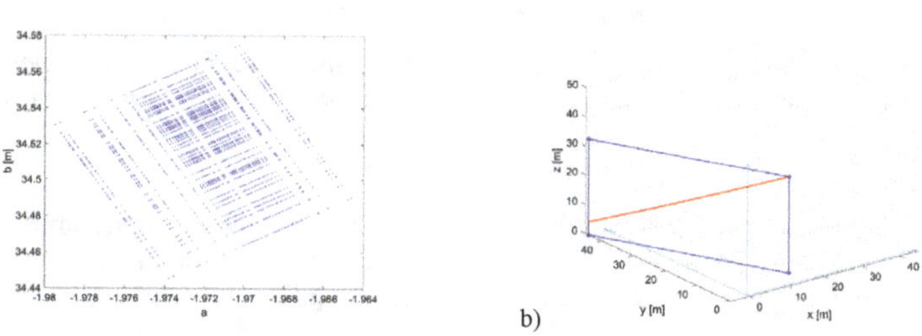

a) b)

Fig. 2. Midplane of the 3D point cloud of cable: a) coefficients a_{ij} and b_{ij}; b) vertical midplane identified (blue) and 3D point cloud model of the cable (red).

By employing nondimensional variables, the generic expression of the cubic function $z(x)$ depends on four unknown coefficients γ_i, according to the formula

$$z(x) = \gamma_0 x^3 + \gamma_1 x^2 + \gamma_2 x + \gamma_3. \tag{5}$$

Therefore, by imposing the geometric boundary conditions $z(x)_{x=0} = 0$ and $z(x)_{x=1} = -\tan\theta$, the cubic expression becomes

$$z(x)=\gamma_0 x^3+\gamma_1 x^2-(\gamma_1+\gamma_0+\tan\theta)x, \tag{6}$$

where the coefficients γ_0 and γ_1 are unknown, but univocally determinable for each pair of laser-scanned points $\mathbf{x}_1=(x_1,z_1)$ and $\mathbf{x}_2=(x_2,z_2)$ lying on the midplane through the formulas

$$\gamma_0=\frac{x_1 z_2(1-x_1)-x_2 z_1(1-x_2)-\Delta(x_1,x_2)}{x_1 x_2(1-x_1)(1-x_2)(x_1-x_2)}, \qquad \gamma_1=\frac{x_2 z_1(1-x_2^2)-x_1 z_2(1-x_1^2)+(x_1+x_2)\Delta(x_1 x_2)}{x_1 x_2(1-x_1)(1-x_2)(x_1-x_2)}, \tag{7}$$

where $\Delta(x_1,x_2)=x_1 x_2(x_1-x_2)\tan\theta$. A subset of $2N_s<N$ points has been selected, namely N_s points $\mathbf{x}_1$ and N_s points $\mathbf{x}_2$ close to the lower and upper thirds of the cable profile, respectively, with the aim of maximizing the accuracy of the coefficient estimates by maximizing the denominators in equations (7). The coefficients γ_0 and γ_1, obtained for each pair of points considered, are represented in the plane (γ_0,γ_1) in Fig. 3. Again, the alignments of the results can be justified by considering that, with respect to a fixed lower third point $\mathbf{x}_1$, the positions of the upper third points $\mathbf{x}_2$ (that are close to each other) determine small variations of the coefficients γ_0 and γ_1, and that the ratio of the first variations is $\delta\gamma_1/\delta\gamma_0=-(x_1+1)$ is constant.

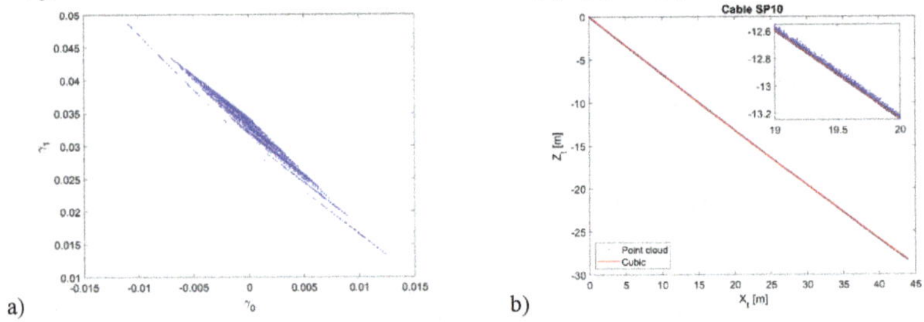

Fig. 3. Geometric interpolation of cable axis: a) coefficients of cubic function, b) overlap of the cubic interpolation with the point cloud model.

The *nonlinear* ordinary differential equation governing the static equilibrium under self-weight of the inextensible cable, considering constant weight per unit natural length w, reads

$$z''=-8\delta\cos\theta\left[1+(z')^2\right]^{1/2}, \qquad \text{with} \quad \delta=(wL)/(8H\cos\theta), \tag{8}$$

where H is the unknown horizontal reaction. It is known that this equation admits the catenary function as exact solution. Since the catenary solution can be difficult to handle, equation (8) can be attacked employing a classic perturbation scheme. Setting the parameter and variable ordering

$$\delta=\varepsilon\delta_1, \qquad \tilde{z}(x)=\varepsilon\tilde{z}_1(x)+\varepsilon^2\tilde{z}_2(x)+\varepsilon^3\tilde{z}_3(x)+\ldots \tag{9}$$

where $\tilde{z}(x)=z(x)-x\tan\theta$ describes the depth of the cable with respect to the chord and $\varepsilon<<1$ is a small auxiliary nondimensional parameter with mere bookkeeping role.

From the physical viewpoint, it is important to note that the ordering entails that the cable weight-to-tension ratio is small ($\delta=O(\varepsilon)$) and the depth of the cable is shallow ($\tilde{z}(x)=O(\varepsilon)$). Introducing the ordering in equation (8), expanding in ε-power series, collecting same-order terms and solving the resulting ordered hierarchy of *linear* ordinary differential equations up to the second order with the boundary conditions $\tilde{z}_i(0)=0$ and $\tilde{z}_i(1)=0$, the solutions are

$$\tilde{z}_1 = 4\delta_1 x(1-x), \quad \tilde{z}_2 = \tfrac{8}{3}\delta_1^2 \sin(2\theta)x(1-x)(1-2x). \tag{10}$$

After reconstruction, the direct comparison of the cubic function (5) and the cubic perturbation solution $z(x) = \tilde{z}(x) + \tan\theta \simeq \tan\theta + \varepsilon\tilde{z}_1(x) + \varepsilon^2\tilde{z}_2(x)$ allows to identify the unknown parameter δ by equating the respective quadratic and cubic coefficients. To circumvent the redundancy, the identification can be based on the quadratic coefficient γ_1, being larger. The equality reads

$$4\delta(1+2\delta\sin(2\theta)) = \gamma_1 \tag{11}$$

and establishes a quadratic equation in the unknown δ, that must be solved for positive values δ_*. The comparison between cubic coefficients, being smaller, can provide an estimate of the identification inaccuracy $\delta_e = -\gamma_0 - \frac{16}{3}\delta_*^2\sin(2\theta)$. Once the parameter δ is determined, the horizontal reaction H and the cable tension $N \simeq \frac{H}{\cos\theta}$ can be assessed by inverting the parameter definition, provided that the cable weight w is known. The mean values (and their variance) obtained with this approach are listed in Tab.1 for seven stays of the Beinasco bridge.

Frequency-based cable axial force identification

The small-amplitude free oscillations of the bridge stays have been investigated according to the linearized undamped dynamics of suspended cables. According to the Irvine cable model [11], [12], which is based on assuming parabolic static profile, linear elastic behavior and negligible longitudinal inertia, the cable axial force can be identified by employing the circular frequency ω_j of j-th out-of-plane modes, according to the direct and inverse (string-like) relations

$$\omega_j = \frac{j\pi\cos\theta}{L}\sqrt{\frac{N}{m}}, \qquad N = \frac{m\omega_j^2 L^2}{j^2\pi^2\cos^2\theta}, \tag{12}$$

where m the uniformly distributed mass density. Accordingly, the j-th frequency is expected to be directly and linearly related to the out-of-plane mode number j in the peak-picking frequency identification based on the Power Spectral Density (PSD) of the experimental signals. Such linear relation is clearly noticeable in the PSD of the camera signal, more than the accelerometer one (Fig. 4), because camera signals allow to identify the fundamental frequencies of the cables, which appeared to be lower than 1 Hz. The mean values (and their variance) obtained with this tension identification approach are listed in Tab. 1 for seven stays of the Beinasco bridge.

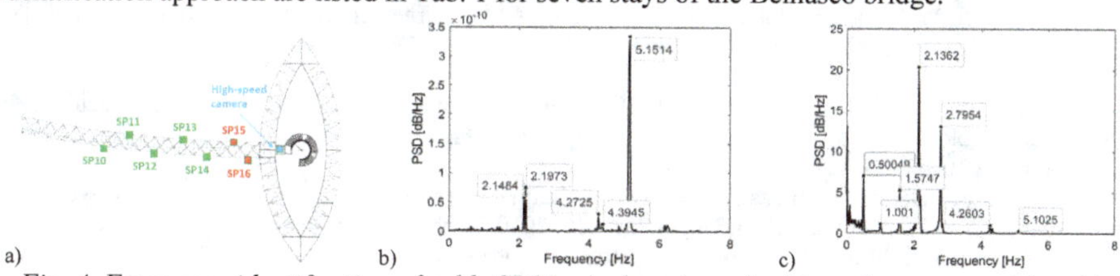

Fig. 4. Frequency identification of cable SP15: a) plan view of cable and camera position; b) PSD of accelerometer signals, c) PSD of image-based displacement signal.

Concluding remarks

The paper investigates the static and dynamic response of the stays supporting a cable-stayed pedestrian bridge, by exploiting experimental data concerning both geometry configuration and vibration response to ambient excitation. By referring to a continuous perfectly flexible model of inclined suspended cable, and by considering the laser-scanned geometry of the stays, the static configuration under self-weight has been consistently described. This description allowed the static (geometry-based) identification of the cable tension, which has resulted in good agreement with the tensions evaluated through the dynamic (frequency-based) output-only identification method, applied to both accelerometer and image-based vibration signals.

Tab. 1. Cable tension from static identification N_s and from dynamic identification N_d.

Cable	γ_0	$\sigma^2_{\gamma_0}$	γ_1	$\sigma^2_{\gamma_1}$	θ	δ_*	$\sigma^2_{\delta_*}$	δ_e	$\sigma^2_{\delta_e}$	N_s [kN]	N_d [kN]
SP10	4.84E-04	1.09E-05	3.22E-02	2.63E-05	0.57	7.94E-03	1.56E-06	-7.91E-04	1.02E-05	97.19	94.89
SP11	-7.35E-03	2.41E-05	5.12E-02	4.64E-05	0.65	1.25E-02	2.63E-06	6.55E-03	2.20E-05	59.39	58.78
SP12	1.35E-02	4.21E-05	1.22E-01	8.32E-05	0.73	2.89E-02	4.18E-06	-1.79E-02	3.44E-05	26.19	26.02
SP13	-3.50E-03	8.55E-05	1.75E-01	1.87E-04	0.83	4.05E-02	8.67E-06	-5.21E-03	6.36E-05	18.39	19.54
SP14	-9.73E-04	2.77E-04	6.14E-02	6.48E-04	0.94	1.49E-02	3.60E-05	-1.56E-04	2.45E-04	54.25	41.68
SP15	-2.06E-02	1.91E-04	2.49E-01	4.68E-04	1.04	5.67E-02	2.01E-05	5.59E-03	1.30E-04	16.32	17.83
SP16	2.95E-02	7.65E-04	7.59E-02	2.12E-03	1.11	1.84E-02	1.15E-04	-3.10E-02	6.58E-04	53.78	44.31

Acknowledgments
This research was in part sponsored by the NATO Science for Peace and Security Programme under grant id. G5924.

References

[1] A.B. Mehrabi, "In-service evaluation of cable-stayed bridges, overview of available methods and findings" *Journal of Bridge Engineering* 11(6), pp. 716–724, 2006. https://doi.org/10.1061/(ASCE)1084-0702(2006)11:6(716)

[2] L. Zhang, G. Qiu, Z. Chen, "Structural health monitoring methods of cables in cable-stayed bridge: A review" *Measurement* 168, id.108343, 2021. https://doi.org/10.1016/j.measurement.2020.108343

[3] M. Lepidi, V. Gattulli, F, Vestroni, F. "Damage identification in elastic suspended cables through frequency measurement" *Journal of Vibration and Control* 15(6), pp.867-896, 2009. https://doi.org/10.1177/1077546308096107

[4] S. Jeong, H. Kim, J. Lee, S.-H. Sim, "Automated wireless monitoring system for cable tension forces using deep learning" *Structural Health Monitoring* 20(4), pp. 1805-1821, 2021. https://doi.org/10.1177/1475921720935837

[5] S.-W. Kim, J.-H. Cheung, J.-B. Park, S.-O. Na, "Image-based back analysis for tension estimation of suspension bridge hanger cables" *Structural Control and Health Monitoring* 27(4), id.e2508, 2020. https://doi.org/10.1002/stc.2508

[6] V. Gattulli, M. Lepidi, F. Potenza, U. Di Sabatino, "Modal interactions in the nonlinear dynamics of a beam-cable-beam" *Nonlinear dynamics* 96(4), pp.2547-2566, 2019. https://doi.org/10.1007/s11071-019-04940-8

[7] V. Gattulli, M. Lepidi, F. Potenza, U. Di Sabatino, "Dynamics of masonry walls connected by a vibrating cable in a historic structure" *Meccanica* 51(11), pp. 2813-2826, 2016. https://doi.org/10.1007/s11012-016-0509-9

[8] C. Gentile, "Deflection measurement on vibrating stay cables by non-contact microwave interferometer" *Ndt & E International* 43(3), pp. 231-240, 2010. https://doi.org/10.1016/j.ndteint.2009.11.007

[9] C. Gentile A. Cabboi, "Vibration-based structural health monitoring of stay cables by microwave remote sensing" *Smart Structures and Systems* 16(2), pp. 263-280, 2015. https://doi.org/10.12989/sss.2015.16.2.263

[10] S. Wangchuk, D.M. Siringoringo, Y. Fujino, "Modal analysis and tension estimation of stay cables using noncontact vision-based motion magnification method" *Structural Control and Health Monitoring* 29(7), id. e2957, 2022. https://doi.org/10.1002/stc.2957

[11] H. M. Irvine, Cable Structures. The MIT Press, 1981.

[12] M. Lepidi V. Gattulli, "Static and dynamic response of elastic suspended cables with thermal effects" *International Journal of Solids and Structures* 49(9), pp. 1103-1116, 2012. https://doi.org/10.1016/j.ijsolstr.2012.01.008

Mechanical modelling of metamaterials and periodic structures

Theoretical and Applied Mechanics - AIMETA 2022 Materials Research Forum LLC
Materials Research Proceedings **26** (2023) 493-498 https://doi.org/10.21741/9781644902431-80

Stability domain and optimal design of a metamaterial made-up of a beam lattice with diagonal cables

Francesco Trentadue[1,a*], Domenico De Tommasi[1,b], Gianluca Caramia[1,c] and Nicola Marasciuolo[1,d]

[1]DICATECh, Politecnico di Bari, 70125 Bari, Italy

[a]francesco.trentadue@poliba.it, [b]domenico.detommasi@poliba.it, [c]gianluca.caramia@poliba.it, [d]n.marasciuolo@studenti.poliba.it

* Corresponding author

Keywords: Metamaterials, Eulerian Stability, Periodic Materials

Abstract. The elastic stability of a plane metamaterial made up of a mesh of orthogonal rods coupled with diagonal cables is studied. The stability domain and the optimal values of the stiffness of the cables are determined.

Introduction

In this work explicit solutions are obtained for the stability domain of a bidimensional periodic metamaterial made up of a mesh of beams and cables. Fig. 1 shows the metamaterial and the chosen periodic cell (RVE). This metamaterial was already studied in [1] in the simpler case of absence of cables by assigning a priori the wave lengths of critical modes.

In [2], by introducing suitable simplifying assumptions, the wave lengths of critical modes were exactly determined by means of Floquet-Bloch theory. The main idea of the paper was to couple periodic patterns of rods with periodic patterns of extremely flexible cables to enhance the stability performance of the obtained metamaterial. Here, we extend our preceding results by considering the more general case in which vertical and horizontal rods have different stiffness.

Mechanical assumptions

The studied metamaterial is shown in Figure 1. In the following analysis we will neglect the axial and shear deformations of rods and denote by J and μJ the inertia moments of horizontal and vertical rods, respectively. Without loss of generality, we assume $\mu < 1$. Further, we assume

$$\frac{J_c}{J} \to 0, \qquad \frac{A_c}{A} \to 0, \tag{1}$$

where J_c and A_c are the moment of inertia and the cross-section areas of diagonal cables and $A = \min\{A_h, A_v\}$ is the minimum cross section area of the rods.

We distinguish (see Figure 1) two types of joints: type 1, connecting four rods and four cables, and type 0, connecting four cables. As a consequence of (1), equilibrium of a type 0 node is identically satisfied and will not be considered in the following.

The chosen RVE is composed of four rods (green in Figure 1) with length l, four internal diagonal cables of length $l/\sqrt{2}$ (red in Figure 1) and four external cables of length $\sqrt{2}l$, whose stiffness is halved since these are shared with the neighboring RVEs.

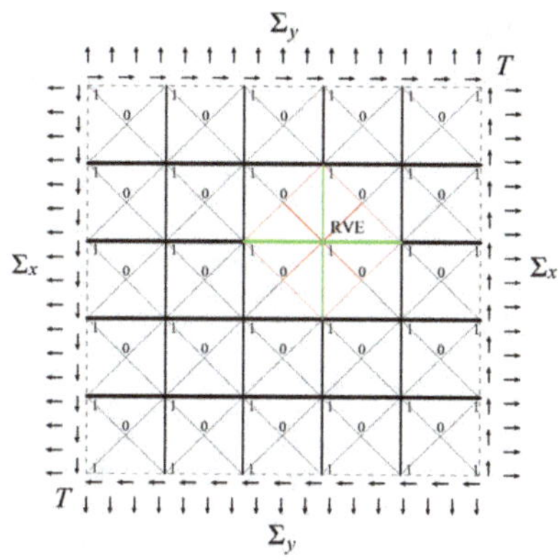

Figure 1: The metamaterial subjected to a generic macrostress

Equilibrium in precritical states

In case of null shear macrostress $(T=0)$, as a consequence of (1), the axial compressive forces in the rods are

$$P_{\Sigma_x} = -\Sigma_x l, \quad P_{\Sigma_y} = -\Sigma_y l, \tag{2}$$

where Σ_x and Σ_y are the traction macrostresses in the horizontal and vertical direction, respectively. In case of pure shear, $(\Sigma_x = \Sigma_y = 0)$ cables in compression immediately buckle (dotted lines in Fig. 2), while cables in tension exert the axial forces N_T (continuous lines in Fig. 2) and the axial compressive forces P_T occur in rods. These are given by

$$N_T = \sqrt{2}\rho|T|l, \quad P_T = \rho|T|l, \tag{3}$$

where ρ and k_e are a distribution factor and a dimensionless measure of elastic cables stiffness (see [2] for details):

$$\rho = \frac{k_e}{k_e + 24\frac{\mu}{1+\mu}}, \quad k_e = \frac{A_c l^2}{\sqrt{2}J}. \tag{4}$$

Here, as usual in Eulerian stability analysis, the mechanical response is assumed linear in precritical states. Then the distribution factor is determined with reference to the unstressed initial state in which the stiffness of cables reduces only to its elastic part k_e.

Under the assumption (1) it is possible to superimpose the two cases above analyzed, to obtain

$$P_x = P_{\Sigma_x} + P_T = -\left(\Sigma_x - \rho|T|\right)l, \quad P_y = P_{\Sigma_y} + P_T = -\left(\Sigma_y - \rho|T|\right)l. \tag{5}$$

Floquet-Bloch analysis

The critical displacement vector field is expressed as (see Fig. 2)

$$\boldsymbol{u}(\boldsymbol{x}) = \{u(\boldsymbol{x}), v(\boldsymbol{x}), \theta(\boldsymbol{x})\}^T \in \mathbb{C}^3 \tag{6}$$

where $\mathbf{x}$ is the node position vector, which is written as

$$\boldsymbol{x} = \zeta_1 l\boldsymbol{i} + \zeta_2 l\boldsymbol{j}, (\zeta_1, \zeta_2) \in \mathbb{Z}^2 . \tag{7}$$

The critical displacement vector is determined according to Floquet-Bloch's theorem

$$\mathbf{u}(\mathbf{x}) = \mathbf{u}(\mathbf{0})e^{i\mathbf{w}\cdot(\zeta_1 l\mathbf{i}+\zeta_2 l\mathbf{j})} = \mathbf{u}(\mathbf{0})e^{i(\omega_1\zeta_1+\omega_2\zeta_2)} . \tag{8}$$

where $\boldsymbol{\omega}$ is a dimensionless wave vector, with $\omega_1, \omega_2 \in (-\pi, \pi)$. Due to the axial inextensibility of rods, only the following cases are feasible:

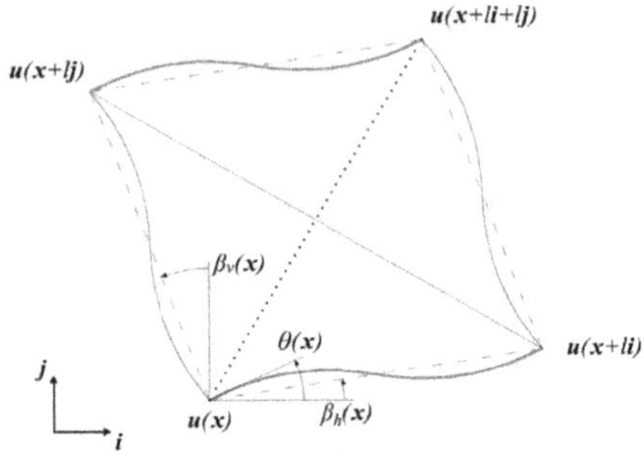

Figure 2: A square mesh of rods and cables

a) no translations of nodes occur: $u(\mathbf{x}) = v(\mathbf{x}) = 0, \ \forall \omega_1, \omega_2 \in (-\pi, \pi)$;

b1) rotations β_h of horizontal rods are null and the horizontal wavelength is infinite ($\omega_1 = 0$);

b2) rotations β_v of vertical rods are null and the vertical wavelength is infinite ($\omega_2 = 0$).

The degrees of freedom of the RVE are chosen as

$$\boldsymbol{\gamma} = \{\beta_h, \beta_v, \theta\}^T = \{\beta_h(\mathbf{x}), \beta_v(\mathbf{x}), \theta(\mathbf{x})\}^T. \tag{9}$$

and its Hermitian stiffness matrix $\mathbf{k}$ is determined as

$$\mathbf{k} = \mathbf{k}_\mathbf{b} + \mathbf{k}_\mathbf{c}, \tag{10}$$

where $\mathbf{k}_\mathbf{b}$ and $\mathbf{k}_\mathbf{c}$ are the stiffness matrices of the rods and cables, respectively. $\mathbf{k}$ depends on both the wave numbers (ω_1, ω_2) and the applied macrostress through the parameters q_x and q_y:

$$q_x = \frac{l}{2}\sqrt{\frac{P_x}{EJ}}, \quad q_y = \frac{l}{2}\sqrt{\frac{P_y}{\mu EJ}}, \tag{11}$$

which can assume either positive real values (compressed rods) or imaginary values (rods in traction). We impose that in stable or critical states the second variation of the total energy is positive semidefinite

$$\frac{1}{2}\boldsymbol{\gamma}^* \mathbf{k}(\boldsymbol{\omega}, q_x, q_y)\boldsymbol{\gamma} \geq 0. \tag{12}$$

The adopted procedure is detailed in [2]. In case a) the stability condition (12) reduces to

$$q_x \cot q_x + \mu q_y \cot q_y \geq 0 \tag{13}$$

and the critical state condition is always attained for $\omega_1 = \omega_2 = \pi$. Figure 3 depicts the corresponding critical mode. Notice that, under the assumption (1), the stability condition (13) is not influenced by the stiffness of cables, which is negligible with respect to those of rods.

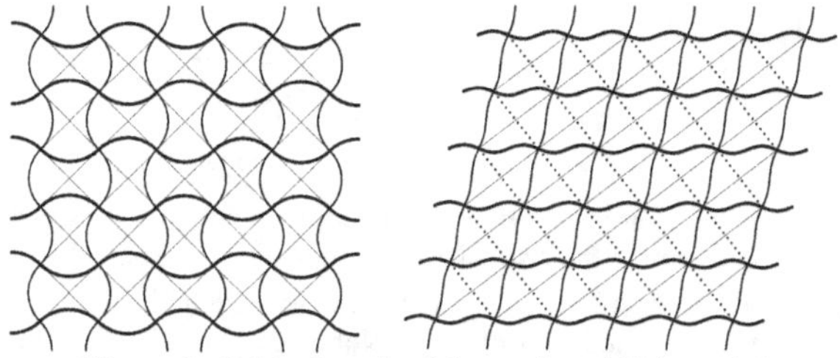

Figure 3: Critical modes I (case a) and II (case b$_1$)

In case b$_1$), from the stability condition (12) we obtain

$$k \geq k_v = \max\left\{8\mu q_y^3 \frac{q_x^2 \cot q_y + \mu q_x q_y \cot q_x - \mu q_y}{q_x q_y\left(q_x \cot q_y + \mu q_y \cot q_x\right) - q_x^2 - \mu q_y^2}, 8\mu q_y^3 \frac{\cot q_y}{q_y \cot q_y - 1}, 0\right\} \quad (14)$$

where $k = k_e + k_T$ is a dimensionless measure of total stiffness of cables, which is sum of an elastic part k_e, already given in (4), and a geometric part

$$k_T = \frac{N_T l^2}{E\sqrt{2}J}. \quad (15)$$

In Figure 3 it is shown the critical mode occurring when $k = k_v$. Similarly, in case b$_2$), from the stability condition (12) we find

$$k \geq k_h = \max\left\{8q_x^3 \frac{\mu q_y^2 \cot q_x + q_x q_y \cot q_y - q_x}{q_x q_y\left(q_x \cot q_y + \mu q_y \cot q_x\right) - q_x^2 - \mu q_y^2}, 8q_x^3 \frac{\cot q_x}{q_x \cot q_x - 1}, 0\right\}. \quad (16)$$

The critical mode III, occurring in case b$_2$) when $k = k_h$ is similar to the mode II rotated by π/2. In absence of cables, the stability conditions in cases b$_1$) and b$_2$) reduce to the conditions

$$\begin{cases} \dfrac{\mu \cot q_x}{q_x} + \dfrac{\cot q_y}{q_y} - \dfrac{\mu}{q_x^2} \geq 0 \\ \dfrac{\mu \cot q_x}{q_x} + \dfrac{\cot q_y}{q_y} - \dfrac{1}{q_y^2} \geq 0 \end{cases} \quad (17)$$

Stability domain

From the above analysis we conclude that the stability domain of the material is determined by the conditions

$$\begin{cases} q_x \cot q_x + \mu q_y \cot q_y \geq 0 \\ k \geq \max(k_h, k_v) = \hat{k}\left(q_x, q_y\right) \end{cases}. \quad (18)$$

In order to obtain a bidimensional representation, we define the dimensionless macrostresses $\tilde{\Sigma}_x$ and $\tilde{\Sigma}_y$

$$\tilde{\Sigma}_x = \frac{\Sigma_x - \rho|T|}{\Sigma_{Ex}}, \quad \tilde{\Sigma}_y = \frac{\Sigma_y - \rho|T|}{\Sigma_{Ey}}, \quad (19)$$

where

$$\Sigma_{Ex} = \frac{\pi^2 EJ}{l^3}, \quad \Sigma_{Ey} = \frac{\pi^2 E\mu J}{l^3}. \quad (20)$$

Therefore, from (11) we get

$$q_x = \frac{i\pi}{2}\sqrt{\tilde{\Sigma}_x}, \qquad q_y = \frac{i\pi}{2}\sqrt{\tilde{\Sigma}_y}, \tag{21}$$

In Figure 4 the boundaries of the stability domains in the plane $\left(\tilde{\Sigma}_x, \tilde{\Sigma}_y\right)$ are shown for different values of μ. Two limit cases are considered: the case in which the cables have a stiffness $k \geq \max(k_h, k_v)$, such to determine the critical mode I, in which the boundaries are the continuous lines with dot marks; the case of absence of cables, in which the boundaries are the inner continuous lines (without dots), deduced by (17). Notice that, in view of (19), to each point of these bidimensional stability domains correspond infinite macrostresses.

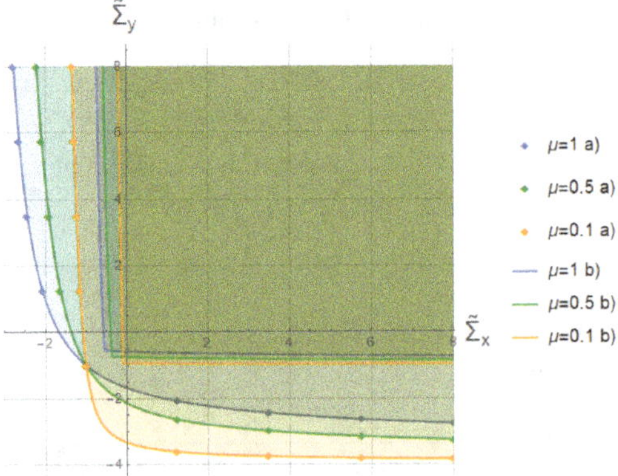

Figure 4: Stability frontiers for different values of μ

In Figure 5, in the case $\mu = 1$, the optimal dimensionless stiffness of cables $\bar{k}\left(\tilde{\Sigma}_x, \tilde{\Sigma}_y\right) = \max(k_h, k_v)$ is represented. This is the minimum value ensuring the stability of the considered macrostress. The minimum value of $\bar{k}$ on the boundary of the stability domain is attained at the point A≡(-1, -1) and is equal to π^2. Among the critical macrostresses corresponding to this point, we have equiaxial compression and pure shear. The maximum value of optimal stiffness is attained at the points (-4, ∞) and (∞, -4) and is equal to $8\pi^2$.

Figure 6 shows the stability boundaries in the three dimensional space $\left(\Sigma_x/\Sigma_{Ex}, \Sigma_y/\Sigma_{Ex}, \mathrm{T}/\Sigma_{Ex}\right)$, for different values of μ, considering an elastic stiffness of the cables $k_e = 8\pi^2$. Notice that in order to preserve the scale, all macrostress components have been divided by the same quantity Σ_{Ex}.

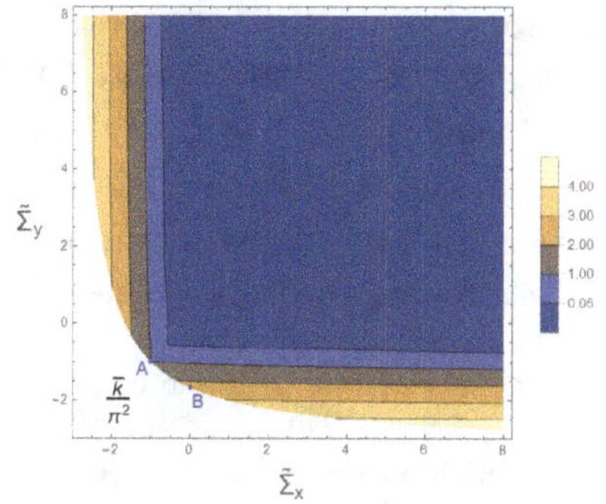

Figure 5: Stability domain and optimal values $\overline{k}$ for $\mu = 1$

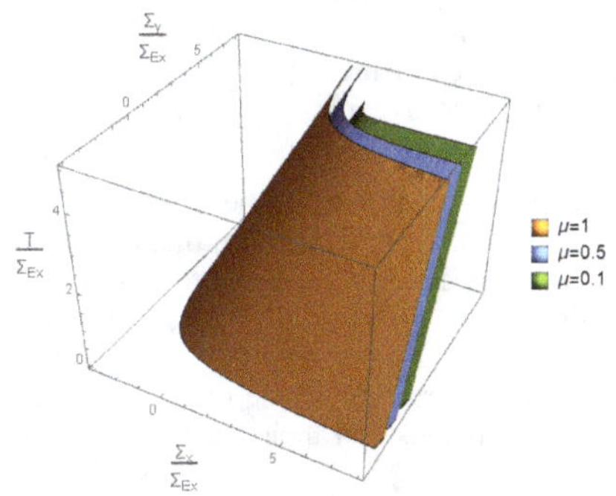

Figure 6: Stability frontiers for $k_e = 8\pi^2$.

Acknowledgements

D. De Tommasi and F. Trentadue have been supported by the Prin Project 2017 J4EAYB 001.

References

[1] B. Haghpanah, J. Papadopoulos, D. Mousanezhad, H. Nayeb-Hashemi, A. Vaziri, «Buckling of regular, chiral and hierarchical honeycombs under a general macroscopic stress state,» *Proceedings of the Royal Society A,* 2014. https://doi.org/10.1098/rspa.2013.0856

[2] F. Trentadue, D. De Tommasi, N. Marasciuolo, «Stability domain and design of a plane metamaterial made up of a periodic mesh of rods with cross-bracing cables,» *Applications in Engineering Science,* vol. 5, 2021. https://doi.org/10.1016/j.apples.2021.100036

[3] D. De Tommasi, G. C. Marano, G. Puglisi, F. Trentadue, «Morphological optimization of tensegrity-type metamaterials,» *Composites Part B: Engineering,* vol. 115, pp. 182-187, 2017. https://doi.org/10.1016/j.compositesb.2016.10.017

[4] D. De Tommasi, G. Puglisi, F. Trentadue, «Elastic response of an optimal tensegrity-type metamaterial,» *Frontiers in Materials,* vol. 6, 2019. https://doi.org/10.3389/fmats.2019.00024

Theoretical and Applied Mechanics - AIMETA 2022 Materials Research Forum LLC
Materials Research Proceedings **26** (2023) 499-504 https://doi.org/10.21741/9781644902431-81

Two-scale asymptotic homogenization of hierarchical locally resonant metamaterials in anti-plane shear conditions

David Faraci[1,a *], Claudia Comi[1,b] and Jean-Jacques Marigo[2,c]

[1] Department of Civil and Environmental Engineering, Politecnico di Milano, Milan, Italy

[2]Laboratoire de Mécanique des Solides, Ecole Polytechnique, Palaiseau, France

[a] david.faraci@polimi.it, [b] claudia.comi@polimi.it, [c] jean-jacques.marigo@polytechinque.edu

Keywords: Homogenization, Local Resonance, Band Gaps

Abstract. Local resonant metamaterials are a class of microstructured man-made material which attenuate the propagation of waves in certain frequency ranges, known as band gaps. In this work, we study through asymptotic homogenization the anti-plane shear wave propagation in metamaterial with a stiff matrix and soft inclusions, periodically distributed, which present a hierarchical geometry. Band gaps of the metamaterial are then analytically predicted by the intervals of frequency in which the effective mass becomes negative.

Introduction

Metamaterials have attracted a lot of interest in dynamic problems due to their peculiar and unusual properties related to wave propagation. A particular class of these materials are the locally resonant metamaterials (LRMs), which are specifically designed to trap the energy of propagating waves by employing the mechanism of local resonance. The frequency intervals in which waves are attenuated are known as band gaps.

Typically, LRMs have a periodic structure, made of one or several phases, hence the Bloch theorem [1,2] can be employed to reconstruct their band structure, including band gaps, in the linear-elastic regime.

In recent years, several researchers propose hierarchical geometries for LRMs to obtain wider band gaps [3,4]. Numerical simulations of a continuum system exhibiting a hierarchical geometry may represent a huge computational burden and analytical predictions of their dynamical properties are mainly restricted to discrete systems of lumped masses and springs.

Two-scale asymptotic homogenization is a promising technique which can be used to study the dynamic effective behaviour of metamaterials [5,6]. This method has been employed to study a periodic LRM made of a connected stiff matrix and soft inclusion in two- and three-phase solids [7,8], and in thin plates [9], allowing the identification of band gaps as the intervals of negative effective dynamic mass.

In this work, we extend the results obtained in [8], to study the anti-plane shear wave propagation in a three-phase hierarchical LRM constituted by a connected matrix with a periodic repetition of coated inclusions which are in turn microstructured, thus realizing a hierarchical LRM. In particular, we consider the presence of several concentrical annular rings of rigid masses, with interposed soft coatings, in each inclusion. The effective mass of the metamaterial, which is frequency-dependent, is derived up to the solution of a tridiagonal linear system. The analytical predictions of band gaps are validated by comparison with those obtained by numerical Bloch-Floquet analyses. Finally, some parametric studies are performed to show the potentiality of the proposed approach for the design and optimization of hierarchical LRMs.

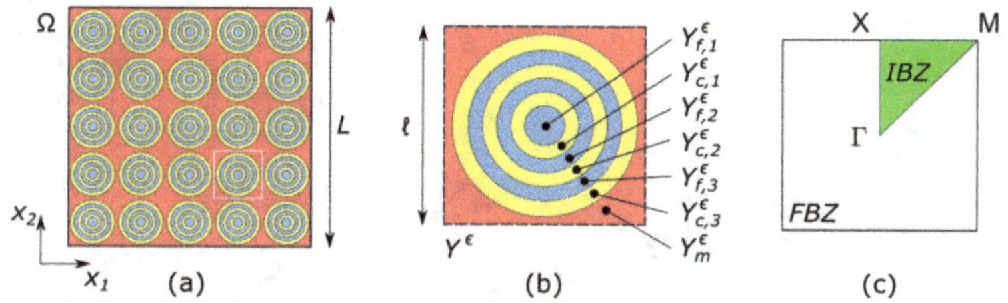

Figure 1: (a) Geometry of the hierarchical metamaterial with its (b) unit cell composed by the matrix (red), coatings (yellow) and rigid masses (blue) in the case $N = 3$. (c) First and Irreducible Brillouin zone of the square lattice.

Two-scale asymptotic homogenization

We consider a three-dimensional heterogeneous linear-elastic media which is characterized by a two-dimensional periodicity. The cross-section Ω of the body, shown in Fig. 1a and having characteristic size L, is made of a connected stiff matrix (m) and several inclusions periodically distributed. Each inclusion presents a hierarchical geometry consisting of N concentric annular rigid masses (f) with interposed soft coatings (c). The unit cell Y^ϵ of the periodic media, depicted in Fig. 1b with its characteristic size ℓ, shows the domains $Y^\epsilon_{f,\alpha}$ and $Y^\epsilon_{c,\alpha}$ of each rigid mass and coating, which are numbered from the innermost to the matrix Y^ϵ_m by $\alpha = 1,\ldots,N$.

Under the hypothesis of separation of scales, i.e., when $\epsilon = \ell/L \ll 1$, the effective behaviour of the metamaterial can be described through asymptotic homogenization.

The anti-plane shear wave propagation problem is governed by the scalar Helmholtz equation

$$\mathrm{div}(\mu^\epsilon \nabla w^\epsilon) + \rho^\epsilon \omega^2 w^\epsilon = 0 \qquad \text{in } \Omega \,, \tag{1}$$

where w^ϵ is the out of plane displacement and ω is the angular frequency of the propagating wave. The terms μ^ϵ and ρ^ϵ in Eq. 1 are the periodically varying shear modulus and mass density of the materials, which are assumed to be μ_m, ρ_m for the matrix, $\epsilon^2\mu_c, \rho_c$ for the coatings and ρ_f for the rigid masses, with μ_m and ρ_m of the same order of magnitude of μ_c and ρ_c, ρ_f.

According to the two-scale homogenization technique, introducing the fast variable $\boldsymbol{y} = \boldsymbol{x}/\epsilon$ and the re-scaled unit cell $Y = Y^\epsilon/\epsilon$, the solution of Eq. 1 is searched in the form

$$w^\epsilon(\boldsymbol{x}) = w^0(\boldsymbol{x}, \boldsymbol{x}/\epsilon) + \epsilon w^1(\boldsymbol{x}, \boldsymbol{x}/\epsilon) + \ldots \,, \tag{2}$$

where the functions $w^i(\mathbf{x},\mathbf{y})$ are defined on $\Omega \times Y$ and are periodic with respect to $\mathbf{y}$. Substituting Eq. (2) into Eq. (1) it is possible to prove that the 0-th order displacement in the matrix depend only on $\mathbf{x}$, i.e., $w^0(\mathbf{x},\mathbf{y}) = W^0_m(\mathbf{x})$ in $\Omega \times Y_m$, see [7] for the full derivation.

Moreover, due to the radial symmetry of coatings and inclusions the masses rigidly translate within the unit cell. That means that $w^0(\mathbf{x},\mathbf{y}) = \psi_\alpha W^0_m(\mathbf{x})$ in $\Omega \times Y_{f,\alpha}$ with $\psi_1, \ldots, \psi_\alpha$ unknown coefficients.

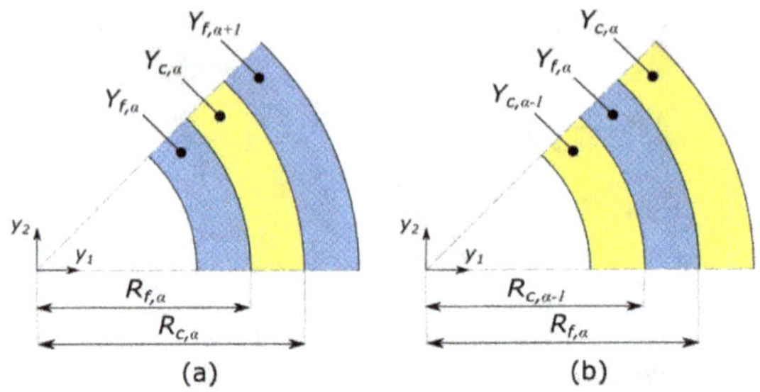

Figure 2: Detail of the circular sector of (a) the α -th coating surrounded by two rigid masses, and (b) the α -th rigid mass surrounded by two coatings.

Motion of the coatings. Restricting Eq. 1 at order δ^0 in the α -th coating, one obtains

$$\mu_c \Delta_y w^0 + \rho_c \omega^2 w^0 = 0 \qquad \text{in } \Omega \times Y_{c,\alpha}, \tag{3}$$

with proper boundary conditions to guarantee the continuity of the solution at interfaces. For $\alpha \neq N$, as shown in Fig. 2a, these BCs read $w^0 = \psi_\alpha W_m^0$ on $\partial Y_{c,\alpha} \cap \partial Y_{f,\alpha}$ and $w^0 = \psi_{\alpha+1} W_m^0$ on $\partial Y_{c,\alpha} \cap \partial Y_{f,\alpha+1}$. Due to linearity the solution of Eq. 3 is given by $w^0(\mathbf{x},\mathbf{y}) = W_m^0(\mathbf{x})\eta_\alpha(\mathbf{y})$ in $\Omega \times Y_{c,\alpha}$. Using polar coordinates, one obtains a solution depending on the radial coordinate r only, in the form

$$\eta_\alpha(r) = \frac{J_0(\lambda r)Y_0(\lambda R_{c,\alpha}) - J_0(\lambda R_{c,\alpha})Y_0(\lambda r)}{J_0(\lambda R_{f,\alpha})Y_0(\lambda R_{c,\alpha}) - J_0(\lambda R_{c,\alpha})Y_0(\lambda R_{f,\alpha})}\psi_\alpha + \frac{J_0(\lambda R_{f,\alpha})Y_0(\lambda r) - J_0(\lambda r)Y_0(\lambda R_{f,\alpha})}{J_0(\lambda R_{f,\alpha})Y_0(\lambda R_{c,\alpha}) - J_0(\lambda R_{c,\alpha})Y_0(\lambda R_{f,\alpha})}\psi_{\alpha+1}. \tag{4}$$

In Eq. 4 J_p and Y_p are the p-order Bessel's functions of first and second kind, $\lambda^2 = \rho_c \omega^2 / \mu_c$, while $R_{c,\alpha}, R_{f,\alpha}$ are the external radius of the α -th coating and inclusion, see Fig. 2a. Note that this solution can be extended also for the N -th coating by setting $\psi_{N+1} = 1$.

Motion of the masses of the inclusions. Up to now, the 0-th order displacement w^0 is determined up to the knowledge of the matrix displacement W_m^0 and of the N constants ψ_α. The latter can be determined by enforcing the global dynamic equilibrium of each rigid mass. For the α -th mass the equilibrium reads

$$\mu_c \int_{\partial Y_{f,\alpha}} \frac{\partial w^0}{\partial r} ds + \rho_f |Y_{f,\alpha}| \omega^2 \psi_\alpha W_m^0 = 0. \tag{5}$$

If $\alpha \neq 1$, i.e., we are not considering the innermost mass (see Fig. 2b), Eq. 5 can be rewritten as

$$f_{\alpha-1}(R_{c,\alpha-1})\psi_{\alpha-1} + \left[g_{\alpha-1}(R_{c,\alpha-1}) - f_\alpha(R_{f,\alpha}) + \frac{\lambda^2 \rho_f}{2\rho_c}(R_{f,\alpha}^2 - R_{c,\alpha-1}^2) \right]\psi_\alpha - g_\alpha(R_{f,\alpha})\psi_{\alpha+1} = 0, \tag{6}$$

where we have introduced the functions

$$f_\alpha(r) = \frac{J_1(\lambda r)Y_0(\lambda R_{c,\alpha}) - J_0(\lambda R_{c,\alpha})Y_1(\lambda r)}{J_0(\lambda R_{f,\alpha})Y_0(\lambda R_{c,\alpha}) - J_0(\lambda R_{c,\alpha})Y_0(\lambda R_{f,\alpha})}\lambda r, \quad g_\alpha(r) = \frac{J_0(\lambda R_{f,\alpha})Y_1(\lambda r) - J_1(\lambda r)Y_0(\lambda R_{f,\alpha})}{J_0(\lambda R_{f,\alpha})Y_0(\lambda R_{c,\alpha}) - J_0(\lambda R_{c,\alpha})Y_0(\lambda R_{f,\alpha})}\lambda r. \tag{7}$$

The dynamic equilibrium of the innermost mass ($\alpha=1$) is expressed again by Eq. 6 with $f_0=g_0=R_{c,0}=R_{f,0}=0$. The system given by Eq. 6 written for $\alpha=1,...,N$ is a tridiagonal system which allows to compute the unknowns ψ_α for each value of λ, i.e., of ω.

Dispersion relation. Taking the average of Eq. 1 at order $\eth^0$ over the unit cell, one obtains the effective equation of motion of the hierarchical metamaterial

$$\mathrm{div}_{\mathbf{x}}(\boldsymbol{\mu}^0\cdot\nabla_{\mathbf{x}}W_m^0)+\rho^0(\omega)\omega^2W_m^0=0 \qquad \text{in } \Omega, \qquad (8)$$

where $\boldsymbol{\mu}^0$ is the positive definite effective stiffness tensor (see [7] for the detail of its derivation) and $\rho^0(\omega)$ is the frequency-dependent effective dynamic mass density, which expression is

$$\rho^0(\omega)=\rho_{st}+\frac{\rho_f}{|Y|}\sum_{\alpha=1}^{N}|Y_{f,\alpha}|(\psi_\alpha-1)+\frac{\rho_c}{|Y|}\sum_{\alpha=1}^{N}|Y_{c,\alpha}|(F_\alpha\psi_\alpha+G_\alpha\psi_{\alpha+1}-1). \qquad (9)$$

In Eq. 9, ρ_{st} is the static mass density and F_α, G_α are constants defined by

$$\rho_{st}=\rho_m\frac{|Y_m|}{|Y|}+\rho_f\frac{|Y_f|}{|Y|}+\rho_c\frac{|Y_c|}{|Y|},$$
$$F_\alpha=\frac{2}{\lambda^2}\frac{f_\alpha(\lambda R_{c,\alpha})-f_\alpha(\lambda R_{f,\alpha})}{R_{c,\alpha}^2-R_{f,\alpha}^2},\quad G_\alpha=\frac{2}{\lambda^2}\frac{g_\alpha(\lambda R_{c,\alpha})-g_\alpha(\lambda R_{f,\alpha})}{R_{c,\alpha}^2-R_{f,\alpha}^2}. \qquad (10)$$

Substituting into Eq. 8 the monochromatic wave $W_m^0(\mathbf{x})=\exp(i\,\mathbf{k}\cdot\mathbf{x})$, where i is the imaginary unit and $\mathbf{k}$ is the wavevector, one obtains the dispersion relation $\rho^0(\omega)\omega^2-\boldsymbol{\mu}^0:\boldsymbol{k}\odot\boldsymbol{k}=0$.

As it can be seen, no real wavevector can satisfy the dispersion relation if the effective mass is negative. This means that the band gaps of the hierarchical metamaterial are identified as the intervals of frequency in which $\rho^0(\omega)<0$.

Example

The design of hierarchical LRM requires, besides the choice of the material and of the shape of the unit cell, also the selection of the number of masses and their distribution. As an example, we will consider a square unit cell, made of an epoxy matrix, rubber coatings and lead masses, see Fig. 1b. Typical values of the materials parameters are reported in Table 1.

***Table1:** Properties of the constituent materials*

	E [MPa]	ν [-]	ρ [kg/m^3]
Matrix (epoxy)	3600	0.370	1180
Coatings (rubber)	0.118	0.469	1300
Rigid masses (lead)	14000	0.420	11340

We keep constant the total mass of coatings and resonant masses and we discuss the effect of changing the inclusions microstructure. Assuming $|Y_f|=0.50\ell^2$ and $|Y_c|=0.13\ell^2$, we obtain different unit cells (Fig. 3) by changing N and assuming that each sub-inclusion has the same mass. In the case of one, two and three rigid masses the effective mass density given by Eq. 9 is plotted against frequency in the left plates of Fig. 4a, 4b and 4c, respectively. Frequencies are

Materials Research Proceedings **26** (2023) 499-504 https://doi.org/10.21741/9781644902431-81

normalized with respect $\omega^* = \sqrt{\mu_c / \rho_c \ell^2}$. The dispersion curves, along the irreducible Brillouin zone of the unit cell (see Fig. 1c), are obtained by the numerical solution of Bloch-Floquet problems and are reported in the right plates of the same figures. The comparison shows the accuracy of the proposed homogenization approach for the prediction of the lowest band gaps (shaded in the figures) in the low-frequency regime.

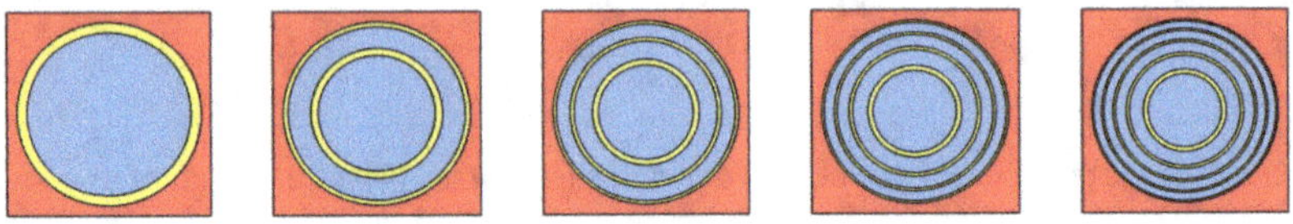

Figure 3: *First five hierarchy level considered.*

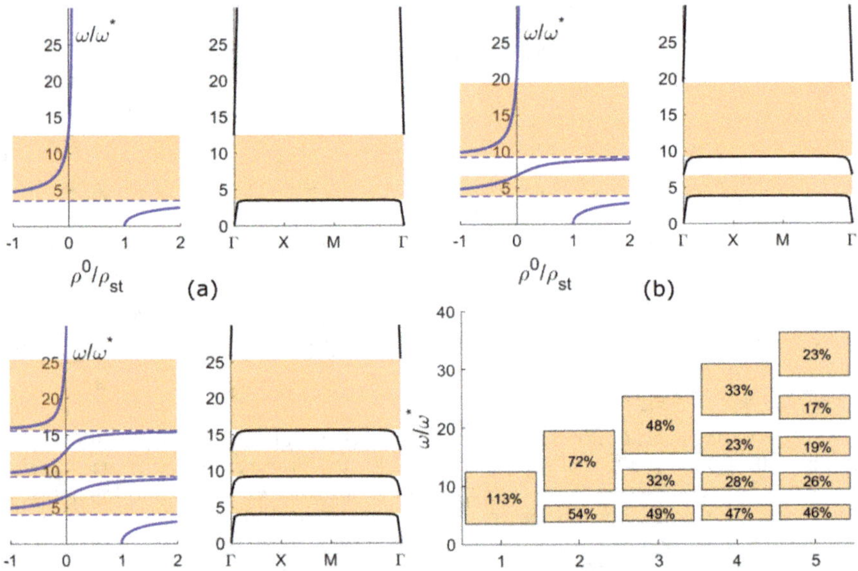

Figure 4: *Effective mass density against frequency (left) and numerical dispersion curves (right) in the case of one (a), two (b) and three (c) rigid masses. (d) Band gaps width as a function of the number of rigid masses.*

Figure 4d shows the width of band gaps for the first five hierarchical geometries with the corresponding gap mid-gap ratio $BG[\%] = 200(\omega_c - \omega_o)/(\omega_c + \omega_o)$, where ω_o and ω_c are the opening and closing frequency of the band gap. It can be observed that an additional band gap appears for each rigid mass considered, while the opening frequency of the first one is almost constant.

Conclusions

In this work, we studied the propagation of anti-plane shear waves in a continuous ternary locally resonant metamaterials, exhibiting a hierarchical mass-in-mass geometry, through asymptotic homogenization. In the simple case of concentric annular coatings and rigid masses we provide the expression of the effective dynamic mass density of the metamaterial up to the solution of a tridiagonal system, which has the same dimension of the number of sub-masses inside the inclusions.

The homogenized mass, which turns out to be frequency-dependent, allows for the prediction of the band gaps as intervals of frequency in which it becomes negative. This result has been validated by comparison with the band gaps obtained through numerical Bloch-Floquet analyses. This allow to perform parametric studies on the hierarchical metamaterial considered, which could be useful in their design phase.

The results obtained in this work can be easily generalized to wave propagation problems in plane strain conditions and for thin plates.

References

[1] C. Kittel. *Introduction to Solid State Physics 7th Ed.* Wiley, 1996.

[2] A. S. Phani, J. Woodhouse,N. A. Fleck. Wave propagation in two-dimensional periodic lattices. *J. Acoust. Soc. Am.*, 119(4):1995-2005, 2006. https://doi.org/10.1121/1.2179748

[3] C. Liu, C. Reina. Broadband locally resonant metamaterials with graded hierarchical architecture. *Journal of Applied Physics*, 123, 095108:1-10, 2018. https://doi.org/10.1063/1.5003264

[4] A. O. Oyelade, T. T. Akano. Graded hierarchical architecture metamaterial in vibration suppression. *UPB Scientific Bulletin, Series D: Mechanical Engineering*, 82(3):41-50, 2020.

[5] J. L. Auriault, G. Bonnet. Dynamique des composites élastiques périodiques. *Arch. Mech.*, 37:269-284, 1985.

[6] R. V. Craster, J. Kaplunov, A. V. Pichugin. High frequency homogenization for periodic media. *Proc. R. Soc. A Math. Phys. Eng. Sci.*, 466(2120):2341-2362, 2010. https://doi.org/10.1098/rspa.2009.0612

[7] C. Comi, J. J. Marigo. Homogenization Approach and Bloch-Floquet Theory for Band-Gap Prediction in 2D Locally Resonant Metamaterials. *Journal of Elasticity*, 139(1):61-90, 2020. https://doi.org/10.1007/s10659-019-09743-x

[8] C. Comi, M. Moscatelli, J. J. Marigo. Two scale homogenization in ternary locally resonant metamaterials. *Materials Physics and Mechanics*, 44(1):8-18, 2020.

[9] D. Faraci, C. Comi, J. J. Marigo. Band Gaps in Metamaterial Plates: Asymptotic Homogenization and Bloch-Floquet Approaches. *Journal of Elasticity*, 148(1):55-79, 2022. https://doi.org/10.1007/s10659-022-09879-3

Theoretical and Applied Mechanics - AIMETA 2022 | Materials Research Forum LLC
Materials Research Proceedings **26** (2023) 505-510 | https://doi.org/10.21741/9781644902431-82

A single-variable approach for layered beams with imperfect interfaces

Ilaria Monetto[1,a*] and Roberta Massabò[1,b]

[1]Department of Civil, Chemical and Environmental Engineering, Univeristy of Genoa, Genova, Italy

[a] ilaria.monetto@unige.it, [b] roberta.massabo@unige.it

Keywords: Single-Variable Formulation, Multi-Layered Beams, Zigzag Theories

Abstract. The first-order shear deformable laminated beam problem is reformulated in terms of a fictitious bending displacement as primal variable. A fourth-order differential equation governs the problem with two additional unknown constants coming from the integration of the longitudinal displacement. With the classical six boundary conditions, the problem is complete and well posed. An isogeometric collocation scheme is developed to solve the problem numerically. The formulation is completely locking-free and satisfies high continuity requirements for the approximation functions. The results for an exemplary structure confirm the validity and the good performance of the method, which is preliminary to the single variable reformulation of a more accurate zig-zag model for laminates with perfect and imperfect interfaces.

Introduction

As is well known, the elastic and thermal mismatch between the layers of composite beams induces complex zig-zag distributions of stresses and displacements in the thickness direction; relative displacements may also arise at the layer interfaces. Alternative approaches to the classical structural theories, which are based on assumptions that make them inadequate, must be used.

The zig-zag theories overcome the drawbacks of equivalent single-layer and layer-wise theories. A global displacement field is enriched by through-the-thickness zigzag functions. This allows to take into account the local effects due to material inhomogenities and to impose stress continuity at the interfaces, while maintaining the number of degrees of freedom low. Originally formulated zigzag theories had some inconsistencies and limitations and were not ideal for the implementation in numerical solution schemes, leading to the development of refined theories, which however also have some important drawbacks (see e.g. [1] for a review).

The validity of a more recent, energetically consistent, multiscale model formulated in [2] was demonstrated for a wide class of problems: plane-strain problems [3]; beams and plates subjected to thermal loadings [4]; propagation of plane-strain harmonic waves [5]; brittle delamination fracture under mode II dominant conditions [6-8].

Our work proposes a novel single-variable approach to the zigzag model in [4], by extending to laminates with and without imperfect interfaces a technique recently developed for homogeneous beams [9,10], as an efficient alternative approach to refined zigzag theories. All details and results will be presented at the conference; here, we describe the methodology and show some preliminary promising results in the framework of classical first-order shear deformation theory, which confirm the validity and efficacy of the method.

Equations governing first-order shear deformation theory for laminates

In this section the equations governing first-order shear deformation theory for laminates are briefly recalled with reference to the multi-layered beam shown in Fig. 1 and having thickness h, unit width in the direction x_1 and length L in the direction x_2. In the system of Cartesian coordinates x_1, x_2, x_3, the plane x_3=0 defines the reference surface of the beam, which is arbitrarily chosen. The beam consists of n layers joined by $n-1$ perfectly bonded interfaces (no relative sliding and transverse displacements are allowed). Each layer is linearly elastic and orthotropic with principal material axes parallel to the geometrical axes; the layers can have different mechanical properties. Coordinate x_3^k defines the interface k, that is the upper surface of the layer k having thickness ${}^{(k)}h$ (with k=1,..,n numbered from bottom to top).

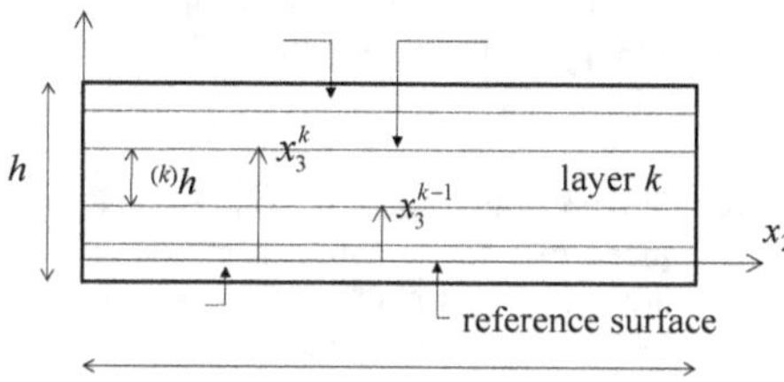

***Fig. 1** Multi-layered beam geometry*

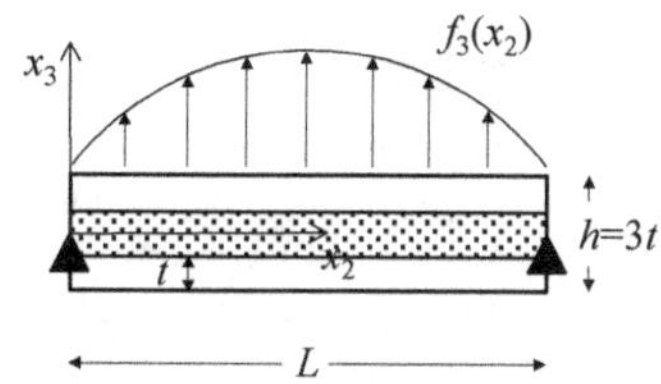

***Fig. 2** Analyzed exemplary beam*

The beam is subjected to boundary conditions and mechanical loads acting on the upper and lower surfaces and on the lateral bounding surface, and applied so as to satisfy plane strain conditions in the plane x_2-x_3 perpendicular to the axis x_1.

Transverse normal stresses are set equal to zero ($\sigma_{33}=0$) and the in plane displacements are assumed as linear functions of the through-the-thickness coordinate. Then, the global equilibrium equations can be written in terms of the global degrees of freedom v_{02}, w_0 and φ_2 (denoting, respectively, the displacements of the reference axis, along the directions x_2 and x_3, and the cross section rotation, about the axis x_1). Using a comma followed by subscripts to denote derivatives with respect to the corresponding coordinates, they are:

$$\begin{aligned}&A_{22}v_{02,22}+B_{22}\varphi_{2,22}+f_2=0\\&B_{22}v_{02,22}+D_{22}\varphi_{2,22}-k_{44}C_{44}^P(\varphi_2+w_{0,2})+f_{2m}=0\ ,\\&-k_{44}C_{44}^P(\varphi_{2,2}+w_{0,22})-f_3=0\end{aligned} \tag{1}$$

with coefficients

$$(A_{22},B_{22},D_{22})=\sum_{k=1}^{n}\int_{x_3^{k-1}}^{x_3^k}{}^{(k)}\overline{C}_{22}(1,x_3,x_3^2)\mathrm{d}x_3 \quad \text{and} \quad C_{44}^P=\sum_{k=1}^{n}\int_{x_3^{k-1}}^{x_3^k}{}^{(k)}C_{44}\mathrm{d}x_3\ , \tag{2}$$

where: for the layer k, ${}^{(k)}\overline{C}_{22}={}^{(k)}(C_{22}-C_{23}C_{32}/C_{33})$ relates local longitudinal normal stress and strain, through ${}^{(k)}\sigma_{22}={}^{(k)}\overline{C}_{22}\,{}^{(k)}\varepsilon_{22}$ under the assumed plane-strain conditions (${}^{(k)}C_{ij}$ with i,j=1,2 being the coefficients of the 6x6 stiffness matrix), and ${}^{(k)}C_{44}$ is the stiffness coefficient relating local transverse shear stress and strain, through ${}^{(k)}\sigma_{23}={}^{(k)}C_{44}2\,{}^{(k)}\varepsilon_{23}$; f_2, f_3, f_{2m} denote distributed longitudinal and transverse global loads (positive if rightward and upward, respectively) and global couples (positive if clockwise) acting on the reference surface. In Eq. (1) a shear correction factor k_{44} is introduced to improve the treatment of shear, given the limitations of the assumed theory to describe the global kinematics. The coefficients A_{22} and D_{22} are the extensional and bending laminate stiffnesses, B_{22} is the coupling stiffness of the laminate and relates bending strain with normal force and vice-versa.

Eq. (1) is a system of coupled differential equations in three unknowns, which is completed by six boundary conditions imposed at the ends x_2=0, L with outward normal $\mathbf{n}=\{0,\mp 1,0\}^{\mathrm{T}}$:

Materials Research Forum LLC
https://doi.org/10.21741/9781644902431-82

$$
\begin{aligned}
v_{02} &= \tilde{v}_{02} \quad \text{or} \quad N_{22}n_2 = [A_{22}v_{02,2} + B_{22}\varphi_{2,2} + w_{0,22}]n_2 = \tilde{N}_{22}n_2 \\
w_0 &= \tilde{w}_0 \quad \text{or} \quad Q_{22}n_2 = k_{44}C^P_{44}(\varphi_2 + w_{0,2})n_2 = \tilde{Q}_{22}n_2 \\
\varphi_2 &= \tilde{\varphi}_2 \quad \text{or} \quad M_{22}n_2 = [B_{22}v_{02,2} + D_{22}\varphi_{2,2}]n_2 = \tilde{M}_{22}n_2
\end{aligned}
\tag{3}
$$

(the tilde defines prescribed values). Once the problem has been solved, the local response for each layer is obtained straightforwardly through local compatibility, constitutive and equilibrium equations.

Single-variable formulation

In this section a single-variable formulation of the problem under consideration is presented, where the system of coupled differential equations (1) is reduced to a single equation in one unknown variable. As discussed in [9,10], different choices are possible for the primal variable. The approach proposed in [10] is followed here and the global transverse displacement w_0 is splitted into two parts: a bending part, w_{0b}, and a shear part, w_{0s}, which are defined as follows

$$
w_0 = w_{0b} + w_{0s} \quad \text{with} \quad \varphi_2 = -w_{0b,2} \quad \text{so that} \quad \varphi_2 + w_{0,2} = w_{0s,2}\,. \tag{4}
$$

Substituting Eq. (4) into the first and second of Eq. (1) yields $v_{02,22}$ and $w_{0s,2}$ as functions of $w_{0b,222}$, which can be derived once with respect to x_2 and introduced into the third of Eq. (1) to obtain a fourth-order differential equation in the one unknown w_{0b}

$$
\left[D_{22} - (B_{22})^2(A_{22})^{-1}\right]w_{0b,2222} = f_3 - B_{22}(A_{22})^{-1}f_{2,2} + f_{2m,2}\,. \tag{5}
$$

When w_{0b} is determined, the global rotation φ_2 follows directly from the second of Eq. (4), whereas the global transverse and longitudinal displacements w_0 and v_{02} are given by , respectively,

$$
\begin{aligned}
w_0 &= w_{0b} - (k_{44}C^P_{44})^{-1}\left\{\left[D_{22} - (B_{22})^2(A_{22})^{-1}\right]w_{0b,22} + \int [B_{22}(A_{22})^{-1}f_2 - f_{2m}]\mathrm{d}x_2\right\}, \\
v_{02} &= B_{22}(A_{22})^{-1}w_{0b,2} - \int\int f_2(A_{22})^{-1}\mathrm{d}x_2\mathrm{d}x_2 + c_5x_2 + c_6
\end{aligned}
\tag{6}
$$

where the integration of v_{02} requires the introduction of two additional arbitrary constants, say c_5 and c_6. To complete the boundary value problem, substituting these results into Eq. (3) leads to the following expressions for the resultant forces and moment in terms of w_{0b}

$$
\begin{aligned}
N_{22}n_2 &= [-\int f_2\mathrm{d}x_2 + A_{22}c_5]n_2 \\
Q_{22}n_2 &= \left\{-[D_{22} - (B_{22})^2(A_{22})^{-1}]w_{0b,222} - B_{22}(A_{22})^{-1}f_2 + f_{2m}\right\}n_2 \\
M_{22}n_2 &= \left\{-[D_{22} - (B_{22})^2(A_{22})^{-1}]w_{0b,22} - B_{22}(A_{22})^{-1}\int f_2\mathrm{d}x_2 + B_{22}c_5\right\}n_2
\end{aligned}
\;. \tag{7}
$$

Eq.s (5)-(7) , together with the second of Eq. (4), represent a single-variable formulation for laminates. All kinematic and static variables are expressed in terms of only one unknown and these expressions do not contain integrals of the unknown funtion. This excludes a priori locking problems and facilitates the employment in discrete numerical solution schemes. Furthermore, analogously to Bernoulli–Euler beam theory, this formulation is rotation-free, but shear deformability is accounted for. For symmetric laminates, when a mid-thickness reference surface x_3=0 is chosen, B_{22}=0 and Eq.s (5)-(7) reduce to those of the single-variable formulation for homogeneous Timoshenko beams in [10], but with f_2 and $f_{2m} \neq 0$. In this case the membrane and bending equilibrium problems are decoupled and the global longitudinal displacement v_{02} is independent of the bending transverse displacement w_{0b}.

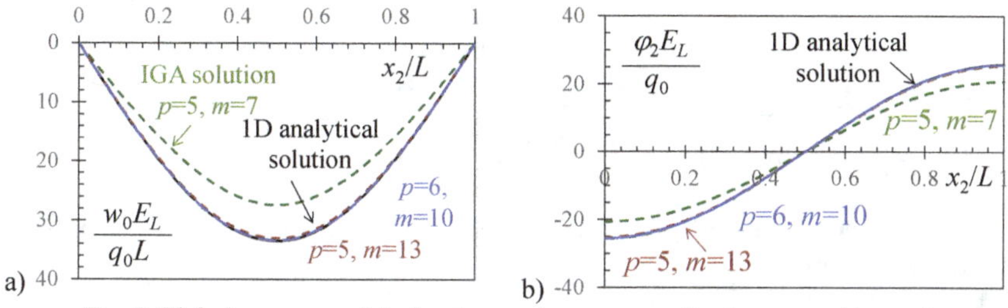

Fig. 3 Global response of the laminate: a) transverse displacement; b) rotation

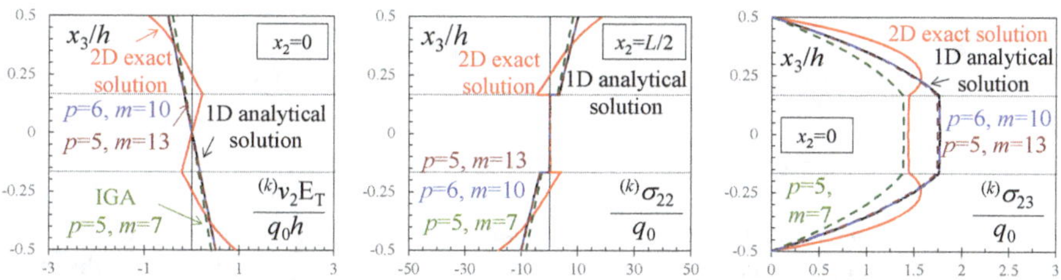

Fig. 4 Local response of the laminate: longitudinal displacement and in-plane stresses

Numerical solution through an isogeometric collocation scheme

In this section we develop an isogeometric collocation scheme (see, e.g., [10] and [11]) to solve numerically the laminate problem governed by the single-variable formulation derived above. In order to do this, the unknown primal variable w_{0b} is approximated by a B-spline curve, that is a linear combination of m piecewise polynomial basis functions of degree p, say $N_{i,p}$ with i=1,..,m, generated from a so-called knot vector and defined recursively starting with piecewise constants:

$$
\begin{aligned}
& w_{0b} \approx \sum\nolimits_{i=1}^{m} N_{i,p}(\xi = x_2 / L)\hat{w}_{0b_i} = \underline{N}\hat{\underline{w}}_{0b} \\
& \text{so that} \quad \frac{\mathrm{d}^k w_{0b}}{\mathrm{d}x_2^k} \approx \sum\nolimits_{i=1}^{m} \frac{\mathrm{d}^k N_{i,p}(\xi)}{\mathrm{d}\xi^k} L^{-k}\hat{w}_{0b_i} = L^{-k}\underline{N}^{(k)}\hat{\underline{w}}_{0b}
\end{aligned}, \tag{8}
$$

where $\underline{N}$ is a row matrix of the B-spline shape functions and $\hat{\underline{w}}_{0b}$ is the vector of control variables which represent the m degreees of freedom of the numerical model (the *mesh*). A knot vector is a set of m+p+1 nondecreasing real numbers (the *knots*); for 1D straight geometries, they represent dimensionless coordinates in the direction x_2, say ξ=x_2/L, and divide the laminate domain (a *patch*) in m+p subintervals (the *knot spans*). The generation of the basis functions starts from a set of m+p piecewise constant functions, which =1 in a single knot span and =0 otherwise, and requires the employment of the recursive formula p times. In order to have interpolatory basis functions at knots located at the ends of the patch, where boundary conditions are imposed, open knot vectors with the first and last knot appearing p+1 times are considered.

Using Eq. (8) into Eq.s (5)-(7) gives global kinematic and static variables in a discrete form, whereas N_{22} is always given by the first of Eq. (7), since it is independent of w_{0b}:

$$
\begin{aligned}
& w_0 \approx \left\{ \underline{N} - (k_{44} C_{44}^P)^{-1} \left[D_{22} - (B_{22})^2 (A_{22})^{-1} \right] L^{-2} \underline{N}'' \right\} \underline{\hat{w}}_{0b} - (k_{44} C_{44}^P)^{-1} \int [B_{22}(A_{22})^{-1} f_2 - f_{2m}] \mathrm{d}x_2, \\
& \varphi_2 \approx -L^{-1} \underline{N}' \underline{\hat{w}}_{0b}, \\
& v_{02} \approx B_{22}(A_{22})^{-1} L^{-1} \underline{N}' \underline{\hat{w}}_{0b} - \iint f_2 (A_{22})^{-1} \mathrm{d}x_2 \mathrm{d}x_2 + c_5 x_2 + c_6, \\
& Q_{22} n_2 \approx \left\{ -[D_{22} - (B_{22})^2 (A_{22})^{-1}] L^{-3} \underline{N}''' \underline{\hat{w}}_{0b} - B_{22}(A_{22})^{-1} f_2 + f_{2m} \right\} n_2, \\
& M_{22} n_2 \approx \left\{ -[D_{22} - (B_{22})^2 (A_{22})^{-1}] L^{-2} \underline{N}'' \underline{\hat{w}}_{0b} - B_{22}(A_{22})^{-1} \int f_2 \mathrm{d}x_2 + B_{22} c_5 \right\} n_2.
\end{aligned}
\tag{9}
$$

To solve the problem, the strong form in Eq. (5) is discretized and evaluated at a set of suitable collocation points, say $\bar{\xi}_i$,

$$
\left[D_{22} - (B_{22})^2 (A_{22})^{-1} \right] L^{-4} \underline{N}''''(\bar{\xi}_i) \underline{\hat{w}}_{0b} = f_3(\bar{\xi}_i) - B_{22}(A_{22})^{-1} f_{2,2}(\bar{\xi}_i) + f_{2m,2}(\bar{\xi}_i). \tag{10}
$$

Following [11], the so-called Greville abscissae related to the fourth derivative space, which corresponds to the order of the differential equation, are here chosen; they are defined as follows

$$
\bar{\xi}_i = \frac{\xi_{i+5} + \xi_{i+6} + .. + \xi_{i+p}}{p-4} \quad \text{for } i = 1,..,m-4. \tag{11}
$$

Eq. (10) gives m–4 equations in the m unknown control variables. The missing equations are obtained by imposing 6 boundary conditions at the ends of the patch ($\bar{\xi}_i = 0,1$) through Eq. (9), which contain the two additional constants c_5 and c_6. To conclude, the discrete problem is governed by a system of m+2 algebric equations in the m+2 unknowns $\underline{\hat{w}}_{0b}$ and c_5, c_6.

Application to an exemplary case and conclusive remarks

The single-variable formulation and related numerical scheme is employed for the solution of the classical problem shown in Fig. 2 and solved in closed form in [12]. A cross-ply laminated beam is simply supported at the ends and subjected to a sinusoidal transverse load $f_3(x_2) = q_0 \sin(\pi x_2 / L)$ ($f_2 = f_{2m} = 0$). The beam consists of n=3 unidirectionally reinforced laminae symmetrically arranged with stacking sequence (0°/90°/0°) and perfectly bonded together. Each lamina has thickness t (h=3t) and is orthotropic with elastic constants E_L, E_T=E_L/25, G_{LT}=E_L/50, G_{TT}=E_L/125, ν_{LT}=ν_{TT}=0.25. A mid-thickness reference surface is chosen.

For the discretization, quintic (p=5) and higher degree (p=6) B-splines are used, since the governing differential equation is of fourth order. Three meshes with different number of degrees of freedom (m=7,13 for p=5 and m=10 for p=6) are considered.

The dimensionless global transverse displacement and rotation, as functions of the dimensionless position along the axis x_2, and local longitudinal displacement, and bending and transverse shear stresses, through the thickness, obtained for k_{44}=1 and L=4h are shown in Fig.s 3 and 4. The numerical results for discretizations p=5 with m=7 (green curves) and m=13 (brown curves), and p=6 with m=10 (blue curves) are presented together with the 1D analytical solution of Eq. (5) (black curves). The analytical and numerical results are in very good agreement, but for the coarsest discretization with p=5 and m= 7 in Fig. 3 and for transverse shear stresses in Fig. 4. In the latter figure the comparison with the 2D exact solution (red curves) allows to conclude that the solution for the highest-order and finest discretizations (p=5, m=13 and p=6, m=10) are statically equivalent, even though locally different, to the 2D exact solution; this is not true for p=5, m= 7. Extension of the technique to the zigzag model in [4], which is in progress, is expected to provide full agreement with the 2D solution.

Acknowledgement
Supported by U.S. ONR and ONR Global, #N62909-21-1-2048.

References

[1] R. Massabò, Effective modeling of interlaminar damage in multilayered composite structures using zigzag kinematic approximations, in: G.Z. Voyiadjis (Ed,) Handbook of Damage Mechanics. Springer, New York, 2020, pp. 1-34. https://doi.org/10.1007/978-1-4614-8968-9_91-1

[2] R. Massabò, F. Campi, Assessment and correction of theories for multilayered plates with imperfect interfaces, Meccanica. 50 (2015) 1045-1071, DOI: 10.1007/s11012-014-9994-x. https://doi.org/10.1007/s11012-014-9994-x

[3] R. Massabò, F. Campi, An efficient approach for multilayered beams and wide plates with imperfect interfaces and delaminations, Composite structures. 116 (2014) 311-324, DOI: 10.1007/s11012-014-9994-x. https://doi.org/10.1007/s11012-014-9994-x

[4] M. Pelassa, R. Massabò, Explicit solutions for multi-layered wide plates and beams with perfect and imperfect bonding and delaminations under thermo-mechanical loading, Meccanica. 50 (2015) 2497-2524. https://doi.org/10.1007/s11012-015-0147-7

[5] R. Massabò, Propagation of Rayleigh-Lamb waves in multilayered plates through a multiscale structural model, Int. J. Solids and Structures. 124 (2017) 108-124. https://doi.org/10.1016/j.ijsolstr.2017.06.020

[6] R. Massabò, Influence of boundary conditions on the response of multilayered plates with cohesive interfaces and delaminations using a homogenized approach, Frattura ed Integrità Strutturale 8 (2014) 230-240. https://doi.org/10.3221/IGF-ESIS.29.20

[7] R. Massabò, H. Darban, Mode II dominant fracture of layered composite beams and wide-plates: a homogenized structural approach, Eng. Fracture Mech. 213 (2019) 280-301. https://doi.org/10.1016/j.engfracmech.2019.03.002

[8] R. Massabò, I. Monetto, Local zigzag effects and brittle delamination fracture of n-layered beams using a structural theory with three displacement variables, Frattura ed Integrità Strutturale. 51 (2020) 275-287. https://doi.org/10.3221/IGF-ESIS.51.22

[9] G. Falsone, D. Settineri, An Euler–Bernoulli-like finite element method for Timoshenko beams, Mechanics Research Communications. 38 (2011) 12-16. https://doi.org/10.1016/j.mechrescom.2010.10.009

[10] J. Kiendl, F. Auricchio, T.J.R. Hughes, A. Reali, Single-variable formulations and isogeometric discretizations for shear deformable beams, Comput. Methods Appl. Mech. Engrg. 284 (2015) 988-1004. https://doi.org/10.1016/j.cma.2014.11.011

[11] J. Kiendl, F. Auricchio, A. Reali, A displacement-free formulation for the Timoshenko beam problem and a corresponding isogeometric collocation approach, Meccanica 53 (2018) 1403-1413. https://doi.org/10.1007/s11012-017-0745-7

[12] N.J. Pagano, Exact solutions for composite laminates in cylindrical bending, J. Compos. Mater. 3 (1969) 398-411. https://doi.org/10.1177/002199836900300304

Microgeometrical design of lightweight bioinspired nacre-like composite materials for wave attenuation tuning

Andrea PRANNO[1,a], Fabrizio GRECO[1,b*], Raimondo LUCIANO[2,c] and Umberto DE MAIO[1,d]

[1]Department of Civil Engineering, Via P. Bucci Cubo 39B, Rende, Italy

[2]Department of Engineering, Parthenope University of Naples, Naples, Italy

[a]andrea.pranno@unicall.it, [b]fabrizio.greco@unical.it, [c]raimondo.luciano@uniparthenope.it, [d]umberto.demaio@unical.it.

Keywords: Bioinspired Composites, Wave Propagation, Instability

Abstract. Numerical simulations give the opportunity for designing periodic microstructured materials by tuning and manipulating their dynamic properties in terms of vibration control by varying their main micro-geometrical parameters. Specifically, the bandgap formation in 2D nacre-like composites characterized by a brick-and-mortar arrangement of stiff platelets and a soft matrix containing periodically arranged cavities was investigated through the Bloch-wave analysis. The numerical outcomes highlighted that, by varying the microstructural topology of the void inclusions and the stiff platelets, enhanced wave absorption capabilities can be obtained providing new opportunities to design periodic lightweight bioinspired composite metamaterials with elastic wave attenuation properties.

Introduction

As engineering requirements have become more strict, composite materials are increasingly being used in many applications [1,2] to meet advanced mechanical, chemical, thermal and vibrational [3,4] requirements. The research community is also constantly interested in developing advanced materials inspired by nature that can reach excellent properties. Thus nacre, also known as the mother of pearl, is among the most examined because of its excellent mechanical properties [5] related to the "brick-and-mortar" pattern characterizing its microstructural arrangement. Due to recent advancements in additive manufacturing, numerous researchers are exploring the high potential for designing nacre-like bioinspired materials with advanced properties finding applications in a wide range of engineering fields [6–8]. Due to their complex microstructure, composite materials are heterogeneous media that are susceptible to various nonlinear phenomena, especially if subjected to large deformation, such as instabilities at the microscopic and macroscopic scale [9] or damage mechanisms involving the microscopic scale such as matrix cracking or debonding of the fiber-matrix interfaces [10–13]. It is widely known that there is a strict correlation between these microscopic mechanisms and macroscopic fracture phenomena such as delamination, interfacial debonding, and multiple crack propagation which are generally considered the most frequent precursors of damage in microstructured composites [14–17]. By adopting hyperelastic constitutive models, numerous advanced numerical modeling techniques, including homogenization [18,19] and multiscale techniques [20,21], have been developed over the past decade to predict the mechanical behavior of composite materials subjected to large deformations and to reduce the computational effort needed to model complex microstructural geometries. Recent research has shown that nacre-like composite materials are capable of influencing wave propagation through their intrinsic periodic microstructure, which has led to an increase in interest in the study of their dynamic and wave propagation properties [22–25]. In this regard, materials with periodic structures, exhibiting physical properties not available in nature such as noise and vibration attenuation, are commonly referred to as metamaterials. They have experienced tremendous growth during the last decade, but they still have a few limitations. In mechanically demanding processes, for example, the mechanical properties of such materials can conflict with the demand for high energy dissipation needed to guarantee the vibration and noise attenuation characteristics. In this work, the Floquet-Bloch analysis is used to investigate the elastic waves propagation in lightened nacre-like composites with alternating stiff platelets and void inclusions in a finite element code (COMSOL Multiphysics 5.6). As a result, we will be able to increase our understanding of this growing area of research by exploring how by varying the main geometrical parameters and the voids inclusion arrangement affect vibration control and band gap phenomena of periodic nacre-like composite materials. Specifically, two void inclusion arrangements were investigated, and a parametric study was conducted by varying the main geometrical parameters, such as void volume fraction, platelets aspect ratio, and the thickness ratio between the horizontal and vertical matrix interfaces. The numerical outcomes have shown that by varying the void inclusion arrangements we are able to provide new opportunities for designing lightened bioinspired composite metamaterials with advanced wave absorption capabilities.

Numerical Results

The 2D unit cells investigated are reported in Fig.1 together with an example of mesh discretization highlighting the main geometrical parameters characterizing the results of the parametric analysis reported in the following. In this work, the matrix interphases are characterized by two different thicknesses b_h and b_v, related to the horizontal and vertical interfaces respectively. The unit cell and the platelets length are denoted by L and L_p, respectively, while the unit cell and the platelets height are denoted by H and H_p, respectively.

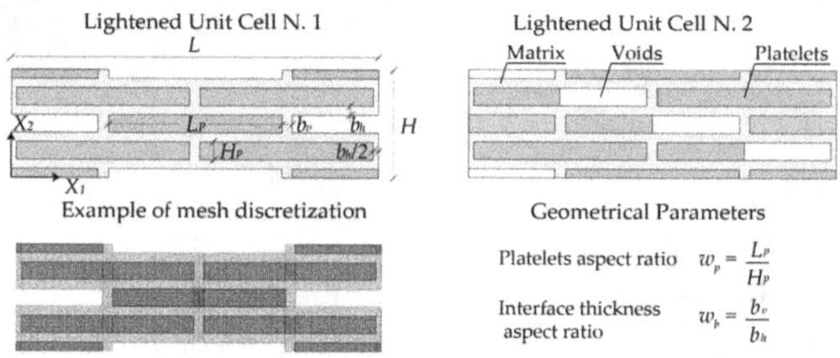

Fig.1 Investigated lightened nacre-like metamaterial with the main geometrical parameters.

The parametric analyses were performed by varying the geometrical parameters in terms of volume fraction $v_f = (8L_pH_p)/[(2L_p + 2b_v)(4H_p + 4b_h)]$, platelets aspect ratio $w_p = L_p / H_p$, with $L_p = 10\mu m$ and interface thickness aspect ratio $w_b = b_v / b_h$. The parameter v_f, ranging from 0.91 to 0.99 , is evaluated considering both the volume occupied by the platelets $v_f^p = (3/4)v_f$ and the one occupied by the voids $v_f^v = (1/4)v_f$. Based on the following neo-Hookean hyperelastic strain energy density function, the constitutive law for each microstructural material phase (matrix and platelets) is defined:

$$W = \frac{1}{2}\mu_{(m,p)}(\text{tr}(C)-3) - \mu_{(m,p)}\ln(J) + \frac{1}{2}\lambda_{(m,p)}\ln(J)^2, \quad (1)$$

Where the material parameters with subscript m and p are related to the matrix and platelets, respectively. Here μ is the initial shear modulus, C is the right elastic Cauchy-Green tensor, J

is the Jacobian of the transformation and λ is the first Lamé parameter which is able to control the material compressibility. With the aim to model the incompressible behavior of material phases $\lambda_{(m,p)}$ was considered equal to $1000\mu_{(m,p)}$. The initial shear modules are equal to $\mu_m = 2.5$ MPa and $\mu_p = k\mu_m$ with k equal to 10000. The Floquet-Bloch analysis was performed by imposing the Floquet-Bloch boundary conditions on the periodic unit cells and by opportunely varying the k-wave vector inside the first Brillouin zone obtaining the corresponding frequency dispersion (additional information about the numerical implementation can be found in the following research works [23,26,27]). Then, by analyzing the obtained dispersion curves the complete bandgap found were stored for each geometrical configuration investigated and then reported in Figures 1 and 2 in order to show the bandgap evolution as a function of the volume fraction parameter for different platelets and interface thickness aspect ratio.

Fig. 2 shows the evolution of the complete bandgaps with reference to the first lightened unit cell as a function of the volume fraction v_f. The figure highlights that numerous bandgaps were obtained for the case in Fig.2a with $w_b = 1$ and that they become slightly wider as the volume fraction increases. Specifically, with $91\% \le v_f \le 95\%$ only two bandgaps were found for the cases with $w_p = 0.5$ and $w_p = 4$ in a range of frequency below 60 [MHz], while with $95\% \le v_f \le 99\%$ we found that numerous bandgaps appear at higher frequency ranges and with different platelets aspect ratios reaching the highest attenuated frequency equal to 250[Mhz].

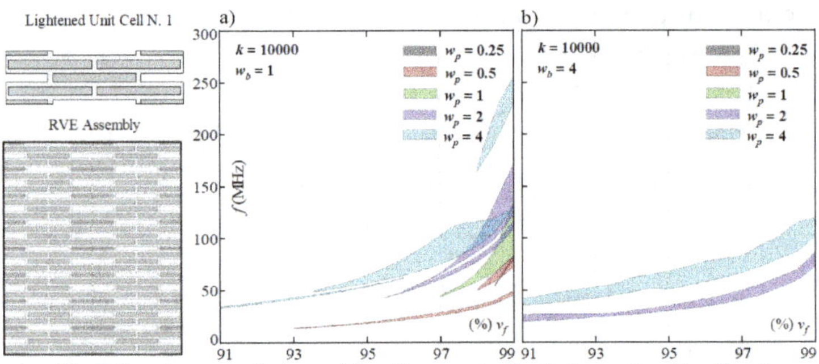

Fig.2 Bandgap evolution with reference to the lightened unit cell n.1

The numerical results with reference to a thickness aspect ratio $w_b = 4$ were reported in Fig.2b, in which it is clear that a higher thickness aspect ratio leads to a change in the wave absorption properties in terms (lowering the attenuated frequency ranges). In fact, only two bandgaps were found for the cases with $w_p = 2$ and $w_p = 4$ reaching the highest attenuated frequency equal to 110 [MHz]. To investigate the influence of different void inclusion arrangements, a second unit cell was proposed considering the same amount of void volume fraction and the obtained results were reported in Fig. 3. Compared with the unit cell n.1, in this case, it is clear that greater bandgaps were observed at lower frequency ranges and that the bandgaps become slightly wider as the volume fraction increases. In Fig.3a the most bandgaps were found for the cases with a volume fraction $94\% \le v_f \le 99\%$, while in Fig.3b the most band gaps were found for $97\% \le v_f \le 99\%$. However, in both Fig.3a and Fig.3b the bandgaps with the highest and widest frequency range were found for the case with a platelets aspect ratio $w_p = 4$ while the bandgaps with the lowest frequency range were found with $w_p = 0.25$. Thus, in general, we observed that a higher volume fraction leads to wider bandgaps with higher attenuated frequency ranges and that a higher platelets aspect ratio leads to higher attenuated frequency ranges.

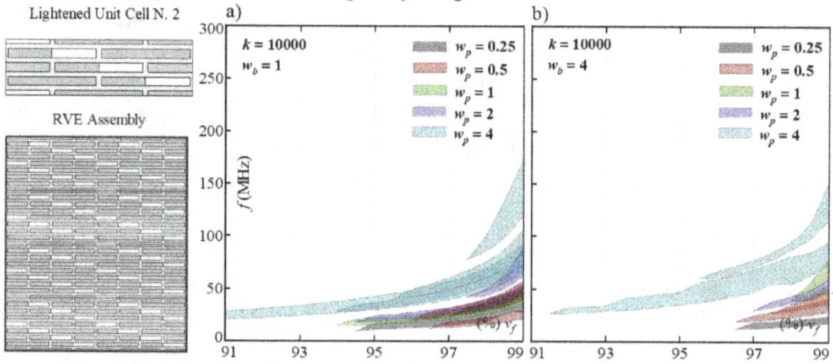

Fig.3 Bandgap evolution with reference to the lightened unit cell n.2

Conclusions

In this study, we investigated the wave propagation properties in two different lightened bioinspired metamaterials by varying the main geometrical parameters. The proposed microstructured metamaterials were lightened by inserting void inclusions in place of some reinforcing platelets in order to promote wave attenuation property. Comprehensive parametric analyses were performed to investigate the influence of the void volume fraction, the platelets aspect and the interface thickness aspect ratios. Based on the numerical results, void inclusions and stiff platelets with different microstructural topologies were found to experience enhanced wave absorption capabilities, thereby providing new opportunities to design periodic lightweight bioinspired composite metamaterials able to attenuate specific frequency ranges of propagating elastic waves.

Acknowledgments

Fabrizio Greco gratefully acknowledge financial support from the Italian Ministry of Education, University and Research (MIUR) National Grant P.R.I.N. 2017 “Multiscale Innovative Materials and Structures” (Project Code 2017J4EAYB). Andrea Pranno gratefully acknowledges financial support from the Italian Ministry of Education, University and Research (MIUR) PON 2014-2020 Azione IV.4 - Rep. N. 1062 of 10/08/2021. Umberto De Maio gratefully acknowledge financial support from the Italian Ministry of University and Research (MUR) National Grant PRIN 2020 “Opportunities and challenges of nanotechnology in advanced and green construction materials” (Project Code 2020EBLPLS).

References

[1] D. Ammendolea, F. Greco, P. Lonetti, R. Luciano, A. Pascuzzo, Crack propagation modeling in functionally graded materials using Moving Mesh technique and interaction integral approach, Compos. Struct. 269 (2021) 114005. https://doi.org/10.1016/j.compstruct.2021.114005

[2] A. Pascuzzo, A. Yudhanto, M. Alfano, G. Lubineau, On the effect of interfacial patterns on energy dissipation in plastically deforming adhesive bonded ductile sheets, Int. J. Solids Struct. 198 (2020) 31–40. https://doi.org/10.1016/j.ijsolstr.2020.04.001

[3] A. Pranno, F. Greco, P. Lonetti, R. Luciano, U. De Maio, An improved fracture approach to investigate the degradation of vibration characteristics for reinforced concrete beams under progressive damage, Int. J. Fatigue. 163 (2022) 107032. https://doi.org/10.1016/j.ijfatigue.2022.107032

[4] F. Greco, D. Gaetano, L. Leonetti, P. Lonetti, A. Pascuzzo, A. Skrame, Structural and seismic vulnerability assessment of the Santa Maria Assunta Cathedral in Catanzaro (Italy): classical and advanced approaches for the analysis of local and global failure mechanisms, Frat. Ed Integrità Strutt. 16 (2022) 464–487. https://doi.org/10.3221/IGF-ESIS.60.32

[5] M.A. Meyers, A.Y.-M. Lin, P.-Y. Chen, J. Muyco, Mechanical strength of abalone nacre: Role of the soft organic layer, J. Mech. Behav. Biomed. Mater. 1 (2008) 76–85. https://doi.org/10.1016/j.jmbbm.2007.03.001

[6] V. Slesarenko, N. Kazarinov, S. Rudykh, Distinct failure modes in bio-inspired 3D-printed staggered composites under non-aligned loadings, Smart Mater. Struct. 26 (2017) 035053. https://doi.org/10.1088/1361-665X/aa59eb

[7] F. Greco, L. Leonetti, A. Pranno, S. Rudykh, Mechanical behavior of bio-inspired nacre-like composites: A hybrid multiscale modeling approach, Compos. Struct. (2019) 111625. https://doi.org/10.1016/j.compstruct.2019.111625

[8] D. Wang, D. Chen, Z. Chen, Recent Progress in 3D Printing of Bioinspired Structures, Front. Mater. 7 (2020) 286. https://doi.org/10.3389/fmats.2020.00286

[9] F. Greco, L. Leonetti, U. De Maio, S. Rudykh, A. Pranno, Macro- and micro-instabilities in incompressible bioinspired composite materials with nacre-like microstructure, Compos. Struct. 269 (2021) 114004. https://doi.org/10.1016/j.compstruct.2021.114004

[10] U. De Maio, F. Greco, L. Leonetti, P. Nevone Blasi, A. Pranno, An investigation about debonding mechanisms in FRP-strengthened RC structural elements by using a cohesive/volumetric modeling technique, Theor. Appl. Fract. Mech. 117 (2022) 103199. https://doi.org/10.1016/j.tafmec.2021.103199

[11] A. Pascuzzo, F. Greco, L. Leonetti, P. Lonetti, A. Pranno, C. Ronchei, Investigation of mesh dependency issues in the simulation of crack propagation in quasi-brittle materials by using a diffuse interface modeling approach, Fatigue Fract. Eng. Mater. Struct. 45 (2022) 801–820. https://doi.org/10.1111/ffe.13635

[12] U. De Maio, D. Cendón, F. Greco, L. Leonetti, P. Nevone Blasi, J. Planas, Investigation of concrete cracking phenomena by using cohesive fracture-based techniques: A comparison between an embedded crack model and a refined diffuse interface model, Theor. Appl. Fract. Mech. 115 (2021) 103062. https://doi.org/10.1016/j.tafmec.2021.103062

[13] A. Pascuzzo, F. Greco, P. Lonetti, D. Ammendolea, Dynamic fracture analysis in quasi-brittle materials via a finite element approach based on the combination of the ALE formulation and M-integral method, Eng. Fail. Anal. (2022) 106627. https://doi.org/10.1016/j.engfailanal.2022.106627

[14] U. De Maio, F. Fabbrocino, F. Greco, L. Leonetti, P. Lonetti, A study of concrete cover separation failure in FRP-plated RC beams via an inter-element fracture approach, Compos. Struct. 212 (2019) 625–636. https://doi.org/10.1016/j.compstruct.2019.01.025

[15] U. De Maio, F. Greco, L. Leonetti, R. Luciano, P. Nevone Blasi, S. Vantadori, A refined diffuse cohesive approach for the failure analysis in quasibrittle materials—part I: Theoretical formulation and numerical calibration, Fatigue Fract. Eng. Mater. Struct. 43 (2020) 221–241. https://doi.org/10.1111/ffe.13107

[16] U. De Maio, F. Greco, L. Leonetti, R. Luciano, P. Nevone Blasi, S. Vantadori, A refined diffuse cohesive approach for the failure analysis in quasibrittle materials—part II: Application to

plain and reinforced concrete structures, Fatigue Fract. Eng. Mater. Struct. 42 (2019) 2764–2781. https://doi.org/10.1111/ffe.13115

[17] U. De Maio, F. Greco, L. Leonetti, P. Nevone Blasi, A. Pranno, A cohesive fracture model for predicting crack spacing and crack width in reinforced concrete structures, Eng. Fail. Anal. 139 (2022) 106452. https://doi.org/10.1016/j.engfailanal.2022.106452

[18] F. Greco, P. Lonetti, R. Luciano, P. Nevone Blasi, A. Pranno, Nonlinear effects in fracture induced failure of compressively loaded fiber reinforced composites, Compos. Struct. 189 (2018) 688–699. https://doi.org/10.1016/j.compstruct.2018.01.014

[19] F. Greco, Homogenized mechanical behavior of composite micro-structures including micro-cracking and contact evolution, Eng. Fract. Mech. 76 (2009) 182–208. https://doi.org/10.1016/j.engfracmech.2008.09.006

[20] F. Greco, L. Leonetti, P. Lonetti, R. Luciano, A. Pranno, A multiscale analysis of instability-induced failure mechanisms in fiber-reinforced composite structures via alternative modeling approaches, Compos. Struct. 251 (2020) 112529. https://doi.org/10.1016/j.compstruct.2020.112529

[21] F. Greco, L. Leonetti, R. Luciano, A multiscale model for the numerical simulation of the anchor bolt pull-out test in lightweight aggregate concrete, Constr. Build. Mater. 95 (2015) 860–874. https://doi.org/10.1016/j.conbuildmat.2015.07.170

[22] Y. Chen, L. Wang, Multiband wave filtering and waveguiding in bio-inspired hierarchical composites, Extreme Mech. Lett. 5 (2015) 18–24. https://doi.org/10.1016/j.eml.2015.09.002

[23] A. Pranno, F. Greco, L. Leonetti, P. Lonetti, R. Luciano, U. De Maio, Band gap tuning through microscopic instabilities of compressively loaded lightened nacre-like composite metamaterials, Compos. Struct. 282 (2022) 115032. https://doi.org/10.1016/j.compstruct.2021.115032

[24] J. Li, V. Slesarenko, S. Rudykh, Auxetic multiphase soft composite material design through instabilities with application for acoustic metamaterials, Soft Matter. 14 (2018) 6171–6180. https://doi.org/10.1039/C8SM00874D

[25] M. Bruggi, A. Corigliano, Optimal 2D auxetic micro-structures with band gap, Meccanica. 54 (2019) 2001–2027. https://doi.org/10.1007/s11012-019-00981-w

[26] N. Arora, J. Li, S. Rudykh, Tunable buckling configurations via in-plane periodicity in soft 3D-fiber composites: Simulations and experiments, Int. J. Solids Struct. 250 (2022) 111711. https://doi.org/10.1016/j.ijsolstr.2022.111711

[27] J. Li, S. Rudykh, Tunable microstructure transformations and auxetic behavior in 3D-printed multiphase composites: The role of inclusion distribution, Compos. Part B Eng. 172 (2019) 352–362. https://doi.org/10.1016/j.compositesb.2019.05.012

Anisotropic behaviours and strain concentration in lattice material evaluated by means of discrete homogenization

Salvatore Gazzo[1, a*], Loredana Contrafatto[1,b], Leopoldo Greco [1,c] and Massimo Cuomo[1,d]

[1] Dipartimento di Ingegneria Civile e Architettura, Università degli Studi di Catania, Italy

[a]salvatore.gazzo@dica.unict.it, [b]loredana.contrafatto@unict.it, [c]leopoldo.greco@unict.it, [d]mcuomo@dica.unict.it

Keywords: Discrete Homogenization, Fibre Network, Anisotropic Material

Abstract. Lattice fibre materials are challenging the standard modelling approaches due to their specific nature that results in peculiar effective behaviours such as extremely anisotropic materials or generalized continuum media. In this context, the aim of this paper is to the determine qualitatively and quantitatively the role of the morphological and mechanical parameters by investigating simple archetypical microstructures. The study is conducted through an up-scaling approach making use of the Homogenization method of discrete periodic media in the framework of a variational approach. The results of the homogenization have been validated comparing the response of the continuum with the response of discrete models.

Introduction

Lattice fibre materials are becoming widespread in engineering applications, thanks to their versatility and their efficient mechanical and physical properties. In addition to conventional fibre materials, widely used in aero-spatial technology and in civil engineering, innovative materials have been introduced by the new manufacturing technologies like additive printing.

The aim of the contribution is to investigate the elastic mechanical behaviour of plane networks (fibre sheets), that present strong directional properties, which can be tuned by suitable design of the network and selection of the fibre mechanical characteristics.

In order to describe the network behaviour, the effective mechanical properties of plane fibre networks have been derived using discrete homogenization [1,2,3]. The resulting equivalent continuum presents strong directional properties, with anisotropy ratios much larger than those that can be found in natural materials.

The results of the homogenization have been validated comparing the response of the continuum with the response of ''discrete'' models, for which each element is modelled as a slender beam, and boundary conditions are accounted for in an exact way.

Brief description of Discrete homogenization technique and some applications on lattice structural types

Discrete homogenization applies to lattices constituted by a finite number of nodes and micro-beam elements. It derives from classical two-scales homogenization, in which the fast (microscopic) variables are substituted by the node numbering. Scale separation hypothesis is assumed, that is, the ratio between the dimension of the unit cell ℓ_c and the dimensions of the network, L is a very small parameter $\varepsilon = \ell_c/L << 1$. Homogeneous Euler–Bernoulli beams are considered for the lattice, for more detaileds on the homogenization method see [3,4].

Two types of networks are examined in this paper, both composed by straight members with rigid connection, representing both biaxial and quadriaxial repetitive cells. The biaxial cell is composed by two families of fibres, as shown in Fig. 1(a); in this case will be considered either orthogonal rectangular cells (so that the fibres form an angle $\alpha = \pi/2$) and skew cell (with an angle between the fibres $0 < \alpha < \pi/2$). The quadriaxial cell is always considered with orthogonal fibres as shown in Fig. 1(b). The elements of the networks are considered as having rectangular cross-sections, with unit depth normal to the plane of the lattice. Plane stress are therefore considered for the homogenized model. Referring to the elements indicated in Fig. 1, the following symbols will be used: l_i is the length of member b_i; I_i is the in-plane moment of inertia of member b_i; A_i is the area of member b_i; h, s will denote the height and width of the members; E_i is the Young Modulus of member b_i; α is the opening angle of the biaxial cell.

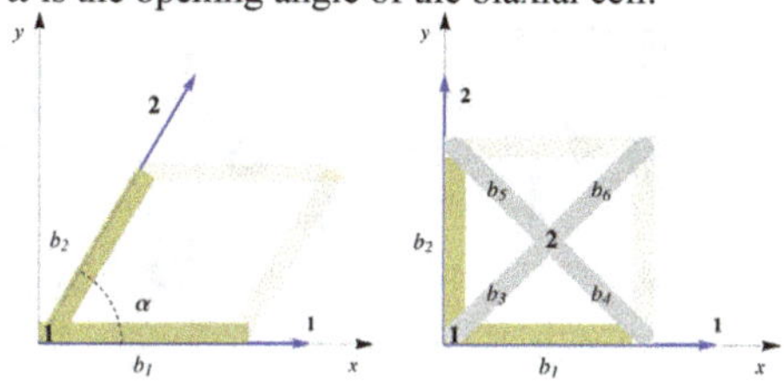

(a) (b)

Fig. 1 Case studies: (a) Skew cell with rigid connection, (b) Rectangular cell braced with rigid connection.

The homogenized constitutive equations of the network constituted by skew repetitive cells with straight fibres forming an angle α between themselves and for the rectangular cell braced with rigid connection can be found in [4,5].

Analysis of the properties of the equivalent materials

The design of lattice structures, like woven fabrics and networks, with controlled anisotropy and mechanical properties is of critical importance for various applications. In order to evaluate the characteristics of anisotropy of networks having as elementary cell one of those examined in the previous sections, we examine the uniaxial stiffness of the network defined as the ratio between an uniaxial stress applied along an axis ξ rotated by an angle θ respect to the x axis (always taken coincident with the direction of fibre 1 and the corresponding strain, see Fig. 2. The ratio $D_{\xi\xi}= \sigma_{\xi\xi}/\epsilon_{\xi}$ denotes the uniaxial stiffness in the direction ξ.

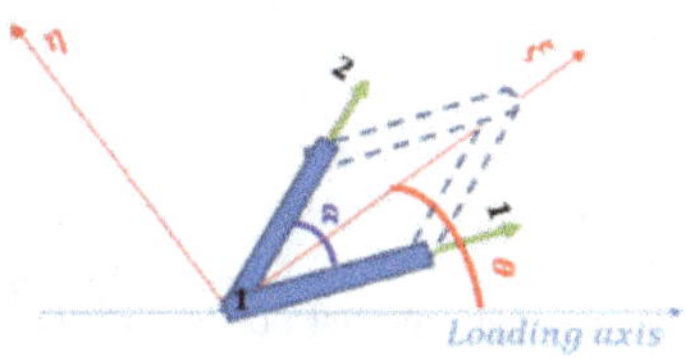

Fig. 2 Definition of the rotated axis ξ.

Extension tests of networks modelled by square biaxial or quadriaxial cells arranged at various angles with respect to the extension axis are considered. The following geometrical and mechanical properties of the fibres have been considered: $l_1 = l_2 = 5$ mm; $h_1 = h_2 = 1$ mm, $s_1 = s_2 = 1$ mm, $E_1 = E_2 = 1600$ Mpa.

Biaxial network

The uniaxial stiffness, normalized with respect to the modulus E_1 of the beams lying in the 1-direction, are represented in polar plots in Fig. 3(a) and 3(b).

Defining the dimensionless parameters $\rho = l_2/l_1$ $\beta = h_2/h_1$ $\gamma = E_2/E_1$ $\mu_1 = l_1/h_1$, the different curves in the plots refer to cells having beams with different length with ρ ranging from 1 to 3, equal section $\beta = 1$, equal modulus $\gamma = 1$, and slenderness of the first set of beams $\mu_1 = 5$.

The material obtained presents a sharp contrast of anisotropy, reaching two order of magnitudes between the directions of the fibres and a direction at 45° with it. For large ratios l_2/l_1 there is also a strong difference between the stiffness along the directions of the fibres.

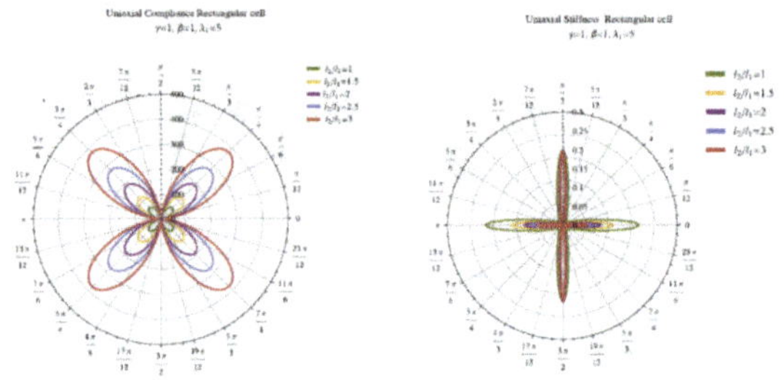

(a) Polar plot of the dimensionless uniaxial compliance $C_{\xi\xi}E_1$

(b) Polar plot of the dimensionless uniaxial stiffness $\frac{D_{\xi\xi}}{E_1}$

Fig. 3 Uniaxial stiffness of a material formed by rectangular reference cells with rigid connections as function of the load directions for different aspect ratio of the cell.

Extension tests are usually performed to get an estimate of the elastic properties of fibre networks. According to ISO standard 13934-1, the sample length-to-width ratio used in laboratory tests for assessing the uniaxial elastic modulus should be equal to 2.

In order to evaluate whether this aspect ratio is appropriate for obtaining stable values of the elastic parameters, the apparent elastic modulus D_{yy} in the axial direction was numerically evaluated as a function of the slenderness of the sample for various inclinations θ of the material axis 1 with respect to the loading direction.

The aspect ratio was varied in the range 0.1 to 10.0. The results are summarized in the plot of Fig. 4(a). All the results are contained within two boundaries, represented by the upper and lower dashed curves, which represent the uniaxial stiffness evaluated analytically under the hypothesis of uniaxial strain state ($\epsilon_{zz} = \gamma_{yz} = 0$, upper curve) and of uniaxial stress state ($\sigma_{zz} = \tau_{yz}$, lower curve).

Quadriaxial network

The results of this section refer to a quadriaxial lattice with orthogonal fibres and rigid connections, as the one examined, in the special case that the internal fibres along the diagonals (usually less stiff) are identical among themselves. The internal node is located at the centre of the cell. The following dimensionless parameters are used: $\gamma_1 = E_2/E_1$; $\gamma_2 = E_{int}/E_1$; $\beta_1 = h_2/h_1$; $\beta_2 = h_{int}/h_1$; $\rho = l_2/l_1$ where E_{int}, h_{int} are the Young modulus and the thickness of the internal fibres.

The length of the internal fibres is $l_{in} = l_1/2\sqrt{1+\rho^2}$. The slenderness of the external fibre 1 is denoted by $\mu_1 = l_1/h_1$. Notice that the slenderness of the beams in the 2-direction and of the internal beams are respectively $\mu_2 = \mu_1\,\rho/\beta$ and $\mu_{int} = \mu_1\,/\beta_2\sqrt{0.25+0.25\rho^2}$.

The case of a square braced cell, $\rho = 1$, with two equal external fibres and $\beta_1 = 1$, $\gamma_1 = \gamma_2 = 1$, is examined first. The slenderness ratio of the external beams has been fixed to $\mu_1 = 5$.

The uniaxial stiffness and compliance are reported in Fig. 5, for different values of β_2 varying from 0.3 to 2. According to whether the internal fibres have a larger thickness than the external ($\beta_2 > 1$), or the opposite, a greater stiffness is obtained at the braces direction (45°), or at 0°. However, in any case, a more isotropic behaviour is obtained with respect to the biaxial lattice.

The effect of the internal fibres is to reduce the contrast of anisotropy. The insertion of diagonal fibres makes the transition of the stiffness from direction 1 to direction 2 more smooth.

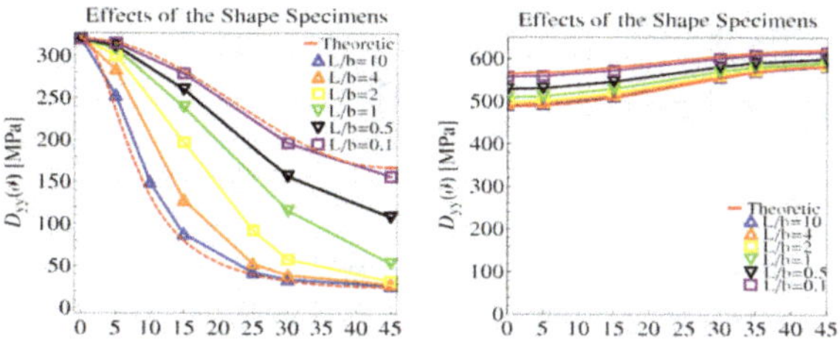

(a) (b)

Fig. 4 Uniaxial stiffness of a material formed by rectangular reference cells with rigid connections as function of the load directions for different aspect ratio of the cell.

The apparent longitudinal elastic modulus as a function of the fibre for aspect ratios of the specimen in the range 0.1 to 10 and for orientations ranging from $\theta = 0°$ to $\theta = 45°$ is presented in Fig. 4(b). Comparing this plot with the one of Fig. 4(a) it is possible to observe that in the case of quadriaxial networks the behaviour of the material is almost independent on the mesh inclination. This effect can be attributed to the stiffness contribution given by the braces.

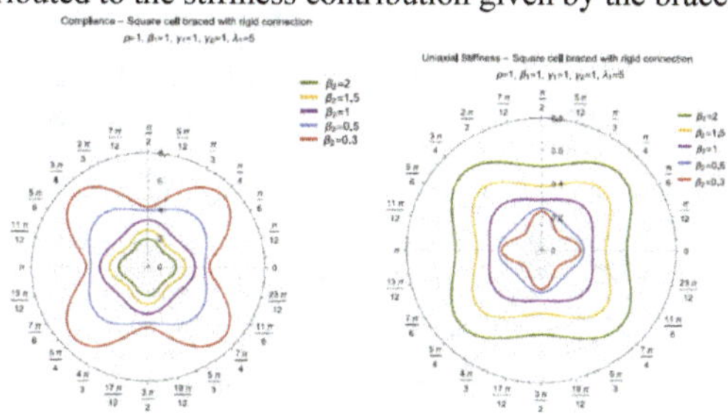

(a) Polar plot of the dimensionless uniaxial compliance $C_{\xi\xi}E_1$ (b) Polar plot of the dimensionless uniaxial stiffness $\frac{D_{\xi\xi}}{E_1}$

Fig. 5 Uniaxial stiffness of a material formed by square braced cell with rigid connections as function of the load directions for different thickness ratios β_2.

Validation

The numerical results obtained using the homogenization model in the simulation of the bias test of the biaxial network with aspect ratio $\rho = 2$ have been compared to the response of a meso-model of the network, realized by means of discrete linear beams.

Four discrete models with a variable number of cells have been employed, corresponding to 20, 40, 80, 160, cells along the longest side of the specimen. For both families of equal fibres the Young modulus $E = 1600$ MPa and slenderness ratio $l_i/h_i = 10$ have been used.

The elastic properties of the homogenized material are $E_{xxxx} = 160$ MPa, $\Gamma = 0.8$ MPa. As the number of elements in the discrete model grows, so as to satisfy the scale separation hypothesis, the results of the discrete and of the homogenized models tend to converge. Fig. 6(a) shows the relative error for the strain energy between the discrete model and a reference value evaluated with the homogenized continuum model, using a very fine mesh. Similar convergence plots refer to the lateral contraction of the specimen, measured at its centre (Fig. 6(b)) and to the shear strain γ_{12} between the fibres directions in the centre of the specimen (Fig. 6(c)).

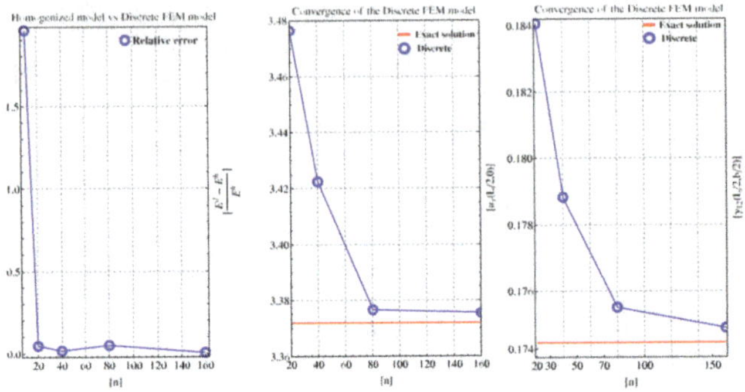

(a) (b) (c)

Fig. 6 Convergence of the results of the discrete models and of the homogenized model as the number of discrete cells increases.

Cylindrical structure

In this section a cylinder whose wall is made by a periodic lattice of biaxial orthogonal fibres with circular cross-section is examined. The discrete and homogenized continuous model will be compared. The homogenised model was defined taking into account both the flexural and the membrane properties simultaneously. The cylindrical structure is shown in the figure 7.

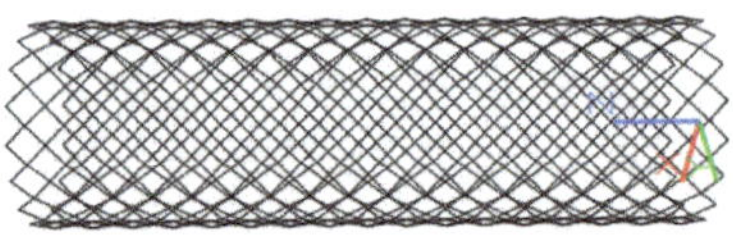

Fig. 7 Cylindrical structure

The geometrical parameters used for the model are: diameter D = 100 mm and height H = 300 mm of the cylindrical structure; dimensions of the repetitive cell l_1 = 4.0 mm e l_2 = 4.0 mm, with rotations of load at 45° of the cell with respect the axis z; diameter of the beams of the cell is d = 0.5 mm. The mechanical properties of the material constituting the lattice are the elastic module E = 1600 N/mm^2 and poisson's ratio ν= 0.15. A line load that tends to uniformly stretch the cylinder has been applied.

At one end of the cylinder translations are prevented, at the other end only axial displacements are allowed. The same simulation using the homogenized continuum has been performed. The vertical displacements of both models, figures 8 and 9, are quite similar. The maximum displacement value is 27.982 mm 28.346 mm respectively for the discrete model and for the homogenized model. The relative error between the maximum displacement obtained for the discrete model and that obtained for the homogenised is:

$$e_r = \frac{|28.346 - 27.982|}{27.982} \times 100 = 1.3\%$$

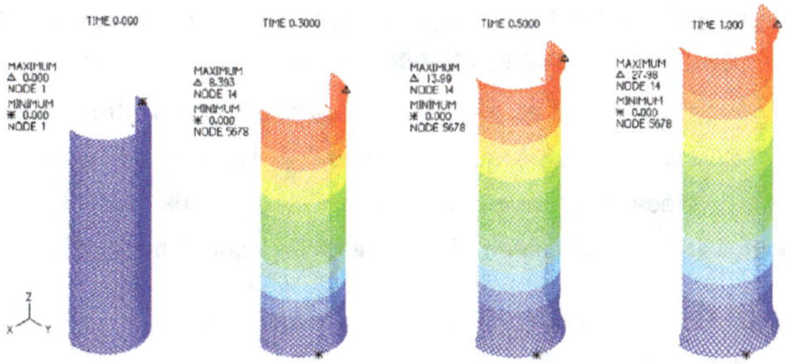

Fig. 8 Band plot of the vertical dislacement for the discrete FEM model for different increment of the load.

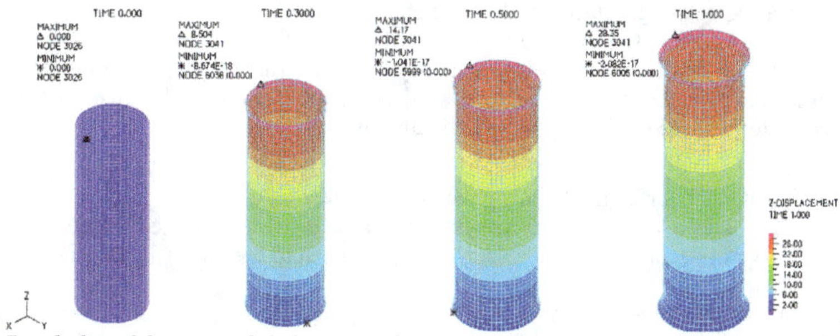

Fig. 9 Band plot of the vertical dislacement for the discrete FEM model for different increment of the load.

References

[1] Abdoul-Anziz, H., Seppecher, P., "Strain gradient and generalized continua obtained by homogenizing frame lattices", Math. Mech. Complex 6, 3, 213-250, (2018). https://doi.org/10.2140/memocs.2018.6.213

[2] dell'Isola, F., Giorgio, I., Pawlikowski, M., Rizzi, N.L., "Large deformations of planar extensible beams and pantographic lattices: heuristic homogenization, experimental and numerical examples of equilibrium", Proc. R. Soc. Lond. Ser. A Math. Phys. Eng. Sci. 472, 472, (2016). https://doi.org/10.1098/rspa.2015.0790

[3] Caillerie, D., Mourad, A., Raoult, A., "Discrete homogenization in graphene sheet modeling", J. Elasticity, 84, 33-68, (2006). https://doi.org/10.1007/s10659-006-9053-5

[4] Gazzo, S., Cuomo, M., Boutin, C., Contrafatto, L., "Directional properties fibre network materials evaluated by means of discrete homogenization", European Journal of Mechanics, A/Solids, 82 (2020). https://doi.org/10.1016/j.euromechsol.2020.104009

[5] Cuomo, M., Boutin, C., Contrafatto, L., Gazzo, S., "Effective anisotropic properties of fibre network sheets", European Journal of Mechanics, A/Solids, 93, 104492, (2022). https://doi.org/10.1016/j.euromechsol.2021.104492

One-dimensional metastructures composed of cables with scatter masses: waves, vibrations and band gaps

Marco Moscatelli[1,2,a *], Claudia Comi[1,b] and Jean-Jacques Marigo[2,c]

[1]Department of Civil and Environmental Engineering, Politecnico di Milano, Milano, Italy

[2]Laboratoire de Mécanique des Solides, École Polytechnique, Palaiseau, France

[a]marco.moscatelli@polimi.it, [b]claudia.comi@polimi.it, [c]jean-jacques.marigo@polytechnique.edu

Keywords: Metamaterials, Cable Dynamics, Wave Attenuation

Abstract. This work analyzes the dynamic behavior of structural elements that can be modelled as taut cables with a discrete array of punctual attached and hanging masses. The propagation of mechanical waves is strongly influenced by the presence of such scatter elements. We found that the problem is governed by a discrete equation, whose solutions depend on the behavior of an equivalent mass density, that varies with frequency. The spectrum of the problem is characterized by the presence of band gaps. This behavior is generally exploited for the design of metastructures. A parametric study of the equivalent mass is finally given.

Introduction

Long and flexible slender elements are typically modelled as cables, i.e. as structures with no flexural rigidity. The dynamics of such systems has been widely studied due to the possible activation of several peculiar mechanisms mainly connected to the intrinsic non-linearities of the problem [1–3].

In most applications, cabling systems present hanging elements that are often periodically repeated. This is the case for instance for the main cables in suspension bridges [4], for cableways [5] and for overhead lines with ball markers or vibrational dissipators [6]. These elements largely influence and modify the dynamic response of the system, acting as scatterers for propagating waves. To show this, we here analyze a simplified problem by studying the propagation of mechanical waves in taut cables with a periodic array of masses, directly attached to the cable or/and hanging to it by means of elastic springs. We focus on the linear dynamics of the system. Specifically, we verified that there exist some intervals of frequencies at which waves cannot propagate through the cable, i.e. band gaps in the spectrum. We thus found that the system can behave as a metastructure.

In this work, we initially derive an equivalent equation governing the linear dynamics of our system and we show that band gaps can indeed appear. We then analyze how these band gaps are modified by a variation of the parameters involved in the problem: this analysis gives useful information for the design phase.

Problem formulation

Let us analyze the propagation of transverse mechanical waves in taut cables. This model can be used to study the behavior of long and slender structural elements, characterized by a low flexural rigidity, that are stretched between two supports positioned at the same height. The cable is here characterized by the presence of a periodic arrangement of pointwise hanging elements, composed of masses m_1 directly attached to the cable and connected to hanging masses m_2 by means of elastic springs k (cf. Fig. 1).

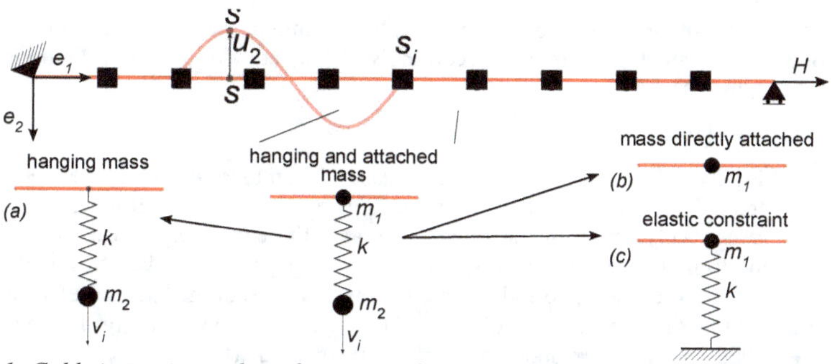

Figure 1: Cable in tension under a force $\mathbf{H}$*, with a periodic distribution of scatter elements composed of masses* $\mathbf{m_1}$ *directly attached to the cable and/or masses* $\mathbf{m_2}$ *hanging through springs* $\mathbf{k}$*. The following limiting cases are shown: in scheme (a) masses are hanging to the cable by means of elastic springs, in scheme (b) masses are directly attached to the cable, and in scheme (c), masses are constrained to the ground by means of elastic springs. Direction* $\mathbf{e_3}$ *is out-of-plane.*

By calling H the pretension applied to the cable, the equilibrium configuration can be approximated to be horizontal when H is much larger than the total weight of cable and attached elements. Specifically, a material point s_0 will be moved to a different position s, during this pretensioning phase, along direction $\boldsymbol{e}_1$ (see Fig. 1). By calling $N_{eq}(s_0)$ the axial force and $\epsilon_{eq}(s_0)$ the static axial strain of the point originally at position s_0, from equilibrium and linear elasticity one has

$$N_{eq}(s_0) = H \quad \text{and} \quad \epsilon_{eq}(s_0) = H/EA, \tag{1}$$

where EA is the axial stiffness of the cable.

Note that, with the current assumptions, the presence of the punctual elements does not influence the static equilibrium of the system. This is not the case when the cable is moving. To show this, let us now take the static equilibrium configuration to be the new reference configuration for the dynamic problem. We call L the distance between the two supports and n the total number of hanging elements. Accordingly, $d = L/(n+1)$ is the distance between two subsequent elements. As we are dealing with a wave propagation problem, we do not consider the boundary conditions.

By studying small oscillations around the equilibrium configuration, the motion problem is governed by three uncoupled equations along the in-plane horizontal and transverse directions and the out-of-plane direction (respectively, directions $\boldsymbol{e}_1$, $\boldsymbol{e}_2$ and $\boldsymbol{e}_3$ in Fig. 1). Specifically, along the transverse direction $\boldsymbol{e}_2$, one has:

$$\ddot{u}(s,t) - c_t^2 u''(s,t) = 0 \quad \forall s \setminus P \tag{2}$$

where we use $(\ddot{\blacksquare})$ and $(\blacksquare'')$ to denote respectively the time and spatial derivatives. In Eq. 2, $c_t^2 = EA(1+\epsilon_{eq})^2/\rho$ is the speed of transverse waves, and we have used P to define the collection of positions $s = s_i$ of the i-th hanging elements. The motion problem is then completed by a set of jump conditions, such that:

$$[\![EA\epsilon_{eq}u']\!](s_i,t) = m_1\ddot{u}(s_i,t) - k\big(v_i(t) - u(s_i,t)\big) \quad with\ 1 < i < n, \tag{3}$$

where $[\![\blacksquare]\!] = (\blacksquare)^+ - (\blacksquare)^-$, with $(\blacksquare)^+$ (resp. $(\blacksquare)^-$) denoting the right (resp. left) limit of $(\blacksquare)$ at s. In the above Eq. 3, we used v_i to denote the vertical displacement of the end point of the spring where the mass m_2 of the i-th element is attached, by making the assumption that these masses can only move in the vertical direction $\boldsymbol{e}_2$. Accordingly, the equation of the vertical motion of the i-th mass m_2 reads as:

$$m_2 \ddot{v}_i + k\big(v_i(t) - u(s_i, t)\big) = 0. \tag{4}$$

Note that Eq. 3 can be also used to model the case when masses $m_1 + m_2$ are directly attached to the cable and the case when the elastic springs are constrained to the ground. For this, one has to consider, respectively, an infinite stiffness k and an infinite mass m_2. Specifically, from Eq. 4, one has for the former case $v_i(t) = u(s_i, t)$, meaning that the system behaves as if masses m_2 were directly attached to the cable and summed to masses m_1. For the latter case, instead, one has $\ddot{v}_i = 0$, meaning that masses m_2 cannot move. These cases are shown in Fig. 1 (schemes (b) and (c)), together with the case corresponding to $m_1 = 0$ (scheme (a)).

By considering that both fields u and v_i are small and harmonically varying in time, such that

$$u(s,t) = \hat{u}\, L\, exp\, \mathrm{i}\omega t \quad \text{and} \quad v_i(s,t) = \hat{v}_i\, L\, exp\, \mathrm{i}\omega t, \tag{5}$$

Eq. 2 can be solved within each interval i-th. The term i is used in relations 5 and in the followings to denote the imaginary unit. By calling $\hat{u}_i$ the transverse displacement of the i-th mass m_1, one obtains

$$\hat{u}(\hat{s}) = \hat{u}_{i-1} cos(\Omega(\mathrm{n}+1)\hat{s} - i + 1) + \frac{\hat{u}_i - \hat{u}_{i-1} cos\,\Omega}{sin\,\Omega} sin(\Omega(n+1)\hat{s} - i + 1) \tag{6}$$

where $\Omega = \omega d/c_t$ and $\hat{s} = s/L$ are respectively dimensionless frequency and coordinate. Using Eq. 6, together with Eq. 3 and 4, one finally finds the following equivalent equation of motion for the points $s_i \in P$:

$$\Delta_i \hat{u} + \mu(\Omega)\hat{u}_i = 0, \quad \text{with} \quad \Delta_i \hat{u} = \hat{u}_{i+1} + \hat{u}_{i-1} - 2\,\hat{u}_i. \tag{7}$$

In the above equation, $\mu(\Omega)$ can be interpreted as a frequency dependent equivalent mass density and it reads

$$\mu(\Omega) = 2(1 - cos\,\Omega) + \left(\Theta_1 + \frac{\hat{k}\Theta_2}{\hat{k} - \Theta_2\Omega^2}\right)\Omega\, sin\,\Omega \tag{8}$$

where Θ_1 and Θ_2 are two mass ratios and $\hat{k}$ is a normalized stiffness of the springs:

$$\Theta_1 = \frac{m_1}{\rho d}, \quad \Theta_2 = \frac{m_2}{\rho d}, \quad \text{and} \quad \hat{k} = \frac{kd}{N_{eq}}. \tag{9}$$

Spectral band gaps

From the derivation in [7], one has the following solutions of Eq. 7:

$$\begin{aligned} \hat{u}_i &= A\, exp(-\mathrm{i}\,\mathrm{K}^* i) + B\, exp(\mathrm{i}\,\mathrm{K}^* i) && \text{for } 0 \le \mu(\Omega) \le 4, \\ \hat{u}_i &= A\, exp(-\mathrm{K}^* i) + B\, exp(\mathrm{K}^* i) && \text{for } \mu(\Omega) < 0, \\ \hat{u}_i &= A\,(-1)^i exp(-\mathrm{K}^* i) + B\,(-1)^i exp(\mathrm{K}^* i) && \text{for } \mu(\Omega) > 4, \end{aligned} \tag{10}$$

where $\mathrm{K}^* \in [0, \pi]$ is a dimensionless wavenumber obtained as:

$$1 - \frac{\mu(\Omega)}{2} = \begin{cases} cos\, K^* & \text{for } 0 \le \mu(\Omega) \le 4, \\ cosh\, K^* & \text{for } \mu(\Omega) < 0, \\ -\,cosh\, K^* & \text{for } \mu(\Omega) > 4. \end{cases} \tag{11}$$

Accordingly, waves propagate freely along the cable when $0 \leq \mu(\Omega) \leq 4$. For either $\mu(\Omega) < 0$ or $\mu(\Omega) > 4$, waves are instead attenuated, as the displacement field is given by a superimposition of exponentials.

The spectrum of the problem governed by Eq. 7 is thus characterized by *band gaps*, i.e. by intervals of frequencies at which wave solutions does not exist.

Parametric study of band gaps

As discussed in the previous section, band gaps are given by those frequencies at which either $\boldsymbol{\mu(\Omega) < 0}$ or $\boldsymbol{\mu(\Omega) > 4}$. Considering for example the case with $\boldsymbol{\Theta_1 = 0}$, $\boldsymbol{\Theta_2 = 1.18}$ and $\boldsymbol{\hat{k}}$ =4 (hanging masses), we report in Fig. 2 the behavior of $\boldsymbol{\mu}$ in function of the frequency $\boldsymbol{\Omega}$. Band gaps are denoted by the green ($\boldsymbol{\mu(\Omega) < 0}$) and red ($\boldsymbol{\mu(\Omega) > 4}$) intervals. Blue intervals correspond to pass bands.

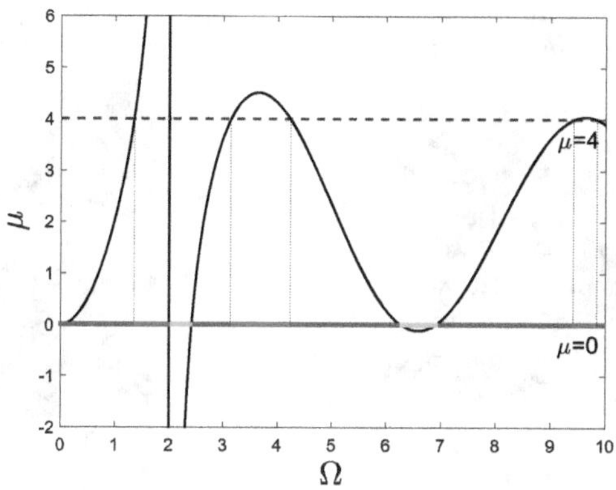

Figure 2: Equivalent mass vs frequency for the system studied in Fig. 5, with $\boldsymbol{\hat{k} = 4}$. *Colors must be interpreted as in Figures 3 to 6: blue is used for pass bands, red and green for band gaps.*

Let us now study how the equivalent mass density $\boldsymbol{\mu(\Omega)}$ is influenced by the parameters given by relations 9. Specifically, we here analyze how band gaps vary in function of them. For this, we show some contour plots (Figures 3 to 6) of the equivalent mass $\boldsymbol{\mu(\Omega)}$ for the following cases:

Fig. 3. Cable with masses m_1 and no hanging mass ($m_2 = 0$).
Fig. 4. Cable with $m_1 = 0$ and $m_2 \to \infty$ (cable lying on a discrete array of elastic constraints).
Fig. 5. Cable with $m_1 = m_2 = m$ ($\Theta_1 = \Theta_2 = \Theta$) and fixed dimensionless spring stiffness $\hat{k}$.
Fig. 6. Cable with hanging masses m_2 ($m_1 = 0$) and varying dimensionless spring stiffness $\hat{k}$.

As before, in the contour plots the conditions $\boldsymbol{\mu(\Omega) < 0}$ and $\boldsymbol{\mu(\Omega) > 4}$ are verified respectively in the green and red regions. The blue areas, instead, correspond to pass bands.

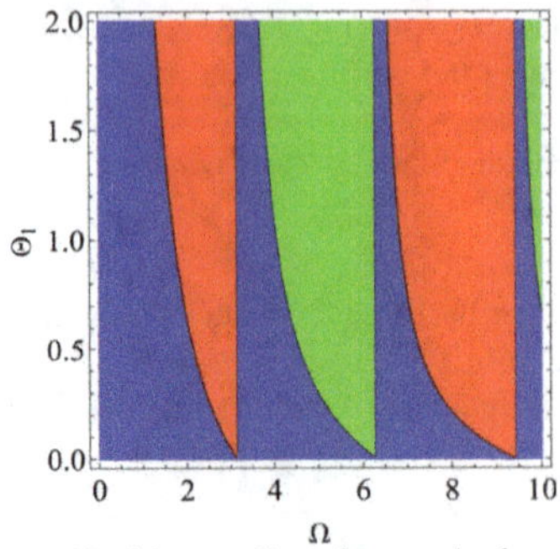

Figure 3: Masses directly attached to the cable for varying mass m1 (m2=0).

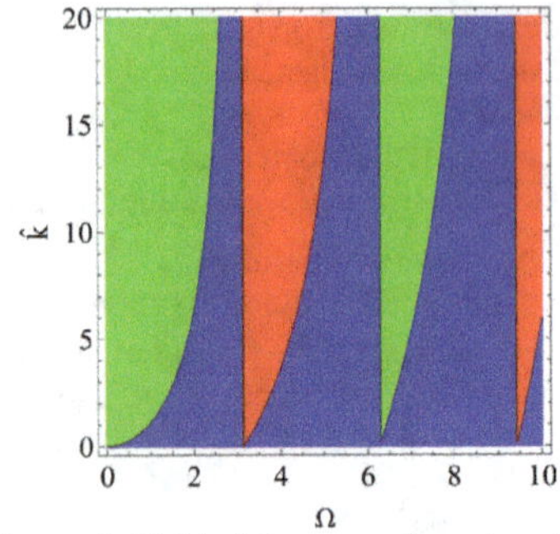

Figure 4: Cable lying on a discrete set of elastic springs for varying stiffness (m1=0).

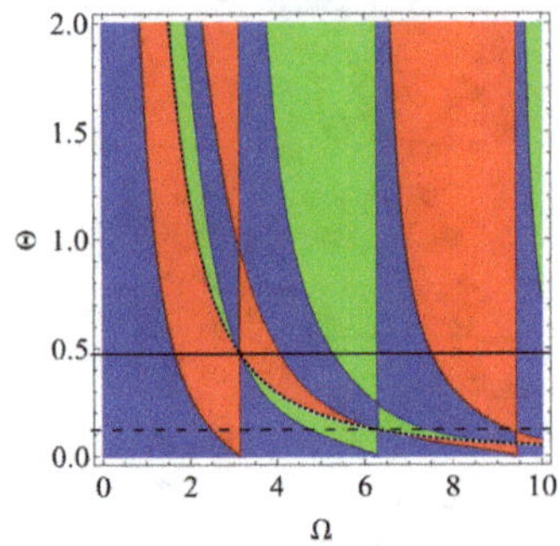

Figure 5: Cable with equal attached and hanging masses for varying mass m1=m2=m. For this plot we used $\widehat{\mathbf{k}} = \mathbf{4.73}$.

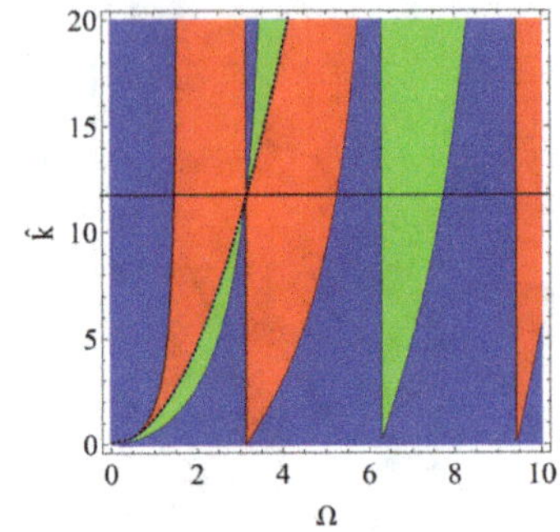

Figure 6: Cable with hanging masses for varying stiffness k (m1=0). For this plot we used $\boldsymbol{\Theta_2} = \mathbf{1.18}$.

Note that the system studied in Fig. 2 corresponds to that used in Fig. 6, by fixing $\hat{k}$ =4. Colors have the same meaning in both Figures.

From Fig. 3, the width of band gaps increases with the mass. For the system analyzed in Fig. 4, their width increases with the stiffness. Most importantly, this latter case is characterized by a cut-off frequency with a band gap opening at zero frequency. Note that the closing frequencies in Fig. 3 and the opening frequencies in Fig. 4 are independent respectively from a variation of attached mass and stiffness. This can be explained by studying the behavior of the equivalent mass μ. For this, let us first rewrite here relation 8 for these two cases

$$\mu(\Omega) = \begin{cases} 2(1 - \cos\Omega) + \Theta_1 \Omega \sin\Omega \\ 2(1 - \cos\Omega) + \hat{k}\frac{\sin\Omega}{\Omega} \end{cases} . \tag{12}$$

From relations 12 one can recognize that at points $j\pi$, with $j \in \mathbb{N}_*$, the first and second terms in the definition of the equivalent mass are both zero independently of Θ_1 and $\hat{k}$.

Let us now comment on Figures 5 and 6. In both cases, the systems used are characterized by the presence of resonant elements (springs and masses m_2) that can locally resonate causing the opening of band gaps. The dotted black curves in these contour plots show how the resonance frequency varies with respect to the varying parameters. Specifically, the equivalent mass is indefinite at resonance and tends to $+\infty$ or $-\infty$ from below or from above (cf. Fig. 2), depending on the case under consideration. Accordingly, band gaps will always appear around the frequency of resonance, as one can check from the contour plots. From Fig. 6, band gaps width clearly increases with the stiffness. The presence of opening and closing frequencies that result to be independent from a variation of mass (Fig. 5) and stiffness (Fig. 6) can be explained by looking at the equivalent mass μ with a reasoning similar to that followed for Figures 3 and 4.

Note that, for the parameters fixed by the black continuous horizontal lines in Figures 5 and 6, the two systems are characterized by a spectrum where the closing frequency of the first band gap is superimposed with the opening frequency of the second band gap: the final band gap becomes thus very wide, resulting in an optimal attenuating behavior. This feature is also valid for the second and third band gaps in the spectrum shown in Fig. 5, as indicated by the black dashed horizontal line.

Conclusions

In this paper we have considered the linear dynamic behavior of a cable with hanging elements. We found that this system behaves as a metastructure, offering the peculiar property that waves cannot freely propagate for some intervals of frequencies, generally known as band gaps. For this, we have derived an equivalent mass density from which we were able to obtain information regarding the influence of the parameters involved in the problem. The result can be of interest as a starting point for more realistic models where the initial sag of the cable is taken into account. This would better idealize the structural applications listed in the introduction of the current work. Moreover, the attenuating capabilities of the system could be exploited for the localization and focusing of mechanical waves, as shown in [8,9].

References

[1] M. Irvine, Cable structures, The Massachusetts Institute of Technology, Cambridge, 1981.

[2] S.A. Nayfeh, A.H. Nayfeh, D.T. Mook, Nonlinear Response of a Taut String to Longitudinal and Transverse End Excitation, Journal of Vibration and Control. 1 (1995) 307–334. https://doi.org/10.1177/107754639500100304

[3] A.H. Nayfeh, H.N. Arafat, C.-M. Chin, W. Lacarbonara, Multimode Interactions in Suspended Cables, Journal of Vibration and Control. 8 (2002) 337–387. https://doi.org/10.1177/107754602023687

[4] F. Gazzola, Mathematical Models for Suspension Bridges, Springer International Publishing, Milan, 2015. https://doi.org/10.1007/978-3-319-15434-3

[5] J.M.W. Brownjohn, Dynamics of an aerial cableway system, Engineering Structures. 20 (1998) 826–836. https://doi.org/10.1016/S0141-0296(97)00113-2

[6] F. Foti, V. Denoël, L. Martinelli, F. Perotti, A stochastic and continuous model of aeolian vibrations of conductors equipped with stockbridge dampers, in: Proceedings of the International Conference on Structural Dynamic , EURODYN, European Association for Structural Dynamics, 2020: pp. 2088–2102. https://doi.org/10.47964/1120.9169.20304

[7] M. Moscatelli, C. Comi, J.J. Marigo, On the dynamic behaviour of discrete metamaterials: From attenuation to energy localization, Wave Motion. 104 (2021) 102733. https://doi.org/10.1016/j.wavemoti.2021.102733

[8] M. Moscatelli, Metamaterials for energy harvesting at small scale, École polytechnique, 2022.

[9] M. Moscatelli, C. Comi, J. Marigo, Attenuation and localization of waves in taut cables with a discrete array of scatter elements, in: Proceedings ECCOMAS, 2022. https://doi.org/10.23967/eccomas.2022.057

Graded meta-waveguides for elastic energy splitting

Luca Iorio[1,a *], Jacopo M. De Ponti[1,b], Raffaele Ardito[1,c] and Alberto Corigliano[1,d]

[1] Dept. of Civil and Environmental Engineering, Politecnico di Milano, Piazza Leonardo da Vinci, 32, 20133 Milano, Italy

[a] luca.iorio@polimi.it, [b]jacopomaria.deponti@polimi.it, [c]raffaele.ardito@polimi.it, [d]alberto.corigliano@polimi.it

Keywords: Metamaterials, Meta-lattices, Wave confinement, Energy Harvesting

Abstract. Engineered metamaterials for precise wave manipulation and control is a new challenge for many applications in the field of mechanical vibration control and energy harvesting. Inspired by the advancement in wave manipulation and localization performance of elastic waveguides thanks to the introduction of resonators, we try to assess if these systems can be applied also into the junctions between beams that make up lattices. The goal is to achieve control over elastic wave propagation and define whether the local resonators are able to enhance the redirection of the wave into a perpendicular beam. Furthermore, evaluations concerning elastic energy localization in the resonators are reported to assess whether these systems are also suitable for energy localization and energy harvesting.

Introduction

Control over wave propagation has attracted growing interest across different realms of physics, with multiple realizations in electromagnetic [1] acoustic [2,3], and elastic systems [4]. Different physical effects have been developed in photonics and researchers have been borrowing those ideas to focus or confine elastic wave energy, such as the creation of elastic lenses [5,6,7,8], cavities [9,10], mirrors [11] or topological modes [12]. Metamaterials devised through the use of local lateral resonators have been recently employed to confine elastic energy for energy harvesting applications or vibration isolation. These resonators filter elastic waves and store elastic energy inside them, efficiently protecting the underlying guide from harmful vibrations [13,14]. More recently, it has been shown how broadband vibration isolation properties and wave redirection effects can be achieved leveraging graded arrays of resonators with spatially varying resonance frequency [15,16]. In the context of wave localization and wave redirection, the idea of implementing novel metalattices is now taking place The underlying idea is to create metalattices that are able, thanks to their microstructure, to guide specific elastic waves, depending on the desired frequency and polarization, into certain predetermined paths in the lattice. The modification of the microstructure of the lattice is able to induce a change in the homogeneous properties at a macro level [17]. As a result, if the variation is properly engineered, lensing effects or in general wave manipulation effects are obtained. The idea now is the opposite: modifying the microstructure of a classic lattice so that the waves are redirected at the micro-structural level. The necessary step to achieve such structures would be to design junctions with peculiar properties that are able to influence the energy redirection between the single beams that compose the junction. To do so, this paper reports some numerical experiments conducted using local resonance systems to partially achieve such junctions. The idea is to engineer lateral resonators that localize and convert waves to achieve control over preferential wave paths.

Analyses

A study of wave manipulation and control over the connection of two infinite waveguides is reported. The aim is to develop new knowledge over wave redirection in metalattices that employ local resonance effects in specific junction points of frame structures to control the flow of elastic energy through space.

Starting from the analysis of a simple T-junction between infinitely long waveguides, the idea is to see if the presence of a resonator is able to single handedly improve the percentage of wave redirected into the perpendicular waveguide. The hypothesis is that, thanks to the localization of energy inside the resonator, the wave emitted by it through anchor losses or other coupling phenomena will be redirected also into the perpendicular waveguide. Then, from these results, rainbow reflection arrays will be implemented to see whether these structures are beneficial for the redirection of the waves.

Materials and methods

First, we evaluate the resonance frequency of the resonator implemented at the junction of the two waveguides with the specific boundary conditions (BCs). In the case of a 13 mm long, 2.5 mm wide, 1.5 mm thick resonator, made of aluminum and attached to the junction point of two infinite waveguides made of aluminum, the resulting resonance frequency is 5800 Hz. At the waveguides boundaries, ALID boundary conditions are implemented to model an infinite domain [18]. Further damping mechanisms are not contemplated. The analyses reported below are all conducted evaluating the displacement field generated by a travelling flexural wave. The numerical results reported are obtained through time domain implicit analyses in COMSOL Multiphysics.

Single resonator at the junction

Fig. (1) reports the geometries analyzed and the results obtained through the comparison between the case with the resonator positioned at the junction point between the waveguides and the junction without the resonator. The displacement field has been normalized with respect to the maximum displacement generated at the input. It shows how the presence of the resonator influences the redirection of the wave at the frequency associated to its resonance. The maximum displacement field in the perpendicular waveguide is enhanced by 43% when the resonator is positioned on the junction. Furthermore, another analysis was conducted to see whether the energy traversing the waveguide is more localized on the resonator in the case of infinite waveguide or the T junction case described above. Fig. (2) shows the setup and the results. It is clear that, for a narrow band signal centered at the resonance frequency of the resonator, the maximum displacement field of the cantilever is enhanced by 134%.

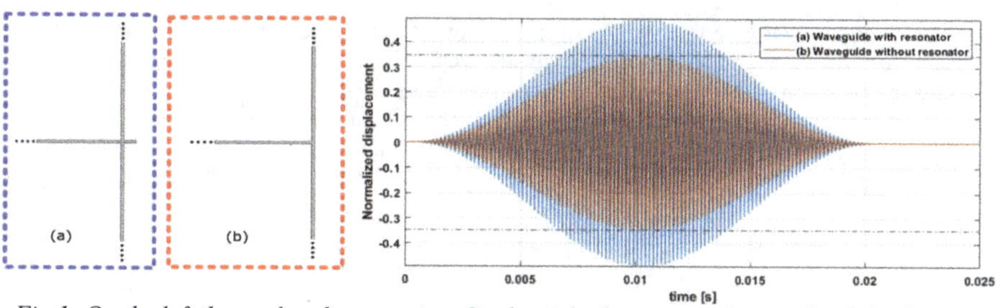

Fig.1: On the left the analyzed geometries. On the right the computed normalized displacement on the perpendicular waveguide for the case of junction with lateral resonator (a), and junction without lateral resonator (b).

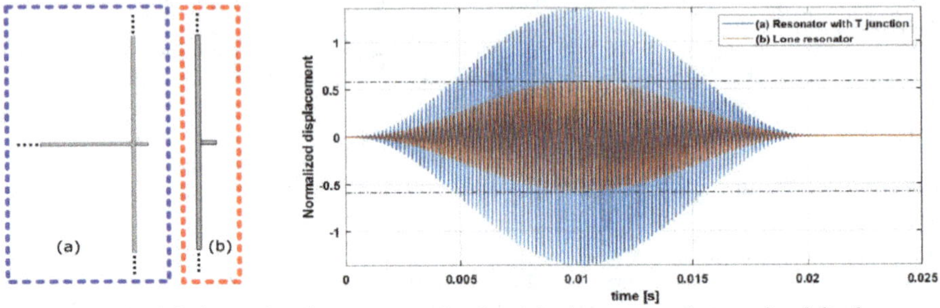

Fig.2: On the left the analyzed geometries. On the right the computed normalized displacement at the resonator's tip for the case of T junction (a), and lone lateral resonator on the waveguide (b).

This is an interesting result for energy localization and energy harvesting application, given that the presence of the junction enhances the energy stored in the lateral resonator.

Asymmetric rainbow at the junction

Having seen how the presence of a single resonator is able to improve wave redirection and energy localization, the next step is to evaluate the effect of placing the resonator inside a rainbow reflection array. This has been done by considering that a rainbow reflection array is able to increase the interaction time between the wave and the target resonator [19]. The goal is to assess whether this system can increase both localization of the wave at the target resonator positioned at the junction point and also redirect more efficiently the energy along the perpendicular waveguide. Before stating the results, it is to be noticed that the resonance frequency of the target resonator is now shifted with respect to the previous analyzed case: this is due to the interaction with the neighbor resonators that pushes down the resonance by 300 Hz. Furthermore, the rainbow reflection configuration is composed of 15 linearly increasing in length and equally spaced resonators so that the target resonator (the 12th) is positioned exactly at the junction point between the two waveguides.

Now the idea is to compare the redirection efficiency of the rainbow structure at the resonance frequency of the target resonator. The analyses are all performed with a narrowband signal generated at the resonance frequency of the target resonator.

Fig. (3) reports the results of the comparison between the case of one single resonator at the junction point (the same one shown before) with the asymmetric rainbow case. The result shows that the rainbow structure is not able to efficiently redirect the wave into the perpendicular waveguide.

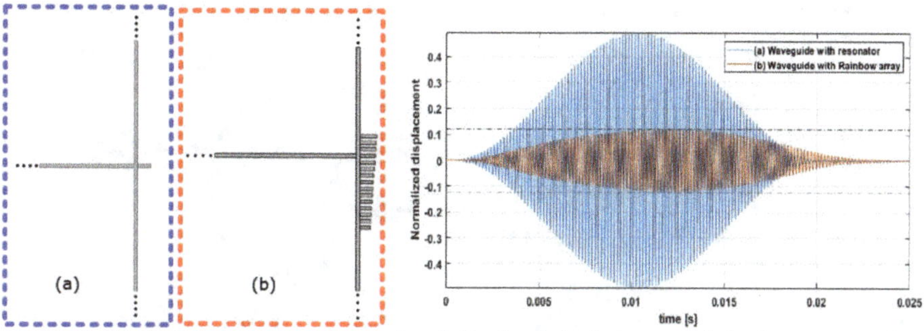

Fig.3: On the left the geometries analyzed. On the right the computed displacement on the perpendicular waveguide for the case of junction with lateral resonator (a), and junction with rainbow reflection array (b).

As for the efficiency of energy localization inside the resonator, Fig. (4) reports that, as stated in previous works [19,20] the enhancement of energy localization is confirmed. However, even if the displacement field of the resonator is enhanced 8 times in the rainbow array, the redirection on the perpendicular waveguide is halved. The energy is simply back scattered towards the input, as it is in the case of simple one dimensional systems.

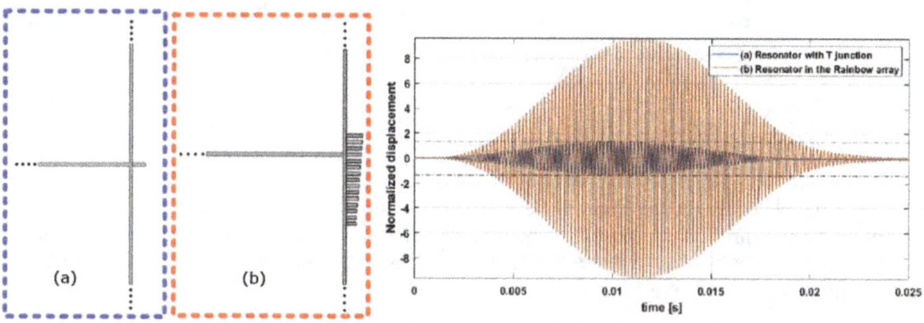

Fig.4: Normalized displacement field of resonators' tip for the case of junction with lateral resonator (a), and junction with rainbow reflection array (b).

The rainbow configuration can be modified both by changing the length of the lateral resonators and the spacing between them. Changing the latter, results in a drastic change in redirection efficiency. For an array with a distance between the resonators in the order of one eight of the wavelength (in the previous case it was 12 times) the redirection is 3% less than that for the single resonator case, while the energy displacement field of the resonator is still maintained at approximately 8 times the one of the single resonator. These results are reported in Fig. (5) and Fig. (6). The reason why the change in distance of the resonator is so important is thought to be linked to the conversion of the travelling wave from flexural to torsional. A torsional wave is better able to be redirected in the perpendicular waveguide. On this topic it has already been shown how a staggered configuration of resonators is able to achieve efficient mode conversion when the distance and the length of the resonators is properly set [21].

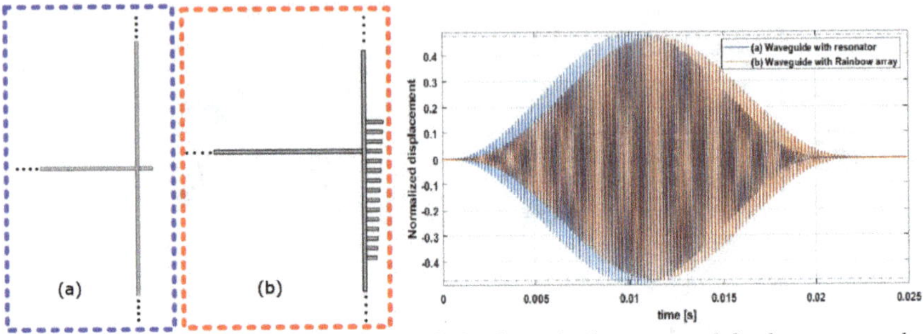

Fig.5: On the left the geometries analyzed. On the right the computed displacement on the perpendicular waveguide for the case of junction with lateral resonator (a), and junction with rainbow reflection array (b).

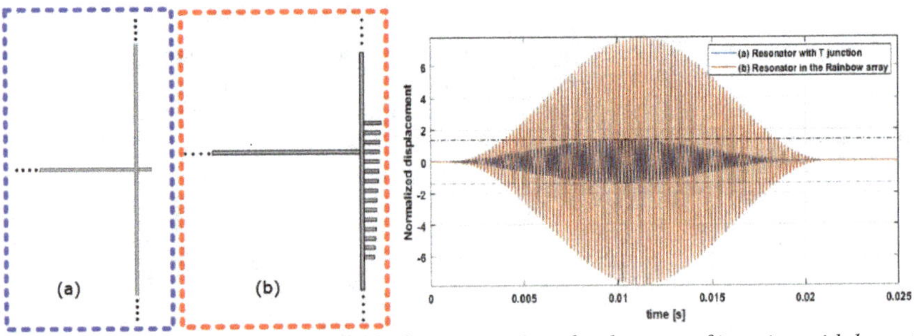

Fig.6: Normalized displacement field of resonators' tip for the case of junction with lateral resonator (a), and junction with rainbow reflection array (b).

In the end, we observe that the resonator and the resonators array are able to influence the wave redirection mainly through conversion of the flexural input wave mode into a propagating torsional mode. Further studies on this topic are necessary to assess whether this wave conversion effect is the key component in wave redirection.

Conclusions

The paper contains some studies aimed at evaluating the effectiveness of implementing different configurations of lateral resonators at 90 degrees junctions of infinite beams. It was showed how the presence of the resonators can be an advantage or disadvantage in terms of wave redirection. It was seen how a single resonator positioned at the junction point is able to partially influence the energy redirection towards the perpendicular waveguide and how its own motion is enhanced by the junction. For what concerns the rainbow reflection system, the results indicate that the single resonator is more effective with respect to the array if the resonators are closely packed together. But for the case of more spaced resonators the two configurations are comparable in wave redirection efficiency. Furthermore, the rainbow array allows for a great enhancement of the displacement of the target resonator. Further numerical and analytical studies must be performed to correctly understand the phenomena that define wave redirection.

References

[1] Pendry, J. B., Holden, A. J., Robbins, D. J. & Stewart, W. Magnetism from conductors and enhanced nonlinear phenomena. IEEE transactions on microwave theory and techniques 47, 2075–2084 (1999). https://doi.org/10.1109/22.798002

[2] Craster, R. V. & Guenneau, S. Acoustic metamaterials: Negative refraction, imaging, lensing and cloaking, vol. 166 (Springer Science & Business Media, 2012). https://doi.org/10.1007/978-94-007-4813-2

[3] Liu, Z. et al. Locally resonant sonic materials. Science 289, 1734–1736 (2000). https://doi.org/10.1126/science.289.5485.1734

[4] Craster, R. & Guenneau, S. World scientific handbook of metamaterials and plasmonics (2017). https://doi.org/10.1142/10642-vol2

[5] Tol, S., Degertekin, F. L. & Erturk, A. Gradient-index phononic crystal lens-based enhancement of elastic wave energy harvesting. Applied Physics Letters 109, 063902 (2016). https://doi.org/10.1063/1.4960792

[6] Tol, S., Degertekin, F. L. & Erturk, A. Phononic crystal luneburg lens for omnidirectional elastic wave focusing and energy harvesting. Applied Physics Letters 111, 013503 (2017). https://doi.org/10.1063/1.4991684

[7] Zareei, A., Darabi, A., Leamy, M. J. & Alam, M.-R. Continuous profile flexural GRIN lens: Focusing and harvesting flexural waves. Applied Physics Letters 112, 023901 (2018). https://doi.org/10.1063/1.5008576

[8] Allam, A., Sabra, K. & Erturk, A. Sound energy harvesting by leveraging a 3D-printed phononic crystal lens. Applied Physics Letters 118, 103504 (2021). https://doi.org/10.1063/5.0030698

[9] Wu, L.-Y., Chen, L.-W. & Liu, C.-M. Acoustic energy harvesting using resonant cavity of a sonic crystal. Applied Physics Letters 95, 013506 (2009). https://doi.org/10.1063/1.3176019

[10] Qi, S., Oudich, M., Li, Y. & Assouar, B. Acoustic energy harvesting based on a planar acoustic metamaterial. Applied Physics Letters 108, 263501 (2016). https://doi.org/10.1063/1.4954987

[11] Carrara, M. et al. Metamaterial-inspired structures and concepts for elastoacoustic wave energy harvesting. Smart Materials and Structures 22, 065004 (2013). https://doi.org/10.1088/0964-1726/22/6/065004

[12] Darabi, A., Ni, X., Leamy, M. & Al`u, A. Reconfigurable Floquet elastodynamic topological insulator based on synthetic angular momentum bias. Science advances 6, (2020). https://doi.org/10.1126/sciadv.aba8656

[13] Gonella, S., To, A. C. & Liu, W. K. Interplay between phononic bandgaps and piezoelectric microstructures for energy harvesting. Journal of the Mechanics and Physics of Solids 57, 621–633 (2009). https://doi.org/10.1016/j.jmps.2008.11.002

[14] Sugino, C. & Erturk, A. Analysis of multifunctional piezoelectric metastructures for low-frequency bandgap formation and energy harvesting. Journal of Physics D: Applied Physics 51, 215103 (2018). https://doi.org/10.1088/1361-6463/aab97e

[15] Colombi, A., Colquitt, D., Roux, P., Guenneau, S. & Craster, R. V. A seismic metamaterial: The resonant metawedge. Scientific reports 6, 1–6 (2016). https://doi.org/10.1038/srep27717

[16] Colombi, A. et al. Elastic wave control beyond band- gaps: shaping the flow of waves in plates and half-spaces with subwavelength resonant rods. Frontiers in Mechanical Engineering 3, 10 (2017). https://doi.org/10.3389/fmech.2017.00010

[17] Aguzzi, G. et al. Octet lattice-based plate for elastic wave control. Scientific reports 12, 1–14 (2022). https://doi.org/10.1038/s41598-022-04900-0

[18] Rajagopal, P., Drozdz, M., Skelton, E. A., Lowe, M. J. & Craster, R. V. On the use of absorbing layers to simulate the propagation of elastic waves in unbounded isotropic media using commercially available finite element packages. NDT & E international 51, 30–40 (2012). https://doi.org/10.1016/j.ndteint.2012.04.001

[19] De Ponti, J. M. et al. Graded elastic metasurface for enhanced energy harvesting. New Journal of Physics 22, 013013 (2020). https://doi.org/10.1088/1367-2630/ab6062

[20] De Ponti, J. M. et al. Experimental investigation of amplification, via a mechanical delay-line, in a rainbow - based metamaterial for energy harvesting. Applied Physics Letters 117, 143902 (2020). https://doi.org/10.1063/5.0023544

[21] De Ponti, J. M. et al. Selective mode conversion and rainbow trapping via graded elastic waveguides. Physical Review Applied 16, 034028 (2021). https://doi.org/10.1103/PhysRevApplied.16.034028

Design of architected materials composed by periodic surfaces

Massimo Cuomo[1], Golshan Farzi[1], Roberto Ruggeri[1], Leopoldo Greco[1]

[1]Department od Civil Engineering and Architecture, University of Catania - Italy

E-mail: mcuomo@dica.unict.it, leopoldo.greco@unict.it

Keywords: Metamaterials, Minimal Surfaces, Large Deformations

Abstract. Architected materials possess extraordinary mechanical properties, that cannot be met by natural materials, either in the static and in the dynamic range. For this reason, the study of metamaterials is a very active field of research. Although lattice architectures are commonly preferred, especially for the relative simplicity of manufacturing, geometries based on thin shells may present considerable advantages. Indeed it has been shown that shell metamaterials, still preserving lightness and versatility, can present excellent stiffness and strength properties, and ability to absorb strain energy by means of very large deformations, that make them useful for the design of shock absorbing elements. In the contribution will be examined a class of shell metamaterials, composed of triple periodic minimal surfaces TPMS, that having zero mean curvature can be in equilibrium with an isotropic state of stress. Two particular surfaces, the Schwarz P and the Schwarz G minimal surfaces, the first centro-symmetric, the second chiral, are examined.

Introduction

Metamaterials represent an innovative approach to the problem of obtaining unusual or extreme physical responses for advanced applications. Since their extreme macroscopic responses depend primarily on the internal low scale pattern, the understanding of how the microstructure topology influences the macroscopic properties is the key-point in the design of new advanced metamaterials.

Metamaterials are usually classified into three-dimensional and surface (planar) materials and are frequently based on elementary geometric patterns.

Periodic cellular materials are characterised by a unit cell that can be translated through them. If the cell is translated in two dimensions, they are designated prismatic cellular materials (e.g., honeycomb), while if the cell is translated in three dimensional periodicity, then they give rise to cellular structures.

A common typology of periodic material is constituted by truss lattices, in which case the unit periodic cell can assume several forms, that have been widely investigated. Hutchinson and Fleck (2006), Thomsen et al. (2018) performed a topological optimization of 2D periodic materials undergoing buckling type instabilities. Moreover, the non-linear response of planar periodic materials has been analyzed by Vigliotti et al. (2014) [6].

Shell lattice structures

AM has enabled the design and manufacturing of cellular structures whose unit cells are composed of plates or shells rather than struts. These lattice structures are commonly described as Triply Periodic Minimal Surfaces (TPMS)-like (though their surfaces do not necessarily have zero mean curvature) and are referred to as "shell lattices".

In this way cellular materials are obtained, that can be closed or open as in natural foams. Manufacture of closed cell plate lattices remains problematic for powder-based AM systems due to the requirement of powder removal (see figure 1, from [1]). Open cells can be obtained exploiting the properties of the geometry of minimal surfaces.

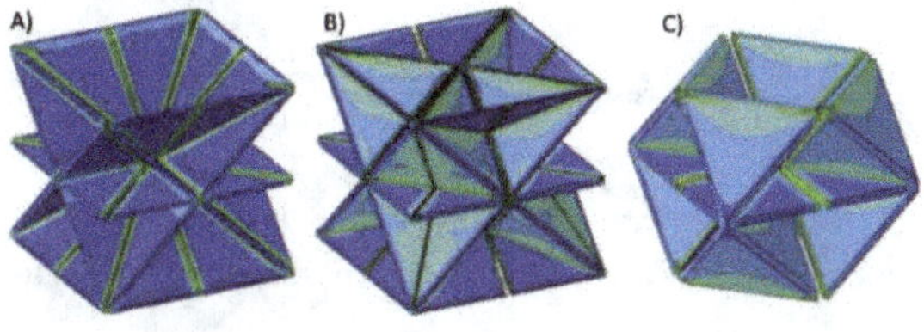

Figure 11. shell lattice cells generated by placing plates on the closest-packed crystals combining simple cubic (SC), BBS and SC-BCC-FCC (B) and SC FCC (C)[1]

Triply periodic minimal surface (TPMS)

A minimal surface is a surface that has a mean curvature equal to zero at every point. When the minimal surface is infinite and periodic in the three independent directions it is called a triply periodic minimal surface (TPMS). TPMS are surfaces that locally minimize surface area for a given boundary [5].

These surfaces have distinctive geometrical characteristics: a minimal surface is smooth everywhere, has no sharp edges or corners, and splits the space into two or more non-intersecting, intertwined and infinite domains that can be repeated periodically in three perpendicular directions.

The first examples of TPMSs were discovered by Schwarz in 1865, followed by his student Neovius in 1883. They described five TPMSs, namely, Schwarz primitive (P), Schwarz diamond (D), Schwarz hexagonal (H), Schwarz crossed layers of parallels, and Neovius (N). The gyroid (G) surface was described by Schoen in 1970, along with another eleven newly discovered TPMSs. In many ways, the G, D, and P surfaces are the most important most commonly observed TPMS structures.

TPMS are widespread in nature. One outstanding example are the cubic membranes, biological membranes formed from lipid bilayer sheets, highly curved. They not only separate the inside and outside of an organism but are also involved in many biological processes, such as selective permeability and energy production.

Single gyroid geometries have been widely discovered in nature, such as alveolar surface of mammalian lung , the prolamellar bodies in plant cells , intracellular cubic membrane in cell organelles , wing scales of green butterflies and exoskeletons of beetles. The SG structure found in the green butterflies is of special interest, exhibiting a complete photonics bandage and polarization dependent optical properties and negative refractions due to inherent chirality.

TPMS can be divided into balanced and unbalanced surfaces. Examples of balanced surfaces are the Gyroid and Primitive surfaces that divide the volume into congruent or interchangeable regions that may (Gyroid) or may not (Primitive) be mirror images of the one another. Unbalanced surfaces such as the I-WP divide the space into labyrinths of unequal morphologies [f14]-[5].

Minimal surfaces can be found by means of form-finding algorithms [cite Bletzinger], However, useful approximations can be obtained using level-set equations derived from a sum defined in terms of the Fourier series [5].

$$\sum_{i=1}^{N} F_i(\boldsymbol{k}_i)\cos(2\pi\boldsymbol{k}_i\cdot\boldsymbol{r}-\boldsymbol{\alpha}_{ki})=t$$

where F is the amplitude, $\boldsymbol{k}_i$ a parameter that defines the relative dimensions of the periodic cell in the three directions, and α a phase parameter. Particularly simple expressions can describe P D and G surfaces. In this paper we focus on gyroid and Schwarz primitive geometries, whose approximated expressions are given by

Schwarz primitive:

$$g_P = \cos\frac{2\pi x}{a} + \cos\frac{2\pi y}{a} + \cos\frac{2\pi z}{a} = t$$

Gyroid

$$g_G = \sin\frac{2\pi x}{a}\cos\frac{2\pi y}{a} + \sin\frac{2\pi y}{a}\cos\frac{2\pi z}{a} + \sin\frac{2\pi z}{a}\cos\frac{2\pi x}{a} = t$$

The constant t is denoted as the level parameter. The best approximation to minimal surfaces is obtained setting t=0.

Varying the value of the level parameter, two families of surfaces are obtained. In the case of P surface, $t \in (-1,1)$, while in the case of the G surface $t \in (-\pi, \pi)$. Figures 2,3 show the surfaces that can be obtained for some values of the level parameter. While for the P surface the sign is critical, it is inessential for the G surface, given its rotational symmetries.

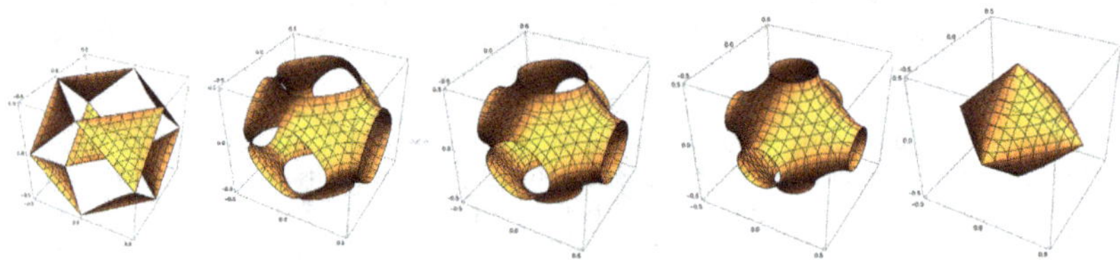

Figure 2. Isolevel surfaces obtained from expression (2), for t=-1, -0.5, 0, 0.5 1. The Schwarz P surface is the one corresponding to t=0.

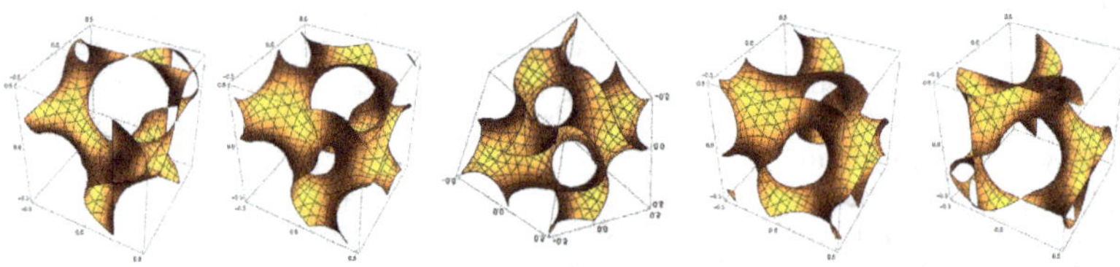

Figure 3. Isolevel surfaces obtained from expression (3)), for t=-1, -0.5, 0, 0.5 1. The Single Gyroid surface is the one corresponding to t=0.

Unit cell for metamaterials can be obtained assigning a thickness to one of the iso-level surfaces. However, this procedure is safe only for very thin surfaces, otherwise there is the risk of intersections. A consistent way for building solid microstructures is obtained using the analytical expressions (2), (3). Since minimal surfaces divide the space in two non connected regions, it is possible to build a so called solid network as the region of space for which $g_G \leq t_1$ (similarly for the P surface). A sheet solid can then be obtained from the intersection of two solid networks with close values of the level parameter. It is observed that replicating the procedure, it is possible to build hierarchical minimal surface unit cells. This aspect will be object of future investigations.

TPMSs are obtained replicating the unit cells in three orthogonal directions. Figure 4 illustrate the procedure. It is interesting to underline that choosing values of the level set parameter close to the extreme values one can approach plate lattices or rod lattices. This possibility makes the present approach very appealing for optimization procedures.

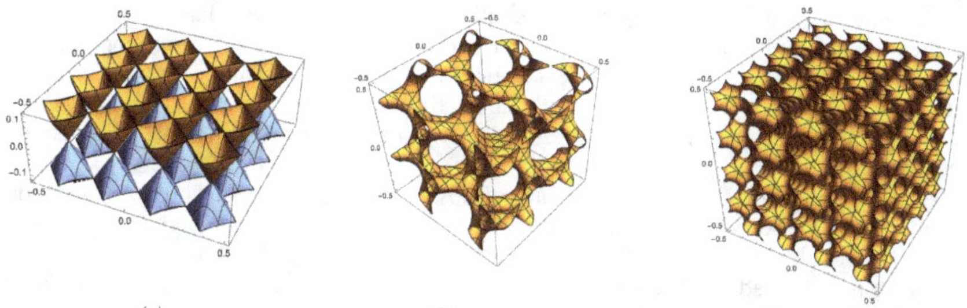

(a) (b) (c)

Figure 4. TPMS obtained as repetition of unit cells. (a) P surface with t=-1, (b) G surface with t=1, (c) G surface with t=-0.25

Among all possible lattices, the unique geometry of the Gyroid minimal surface and its related morphologies makes Gyroid structured materials a fascinating subject across many disciplines including biology, physics, optics and topological photonics. Gyroids, like most biomimetic designs, are mechanically robust as evolution of self-assembly leads to mechanically stable structures.

Gyroids present cubic symmetry and chirality. They bisect space into a pair of 3D labyrinths, one left-handed (LHD) and one right-handed (RHD). These channels resemble a spiral-like shape and are indeed chiral [F20]. With the left- and right-handed channels having opposing handedness, the infinitely periodic gyroid surface has zero net geometric chirality due to equal amounts of LHD and RHD curvature

Physical and Mechanical properties of TPMS-Gyroid structure in the linear range

From an engineering point of view, the main properties that need to be determined in order to investigate the linear response of the material are density and stiffness. Next some preliminary observations about them will be outlined. Similar results can be found in [..] for the density, and in [Maskery] for the elastic properties.

Figure 5a shows a plot of the relative density for a P sheet network as a function of the level parameter. The relative density is defined as the volume occupied by the sheet with respect to the volume of the bounding cube. Three different values of the thickness are considered, viz., $h/L = 0.05, 0.1, 0.15$, with L the size of the bounding cube. The density is relatively independent from the level parameter, and of its order of magnitude can be approximated as 1.5 h.

A similar result is found for the gyroid, also in this case the relative density is almost constant with the level parameter.

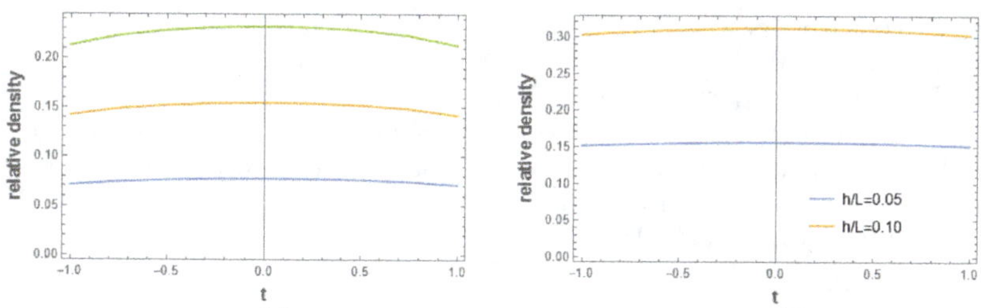

(a) (b)

Figure 5 Relative density of sheet networks (a) P surface (b) G surface

Materials Research Proceedings **26** (2023) 535-539
https://doi.org/10.21741/9781644902431-87

Many investigations have focused on the compressive behaviour of TPMS. Maskery et al.. Wang et al. Yan et al. A stable and progressive buckling collapse is observed during compression, and absorbed energy and crushing load are affected by relative density. Li et al. conducted a comparison of gyroid structures with different lattices at quasi-static loading Ref. Their results highlighted that the gyroid lattice structures with the relative density of 20% and 30% absorb more energy per unit mass than the other structures, indicating the potential application in protective structures. [Abueidda]

Maskery et al.[The deformation and elastic anisotropy of a new gyroid-based honeycomb made by laser sintering I. Maskery*, I.A. Ashcroft] and Abueidda et al. [Effective conductivities and elastic moduli of novel foams with triply periodic minimal surfaces Diab W. Abueidda a,b, Rashid K. Abu Al-Rub a,*, Ahmed S. Dalaq a, Dong-Wook Lee a, Kamran A. Khan c, Iwona Jasiuk] among others studied the mechanical performance of three TPMS structures: Primitive, Gyroid, and Diamond. They showed that the Primitive structure has higher elastic modulus than the other two TPMS designs. The Primitive type also exhibits strut stretching and buckling, as opposed to the bending dominated deformation modes observed in other TPMS.

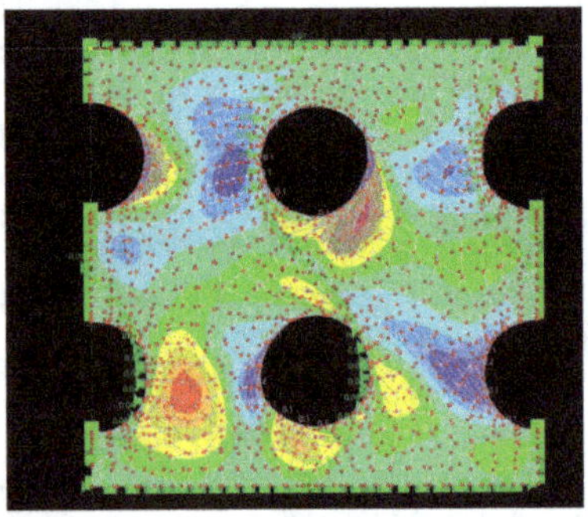

Figure 7 shows the horizontal displacements of a unit gyroid cell to uniaxial stretch. At odd with P cells, the gyroid tends to rotate in addition to deform. This aspect has been not observed in previous studies, indicating the need of using non standard material models for interpreting the behaviour of lattice gyroid metamaterials, opening the way to very interesting applications.

References

[1] Ruggeri, R., “Design of tissue engineering scaffolds based on hyperbolic surfaces: Structural numerical evaluation”, Medical Engineering & Physics 36 1033–1040 (2014.) https://doi.org/10.1016/j.medengphy.2014.05.006

[2] X. Guo, J. Ding, X. Li, . Qu, X. Song, J. Y. Hsi Fuh, W. F. Lu, W. Zhai, “Enhancement in the mechanical behaviour of a Schwarz Primitive periodic minimal surface lattice structure design, International Journal of Mechanical Sciences 216 (2022) 106977. https://doi.org/10.1016/j.ijmecsci.2021.106977

Novel stochastic dynamics methodologies & signal processing techniques for civil engineering applications

Theoretical and Applied Mechanics - AIMETA 2022 Materials Research Forum LLC
Materials Research Proceedings **26** (2023) 543-548 https://doi.org/10.21741/9781644902431-88

A distributed analysis of vibration signals for leakage detection in Water Distribution Networks

Gabriele Restuccia[1,2,a], Ilenia Tinnirello[1,2,a], Fulvio Lo Valvo[1,a], Giacomo Baiamonte[1,a], Domenico Garlisi[1,2,a*], and Costantino Giaconia[1,2,a]

[1]University of Palermo, IItaly –

[2]CNIT, Parma, Italy

[a]{name}.{surname}@unipa.it

Keywords: Smart Water Distribution Network (SWDN), LoRaWAN, Vibrations Sensor

Abstract. It is well known that Water Distribution Networks (WDNs) are very inefficient and, in Italy, 40% of water is lost during distribution. In this paper, we present a solution for detecting leakages in WDNs, based on three main components: i) an innovative sensing element to be deployed at the sensor nodes, which analyses vibrations in the acoustic range for classifying external noise sources, induced by water leakages, by means of suitable machine learning techniques; ii) an Internet of Things (IoT) system of sensors, deployed at the junctions of the WDNs, for comparing the measurements collected at different critical points of the network; iii) a machine learning algorithm for processing the data. After the definition of the WDN structure, we introduce some numerical simulation tools suitable for studying our system and modeling the proposed sensing solution. Given the geometry, physical properties (pipe lengths, diameters, roughness, reservoir shapes and levels, pump and valve characteristic curves) and nodal demands, the simulation tool is able to compute leakages in pipes or nodes over time. In parallel, we simulate our IoT system coupled to the WDN, by logging partial information about the WDN status, which corresponds to the demand readings at the edge nodes or at some junction nodes, together with the (optional) measurements of the deployed sensing elements. On the basis of this data, we analyze the possibility of identifying the leakages in the network, even without knowing the exact or complete topology of the WDN. Our solution exploits different machine learning techniques devised to indirectly retrieve topological information, by correlating the balance of the flows as the water demand varies over time.

Introduction

A Water Supply System (WSS) is the infrastructure that connects water suppliers and customers to provide water to households and businesses. A conventional WSS comprises a water source, a water treatment, a pumping station, a Water Distribution Network (WDN), and finally, the end users. Over the past 100 years, the global water consumption has increased sixfold and this trend has continued at a rate of about 1% per year due to population growth and to the economic development [1]. According to the current literature, the world may face a global water deficit of 40% by 2030. At the same time, massive amounts of water are wasted because of leakages in (aging) water distribution infrastructures. According to the more recent statistics of the European associations of water services (EurEau), the mean value for water wasted is about 26% in Europe [2], but such a value can be even higher than 50% in some networks. In this context, it is of paramount interest to exploit ICT solutions for improving the monitoring and management of WSSs. Indeed, most water utilities address the leakage control in a passive way; leaks are repaired only when they are visible. A standard way to facilitate the leakage control is to partition the network into District Metered Areas (DMAs), where the flow and the pressure at the inlet are monitored [3]. Recently, classifiers have been proposed to analyze the residuals in [4]. In this paper, we focus on the distribution network of the WSS and, in particular, on the problem of leakage detection and localization. We consider three main technical solutions for addressing this problem: (i) the utilization of embedded devices to be deployed at the nodes and along the transportation pipes, for collecting demand measurements and acquiring vibration signals due to the water flows; (ii) a local form of intelligence, running at the gateway nodes collecting information about the vibration signals, for removing noise sources (such as the vibration signals caused by a vehicle on the road) not related to water leakages; (iii) an high-level aggregation scheme, responsible of mapping the data collected from the whole network or DMA (demand, pressure, presence of leakage) into a per-node estimate of presence/absence of leakages. For designing our scheme, we used a small-scale testbed, devised to optimize the local intelligence for filtering the vibration measurements, and a well known WSS simulation tool, called EPANET, for large-scale analysis. The simulator has been extended for emulating the innovative sensing system based on acoustic signals. Our solution has some interesting practical implications. First, it does not require a complete knowledge of the WSS topology and any calibration of an hydraulic model. Second, it only needs data from normal operation behavior. The remainder of the paper is organized as follows. An analysis of the state of the art can be found in Section 2. The design of the vibration sensor is presented in Section 3, together with the experimental results in the local testbed. In Section 4 we provide performance results on the considered large scale scenarios, and discuss the results of our simulations. Finally, some conclusions are drawn in Section 5.

Related work

A leakage detection technique using acoustic emissions is based on the principle that water flows due to losses passing through a perforation in the pipe create an acoustic signal [5]. When a leak occurs, acoustic sensors installed outside the pipe, track and detect the acoustic signal as it propagates along the pipeline. Acoustic signals are greatly influenced by the distance between the perforation and the sensor, as well as by the background noise from the environment and those produced by a pipe burst. Therefore, their utilization requires some signal processing techniques, which can be expensive in large-scale networks. An alternative approach for monitoring WDNs is the usage of Micro Electro-Mechanical System (MEMS). Indeed, MEMS are low-cost solutions, with a relatively small dimension and low weight, which make them suitable for identifying leakages by means of vibration signals. MEMS accelerometers can have a range of acceleration from 0.5g to 200g (being "g" the gravity acceleration) and a bandwidth spacing from 10 Hz to order of tens kHz, which make possible the analysis of most type of vibration signals in the WDN field. In some cases, sensing can exploit both acoustic and vibration signals [6]. In [7], a method to identify leaks is proposed for identifying leakages even when they are far from the sensing points (blind spots), while [8] proposes models of buried pipelines to estimate wave velocities. In our work, we propose a small footprint of the sensing devices, assuming that they can be installed inside a pipe during repairs with a no-dig technology; in perspective, we can envision a scenario where pipes are continuously monitored against water leaks. Apart from the sensing and noise filtering problem, several works focus on data aggregation for leak localization. For this problem, black-box models based on Machine Learning (ML) are becoming a dominant solution. Different approaches, such as Support Vector Machines (SVMs), k-NN classifiers or multi-layer neural networks have been investigated in [9] [10] [11], respectively. We also exploit ML for aggregating the data collected by the sensors deployed within the network.

Leakage detection using vibration signals and AI on board

In this section, we describe the design and implementation of our sensing element for water leakages, based on accelerometer measurements and on embedded processing. We chose a commercial device by STMicroelectronics, with a low noise MEMS-based accelerometer IIS3DWB, featuring a programmable full-scale measurement and set up to measure in a range of

Materials Research Proceedings **26** (2023) 543-548 https://doi.org/10.21741/9781644902431-88

±2g in each of the three axes. The system works with a sampling rate of 26.71 kHz, thus providing a new measurement every 37μs. Acquired data are digitally converted with a 16-bit Analog to Digital Converter (ADC) integrated into the accelerometer, reaching a nominal resolution of about 61μg. Since terrestrial gravity, affecting vertical direction, contains no information related to pipe vibrations, it's possible to remove the terrestrial acceleration component to maximize precision. The sensor is interfaced with a low power 32-bit microcontroller, namely STM32L4R9ZI by STMicroelectronics as well. The microcontroller is responsible of running the processing tasks for extracting some relevant features from a temporal sequence of measurements and classifying the sequence as a leakage/non-leakage event. Indeed, as discussed in [12] by Youngseok et al., the spectrum of the vibration signal in case of regular functioning of the pipe is quite different from the one acquired during the leakage: in presence of leakages, there are some frequential components in the range from 400 Hz to 5 kHz, according to the entity of the leakage and the dimension and material of the pipe. The amplitude of the sensed vibration varies based on the distance between the leakage and the sensor and it can change depending on the amount of wasted water. Since the vibrations are in the acoustic range, we designed a spectrum analysis scheme based on typical solutions for speech recognition: More into details, we extracted the Mel Frequency Cepstral Coefficients (MFCCs) [13] of different groups of measurements. We organized consecutive samples of the accelerometer measurements into temporal windows of 0.46 s (corresponding to 6 groups of 2048 measurement samples). The sequence is divided into 9 sub-windows of 2048 measurements, partially overlapping, and each window is converted in 40 MFCCs with a float precision.

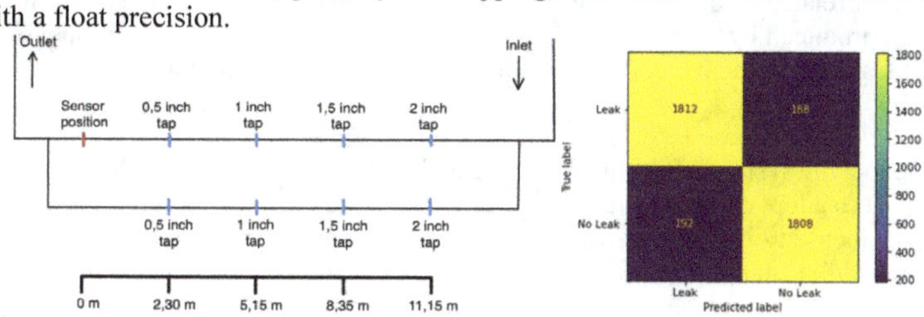

Figure 1 - Scheme of the used pipe system (left), and confusion matrix of the model performance (right).

The MFFCs matrix of 40x9 coefficients is then used as an input of a multilayer perceptron network, with two 256-neurons fully connected layers by a flattening layer. In order to classify the input as affected by a leakage, the output of the algorithm produces a number between 0 and 1 which represents the probability of a regular working condition of the sensed pipe. For the purpose of classification, outputs greater than or equal to 0.5 have been classified as "no leak" while outputs less than 0.5 are considered "leak".

Smart pipe performance evaluation in real testbed scenario

For training our model, we collected some experimental data by exploiting a small-scale testbed with real pipes and taps to be used as controlled leakages. The taps are different in diameter and distance from the sensor, as depicted in Figure 1(left). During data acquisition, the electronic system has been placed on the pipe and various scenarios, in terms of taps opened over time, have been recorded as a ground-truth, together with the relevant measurements' traces. The collected data has been split in training, validation and test set, and then used to train the model and evaluate its performances. The tested topology produced a model that is able to correctly classify samples extracted from a leaking condition 90.4% of the time and 90.6% of the time in the case of a regular functioning pipe, as shown in the confusion matrix represented in the rightmost part of Figure 1.

After the evaluation, the model has been tested on the complete recorded scenarios, even with unseen scenarios, showing promising results. The performance of the algorithm can be improved by introducing some memory effects: indeed, in case of leakages, it is very likely that the leakage will last for long time intervals and therefore consecutive outputs can be averaged for filtering sporadic classification errors.

Leakage detection in large scale scenario

For analyzing the possibility of identifying leakage events in real WDNs, we worked on simulation, where it is relatively easily to build a complex topology of a water grid, comprised of a number of links connected together to form loops or branches. Each link may contain one or more pipes (with different diameters) connected in series. A pipe is a segment of a link that has a constant flow with a constant diameter, and no branches. Two or more pipes are joined by node elements. Water can leave the network at these nodes, where we also assume that smart readings of water demand are possible. For simulating a realistic scenario, we made some assumptions on the sensors and meters available in the system, as well as on the communication networks deployed for transporting the data to a central aggregation server (working at different scales, including a district-level scale). These assumptions are used for filtering the whole trace produced by the simulator (which characterizes the complete view of the system) into a set of measurements visible to the leakage detection system. The system works on the basis of two ML models: a low level one, which only runs on the sensors for processing acceleration measurements and identifying leakage events related to the node where the sensor is deployed; a high-level one, which uses all the readings produced by smart meters and sensors and flow balancing conditions for identifying the overall leakages in the network (even where sensors are not available).

Communication Network

We assume that the system works thanks to IoT technologies. According to the WDN architecture, we can expect to adopt cellular technologies, or technologies working in unlicensed bands. For our analysis, we focused on the usage of LoRaWAN [14], a wireless technology offering low-cost, low-power, and long-range communications (up to a few kilometers). The architecture is based on a start of star deployment, where End Devices (EDs) transmit data to multiple nearby Gateways (GWs) deployed in their covered area, which in turns forwards frames to the Network Server (NS) and eventually to the corresponding IoT applications server. We assume that smart readers and sensing elements act as EDs, equipped with a LoRaWAN interface, and that at least one GW is able to collect the measurements provided by the EDs.

Water system simulator

Our work is based on the Water Network Tool for Resilience (WNTR) [15]. The tool is a Python package designed to simulate and analyze the resilience of water distribution networks based on EPANET, an open source software for modeling hydraulic and quality dynamics of a WDN) [16]. WNTR was developed to extend the capabilities of EPANET and simulate the dynamics of water flows across pipelines, taking into account bulk flows and pipe wall reactions, as well as the availability of water sources and reservoirs. WNTR has an application programming interface (API) that is flexible and allows the configuration of the network topology and the scheduling of disruptive incidents and recovery actions. We use the WNRT simulator to generate a suitable dataset to feed and test the ML approach. On one side, the simulator generates a complete trace, with the status of each node over time, by recording a log file containing the following fields: *timestamp, nodeID, demand, head (elevation + pressure head), pressure, has_leak, smart_pipe*. On the other side, we filter a subset of data corresponding to smart meters and sensors, together with the real status of nodes, for training a ML model whose input is the total set of measurements in a given time window, and whose output is a vector with the status leakage/non-leakage of each node. An interesting feature of our model is the lack of input data related to the network topology. Indeed, topological information are indirectly retrieved thanks to the balancing of water flows over time, in presence of time varying water demands. For testing our approach, we used some pre-defined network models offered by the Open Water Analytics's community public repository. Specifically, we considered three different sub-networks [17], named A, B, and C, with an increasing level of complexity (in terms of the total amount of nodes). At edge nodes, we also defined a set of demand-patterns, which represent water consumption at different timesteps. We also considered an interval of one hour as minimum time interval for changing the demand patterns. In one half of the nodes, we added the possibility of adding leakage events. We then run simulations for each network topology and for different demand-patterns, lasting a total duration of one week or one month. Pipes with leaks are chosen from a pseudo-random function in the simulation script (coded in Python) and feed into the entire simulation's time intervals.

Table 1 - Parameters of three considered network.

Network	Total network demand [gal/min]	Leak demand [%]	# of nodes	# of reservoirs	# of nodes with leakage	# of mart pipes
A	0.00333947	43.45	83	1	41	20
B	0.13837243	48.74	1038	1	518	259
C	0.30645519	42.38	2099	2	1048	524

Table 1 shows the list of the parameters present in each WDN network used to generate the dataset. For each network, we report the total amount of the demand from the nodes and the percentage of nodes involved in water leakages. The table also summarizes the total number of network nodes, the total number of reservoirs, the number of nodes with leakages and the number of installed smart pipes (equipped with the sensing element working on acceleration data). We represent our use case as a Supervised Classification problem in which we try to predict if there is a leakage on the pipe. From the simulation trace, we extracted our training set with the following features: *demand, head, pressure, smart_pipe;* this latter feature can assume three different values: 0, if the vibration sensor is present and a leakage is not detected; 1, if the vibration sensor is present and the leakage is detected; 2 when there is no sensor. Since this is a classification problem, our test set also contains the status of each node, in the boolean variable *has_leak.*

Performance results

In order to choose the most adequate classification model, we applied different techniques and confronted them. We tested the following classificators: KNeighborsClassifier, LinearSVM, RBFSVM, DecisionTree, RandomForest, AdaBoostClassifier, GaussianNaiveBayes. Models have been compared in terms of accuracy on the node classification. We analyzed accuracy results for the trained model, in the three different network topologies (A, B and C), when only one week of data is available and without the usage of vibration sensors. Accuracy values are calculated after the fitting and prediction phases and after the confusion matrices generation. Specifically, we extract the *True Positives (TP)*, *True Negatives (TN)*, *False Positives (FP)* and *False Negatives (FN)* values and calculate the accuracy as seen here: (TP + TN) / (TP + TN + FP + FN). Each classificator is evaluated through a series of k={7,4} cross-folds. The maximum accuracy **(about 70%)** is reached for **network A**, which has a lower number of nodes and therefore a simpler topology to be inferred. The presence of the vibration sensor in all the considered networks, improve the accuracy of the system about the **10%**. Finally, we analyze the impact of a longer data trace, corresponding to a whole month of data, for training the system equipped with the smart pipes. We note that longer traces are able to solve most of topological ambiguities and achieving very high accuracy (as high as **98%** for a model based on Decision Trees) even for the more complex network topologies B and C.

Conclusion

This work presented a system architecture and a data-drivn approach to detect leakages in WDN. A LoRaWAN IoT network is considered to collect information at WDN nodes, where water demands are measured over time. Moreover, in a sub-set of nodes we also assume to deploy innovative sensors, equipped with embedded intelligence, for processing accelerometer measurements and identify leakage events. We demonstrate that demand measurements over time can be exploited for retrieving topological information of the network thanks to the balance conditions among flows. Starting from a realistic trace of water demand measurements and optional smart sensor readings, a model can be trained for identifying leakage events at each network node. Although our model has been trained and validated by exploiting numerical simulation, we expect that results can be easily generalized to real deployments.

Acknowledgments

This work has been partially financed by the Project “TiSento” (Azione 1.1.5. - POC Sicilia 2014/2020 Asse 1 - PO FESR 2014/2020).

References

[1] WWAP: World Water Assessment Programme, The United Nations world water development report 2020: water and climate change:WWDR2020,978-92-3-100371-https://unesdoc.unesco.org/notice?id=p::usmarcdef_0000372985

[2] Europe’s Water in Figures An overview of the European drinking water and waste water sectors 2021 edition EurEau ISBN 978-2-9602226-3-0.

[3] Puust, R.; Kapelan, Z.; Savi´c, D.A.; Koppel, T. A review of methods for leakage management in pipe networks. Urban Water J. 2010, 7, 25–45. https://doi.org/10.1080/15730621003610878

[4] Khulief, Y.; Khalifa, A.; Mansour, R.; Habib, M. Acoustic Detection of Leaks in Water Pipelines Using Measurements inside Pipe. J. Pipeline Syst. Eng. Pract. 2012, 3, 47–54. https://doi.org/10.1061/(ASCE)PS.1949-1204.0000089

[5] Adedeji, Kazeem & Hamam, Yskandar & Abe, B.T. & Abu-Mahfouz, Adnan. (2017). Towards Achieving a Reliable Leakage Detection and Localization Algorithm for Application in Water Piping Networks: An Overview. IEEE Access. 5. 20272 - 20285. https://doi.org/10.1109/ACCESS.2017.2752802

[6] Fuchs, H.; Riehle, R. Ten years of experience with leak detection by acoustic signal analysis. Appl. Acoust. 1991, 33, 1–19. https://doi.org/10.1016/0003-682X(91)90062-J

[7] Yang, J.; Wen, Y.; Li, P. Leak location using blind system identification in water distribution pipeline. J. Sound Vib. 2008, 310, 134–148. https://doi.org/10.1016/j.jsv.2007.07.067

[8]. Muggleton, J.; Brennan, M.; Pinnington, R. Wavenumber prediction of waves in buried pipes for water leak detection. J. Sound Vib. 2002, 249, 939–954. https://doi.org/10.1006/jsvi.2001.3881

[9]. Ferrandez-Gamot, L.; Busson, P.; Blesa, J.; Tornil-Sin, S.; Puig, V.; Duviella, E.; Soldevila, A. Leak Localization in Water Distribution Networks using Pressure Residuals and Classifiers. IFAC-PapersOnLine 2015, 48, 220–225. https://doi.org/10.1016/j.ifacol.2015.09.531

[10]. Wachla, D.; Przystalka, P.; Moczulski, W. A Method of Leakage Location in Water Distribution Networks using Artificial Neuro-Fuzzy System. IFAC-PapersOnLine 2015, 48, 1216–1223. https://doi.org/10.1016/j.ifacol.2015.09.692

[11]. Verde, C.; Torres, L. Modeling and Monitoring of Pipelines and Networks: Advanced Tools for Automatic Monitoring and Supervision of Pipelines; Springer: Berlin, Germany, 2017. https://doi.org/10.1007/978-3-319-55944-5

[12] Kim, Youngseok & Jung, Haewook & Ryou, Jaesuk & Choi, Jaehyuk. (2021). A Basic Experimental Study on Analysis of Leak Signal and Monitoring Method for Water Supply Pipe. Applied Sciences. 11. 2097. https://doi.org/10.3390/app11052097

[13] Ittichaichareon, C., Suksri, S., & Yingthawornsuk, T. (2012, July). Speech recognition using MFCC. In International conference on computer graphics, simulation and modeling (Vol. 9).

[14] LoRa Alliance Technical Committee, “LoRaWAN 1.1 Spec.” https://lora-alliance.org/resource-hub/lorawantm-specification-v11, 2017.

[15] Klise, K.A.; Bynum, M.; Moriarty, D.; Murray, R. A software framework for assessing the resilience of drinking water systems to disasters with an example earthquake case study. Environ. Model. Softw. 2017. https://doi.org/10.1016/j.envsoft.2017.06.022

[16] Jun, L.; Guoping, Y. Iterative Methodology of Pressure-Dependent Demand Based on EPANET for Pressure-Deficient Water Distribution Analysis. J. Water Res. Plan. Manag. 2013, 139, 34–44. https://doi.org/10.1061/(ASCE)WR.1943-5452.0000227

[17] https://github.com/gabrielication/py_epanet2csv/tree/main/networks

Theoretical and Applied Mechanics - AIMETA 2022 Materials Research Forum LLC
Materials Research Proceedings **26** (2023) 549-554 https://doi.org/10.21741/9781644902431-89

Fractional differential equations under stochastic input processes handled by the improved pseudo-force approach

Alba Sofi[1,a*], Giuseppe Muscolino[2,b] and Mario Di Paola[3,c]

[1]Department of Architecture and Territory, University "Mediterranea" of Reggio Calabria, Via dell'Università 25, 89124 Reggio Calabria, Italy

[2]Department of Engineering, University of Messina, Villaggio S. Agata, Messina, Italy

[3]Department of Engineering, University of Palermo, Viale delle Scienze Ed. 8, 90128 Palermo, Italy

[a]alba.sofi@unirc.it, [b]gmuscolino@unime.it, [c]mario.dipaola@unipa.it

*corresponding author

Keywords: Fractional Differential Equations, Stochastic Processes, Step-By-Step Integration

Abstract. This paper presents a step-by-step procedure for the numerical integration of the fractional differential equation governing the response of a single-degree-of-freedom (SDOF) system with fractional derivative damping. The procedure is developed by extending the *improved pseudo-force method* proposed by the second author for the numerical integration of classical differential equations. To this aim, the *Grünwald–Letnikov* approximation of the fractional derivative is adopted. The proposed numerical procedure is exploited to compute response statistics of a SDOF system subjected to stochastic excitation by applying classical *Monte Carlo Simulation.*

Introduction

Within a deterministic setting, fractional differential equations are usually solved by using the step-by-step *Grünwald–Letnikov* (*GL*) procedure [1,2]. The main issue of this approach is that at each new time step a new term appears in the summation representing the *GL* approximation of the fractional derivative (long tail memory) [2]. Moreover, in order to achieve good accuracy, a very small time step is necessary. This implies an increase of the computational effort which becomes prohibitive in stochastic analysis, especially when *Monte Carlo simulation* (*MCS*) is applied.

In 1996, Muscolino [3] proposed an *improved pseudo-force method* (*IPFM*) for evaluating the solution of linear and non-linear classical differential equations by a step-by-step procedure. This method requires two main steps: i) to consider some (linear or non-linear) terms of the differential equation, depending on the structural response at the current time instant, as pseudo-forces; ii) to accordingly modify the so-called *fundamental* (or *transition*) *matrix* as well as the forcing terms involved in the step-by-step procedure. In this way, it is possible to increase the size of the time step of more than one order of magnitude with respect to classical step-by-step integration schemes, like the *finite difference method* (*FDM*).

In this study, the *IPFM* is revisited and properly adapted to perform the numerical integration of the differential equation governing the response of a single-degree-of-freedom (SDOF) system with fractional derivative damping. Then, the *IPFM* is applied to evaluate response statistics of fractionally damped SDOF systems subjected to stochastic excitation by means of classical *MCS*. The *IPFM* can be extended to multi-DOF systems as well as to non-linear problems.

Grünwald–Letnikov definition of fractional derivative

The *fractional calculus* started with the definition of the *Riemann–Liouville* (*RL*) fractional integral. After that, several definitions for the fractional-order derivative were introduced. Under appropriate conditions, such definitions are equivalent for a wide class of functions [1]. A very useful definition for engineering applications is the *Grünwald–Letnikov* (*GL*) representation of the fractional derivatives, given as:

$$ {}_{a}^{GL}\mathrm{D}_{t}^{\beta}\langle f(t)\rangle = \lim_{N\to\infty}\left\{\left(\frac{t-a}{n}\right)^{\beta}\frac{1}{\Gamma(-\beta)}\sum_{r=0}^{n-1}\frac{\Gamma(r-\beta)}{\Gamma(r+1)}f\left(t-r\frac{t-a}{n}\right)\right\} \tag{1} $$

where $\Gamma(\bullet)$ is the Euler's gamma function and ${}_{a}^{GL}\mathrm{D}_{t}^{\beta}\langle\bullet\rangle$ is the *GL* fractional time derivative operator. Notice that Eq. (1) holds if $f(t)$ is continuous and differentiable up to the order $n-1$. Unfortunately, explicit expressions of the *GL* fractional derivative are usually unavailable, so that numerical procedures are needed. To this aim, the time interval $[a=0,\ t_n]$ is subdivided into small intervals of equal length Δt such that $t_0=0,\ t_1=\Delta t, t_2=2\Delta t,\dots,t_j=j\Delta t,\dots,\ t_n=n\Delta t$ are the subdivision times. To numerically solve Eq. (1), the *GL* approximation of the *GL* fractional operator based on finite differences [2] can be adopted:

$$ {}_{0}^{GL}\mathrm{D}_{t}^{\beta}\langle f(t_n)\rangle \cong \frac{1}{(\Delta t)^{\beta}}\sum_{j=1}^{n}\lambda_j(\beta)\, f(t_{n+1-j}) \tag{2} $$

where $\lambda_j(\beta)$ may be easily evaluated in recursive form:

$$ \lambda_1(\beta)=1, \lambda_2(\beta)=-\beta,\dots, \lambda_j(\beta)=\left(\frac{j-2-\beta}{j-1}\right)\lambda_{j-1}(\beta),\dots,\ j=3,4,\dots,n. \tag{3} $$

Note that, the *GL* approximation in Eq. (2) is asymptotically and absolutely stable. Since for $\beta\to 1$ the results reduce to those of the classical Euler methods, the *GL* approximation may be viewed as an extension of the classical explicit and implicit Euler methods [2].

Improved pseudo-force step-by-step integration procedure

In the framework of stochastic dynamics, the most general procedure for evaluating the response of linear and nonlinear structural systems is the *Monte Carlo Simulation* (*MCS*) method. The main advantage of this method is the ability to obtain sufficiently accurate results for any problem for which a deterministic, analytical or numerical, solution is available. Therefore, *MCS* method is currently the only tool available to solve the widest range of stochastic problems involving non-linearities of various kinds, for which there are no analytical solutions, as well as to validate approximate solutions. The main drawback of *MCS* is the heavy computational burden needed to obtain statistically meaningful solutions, which, in some cases, can involve particularly long computation times. This drawback is almost insurmountable for devices having long memory leading to fractional differential equations with stochastic excitations. Indeed, the solution of a problem by *MCS* involves the following three main steps: *i*) *Generation of Samples* - the number of samples depends on the purposes of the analysis and the precision required in determining response statistics; *ii*) *Deterministic Dynamic Analysis* - for each sample of the forcing process, the corresponding sample of the response process is determined by means of the methods of deterministic dynamics; iii) *Evaluation of Response Statistics* - the various probabilistic characteristics of the response process of interest are estimated "downstream" by means of a statistical analysis.

In this section, a very efficient and accurate method, based on the so-called *improved pseudo-force method* (*IPFM*) [3] to evaluate the response of fractional oscillators is proposed. To illustrate the proposed method, let us consider the following fractional differential equation ruling the motion of a SDOF system:

$$\ddot{x}(t)+c_{\beta}\,{}_{0}^{GL}\mathrm{D}_{t}^{\beta}\langle x(t)\rangle+\omega_0^2 x(t)=p(t) \tag{4}$$

where ω_0 is the natural circular frequency, c_β is the fractional viscoelastic factor, and $p(t)$ is the forcing function which may be the generic sample of a stochastic process. In order to numerically solve Eq. (4), first the state variable vector $\mathbf{y}(t)=\left[x(t)\;\dot{x}(t)\right]^{\mathrm{T}}$ is introduced. Then, Eq. (4) is rewritten in terms of state variables as follows:

$$\dot{\mathbf{y}}(t)=\mathbf{D}\,\mathbf{y}(t)+\mathbf{v}_{\beta}\,{}_{0}^{GL}\mathrm{D}_{t}^{\beta}\langle x(t)\rangle+\mathbf{v}_{p}\,p(t) \tag{5}$$

where

$$\mathbf{D}=\begin{bmatrix}0 & 1\\ -\omega_0^2 & 0\end{bmatrix};\quad \mathbf{v}_{\beta}=\begin{Bmatrix}0\\ -c_{\beta}\end{Bmatrix};\quad \mathbf{v}_{p}=\begin{Bmatrix}0\\ 1\end{Bmatrix}. \tag{6}$$

The solution of Eq. (5) can be formally written as:

$$\begin{aligned}\mathbf{y}(t)&=\mathbf{\Theta}(t-t_0)\mathbf{y}_0+\int_{t_0}^{t}\mathbf{\Theta}(t-\tau)\left[\mathbf{v}_{\beta}\,{}_{0}^{GL}\mathrm{D}_{\tau}^{\beta}\langle x(\tau)\rangle+\mathbf{v}_{p}\,p(\tau)\right]\mathrm{d}\tau\\ &=\mathbf{\Theta}(t-t_0)\mathbf{y}_0+\int_{t_0}^{t}\mathbf{\Theta}(t-\tau)\,\mathbf{f}_{\beta}(\tau)\,\mathrm{d}\tau\end{aligned} \tag{7}$$

where the *transition matrix* of the undamped system and the *pseudo-force vector* are introduced:

$$\mathbf{\Theta}(t)=\exp\left(\mathbf{D}t\right)=\begin{bmatrix}\cos(\omega_0 t) & \dfrac{1}{\omega_0}\sin(\omega_0 t)\\ -\omega_0\sin(\omega_0 t) & \cos(\omega_0 t)\end{bmatrix};\quad \mathbf{f}_{\beta}(t)=\mathbf{v}_{\beta}\,{}_{0}^{GL}\mathrm{D}_{t}^{\beta}\langle x(t)\rangle+\mathbf{v}_{p}\,p(t). \tag{8}$$

It has to be emphasized that in Eq. (8), $p(t)$ is the known external force, while the term ${}_{0}^{GL}\mathrm{D}_{t}^{\beta}\langle x(t)\rangle$, which depends on the response $x(t)$, is an a priori unknown quantity.

In order to evaluate the integral appearing in Eq. (7), formal solution of the differential Eq. (5) where a fractional derivative appears, a step-by-step procedure is adopted. To this aim, first the time axis is subdivided into small intervals of equal length Δt, with $t_0=0, t_1,\ldots,t_{n-1},t_n,t_{n+1},\ldots$ denoting the subdivision times. Then, the pseudo-force vector, $\mathbf{f}_{\beta}(t)$, in Eq. (8) is assumed piecewise linear in each time interval. Finally, by adopting the *GL* approximation of fractional derivative given by Eq. (2), the following relationships can be written:

Materials Research Forum LLC
https://doi.org/10.21741/9781644902431-89

$$
\begin{aligned}
&\mathbf{v}_\beta \, {}_0^{GL}\mathrm{D}_t^\beta \langle x(t_n)\rangle = \frac{\mathbf{V}_\beta}{(\Delta t)^\beta}\sum_{j=1}^{n}\lambda_j(\beta)\,\mathbf{y}(t_{n+1-j});\\
&\mathbf{v}_\beta \, {}_0^{GL}\mathrm{D}_t^\beta \langle x(t_{n+1})\rangle = \frac{\mathbf{V}_\beta}{(\Delta t)^\beta}\sum_{j=1}^{n+1}\lambda_j(\beta)\,\mathbf{y}(t_{n+2-j})\\
&\qquad = \frac{1}{(\Delta t)^\beta}\mathbf{V}_\beta\,\mathbf{y}(t_{n+1}) + \frac{\mathbf{V}_\beta}{(\Delta t)^\beta}\sum_{j=2}^{n+1}\lambda_j(\beta)\,\mathbf{y}(t_{n+2-j})
\end{aligned}
\tag{9}
$$

where $\mathbf{V}_\beta$ is a matrix defined in Eq. (11). By substituting Eq. (9) into Eq. (7), and according to the assumption that the pseudo-force vector $\mathbf{f}_\beta(t)$ is linear in each time step, the following recursive equation, numerical solution of Eq. (5), can be written [3]:

$$
\begin{aligned}
\mathbf{y}(t_{n+1}) &= \boldsymbol{\Theta}(\Delta t)\mathbf{y}(t_n) + \boldsymbol{\gamma}_0(\Delta t)\left[\frac{\mathbf{V}_\beta}{(\Delta t)^\beta}\sum_{j=1}^{n}\lambda_j(\beta)\mathbf{y}(t_{n+1-j}) + \mathbf{v}_p\,p(t_n)\right]\\
&+\boldsymbol{\gamma}_1(\Delta t)\left[\frac{1}{(\Delta t)^\beta}\mathbf{V}_\beta\,\mathbf{y}(t_{n+1}) + \frac{\mathbf{V}_\beta}{(\Delta t)^\beta}\sum_{j=2}^{n+1}\lambda_j(\beta)\mathbf{y}(t_{n+2-j}) + \mathbf{v}_p\,p(t_{n+1})\right]
\end{aligned}
\tag{10}
$$

where $\boldsymbol{\Theta}(\Delta t)$ is the *transition matrix* of the undamped SDOF system and:

$$
\begin{aligned}
&\mathbf{V}_\beta = \begin{bmatrix} 0 & 0 \\ -c_\beta & 0 \end{bmatrix};\ \mathbf{D}^{-1} = \begin{bmatrix} 0 & -\dfrac{1}{\omega_0^2} \\ 1 & 0 \end{bmatrix};\ \mathbf{L}(\Delta t) = \left[\boldsymbol{\Theta}(\Delta t) - \mathbf{I}_2\right]\mathbf{D}^{-1};\\
&\boldsymbol{\gamma}_0(\Delta t) = \left[\boldsymbol{\Theta}(\Delta t) - \frac{1}{\Delta t}\mathbf{L}(\Delta t)\right]\mathbf{D}^{-1};\ \boldsymbol{\gamma}_1(\Delta t) = \left[\frac{1}{\Delta t}\mathbf{L}(\Delta t) - \mathbf{I}_2\right]\mathbf{D}^{-1}.
\end{aligned}
\tag{11}
$$

Equation (10) gives the exact response if the forcing term is piecewise liner. When the latter condition is not strictly verified, Eq. (10) provides much more accurate solutions than all other numerical procedures proposed in literature, with the same integration step.

The main idea of the *IPFM* [3] consists of moving all terms containing the unknown vector $\mathbf{y}(t_{n+1})$ from the right- to the left-hand side, so that, after some algebra, the following step-by-step procedure is obtained:

$$
\begin{aligned}
\mathbf{y}(t_{n+1}) &= \tilde{\boldsymbol{\Theta}}(\Delta t)\mathbf{y}(t_n) + \tilde{\boldsymbol{\gamma}}_0(\Delta t)\left[\frac{\mathbf{V}_\beta}{(\Delta t)^\beta}\sum_{j=1}^{n}\lambda_j(\beta)\,\mathbf{y}(t_{n+1-j}) + \mathbf{v}_p\,p(t_n)\right]\\
&+\tilde{\boldsymbol{\gamma}}_1(\Delta t)\left[\frac{\mathbf{V}_\beta}{(\Delta t)^\beta}\sum_{j=2}^{n+1}\lambda_j(\beta)\,\mathbf{y}(t_{n+2-j}) + \mathbf{v}_p\,p(t_{n+1})\right]
\end{aligned}
\tag{12}
$$

where

$$
\begin{aligned}
&\tilde{\boldsymbol{\Theta}}(\Delta t) = \left[\mathbf{I}_2 - \frac{1}{(\Delta t)^\beta}\boldsymbol{\gamma}_1(\Delta t)\mathbf{V}_\beta\right]^{-1}\boldsymbol{\Theta}(\Delta t);\quad \tilde{\boldsymbol{\gamma}}_0(\Delta t) = \left[\mathbf{I}_2 - \frac{1}{(\Delta t)^\beta}\boldsymbol{\gamma}_1(\Delta t)\mathbf{V}_\beta\right]^{-1}\boldsymbol{\gamma}_0(\Delta t);\\
&\tilde{\boldsymbol{\gamma}}_1(\Delta t) = \left[\mathbf{I}_2 - \frac{1}{(\Delta t)^\beta}\boldsymbol{\gamma}_1(\Delta t)\mathbf{V}_\beta\right]^{-1}\boldsymbol{\gamma}_1(\Delta t).
\end{aligned}
\tag{13}
$$

The step-by-step procedure (12) gives the numerical solution $\mathbf{y}(t_{n+1})$ of the set of fractional

 Materials Research Forum LLC
https://doi.org/10.21741/9781644902431-89

differential equations (5) at time step t_{n+1}. It has been proved that this procedure is unconditionally stable [3].

The use of the *GL* approximation (2) in the framework of step-by-step integration procedures may be time-consuming. Indeed, due to the memory of the response, the summation in Eq. (3) involves a larger number of terms as time increases. This implies that computational times become prohibitive, especially when *MCS* is applied to perform the stochastic dynamic analysis of structures with fractional damping. To cope with this issue, the decreasing feature of the absolute value of the weights $\lambda_j(\beta)$ (see Eq. (3)) as the number of the time step increases (see e.g. [4,5]) may be exploited to truncate the *GL* approximation. In this way, the calculation at each time step needs not to go back to the beginning of the motion [4], i.e.:

$$\mathbf{v}_\beta \, {}_0^{GL}\mathsf{D}_t^\beta \langle x(t_n) \rangle = \begin{cases} \dfrac{\mathbf{V}_\beta}{(\Delta t)^\beta} \displaystyle\sum_{j=1}^{n} \lambda_j(\beta) \, \mathbf{y}(t_{n+1-j}) & \text{if } n \le n_T \\ \dfrac{\mathbf{V}_\beta}{(\Delta t)^\beta} \displaystyle\sum_{j=1}^{n_T} \lambda_j(\beta) \, \mathbf{y}(t_{n+1-j}) & \text{if } n > n_T \end{cases} \tag{14}$$

where n_T is the number of time instants that must be considered in the analysis to achieve good accuracy. Notice that the truncation can be applied if and only if $\beta < 1$.

Numerical application and discussion

The accuracy of the proposed *IPFM* is assessed by analyzing the response of a SDOF system with fractional damping characterized by the following parameters: $\omega_0 = 1.0$ rad/s and $c_\beta = 1.0$. Different values of the fractional derivative order β are considered. It is assumed that the SDOF system is subjected to a Gaussian white noise $W(t)$ with Power Spectral Density $S_0 = 1/2\pi$. The generic sample of the excitation is generated by means of the well-known spectral representation. The corresponding sample of the response is obtained by applying the proposed *IPFM* with ($n_T = 500$) and without truncation assuming a time step $\Delta t = 0.03$ s. For comparison purposes, the response obtained by applying the finite difference method (*FDM*) with a time step $\Delta t = 0.0001$ s is assumed as reference solution.

Figure 1a displays the time-history of the generic sample of the displacement $x(t)$ for $\beta = 0.2$. The responses provided by the *IPFM* and the *FDM* with a time step $\Delta t = 0.03$ s are contrasted with the reference solution. It is observed that the proposed *IPFM* is much more accurate and efficient than the *FDM* as it yields a solution in excellent agreement with the reference one by using a much larger time step. The enlargement in Fig. 1b shows that the *IPFM* achieves a very good match with the reference solution also when the *GL* approximation of the fractional derivative is truncated to the first $n_T = 500$ terms. This implies a drastic reduction of the computational effort. Numerical investigations, omitted for brevity, have shown that the same degree of accuracy is achieved for other values of β.

Figure 2 shows the time-history of the variance of the displacement provided by *MCS* ($N_s = 5000$ samples) along with the exact steady-state value for $\beta = 0.2$ and $\beta = 0.5$. It can be seen that a very accurate prediction of the displacement variance is provided by the *IPFM* with a time step $\Delta t = 0.03$ s even when a truncation step $n_T = 500$ is considered. Conversely, the *FDM* with $\Delta t = 0.03$ s highly underestimates the variance of the response. A smaller time step is needed to improve the accuracy of the *FDM*.

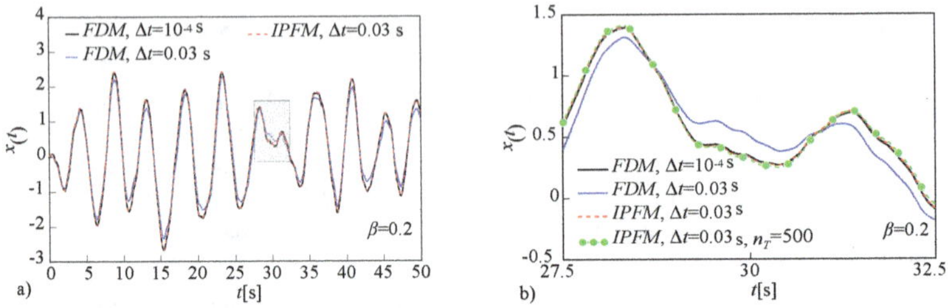

Figure 1. a) Sample of the displacement provided by the FDM and the IPFM; b) enlargement showing the accuracy of the IPFM with truncation ($\beta = 0.2$).

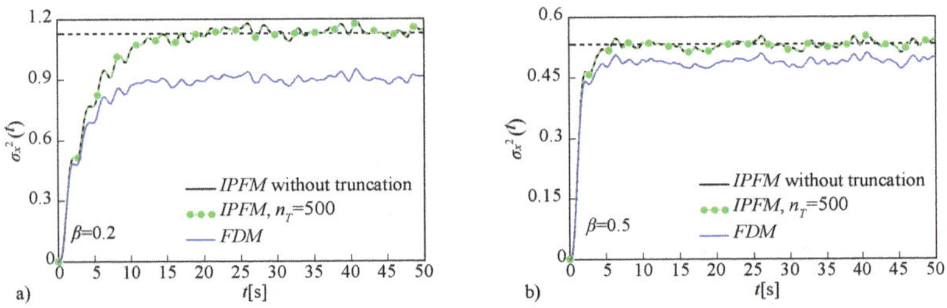

Figure 2. Variance of the displacement provided by MCS ($N_s = 5000$ samples) along with the exact steady-state value (black dashed line): a) $\beta = 0.2$ and b) $\beta = 0.5$ ($\Delta t = 0.03$ s).

Numerical results demonstrate that the *IPFM* allows a drastic enhancement of the computational efficiency of *MCS* since it is able to provide accurate estimates of the response using much larger time steps than the classical *FDM*. The computational burden can be further reduced by truncating the *GL* approximation of the fractional derivative.

References

[1] I. Podlubny, Fractional Differential Equations. An Introduction to Fractional Derivatives, Fractional Differential Equations, to Methods of their Solution and Some of their Applications, Academic Press, San Diego, CA, 1999.

[2] R. Scherer, S.L. Kalla, Y.Tang, J. Huangc, The Grünwald–Letnikov method for fractional differential equations, Comput. Math. Appl. 62 (2011) 902–917. https://doi.org/10.1016/j.camwa.2011.03.054

[3] G. Muscolino, Dynamically modified linear structures: deterministic and stochastic response. J. Eng. Mech. 122(11) (1996), 1044-1051. https://doi.org/10.1061/(ASCE)0733-9399(1996)122:11(1044)

[4] J. Xu, J. Li, Stochastic dynamic response and reliability assessment of controlled structures with fractional derivative model of viscoelastic dampers, Mech. Syst. Sig. Process. 72-73(2016) 865–896. https://doi.org/10.1016/j.ymssp.2015.11.016

[5] M.P. Singh, T.-S. Chang, H. Nandan, Algorithms for seismic analysis of MDOF systems with fractional derivatives, Eng. Struct. 33(8) (2011) 2371 – 2381. https://doi.org/10.1016/j.engstruct.2011.04.010

Theoretical and Applied Mechanics - AIMETA 2022 | Materials Research Forum LLC
Materials Research Proceedings **26** (2023) 555-560 | https://doi.org/10.21741/9781644902431-90

Stochastic analysis of double-skin façades subjected to imprecise seismic excitation

Federica Genovese[a], Alba Sofi [b *]

Department of Architecture and Territory, University "Mediterranea" of Reggio Calabria, Via dell'Università 25, 89124 Reggio Calabria, Italy

[a]federica.genovese@unirc.it, [b]alba.sofi@unirc.it

Keywords: Double-Skin Façades, Interval Analysis, Imprecise Power Spectral Density Function, Passive Control

Abstract. Double-skin façades (DSFs), widely used in buildings to provide specific thermal efficiency, acoustic isolation, and weather resistance properties, have been recently used as passive control systems. The present study focuses on the stochastic analysis of a shear-type frame equipped with a DSF subjected to ground motion acceleration modelled as a zero-mean stationary Gaussian random process fully characterized by an imprecise power spectral density function i.e., with interval parameters. The influence of imprecision of the seismic excitation on the performance of the DSF is investigated by evaluating the bounds of the interval reliability function for a selected displacement process of the frame structure, in the framework of the classical first-passage problem.

Introduction

A double-skin façade (DSF) is a multi-layered structure composed of a second "skin" placed in front of a regular building façade. DSFs are becoming increasingly popular for their capability of improving the energy performance and the aesthetics of buildings. Recently, the application of DSFs as vibration absorbers has been investigated assuming either deterministic [1] or stochastic excitation [2]. Previous studies mainly focused on the optimal design of the DSF by varying the layout of panels and design parameters such as the stiffness of links to the building.

The present study addresses the stochastic analysis of a shear-type frame equipped with a DSF subjected to ground motion acceleration taking into account epistemic uncertainties affecting the excitation. Following Pipitone et al. [2], the DSF is modelled as a set of independent panels, each one described as a mass lumped system connected to the main structure by elastic springs at the floor level. Within the *strong motion* phase, seismic excitation is modelled as a zero-mean stationary Gaussian process, fully characterized by an imprecise power spectral density (IPSD) function, recently proposed by Muscolino et al. [3]. The three parameters characterizing the assumed spectral model are described as interval variables by applying the so-called *Improved Interval Analysis* [4]. The bounds of the spectral parameters are estimated through the analysis of a set of accelerograms recorded on rigid soil deposits. Since the IPSD function has an interval nature, response statistics of the seismically excited system are described by intervals. To assess structural safety, the IPSD function of ground motion acceleration is incorporated into the formulation of the classical first-passage problem and the bounds of the interval reliability function for a selected response process are estimated.

Problem formulation

Let us consider a combined system (see Fig. 1) consisting of a $s-$storey shear-type frame (primary system) equipped with a DSF (secondary system) subjected to seismic excitation modelled as a zero-mean stationary Gaussian random process fully characterized by an IPSD i.e., with interval parameters. The primary system is characterized by storeys having equal mass m, lateral stiffness k and inter-storey height h. A constant viscous damping ratio ζ_0 is assumed for all modes of vibration. Following Pipitone et al. [2], the DSF is modelled as a set of N independent panels connected to the main structure by elastic springs at the floor level. The generic panel is modelled as a system composed by j lumped masses equally spaced by $h/2$. Each lumped mass has two degrees-of-freedom (DOFs), namely the horizontal displacement $x_j(t)$ and the rotation $\theta_j(t)$ (see Fig. 1). The flexural stiffness of the panels k_P and the stiffness of the springs are assumed proportional to the lateral stiffness k of the frame through the dimensionless coefficients υ and α, respectively. The same viscous damping ratio ζ_P is assumed for all the N panels.

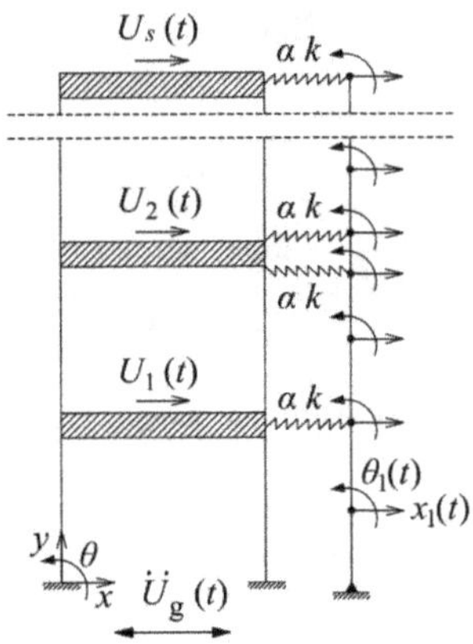

Figure 1: Shear-type primary structure with double-skin façade.

After a static condensation of the rotational DOFs of the panels, the equations of motion of the coupled system subjected to imprecise seismic excitation can be written as:

$$\mathbf{M}\ddot{\mathbf{U}}^I(t)+\mathbf{C}\dot{\mathbf{U}}^I(t)+\mathbf{K}\mathbf{U}^I(t)=-\mathbf{M}\boldsymbol{\tau}\ddot{U}_g^I(t) \tag{1}$$

where a dot over a variable denotes differentiation with respect to time t; $\mathbf{U}^I(t)$ is the interval vector random process that collects the n dynamically significant nodal displacements, with the apex I denoting interval quantities; $\ddot{U}_g^I(t)$ is the ground motion acceleration characterized by an IPSD function; $\boldsymbol{\tau}$ is the n-vector listing the influence coefficients; $\mathbf{M}$ and $\mathbf{K}$ are the $n\times n$ mass and stiffness block matrices of the coupled structure:

$$\mathbf{M}=\begin{bmatrix}\mathbf{M}_0 & \mathbf{O} & \cdots & \mathbf{O}\\ \mathbf{O} & \mathbf{M}_1 & \cdots & \mathbf{O}\\ \vdots & \vdots & \ddots & \vdots\\ \mathbf{O} & \mathbf{O} & \cdots & \mathbf{M}_N\end{bmatrix};\quad \mathbf{K}=\begin{bmatrix}\mathbf{K}_0+\mathbf{K}_{S0} & \mathbf{K}_{01} & \cdots & \mathbf{K}_{0N}\\ \mathbf{K}_{01}^{\mathrm{T}} & \mathbf{K}_1+\mathbf{K}_{S1} & \cdots & \mathbf{O}\\ \vdots & \vdots & \ddots & \vdots\\ \mathbf{K}_{0N}^{\mathrm{T}} & \mathbf{O} & \cdots & \mathbf{K}_N+\mathbf{K}_{SN}\end{bmatrix} \tag{2}$$

where $\mathbf{M}_0$ and $\mathbf{K}_0$ are the mass and stiffness matrices of the primary structure; $\mathbf{M}_i$ and $\mathbf{K}_i$ are the mass and stiffness matrices (after static condensation) of the i-th panel; $\mathbf{K}_{S0}$ represents the increment of the stiffness matrix of the primary structure due to the secondary system; $\mathbf{K}_{Si}$ and $\mathbf{K}_{0i}$ contain the stiffness of the elastic springs at each storey level; $\mathbf{O}$ is a zero matrix of appropriate dimensions. The primary building and the panels are individually considered as classically damped subsystems. In Eq. (1), the damping matrix $\mathbf{C}$ of the coupled system is defined as:

$$\mathbf{C} = \mathbf{\Gamma}^{-1}\,\mathbf{\Xi}\,\mathbf{\Gamma}^{-\mathrm{T}} \tag{3}$$

where $\mathbf{\Gamma}$ is a convenient transformation matrix while $\mathbf{\Xi}$ is the modal damping matrix.

Imprecise power spectral density function

The model of the IPSD function of ground motion acceleration recently proposed in [3] is assumed:

$$G^I_{\ddot{U}_g}(\omega) = \left(\sigma^2_{\ddot{U}_g}\right)^I \beta^I_0 \left(\frac{\omega^2}{\omega^2 + \left(\omega^I_H\right)^2}\right)\left(\frac{\left(\omega^I_L\right)^4}{\omega^4 + \left(\omega^I_L\right)^4}\right)\frac{\rho_0}{\pi}\left[\frac{1}{\left(\rho^I_0\right)^2 + (\omega + \Omega^I_0)^2} + \frac{1}{\left(\rho^I_0\right)^2 + (\omega - \Omega^I_0)^2}\right] \tag{4}$$

where β^I_0 is defined so as to ensure that the interval stochastic process $\ddot{U}^I_g(t)$ possesses variance $\left(\sigma^2_{\ddot{U}_g}\right)^I$, while $\omega^I_L = \Omega^I_0 + 0.8\,\rho^I_0$ and $\omega^I_H = 0.1\,\Omega^I_0$ are, respectively, the interval frequency control of the second order low-pass and first order high-pass butterworth filters, that in turn depend on the interval predominant circular frequency Ω^I_0 and frequency bandwidth ρ^I_0 of the filtered stationary process [5]. Since the one-sided PSD function in Eq. (4) depends linearly on $\left(\sigma^2_{\ddot{U}_g}\right)^I$, this variable could be set a posteriori.

In Eq. (4), the generic interval variable z^I_i is expressed by using the *Improved Interval Analysis* [4] as follows:

$$z^I_i = \left[\underline{z}_i, \overline{z}_i\right] \equiv z_{\mathrm{mid},i}(1+\delta^I_i) = z_{\mathrm{mid},i}(1+\Delta\delta_i \hat{e}^I_i) \tag{5}$$

where the symbols $\underline{z}_i$ and $\overline{z}_i$ denote the lower bound (*LB*) and the upper bound (*UB*) of the interval, respectively; $\hat{e}^I_i = [-1,1]$ is the so-called *extra unitary interval* (*EUI*) associated with the i-th interval variable. In Eq. (5), $z_{\mathrm{mid},i}$ and $\Delta\delta_i$ are the *midpoint value* (*mid*) and the normalized *deviation amplitude* of z^I_i, given, respectively, by:

$$z_{\mathrm{mid},i} = \frac{\underline{z}_i + \overline{z}_i}{2}; \quad \Delta\delta_i = \frac{\Delta z_i}{z_{\mathrm{mid},i}} = \frac{\overline{\delta}_i - \underline{\delta}_i}{2} > 0 \tag{6}$$

where $\Delta z_i = (\overline{z}_i - \underline{z}_i)/2$ is the *deviation amplitude* (*dev*) of z^I_i. In Eq. (5), $\delta^I_i = \Delta\delta_i \hat{e}^I_i$ denotes the dimensionless interval fluctuation around $z_{\mathrm{mid},i}$ such that $\Delta\delta_i < 1$.

Interval reliability analysis

The interval displacement vector ruled by Eq. (1) is described by an interval zero-mean stationary Gaussian random vector process, completely characterized in the frequency domain by the knowledge of the interval one-sided PSD function matrix, given by:

$$\mathbf{G}^I_{\mathrm{U}}(\omega) = \mathbf{H}^*(\omega)\mathbf{p}\mathbf{p}^{\mathrm{T}}\mathbf{H}^{\mathrm{T}}(\omega)G^I_{\ddot{U}_g}(\omega); \;\; \mathbf{H}(\omega) = \left[\mathbf{K} - \omega^2\mathbf{M} + \mathrm{j}\omega\mathbf{C}\right]^{-1}; \;\; \mathbf{p} = -\mathbf{M}\boldsymbol{\tau} \tag{7}$$

where $\mathbf{H}(\omega)$ is transfer matrix; $\mathrm{j} = \sqrt{-1}$; the asterisk means complex conjugate.

To perform structural safety assessment under seismic excitation, the IPSD function is herein incorporated into the formulation of the classical first-passage problem. Let U_h^I be the interval displacement process of interest and $U_{\max,h}^I(T)=\max_{0\le t\le T}\left|U_h^I(t)\right|$ the associated *extreme value* process. Adopting the Vanmarcke's failure criterion [6], the *interval cumulative distribution function* (ICDF) or *interval reliability function* $L_{U_{\max,h}}^I(b,T)$ can be expressed as [3]:

$$\begin{aligned}L_{U_{\max,h}}^I(b,T)&\equiv L_{U_{\max,h}}\left(b,T;\Omega_0^I,\rho_0^I,(\sigma_{\ddot{U}_g}^2)^I\right)=\Pr\left[U_{\max,h}^I(T)\le b\right]\\&\cong\exp\left[-T\frac{1}{\pi}\sqrt{\frac{\tilde{\lambda}_{2,U_h}^I}{\tilde{\lambda}_{0,U_h}^I}}\left[\frac{1-\exp\left(-b\left(\tilde{\delta}_{U_h}^I\right)^{1.2}\sqrt{\frac{\pi}{2(\sigma_{\ddot{U}_g}^2)^I\,\tilde{\lambda}_{0,U_h}^I}}\right)}{\exp\left(\frac{b^2}{2(\sigma_{\ddot{U}_g}^2)^I\,\tilde{\lambda}_{0,U_h}^I}\right)-1}\right]\right]\end{aligned} \tag{8}$$

where $\tilde{\lambda}_{i,U_h}^I$ $(i=0,1,2)$ are the interval spectral moments of the response process $\tilde{U}_h^I(t)=U_h^I(t)/\sigma_{\ddot{U}_g}^I$ under unit variance seismic acceleration; $\tilde{\delta}_{U_h}^I$ is the *interval bandwidth parameter*; b represents the barrier level; T is the observation time. The *LB* and *UB* of the ICDF can be evaluated by performing global optimization for each value of the barrier level b under the constraint that the uncertain spectral parameters range within the pertinent intervals. Alternatively, accurate estimates of the bounds of the ICDF can be efficiently obtained as [3]:

$$\begin{aligned}\underline{L}_{U_{\max,h}}(b,T)&=L_{U_{\max,h}}\left(b,T;\overline{\sigma}_{\ddot{U}_g}^2,\overline{\tilde{\lambda}}_{0,U_h},\overline{\tilde{\lambda}}_{1,U_h},\overline{\tilde{\lambda}}_{2,U_h}\right);\\\overline{L}_{U_{\max,h}}(b,T)&=L_{U_{\max,h}}\left(b,T;\underline{\sigma}_{\ddot{U}_g}^2,\underline{\tilde{\lambda}}_{0,U_h},\underline{\tilde{\lambda}}_{1,U_h},\underline{\tilde{\lambda}}_{2,U_h}\right)\end{aligned} \tag{9}$$

where $\underline{\tilde{\lambda}}_{i,U_h}$ and $\overline{\tilde{\lambda}}_{i,U_h}$ are the bounds of the interval spectral moments $\tilde{\lambda}_{i,U_h}^I$ $(i=0,1,2)$. Since the monotonic behaviour is not guaranteed, such bounds might correspond to intermediate values of the interval spectral parameters Ω_0^I and ρ_0^I, and can be evaluated by optimization.

Numerical application and discussion

A six-storey shear-type frame equipped with a DSF consisting of two independent panels is considered [2]. The primary building, having a fundamental period of vibration $T_1=0.582\,\text{s}$, is characterized by the following parameters: floor mass $m=20000\,\text{kg}$, lateral stiffness $k=4\times10^7\,\text{N/m}$; inter-storey height $h=3.32\,\text{m}$; constant viscous damping ratio $\zeta_0=0.02$ for all modes of vibration. The total mass of the two panels is assumed as 10% of the mass of the primary building [2]. For the two panels, $\upsilon=0.392$ and $\zeta_P=0.13$ are considered. Different values of the stiffness of the links to the frame αk are considered such that $\alpha_{\min}\le\alpha\le\alpha_{\max}$, with $\alpha_{\min}=1.72\times10^{-3}$ and $\alpha_{\max}=5.16\times10^{-3}$. The value $\alpha_0=3.44\times10^{-3}$ is assumed as the nominal one. The displacement of the first floor of the primary structure, $U_1^I(t)$, is selected as response quantity of interest. The bounds of the interval parameters Ω_0^I, ρ_0^I, $\left(\sigma_{\ddot{U}_g}^2\right)^I$ entering the IPSD function in Eq. (4) are estimated by analysing 10 site-compatible accelerograms, recorded on rigid soil deposits, downloaded from PEER [8] and Engineering Strong Motion [9] database. All the selected accelerograms are scaled to the target peak ground acceleration value $a_g = 3.33$ m/s^2. Table 1 lists the *LB*, *UB*, *mid* and percentage ratio (*dev/mid*)% of the spectral parameters.

Table 1: Main characteristics of the interval spectral parameters.

Parameter	*LB*	*UB*	*mid*	*(dev/mid) %*
$\sigma^2_{\ddot{U}_g}$ $[\mathrm{m}^2/\mathrm{s}^4]$	0.64	1.47	1.06	39.09
Ω_0 [rad/s]	23.91	45.22	34.57	30.82
ρ_0[rad/s]	12.10	22.56	17.33	30.17

The realizations of the IPSD function $\tilde{G}^I_{\ddot{U}_g}(\omega) = G^I_{\ddot{U}_g}(\omega)/\left(\sigma^2_{\ddot{U}_g}\right)^I$ pertaining to the extreme values of the interval parameters Ω^I_0 and ρ^I_0 along with the nominal spectrum are reported in Fig. 2a. The ratio $\tilde{J}_{U_1} = \tilde{\sigma}^2_{U_1,c} / \tilde{\sigma}^2_{U_1,u}$ between the variance of the displacement $\tilde{U}^I_1(t) = U^I_1(t)/\sigma^I_{\ddot{U}_g}$ of the primary structure with and without the DSF [2] is herein taken as representative of structural performance under seismic excitation. The subscripts "c" and "u" stand for controlled and uncontrolled. Fig. 2b displays the ratio $\tilde{J}_{U_1}$ versus the coefficient α for all possible combinations of the endpoints of the interval spectral parameters Ω^I_0 and ρ^I_0 as well as for the nominal values. The value of the dimensionless coefficient α which minimizes $\tilde{J}_{U_1}$ is nearly $\alpha_M = 0.00392$ in all the considered cases. Furthermore, over the whole range of α, the smallest value of the ratio $\tilde{J}_{U_1}$ pertains to $\overline{\Omega}_0$ and $\underline{\rho}_0$ which indeed provide the *LB*, $\underline{\tilde{\lambda}}_{0,U_1}$, of $\tilde{\lambda}^I_{0,U_1}$.

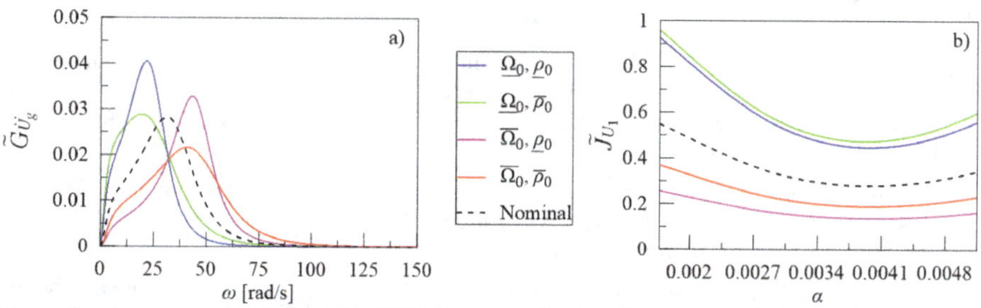

Figure 2: a) some realizations of the IPSD function of ground motion acceleration; b) associated ratios between the variance of $\tilde{U}^I_1(t) = U^I_1(t)/\sigma^I_{\ddot{U}_g}$ with and without DSF versus α.

Fig. 3a shows the *LB* and *UB* of the ICDF, $L^I_{U_{\max,1}}(b,T)$, of the *extreme value* process $U^I_{\max,1}(T)$, along with the nominal solution for $T = 30$s and $\alpha = \alpha_0$. Notice that the proposed bounds (Eq. (9)) are in excellent agreement with the "Exact" ones obtained by applying the scanning method. As expected, neglecting the uncertainties affecting the main parameters of the IPSD function of ground motion acceleration may lead to serious overestimation of the safety level. To ensure a conservative design, the worst-case scenario, corresponding to the *LB* of the ICDF, needs to be considered [3]. To assess the influence of the DSF on structural reliability, in Fig. 3b the bounds of the ICDF, $L^{I}_{U_{\max,1}}(b,T)$, provided by Eq. (9) for $\alpha = \alpha_{\min}, \alpha_M, \alpha_{\max}$ are contrasted with the ones pertaining to the frame without DSF. It can be noticed that the DSF significantly improves the seismic performance of the primary structure. In particular, the DSF with link stiffness $k\alpha_M$ ensures the highest safety level.

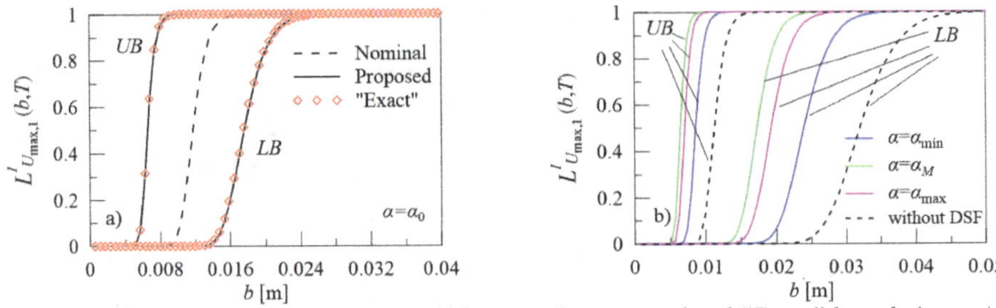

Figure 3: ICDF of $U^{I}_{\max,1}(T)$: a) comparison between the proposed and "Exact" bounds ($\alpha = \alpha_0$); b) proposed bounds for different configurations.

References

[1] K.S. Moon, Tall building motion control using double skin façades, J. Archit. Eng. 15(3), (2009) 84–90. https://doi.org/10.1061/(ASCE)1076-0431(2009)15:3(84)

[2] G. Pipitone, G. Barone, A. Palmeri, Stochastic design of double-skin façades as seismic vibration absorbers, Adv. Eng. Softw. 142, 102749 (2020). https://doi.org/10.1016/j.advengsoft.2019.102749

[3] G. Muscolino, F. Genovese, A. Sofi, Reliability Bounds for Structural Systems Subjected to a Set of Recorded Accelerograms Leading to Imprecise Seismic Power Spectrum, ASCE-ASME J. Risk Uncertain. Eng. Syst. A: Civ. Eng. 8(2), 04022009 (2022). https://doi.org/10.1061/AJRUA6.0001215

[4] G. Muscolino, A. Sofi, Stochastic Analysis of Structures with Uncertain-but-Bounded Parameters via Improved Interval Analysis, Probab. Eng. Mech. 28, (2012) 152-163. https://doi.org/10.1016/j.probengmech.2011.08.011

[5] G. Muscolino, F. Genovese, G. Biondi, E. Cascone, Generation of fully non-stationary random processes consistent with target seismic accelerograms. Soil Dyn. Earthq. Eng. 141, 106467 (2021). https://doi.org/10.1016/j.soildyn.2020.106467

[6] E.H. Vanmarcke, On the distribution of the first-passage time for normal stationary random processes, J. Appl. Mech. 42 (1), (1975) 215–220. https://doi.org/10.1115/1.3423521

[7] T.D. Ancheta, R.B. Darragh, J.P. Stewart, E. Seyhan, W.J. Silva, B.S.J. Chiou, K.E. Wooddell, R.W. Graves, A.R. Kottke, D.M. Boore, T. Kishida, and J.L. Donahue, PEER NGA-West2 Database (2013). https://doi.org/10.1193/070913EQS197M

[8] L. Luzi, G. Lanzano, C. Felicetta, M.C. D'Amico, E. Russo, S. Sgobba, F. Pacor, ORFEUS Working Group 5, Engineering Strong Motion Database (ESM), V.2.0 (2020).

Theoretical and Applied Mechanics - AIMETA 2022 Materials Research Forum LLC
Materials Research Proceedings **26** (2023) 561-566 https://doi.org/10.21741/9781644902431-91

Digital simulation of multi-variate stochastic processes

Salvatore Russotto[1, a *], Mario Di Paola[1, b] and Antonina Pirrotta[1, c]

[1]Dipartimento di Ingegneria, Università degli Studi di Palermo, Viale delle Scienze – Ed.8, 90128, Palermo

[a]salvatore.russotto01@unipa.it, [b]mario.dipaola@unipa.it, [c]antonina.pirrotta@unipa.it

Keywords: Stochastic Processes, Wind Field Velocity, Monte Carlo Simulation, Reduction of the Computational Burden

Abstract. Stochastic dynamic analysis of linear or nonlinear multi-degree-of-freedom systems excited by multi-variated processes is usually conducted by using digital Monte Carlo (MC) simulation. Since in structural systems few modal shapes contribute to the response in the nodal space, the computational burden of MC simulation is mainly related to the digital simulation of the input process. Usually, the generation of multi-variated samples of Gaussian input process is performed with the aid of the Shinozuka formula. However, since in this procedure the stochastic process is given as a summation of waves with random amplitude amplified by the square root of the power spectral density, the randomness is due to a random phase angle of each wave, therefore a very large number of waves is required to reach the Gaussianity, i.e. the process is only asymptotically stable. Moreover, the computational burden increases in case of multi-variated processes. The paper aims to drastically reduce the generation time of the input process through the use of a two-step procedure. In the first step, by using the Priestley formula, each wave is normally distributed. This first aspect allows to drastically reduce the computational effort for the mono-variate process since few waves are sufficient to reach the Gaussianity. In the second step, the multi-variate process is reduced as a summation of independent fully coherent vectors if the quadrature spectrum (q-spectrum) can be neglected. An application of digital simulation of the wind velocity field is discussed to prove the efficiency of the proposed approach.

Introduction

Monte Carlo (MC) simulation is commonly used in the stochastic dynamic analysis of linear or nonlinear multi-degree-of-freedom systems excited by multi-variated processes [1,2]. It consists, initially, in the generation of samples of an input process characterized in probabilistic setting; then the output process is calculated by performing a deterministic analysis [1, 2]. Finally, a probabilistic analysis of the results is performed [1, 2].

As far as the input process generation are concerning it can be used for different purposes [3-7] and, if the input process is Gaussian, then it is fully characterized in probabilistic setting by its Power Spectral Density (PSD). The generation of samples of a Gaussian multi-variate input process can be performed in different ways; such an example Auto Regressive (AR) or Auto Regressive Moving Average (ARMA) techniques can be used for this purposes [1,8]. Although AR or ARMA techniques lead to good results, these kind of techniques requires a very high computational burden [1]. Nowadays, the generation of samples of the input process is usually performed through the use of the Shinozuka's formula [9] that consists in a summation of harmonic waves each of which has a different frequency and a random phase angle. The latter is the realization of a random variable uniformly distributed in the range $[0 - 2\pi]$. However, since the phases are uniformly distributed random variables, a high number of waves is required to reach the Gaussianity [1,8]. For this reason, the Priestley's formula is preferable [8]. It is similar to the Shinozuka's formula but, in this case, each harmonic wave is multiplied by the realization of a zero-mean complex random variable having normal distribution. This difference, drastically reduces the number of waves required to reach the Gaussianity since the random variables in the Priestley's formula have normal distribution unlike the random phases present in the Shinozuka's formula that are uniformly distributed.

A multi-variate stochastic process can be decomposed as a summation of fully coherent vectors [1,8]. This decomposition can be performed by using different method to decompose the PSDs matrix. One of these ways has been consists of decomposing it into the basis of the eigenvectors of the PSDs matrix [1,8]. This way is preferable to the decomposition of the PSDs matrix into the product between two triangular matrices not only because it is faster than the second one, but also because, in the wind field simulation, it has a strong physical significance. In fact, the eigenvalues are related to the power of the process that will be generated, while the eigenvectors, usually called *blowing mode shapes*, represent the modal shapes associated with the wind field velocity [1,8].

In this paper an innovative method for the simulation of multi-variate stationary processes is proposed. It exploits the advantages deriving from the use of the Priestley's formula that allows to reach the Gaussianity with a little number of harmonics. Moreover, the decomposition of the PSDs matrix, performed by using its eigenvalues and its eigenvectors, is used to select only the first eigenvalue and the first eigenvector that give the major contribution to the total input process. However, since the contribution of the other eigenvalues and eigenvectors can be not negligible, a simple technique to preserve the variance of the input process is adopted. The proposed method allows to generate a multi-variate stochastic process with good precision and it is faster than the methods commonly used.

In order to prove the reliability of the proposed method a numerical simulation concerning the generation of multi-variate wind field velocity has been performed and the results obtained are discussed in detail.

Proposed Method

Let $\mathrm{V}(x,y,z;t)$ be the velocity field. It can be expressed as [1,8]

$$\mathrm{V}(x,y,z;t)=\bar{V}(z)+V(x,y,z;t) \tag{1}$$

in which $\bar{V}(z)$ represents the mean value and $V(x,y,z;t)$ is a zero-mean stationary process that represents the fluctuating component of the wind speed. The mean value $\bar{V}(z)$ can be expressed by the logarithmic law $\bar{V}(z)=u_*\ln(z/z_0)/k$ in which k is the Von Karman's constant, z_0 is the roughness length and u_* is the shear velocity [1,8]. In this paper, the mean value $\bar{V}(z)$ is expressed, without loss of generality, as the logarithmic law

$$\bar{V}(z)=\bar{V}^{(10)}\frac{\ln(z+1)}{\ln(11)} \tag{2}$$

being $\bar{V}^{(10)}$ the wind velocity at $z=10\mathrm{m}$.

The one-side PSD of $V_i=V(x_i,y_i,z_i;t)$ and $V_j=V(x_j,y_j,z_j;t)$ can be calculated, considering that its imaginary part can be neglected [1,10], as [1,8]

$$G_{V_iV_j}(\omega)=\sqrt{G_{V_iV_i}(\omega)G_{V_jV_j}(\omega)}\exp(-f_{ij}(\omega)) \tag{3}$$

in which $G_{V_iV_i}(\omega)$ is the auto-PSD of the process V_i and $f_{ij}(\omega)$ is expressed as [1,8]

$$f_{ij}(\omega)=|\omega|\frac{\sqrt{C_y^2(y_i-y_j)^2+C_z^2(z_i-z_j)^2}}{2\pi(\bar{V}(z_i)+\bar{V}(z_j))}. \tag{4}$$

In Eq.(4) C_y and C_z are decay coefficients that must be determined experimentally [1,8]. The auto-PSD $G_{V_jV_j}(\omega)$ can be calculated by using the Solari's formula [11], i.e.

$$G_{V_jV_j}(\omega)=\frac{6.868\sigma_V^2\gamma_j(\omega)L_V/z_j}{\frac{\omega}{2\pi}\left(1+10.302\gamma_j(\omega)L_V/z_j\right)^{\frac{5}{3}}} \tag{5}$$

in which L_V is the integral length scale of the turbulence, σ_V^2 is the variance of the longitudinal component of the velocity fluctuations and $\gamma_j(\omega)$ is the Monin coordinate expressed as [1,8]

$$\gamma_j(\omega)=\frac{\omega z_j}{2\pi\bar{V}(z_j)}. \tag{6}$$

Defining the multi-variate vector process $\mathbf{V}(t)=[V_1\ \ V_2\ \ \dots\ \ V_n]^T$, its PSDs matrix, denoted as $\mathbf{G}_\mathrm{V}(\omega)$, can be calculated through the use of Eq.(3) and then $\mathbf{G}_\mathrm{V}(\omega)$ can be decomposed as [1,8]

$$\mathbf{G}_\mathrm{V}(\omega)=\mathbf{\Psi}(\omega)\mathbf{\Lambda}(\omega)\mathbf{\Psi}^T(\omega) \tag{7}$$

in which $\mathbf{\Psi}(\omega)$ is a matrix whose columns are the eigenvectors of $\mathbf{G}_\mathrm{V}(\omega)$ while $\mathbf{\Lambda}(\omega)$ is a diagonal matrix that contains the eigenvalues of $\mathbf{G}_\mathrm{V}(\omega)$.
The vector process $\mathbf{V}(t)$ can be decomposed into a summation of multi-variate fully coherent normal vectors $\mathbf{Q}_j(t)$ in the form [1,8]

$$\mathbf{V}(t)=\sum_{j=1}^{n}\mathbf{Q}_j(t) \tag{8}$$

and each vector $\mathbf{Q}_j(t)$ can be generated through the use of the Priestley's formula, taking into account the decomposition of the PSDs matrix in Eq.(7), as [1,8]

$$\mathbf{Q}_j(t)=\sum_{r=1}^{N}\mathbf{\Psi}_j(\omega_r)\sqrt{2\Lambda_j(\omega_r)\Delta\omega}P_r^{(j)} \tag{9}$$

in which $\mathbf{\Psi}_j(\omega_r)$ is the j-th eigenvector of $\mathbf{G}_\mathrm{V}(\omega_r)$, $\Lambda_j(\omega_r)$ is the j-th eigenvalue of $\mathbf{G}_\mathrm{V}(\omega_r)$, N is the number of harmonics waves considered in the summation, $N\Delta\omega$ is the cut-off frequency and $P_r^{(j)}$ is the r-th realization of a zero-mean normal complex random variable obeying the condition $E\left[P_s^{(i)}P_r^{(j)*}\right]=\delta_{ij}\delta_{sr}$. Eq.(9) can be expressed in its real form as [1,8]

$$\mathbf{Q}_j(t)=\sum_{r=1}^{N}\mathbf{\Psi}_j(\omega_r)\sqrt{2\Lambda_j(\omega_r)\Delta\omega}\left(R_r^{(j)}\cos(\omega_r t)+I_r^{(j)}\sin(\omega_r t)\right) \tag{10}$$

being $R_r^{(j)}$ and $I_r^{(j)}$, respectively, the real part and the imaginary part of $P_r^{(j)}$.

The use of Eq.(10) allows to generate samples of the fully coherent vector $\mathbf{Q}_j(t)$ faster than the Shinozuka's formula. In fact, since $P_r^{(j)}$ is normally distributed, a small number of harmonic waves can be considered in the summation and thus the computational burden can be drastically reduced. The time needed for the computation of the process $\mathbf{V}(t)$ can still be significantly reduced taking into account only $j=1$ in Eq.(8). However, if only the first terms of the summation in Eq.(8) is considered, then the variance of the process $V_i(t)$ is not well estimated and thus, to have good results, the variance has to be preserved. To do this, firstly, the exact value of the variance of $V_i(t)$ can be simply calculated as

$$\sigma_{V_i}^2 = \sum_{r=1}^{N} G_{V_i V_i}(\omega_r)\Delta\omega \tag{11}$$

and then the vector process $\mathbf{V}(t)$ can be approximated as

$$\mathbf{V}(t) \approx \tilde{\mathbf{Q}}_1(t) \tag{12}$$

in which the elements $\tilde{Q}_i(t)$ of the vector $\tilde{\mathbf{Q}}_1(t)$ are defined as

$$\tilde{Q}_i(t) = \frac{1}{\sigma_{Q_{i1}}} \sum_{r=1}^{N} \Psi_{i1}(\omega_r)\sqrt{2\Lambda_1(\omega_r)\Delta\omega}P_r^{(1)}\sqrt{\sigma_{V_i}^2} \tag{13}$$

being $\sqrt{\sigma_{V_i}^2}$ the standard deviation of $V_i(t)$ and $\sigma_{Q_{i1}}$ the standard deviation of the i-th row of the vector $\mathbf{Q}_1(t)$. The vector $\mathbf{Q}_1(t)$ can be simply obtained considering $j=1$ in Eq.(10).

In this way, the computational burden required is drastically reduced since only the first blowing modal shape is considered. Particularly the time required to generate the process $\mathbf{V}(t)$ by using the proposed method can be expressed as a percentage of the time required to generate the process $\mathbf{V}(t)$ considering all the terms in the summation present in Eq.(8). This percentage is labelled as $T_*\%$ and can be expressed in the form

$$T_*\% \approx 100/n\,. \tag{14}$$

From Eq.(14) it is clear that the proposed method allows to drastically reduce the time required to generate a stationary Gaussian multi-variate stochastic process.

Numerical Simulation

To assess the reliability of the proposed method a numerical simulation has been performed considering three different points, $P_1(x_1,y_1,z_1)$; $P_2(x_2,y_2,z_2)$ and $P_3(x_3,y_3,z_3)$, having $x_1=x_2=x_3$, $y_1=y_2=y_3$, $z_1=10$ m, $z_2=20$ m and $z_3=30$ m. The velocity $\bar{V}^{(10)}$ is equal to 22 m/s ,while L_V, σ_V^2 and C_z has been assumed as unitary values. In this simulation 1000 samples of the process $\mathbf{V}(t)$ have been generated. Each sample has a duration of 100 s discretized with a sampling frequency of 100 Hz. The components of PSDs matrix of the process $\mathbf{V}(t)$ have been calculated by using Eq.(3) and are represented in Fig.1.

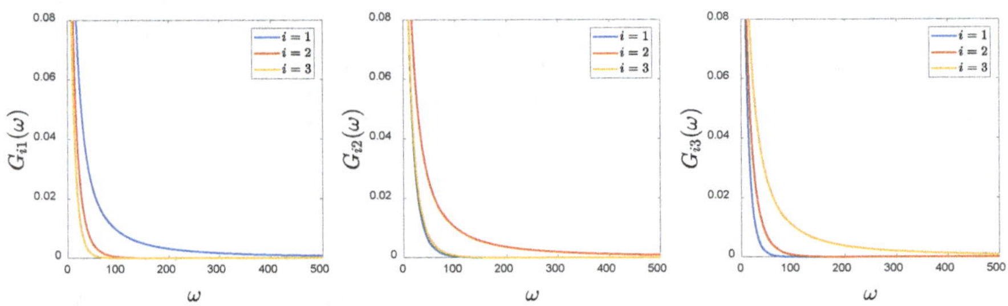

Figure 1 – Components of the PSDs matrix of the process $\mathbf{V}(t)$

From Fig.1 it is clear that the cross-PSDs decrease if the distance between the point increases. Moreover, all the components of the PSDs matrix decrease when the frequency increases.

The eigenvalues and the eigenvectors of the PSDs matrix have been calculated and are reported in Fig.2.

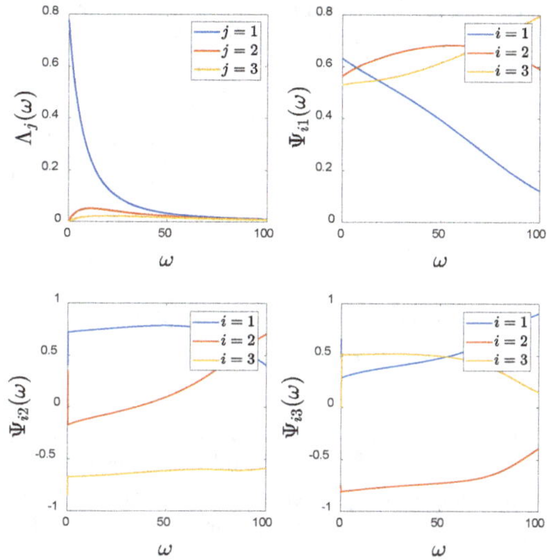

Figure 2 – Eigenvalues ($\Lambda_j(\omega)$) and components of the eigenvectors ($\Psi_{ij}(\omega)$) of $\mathbf{G_V}(\omega)$

From Fig.2 can be observed that the eigenvalues have the same behavior of the PSDs matrix, i.e. they decrease when the frequencies increases. Moreover, from Fig.2, it can be seen that the first eigenvalue gives the major contribution to the total vector process.

Once that the PSDs matrix of the process $\mathbf{V}(t)$ and its eigenvalues and eigenvectors have been calculated, Eqs.(12, 13) have been used to generate the process $\mathbf{V}(t)$.

Since Priestley's formula has already been applied in the literature to reduce the computational burden required to generate a process $\mathbf{V}(t)$ [1,8], the main innovation of this paper is the truncation of the summation in Eq.(8) linked to the way suggested to preserve the variance. For this reason, in this numerical simulation, two processes have been generated considering the same number of waves: the first one has been obtained by using Eqs.(8,10) without any truncation, while the second one has been generated through the use of the proposed method, i.e. by using Eqs.(12,13). In Fig.3 few second of a sample of the two generated processes are reported.

From Fig.3 it is clear that the proposed method leads to good results. Moreover, the discrepancies between the variances of the processes calculated by using Eq.(12) and the variances of the processes calculated by using Eq.(8) are computed as

$$\varepsilon_j\% = 100\frac{\int_0^{t_f}\left|\sigma^2_{V_j}(t)-\sigma^2_{\tilde{Q}_j}(t)\right|dt}{\int_0^{t_f}\left|\sigma^2_{V_j}(t)\right|dt} \tag{15}$$

The results obtained from Eq.(15) are reported in Tab.1

Table 1 – Discrepancies between the variances of the processes $V_i(t)$ generated by using Eq.(8) and the variances of the processes $V_i(t)$ generated by using Eq.(12)

$\varepsilon_1\%$	$\varepsilon_2\%$	$\varepsilon_3\%$
3.5464	3.0232	2.5727

From these results it is clear that the proposed method can be considered reliable.

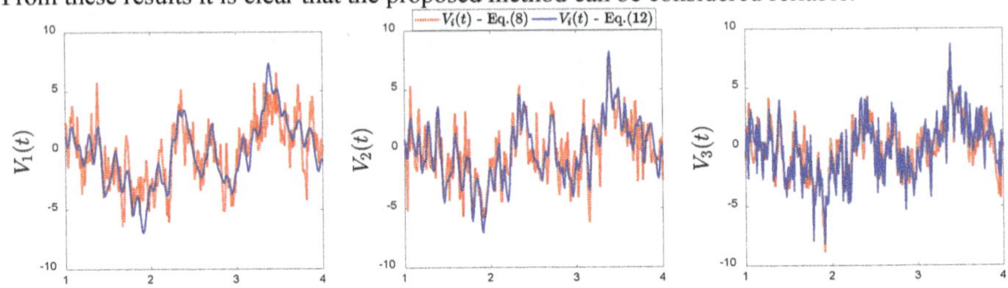

Figure 3 – Comparison between the processes $V_i(t)$ generated by using Eq.(8) and the processes $V_i(t)$ generated by using Eq.(12)

The time taken to generate 1000 samples of the process $\mathbf{V}(t)$ by using Eq.(8) is equal to 176 min, while the time taken to generate the same number of samples by using Eq.(12) is equal to 59.33 min, thus, according to Eq.(14), $T.\% = 33.71\% \approx 1/n$. Furthermore, the computational burden can still be reduced by exploit the advantages deriving from the use of the Priestley's formula. To this regard, the PSDs in Fig.1 and its eigenvalues reported in Fig.2 can help to choose the cut-off frequency $N\Delta\omega$.

Conclusions

In this paper an innovative method to simulate multi-variate stochastic processes is proposed. It uses the Priestley formulation since it is more performant than the Shinozuka's one for the generation of Gaussian processes. The computational burden required to generate a multi-variate stochastic processes is drastically reduced taking into account only the first blowing mode shape provided that the variance of the process is preserved. The numerical simulations performed show that the proposed method is reliable, thus it can be used to drastically reduce the computational burden required to generate multi-variate stochastic processes.

References

[1] M. Di Paola, I. Gullo, Digital generation of multivariate wind field processes, Probabilistic Eng. Mech. 16 (2001) 1-10. https://doi.org/10.1016/S0266-8920(99)00032-6

[2] N.T. Thomopoulos, Essentials of Monte Carlo Simulation, first ed., Springer, New York, 2013. https://doi.org/10.1007/978-1-4614-6022-0

[3] M. Di Paola, S. Russotto, A. Pirrotta, Self-similarity and response of fractional differential equations under white noise input, Prob. Eng. Mech. 70 (2022) 103327. https://doi.org/10.1016/j.probengmech.2022.103327

[4] S. Russotto, A. Di Matteo, A. Pirrotta, An Innovative Structural Dynamic Identification Procedure Combining Time Domain OMA Technique and GA, Buildings 12(7) (2022) 963. https://doi.org/10.3390/buildings12070963

[5] S. Russotto, A. Di Matteo, C. Masnata, A. Pirrotta, OMA: From research to engineering applications, Lect. Notes Civ. Eng. 156 (2021) 903-920. https://doi.org/10.1007/978-3-030-74258-4_57

[6] A. Di Matteo, C. Masnata, S. Russotto, C. Bilello, A. Pirrotta, A novel identification procedure from ambient vibration data, Meccanica 56 (2021) 797-812. https://doi.org/10.1007/s11012-020-01273-4

[7] M. Di Paola, A. Pirrotta, Non-linear systems under impulsive parametric input, Int. J. Non Linear Mech. 34(5) (1999) 843-851. https://doi.org/10.1016/S0020-7462(98)00057-2

[8] M. Di Paola, Digital simulation of wind field velocity, J. Wind. Eng. Ind. Aerodyn. 74-76 (1998) 91-109. https://doi.org/10.1016/S0167-6105(98)00008-7

[9] M. Shinozuka, Simulation of multivariate and multidimensional random processes, J. Acoust. Soc. Amer. 49(1) (1971) 357-367. https://doi.org/10.1121/1.1912338

[10] E. Simiu, R.H. Scanlan, Wind effects on structures, Wiley, New York, 1978.

[11] G. Solari, Gust Buffeting. I: Peak Wind Velocity and Equivalent Pressure, J. Struct. Eng. 119(2) (1993) 365-382. https://doi.org/10.1061/(ASCE)0733-9445(1993)119:2(365)

Theoretical and Applied Mechanics - AIMETA 2022 Materials Research Forum LLC
Materials Research Proceedings **26** (2023) 567-572 https://doi.org/10.21741/9781644902431-92

Identification of tie-rod properties in monumental buildings under uncertainty

Chiara Pepi[1,a *], Mircea D. Grigoriu[2,b] and Massimiliano Gioffrè [1,c]

[1]Dept. of Civil and Environmental Engineering, Univ. of Perugia via G. Duranti 93, 06125 Perugia, Italy

[2]Dept. of Civil and Environmental Engineering, Cornell University Hollister Hall, Ithaca, New York 14853, USA

[a]chiara.pepi@unipg.it, [b]mdg12@cornell.edu, [c]massimiliano.gioffre@unipg.it

Keywords: Uncertainty Quantification, Structural Identification, Stochastic Reduced Order Models, Structural Health Monitoring, Tie-Rods in Monumental Buildings

Abstract. Tie-rods are simple structural elements that have been commonly used throughout the centuries to improve the stability of old masonry buildings. Measuring the dynamic vibrations of such elements can give important information on the structural performance variation with time if proper system models can be calibrated taking into account all the relevant sources of uncertainties. This paper poses and solves the inverse problem of quantifying the uncertainty in the mechanical properties, tensile axial force and boundary conditions of tie-rods from measurements of their experimental frequencies of vibrations. It is assumed that the probabilistic models of the random entries of the problem, i.e., the mass per unit length, the bending stiffness, the tensile force and the boundary conditions, are known up to some parameters which need to be estimated. Both the physical and probability spaces are discretized using finite dimensional (FD) models, i.e., deterministic functions of time and/or space and finite numbers of random variables. Monte Carlo simulation is finally used to completely characterize the uncertain features obtaining the best estimation of all the unknown features such as mass per unit length, stiffness, boundary conditions and tensile force that give the target free vibration modal parameters, i.e. natural frequencies.

Introduction

The accurate estimation of the axial tensile force in tie-rods is one of the concerns when dealing with the performance assessment of old masonry structures. The problem is complex and it is yet far from having a unique solution. Tie-rods are simple structural elements that, when placed at the arches or vaults springings, contribute to the construction stability balancing the horizontal thrusts. Monitoring the changes in time of the tensile axial force in these elements can give information on possible mechanical failures causing a new internal forces distribution among the structural elements. As an example, cracks in the masonry walls or differential settlements, which can compromise the stability and the integrity of the whole building, can initiate this process of change in the internal forces in the structural elements, including the tie-rods.

Several approaches have been proposed in the literature to estimate the axial tensile force, starting from the assumption to consider the tie-rod as a vibrating wire and measuring its first modal frequency [1]. Among all the proposed approaches it is worth recalling the first works where non-destructive static and/or dynamic tests were used [2-7] up to the most recent contributions where the combined use of experimental measures of the frequency response funct-

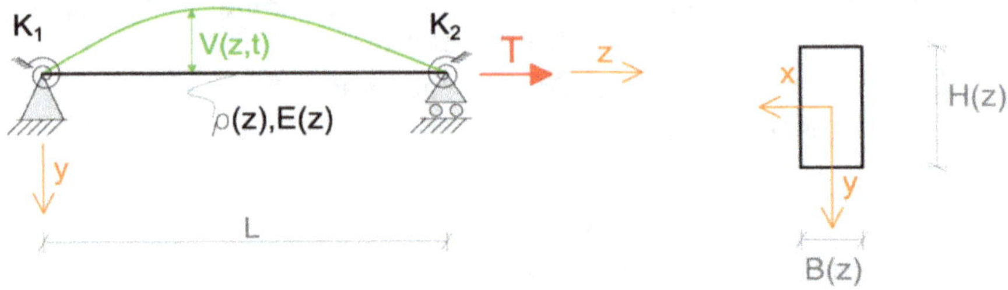

Figure 1. Structural model for the tie-rod.

ions and numerical solutions of the structural problem with unknown parameters gave interesting results [8-12].

The axial tensile force in tie-rods estimation is a classical inverse problem that can be ill-posed or ill-conditioned, depending on the available information, and the majority of the solution approaches proposed in the literature are set in a deterministic environment. On the other hand most of the geometrical and mechanical parameters, together with the boundary conditions are significantly affected by uncertainties, making the estimation problem even more difficult. A recent work investigated the influence of significant parameters on the evaluation of axial load through the application of frequency identification methods using the General Polynomial Chaos Expansion (gPCE) to perform sensitivity analysis [13].

This paper proposes a first approach to identify the random features of all the tie-rods uncertain quantities using numerical structural analysis and Monte Carlo simulation. First, initial assumption on the random models are proposed. Second, samples of all the involved tie-rod random parameters are generated and the corresponding numerical solutions of the tie-rod free vibration problem are obtained. Third, a suitable error function between the numerical natural frequencies and the target experimental values is minimized changing the main features of the assumed random models. The optimal solution is finally used to generate samples of all the involved uncertain tie-rod characteristics and to describe their main random properties. The proposed approach provides the tensile axial force probability density function estimation.

Problem definition

The structural model of the tie-rod considered in this paper is summarized in Fig. 1. The tie-rod deflection is a vector-valued function of time and space which depend on the cable mechanical properties and tensile force T. In this paper, for simplicity, the projection $\{V(z;t);\ 0 \leq z \leq l;\ t \geq 0\}$ of the deflection on the $y-z$ plane will be considered. $V(z;t)$ is a real-valued function of time and space which satisfies the partial differential equation

$$\frac{\partial^2}{\partial z^2}\left[EI(z)\,\frac{\partial^2 V(z,t)}{\partial z^2}\right] + T\,\frac{\partial^2 V(z,t)}{\partial z^2} = R(z)\,\frac{\partial^2 V(z,t)}{\partial t^2},\ 0 < z < l \qquad (1)$$

with arbitrary initial conditions, $V(z, 0)$ and $\dot{V}(z, 0)$, and the boundary conditions,

$$V(0,t)=0, \qquad V(l,t)=0$$
$$EI(z)\,\frac{\partial^2 V(0,t)}{\partial z^2}-K_1\frac{\partial\ V(0,t)}{\partial z}=0 \tag{2}$$
$$EI(z)\,\frac{\partial^2 V(l,t)}{\partial z^2}-K_2\frac{\partial\ V(l,t)}{\partial z}=0$$

where $EI(z)$ denotes the flexural stiffness, $R(z) = \rho(z)A(z)$ is the mass per unit length, $\rho(z)$ is the mass density, $A(z)$ is the cross-section area, and K_i, $i = 1, 2$, denote the stiffnesses of the end rotational springs. The solution $V(z, t)$ of this equation describes the free vibration of the tie-rod in Fig. 1. It depends on the beam properties described by the parameters K_1, K_2 and T and the functions $EI(z)$ and $R(z)$.

The objective of this study is to identify the tie-rod mechanical properties from experimental measurements giving natural frequencies and corresponding modal shapes. This problem does not admit an analytical solution even in the deterministic setting. The uncertainty in the beam mechanical properties is captured by the random variables K1, K2 and T and the random fields EI(z) and R(z). The main task in the stochastic setting is to identify the probability laws of the random variables K1, K2 and T and of the random fields EI(z) and R(z).

Assumptions

The random entries of Eq. 1 must satisfy physics constraints, i.e., the samples of the random variables K_1, K_2 and T and the samples of the random fields $EI(z)$ and $R(z)$ must take values in bounded intervals of $(0, \infty)$. These constraints impose restrictions which are used to partially specify probabilistic models of the random elements in Eq. 1.

The case study of this work is one of the four tie-rods in Santa Maria della Consolazione Temple, one of the most important monumental buildings in Umbria Region, Central Italy, located in the South West Area of Todi's city. Laser vibrometer and radar interferometer experimental vibration measurements where performed with the aid of drones (Unmanned Arial Vehicles) because of the inaccessibility of the tie-rods [14]. The following probabilistic models are assumed:

- *Cross-section dimensions, $B(z) = H(z) = 7$ cm,* can be considered constant along the tie-rod. It follows that the cross-section area, $A(z) = A$, and second moment, $I(z) = I$, are also constant.
- *Rotational springs* K_1 and K_2 are independent Beta random variables with known support $[0,\ 2\text{x}10^6]$ N m and unknown shape parameters (p_i,q_i), taking values in the bounded triangles $J_i=\{(u,v) : 1 \le u \le 8,\ 0 \le v \le u\}$, $i = 1, 2$.
- *Tension* T is a Beta random variable with specified shape parameters $p = q = 8$ but unknown mean μ_T, and support $[\mu_T - \alpha,\ \mu_T + \alpha]$, $0 \le \alpha \le \mu_T$, , i.e., $T = \mu_T + 2\,\alpha\,(X - 1/2)$, where X denotes the standard Beta variable with support [0, 1] and $p = q = 8$.
- *Mass per unit length* $R(z) = F_R \circ\Phi\,(G_R(z))$ is a Beta translation random field whose image $G_R(z)$ is a homogeneous Gaussian field with zero-mean, unit-variance and correlation function $E[G_R(z)\,G_R(z+y)] = \exp - \lambda_R\,|y|$, $\lambda_R > 0$. The map $G_R(z) \rightarrow R(z)$ is defined by the marginal distribution F_R of $R(z)$ and the distribution Φ of the standard Gaussian variable. It is assumed that the support of F_R is $[0.8,\ 1.2]\ \mu_\rho$, where $\mu_\rho = 7500\text{kg/m}^3$, the shape parameters (p_R, q_R) of F_R are unknown with values in $[1, 8]^2$ and the decay parameter λ_R is in [0, 2].
- *Joung modulus* $E(z) = F_E \circ \Phi\ G_E(z)$ is assumed to be a Beta translation random field with the same form as $R(z)$ but independent of this field. Its samples take values in the range $[0.8,\ 1.2]\ \mu_E$, where $\mu_E = 1.8\text{x}10^{11}$ Pa, the shape parameters (p_E, q_E) of the Beta distribution take values in $[1, 8]^2$ and the decay parameter λ_E of the correlation function of $G_E(z)$ is in [0, 2].

The ranges of the unknown parameters are selected after a preliminary sensitivity analysis solving the direct problem in Eq. 1 with the finite element (FE) approach.

Since the probabilistic models of the random variables $K1$, $K2$ and T and the random fields $R(z)$ and $E(z)$ are mutually independent, samples of these random elements can be generated by standard algorithms, provided that the unknown parameters of their models are specified. The discretization of the probability space requires to approximate the random functions $R(z)$ and $E(z)$ in the definition of the stochastic problem, which constitute uncountable families of random variables indexed by time and/or space, by finite dimensional (FD) models, i.e., deterministic functions of time and/or space and finite numbers $d < \infty$ of random variables. In this case study, samples of the Gaussian image of $R(z)$ are generated from the recurrence formula $G(z + \Delta z) = (1 - \lambda_G \Delta z)\ G(z) + (2\ \lambda_G\ \Delta z)^{1/2}\ N_z$ with the initial condition $G(0) \sim N(0,\ 1)$, where $\Delta z > 0$, $N(0,\ 1)$ is a standard Gaussian variable and $\{N_z\}$ are independent $N(0,\ 1)$ variables which are independent of $G(0)$ [15].

Identification of unknown model parameters

Denote by $(v_1(\xi,\omega), v_2(\xi,\omega),...)$ the modal frequencies of the system in Eq. 1 obtained from a sample ω of the random variables K_1, K_2 and T and the random fields $R(z)$ and $E(z)$ corresponding to a given value of the vector $\xi = (p_1, q_1, p_2, q_2, \mu_T, \alpha, p_R, q_R, \lambda_R, p_E, q_E, \lambda_E)$ of unknown parameters of the probabilistic models of these random elements. Samples $(v_1(\xi,\omega), v_2(\xi,\omega),...)$ of the modal frequencies are delivered by the FE numerical solution of Eq. 1.

Denote by $(v_{exp,1},\ v_{exp,2}, \ldots)$ the experimental values of the modal frequencies. The mean square error $E[(v_q(\xi) - v_{exp,q})^2]$, $q = 1, 2, \ldots$, is used to quantify the discrepancy between calculated and measured modal frequencies and can be estimated from samples $(v_1(\xi,\ \omega),\ v_2(\xi,\ \omega),...)$ of the modal frequencies. The discrepancy between the first $\bar{q}$ calculated and measured modal frequencies is measured by the objective function

$$\varepsilon(\xi) = \sum_{q=1}^{\bar{q}} w_q E[(v_q(\xi) - v_{exp,q})^2] \qquad \bar{q} = 1,2,\ldots \tag{3}$$

where $\{w_q\}$ are weighting factors. The optimal value ξ_{opt} of ξ minimizes the objective function, i.e., $\varepsilon(\xi_{opt}) = \min_{\varepsilon} \{\varepsilon(\xi)\}$.

Standard genetic algorithms are used to find ξ_{opt}, i.e., identify the optimal models of the random elements in the definition of Eq. 1, by a two step procedure. First, the initial vector of parameters $\xi_0 = \hat{\xi}_0$ (first row of Table 1) using, as much as possible, the information obtained from a site survey are set. Second, the objective function in Eq. 3 is minimized by genetic algorithms varying the number of samples, n_s of the random variables K_1, K_2 and T and the random fields $R(z)$ and $E(z)$ corresponding to a given value of the vector ξ of unknown parameters of the probabilistic models, i.e. $n_s = [10,\ 50,\ 100]$, between the limits fixed in rows two, ξ_{min}, and three ξ_{max} of Table 1.

Table 1. Values of the parameter vector ξ and corresponding $\varepsilon(\xi_{opt})$.

	p_1	q_1	p_2	q_2	μ_T 10^5 N	α 10^5 N	p_R	q_R	λ_R	p_E	q_E	λ_E	$\varepsilon(\xi_{opt})$
$\hat{\xi}_0$	5	1.5	5	1.5	2	0.30	2	3	0.3	3	2	0.3	—
ξ_{min}	1	1.0	1	1.0	0	0.05	1	1	0.0	1	1	0.0	—
ξ_{max}	8	8.0	8	8.0	5	0.90	8	8	2.0	8	8	2.0	—
ξ_{opt10}	5.9	2.8	5.6	2.2	4.2635	0.6051	3.7	5.4	1.7	4.1	3.0	1.3	0.0222
ξ_{opt50}	6.3	3.3	6.2	3.3	4.2823	0.6033	3.5	5.2	1.8	4.7	3.5	1.7	0.0215
ξ_{opt100}	7.2	3.0	6.1	2.8	4.1027	0.6206	3.1	6.5	1.9	4.8	4.7	1.5	0.0239

Table 2. First two tie-rod natural frequencies statistics estimated from n_s = 10000 samples.

	$v_{exp,1} = 4.4920$				$v_{exp,2} = 9.4320$			
	$E[v_1(\xi_{opt},\omega)]$	cov	γ_3	γ_4	$E[v_2(\xi_{opt},\omega)]$	cov	γ_3	γ_4
ξ_{opt10}	4.4613	0.0191	-0.0324	2.8871	9.4321	0.0178	-0.0324	2.9338
ξ_{opt50}	4.4613	0.0190	-0.0264	2.9221	4.4280	0.0176	-0.0264	2.8619
ξ_{opt100}	4.4629	0.0191	0.0126	2.8636	4.4394	0.0177	0.0126	2.8743

Results

Using the described assumptions, the objective function in Eq. 3, with $\bar{q} = 2$ ($v_{exp,1} = 4.492$, $v_{exp,2} = 9.432$) and $w_q = [1, 1]$, is minimized considering the optimum reached when the variation in the objective function in Eq. 3 using the last 50 populations is lower than 0.01. The results of the optimization are listed in Table 1, that shows the optimal parameter vectors, ξ_{opt10}, ξ_{opt50}, ξ_{opt100}. Table 2 shows the estimated first four statistical moments for the two natural frequencies $v_1(\xi_{opt}, \omega)$ and $v_2(\xi_{opt}, \omega)$ based on n_s = 10000 samples of the involved random variables K_1, K_2, T, and fields, $R(z)$, $E(z)$, corresponding to each of the optimal vectors $\xi_{opt_{ns}}$. Given the limited space available the estimated probability density functions of the tensile axial force and the estimated numerical natural frequencies are not shown.

Two main findings can be highlighted. First, Table 1 shows slight variations on the components of the optimal parameter vectors $\xi_{opt_{ns}}$ obtained with different values of $n_s \in \{10, 50, 100\}$. These variations seem to be related to the inherent statistical variability of the proposed method. Second, this apparent variability in the optimal results is not reflected in the estimated random features of both tensile force and natural frequencies. The coefficient of variation is about 3% and lower than 2% (Table 2) for the axial tensile force and natural frequencies, respectively.

References

[1] S. Mastrodicasa, Dissesti statici delle strutture edilizie, Hoepli, Milan, 1974.

[2] T. Livingston, J.G. Be´liveau, D.R. Huston, Estimation of axial load in prismatic members using flexural vibrations, Journal of Sound and Vibration, 179(5), 899–908, 1995. https://doi.org/10.1006/jsvi.1995.0060

[3] M.L. Beconcini, A practical method for the tensile axial force identification of ancient tie-rods, Costruire in laterizio, 54, 299-301, 1996.

[4] S. Sorace, Parameter models for estimating in-situ tensile force in tie-rods, Journal of Engineering Mechanics, ASCE, 122(9), 818–825, 1996. https://doi.org/10.1061/(ASCE)0733-9399(1996)122:9(818)

[5] S. Briccoli Bati,U. Tonietti, Experimental methods for estimating in situ tensile force in tie-rods, Journal of Engineering Mechanics, ASCE, 127(12), 1275–1283, 2001. https://doi.org/10.1061/(ASCE)0733-9399(2001)127:12(1275)

[6] N. Tullini N, F. Laudiero, Evaluation of the axial force in prismatic beams by means of modal flexural parameters, in: XVI Congress of the Italian Association of Theoretical and Applied Mechanics - AIMETA 2003, Ferrara, IT, September 9-12, 2003.

[7] S. Lagomarsino, C. Calderini, The dynamical identification of the tensile force in ancient tie-rods, Engineering Structures, 27(6), 846–856, 2005. https://doi.org/10.1016/j.engstruct.2005.01.008

[8] M. Amabili, S. Carra, L. Collini, R. Garziera, A. Panno, Estimation of tensile force in tie-rods using a frequency-based identification method, Journal of Sound and Vibration, 329(11), 2057–2067, 2010. https://doi.org/10.1016/j.jsv.2009.12.009

[9] H. Luong, L.F. Ramos, R. Aguilar, Identification of the tensile force in tie-rods of historical constructions, Civil Engineering Topics, Volume 4, 71–81, 2011. https://doi.org/10.1007/978-1-4419-9316-8_6

[10] N. Tullini, G. Rebecchi, F. Laudiero, Reliability of the tensile force identification in an- cient tie-rods using one flexural mode shape, International Journal of Architectural Heritage, 13(3), 402–410, 2019. https://doi.org/10.1080/15583058.2018.1563227

[11] C. Gentile, C. Poggi, A. Ruccolo, M. Vasic, Vibration-based assessment of the tensile force in the tie-rods of the Milan Cathedral, International Journal of Architectural Heritage, 13(3), 411–424, 2019. https://doi.org/10.1080/15583058.2018.1563235

[12] D. Camassa, A. Castellano, A. Fraddosio, G. Miglionico, M.D. Piccioni, Dynamic identification of tensile force in tie-rods by interferometric radar measurements, Applied Sciences, 11(8), 3687, 2021. https://doi.org/10.3390/app11083687

[13] A. De Falco, C. Resta, G. Sevieri, Sensitivity analysis of frequency-based tie-rod axial load evaluation methods, Engineering Structures, 229, 111568, 2021. https://doi.org/10.1016/j.engstruct.2020.111568

[14] M. Gioffrè, N. Cavalagli, C. Pepi, M. Trequattrini, Laser doppler and radar interferometer for contactless measurements on unaccessible tie-rods on monumental buildings: Santa Maria della Consolazione Temple in Todi, Journal of Physics Conf. Series, 778, 2017. https://doi.org/10.1088/1742-6596/778/1/012008

[15] M. Grigoriu, Stochastic Systems. Uncertainty Quantification and Propagation. Springer Series in Reliability Engineering. Springer, London Heidelberg NY Dordrecht, 2012. https://doi.org/10.1007/978-1-4471-2327-9

Open issues on procedures and methodologies for the vibration-based monitoring and dynamic identification of historic constructions

Materials Research Forum LLC
https://doi.org/10.21741/9781644902431-93

Post-earthquake continuous dynamic monitoring of the twin belfries of the Cathedral of Santa Maria Annunziata of Camerino, Italy

Gianluca Standoli[1,a], Francesco Clementi[1,b *], Carmelo Gentile[2,c], Stefano Lenci[1,d]

[1] Department of Civil and Building Engineering, and Architecture, Polytechnic University of Marche, Ancona, Italy

[2] Department of Architecture, Built Environment and Construction Engineering, Technical University of Milan, Milan, Italy

[a]g.standoli@univpm.it, [b] francesco.clementi@univpm.it, [c]carmelo.gentile@polimi.it, [d]s.lenci@univpm.it

Keywords: Masonry Towers, Structural Health Monitoring, MEMS, Weather Conditions

Abstract. In this paper the results of over two years of continuous monitoring of the twin bell towers of the Cathedral of Camerino (Italy) are presented. The monitoring activity target is the evaluation of the dynamic behavior of the twin belfries after the damages occurred during the seismic events of 2016 and the application of fast securing. The experimental data are acquired in continuous using four triaxial MEMS accelerometers, two for each structure, on two opposite corners of the bell cells. Data processing is managed with an automatic system which elaborates the signals and executes Operational Modal Analysis to track the modal characteristics of the structures and their evolution in time. Correlation with environmental factors allows to discern the effect of climatic conditions on the variations of dynamics.

Introduction

The preservation of masonry Cultural Heritage (CH) buildings has become one of the main goals for civil engineering [1-3]. Over the last decades, due to their vulnerability to natural hazards, i.e. earthquakes, different methodologies have been developed to assess their health status and consequently design intervention plans. Among the various Non-Destructive Tests (NDT) methods available, one of the techniques that mostly attracts the interest of researchers is the Ambient Vibration Testing (AVT) [4], which consists in monitoring the environmental and anthropogenic vibrations induced on the structure in its operating conditions [5-7], and finally extracting the main dynamic parameters (frequencies, modal shapes, and damping factors) [8, 9], after filtering of the non-structural effects [1,10-12].

A particular application of this technique is the system, composed of four MEMS accelerometers, installed on the 4th February 2020 in the bell towers of the Cathedral of Santissima Maria Annunziata of Camerino. This monument suffered severe damage during the seismic sequence of 2016 that stroked Central Italy, leading to reinforcement interventions that made the church an iconic case study for the development of a continuous monitoring system. In the paper, the first results of the modal parameters tracking are presented, automatically executed [13] with a self-made Matlab© script, on the vibrations data acquired for the twin bell towers during over a year of monitoring activity.

The Case Study: Camerino Cathedral Bell Towers

The city of Camerino (Fig. 1a) arises on a hill (661 m a.m.s.l.) of the Macerata Province, in the Marche region (Italy). In the main square is locate the Cathedral (Fig. 1b), nowadays known as Santissima Annunziata of Camerino, originally built during the V century d.C. as a Roman basilica dedicated to San Giuseppe. Starting from XIII century, according to the historical documentation, the structure is subject to great alterations because of the Swabian devastations and the evolution of the economy and demography of the city. The renovation culminated with the restoration works following the complete collapse of the bell tower caused by an earthquake in 1279. New restyling operations started between 1748-1749, when the XIII century façade was completely rebuilt with baroque canons.

Figure 1. Localization of the city of Camerino (a) and a view of Santissima Maria Annunziata Cathedral (b)

Due to the 1799 seismic events, the building was destroyed, thus re-building works started under the direction of Andrea Vici. This project contemplated the complete reconstruction of the church with a Latin-cross planimetry and the expansion of the square in the front.

After a stop, due to the French occupation in 1807, the restoration works were resumed 10-20 years later and the Cathedral assumed the shape we observe nowadays, with the two robust symmetrical belfries and the front gallery. The works finished in 1832, when the church was consecrated to Santissima Annunziata.

Other earthquakes occurred in the following years (i.e., 1873, 1897, 1979) and only after the seismic events of 1997 steel curbs were installed in the roof.

The geometrical and material surveys of the Church were operated through laser scanning technology, which allowed to extract the planimetry, the lateral views and the sections of the building. From the right aisle, it is possible to access the relative tower, otherwise the left tower (Fig. 1) is accessible through a door from the outside. The towers are symmetrical, with a peak height of 40.8 meters above the countryside level, with an irregular octagonal cross-section and a planimetric footprint of around 7.40 x 6.92 m^2. The belfries floors stand at around 25 m of altitude and present four vast arch opening on the sides (Fig. 1b).

Several types of masonry texture [14], the presence of opening and different heights of the foundations constitutes vulnerability elements which contributed to reach the observed level of damage.

Ambient Vibration Testing

The continuous monitoring system is composed of four triaxial MEMS accelerometers, installed in pairs on two opposite corners of the towers, at a height of around 4 meters above the bell cell floors (Fig. 2). All the sensors are connected in chain, assuring the synchronization of the measures. Data are acquired for 20 minutes every hour, with a sampling frequency of 200 Hz. When the continuous acquisition process stops, acquisition on trigger activates and, in case of events, registrations start with sampling frequency of 1000 Hz and duration of 90 seconds of pre- and post-triggering.

Materials Research Proceedings **26** (2023) 575-580
https://doi.org/10.21741/9781644902431-93

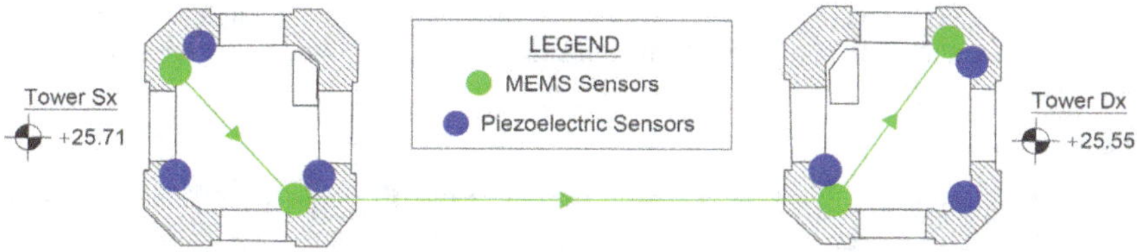

Figure 2. Sensors' layouts: respectively the continuous monitoring system (in green) and the short-time monitoring system (in blue).

Environmental data, concerning external temperature, external humidity, and wind speed, are collected through a weather station positioned in the proximity of the structure, available for consultation on the site http://app.protezionecivile.marche.it/sol/info.sol?lang=it.

In order to assess the initial conditions of the towers at the start of the long-time monitoring process, a short-time acquisition has been made, using 18 monoaxial piezoelectric accelerometers (Fig. 2), with a sensitivity of 10 V/g, fixed in groups of three sensors in six corners of the bell cell (in proximity of the MEMS sensors). This acquisition lasted 40 minutes with a sampling frequency of 1000 Hz.

Modal identification

Data are elaborated automatically through a self-made script developed in Matlab© environment. The towers data are analyzed separately, to highlight the difference in dynamic behavior including the influence of temporary (safety provisions) reinforcement interventions.

The script firstly pre-processes the short-time acquisition, applying de-trending, low-pass filtering and then decimating data in the range of 0-12.5 Hz, which is the one of interest for this type of structure. Then modal parameters identification is operated through automatic Enhanced Frequency Domain Decomposition (EFDD) [15] and Stochastic Subspace Identification - Covariance Based (SSI-Cov) methodology [16] for both towers, and the resulting modal frequencies and mode shapes (Fig. 3) become the targets for modal tracking over the continuous monitoring data.

From the identification operated over the first dataset (Table 1), it is observable that the two towers, have similar in terms of modal frequencies values. In Fig. 3 is reported the modal shapes of only the right tower for brevity reasons.

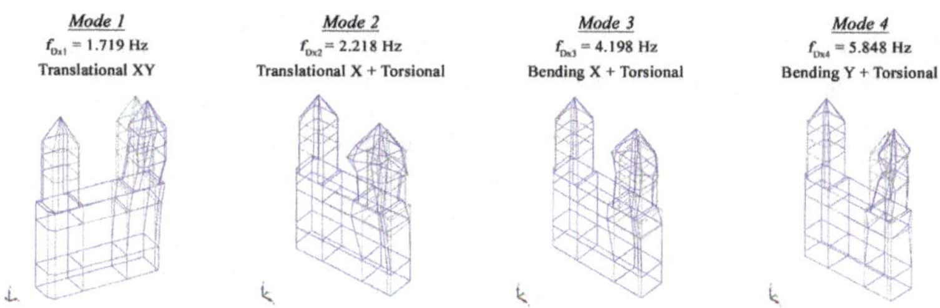

Figure 3. Target frequencies and mode shapes obtained from short-time monitoring data analysis for the right tower

Acceleration files stored through MEMS sensors are initially pre-processed, repeating the operations described for the short-time data. The parameters extraction process is instead operated recurring to two clustering levels: in the first step, hard validation, and minimization of a two terms objective function in Eq. 1 [17] which considers both frequencies variation and MAC values [18] between consecutives model orders; the second step starts from grouping the parameters corresponding to the same modes and then comparing these ones to the target ones, selecting the values with the best fitting.

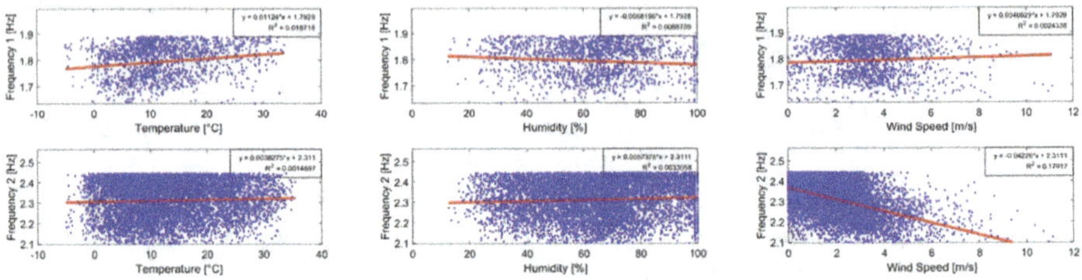

Figure 4. Example of interpolation of environmental data (Temperature, Humidity, Wind Speed) with the first two frequencies of the right tower

Automatically identified parameters are reported and their mean values are compared with the target ones in Table 1.

Table 1. Comparison between target and median automatically identified parameters

Mode	Left Tower			Right Tower			
	f_{Target} [Hz]	f_{Mean} [Hz]	$\Delta f_{Target\text{-}Tracked}$ [%]	f_{Target} [Hz]	f_{Mean} [Hz]	$\Delta f_{Target\text{-}Tracked}$ [%]	$\Delta f_{L\text{-}R}$ [%]
1	1.805	1.904	5.20	1.719	1.815	5.29	4.90
2	2.318	2.310	0.35	2.218	2.331	4.85	0.90
3	4.218	4.287	1.61	4.198	4.103	2.32	4.48
4	5.906	5.913	0.12	5.848	5.872	0.41	0.70

$$\frac{|f_i - f_{Target}|}{f_{Target}} + (1 - MAC_{i-Target}) \leq 0.05 \tag{1}$$

By overlapping the data before and after environmental effect removal, an interesting phenomenon is highlighted in the fourth modes of both towers: the occurrence of a freezing condition on 15th of February 2021 (day when the minima temperatures of the year were registered), due to the freezing of the water particles contained in masonry micropores, testified by a sudden increment in towers rigidity [5]. In Figure 5, for brevity issues, the results of only the right tower are reported.

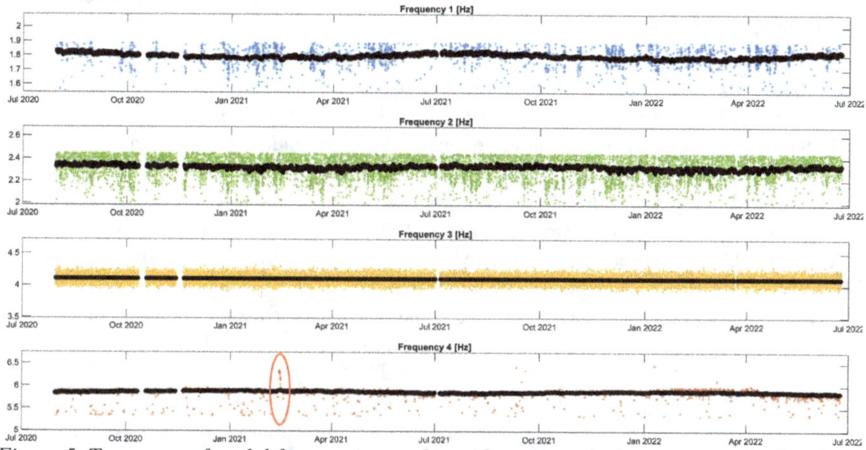

Figure 5. Two years of modal frequencies tracking after removal of environmental effects. The red circle highlights the freezing condition.

Summary

The paper shows an application of continuous structural monitoring by a particular type of accelerometric sensors, i.e. MEMS. Through a Matlab© script it was possible to identify the dynamic characteristics associated to the first four modes of the two twin towers of Santissima Maria Annunziata Cathedral in Camerino, a historical building strongly damaged by the seismic events of 2016-'17, and then subdued to temporary reinforcement interventions.

The procedure produced interesting results, allowing to understand the differences in the dynamic behavior of the two towers, despite their symmetry. Moreover, it was possible to prove the efficiency of MEMS sensors for continuous monitoring activity.

References

[1] M.G. Masciotta, J. C. A. Roque, L. F. Ramos, and P. B. Lourenço, A multidisciplinary approach to assess the health state of heritage structures: The case study of the Church of Monastery of Jerónimos in Lisbon, Constr. Build. Mater. 116, 169–187 (2016). https://doi.org/10.1016/j.conbuildmat.2016.04.146

[2] F. Magalhães, A. Cunha, and E. Caetano, Vibration based structural health monitoring of an arch bridge: From automated OMA to damage detection, Mech. Syst. Signal Process. 28, 212–228 (2012). https://doi.org/10.1016/j.ymssp.2011.06.011

[3] F. Clementi, A. Ferrante, E. Giordano, F. Dubois and S. Lenci, Damage assessment of ancient masonry churches stroked by the Central Italy earthquakes of 2016 by the non-smooth contact dynamics method, Bull Earthquake Eng. 18, 455–486 (2020). https://doi.org/10.1007/s10518-019-00613-4

[4] A. Pierdicca, F. Clementi, A. Fortunati, S. Lenci, Tracking modal parameters evolution of a school building during retrofitting works, Bull Earthquake Eng. 17(2), 1029-1052 (2019). https://doi.org/10.1007/s10518-018-0483-9

[5] F. Ubertini, G. Comanducci, N. Cavalagli, A. Laura Pisello, A. Luigi Materazzi, and F. Cotana, Environmental effects on natural frequencies of the San Pietro bell tower in Perugia, Italy, and

their removal for structural performance assessment, Mech. Syst. Signal Process. 82, 307–322 (2017). https://doi.org/10.1016/j.ymssp.2016.05.025

[6] A. Deraemaeker, E. Reynders, G. De Roeck, J. Kullaa, Vibration-based structural health monitoring using output-only measurements under changing environment, Mech. Syst. Signal Process. 22, 34–56 (2008). https://doi.org/10.1016/j.ymssp.2007.07.004

[7] A. Saisi, C. Gentile, M. Guidobaldi, Post-earthquake continuous dynamic monitoring of the Gabbia Tower in Mantua, Italy, Constr. Build. Mater. 81, 101-112 (2015). https://doi.org/10.1016/j.conbuildmat.2015.02.010

[8] W. Torres, J. L. Almazán, C. Sandoval, and R. Boroschek, Operational modal analysis and FE model updating of the Metropolitan Cathedral of Santiago, Chile, Eng. Struct.143, 169–188 (2017). https://doi.org/10.1016/j.engstruct.2017.04.008

[9] L. F. Ramos, G. De Roeck, P. B. Lourenço, and A. Campos-Costa, Damage identification on arched masonry structures using ambient and random impact vibrations, Eng. Struct. 32, 146–162 (2010). https://doi.org/10.1016/j.engstruct.2009.09.002

[10] C. Gentile, M. Guidobaldi, and A. Saisi, One-year dynamic monitoring of a historic tower: damage detection under changing environment, Meccanica 51, 2873–2889 (2016). https://doi.org/10.1007/s11012-016-0482-3

[11] P. F. Giordano, F. Ubertini, N. Cavalagli, A. Kita, and M. G. Masciotta, Four years of structural health monitoring of the San Pietro bell tower in Perugia, Italy: two years before the earthquake versus two years after, Int. J. Mason. Res. Innov. 5, 445–467 (2020). https://doi.org/10.1504/IJMRI.2020.111797

[12] R. Ceravolo, G. Coletta, G. Miraglia, and F. Palma, Statistical correlation between environmental time series and data from long-term monitoring of buildings Statistical correlation between environmental time series and data from long-term monitoring of buildings, Mech. Syst. Signal Process. 152 (2021). https://doi.org/10.1016/j.ymssp.2020.107460

[13] E. García-Macías and F. Ubertini, MOVA/MOSS: Two integrated software solutions for comprehensive Structural Health Monitoring of structures, Mech. Syst. Signal Process. 143 (2020). https://doi.org/10.1016/j.ymssp.2020.106830

[14] G.P. Salachoris, E. Magagnini, F. Clementi, Mechanical characterization of "Scaglia Rossa" stone masonry through experimental and numerical analyses," Constr. Build. Mater., 303 (2021). https://doi.org/10.1016/j.conbuildmat.2021.124572

[15] N. J. Jacobsen, P. Andersen, and R. Brinker, Using Enhanced Frequency Domain Decomposition as a Robust Technique to Harmonic Excitation in Operational Modal Analysism, in ISMA2006: International Conference on Noise and Vibration Engineering, Katholieke Universiteit, (2006).

[16] B. Peeters and G. De Roeck, Reference-Based Stochastic Subspace Identification for Output-Only Modal Analysis, Mech. Syst. Signal Process. 13, 855–878 (1999). https://doi.org/10.1006/mssp.1999.1249

[17] G. Standoli, G.P. Salachoris, M.G, Masciotta, F. Clementi, Modal-based FE model updating via genetic algorithms: Exploiting artificial intelligence to build realistic numerical models of historical structures, Constr. Build. Mater., 303 (2021). https://doi.org/10.1016/j.conbuildmat.2021.124393

[18] M. Pastor, M. Binda, and T. Harčarik, Modal Assurance Criterion, Procedia Eng. 48, 543–548, (2012). https://doi.org/10.1016/j.proeng.2012.09.551

Theoretical and Applied Mechanics - AIMETA 2022 Materials Research Forum LLC
Materials Research Proceedings **26** (2023) 581-586 https://doi.org/10.21741/9781644902431-94

Deep learning for structural health monitoring: An application to heritage structures

Fabio Carrara, Fabrizio Falchi, Maria Girardi*, Nicola Messina, Cristina Padovani and Daniele Pellegrini

Institute of Information Science and Technologies "A. Faedo", ISTI-CNR, Via Moruzzi 1, 56124 Pisa, Italy

* maria.girardi@isti.cnr.it

Keywords: Heritage Structures, Anomaly Detection, Deep Learning

Abstract. In this work, we employ recent deep learning techniques for time-series forecasting to inspect and detect anomalies in the large dataset recorded during a long-term monitoring campaign conducted on the San Frediano bell tower in Lucca. We frame the problem as an unsupervised anomaly detection task and train a Temporal Fusion Transformer to learn the normal dynamics of the structure. We then detect the anomalies by looking at the differences between the predicted and observed frequencies.

Introduction

Heritage structures are threatened worldwide by aging, material deterioration, environmental actions, and extreme meteorological events due to climate changes, and therefore they need maintenance and restoration. Structural health monitoring (SHM) provides a relatively inexpensive tool for promptly assessing the structural conditions, planning, and controlling maintenance interventions.

The availability of high-sensitive instrumentation and extensive data sets from long-term monitoring protocols opened new issues about data analysis, particularly regarding the implementation of automatic anomaly detection and early warning procedures. In this data-driven context, exploiting artificial intelligence (AI) represents a significant opportunity. Machine learning (ML) algorithms originated from the broader domain of AI and have recently had a significant diffusion, with applications in many research and industrial areas. ML covers a wide range of algorithms that recognize patterns and build regression models for large multi-source heterogeneous datasets: this approach naturally fits with data collected by SHM systems [1].

Bridges' construction and maintenance represent the classical application fields for automated vibration monitoring. Many papers regarding algorithms for operational modal analysis and damage detection in the domain of ML are available in the literature [2], [3], [4]. The application of ML to the preservation of architectural heritage is relatively recent [5].

In recent years, Deep Learning (DL) techniques have become the state-of-the-art for processing sequential data, like text or audio. The recent astonishing advancements in this field brought important innovations also in the context of time-series processing, obtaining interesting results in time-series forecasting, classification, and anomaly detection [6, 7]. The key to the success of DL in these tasks can be attributed to some neural architectures, such as the Recurrent Neural Networks or the recently introduced Transformer Networks [8]. Despite the fast and pervasive development of DL techniques for time-series processing, relatively few works [9, 10] tried to apply this promising technology to monitor heritage structures and find possible anomalies.

In this paper, we propose to use a recently developed Transformer Network [11] to reveal possible anomalies from the data recorded by the high-sensitive instrumentation installed on heritage structures. We analyze the San Frediano bell tower in Lucca [12], subject to a long-term vibration monitoring campaign from October 2015 to November 2017. The Transformer network is trained to predict the main natural frequencies of the tower starting from some environmental data and the frequencies observed in the very recent past. Anomalous events are then found by comparing how much the actual frequencies deviate from the predicted ones. The data analysis shows promising results, highlighting anomalous events like the Amatrice earthquake that happened on 24 Aug 2016 or the Santa Croce celebrations on 13 Sep 2016.

Description of the algorithm

The proposed approach belongs to a set of techniques that pertain to the domain of *unsupervised anomaly detection*. Specifically, a network is trained to understand the normality patterns present in some data and then, at inference time, the more significant deviations from the learned normality patterns are considered anomalies. In order to learn the concept of *normality* during the training process, the network is usually trained to predict one or more samples in the future, given observations from the recent past. In this way, the network learns to approximate the normal dynamics of the system under observation. In this work, we use a state-of-the-art deep neural network for learning the structure's dynamics directly from data. This network is called Temporal Fusion Transformer (TFT) [11]. In the following two paragraphs, we discuss how the TFT can predict the structure dynamics while characterizing its uncertainty and how this output can be used for anomaly detection.

The Temporal Fusion Transformer. In heritage structure monitoring, data are usually composed of post-processed sensor measurements from the monitored structure, plus some environmental data, like temperature, wind speed, rain, or humidity. The core part of the algorithm is constituted by the network, which predicts the structure's dynamic evolution, given the sensor and environmental data from the recent past. In this study, we use TFT [11], one of the state-of-the-art Transformer Networks for time-series forecasting. Apart from using state-of-the-art attentive modules for processing time-series, the TFT can also estimate the uncertainty of the prediction. This output is fundamental during the anomaly detection phase: we can quickly know whenever the observed actual values fall inside the predicted confidence interval and, if not, easily quantify the deviation from these margins. In the light of this, the system that we develop can be formalized as follows:

$$X_{T+1} = TFT(x_1, x_2, \ldots, x_T;\ u_1, u_2, \ldots, u_T), \tag{1}$$

where $\boldsymbol{x}_t$ is a vector encoding the values measured from the instrumentation in the structure at time t, $\boldsymbol{u}_t$ is a vector encoding the environmental (external) factors like temperature, wind speed, rain, or humidity at time t, while the output $\mathbf{X}_{T+1}$ is a random variable encoding the predicted sensor data at time $T+1$. The model characterizes this output random variable by predicting its mean value and 1, 10, 25, 50, 75, 90, 99 percentiles. The training procedure can be treated as a regression problem, where the objective consists of estimating the parameters of the TFT model given some training data. In TFT, the optimization is performed using a quantile loss, which estimates the aleatoric uncertainty of the data. More details in the original TFT paper [11].

Anomaly Detection. Once the TFT has been trained on some non-anomalous data, it should have learned the *normal* dynamics of the given system to some extent. Therefore, it should be able to predict the next system state while also quantifying the uncertainty of this prediction. At this point, we can search for anomalous events in a time interval never seen from the network at training time. Given the aleatoric prediction $\mathbf{X}_{T+1}$ and the actual observed state $\boldsymbol{x}_{T+1}$, we can define the state $\boldsymbol{x}_{T+1}$ as *anomalous* if it is an outlier with respect to the estimated distribution of the random variable $\mathbf{X}_{T+1}$. More formally, we can define $\boldsymbol{x}_{T+1}$ as an outlier – or an anomalous sample – by checking if the following condition is NOT satisfied:

$$\pi_{100-p}\,(X_{T+1}) \leq x_{T+1} \leq \pi_p\,(X_{T+1}), \tag{2}$$

where $\pi_p\,(\mathbf{X})$ represents the p-percentile of the random variable **X**. In our experiments, we use p = 99. This procedure is repeated over a sliding window having width T over the whole observation period so that we can predict the presence of anomalies at each timestep t given the sensor and environmental data from the timesteps $\{t\text{-}1,\ t\text{-}2,\ ...,\ t\text{-}T\text{-}1\}$.

Case study: The San Frediano bell tower

The San Frediano bell tower in the historic center of Lucca has been subjected to a long-term vibration monitoring campaign from October 2015 to November 2017 [12]. Data from the first year of monitoring (28 Oct 2015 – 13 Aug 2016) are used in the paper to train and validate the TFT model. They were measured by four seismic stations placed alongside the tower's height, each equipped with a tri-axial velocimeter and a 24-bit digitizer of the SARA Electronic Instruments firm. Data were sampled at 100 Hz and continuously acquired over the monitoring period. The training dataset was built by processing data via the covariance-driven Stochastic Subspace Identification technique to extract the first five natural frequencies from the tower's response. We organized data in hourly packages; therefore, the training set consisted of hourly samples of the five frequencies. Together with the frequencies, we considered other environmental variables in the model: air temperature, rainfall and air humidity, wind speed (average and peak values), and wind direction. Furthermore, we conditioned the model on the temporal fingerprints of the observations to better capture periodical patterns at different time scales. Specifically, we conditioned the model on the hour (1-24), the day (1-30), and the month (1-12).

We run the trained model on a time interval not used during the training phase. Specifically, we considered the period from 19 Aug 2016 to 16 Oct 2016, with a window width T = 96 hours (4 days). The results of the analysis are shown in Fig. 1 and Fig.2, reporting the predictions of the model and the observed frequencies. Every prediction is accompanied by the estimated uncertainty, represented by the area between the 1st and 99th percentiles at every prediction timestep. Vertical bars show the time locations when at least one of the frequencies violates the condition expressed in Eq. 2. The color intensity of these lines depends on how much the actual observed value falls outside the estimated uncertainty margins. The final color intensity (yellow low - red high) is obtained by summing the anomaly contributions from each frequency, using the inverse of the frequency as a weight, to attenuate the contribution from higher frequencies. Fig. 1 highlights two major anomalous events, on 24 Aug and 13 Sep 2016. The former event corresponds to the Amatrice earthquake that struck central Italy with a 6.0 magnitude and, despite the distance of 400 km between Amatrice and Lucca, was clearly detected by the sensors installed on the tower and induced vibrations in the same order of magnitude as those caused by the bells' swinging. The latter anomaly corresponds to the Santa Croce celebrations and relates to the presence of many people moving in the town and affecting the tower's vibrations. The algorithm can also detect the vibrations induced by the swinging bells. This anomaly involves the second frequency (related to the swinging direction) and corresponds to the main religious ceremonies held in the Cathedral, particularly on Saturday afternoon and Sunday morning.

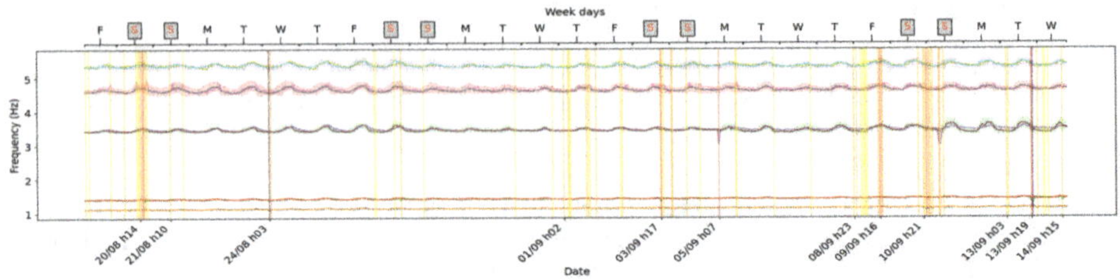

Figure 1: First five predicted and experimental (observed) frequencies, in the period 19 Aug 2016 - 14 Sep 2016.

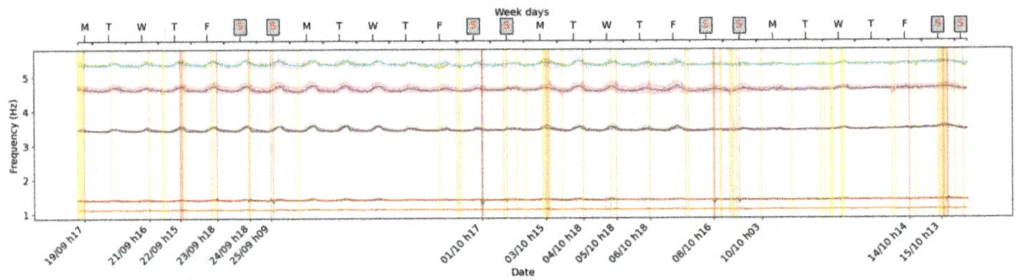

Figure 2: First five predicted and experimental (observed) frequencies, in the period 19 Sep 2016 - 16 Oct 2016.

The anomalous events detected by the TFT procedure are better emphasized in Fig. 3, reporting the observed and predicted values of the first two frequencies from 20 to 26 August 2016, together with the confidence interval evaluated by the algorithm and reported in a dashed line. The anomalies caused by the swinging bells on Saturday (20 Aug, h 17:00) and Sunday (21 Aug, h 10:00 and h 12:00) and by the Amatrice earthquake (24 Aug, h 3:00) are highlighted in the figure; the first two affect mainly the second frequency of the tower, while the effects of the earthquake are evident in the whole dynamic response of the structure.

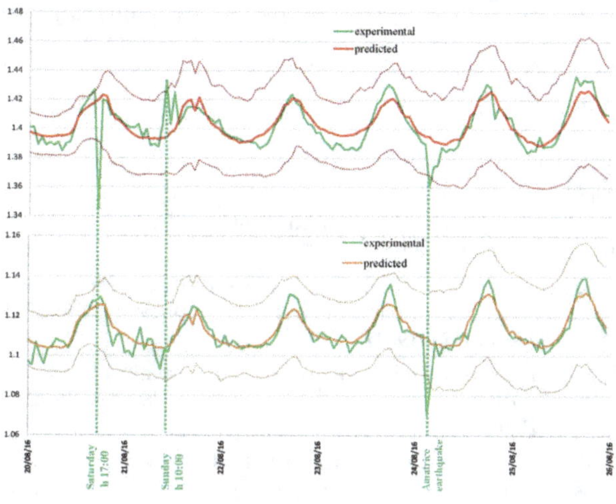

Figure 3: First two predicted and experimental (observed) frequencies, 20-26 Aug 2016.

Conclusions

In this paper, we explored the potential of state-of-the-art attentive deep neural networks for monitoring heritage structures and spotting possible anomalies. We employed the Temporal Fusion Transformer to learn the normal dynamics of the structure. The tower's dynamic behavior was encoded as its first five natural frequencies, extracted through the covariance-driven Stochastic Subspace Identification technique. The predicted distributions of the frequencies were then used to spot outliers in the test data. This preliminary work showed promising results in the case study of the San Frediano bell tower in Lucca, where we were able to correctly identify some important events, like the Amatrice earthquake and the Santa Croce celebrations.

Further investigations are necessary to confirm the good performance and reliability of the adopted method. A comparison between the anomaly detection techniques used in the literature and the TFT model will be carried out in a future work, where the performance of the Transformer network will be tested on artificial damage scenarios. Furthermore, it would be interesting to characterize the found anomalies by inspecting and clustering the hidden representations learned by the model. This characterization would be helpful, for example, for distinguishing a celebration event happening near the tower from an unexpected structural failure.

Acknowledgments

This research has been supported by the Fondazione Cassa di Risparmio di Lucca (SOUL project, 2019-2022), and by the INAROS project (CNR4C program, CUP B53D21008060008). This support is gratefully acknowledged.

References

[1] Y. Bao, Y., H. Li, Machine learning paradigm for structural health monitoring, Structural Health Monitoring 20(4) (2021) 1353-1372. https://doi.org/10.1177/1475921720972416

[2] C.R. Farrar, K. Worden, Structural health monitoring: a machine learning perspective, John Wiley & Sons, 2012. https://doi.org/10.1002/9781118443118

[3] J. Lim, S. Kim, H. Kim, Using supervised learning techniques to automatically classify vortex-induced vibration in long-span bridges, Journal of Wind Engineering and Industrial Aerodynamics 221 (2022) 104904. https://doi.org/10.1016/j.jweia.2022.104904

[4] A.J. Hughes, L.A. Bull, P. Gardner, R.J. Barthorpe, N. Dervilis, K. Worden, On risk-based active learning for structural health monitoring, Mechanical Systems and Signal Processing 167 (2022) 108569. https://doi.org/10.1016/j.ymssp.2021.108569

[5] M. Mishra, Machine-learning techniques for structural health monitoring of heritage buildings: A state-of-art review and case studies, Journal of Cultural Heritage 47 (2021) 227-245. https://doi.org/10.1016/j.culher.2020.09.005

[6] F. Carrara, P. Elias, J. Sedmidubsky, P. Zezula, LSTM-based real-time action detection and prediction in human motion streams, Multimedia Tools and Applications 78 (2018) 19: 27309-27331. https://doi.org/10.1007/s11042-019-07827-3

[7] B. Lim, S. Zohren, Time-series forecasting with deep learning: a survey. Philosophical Transactions of the Royal Society A 379 (2194) (2021) 20200209. https://doi.org/10.1098/rsta.2020.0209

[8] A. Vaswani, N. Shazeer, N. Parmar, J. Uszkoreit, L. Jones, A.N. Gomez, L. Kaiser, I. Polosukhin, Attention is all you need, Advances in neural information processing systems (2017).

[9] P. Seventekidis, D. Giagopoulos, A. Arailopoulos, O. Markogiannaki, Structural Health Monitoring using deep learning with optimal finite element model generated data, Mechanical Systems and Signal Processing 145 (2020) 106972. https://doi.org/10.1016/j.ymssp.2020.106972

[10] H.V. Dang, H. Tran-Ngoc, T.V. Nguyen, T. Bui-Tien, G. De Roeck, H.X. Nguyen, Data-driven structural health monitoring using feature fusion and hybrid deep learning, IEEE Transactions on Automation Science and Engineering 18(4) (2020) 2087-2103. https://doi.org/10.1109/TASE.2020.3034401

[11] B. Lim, S.O. Arık, N. Loeff, N., T. Pfister, Temporal fusion transformers for interpretable multi-horizon time series forecasting, International Journal of Forecasting 37(4) (2021) 1748-1764. https://doi.org/10.1016/j.ijforecast.2021.03.012

[12] R.M. Azzara, G. De Roeck, M. Girardi, C. Padovani, D. Pellegrini, E. Reynders, The influence of environmental parameters on the dynamic behaviour of the San Frediano bell tower in Lucca, Engineering Structures 1568 (2018) 175-187. https://doi.org/10.1016/j.engstruct.2017.10.045

Theoretical and Applied Mechanics - AIMETA 2022 Materials Research Forum LLC
Materials Research Proceedings **26** (2023) 587-592 https://doi.org/10.21741/9781644902431-95

Application of OMA technique to masonry slender towers: FEM updating and sensitivity analysis

Davide Li Rosi[1, a *], Loredana Contrafatto[1, b], Salvatore Gazzo[1, c], Leopoldo Greco[1, d] and Massimo Cuomo[1, e]

[1]Department of Civil Engineering and Architecture, University of Catania, Italy

[a]davide.lirosi@phd.unict.it, [b]loredana.contrafatto@unict.it, [c]salvatore.gazzo@dica.unict.it, [d]leopoldo.greco@unict.it, [e]mcuomo@dica.unict.it

Keywords: OMA, Masonry, Bell Tower

Abstract. Bell towers are masonry slender structures prone to damage caused by vibrations due to human activities, such as motor-vehicle traffic and operating machines in building sites, to bells' ringing and, of course, vulnerable to strong vibrations induced by earthquakes. Several studies with Operational Modal Analysis applications to masonry historical slender towers demonstrate that their modal properties are typical and recurrent and depend mainly on the geometric features determining mass distribution and on the boundary conditions due to adjacent buildings and subjacent foundation soil. In the paper a sensitivity analysis of modal frequency obtained by FEM eigenvalue analysis to the soil modelling approach is carried out. The case study is the historical bell tower of San Giuseppe in Aci Castello church. The natural frequencies of the tower, identified through OMA procedure are considered for comparison.

Introduction

Italy has a rich architectural heritage which belong to thousands of years of history. Among the architectural monuments, one of the most iconic categories is represented by bell towers, typical landmarks of the Italian landscape since the Middle Ages. Depending on their geometric and architectural features, bell towers are masonry slender structures characterised by intrinsic vulnerability to dynamic loads. Therefore, complete dynamic identification coupled with a calibrated numerical model is a major issue as a starting point for seismic vulnerability evaluation, damage prediction and preliminary effectiveness assessment of structural interventions.

Operational Modal Analysis (OMA) technique appears a proper method to perform dynamic identification [1], i.e. to experimentally determine modal frequencies, modal shapes and modal damping of a structure through to environmental vibrations. OMA methods present several advantages such as non-destructiveness, cost effectiveness, lightness, and portability of the required instrumentation and, above all, the possibility to measure values of the accelerations in the actual service conditions.

Several studies about masonry historical towers have been published showing that their modal shapes and frequencies are typical and recurrent and depend mainly on height and width of the structure, that is on the geometric features determining mass distribution, on the boundary conditions due to adjacent buildings [2, 3] and foundation soil characteristics [4, 5, 6].

OMA results allow the updating of input parameters of FEM analysis, such as elastic modulus and density, or the boundary conditions, improving the accuracy of the numerical model in representing the actual dynamic behavior of the structure [7].

In this paper the importance of the foundation soil modelling is highlighted with regard to the numerical evaluation of modal frequencies, especially in the case of soil characterised by low stiffness values. A refined FEM modelling of a historical masonry tower taking into account soil-structure interaction effects was implemented after the architectural relief and the dynamic identification through OMA techniques. The practical application concerns the bell tower of the church of San Giuseppe in the town of Aci Castello, (Eastern Sicily, Italy). Three different soil modelling approaches were followed to refine the model: fixed-base, Winkler model with linear springs, complete 3D model of the layered soil. Different arrangements of the tower foundation were tested. Mechanical and physical properties used as input parameters for the FEM modelling of the soil beneath the church were obtained through previous geological and geophysical surveys [8]. The eigenvalue analyses were carried out with Midas FEA NX Software [9]. The resulting values of modal frequencies were compared to those obtained through the experimental OMA measurements, to evaluate the influence and the accuracy of different soil modelling approaches.

Description of the case study

The bell tower of the church of San Giuseppe was chosen as a case study. The church is located in Aci Castello, a municipality of the Metropolitan City of Catania, Sicily (Italy), and was probably built around 1748 in the location in which another ecclesiastical building dedicated to Sant'Agata may has already existed previously, destroyed by an earthquake in 1547 [10]. According to this opinion, the bell tower may belong to a previous construction phase with respect to the church. This would be confirmed by architectural features of the tower and allows to suppose that there is no effective link between the two structures, so that they could be considered detached. The tower can therefore be studied as isolated.

The location of the church and the tower with respect to the coastal cliffs is highlighted in Fig. 1a. The tower stands at the south-eastern boundary of the town, about 25 meters away from a about 6-metres-high coastal cliff characterised by near-vertical rock faces of jointed volcanic rock, with a massive to vesicular intact rock texture [11]. A crypt, used in the past as a burial site, lays beneath the church with the same extension of the nave. The original tapering pyramidal masonry spire on the top of the tower was demolished after 1908 earthquake of Messina; nowadays, there is only a simple four-pitched roof with clay tiles on wooden frame.

An architectural survey was carried out to obtain geometrical and material information about the tower, shown in Fig. 1b. The tower has rectangular section, with sides of about 3.8 m and 4.8 m at the bottom, reducing to 3.3 m and 3.8 m from 2.5 m to the top bell chamber, for a total height of 15.4 m (excluding the tiles roof). Fig. 1c reports the layout of the tower.

(a) (b) (c)

Figure 1. Church of San Giuseppe in Aci Castello (Catania, Italy): (a) View of the area, with the location of the church; (b) The church and (c) Plane of the ground floor of the bell tower.

Methodological approach

In order to experimentally determine modal parameters of the bell tower, an Operational Modal Analysis based on the Frequency Domain Decomposition (FDD) method was previously carried out. The estimated frequencies were: f_1 = 3.94 Hz, associated to a flexural mode in the East/West direction, with a small translation component in the North/South direction and negligible torsional rotation; f_2 = 4.50 Hz, associated to a flexural mode in the North/South direction, with a translation component in the East/West direction and negligible torsional rotation; f_3 = 9.78 Hz, associated to a torsional mode, with a translation component in the North/South direction.

Materials Research Forum LLC
https://doi.org/10.21741/9781644902431-95

The mechanical characteristics of the masonry are analogous to that of the church both in terms of layout and materials. Masonry walls are made up by lavic rubble, clay tiles fragments and lime mortar; the thickness is between 0.65 m and 0.75 m. The mechanical parameters, derived from literature data for similar materials and updated after preliminary analysis have been set at: Young Modulus E = 2800 MPa, Poisson ratio $\nu = 0.3$, unit weight $\gamma = 17$ kN/m^3 [12].

In a previous *in situ* experimental campaign of measurements [8], Multichannel Analysis of Surface Waves (MASW) and seismic tomography tests were carried out to map and identify the soil in the tower construction site. Measurements of the profiles of shear waves and compression waves, V_s and V_p, up to a depth of 27 m were also conducted. Six soil layers with different mechanical characteristics were identified. A soft surface soil layer was found, laying on layers of fractured ancient lava of increasing stiffness with the depth. The soft layer consists of filling material used for leveling the foundation plane of the church. Density (ρ) and dynamic elastic modulus (E_{dyn}) for each soil layer were derived from theoretical correlations. In Table 1 soil physical and mechanical parameters are summarised.

Table 1. Soil parameters.

Soil layer n.	Depth [m]	S-Waves velocity V_s [m/s]	P-Waves velocity V_p [m/s]	Density ρ [g/cm^3]	Dynamic elastic modulus E_{dyn} [MPa]	ν [-]
1	0.00	355.06	1684.12	1.85	689	0.48
2	3.90	351.46	1680.12	1.84	671	0.48
3	7.06	446.68	1785.82	1.88	1100	0.47
4	13.27	525.04	1872.79	1.90	1527	0.46
5	21.85	621.05	1979.36	1.94	2163	0.45
6	27.31	705.05	2072.61	1.96	2795	0.43

A 3D FEM model of the tower was built in Midas FEA NX. A tetrahedral 4-node elements discretization was defined, with maximum element size 300 mm. Masonry was modeled as a linearly elastic isotropic material. A static dead load simulates the self-weight of the roof. Different boundary conditions were considered at the base of the tower. Footing structure and foundation soil were modeled to refine the model and analyse the effect of the soil modeling on the results of the eigenvalue analysis. Firstly, a fixed-base scheme, with hinges directly applied to each node at the bottom surface of the tower walls was analysed. Fig. 2 shows the FEM model.

Then, the fixed-base idealization, characterised by bilateral contraints, was removed. A FEM modeling of the layered soil was built, according to data in Table 1, up to the depth of 30 m. The linear elastic isotropic material model was used. Three different widths (50, 100, 200 m) of the volume of soil were considered, as shown in Fig. 3a.

Materials Research Proceedings **26** (2023) 587-592 https://doi.org/10.21741/9781644902431-95

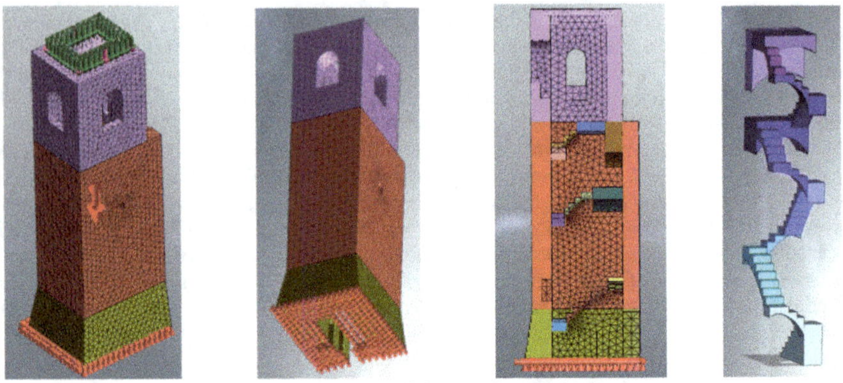

Figure 2. FEM model on Midas FEA NX.

The soil was simplified as horizontally layered, with each layer discretized in tetrahedral 4-nodes elements (Fig. 3b). The bottom nodes were constrained with hinges, while horizontal-axis roller boundary conditions were assigned to lateral boundaries. In the 200 m-side model, to simulate the presence of the coastal cliff (Fig. 3c) lateral nodes were not constrained along the height of the escarpment (Fig. 3d).

(a) (b) (c) (d)

Figure 3. (a) section detail of the layered soil FEM model (b) squared 50-, 100- and 200-meters side volumes of soil modeled in 3D FEM analysis and contour lines map. The dotted line highlights the coastal cliff; (c) The rocky front and (d) 200 m-side layered soil FEM model with B.C..

Finally, the Winkler's idealization of the soil was considered. Linearly elastic independent springs were introduced in the model, with stiffness given by Eq. 1:

$$k = \frac{E_{dyn}\cdot\frac{(1-\nu)}{(1+\nu)\cdot(1-2\nu)}}{1.5\cdot B}\ [\mathrm{N/mm^3}] \quad (1)$$

Two different foundation type were considered: a strip foundation 1.5 m deep; a slightly protruding slab foundation 1.5 m deep, with 5.3x4.3 m footprint area. Springs were applied to the three cases: directly to bottom nodes of the tower model, with $k = 2.91$ [N/mm^3]; to the bottom nodes of the strip foundation, with $k = 2.26$ [N/mm^3]; to the bottom nodes of the slab, with $k = 1.4$ [N/mm^3].

Results, discussion and conclusions

The eigenvalue analysis of the San Giuseppe's Bell Tower was performed with Midas FEA NX software. The three dominant natural frequencies were identified. The first and the second one are related to bending modes, associated to the Y-axis (i.e. North-South direction) and X-axis (i.e. East-West direction); the third one is related to a torsional mode around the vertical axis, Z.

Fig. 4a shows a comparison between the numerical modal frequencies and the experimental frequencies obtained from OMA. As expected, the higher the stiffness of the constraint configuration, the higher are the frequency values. The role of the surface soft layer is denounced by the lowering of the frequencies compared to the model on a fixed base. This reduction is more pronounced when a foundation body is present between the base of the tower and the soil, both in the case of the simplified Winkler models (W2, W3 in Fig. 4a) and in the case of the complete layered model (50, 100, 200 in Fig. 4a). In the picture 4a, in the cases of layered soil the results related to the slab type foundation are reported, the other solutions being basically the same. This last statement is confirmed in Fig. 4b where the comparisons are made considering the 50x50x30 m layered soil volume varying the foundation arrangement.

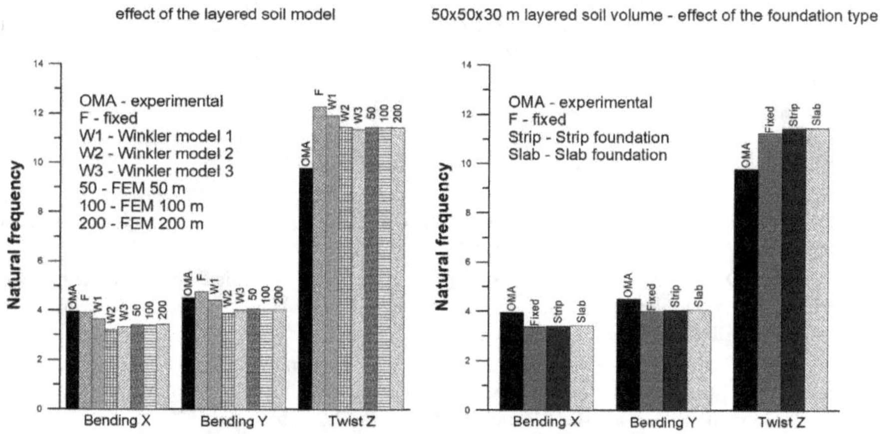

Figure 4. Comparison between FEA main numerical frequencies and OMA identified frequencies (a) varying the soil model and (b) varying the foundation type.

It is clear that in the study case it is sufficient to consider the smallest volume of soil and that, whatever the model of the foundations is considered, the Winkler hypothesis (W2, W3) leads to lower frequencies than those obtained by modeling the layered soil. It should be noted that the frequencies related to the bending modes are underestimated w.r.t. OMA frequencies, except in the Y direction in the case of the fixed-base tower model. On the contrary, the torsional frequency is always overestimated. The reasons for these differences can be found in two different factors. The first concerns the mass and stiffness properties of the masonry walls, which are derived from literature data. Provided that the geometric model is exact, in the absence of experimental data the masonry mechanical characteristics updating is therefore necessary. The second factor concerns the high torsional stiffness of the tower, denounced by overestimated numerical prediction of the twist eigenvalue. This factor can be attributed to the presence in the FEM model of helicoidal-like staircase inside the tower, shown in Fig. 2. The staircase was modeled perfectly connected to the perimeter walls and with mechanical properties identical to those of the load-bearing walls. Both hypotheses are unrealistic and can be considered as a source of the overestimation of the torsional frequency.

The overall results prove the need for a high level of knowledge in order to correctly identify the dynamic properties numerically [13]. Furthermore, when experimental measurements of the mechanical properties of the masonry are not available, the need of the FEM model updating is evident, by adjusting the properties of mass and stiffness. In the presence of soft foundation soils, the model should also include an accurate stratigraphy.

References

[1] G. Milani, F. Clementi, Advanced Seismic Assessment of Four Masonry Bell Towers in Italy after Operational Modal Analysis (OMA) Identification, Int. J. Archit. Herit, 15:1 (2021) 157-186. https://doi.org/10.1080/15583058.2019.1697768

[2] M. Shakya, H. Varum, R. Vicente, A. Costa, Seismic vulnerability assessment methodology for slender masonry structures, Int. J. Archit. Herit., 12:7-8 (2018) 1297-1326. https://doi.org/10.1080/15583058.2018.1503368

[3] M. Valente, G. Milani, Effects of Geometrical Features on the Seismic Response of Historical Masonry Towers, J. Earthq. Eng., 22 (2018) 2-34. https://doi.org/10.1080/13632469.2016.1277438

[4] A. Shabani, M. Feyzabadi, M. Kioumarsi, Model updating of a masonry tower based on operational modal analysis: The role of soil-structure interaction, in Case Studies in Construction Materials 16 (2022) 1-18. https://doi.org/10.1016/j.cscm.2022.e00957

[5] S. Biondi, G. Camata, E. Candigliota, E. Spacone, C. Valente, Identificazione strutturale di una torre in muratura per la messa a punto del modello numerico, Conference paper XIII Convegno ANIDIS L'Ingegneria Sismica in Italia ANIDIS, Bologna (2009).

[6] G. Torelli, D. D'Ayala, M. Betti, G. Bartoli, Analytical and numerical seismic assessment of heritage masonry towers, Bulletin of Earthquake Engineering 18 (2020) 969-1008. https://doi.org/10.1007/s10518-019-00732-y

[7] A. Cabboi, C. Gentile, A. Saisi, From continuous vibration monitoring to FEM-based damage assessment: Application on a stone-masonry tower, Construction and Building Materials, 156 (2017) 252-265. https://doi.org/10.1016/j.conbuildmat.2017.08.160

[8] S. Imposa, M. Cuomo, L. Contrafatto, S. Mineo, S. Grassi, D. Li Rosi, M.S. Barbano, G. Morreale, M. Galasso, G. Pappalardo, Engineering geological and geophysical studies supporting FEA of historical buildings after dynamic identification, submitted to Bulletin of Engineering Geology and the Environment (2022).

[9] MIDAS FEA NX (2021), https://www.midasoft.com/

[10] S. Castorina, Sotto il piano del castello. Quattromila anni di presenza umana nella benna dell'escavatore, in Agorà IX (3) (2002), 16-20.

[11] S. Branca, M. Coltelli, G. Groppelli, F. Lentini, Geological map of Etna volcano, 1:50,000 scale. Italian Journal of Geosciences, 130:3 (2011) 265-291. https://doi.org/10.3301/IJG.2011.15

[12] D. Liberatore, L. Gambarotta, G.C. Beolchini, L. Binda, G. Magenes, S. Cocina, E. Lo Giudice, S. Scuderi, Tipologie Edilizie In Muratura del Comune Di Catania, in: D. Liberatore, (ed.), Progetto Catania: indagine sulla risposta sismica di due edifici in muratura, Gruppo Nazionale per la Difesa dai Terremoti, (2000), 3-22.

[13] L. Contrafatto, A. Lo Faro, S. Gazzo, A. Purrazzo, Factors affecting the seismic analysis of historical masonry structures: A Case on single nave church damaged during the 2009 L'Aquila earthquake, Journal of Architectural Engineering (2022) 10.1061/(ASCE)AE.1943-5568.0000556. https://doi.org/10.1061/(ASCE)AE.1943-5568.0000556

Theoretical and Applied Mechanics - AIMETA 2022
Materials Research Proceedings **26** (2023) 593-598

Materials Research Forum LLC
https://doi.org/10.21741/9781644902431-96

A proposal of classification for machine-learning vibration-based damage identification methods

Francesca Marafini[1, a *], Michele Betti[1,b], Gianni Bartoli[1,c], Giacomo Zini[1,d], Alberto Barontini [2,e], Nuno Mendes [2,f]

[1]Department of Civil and Environmental Engineering (DICEA), Florence, Italy

[2]ISISE, Department of Civil Engineering, University of Minho, Guimarães, Portugal

[a]francesca.marafini@unifi.it [b]michele.betti@unifi.it [c]gianni.bartoli@unifi.it, [d]giacomo.zini@unifi.it, [e]albe.barontini@gmail.com, [f]nunomendes@civil.uminho.pt

Keywords: Machine Learning, Damage Identification, Vibration-Based SHM

Abstract. Recent advances in computing power and sensing technology led to a significant evolution of Structural Health Monitoring (SHM) techniques, transforming SHM into a "Big Data" problem. The use of data-driven approaches for damage identification purposes, specifically Machine Learning (ML) methods, has gained popularity. ML can help at various levels of the SHM process: to pre- and post-process input data, extract damage sensitive features, and operate pattern recognition in measured data and output valuable information for damage identification. In this paper, the role of ML in SHM applications is discussed together with a new scheme for classifying ML applications in SHM, especially focusing on vibration-based monitoring, given its consolidated theoretical base. Finally, the implications of the application of these methods to historic structures are discussed, with a brief account of existing case studies. The proposed classification is exemplified using the most recent studies available in the literature on cultural heritage structures.

Introduction

Structural Health Monitoring (SHM), as the process of implementing strategies for Damage Identification (DI) [6], is an interdisciplinary field which has been successfully investigated over the last decades. Meanwhile, advancements in computational power and data science have opened new avenues for the development of data-driven approaches for SHM and Machine-Learning (ML) [7]. The research effort in this direction led to a literature explosion in ML, with the number of papers published on the topic rapidly increasing in the last 20 years (Figure 1).

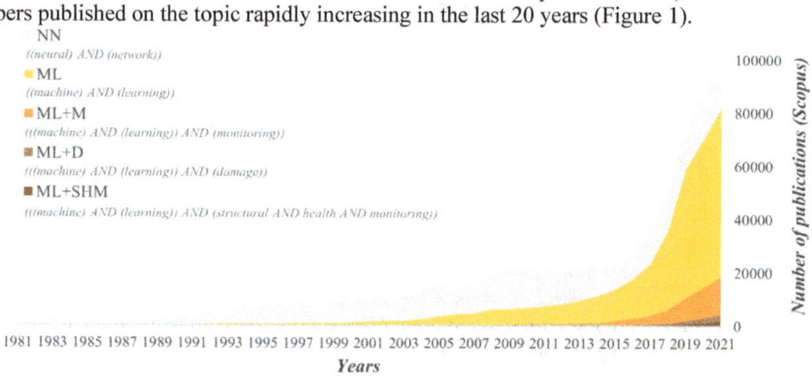

Figure 1 – Number of publications per year in the last two decades. Research operated on Scopus for words in Title, Keywords and Abstract.

Since the knowledge and methodologies on Data Science (DS) and Artificial Intelligence (AI) are increasingly being transferred to Civil Engineering, the need for a shared, clear glossary and framework is arising, to allow professionals and researchers to explore all capabilities of ML algorithms and connect transversal topics in the growing interdisciplinarity of SHM. Numerous published reviews have successfully formulated guidelines to approach the state-of-the-art research on ML applications to SHM [2], [3], [4], [8], [9], [10], [11]. However, differences in the classification of existing studies can still be found. The objectives of this paper are: (i) to provide basic notions to approach the study of ML applications to SHM, focusing on the use of vibration signatures, (ii) to propose a new classification methodology to operate a review of existing studies, and (iii) to briefly discuss peculiarities related to historic structures in the framework of ML applications to data-driven and vibration-based SHM.

Definitions

ML was introduced, as a subset of AI, to overcome the limitations of knowledge-based approaches [2]. ML algorithms "learn" systematically from a sufficient amount of data without the use of explicit programming [13]. The construction of a ML model entails the presence of input data, commonly divided into training, validation, and testing sets. The datasets are kept independent to ensure a correct assessment of the prediction accuracy over the validation set, preventing an overfitting of the model against the training set [12]. After repeated training, once the model is optimized, the testing set is fed to the model to operate a check against new data.

The ML training process is called *supervised* or *unsupervised*, based on the type of training data, which can be labelled or unlabelled, respectively. If both labelled and unlabelled data are used, the process is called *semi-supervised* ML. Moreover, in *reinforcement* learning, the use of unlabelled data is accompanied by agents that positively correct predictions in a trial-and-error process, reducing the requirements of training data. The availability and type of data is a key element in the choice of ML model.

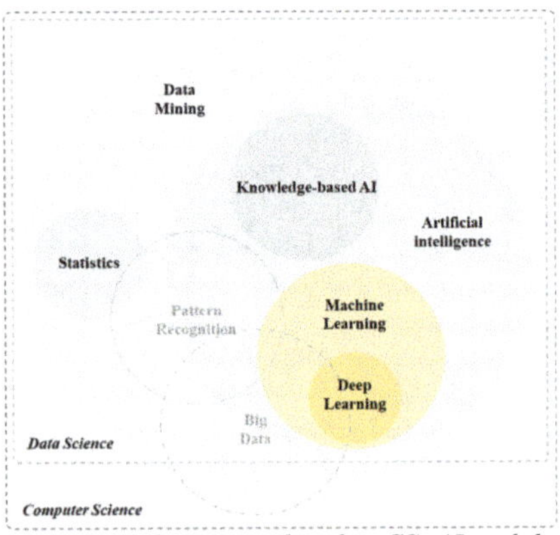

Figure 2 – Relationship between techniques related to CS, AI and data mining, the process of extracting useful knowledge and information from the bases of data. Pattern recognition is not a methodology or a technique [10], it is the problem of discovering automatically irregularities in data through the use of computer algorithms [13]. The expression Big Data is now often used to indicate a "field" of CS [14]. (Image adapted from [14]).

Basic ML algorithms require the conversion of data in a fixed number of features. Big Data will often present higher sparsity requiring more features to describe it [1], lowering model reliability and statistical significance. The higher the number of features, the larger the data requirements of

Materials Research Proceedings **26** (2023) 593-598
https://doi.org/10.21741/9781644902431-96

the ML algorithms, hence the so-called *curse of dimensionality* [12]. Discarding redundant information is often applied to overcome this issue. Moreover, to avoid the handcrafting of features in complex applications, Deep Learning (DL) methodologies were developed, to operate the feature selection process autonomously. DL algorithms explain high-level and abstract features as a hierarchy of simple and low-level learned features, ultimately reducing the dimension of feature vectors [2]. Relationships between AI, ML, DL, etc. are shown graphically in Figure 2.

The ML model is then finally used to solve a specific learning problem, such as classification, regression or prediction, clustering or density estimation problems [2]. The output of the application of such processes can yield both true and false "predictions".

ML in the SHM process

In data-driven vibration-based SHM, ML applications can be found at different levels of the process [11]. The most relevant are feature extraction/selection, dimensionality reduction and discrimination of the effect of environmental and operational variabilities (EOVs), statistical pattern recognition (SPR) and, finally, DI.

In an SPR framework, detecting the presence of damage means distinguishing between an initial "healthy" or undamaged state and a damaged state [12]. Detection problems represent the largest studied level of the DI hierarchy [15], comprised of detection, localization, assessment, quantification and prognosis. If operated unsupervised, damage detection is referred to as novelty/anomaly or outlier detection, a long-established statistical technique that can also be addressed through ML inference. Classification between damaged/undamaged is mostly operated, so much that ML algorithms are often called simply classifiers [2].

ML applications are steadily gaining recognition as viable techniques to operate data-driven vibration-based SHM. Several challenges are still to be addressed. The lack of data with damage is a long-standing obstruction to a more widespread use of ML. Over time viable solutions are being identified, by creating thresholds, synthetically generating data with damage and operating experiments. At the forefront of this research are the studies on Population-Based SHM (PBSHM), seeking to group similar structures in populations and using transfer-learning to overcome the lack of damaged data [20].

The selection of *damage sensitive* features, namely factors that make explicit the damage pattern to be learned from data [6], is still an ongoing research topic. Finally, going beyond the detection stage with unsupervised techniques is still a challenge, and damage prognosis is still achieved only when the physics of damage progression is included in a hybrid data/model-driven approach. Physics-informed ML applications are working towards this goal [21].

Proposed classification scheme for Machine-Learning Application to SHM

Considering the complexity of the subject and the interdisciplinarity necessary to approach the state-of-the-art research on ML applications, a new classification scheme for the development of a detailed review is proposed herein. The methodology encompasses a bottom-up strategy from a civil engineering perspective while gathering the necessary information to draw relevant conclusions from the analysed literature, in terms of recurrence of the methodologies, performance, success rate, etc.

According to the proposed classification, the level of the damage hierarchy reached is identified first. Then, a distinction is operated among three aspects: the type of extracted features (i.e., modal parameters, or statistical parameters in Autoregressive models etc.), the metrics used as a novelty/damage index (i.e., a distance metrics like the Mahalanobis distance), and the model built to highlight the presence of an underlying pattern (i.e., a simple Artificial Neural Network (ANN)).

To summarise, four elements are at the core of the classification, which should be used to operate a clear distinction in applications, as follows.

- The level of the damage hierarchy.

- The type of features.
- The type of SPR/ML method.
- The damage index/indices.

It is important to notice how ML could be used for one or both feature extraction/selection and SPR, but also at other stages of the SHM process, like data pre-processing or performance optimization. Moreover, the identification of the adopted damage index or indices is not highlighted in existing reviews, while it could show how the same index could apply to different extracted features.

Additionally, other information on the input data should be part of the classification if available. The methodology may have been tested on data from various sources, numerical, experimental or real data. The number and type of sensors may play a role in the quality and significance of the data obtained with respect to the structure analysed. Feature extraction and selection techniques should be added to the classification as well. Together with the number and types of features, the classification should specify whether they are considered as stochastic variables allowing for uncertainty estimation, and which technique was used, for example Gaussian Processes or Bayesian statistical inference. Also, if a dimensionality reduction technique was employed it should be noted, even if it is not a ML one. These techniques are used to reduce the size of input parameters without losing the information content of the data, for example Principal Component Analysis (PCA), and they can be instrumental in the overall performance of the application.

The algorithm for SPR is also to be specified further, in terms of architecture family, distinguishing for example between instance-based and clustering algorithms. The same stage of the damage hierarchy could be addressed with a different learning problem depending on the structure and the available data, for example the type of damage can be classified or clustered into several known damage scenarios or mechanisms. Strictly connected to the learning problem chosen, the nature of the output (Boolean, cluster, etc.) should also be identified. Finally, specific information on whether and how the performance of the algorithm was evaluated in the study should be included, as they could give insights into the validity of the methodology. Critical considerations on the pros and cons of the use of ML in the analysis application could be added, if provided by the paper authors. The classification should be complemented with all the necessary information to reference authors and publications. An example of two of the recent ML applications to data-driven vibration-based SHM of historic structures is provided in Table 1.

ML applications on SHM for Historic Structures

SHM operated on historic structures present a series of peculiarities which make the application of a data-driven approach and ML techniques even more challenging than in civil structures. Architectural heritage often encompasses complex structures in terms of geometry and mechanical behaviour. Materials are often heterogeneous and behave nonlinearly, with strict dependence on environmental conditions. Operating manual mapping of damages on these structures is even more expensive and time-consuming than it is on civil structures, given the ancient designs and construction techniques, and the fact that damage is often hidden to visual inspection.

In historic structures, damage can present itself at a global and a local level, potentially with equal relevance. One challenge is to detect relevant sudden changes in the state of the structure in real-time, to trigger consequent inspections and controls. Another is the question of identifying trends of accumulation of damage over extended periods of time, filtering out the effect of environmental and operational parameters, learning how they factor in, in the different damage mechanisms.

Many uncertainties arise when model-based approaches are pursued to monitor historic structures. SHM approaches based on dynamic identification are among the most widespread used techniques, given their strong theoretical base and the direct interpretability of the output.

Nonetheless, their support for engineering decisions is still limited, given the *global* nature of the output and the often *local* nature of the damage [5]. To date, a replicable and generalized approach to SHM of historic structures is yet to be reached.

ML methods are slowly starting to gain traction also for SHM of historic structures, on different scales, tackling a variety of problems [17], but a lot of work still needs to be done in this direction.

Table 1 – Classification of two ML applications to data-driven vibration-based SHM according to the proposed scheme.

Year of publication: 2018 **Authors:** Marrongelli et al. [18]	**Title:** *An artificial intelligence strategy to detect damage from response measurements: application to an ancient tower.* **Published in:** MATEC Web of Conferences 211, 21002 (2018) VETOMAC XIV
Damage level: Level 1 – Damage detection	
Features: Modal parameters – 5 natural frequencies	
Statistical Pattern Recognition Model: Support Vector Machines (SVMs)	
Damage index: SVM probability	
Input ***Data source:*** real data ***Type of structure***: ancient tower ***Material:*** brick masonry ***Sensors:*** 3 piezoelectric triaxial accelerometers positioned along the tower height, 1 temperature sensor	**Methodology** ***Feature extraction technique:*** covariance-driven Stochastic Subspace Identification (SSI-Cov) ***Features:*** 5 natural frequencies, alone and combined with value of temperature ***Algorithm architecture:*** Instance-based ***Learning problem:*** Classification ***Objective:*** seismic damage detection ***Performance evaluation:*** N/A
Comments: Uncoupling frequency and temperature input yields better results.	

Year of publication: 2017 **Authors:** Chalouhi et al. [19]	**Title:** *Damage detection in railway bridges using Machine Learning: application to a historic structure.* **Published in:** X International Conference on Structural Dynamics, EURODYN17
Damage level: Level 1 – Damage detection	
Features: Acceleration signals	
Statistical Pattern Recognition Model: Artificial Neural Network (ANNs) and Gaussian Processes (GP)	
Damage index: Normalized prediction error between the real and simulated data	
Input ***Data source:*** real and numerically simulated data ***Type of structure***: historic bridge ***Material:*** cast iron deck and arches and brick masonry piers ***Sensors:*** 21 MEMS accelerometers, 3 for each bridge section, 2 air temperature sensors	**Methodology** ***Feature extraction technique:*** Selection of acquired acceleration signals at each train passage ***Features:*** Acceleration signals ***Algorithm architecture:*** Neural Network and Gaussian Process ***Learning problem:*** Classification ***Objective:*** anomaly detection against predicted values ***Performance evaluation:*** Receiver Operating Characteristics (ROC)
Comments: Detection reliability decreases with the number of train passages.	

Conclusions

The proposed classification is intended as a base for a future review work, aimed at critically examining the existing methods, applying them to benchmark case studies and providing meaningful comparisons in terms of computational cost and accuracy. Future work is aimed at identifying key factors in the successful applicability of ML techniques to SHM of historic structures.

References

[1] C. Cremona and J. Santos, "Structural health monitoring as a big-data problem," *Struct. Eng. Int.*, vol. 28, no. 3, pp. 243–254, 2018. https://doi.org/10.1080/10168664.2018.1461536

[2] O. Avci, O. Abdeljaber, S. Kiranyaz, M. Hussein, M. Gabbouj, and D. J. Inman, "A review of vibration-based damage detection in civil structures: From traditional methods to Machine Learning and Deep Learning applications," *Mech. Syst. Signal Process.*, vol. 147, p. 107077, 2021. https://doi.org/10.1016/j.ymssp.2020.107077

[3] H.-T. Thai, "Machine learning for structural engineering: A state-of-the-art review," *Structures*, vol. 38, no. December 2021, pp. 448–491, 2022. https://doi.org/10.1016/j.istruc.2022.02.003
[4] X. W. Ye, T. Jin, and C. B. Yun, "A review on deep learning-based structural health monitoring of civil infrastructures," *Smart Struct. Syst.*, vol. 24, no. 5, pp. 567–585, 2019. https://doi.org/10.12989/sss.2019.24.5.567.
[5] M. Mishra, "Machine learning techniques for structural health monitoring of heritage buildings: A state-of-the-art review and case studies," *J. Cult. Herit.*, vol. 47, pp. 227–245, 2021. https://doi.org/10.1016/j.culher.2020.09.005
[6] K. Worden, C. R. Farrar, G. Manson, and G. Park, "The fundamental axioms of structural health monitoring," *Proc. R. Soc. A Math. Phys. Eng. Sci.*, vol. 463, no. 2082, pp. 1639–1664. https://doi.org/10.1098/rspa.2007.1834
[7] R. J. Barthorpe, "On Model and Data Based Approaches to Structural Health Monitoring," 2010.
[8] Y. Hamishebahar, H. Z. Li, and H. Guan, "Application of machine learning algorithms in structural health monitoring research," *Lect. Notes Civ. Eng.*, vol. 101, pp. 219–228, 2021. https://doi.org/10.1007/978-981-15-8079-6_21
[9] A. Malekloo, E. Ozer, M. AlHamaydeh, and M. Girolami, *Machine learning and structural health monitoring overview with emerging technology and high-dimensional data source highlights*, vol. 0, no. 0. 2021. https://doi.org/10.1177/14759217211036880
[10] H. Salehi and R. Burgueño, "Emerging artificial intelligence methods in structural engineering," *Eng. Struct.*, vol. 171, no. April, pp. 170–189, 2018. https://doi.org/10.1016/j.engstruct.2018.05.084
[11] D. A. Tibaduiza Burgos, R. C. Gomez Vargas, C. Pedraza, D. Agis, and F. Pozo, *Damage identification in structural health monitoring: A brief review from its implementation to the use of data-driven applications*, vol. 20, no. 3. 2020. https://doi.org/10.3390/s20030733
[12] C. R. Farrar and K. Worden, "Structural Health Monitoring: A Machine Learning Perspective," *Struct. Heal. Monit. A Mach. Learn. Perspect.*, 2012. https://doi.org/10.1002/9781118443118
[13] C. M. Bishop, "Pattern Recoginiton and Machine Learning," *Inf. Sci. Stat.*, p. 738, 2006, Accessed: Jul. 03, 2022.
[14] M. Gordan *et al.*, "State-of-the-art review on advancements of data mining in structural health monitoring," *Meas. J. Int. Meas. Confed.*, vol. 193, no. February, 2022. https://doi.org/10.1016/j.measurement.2022.110939
[15] A. Rytter, "Vibrational Based Inspection of Civil Engineering Structures.," Aalborg University, 1993.
[16] D. M. Bourdalos, I. A. Iliopoulos, and J. S. S. B, "On the Detection of Incipient Faults in Rotating Machinery Under Different Operating Speeds Using Unsupervised Vibration-Based Statistical Time Series," no. Cm, pp. 287–296. https://doi.org/10.1007/978-3-031-07258-1_30
[17] F. Clementi, A. Formisano, G. Milani, and F. Ubertini, "Structural Health Monitoring of Architectural Heritage: From the past to the Future Advances," *Int. J. Archit. Herit.*, vol. 15, no. 1, pp. 1–4, 2021. https://doi.org/10.1080/15583058.2021.1879499
[18] G. Marrongelli, R. Finotti, C. Gentile, and F. Barbosa, "An artificial intelligence strategy to detect damage from response measurements: application on an ancient tower," *MATEC Web Conf.*, vol. 211, p. 21002, Oct. 2018. https://doi.org/10.1051/matecconf/201821121002
[19] E. K. Chalouhi, I. Gonzalez, C. Gentile, and R. Karoumi, "Damage detection in railway bridges using Machine Learning: application to a historic structure," *Procedia Eng.*, vol. 199, pp. 1931–1936, Jan. 2017. https://doi.org/10.1016/j.proeng.2017.09.287
[20] J. Gosliga, P. A. Gardner, L. A. Bull, N. Dervilis, and K. Worden, "Foundations of Population-based SHM, Part {II}: Heterogeneous populations--Graphs, networks, and communities," *Mech. Syst. Signal Process.*, vol. 148, p. 107144, 2021. https://doi.org/10.1016/j.ymssp.2020.107144
[21] E. J. Cross, S. J. Gibson, M. R. Jones, D. J. Pitchforth, S. Zhang, and T. J. Rogers, "Physics-Informed Machine Learning for Structural Health Monitoring," *Struct. Integr.*, vol. 21, pp. 347–367, 2022. https://doi.org/10.1007/978-3-030-81716-9_17

Influence of a hemp biocomposite reinforcement on masonry vaults dynamic response

Massimiliano Gioffrè[1,a], Giacomo Navarra[2,b], Nicola Cavalagli[1,c], Francesco Lo Iacono[2,d], Roberta Scungio[1,e], Vittorio Gusella[1,f] and Chiara Pepi[1,g*]

[1]Department of Civil and Environmental Engineering, University of Perugia, via G. Duranti, 93, 06125, Perugia, Italy

[2]Facoltà di Ingegneria e Architettura, University of Enna Kore, Cittadella Universitaria, Enna, Italy

[a]massimiliano.gioffre@unipg.it, [b]giacomo.navarra@unikore.it, [c]nicola.cavalagli@unipg.it, [d]francesco.loiacono@unikore.it, [e]roberta.scungio@studenti.unipg.it, [f]vittorio.gusella@unipg.it, [g]chiara.pepi@unipg.it

Keywords: Masonry Vaults, Biocomposite Reinforcement, Dynamic Behavior, Shaking Table Test

Abstract. In this paper the effectiveness of the application of a hemp biocomposite reinforcement at the extrados of a masonry barrel vault was investigated. To this aim, two specimens of unreinforced (UV) and reinforced (RV) masonry vaults were built having the same geometry with brickwork material and subjected to shaking table seismic tests. The comparison of the dynamic behavior between the two specimens was carried out by increasing the intensity of the seismic input, which was simultaneously applied on both the structures. The evaluation of the progressive damage induced by the seismic input was observed by visual inspections and natural frequencies decay estimated by experimental dynamic identification through the application of Operational Modal Analysis (OMA) techniques.

Introduction

Masonry vaults are common structural elements of the Italian architectural heritage, that in some cases are significantly vulnerable to seismic loads. For this reason, strengthening measures are often required to prevent damages or even collapses [1]. In the last decade, the use of traditional reinforcement techniques has been progressively replaced by Fiber Reinforced Polymers (FRP) bonded at the vault extrados/intrados. However, this approach has shown drawbacks mainly associated to the polymeric nature of matrix and physical compatibility between the matrix and the masonry support. Consequently, FRP strengthening evolved to innovative composite materials obtained by cement based matrix, i.e. Fiber Reinforced Cementitious Matrix [2,3]. Moreover, within the growing environmental sustainability issues, the scientific research is focusing on the development of more sustainable composite materials, based on vegetal fibers and natural matrix [4,5].

This paper presents the results of an experimental campaign aimed at verify the effectiveness of a biocomposite reinforcement applied on the extrados of a barrel vault in the increase of its strength. To this aim, two full-scale masonry barrel vaults were constructed with the same geometry and mechanical properties: one was unreinforced and the latter was reinforced at the extrados by continuous bidirectional hemp ropes bonded in a layer of cocciopesto matrix (i.e. a mix of ground clay bricks and organic binder). The two models were built and tested in the 3D shaking table of the Laboratory of Earthquake Engineering and Dynamic Analysis (LEDA) at Enna Kore University. Each vault was subjected to seismic and dynamic identification tests: the seismic load was given with increasing amplitude at the base of the two models while the dynamic identification tests were carried out after each seismic sequence in order to evaluate the variation of the dynamic properties, testifying structural damage.

The strengthening effectiveness was assessed by comparing the damages detected on the unreinforced and reinforced masonry vaults. Results showed that the proposed bio composite strengthening material is effective for improving the seismic performance of masonry vaults.

Description of the masonry vault specimens

The masonry vaults specimens were built in full scale, following the dimensions indicated in previous experiments made by [3]. One of them consists only by brickwork masonry (UV), while the other is strengthened with an hemp biocomposite reinforcement at the extrados (RV). More in detail, referring to the nominal dimensions, the span of the vaults is 291.6 cm and the rise 114.1 cm. Considering the thickness of the abutments, the total length of the structures at the base is 384.5 cm and the width in the out-of-plane direction is 207 cm. Both the specimens were built using standard bricks having dimensions of 5.5×12×25 cm, so that the thickness of the vaults is of 12 cm. An M15 class of mortar was used in the construction of masonry material with joints of about 1.5 cm in thickness.

The reinforcement system of the strengthened masonry vault is characterized by a biocomposite matrix made by a mixture of cocciopesto and resin in which a rectangular mesh net of hemp ropes is included. The composite material is then characterized by a significant tensile strength given by the hemp ropes and a particular elasto-plastic behavior given by the matrix in compression and bending loads [4]. The reinforcement was applied on the vault with a constant thickness of about 3 cm. In Figure 1 the two specimens fixed on the shaking tables of LEDA Laboratory of Enna Kore University are shown.

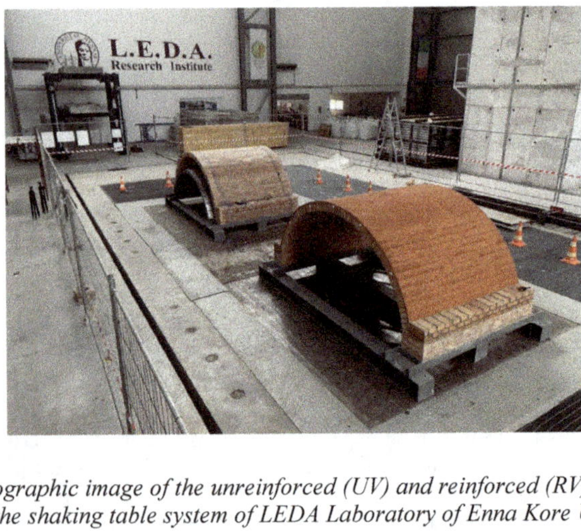

Figure 1. Photographic image of the unreinforced (UV) and reinforced (RV) masonry vaults fixed on the shaking table system of LEDA Laboratory of Enna Kore University.

Experimental setup

In order to investigate the effectiveness of the proposed reinforcement technique during earthquakes, an extensive experimental campaign was carried out on the 3D shaking table system at the Laboratory of Earthquake engineering and Dynamic Analysis (L.E.D.A.) Research Institute of "Kore" University [6]. This system consists of two identical 6 degrees of freedom (DOFs) 4×4

Materials Research Proceedings **26** (2023) 599-604 https://doi.org/10.21741/9781644902431-97

m shaking tables that can operate separately and simultaneously, and with asynchronous or synchronous motion.
The two specimens were equipped with two measurement systems. The first consists of MEMS accelerometers of 1 V/g in sensitivity deployed on the structures with two different configurations. In Configuration 1 the two vaults were instrumented with the same setup as illustrated in Figure 2(a) using 22 sensors, while Configuration 2, characterized by 28 sensors deployed as illustrated in Figure 2(b), was used only for the RV for high values of dynamic loads. The second measurement system is an optical-based equipment, through which the displacement time histories of specific markers installed on the vaults have been recorded by means of an innovative three-dimensional motion tracking. It consists of a set of eight high-speed, high-resolution cameras operating in the infrared field and reflective markers placed appropriately on the two vaults.

The dynamic input load was the acceleration records acquired by the seismic station of Norcia (Central Italy) during the earthquake of October 30th in 2016: both horizontal and vertical measured acceleration time histories (Figure 3) were used as input for the shaking table system. The choice of this seismic input was due to its peculiarity in the strong vertical acceleration time history, which causes significant damages in the city of Norcia and other neighboring towns [7,8].

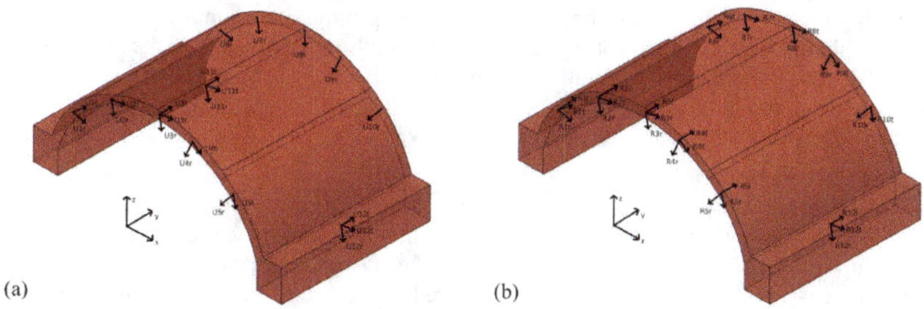

Figure 2. MEMS accelerometers layout installed on the specimens according the Configuration 1 (UV and RV) and the Configuration 2 (RV).

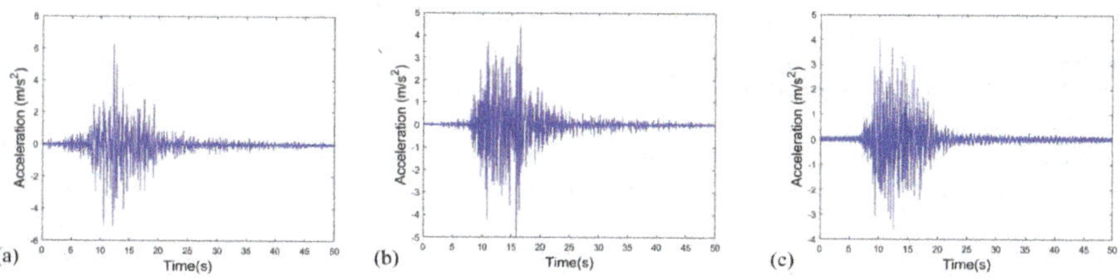

Figure 3. Acceleration time histories recorded by NOR seismic station on October 30th of 2016 used as seismic input during the tests.

Mode 1 Mode 2 Mode 3

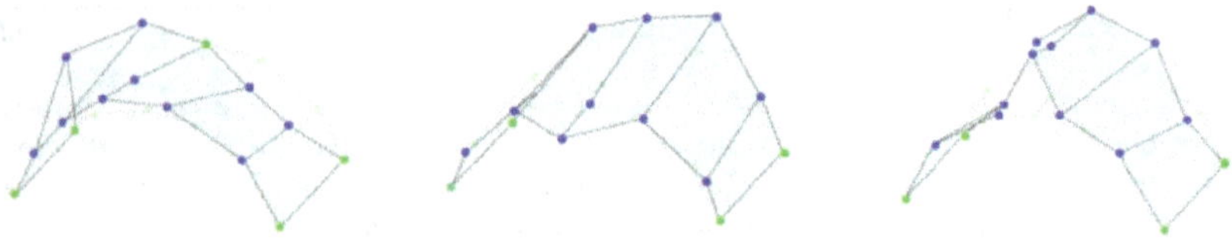

f_1^{UV}=18.36 Hz
f_1^{RV}=19.14 Hz

f_2^{UV}=25.00 Hz
f_2^{RV}=27.54 Hz

f_3^{UV}=37.40 Hz
f_3^{RV}=37.80 Hz

Figure 4. First natural frequencies and related mode shapes identified for UV and RV speciment in undamaged condition.

The input load was applied simultaneously to both the specimens thanks to the laboratory facilities, which allow to use coupled shaking tables. The seismic tests were performed scaling the amplitude of Norcia acceleration time histories by different Scale Factors (SF) ranging from 0.2 to 0.4 for the UV and from 0.2 to 2.30 for the RV, with progressive steps of 0.1 SF. Between each step of the increasing seismic load, Ambient Vibration Tests were carried out through operational modal analysis techniques to identify and localize the progressive damage developed in the structures. It should be noted that, taking into account both the added mass and the stiffness of the composite, the reinforcement system did not increase significantly the first natural frequencies of the vault in undamaged condition (Figure 4), so that the energy amount transmitted to the structure can be considered almost the same for both UV and RV.

Analysis of the results

As described in the previous Section, the seismic tests were performed by using two configurations of the sensors layout. In Configuration 1 the deployment of the sensors was equal for both the UV and RV. With this configuration, only two values of the seismic input SF were used, i.e. 0.2 and 0.4. After the first dynamic test no damages occurred on both the specimens, while the second dynamic test induced the activation of a kinematic 4-hinges mechanism on the UV specimen. However, after the rocking motion, the UV did not reach the collapse. In Figure 5 the position of the four hinges is indicated in the front and behind sides of the vault (Figures 5(a) and 5(b)), while Figure 5(c) and 5(d) show the development of the hinges through the width of the vaults, which clearly appear as cylindrical hinges. At the same step, the RV specimen did not show any significant damage.

After the achievement of the critical equilibrium condition for the UV, the test continued only for the RV specimen, with a sensor deployment described by the Configuration 2. By increasing the intensity of the seismic input, the first damages occurred on the structures at a value of SF equal to 2.20, with mortar detachments at the intrados. During the dynamic motion of the further step (i.e. SF equal to 2.30) significant damages occurred: a series of cylindrical hinges activated along the development of the vault and a sliding failure at the right impost (Figure 6(a)); sliding and debonding failure at the crown of the vaults and cylindrical hinge at the left impost (Figure 6(b)); opening of other hinges and increasing of the developed damages (Figures 6(c) and 6(d)).

The results of the AVTs carried out after each step of dynamic tests highlighted a decrease of the natural frequency due to the damage of about 45%, 30% and 10% for the first three modes of UV specimen and of about 65%, 55% and 35% for the first three modes of RV specimen, but, at the same time, with an increase of strength of about 5 times.

Materials Research Forum LLC
https://doi.org/10.21741/9781644902431-97

Figure 5. Damage configuration developed on the UV specimen after the application of a seismic input with SF equal to 0.4.

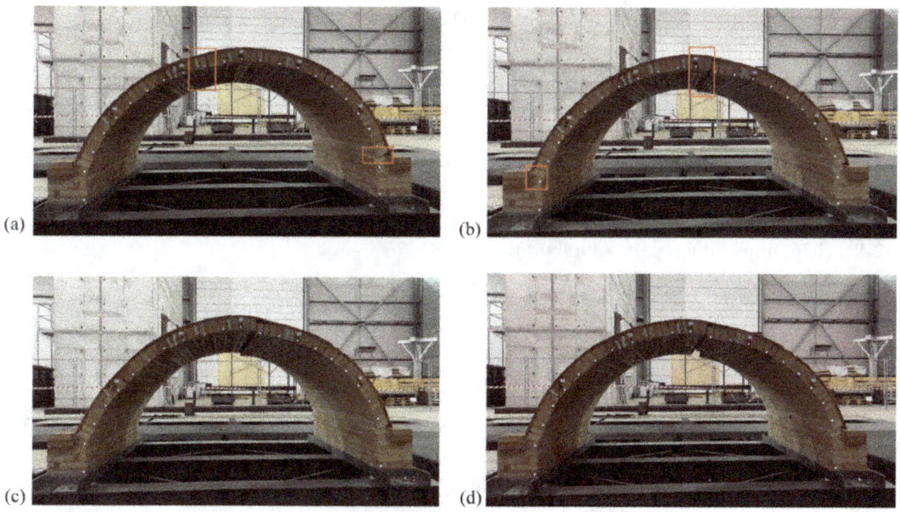

Figure 6. Damage configuration developed on the RV specimen after the application of a seismic input with SF equal to 2.3.

Conclusions

In this work the effectiveness of an hemp biocomposite reinforcement applied on a masonry barrel vault was investigated by means of shaking table tests carried out on two full scale specimens: unreinforced (UV) and reinforced (RV) masonry vaults. The reinforcement consists of hemp ropes included into a composite matrix made by cocciopesto and resin material. The seismic input was applied simultaneously on both the structures with a progressive increase of the intensity. The results highlighted a significant increase in strength of the RV specimen, on which significant damages occurred with the application of a scale factor of 2.30 to the seismic input, if compared to the UV specimen, in which a kinematic mechanism occurred with the application of a scale factor of 0.40 to the seismic input. Moreover, by means of visual inspections and frequency decay estimation as indicator of damage parameter, it was observed that UV specimen was characterized by very localized damages with small intensity, which essentially correspond to the activation of hinges, while RV specimen shown a more spread damage pattern, in some case with great intensity and in large displacements, but without reaching the collapse.

References

[1] Pepi, C., Cavalagli, N., Gusella, V., & Gioffrè, M. (2021). An integrated approach for the numerical modeling of severely damaged historic structures: Application to a masonry bridge. Advances in Engineering Software, 151, 102935. https://doi.org/10.1016/j.advengsoft.2020.102935

[2] M.R. Valluzzi, M. Valdemarca, C. Modena, Behavior of brick masonry vaults strengthened by FRP laminates, J. Compos. Constr. 5(3) (2001) 163-169. https://doi.org/10.1061/(ASCE)1090-0268(2001)5:3(163)

[3] V. Giamundo, G.P. Lignola, G. Maddaloni, A. Balsamo, A. Prota, G. Manfredi, Experimental investigation of the seismic performances of IMG reinforcement on curved masonry elements, Compos. Part B-Eng. 70 (2015) 53-63. https://doi.org/10.1016/j.compositesb.2014.10.039

[4] M. Gioffré, A. Vincenzini, N. Cavalagli, V. Gusella, M.A. Caponero, A. Terenzi, C. Pepi, A novel hemp-fiber bio-composite material for strengthening of arched structures: Experimental investigation, Constr. Build. Mater. 308 (2021) 124969. https://doi.org/10.1016/j.conbuildmat.2021.124969

[5] Liberotti, R., Cluni, F., Faralli, F., & Gusella, V. (2022). Natural FRCM and Heritage Buildings: Experimental Approach to Innovative Interventions on "Wall Beams". Buildings, 12(8), 1076. https://doi.org/10.3390/buildings12081076

[6] M. Fossetti, F. Lo Iacono, G. Minafò, G. Navarra, G. Tesoriere, A new large scale laboratory: The LEDA research centre (Laboratory of Earthquake engineering and Dynamic Analysis), International Conference on Advances in Experimental Structural Engineering (2017) 2017-September, 699–717.

[7] M. Gioffrè, N. Cavalagli, V. Gusella, C. Pepi, Confined vs. unreinforced masonry: Construction and shaking table tests of two-storey buildings, Constr. Build. Mater. 333 (2022) 126961. https://doi.org/10.1016/j.conbuildmat.2022.126961

[8] Pepi, C., Cavalagli, N., Gusella, V., & Gioffrè, M. (2021). Damage detection via modal analysis of masonry structures using shaking table tests. Earthquake Engineering & Structural Dynamics, 50(8), 2077-2097. https://doi.org/10.1002/eqe.3431

Modeling and analysis of nanocomposites and small-scale structures

Theoretical and Applied Mechanics - AIMETA 2022
Materials Research Forum LLC
Materials Research Proceedings **26** (2023) 607-612
https://doi.org/10.21741/9781644902431-98

Multiscale failure analysis of fiber-reinforced composite structures via a hybrid cohesive/volumetric nonlinear homogenization strategy

Daniele Gaetano[1, a], Fabrizio Greco[1,b *], Lorenzo Leonetti[1,c], Paolo Nevone Blasi[1,d] and Arturo Pascuzzo[1,e]

[1]Department of Civil Engineering, University of Calabria, Rende, Italy

[a]daniele.gaetano@unical.it, [b]fabrizio.greco@unical.it, [c]lorenzo.leonetti@unical.it, [d]paolo.nevoneblasi@unical.it, [e]arturo.pascuzzo@unical.it

Keywords: Cohesive/Volumetric Multiscale Modeling, Failure Analysis, Fiber-Reinforced Composite Structures

Abstract. In this work, a novel multiscale model for softening periodic microstructures is proposed, relying on a nonlinear homogenization method combined with a cohesive/volumetric finite element model. This strategy is able to overcome the mesh sensitivity issues usually experienced by purely volumetric homogenization techniques in presence of strain localization. As the main ingredient of the proposed approach, a microscopically informed traction-separation law for the embedded interfaces is extracted, starting from the homogenized bulk behavior obtained for a suitably chosen Repeating Unit Cell (RUC) subjected to different macro-strain paths. The present approach has been fully validated by performing several numerical simulations of the main damage phenomena experienced by fiber-reinforced composite structures, with special reference to transverse micro-cracking. Finally, to investigate the reliability and the accuracy of the proposed model, a comparison with direct simulations performed on fully meshed specimens has been presented, in terms of both load-displacement curves and associated crack patterns.

Introduction

Composite materials are nowadays being frequently used in a wide variety of engineering fields due to their superior mechanical properties compared to conventional single-phase materials. Their exceptional capability to face extreme loads is essentially due to their peculiar microstructural arrangements. Nevertheless, such engineered materials are prone to distinctive failure modes, such as delamination, matrix cracking, fiber/matrix debonding, fiber breakage and buckling [1].

Fully microscopic models are usually required for accurately simulating all these failure mechanisms, although being unpractical for large-scale problems due to the huge computational cost [2]. As a consequence, more effective strategies have been widely used in the technical literature. Most of these strategies are based on micromechanical and/or multiscale models [3-7].

However, traditional micromechanical approaches are not appropriate when strain localization is likely to occur. This is essentially due to the ill-posedness of the resulting macroscopic boundary value problem. To overcome such a drawback, more advanced approaches have been proposed, including higher-order [8], coupled-volume [9], micropolar [10,11], and continuous-discontinuous [12-16] homogenization. Most of these approaches have been used in the spirit of the FE^2 methods [17], which are often too costly, especially if applied to real-life structural applications.

In this work, a novel continuous/discontinuous multiscale model for periodic microstructures is proposed, based on a hybrid cohesive/volumetric hierarchical homogenization, in which the micro- and macro-scales are only one-way coupled. In particular, two independent homogenized constitutive responses are extracted: (i) a homogenized anisotropic damage evolution law, valid up to the occurrence of strain localization; and (ii) a homogenized mixed-mode traction-separation law. The main advantage of the present multiscale approach relies in the possibility to derive the overall mechanical response in a very efficient manner, by performing off-line computations.

The present approach is validated by performing failure analyses of fiber-reinforced composite beams experiencing transverse micro-cracking, and by comparing the related numerical results with those arising from direct simulations performed on fully meshed specimens.

Theoretical formulation of the cohesive/volumetric homogenization approach

A discretized 2D macroscopic solid Ω_M is considered here (the subscript M denoting the macro-scale), subjected to external tractions $\bar{\boldsymbol{t}}_M$ and imposed displacements $\bar{\boldsymbol{u}}_M$ on its Neumann and Dirichlet boundaries, respectively (see Fig. 1a). This solid is damageable, so that fracture is represented via a Diffuse Interface Modeling (DIM) approach, according to which cohesive interfaces, denoted by Γ_M^{coh}, are placed along all the internal mesh boundaries (for additional details see, for instance, [18-23]). Given the periodic nature of the considered microstructure, a suitable Repeating Unit Cell (RUC) is defined to microscopically derive the bulk and interface constitutive relations, which are valid before and after strain localization, respectively (see Figs. 1b and 1c).

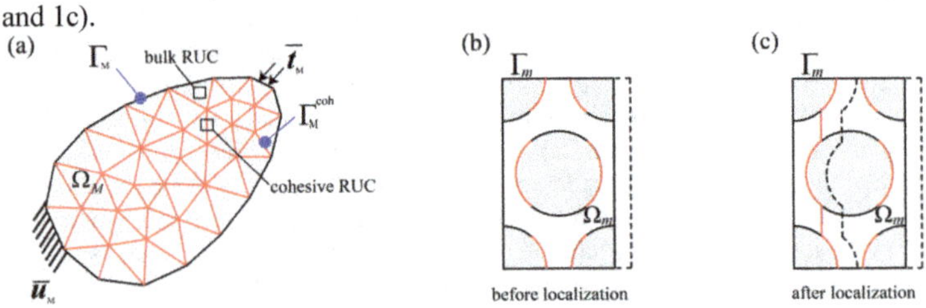

Fig. 1. Schematic of the cohesive/volumetric homogenization approach: (a) homogenized macroscopic problem; (b) bulk microscopic problem; (c) cohesive microscopic problem.

Under the assumptions of scale separation and local periodicity, the macro-stress and strain fields, indicated by $\boldsymbol{\sigma}_M$ and $\boldsymbol{\varepsilon}_M$ respectively, are defined in a standard manner as functions of boundary tractions $\boldsymbol{t}_m$ and displacements $\boldsymbol{u}_m$:

$$\boldsymbol{\sigma}_M = \frac{1}{|\Omega_m|}\int_{\Gamma_m} \boldsymbol{t}_m \otimes \boldsymbol{x}_m dS, \qquad \boldsymbol{\varepsilon}_M = \frac{1}{|\Omega_m|}\int_{\Gamma_m} \boldsymbol{u}_m \otimes_s \boldsymbol{n}_m dS. \tag{1}$$

In Eq. (1), Ω_m and Γ_m represent the RUC and its external boundary, respectively, whereas $\boldsymbol{x}_m$ is a generic material point inside the RUC, and $\boldsymbol{n}_m$ is the outer unit normal at the RUC boundary. The micro-to-macro transition based on the use of Eq. (1) is no longer objective if applied after strain localization has appeared. The resulting localization band, whose width goes to zero in a cohesive model, is described as a zero-thickness interface, equipped with a homogenized traction-separation law, which can be extracted from the homogenized bulk response by using the following relations:

$$t_M^{coh} = \sigma_M n_M, \quad \frac{1}{h}[\![u_M]\!] \otimes_s n_M = \varepsilon_M. \tag{2}$$

$\boldsymbol{n}_M$ being the normal to the macro-crack (supposed to be a priori known coherently with the Diffuse Interface Model adopted at the macro-scale), and h is the RUC size measured parallel to $\boldsymbol{n}_M$. The homogenized traction-separation law, as computed via Eq. (2), is subsequently depurated from the hardening contribute, associated with the early crack propagation prior to strain localization.

Numerical implementation of the proposed multiscale model

The proposed hybrid cohesive/volumetric nonlinear homogenization has been numerically implemented by considering three operational steps:

1. Computation of the undamaged elasticity tensor by means of the solution of linear boundary value problems for the RUC subjected to different pure macro-strain paths.
2. Computation of the homogenized damage evolution law, extracted from the stress-strain relations obtained via a nonlinear bulk homogenization along different loading paths.
3. Computation of the homogenized traction-separation law, obtained by projecting the previous nonlinear stress-strain relations along $\boldsymbol{n}_M$ (which varies depending on the actual orientation of the embedded cohesive interfaces for the macro-scale problem).

Numerical experiments: three-point bending test on a fiber-reinforced beam

The case study considered here for validation purposes is the three-point bending test performed on a fiber-reinforced composite beam, analyzed by Canal et al. [24]. Here, plane strain is assumed. The composite beam, whose geometry is depicted in Fig. 2a (with thickness of 2 mm), possesses a periodic microstructure, made with an epoxy matrix and glass fibers, having diameter of 15 μm and volume fraction of 54%, placed in a hexagonal arrangement. The elastic parameters (Young's modulus and Poisson's ratio) of the two constituents are listed in Table 1. Matrix cracking is described by an isotropic damage model with linear softening, having a tensile strength of 75 MPa and a fracture energy of 200 N/m, whereas fiber/matrix debonding is described by a mixed-mode cohesive zone model with linear softening, whose parameters are reported in Table 2.

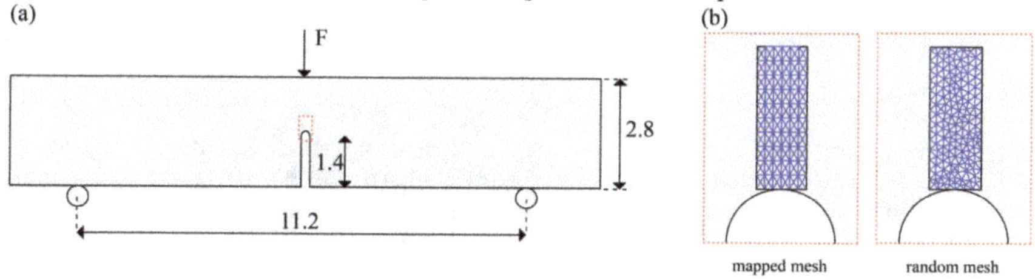

Fig. 2. Three-point bending test configuration: (a) geometry and boundary conditions of the macroscopic specimen (dimensions in mm); (b) detail of the mesh configurations within the damageable zone.

Table 1. Elastic parameters of the constituents.

Component	E [GPa]	ν [-]
Matrix	3.5	0.35
Fiber	74.0	0.20

Table 2. Elastic and inelastic parameters of the fiber/matrix interfaces.

$K_n = K_s$ [N/mm^3]	σ_{nc} [MPa]	σ_{sc} [MPa]	G_{Ic} [N/m]	G_{IIc} [N/m]
1.00×10^8	50.0	75.0	150	150

The proposed multiscale model has been used to derive the nonlinear structural behavior of this composite beam. According to the steps reported in the previous section, the undamaged elasticity tensor, as well as the homogenized damage evolution and traction-separation laws for the given microstructure have been computed. Then, a macro-scale analysis of the beam, here referred to as Multiscale Numerical Simulation, is performed by using the previous databases as material inputs. The macro-scale response is reported in terms of the applied force, F, as a function of the Crack Mouth Opening Displacement, CMOD. With the aim of investigating the influence of the mesh topology, two discretizations have been considered, i.e., a cross-triangle mapped and a Delaunay (random) mesh, shown in Fig. 2b. For validation purposes, the results of these multiscale analyses, or Multiscale Numerical Simulations (MNSs), have been compared with those obtained via a

Direct Numerical Simulation (DNS), in which the microstructural details are explicitly modeled. As shown in Fig. 3a, the MNS results obtained with the mapped mesh are almost coincident with the (reference) DNS results. Conversely, a little overestimation (of about 3%) of the peak load is found for the unstructured mesh (see Fig. 3b) This is not due to the homogenization scheme, but rather to the artificial crack path tortuosity induced by the DIM approach used at the macro-scale.

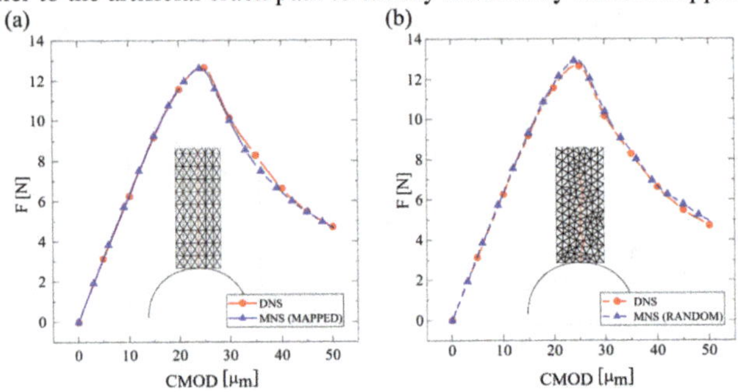

Fig. 3. Comparison between direct and multiscale simulations by using mapped (a) and random (b) meshes.

Conclusions

In this work, a novel hybrid cohesive/volumetric multiscale model is presented, to be adopted for the accurate failure simulation in composite structures. Such a model is based on a diffuse interface approach used in synergy with a continuous/discontinuous nonlinear homogenization scheme. Its main advantage is its greater efficiency with respect to most of the existing multiscale approaches, being related to derivation of homogenized bulk and interface responses via off-line computations.

As a validation step, the present model has been applied for analyzing the failure of a fiber-reinforced beam experiencing transverse cracking under Mode-I fracture conditions. Furthermore, the comparison with a fully detailed model has demonstrated that the proposed numerical strategy preserves a high accuracy, in terms of both load-carrying capacity and average cracking pattern.

A possible improvement of the present model could be the incorporation of plasticity [25] as well as the adoption of more efficient crack models based on moving mesh approaches [26-28].

Acknowledgements

Fabrizio Greco and Paolo Nevone Blasi gratefully acknowledge financial support from the Italian Ministry of Education, University and Research (MIUR) under the P.R.I.N. 2017 National Grant "Multiscale Innovative Materials and Structures" (Project Code 2017J4EAYB; University of Calabria Research Unit). Lorenzo Leonetti and Arturo Pascuzzo gratefully acknowledge financial support from the Italian Ministry of Education, University and Research (MIUR) under the P.R.I.N. 2020 National Grant "Opportunities and challenges of nanotechnology in advanced and green construction materials" (Project Code 2020EBLPLS; University of Calabria Research Unit).

References

[1] F. Greco, P. Lonetti, R. Zinno, An analytical delamination model for laminated plates including bridging effects, Int. J. Solids Struct. 39(9) (2002) 2435-2463. https://doi.org/10.1016/S0020-7683(02)00118-X

[2] M. Pepe, M. Pingaro, P. Trovalusci, E. Reccia, L. Leonetti, Micromodels for the in-plane failure analysis of masonry walls: Limit analysis, FEM and FEM/DEM approaches, Frat. ed Integrita Strutt. 14(51) (2020) 504-516. https://doi.org/10.3221/IGF-ESIS.51.38

[3] T. Belytschko, J.-H. Song, Coarse-graining of multiscale crack propagation, Int. J. Numer. Meth. Engng. 81 (2010) 537-563. https://doi.org/10.1002/nme.2694

[4] F. Greco, L. Leonetti, R. Luciano, A multiscale model for the numerical simulation of the anchor bolt pull-out test in lightweight aggregate concrete, Constr. Build. Mater. 95 (2015) 860-874. https://doi.org/10.1016/j.conbuildmat.2015.07.170

[5] F. Greco, L. Leonetti, R. Luciano, P. Nevone Blasi, An adaptive multiscale strategy for the damage analysis of masonry modeled as a composite material, Compos. Struct. 153 (2016) 972-988. https://doi.org/10.1016/j.compstruct.2016.06.066

[6] F. Greco, L. Leonetti, U. De Maio, S. Rudykh, A. Pranno, Macro- and micro-instabilities in incompressible bioinspired composite materials with nacre-like microstructure, Compos. Struct. 269 (2021) 114004. https://doi.org/10.1016/j.compstruct.2021.114004

[7] A. Pranno, F. Greco, L. Leonetti, P. Lonetti, R. Luciano, U. De Maio, Band gap tuning through microscopic instabilities of compressively loaded lightened nacre-like composite metamaterials, Compos. Struct. 282 (2022) 115032. https://doi.org/10.1016/j.compstruct.2021.115032

[8] V.G. Kouznetsova, M.G.D. Geers, W.A.M. Brekelmans, Multi-scale second-order computational homogenization of multi-phase materials: a nested finite element solution strategy, Comput. Methods Appl. Mech. Eng. 193(48-51) (2004) 5525-5550. https://doi.org/10.1016/j.cma.2003.12.073

[9] I.M. Gitman, H. Askes, L.J. Sluys, Coupled-volume multi-scale modelling of quasi-brittle material, Eur. J. Mech. A Solids. 27(3) (2008) 302-327. https://doi.org/10.1016/j.euromechsol.2007.10.004

[10] P. Trovalusci, M.L. De Bellis, M. Ostoja-Starzewski, A. Murrali, Particulate random composites homogenized as micropolar materials, Meccanica 49 (2014) 2719-2727. https://doi.org/10.1007/s11012-014-0031-x

[11] M. Tuna, L. Leonetti, P. Trovalusci, M. Kirca, 'Explicit' and 'implicit' non-local continuous descriptions for a plate with circular inclusion in tension, Meccanica 55(4) (2020) 927-944. https://doi.org/10.1007/s11012-019-01091-3

[12] T.J. Massart, R.H.J. Peerlings, M.G.D. Geers, An enhanced multi-scale approach for masonry wall computations with localization of damage, Int. J. Numer. Methods Eng. 69(5) (2007) 1022-1059. https://doi.org/10.1002/nme.1799

[13] T. Belytschko, S. Loehnert, J.-H. Song, Multiscale aggregating discontinuities: A method for circumventing loss of material stability, Int. J. Numer. Methods Eng. 73(6) (2008) 869-894. https://doi.org/10.1002/nme.2156

[14] V.P. Nhuyen, O. Lloberas-Valls, M. Stroeven, L.J. Sluys, On the existence of representative volumes for softening quasi-brittle materials – A failure zone averaging scheme, Comput. Methods Appl. Mech. Eng. 199(45-48) (2010) 3028-3038. https://doi.org/10.1016/j.cma.2010.06.018

[15] C.V. Verhoosel, J.J.C. Remmers, M.A. Gutiérrez, R. de Borst, Computational homogenization for adhesive and cohesive failure in quasi-brittle solids, Int. J. Numer. Methods Eng. 83(8-9) (2010) 1155-1179. https://doi.org/10.1002/nme.2854

[16] V.P. Nhuyen, O. Lloberas-Valls, M. Stroeven, L.J. Sluys, Homogenization-based multiscale crack modelling: From micro-diffusive damage to macro-cracks, Comput. Methods Appl. Mech. Eng. 200(9-12) (2011) 1220-1236. https://doi.org/10.1016/j.cma.2010.10.013

[17] F. Feyel, A multilevel finite element method (FE2) to describe the response of highly non-linear structures using generalized continua, Comput. Methods Appl. Mech. Eng. 192(28-30) (2003) 3233-3244. https://doi.org/10.1016/S0045-7825(03)00348-7

[18] U. De Maio, D. Cendón, F. Greco, L. Leonetti, P. Nevone Blasi, J. Planas, Investigation of concrete cracking phenomena by using cohesive fracture-based techniques: A comparison between an embedded crack model and a refined diffuse interface model, Theor. Appl. Fract. Mech. 115 (2021) 103062. https://doi.org/10.1016/j.tafmec.2021.103062

[19] F. Greco, D. Gaetano, L. Leonetti, P. Lonetti, A. Pascuzzo, A. Skrame, Structural and seismic vulnerability assessment of the Santa Maria Assunta Cathedral in Catanzaro (Italy): classical and advanced approaches for the analysis of local and global failure mechanisms, Frat. ed Integrita Strutt. 16(60) (2022) 464-487. https://doi.org/10.3221/IGF-ESIS.60.32

[20] A. Pranno, F. Greco, P. Lonetti, R. Luciano, U. De Maio, An improved fracture approach to investigate the degradation of vibration characteristics for reinforced concrete beams under progressive damage, Int. J. Fatigue 163 (2022) 107032. https://doi.org/10.1016/j.ijfatigue.2022.107032

[21] U. De Maio, F. Greco, L. Leonetti, P. Nevone Blasi, A. Pranno, A cohesive fracture model for predicting crack spacing and crack width in reinforced concrete structures, Eng. Fail. Anal. 139 (2022) 106452. https://doi.org/10.1016/j.engfailanal.2022.106452

[22] U. De Maio, F. Greco, L. Leonetti, P. Nevone Blasi, A. Pranno, An investigation about debonding mechanisms in FRP-strengthened RC structural elements by using a cohesive/volumetric modeling technique, Theor. Appl. Fract. Mech. 117 (2022) 103199. https://doi.org/10.1016/j.tafmec.2021.103199

[23] A. Pascuzzo, F. Greco, L. Leonetti, P. Lonetti, A. Pranno, C. Ronchei, Investigation of mesh dependency issues in the simulation of crack propagation in quasi-brittle materials by using a diffuse interface modeling approach, Fatigue Fract. Eng. Mater. Struct. 45(3) (2022) 801-820. https://doi.org/10.1111/ffe.13635

[24] L.P. Canal, C. González, J. Segurado, J. LLorca, Intraply fracture of fiber-reinforced composites: Microscopic mechanisms and modeling. Compos. Sci. Technol. 72(11) (2012) 1223-1232. https://doi.org/10.1016/j.compscitech.2012.04.008

[25] A. Pascuzzo, A. Yudhanto, M. Alfano, G. Lubineau, On the effect of interfacial patterns on energy dissipation in plastically deforming adhesive bonded ductile sheets, Int. J. Solids Struct. 198 (2020) 31-40. https://doi.org/10.1016/j.ijsolstr.2020.04.001

[26] D. Ammendolea, F. Greco, P. Lonetti, R. Luciano, A. Pascuzzo, Crack propagation modeling in functionally graded materials using Moving Mesh technique and interaction integral approach, Compos. Struct. 269 (2021) 114005. https://doi.org/10.1016/j.compstruct.2021.114005

[27] F. Greco, D. Ammendolea, P. Lonetti, A. Pascuzzo, Crack propagation under thermo-mechanical loadings based on moving mesh strategy, Theor. Appl. Fract. Mech. 114 (2021) 113033. https://doi.org/10.1016/j.tafmec.2021.103033

[28] A. Pascuzzo, F. Greco, P. Lonetti, D. Ammendolea, Dynamic fracture analysis in quasi-brittle materials via a finite element approach based on the combination of the ALE formulation and M-integral method, Eng. Fail. Anal. (2022) 106627. https://doi.org/10.1016/j.engfailanal.2022.106627

Theoretical and Applied Mechanics - AIMETA 2022 Materials Research Forum LLC
Materials Research Proceedings **26** (2023) 613-618 https://doi.org/10.21741/9781644902431-99

Line element-less method (LEM) for arbitrarily shaped nonlocal nanoplates: exact and approximate analytical solutions

DI MATTEO Alberto[1, a *] and PIRROTTA Antonina[1,b]

[1]Dipartment of Engineering, University of Palermo, Viale delle Scienze Ed. 8, 90128, Palermo, Italy

[a]alberto.dimatteo@unipa.it, [b]antonina.pirrotta@unipa.it

* please mark the corresponding author with an asterisk

Keywords: Kirchoff Plate, Nonlocal Eringen Model, Harmonic Polynomials, Line Element-Less Method, Meshfree Method

Abstract. This paper presents an innovative procedure for the analysis of nonlocal plates with arbitrary shape and various boundary conditions. In this regard, the Eringen's nonlocal model is used to capture small length scale effects. The proposed procedure, referred to as Line Element-Less Method (LEM), is a completely meshfree approach requiring the evaluations of simple line integrals along the plate boundary parametric equation. Further, the deflection function is represented by a series expansion is terms of harmonic polynomials whose coefficients are found by performing variations of appropriately introduced functionals, leading to a linear system of algebraic. Notably, the proposed procedure yields approximate analytical solutions for general shapes and boundary conditions, and even exact solutions for some plate geometries.

Introduction

The mechanical behavior of most structures at the nanoscale is typically size dependent, and classical approaches of continuum mechanics cannot capture this peculiar characteristic. Therefore, more sophisticated continuum theories have been introduced, and several different models have been developed [1-4].

One of the most widely adopted is the nonlocal elasticity theory, firstly introduced by Eringen [5-6], in which the stress at some reference point is assumed to be a function of the strain field at every point in the body, and the size-effect feature is captured in the model through an additional material parameter generally referred to as "the nonlocal parameter". This has paved the way for the application of Eringen's nonlocal elasticity theory in a plethora of studies involving the mechanical behavior of structural systems at the nanoscale, mostly related to nanobeams [7-9].

Recently, Eringen's nonlocal elasticity theory has been also used for the analysis on nanoplates. Initial contributions in [9-11] treated both isotropic linear elastic Kirchhoff plates as well as nonlinear plates considering higher-order shear deformation theory. Notably, most of these studies have focused on plates with simple rectangular shape. In this regard, the classical Navier's or Levy's approaches have been used in [10, 12]. Few other studies have focused on the analysis of nonlocal plates of different shapes [13], with most contributions mainly related to circular shapes [14, 15].

On this base, this study deals with the bending response of micro and nanoscale Kirchhoff plate, using Eringen's nonlocal theory, and considering arbitrary geometries and general boundary conditions. Specifically, the so-called Line Element-less Method (LEM) [16-25], is here extended to determine the deflection and bending moments of nonlocal plates subjected to transversal loads. Notably, the proposed procedure only requires the solution of simple line integrals of harmonic polynomials with unknown coefficients, along the boundary parametric equation and, eventually, the solution of a set of linear algebraic equations for these unknown terms. The LEM is completely element-free, since it does not require any discretization, be it in the domain or in the boundary, and it also differs from other so-called meshfree procedure since the expansion coefficients are not determined by collocation. Notably, this procedure yields approximate analytical solutions for generally shaped nonlocal plates, and even exact closed-form solution for some geometries and boundary conditions. These aspects clearly represent attractive features of the proposed procedure, especially with respect to other meshfree methods that are of numerical nature only. In this regard, several applications will be discussed, assessing the simplicity and accuracy of the considered approach.

Preliminary Concepts on Nonlocal Plate in Bending

Consider a homogeneous, isotropic, linear elastic Kirchhoff plate of arbitrary shape with contour Γ, domain Ω, and uniform thickness h, subjected to a transverse distributed load $q(x,y)$, and satisfying the Eringen's nonlocal model [5]. The plate is characterized by the modulus of elasticity E, Poisson's ratio ν, and nonlocal parameter $\lambda = (l e_0)^2 \geq 0$, where l is an internal characteristic length, whereas e_0 is the small length scale coefficient. Note that when $\lambda = 0$ the classical local Kirchhoff plate is obtained. The corresponding biharmonic governing differential equation for bending of the Eringen's nonlocal plate in terms of deflection function can be written as

$$D\left(\frac{\partial^2}{\partial x^2}+\frac{\partial^2}{\partial y^2}\right)\left(\frac{\partial^2 w}{\partial x^2}+\frac{\partial^2 w}{\partial y^2}\right)=\left[q-\lambda\left(\frac{\partial^2 q}{\partial x^2}+\frac{\partial^2 q}{\partial y^2}\right)\right] \tag{1}$$

where $D = Eh^3/12(1-\nu^2)$ is the plate flexural rigidity. Notably, as it can be seen in Eq. (1), the only difference with respect to classical plate differential equation (local model) stands in the additional term $\lambda\nabla^2 q(x,y)$ at the right-hand-side of Eq. (1), which is therefore an inherent effect due to the employed Eringen's nonlocal model.

Further, introduce the Marcus' moment sum $M(x,y)$, defined as [26, 27] $M(x,y) = (M_x + M_y)/(1+\nu)$, where $M_x(x,y)$ and $M_y(x,y)$ are the bending moments, Eq. (1) can be decomposed into two Poisson's equations as

$$\begin{aligned} &\nabla^2 M(x,y) = -q(x,y) \\ &D\nabla^2 w(x,y) = -(1-\lambda\nabla^2)M(x,y) \end{aligned} \tag{2.a, b}$$

Note that, in this manner, the solution of the plate problem Eq. (1) reduces to the integration, in sequence, of the two Poisson differential equations Eqs. (2), respectively, which is sometimes preferred depending upon the method of solution employed. As far as the boundary conditions (BCs) are concerned, denote as **n** the outward unit normal vector at a point of a generic curvilinear edge of the contour Γ. Thus, the BCs for the simply support curvilinear edge can be assumed as those of the classical local plate, that is $w(x,y)=0$ and $M_n(x,y)=0$, where $M_n(x,y) = n_x^2 M_x + n_y^2 M_y + 2 n_x n_y M_{xy}$ denotes the normal bending moment applied at the edge, and n_x and n_y are the components of the unitary vector **n** along the x and y axes, respectively. Note that analogous expression can be found for different BCs in [25].

Line Element-Less Method for nonlocal plate analysis

In this section LEM approach is introduced for the analysis of nonlocal Kirchhoff plates of general shape, subjected to a transversal load $q(x,y)$. Specifically, based on the previous studies on the use of the LEM [16-25] and considering Eqs. (2), $M(x,y)$ and $w(x,y)$ can be expressed in terms of the so-called harmonic polynomials P_k and Q_k, generally defined recursively as.

$$\begin{aligned} P_k(x,y) &= P_{k-1}x - Q_{k-1}y \\ Q_k(x,y) &= Q_{k-1}x - P_{k-1}y \end{aligned} \tag{3.a, b}$$

which are valid for $k > 0$, and with $P_0 = 1$ and $Q_0 = 0$.

Based on the above relations, a solution of Eq. (2.a) can be obtained expressing the moment sum function in terms of harmonic polynomials, plus a particular solution of the Poisson equation Eq. (2.a), namely $M_c(x,y)$. That is

$$M(x,y) \cong \sum_{k=0}^{p} a_k P_k(x,y) + \sum_{k=1}^{p} b_k Q_k(x,y) + M_c(x,y) \tag{4}$$

where a_k and b_k are $(2p+1)$ unknown coefficients to be determined, and p is the truncation limit of the series expansion. Note that, for the typical case of a uniformly distributed load $q(x,y) = q_0$, the following expressions of $M_c(x,y) = -q_0(x^2+y^2)/4$. As far as the unknown coefficients in Eq. (4) are concerned, the $(2p+1)$ values of a_k and b_k can be determined appropriately imposing the specified BCs of the plate. In this context, it is convenient to address the case of polygonal plates with simply-supported edges, while the generalization to arbitrary shaped plates is reported in [25]. Specifically, in this case, the moment sum function must be zero along the entire contour of the polygonal plate; thus, $M_n(x,y) = M(x,y) = 0$ in Γ.

Therefore, the unknown coefficients (a_k, b_k) in Eq. (4) can be evaluated applying a minimization procedure on the closed contour path integral of the squared moment sum function; that is $\Psi(a_k,b_k) = \oint_\Gamma [M(x,y)]^2\, d\gamma$. Thus, performing variations of the aforementioned functional with respect to (a_k, b_k), yields a linear algebraic system of $(2p+1)$ equations in the unknowns (a_k, b_k). In this manner, once the coefficients are determined, the moment sum function can be found through Eq. (4).

Next, the deflection function $w(x,y)$ can be obtained solving Eq. (2.b). In this regard, following a similar approach, a solution of this equation can be sought assuming $w(x,y)$ as the sum of harmonic polynomials, and a particular solution of the Poisson equation Eq. (2.b), namely $w_c(x,y)$; that is

$$w(x,y) \cong \sum_{k=0}^{m} c_k P_k(x,y) + \sum_{k=1}^{m} d_k Q_k(x,y) + w_c(x,y) \tag{5}$$

where c_k and d_k are $(2m+1)$ unknown coefficients to be determined, and m is the truncation limit of the series expansion. Again, note that the particular solution $w_c(x,y)$ can be evaluated applying the procedure in [25], considering the term at the right-hand side of Eq. (2.b).

As far as the unknown coefficients in Eq. (5) are concerned, the $(2m+1)$ values of c_k and d_k are determined appropriately imposing the BCs. In this regard, considering the case of a simply-supported plate, the coefficients (c_k, d_k) can be found minimizing the closed contour path integral of the squared deflection function; that is $\Theta(c_k, d_k) = \oint_\Gamma [w(x,y)]^2\, d\gamma$. Next, introducing Eq. (5) into the previous functional, and performing variation with respect to the unknown coefficients leads to an algebraic linear system in terms of the unknowns (c_k, d_k). Solution of the this set of $(2m+1)$ equation yields the sought deflection function of the plate $w(x,y)$ through Eq. (5).

Numerical Applications

In this section, the proposed LEM is applied to some nonlocal plate configurations, considering various shapes and boundary conditions. Specifically, the proposed method is employed for the analysis of different well-known examples, showing the ability of the approach to yield even exact solutions. In this regard, a triangular simply-supported plate and a simply-supported circular plate are considered.

Specifically, firstly consider the case of an equilateral triangular shaped plate with length side $2\sqrt{3}l$ under a uniformly distributed load q_0 and with simply-supported edges. Applying the previously described procedure, the obtained closed-form expression of the deflection function is

$$w(x,y) = \frac{q_0}{192 l D}(l-y)\left[(2l-y)^2 - 3x^2\right]\left(4l^2 - x^2 - y^2 + 16\lambda\right) \tag{6}$$

Notably, Eq. (6) reverts to the classical solution of the Krichoff local plate for $\lambda = 0$. In this regard, deflection profile for $x = 0$ is shown in Fig. 1(a) for different values of nonlocal parameters.

Next, consider next the case of a circular plate of radius r under a uniformly distributed load q_0. Note that, in this case, the equilibrium equations of axisymmetric bending of circular nanoplates can be more simply written in polar coordinates as in [15]. Then, applying the previously described procedure, the closed-form expression of the deflection function is given as

$$w(x,y) = \frac{q_0\left(r^2 - x^2 - y^2\right)\left[-\left(x^2+y^2\right)(1+\nu) + r^2(5+\nu) + 8\lambda(3+\nu)\right]}{64 D(1+\nu)} \tag{7}$$

Notably, again Eq. (7) reverts to the classical solution of the Krichoff local plate for $\lambda = 0$. In this regard, the deflection profile for $x = 0$ is given in Fig. 1(b) for different values of nonlocal parameter.

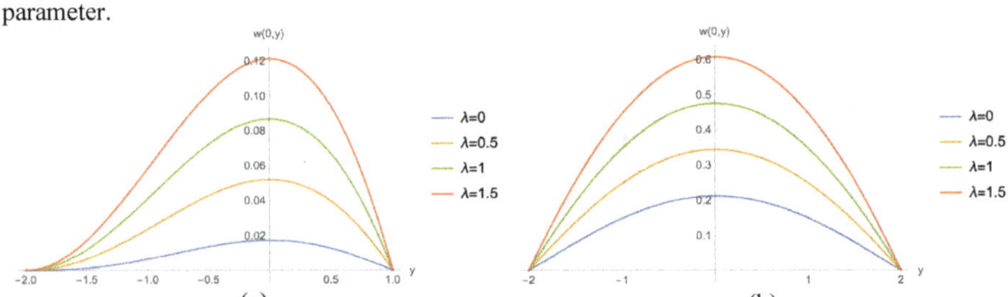

(a) (b)

Fig. 1: Profile of $w(0,y)$ for different value of λ. (a) Simply-supported Triangular Plate; (b) Simply-supported Circular Plate.

Summary

In this paper, the so-called Line Element-Less Method (LEM) has been proposed for the analysis of nonlocal plates with arbitrary shape. Specifically, the classical Kirchhoff plate model has been assumed, employing the well-known Eringen's nonlocal elasticity theory to capture small length scale effects. The method is based on an expansion of the deflection function in terms of harmonic polynomials, whose expansion coefficients can be easily found solving a linear system of algebraic equations. Notably, the entire procedure proves to be entirely mesh-free, since it only requires the definition of simple line integrals which appear in appropriately introduced functionals. These functionals are employed to take into account the pertinent plate boundary conditions (BCs). It is worth mentioning that, the proposed approach yields approximate analytical solutions for general plate shapes and BCs, while exact closed form expressions of the deflection functions of nonlocal plates can be found for particular shapes.

References

[1] A.C. Eringen, E.S. Suhubi, Nonlinear theory of simple micro-elastic solids-I. Int. J. Eng. Sci. 2 (1964)189-203. https://doi.org/10.1016/0020-7225(64)90004-7

[2] M. Di Paola, A. Pirrotta, M. Zingales, Mechanically-based approach to non-local elasticity: variational principles. Int. J. Solids Struct. 47 (2010) 539–548. https://doi.org/10.1016/j.ijsolstr.2009.09.029

[3] M. Di Paola, G. Failla, A. Pirrotta, A. Sofi, M. Zingales. The mechanically based non-local elasticity: an overview of main results and future challenges. Phil. Trans. R. Soc. A. (2013) 37120120433. https://doi.org/10.1098/rsta.2012.0433

[4] F.P. Pinnola, S.A. Faghidian, R. Barretta, F. Marotti de Sciarra. Variationally consistent dynamics of nonlocal gradient elastic beams. Int. J. Eng. Sci. 149 (2020) 103220. https://doi.org/10.1016/j.ijengsci.2020.103220

[5] A.C. Eringen, D.G.B. Edelen. On nonlocal elasticity. Int. J. Eng. Sci. 10 (1972) 233–248. https://doi.org/10.1016/0020-7225(72)90039-0

[6] A.C. Eringen. Linear theory of nonlocal elasticity and dispersion of plane waves. Int. J. Eng. Sci. 10 (1972) 425–435. https://doi.org/10.1016/0020-7225(72)90050-X

[7] J. Peddieson, G.R. Buchanan, R.P. McNitt. Application of nonlocal continuum models to nanotechnology. Int. J. Eng. Sci. 41 (2003) 305–312. https://doi.org/10.1016/S0020-7225(02)00210-0

[8] R. Barretta, L. Feo, R. Luciano, F. Marotti de Sciarra. Application of an enhanced version of the Eringen differential model to nanotechnology. Compos. B Eng. 96 (2016) 274–280. https://doi.org/10.1016/j.compositesb.2016.04.023

[9] J.N. Reddy. Nonlocal nonlinear formulations for bending of classical and shear deformation theories of beams and plates. Int J Eng Sci. 48 (2010) 1507-1518. https://doi.org/10.1016/j.ijengsci.2010.09.020

[10] P. Lu, P.Q. Zhang, H.P. Lee, C.M. Wang, J.N. Reddy. Nonlocal elastic plate theories. Proc. R. Soc. Lond. A Math. Phys. Eng. Sci. 463 (2007) 3225-3240. https://doi.org/10.1098/rspa.2007.1903

[11] R. Ansari, J. Torabi, A. Norouzzadeh. Bending analysis of embedded nanoplates based on the integral formulation of Eringen's nonlocal theory using the finite element method, Physica B. 534 (2018) 90-97. https://doi.org/10.1016/j.physb.2018.01.025

[12] T. Aksencer, M. Aydogdu. Levy type solution method for vibration and buckling of nanoplates using nonlocal elasticity theory. Physica E. 43 (2011) 954–959. https://doi.org/10.1016/j.physe.2010.11.024

[13] H. Irschik, R. Heuer. Analogies of simply supported nonlocal Kirchhoff plates of polygonal planform. Acta Mech. 229 (2018) 867-879. https://doi.org/10.1007/s00707-017-2005-2

[14] J.W. Yan, L.H. Tong, C. Li, Y. Zhu, Z.W. Wang. Exact solutions of bending deflections for nano-beams and nano-plates based on nonlocal elasticity theory. Compos. Struct. 25 (2015) 304-313. https://doi.org/10.1016/j.compstruct.2015.02.017

[15] W.H. Duan, C.M. Wang. Exact solutions for axisymmetric bending of micro/nanoscale circular plates based on nonlocal plate theory. Nanotechnology. 18 (2007) 385704. https://doi.org/10.1088/0957-4484/18/38/385704

[16] G. Battaglia, A. Di Matteo, G. Micale, A. Pirrotta. Arbitrarily shaped plates analysis via Line Element-Less Method (LEM). Thin-Walled Struct. 133 (2018) 235-248. https://doi.org/10.1016/j.tws.2018.09.018

[17] G. Battaglia, A. Di Matteo, G. Micale, A. Pirrotta. Vibration-based identification of mechanical properties of orthotropic arbitrarily shaped plates: Numerical and experimental assessment. Compos. Pt. B-Eng. 150 (2018) 212-225. https://doi.org/10.1016/j.compositesb.2018.05.029

[18] A. Pirrotta, C. Bucher. Innovative straight formulation for plate in bending, Comput. Struct. 180 (2017) 117-124. https://doi.org/10.1016/j.compstruc.2016.01.004

[19] A. Pirrotta, C. Proppe. Extension of the line element-less method to dynamic problems. Meccanica. 55 (2020) 745–750. https://doi.org/10.1007/s11012-019-01120-1

[20] M. Di Paola, A. Pirrotta, R. Santoro. Line Element-less Method (LEM) for beam torsion solution (Truly no-mesh method). Acta Mech. 195 (2008) 349-63. https://doi.org/10.1007/s00707-007-0557-2

[21] M. Di Paola, A. Pirrotta, R. Santoro. De Saint-Venant flexure-torsion problem handled by Line Element-less Method (LEM). Acta Mech. 217 (2011) 101-118. https://doi.org/10.1007/s00707-010-0376-8

[22] A. Pirrotta. LEM for twisted re-entrant angle sections. Comput Struct. 133 (2014) 149-155. https://doi.org/10.1016/j.compstruc.2013.11.015

[23] G. Barone, A. Pirrotta, R. Santoro. Comparison among three boundary element methods for torsion problems: CPM, CVBEM, LEM. Eng. Anal. Bound. Elem. 35 (2011) 895–907. https://doi.org/10.1016/j.enganabound.2011.02.003

[24] R. Santoro. Solution of de Saint Venant flexure-torsion problem for orthotropic beam via LEM (Line Element-less Method). Eur J Mech A-Solids. 30 (2011) 924-939. https://doi.org/10.1016/j.euromechsol.2011.06.003

[25] A. Di Matteo, M. Pavone, A. Pirrotta. Exact and approximate analytical solutions for nonlocal nanoplates of arbitrary shapes in bending using the line element-less method. Meccanica. 57 (2022) 923-941. https://doi.org/10.1007/s11012-021-01368-6

[26] E. Murtha-Smith. Plate analogy for the torsion problem. J Eng Mech. 116 (1990) 1-17. https://doi.org/10.1061/(ASCE)0733-9399(1990)116:1(1)

[27] H. Irschik. Analogies between bending of plates and torsion problem. J Eng Mech. 117 (1991) 2503-2508. https://doi.org/10.1061/(ASCE)0733-9399(1991)117:11(2503)

Materials Research Proceedings **26** (2023) 619-624 https://doi.org/10.21741/9781644902431-100

Perturbations for vibration of nano-beams of local/nonlocal mixture

Ugurcan Eroglu[1, a] and Giuseppe Ruta[2,3,b *]

[1]Department of Mechanical Engineering, MEF University, Istanbul, Turkey

[2]Department of Structural and Geotechnical Engineering, Università 'La Sapienza', Roma, Italy

[3]National Group for Mathematical Physics, Roma, Italy

[a]erogluug@mef.edu.tr, [b]giuseppe.ruta@uniroma1.it, * corresponding author

Keywords: Perturbation, Nonlocal Elasticity, Free Vibration

Abstract. Here we extend the perturbation approach, previously presented in the literature for Eringen's two-phase local/nonlocal mixture model, to free vibration of purely flexible beams. In particular, we expand the eigenvalues and the eigenvectors into power series of the fraction coefficient of the non-local material response up to 2nd order. We show that the family of 0th order bending couples satisfy the natural and essential boundary conditions of the 1st order; hence, the 1st order solution can conveniently be constructed using the eigenspace of the 0th order with no necessity of additional conditions. We obtain the condition of solvability that provides the incremental eigenvalue in closed form. We further demonstrate that the 1st order increment of the eigenvalue is always negative, providing the well-known softening effect of long-range interactions among the material points of a continuum modelled with Eringen's theory. We examine a simply supported beam as a benchmark problem and present the incremental eigenvalues in closed form.

Introduction

Structures with comparable internal and external length scales can be modelled by suitable quasi-continuum models, for which the distances shorter than their scale parameters have no physical meaning. Among many well-established models, we focus on Eringen's strain driven model, due to the efforts of Eringen and co-workers [1,2]. The application of this model to finite domains, however, needs additional mathematical conditions (the meaning of which is dubious) for a solution to exist in a certain form [3]. This led to criticisms on the validity of the material model itself, despite its strong mathematical and philosophical foundations [4].

Here we use the perturbation approach proposed in [5] to investigate free vibration of purely flexible beams composed of a local/nonlocal mixture. We get a hierarchy of equations that at the 0th order match with the well-known ones of local elastic beams. We show that the family of 0th order bending couples satisfy the natural and essential boundary conditions of 1st order; hence, the 1st order solution can conveniently be built using the 0th order eigenspace with no need of additional conditions. Exploiting the eigenfunctions orthogonality, we obtain a condition of solvability that provides the incremental eigenvalues in closed form and proves that the 1st order increment of the eigenvalue is always negative, yielding the well-known softening effect of long-range actions among the material points of a continuum according to Eringen's theory.

Direct 1-D Beam Model

We fix an origin and a Cartesian coordinate frame xyz, equipped with ortho-normal base vectors $\{\boldsymbol{i}, \boldsymbol{j}, \boldsymbol{k}\}$, in the 3D Euclidean ambient space. The reference configuration of a beam is defined as a collection of equal plane cross-sections attached through their centroid to a portion of the z-axis of length l, called beam axis. Another configuration is described by the translation of the cross-sections centroids, represented by the vector field $\mathbf{u}(z)$; and by the cross-sections rotation,

 Materials Research Forum LLC
https://doi.org/10.21741/9781644902431-100

represented by the orthogonal tensor field **R**(z). A suitable difference between these two configurations provides the following strain measures

$$\mathbf{E} = \frac{d\mathbf{R}}{dz}\mathbf{R}^{\mathrm{T}}, \qquad \mathbf{e} = \frac{d}{dz}(z\mathbf{k} + \mathbf{u}) - \mathbf{R}\mathbf{k}, \tag{1}$$

where **E** is a skew-symmetric tensor field of the curvature of the beam in the actual configuration and **e** provides shearing and elongation of the beam axis.

We are interested in planar motion; that is, the forces and couples acting on the beam are in the yz plane, and the deformed beam axis is also in the yz plane. Therefore, the components w, v of **u**(z) along z and y respectively are nonzero, and the only nonzero rotation angle Ω of the cross-sections is about the x-axis. Usual assumptions of small-amplitude displacements and rotations lead to following linearized strain measures

$$\varepsilon = \frac{dw}{dz}, \qquad \gamma = \frac{dv}{dz} + \Omega, \qquad \chi = \frac{d\Omega}{dz}, \tag{2}$$

where ε, γ, and χ stand for axial elongation, shearing strain, and bending curvature, respectively, while the displacement components and rotation now describe their first-order increments.

The balance in the actual shape, in the absence of the so-called *nonlocal residuals*, read

$$\frac{d\mathbf{n}}{dz} + \mathbf{q} = \mathbf{0}, \qquad \frac{d\mathbf{m}}{dz} + \frac{d(z\mathbf{k} + \mathbf{u})}{dz} \times \mathbf{n} + \mathbf{t} = \mathbf{0}, \tag{3}$$

where **n**,**m** are inner force and couple, while **q**,**t** stand for the external force and couple densities, respectively. When the couples and forces lie in the yz plane, Eq. (3) lead to the following scalar incremental balance equations

$$\frac{dN}{dz} + q_z = 0, \qquad \frac{dT}{dz} + q_y = 0, \qquad \frac{dM}{dz} - T + t_x = 0, \tag{4}$$

where: N, T, M are the increments of the inner normal force, shear force, and bending couple, respectively; q_z, q_y, t_x are the increments of the external actions reduced to the beam axis in the direction indicated by the subscripts.

The linear elastic problem is closed by the constitutive equation, which relates strain measures and dual work-conjugate internal actions. Here we assume that the beam is purely flexible, and the material obeys Eringen's model of a two-phase local/nonlocal mixture:

$$M = B[(1-\xi)\chi + \xi K * \chi], \tag{5}$$

where B is the bending stiffness of the cross-section and ξ is the nonlocal portion of the material response; the latter is modelled by the kernel function K, accounting for long-range interactions among material points. We consider an exponential kernel in the following form

$$K * f = \int_0^L K(\zeta, z) f(\zeta) dz, \qquad K(\zeta, z) = \frac{1}{2\kappa} e^{-\frac{|z-\zeta|}{\kappa}}, \qquad \int_{-\infty}^{\infty} K(\zeta, z) dz = 1 \tag{6}$$

where κ is called nonlocal parameter and is a rough measure of a 'radius of activity' or 'radius of extinction' of long-range interactions.

Transverse Vibration

Let the only non-zero component of external action be transverse inertia, that is,

$$q_z = t_x = 0, \qquad q_y = -m\ddot{v}, \tag{7}$$

m being the mass of the beam per unit length of its axis and over-dots denote time derivatives. If the fields of interest are harmonic in time with angular frequency ω, we choose to use the same notation to indicate their spatial part only, and define the following nondimensional quantities:

$$\bar{z} = \frac{z}{L}, \qquad \bar{v} = \frac{v}{L}, \qquad \bar{\kappa} = \frac{\kappa}{L}, \qquad \bar{T} = \frac{TL^2}{B}, \qquad \bar{M} = \frac{ML}{B}, \qquad \bar{\lambda} = \frac{m\omega^2 L^2}{B}. \tag{8}$$

For the ease of notation, the overbars will be omitted, except when confusion may arise.

With the assumptions on the initial and current shape of the beam, the transverse and axial motions are uncoupled. We are interested only in the transverse motion that is quantified by: the transverse displacement component v; the cross-section rotation angle Ω; its dual work conjugate, the bending couple M; and, for balance, the shear force T. The corresponding four 1st order integral-differential equations may be reduced into a single one of 4th order after trivial operations

$$(1-\xi)f^{IV} + \xi\left(K * f^{(i)}\right)^{(4-i)} - \lambda f = 0, \quad \begin{matrix} f = v \Rightarrow i = 1, & f = T \Rightarrow i = 3, \\ f = \Omega \Rightarrow i = 2, & f = M \Rightarrow i = 4; \end{matrix} \tag{9}$$

where the superscript in the parentheses indicates the order of the spatial derivative.

The usual boundary conditions are listed in Table 1.

Table 1. *Natural and essential boundary conditions for particular selection of f.*

	pin	clamp	free
v	$f = 0$, $(1-\xi)f'' + \xi K * f'' = 0$	$f = 0$, $f' = 0$	$(1-\xi)f'' + \xi K * f'' = 0$ $(1-\xi)f''' + \xi(K * f'')' = 0$
Ω	$(1-\xi)f' + \xi K * f' = 0$ $(1-\xi)f''' + \xi(K * f')'' = 0$	$(1-\xi)f''' + \xi(K * f')'' = 0$ $f = 0$	$(1-\xi)f' + \xi K * f' = 0$ $(1-\xi)f'' + \xi(K * f')' = 0$
T	$f' = 0$ $(1-\xi)f''' + \xi K * f''' = 0$	$f' = 0$ $f'' = 0$	$f = 0$ $(1-\xi)f''' + \xi K * f''' = 0$
M	$f = 0$ $f'' = 0$	$f'' = 0$ $f''' = 0$	$f = 0$ $f' = 0$

It is notable that when the bending couple is used as the unknown field in Eq. 9, none of the classical boundary conditions include the convolution.

Perturbation with respect to the Nonlocal Fraction

If, as physically reasonable, all the quantities of interest depend on the nonlocal fraction ξ of the material response, we can approximate them by their ξ-power series expansions about a given value ξ_0

$$f \cong \sum_{j=0}^{n} \frac{(\xi-\xi_0)^j}{j!} f_j, \quad \lambda \cong \sum_{j=0}^{n} \frac{(\xi-\xi_0)^j}{j!} \lambda_j, \quad f_j = \left.\frac{\partial f}{\partial \xi}\right|_{\xi=\xi_0}. \tag{10}$$

A crucial choice for our aim is $\xi_0 = 0$, corresponding to a purely local elastic response. Thus, all quantities of interest are evaluated as if the beam were local. The expansions in Eq. 10 are fully reliable for 'small' values of ξ, i.e., $\xi \cong 0.1$, but are expected to give satisfactory results also for 'moderate' values of ξ, i.e., $\xi \cong 0.3 - 0.5$; in our applications we will give a numerical example.

Inserting Eq. 10 into Eq. 9 provides a hierarchy of equations for different orders of ξ:

0th order:
$$f_0^{IV} - \lambda_0 f_0 = 0, \tag{11}$$

1st order:
$$f_1^{IV} - \lambda_0 f_1 = f_0^{IV} - \left(K * f_0^{(i)}\right)^{(4-i)} + \lambda_1 f_0, \tag{12}$$

2nd order:
$$f_2^{IV} - \lambda_0 f_2 = 2f_1^{IV} - 2\left(K * f_1^{(i)}\right)^{(4-i)} + 2\lambda_1 f_1 + \lambda_2 f_0, \tag{13}$$

or, in general,

$$Df_0 = 0, \qquad Df_k = b_k, \qquad k = 1,2,3,\ldots \tag{14}$$

for which the usual pattern of perturbation expansions is apparent: that is, the differential operator $D(\cdot) = d^4(\cdot)/dz^4 - \lambda_0(\cdot)$ is the same at all orders and the 'forcing' terms on the right side depend on the solutions of previous orders. It is crucial to remark is that we turned the integral-differential system into a set of differential equations by perturbing the unknown field and the eigenvalues about the local problem, a solution of which we know to exist and be unique.

The boundary conditions for different ξ-orders are provided in Tables 2-3.

Table 2. *0th order boundary conditions for particular selections of f.*

	pin	clamp	free
v	$f_0 = 0,$ $f_0'' = 0$	$f_0 = 0,$ $f_0' = 0$	$f_0'' = 0$ $f_0''' = 0$
Ω	$f_0' = 0$ $f''' = 0$	$f_0''' = 0$ $f_0 = 0$	$f_0' = 0$ $f_0'' = 0$
T	$f_0' = 0$ $f_0''' = 0$	$f_0' = 0$ $f_0'' = 0$	$f_0 = 0$ $f_0''' = 0$
M	$f_0 = 0$ $f_0'' = 0$	$f_0'' = 0$ $f_0''' = 0$	$f_0 = 0$ $f_0' = 0$

Table 3. *1st order boundary conditions for particular selections of f.*

	pin	clamp	free
v	$f_1 = 0,$ $f_1'' = f_0'' - K * f_0''$	$f_1 = 0,$ $f_1' = 0$	$f_1'' = f_0'' - K * f_0''$ $f_1''' = f_0''' - (K * f_0'')'$
Ω	$f_1' = f_0' - K * f_0'$ $f_1''' = f_0''' - (K * f_0')''$	$f_1''' = f_0''' - (K * f_0')''$ $f_1 = 0$	$f_1' = f_0' - K * f_0'$ $f_1'' = f_0'' - (K * f_0')'$
T	$f_1' = 0$ $f_1''' = f_0''' - K * f_0'''$	$f_1' = 0$ $f_1'' = 0$	$f_1 = 0$ $f_1''' = f_0''' - K * f_0'''$
M	$f_1 = 0$ $f_1'' = 0$	$f_1'' = 0$ $f_1''' = 0$	$f_1 = 0$ $f_1' = 0$

We see that the 1st order equations admit nonhomogeneous boundary conditions even though the corresponding 0th order equations have homogeneous boundary conditions.

Constructing a Solution

We define the scalar product for continuously differentiable functions h, g that satisfy the 0th order boundary conditions and have support (0,1) as

$$\langle h, g\rangle = \int_0^1 h(z)g(z)dz \tag{15}$$

The differential operator D is self-adjoint with respect to the scalar product defined in Eq. 15,

$$\langle Dh, g\rangle = \langle h, Dg\rangle. \tag{16}$$

We can also show that the family f_{0i} of solutions to Eq. 11, associated with the eigenvalue λ_{0i}, are orthogonal to each other; that is,

$$\langle f_{0i}, f_{0j}\rangle = 0, \qquad i \neq j. \tag{17}$$

Recalling that the boundary conditions for higher-order equations are homogeneous only if $f = M$, we choose the bending couple as unknown field for convenience in mathematical operations. Indeed, it is possible to write the bending couple by the following eigenfunction expansion

$$M_{ki} = M_{0i} + \sum_{i \neq j} \alpha_{ij} M_{0j}, \qquad k = 1,2,3, \dots \tag{18}$$

for which the boundary conditions are satisfied; α_{ij} are the constants providing the contribution of the jth mode onto the ith mode of the 0th order solution.

Incremental Eigenvalues

Inserting Eq. 18 into Eq. 12 provides

$$\sum_{i\neq j}\alpha_{ij}\lambda_{0j}M_{0j}-\lambda_{0i}\sum_{i\neq j}\alpha_{ij}M_{0j}=\lambda_{0i}M_{0i}-\lambda_{0i}K*M_{0i}+\lambda_{1i}M_{0i} \tag{19}$$

multiplying both sides by M_{0i} and integrating over the domain leads to

$$\int_0^1(\lambda_{0i}M_{0i}M_{0i}-\lambda_{0i}(K*M_{0i})M_{0i}+\lambda_{1i}M_{0i}M_{0i})dz=\langle M_{0i},b_1\rangle=0, \tag{20}$$

that is, the 0th order solution for the ith mode shall be orthogonal to the forcing term of the 1st order equation for the corresponding mode. This is the Fredholm compatibility (solvability) condition, which we can solve for the incremental eigenvalue

$$\lambda_{1i}=\left(\frac{\langle K*M_{0i},M_{0i}\rangle}{\langle M_{0i},M_{0i}\rangle}-1\right)\lambda_{0i} \tag{21}$$

For a bounded and integrable function g with a compact support S the Lp norm is

$$\|g\|_p=\left(\int_S|g|^p dS\right)^{1/p}. \tag{22}$$

Young's Convolution Identity reads

$$\|f*g\|_r\leq\|f\|_p\|g\|_q,\qquad 1/p+1/q=1/r+1. \tag{23}$$

Then, we can write

$$\|K*M_{0i}\|_2\leq\|K\|_1\|M_{0i}\|_2,. \tag{24}$$

Since the kernel function K is a positive symmetric radial function, and considering Eq. 6-3, it is

$$\int_S K(\zeta,z)dz=\int_S|K(\zeta,z)|dz;\qquad \int_{-\infty}^{\infty}K(\zeta,z)dz=1\Rightarrow\int_0^1 K(\zeta,z)dz=\|K\|_1\leq 1, \tag{25}$$

which, along with Eq. 24, leads to

$$\|K*M_{0i}\|_2\leq\|M_{0i}\|_2,. \tag{26}$$

Referring to Hölder's Inequality

$$|\langle f*g\rangle|\leq\|f\|_2\|g\|_2, \tag{27}$$

we write

$$\int_0^1 M_{0i}(K*M_{0i})dz\leq\left|\int_0^1 M_{0i}(K*M_{0i})dz\right|\leq\left(\int_0^1 M_{0i}^2dz\right)^{1/2}\left(\int_0^1|K*M_{0i}|^2dz\right)^{1/2}, \tag{28}$$

which, along with Eq. 26, gives

$$\int_0^1 M_{0i}(K*M_{0i})dz\leq\int_0^1 M_{0i}M_{0i}dz\Rightarrow\frac{\int_0^1 M_{0i}(K*M_{0i})dz}{\int_0^1 M_{0i}M_{0i}dz}=\frac{\langle K*M_{0i},M_{0i}\rangle}{\langle M_{0i},M_{0i}\rangle}\leq 1. \tag{29}$$

This is a notable result indicating that the incremental eigenvalue is always negative, which is a feature ascribable to the well-known softening behavior of Eringen's model.

An Example: Simply supported beam

For a simply supported beam, the well-known local (0th order) solution reads

$$M_{0i}=C\sin i\pi z,\qquad \lambda_{0i}=i^4\pi^4 \tag{31}$$

where C is found by a suitable normalization.

Inserting Eq. 30 into Eq. 21 provides the slope of the $\lambda-\xi$ curve at $\xi=0$, that is, a linear approximation of the dependance of the eigenvalues on the non-local fraction coefficient. Further, what is most interesting is that such approximation is given in a closed form as follows

$$\lambda_{1i} = -\frac{\pi^6\kappa^2 i^6\{1+\kappa[i^2\pi^2\kappa+(-1)^i 2e^{-1/\kappa}-2]\}}{(1+i^2\pi^2\kappa^2)^2} \tag{32}$$

Table 4. *Comparison of eigenvalues for different parameters.*

	$\kappa = 0.05$		$\kappa = 0.1$	
	$\xi = 0.25$	$\xi = 0.5$	$\xi = 0.25$	$\xi = 0.5$
[5]	9.84255	9.81485	9.77714	9.67978
[6]	9.84255	9.81485	9.77714	9.67978
Present	9.84276	9.81584	9.77854	9.68661

It is apparent in Table 4 that our perturbation approach can reply almost exactly the results found in the literature. It must be remarked that our linear perturbation approach is reliable even for moderate non-local fractions, which is not to be taken for granted in advance. That is, even though perturbations are usually reliable only for 'small' values of the perturbation parameter, in this case, as advanced previously, our perturbation yields reliable results also for 'non-small' values of ξ, i.e., $\xi = 0.5$.

Conclusions

By a perturbation approach we turned the integral-differential field system for beams composed of a two-phase local/nonlocal mixture into a hierarchy of bulk equations completed by the usual boundary conditions of local elasticity. Non-triviality of solutions and solvability conditions at successive steps of the hierarchy led to closed-form solutions for incremental natural angular frequencies of transverse natural vibration. Using well-known basic identities and inequalities of functional analysis we showed that the incremental eigenvalue is always negative, providing the well-known softening effect of Eringen's theory. The closed-form expressions can be of interest in the modelling and identification of nanomaterials.

References

[1] Eringen A.C., "Linear theory of nonlocal elasticity and dispersion of plane waves", International Journal of Engineering Sciences 10 (1972), 425–435. https://doi.org/10.1016/0020-7225(72)90050-X

[2] Eringen A.C., Edelen D., "On nonlocal elasticity", International Journal of Engineering Sciences 10 (1972), 233–248. https://doi.org/10.1016/0020-7225(72)90039-0

[3] Romano G., Barretta R., Diaco M., Marotti de Sciarra F., "Constitutive boundary conditions and paradoxes in nonlocal elastic nanobeams", International Journal of Mechanical Sciences 121 (2017), 151–156. https://doi.org/10.1016/j.ijmecsci.2016.10.036

[3] Eringen A.C., Nonlocal Continuum Field Theories. Springer (2002).

[4] Eroglu U., "Perturbation approach to Eringen's local/nonlocal constitutive equation with applications to 1-D structures", Meccanica, 55 (2020), 1119-1134. https://doi.org/10.1007/s11012-020-01145-x

[5] Fernández-Sáez J., Zaera R., Vibrations of Bernoulli-Euler beams using the two-phase nonlocal elasticity theory, International Journal of Engineering Science 119 (2017), 232–248. https://doi.org/10.1016/j.ijengsci.2017.06.021

[6] Fakher M., Hosseini-Hashemi S., Nonlinear vibration analysis of two-phase local/nonlocal nanobeams with size-dependent nonlinearity by using Galërkin method, Journal of Vibration and Control 27(3-4) (2021), 378–391. https://doi.org/10.1177/1077546320927619

Reaction-diffusion-drift equations and gradient flows in mechanics and continuum physics

Theoretical and Applied Mechanics - AIMETA 2022 Materials Research Forum LLC
Materials Research Proceedings **26** (2023) 627-631 https://doi.org/10.21741/9781644902431-101

An effective strategy to transform second-gradient equilibrium equations from the Eulerian to the Lagrangian configuration

Roberto Fedele[1,a *], Francesco dell'Isola[2,b], Pierre Seppecher[3,c], Simon R. Eugster[4,d]

[1] Department of Civil and Environmental Engineering (DICA), Politecnico di Milano, Milan, Italy

[2] Dipartimento di Ingegneria Civile, Edile-Architettura e Ambientale (DICEAA) and International Research Center on Mathematics and Mechanics of Complex Systems, University of L'Aquila, L'Aquila, Italy

[3] Institut de Mathématiques de Toulon, Université de Toulon, Toulon, France

[4] Institute for Nonlinear Mechanics, University of Stuttgart, Stuttgart, Germany

[a] roberto.fedele@polimi.it , [b] francesco.dellisola@univaq.it , [c] pierre@seppecher.fr , [d] eugster@inm.uni-stuttgart.de

Keywords: Continuum Mechanics, Second-Gradient Materials, Nonstandard Boundary Conditions, Variational Approach

Abstract. In this communication the problem of transforming the equilibrium equations from the Eulerian to the Lagrangian form is discussed with reference to materials governed by second-gradient energy densities. In particular, novel theoretical achievements are outlined, which represent intermediate steps to attain the purpose: the transformation of edge vectors and of complementary orthogonal projectors over the boundary surface; a novel formula based on the divergence theorem for curved surfaces with boundary, relating material and spatial expressions; a remarkable relationship between Lagrangian and Eulerian (hyper-)stress tensors of different orders.

Introduction

Higher gradient materials constitute a wide class of materials for which the stored energy density depends not only on the deformation gradient but also on its higher order derivatives. In the last decade, such kind of mathematical models have attracted an increasing number of researchers in continuum and computational mechanics. This circumstance is mainly due to the fact that higher gradient modelling is capable of describing complex phenomena which cannot be predicted by the conventional Cauchy approach. Among others, we can mention size effects taking into account characteristic length scales, boundary layers, corner and surface effects, which are crucial in all those scenarios in which the separation of scales is not sharp. Moreover, higher gradient materials admit "exotic" loading which cannot be sustained by a Cauchy medium, such as double forces, expending work under the normal derivative of the virtual placement, edge or wedge loading (see e.g. [1-2]): these generalized forces represent versatile tools when investigating surface tension in fluids or other interface issues, wave propagation in crystals, fiber nets interacting with the matrix in composites and more in general the mechanical behaviour of the so called metamaterials.

In this study a strategy is proposed, to transport the governing equations for second-gradient materials from the Eulerian to the Lagrangian form. Such a study revealed important differential geometric features of the equilibrium problem, and is expected to play an important role for the formulation and the implementation of advanced mechanical theories.

Variational approach

In the present approach, the deformation process of a continuum body is described as a bijection between two configurations, namely a reference configuration $\Omega_{å} \subseteq \mathsf{R}^3$, referred to as material or Lagrangian, and another one, usually referred to as spatial or Eulerian $\Omega \subseteq \mathsf{R}^3$. Such a map is continuous and differentiable, being its inverse also continuous and differentiable, and is denoted by the symbol $\boldsymbol{\chi}: \mathbf{X} \in \Omega_{å} \to \mathbf{x} \in \Omega$. It represents a diffeomorphism between submanifolds with boundary, for which we have $J = \det(\mathbf{F}) > 0$, where symbol $\mathbf{F} = \partial\boldsymbol{\chi} / \partial\mathbf{X}$ denotes the deformation gradient. We assume moreover that the (differential and topological) boundary of the above domains is constituted of the union of regular *faces*, having in common two by two parts of their border, referred to as *edges*, which represent discontinuity loci for the face normals. In this study we consider an energy density depending on the first and second gradient of the placement map, namely $\mathrm{W}(\mathbf{X}, \mathbf{F}, \nabla\mathbf{F})$, see [1]. The objectivity of such an energy is guaranteed by prescribing the dependence on the right Cauchy-Green tensor $\mathbf{C} = \mathbf{F}^T\mathbf{F}$. The equilibrium configuration corresponds to the placement minimizing the total potential energy functional [3], namely

$$\hat{\boldsymbol{\chi}} = \arg\min_K \left\{ \mathsf{E}^{\mathrm{TOT}}(\boldsymbol{\chi}) = \int_{\Omega_{å}} \mathrm{W}\left(\mathbf{F}, \nabla\mathbf{F}\right) d\Omega_{å} - \mathsf{E}^{\mathrm{EXT}}(\boldsymbol{\chi}, \nabla\boldsymbol{\chi}) \right\} \tag{1}$$

Symbol K denotes herein an admissible set of functions sufficiently regular (also as for their trace) which incorporate the essential boundary conditions on the placement map and its normal derivative. According to a variational approach, the stationarity condition follows by imposing the first variation of the above functional to vanish. As for the inner virtual work, one obtains

$$\begin{aligned}
&\delta \int_{\Omega_{å}} \mathrm{W}\left(\mathbf{F}, \nabla\mathbf{F}\right) d\Omega_{å} = \int_{\Omega_{å}} \frac{\partial \mathrm{W}}{\partial \mathbf{F}} : \delta\mathbf{F} + \frac{\partial \mathrm{W}}{\partial \nabla\mathbf{F}} \vdots \delta\left(\nabla\mathbf{F}\right) d\Omega_{å} = \\
&= \underbrace{\int_{\Omega_{å}} \frac{\partial \mathrm{W}}{\partial F^i_{\ A}} \delta F^i_{\ A} \, d\Omega_{å}}_{=\delta\mathsf{E}_I^{\mathrm{DEF}}} + \underbrace{\int_{\Omega_{å}} \frac{\partial \mathrm{W}}{\partial F^i_{\ AB}} \delta F^i_{\ AB} \, d\Omega_{å}}_{=\delta\mathsf{E}_{II}^{\mathrm{DEF}}} = \\
&= \int_{\Omega_{å}} P_i^A \, \delta F^i_{\ A} \, d\Omega_{å} + \int_{\Omega_{å}} P_i^{AB} \, \delta F^i_{\ AB} \, d\Omega_{å} \,;
\end{aligned} \tag{2}$$

In the above equation the parts of the energy variation relevant to the first and second gradient were marked by subscript *I* and *II*. As a key ingredient of the present formulation for second-gradient materials, the inner work depends not only on the second-rank Piola stress tensor, denoted above by P_{1i}^A, but also on a third-rank tensor P_{2i}^{AB}, referred to as hyper-stress tensor (or double-stress), see [2]. The present variational approach (and the more general principle of the virtual work which does not require any constitutive assumptions, see [4]) leads naturally to nonstandard boundary conditions and allows one to specify the admissible classes of external actions not known a priori. By the reiterated application of the integration by parts and of the divergence theorem extended to curved surfaces with boundary, we obtain a representation of the inner virtual work as the sum of diverse terms, including a novel surface action expending work versus the normal derivative of the virtual placement, an edge term, and contact pressures over the surface in which the linear dependence on the normal (according to the Cauchy's postulate) is superseded by the sum of nonlinear expressions involving the product of normals and of their derivatives: in particular, the dependence on the local mean curvature is made explicit, see [5].

Work-conjugate variables

The above formulation, based on the definition of a suitable energy density, is truly Lagrangian. However, the equilibrium problem can be formulated in an abstract setting by a proper choice of work-conjugate variables: then the governing equations assume the same form in the Lagrangian and in the Eulerian configuration. In fact, after computing the Eulerian counterparts of the virtual placement gradients, namely

$$\begin{aligned} \delta D^i_j &= \frac{\partial}{\partial x^j}\delta\chi^i(\mathbf{x}) = \frac{\partial \delta\chi^i}{\partial X^A}\frac{\partial X^A}{\partial x^j} = \delta F^i_A\left(\mathbf{F}^{-1}\right)^A_j ; \\ \delta D^i_{kj} &= \frac{\partial^2}{\partial x^j x^k}\delta\chi^i(\mathbf{x}) = \frac{\partial}{\partial x^k}\left(\frac{\partial}{\partial x^j}\delta\chi^i(\mathbf{x})\right) = \\ &= \frac{\partial}{\partial x^k}\left(\mathbf{F}^{-1}\right)^A_j \delta F^i_A + \left(\mathbf{F}^{-1}\right)^A_j \delta F^i_{AB}\left(\mathbf{F}^{-1}\right)^B_k ; \end{aligned} \tag{3}$$

equating the expressions of the Lagrangian and of the Eulerian inner virtual work, one obtains remarkable relationships between the Eulerian and Lagrangian hyper-stress tensors [6], namely

$$\begin{aligned} T^j_{1i} &= J^{-1} P^A_{1i} F^j_A + J^{-1} P^{AB}_{2i} F^j_{AB} \\ T^{jk}_{2i} &= J^{-1} P^{AB}_{2i} F^j_A F^k_B \end{aligned} \tag{4}$$

These relationships highlight a top-down structure typical of higher gradient modelling: in fact, the Lagrangian hyperstress affects both lower order Eulerian tensors, whilst the Piola stress tensor affects only the Cauchy stress T^j_{1i}.

Edge vectors

In the classical treatises of continuum mechanics a few formulae were already available to transform vectors defined over the boundary faces from the Lagrangian to the Eulerian configuration, such as those concerning the contravariant tangent vector and the covariant normal. However, the authors proposed novel relationships for the covariant and contravariant form of the border normal, namely the normal to the border edge which is orthogonal to the tangent and belongs to the face tangent plane. By assuming as ansatz an affine function of the Lagrangian variable, and exploiting the above mentioned orthogonality conditions the following relationship was derived for the contravariant form of the edge normal

$$b^r = \left\{ F^r_R B^R - \frac{\left(g_{rs} F^r_R B^R F^s_S T^S\right)}{\left(g_{rs} F^r_R T^R F^s_S T^S\right)} F^r_R T^R \right\} \frac{\| \mathbf{FT} \|}{\| J\mathbf{F}^{-T}\mathbf{N} \|} ; \tag{5}$$

An analogous formula can also be provided for the covariant form of the edge normal, see [6]. It is worth emphasizing that such transformations are not unique and alternative expressions may exist: not necessarily they are available in closed form for both the contravariant and covariant representation of the same vector. The above formula can be regarded as an application of Gram-Schmidt orthonormalization procedure, where a key role is played by the pull-back metric tensor.

Transport of surface projectors

As well known, at each point of a surface a pair of complementary orthogonal projectors can be defined, referred to as normal and tangential, and denoted by symbols $[M_\perp]^S_V = N^S N_V$ and $[M_\parallel]^S_V = \delta^S_V - N^S N_V$ respectively, apt to project any vector of the space environment onto the normal or the tangent space at that point. We provided effective transformation formulae for such projectors from the Lagrangian to the Eulerian configuration. For the normal projector, exploiting the transformation rule for the covariant normal we get

$$
\begin{aligned}
&[m_\perp]^r_s = n^r n_s = g^{rt} \frac{\left(\mathbf{F}^{-1}\right)^Q_t N_Q}{\| \mathbf{F}^{-T}\mathbf{N} \|} \frac{\left(\mathbf{F}^{-1}\right)^V_s N_V}{\| \mathbf{F}^{-T}\mathbf{N} \|} = \\
&= \frac{g^{rt}\left(\mathbf{F}^{-1}\right)^Q_t g_{QS}}{\| \mathbf{F}^{-T}\mathbf{N} \|^2}\left[N^W N_V\right]\left(\mathbf{F}^{-1}\right)^V_s = \\
&= \frac{g^{rt}\left(\mathbf{F}^{-1}\right)^Q_t g_{QS}}{\| \mathbf{F}^{-T}\mathbf{N} \|^2}[M_\perp]^S_V\left(\mathbf{F}^{-1}\right)^V_s;
\end{aligned} \tag{6}
$$

whilst a more complex relationship is met for the tangential projector. It is worth underlying that the Eulerian tangential projector depends on both the Lagrangian projectors: hence, an Eulerian vector normal to an Eulerian surface when transported to the Lagrangian configuration is expected to possess non-vanishing components in both the normal and the tangent space at the corresponding point.

Divergence theorem revisited

In the present approach recourse is made to the divergence theorem formulated for submanifolds with boundary, which represents an important result of differential geometry, e.g. see [6]. Such a theorem was revisited, providing a novel relationship between spatial and material expressions extremely useful for the analytical developments, namely

$$
\begin{aligned}
&[m_\|]^c_a \frac{\partial}{\partial x^c}([m_\|]^a_b w^b)\|J\boldsymbol{F}^{-T}\boldsymbol{N}\| = \\
&= [M_\|]^A_S \frac{\partial}{\partial X^A}(\|J\boldsymbol{F}^{-T}\boldsymbol{N}\|(\boldsymbol{F}^{-1})^R_a([m_\|]^a_b w^b)[M_\|]^S_R);
\end{aligned} \tag{7}
$$

In a sense, this formula generalizes Piola's bulk transformation, see [4].

Closing remarks and future prospects

In this study the transport from the Eulerian to the Lagrangian configuration of the equilibrium equations was addressed for the first time with reference to second-gradient modelling. To attain the purpose, novel theoretical results were achieved, which represent intermediate steps and however exhibit a general interest. The methodology proposed above can be easily extended to higher-order gradient materials (see e.g. [7]), possibly enriched by damage and plasticity. Moreover the novel results are expected to play a role for advanced mechanical theories and their implementation (see e.g. [8]), possibly concerning fracture propagation and contact mechanics.

References

[1] N. Auffray, F. dell'Isola, V. Eremeyev, A. Madeo, G. Rosi, Analytical continuum mechanics à la Hamilton-Piola least action principle for second gradient continua and capillary fluids, Math. Mech. Solids. 20(4) (2015) 375-417. https://doi.org/10.1177/1081286513497616

[2] C. Polizzotto, A note on the higher order strain and stress tensors within deformation gradient elasticity theories: Physical interpretations and comparisons, Int. J. Solids Struct. 90(4) (2016) 375-417. https://doi.org/10.1016/j.ijsolstr.2016.04.001

[3] R. Fedele, Piola's approach to the equilibrium problem for bodies with second gradient energies. Part I: First gradient theory and differential geometry, Continuum Mech. Thermodyn. 34 (2022) 445–474. https://doi.org/10.1007/s00161-021-01064-6

[4] S.R Eugster, F. dell'Isola, R. Fedele, P. Seppecher, Piola transformations in second-gradient continua, Mech. Research Commun. 120 (2022) 103836. https://doi.org/10.1016/j.mechrescom.2022.103836

[5] R. Fedele, Approach à la Piola for the equilibrium problem of bodies with second gradient energies. Part II: Variational derivation of second gradient equations and their transport, Continuum Mech. Thermodyn. 34 (2022). https://doi.org/10.1007/s00161-022-01100-z

[6] G. Capobianco, S.R. Eugster, On the divergence theorem for submanifolds of Euclidean vector spaces within the theory of second-gradient continua, Z. Angew. Math. Phys. (2022). https://doi.org/10.1007/s00033-022-01718-8

[7] R. Fedele, Third gradient continua: nonstandard equilibrium equations and selection of work conjugate variables, Math Mech Solids (2022). https://doi.org/10.1177/10812865221098966

[8] Reiher JC, Giorgio I and Bertram A. Finite-element analysis of polyhedra under point and line forces in second-strain gradient elasticity. J. Eng. Mech. 143(2) (2017) 04016112. https://doi.org/10.1061/(ASCE)EM.1943-7889.0001184

New frontiers in multibody systems vibration analysis

A simple tool to forecast the natural frequencies of thin-walled cylinders

Marco Cammalleri[1,a *], Antonella Castellano[1,b] and Marco Abella[1,c]

[1]Department of Engineering, University of Palermo, Palermo, Italy

[a]marco.cammalleri@unipa.it, [b]antonella.castellano@unipa.it, [c]marco.abella@you.unipa.it

Keywords: Circular Cylindrical Shell, Natural Frequencies, Eigenvalue Problem

Abstract. This paper presents an approximate method to predict the natural frequencies of thin-walled cylinders. By taking inspiration from a previous work of one of the authors, the starting point of the proposed approach is a proper construction of reasonable eigenfunctions. However, a new simple tool based on the principle of virtual work has been developed to estimate the natural frequencies and the amplitude of vibration without complex numerical resolution. Moreover, the applicability of the model is extended to all the most common constraint conditions. The identification of the natural frequencies of a continuous cylinder is reduced to an eigenvalue problem based on a matrix whose elements depend only on the geometric characteristics of the cylinder, the mechanical properties of the material and known numerical parameters. The latter are precalculated for given boundary conditions, covering clamped or pinned end constraints. Although the proposed formulation can address any constraints combination, only a pinned-pinned cylinder is analyzed here for brevity. The reliability of the model was tested against FEM analysis results. These comparisons showed that the maximum error versus the exact solutions for the lowest natural frequency is around 2% for all the mode shapes of the pinned-pinned case, offering an excellent trade-off between accuracy and ease of use.

Introduction

The characterization of the vibratory behaviour of thin-walled cylinders attracts researchers' interest due to the wide use of shells as structural elements in several engineering fields. In particular, predicting their natural frequency is crucial to prevent severe faults during the manufacturing process and the regular use, when time-varying forces often stress these elements.

Due to the continuous nature of thin-walled cylinders, studying their free vibrations is far more complex than a discrete multi-degree-of-freedom system. The integration of the partial differential equations rarely leads to an exact closed-form solution, which, however, is rather convoluted [1]. More frequently, the resolution is achieved by finite element analysis, numerical approaches [1–7] or simplified analytical models [8–12]. Nonetheless, the finite element method (FEM) may require a convergence analysis and lacks intelligibility. On the other hand, advanced numerical techniques enable the resolution of highly accurate models but might be challenging to program, while the introduction of simplifying assumptions allows an analytical solution to the problem at the expense of accuracy.

In contrast, the novel model presented here combines good accuracy with ease of use. It takes inspiration from [9], which provided the natural frequencies by a simple sequence of explicit algebraic equations without the need for complex or iterative numerical resolution. Starting from Love's theory for thin-walled cylinders modified by Reissner and simplified by Donnell's assumptions, the dynamic equilibrium equations are derived as functions of displacements. Then, Hamilton's principle is applied. The assumption of reasonable eigenfunctions, similarly to Rayleigh's method, enables a fast-solving procedure based on the resolution of a cascade of simple algebraic equations. Nonetheless, this method applies only to clamped-clamped cylinders and involves two different sets of eigenfunctions depending on the mode shape order.

In this paper, a reformulation of [9] is proposed. In particular, the equilibrium, compatibility and constitutive equations are the same. Nevertheless, different eigenfunctions are hypothesized at the beginning and the principle of virtual work is used. Moreover, the cascaded algebraic resolutive approach is converted to an eigenvalue problem. The new procedure leads to a faster resolution that can be easily adapted to any constraint condition, not just the clamped-clamped one. By way of example, this paper present only the results for a pinned-pinned cylinder. The reliability of the proposed method is tested by a comparison against FEM analysis results. A maximum error of 2% has been obtained, proving that the model is effective and efficient.

Method

Given a thin-walled circular cylinder having a finite length l, constant thickness h and mean radius a consisting of a material having a density ρ, Young's modulus E and Poisson's ratio ν, Fig. 1 shows the orthogonal local reference system consisting of longitudinal direction x, circumferential direction s and radial direction r.

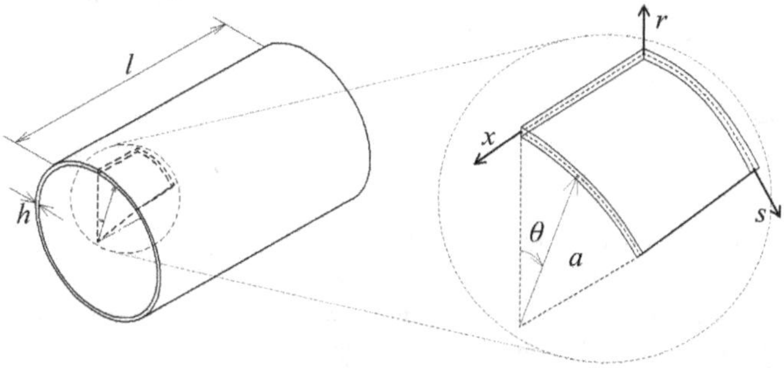

Figure 1. Geometry and local reference system of a thin-walled cylinder.

On the basis of Love's theory [13] modified by Reissner [14] and Donnell's assumptions [15], the dynamic equilibrium equations and the compatibility equations are derived. Then, the constitutive equations are used to express the internal forces and moments as functions of the deformations. After substituting the compatibility equations into these latter equations, forces and moments are obtained as functions of displacements. Then, these forces and moments are introduced into the dynamic equilibrium, and the following equations of motion are obtained:

$$\begin{cases} K\left(\frac{\partial^2 u_x}{\partial x^2}+\frac{1-\nu}{2a^2}\frac{\partial^2 u_x}{\partial \theta^2}+\frac{1+\nu}{2a}\frac{\partial^2 u_s}{\partial x \partial \theta}+\frac{\nu}{a}\frac{\partial u_r}{\partial x}\right)=\rho h\frac{d^2 u_x}{dt^2}, \\ K\left(\frac{1}{a^2}\frac{\partial^2 u_s}{\partial \theta^2}+\frac{1-\nu}{2}\frac{\partial^2 u_s}{\partial x^2}+\frac{1+\nu}{2a}\frac{\partial^2 u_x}{\partial x \partial \theta}+\frac{1}{a^2}\frac{\partial u_r}{\partial \theta}\right)=\rho h\frac{d^2 u_s}{dt^2} \\ K\left[-\frac{\nu}{a}\frac{\partial u_x}{\partial x}-\frac{1}{a^2}\frac{\partial u_s}{\partial \theta}-\frac{u_r}{a^2}-\frac{h^2}{12}\left(\frac{\partial^4 u_r}{\partial x^4}+\frac{1}{a^4}\frac{\partial^4 u_r}{\partial \theta^4}+\frac{2}{a^2}\frac{\partial^4 u_r}{\partial x^2 \partial \theta^2}\right)\right]=\rho h\frac{d^2 u_r}{dt^2} \end{cases} \quad (1)$$

where $K=\frac{Eh}{1-\nu^2}$. The mathematical procedure performed to obtain Eq. 1 is omitted for brevity but can be found in [9]. Nevertheless, if in [9] the equations of motion were used to apply Hamilton's principle, here are introduced in the principle of virtual work (Eq. 2), according to which the virtual work δW of all forces applied to the system, including the inertial actions, is zero for any virtual displacements δu_x, δu_s, and δu_r that meet the constraints.

$$\delta W = aK\int_0^{2\pi}\int_0^l\left\{\left[\frac{\partial^2 u_x}{\partial x^2}+\frac{1-\nu}{2a^2}\frac{\partial^2 u_x}{\partial \theta^2}+\frac{1+\nu}{2a}\frac{\partial^2 u_s}{\partial x \partial \theta}+\frac{\nu}{a}\frac{\partial u_r}{\partial x}-\frac{1-\nu^2}{E}\rho\frac{d^2 u_x}{dt^2}\right]\delta u_x+\right.$$

$$+\left[\frac{1}{a^2}\frac{\partial^2 u_s}{\partial\theta^2}+\frac{1-\nu}{2}\frac{\partial^2 u_s}{\partial x^2}+\frac{1+\nu}{2a}\frac{\partial^2 u_x}{\partial x\partial\theta}+\frac{1}{a^2}\frac{\partial u_r}{\partial\theta}-\frac{1-\nu^2}{E}\rho\frac{d^2 u_s}{dt^2}\right]\delta u_s+ \qquad (2)$$
$$-\left[\frac{\nu}{a}\frac{\partial u_x}{\partial x}+\frac{1}{a^2}\frac{\partial u_s}{\partial\theta}+\frac{u_r}{a^2}+\frac{h^2}{12}\left(\frac{\partial^4 u_r}{\partial x^4}+\frac{1}{a^4}\frac{\partial^4 u_r}{\partial\theta^4}+\frac{2}{a^2}\frac{\partial^4 u_r}{\partial x^2\partial\theta^2}\right)+\frac{1-\nu^2}{E}\rho\frac{d^2 u_r}{dt^2}\right]\delta u_r\}dxd\theta=0$$

Given the arbitrariness of the virtual displacements δu_x, δu_s and δu_r Eq. 2 can be respected only if each of the three addends is null.

Similarly to [9], the assumption of reasonable displacements u_x, u_s and u_r as eigenfunctions of the problem of free vibrations of cylindrical shells enables a simpler approximated mathematical treatment. Free vibrations of a thin-walled circular cylinder consists of n circumferential waves and m longitudinal half-waves. Therefore, each mode shape is characterized by a pair of values of n and m. Circumferential waves are independent from the boundary conditions, unlike longitudinal half-waves which, instead, depend on them, similarly to the transverse vibrations of beams subjected to the same constraints. Thus, the same solutions hypothesized in [9], properly chosen to respect orthogonality, are also considered here:

$$\begin{cases} u_x = A_x\frac{d}{dx}f_r(x)cos(n\theta)cos\omega t \\ u_s = A_s f_r(x)\,sin(n\theta)\,cos\omega t \\ u_r = A_r f_r(x)cos(n\theta)cos\omega t \end{cases} \qquad (3)$$

where $f_r(x)$ is the eigenfunction of the beam subjected to the same constraints of the cylinder under analysis. For example, for a pinned-pinned beam, $f_r(x)=\sin\beta_i l\frac{x}{l}$, where $\beta_i l$ are the roots of the related frequency equation $\sin\beta_i l=0$. However, the formulation proposed in the following is highly general and thus can be extended to any boundary conditions.

By normalizing Eq. 2 by the cylinder length l and considering Eq. 3, a three-equation system is obtained:

$$\begin{cases} \int_0^1\left[\left(\frac{\partial^2 u_x}{\partial X^2}+\frac{1-\nu}{2\alpha^2}\frac{\partial^2 u_x}{\partial\theta^2}+\frac{1+\nu}{2\alpha}\frac{\partial^2 u_s}{\partial X\partial\theta}+\frac{\nu}{\alpha}\frac{\partial u_r}{\partial X}+\Delta\,u_x\right)\delta u_x\right]dX=0 \\ \int_0^1\left[\left(\frac{1}{\alpha^2}\frac{\partial^2 u_s}{\partial\theta^2}+\frac{1-\nu}{2}\frac{\partial^2 u_s}{\partial X^2}+\frac{1+\nu}{2\alpha}\frac{\partial^2 u_x}{\partial X\partial\theta}+\frac{1}{\alpha^2}\frac{\partial u_r}{\partial\theta}+\Delta\,u_s\right)\delta u_s\right]dX=0 \\ \int_0^1\left\{\left[\frac{\nu}{\alpha}\frac{\partial u_x}{\partial X}+\frac{1}{\alpha^2}\frac{\partial u_s}{\partial\theta}+\frac{u_r}{\alpha^2}+\frac{\eta^2}{12}\left(\frac{\partial^4 u_r}{\partial X^4}+\frac{1}{\alpha^4}\frac{\partial^4 u_r}{\partial\theta^4}+\frac{2}{\alpha^2}\frac{\partial^4 u_r}{\partial X^2\partial\theta^2}\right)-\Delta\,u_r\right]\delta u_r\right\}dX=0 \end{cases} \qquad (4)$$

where $X=\frac{x}{l}$, $\Delta=\frac{1-\nu^2}{E}\rho l^2\omega^2$, $\alpha=\frac{a}{l}$ and $\eta=\frac{h}{l}$. By substituting Eq. 3 in Eq. 4, the following matrix formulation is derived:

$$(\bar{\bar{D}}-\Delta\bar{\bar{I}})\{A\}=\{0\} \qquad (5)$$

where $\{A\}=\{A_x;A_s;A_r\}$ is the unknown vector containing the displacements amplitudes in the three directions, $\bar{\bar{I}}$ is the identity matrix and $\bar{\bar{D}}$ is the following matrix:

$$\bar{\bar{D}}=\begin{bmatrix} -\frac{I_{13}}{I_{11}}+\frac{1-\nu}{2\alpha^2}n^2 & -n\frac{1+\nu}{2\alpha} & -\frac{\nu}{\alpha} \\ \frac{1+\nu}{2}\frac{n}{\alpha}\frac{I_{02}}{I_{00}} & -\frac{1-\nu}{2}\frac{I_{02}}{I_{00}}+\frac{n^2}{\alpha^2} & \frac{n}{\alpha^2} \\ \frac{\nu}{\alpha}\frac{I_{02}}{I_{00}} & \frac{n}{\alpha^2} & \frac{1}{\alpha^2}+\frac{\eta^2}{12}\left(\frac{I_{04}}{I_{00}}+\frac{n^4}{\alpha^4}-\frac{2}{\alpha^2}n^2\frac{I_{02}}{I_{00}}\right) \end{bmatrix} \qquad (6)$$

where $I_{13} = \int_0^1 f_r^1(X)\, f_r^3(X)\, dX$, $I_{11} = \int_0^1 f_r^1(X)\, f_r^1(X)\, dX$, $I_{00} = \int_0^1 f_r(X)\, f_r(X) dX$, $I_{02} = \int_0^1 f_r(X) f_r^2(X)\, dX$ and $I_{04} = \int_0^1 f_r(X) f_r^4(X)\, dX$; $f_r^k(X)$ is the *k*-order derivative of $f_r(X)$. Thanks to the normalization by the cylinder length, these integrals may be evaluated a priori without knowing the actual cylinder dimension. Thus, they need to be calculated only once for any constraint condition and then can be exploited for the free vibrations analysis of any cylinder subjected to the same constraints. Lastly, as it is apparent from Eq. 5, the natural frequency of any thin-walled cylinder can be easily calculated by solving the eigenvalue problem of the matrix $\overline{\overline{D}}$. From the three eigenvalues Δ_1, Δ_2 and Δ_3, the natural frequency is obtained as follows:

$$f_i = \frac{1}{2\pi}\sqrt{\frac{E\Delta}{(1-\nu^2)\rho l^2}} \qquad \text{for } i = 1, 2, 3 \tag{7}$$

To sum up, the proposed analysis of free vibrations of cylindrical shells involves the initial identification of the eigenfunction describing the transverse free vibrations of a beam subjected to the same constraints of the cylinder. Then, a pair of values for *m* and *n* is selected, and the matrix $\overline{\overline{D}}$ is populated. The natural frequencies are evaluated from the eigenvalues of the matrix $\overline{\overline{D}}$, while its eigenvectors contain the displacements amplitude ratios of each modal shape.

Results

Table 1 shows the frequency f_1, f_2 and f_3 for a pinned-pinned cylinder characterized by $a = 76$mm, $l = 305$mm, $h = 0.254$ mm, $\rho = 7833$ kg/m^3, $E = 207$ kN/mm^2, $\nu = 0.3$. For brevity, only results for $m \leq 3$ and $n \leq 8$ are reported. Nonetheless, they are sufficient to notice that f_1 is lower by un order of magnitude than f_2 and f_3; thus, it is the frequency related to the highest risk of the arising of redundancy condition. Moreover, f_1 shows a minimum for fixed *m*, which occurs for a higher value of *n* if *m* increases, while f_2 and f_3 are monotonically increasing by both *n* and *m*.

Table 1. Natural frequencies for m ≤ 3 and n ≤ 8.

	$m = 1$			$m = 2$			$m = 3$		
n	f_1 [Hz]	f_2 [Hz]	f_3 [Hz]	f_1 [Hz]	f_2 [Hz]	f_3 [Hz]	f_1 [Hz]	f_2 [Hz]	f_3 [Hz]
1	2,886	10,009	17,209	6,445	14,059	21,944	8,556	17,890	29,228
2	1,265	14,889	26,403	3,677	18,061	29,947	5,821	21,647	35,508
3	656	20,923	36,632	2,178	23,194	39,414	3,896	26,222	43,820
4	423	27,319	47,298	1,396	29,012	49,558	2,681	31,491	53,176
5	372	33,847	58,183	985	35,177	60,072	1,931	37,225	63,123
6	431	40,433	69,190	792	41,523	70,807	1,477	43,249	73,434
7	551	47,048	80,272	751	47,971	81,681	1,227	49,455	83,983
8	705	53,681	91,403	819	54,481	92,649	1,132	55,780	94,694

Table 2 shows the amplitude ratios only for $m \leq 3$ and $n = 4$, but similar trends are obtained for other combinations of *m* and *n*. The predominant amplitude at the lowest natural frequency f_1 is A_r, so the associated motion is mostly radial (transverse mode of vibration). Conversely, at frequencies f_2 and f_3, A_x and A_s prevail, respectively; thus, the associated modes are called longitudinal and circumferential.

Table 2. Amplitude ratios for $m \leq 3$ and $n = 4$.

	$m = 1$		$m = 2$		$m = 3$	
	A_x/A_r	A_s/A_r	A_x/A_r	A_s/A_r	A_x/A_r	A_s/A_r
f_1	0.014	0.252	0.011	0.255	0.008	0.249
f_2	3.697	1.897	1.020	2.155	0.521	2.561
f_3	0.240	4.097	0.259	4.380	0.288	4.822

The accuracy of the results was tested by comparing them against those derived from a FEM analysis carried out in Ansys 2022, where a thin-walled circular cylinder was analyzed using 36,433 SHELL181 linear elements. Fig. 3 and Table 3 reveal a high correspondence between the FEM results and those provided by the proposed model, with a maximum error of 2.07% for $m = 1$ and $n = 6$. The comparison considers only the frequency f1 for brevity since it is the lowest.

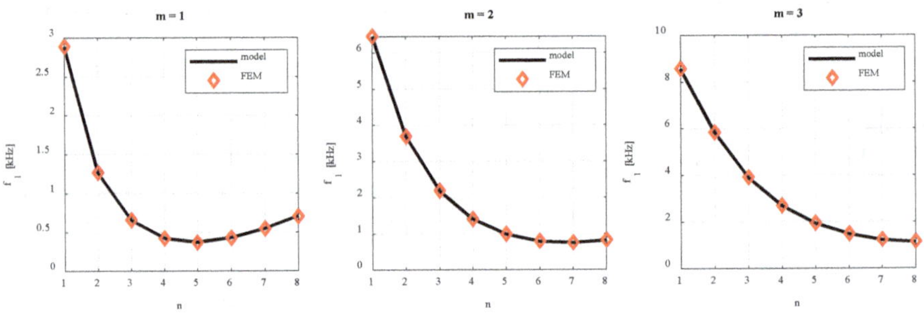

Figure 2. Comparison between the natural frequency f_1 obtained by the model and FEM analysis ($m \leq 3$ and $n \leq 8$).

Table 3. Percentage error on the f_1 frequency for $m \leq 8$ and $n \leq 8$.

	$m = 1$	$m = 2$	$m = 3$	$m = 4$	$m = 5$	$m = 6$	$m = 7$	$m = 8$
$n = 1$	-0.0004	-0.0103	-0.0181	-0.0212	-0.0225	-0.0228	-0.0229	-0.0229
$n = 2$	0.0106	-0.0254	-0.0427	-0.0550	-0.0628	-0.0675	-0.0704	-0.0723
$n = 3$	0.1934	-0.0336	-0.0715	-0.0957	-0.1137	-0.1269	-0.1365	-0.1435
$n = 4$	0.9627	0.0093	-0.0921	-0.1365	-0.1681	-0.1927	-0.2120	-0.2222
$n = 5$	2.0088	0.1821	-0.0822	-0.1683	-0.2199	-0.2591	-0.7405	-0.3167
$n = 6$	2.0702	0.5078	-0.0131	-0.1776	-0.2621	-0.3210	-0.3678	-0.9082
$n = 7$	1.5567	0.7645	0.1176	-0.1527	-0.2864	-0.3729	-0.4387	-0.4930
$n = 8$	1.0193	0.7233	0.2303	-0.1016	-0.2895	-0.4100	-0.4991	-0.5714

Conclusions

Starting from the standard equations for modeling equilibrium, deformations and displacements of thin-walled cylinders, this paper used the principle of virtual work to reduce the calculation of the natural frequencies to an eigenvalue problem, thanks to the simplifying assumption of eigenfunctions deduced from the beam theory. The results for a pinned-pinned cylindrical shell were compared to those obtained by a FEM analysis, reporting a maximum error of 2%. Thus, the novel model combines good accuracy with ease of use, being ideal for preliminary investigation of the resonance condition of shell structures. Moreover, it can be potentially extended to any boundary conditions: clamped, pinned, and free end cylinders can be addressed by a unique formulation. Future works will test the model accuracy with other constraints combinations.

References

[1] Y. Xing, B. Liu, T. Xu, Exact solutions for free vibration of circular cylindrical shells with classical boundary conditions, Int. J. Mech. Sci. 75 (2013) 178–188. https://doi.org/10.1016/j.ijmecsci.2013.06.005

[2] Y. Dubyk, I. Orynyak, O. Ishchenko, An exact series solution for free vibration of cylindrical shell with arbitrary boundary conditions, Sci. J. Ternopil Natl. Tech. Univ. 89 (2018) 79–88. https://doi.org/10.33108/visnyk_tntu2018.01.079

[3] Cammarata, A., Sinatra, R., & Maddio, P. D. (2019, July). An extended Craig-Bampton method for the modal analysis of mechanisms. In IFToMM World Congress on Mechanism and Machine Science (pp. 3329-3338). Springer, Cham. https://doi.org/10.1007/978-3-030-20131-9_329

[4] Cammarata, A., Sinatra, R., & Maddìo, P. D. (2019). Static condensation method for the reduced dynamic modeling of mechanisms and structures. Archive of Applied Mechanics, 89(10), 2033-2051. https://doi.org/10.1007/s00419-019-01560-x

[5] M. Cammalleri, E. Pipitone, S. Beccari, G. Genchi, A Mathematical Model for the Prediction of the Injected Mass Diagram of a S.I. Engine Gas Injector, Journal of Mechanical Science and Technology, Springer vol. 11 (2013) pagg. 3253-3265. https://doi.org/10.1007/s12206-013-0848-6

[6] Cammalleri M., Castellano A., Analysis of hybrid vehicle transmissions with any number of modes and planetary gearing: kinematics, power flows, mechanical power losses. Mechanism and Machine Theory 162 (2021) 104350. https://doi.org/10.1016/j.mechmachtheory.2021.104350

[7] S.M.R. Khalili, A. Davar, K. Malekzadeh Fard, Free vibration analysis of homogeneous isotropic circular cylindrical shells based on a new three-dimensional refined higher-order theory, Int. J. Mech. Sci. 56 (2012) 1–25. https://doi.org/10.1016/j.ijmecsci.2011.11.002

[8] C. Wang, J.C.S. Lai, Prediction of natural frequencies of finite length circular cylindrical shells, Appl. Acoust. 59 (2000) 385–400. https://doi.org/10.1016/S0003-682X(99)00039-0

[9] M. Cammalleri, A. Costanza, A closed-form solution for natural frequencies of thin-walled cylinders with clamped edges, Int. J. Mech. Sci. 110 (2016) 116–126. https://doi.org/10.1016/j.ijmecsci.2016.03.005

[10] F. Sorge, M. Cammalleri, Control of hysteretic instability in rotating machinery by elastic suspension systems subject to dry and viscous friction, J. Sound Vib. 329 (2010) 1686. https://doi.org/10.1016/j.jsv.2009.12.007

[11] F. Sorge, M. Cammalleri, On the beneficial effect of rotor suspension anisotropy on viscous-dry hysteretic instability, Meccanica. 47 (2012) 1705–1722. https://doi.org/10.1007/s11012-012-9549-y

[12] A. Kumar, S.L. Das, P. Wahi, Effect of radial loads on the natural frequencies of thin-walled circular cylindrical shells, Int. J. Mech. Sci. 122 (2017) 37–52. https://doi.org/10.1016/j.ijmecsci.2016.12.024

[9] A.E.H. Love, XVI. The small free vibrations and deformation of a thin elastic shell, Philos. Trans. R. Soc. London. 179 (1888) 491–546. https://doi.org/10.1098/rsta.1888.0016

[10] E. Reissner, A New Derivation of the Equations for the Deformation of Elastic Shells, Am. J. Math. 63 (1941) 177. https://doi.org/10.2307/2371288

[11] L.H. Donnell, Beams, plates and shells, McGraw-Hill, New York, 1976.

Multibody dynamics modeling of drivetrain components: On the caged-roller dynamics of centrifugal pendulum vibration absorbers

Mattia Cera[1,a*], Marco Cirelli[2,b], Luca D'Angelo[1,c], Ettore Pennestrì[1,d] and Pier Paolo Valentini[1,e]

[1]Department of Enterprise Engineering, University of Rome Tor Vergata, Rome, Italy

[2]Department of Mechanical Engineering, University Niccolò Cusano, Rome, Italy

[a]mattia.cera@uniroma2.it, [b]marco.cirelli@unicusano.it [c]luca.dangelo.02@alumni.uniroma2.eu, [d]pennestri@mec.uniroma2.it, [e]valentini@ing.uniroma2.it

Keywords: Centrifugal Pendulum Vibration Absorbers, Multibody Dynamics Contact and Friction Models, Caged-Roller Joint, Higher Path Curvature Analysis

Abstract. Centrifugal pendulum absorbers are passive dampers mainly employed nowadays to attenuate torsional vibrations in modern drivetrains to reduce fuel consumption and CO2 emissions. The absorber is linked to the drivetrain by means of a higher kinematic joint composed of slots and rollers, termed *caged-roller joint*. This work aims to investigate the contact between the rollers and the slots through multibody dynamics simulations. As a result, the sliding between the profiles, usually neglected in the design model of the caged-roller joint, is assessed and an estimate of the power loss is provided.

Introduction

Current trends for internal combustion engines (ICEs) suggest operation just above the idle-speed range to reduce fuel consumption. This generates low-frequency and high-amplitude oscillations. On the other hand, to avoid a drop in performance, high driving torque levels should be ensured. modern drivetrains require more effective torsional vibration isolation. The class of centrifugal dampers constitutes a widespread industrial solution adopted by many car manufacturers [1]. Centrifugal pendulum vibration absorbers (CPVAs) are order-tuned passive devices [2,3]. The absorbers counteract the torsional disturbance through their oscillations along a prescribed path within a centrifugal force field. The distinctive feature of CPVAs is that the absorber frequency of oscillation matches, for any rotor speed, the one of the torque to be damped. This is realized by selecting the absorber center of mass (COM) path curvature according to a design equation termed *tuning condition.* The classical approaches for the study of CPVAs nonlinear dynamics (*e.g.* [4-10]) are based on perturbation methods. To offer holistic modeling of these devices, the dynamic behavior of this device is herein explored through multibody techniques (*e.g.* [11]). The damper architecture herein portrayed is the parallel bifilar pendulum, where the absorbers are constrained to translate w.r.t. the rotor by two higher kinematic pairs, each constituted by a roller mating with two slots, denoted as caged-roller joint [12-14]. The ideal design of the slot profiles, needed as a constraint to enforce the tuning condition, requires the assumption of pure-rolling between the mating elements. However, this condition heavily depends on normal contact force amplitude throughout the absorbers oscillations. Hence, this paper focuses on analyzing the sliding and the power losses estimation due to the contact between the roller and the slot by means of multibody dynamics simulations.

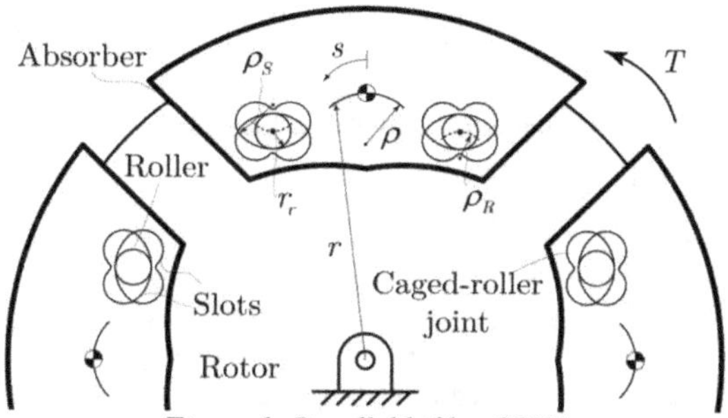

Figure 1: Parallel bifilar CPVA

Damper device modeling

The model kinematic structure is represented in Figure 1 in the reference configuration. Four absorbers of total mass m are linked to the rotor, characterized by a moment of inertia I through eight caged-roller joints. Each one of these joints is composed of a roller and two slots carved on the rotor and on the absorbers, respectively. A 2D circle-to-curve contact is established between the conjugate profiles[1]. The rotor is free to rotate w.r.t. the ground with an initial speed Ω and is subjected to the cubic ramp harmonic torque disturbance:

$$T(t)=T_0\left[3\left(\frac{t}{t_f}\right)^2-2\left(\frac{t}{t_f}\right)^3\right]\sin n\Theta(t)\,, \tag{1}$$

where Θ is the rotor rotation angle, n is the harmonic order to be damped and t_f is the end-time of the simulation. The CPVA conceptual design is hinged on the following requirements:

- the tuning condition governs the absorber COM path curvature for small oscillations [2-4]:

$$\rho_0=\frac{r_0}{n^2\left(1+\sigma\right)+1}\,, \tag{2}$$

  where ρ_0 is the absorber COM radius of curvature, r_0 denotes the rotor-absorber COM center distance, the subscript (0) indicates that the variable is evaluated at the reference position, and σ is the detuning, namely the difference between the disturbance and the absorber tuned orders.

- the damper dynamic behavior for large oscillations is controlled by the full absorber COM path. This trajectory, in intrinsic coordinates (ρ, s), is usually selected from the epicycloid family of curves [4-7]:

$$\rho(s)=\rho_0+\lambda_2 s^2\,, \tag{3}$$

where s is the absorber curvilinear abscissa and λ_2 is the second curvature ratio of the path $\rho(s)$, which is constant for epicycloids[2]. Since the absorbers undergo a relative translation w.r.t. the rotor, the roller center relative paths (ρ_R, s_R), are obtained by halving Eq.(3):

[1] Contact force models for multibody systems have been analyzed e.g. by Flores [15] and Pereira et al. [16].
[2] For further details on higher path curvature analysis and CPVA design procedures, refer to [4-9,14,17]

Materials Research Forum LLC
https://doi.org/10.21741/9781644902431-103

$$s_R = \frac{s}{2}, \quad \rho_R = \frac{\rho}{2}, \tag{4}$$

Then, the rotor and absorber slots paths are identical and can be computed as parallel curves of the roller paths (ρ_S, s_S) [14,17]:

$$s_S = s_R\left(1 + \frac{r_r}{\rho_R}\right), \quad \rho_S = \rho_R + r_r, \tag{5}$$

where r_r is the roller radius.

The roller-slot contact detection is based on an algorithm with a pre-search to identify contact zones and a detailed-search to find the penetration depth [19]. The contact normal behavior is governed by the Johnson model [16] for the elastic contribution:

$$\delta = \frac{2P(1-\nu^2)}{\pi E}\left[\log\left(\frac{4\pi E\Delta R}{2P(1-\nu^2)}\right) - 1\right], \tag{6}$$

where δ is the penetration, P is the load per unit length, $\Delta R = (\rho_S - r_r)|_{s=0} = \frac{\rho_0}{2}$, and (E, ν) are the elastic properties of the steel. This model is fitted with MATLAB fit routine to compute the normal force as follows:

$$F_n = k\delta^{m_1}\left(1 + \frac{3}{2}\alpha\dot{\delta}\right), \tag{7}$$

where the damping term $\left(\frac{3}{2}\alpha k\delta^{m_1}\dot{\delta}\right)$ is expressed according to Hunt and Crossley [19] model.

On the other hand, the tangential behavior follows a stick-slip dry friction portrayed in the investigations of Cha et al. [20] and Pennestrì et al. [21]:

$$\begin{aligned} F_t &= \left[(1-\beta)\mu_\Delta + \mu_v\right]F_n, & v < v_s \text{ (static friction)} \\ F_t &= -\text{sgn}(\mu)F_n, & v \geq v_s \text{ (dynamic friction)} \end{aligned} \tag{8}$$

This model, and in particular the terms β, μ_Δ, μ_v, μ, depend on five parameters, namely: μ_s, μ_d (static and friction coefficients), v_s, v_d (static and dynamic threshold velocity) and Δ_{max} (max stiction deformation). For the further details, refer to the cited works. The data of the system's geometry and inertia, as well as the contact parameters, are reported in Table 1.

Table 1: systems data and parameters

n	σ	λ_2	Ω [rpm]	m [kg]	r_0 [m]	T_0 [Nm]	I [kg m^2]
2	0.01	-0.84	1000	0.3	0.06	21	0.04
k [N/m^m1]	m_1	α	v_s [m/s]	v_d [m/s]	μ_s	μ_d	Δ_{max}
6.874e9	1.072	0.08	0.15	0.1	0.05	0.04	0.1

Results analysis

The first result needed to assess the CPVA dynamics behavior is the relationship between the torque amplitude and the absorber COM oscillation amplitude, depicted in the plot of Figure 2.

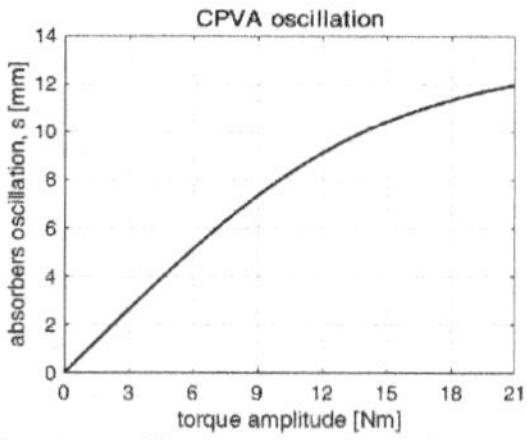

Figure 2: absorber oscillation amplitude .vs. torque amplitude.

For low torque levels, the device response is quasi-linear, whereas for large oscillations the resulting behavior is hardening. This feature depends on the selected value of the curvature ratio λ_2. Then, in Figure 3 are represented the friction coefficient and sliding velocity curves with respect to the torque amplitude. In particular, the gray line is for the interaction between the absorber slot and the roller, whereas the black line is for the rotor slot-roller contact. Moreover, the dashed lines indicate the threshold values v_s, v_d, μ_s, μ_d, respectively.

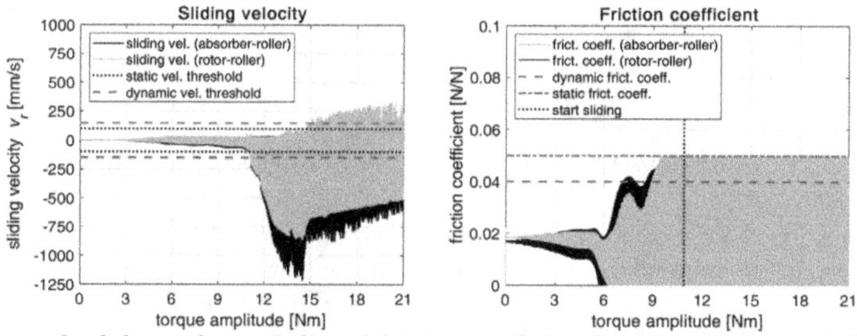

Figure 3: sliding velocity (left) and friction coefficient (right) .vs. torque amplitude .

From these plots, the following considerations arise:

- For low torque values, the sliding velocity is close to zero, which means that the pure-rolling condition is satisfied completely, due to stiction.
- For medium torque levels, there is a finite sliding velocity, but its value is less than the threshold velocity. Hence, the rolling condition is still valid.
- For high torque amplitudes, the sliding velocity is greater than the threshold and the friction coefficient reaches the dynamic value. In these cases, there is slipping between the roller and the slot.

However, observing Figure 5, which is the zoom versions of Figure 3, the slipping is only localized at the end of the time period of oscillation.

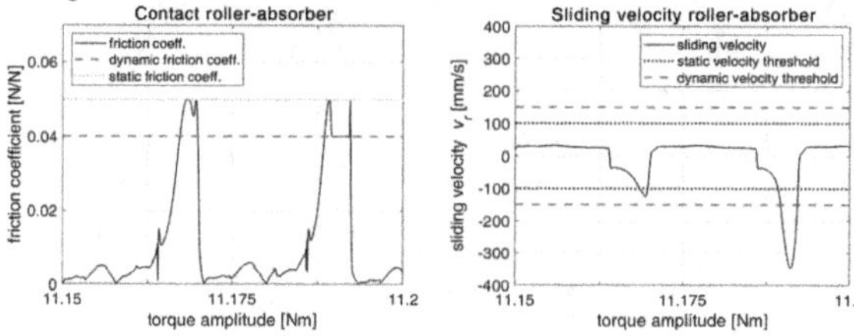

Figure 5: sliding velocity and friction coefficient .vs. torque amplitude (zoom version)

To assess the power losses due to the roller-slot slipping, the ratio between friction power loss of all the caged-roller joints and disturbance power *FP/DP*:

$$\frac{FP}{DP}=\frac{\int_{-\pi}^{\pi}\sum_{i=1}^{8}F_t v_r \mathrm{d}\Theta}{\int_{-\pi}^{\pi}T\Omega\sin n\Theta\mathrm{d}\Theta} \tag{9}$$

averaged over a period is computed. This ratio is plotted with respect to the torque amplitude in Figure 6. As expected, since the slipping is very limited, the power loss is negligible even for large oscillations.

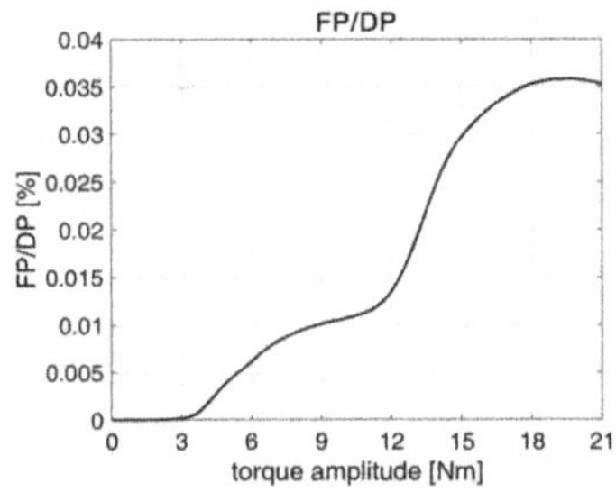

Figure 6: ratio *FP/DP* .vs. torque amplitude

Conclusions

In this investigation, the internal dynamics of the caged-roller joints of a centrifugal pendulum is analyzed by means of the multibody environment. In the caged-roller joints, the pure-rolling between the mating elements is required. However, as observed in the multibody dynamics simulations, there could be slipping for large absorber oscillations. This occurs mostly for low rotor speed, when the centrifugal force field amplitude is not enough to guarantee a sufficient normal force between the roller and the slot throughout the entire oscillation. To assess the power losses associated with this phenomenon, the ratio between the friction and the disturbance powers is computed. From this comparison, the friction power losses are negligible for any oscillations.

References

[1] A. Kooy and R. Seebacher. Best-in-class dampers for every driveline concept. Schaeffler Symposium, (2018)

[2] J. P. Den Hartog. Mechanical vibrations, fourth edition. J. R. Aeronaut. Soc. 61 (554) (1957).

[3] H.H. Denman, Tautochronic bifilar pendulum torsion absorbers for reciprocating engines, J. Sound Vib. 159 (2) (1992). https://doi.org/10.1016/0022-460X(92)90035-V

[4] C.P. Chao, S.W. Shaw, C.T. Lee, Stability of the unison response for a rotating system with multiple tautochronic pendulum vibration absorbers, J. Appl. Mech. 64 (1) (1997) 149–156. https://doi.org/10.1115/1.2787266

[5] A.S. Alsuwaiyan and S.W. Shaw. Performance and dynamic stability of general-path centrifugal pendulum vibration absorbers. J. Sound Vib., 252(5) (2002). https://doi.org/10.1006/jsvi.2000.3534

[6] J. Mayet, H. Ulbrich, First-order optimal linear and nonlinear detuning of centrifugal pendulum vibration absorbers, J. Sound Vib. 335 (2015) 34–54. https://doi.org/10.1016/j.jsv.2014.09.017

[7] M. Cirelli, M. Cera, E. Pennestrì, P.P. Valentini, Nonlinear design analysis of centrifugal pendulum vibration absorbers: an intrinsic geometry-based framework, Nonlinear Dynam. 102 (3) (2020) 1297–1318. https://doi.org/10.1007/s11071-020-06035-1

[8] M. Cera, M. Cirelli, E. Pennestrì, and P. P. Valentini. Design analysis of torsichrone centrifugal pendulum vibration absorbers. Nonlinear Dynam. 104 (2) (2021) 1023–1041. https://doi.org/10.1007/s11071-021-06345-y

[9] M. Cera, M. Cirelli, E. Pennestrì, P.P. Valentini, Nonlinear dynamics of torsichrone CPVA with synchroringed form closure constraint, Nonlinear Dynam.105 (3) (2021) 2739–2756. https://doi.org/10.1007/s11071-021-06732-5

[10] E.R. Gomez, I.L. Arteaga, L. Kari, Normal-force dependent friction in centrifugal pendulum vibration absorbers: Simulation and experimental investigations, J. Sound Vib. 492 (2021). https://doi.org/10.1016/j.jsv.2020.115815

[11] O. Bauchau, J. Rodriguez, and S.Y. Chen. Modeling the bifilar pendulum using nonlinear, flexible multibody dynamics. J. American Helicopter Society 48 (2003). https://doi.org/10.4050/JAHS.48.53

[12] M. Cera, M. Cirelli, E. Pennestrì, P.P. Valentini, The kinematics of curved profiles mating with a caged idle roller - higher-path curvature analysis, Mech. Mach. Theory 164 (2021). https://doi.org/10.1016/j.mechmachtheory.2021.104414

[13] J. Mayet, Effective and robust rocking centrifugal pendulum vibration absorbers, J. Sound Vibr. 527 (2022). https://doi.org/10.1016/j.jsv.2022.116821

[14] M. Cera, M. Cirelli, E. Pennestrì, P.P. Valentini, Design and comparison of centrifugal dampers modern architectures: The influence of roller kinematics on tuning conditions and absorbers nonlinear dynamics, Mech. Mach. Theory 174 (2022). https://doi.org/10.1016/j.mechmachtheory.2022.104876

[15] P. Flores, Contact mechanics for dynamical systems: a comprehensive review. Mult. Syst. Dynam., 54(2) (2021). https://doi.org/10.1007/s11044-021-09803-y

[16] C. M. Pereira, A. L. Ramalho, J. A. Ambrósio, A critical overview of internal and external cylinder contact force models, Nonlinear Dynam. 63 (2011). https://doi.org/10.1007/s11071-010-9830-3

[17] M. Cera, E. Pennestrì, The mechanical generation of planar curves by means of point trajectories, line and circle envelopes: a unified treatment of the classic and generalized Burmester problem, Mech. Mach. Theory 142 (2019). https://doi.org/10.1016/j.mechmachtheory.2019.103580

[18] F. Freudenstein, L.S. Woo, On the curves of synthesis in plane instantaneous kinematics, in: M. Hetényi, W.G. Vincenti (Eds.), Applied Mechanics. International Union of Theoretical and Applied Mechanics, Springer (1969)

[19] J. Choi, H. S. Ryu, C. W. Kim, and J. H. Choi. An efficient and robust contact algorithm for a compliant contact force model between bodies of complex geometry. Mult. Syst. Dynam., 23(1) (2009). https://doi.org/10.1007/s11044-009-9173-3

[20] H. Y. Cha, J. Choi, H. S. Ryu, and J. H. Choi. Stick-slip algorithm in a tangential contact force model for multi-body system dynamics. J. Mech. Science and Technology, 25(7) (2011). https://doi.org/10.1007/s12206-011-0504-y

[21] E. Pennestrì, V. Rossi, P. Salvini, and P. P. Valentini. Review and comparison of dry friction force models. Nonlinear Dynam., 83(4) (2015). https://doi.org/10.1007/s11071-015-2485-3

Materials Research Proceedings **26** (2023) 647-652
https://doi.org/10.21741/9781644902431-104

Modal analysis of a four-bar linkage MEMS microgripper with co-operative electrostatic actuation

Andrea Rossi[1,a,*], Nicola Pio Belfiore[1,b]

[1]Department of Industrial, Electronic and Mechanical Engineering, Roma Tre University, Via della Vasca Navale 79, 00146 Rome, Italy

[a]andrea.rossi@uniroma3.it, [b]nicolapio.belfiore@uniroma3.it

Keywords: MEMS, Comb Drive Actuators, Modal Analysis

Abstract. MEMS-Technology based microgrippers have been recently used in different fields of applications. These microsystems can be actuated by means of electrostatic actuators, such as linear or rotary comb drives, but the robustness and feasibility of such components in static and as well as dynamic conditions still raises some concerns. In order to contribute to fill this gap, the dynamic properties of a MEMS-Techology Based silicon microgripper, based on four-bar linkage, with co-operative comb-drives, are here numerically analysed. The analysis of the vibrations is essential in many MEMS applications since vibrations may lead to significant positioning errors or undesirable contacts between the anchored and floating fingers of the comb-drives. The present investigation aims to assess the critical modes of a MEMS microgripper in order to ascertain the possibility of impact between the fixed and moving fingers of the comb-drives. The relative displacements between the anchored and the floating fingers are tolerable only if the center of the relative rotation is coincident with the center of the conjugate profiles. Hence, the nature of the relative motion for the first vibration modes has been assessed by means of Finite Element Analysis (FEA) in order to avoid operational issues.

Introduction

The variety of flexure hinges [1] and the actual technological capabilities in MEMS fabrication (micro electro mechanical system) enabled the development of new microsystems for the manipulation at the microscale. The recent progress in micro and nano machining gave rise to the deployment of some multi-hinge and multi-DoF (Degrees of Freedom) MEMS-Technology based microsystem. The crucial issue regards the design and fabrication of lumped flexure hinges that are also the smallest structural elements in the device. New microsystems equipped with Conjugate Surface Flexure Hinges (CSFHs) were conceived and fabricated [2] to improve their positioning accuracy. A CSFH is a peculiar flexure that consists of a curved beam, which provides compliance, together with a portion of conjugate-profiles. It can be shown that parasitic deformations are minimized when the curved beam elastic weights center is overlapping with the center of the conjugate profiles. The CSFH was adopted in several microdevices, such as in micromanipulators [3], micro mechanisms, grippers [4], microtribometers, etc. These microsystems can be actuated by means of electrostatic actuators, such as linear or rotary comb drives, but the robustness and feasibility of such components in static and as well as dynamic conditions still raises some concerns. A MEMS-technology based device may involve dimensions ranging from a few microns to millimeters. In this work, a monolithic microgripper operated by means of electrostatic rotary comb-drives has been considered. Fig. 1c shows the overall geometry of the microgripper that consists of two tip points, 8 CSFHs, 4 comb-drives with electric pads. The grasping task can be obtained by moving the jaws that are attached to the coupler links of the two 4-bar linkage structures that correspond to the pseudo-rigid body equivalent mechanism (PRBM) [5] used to synthetize the compliant mechanism. Since the jaw is attached to the coupler link, the jaw instantaneous rotation axis can be conveniently predefined by the designer. The input link is driven

Materials Research Proceedings **26** (2023) 647-652 https://doi.org/10.21741/9781644902431-104

by a pair of co-operating rotary comb-drives supplied with the same voltage. The design requirements suggested the use of monocrystalline silicon as a structural material since it offers mechanical and electrical properties with very good performance. A Silicon On Insulator (SOI) wafer with the following size has been considered: 40 μm thick device layer, 3-4 μm silicon oxide layer, handle layer with a thickness of 400 μm to ensure high structural mechanical reliability.

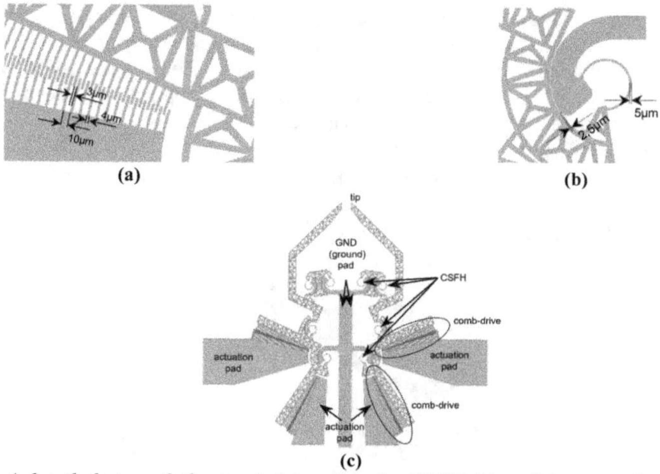

***Figure 1** A detailed view of: the comb drives (a), the CSFH (b) and the overall geometry of the proposed four-bar linkage microgripper with co-operative electrostatic actuation.*

It is worth noting that the structural reinforcement is provided in the elements that do not show any motion during the actuation, i.e. the pads and the central frame structure for the CSFHs (Fig. Ec). The device structure consists of elements with heterogeneous dimensions and, among these, some parts may provide functional constraints. For instance, the comb-drive fingers and the gap between non-moving and moving fingers should be limited to a few microns (see Figure 1a,b).

More details regarding the considered CSFH flexures and comb-drives are listed in Table 1.

Table 1

Component	Label	Value
Finger	Width	4 μm
	Out-of-plane thickness	40 μm
	Distance	10 μm
	Finger clearance	3 μm
SOI wafer	Handle layer	400 μm
	Device layer thickness	40
Overlapping	Initial angle	2°
	Rotor-stator finger gap	3 μm
CSFH	Curved beam width	5 μm
	Curved beam length	252 μm
	Curved beam thickness	40 μm
	Conjugate surfaces clearance	2,5 μm
	Curvature radius	62,5 μm

Numerical microgripper modal analysis

The versatility of the CSFH is appealing for several MEMS applications such as microsurgery, biological tissue manipulation, etc. The modal analysis turns out to be insightful on how the device dynamic operation may be affected. As a matter of fact, vibrations may substantially affect positioning accuracy or unwanted stator-to-rotor fingers impact that result in hazardous short circuit or mechanical damage [6]. On the other hand, vibration may enhance the releasing of micro objects or the assessment of the biological soft tissues mechanical properties [7]. In this work, the eigenmodes have been conveniently analyzed, via Finite Element Analysis, by implementing a COMSOL Multiphisycs code. Symmetry boundary conditions have been conveniently exploited to reduce the computational costs, so only one-half of the microgripper has been considered (Fig. 2). The considered mechanical boundary conditions are reported below:

- the non-moving fingers are anchored (A);
- the moving fingers and the remaining gripper links (B) are free to move;
- the frame link is fixed (C);
- the symmetric boundary condition is applied to the surface (D).

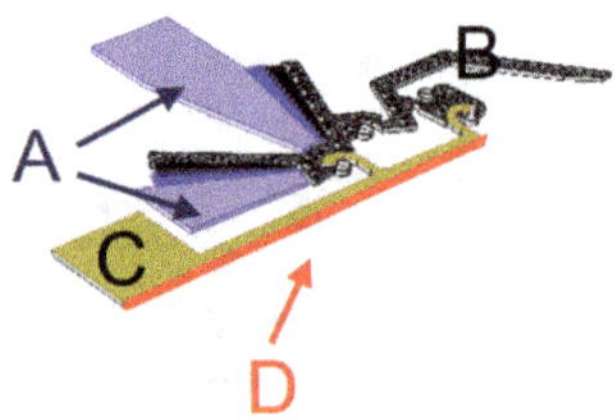

Figure 2 *Boundary conditions adopted in FEA simulations: anchored non-moving fingers (A), the moving fingers and the remaining gripper links are free to move(B), the frame link is fixed (C), the surface (D) represents the symmetric boundary condition.*

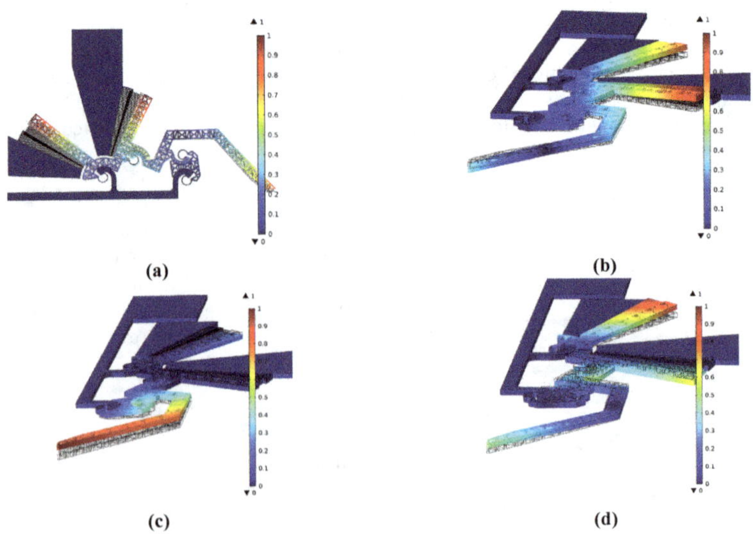

(e) (f)

Figure 3 *First six eigenmodes of the proposed microdevice: I mode (a), II mode (b), III mode (c), IV mode (d), V mode (e) and VI mode (f).*

The first six eigenmodes and natural frequencies are respectively reported in Figure 3 and in Table 2.

Table 2

Mode	Frequency (kHz)	Critical	Plane	Rotation axes
I	1,45	no	in	3
II	2,00	no	out	1
III	2,70	no	out	1
IV	3,67	no	out	2
V	6,09	yes	in	3
VI	6,98	no	out	2

The relative displacements between the anchored and the comb-drives moving fingers are acceptable whenever they correspond to a relative rotation and its center overlapping with the conjugate profiles center. Thereby, it is crucial to evaluate the relative motion nature of all the vibration modes that may be excited during the operational stages. In particular, the first and fifth modes present radial displacements for the fingers and therefore finger contact appears to be theoretically possible.

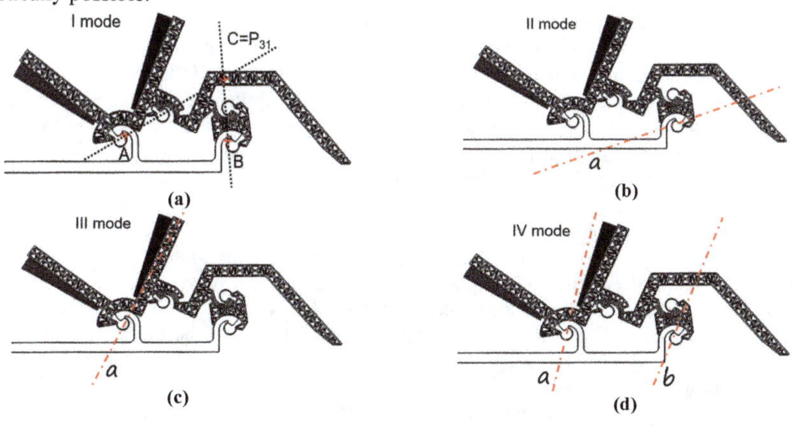

(a) (b) (c) (d)

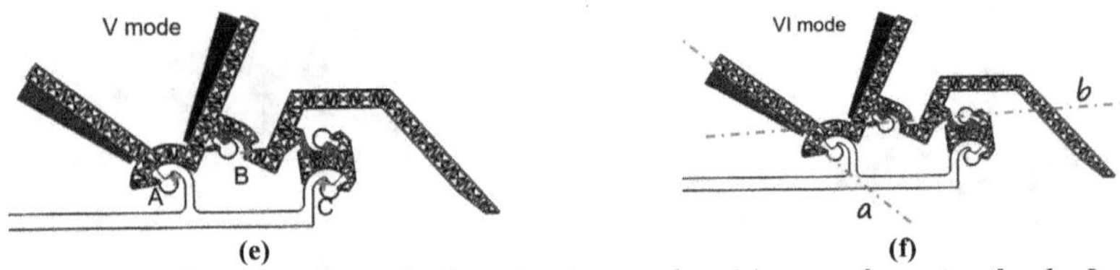

Figure 4 *Axes of rotation for to the first six eigenmodes: (a) axes of rotation for the I mode (the axes are normal to the xy plane, red cross marks), (b) out-plane rotation axis for the II mode (red line), (c) out-plane rotation axis for the III mode (red line), (d) out-plane rotation axis for the IV mode (red lines), (e) in-plane rotation axes for the V mode (red cross marks), (f) out-plane rotation axes for the VI mode (red lines); black dotted lines represent geometric properties of the PRBM, while red lines represent the nodal lines.*

FEA results have been used to identify the minimum displacement areas for each vibration mode. Such zones can be easily identified and the structural behaviour can be physically interpreted by tracing the nodal rotation axes. When the first mode is excited (Figs. 3a, 4a) three elements of the structure act as pseudo-rigid bodies, i.e. the coupler and the two rockers links. Motion is provided by the four flexure hinges and all the mechanism elements are characterized by an in-plane motion. The nodal rotation axes are normal to the *xy* plane (red crosses), thereby their intersections with the mechanism plane are identified by the points A, B and C depicted in Fig. 4a. It is worth noting that the third axis of rotation (c) passes through the instantaneous center of velocity of the coupler with respect to the frame link (P_{31}). Hence, the first mode motion is compatible with the microsystem design, so the involved deformations are compatible with the comb-drives geometry since the relative rotation axes are practically coincident with the CSFH rotation axes and no stator-to-rotor fingers contact is implied. The second mode analysis (Figs. 3b, 4b) shows that the whole mechanism behaves as a plate that rotates around the axis (a) (red line) passing through the center of the CSFH that links the frame to the second rocker (which has no comb-drive mounted on). This axis lies in the *xy* plane so out-of-plane vibrations are entailed. However, the moving fingers are all located on one side of the axis, thereby their motion acts along the perpendicular to the gap between the fingers. As a consequence, the fingers contact is still unlikely. The third mode (Figs. 3c, 4c) is similar to the second but its nodal axis passes through the center of the hinge that joins the frame to the rocker equipped with the comb-drives (red line) . The nodal axis lies in the mechanism plane and the moving fingers are still located on one side of this axis, hence fingers contact seems to be unlikely. The modal analysis for the fourth mode (Fig. 3d) reveals that the microgripper inflects around two parallel axes a and b depicted in Fig. 4d (red lines). Such axes lies in the *xy* plane, then the displacements will occur out-of-plane. The system behaviour is similar to a flexible plate which oscillates around a and b nodal axes, resulting in three different zones. For instance, if the central area inflects downwards, the lateral zones will inflect upwards and vice versa. From the perspective of preventing contact between moving and non-moving fingers, the fifth mode (Figs. 3e, 4e) represents the most critical one. In fact, the system behaves, approximately, as a pseudo-rigid plate that rotates around an axis (c). Such axis passes through the point C and is located within the internal zone of the mechanism, as depicted in Fig. 4e (red crosses). Unfortunately, such motion can be rather detrimental for the comb drives since the moving fingers do not rotate about the CSFH centers. As a consequence, the fingers will no more travel along the natural span of the fixed fingers gap, therefore the moving fingers may collide with the fixed ones. In such a situation, the moving fingers undergoes a radial displacement component since the rotation does not occur around the original center of rotation. The sixth mode has a rather complicated shape and three nodal axes can be identified (red lines in Figs. 3f,4f). The axes lies all in the mechanism plane. The microgripper acts as a flexible plate that inflects around the nodal lines and out-of-plane displacements are entailed, but no critical configuration is expected for the comb-drives. Since fingers contact can be expected only when the fifth mode is excited, its occurrence has a limited impact on the microgripper dynamic operations.

Conclusions

The present investigation has shown that the proposed microgripper may experience contact issues between moving and non-moving fingers of comb-drives when, during its dynamic operation, the fifth mode is excited. FEA has been exploited to identify the main modes of vibration, whose examination is required to prevent critical configurations that happen when there is a contact among moving and non-moving fingers of comb-drives. The modal analysis confirmed that the proposed microgripper unlikely undergoes to critical configurations, providing significant reliability during dynamic operations and it is promising for biomedical and soft tissue manipulation applications.

References

[1] Lobontiu, N.; Garcia, E.; Canfield, S. Torsional stiffness of several variable rectangular cross-section flexure hinges for macro-scale and MEMS applications. Smart Mater. Struct. 2004, 13, 12–19. https://doi.org/10.1088/0964-1726/13/1/002

[2] Belfiore, N.P.; Broggiato, G.B.; Verotti, M.; Balucani, M.; Crescenzi, R.; Bagolini, A.; Bellutti, P.; Boscardin, M. Simulation and construction of a mems CSFH based microgripper. Int. J. Mech. Control 2015, 16, 21–30.

[3] Balucani, M.; Belfiore, N.;Crescenzi, R.; Verotti, M. The development of a MEMS/NEMS-based 3 D.O.F. compliant micro robot. Int. J. Mech. Control 2011, 12, 3–10. https://doi.org/10.1109/RAAD.2010.5524590

[4] Botta F, Rossi A, Belfiore NP. A Feasibility Study of a Novel Piezo MEMS Tweezer for Soft Materials Characterization. Applied Sciences. 2019; 9(11):2277. https://doi.org/10.3390/app9112277

[5] Howell, L.L.; Midha, A. A method for the design of compliant mechanisms with small-length flexural pivots. J. Mech. Des. Trans. ASME 1994, 116, 280–290. https://doi.org/10.1115/1.2919359

[6] Demaghsi, H.; Mirzajani, H.; Ghavifekr, H.B. A novel electrostatic based microgripper (cellgripper) integrated with contact sensor and equipped with vibrating system to release particles actively. Microsyst. Technol. 2014, 20, 2191–2202. https://doi.org/10.1007/s00542-013-1989-3

[7] Valtorta, D.; Mazza, E. Dynamic measurement of soft tissue viscoelastic properties with a torsional resonator device. Med. Image Anal. 2005, 9, 481–490. https://doi.org/10.1016/j.media.2005.05.002

Theoretical and Applied Mechanics - AIMETA 2022 Materials Research Forum LLC
Materials Research Proceedings **26** (2023) 653-658 https://doi.org/10.21741/9781644902431-105

Interface reduction in flexible multibody systems

CAMMARATA Alessandro[1,a*], MADDÌO Pietro Davide[1,b] and SINATRA Rosario[1,c]

[1]Dipartimento di Ingegneria Civile e Architettura, Università degli Studi di Catania, Via S. Sofia 64, 95125, Catania, Italia

[a]alessandro.cammarata@unict.it, [b]pietro.maddio@unict.it, [c]rosario.sinatra@unict.it

Keywords: Floating Frame of Reference, Multipoint Constraints, Reference Conditions

Abstract. The problem of imposing the reference conditions in a floating frame of reference formulation is coupled with the necessity to reduce the interfaces to virtual nodes required to define the multibody joints. Two methods are implemented for rigid and interpolation multipoint constraints, and the reference condition matrix is derived employing all the interface dofs. The case study of a slider-crank mechanism is discussed to show how different sets of reference conditions can modify the system's dynamics.

Introduction

Interface reduction is a recurring problem in substructuring and model reduction theory [1]. In flexible multibody dynamics, the interfaces are primitive geometric features employed to form a joint. In practice, what is done is to individually reduce each interface to a single virtual node, usually a not collocated node outside the volume of the body. This reduction occurs through two types of multipoint constraints (MPCs): the rigid multipoint constraint, usually referred to as the RBE2 element, and the interpolation multipoint constraint, usually referred to as the RBE3 element [2]. Subsequently, the virtual nodes of the two interfaces are linked through kinematic constraints necessary to define a joint. In [3], the authors raised the problem of the scarce use of RBE3 in multibody simulations and identified the disappearance of the dependent coordinates, operated by FE software after eliminating the multipoint constraints, as one of the possible causes. The method proposed in [3] has practical implications to be used in commercial FE software. Still, it neglects essential aspects related to the presence of MPCs and generic reference conditions (RCs) within the Floating Frame of Reference Formulation (FFRF) [4].

Here, a different approach is presented that is perfectly integrated inside the FFRF working with every RCs. Both types of MPC are treated, and the interpolation MPC exploits all interface DOFs without the need to select dependent nodes or to introduce selection criteria. MPCs are directly connected to the reference conditions necessary to define the floating frame correctly. This issue has only been marginally addressed in the literature without providing a general treatment for any RC.

The paper is organized as follows. First, the mathematical background of the FFRF and the role of the RCs are introduced. Then, the method to apply the reference conditions to the virtual nodes of rigid and interpolation MPCs is provided. The reference condition matrix is obtained in explicit form for both cases. The case study of a flexible crank is analyzed, and three different sets of RCs have been applied to the interface virtual nodes of the component. Finally, the crank is assembled with a flexible connecting rod and a rigid piston to simulate a single cylinder of an internal combustion engine.

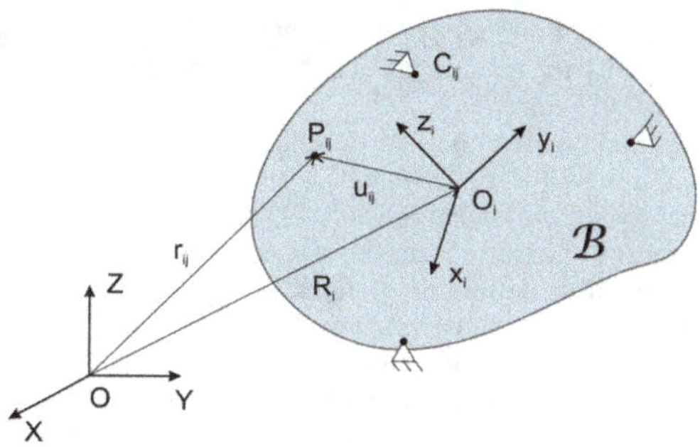

Figure 1: Floating frame of reference nomenclature and reference conditions

Background

The Floating Frame of Reference formulation describes the motion of a flexible body composing the gross motion of a particular frame, i.e., the floating frame and the local deformation with respect to the floating frame. In the finite element theory, the shape function necessary to describe the elastic behavior of a deformable body must be able to represent the rigid-body motion. Otherwise, the body cannot satisfy the *objectivity* property. The rigid-body motion provided by the shape function can be a duplicate or interfere with the gross motion of the floating frame. A set of linear constraints must be applied at some points C_{ij} to remove the rigid-body motion. These constraints are referred to as the *reference conditions* (RCs) and define the nature of the floating frame.

Denoting with $\boldsymbol{R}_i$ and $\boldsymbol{A}_i$ the position and the rotation matrix of the floating frame ($x_i y_i z_i$) with respect to the inertial frame (XYZ), and with $\boldsymbol{u}_{ij}$ the position of a generic point P_{ij} with respect to the floating frame, we write

$$\boldsymbol{r}_{ij} = \boldsymbol{R}_i + \boldsymbol{A}_i\bar{\boldsymbol{u}}_{ij} = \boldsymbol{R}_i + \boldsymbol{A}_i\boldsymbol{S}_{ij}(\boldsymbol{q}_0 + \boldsymbol{B}_2\boldsymbol{q}_f) \tag{1}$$

where $\boldsymbol{S}_{ij}$ is the shape matrix, $\boldsymbol{q}_0$ and $\boldsymbol{q}_f$ respectively are the vectors of elastic coordinates in the undeformed and deformed configuration and $\boldsymbol{B}_2$ is the reference condition matrix that removes the rigid-body motion from the elastic displacements. It is noteworthy that the gross motion defined by $\boldsymbol{R}_i$ and $\boldsymbol{A}_i$ is not, in general, a rigid-body motion. Only when the RCs impose a fixed constraint at point O_i the gross motion becomes a rigid-body motion. The RCs can be applied at any point of the body, respecting that the final structure is isostatic or hyperstatic. Although this is the only prescription imposed on the RCs, however, further advice based on the experience is advisable:

1. the RCs should be a subset of the multibody joints, meaning that the body should deform following shapes allowed by the actual joints.
2. different RCs yield different results and should be experimentally validated.

The first advice comes from a recent paper [5] in which it is demonstrated that the free-free or mean-axis RCs, in particular system layouts, do not satisfy the mechanical joints. If the RCs must respect the mechanical joints, they must be applied on the same nodes used to define those joints. These nodes can be points of the structure, as in the case of beam or plate elements, or they can be virtual points, as in most cases where three-dimensional elements are used.

The second piece of advice comes from the evidence that different RCs create different component modes of the reduced system and eventually modify the elastic response of the system.

Materials Research Forum LLC
https://doi.org/10.21741/9781644902431-105

This probably stems from the finiteness of the component mode set employed to obtain the reduced-order model. In [6-10], the authors investigated this problem by providing different planar and spatial examples based on beams.

In summary, it would be convenient to use different sets of RCs. Advisable FFR-based methods must be able to apply the RCs to any physical or virtual point of the structure to create the component modes necessary for the dynamic analysis [11]. In FEA, this issue is solved using the rigid and the interpolation multipoint constraints (MPC), often referred to as RBE2 and RBE3 elements, respectively. In flexible multibody systems, as for structures, the RBE elements are also employed to create multibody joints. In RBE2, a group of dependent nodes follows the rigid body displacements of a single independent node. In RBE3, the displacement of a given dependent node is calculated using the displacements of a group of independent nodes.

Methodology

Given a component discretized into FE, let *B*, *I*, and *V* be the sets of boundary, internal, and virtual nodes, respectively. All interface nodes belong to *B*, while *V* contains the virtual nodes necessary to create an MPC; the remaining nodes belong to *I*. The virtual nodes are essential creating an MPC and can be collocated, that is, physical nodes of the mesh or non-collocated nodes, i.e., nodes not belonging to the body's volume. Then, the RBE2 element can be described in terms of the mentioned sets:

$$\boldsymbol{F}: B \longrightarrow V, \boldsymbol{q}_B = \boldsymbol{F}\boldsymbol{q}_V \text{ (RBE2)} \tag{1}$$

Fort he and RBE3 element, it follows

$$\boldsymbol{G}: V \longrightarrow B, \boldsymbol{q}_V = \boldsymbol{G}\boldsymbol{q}_B \text{ (RBE3)} \tag{2}$$

where $\boldsymbol{F}$ and $\boldsymbol{G}$ are linear functions of the independent nodes whose expressions are reported in [2, 3]. The vectors $\boldsymbol{q}_B$ and $\boldsymbol{q}_V$ contain the displacements of nodes belonging to *B* and *V*, respectively. Usually, dim(*V*) < dim(*B*), and this explains the limited use of RBE3 in multibody applications. While imposing the RCs on the virtual nodes of the RBE2 elements is often immediate, doing the same with the RBE3 element needs some tricks. Applying the RCs on the virtual nodes of *V*, the RCs can be expressed through the following linear constraint equations

$$\boldsymbol{D}\boldsymbol{q}_V = \boldsymbol{0} \tag{3}$$

where $\boldsymbol{D}$ is a matrix containing the coefficients of these equations.

RBE2 element

By introducing Eq.(1) into Eq.(3), we derive

$$\boldsymbol{D}\boldsymbol{F}^{\dagger}\boldsymbol{F}\boldsymbol{q}_V = \boldsymbol{0} \rightarrow \boldsymbol{D}\boldsymbol{F}^{\dagger}\boldsymbol{q}_B = \boldsymbol{0} \rightarrow \boldsymbol{B}_2 = null(\boldsymbol{D}\boldsymbol{F}^{\dagger}) \tag{4}$$

where $\boldsymbol{F}^{\dagger}$ is the generalized inverse of $\boldsymbol{F}$ and $\boldsymbol{B}_2$ is the matrix of the RCs such that $\boldsymbol{q}_B = \boldsymbol{B}_2\boldsymbol{\gamma}$ being $\boldsymbol{\gamma}$ a reduced set of independent elastic parameters.

Usually, this procedure is not needed as $\boldsymbol{B}_2$ can be directly found by removing the constrained dof from $\boldsymbol{q}_V$, i.e.

$$\boldsymbol{q}_V = \boldsymbol{B}_2\boldsymbol{q}_V^* \tag{5}$$

where $\boldsymbol{q}_V^*$ is the reduced set of independent elastic coordinates. Exploiting Eq.(5), it is derived that

$$\boldsymbol{q}_B = \boldsymbol{F}\boldsymbol{B}_2\boldsymbol{q}_V^* \rightarrow \boldsymbol{T} = \boldsymbol{F}\boldsymbol{B}_2 \tag{6}$$

in which $\boldsymbol{T}$ is the transformation matrix that contains both the RBE2 and RCs.

Materials Research Forum LLC
https://doi.org/10.21741/9781644902431-105

RBE3 element

By introducing Eq.(2) into Eq.(3), we have

$$\boldsymbol{DGq}_B = \boldsymbol{0} \rightarrow \boldsymbol{q}_B = \boldsymbol{B}_2\boldsymbol{\rho},\ \boldsymbol{B}_2 = null(\boldsymbol{DG}) \tag{7}$$

By substituting into Eq.(2), we obtain $\boldsymbol{q}_V = \boldsymbol{GB}_2\boldsymbol{\rho}$. In this case, the direct substitution carried out in the RBE2 element is not possible, and the final transformation matrix $\boldsymbol{T} = \boldsymbol{GB}_2$ maps a reduced set of independent elastic parameters into the virtual node displacement vector $\boldsymbol{q}_V$.

Numerical simulation

The method presented in the previous sections is applied to a slider-crank mechanism. In Figure 2, the crank is modeled considering tetrahedrons. The interface nodes of set B are highlighted using red dots. The three interfaces refer to the revolute joints employed to connect the crankshaft to the frame (B_1 and B_3) and the revolute joint between the crank and the connecting rod (B_2). Only RBE3 elements are employed, and a virtual node is defined at the geometric center of each interface. Then, the three virtual nodes are used to determine the revolute joints. To define $\boldsymbol{B}_2$ as in Eq.(7), a set of RCs must be imposed by means of the matrix $\boldsymbol{D}$ of Eq.(3). Here, we consider three sets of RCs as reported in Tab. 1. All sets lead to an isostatic structure.

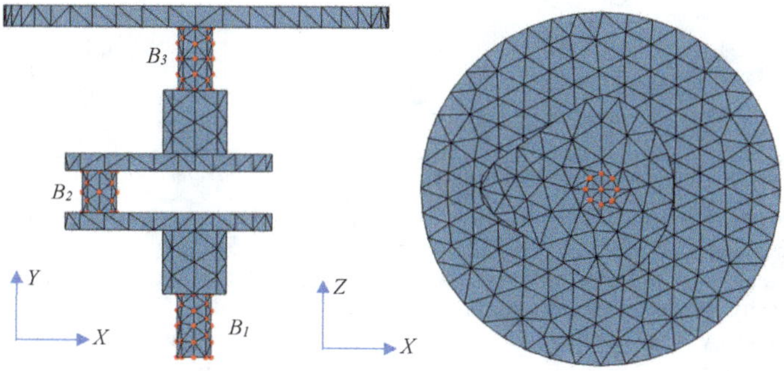

Figure 2: Crank layout and interface nodes highlighted using red dots.

Table 1: Three sets of reference conditions. T and R stand for forbidden translations and rotations.

RCs	Interface B_1	Interface B_2	Interface B_3
Set 1	Spherical (TxTyTz)	-	Universal (TxTyTzRy)
Set 2	Spherical (TxTyTz)	Translation (Tz)	Translations (TxTz)
Set 3	-	Fixed (TxTyTzRxRyRz)	-

The three sets of RCs are used to define the normal modes, i.e., eigenmodes of the component constrained using the RCs [4]. Retained sets of normal modes serve as a basis for reducing the elastic coordinates. The final set of coordinates includes only gross motion coordinates and the modal amplitudes corresponding to the retained normal modes. As can be shown in Fig. 3, the modes and their corresponding frequencies change, passing from one set to another. As already remarked, this modifies the dynamics of the system.

To observe how the elastic field modifies the dynamics of a system, we assembled a slider-crank mechanism simulating an internal combustion engine with a single-cylinder. The connecting rod has been modeled using tetrahedrons and simply supported RCs, while the cylinder has been modeled as a rigid body. An external force arising from the in-cylinder pressure shown in Fig. 4

has been applied to the piston, as detailed in [12]. Two torques simulating the starter electric motor and the external load applied to the crankshaft have also been included in the model. All components and the dynamics simulation have been performed using the commercial software Matlab®. The implicit *generalized-alpha method* has been employed to integrate the equations of motion.

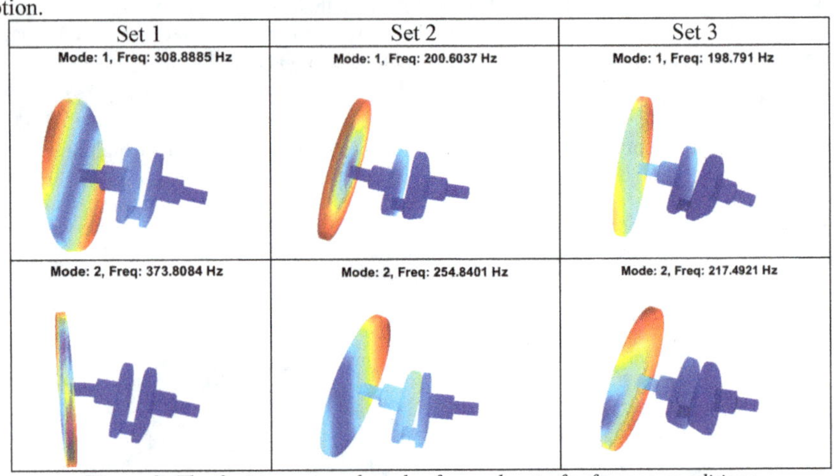

Figure 3: The first two normal modes for each set of reference conditions.

As can be observed in Fig. 4, after a transitory phase in which the system accelerates, the angular speed of the crankshaft stabilizes. The oscillations around the regime speed are typical of a single-cylinder system due to unbalanced forces during the rotation. Comparing the three sets of RCs, the angular velocity of the crankshaft starts to differ for high speeds. This effect comes from the different frequency ranges of the three normal mode sets employed in the reduction process.

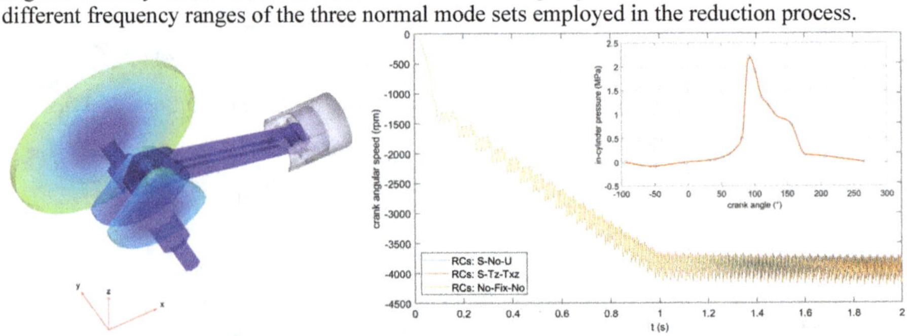

Figure 4: Simulation snapshot of the slider-crank system, in-cylinder pressure, and angular speed of the crankshaft considering the three sets of RCs of Table 1.

Conclusions

This paper presents a method to combine interface reduction and reference conditions to create component modes to be used for the dynamics of flexible multibody systems. The procedure is applied to both rigid and interpolation multipoint constraints. For the latter, all interface dofs are employed without resorting to any selection criteria of the independent dofs. The case study of a slider-crank mechanism simulating an internal combustion engine with a single cylinder is provided to demonstrate how different sets of reference conditions can influence the dynamics of a complex multibody system.

References

[1] Allen, M. S., Rixen, D., Van der Seijs, M., Tiso, P., Abrahamsson, T., & Mayes, R. L. Substructuring in engineering dynamics. Springer International Publishing, 2020. https://doi.org/10.1007/978-3-030-25532-9

[2] M. Nastran, Basic dynamic analysis user's guide, MSC. Software Corporation. USA 546, 2004.

[3] G. H. Heirman, W. Desmet, Interface reduction of flexible bodies for efficient modeling of body flexibility in multibody dynamics, Multibody System Dynamics 24 (2) (2010) 219–234. https://doi.org/10.1007/s11044-010-9198-7

[4] Shabana, A. Dynamics of multibody systems. Cambridge university press, 2020. https://doi.org/10.1017/9781108757553

[5] Shabana, A.A., Wang, G., Kulkarni, S., Further investigation on the coupling between the reference and elastic displacements in flexible body dynamics, J.Sound Vib. 427 (2018) 159–177. https://doi.org/10.1016/j.jsv.2018.02.054

[6] Cammarata, A., Pappalardo, C.M., On the use of component mode synthesis methods for the model reduction of flexible multibody systems within the floating frame of reference formulation, Mech. Syst. Signal Process. 142 (2020) 106745. https://doi.org/10.1016/j.ymssp.2020.106745

[7] Cammarata, A., Sinatra, R., Maddio, P.D., A two-step algorithm for the dynamic reduction of flexible mechanisms, Mechanisms and Machine Science, 66, (2019), 25–32. https://doi.org/10.1007/978-3-030-00365-4_4

[8] Tanev, T.K., Cammarata, A., Marano, D., Sinatra, R. Elastostatic model of a new hybrid minimally-invasive-surgery robot, 2015 IFToMM World Congress Proc., IFToMM, (2015).

[9] Huang, Z., Xi, F., Huang, T., Dai, J.S., Sinatra, R. Lower-mobility parallel robots: Theory and applications, Advances in Mechanical Engineering, 2010, n. 927930, (2010). https://doi.org/10.1155/2010/927930

[10] Cammarata, A., J. Angeles, and R. Sinatra. "The dynamics of parallel Schönflies motion generators: the case of a two-limb system." Proceedings of the Institution of Mechanical Engineers, Part I: Journal of Systems and Control Engineering 223.1 (2009): 29-52. https://doi.org/10.1243/09596518JSCE623

[11] Cammarata, A., Sinatra, R., and Maddio, P. D., Interface reduction in flexible multibody systems using the Floating Frame of Reference Formulation. Journal of Sound and Vibration, 523, 116720, (2022). https://doi.org/10.1016/j.jsv.2021.116720

[12] Mauro, S., Şener, R., Gül, M. Z., Lanzafame, R., Messina, M., and Brusca, S., Internal combustion engine heat release calculation using single-zone and CFD 3D numerical models. International Journal of Energy and Environmental Engineering, 9(2), (2018), 215-226. https://doi.org/10.1007/s40095-018-0265-9

Theoretical and Applied Mechanics - AIMETA 2022 Materials Research Forum LLC
Materials Research Proceedings **26** (2023) 659-664 https://doi.org/10.21741/9781644902431-106

Dynamic analysis of lightweight gears through multibody models with movable teeth

Marco Cirelli [1,a*], Alessio Cellupica [2,b], Mattia Cera [2,c], Oliviero Giannini [1,d], Pier Paolo Valentini [2,e], Ettore Pennestrì [2,f]

[1] Department of mechanical engineering, University Niccolò Cusano, Rome, Italy

[2] Department of enterprise engineering, University of Rome Tor Vergata, Rome, Italy

[a]marco.cirelli@unicusano.it, [b]alessio.cellupica@alumni.uniroma2.eu, [c]mattia.cera@uniroma2.it, [d]oliviero.giannini@unicusano.it, [e]valentini@ing.uniroma2.it, [f]pennestri@mec.uniroma2.it

Keywords: Lightweight Gears, Pseudo Rigid Teeth, Contact

Abstract. This paper aims the study of lightweight gears through a novel approach based on a multibody model contact-based with pseudo-rigid teeth. This method considers the teeth as pseudo-rigid bodies attached to the main body through revolute joints and torsion springs with precomputed stiffness. Thanks to this approach, the lightening is considered by varying the value of the stiffness parameter. To validate the method, the first step compares the Transmission Error computed with this novel approach with the Finite Element method in quasi-static operating conditions. The second step highlights the behaviour of the Transmission Error, increasing the rotation speed.

Introduction

Lightweight gears (LGs) represent one of the most common design solutions to increase the efficiency of the transmission system. This approach is based on the reduction of inertia properties that generates a decrease in the potential energy dissipated in the transient phase. Lightening is produced by reducing the gear rim thickness or by introducing holes on the gear rim. Recently many researchers have found solutions to consider the effects on the transmission error (TE) introduced by lightening. In reference [1], the dynamic behaviour of LG is investigated through a one degree-of-freedom (DOF) considering a time depending on mesh stiffness (MS) computed through the Finite Element (FE) method. This approach is well documented in the literature and remains widely adopted at the design level. In reference [2], the TE and the strain analysis of LGs are obtained by using a hybrid FE-analytical gear contact model. The weakness of the analytical approach is that the MS needs be previously computed through FE models. For this reason, a multibody model contact-based with pseudo-rigid tooth (MUBOCO-PR) has been introduced [3, 4]. This method considers the teeth as pseudo-rigid bodies attached to the main body through revolute joints and torsion springs having a precomputed stiffness. In this way, the flexibility of the tooth and the gear body can be condensed into the torsion spring. This approach has been already adopted for gears having tip relief modifications [5]. The main purpose of this investigation is to show how this method can be extended to the study of the LGs. To simplify the interpretation of the numerical results, and consistently with the findings reported in [6], it is introduced the definition of the transmission error (TE) as "*the deviation in position of the driven gear and the position it would occupy if the gear drive were perfectly conjugate*". Using the subscript s 1 and 2 to denote the driving and the driven gear, respectively, the TE along the line of action in an ordinary gear train is expressed as

$$TE(t) = r_{b1}\theta_1(t) + r_{b2}\theta_2(t) \tag{1.1}$$

for the *i*-th gear ($i=1,2$), $\theta_i(t)$ is the angular position, measured from the nominal position, and r_{bi} the base radii, respectively.

If the TE is measured in static conditions, it can be defined as Static transmission error (STE). Conversely, if it is evaluated under operating conditions, it is termed dynamic transmission error (DTE). While the STE depends on compliance contributions of the system and geometrical properties of the profile, the dynamic response involves phenomena such as impacts, nonlinear effects, and vibrations. The paper is organized as follows: in the first part, a general methodology to generate a MUBOCO-PR model is presented. Then, STE numerical results have been compared with FE model in quasi-static conditions.

Method

MUBOCOF-PR, considers each tooth as an individual body elastically connected to the gear hub through a specific elastic joint, considering the elastic contributions that represent the source of excitation of the system. In particular, the rim body and the teeth compliances are condensed into specific spring joints, while the contact compliance is assigned to the contact stiffness. In this way, a suitable separation of all the described contributions can be achieved.

The parameters to be specified are load conditions, stiffness, and damping for the spring at each tooth joint and the parameters for the multibody contact formulation. The generalized spring force is computed by means of the following equation:

$$T_{junction} = -k_{\varphi}(\varphi-\varphi_0)^{n_{k\varphi}} - c_{\varphi}\dot{\varphi}^{n_{c\varphi}} + T_0 \tag{1.2}$$

where k_{φ} and c_{φ} are the spring stiffness and damping coefficients, respectively; $n_{k\varphi}, n_{c\varphi}$ the exponents of the relative rotation and angular velocity, respectively; T_0 the free length spring torque; φ the angle of the current rotation; φ_0 the free angle. Assuming T_0=0 and φ_0=0, only four parameters, namely $k_{\varphi}, c_{\varphi}, n_{k\varphi}, n_{c\varphi}$, need to be identified.

The contact formulation adopted in this work is based on a double detection phase: pre-search to identify contact zones and a detailed search to find the penetration depth of the contact regions [7]. A 2D contact model without friction has been used to estimate the penetration depth with a feasible accuracy. The contact force generated at the contact point is based on a penalty contact force [8, 9]. Contact force can be calculated using the following relationship:

$$F_n = k_{con}\delta^{m_1} + c_{con}\frac{\dot{\delta}}{|\dot{\delta}|}|\dot{\delta}|^{m_2}\,\delta^{(m_3)} \tag{1.3}$$

where δ is the penetration, $\dot{\delta}$ is the penetration speed, k_{con} and c_{con} are the stiffness and the damping coefficients, respectively, m_1, m_2 and m_3 are the stiffness, the damping, and the indentation exponents, respectively. With this formulation, there are five parameters to be identified, namely $k_{con}, c_{con}, m_1, m_2, m_3$. In conclusion, nine parameters have to be defined, *i.e.* four regarding the rotational spring and five regarding the contact formulation, respectively. These parameters are divided into two groups: the first one contains the parameters linked to the stiffness characteristics ($k_{\varphi}, n_{k\varphi}$, k_{con}, m_1); the second one includes the parameters relating to the damping properties (c_{φ}, $n_{c\varphi}, c_{con}, m_2, m_3$). The first set influences the solution both in quasi-static conditions and dynamic, and the second set is relevant mainly in dynamic simulations.

Modifications for lightweight can be done essentially in two different ways: creating holes in the rim of the gear or reducing the gear thickness of the rim. Figure 1 reports the main geometrical parameters associated with a lightening solution. In particular, in the case of gear lightening by holes, the following geometric parameters can be identified: the number of holes N, the radius of the holes R_h and the radius of the circumference of the hole R_{ch} (Figure 1a). The rim thickness reduction, for generic spur gear, can be synthesized with four additional dimensions: t_2 is gear width, t_1 is section thickness, l_1 the crown height, l_2 the central height and l_3 the rim height (Figure 1b).

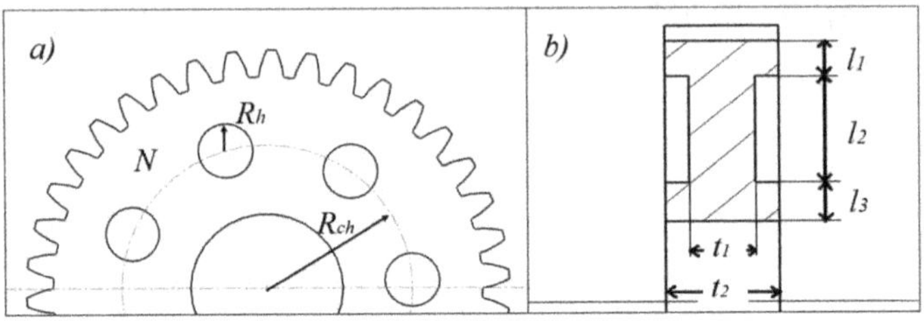

Figure 1 Lightening parameters

Numerical Example

A numerical example was carried out adopting two gears having identical macro geometry properties but different lightening conditions. In particular, the driving gear is considered without lightening conditions, while the driven gear is characterized both by reduction of the body thickness and by the presence of holes. Table 1 shows the geometric characteristics of both gears.

Table 1 Gear geometrical properties: A) general parameters B) lightning parameters

A)	Gear A	Nomenclature	UoM
m	1.5	*Module*	mm
Z	67	*Number of Teeth*	
d_p	100.5	*Pitch Circle Diameter*	mm
ϕ_p	20	*Pressure angle*	deg
a	1	*Addendum Coefficient*	mm
b	1.25	*Dedendum Coefficient*	mm
c	0.07	*Backlash*	mm
t_2	17	*Tooth Width*	mm
R_{fill}	0.45	*Fillet tooth radius*	mm

B)	Lightening modifications		
N	6	*Number of holes*	
R_h	6	*Hole radius*	mm
R_{ch}	32.36	*Position of holes*	mm
l_1	6.075	*External length*	mm
l_2	23.25	*Length of reduction*	mm
l_3	7.425	*Internal length*	mm
t_1	10	*Thickness reduction*	mm

In the example, the gears have been connected to the ground using revolute joints at the centres of the gears. A ramp motion is prescribed to the driving gear, while a similar ramp is adopted for the resistant torque applied to the driven gear to avoid impacts. Figure 2 shows the setup of the model. Contact stiffness parameters (k_{con} and m_1) are identified through a fit of the results from Weber's analytical solution [10]. Contact stiffness is not influenced by the lightening because it depends only on the geometrical profile and material of the gear. In this simulation $k_{con} = 5.64 \cdot 10^6$ N/mm; $m_1 = 1.24$.

Figure 2 Numerical example setup

Spring coefficients of the unmodified gear are computed by adopting the methodology developed in [3]. In this simulation $k_\phi = 1.06 \cdot 10^6$ Nmm/rad; $n_1 = 1.1$. The spring coefficients of the lightened gear are computed with the same method, but the displacement of the tooth is obtained for each tooth through a set of FE models in which the gear has been considered fixed in the inner hole and loaded at the primitive circle. The results of the stiffness computation are reported in Figure 3. In particular, Figure 3a shows the adopted tooth enumeration, while Figure 3b reports the stiffness computed for each tooth. An interesting behaviour can be observed from this first FE result: the stiffness is higher in teeth positioned over the holed. It is apparently an unintuitive result but can be explained in the stress contour. The most stressed zone is not the portion under the tooth but the zone near the fillet radius.

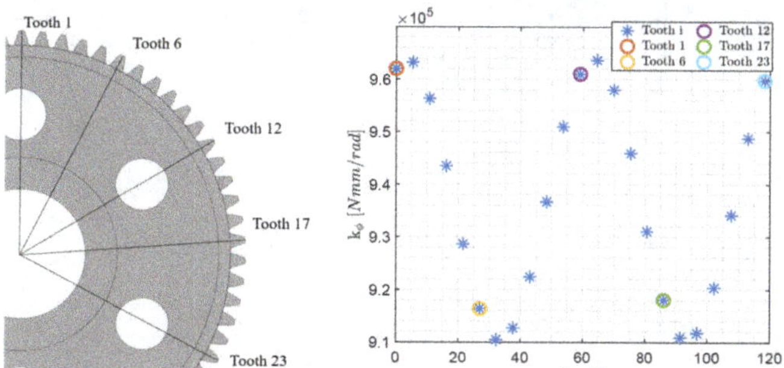

Figure 3 a) Tooth enumeration b) stiffness computed for each single tooth

Quasi-static results

For the quasi-static simulation, the ramp of motion starts from 0 rad/s at the time 0 s, and reaches 1.0 rad/s at time 0.001 s. The same ramp is adopted for the resisting torque applied to the driven gear: 0 Nm at 0 s, end 50 Nm at 0.001 s. The simulation end time is set to admit a 120 deg of gear rotation to highlight the effect of the holes on the STE. The same simulation conditions have been adopted in the reference model based on FE model, composed from 2D 4-side elements, under the assumption of planar strain. The contact formulation adopted for FE models, is a 2D contact between involute profiles based on the concept of hard contact. As the result, a comparison among FE models and MUBOCOF models is reported in Figure 4. In particular, Figure 4a reports the STE computed from 0 to 120 deg to highlight the effects of the holes, while Figure 4b highlights a magnification of the STE from 20 to 40 deg to better underline the comparison. The effects of the holes can be replicated in a satisfactory approximation. The only difference between FE and MUBOCOF is the amplitude of the STE. It can be attributed to an overestimation of the stiffness when two couple of the teeth are in contact. In fact, MUBOCOF does not consider the coupling effect in deformation due to the foundation.

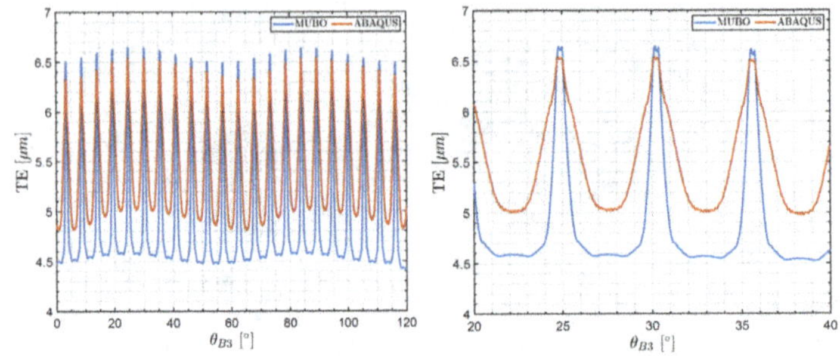

Figure 4 a) STE computed for 120 deg; b) STE magnification

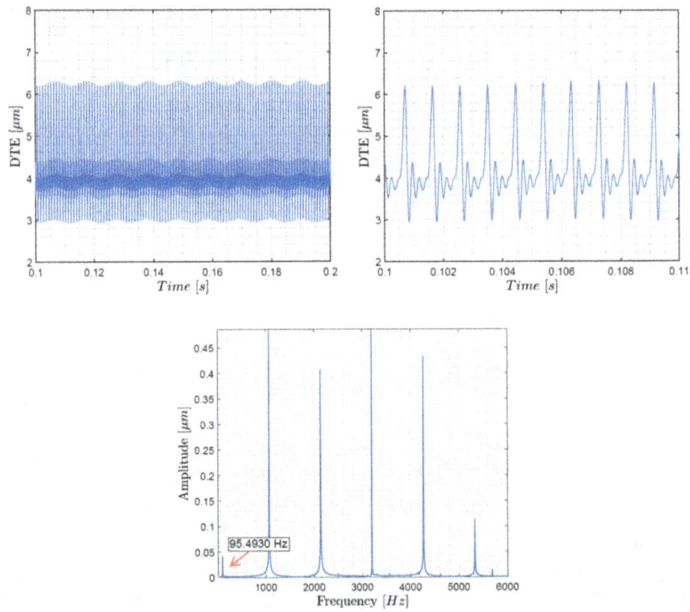

Figure 5 a) DTE signal; b) Magnification of DTE; c) Fast Fourier Transform of DTE

Dynamic analysis

In the dynamic simulation, all the parameters have been maintained except the ramp for the motion and the applied torque. The ramp of motion begins from 0 rad/s at the time 0 s, and reaches 100 rad/s at time 0.01 s. The ramp for the resisting torque applied to the driven gear is 0 Nm at 0 s, end 50 Nm at 0.01 s. Figure 5 summarizes the DTE results. The total simulation time is set to 0.2 s. Figure 5a reports the DTE signal in which the transient phase is avoided. It can be observed the vibration component generated from the holes is maintained. Moreover, the additional vibration component due to the oscillation of the teeth is introduced (Figure 5a). A very interesting plot is reported in Figure 5c in which the Fast Fourier Transform (FFT) of the DTE signal is reported. It can be seen how the first peak represents the vibration component induced by the holes.

Conclusions

In this paper, a preliminary study of the LGs through a MUBOCO-PR is performed. TE results in quasi-static conditions demonstrate a good agreement with the FE method. STE results highlight a slight difference in STE value when multiple teeth are in contact. The effects due to the variation of the stiffness with respect to the angular position are correctly identified. Dynamic analysis shows how the vibration induced from the flexibility of the tooth generates an additional oscillating component and a dynamic amplification of the TE.

This research is funded by MUR through the PRIN *"Innovative contact-based multibody model for noise and vibration prediction in high-performance gears"*. Code project: 202022Y4N5.

References

[1] S. Shweiki, A. Palermo and D. Mundo, "A Study on the Dynamic Behaviour of Lightweight Gears," *Hindawi Shock and Vibration,* vol. 2017, p. 12, 2017. https://doi.org/10.1155/2017/7982170

[2] S. Shweiki, A. Rezayat, T. Tamarozzi and D. Mundo, "Transmission Error and strain analysis of lightweight gears by using a hybrid FE-analytical gear contact model," *Mechanical Systems and Signal Processing,* pp. Volume 123, Pages 573-590, 2019. https://doi.org/10.1016/j.ymssp.2019.01.024

[3] M. Cirelli, P. P. Valentini and E. Pennestrì, "A study of the non-linear dynamic response of spur gear using a multibody contact based model with flexible teeth," *Journal of Sound and Vibration,* 2019. https://doi.org/10.1016/j.jsv.2019.01.019

[4] M. Cirelli, P. P. Valentini and E. Pennestrì, "Multibody dynamics of gear pairs: comparison among different models," in *Proceedings of the 8th ECCOMAS Thematic Conference on MULTIBODY DYNAMICS*, Prague, 2017.

[5] M. Cirelli, O. Giannini, P. P. Valentini and E. Pennestrì, "Influence of tip relief in spur gears dynamic using multibody models with movable teeth," *Mechanism and Machine Theory,* vol. 152, no. 103948, 2020. https://doi.org/10.1016/j.mechmachtheory.2020.103948

[6] S. Harris, "Dynamic loads on the teeth of spur gears," *Proceedings of the Institution of Mechanical Engineering,* vol. 172, p. 87, 1958. https://doi.org/10.1243/PIME_PROC_1958_172_017_02

[7] J. Choi, H. S. Ryu, C. W. Kim and J. H. Choi, "An efficient and robust contact algorithm for a compliant contact force model between bodies of complex geometry," *Multibody Syst Dyn,* vol. 23, p. 99–120, 2010. https://doi.org/10.1007/s11044-009-9173-3

[8] J. Choi, S. Rhim and J. H. Choi, "A general purpose contact algorithm using a compliance contact force model for rigid and flexible bodies of complex geometry," *International Journal of Non-Linear Mechanics,* vol. 53, pp. 12-23, 2013. https://doi.org/10.1016/j.ijnonlinmec.2013.01.017

[9] M. Machado, P. Moreira, P. Flores and H. M. Lankarani, "Compliant contact force models in multibody dynamics Evolution of the Hertz contact theory," *Mechanism and Machine Theory,* vol. 53, p. 99–121, 2012. https://doi.org/10.1016/j.mechmachtheory.2012.02.010

[10] C. Weber, "The Deformations of Loaded Gears and The Effect on Their Load-Carrying Capacity," *British Dept. of Scientific and Industrial Research,* vol. Report No.3, 1949.

Theoretical and Applied Mechanics - AIMETA 2022 Materials Research Forum LLC
Materials Research Proceedings **26** (2023) 665-670 https://doi.org/10.21741/9781644902431-107

Surface error correction of a mesh deployable reflector

MADDIO Pietro Davide[1, a *], SALVINI Pietro[2,b], SINATRA Rosario[1,c] and CAMMARATA Alessandro[1,d]

[1]University of Catania, DICAR, Via S. Sofia 64, 95123 Catania, Italy

[2]University of Rome "Tor Vergata", Department of Enterprise Engineering, Via del Politecnico 1, 00133 Rome, Italy

[a]pietro.maddio@unict.it, [b]salvini@uniroma2.it, [c]rosario.sinatra@unict.it, [d]alessandro.cammarata@unict.it

Keywords: Form-Finding, Deployable Reflector, Optimization Problem

Abstract. Large Deployable Reflector (LDR) systems are commonly used as mesh reflectors for large aperture space antennas in aerospace applications since they provide affordability while guaranteeing at the same time a high gain and a high directivity. To improve the surface accuracy several methods have been studied, most of which measure the distance between the cable-net system that forms the reflector surface and the desired paraboloid. In this paper we want to improve the reflector's ability to convey a greater concentration of reflected rays in the direction of the feed. To deal with this issue, a numerical optimization algorithm has been proposed.

Introduction

Applications that make use of satellite communications are now widely spread, both in civil and military fields. Bear in mind, for example, about the analysis of images for the study of terrestrial morphology, or in the meteorological field for weather forecasts, or even about studies on climate change. In the civil field, then, almost all mobile devices, thus far, are equipped with a GPS module for tracking the position. Large Deployable Reflectors (LDR) are the main types of reflectors used for most of the above mentioned fields. LDR systems are commonly used as mesh reflectors for large aperture space antennas in aerospace applications since they provide affordability while guaranteeing at the same time a high gain and a high directivity.

The key features that characterize the geometry of LDR systems are closely connected with volume constraints of launch vehicles, mainly because of budget problems [1]. Deployable mesh reflectors are composed of rigid bodies, deformable components, mechanical joints, and control actuators [2-4] which allows for achieving a complete transition between the initial stowed configuration to the final deployed configuration.

The fundamental problems for the correct functioning of an LDR system are, therefore, the proper deployment of the folding mechanism and the form-finding of the cable-net which serves as support for the metal mesh. To maintain excellent reflective qualities and meet the prescribed bandwidth requirements, the reflector surface must be as close as possible to the shape of a paraboloid. Most of the methods used in the literature define the best surface of the reflector as the one passing through the nodes of the cable system of the front net or through the centroid of each triangular facet [5-10]. In this case, the RMS error depends on the distance between the nodes of the front net with respect to the desired working surface. Agrawal et al. [11] examined the RMS error between the best-fit paraboloid with flat facets and a sphere. Deng et al. [12] defined the RMS error as the distance between the desired working surface and the nodes of the front net. A similar interpretation is given by Morterolle et al. [13], where the z-direction distance between the desired working surface and a facet of each triangular facet is used to calculate the surface accuracy. In this work, however, we want to focus on the amount of energy that hits the feed, thus investigating the best topology of the net that guarantees a greater concentration of the incident rays directed towards the focus of the paraboloid. This paper is organized as follows. At first, the receiver deviation error is introduced, both in the two-dimensional and in the three-dimensional model. Hence the optimization model is presented, along with the results. Finally, the conclusions are reported.

Receiver deviation error

Figure 1 shows a polygonal chain in blue, whose vertices pass through the parabola drawn in red and having the following equation:

$$y = \frac{x^2}{4F} \tag{1}$$

where *F* is the focus of the parabola.

According to what is described in the literature, the polygonal chain would represent the ideal condition for the reflector surface as its vertices are located exactly on the parabola. In fact, if we consider the RMS error as

$$e_{RMS} = \sqrt{\frac{1}{n}(d_1^2 + d_2^2 + \ldots + d_n^2)} \tag{2}$$

where d_n represents the distance between each vertex of the polygonal and the parabola, we obtain a value equal to zero. Despite this result, the most representative value that guarantees the reflector a good frequency band is given by the contribution of each reflected ray that intercepts the feed located at the point *F*. As shown in Fig. 2 (left), taking into consideration a prime-focus antenna, where the focus is positioned in the center of the reflector, the incident rays parallel to the *y*-axis are reflected on the surface of the reflector and then directed towards the feed.

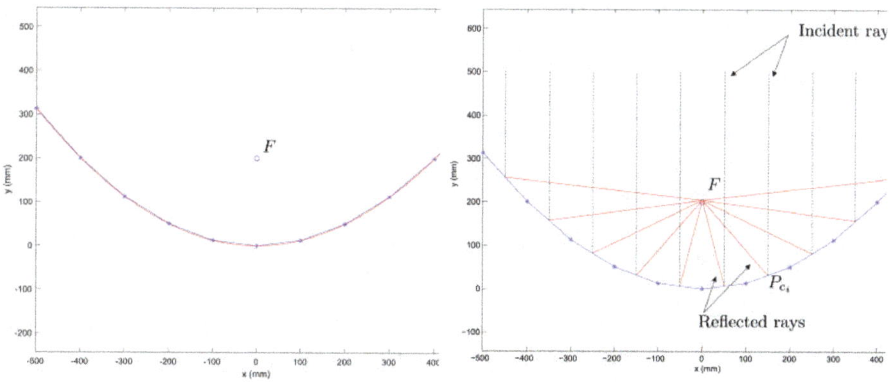

Fig. 1: Polygonal chain and parabola (left); Incident and reflected rays (right).

The point P_{ci} represents the centre of the *i*-th line segment of the polygonal. Figure 2 shows an enlargement around the feed, which highlights how the incident rays coming from the points are not directed exactly on the focus *F*, but are distant from it by a quantity d_i. To determine this quantity, we need to calculate the minimum distance between the focus and each reflected ray coming from all line segments of the polygonal chain.

By keeping the aperture width of the parabola fixed, it is easy to understand how by increasing the number of line segments of the polygonal, the RMS error decreases exponentially, as the approximation of the parabola improves. Theoretically, the ideal configuration occurs when the number of line segments tends to infinity, but this information collides with the manufacturing limits, as it is not possible to make a cable-net with an infinitesimally small cable length.

What described for the two-dimensional model can be even extended in the three-dimensional model. Unlike the 2D model, the polygonal chain is replaced by the surface reflector consisting of the cable-net system, whose nodes are located on the paraboloid surface. Here, for the RMS error evaluation, the vertices of the triangular facets and their centroid are considered.

Using the equation to determine the minimum distance of the reflected rays from the focus of the paraboloid and considering a prime-focus antenna, we obtain what is depicted in Fig. 3. This figure shows in blue the points of minimum distance calculated by the reflected rays coming from the vertices of each triangle, while in red those coming from the centroids. As for the 2D model, also in this case it is necessary to associate a weight according to the different contributions of the rays. At the centroid a weight of 1/2 is assigned, while at the three vertices of the triangle a weight of 1/6 is assigned, since their contribution needs to be split equally between them. In the 3D model, by adding the weights, , the Eq. (2) changes as follows:

$$e_{RMS} = \sqrt{\frac{(w_1 d_1)^2+(w_2 d_2)^2+\cdots+(w_n d_n)^2}{3n}} \tag{3}$$

Correction algorithm

To improve the performance of the reflector, the best geometric figure formed by contiguous and flat triangular facets needs to be found.

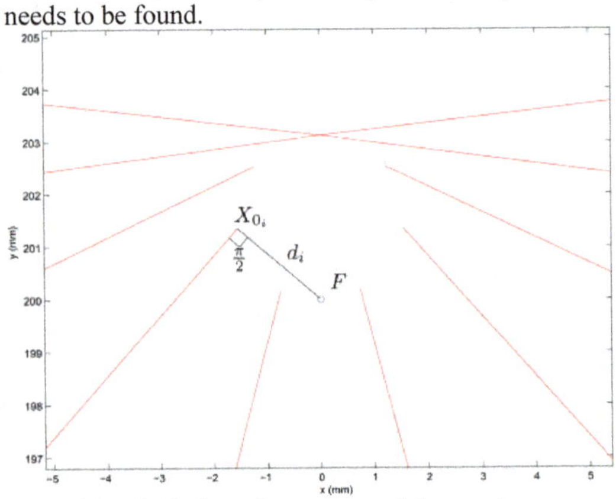

Fig. 2: Reflected rays around the receiver.

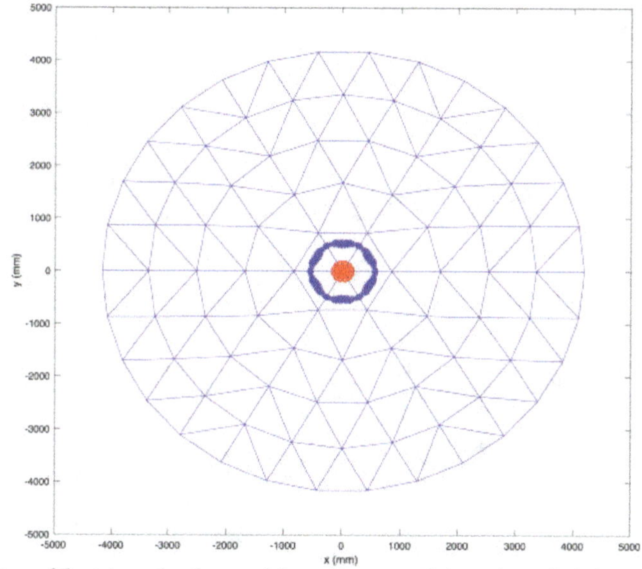

Fig. 3: Top view of the triangular facets of the cable-net and the points of minimum distance of each reflected ray for a prime-focus reflector.

The geometry of the above said paraboloid can be achieved thanks to the formulation of an optimization algorithm which, by varying the position of the nodes of the cable-net, minimizes the distance of each reflected ray with respect to the focus of the paraboloid.
The optimization problem can be defined as follows:

$$\begin{aligned} &\text{Find} \quad \mathbf{x}, \mathbf{y}, \mathbf{z} \\ &\min \quad e_{RMS} \end{aligned} \tag{4}$$

where **x,y,z** represent the coordinate of the internal nodes of the cable-net, since the nodes located on the perimeter remain fixed in their initial position, while e_{RMS} is the RMS error defined by Eq. (3). By applying this algorithm to an offset reflector with a parent paraboloid diameter of 6 meters, it is possible to appreciate how the cable-net topology changes according to the minimum distance between each reflected ray and the receiver.

Figure 4, on the left, shows the initial topology of the cable-net of the offset reflector. On the left side the location of the points of minimum distance of the reflected rays with respect to the focus is shown. In particular, the blue points are closer to the focus as they represent those coming from the centroids of each triangle of the cable-net, while those in red from the three vertices. By applying the optimization problem described above, the result shown in the right side of Fig. 4 is obtained. The greater concentration of points around the focus demonstrates the effectiveness of the algorithm, and this confirms how the optimized net topology concentrates the rays reflected by the reflector in a smaller area of the illuminator.

In Fig. 5 it is possible to better evaluate a comparison between the two obtained results.

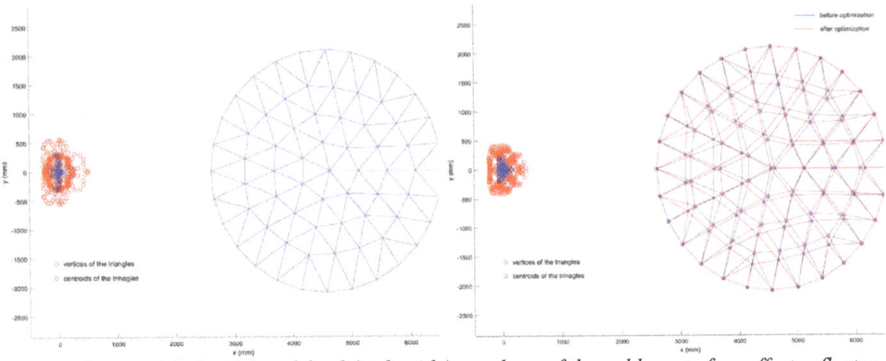

Fig. 4: The initial (left side) and final (right side) topology of the cable-net of an offset reflector and the points of minimum distance with respect to the focus.

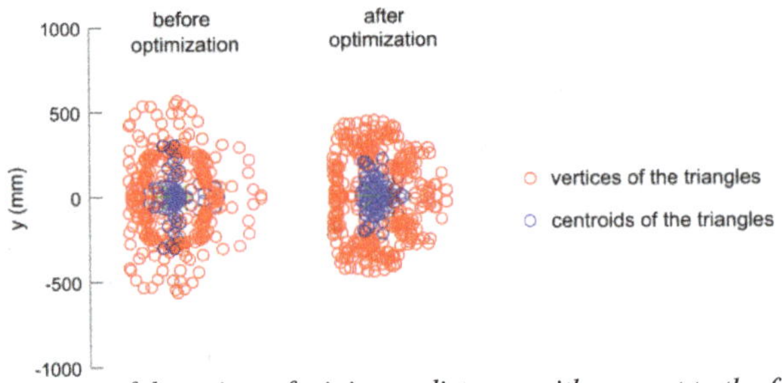

Fig. 5: An enlargement of the points of minimum distance with respect to the focus before and after optimization.

Conclusions

The surface accuracy of mesh reflectors is the main objective to be achieved in order to guarantee the prescribed band-width requirements. In this work, a numerical optimization algorithm able to ensure a greater concentration of reflected rays around the focus has been developed. The novelty introduced in this paper concerns the ability to optimize the reflecting surface according to the quantity of reflected rays directed towards the illuminator, unlike the most used method which measures the distance of the nodes of the net with respect to the ideal paraboloid. This approach provides a direct measure of the error of all incoming electromagnetic rays missing the receiver. One operating condition has been addressed: the case of the offset reflector. In this case, an unconstrained numerical optimization problem was adopted as the nodes were fixed on the external circumference and the only variables were represented by the internal nodes, which are free to move over the entire surface of the reflector. The numerical results confirm the goodness of the proposed method, modifying the topology of the cable-net in order to reduce the mean square error around the feed of the reflector.

References

[1] L. Puig, A. Barton, N. Rando. A review on large deployable structures for astrophysics missions. Acta Astronautica 67.1-2 (2010): 12-26. https://doi.org/10.1016/j.actaastro.2010.02.021

[2] A. Cammarata, M. Lacagnina, R. Sinatra. Closed-form solutions for the inverse kinematics of the Agile Eye with constraint errors on the revolute joint axes. 2016 IEEE/RSJ International Conference on Intelligent Robots and Systems (IROS). IEEE, 2016. https://doi.org/10.1109/IROS.2016.7759073

[3] A. Cammarata, R. Sinatra, A. Rigano, M. Lombardo, P.D. Maddio. Design of a large deployable reflector opening system. Machines, 8(1), (2020). 7. https://doi.org/10.3390/machines8010007

[4] A. Cammarata, R. Sinatra, P.D. Maddio. A two-step algorithm for the dynamic reduction of flexible mechanisms. In IFToMM Symposium on Mechanism Design for Robotics (pp. 25-32). (2018). Springer, Cham. https://doi.org/10.1007/978-3-030-00365-4_4

[5] Z. Huang, F. Xi, T. Huang, J. S. Dai, R. Sinatra. Lower-mobility parallel robots: theory and applications. Advances in Mechanical Engineering 2 (2010): 927930. https://doi.org/10.1155/2010/927930

[6] S. Yuan, B. Yang, H. Fang. The Projecting Surface Method for improvement of surface accuracy of large deployable mesh reflectors. Acta Astronautica 151 (2018): 678-690. https://doi.org/10.1016/j.actaastro.2018.07.005

[7] S. Yuan, B. Yang, H. Fang. Direct root-mean-square error for surface accuracy evaluation of large deployable mesh reflectors. AIAA SciTech 2020 Forum. 2020. https://doi.org/10.2514/6.2020-0935

[8] S. Yuan, B. Yang, H. Fang. Improvement of surface accuracy for large deployable mesh reflectors. AIAA/AAS Astrodynamics Specialist Conference. 2016. https://doi.org/10.2514/6.2016-5571

[9] Y. Tang, T. Li, Z. Wang, H. Deng. Surface accuracy analysis of large deployable antennas. Acta Astronautica 104.1 (2014): 125-133. https://doi.org/10.1016/j.actaastro.2014.07.029

[10] A. Cammarata, R. Sinatra, R. Rigato, P.D. Maddio. Tie-system calibration for the experimental setup of large deployable reflectors. Machines 7.2 (2019): 23. https://doi.org/10.3390/machines7020023

[11] P. Agrawal, M. Anderson, M. Card. Preliminary design of large reflectors with flat facets. IEEE transactions on antennas and propagation 29.4 (1981): 688-694. https://doi.org/10.3390/machines7020023

[12] H. Deng, T. Li, Z. Wang, X. Ma. Pretension design of space mesh reflector antennas based on projection principle. Journal of Aerospace Engineering 28.6 (2015): 04014142. https://doi.org/10.1061/(ASCE)AS.1943-5525.0000483

[13] S. Morterolle, B. Maurin, J. Quirant, C. Dupuy. Numerical form-finding of geotensoid tension truss for mesh reflector. Acta Astronautica 76 (2012): 154-163. https://doi.org/10.1016/j.actaastro.2012.02.025

Theoretical and Applied Mechanics - AIMETA 2022 Materials Research Forum LLC
Materials Research Proceedings **26** (2023) 671-676 https://doi.org/10.21741/9781644902431-108

Multibody simulations of a distributed-compliance helical transmission joint for largely misaligned shafts

Sorgonà Orazio[1,a *], Giannini Oliviero[1,b], Cirelli Marco[1,c] Pier Paolo Valentini [2,d]

[1] University of Rome Niccolò Cusano, via Don Carlo Gnocchi 3, 00166 Rome (RM), Italy

[1] University of Rome Tor Vergata, via del Politecnico 1, 00133 Rome (RM), Italy

[a *] orazio.sorgona@unicusano.it, [b] oiviero.giannini@unicusano.it, [c] marco.cirelli@unicusano.it, [d] valentini@ing.uniroma2.it

Keywords: Transmission Joints, Compliant Mechanisms, Machined Springs, Flexible Multibody Simulations

Abstract. The manuscript presents the development of a flexible multibody model to study the behavior of a distributed-compliance helical transmission joint. This kind of joint has not already been studied in literature and its simulation challenges with complexities. The flexible multibody approach breaks through these and provides some interesting results. It's exposed first upon the kinematic behavior of the joint; then a stress analysis is provided; and lastly the results of dynamical tests are shown.

Introduction

Transmission joints allow the transfer of mechanical power between generally misaligned rotating shafts [4,6,7,8]. Their design confronts Applied Mechanics with two main tasks, it is:

a) the regularity of the transmission of motion, i.e. obtaining or approximating the optimal transmission ratio of the angular velocities, which is equal to one.
b) the power efficiency, i.e. to reduce the passive work of friction actions. Contact forces and consequently friction tend to be large, due to the necessity of transmitting a high torque through a minimized radial extension.

Concerning both points, rigid-link transmission joints are moreover extremely sensitive to construction and assembly errors [6,8]. Relevant lubrication is needed, and wearing deteriorates the quality of the kinematic operation. A good solution for the task (b) may be the adoption of compliant mechanisms as transmission joints. These kinds of mechanisms work in fact without contacts, save when they are specifically designed, for they are monolithic deformable structures, or made of integrated parts, in which the deformation of some of their portions allows for the desired displacement of others [1,6]. This anyway makes them intrinsically load-sensitive, and so their fitness to task (a) needs to be evaluated. Severe fatigue issues may arise and, further, the dynamics of the compliant-jointed system will exhibit critical frequencies in which resonating effects are produced.

Compliant kinematic-equivalent joints. Compliant transmission joints had insofar been designed generally to reproduce rigid-links kinematic pairings [2,3,5]. In [2] there are examples of several flexures to reproduce revolute joints. In [5] torsional prisms are used to simulate the functional couplings of a Hooke-Joint.

Compliant helical joint (CHJ). Another kind of compliant joint is not based on some rigid-links ideal or actual counterpart but on helical machined springs [9]. The structural concept of CHJ's is to join a high torsional stiffness with a low flexural. These kinds of devices may work as couplings for small axes misalignments, or transmission joints for large, either angular, parallel, or skewed. The manufacturing of machined springs enjoys all the common advantages for compliant mechanisms, it is no necessity of assembly thus integration with other elements, and high

Materials Research Proceedings 26 (2023) 671-676 https://doi.org/10.21741/9781644902431-108

precision. It has some advantages even with the common manufacturing of extruded wire springs, while the machined coil has negligible residual stress. CHJ's are produced and can be purchased. To the knowledge of the authors, there is no existing literature on the subject, despite this kind of compliant joint is interesting from a mechanical point of view, and possibly economical too.

Cases of study

Multibody simulations. Multibody dynamics simulations represent a very useful approach to study transmission joints because thanks to this approach it is possible to consider the behaviour of the joint in several operating conditions. [10].

In this case, a sample CHJ is modelled in CAD and employed in full-flexible multibody simulations. In these, the coil is treated as a flexible component discretized as FEM, and it's analyzed along with the computation of the general multibody dynamics. The purpose is to highlight the basic features of a generic CHJ about its kinematic, dynamic, and structural performances. At this earlier stage of the study the influences of variations in the coil geometry had not been investigated.

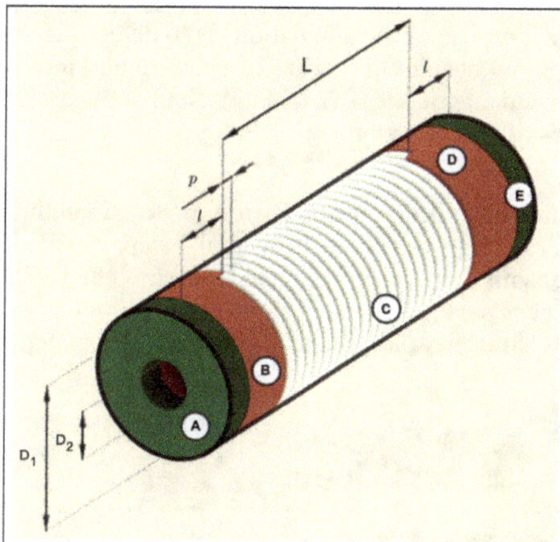

Fig. 1. CAD model of the sample CHJ.

Table 1. Nomenclature and sizes.

A	Input bearing	D_1	19,2 [mm]
B	Input rigid end	D_2	6,35 [mm]
C	Flexible coil	l	8,50 [mm]
D	Output rigid end	p	1,98 [mm]
E	Output bearing	L	39,6 [mm]

Model description. The sample CHJ is featured and sized according to a joint of the kind at [11]. This is an inox-steel single coil and single coil-start monolith, actually designed for angularly misaligned shafts. Two rigid ends are integral to the coil and the two input and output shafts bearings are added in the model.

Simulations environment and setups. In a full-flexible multibody simulation environment, the helical coil is modelled in shell Quad4 elements. With reference to Fig.2, the bearings are joined by a cylindrical pair at their middle point along the line of the aligned axes, with allowed perpendicular translations S and rotations R. It has to be noticed that this is a *particular* condition of kinematic constraints, although not unrealistic. That is, a more complete study on CHJ's shall account for different conditions too, like for instance possibility for one or both of the rigid ends of the joint to slide along the respective shaft. The model is completed with a massive body M integral to the output rigid end, with parametric moments of inertia J in the direction of the output axis.

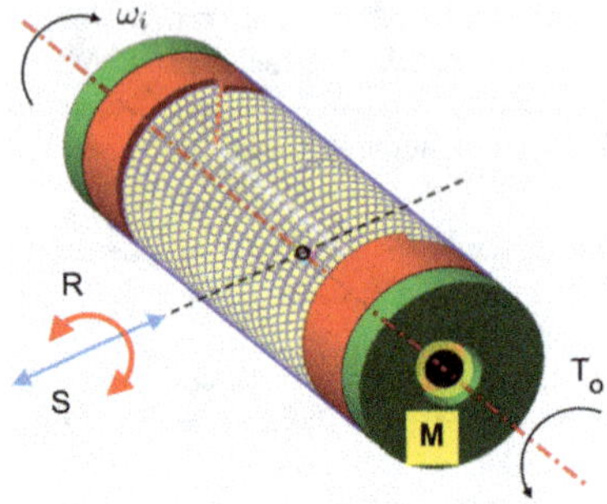

Fig.2 Flexible Multibody modelling and simulations setup.

Tests are generally with skewed misalignments, even if the sample joint is declared to be specifically designed for angular ones. In fact, the purpose of the simulations is to detect general tendencies, apart from the definition of some optimal designs. Tests are of quasi-static kinds to measure the error of transmission and stress peaks and magnitudes of oscillation; and of dynamical kind to find the resonating frequencies and factors of amplification.

Results

Quasi-static tests had been carried out varying angular misalignments R from 0° to 45° with 9° step; and translational S by multiples of $r = D_1/2$, ranging from 0,0 to 1,0 with 0,2 step.

Transmission error. The main results on transmission error, calculated on smoothed data, demonstrate that: i) it depends *only* on misalignment parameters; ii) is anyway very limited, it is $O(10^{-3})$. This from an engineering point of view confirms the producers' claim of an *effective constant-velocity* performance of these joints.

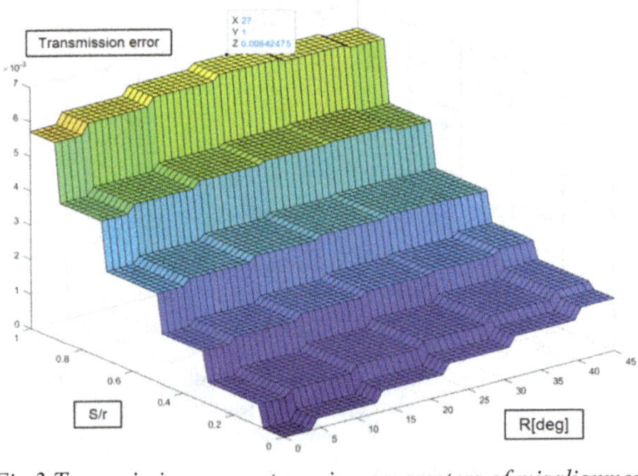

Fig.3 Transmission error at varying parameters of misalignment.

Stress distributions and severities. The behavior of the coil at different kinds of misalignment can be appreciated in Fig. 4 *a) b) c)*. These show the distributions of Von Mises stress at the extreme points of pure rotation, of pure translation and in the major skewing, with an applied output torque T_o=500 Nmm.

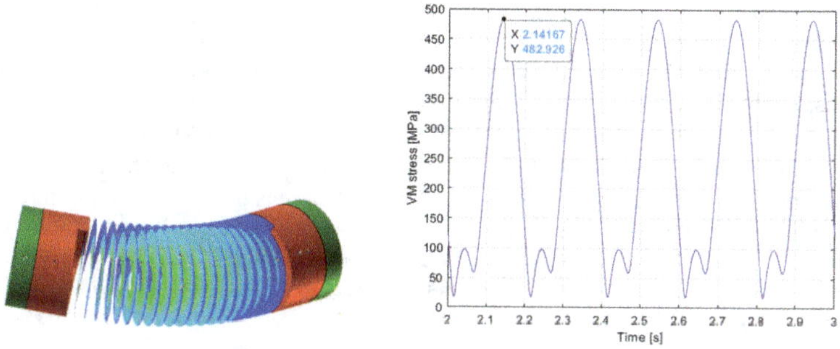

a) R=45°, S/r = 0;

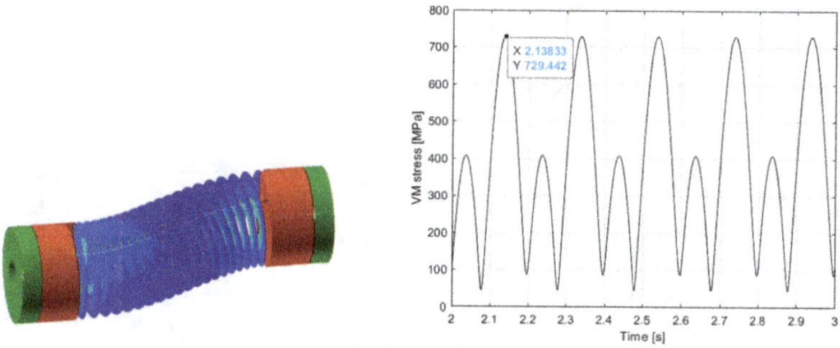

b) R=0, S/r = 1;

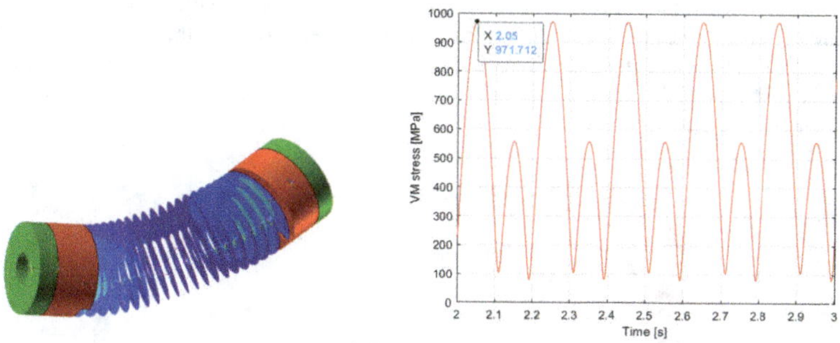

c) R=45°, S/r = 1;

Fig.4 Von Mises stress distributions in the coil and developments at more stressed points in different conditions of misalignment.

In all the cases the distributions remain quite constant in the control volume. In the cases of pure rotation stress increases axially from the ends to the middle; in the cases of pure translations instead, it decreases axially from the ends to the middle; in skewed misalignments it has a hybrid behaviour. In all the cases stress increases radially from the external to the internal edge of the coil. This is consistent with the theory of beams with high curvatures, that predicts a non-homogeneous distribution of stiffness on the cross-sections, and more specifically increasing toward the centre of curvature. This implies a greater stress on the same deformations.

Dynamic tests. Resonating frequencies had been obtained through simulations at different values of misalignments, torque, and moments of inertia J. A nonlinear behavior had been detected and it's shown in Fig.5 on normalized variables. The values f_0 of the frequencies at aligned axes seem to scale in inverse ratio with $\sqrt{J}$ as in the linear theory. These values decrease with the translational misalignments, and not with the rotational *if* there is no applied output torque. If the torque is applied a greater decrease it's observed with the translational, and then with the rotational too. For this behavior it had been hypothized a law which is *trilinear* in the parameters of misalignment and output torque. Coefficients had been established on the data obtained for J=400[kgmm^2], and the hypothetical laws thus reads as:

$$f = \frac{20}{\sqrt{J}}\left\{18.5 - \frac{S}{r}\left[2.7 + \frac{T_o}{500}\left(1.44 + 0.24\frac{R}{45°}\right)\right]\right\},$$

where R is expressed in degrees.

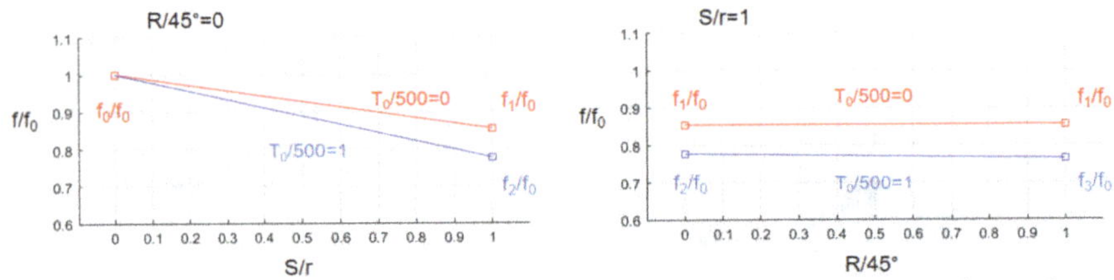

Fig.4 Adimensional variations of the frequencies of resonance in dependence of: (left) translational misalignments, and output torque; (right) translational misalignments, output torque, and rotational misalignments.

The formula gives some good predictions, the results of some of which are shown in Fig.6 in terms of the difference between output and input angular velocities. In the case on the right side the prediction of 5.29 Hz is not truly remote from the actual resonance, which had been detected at ab. 5.50 Hz. The error is thus less than 4%. A more refined model may be developed in further work.

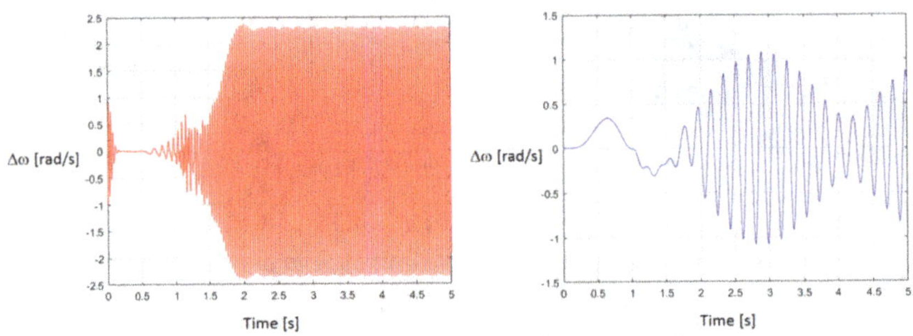

Fig.6 Validation of predictions on frequencies of resonance: (left) J=100.0 [kgmm^2], S/r=1.0, R=45°, f = 30.9 [Hz];

(right) J=3600 [kgmm^2], S/r=0.6, R=45°, f= 5.29 [Hz].

Conclusions

This paper deals with study of a distributed compliance helical transmission joint through a methodology of flexible multibody simulations. The compliance of the coils is considered by the adoption of a shell FEA model of same thickness of the reference geometry, in order to speed up the simulations. The results demonstrate that the transmission error is very limited and depends only on misalignment parameters. Moreover, the stress distribution appears to agree with the stress theory of deformed beams. Dynamic behavior exhibits some nonlinearity, expressed in a first hypothesis as a multilinear relation.

References

[1] Howell, L. L., "Compliant mechanisms", John Wiley & Sons,(2001).

[2] Ling, M., Howell, L.L., Cao, J., Chen, G., "Kinetostatic and Dynamic Modeling of Flexure-Based Compliant Mechanisms: A Survey" Applied Mechanics Reviews, Vol.72 (2020). https://doi.org/10.1115/1.4045679

[3] Machekposhti, D. F., Tolou, N., and Herder, J. L., "A Review on Compliant Joints and Rigid-Body Constant Velocity Universal Joints Toward the Design of Compliant Homokinetic Couplings." ASME. J. Mech. Des., 137(3): 032301(12 pages). (2015) https://doi.org/10.1115/1.4029318

[4] Machekposhti, D. F., Tolou, N., and Herder, J. L. (November 9, 2017). "A Fully Compliant Homokinetic Coupling." ASME. J. Mech. Des., 140(1): 012301(9 pages), (2018). https://doi.org/10.1115/1.4037629

[5] Schmelz, F., Seherr-Thoss, C. H.-C., and Aucktor, E., "Universal jointed driveshafts for transmitting rotational movements". In Universal Joints and Driveshafts. Springer, pp. 1–28, (1992). https://doi.org/10.1007/978-3-662-02746-2_1

[6] Tanik, C.M., Parlaktaş, V., Tanik, E., Kadioğlu, S., "Steel compliant Cardan universal joint", Mechanism and Machine Theory, 92, pp. 171-183, (2015). https://doi.org/10.1016/j.mechmachtheory.2015.05.009

[7] Yin, L., Ananthasuresh G. K., "Design of Distributed Compliant Mechanisms", Mechanics Based Design of Structures and Machines, 31(2), pp. 151-179, (2003). https://doi.org/10.1081/SME-120020289

[8] Valentini, P., " Effects of the dimensional and geometrical tolerances on the kinematic and dynamic performances of the rzeppa ball", Proceedings of the Institution of Mechanical Engineers, Part D: Journal of Automobile, 228(1), pp. 37-49, (2013). https://doi.org/10.1177/0954407013505745

[9] Cirelli, M. , Giannini, O., Cera, M., De Simoni, F., Valentini, P.P., Pennestrì, E., " The mechanical efficiency of the Rzeppa transmission", Mechanism and Machine Theory, vol 164 (2021). https://doi.org/10.1016/j.mechmachtheory.2021.104418

[10] Cirelli, M., Rossi, V., Valentini, P.P. and Pennestrì, E. 'A dynamic model of a Cardan joint to evaluate the effect of elasticity and manufacturing errors', Int. J. Vehicle Performance, Vol. 7, Nos. 1/2, pp.136–155 (2021). https://doi.org/10.1504/IJVP.2021.113423

[11] Perez, C. Calderon, J.D.D., On the general characterization of machined springs and their manufacturing process; Mechanical Engineering Congress and exposition IMECE, San Diego, California, USA (2012). https://doi.org/10.1115/IMECE2013-63402

[12] Mariti L. Belfiore, N.P. Pennestrì, E. Valentini, P.P. Comparison of solution strategies for multibody dynamics equations. International Journal for numerical methods in Engineering. Nov, 18,88(7): 637-56 (2011). https://doi.org/10.1002/nme.3190

[13] Information on https://helicalcouplings.com/helical-flexured-u-joints-2.aspx

Mechanics of renewable energy systems

Theoretical and Applied Mechanics - AIMETA 2022 | Materials Research Forum LLC
Materials Research Proceedings **26** (2023) 679-684 | https://doi.org/10.21741/9781644902431-109

Combining pendulum and gyroscopic effects to step-up wave energy extraction in all degrees of freedom

Giuseppe Giorgi[1,a]*, Fabio Carapellese[1,b], Mauro Bonfanti[1,c], and Sergej Antonello Sirigu[3,d]

[1] Marine Offshore Renewable Energy Lab (MOREnergy Lab), Department of Mechanical and Aerospace Engineering (DIMEAS), Politecnico di Torino, Italy

[a]giuseppe.giorgi@polito.it, [b]fabio.carapellese@polito.it, [c]mauro.bonfanti@polito.it, [d]sergej.sirigu@polito.it

*corresponding author

Keywords: Wave Energy, Gyroscopic Effects, Inertial Effects

Abstract. The fight against the global threat of climate change requires, among other actions, to increase the penetration of renewable energy technologies and diversify the energy mix in order to support a resilient energy system that can reach net-zero greenhouse gas emissions. Offshore energy is expected to drive the energy transition, with wave energy having the major role to provide a reliable baseload and reduce the need for storage; however, its techno-economic feasibility requires reduction of costs and increase of energy conversion efficiency. This paper tackles a fundamental innovation of a device's working principle which, jointly exploiting pendulum and gyroscopic effects, steps-up the overall conversion efficiency in real operational conditions. A recent patent proposes a technological solution that conveniently combines pendulum and gyroscopic effects in order to effectively exploit motion also outside the plane, namely in the three-dimensional space and from all degrees of freedom (DoFs). This paper tackles the endeavour of the analytical formulation of the electro-mechanical conversion system dynamics, considering at first the fully-nonlinear equation of motion, obtained through a Lagrangian approach. Consequently, incremental simplifications are applied to accommodate practical application, based on the study on the relative importance of each term in the equation of motion. Furthermore, preliminary results are produced and discussed, comparing the behaviour in response to 3-DoF to 6-DoF exploitation.

Introduction

The European Commission has set ambitious carbon neutrality targets, reaching net-zero greenhouse gas emission by 2050, as defined in the Green Deal Communication [1]; moreover, the key enabling role of ocean energy (tidal and wave energy) and offshore wind energy has been expressively declared by the European Commission [2], including quantitative targets for the envisioned installed capacity: 1 GW installed capacity by 2030 and 40 GW by 2050 in the European Union waters for ocean energy devices, and 60 GW by 2030 and 300 GW by 2050, for fixed and floating wind turbines. A techno-economic improvement is mandatory for both offshore wind and ocean technologies to become viable on a such a large and systemic scale, which requires further developing of floating wind substructure and the related installed turbine, via techno-economic optimization and numerical modelling [3]; likewise, leveraging the complementarity between wave and wind energy (load profiles, low statistical correlation [4], and cost sharing synergies [5]), it is also crucial to flank the offshore wind development with more efficient wave energy converters (WECs) which, albeit a smaller total capacity, provide a more stable and reliable baseload to reduce production variability and need for storage. Increasing WECs efficiency is a complex and multidisciplinary task, relying on a variety of energy-maximisation techniques, e.g.

via model-predictive control [6], extremum-seeking [7], or time-varying control [8], time-effective estimation [9] and forecasting [10] algorithms, data-driven improved accuracy of numerical models of fluid-structure interactions [11] by means of the identification of grey-parameters in low-complexity models [12], mid-fidelity partially-nonlinear models [13], or direct implementation of fully-nonlinear Navier-Stokes fluid-dynamics equations [14]. Nevertheless, such models should preserve the inherent passive characteristics of the system [15], while remaining able to articulate typical instabilities of floating structures, e.g. parametric resonance [16] or yaw instability [17]. Finally, due to the resulting potential complexity of such nonlinear models, it is often convenient to apply model order reduction techniques to achieve real-time computation, required for practical implementation [18].

All of the previous techniques are applicable to arbitrary WEC concepts; however, more fundamental routs to increase energy efficiency is to operate on the underlying conversion principle [19]. A novel device is indeed proposed in this paper and in a recent patent, based on the idea to notionally merge two already successful devices, both using inertial coupling between the motion of a floating hull and the inner electromechanical system: the ISWEC [9] (Inertial Sea Wave Energy Converter) and the PeWEC [20] (Pendulum Wave Energy Converter). Both technologies perform best when the incoming wave direction is along the hull main axis, and/or the spreading factor of the wave is infinite (i.e. all spectral components are travelling in the same direction) [21], so the WEC motion remains in a plane. The herein discussed technological solution conveniently combines gyroscopic and pendulum effects to expand the operational space of the WEC also outside the plane, namely to extract energy from all degrees of freedom (DoFs). The content of the paper is the analytical formulation of the electromechanical conversion system dynamics, starting from the fully-nonlinear equation of motion, obtained through a Lagrangian approach; consequently, a sensitivity analysis explores the system's most influential mechanical parameters. Such results constitute the first stepping stone towards a holistic techno-economic analysis, based on the results presented in this paper.

Mechanical device and its mathematical modelling

This section introduces the underlying mechanical innovation of the analysed system and presents the related mathematical model. The technology is composed of a flywheel rotating around an axis φ, hosted within a gimbal and vertical at static equilibrium. The gimbal is supported by a structure able to rotate around an axis ε, normal to φ. The main innovation with respect to the ISWEC system is to displace the middle symmetry plane of the flywheel by a distance λ from the axis ε; in this way, the arm λ introduces the pendulum behaviour while the gyroscopic action is obtained by the flywheel rotation $\dot{\varphi}$. Both torques are parallel to the ε axis, so a potentially constructive interference may be obtained. A power take-off unit converts the mechanical energy into electrical

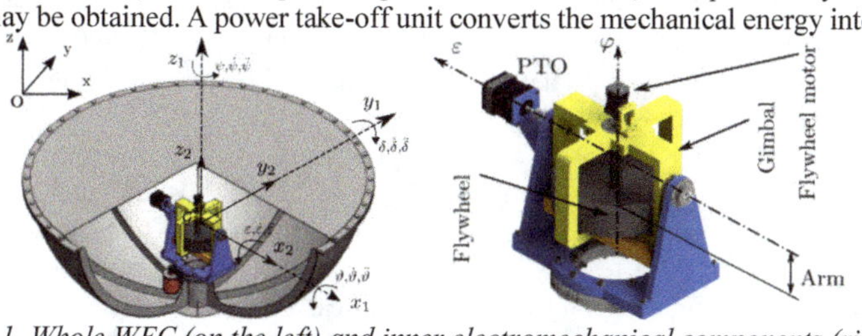

Figure 1. Whole WEC (on the left) and inner electromechanical components (right)

Figure 1 shows a CAD impression of such technology, and a possible installation within a sealed floating hull, loosely moored to the seabed to enable rotations around the horizontal axis. Considering the inertial frame $O(x, y, z)$, the pose of both the gimbal and the flywheel are computed by means of a set of linear differential equations, assuming small oscillations to make

Materials Research Forum LLC
https://doi.org/10.21741/9781644902431-109

the use of linear mapping reasonable. The system motion is described with $n = 6$ degrees of freedom (DoFs), so that the vector of generalized coordinates $X(t) \in R^n$ is defined as:

$$\mathrm{X(t)} = [\mathrm{x(t)}, \mathrm{y(t,)}, \mathrm{z(t)}, \theta(\mathrm{t}), \delta(\mathrm{t}), \varepsilon(\mathrm{t})]^T \tag{1}$$

which refer to the surge, sway, and heave displacements, and the roll and pitch rotations of the floater; in addition, $\varepsilon(t)$ is the pendulum rotation with respect to the precession axis. Note that no yaw motion is considered, since the axisymmetric floater is not affected by yaw hydrodynamic torques. Thanks to the linear assumptions, equation of motion is derived through the superposition principles, combing both the elastic effect due to the flywheel mass m_f and the gyroscopic effect. It is further assumed that the centres of mass of the floater and the gimbal coincide, both placed on the precession axis. The corresponding inertia matrix are defined as $I_{g,\mathcal{G}} = \mathrm{diag}(I_{g,x}, I_{g,y}, I_{g,z})$ and $I_{f,\mathcal{G}} = \mathrm{diag}(I_{f,x}, I_{f,y}, I_{f,z})$, where $\{I_g, I_f\} \subset R^{3\times3}$ are the inertia matrices of the gimbal and flywheel, respectively. Then, the system of differential equations describing the mechanical coupling between pendulum and hull become $M_w\ddot{X}(t) + C_wX(t) + K_wX(t) = \mathcal{F}_{ext}(t)$, with:

$$C_w = \begin{bmatrix} 0 & 0 & 0 & 0 & 0 & 0 \\ 0 & 0 & 0 & 0 & 0 & 0 \\ 0 & 0 & 0 & 0 & 0 & 0 \\ 0 & 0 & 0 & 0 & 0 & 0 \\ 0 & 0 & 0 & 0 & 0 & -\mathrm{I_{f_z}}\dot{\varphi} \\ 0 & 0 & 0 & \mathrm{I_{f_z}}\dot{\varphi} & 0 & 0 \end{bmatrix}, K_w = \begin{bmatrix} 0 & 0 & 0 & 0 & 0 & 0 \\ 0 & 0 & 0 & 0 & 0 & 0 \\ 0 & 0 & 0 & 0 & 0 & 0 \\ 0 & 0 & 0 & \mathrm{m_f g l_f} & 0 & \mathrm{m_f g l_f} \\ 0 & 0 & 0 & 0 & \mathrm{m_f g l_f} & 0 \\ 0 & 0 & 0 & \mathrm{m_f g l_f} & 0 & \mathrm{m_f g l_f} \end{bmatrix}$$
$$M_w = \begin{bmatrix} \mathrm{m_f + m_g} & 0 & 0 & 0 & -\mathrm{m_f l_f} & 0 \\ 0 & \mathrm{m_f + m_g} & 0 & \mathrm{m_f l_f} & 0 & \mathrm{m_f l_f} \\ 0 & 0 & \mathrm{m_f + m_g} & 0 & 0 & 0 \\ 0 & \mathrm{m_f l_f} & 0 & \mathrm{I_{f_x} + I_{g_x}} & 0 & \mathrm{I_{f_x} + I_{g_x}} \\ -\mathrm{m_f l_f} & 0 & 0 & 0 & \mathrm{I_{f_y} + I_{g_y}} & 0 \\ 0 & \mathrm{m_f l_f} & 0 & \mathrm{I_{f_x} + I_{g_x}} & 0 & \mathrm{I_{f_x} + I_{g_x}} \end{bmatrix} \tag{2}$$

where m_g is the gimbal mass. Considering that the gyropendulum motion is controlled by the torque $\tau_\varepsilon(t)$, the forcing vector $\mathcal{F}_{ext}(t) \in R^n$ is defined as $\mathcal{F}_{ext}(t) = [\mathbf{0}_{1\times n}, \tau_\varepsilon(t)]^T$. The changing in time of position and orientation of the body-fixed reference frame and the potential energy variation result in inertial, elastic and gyroscopic forces that induce the gyropendulum oscillation. Therefore, energy from the wave forces induces the movement of the floater, hence the oscillation of the PTO axis via the gyroscopic effect and/or the restoring force due do the eccentric mass. Fluid-structure interactions are modelled using linear potential flow theory, based on the assumptions of inviscid fluid and incompressible and irrotational flow. The floater-only response with respect to its generalized coordinates in Fourier transform form $X_f(\omega) = [\hat{x}(\omega), \hat{y}(\omega), \hat{z}(\omega), \hat{\vartheta}(\omega), \hat{\delta}(\omega)]^T$ is defined by the following frequency-domain equation:

$$-\omega^2 M_f X_f(\omega) = F_r(\omega) + F_{exc}(\omega) + F_h(\omega) + F_p(\omega) \tag{3}$$

where $M_f \in R^{5\times5}$ is the floater inertia matrix, $F_r(\omega) \in R^{5\times1}$ is the radiation force, $F_h = K_h X_f(\omega) \in R^{5\times1}$ is the hydrostatic restoring force, proportional to the hydrostatic stiffness matrix $K_h X \in R^{5\times5}$, $F_{exc}(\omega) \in R^{5\times1}$ is the wave exiting force and $F_c(\omega) \in R^{5\times1}$ is the reaction force generated by the dynamic coupling of the mechanical system. Finally, it is now possible to combine the two sub-systems (inner mechanics and wave-structure system) into the WEC entire device dynamics:

$$-\omega^2\left(M_w + \widetilde{M_f} + \tilde{A}(\omega)\right)X(\omega) + j\omega\tilde{B}(\omega)X(\omega) + \left(K_w + \widetilde{K_h}\right)X(\omega) = F_e(\omega) \tag{4}$$

$$\widetilde{M_f} = \begin{bmatrix} M_f & \mathbf{0}_{5\times1} \\ \mathbf{0}_{1\times5} & \mathbf{0}_{1\times1} \end{bmatrix}, \quad \widetilde{K_h} = \begin{bmatrix} K_h & \mathbf{0}_{5\times1} \\ \mathbf{0}_{1\times5} & \mathbf{0}_{1\times1} \end{bmatrix} \tag{5}$$

where, $X(\omega)$ is the vector containing the Fourier transforms of the generalized displacements $X(t)$, $\widetilde{M_f}$ and $\widetilde{K_h}$ are the mass and stiffness matrices defined as block matrices, and $\tilde{A}(\omega)$ and $\tilde{B}(\omega)$ are the added mass and the radiation damping matrices with respect to the generalized coordinates, and $F_e(\omega)$ is the excitation forces vector. The excitation forces vector $F_e(\omega)$ can be defined as $F_e(\omega) = [F_{exc}(\omega), \tau_\varepsilon(\omega)]^T$, where $\tau_\varepsilon(\omega)$ is the actual control action on the gyropendulum axis ε. The control is actuated by the PTO system that exerts a damping torque $\tau_\varepsilon(t)$ proportional to the pendulum swinging velocity $\dot{\varepsilon}(t)$. In this work, the simulations are performed considering a frequency domain model, whose steady-state response under stochastic loads can be generally determined on the basis of the transfer function of the system. The related complex form of transfer function is:

$$H(\omega) = \left[-\omega^2\left[M_w + \widetilde{M_f} + \tilde{A}(\omega)\right] + j\omega\tilde{B}(\omega) + K_w + \widetilde{K_h}\right]^{-1} \tag{6}$$

The frequency domain representation of eq. (6) is used as a basis for a spectral-domain description of the entire energy conversion process, which considers a probabilistic representation of the waves and of the model response as input and output respectively. The free surface elevation is defined as a discretised signal with N fixed time steps (Δt). According to [22], a sea wave can be modelled using a Random Amplitude Scheme: considering the finite realization length $T = N\Delta t$, the discrete sequence of simulated free surface elevation $\eta(t_i)$ with $t_i = i\Delta t$ can be represented as: $\eta(t_i) = \sum_{k=1}^{N} a_k \cos(\omega_k t_i + \phi_k)$, with randomly chosen phases (ϕ_k) from a uniform distribution $\mathcal{U}(0,2\pi)$, while the amplitudes (a_k) follow a Rayleigh distribution $\mathcal{R}(\sigma)$ with variance $\sigma = 2S_\eta(\omega_k)\Delta\omega$. According to the Gaussian closure assumption, the WEC can be simulated as a linear Gaussian process, as the Gaussian process that drives the system: being the system linear, its steady-state response to a sum of orthogonal frequency components is the sum of the responses to each of the frequency components, with phase shift $\Delta\phi_k$. The calculation of the mean absorbed power ($P_{abs,avg}$), which is the main variable of interest, is non-deterministic and is a function of the sea-state conditions and of the control action c_{PTO} applied on the WEC. Under the hypothesis of a large number of realizations, the expected value of $P_{abs,avg}$, denoted by $E(P_{abs,avg})$, is defined as $E(P_{abs,avg}) = -c_{PTO} \sum_{k=1}^{N} S_{\dot{\varepsilon}\dot{\varepsilon}}(\omega_k)\Delta\omega$ [22], where $S_{\dot{\varepsilon}\dot{\varepsilon}}$ refers to the power spectral density function of the DoF related to power extraction. It is worth noticing that, having considered the power output a Gaussian process, $E(P_{abs,avg})$ is also the mean value $\mu_{P_{abs,avg}}$ of the distribution.

Results and conclusions

Based upon the linear model presented in the previous section, the dynamic of the system can be studied directly at the PTO DoF applying the impedance matching theory introduced in [20]. Let us recall the principle of maximum power transfer that is obtained applying a control action equal to the inverse of the impedance of the equivalent circuit related to the model stated in Eq (2).

Based upon the device characteristic, a sensitivity analysis of the precession axis equivalent force is performed, varying both the flywheel speed and the wave direction. The analysis is computed considering an irregular input wave, with a unitary amplitude and varying the energetic period of the input wave, selected between 3s and 10s. Figure 2 presents an appraisal of the equivalent force at the PTO axis $\hat{F}_{w,swingo}$ varying the flywheel angular velocity. The extremum scenario are considered, where the wave impacts the floater at 0° and 90°, with a third intermediate operating condition of 45°. We can notice that the gyropendulum mechanism, mounted in the SWINGO device, allows the increment of the total force acting on the power take off DoF with respect to the gyroscope mechanism. Such an improvement ranges from the 70% to the 10% considering the overall set of operating condition. Moreover, we notice that the greatest improvement happens when the flywheel velocity $\dot{\varphi} = 0$, then the percentage gradually decreases, increasing the flywheel velocity. We state that even if the gyropendulum system can improve the equivalent force acting on the PTO DoF, for a wide set of operating condition, the highest advantages set up in term of input force is achieved for zero velocity of the flywheel.

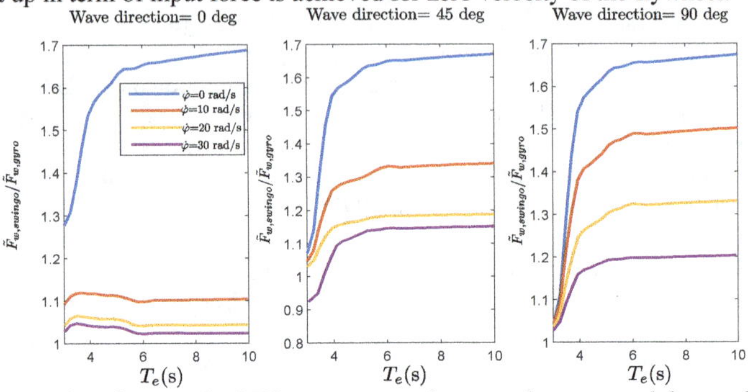

Figure 2 Equivalent force at the PTO: comparison between the gyropendulum and simple gyroscope.

In conclusion, in this work we have introduced a new concept of wave energy converter based on both pendulum and gyroscopic device. The resulting mechanism is the gyropendulum system, that because of its characteristic can be activated with respect to the ε-axis independently from the incoming wave direction. Such a device has the peculiarity of an eccentric flywheel mounted at a distance l_f from the precession axis. Applying the impedance matching principle, the dynamic can be mapped to the PTO axis and then the equivalent system depends on an equivalent force $\widetilde{F_w}$ and impedance $I(j\omega)$. Then we have demonstrated that the gyropendulum system allows to project more input force on the PTO axis than a simple gyroscopic configuration. Such an improvement of input power ranges from 70 to 20%, accordingly to the flywheel speed.

References

[1] European Commission, The European Green Deal, EUR-Lex – EN, COM/2019/640, 2019.

[2] European Commission, An EU strategy to harness the potential of offshore renewable energy (ORE) for a climate neutral future, EUR-Lex - 52020DC0741 – EN, COM/2020/741, 2020.

[3] Ghigo, A., Cottura, L., Caradonna, R., Bracco, G., Mattiazzo, G.: Platform Optimization and Cost Analysis in a Floating Offshore Wind Farm. Jour. of Mar. Sci. and Eng. 2020, 8, 835. https://doi.org/10.3390/jmse8110835

[4] Kardakaris, K., Boufidi, I., Soukissian, T.: Offshore Wind and Wave Energy Complementarity in the Greek Seas Based on ERA5 Data. Atmosphere 2021, Vol. 12, Page 1360, 1360 (2021). https://doi.org/10.3390/atmos12101360

[5] Fenu, B., Attanasio, V., Casalone, P., Novo, R., Cervelli, G., Bonfanti, M., Sirigu, S.A., Bracco, G., Mattiazzo, G.: Analysis of a Gyroscopic-Stabilized Floating Offshore Hybrid Wind-Wave Platform. Journal of Marine Science and Eng., 8, pp. 439 (2020). https://doi.org/10.3390/jmse8060439

[6] Amann, K.U., Magaña, M.E., Sawodny, O.: Model Predictive Control of a Nonlinear 2-Body Point Absorber Wave Energy Converter With Estimated State Feedback. IEEE Transactions on Sustainable Energy 6(2), 336–345 (2015). https://doi.org/10.1109/TSTE.2014.2372059

[7] Parrinello, L., Dafnakis, P., Pasta, E., Bracco, G., Naseradinmousavi, P., Mattiazzo, G., Bhalla, A.P.S.: An adaptive and energy-maximizing control optimization of wave energy converters using an extremum-seeking approach. Physics of Fluids 32, (2020). https://doi.org/10.1063/5.0028500

[8] Giorgi, G., Faedo, N.: Performance enhancement of a vibration energy harvester via harmonic time-varying damping: A pseudospectral-based approach. Mechanical Systems and Signal Processing 165, 108331 (2022). https://doi.org/10.1016/j.ymssp.2021.108331

[9] Bonfanti, M., Hillis, A., Sirigu, S.A., Dafnakis, P., Bracco, G., Mattiazzo, G., Plummer, A.: Real-Time Wave Excitation Forces Estimation: An Application on the ISWEC Device. Journal of Marine Science and Engineering 2020, 8, pp. 825 (2020). https://doi.org/10.3390/jmse8100825

[10] Li, G.,Weiss, G., Mueller, M., Townley, S., Belmont, M.R.:Wave energy converter control by wave prediction and dynamic programming. Renewable Energy 48, 392–403 (2012). https://doi.org/10.1016/j.renene.2012.05.003

[11] Dafnakis, P., Bhalla, A.P.S., Sirigu, S.A., Bonfanti, M., Bracco, G., Mattiazzo, G.: Comparison of wave–structure interaction dynamics of a submerged cylindrical point absorber with three degrees of freedom using potential flow and computational fluid dynamics models. Physics of Fluids 32(9), 093307 (2020). https://doi.org/10.1063/5.0022401

[12] Fontana, M., Casalone, P., Sirigu, S.A., Giorgi, G., Bracco, G., Mattiazzo, G.: Viscous Damping Identification for a Wave Energy Converter Using CFD-URANS Simulations. Journal of Marine Science and Engineering, 355, 8(5) (2020). https://doi.org/10.3390/jmse8050355

[13] Giorgi, G., Sirigu, S., Bonfanti, M., Bracco, G., Mattiazzo, G.: Fast nonlinear Froude–Krylov force calculation for prismatic floating platforms: a wave energy conversion application case. Jour. of Ocean Eng. and Mar. En. 7(4), 439–457 (2021). https://doi.org/10.1007/s40722-021-00212-z

[14] Casalone, P., Dell'Edera, O., Fenu, B., Giorgi, G., Sirigu, S.A., Mattiazzo, G.: Unsteady RANS CFD Simulations of Sailboat's Hull and Comparison with Full-Scale Test. Journal of Marine Science and Engineering 2020, 8 (6), pp. 394 (2020). https://doi.org/10.3390/jmse8060394

[15] Faedo, N., Peña-Sanchez, Y., Carapellese, F., Mattiazzo, G., Ringwood, J.V.: LMI-based passivation of LTI systems with application to marine structures. IET Renewable Power Generation 15(14), 3424–3433 (2021). https://doi.org/10.1049/rpg2.12286

[16] Habib, G., Giorgi, G., Davidson, J.:Coexisting attractors in floating body dynamics undergoing parametric resonance. Acta Mec. 2351-2367(2022). https://doi.org/10.1007/s00707-022-03225-3

[17] Giorgi, G., Davidson, J., Habib, G., Bracco, G., Mattiazzo, G., Kalmár-Nagy, T.: Nonlinear Dynamic and Kinematic Model of a Spar-Buoy: Parametric Resonance and Yaw Numerical Instability. Jour. of Mar. Sci. and Eng., 8 (7), pp 504, (2020). https://doi.org/10.3390/jmse8070504

[18] Faedo, N., Dores Piuma, F.J., Giorgi, G., Ringwood, J.V.: Nonlinear model reduction for wave energy systems: a moment-matching-based approach. Nonlinear Dynamics 102(3), 1215–1237 (2020). https://doi.org/10.1007/s11071-020-06028-0

[19] Ermakov, A., Ringwood, J.V.: Rotors for wave energy conversion—Practice and possibilities. IET Renewable Power Generation 15(14), 3091–3108 (2021). https://doi.org/10.1049/rpg2.12192

[20] Faedo, N., Carapellese, F., Pasta, E., Mattiazzo, G.: On the principle of impedance-matching for underactuated wave energy harvesting systems. Applied Ocean Research 118, 102958 (2022). https://doi.org/10.1016/j.apor.2021.102958

[21] Amini, E., Golbaz, D., Amini, F., Nezhad, M.M., Neshat, M., Garcia, D.A.: A parametric study of wave energy converter layouts in real wave models. Energies 13(22), 1–23 (2020). https://doi.org/10.3390/en13226095

[22] Mérigaud, A., Ringwood, J.V.: Free-Surface Time-Series Generation for Wave Energy Applications. IEEE Journal of Oceanic Engineering 43(1), 19–35 (2018). DOI 10.1109/JOE.2017.2691199. https://doi.org/10.1109/JOE.2017.2691199

Theoretical and Applied Mechanics - AIMETA 2022 | Materials Research Forum LLC
Materials Research Proceedings **26** (2023) 685-690 | https://doi.org/10.21741/9781644902431-110

On the manipulation of the magnetic forces for improving the contactless plucking in piezoelectric vibration energy harvesters

Michele Rosso[1,a*], Filippo Pietro Perli[1,b], Alberto Corigliano[1,c] and Raffaele Ardito[1,d]

[1]Department of Civil and Environmental Engineering, Politecnico di Milano, Piazza Leonardo da Vinci 32, 20133 Milano, Italy

[a]michele.rosso@polimi.it, [b]filippopietro.perli@mail.polimi.it, [c]alberto.corigliano@polimi.it, [d]raffaele.ardito@polimi.it

Keywords: Piezoelectric Vibration Energy Harvesting, Magnetic Plucking, Nonlinear Dynamics, Advanced Frequency Up-Conversion

Abstract. We propose a shielding technique for the magnetic field localization around permanent magnets (PMs) for sharpening the magnetic force-displacement curve in the frequency up-conversion (FuC) of piezoelectric vibration energy harvesters (PVEHs). We present a concept with theoretical formulation, computational analyses, and experimental validation that confirms the supposed principle. The numerical study on a PVEH shows that with the shielding, for varying actuation velocity, high values of peak power (50 mW – 150 mW) can be reached at low speeds (e.g. 0.5 m/s – 2.6 m/s). The unshielded device exhibits a good behavior for velocities over 3 m/s. This result makes the technique useful for the design of FuC mechanisms depending to its operating velocity.

Introduction

The manipulation of the magnetic field has been largely studied in the last decades because of immediate practical applications (e.g. electromagnetic induction heat treatments). The basic idea to manipulate the magnetic field is to exploit a magnetic flux concentrator (MFC) which is a piece of soft ferrous material with high magnetic permeability. As a consequence, the field lines are forced to follow a specific path in the surrounding space and they remain free to expand where there is not high permeability as shown in Fig. 1a.

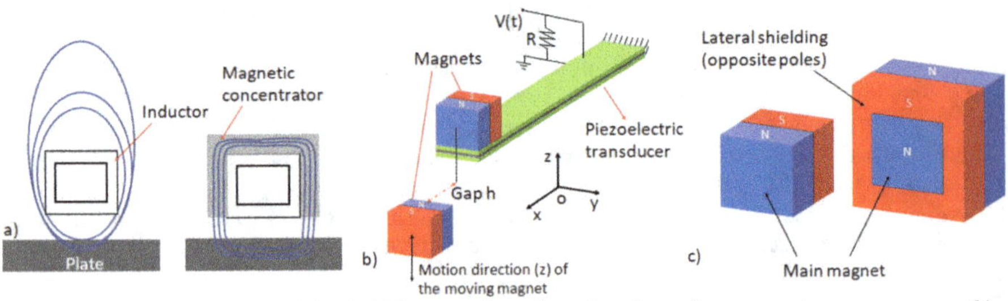

Figure 1. a) Illustration of the field lines path with and without the magnetic concentrator, b) proposed shielding layout of a PM with indication of the poles, c) PVEH with magnetic FuC

The approach can be used to control rapid temperature cycling (RTC) technologies as made by Mrozek et al [1], or to improve the gap-to-gap induction heating as proposed by Wen et al [2]. In the micro-electro-mechanical systems (MEMS) industry the MFCs are used to increase the sensitivity in Hall-effect based transducers [3]. The magnetic field can be also manipulated with high-rate triode sputtering of patterned thin films as made by Chigirinsky et al [4]. In this work we propose a technique of concentration with an innovative application on the contactless frequency up-conversion in PVEHs. The motivation is the fact that it is not always possible to have a real impulsive phenomenon [5] in plucking of transducers. To have an efficient mechanism also at a very low frequency or input velocities such in case of the human motion (e.g. 0.1-5 Hz), it is possible to tune the shape of the magnetic force by means of additional PMs.

Shielding technique

The magnetic plucking mechanism, as proposed in [6], is composed of a piezoelectric transducer equipped with a PM that magnetically interacts with another dynamical systems also equipped with a PM, Fig. 1b. In this framework the goal is the sharpening of the magnetic force with respect to the relative distance between them. The idea is to capture the field lines with additional PMs with an inverse poles orientation as indicated in Fig. 1c. The shielding material is put only on the sides of the central (main) magnet that contain both the poles (4 over 6 faces in a parallelepiped case). Both the main magnet and the material around it are made of Neodymium-Boron-Iron alloy (NdFeB). An important point is that the technique remains conceptually valid if either both or only one of the involved main magnets are shielded.

Mathematical modeling

In this section we present the modelling of the magnetostatics problem and the PVEH reduced order model.

Magnetostatics problem

In absence of electrical current, the first Maxwell's equation [7] in a domain Ω is:

$$\nabla \times \mathbf{H}=\mathbf{0} \qquad \text{in } \Omega \tag{1}$$

where **H** is the magnetic field density vector. The problem is conservative, and it is possible to define a scalar magnetic potential φ_M inside the domain:

$$\mathbf{H}=-\nabla\varphi_M \qquad \text{in } \Omega \tag{2}$$

Then we assume a linear constitutive law between the magnetic field **H** and the magnetic flux density **B** also by considering the magnetization vector **M** of the PM:

$$\mathbf{B}=\mu_0(\mathbf{H}+\mathbf{M}) \qquad \text{in } \Omega \tag{3}$$

where μ_0 is the permeability of the vacuum equal to 4πe-7 N/A^2. It is then possible to write an equation of magnetic flux conservation by means of the Gauss' Law:

$$\nabla\cdot\mathbf{B}=\mathbf{0} \qquad \text{in } \Omega \tag{4}$$

Eqs. (1)-(4) must be considered together with the boundary conditions (BCs). A Neumann BC, also called insulation equation, can be put on symmetry planes where the magnetic field is tangential to the plane:

$$\mathbf{n}\cdot\mathbf{B}=0 \qquad \text{on} \qquad \partial\Omega_{ins} \tag{5}$$

where **n** is the normal outward unit vector and $\partial\Omega_{ins}$ is the insulated surface. In case the magnetic field is orthogonal to the boundary, a constant value of the magnetic potential leads to impose a Dirichlet BC:

$$\varphi_M=\bar{\varphi} \qquad \text{on} \qquad \partial\Omega_{pot} \tag{6}$$

where $\bar{\varphi}$ is the prescribed magnetic potential value at the boundary $\partial\Omega_{pot}$. The magnetic force is then computed by integrating over the surface of the permanent magnet the Maxwell's stress tensor T.

$$\mathbf{F}_{MAG} = \int_{\partial\Omega_{MAG}} \mathbf{n}\, T\, dS = -\frac{1}{2}\int_{\partial\Omega_{MAG}} [\mathbf{n}(\mathbf{H}\cdot\mathbf{B}) + (\mathbf{n}\cdot\mathbf{H})\mathbf{B}^T]\, dS \qquad \text{on} \qquad \partial\Omega_{MAG} \tag{7}$$

The problem is then solved via the Finite Element Method (FEM) through COMSOL Multiphysics® with serendipity quadratic FEs.

Piezoelectric vibration energy harvester

The typical PVEH is a layered cantilever beam as represented in Fig. 1b. In this work we assume linear behavior of the material [8,9]. The modelling of the beam is carried out by considering only the so called 31-mode and through a lumped-parameter approach [10] with one degree-of-freedom (dof) both for the displacement field and the voltage. The electromechanical equations of motion are derived by using the Euler-Lagrange equations obtaining the following differential system that has been implemented in a MATLAB© code:

$$\begin{cases} m\ddot{W}(t) + c_m\dot{W}(t) + k_L W(t) - \theta V(t) = f_{ext} \\ C\,V(t) + \theta\ddot{W}(t) + V(t)/R = 0 \end{cases} \tag{9}$$

where W is the tip displacement of the cantilever, V the voltage output from the piezo. m is the inertial term, c_m the linear damping coefficient, k_L the linear stiffness, C the capacitance of the piezo and θ the piezoelectric coupling coefficient. f_{ext} is the forcing function, and it includes the magnetic force.

Simulation

Magnetic force

To numerically investigate the shielding technique, we simulate the magnetic force between two magnetic systems as depicted Fig. 1c, in repulsive configuration with both or only one shielded magnets. The main magnet is a cube with a side length of 3 mm and a magnetization **M** equal to 1.32 T. The cover is realized with the same material and magnetization. In the case in which both magnets are shielded, we consider two values as cover thickness (0.5 mm, 1.0 mm) and four values of gap distance h (0.5, 1.0, 1.5, 2.0 mm). In the case in which only one of the cubic PMs is shielded, we consider the cover thickness of 1 mm and two values of gap distance (1.0 and 1.5 mm). We compare all the cases with the layout without shielding. According to the reference system of Fig. 1b we are interested in the z-component of the magnetic force with respect to the relative distance on the same axis. All the plots in the Figs. 2 and 3 show that, using the shielding on one or both PMs, it is possible to tighten the force-distance curves. The cover induces an inversion in sign in such way that the gradient of the force is modified, and the sharpening is the result.

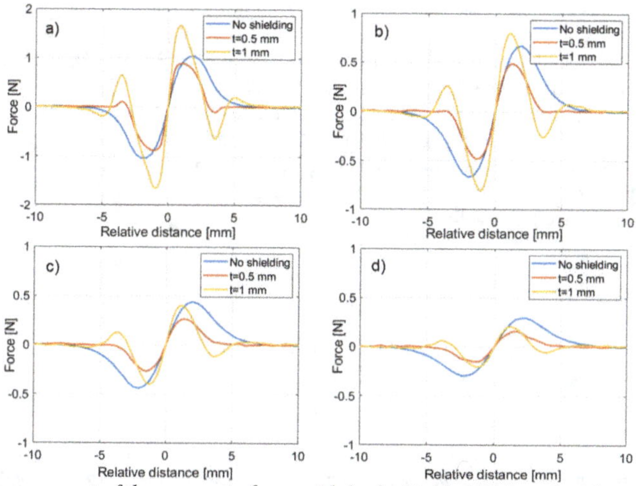

Figure 2. z-component of the magnetic force with both shielded PMs and without shielding for different gaps: a) 0.5 mm, b) 1.0 mm, c) 1.5 mm d) 2.0 mm

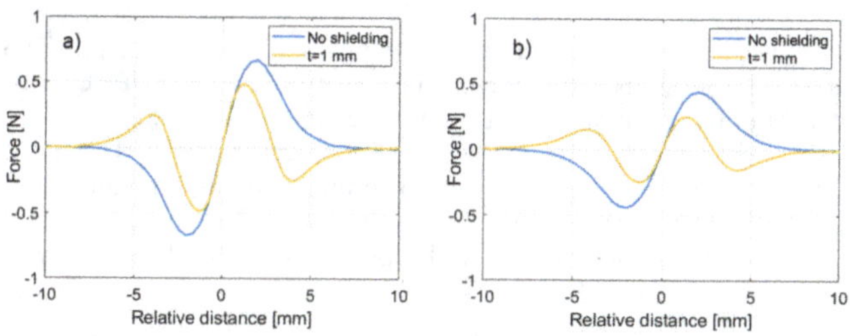

Figure 3. z-component of the magnetic force with only one shielded PM and without shielding for different gaps: a) 1.0 mm, b) 1.5 mm

Magnetically plucked piezoelectric energy harvester

In this section we use the computed magnetic force to analyze the plucking mechanism on a PVEH. The data of the cantilever are summarized in the Table 1.

Material	ρ [kg/m³]	E [GPa]	n [-]	d_{31} [pC/N]	ε_{33}^s[-]	t[μm]	Width [mm]	Length [mm]
Titanium	4500	115	0.3	-	-	65	1.5	15
PZT	7500	60	0.3	212	2000	280 per layer (series)	1.5	15

Table 1. Physical parameters and geometry of the bimorph

We simulate the response of the cantilever by solving the differential system of Eq. (9) by using the presented magnetic forces in repulsive configuration as f_{ext}. We look at the peak of power output for varying velocity of the interaction in the range 0.1-10 m/s (i.e. velocity of the moving magnet in Fig. 1c supposed at constant velocity) with a value of the resistor $R = 100$ kΩ. The instantaneous power is computed trough the Joule's Law:

$$P = V^2/R \tag{10}$$

To compare two oscillators working at the same frequency of 366.06 Hz, we use in the layout without shielding two magnets with dimensions 5 x 5 x 3 mm^3 and in the case with the shielding a cubic main magnet with a length side of 3 mm and a cover thickness of 1 mm to get the same global dimensions. The results in Fig. 3b show that the adoption of shielding induces a huge increment of peak power in the low-speed range (0.1-2.6 m/s), that is typical for the human movements. On the contrary, the case without shielding is connected with the absolute maximum power, reached for quite large velocity (4.5 m/s).

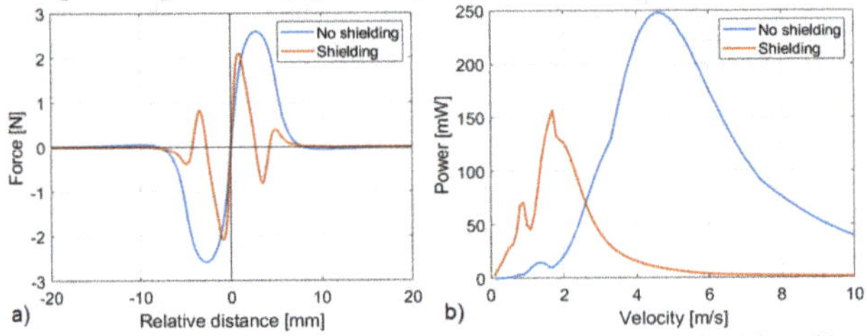

Figure 4. a) z-component of the magnetic force with and without shielding b) peak power for different velocities of the seismic system with a resistor $R = 100k\Omega$

Experimental validation

We present in this section an experimental validation of the concept in case only one of the couple of interacting PMs is shielded for gap values of 1.0 mm and 1.5 mm. The details of the setup are summarized in [11]. The supposed principle is confirmed because by considering only the experimental curves in the plots of Fig.5, the shielded layout (S) shows a sharper behavior with respect the unshielded case (NS) in terms of force-distance curve. On the other side, the comparison between numerical results and experiments with shielding (S) shows some discrepancies in the inversion of sign which may be the subject of further studies and works. Anyway, the physics of the phenomenon is well captured as well as the peak force.

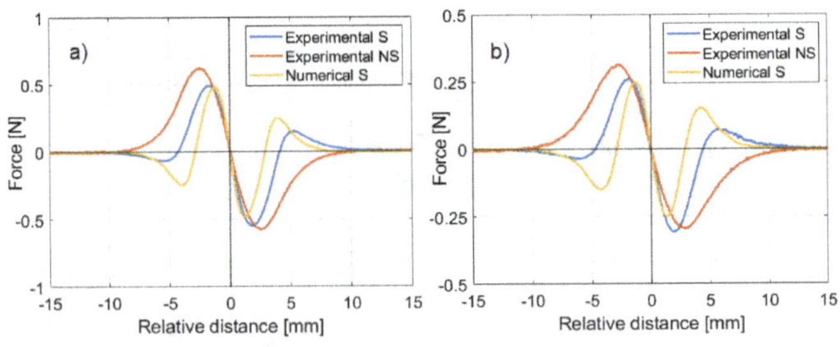

Figure 5. z-component of the magnetic force for the case of a) gap 1.0 mm and b) 1.5 mm. NS: no shielding, S: with shielding

Conclusions

In this work we studied a strategy to sharpen the force-displacement magnetic curve between PMs through a specific arrangement of the magnetization of additional polarized ferromagnetic material to a main magnet. The aim is to realize a dynamical load as impulsive as possible for PVEH systems. We proposed a concept, a theoretical formulation, and a computational study together with an experimental validation that confirms the principle. The numerical investigation applied to a PVEH case study shows that the technique can be used to set the performance of the harvester depending on the velocity of the magnetic interaction. The shielded harvester obtains a larger amount of power (until 150 mW) than the unshielded one in a very low speed range (0.1-2.6 m/s). The technique can be also applied in a more general context of the mechanics (e.g. actuation and sensing) and this opens new motivations to improve the modelling and to implement this technique also at the MEMS scale.

References

[1] K. Mrozek, P. Muszyński, P. Poszwa, Application of Magnetic Concentrator for Improvement in Rapid Temperature Cycling Technology, Polymers 13 (2021), 91. https://doi.org/10.3390/polym13010091

[2] H. Wen, X. Zhang, H. Ye, Y. Han, Research on the mechanism of magnetic flux concentrator in the gap-to-gap induction heating of wing power gear, International Journal of Thermal Sciences, 168 (2021) https://doi.org/10.1016/j.ijthermalsci.2021.107055

[3] Ed. Ramsden, Hall-Effect Sensors, second edition, Newnes, Burlington, MA, USA 2006. https://doi.org/10.1016/B978-075067934-3/50002-8

[4] S. Chigirinsky, M. Kustov, N. Dempsey, C. Ndao, R. Grechishkin, Calculations and measurements of the magnetic field of patterned permanent magnetic films for lab-on-chip applications. Rev. Adv. Mat. Sci. (2009) 85-91.

[5] P. Janphuang, R. A. Lockhart, D. Isarakorn, S. Henein, D. Briand, and Nico F. de Rooij, Harvesting energy from a rotating gear using an AFM-like MEMS piezoelectric frequency up-converting energy harvester" Journal of Microelectromechanical Systems, 24 (2015), 742-754. https://doi.org/10.1109/JMEMS.2014.2349794

[6] P Pillatsch, E M Yeatman and A S Holmes, Magnetic plucking of piezoelectric beams for frequency up-converting energy harvesters, Smart Mater. Struct. 23 (2014) 025009 (12pp). https://doi.org/10.1088/0964-1726/23/2/025009

[7] J. Jin, The Finite Element Method in Electromagnetics, second edition, John Wiley & Sons, New York, USA 2002.

[8] A. Erturk, D.J. Inman, Piezoelectric Energy Harvesting, John Wiley & Sons Ltd, New York, USA (2011). https://doi.org/10.1002/9781119991151

[9] Standards Committee of the IEEE Ultrasonics, Ferroelectrics, and Frequency Control Society. IEEE, Standard on Piezoelectricity, IEEE, New York (1987)

[10] R. Ardito, A. Corigliano, G. Gafforelli, C. Valzasina, F. Procopio and R. Zafalon, Advanced model for fast assessment of piezoelectric micro energy harvesters, Front. Mater., 19 April 2016. https://doi.org/10.3389/fmats.2016.00017

[11] M. Rosso, A. Corigliano, R. Ardito, Numerical and Experimental Evaluation of the Magnetic Interaction for Piezoelectric Vibration Energy Harvesters, Meccanica 2022, 57, 1139–1154. https://doi.org/10.1007/s11012-022-01481-0

Theoretical and Applied Mechanics - AIMETA 2022 Materials Research Forum LLC
Materials Research Proceedings **26** (2023) 691-696 https://doi.org/10.21741/9781644902431-111

First offshore windfarm in the Mediterranean Sea - Italy

Luigi Severini[1, a *], Alessandro Severini[1,b], Sara Bray[3,c] and Simona Capozza[1,d]

[1] via Solito 85, Taranto, Italy

[a] manager@studioseverini.eu, [b] a.severini@studioseverini.eu, [c] s.bray@nicetechnology.eu, [d] s.capozza@studioseverini.eu

Keywords: Offshore Wind Farms, Wind Turbines, Mediterranean Sea

Abstract. The global demand for renewable energy is expanding in the attempt to mitigate the impacts of climate change. Offshore wind energy is, alongside with other "green" forms of energy, at the core of seeking carbon-neutrality of various countries. Based on the International Energy Agency, offshore wind energy could become the main source of power generation in Europe by 2042 [1]. Consequently, interest in offshore wind power is on the rise. By 2020, already 25 GW had been installed in Europe, and according to the latest projections by Wind Europe, 450 GW could be deployed by 2050 [2]. The article will present in detail the project of the offshore wind farm built a few months ago near the Taranto harbor. The project was conceived and carried out entirely by the iLStudio and Nicetechnology srl. In particular, not only the design phases but also the bureaucratic and planning process of the plant will be described, highlighting how long it took to get to its complete realization (14 years). At present we have become pioneers in the development of this technology.

Introduction

Climate change is the challenge of our time. Carbon dioxide levels in the atmosphere are higher than they have been for the last 800.000 years and are increasing, with an almost linear correlation with the average increase in our planet's surface temperature.

The solution clearly passes through the decarbonization process: the energy transition towards a massive use of renewable energy sources is underway, but further progress is needed to trigger positive effects on climate change and at the same time satisfy the growing energy demand.

The EU Strategy [3] for offshore renewable energy has set as a goal the installation of a capacity of 60 GW of wind energy by 2030 and 340 GW by 2050. Directive 2018/2001/EU (RED II) [4] provides for the gross consumption of renewable energy to be equal to (at least) 32% of the energy mix. According to statistics published by the World Wind Energy Association (WWEA) [5], the overall power generation capacity through wind farms currently exceeds 840 GW.

Considering that for 60% of the population wind power represents the most convenient option for producing electricity, the importance of this sector for the energy transition and sustainable development is undeniable.

In this context, building wind farms that are efficient and respectful of the environment represents a fundamental challenge for the near future.

Making wind farms sustainable means not only eliminating harmful emissions, but also protecting biodiversity and natural habitats.

For what concerns offshore wind sector the PNIEC (Piano Nazionale Integrato per l'Energia e il Clima) [6] sets a target of 900 MW for 2030.

Currently, the only existing offshore wind farm in the Mediterranean Sea is the one built in the external harbour of Taranto. The idea of exploiting the wind resource offered by the Mediterranean Sea was born from the intuition of the Eng. Luigi Severini. It should be noted that over 20 offshore wind farm projects were rejected and did not reach the end of the procedural process.

Materials Research Proceedings **26** (2023) 691-696 https://doi.org/10.21741/9781644902431-111

The offshore wind farm consists of 10 turbines of 3 MW each, for a total nominal power of 30 MW.

Following, Chapter 2 will describe in detail the Taranto offshore wind farm project, with particular reference to the geotechnical and structural characteristics of the windfarm. Chapter 3 will describe the bureaucratic process that led to the final construction of the park and, in the end, the conclusions and future developments in this sector.

Taranto Offshore wind farm project: design process

The Taranto wind farm was inaugurated on 21 April 2022.

Located in Italy's Apulia Region near the Taranto harbour, the 30 MW offshore wind farm will provide about 58.000 MWh of electricity per year, equal to the annual requirement of 60 thousand people. In environmental terms, it means that, over the 25-year life, it will save approximately 730.000 tons of CO_2.

Figure 1 Taranto Offshore Wind Farm [ph.: Renexia]

The windfarm layout creates a design conforming to the site's current infrastructure templates.

The windfarm is divided into two subgroups: the first group of turbines is located at a distance of about 300 m from the coast and is parallel the multi-sector pier, while the second one is positioned behind the breakwater and has a triangular layout (Fig. °1). This area was selected on the basis of careful studies, in consideration of the available wind resource, the presence of regulatory, urban and environmental constraints, the distance from the coastline, the characteristics and depth of the seabed and the possibility of connection to the national electricity grid.

The height of the wind towers from the center of the rotor to mid-sea level is about 100 m; the submerged part of the tower varies from 4 m to 18 m; the foundation reaches a depth of about 30 m from the seabed.

The energy produced by each low voltage wind turbine is transformed to 33 kV by the transformer present in the turbine itself and transported to the base of the tower by internal cables. Subsequently, the energy is transported through submarine cables to the junction box which connects it to the onshore cables and then to the transformer substation, located onshore, that transforms the produced energy at the distribution voltage of 150 kV (Fig. °2). Electricity production will be used to fuel all port activities as well, allowing their energy transition.

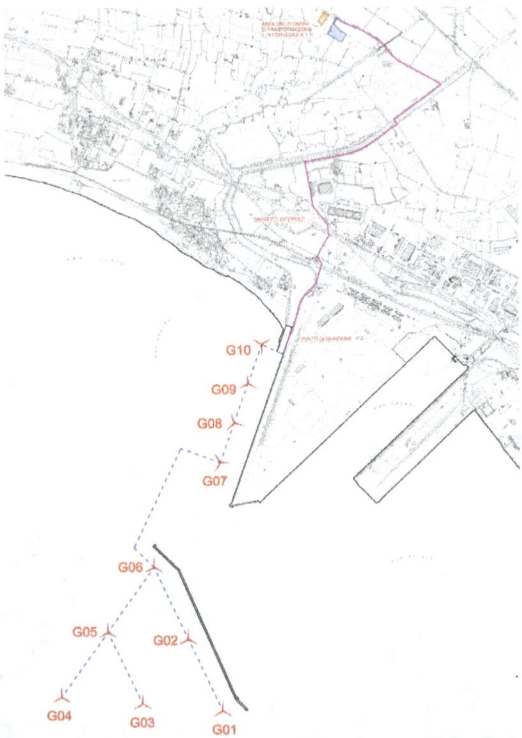

Figure 2 Taranto offshore windfarm layout [iLStudio]

During all the design stages, there were many challenges to be addressed and overcome. First of all, the design of monopile foundations embedded into an over-consolidated clay (grey-blue clay). This type of soil is still not well investigated for what concern the soil-foundation interaction.

A geognostic campaign and laboratory geotechnical analysis were carried out in order to obtain geophysical, geomorphological and geotechnical characterization of the site.

Consequently, many national/international standards had been followed and many FEM analyses had been carried out regarding the most suitable methodology for the monopiles design (i.e. p-y curve method [7]). All analyzes and software used were in source. Besides, it was designed the direct coupling [8] between the monopile flange and the wind turbine flange without using the transition piece.

Further issues concerned the design of the secondary steel (service platform, boat landing, etc.) and the electrical cables installation which had to be adapted to the innovative configuration of entire structure.

The foundations of the wind turbines consist of steel monopiles. All the monopiles have an external diameter of 4,5 m and a plate thickness ranging from a minimum of 50 mm at toe to a maximum of 85 mm. The monopiles were installed with a suitable hydraulic hammer.

The service platform and related accessory structures (stairs, boat landing, etc.) were also designed by performing finite element simulations using SOLIDWORKS. Then, the platform and the other structures were installed on the top of the monopile. The entire installation process was carried out by Van Oord with Van Oord's offshore installation vessel MPI Resolution with a variety of equipment on board including a hammer, sea fastening, upend tool and lifting equipment.

The dimensioning of the foundations was carried out considering the geotechnical data obtained from the geognostic campaigns carried out on site.

The design loads relating to the wind turbine were supplied by the turbine manufacturer (MingYang Smart Energy), who assessed, for the tower - foundation assembly, the loads acting according to the IEC 61400 deriving from the operating conditions, from the maximum design wave, from currents, from impact with ships and from possible seismic interactions. The calculation was performed with reference to the Technical Standards for Construction NTC 2018 (D.M. January 17, 2018) and the International Standards IEC 61400.

The numerical analyzes for the pile-soil interaction were processed with MIDAS GTS NX a geotechnical software while the mechanical structural analyzes of the monopiles and secondary steel were processed with finite element software (FEM) SOLIDWORKS (v.2017).

The design and verification of foundations as well as the management of the entire project over all these years has been carried on by iLStudio with the support of Nicetechnology s.r.l.

An overview of the entire bureaucratic and management process that led, after more than ten years, to the construction of the first offshore wind farm in the Mediterranean is presented below.

Taranto offshore wind farm project: bureaucratic and managerial overview

After 14 years of waiting and various vicissitudes (corporate, industrial, administrative and judicial vicissitudes in front of the TAR (Regional Administrative Court)), Taranto inaugurated the first Italian offshore wind farm on 21 April. This project was done by Renexia, a company of Toto holding. The investment is 82 million euros.

The offshore wind farm will supply not only energy to the local population but also to the port of Taranto. An agreement already signed provides for the sale of at least 10% of the energy produced for a quantity of not less than 220 MWh per year. But this park also aims to supply, through electrolysis, green hydrogen to the former Ilva steel plant and to the Eni refinery.

The 14 years it took the Taranto project to complete were also an opportunity to reiterate, once again, that investments in renewable energy sources must be accelerated if you don't want the energy transition to stay on paper.

It represents the first example of offshore wind farm fully endorsed by the Italian Institutions and its realization allows the national and local industrial system to acquire an important primacy of knowledge and skills about the construction of such installations.

The bureaucratic procedure for obtaining the State of Property Concession involved a considerable amount of time during which purely administrative issues are intertwined with purely economic-design issues. A brief description of the entire procedure is shown below:

The State Property Concession of the project area has been formally requested to the Port Authority on 2008. After this date, a period of eight years has taken place during which further authorizations have been requested for the final grant of the Concession.

The required Authorization are as follows:

- The Environmental Compliance Decree, with VIA / VAS procedure, was issued by the Minister of Environment in agreement with Minister of Cultural Heritage and Activities on 24.07.2012;
- The Single Authorization for the construction and operation of the plant was issued by the Ministry of Infrastructure and Transport by Order No. 1 of 27.06.2013.

- After obtaining the previous Authorizations, the State of Property Concession has been granted on 03/11/2016.
- The favorable opinion expressed by MIBAC (Ministry of Cultural Heritage and Activities and Tourism) is also peculiar: "... it is possible to share the methodological setting of the project ... structures should not compromise the recognition elements of sites but they have to introduce new perceptual values through non-random projects capable of introducing new forms of spatial relationship with precise alignments and compositional devices. ... the localization of wind turbines ... capable of creating an evocative layout ... is not definable as the pathological alteration of existing visuals ...At this project goes the merit of having correctly identified the maximum number of site-compatible wind turbines ... unlike how often it can be found in other offshore wind power projects ... out of any relationship with the existing ....In this particular case, the interference generated by the new wind farm is not dissonant with respect to the present condition of the landscaping interest of the area, whose values are still preserved. "

Furthermore, during these years the supplier of the turbines, previously appointed, has been replaced (due to the insolvency of the company in 2019). This led to additional times, before the actual construction of the windfarm, resulting from the choice of a new supplier by proceeding with a specific request for a variant to the Ministry of the Environment (for dimensional changes of the new turbines).

After having obtained a positive opinion from the Ministry of the Environment, it was possible to continue with the management of the project from the logistical / implementation point of view.

The case of Taranto is unfortunately only the tip of the iceberg. If we want to close all fossil fuel power plants by 2035, we must push on the development of renewables in Italy, build many wind farms and we must overcome the no of the Superintendencies, Regions, local administrations, local committees and environmental associations.

Conclusions

Wind-based renewable energy sources represent one of the key technologies for economic recovery and the creation of millions of jobs, towards a fully sustainable future.

Studio Severini with the support of Nicetechnology s.r.l. has managed and fully carried out all the activities necessary for the complete realization of the of the first offshore wind farm in the Mediterranean Sea: bureaucratic procedure, foundations design, power transmission grid design, structural and geotechnical analysis, development of innovative design solutions for secondary steel, material research and contact with suppliers.

Every challenge was overcome in order to create an offshore wind power plant that could be integrated in the best possible way within the surrounding environment, which is already heavily harmed by industrial pollution.

It is a symbolic case of the red tape in Italy. The project proposal was presented in 2008 (14 years ago). The Superintendence succeeded in giving a negative opinion due to the visual impact generated (in front of it there are the chimneys of the former Ilva, the Eni refinery, the cement factory and the cranes of the industrial port).

We need to increase the installation speed of renewable energy plants tenfold - primarily wind by 2025. The Planet can no longer wait.

Wind farms make it possible to make great strides towards the definitive exit from dependence on fossil fuels, without causing impacts on the environment and biodiversity.

Future aims will be to investigate in detail the most promising areas and promote the developing new projects in selected sites in Italy (ex. floating offshore windfarm). The objective is to become pioneers in the national offshore wind sector, opening the way for the Italian growth of this type of technology.

References

[1] Freeman K, Frost C, Hundleby G, Roberts A, Valpy B, Holttinen H, et al. Our energy, our future. How offshore wind will help Europe go carbon neutral. WindEurope, 2019.

[2] Ramirez L, Fraile D, Brindley G. Offshore wind in Europe, key trends and statistics 2020, 2020.

[3] European Commission, "An EU Strategy to harness the potential of offshore renewable energy for a climate neutral future", Bruxelles, 19.11.2020, page 2, 2020.

[4] European Parliament, Council of the European Union, "DIRECTIVE (EU) 2018/2001 OF THE EUROPEAN PARLIAMENT AND OF THE COUNCIL of 11 December 2018 on the promotion of the use of energy from renewable sources", Bruxelles, 11/12/2018, page 2, 2018.

[5] https://wwindea.org/world-market-for-wind-power-saw-another-record-year-in-2021-973-gigawatt-of-new-capacity-added/

[6] Ministero dello Sviluppo Economico, Ministero dell'Ambiente e della Tutela del Territorio e del Mare, Ministero delle Infrastrutture e dei Trasporti, "PIANO NAZIONALE INTEGRATO PER L'ENERGIA E IL CLIMA", Italia, 12/2019, page 57, 2019.

[7] Reese, L., Cox, W., and Koop, F., "Field testing and analysis of laterally loaded piles in stiff clays" in Proc. 7th Offshore Technology Conference, Houston, TX., Volume 2, OTC 2312, pp. 671–690, (1975). https://doi.org/10.4043/2312-MS

[8] Severini, A., "SISTEMA DI CORREZIONE AD ALTA PRECISIONE DELLA VERTICALITA' DELLE TORRI EOLICHE MEDIANTE ELEMENTI DI CONNESSIONE A FLANGE REGOLABILI IN SITO" Domanda di Brevetto per invenzione industriale numero 102016000083638, Italia, 11/02/2018, (2018).

Advances in mathematical modeling and experimental techniques for quantification and prediction of fluid dynamic noise

Multi-modal noise generation in low Mach number orifice plates: An experimental investigation

Luca Nicola QUARONI[1, a *], Islam RAMADAN[2], Simon RAMPNOUX[2], Stefano MALAVASI[1], Emmanuel PERREY-DEBAIN[2] ,

[1]Department of Civil and Environmental Engineering, Politecnico di Milano, Piazza Leonardo da Vinci, Milan (20133), Italy

[2]Laboratoire Roberval, Université de Technologie de Compiègne, Centre de recherche Royallieu – CS 60 319 – 60 203 Compiègne Cedex, France

[a]lucanicola.quaroni@polimi.it

*corresponding author

Keywords: Duct Acoustics, Orifice Plates, Higher-Order Acoustic Modes

Abstract. An experimental campaign on aerodynamic noise generation by orifice plates in rectangular ducts for low Mach number flows is carried. The test plant allows for the measurement of the total emitted acoustic power both upstream and downstream of the orifice plates for varying mass flow rates through the 2N-Port method. A modal decomposition for the first eight propagating higher-order acoustic modes is also performed. Keeping the free flow area and the plates' thickness constant, the tests allowed for a study of the influence of the number of circular orifices on the generated acoustic power as well as its modal composition. It is found that the orifices' number does not influence the overall pressure drop caused by the presence of the obstacle. The generated noise on the contrary is greatly affected by such a parameter both for the plane-wave frequency range and especially so in the multi-modal one.

Introduction

Heating, Ventilation and Air Conditioning (HVAC) systems play a fundamental role in ensuring a safe and comfortable environment in both buildings and vehicles. Part of the requirements for such systems is that they generate as less noise and vibrations as possible during operation. Amongst the various components concurring to the production of noise (e.g., fans, compressors…), flow singularities in ducts are one of the most common sources [1]. The flow is usually in the turbulent incompressible regime with very low Mach numbers ($Ma < 0.1$). In undisturbed flows (e.g., sufficiently long straight ducts), sound is aerodynamically generated by the turbulent boundary layer, and it is propagated inside the duct [2]. Whenever a singularity is present (e.g., valves, bends) intense turbulence is produced in its vicinity as the flow detaches from the ducts' walls. Part of the turbulence kinetic energy is converted into sound, which reaches largely higher values than in the case of undisturbed flows [3]. In unconfined low Mach number flows, previous studies have shown that in the presence of fixed surfaces noise generation can be predicted once the pressure fluctuations on such surfaces are known [4]. This amounts to equating the noise generation mechanism to a distributed dipole source. Addition of confinement (as in ducts) further modifies such noise generation mechanism. In particular, the fact that sound propagates in higher-order (or transversal) modes from a given cut-on frequency onwards fundamentally changes the propagation mechanisms compared to unconfined free-field conditions [5]. Obstacles perpendicular to the flow have received much attention due to their relatively simple geometry as well as their widespread use in duct systems (e.g., spoilers, orifice plates). A successful theory has been developed for aerodynamic noise prediction in the plane-wave frequency range by assuming that the pressure fluctuations over the obstacle's surface be proportional to the total drag force acting on it [6,7]. Attempts at models involving higher-order acoustic modes have been however somewhat less conclusive. Recent experimental work has shed more light on the phenomenon by directly measuring the fluctuating pressures over both sides of a circular orifice plate perpendicular to the flow [8]. It was found that the intensity of the pressure fluctuations was highest in the immediate vicinity of the orifice's edge, after which an exponential decrease was observed.

In the present paper, the results of an experimental investigation on noise production by ducted orifice plates are presented. The tests were performed at the Roberval Laboratory of the University of Technology of Compiègne (France), where an experimental bench allows for the measurement of the total internal aerodynamically generated acoustic power as well as its modal decomposition in the propagating higher-order acoustic modes through the 2N-port method [9]. The remainder of the paper is structured as follows. The basic theory of sound propagation in ducts and of aerodynamic sound production in ducts is first recalled. The experimental test plant is then briefly described, together with the tested orifice plate geometries. Finally, the results of the tests on the plates are discussed and some conclusions are drawn.

Theory

Acoustic pressure waves in ducts travel with a constant amplitude distribution over the cross-section until a particular "cut-on" frequency is reached. Above that, waves with a non-uniform sectional pressure distribution propagate alongside the plane wave form. The cut-on frequencies are usually referred to as f_{mn} where the subscripts m and n refer to the number of nodal lines along a particular direction depending on the type of cross section. For an obstacle inserted perpendicularly to a ducted low Mach number flow, the Fourier transform of the aerodynamically generated acoustic pressure in a point $(x_1, x_2. x_3)$ can be expressed as [10]:

$$\hat{p}(x_1, x_2, x_3) = \sum_m \sum_n \frac{1}{2A} \rho_0 \hat{c}_{mn} \psi_{mn}(x_1, x_2) e^{-i\hat{k}_{mn}x_3} \hat{Q}_{mn} \tag{1}$$

where A is the duct's area, ρ_0 is the density of the medium at rest, $\hat{c}_{mn} = 2\pi f / k_{mn}$ is the modal phase velocity, ψ_{mn} is the modal shape function of the duct, $\hat{k}_{mn}$ is the axial modal wave number and $\hat{Q}_{mn}$ is a coefficient which takes into account the coupling between the fluctuating force spatial distribution and the ducts' modal shape function. The expression of such a coefficient for the present case is [6]:

$$\hat{Q}_{mn} = \frac{1}{\hat{c}_{mn}} \int \int_{A_s} \hat{f}_3 \, (x_1', x_2') \psi_{mn}(x_1', x_2') e^{i\hat{k}_{mn}x_3} dA \tag{2}$$

where $\hat{f}_3$ is the Fourier transform of the net fluctuating perpendicular force per unit area and per unit density over the obstacle's surface A_s. This expression shows that the value of the coefficient $\hat{Q}_{mn}$, and therefore of $\hat{p}$ through (1), increases the more the distribution of $\hat{f}_3$ follows that of the modal shape function ψ_{mn}.

Experimental setup

The experimental set-up for the results discussed in the present paper was conceived for the acoustic characterization and the estimation of noise emissions of HVAC components through the 2N-port method. Such experimental plant was used in several other publications on the subject (e.g., [2]) and the reader is referred to [9] for a more complete description both of the plant and of the 2N-port method.

The setup consists of a duct of rectangular cross section of width b equal to 200 mm and height h of 100 mm. Its total length is 4.1 m and anechoic terminations on both sides allow for a removal of end reflections starting from a frequency of 100 Hz. For such a cross section and neglecting the effect of the flow, the cut-on frequencies are computed from $f_{mn} = 1/c\sqrt{(m\pi/b)^{\wedge}2 + (n\pi/h)^{\wedge}2}$. The modal shape functions are instead given by $\psi_{mn} = N_{mn}\cos(m\pi x_1/b)\cos(n\pi x_2/h)$ with N_{mn} a normalizing factor ensuring that the average squared value of ψ_{mn} is 1 over the duct's cross section. The cut-on frequencies in the range of study [200 Hz, 3200 Hz] for the present configuration are reported in Table 1.

Table 1 - Cut-on frequencies

Mode (m,n)	(0,0)	(1,0)	(2,0)	(0,1)	(1,1)	(2,1)	(3,0)	(3,1)
f_{mn} **[Hz]**	0	849	1717	1717	1920	2428	2575	3095

Air is supplied to the system through a centrifugal fan powered by a variable-speed electric motor connected to a frequency handle for flow regulation. To reduce the disturbances by the machine, an in-line silencer is placed before a stabilization tank which is itself needed to compensate for the fluctuations of the incoming flow. The mass flowrate is then measured through a Venturi-like flow meter. Static pressure and temperature are also recorded at the stabilization tank through a pressure transducer and a Pt-100 RTD probe respectively. A total of 96 condenser microphones of 1/4" diameter is flush mounted along the duct between the two anechoic terminations in couples of twelve along a given cross-section, four of which upstream and four downstream of the mounted obstacles.

Three different orifice plate geometries have been experimentally tested. The flow to duct area ratio A_{flow}/A and the plates' thickness have been kept as constant for all three cases and equal to 0.25 and 8 mm respectively. The shape of the applied orifices was chosen as circular to avoid corners; all are sharp-edged. Table 2 reports the geometry of the tested orifices as well as their identification strings and the diameters of the applied holes d_h.

Table 2 - Orifice plate geometries

Plate ID	A1	A2	A4
Geometry			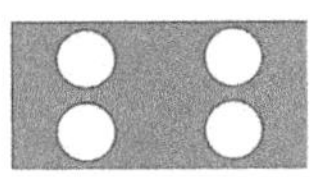
d_h **[mm]**	80	56	40

Results

A preliminary measurement campaign was performed to characterize the pressure drop Δp_R caused by insertion of the orifice plates into the duct. This was necessary to understand whether differences in noise emissions could be due to such a factor, as previous literature highlights its importance. The pressure drop was measured varying the mass flowrate and keeping the outlet pressure at atmospheric level as the duct discharges into the outside environment. Figure 1 reports a dimensional and a nondimensional plot of the pressure drop due to the added resistance as a function of the mass flowrate. The nondimensionalization of Δp_R was made by computing the total drag force acting on the plate as $\bar{F}_3 = \Delta p_R A$ and expressing it in terms of a drag coefficient C_d as $\bar{F}_3 = 0.5\, \rho_0 U^2 C_d A_S$ where U is the average velocity of the flow in the axial direction. The Reynolds' number is defined as $Re = \rho_0 U D_h / \mu$ with D_h the hydraulic diameter of the duct ($D_h = 4A/P$, with P the duct's cross section perimeter). The pressure drop (or the drag coefficient) shows a weak dependence on the geometry of the plates having kept constant the flow area and the plates' thickness, with the A2 plate indicating a slightly higher resistance to the flow.

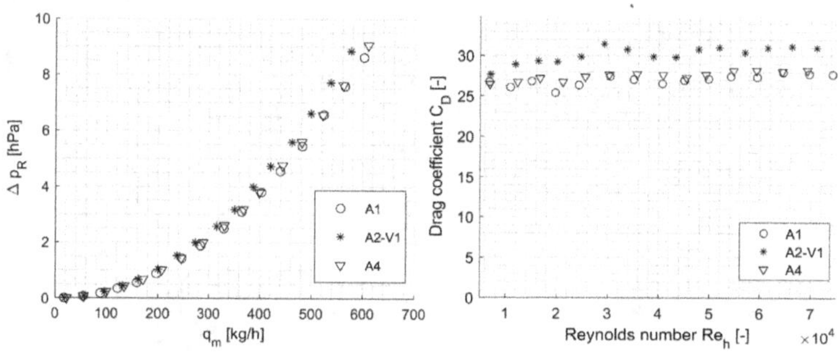

Figure 1 - (left) Pressure drop versus mass flowrate and (right) drag coefficient versus Reynolds' number.

The sound power level (SWL) in fine bands (Δf = 1.76 Hz) produced by the three orifice plates at a section 15 cm upstream of it for a mass flowrate of 505 kg/h (corresponding to a maximum local Mach number of ≈ 0.1) is reported in Figure 2. Differences of more than 10 dB are observed between, with the A2 plate producing the lowest noise and the A4 the highest outside of the plane-wave range. The A1 plate is particularly sensitive to the onset of certain higher-order modes (e.g. (2,0) and (1,0) modes). The modal decomposition of the SWL for the same tests resulting from application of the 2N-port method are reported in Figure 3. The amplitude of a given higher-order mode (m,n) in comparison to the corresponding plane-wave mode (0,0) depends greatly on the orifice plate geometry. The (2,0) mode for example has a higher value than its plane-wave counterpart for the A1 plate, a slightly lower value for the A2 plate and a much lower value for the A4 plate. Again, the (1,0) mode displays a relative amplitude with respect to (0,0) lower for the A1 geometry and essentially equal for the two plates A2 and A4.

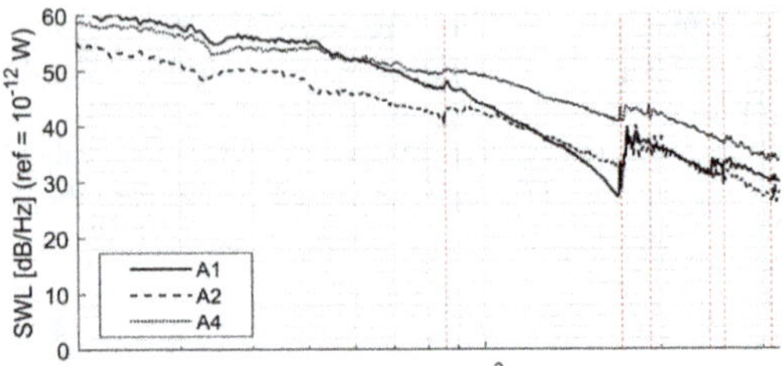

Figure 2 - Total upstream sound power level at 505 kg/h for the three geometries tested. Cut-on frequencies are highlighted as dashed red lines.

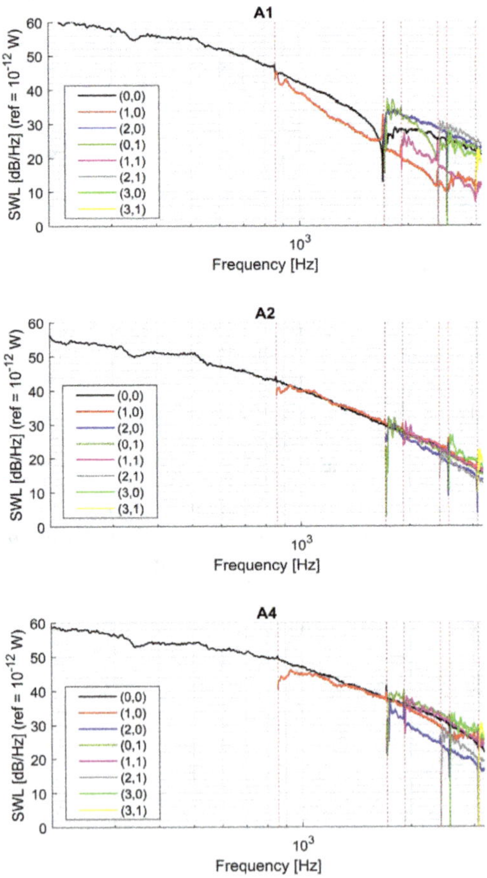

Figure 3 – Modal decomposition into propagating higher-order acoustic modes of the sound power level (SWL) for the different orifice plates tested at 505 kg/h. Cut-on frequencies are highlighted as dashed red lines.

Materials Research Forum LLC
https://doi.org/10.21741/9781644902431-112

Conclusions

In the present short paper, it has been shown that keeping a fixed flow area and plate thickness, the pressure drop caused by a low Mach number flow through circular-shaped sharp-edged orifices in a rectangular duct does not vary appreciably if the number of the holes is changed. On the contrary, the total emitted acoustic power is very sensitive to such parameter. A modal decomposition of such acoustic power into the first eight propagating higher-order acoustic modes has then been performed. It has been found that the contribution of a given higher-order acoustic mode (m,n) relative to the plane-wave mode (m,n) greatly varies depending on the plate geometry.

References

[1] Kårekull, Oscar. “Predicting Flow-Generated Noise from HVAC Components.” PhD. thesis, Engineering Sciences, KTH Royal Institute of Technology, 2015.

[2] David, Antoine, Florian Hugues, Nicolas Dauchez, and Emmanuel Perrey-Debain. 2018 “Vibrational Response of a Rectangular Duct of Finite Length Excited by a Turbulent Internal Flow.” Journal of Sound and Vibration 422: 146–60. https://doi.org/10.1016/j.jsv.2017.11.052

[3] Norton, Micheal, Karczub Denis. Fundamentals of Noise and Vibrations for Engineers. 2nd ed. Cambridge University Press, 2003. https://doi.org/10.1017/CBO9781139163927

[4] Curle, N., and M. J. Lighthill. “The Influence of Solid Boundaries upon Aerodynamic Sound.” Proceedings of the Royal Society of London. Series A. Mathematical and Physical Sciences 231, no. 1187 (1955): 505–14. https://doi.org/10.1098/rspa.1955.0191

[5] Davies, H. G., and J. E. Ffowcs Williams. “Aerodynamic Sound Generation in a Pipe.” Journal of Fluid Mechanics 32, no. 4 (June 18, 1968): 765–78. https://doi.org/10.1017/S0022112068001011

[6] Nelson, P.A., and C.L. Morfey. “Aerodynamic Sound Production in Low Speed Flow Ducts.” Journal of Sound and Vibration 79, no. 2 (November 1981): 263–89. https://doi.org/10.1016/0022-460X(81)90372-2

[7] Oldham, D.J., and A.U. Ukpoho. “A Pressure-Based Technique for Predicting Regenerated Noise Levels in Ventilation Systems.” Journal of Sound and Vibration 140, no. 2 (July 1990): 259–72. https://doi.org/10.1016/0022-460X(90)90527-7

[8] Tao, Fuyang, Phillip Joseph, Xin Zhang, Oksana Stalnov, Matthias Siercke, and Henning Scheel. “Investigation of the Sound Generation Mechanisms for In-Duct Orifice Plates.” The Journal of the Acoustical Society of America 142, no. 2 (August 2017): 561–72. https://doi.org/10.1121/1.4996459

[9] Trabelsi, Hassen, Nicolas Zerbib, Jean-Michel Ville, and Félix Foucart. 2011. “Passive and Active Acoustic Properties of a Diaphragm at Low Mach Number: Experimental Procedure and Numerical Simulation.” European Journal of Computational Mechanics 20(1–4): 49–71. https://doi.org/10.3166/ejcm.20.49-71

[10] Doak, P.E. “Excitation, Transmission and Radiation of Sound from Source Distributions in Hard-Walled Ducts of Finite Length (I): The Effects of Duct Cross-Section Geometry and Source Distribution Space-Time Pattern.” Journal of Sound and Vibration 31, no. 1 (November 1973): 1–72. https://doi.org/10.1016/S0022-460X(73)80249-4

Advanced process mechanics

Materials Research Proceedings **26** (2023) 707-712 https://doi.org/10.21741/9781644902431-113

Milling of alumina-based ceramic foams: tool material effects

Giovanna Rotella[1], Maria Rosaria Saffioti[2,*], Michela Sanguedolce[2], Flaviano Testa[3], Luigino Filice[2], Fabrizio Micari[4]

[1]Department of Engineering for Innovation, University of Salento, LE 73100, Italy

[2]Department of Mechanical, Energy and Management Engineering, University of Calabria, CS 87036, Italy

[3]Department of Computer Engineering, Modeling, Electronics and Systems Engineering, University of Calabria, CS 87036, Italy

[4]Department of Engineering, University of Palermo, Viale delle Scienze, Palermo, PA 90128, Italy

* Maria Rosaria Saffioti. Tel.: +39 0984 494637; E-mail address: mariarosaria.saffioti@unical.it

Keywords: Milling, Porous Ceramics, Surface Processing

Abstract. Porous ceramics are recognized to have the potential to be used for a wide variety of industrial applications, such as catalyst support and molten metal filter. It is a common practice to adjust the shape of ceramics in a pre-sintering stage in order to obtain various configurations but some limitations due to premature failure of the product, marked tool wear or inhomogeneous shrinkage during processing stages may arise, whenever a complex shape is requested. Almost the same happens in the case of post-sintering stage ceramics machining leading, in most cases, to an inefficient process characterized by a severe tool wear and poor final product quality. Such problems, related to tool material, material removal modes and hence process parameters represent a huge limitation for many industrial applications of the above materials. Milling of alumina-based ceramic foams in a sintered state is the focus of the present work. An experimental campaign has been carried out at fixed axial depth of cut, feed rate, number of passes, spindle speed and under flood lubrication regime varying tool material (aluminum oxide-based, diamond-coated). Tool wear mechanisms and final product quality have been investigated through surface analysis in order to verify the workability of alumina based ceramic foams and the related tool wear.

Introduction

Porous ceramics are classified as advanced materials, able to supply additional functionalities with respect to conventionally used ones; these materials cover a wide range of structures based on different morphologies and composition [1, 2]. High porosity materials can be used for several applications due to their elevated surface-volume ratio; some applications may require high level of refractoriness, creep and corrosion resistance, as properties belonging to ceramic materials.

For the ceramic foams production several processes are available, some of them involve other materials (such as polymer) as precursor, in order to provide the lattice shape after sintering [3]. Usually, to change the structure of the obtained ceramic foams, it is possible to take action on precursor shape or working on material ahead of sintering [4]. Nowadays, many studies are being carried out about new sustainable production processes detection and on complex shape development [5], also involving additive manufacturing [6], to avoid premature product failure during the production process and inhomogeneous shrinkage in sintering phase [7, 8]. Several studies are referred to conventional and non-conventional machining methods of bulk ceramic materials [9] while there is still a lack of knowledge concerning the machinability of porous ceramic materials. These latter are considered difficult to machine, since they lead to an accelerated tool wear [10], together with a poor final surface quality.

It is extremely important to find innovative, cheap and easy to perform production methods in order to guarantee high performance, reliable, long lifetime ceramic filters. At the same time, it is necessary to optimize resources minimizing the maintenance costs and downtime arising from failure of a filter element ensuring, at the same time, the compliance with the environmental constraints.

It is worth noting that effective machining of ceramic foams has not been yet assessed. Most commonly used machining techniques for bulk ceramics involve conventional methods (e.g. abrasive wheel cutting), non-conventional machining (e.g. wire electrical discharge machining) and hybrid machining (e.g. hybrid laser waterjet machining) and the most commonly used tools are diamond based ones. However, studies based on these techniques [9] demonstrated that the machinability of bulk ceramics still represents an issue for both tool wear and surface integrity of the obtained components.

This paper presents the evaluation of the machinability of alumina based ceramic foams by means of spherical end milling process using different tools materials and under flood lubrication method.

Materials and Methods

The samples under investigation are cylindrical alumina-based ceramic foams, with a diameter of 30 mm, height of 30 mm and pore density of 30 pores per inch (ppi). They have been provided by Lanik s.r.o. and have the following chemical composition: Al_2O_3 84.0%, SiO_2 14.0%, MgO 0.8%, Other 1.2%. The foams, produced via the replica process [8, 11], consist of a network of randomly-oriented dodecahedral-shaped cells interconnected through struts (Fig. 1).

Two different tool materials (Fig. 2) have been selected for the experimental campaign on spherical end milling process: vitrified hard bond pink aluminum oxide (average grit size 120 μm) and electroplated diamonds on a metal substrate (average grit size 118 μm). Tool shank and head diameter were respectively 3 mm and 6 mm for both tools. The experimental campaign has been carried out varying tool material and at fixed axial depth of cut, feed rate, number of passes, spindle speed and under flood lubrication regime.

Preliminary milling tests have been performed in order to choose the set of parameters for the experimental tests, starting from values recommended by the manufacturers for similar bulk materials. However, it is worth noting that there are no available recommendations for milling ceramic foams. The factors considered for experimental campaign are reported in Table 1. After machining, samples and tools have been inspected under a Scanning Electron Microscope in order to analyze the surface characteristics prior and after processing and the tool wear mechanisms.

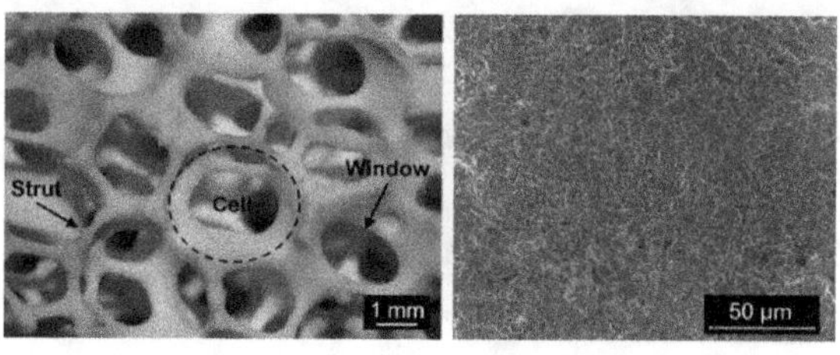

Fig. 1 – As received alumina based ceramic foam (a) foam structure, (b) SEM image of the as received surface.

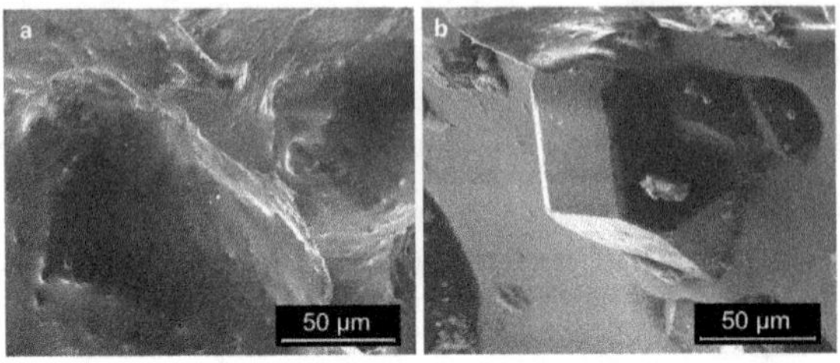

Fig. 2 – SEM images of as received pink aluminum oxide (a) and diamond coated (b) tools.

Table 1. Design of experiments for the experimental campaign.

Factors	
Tool material	Aluminum oxide, Diamond coated
Lubrication conditions	Flood
Spindle speed [RPM]	10000
Number of Passes	2
Depth of Cut [mm]	0.5
Feed Rate [mm/min]	50

Results and Discussion

Tool wear mechanisms

Tool wear is a direct expression of tool life and it is strictly related to machined surface and subsurface quality. Fig. 3a shows the low magnified (100x) surfaces of aluminum oxide tools in which the damage is immediately recognizable. Also, adhesive wear takes place during the process since a certain amount of adhered foam material can be easily found on tool surface.

The phenomena of flattening and materials mixing can be attributed to tool and sample material affinity in terms of composition and mechanical properties. In particular, adhesion is generally reduced by flood lubrication which results, by nature, into an efficient lubrication effect.

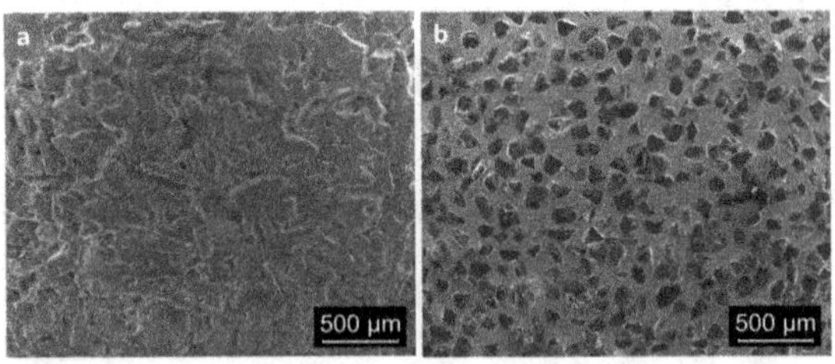

Fig. 3 – Aluminum oxide (a) and diamond coated (b) tools after milling at different magnification.

Concerning diamond coated tools, damage is not easily recognizable at 100x magnification, as shown in Fig. 3b.

Fig. 4 shows a detailed view of tool surfaces after milling process. Fig. 4a has been taken using Backscattered Electrons (BSE) mode in order to highlight the presence of different materials, Fig. 4c shows a void left by a diamond being pulled out while Figure 4d displays a damaged diamond.

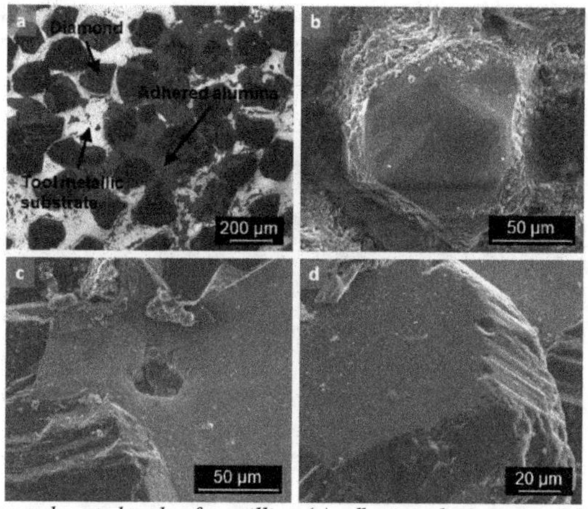

Fig. 4 – Details of diamond coated tools after milling (a) adhesion phenomenon, BSE image, (b) adhesion phenomenon, (c) diamond pull out, (d) diamond fracture.

Machined surface analysis

The presented process inevitably involves both brittle and ductile fractures due to the multitude of scratches and their relative interactions when abrasive grains cut into ceramic specimens. The machined surface results ideally deformed similarly to that left by an indentation process but other mechanisms can also take place, such as adhesion between tool/workpiece surfaces, mainly due to chemical affinity. Also, excessive local forces can generate different defects like chips, fissures, cracks or flaws. The above issues result exacerbated by the geometric characteristics of the workpiece under investigation.

When aluminum oxide tools are employed, the adhesion effect evidently influences the fracture mechanism, as shown in Fig. 5. In fact, the chemical affinity between the foam and the tool brings to machined surfaces characterized by a higher content of brittle fractures. The analysis reported herein also confirms the presence of adhesive wear on the tool, as presented in the previous section.

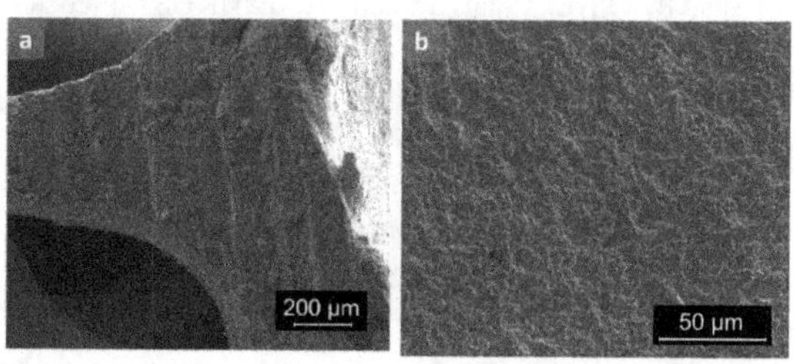

Fig. 5 – Details of ceramic foam surface after milling with aluminum oxide tools.

Diamond coated tools result into a more efficient cutting process since less adhesion effect is present. Fig. 6 clearly show both ductile and brittle fracture areas, which are typical of the process. The diamond abrasive grains exposed on the lip of the milling tool are of random distribution, with a consequent difference in the exposure highness of each diamond.

The overall machining is achieved by the combined actions of what can be thought as numerous micro-cutting edges with locally different depths of cut, causing areas of both ductile and brittle fracture when machining ceramics.

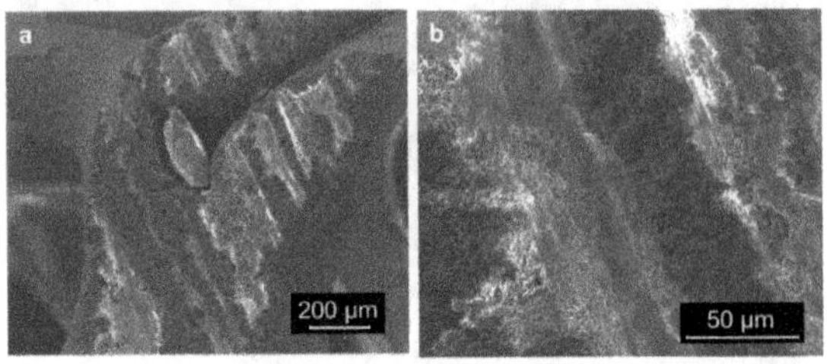

Fig. 6 – Details of ceramic foam surface after milling with diamond coated.

Conclusions

This paper presents an analysis of workability of alumina based ceramic foams. In particular, such materials combine the issues of machining conventional ceramics with those related to machinability of geometrically complex workpieces. The experimental campaign involved the spherical end milling of the workpieces at varying cutting tool materials.

The overall results highlighted the tendency of such process to fast wearing the tools and the combination of ductile and brittle deformation mode on the machined surfaces.

The results obtained allow to state that ceramic foams can be machined by spherical end milling in order to try to obtain a variety of complex shapes and that diamond coated tools are, up to now, the best available choice. It should be noted that the overall process needs to be optimized in order to define a workability window able to minimize the tool wear and maximize the surface quality.

References

[1] A. Salomon, M. Dopita, M. Emmel, S. Dudczig, C.G. Aneziris, D. Rafaja, Reaction mechanism between the carbon bonded magnesia coatings deposited on carbon bonded alumina and a steel melt. J Eur Ceram Soc (2015); 35: 795–802. https://doi.org/10.1016/j.jeurceramsoc.2014.09.033

[2] T. Granato, F. Testa, R. Olivo, Catalytic activity of HKUST-1 coated on ceramic foam. Microporous Mesoporous Mater (2012); 153: 236–246. https://doi.org/10.1016/j.micromeso.2011.12.055

[3] A.R. Studart, U.T. Gonzenbach, E. Tervoort, L.J. Gauckler, Processing routes to macroporous ceramics: A review, J. Am. Ceram. Soc. 89 (2006) 1771–1789. https://doi.org/10.1111/j.1551-2916.2006.01044.x

[4] M. V. Twigg, J.T. Richardson, Fundamentals and applications of structured ceramic foam catalysts, Ind. Eng. Chem. Res. 46 (2007) 4166–4177. https://doi.org/10.1021/ie061122o

[5] G. V. Franks, C. Tallon, A.R. Studart, M.L. Sesso, S. Leo, Colloidal processing: enabling complex shaped ceramics with unique multiscale structures, J. Am. Ceram. Soc. 100 (2017) 458–490. https://doi.org/10.1111/jace.14705

[6] G. Liu, Y. Zhao, G. Wu, J. Lu, Origami and 4D printing of elastomer-derived ceramic structures, Sci. Adv. 4 (2018) 1–11. https://doi.org/10.1126/sciadv.aat0641

[7] B. Su, S. Dhara, L. Wang, Green ceramic machining: A top-down approach for the rapid fabrication of complex-shaped ceramics. J Eur Ceram Soc (2008); 28: 2109–2115. https://doi.org/10.1016/j.jeurceramsoc.2008.02.023

[8] P. Colombo, H.P. Degischer, Highly porous metals and ceramics, Mater. Sci. Technol. 26 (2010) 1145–1158. https://doi.org/10.1179/026708310X12756557336157

[9] R. Rakshit, A.K. Das, A review on cutting of industrial ceramic materials. Precis Eng (2019); 59: 90–109. https://doi.org/10.1016/j.precisioneng.2019.05.009

[10] G. Rotella, M. Sanguedolce, M.R. Saffioti, L. Filice, F. Testa, Strategies for Shaping of Different Ceramic Foams, Procedia Manuf. 47 (2020) 493–497. https://doi.org/10.1016/j.promfg.2020.04.345

[11] P. Colombo, Conventional and novel processing methods for cellular ceramics, Philos. Trans. R. Soc. A Math. Phys. Eng. Sci. 364 (2006) 109–124. https://doi.org/10.1098/rsta.2005.1683

Progresses in multi-materials billet manufacturing out of metal scraps through friction stir consolidation

Abdul Latif[1, a], Giuseppe Ingarao[1,b*], Livan Fratini[1,c]and Fabrizio Micari[1,d]

[1]Department of Engineering, University of Palermo, Viale delle Scienze, Palermo, 90128,Italy

[a]abdul.latif@unipa.it, [b*] giuseppe.ingarao@unipa.it , [c]livan.fratini@unipa.it,
[d]fabrizio.micari@unipa.it

Keywords: Friction Stir Consolidation, Solid State Technique, Recycling and Upcycling, Functionally Graded Material, Aluminum Alloys

Abstract. The evolution of Friction Stir Consolidation (FSC) from recycling towards upcycling technique proved to be one of the excellent solid-state methods for manufacturing functionally graded billets. Multi-material Functional Graded Materials (FGMs) represent a novel class of materials characterized by a gradual change in properties and functions which can be tailored to enhance components performance. Manufacturing techniques play a critical role in achieving the designed compositional and microstructural distribution. Specifically, FSC allows the manufacturing of FGM billets out of metallic chips; the mixing of different metallic chips offers mutually exclusive mechanical properties like high hardness and good ductility in a single FSC billet with excellent formability. The present research further explores some challenges while combining dissimilar aluminum alloys chips in different percentages and spatial order, especially in the radial direction. The mechanical and metallurgical properties were assessed through the Vickers hardness measurements and microstructure analysis. The results revealed that new strategies are needed for a better exploitation of FSC as a solid-state method for fabrication of Functionally Graded Material.

Introduction

The demand for aluminum is rapidly increasing due to its growing application. The accelerating consumption is putting immense pressure on industries to increase the production rate [1]. Roughly 100 million metric tons of aluminum are currently produced per year. However, per 1 ton of aluminum production from the primary source, 12-16 tons of greenhouse gas (GHG) are produced. Almost 35 % of the aluminum demand is met by recycling aluminum scraps [2]. The conventional recycling route skips many complex steps of aluminum extraction from the primary source, and therefore it is a highly energy efficient process. However, this method has further limitations, especially during the recycling of aluminum machining chips. Due to their high surface-to-volume ratio, these scraps are prone to oxidation, causing permanent material loss during the melting process. Therefore, the researchers turned to solid-state recycling (SSR) techniques.

SSR methods directly transform metal scraps into finished or semi-finished billet through mechanical means [3]. Recently, SSR processes have proved their energy and resource efficiency [4]. The new frontier of SSR processes could be their application for manufacturing multi-material functional graded semi-finished products. At the same time, there is an urgent need to find manufacturing solutions for obtaining multi-material components. So far, aluminum alloys have been successfully joined with magnesium alloys in bimetallic sheet applications [5]. Graded components are nowadays crucial for enhancing the performance of the product under mechanical, electrical, and environmental angles [6].

The authors have recently successfully applied FSC for producing multi-material based FGMs [7]. FSC has two main steps: compaction and consolidation. During compaction, chips or powder are pressed in a hollow die chamber by applying a specific load through a cylindrical tool. Then compacted materials are further pressed and stirred through high the tool's downward force and rotational speed during the consolidation phase. The authors, in their previous research, have successfully obtained bimetallic billet in a multimaterial approach starting from AA 7075 and AA 2011-T3 chips [7]. Further, a discrete change in hardness profile was noticed, a property that is found in discontinue FGM. In the present research, chips of aluminum alloys AA 7075 and AA 2011-T3 were mixed at different proportions to get a continuous graded hardness profile like continuous FGM. Further, attempts were made to develop multi-materials based shell-core billet by combining the chips along the radial direction of the billet. The purpose was to extend the capabilities of the FSC process in manufacturing FGM. The quality of both continuous FGM and shell-core billets was evaluated by comparing the results with the previous study [7]. Hardness and microstructure are considered the main output criteria.

Material and methods

Material and process set-up

Aluminum alloys AA 2xxx and AA 7xxx were considered in the current studies because they are popular alloys in the transport and aerospace manufacturing industries where a good balance between the product in service and scraps [8] exists. The machining scraps were obtained from aluminum alloys AA 7075 and AA 2011-T3 through turning and milling operations, respectively. The scraps were cleaned by submerging them in acetone for 30 minutes. Mono-material billets were developed from chips of both AA 7075 and AA 2011-T3. Then chips of both alloys were combined under proportion (Fig. 1a) and spatial order (Fig .1b) to develop a bi-material billet. However, the overall mass of input material was 15g (constant) in all experiments. First, the chips were loaded in a cylindrical die with a nominal diameter of 25.4 mm and then compacted at 5 kN force by an H13 steel cylindrical tool with a 25 mm diameter. The die and pressing tool system was integrated with ESAB-LEGIO (Fig. 1a), a dedicated friction stir welding machine. The final consolidation step was applied at 20 kN with a high tool rotational speed of 1500 rpm.

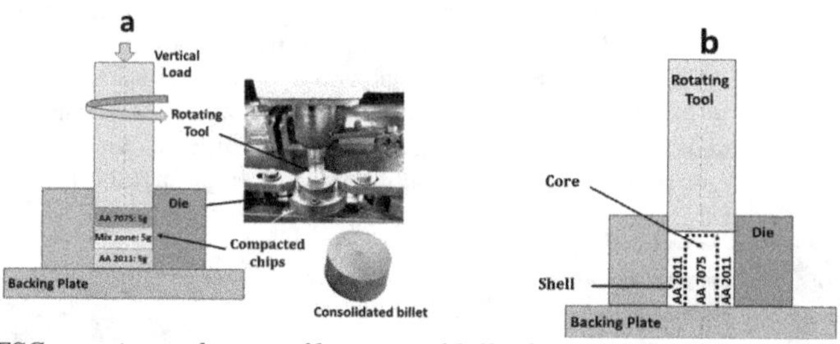

Figure 1. *FSC experimental setup of bi-material billet for (a) mixing chips at various proportion in the mid layer, and (b) chips combining for shell-core manufacturing*

Measured output

The formability of the FSC billet was evaluated with a forging test under cold conditions to turn the billet into near-net shape parts. The forging die and tool were coupled with Galdabini hydraulic press tensile testing machine (Fig. 2), and billets were reduced to 40% of the initial height.

The hardness was evaluated through the Vickers hardness measurement. A load of 49 N (5 kg) was used for 15 seconds. Due to process symmetry, the FSC billet was sectioned along the longitudinal axis, and only four lines were selected along the longitudinal direction, as shown in Fig. 3. Keller's reagent was also applied to the cross-section to examine the microstructure of the samples.

Materials Research Forum LLC
https://doi.org/10.21741/9781644902431-114

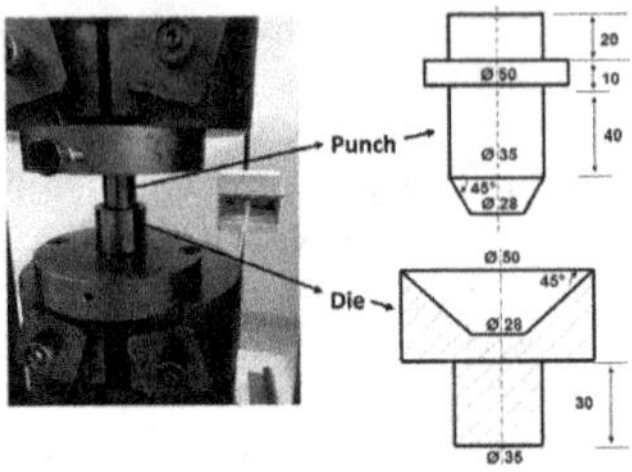

Figure 2. Forging tool and die setup integrated with Galdabini tensile testing machine

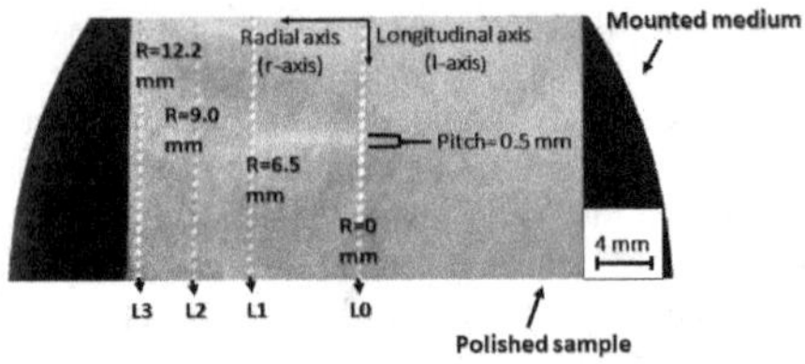

Figure 3. Sample for Vickers hardness measurement

Results and discussion

Hardness of the typical FSC bi-material AA 7075/AA 2011-T3 billet

Two factors were found that caused variation in the hardness of the bi-material billet. First was the material aspect, as material retained mechanical properties their properties in the case of bi-material billet [7]. Thus, the part of bi-material that consisted of soft material (AA 2011) showed lower hardness comparing the part that consisted of hard material (AA 7075). Second, a typical FSC mono-material billet is characterized by decaying hardness values along the longitudinal direction and radial direction due to different strain rates and thermal gradients during the FSC process.

Forgeability characteristics of bi-material AA 7075/AA 2011-T3 billet

Analyzing the cross-section of the forged section, a defect-free sample was observed for mono-material AA 2011 billet that showed good forgeability due to its good ductility. However, many defects in the form of debonding chips occurred near the external surface for the AA 7075 billet section (Fig. 4). The reason is that AA 7075 is a poorly ductile material. While combining AA 7075 and AA 2011 in a bi-material billet, the sample showed good forgeability without any defect. The reason was that AA 7075, exploited the good ductility of AA 2011 in the case of bi-material billet. Further, changes due to forging were analyzed by comparing the hardness profile of non-forged and forged bi-material. For the sake of better understanding, only hardness along one line (L0) was considered, as shown in Fig 5. Forged bi-material billet showed an increase in hardness due to strain hardening and a reduction in height compared to unforged billet.

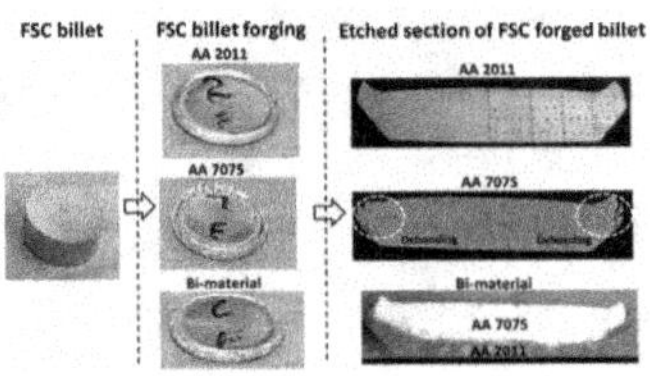

Figure 4. Forged and etched samples of mono and bi-materials billets

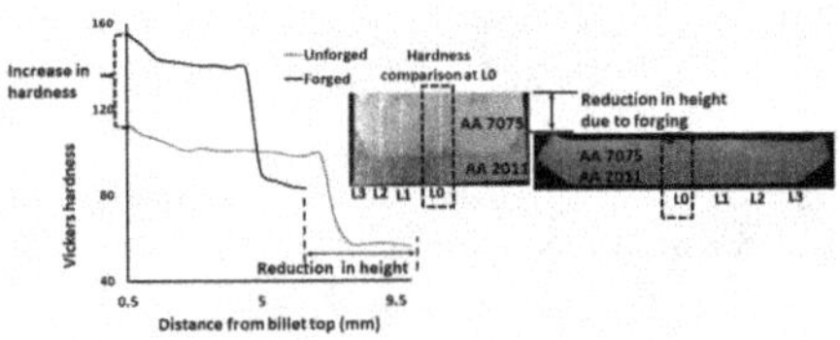

Figure 5. Hardness profile comparison at L0 for bi-material forged and unforged billets

Chips mixing of two dissimilar aluminum alloys in the middle zone of the billet

FSC process mechanics and material aspects are the main factors that cause variation in hardness. Comparatively changing material is more dominant than the FSC process mechanics factor. However, the material change causes abrupt variation in hardness and leads to fabricating billet with a hardness profile similar to a discontinuous functionally graded material. Therefore, attempts were made to control hardness based on material mixing and obtain a billet that possessed properties like continuous graded materials. Therefore, a three layers multi-material billet was developed, as shown in Fig. 6a. Each layer has a constant mass of 5g. AA 7075 chips were placed in the top layer, AA 2011-T3 existed in the bottom layer, while in the middle layer, both alloys were mixed at different proportions. For the sake of better understanding, the percentage of AA 2011 in the middle layer was considered a reference. An abrupt drop from higher to lower hardness value was found for 100% (Fig. 6b) and 0% (Fig. 6f) percent AA 2011 composition in the middle layer. In the case of 75 % and 25% the fluctuation trend was noticed in the transition zone, but still, the hardness profile was more inclined towards AA 2011 and AA 7075, respectively. For 50 %, a balance fluctuated hardness profile between AA 2011 and AA 7075 was observed. These results show that continuous hardness variation is not possible based on changing the proportion. The fact is also visible from the etched section (Fig. 7). Actually, the two alloys do not mix with each other to get a uniform solution because FSC is a solid-state process, and the process temperature does not reach to the melting point of aluminum alloys.

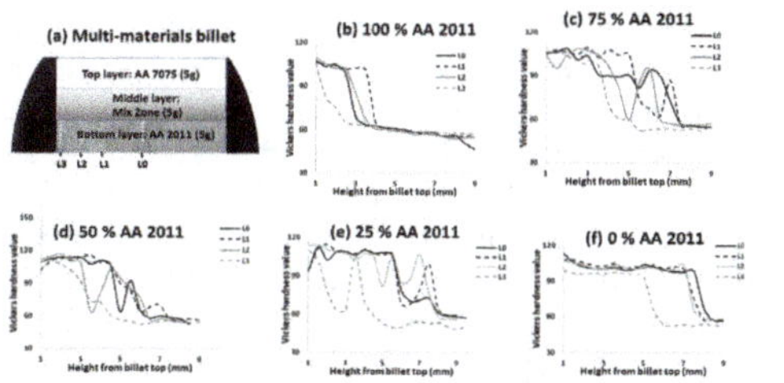

Figure 6. *Three layers multi-material billet (a) schematic, and hardness profile with percentage distribution of AA 2011-T3 in the middle layer (b) 100%, (c) 75 %, (d) 50%, (e) 25 %, and (f) 0%*

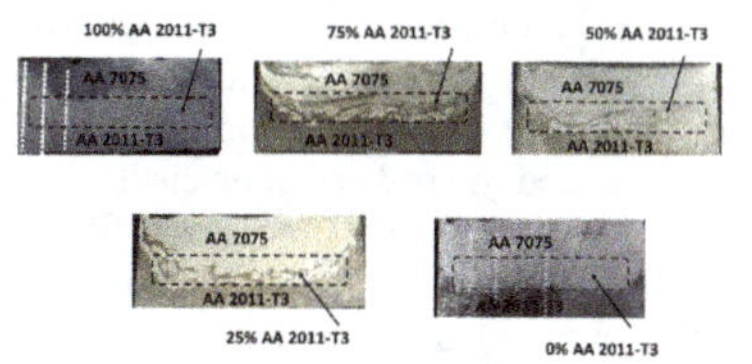

Figure 7. *Eetched section for bi-material billet with 100%, 75 %, 50%, 25 %, and 0% AA 2011-T3 distribution in the middle layer*

Bi-material billet in radial direction

Two different single-step strategies were applied to manufacture multi-material shell-core cylindrical billet from AA 7075 and AA 2011 chips.

Method I

In this method, a hollow flexible thin wall plastic cylinder was placed in the die. The plastic cylinder had a diameter less than the diameter of the die and was concentric with the die. AA 7075 chips were loaded inside the plastic cylinder and surrounded by AA 2011 chips (Fig. 8a). The plastic cylinder was carefully removed after chips compaction. Then the whole charge was consolidated. Keller's reagent was applied to analyze the billet section (Fig. 8b) as this method was previously proved to effectively differentiate the boundary between different aluminum alloys in multi-material billet [7]. Problems like the mixing of chips at the top zone of the billet and an irregular boundary line between the shell and core were found. It is assumed that problems were caused by material flow due to unwanted backward material extrusion, which will be discussed in the following section. Further, this strategy also involves manual steps. So, the next strategy was tested.

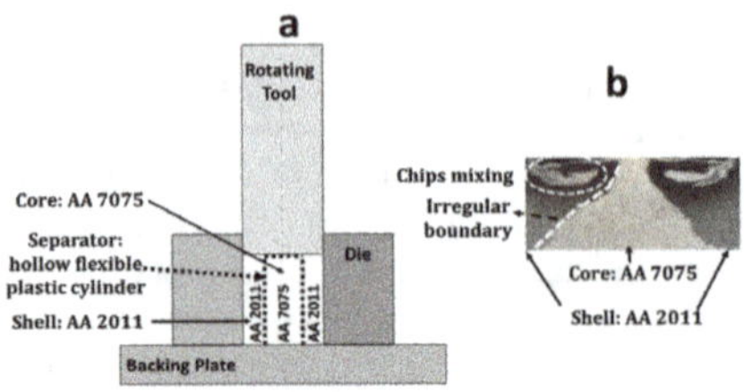

Figure 8. *(a) Method 1 schematic presentation for fabricating of shell-core multi-material billet, and (b) etched section of the shell-core multi-material FSC billet*

Method II

In the second strategy (Fig. 9), a solid cylindrical rod was placed inside the die, and the gap between the rod and the die was filled with AA 7075 chips (shell). Then hollow tool was used to press the chips AA 7075. The purpose of the rod was to develop a hollow shell of AA 7075 and provide support to the internal wall support of the compacted step. The surface of the supporting rod was also very well polished so that chips of the shell did not stick during the compression phase, and the rod could be easily removed. In the next step, cylindrical support was removed, and the cylindrical cavity was then filled with chips of AA 2011 for core development. A tool with a diameter comparable to the diameter of the core was used for chips compaction of the core. Finally, the whole charge was consolidated through the FSC tool. On analyzing the etched cross-section (Fig. 10) of the multi-material core-shell billet, again irregular boundary, chips mixing, and further non-symmetrical (along the longitudinal axis) core material distribution were found. These problems were caused by problems discussed in the previous method.

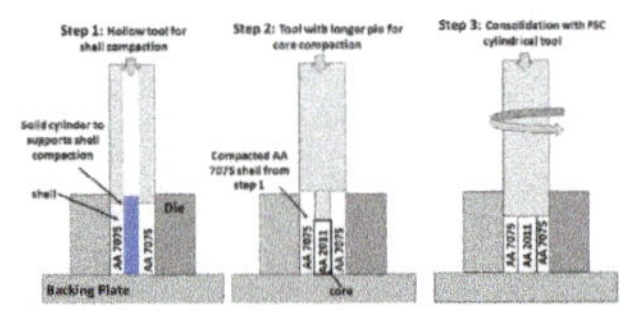

Figure 9*. Method 2 schematic presentation for fabricating of shell-core multi-material billet*

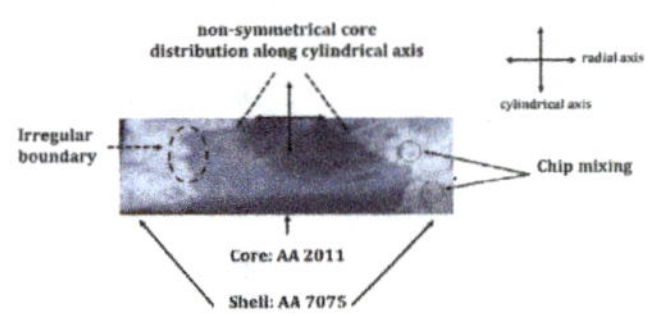

Figure 10. *Method 2 etched section ofthe shell-core multi-material FSC billet*

Additional challenges in manufacturing FSC multi-material billet

During the FSC process, unwanted material flow within the clearance between the tool and the die occurred. It is assumed that by proper selection of process parameters or at a higher technology readiness level (TRL), this unwanted material flow can be controlled [7].

Besides, it is challenging to numerically model a bi-material billet. As assigning two materials to a single billet is possible through phase transformation in Deform 3D software, but it is very difficult to obtain a discrete boundary between materials, even choosing a very refined mesh size.

Conclusion

FSC stir consolidation is an efficient technique to develop multi-material based FGM with excellent control on mechanical properties. Based on the previous discussion, the following conclusion can be drawn.

1. Mixing of different alloy chips at any proportion does not lead to develop continuous FGM. FSC is solid-state process and therefore, materials do not join to mix and form a uniform solution.
2. Multi-material shell-core billet is challenging to develop through a single-step FSC process.

3. Backward extrusion is the drawback of the FSC and can be controlled with a proper selection of key process parameters or at a higher technology readiness level.
4. The existing numerical model needs improvement to bring it closer to the real FSC process for proper modelling.

The future potential research in FSC could concern: 1/advanced mechanical and metallurgical characterization, 2/ numerical modelling and environmental assessment through Life Cycle Assessment methods.

References

[1] T.G. Gutowski, S. Sahni, J.M. Allwood, M.F. Ashby, E. Worrell, The energy required to produce materials: constraints on energy-intensity improvements, parameters of demand, Philosophical Transactions of the Royal Society A: Mathematical, Physical and Engineering Sciences 371.1986 (2013): 20120003. https://doi.org/10.1098/rsta.2012.0003

[2] D. Raabe et al., Making sustainable aluminum by recycling scrap: The science of "dirty" alloys, Progress in Materials Science. (2022): 100947. https://doi.org/10.1016/j.pmatsci.2022.100947

[3] B. Wan, W. Chen, T. Lu, F. Liu, Z. Jiang, M. Mao, Review of solid state recycling of aluminum chips, Resources, Conservation and Recycling.125 (2017):37-47. https://doi.org/10.1016/j.resconrec.2017.06.004

[4] J.R. Duflou, A.E. Tekkaya, M. Haase, T. Welo, K. Vanmeensel, K Kellens, W Dewulf, D. Paraskevas, Environmental assessment of solid state recycling routes for aluminium alloys: can solid state processes significantly reduce the environmental impact of aluminium recycling?. CIRP Annals (2015) 64(1) pp 37-40. https://doi.org/10.1016/j.cirp.2015.04.051

[5] J.H. Bae, A.P. Rao, K.H. Kim, N.J. Kim, Cladding of Mg alloy with Al by twin-roll casting Scripta Materialia (2011) 64(9) pp 836-9. https://doi.org/10.1016/j.scriptamat.2011.01.013

[6] C. Zhang, F. Chen, Z. Huang, M. Jia, G. Chen, Y. Ye, E.J. Lavernia, Additive manufacturing of functionally graded materials: a review Materials Science and Engineering A (2019) 764 p 138209. https://doi.org/10.1016/j.msea.2019.138209

[7] A. Latif, G. Ingarao, L. Fratini, Multi-material based Functionally Graded billets manufacturing through Friction Stir Consolidation of aluminum alloys chips CIRP ANNALS, in press (2022). https://doi.org/10.1016/j.cirp.2022.03.035

[8] J.M. Allwood, J.M. Cullen, M.A. Carruth, D.R. Cooper, M. McBrien, R.L. Milford, M.C. Moynihan, A.C. Patel, Sustainable materials: with both eyes open, Vol. 2012, Cambridge, UK: UIT Cambridge Limited, 2012.

Materials Research Proceedings **26** (2023) 719-724 https://doi.org/10.21741/9781644902431-115

Process mechanics in continuous friction stir extrusion process of aluminum alloy

Gianluca Buffa[1, a], Davide Campanella[1, b *], Livan Fratini[1, c], Adnan Muhuammed[1, d] and Fabrizio Micari[1, e]

[1]Department of Engineering, University of Palermo Italy

[a]gianluca.buffa@unipa.it, [b]davide.campanella@unipa.it, [c]livan.fratini@unipa.it, [d]muhammad.adnan06@community.unipa.it [e]fabrizio.micari@unipa.it

Keywords: Recycling, Metals, Friction Stir Extrusion

Abstract. Friction Stir Extrusion (FSE) is an innovative direct-recycling technology developed for metal machining chips. A rotating die is plunged into a cylindrical chamber containing the material to be recycled during the procedure. The die's stirring action causes strong bonding, enabling for the back extrusion of a full dense rod. One of the technology's major flaws is the process' discontinuity, which limits the volume of extrudes to the chamber's capacity. Based on the previous experience of some of the authors have designed a machine tool for continuous solid-state recycling of metal chips with the aims to overcome the limitations of the discontinuous process which makes it potentially appealing to industry. In this work, Friction Stir Extrusion process is defined, the set-up machines tool is presented, and the experimental findings of case studies are shown and analyzed. Finally, with the goal the mechanics of the FSE continuous process, a numerical model was created starting from the experimental results obtained by discontinuous process.

Introduction

One of the main issues facing the current era is the development of environmentally sustainable industrial techniques. Numerous international agreements, including the Kyoto Protocol, oblige signatory nations to gradually cut their emissions of greenhouse gases over the following few years. It's interesting to note that the production of raw materials contributes significantly to these annual worldwide emissions. According to a study by Worrell et al. [1], five materials in particular steel, cement, paper, aluminum alloys, and polyethylene plastics have the greatest energy impact. For instance, cement has a very cheap production cost, but because it is a material that is utilized extensively, its impact is far from insignificant. Thus according to research by Worrel et al., around 20% of the world's CO_2 emissions are released into the atmosphere during the manufacture of materials. It is predicted that demand for aluminum will expand exponentially and at a faster rate than that of the other materials indicated above, based on the trend in demand seen in previous years. It is vital to put procedures in place that will extend the material's usable life as much as feasible in order to reduce the subsequent rise in emissions caused by the rising demand for the material. Since it has several technological and environmental advantages over other methods, including as large energy savings compared to original production, recycling is currently the most popular method for dealing with metals. Recycling can result in primary energy savings of up to 90% for aluminum alloys, as well as for magnesium and titanium [2]. The savings for steel are a little less, at 69 percent. [3], the demand for aluminum is anticipated to increase by a factor of 2.6 to 3.5 between 2005 and 2050, while the demand for steel is anticipated to increase by a factor of 1.8 to 2.2. Due to the circular economy is become more popular [4] the most popular method for recycling metals now is conventional recycling, which has numerous advantages over alternative technologies in terms of both technology and the environment. Recycling is significantly more cost-effective from an energy consumption point of view when it comes to light alloys, where main energy savings of up to 90% can be made [5]. Furthermore, metal scrap frequently consists of sheets from finishing after forming procedures and chips from machining operations. The type of scrap is categorized as "process scrap" or "new scrap" [6] according to European standards and is made up of leftover materials from the entire production chain of the product, from its creation to its sale to the consumer. Because of its high surface area to volume ratio and the presence of oxides and/or other impurities on its surface, this form of scrap is among the most difficult to recycle. Due to this, material losses of up to 15-20% can be attained when recycling light alloys using annealed scrap metal [7]. Even "old scraps" from used products can be transformed into "process scraps" through the application of the proper cleaning, sorting, and cutting procedures reducing the melting energy cost. According to Samuel's study [8], one of the first initiatives using this type of processing led to the recycling of 96% of the processed scrap with very little loss. A revolutionary process called friction stir extrusion (FSE), which is based on the FSW technology, has been created. Its distinctive characteristic is that frictional heat is also provided to the extrusion process in addition to the softening of materials. The FSE process, which is widely used in the creation of high-quality wires or rods from metal powders, chips, or billet, has recently been presented as a direct conversion technique of recycling. The limited length of wire that may be produced in a single operation using the FSE technique is one of its key drawbacks. This problem can be attributed to both the inherent discontinuity of the operation and the design of the extrusion machine itself. Most of the equipment used by researchers is an adaptation of milling machines, much as it was during the early development of FSW. In this paper a process design approach was carried out through numerical analyses based on FEM models, aimed to define the conditions for a continuous extrusion process. In particular, after preliminary analyses carried out using a Von Mises plastic yield surface of the material, the Shyma-Oyane models was used, modelling the incoherent chips as a porous material, to follow the density evolution during the stage of the process and to get closer to the transient condition of the chips.

Experimental setups

Some of the authors of this research conducted FSE experimental campaigns on both aluminum alloys [9-10] and magnesium alloys, such as AZ31B, [11] to examine the effects of the key process parameter and create a specialized numerical model to study the mechanics of the process[12]. An ESAB LEGIO 3ST machine was used for the experiments. This machine was developed for FSW processes and allow to control the vertical force on the tool. This capability resulted to be particularly relevant since the extrusion force is one of the most important technological parameters to be controlled during FSE. The tools used for the experiments were a hollow chamber and a specifically designed rotary die mounted on the machine chuck. The extruded wire was housed inside the mandrel and successively extracted in between the extrusions. The die presented a conical shaped shoulder to guide the material flow toward the extrusion channel. The grasping of the die during the trials was a recurrent problem caused by the plasticized material flowing into the space between the die and the chamber itself, so the chamber was created as a detachable tool. This experimental setup enabled the production of sound wire using both magnesium and aluminum alloys. The following reference experiments were run with the fixture shown. As shown in Fig. 3, the Esab Legio 3ST machine could manufacture wires up to a maximum length of 70 mm (2 pieces), 40 mm (1 piece), and 30 mm (3 pieces). During a preliminary campaign that also enabled fine-tuning of the tool design with an emphasis on the extension in length of the extrusion channel, the process parameters for the extrusion (R = 900 rpm F = 22 kN) were chosen. Three times were given to each set of procedure parameters, and each time, samples were taken for further study.

Materials Research Forum LLC
https://doi.org/10.21741/9781644902431-115

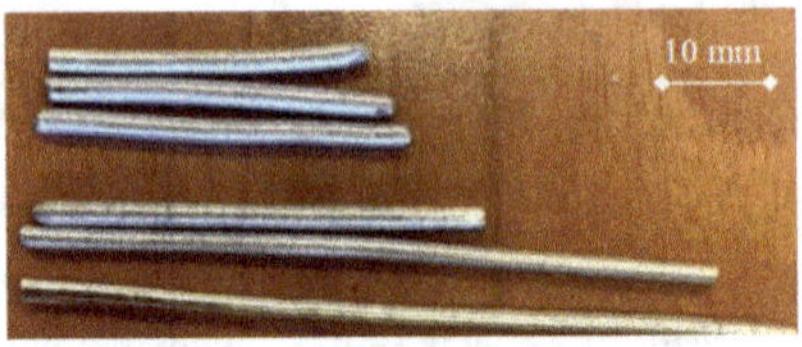

Fig. 2 Extruded wire at 900 rpm and 22 kN with an extrusion ratio equal to 8

Numerical models

Such activity as mentioned was carried out through a numerical simulation-based study was carried out using the Deform software characterized by a lagrangian implicit solver able to carry out thermo-mechanical 3D analyses whose governing equation have been discussed in detail by Tekkaya [13]. The software was used with the aim to investigate the effects of the most relevant process parameters and to define an actual process window able to obtain sound continuous wires. In Fig. 3 sketches of some parts of the machine, designed through the numerical simulation are shown. The tools have been simulated as rigid parts by attributing as material AISI-H-13 present in the Deform library. While the chip, considered a coherent workpiece, was imported as a deformable element, attributing it as material AA2024 (20-550 °C). The geometry of the cochlea has also been optimized with the aid of simulation. To obtain the extrusion, two different mesh windows were created which allowed local refinement in critical points thus avoiding the need for recurrence to an excessive number of elements along the extrusion mouth and in the compacting area of the material. The single block cochlea, the container in which it rotates, with static walls of the extrusion chamber, and the rotating extrusion die were modelled as rigid bodies and meshed with 27778, 19915 and 14175, elements, respectively, to solve the thermal problem. The workpiece, representing the AA2024 aluminium alloy compacted chips, was modelled as a unique body and meshed with 40408 tetrahedral elements of variable size. It is assumed that the chips have already fulfilled the whole screw thread and the extrusion chamber. The cochlea rotated around its longitudinal axis and was considered fixed in its axial position; the extrusion chamber, together with the container, was fixed in space, while the extrusion die rotated around its longitudinal axis and was fixed in its axial position. No addition of chips was considered. An experimental campaign was carried out aimed at knowing the best process conditions. It was decided to perform a velocity control on the auger and matrix, thus reducing the computational difficulties that would have arisen in the case of a force control. A speed of 1600 rpm was set for the axial rotation of the die, and a rotational speed of 100 rpm for the screw; with all elements fixed in space. Having defined the movements of the various elements, the contacts between them were generated by establishing the friction conditions. A friction coefficient of $\mu = 0.7$ was imposed in the contact elements between chip and die.

Results

Several problems were encountered: the breakage of the plastic material during the process due to the shear action of the cochlea, that, as the rotation speed increased, determined the fact that the shear strength limit of the material was locally reached, the contacts between the tool and the material that can be estimated from the load trend (peaks on the load). The loss of contacts led to a reduction in temperature as well as pressure. To obtain an effective simulation a strong increase of the number of elements here and there in the mesh was necessary: in this way in the critical points two mesh windows (as shown in Fig. 4c) were used. Furthermore, a frequent remeshing was adopted due to the strong strain level and distortion of the elements. Once the model was set up, a proper calibration was needed: the latter was carried out through an inverse approach starting from experimental evidence using temperature history data derived from the discontinuous FSE process, where a proper thermocouple was placed in the scrap metal chamber. For the tuning phase of the simulation, therefore, an analysis was made using the temperature acquired in the scrap metal chamber. To proceed with the actual tuning referred to the new fixture, 4 temperatures loci were tacked on the basis of the positions of the four thermocouples placed in the actual machine. In the Fig. 9 show the comparison between the numerically calculated temperatures in positions 1 and 2.

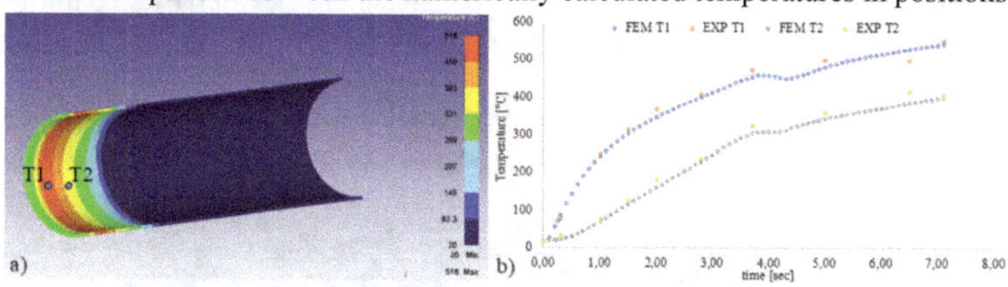

Fig. 9 (a) Position of the thermocouples in the scrap metal chamber and (b) comparison between temperature acquired by thermocouples T1 and T2 and numerical prediction - R=900 rpm, F=22 kN

Using the point tracking was possible to perform a study on the flow of the material generated during the process. Fig. 10 show the positioning of these points.

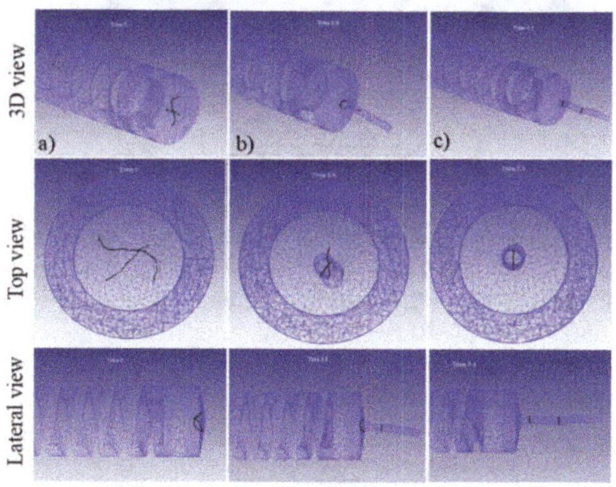

Fig. 10 Point tracking positions in three different steps: (a) initial, b)-c) out of the extrusion channel at different time – R = 900 rpm, F = 22 kN

Once the simulation of the process was established, the material modeled by Shyma-Oyane yield surface was used, with an initial density of 0.8. The process therefore remains the same with the only new filed variable being the porosity of the material. It has been observed that the extrusion mechanics is activated only after reaching the full density (1) of the material (Fig. 11)

Numerical instabilities were observed due to the quite low value of the material density with some sudden variations of temperature due to local losses of contact. In Fig. 11 (a and b) are shown difference noted in the achievement of the unitary value of the density, compared to what happens in the traditional process [14]. In such continuous process an intermediate compacted layer of the material is obtained depending on the actual distance between the cochlea and the scrap metal chamber during the process. This observation led to the conclusion that the process must consist of the following three distinct phases:

Materials Research Proceedings **26** (2023) 719-724 https://doi.org/10.21741/9781644902431-115

I. Pre-compacting by cochlea with zero tool rotation: in such stage just the axial motion of the cochlea is acted determining a sort of upsetting process mechanics and the consequent compacting of the chips.
II. Under a constant load by the cochlea, the tool starts rotating with established rotational speed: the axial force by the cochlea increases step by step until reaching the imposed value (process parameter); (at this stage, the process becomes a load controlled one).
III. Locking the advancement of the cochlea and starting the rotation of the cochlea to intake new scrap metal maintaining thus a constant pressure in the chamber.

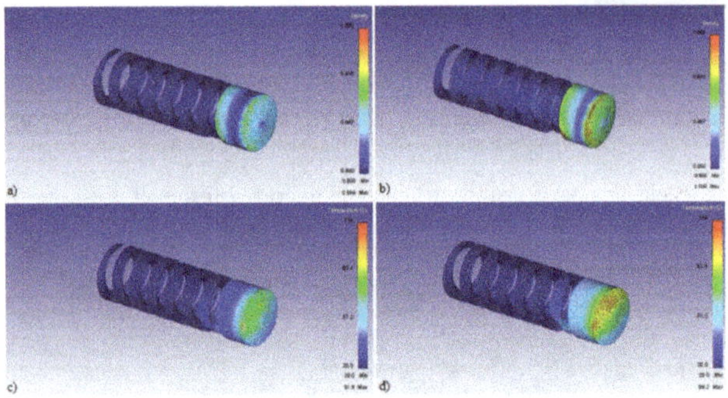

Fig. 11 Relative density and temperature at a)-c) 1.5 sec and b)-d) 2.5 sec

The contact between the chip and the tool is also linked to the inclination of the thread, therefore there is a point of maximum pressure and a minimum point, a hypothesis in the verification phase provides for a linear displacement of the auger, by means of application of a force towards the tool, in order to self-regulate the pressure in the chamber avoiding lack of pressure phenomena which contribute to the interruption of the material plastic flow directed towards the tool and then the interruption of the consolidated material extrusion. Once the obtained numerical data will be verified with the experimental ones, it will then be possible to proceed with the full process engineering.

Conclusion

The study conducted demonstrates that Friction Stir Extrusion technology operated continuously represents a state-of-the-art solution for the primary recycling of light alloys, such as aluminum. The continuous process represents a very viable alternative to more classical material remelting methods and an excellent alternative to the Solid-State Recycling landscape. The numerical analyses carried out show how the proposed FEM model fully represents the reality of the process, being able to provide that information which is not directly measurable by the experimental test. It has been shown how the model is able to help the experimental campaign by providing useful data such as: output parameters for the achievement of the solid bonding phenomenon such as, the temperature distribution, the strain rate trend and thus the material flow; but also input parameters necessary for the proper operation of the process such as the force trend required for the extrusion and the rotation speeds of the moving elements. The solution, identified in the continuous extrusion process, allows recycling to be carried out directly "on site". An additional advantage of the application of continuous FSE is that it overcomes the geometric limitations associated with its "single block" version. Finally, it is noted that the developed FEM model is a first attempt model, which has shown encouraging and plausible results, but to become an industrial design tool it will have to be calibrated appropriately. A study should, in the future, be carried out as parameters such as friction, conductivity coefficients and mesh sensitivity.

References

[1] Gutowski, T. G., Sahni, S., Allwood, J. M., Ashby, M. F., & Worrell, E. (2013). The energy required to produce materials: Constraints on energy-intensity improvements, parameters of demand. https://doi.org/10.1098/rsta.2012.0003

[2] Ingarao, G. (2017). Manufacturing strategies for efficiency in energy and resources use: The role of metal shaping processes. Journal of Cleaner Production, 142, 2872–2886. https://doi.org/10.1016/j.jclepro.2016.10.182

[3] European Aluminum Association, 2013. Environmental Profile Report for the European Aluminium Industry.

[4] The energy required to produce materials: Constraints on energy-intensity improvements, parameters of demand. Philosophical Transactions of the Royal Society A: Mathematical, Physical and Engineering Sciences, 371(1986).

[5] Vargel, C., 2004. Corrosion of Aluminium. Elsevier. https://doi.org/10.1016/B978-008044495-6/50012-4

[6] Simon, L., Moraes, C. A. M., Modolo, R. C. E., Vargas, M., Calheiro, D., & Brehm, F. A. (2017). Recycling of contaminated metallic chip based on eco-efficiency and eco-effectiveness approaches. Journal of Cleaner Production, 153, 417–424. https://doi.org/10.1016/j.jclepro.2016.11.058

[7] Grayson J., 2017. Reducing Melt Loss and Dross Generation.

[8] Baffari D, Buffa G, Campanella D, Fratini L, Micari F. 2014. Friction based Solid State Welding Techniques for Transportation Industry Applications. https://doi.org/10.1016/j.procir.2014.06.125

[9] D. Baffari, A.P. Reynolds, X. Li, L. Fratini, Influence of processing parameters and initial temper on friction stir extrusion of 2050 aluminum alloy, Journal of Manufacturing Processes, 28 (2017) 319–325. https://doi.org/10.1016/j.jmapro.2017.06.013

[10] D. Baffari, G. Buffa, D. Campanella, L. Fratini, Al-SiC metal matrix composite production through friction stir extrusion of aluminum chips, Procedia Engineering, 207 (2017) 419–424. https://doi.org/10.1016/j.proeng.2017.10.798

[11] D. Baffari, G. Buffa, D. Campanella, L. Fratini, A.P. Reynolds, Process mechanics in friction stir extrusion of magnesium alloys chips through experiments and numerical simulation, Journal of Manufacturing Processes, 29 (2017) 41–49. https://doi.org/10.1016/j.jmapro.2017.07.010

[12] D. Baffari, G. Buffa, L. Fratini, A numerical model for wire integrity prediction in friction stir extrusion of magnesium alloys, Journal of Materials Processing Technology, 247 (2017) 1–10. https://doi.org/10.1016/j.jmatprotec.2017.04.007

[13] Tekkaya AE. Simulation of Metal Forming Processes 2000:251–302. https://doi.org/10.1007/978-3-662-04013-3_7

[14] Baffari Dario; Buffa Gianluca; Fratini Livan (2017). A numerical model for Wire integrity prediction in Friction Stir Extrusion of magnesium alloys. Journal of Materials Processing Technology, 247, 1. https://doi.org/10.1016/j.jmatprotec.2017.04.007

Keyword Index

www.ingramcontent.com/pod-product-compliance
Lightning Source LLC
LaVergne TN
LVHW052204200926
848481LV00001B/26
* 9 7 8 1 6 4 4 9 0 2 4 2 4 *